ANALOG CIRCUIT DESIGN

DISCRETE AND INTEGRATED

ANALOG CIRCUIT DESIGN

DISCRETE AND INTEGRATED

Sergio Franco
San Francisco State University

Dedication

In Memory of My Parents

Luigia Braidotti and Luigi Franco

Contents

Preface

This textbook is intended for EE majors envisioning industrial careers in analog electronics. Analog integrated-circuit designers, product/process/reliability engineers, test/test-development engineers, and analog applications/marketing/customer-support engineers are always in great demand. The book is the result of my teaching experience at San Francisco State University, where over the years I have contributed to the formation of hundreds of students now gainfully employed in Silicon Valley in a wide range of analog positions. Here are three important features of this book:

- Both *bipolar* and *CMOS* technologies are covered. Even though digital electronics is dominated by CMOS technology, analog electronics relies on both CMOS and bipolar, the latter being the technology of choice in high-quality analog circuits as well as a fundamental part of BiCMOS technology.
- Both *discrete* and *integrated* designs are covered. Though nowadays the ultimate form of an analog system is likely to be of the integrated-circuit type, testing and applications often require ancillary functions such as conditioning and interfacing that are best realized with ad-hoc discrete designs. (Anyone familiar with the work by recognized leaders in analog applications/testing/instrumentation like Jim Williams and Robert Pease will agree to this.) In this respect, BJTs are available in a wide selection of off-the-shelf discrete types to serve a variety of needs, including practical experimentation in the lab. Moreover, for pedagogical reasons it is convenient to cover simple discrete circuits before tackling the more complex integrated circuits.
- *Semiconductor theory* is presented in sufficient depth to reflect the daily needs of a practicing engineer in industry. Every analog function is inextricably rooted on a physical phenomenon, so analog engineers, particularly IC designers and product/process/reliability engineers, need to be conversant with the physics of semiconductors in order to function optimally.

BOOK ORGANIZATION

The book is organized into two parts:

- The first part covers (1) *Diodes*, (2) *BJTs*, and (3) *MOSFETs*; as such, it is suited for a *first course in electronics*, typically at the junior level. The material is sequenced according to the technological evolution of electronics. However, the instructor who wishes to change the order of coverage of BJTs and MOSFETs can easily do so, as the two chapters are autonomous. Regardless of the order, the last chapter in the sequence can be covered much faster than the previous one as the student has already faced the challenges of dc biasing and large/small signal modeling.

- The second part covers (4) *Analog IC Building Blocks*, (5) *Representative Analog ICs*, (6) *Frequency and Time Responses*, and (7) *Negative Feedback, Stability*, and *Noise*. This part is suited for a *junior/graduate-level course* in analog IC *analysis and design*. In this part, BJTs and MOSFETs are mostly covered *side by side* so as to offer a unified treatment emphasizing similarities while recognizing inherent differences.

Each chapter provides a fairly comprehensive coverage of its title subject, so chapters are of necessity long (>100 pages each). The overall amount of material exceeds that of a typical two-semester or four-quarter course sequence, so the instructor has considerable leeway in material selection. Also, the author feels that the student needs to fully master low-frequency circuits before proceeding to the more challenging frequency and time responses of Chapter 6. But, the instructor can modify the coverage by skipping suitable topics from Part I in favor of selected topics from Chapter 6.

As mentioned, the first part focuses on basic transistor circuits emphasizing the more traditional *discrete design* approach. Pedagogically, it makes sense to study single-transistor circuits before progressing to multi-transistor systems—so much the better if this part is accompanied by a lab, where it is easier to investigate simpler circuits. In fact, at this level, a lab (complete with smoke!) is likely to offer a far more valuable learning environment than a computer-simulated one. A notorious drawback of discrete circuits is the need for coupling/bypassing capacitors, which introduce an element of additional distraction as it takes time for the student to develop full confidence with these capacitance functions. Cognizant of this, I have tried to demystify capacitors via detailed visual examples (see Fig. 2.55, p. 181, and Fig. 3.60, p. 302).

Following discrete circuits, the book progresses to *integrated circuits*. The integrated and discrete approaches are contrasted in Section 4.1; also, an intuitive borderline between the two is discussed in Exercise 4.3, p. 357. The second part progresses in complexity as well as sophistication from building blocks, to representative analog ICs, to IC dynamics, and finally to IC operation in negative-feedback, along with stability considerations, frequency compensation, and noise. This part is intended for IC designers but also for all other categories of engineers involved in fabrication, test, and applications. Application engineers, by far the largest group, need a working familiarity both with the technology (in order to make educated selections) and with the IC's inner functioning (in order to optimize its application). The book's aim is to promote a balance between the ability to *design on chip* and the ability to *design on board*.

The book website features a solutions manual and PowerPoint lecture slides for instructors as well as a list of helpful web links and errata. The author appreciates being notified of any possible errata.

MOTIVATION

After having experimented with a number of other textbooks, I decided to write my own in response to a number of student concerns and also to implement my own ideas on how to better serve the needs of our graduates, who generally pursue careers in industry. Following are the most common student concerns:

- *Need to see lots of examples, especially of the type practicing engineers face daily.* I have painstakingly thought out each in-text example and end-of-chapter problem to serve two broad needs: to help the student develop a *feel* for the orders of magnitude of the quantities under study (see, for instance, Example 1.8, p. 42), and to deepen student *understanding* by following a circuit's evolution through different states or through increasing levels of complexity (see Figs. 1.18–20, pp. 16–17). In this respect, I made a concerted effort to develop a *systematic problem-solving methodology* emphasizing *thinking* and *physical intuition* as opposed to rote calculations. For it is in physical understanding, not in mathematical manipulations or computer simulations, that is rooted the design creativity the student will be called to exercise on the job. The book contains valuable rules-of-thumb working engineers use daily (see pp. 50, 52–55). Whenever possible, the student is reminded to use intuition and physical insight to anticipate what to *expect* from calculations or computer simulations, and to *check* results against physical substance (for physical insight see Example 6.5, p. 583, and Fig. 793, p. 787).
- Incorporate *SPICE simulation within the text.* SPICE has been integrated throughout the text both as a *pedagogical aid* to confer more immediacy to a new concept (see Fig. 4.66, p. 424), and as a *validation tool* for hand calculations. If unreasonable discrepancy is found between calculations and simulations, the student is challenged to account for possible causes (see Example 5.2, p. 490). Finally, SPICE is used to bring out nuances that would be too complex for hand calculations (see Example 6.11, p. 602). Nowadays a plethora of SPICE versions are available. Rather than committing to a particular version, I have decided to keep schematic capture circuits simple enough for students to set them up in their preferred SPICE version in a matter of minutes.
- *Provide practical exposure to basic semiconductor concepts.* The majority of graduates from my own institution (a state university) pursue industrial careers spanning a wide range of positions from IC designers to product and reliability engineers, test and test-development engineers, and application and customer-support engineers, where a broad background is far more desirable than a narrow specialization. A basic understanding of semiconductor principles is an integral component of such a background, especially for future product and reliability engineers.
- *Make generous use of figures* to comply with today's visually oriented learning trends. Most figures consist of two or more components placed side by side to visualize different facets of the same concept, be they different states of a circuit, or models, or time frames, or cause-effect relationships (see Fig. 1.59, p. 69). Also, the most relevant formulas intervening in a given analytical process have

been boxed for easy visualization, especially when the student crams for quizzes and tests. When appropriate, entire groups of formulas have been tabulated to facilitate their comparison (see Fig. 3.50, p. 289).

I tried to address the above concerns by returning to an essential, no-frills, no-distractions textbook format. Each chapter starts with a brief historical background and motivational framework, followed by a brief outline of the topics to be covered, followed by the chapter body proper. It concludes with a variety of carefully thought-out problems emphasizing intuition and physical insight.

THE CONTENTS AT A GLANCE

Chapter 1 starts out with the *ideal diode* as a vehicle for introducing the student to nonlinear circuit analysis and applications. This is followed by a review of the *operational amplifier* to pave the way for additional applications of diodes and, later, transistors. It is now time to introduce the student to the most common physical device approximating the diode function, the *pn* junction. After an intuitive review of *semiconductors*, the *pn* junction is discussed in proper detail, using *rules of thumb* to highlight those practical aspects that engineers use daily on the job. Working familiarity with the *pn* junction is crucial for the understanding of transistor physics in the following two chapters. Finally, various *popular diode applications* are discussed, often using PSpice as a pedagogical aid to enhance understanding.

Chapter 2 introduces the *bipolar junction transistor* (BJT) as a technological (and historical) evolution of the *pn* junction. Mirroring Chapter 1, we start out with the *physical structure* of the BJT, followed by the derivation of its *i-v characteristics*, the development of *large- and small-signal models*, *dc biasing*, and finally the analysis and design of *single-transistor amplifiers* and *buffers*. The common-emitter configuration is presented as the natural realization of voltage amplification, whereas the common-collector and common-base configurations serve most naturally as voltage and current buffers, respectively. Great emphasis is placed on the role of the BJT as a *resistance transformation device* (which actually provided the basis for its very name.) The transformation equations are conveniently tabulated for easy reference in later chapters.

Chapter 3 covers the MOSFET in similar fashion as the BJT of Chapter 2. However, the two chapters are kept independent of each other, so the order of coverage can be interchanged if desired. The chapter begins with a detailed discussion of the physical basis of the *native threshold* for the benefit of those students who will pursue careers as product, process, and reliability engineers. Next the MOSFET's *i-v characteristics* are derived, followed by the development of *large- and small-signal models*, *dc biasing*, and finally the analysis and design of *single-transistor amplifiers* and *buffers*. The common-source configuration is presented as the natural realization of voltage amplification, whereas the common-drain and common-gate configurations serve most naturally as voltage and current buffers, respectively. The chapter also covers the *CMOS inverter* and basic CMOS *logic gates* so as to provide a more balanced treatment for the benefit of computer engineering majors (see the PSpice noise-margin illustration of Fig. 3.44, p. 281).

Chapter 4 brings the student to a higher level of circuit sophistication by introducing the analog IC *building blocks* in widest use today. *Cascode* configurations, *differential amplifiers*, *current mirrors* of all types, *active loads*, and *push-pull output stages* are treated in proper detail in anticipation of their utilization in the following chapter. Whenever possible, BJTs and MOSFETs are covered side by side so as to present the reader with a *uniform treatment* and thus save space as well as effort.

Chapter 5 puts to use the blocks of Chapter 4 in the design of a *representative mix* of analog ICs in both bipolar and CMOS technologies, namely: *high-gain amplifiers* such as op amps, voltage comparators, and fully-differential op amps; *voltage and current references* such as bandgap references; *current-mode ICs* such as transconductors, OTAs, and current-feedback amplifiers; and, finally, *switched-capacitor circuits*.

Chapter 6 deals with the *frequency and time responses* of individual devices all the way up to complex circuits such as the ICs of Chapter 5. Frequency analysis relies on the *Miller approximation* as well as the *open-circuit time-constant technique*. The switching times of *pn* diodes and BJTs, unjustifiably ignored by current textbooks in spite of their enduring industrial relevance, are investigated via *charge-control analysis* emphasizing physical insight. Also covered are the *switching times of CMOS gates* for the benefit of computer engineering majors. This is a chapter in which in-text PSpice is put to frequent use as a verification tool for hand calculations.

Chapter 7 starts out with a comprehensive treatment of *negative feedback* as applied to the electronic circuits of all previous chapters, from single-transistor stages all the way to op amps. Both *two-port* and *return-ratio analyses* are presented and compared via a variety of carefully thought out examples. Also *Blackman's impedance formula* and *injections methods* are presented in a practical manner. The chapter proceeds to the subject of *stability* and *frequency-compensation* for op amps, both bipolar and CMOS (here again PSpice proves a most useful pedagogical tool). The chapter concludes with the study of noise in integrated circuits. After an introduction to basic noise properties, analytical tools, and noise types, the noise models of diodes and transistors are discussed. Finally, noise analysis is applied to representative circuit configurations such as op amp circuits and differential pairs.

A WORD OF ADVICE TO THE STUDENT

Your electronics courses provide the foundation for your career in EE. The objective of these courses is not only to introduce you to new devices such as *diodes* and *transistors*, but also to help you establish a *thinking style* and develop a *problem-solving methodology* that are unique to this challenging but most interesting field. Cognizant of the fact that a large proportion of EE graduates end up working in industry, I have emphasized those practical aspects that are of relevance in today's industrial milieu. Whether you pursue a career as an IC designer, a product engineer, a test or test-development engineer, or an applications or customer support engineer, the basic material of your first electronics courses will always resurface in a variety of ever-changing situations, so you may wish to invest far more time and effort in this one course than you would normally do—the benefits will be quite rewarding.

Even though diodes and transistors are *highly nonlinear* devices, special techniques have been developed for their analysis, which draw quite heavily from those covered in linear-circuits courses. Far from being a waste of time, the analytical tools learned in these prerequisite courses will be put to heavy use also in the study of electronics. Specifically, *Ohm's law*, *Kirchhoff's laws* (KVL and KCL), *Nodal/ Loop Analysis*, *Thévenin's/Norton's theorems*, the *Superposition principle*, and the *Op Amp Rule* will continue to be valuable analytical tools as we venture into the exciting realm of electronics.

Electronics, like any other branch of engineering, deals with the *physical reality* of its devices and systems. We use mathematics as a *tool* to understand or predict their operation as well as to design new ones, and computer simulation as a verification tool. Any conceptual derivation or prediction must ultimately be checked against physical substance and never be taken for granted on its own. The use of physical reasoning to corroborate any conceptual process, be it a mathematical derivation or a computer simulation, is at the very core of this entire course sequence.

Beside proficiency in linear circuit analysis techniques, the student is expected to possess a working knowledge of basic calculus concepts, such as slope and area under a curve, as well as basic electrostatics concepts such as Gauss's theorem and the relations between electric field and potential. Also, the ability to perform circuit simulations via PSpice, as learned in prerequisite circuits courses and labs, will prove extremely useful for checking the results of hand analysis.

ACKNOWLEDGEMENTS

A number of reviewers provided detailed commentaries and many valuable suggestions. I tried to implement their recommendations whenever I could, but in the presence of conflicting viewpoints, I had to draw a line and pursue my own. To all who provided feedback, my sincere thanks. I would especially like to acknowledge Stephen Hubbard, Clemson University; Santosh Pandey, Iowa State University; and Donna Ginger Yu, North Carolina State University. Finally, I wish to express my gratitude to Diana May, my wife, for her encouragement and steadfast support.

Sergio Franco
San Francisco State University

Diodes and the *pn* Junction

Chapter Outline

The diode is the most basic electronic device. In fact, its invention, over a century ago, is credited with ushering in the era of *electronics*. Like the resistor, the diode comes with two terminals; however, unlike the bidirectional resistor, the diode carries current only in *one direction*. For a qualitative understanding of how this can happen, think of the early diodes, which were of the vacuum-tube type. A vacuum diode had an incandescent filament called a *cathode*, acting as a copious source of free electrons, and a plate called an *anode*, to control current flow. Applying a *positive* voltage to the anode relative to the cathode would attract the negatively charged electrons and thus *sustain electron flow* from cathode to anode. Conversely, applying a *negative* voltage to the anode would repel the electrons and thus *inhibit electron flow*. By a hydraulic analogy, the diode can be visualized as a *one-way valve*.

The vacuum-tube diode was invented by John A. Fleming in 1904. Just two years later, in 1906, Greenleaf W. Pickard invented an alternative diode type by forming a point contact to a slab of silicon, thus creating the first solid-state electronic device. However, it took half a century for the semiconductor industry to become a commercial reality, so the first half of the twentieth century was dominated by vacuum-tube electronics.

Nowadays diodes are made of *semiconductor materials*, with dramatic advantages over their vacuum-tube counterparts in terms of miniaturization, reliability, power consumption and cost. Specifically, the most common diode today is the silicon *pn* junction, though other types of materials and junctions are also in use. The *pn* junction plays a central role in microelectronics in that not only does it provide the diode function at the basis of the above applications, but it is also at the basis of the *bipolar junction transistor* (BJT), the *junction field-effect transistor* (JFET), and other semiconductor devices such as the *silicon-controlled rectifier* (SCR). The *pn* junction is also present in the *metal-oxide-semiconductor field-effect transistor* (MOSFET), the device most widely used in today's microelectronics products. Furthermore, in its reverse-biased form, the *pn* junction is used to *isolate* from each other different devices coexisting on the same semiconductor chip.

The student who comes from prerequisite courses in linear circuits will immediately find that diodes, like the transistors to follow, are highly *nonlinear* devices. Mercifully, a number of techniques have been developed for the analysis of nonlinear devices that draw quite heavily from those covered by linear-circuits courses. Far from being a waste of time, the analytical tools learned in circuits courses will also be put to heavy use in the study of electronics. Specifically, *Ohm's Law*, *Kirchhoff's Laws* (KVL and KCL), the *Voltage/Current Divider Rules*, *Thévenin's/Norton's Theorems*, and the *Superposition Principle* will continue to be our precious analytical tools as we venture into the exciting realm of electronic devices and systems.

CHAPTER HIGHLIGHTS

The chapter begins with the *ideal diode*, a concept designed to develop a basic feel for diode behavior as well as introduce the student to nonlinear circuit analysis techniques, the backbone of all subsequent electronics. The applications covered are *diode rectifiers*, *diode logic gates*, *voltage clamps*, *piecewise-linear function generators*, *peak detectors*, *dc restorers*, and *voltage multipliers*.

Next, we review the basic operational amplifier principles of prerequisite circuits courses because diodes (and later, transistors) offer a fertile application ground for op amps. The first diode-op-amp application to be considered is *full-wave rectification*, but others will follow as we proceed.

As mentioned, today diodes are made of semiconductors, so the next objective is the study of basic *pn* junction theory. After a review of the semiconductor basics normally covered in prerequisite physics courses, the chapter develops an intuitive discussion of the *pn* junction, highlighting those practical aspects (rules of thumb) that form the working knowledge of the electronics engineer in the modern industrial milieu. Whether the student will pursue a career as an IC designer, or a product, process, or reliability engineer, or as a test or applications engineer, the *pn* junction will always resurface in a variety of situations, so it is only appropriate that we address it in some depth. All the student needs to remember from basic physics is Gauss's Theorem as well as the relationship between electric field and potential,

$$\frac{dE}{dx} = \frac{\rho(x)}{\varepsilon_{si}} \qquad E = -\frac{dv(x)}{dx}$$

The outcome of *pn* junction theory is the concept of a *real-life diode*, a device that in spite of its departure from the ideal diode is still analyzed via suitable linearizing techniques. This introduces the student to the *large-signal model* as well as the *small-signal model* for the diode, models that will be expanded further in the following chapters when we study transistors.

The last part of the chapter applies the above models to the study of a number of widely used practical circuits such as *rectifiers, voltage references, basic non-linear op amp circuits,* and *dc power supplies.* The Appendix discusses the parameters intervening in the diode model used by SPICE.

Most likely the chapter covers more material than feasible in a typical junior course. But, the instructor can easily skip selected topics, such as the sections on semiconductor theory, especially if these topics are covered in alternative courses. I have written this chapter with the aim to have all (or almost all) pertinent diode material in one place.

1.1 THE IDEAL DIODE

The diode is a two-terminal device designed to conduct current in one direction only. Unlike the resistor, which conducts in either direction, the diode carries current only from the terminal called *anode* (*A*) to the terminal called *cathode* (*C*). Its circuit symbol, shown in Fig. 1.1*a*, uses an arrowhead to signify this directionality. The voltage across the diode is defined as positive at the anode and negative at the cathode, thus conforming to the *passive sign convention* of other popular devices such as resistors.

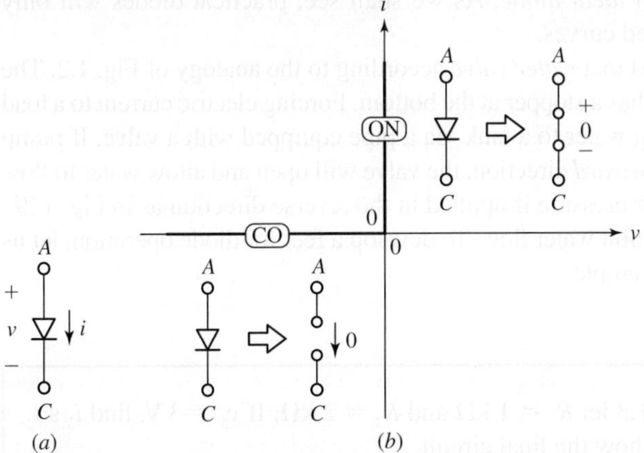

FIGURE 1.1 (*a*) Circuit symbol and sign convention for the diode. (*b*) Ideal-diode *i-v* characteristic and diode models in the *on* (ON) region and in the *cut-off* (CO) region of operation.

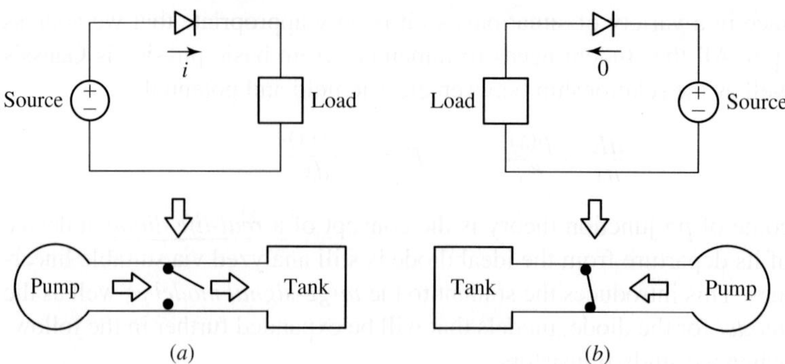

FIGURE 1.2 Valve analogy of a diode: (a) forward operation and (b) reverse operation.

When invited to draw current in the direction of its arrowhead ($i > 0$), also called the *forward* (F) direction, a diode will eagerly conduct the given current by acting as a *short circuit* ($v = 0$). In this case the diode is said to be *forward biased*, or also to be *on* (ON). However, if we try to force current in the opposite direction, also called the *reverse* (R) direction, the diode will stubbornly oppose current flow by acting as an *open circuit* ($i = 0$). The diode is now said to be *reverse biased*, or also to be *cut off* (CO). When cut off the diode will sustain whatever voltage ($v < 0$) is imposed by the surrounding circuitry.

Figure 1.1*b* shows the *i-v* characteristic of the diode, which we express mathematically as

$$v = 0 \qquad \text{for } i > 0 \tag{1.1a}$$
$$i = 0 \qquad \text{for } v < 0 \tag{1.1b}$$

Also shown next to the curves are the diode models (a short circuit and an open circuit) corresponding to the two modes of operation. A device with the characteristic shown is referred to as an *ideal diode*. As we shall see, practical diodes will only approximate these idealized curves.

A diode can be likened to a *water valve* according to the analogy of Fig. 1.2. The valve hinges at the top and has a stopper at the bottom. Forcing electric current to a load via a diode is like pumping water to a tank via a pipe equipped with a valve. If pump pressure is applied in the *forward* direction, the valve will open and allow water to flow as in Fig. 1.2*a*. However, if pressure is applied in the reverse direction as in Fig. 1.2*b*, the valve will close and inhibit water flow. To develop a feel for diode operation, let us consider our first circuit example.

EXAMPLE 1.1

(a) In the circuit of Fig. 1.3 let $R_1 = 1\ \text{k}\Omega$ and $R_2 = 2\ \text{k}\Omega$. If $v_S = 3$ V, find i_S so that D draws 1 mA. Show the final circuit.

(b) If $i_S = 3$ mA, find v_S so that D drops 2 V. Show the circuit.

(c) If $i_S = 2$ mA and $v_S = 6$ V, find R_1 and R_2 so that D operates at the origin of the *i-v* plane, where $v = 0$ and $i = 0$.

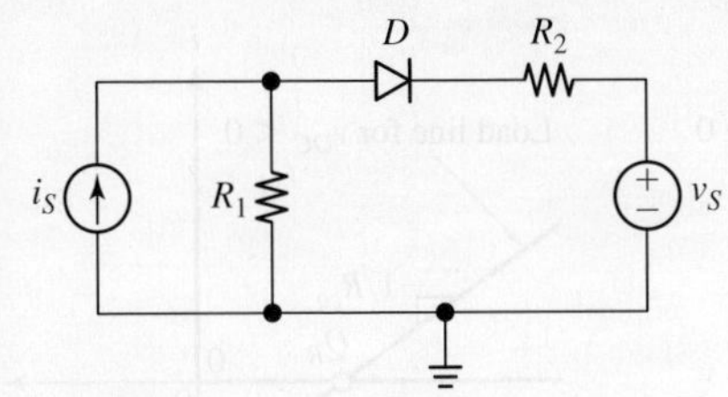

FIGURE 1.3 Circuit of Example 1.1.

Solution

(a) In conduction D acts as a short-circuit, so $v_A = v_C$. By Ohm's law and KVL, $v_A = v_C = (2\ \text{k}\Omega) \times (1\ \text{mA}) + (3\ \text{V}) = 5\ \text{V}$, causing the 1-k$\Omega$ resistance to draw $(5\ \text{V})/(1\ \text{k}\Omega) = 5\ \text{mA}$, flowing downward. KCL gives $i_S = (1\ \text{mA}) + (5\ \text{mA}) = 6\ \text{mA}$, so the situation is as in Fig. 1.4a.

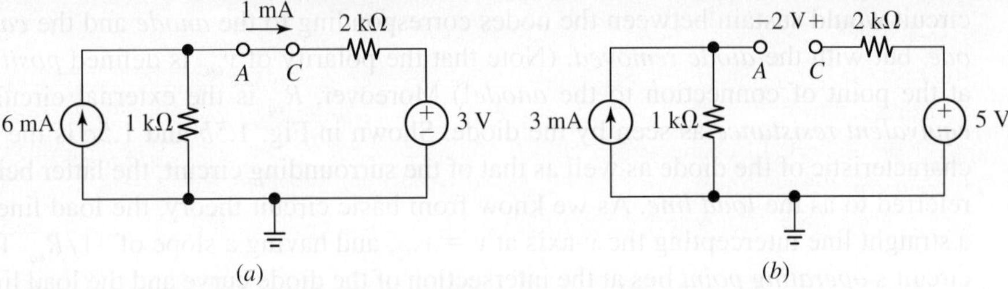

(a)   (b)

FIGURE 1.4 Circuit solutions to Example 1.1.

(b) In cutoff D acts as an open circuit, so the 2-kΩ resistance drops 0 V. By Ohm's law, $v_A = 3 \times 1 = 3$ V. By KVL, the voltage at the cathode is $v_C = v_A + 2 = 3 + 2 = 5$ V, so $v_S = v_C = 5$ V as shown in Fig. 1.4b.

(c) $i = 0 \Rightarrow v_C = v_S = 6$ V; $v = 0 \Rightarrow v_A = v_C = 6$ V and $R_1 = v_A/i_S = 6/2 = 3$ kΩ. As long as D is cut off, the value of R_2 is irrelevant. R_2 has an impact only when D is on.

Finding the Operating Mode of a Diode

Even though it consists of two straight segments, the diode characteristic is *nonlinear*. (In fact, it is said to be *piecewise linear*.) Yet, as demonstrated by Example 1.1, we can still apply the analytical techniques learned in linear-circuits courses because at any given time the diode operates *only in one* of its two possible modes (either ON or CO), where it admits a model (open or short circuit) that is indeed linear. Thus, to carry out our analysis, we only need to determine which of the two modes the diode happens to be operating in at a given time.

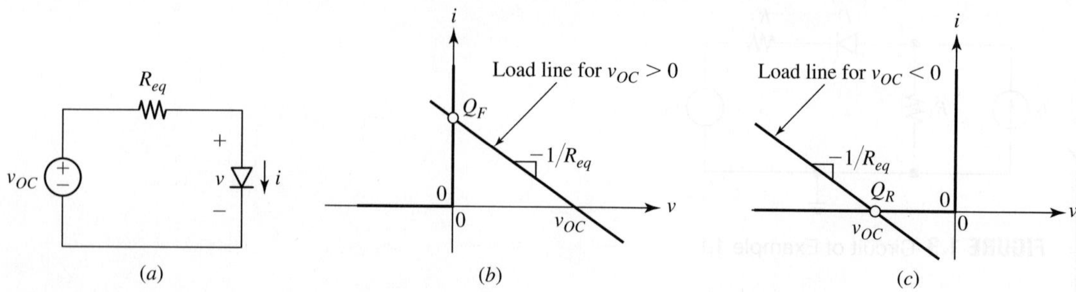

FIGURE 1.5   (a) Basic diode circuit, and (b), (c) graphical method to find the operating point.

There are many situations in which the diode is embedded in a *linear circuit*, indicating that we can simplify our analysis significantly if we replace the surrounding circuitry with its *Thévenin equivalent*. After doing so we end up with the basic situation of Fig. 1.5a, where v_{OC} is the *open-circuit voltage* that the external circuit would sustain between the nodes corresponding to the *anode* and the *cathode,* but with the *diode removed.* (Note that the polarity of v_{OC} is defined *positive* at the point of connection to the *anode!*) Moreover, R_{eq} is the external circuit's *equivalent resistance* as seen by the diode. Shown in Fig. 1.5b and 1.5c is the *i-v* characteristic of the diode as well as that of the surrounding circuit, the latter being referred to as the *load line.* As we know from basic circuit theory, the load line is a straight line intercepting the *v*-axis at $v = v_{OC}$, and having a slope of $-1/R_{eq}$. The circuit's *operating point* lies at the intersection of the diode curve and the load line, where the diode and the surrounding circuit share the *same* voltage and current. It is apparent that

- If $v_{OC} > 0$, the operating point (Q_F) lies on the ON segment, where the diode is forward biased and thus acts as a short circuit to yield

$$v = 0 \qquad i = \frac{v_{OC}}{R_{eq}} \, (> 0) \tag{1.2a}$$

- Conversely, if $v_{OC} < 0$, the operating point (Q_R) lies on the CO segment, where the diode is reverse biased and thus acts as an open circuit to yield

$$i = 0 \qquad v = v_{OC} \, (< 0) \tag{1.2b}$$

Let us illustrate via an actual example.

EXAMPLE 1.2   (a) Find v and i in the circuit of Fig. 1.6a if $v_S = 12$ V, $R_1 = 10$ kΩ, $R_2 = 30$ kΩ, and $R_3 = R_4 = 15$ kΩ.

(b) Repeat, but with R_2 lowered from 30 kΩ to 2.0 kΩ.

(c) What happens if we reverse the diode's direction in part (a)? In part (b)?

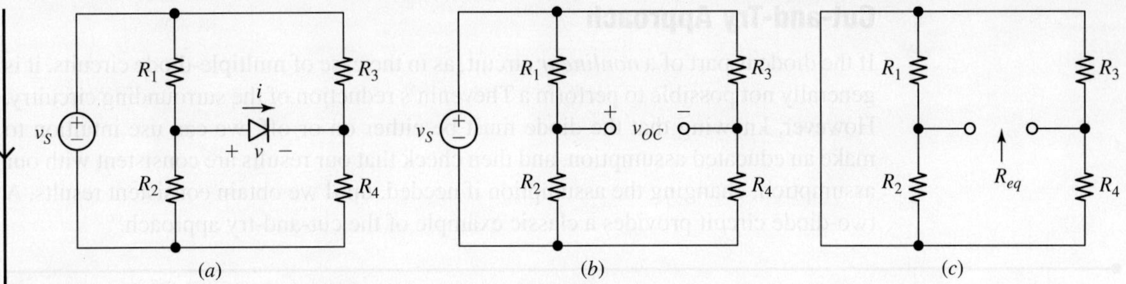

FIGURE 1.6 (a) Circuit of Example 1.2. (b), (c) Subcircuits to find the Thévenin's equivalent of the network surrounding the diode. Note that the diode has been removed.

Solution

(a) Removing the diode yields the subcircuit of Fig. 1.6b, where the voltage divider rule gives

$$v_{OC} = \left(\frac{R_2}{R_1 + R_2} - \frac{R_4}{R_3 + R_4} \right) v_S$$

Substituting the given component values gives $v_{OC} = 9 - 6 = 3$ V. Since $v_{OC} > 0$, the diode will be on, acting as a short circuit with $v = 0$. To find i, we need R_{eq}. To this end, we suppress the voltage source to obtain the subcircuit of Fig. 1.6c. By inspection,

$$R_{eq} = (R_1//R_2) + (R_3//R_4)$$

Substituting the given component values gives $R_{eq} = 15$ kΩ. Consequently, by Eq. (1.2a), $i = 3/15 = 0.2$ mA. The situation is shown in Fig. 1.7a, where the student is encouraged to find all other voltages and currents in the circuit as a check.

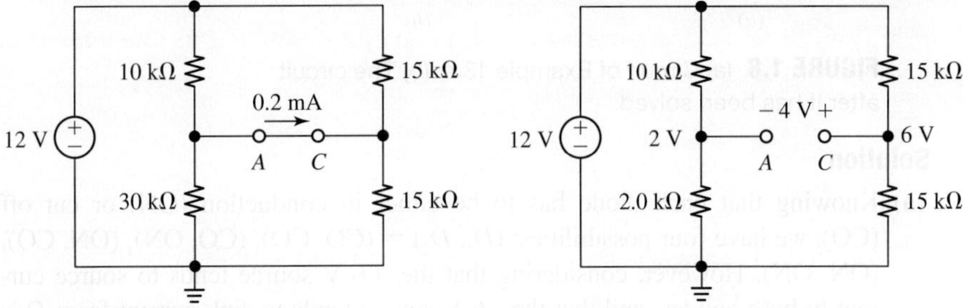

FIGURE 1.7 Circuit of Example 1.2a, where the diode is *on*, and Example 1.1b, where the diode is *off*.

(b) With R_2 lowered to 2.0 kΩ we get $v_{OC} = 2 - 6 = -4$ V. Since we now have $v_{OC} < 0$, the diode is off, acting as an open circuit with $i = 0$. The situation is depicted in Fig. 1.7b.

(c) Reversing the diode direction so that the anode is now at the right and the cathode is at the left will cause the diode to be *off* in part (a) and *on* in part (b). Thus, in part (a) the diode current is zero, and the diode voltage is -3 V. In part (b) we get $R_{eq} = (10//2) + (15//15) = 55/6$ kΩ, $v = 0$ V, and $i = 4/(55/6) = 24/55$ mA, now flowing toward the left.

Cut-and-Try Approach

If the diode is part of a *nonlinear* circuit, as in the case of multiple-diode circuits, it is generally not possible to perform a Thévenin's reduction of the surrounding circuitry. However, knowing that the diode must be either on or off, we can use intuition to make an educated assumption, and then check that our results are consistent with our assumption, changing the assumption if needed, until we obtain consistent results. A two-diode circuit provides a classic example of the cut-and-try approach.

EXAMPLE 1.3

(a) The circuit of Fig. 1.8*a* is powered by a dual ±6-V dc power supply system, which we show in concise form (that is, without drawing the actual dc voltage sources) in order to reduce cluttering in the circuit diagram. Find the current I_D and voltage V_D for each diode.

(b) Repeat, but with the resistances interchanged with each other.

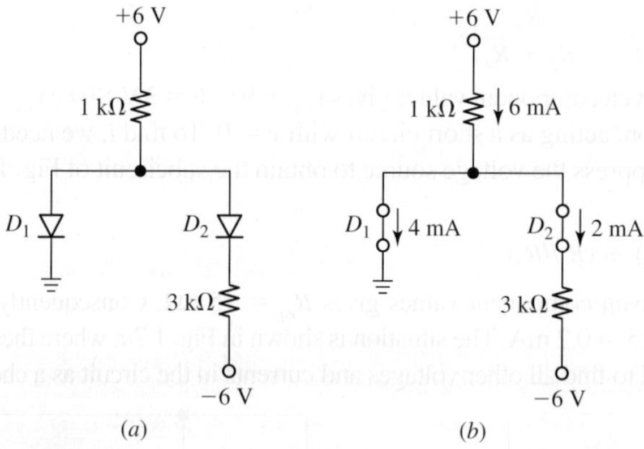

(*a*)   (*b*)

FIGURE 1.8 (a) Circuit of Example 13.*a*. (b) The circuit after it has been solved.

Solution

(a) Knowing that each diode has to be either in conduction (ON) or cut off (CO), we have four possibilities: (D_1, D_2) = (CO, CO), (CO, ON), (ON, CO), (ON, ON). However, considering that the +6-V source tends to source current to both anodes, and that the −6-V source tends to sink current from D_2's cathode, it seems reasonable to assume that *both* diodes be on, or (D_1, D_2) = (ON, ON). Thus, replacing each diode with a short circuit, we end up with the situation of Fig. 1.5*b*, where the diode voltages are

$$V_{D_1} = V_{D_2} = 0 \text{ V}$$

Moreover, applying first Ohm's law and then KCL we have

$$I_{D_2} = I_{3\,k\Omega} = \frac{0 - (-6)}{3} = 2 \text{ mA} \qquad I_{D_1} = I_{1\,k\Omega} - I_{D_2} = \frac{6 - 0}{1} - 2 = 4 \text{ mA}$$

Both diodes carry current in the forward direction, so our results are consistent with our initial assumption, and we are done.

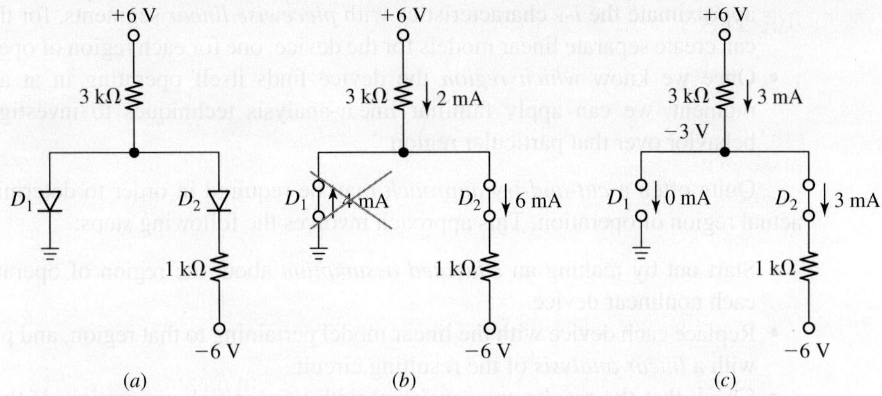

FIGURE 1.9 (a) Circuit of Example 1.3b. (b) First attempt (wrong), and (c) second attempt (successful).

(b) Swapping the resistances results in the circuit of Fig. 1.9a. We may again assume the state $(D_1, D_2) = $ (ON, ON) and refer to the circuit of Fig. 1.9b for our calculations. By Ohm's law, $I_{3\,k\Omega} = 6/3 = 2$ mA and $I_{1\,k\Omega} = [0 - (-6)]/1 = 6$ mA. Then, to satisfy KCL, D_1 would have to supply 4 mA flowing upward, that is, in the reverse direction. But this is impossible, indicating that the assumption $(D_1, D_2) = $ (ON, ON) was wrong. Yet, we note that D_2 must be ON because the negative supply draws current from its cathode. The only plausible assumption is thus $(D_1, D_2) = $ (CO, ON), as shown in Fig. 1.9c. We now have

$$I_{D_1} = 0 \qquad I_{D_2} = \frac{6 - (-6)}{3 + 1} = 3 \text{ mA}$$

By KVL, the voltage at the node common to the anodes is $6 - (3 \times 3) = -3$ V, indicating that D_1 is indeed reverse biased, as assumed at the onset of our second try. So, $V_{D_1} = -3$ V, and $V_{D_2} = 0$.

Exercise 1.1

Find I_D and V_D for each diode in the circuit of Fig. 1.9a if (a) the direction of D_1 is reversed while D_2 is as shown; (b) the direction of D_2 is reversed while D_1 is as shown; (c) the direction of each diode in Fig. 1.9a is reversed.

Ans. (a) $I_{D_1} = 4$ mA, $V_{D_1} = 0$, $I_{D_2} = 6$ mA, $V_{D_2} = 0$; (b) $I_{D_1} = 2$ mA, $V_{D_1} = 0$, $I_{D_2} = 0$, $V_{D_2} = -6$ V; (c) $I_{D_1} = I_{D_2} = 0$, $V_{D_1} = -6$ V, $V_{D_2} = -12$ V.

Concluding Observations

Even though the diode is the simplest electronic element, it offers a glimpse into important features that are common to other more complex electronic devices, such as transistors to be studied later.

- The diode is a classic example of a *nonlinear* circuit element.
- Strictly speaking, the analysis techniques learned in linear-circuits courses cannot be applied to nonlinear elements. However, every attempt should be made to

approximate the *i-v* characteristics with *piecewise linear* segments, for then we can create separate linear models for the device, one for each region of operation.

- Once we know *which region* the device finds itself operating in at a given moment, we can apply familiar linear-analysis techniques to investigate its behavior over that particular region.

Quite often a *cut-and-try approach* may be required in order to determine the actual region of operation. This approach involves the following steps:

- Start out by making an *educated assumption* about the region of operation of each nonlinear device.
- Replace each device with the linear model pertaining to that region, and proceed with a *linear analysis* of the resulting circuit.
- Check that the results are *consistent* with your initial assumption. If they are, we are finished. If they aren't, we need to change our initial assumption until consistent results are obtained.

We shall encounter plenty of examples as we proceed. Lastly, it must be said that computer simulation, such as PSpice, is a powerful tool not only to corroborate any assumptions that we make, but also to provide refinements and details that a piecewise linear approximation of necessity often misses.

1.2 BASIC DIODE APPLICATIONS

Having introduced the fundamentals of diode behavior and diode circuit analysis, we are now ready to examine some of the most common diode applications.

Rectifiers

Figure 1.10*a* depicts one of the most popular diode applications, namely, *signal rectification*. During the *positive* alternations of v_I the diode is on and the circuit gives $v_O = v_I$ (see Fig. 1.10*b*). During the *negative* alternations of v_I the diode is cut off, and the circuit gives $v_O = 0$ V (see Fig. 1.10*c*). Figure 1.11*a* shows the response v_O to a sinusoidal input v_I. We say that the circuit passes on to the load R only the *positive* portions of the input waveform, while the negative portions are *inhibited*. We observe that while v_I alternates in polarity and averages out to zero, v_O is *unipolar* and thus exhibits a nonzero average, or a *nonzero dc component*. For this reason the circuit is referred to as *rectifier*. Its behavior can be visualized also via the *voltage*

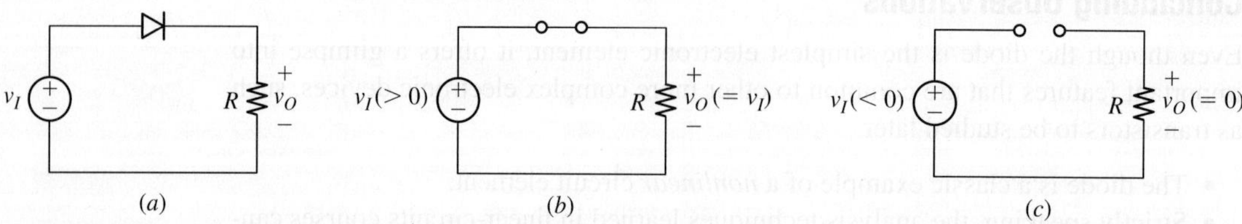

FIGURE 1.10 (a) *Half-wave rectifier* and its equivalent circuits for (b) $v_I > 0$ and (c) $v_I < 0$.

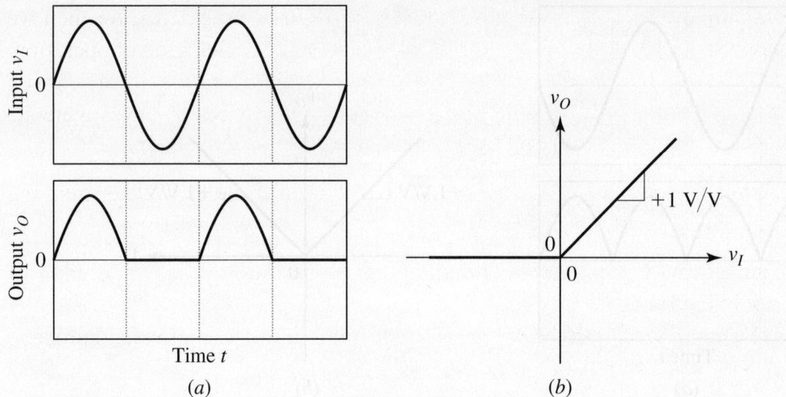

FIGURE 1.11 Illustrating *half-wave rectification* via (a) the *input/output* waveforms and (b) the *VTC*.

transfer curve (VTC), representing the plot of v_O versus v_I. This curve, expressed mathematically as

$$v_O = v_I \quad \text{for } v_I > 0 \qquad \text{(1.3a)}$$
$$v_O = 0 \quad \text{for } v_I < 0 \qquad \text{(1.3b)}$$

is shown in Fig. 1.11b. Needless to say, the presence of the diode results in a *nonlinear* VTC. The student can easily verify that *reversing* the diode interconnection in Fig. 1.10a results in a rectifier that passes only the *negative* portions of the input waveform.

Since it passes only half the input wave, the circuit of Fig. 1.10a is called a *half-wave rectifier*. By contrast, a *full-wave rectifier* passes *both* halves of the input wave, one unchanged and the other inverted. Also called *absolute-value circuit*, it is implemented using the four-diode arrangement of Fig. 1.12a, known as a *diode bridge*. To understand its operation, consider first the case $v_I > 0$, then the case $v_I < 0$.

- For $v_I > 0$ we expect the input source v_I to *source* current to D_1's anode and to *sink* current from D_4's cathode, causing both diodes to be on, as depicted in Fig. 1.12b. Under these circumstances we have $v_O = v_I \, (> 0)$. Moreover, we note that both D_2 and D_3 are reverse biased by the amount v_O, as also shown in Fig. 1.12b. The current loop is thus *source* $\rightarrow D_1 \rightarrow R \rightarrow D_4 \rightarrow$ *source*.

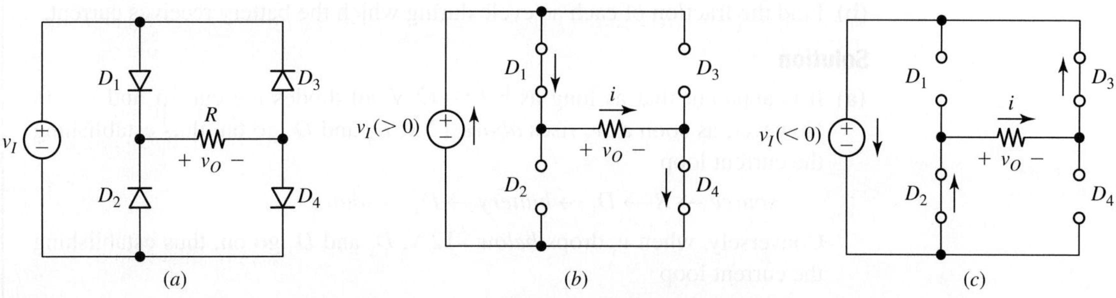

FIGURE 1.12 (a) Full-wave rectifier and its equivalent circuits for (b) $v_I > 0$ and (c) $v_I < 0$.

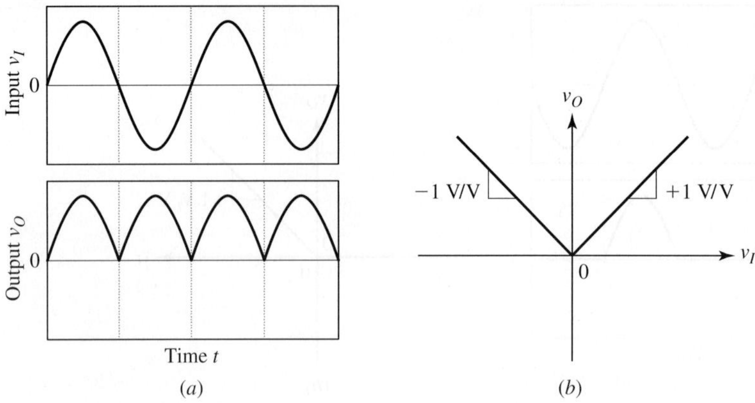

FIGURE 1.13 Illustrating full-wave rectification via (a) the input/output waveforms and (b) the *VTC*.

- For $v_I < 0$ we expect the input source v_I to *sink* current from D_3's cathode and to *source* current to D_2's anode, causing both diodes to be on, as depicted in Fig. 1.12c. We now have $v_O = -v_I$, or $v_O > 0$ because $v_I < 0$. Moreover, both D_1 and D_4 are now reverse biased, as also shown in Fig. 1.12c. The current loop is now *source* → $D_2 \rightarrow R \rightarrow D_3 \rightarrow$ *source*. We observe that the current through R flows toward the right in *both* cases, confirming the absolute-value function mentioned earlier.

Figure 1.13a shows the response v_O to a sinusoidal input v_I, while Fig. 1.13b shows the VTC, which we express concisely as

$$v_O = |v_I| \tag{1.4}$$

Rectifier circuits find application in power electronics, communications, and instrumentation.

EXAMPLE 1.4   (a) Shown in Fig. 1.14 is a simple—if not overly efficient—circuit for charging a 12-V car battery. Assuming v_S is an ac source with a 24-V peak amplitude obtained from the household ac power via a step-down transformer, sketch and label v_S as well as the battery current i.

(b) Find the fraction of each ac cycle during which the battery receives current.

Solution

(a) It is apparent that as long as $|v_S| \leq 12$ V, all diodes are cut off and $i = 0$. However, as soon as v_S rises *above* 12 V, D_1 and D_4 go on, thus establishing the current loop

$$source \rightarrow R \rightarrow D_1 \rightarrow battery \rightarrow D_4 \rightarrow source$$

Conversely, when v_S drops *below* –12 V, D_2 and D_3 go on, thus establishing the current loop

$$source \rightarrow D_2 \rightarrow battery \rightarrow D_3 \rightarrow R \rightarrow source$$

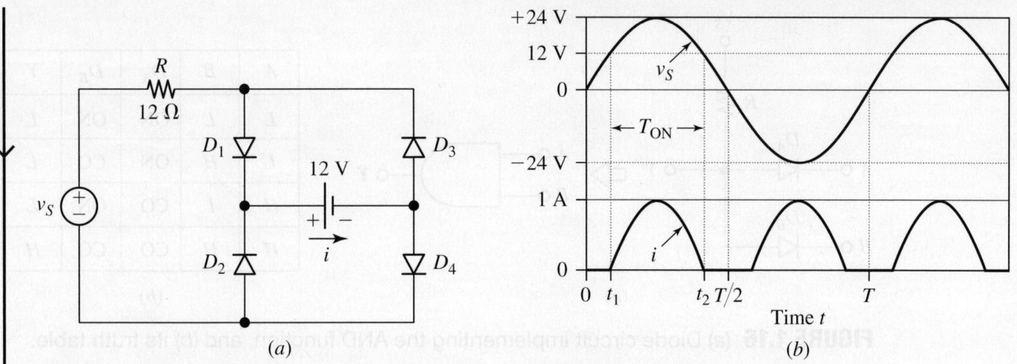

FIGURE 1.14 (a) Simple battery charger, and (b) voltage and current waveforms.

In either case, we have

$$i = 0 \quad \text{for} \quad |v_S| \le 12\,\text{V} \qquad i = \frac{v_S - 12\,\text{V}}{R} \quad \text{for} \quad |v_S| > 12\,\text{V}$$

The peak value of the current is $(24 - 12)/12 = 1$ A. The waveforms are shown in Fig. 1.14b.

(b) During the first cycle, i starts to flow into the battery at the instant t_1 such that

$$24 \sin\frac{2\pi t_1}{T} = 12$$

or $t_1 = T/12$, and stops conducting at $t_2 = T/2 - T/12$. Consequently, the conduction interval is $T_{\text{ON}} = t_2 - t_1 = T/3$, or $T_{\text{ON}} = (2/3)(T/2)$, where $T/2$ is the period of the current waveform. In summary, the battery receives current during $2/3$ of the time.

Diode Logic Gates

Figures 1.15a and 1.16a show how diodes can be used to implement elementary functions of the logic type, which are at the basis of digital systems. The inputs A and B and the output Y are binary-valued voltages that can be either low (L), such as 0 V, or high (H), such as the power-supply voltage V_S (typically, 5 V). Circuit behavior is best understood by examining the rows of the accompanying tables, one at a time.

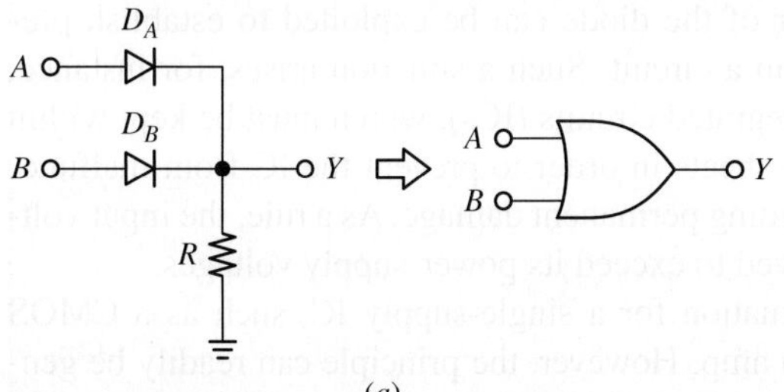

A	B	D_A	D_B	Y
L	L	CO	CO	L
L	H	CO	ON	H
H	L	ON	CO	H
H	H	ON	ON	H

(a)   (b)

FIGURE 1.15 (a) Diode circuit implementing the OR function, and (b) its truth table.

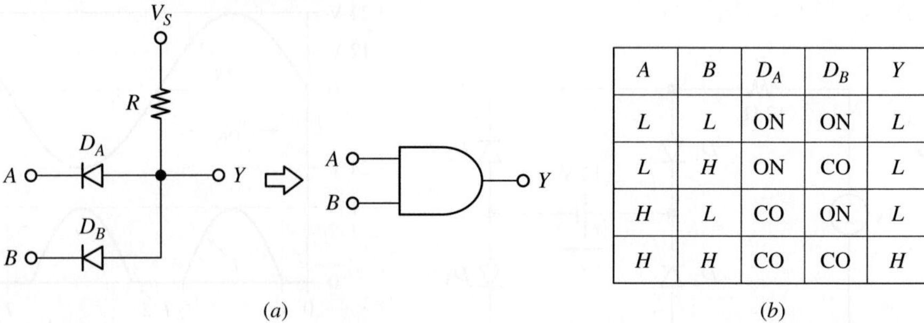

FIGURE 1.16 (a) Diode circuit implementing the AND function, and (b) its truth table.

- As long as both inputs A and B are low (0 V) in Fig. 1.15a, both diodes are off, so all voltages and currents are zero, and Y is low. However, if we drive at least one input high (5 V), we will be sourcing current to the anode of the corresponding diode, turning it on and thus pulling Y also high, as illustrated in Fig. 1.15b. We summarize by saying that Y is high if A *or* B *or both* are high. This logic function, aptly called the OR function, is identified by the logic-gate symbol next to the circuit itself, in Fig. 1.15a. The table of Fig. 1.15b summarizes its behavior and is called the *truth table*.
- As long as at least one input is low (0 V) in Fig. 1.16a, the corresponding diode will be on as we are sinking current from its cathode. So, Y is also low. Only if both A and B are high will both diodes be cut off, causing the current through R to drop to zero. With zero current, the voltage drop across R is also zero, and we say that R *pulls* Y *to* V_S, or high, as illustrated in Fig. 1.16b. This state of affairs is summarized by saying that Y is high if A *and* B are high. This logic function, aptly called the AND function, is identified by the logic-gate symbol next to the circuit itself, in Fig. 1.16a. The table of Fig. 1.16b summarizes its behavior and is called the *truth table*.

Each gate can readily be expanded to handle more than just two inputs by interconnecting additional diodes, as needed. The above gates, though certainly useful, are not sufficient to build a complete digital system because we also need, among others, the *inversion* function. This requires a *transistor*, as we shall study in the next two chapters.

Voltage Clamps

The unidirectional-switch behavior of the diode can be exploited to establish prescribed limits for certain voltages in a circuit. Such a situation arises, for instance, in connection with the inputs to integrated circuits (ICs), which must be kept within limits as recommended in the data sheets in order to prevent the IC from malfunctioning or, even worse, from undergoing permanent damage. As a rule, the input voltages to an IC should never be allowed to exceed its power-supply voltages.

Figure 1.17a illustrates the situation for a single-supply IC, such as a CMOS digital circuit or a single-supply op amp. However, the principle can readily be generalized to multiple-supply systems, such as dual-supply op amps. With reference to Fig. 1.17a, we readily see that as long as the *external* input v_I lies within the range $0 \le v_I \le V_S$, both diodes are off, so the signal *right* at the IC's *input pin* is $v_{IC} \cong v_I$

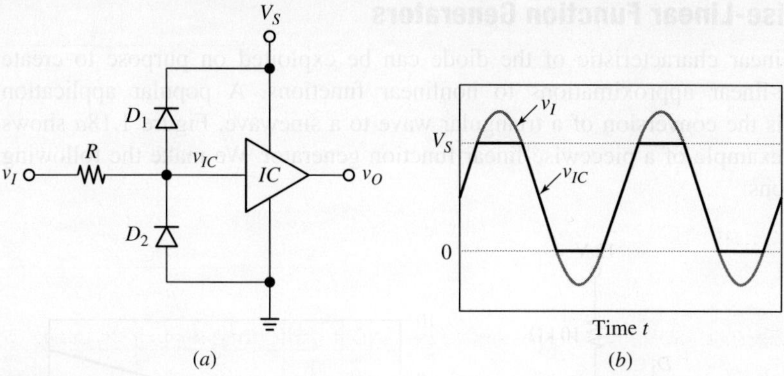

FIGURE 1.17 (a) Using diode clamps to limit the input voltage range of an IC. (b) Waveforms.

(assuming the IC draws negligible input current, which is indeed the case with CMOS logic circuits as well as op amps.) However, should v_I exceed the supply voltage V_S, either because of an oversight by the user or because of interference noise superimposed upon v_I itself, D_1 will go on, thus establishing a short between v_{IC} and V_S. We say that D_1 *clamps* v_{IC} at V_S. By similar reasoning, should v_I drop below ground potential, D_2 will *clamp* v_{IC} at 0 V. As depicted in Fig. 1.17b, the diodes protect the IC against possible input overdrive situations by *limiting* the input-pin voltage within the range

$$0 \leq v_{IC} \leq V_S$$

For this reason, a diode clamp is also referred to as a *limiter*, and since it clips the portions of the input waveform falling outside the allowed range, it is also called a *clipper*.

The need for protection against input overdrives is so important and so common that many ICs come already equipped with internal diode-clamping networks to relieve the user from this worry. In this respect it is worth mentioning the MOSFET as a device particularly sensitive to input overvoltages when handled by humans. Since its gate terminal is the plate of an extremely tiny capacitor, any electrostatic charge that may have accumulated on the body of the user will be transferred to this capacitor during manual contact, thus leading to potentially high voltages (as per $V = Q/C$) that are likely to destroy the capacitor's dielectric. However, in the presence of suitable diode clamps, the body of the handler will discharge through one of the diodes, saving the device from assured destruction.

Exercise 1.2

In the circuit of Fig. 1.17 let $R = 10$ kΩ and $V_S = 5$ V. Assuming the IC draws no current at its input terminal, find the current through R (magnitude and direction) for the following values of v_I: (a) $= 1$ V, (b) 8 V, (c) -2 V, (d) 4.5 V. (e) If it is found that R draws 0.25 mA flowing toward the right, what do you conclude about v_I? (f) What if R draws 0.5 mA flowing toward the left? (g) What if the current through R is zero?

Ans. (a) 0 mA; (b) 0.3 mA ($\rightarrow$); (c) 0.2 mA ($\leftarrow$); (d) 0 mA; (e) $v_I = 7.5$ V; (f) $v_I = -5$ V; (g) $0 \leq v_I \leq 5$ V.

Piecewise-Linear Function Generators

The nonlinear characteristic of the diode can be exploited on purpose to create piecewise-linear approximations to nonlinear functions. A popular application example is the conversion of a triangular wave to a sinewave. Figure 1.18*a* shows a simple example of a piecewise-linear function generator. We make the following observations:

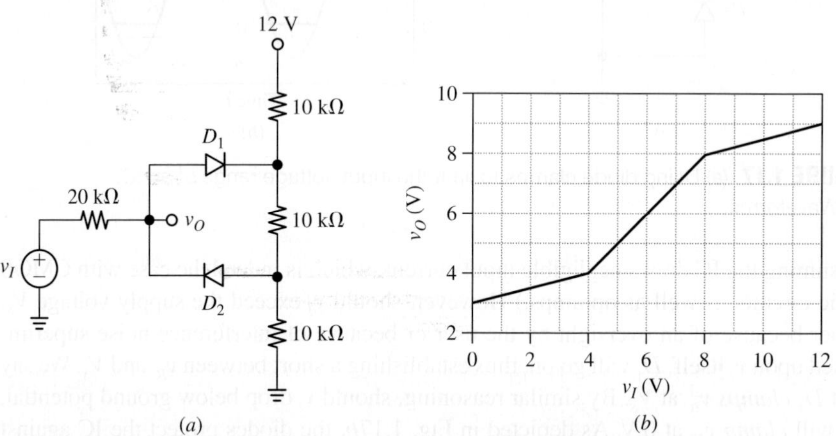

FIGURE 1.18 (*a*) Piecewise-linear function generator example, and (*b*) its VTC.

- With both diodes off, the three 10-kV resistors partition the 12-V supply voltage into three equal voltage drops, thus giving 4 V and 8 V, respectively. This is illustrated in Fig. 1.19*a*, where we observe that both diodes are off as long as the input is kept within the range 4 V $< v_I <$ 8 V. Since no current flows through the 20-kΩ resistor, the latter drops 0 V, so the circuit gives

$$v_O = v_I \qquad \text{for } 4 \text{ V} < v_I < 8 \text{ V} \tag{1.5a}$$

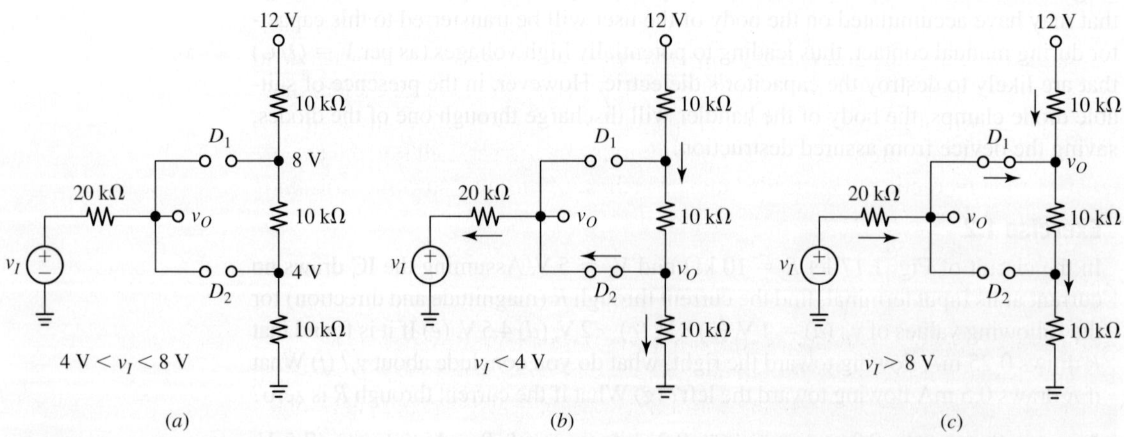

FIGURE 1.19 Equivalent circuits of the function generator of Fig. 1.18 for different input conditions.

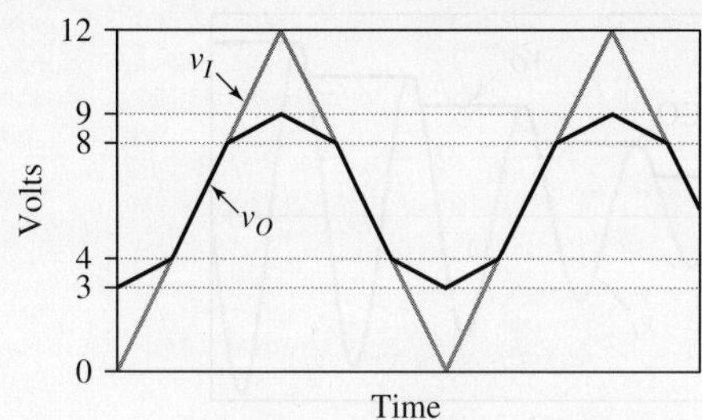

FIGURE 1.20 Illustrating
the waveshaping effect of
the piecewise-linear circuit
of Fig. 1.18a.

- If we lower v_I *below* 4 V, D_2 goes on while D_1 remains off, resulting in the situation of Fig. 1.19b. By KCL we have

$$\frac{12 - v_O}{10 + 10} = \frac{v_O - v_I}{20} + \frac{v_O}{10}$$

which is readily solved for v_O to give

$$v_O = 0.25v_I + 3 \text{ V} \qquad \text{for } v_I < 4 \text{ V} \tag{1.5b}$$

- If we raise v_I *above* 8 V, D_1 goes on while D_2 remains off, resulting in the situation of Fig. 1.19c. Applying again KCL, we get

$$\frac{v_I - v_O}{20} + \frac{12 - v_O}{10} = \frac{v_O}{10 + 10}$$

which is readily solved for v_O to give

$$v_O = 0.25v_I + 6 \text{ V} \qquad \text{for } v_I > 8 \text{ V} \tag{1.5c}$$

Figure 1.18b shows the circuit's VTC. Clearly, we have *three* separate regions of operation. Within each region, the VTC is a straight segment whose analytical expression is obtained by drawing the corresponding equivalent circuit, as per Fig. 1.19, and subjecting it to elementary linear analysis techniques. This is typical of diode circuits.

Figure 1.20 shows the waveshaping effect of the circuit of Fig. 1.18 upon a triangular wave. Here, the aim is to compress the top and the bottom of the triangle to approximate a sine. Clearly, the present example provides a rather crude approximation, but it is not hard to imagine that we can improve it by using additional segments. This is normally accomplished with *pn* junction diodes. As we shall see, the characteristic of the *pn* diode has a rounded knee which helps ensure a smoother transition from one segment to the next of the VTC.

Peak Detectors

Additional interesting applications arise if a diode D is partnered by a capacitor C. As our first example we consider the circuit of Fig. 1.21a, known as a *peak detector*. To understand its operation, refer to the waveforms of Fig. 1.21b, where C is assumed initially discharged. As v_I swings positive, current will be sourced to the anode, forcing D to conduct and thus to charge up C. Since D acts as a short circuit, v_O simply follows v_I.

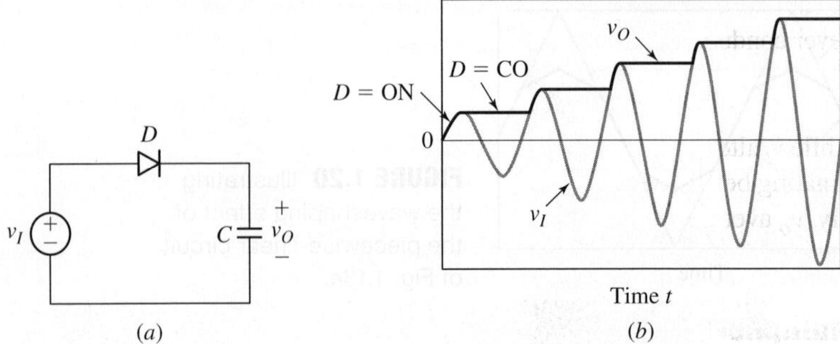

FIGURE 1.21 (*a*) Peak detector, and (*b*) illustrative waveforms.

Once v_I peaks out, the anode will start swinging negative relative to the cathode, turning D off and thus leaving C to hold the previously acquired voltage, whose value coincides with that of the previous input peak. This *memory action* by C will persist until v_I rises again to a new, higher peak. When this happens, D will go on again, charging up C to this new peak.

It is apparent that the diode's directionality causes the capacitor voltage to experience only *increases*, this being the reason why the circuit shown is said to be a *positive* peak detector. Reversing the diode interconnection in the circuit will cause the capacitor voltage to experience only *decreases*, thus yielding a *negative* peak detector. Peak detectors are at the basis of demodulators in the detection of audio signals in *amplitude-modulation* (AM) radio receivers.

The Clamped Capacitor or Dc Restorer

In the circuit of Fig. 1.22*a* the diode provides a clamping function that prevents v_O from ever going negative. To understand circuit operation, refer to the waveforms of Fig. 1.22*b*, where the input is a sine wave alternating between $+V_m$ and $-V_m$, and C is assumed initially discharged. During the initial positive alternation of v_I we simply have $v_O = v_I$ as the voltage across C is zero and the diode is off. However, as v_I goes *below* 0 V for the first time, the input source will sink current from D's cathode via C, thus turning D on and forcing C to charge. As v_I reaches its negative peak of $-V_m$ at $t = t_2$,

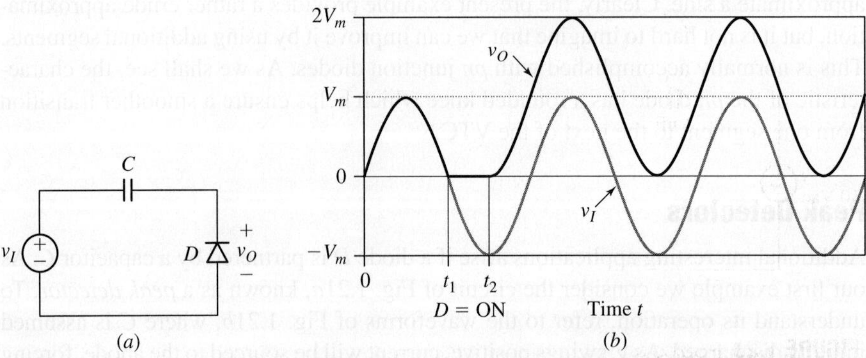

FIGURE 1.22 (*a*) Clamped capacitor, and (*b*) illustrative waveforms.

C will have acquired a voltage equal to V_m, positive at the right plate. After $t = t_2$, the diode will never conduct again, and C will retain the voltage acquired last, thus giving

$$v_O = v_I + V_m \qquad (1.6)$$

for $t \geq t_2$. While v_I alternates between $-V_m$ and $+V_m$, and thus averages out to zero, v_O ends up alternating between 0 and $+2V_m$ because of the voltage offset provided by C. Consequently, v_O averages out to $+V_m$, and because of this nonzero dc component at its output, the circuit is also referred to as a *dc restorer*.

Voltage Multipliers and PSpice Simulation

An interesting property of the clamped-capacitor circuit of Fig. 1.22*a* is that it provides a positive output peak that is *twice* that of the input. This indicates that if we feed the output of the clamped capacitor circuit to a peak detector, we can synthesize a dc voltage of value $+2V_m$, that is, of twice the amplitude of the input! To better understand the behavior of this composite circuit, we simulate it via PSpice. Aptly called a *voltage doubler*, the circuit is shown in Fig. 1.23 for the case of a 1-kHz, 10-V sinusoidal input. The circuit utilizes pseudo-ideal diodes, whose PSpice models have been created by editing one of the already existing diode models available in PSpice's library, and then setting the parameters I_s and n to the values shown in the table (more on this in Appendix 1A).

The various waveforms are displayed in Fig. 1.24, where we observe that after a transient situation lasting less than ten cycles, the output settles to the dc value

$$v_2 \to 2V_m \qquad (1.7)$$

or $v_2 = 20$ V in the present example. The reason it takes several cycles to attain the desired steady-state situation is that C_1 is being *loaded* by C_2. If the circuit were implemented with $C_1 \gg C_2$, steady state would be achieved essentially at the second positive peak of v_I. However, with equal capacitors, which is how the circuit is usually implemented, the charge accumulated by C_1 at each positive peak of v_I redistributes equally between C_1 and C_2, causing a reduction in v_1 which, though significant at the first peak of v_I, becomes progressively less relevant with each subsequent peak. One can prove (see Exercise 1.3), that the values attained by v_2 at each peak of v_I are 5 V, 12.5 V, 16.25 V, 18.125 V. . . .

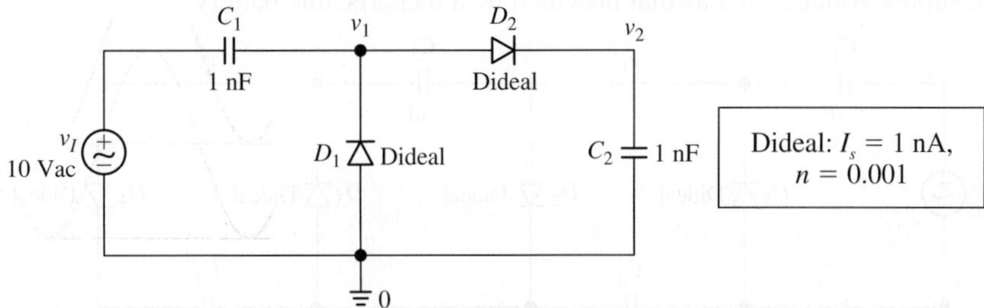

FIGURE 1.23 PSpice circuit simulating a *voltage doubler* based on pseudo-ideal diodes. The parameters of the PSpice diode model are shown inside the box at the right.

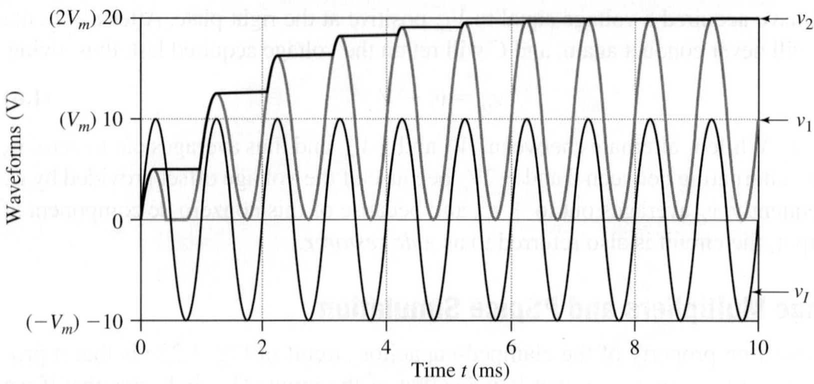

FIGURE 1.24 Waveforms for the PSpice circuit of Fig. 1.23.

Exercise 1.3

Denoting the value of v_2 following the k-th peak of v_1 as $v_2(k)$, use the charge conservation principle to show that $v_2(k)$ is related to the value $v_2(k-1)$ following the previous peak as

$$v_2(k) = 0.5v_2(k-1) + 10 \text{ V}$$

$k = 2, 3, 4 \ldots$, and $v_2(1) = 5$ V. At which peak of v_1 does v_2 come within about 10% of 20 V? 1% of 20 V?

Ans. For 10%, $k = 3$. For 1%, $k = 6$.

The principle at the basis of the voltage doubler can be generalized to achieve higher voltage-multiplication factors. Figure 1.25 shows a *voltage quadrupler,* whose waveforms are displayed in Fig. 1.26. We see again that after a transient situation lasting a certain number of cycles the output v_4 eventually settles to the dc value

$$v_4 \to 4V_m \tag{1.8}$$

or $v_4 = 40$ V in the present example. The reader is encouraged to trace through each waveform in detail to develop a feel for the workings of this clever circuit.

Voltage multipliers find application in integrated circuits, where it is desired to synthesize specific voltages to bias different on-chip circuits, starting out with a single supply voltage such as that provided by a rechargeable battery.

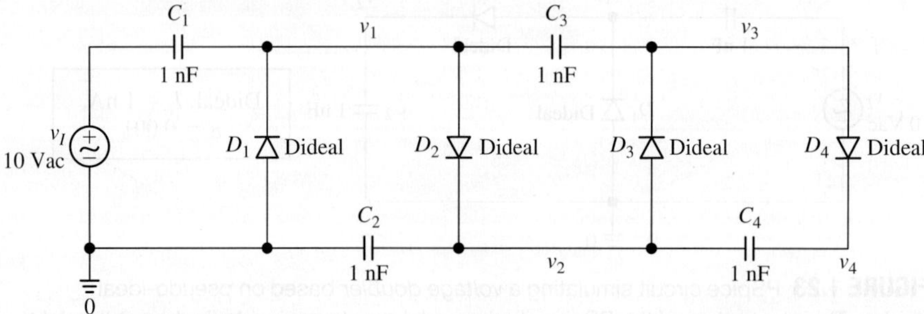

FIGURE 1.25 Voltage quadrupler.

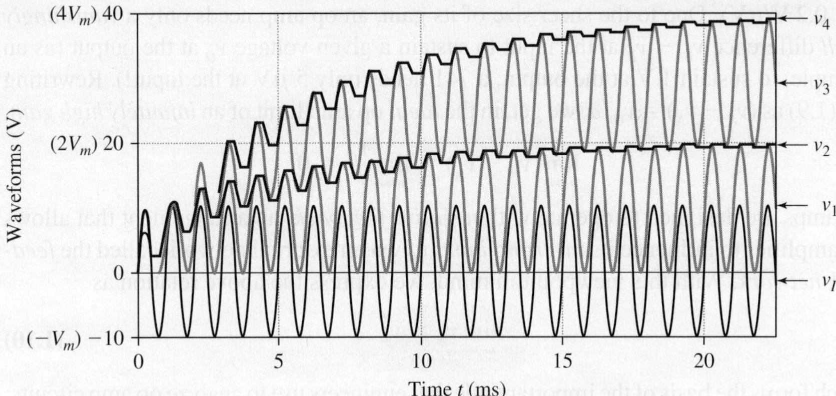

FIGURE 1.26 Waveforms for the voltage quadrupler of Fig. 1.25.

1.3 OPERATIONAL AMPLIFIERS AND DIODE APPLICATIONS

As we move along we shall find that the range of applications for diodes and, later, for transistors can be expanded significantly if we team up these devices with the operational amplifier, or op amp for short. Recall from prerequisite circuit courses that an op amp is a *high-gain voltage amplifier* that accepts two inputs, called the *inverting* input v_N and the *noninverting* input v_P, and yields an output v_O such that

$$v_O = a(v_P - v_N) \tag{1.9}$$

where a (in V/V) is the *voltage gain* (see Fig. 1.27a). In order to function, an op amp needs to be powered. Figure 1.27a shows dual power supplies of $\pm V_S$, but a single supply is also common. (To reduce cluttering, it is customary to omit showing the supplies explicitly.) Physically, an op amp cannot swing v_O above $+V_S$ or below $-V_S$. Overdriving it will simply cause v_O to *saturate* at some voltage V_{OH} in the vicinity of $+V_S$ or at some voltage V_{OL} in the vicinity of $-V_S$, so Eq. (1.9) holds only so long as we confine v_O within the range $V_{OL} < v_O < V_{OH}$, aptly called the *linear output range*. Figure 1.27b shows the op amp's voltage transfer curve (VTC).

As we move along we shall find that the higher the voltage gain a, the better (as an example, the venerable 741 op amp has $a = 200{,}000$ V/V, also expressed as

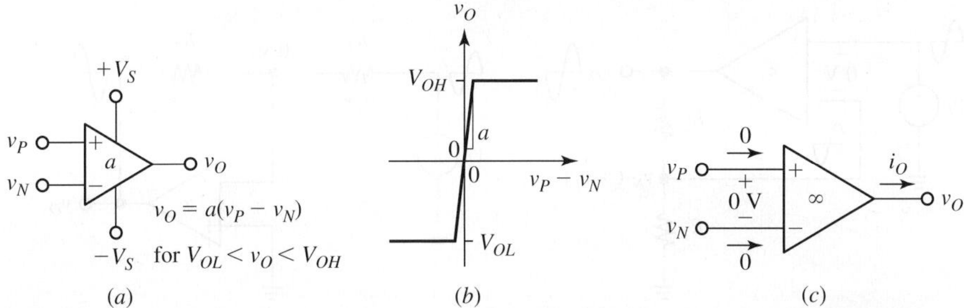

FIGURE 1.27 (a) Op amp symbol and labels. (b) The gain a is the slope of the voltage transfer curve (VTC). (c) The input voltage of an *ideal op amp* approaches 0 V. Moreover, it draws no input currents.

$a = 0.2$ V/μV). Due to the sheer size of its gain, an op amp needs only a *vanishingly small* difference $v_P - v_N$ at the input to sustain a given voltage v_O at the output (as an example, to sustain 1 V at the output, a 741 needs only 5 μV at the input!). Rewriting Eq. (1.9) as $(v_P - v_N) = v_O/a$, we get, in the *ideal* op amp limit of an *infinitely high gain*,

$$\lim_{a \to \infty} (v_P - v_N) = \lim_{a \to \infty} \frac{v_O}{a} = 0$$

Op amps are designed to operate with *negative feedback*, an arrangement that allows the amplifier to influence its *inverting input* v_N via an external network called the *feedback network*. With this viewpoint in mind, we express the above relation as

$$\lim_{a \to \infty} v_N = v_P \tag{1.10}$$

which forms the basis of the important rule that engineers use to analyze op amp circuits:

Op Amp Rule: When given the ability to influence its own input v_N via negative feedback, an ideal op amp will output whatever voltage v_O and current i_O it takes to force v_N to *track* v_P. Moreover, the op amp will do this without drawing any current at either input pin (see Fig. 1.27c).

Let us put to use this rule to review the most popular op amp circuits.

Basic Op Amp Circuits

The most popular op amp circuits are the *noninverting amplifier*, the *inverting amplifier*, the *summing amplifier*, and the *voltage buffer*.

- The **Noninverting Amplifier**: In the circuit of Fig. 1.28a the op amp influences v_N via the voltage divider made up of R_1 and R_2 to give

$$v_N = \frac{R_1}{R_1 + R_2} v_O = \frac{v_O}{1 + R_2/R_1}$$

By the op amp rule we have $v_N = v_P (= v_I)$. Consequently, replacing v_N with v_I in the above expression and solving for the ratio v_O/v_I gives

$$A = \frac{v_O}{v_I} = 1 + \frac{R_2}{R_1} \tag{1.11}$$

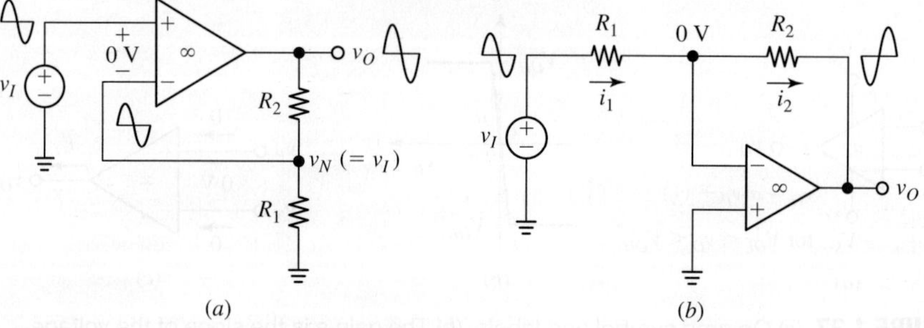

FIGURE 1.28 (a) Noninverting and (b) inverting op amp configurations.

where A represents the gain of the entire circuit (not to be confused with the gain $a\ (\to \infty)$ of the basic op amp). Since v_O has the *same* polarity as v_I the circuit is referred to as *noninverting amplifier*.

- The **Inverting Amplifier**: In the circuit of Fig. 1.28b the op amp influences v_N via the *feedback resistance* R_2, and since the op amp rule implies $v_N = v_P = 0$, we refer to node v_N as a *virtual ground*. Any current injected by v_I via R_1 is removed by v_O via R_2, or $i_1 = i_2$. Using Ohm's law,

$$\frac{v_I - 0}{R_1} = \frac{0 - v_O}{R_2}$$

Solving again for the ratio v_O/v_I gives

$$A = \frac{v_O}{v_I} = -\frac{R_2}{R_1} \tag{1.12}$$

Since the polarity of v_O is *opposite* to that of v_I the circuit is called an *inverting amplifier*.

- The **Summing Amplifier**: By the op amp rule, the inverting-input node in Fig. 1.29 is at virtual ground. This node is called a *summing junction* because it sums the currents coming from the input sources v_1 and v_2 and diverts this sum to the output node v_O to give

$$\frac{v_1 - 0}{R_1} + \frac{v_2 - 0}{R_2} = \frac{0 - v_O}{R_3}$$

Solving for v_O,

$$v_O = -\left(\frac{R_3}{R_1}v_1 + \frac{R_3}{R_2}v_2\right) \tag{1.13}$$

If $R_1 = R_2$, the circuit gives $v_O = -(R_3/R_1)(v_1 + v_2)$ and is aptly called a *summing amplifier*.

- The **Voltage Buffer**: Letting $R_2 = 0$ and $R_1 = \infty$ in the circuit of Fig. 1.28a turns it into a *unity-gain amplifier* ($A = 1$ V/V). Its main application is as a *voltage buffer* to eliminate inter-stage loading. As an example, consider Fig. 1.30a, where a signal

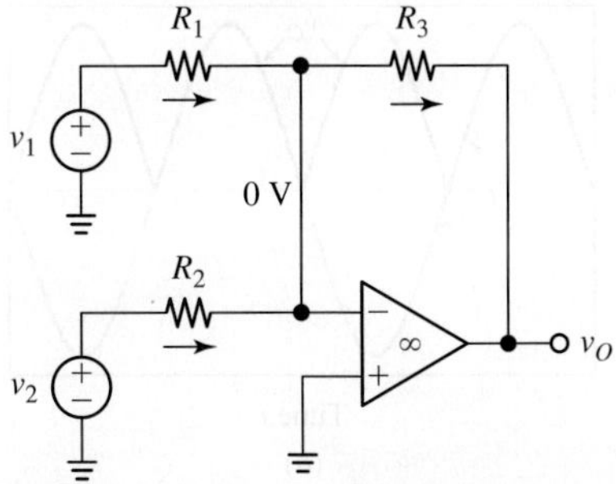

FIGURE 1.29 Summing amplifier.

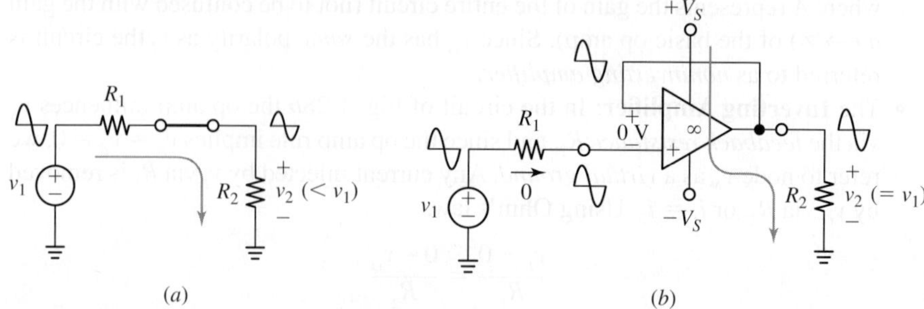

FIGURE 1.30 Using a unity-gain voltage buffer to eliminate loading (current polarities shown for $v_1 > 0$).

source v_1 with internal resistance R_1 is to be fed to a load R_2. If we connect the source to the load via a plain wire, R_2 will form a voltage divider with R_1, giving $v_2 = v_1/(1 + R_1/R_2)$. Clearly v_2 is *less* than v_1, a situation referred to as *loading* and stemming from the fact that R_2 draws current via R_1, so there is voltage *loss* across R_1.

However, if we couple the source to the load via a buffer as in Fig. 1.30*b*, there will be no voltage drop across R_1 because the op amp is designed to draw no input currents. Consequently the buffer eliminates loading to give $v_2 = v_1$. Of course R_2 does draw current, but this is supplied by the op amp, which in turn draws it from the power supply $+V_S$ rather than from v_1. (The illustration refers to the case $v_1 > 0$; for $v_1 < 0$ the load current will flow from R_2 through the op amp to $-V_S$.)

The above configurations show up so often, either on their own or as subcircuits of more complex systems, that we shall apply Eqs. (1.10) through (1.13) quite frequently.

Our First Diode/Op-Amp Circuit

Having reviewed op amp basics, we are now ready to investigate our first diode/op-amp circuit, shown in Fig. 1.31*a*. Where do we begin? As a rule, start out using *simple inspection* to see if you can identify *already familiar subcircuits*, and then build up your understanding from there, one step at a time. In the present case we observe that D_2 and R_3 form a *half-wave rectifier* of the type of Fig. 1.10*a*, so we can follow the line of reasoning developed there and analyze the cases $v_I > 0$ and $v_I < 0$ separately.

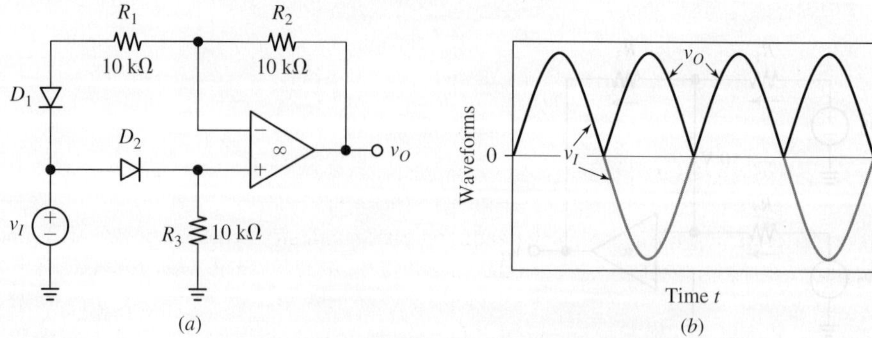

FIGURE 1.31 (a) Our first diode/op-amp circuit, and (b) its input and output waveforms.

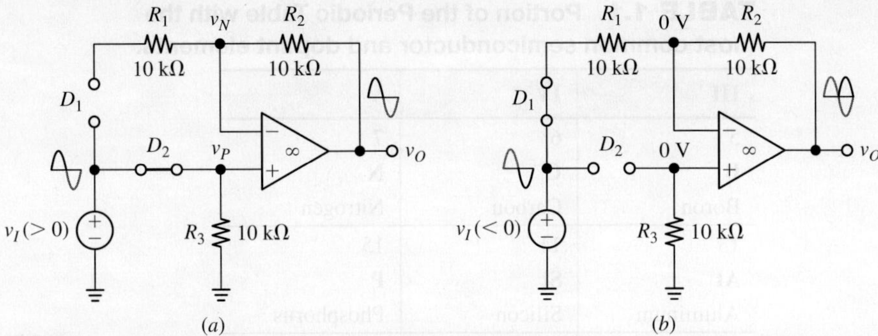

FIGURE 1.32 Redrawing the circuit of Fig. 1.31a for (a) $v_I > 0$ and (b) $v_I < 0$.

- For $v_I > 0$, D_2 is on, making $v_P = v_I$. But, by the op amp rule we have $v_N = v_P$, and thus $v_N = v_I$. This causes D_1 to be off, as pictured in Fig. 1.32a. In the absence of any current through R_2 the op amp gives $v_O = v_N$, so

$$v_O = v_I \tag{1.14a}$$

- For $v_I < 0$, D_2 is off, making $v_P = 0$ and thus, by the op amp rule, $v_N = 0$. D_1 is now on, as pictured in Fig. 1.32b, and the op amp acts as an *inverting amplifier* to give

$$v_O = -\frac{R_2}{R_1}v_I = -\frac{10}{10}v_I = -v_I \tag{1.14b}$$

- We combine the two expressions by writing

$$v_O = |v_I| \tag{1.15}$$

and by stating that the circuit is a *full-wave rectifier*. The function is similar to that provided by the circuit of Fig. 1.12a and graphed in Fig. 1.13. There is a difference, however: in the circuit of Fig. 1.12a the rectified signal appears across a *floating load*, whereas in the op amp version of Fig. 1.31a the rectified signal is referenced to ground and as such it can be applied to a *grounded load*.

Exercise 1.4

Find v_O in the circuit of Fig. 1.31a if (a) R_3 is doubled; (b) R_2 is doubled; (c) R_1 is doubled; (d) the direction of each diode is reversed; (e) only the direction of D_1 is reversed.

Ans. (a) $v_O = |v_I|$; (b) $v_O = v_I$ for $v_I > 0$; $v_O = -2v_I$ for $v_I < 0$; (c) $v_O = v_I$ for $v_I > 0$, $v_O = -0.5v_I$ for $v_I < 0$; (d) $v_O = -|v_I|$; (e) $v_O = v_I$ for $v_I > 0$, $v_O = 0$ for $v_I < 0$.

1.4 SEMICONDUCTORS

The material in widest use by the semiconductor industry today is *silicon* (Si), an element of Group IV of the periodic table of elements (see the portion shown in Table 1.1). The atoms of Group IV elements possess *four* electrons in their outer electron shell, also called the *valence band*. Each atom shares these four electrons with its four nearest neighbor atoms to form *covalent bonds*. These bonds keep atoms

TABLE 1.1 Portion of the Periodic Table with the most common semiconductor and dopant elements.

III	IV	V
5	6	7
B	**C**	**N**
Boron	Carbon	Nitrogen
13	14	15
Al	**Si**	**P**
Aluminum	Silicon	Phosphorus
31	32	33
Ga	**Ge**	**As**
Gallium	Germanium	Arsenic
49	50	51
In	**Sn**	**Sb**
Indium	Tin	Antimony

bound to fixed positions of an orderly spatial structure known as *crystal lattice*. This is depicted in the two-dimensional rendition of Fig. 1.33*a*. The number of silicon atoms per unit volume (atoms/cm^3), also called *atomic density*, is

$$N_{si} = 5 \times 10^{22} \text{ atoms/cm}^3 \tag{1.16}$$

Due to thermal agitation, a covalent bond may break on occasion, freeing an *electron* which then becomes available for conduction. Consequently, the parent atom is said to be *ionized*. As we know, the electron charge is $-q$, where

$$q = 1.602 \times 10^{-19} \text{ C} \tag{1.17}$$

Once an electron moves away from a covalent site, it leaves behind a vacancy having charge $+q$, as shown in Fig. 1.33*b*. An electron from another covalent bond may fill this vacancy, leaving in turn behind another vacancy at the covalent site of origin.

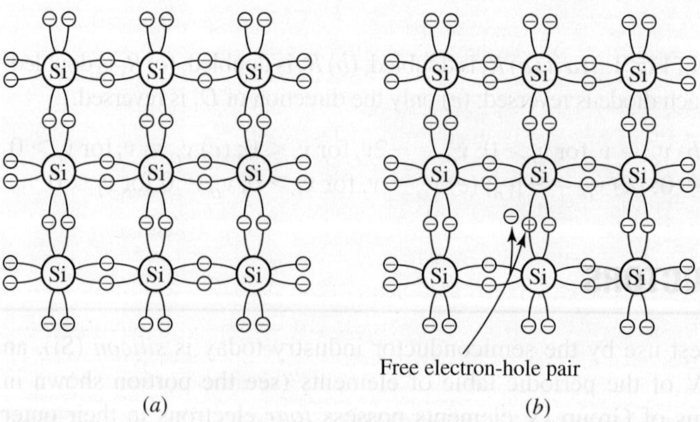

FIGURE 1.33 (a) Pure silicon. (b) Creation of an electron-hole pair by thermal agitation.

As this process repeats itself, we are in effect witnessing the motion of positively charged vacancies, or *holes*, through the crystal.

Just as thermal agitation results in the creation of free electron-hole pairs, an electron and a hole may recombine and thus disappear from the pool of free charges. The recombination rate is proportional to the number of free electron-hole pairs available, which in turn is a strong function of temperature. In thermal equilibrium, the recombination rate equals the generation rate, resulting in an equilibrium *electron concentration* (or *density*) n (electrons/cm^3) and *hole concentration* (or *density*) p (holes/cm^3) such that

$$n = p = n_i \tag{1.18}$$

where n_i is called the *intrinsic concentration*. For silicon, n_i is such that

$$n_i^2(T) = BT^3 e^{-V_{G0}/V_T}\ \text{cm}^{-6} \tag{1.19}$$

where T is absolute temperature, in K, B is a suitable constant, $V_{G0} = 1.205$ V is the *bandgap voltage* for silicon, and

$$V_T = \frac{kT}{q} \tag{1.20}$$

is a temperature-dependent scaling factor, in V, often arising in semiconductor physics and called the *thermal voltage*. Here q is the electron charge and $k = 1.381 \times 10^{-23}$ J/K is Boltzmann's constant. At room temperature ($T = 300$ K), we have $V_T = 25.86$ mV $\cong 26$ mV.

For silicon, Eq. (1.19) takes on the form

$$n_i^2(T) = 1.5 \times 10^{33} T^3 e^{-14,028/T}\ \text{cm}^{-6} \tag{1.21}$$

We observe that n_i is a strong function of temperature. At $T = 300$ K we have $n_i^2 = 2 \times 10^{20}$ cm^{-6}, or $n_i = 1.4 \times 10^{10}$/cm^3, indicating that only one in about 36×10^{12} silicon atoms is ionized. By contrast, in a good conductor each atom contributes one or more electrons to conduction. It is apparent that at room temperature pure silicon is not much of a conductor.

Doping

The electric properties of an element of Group IV can be altered dramatically by replacing some of its atoms with atoms of elements from the adjacent groups. For instance, replacing an atom of silicon with one of *phosphorous* (P), which belongs to Group V and thus possesses *five* electrons in its outer shell, results in the situation of Fig. 1.34a. Four of the five electrons will go into forming covalent bonds, just like those of a silicon atom would; the fifth electron, due to thermal agitation, will wander throughout the crystal, no longer belonging to any particular atom and thus being free and available for conduction. Group V elements are referred to as *donors* in that each atom contributes, or donates, an electron to the silicon crystal. By contrast, replacing an atom of silicon with one of *boron* (B), which belongs to

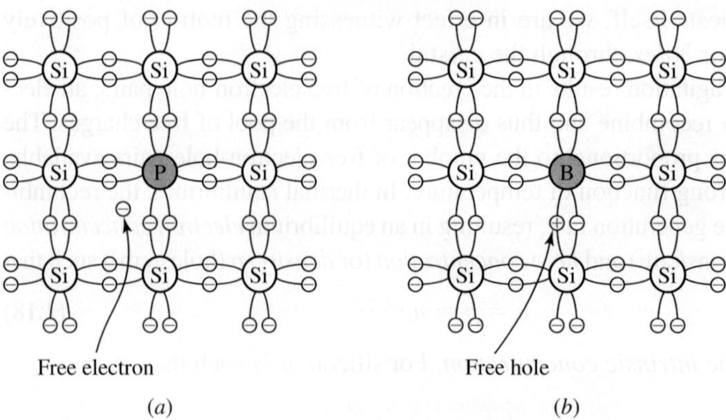

FIGURE 1.34 Silicon with (a) a *donor* atom, and (b) an *acceptor* atom.

Group III and thus possesses only *three* electrons in its outer shell, will lead to the situation of Fig. 1.34*b*. Here, the lack of a fourth electron results in a hole, and since holes in turn accept electrons when they recombine, Group III elements are referred to as *acceptors*.

The replacement of silicon atoms with donor or acceptor atoms is called *doping*. Since doped silicon is no longer pure, donor and acceptor atoms are collectively referred to as *impurities*. By doping silicon with a sufficient number of impurities, we can turn it into a good conductor—hence the reason for calling it a *semiconductor*. Doping can be achieved in more than one way. With *solid-state diffusion*, the doping material is deposited on a selected area of the silicon crystal, which is then placed in a high temperature furnace to force the impurity atoms to penetrate and diffuse into the underlying region of the crystal. With *ion implantation*, the silicon crystal is bombarded with ions of the desired impurity material, which then remain embedded in the crystal. Doped silicon can also be *grown directly as a crystal*, starting with a suitable mixture of silicon and impurity atoms of the desired type.

The donor and acceptor *concentrations* (atoms/cm^3), also called doping *densities*, are denoted as N_D and N_A, respectively. Depending on the particular requirements, doping doses may range from as low as 10^{14} atoms/cm^3 to as high as 10^{21} atoms/cm^3, that is, practical impurity densities are always *much higher* than the room-temperature intrinsic electron-hole concentration ($n_i = 1.4 \times 10^{10}$/cm^3). Consequently, at room temperature, silicon doped with *donor* impurities, also called *n-type silicon*, has $n >$ N_D, while silicon doped with *acceptor* impurities, also called *p-type silicon*, has $p > N_A$.

Regardless of the type and amount of doping, the electron and hole concentrations satisfy at all times the **mass-action law**,

$$n \times p = n_i^2 \qquad (1.22)$$

or $n \times p = 2 \times 10^{20}$ cm^{-6} at room temperature ($T = 300$K). Consequently, in *n-type* silicon we have

$$n \cong N_D \qquad p \cong \frac{n_i^2}{N_D} \qquad (1.23a)$$

while in *p-type* silicon we have

$$p \cong N_A \qquad n \cong \frac{n_i^2}{N_A} \qquad\qquad (1.23b)$$

To put numbers in perspective, suppose some *n*-type silicon has been doped with $N_D = 10^{16}/cm^3$. Then, $n \cong 10^{16}/cm^3$ and $p \cong 2 \times 10^{20}/10^{16} = 2 \times 10^4/cm^3$, indicating a material much richer in electrons and much poorer in holes than intrinsic silicon. Since $n \gg p$, electrons in *n*-type silicon are aptly called *majority charge carriers*, and holes *minority charge carriers*. Conversely, a *p*-type silicon with $N_A = 10^{18}/cm^3$ would have $p \cong 10^{18}/cm^3$ and $n \cong 2 \times 10^2/cm^3$. Since now $p \gg n$, the majority carriers in *p*-type silicon are holes, and the minority carriers are electrons.

It is understood that the thermal generation of electron-hole pairs continues to take place just as in the intrinsic case. However, with such an abundance of majority carriers, the chance of recombining for minority carriers is now much higher, this being the reason for their much reduced concentration. In fact, the balance between the two charge types is governed by the mass-action law!

We note that the designations *n-type* and *p-type* identify only the type of majority carriers in the given material. They should not mislead the reader into regarding *n*-type material as negatively charged, or *p*-type material as positively charged! Regardless of the doping type, the material remains at all times *neutrally charged* because for every charge that has been freed there is the charge of the ionized atom left behind, which is of the opposite polarity. Figure. 1.35 depicts *n* and *p* for three significant cases.

Drift and Diffusion Currents

There are two types of conduction mechanisms in semiconductors, often coexisting: *drift* and *diffusion*.

- **The Drift Current:** To discuss the drift mechanism, refer to Fig. 1.36*a*, top, where a slab of *p*-type material is assumed to be immersed in an electric field of strength E (in V/cm). Such a field can be produced by connecting an external battery across the slab. If the slab is homogeneous, as assumed here, the

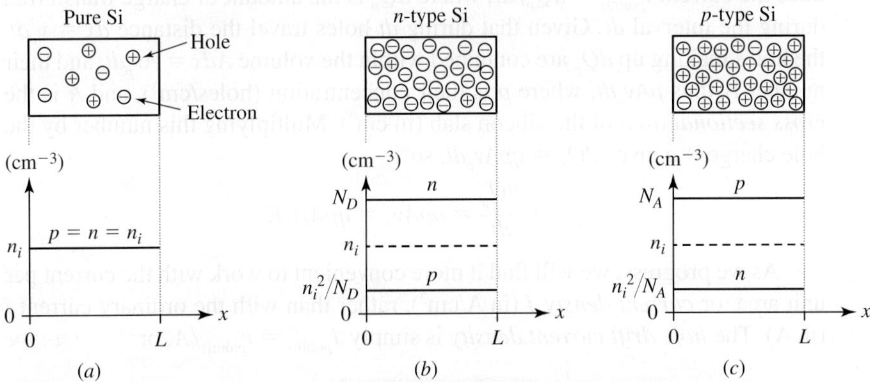

FIGURE 1.35 Mobile-charge concentrations in a slab of (a) *pure* silicon, (b) *n-type* silicon, and (c) *p-type* silicon. (The vertical scales are logarithmic.)

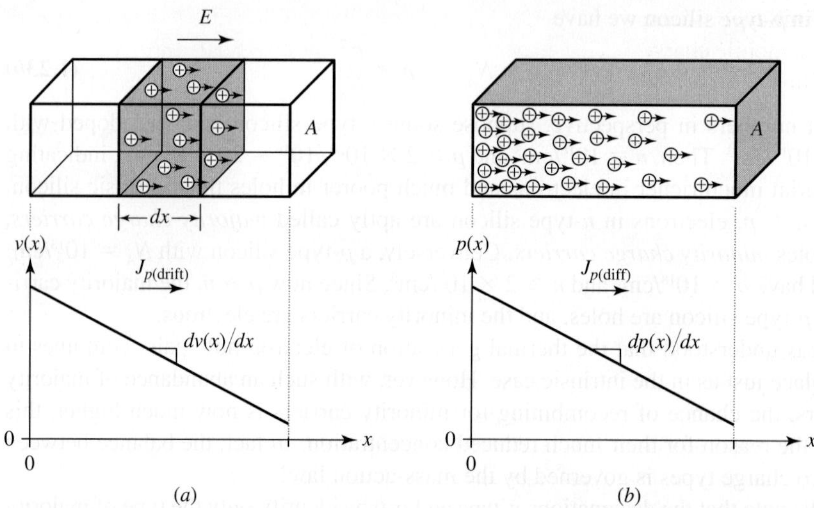

FIGURE 1.36 Illustrating (a) current *drift* and (b) current *diffusion* for the case of holes.

potential $v(x)$ will vary linearly across the slab, as shown at the bottom. From basic physics we know that field and potential are related as

$$E = -\frac{dv(x)}{dx} \tag{1.24}$$

Consequently, E will be constant throughout the homogeneous slab.

Now, under the accelerating effect of E, holes will drift in the same direction as the field and achieve an *average drift velocity* v_p (in cm/s) which is linearly proportional to the field strength,

$$v_p = \mu_p E \tag{1.25a}$$

Aptly called the *hole mobility*, μ_p (in cm^2/Vs) gives a measure of the average drift velocity acquired by holes for a given applied field. As holes drift, they produce the current $i_{p(\text{drift})} = dQ_p/dt$, where dQ_p is the amount of charge transferred during the interval dt. Given that during dt holes travel the distance $dx = v_p dt$, the holes making up dQ_p are contained within the volume $A dx = A v_p dt$, and their number is thus $pAv_p dt$, where p is their concentration (holes/cm^3), and A is the *cross sectional area* of the silicon slab (in cm^2). Multiplying this number by the hole charge $+q$ gives $dQ_p = qpAv_p dt$, so

$$i_{p(\text{drift})} = \frac{dQ_p}{dt} = qpAv_p = qpA\mu_p E$$

As we progress, we will find it more convenient to work with the current per unit area, or *current densiy J* (in A/cm^2), rather than with the ordinary current i (in A). The *hole drift current density* is simply $J_{p(\text{drift})} = i_{p(\text{drift})}/A$, or

$$J_{p(\text{drift})} = qp\mu_p E \tag{1.25b}$$

Similar considerations hold for a slab of *n*-type material, except that in this case the mobile charges are electrons, whose concentration and mobility are n and μ_n. Thus, the *average drift velocity* of electrons is

$$v_n = \mu_n E \qquad (1.26a)$$

where μ_n (in cm²/Vs) is the *electron mobility*, and the *electron drift current density* is

$$J_{n(\text{drift})} = qn\mu_n E \qquad (1.26b)$$

Equation (1.26*b*) is also at the basis of metals and ordinary conductors such as composition resistors, where the only type of mobile charges available are electrons. Both equations indicate the necessary ingredients for good conduction: *high concentration* along with *high mobility*.

- **The Diffusion Current**: The other mechanism for charge-carrier motion in semiconductors is diffusion—a mechanism not found in ordinary conductors. As we progress, we shall see that in semiconductor devices it is possible to establish and continuously maintain *non-uniform profiles* for the mobile charge densities. Figure 1.36*b* shows an example of a *linear* density profile, such as that found in the base region of a forward-biased bipolar junction transistor. As holes wander randomly because of thermal agitation, they tend to *diffuse* from regions of higher density to regions of lower density, or toward the right in our example. This is similar to particles of smoke diffusing from the smoking area to the rest of a room. If holes are continuously injected to the left while being removed from the right, a sustained current flow will result. The *hole diffusion current density* is

$$J_{p(\text{diff})} = -qD_p\frac{dp(x)}{dx} \qquad (1.27a)$$

where D_p is the *hole diffusivity* (in cm²/s). The negative sign stems from the fact that holes flow in the direction of the *negative gradient* in p. Likewise, the *electron diffusion current density* is

$$J_{n(\text{diff})} = qD_n\frac{dn(x)}{dx} \qquad (1.27b)$$

where D_n is the *electron diffusivity* (in cm²/s), and the sign is now positive due to the negative charge of electrons.

We note strong similarities between the expressions for the drift and diffusion currents. In fact, substituting Eq. (1.24) into Eqs. (1.25*b*) and (1.26*b*) gives

$$J_{p(\text{drift})} = -qp\mu_p\frac{dv(x)}{dx} \qquad J_{n(\text{drift})} = qn\mu_n\frac{dv(x)}{dx}$$

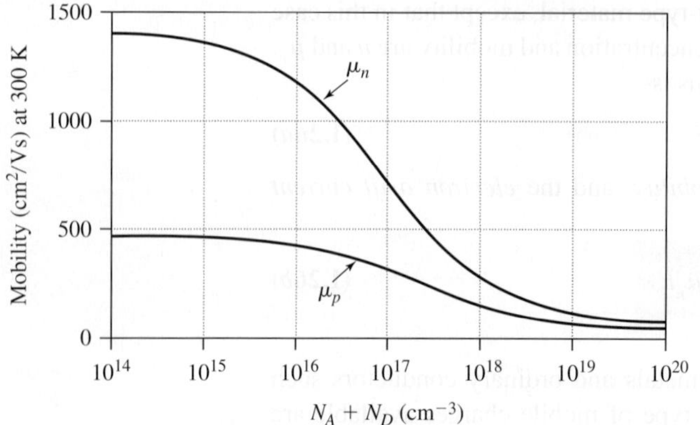

$$\mu_n = 68 + \frac{1346}{1 + \left(\dfrac{N_A + N_D}{9.2 \times 10^{16}}\right)^{0.71}} \; \text{cm}^2/\text{Vs}$$

$$\mu_p = 45 + \frac{427}{1 + \left(\dfrac{N_A + N_D}{2.2 \times 10^{17}}\right)^{0.72}} \; \text{cm}^2/\text{Vs}$$

FIGURE 1.37 Room-temperature dependence of the mobilities on the total doping density, and empirical formulas for their calculation for the case of donor atoms of *phosphorous* and acceptor atoms of *boron*.

which are even closer in form to Eqs. (1.27*a*) and (1.27*b*). These equations indicate that:

- To sustain a given current we need a *gradient* (a *voltage gradient* to sustain *drift*, a *density gradient* to sustain *diffusion*).
- Charge flow is in the direction of a *decreasing* gradient.
- The diffusivities D_p and D_n play a similar role to the mobilities μ_p and μ_n in that each offers a measure of how much current stems from a given gradient. Indeed, it turns out that diffusivities and mobilities are related by Einstein's relations

$$\frac{D_n}{\mu_n} = \frac{D_p}{\mu_p} = V_T \tag{1.28}$$

where $V_T \cong 26$ mV is the thermal voltage of Eq. (1.20).

Mobilities and diffusivities are greatest when silicon is pure, but *decrease* with *doping* as well as *temperature*. Figure 1.37 shows the dependence of μ_n and μ_p on the *total* doping density $(N_A + N_D)$, at room temperature. The higher mobility (by a factor of two to three) exhibited by electrons compared to holes is the primary reason why *n*-type materials are generally preferred over *p*-type materials, particularly in the fabrication of devices intended for high-speed operation.

Lastly, it must be pointed out that the linear relationships between velocities and electric field, expressed as $v_n = \mu_n E$ and $v_p = \mu_p E$, hold only up to a certain field strength, typically on the order of 5 kV/cm. Past this limit, electron and hole velocities saturate at approximately 10^7 cm/s. Aptly called *velocity saturation*, this phenomenon sets an upper limit on the speed of operation of semiconductor devices such as MOSFETs.

An Example: The Integrated-Circuit Diode

Figure 1.38 illustrates the most basic steps involved in the fabrication of the *pn junction diode*, a semiconductor device at the basis of most other integrated-circuit (IC) devices. Starting out with a lightly doped *n*-type slab, such as $N_D = 10^{15}/\text{cm}^3$, a localized boron diffusion is made to create a *p*-type region. Clearly, in order to overcome

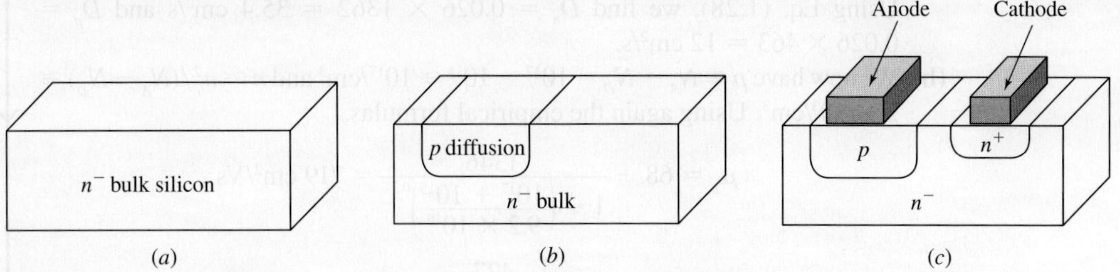

FIGURE 1.38 Basic fabrication steps of an IC diode: (*a*) starting material, (*b*) *p*-type diffusion, and (*c*) provision for its connection to the outside.

the existing *n*-type nature of this region, the acceptor density N_A must *exceed* the existing donor density N_D there. Then, in this region we have

$$p \cong N_A - N_D \qquad n \cong \frac{n_i^2}{N_A - N_D}$$

An additional diffusion is made to create a heavily doped *n*-type region to ensure an ohmic contact between the *n*-type slab and a metal (more on this later in the chapter), and finally metal depositions are made to allow for the interconnection of the device to external circuitry.

The dimensions of the above device are in the range of *micrometers* ($1\ \mu\text{m} = 10^{-6}\text{m}$). Such tiny dimensions allow for the simultaneous fabrication of a very large number of devices on the same wafer. To prevent different devices from interfering with each other, we must keep them electrically isolated from each other. Interestingly enough, a popular way of achieving isolation is through additional reverse-biased *pn* junctions, a subject that we will illustrate in greater detail when studying the fabrication of transistors. The interested student is encouraged to search the Web for videos and articles illustrating the fascinating subject of integrated circuit fabrication.

Find the electron and hole concentrations n and p as well as the mobilities μ_n and μ_p, and the diffusivities D_n and D_p in the three regions of the structure of Fig. 1.38, assuming the following doping densities:

EXAMPLE 1.5

(a) n^--type bulk: $N_D = 10^{15}$ phosphorous atoms/cm^3
(b) p-type diffusion: $N_A = 10^{17}$ boron atoms/cm^3
(c) n^+-type diffusion: $N_D = 10^{20}$ phosphorous atoms/cm^3

Solution

(a) We have $n \cong N_D = 10^{15}/\text{cm}^3$ and $p \cong n_i^2/N_D = 2 \times 10^{20}/10^{15} = 2 \times 10^5/\text{cm}^3$. Using the empirical formulas of Fig. 1.37, we find

$$\mu_n = 68 + \frac{1346}{1 + \left(\dfrac{10^{15}}{9.2 \times 10^{16}}\right)^{0.71}} = 1362\ \text{cm}^2/\text{Vs}$$

$$\mu_p = 45 + \frac{427}{1 + \left(\dfrac{10^{15}}{2.2 \times 10^{17}}\right)^{0.72}} = 463\ \text{cm}^2/\text{Vs}$$

Using Eq. (1.28), we find $D_n = 0.026 \times 1362 = 35.4$ cm²/s and $D_p = 0.026 \times 463 = 12$ cm²/s.

(b) We now have $p \cong N_A - N_D = 10^{17} - 10^{15} \cong 10^{17}$/cm³ and $n \cong n_i^2/(N_A - N_D) \cong 2 \times 10^3$/cm³. Using again the empirical formulas,

$$\mu_n = 68 + \frac{1346}{1 + \left(\dfrac{10^{17} + 10^{15}}{9.2 \times 10^{16}}\right)^{0.71}} = 719 \text{ cm}^2/\text{Vs}$$

$$\mu_p = 45 + \frac{427}{1 + \left(\dfrac{10^{17} + 10^{15}}{2.2 \times 10^{17}}\right)^{0.72}} = 317 \text{ cm}^2/\text{Vs}$$

Moreover, $D_n = 0.026 \times 719 = 18.7$ cm²/s and $D_p = 8.3$ cm²/s.

(c) We now have $n \cong 10^{20} + 10^{15} \cong 10^{20}$/cm³, $p \cong 2$/cm³; $\mu_n = 78$ cm²/Vs, $D_n = 2$ cm²/s; $\mu_p = 50$ cm²/Vs, and $D_p = 1.3$ cm²/s.

1.5 THE *pn* JUNCTION IN EQUILIBRIUM

When a *p*-type and an *n*-type region are joined together, they are said to form a *pn junction*. Even though in practice they are fabricated contiguously as exemplified in Fig. 1.38, from a pedagogical viewpoint it is convenient to consider two slabs that have been fabricated separately and are brought into contact with each other subsequently, in the manner depicted in Fig. 1.39, top. To develop a numerical feel for the various quantities involved, we shall work with the following doping concentration example:

$$N_A = 10^{18}/\text{cm}^3 \qquad N_D = 10^{16}/\text{cm}^3 \tag{1.29}$$

Assume donor atoms of phosphorous and acceptor atoms of boron, so we can use the formulas of Fig. 1.37, when necessary. Using the subscript zero to identify equilibrium concentrations, we exploit Eq. (1.23b) to find the *p*-side hole and electron concentrations

$$p_{p0} \cong N_A = 10^{18}/\text{cm}^3 \qquad n_{p0} \cong \frac{n_i^2}{N_A} = 2 \times 10^2/\text{cm}^3 \tag{1.30a}$$

and we exploit Eq. (1.23a) to find the *n*-side electron and hole concentrations

$$n_{n0} \cong N_D = 10^{16}/\text{cm}^3 \qquad p_{n0} \cong \frac{n_i^2}{N_D} = 2 \times 10^4/\text{cm}^3 \tag{1.30b}$$

Once the two slabs are brought into contact, holes will diffuse from the *p*-side, where they are highly concentrated (10^{18} holes/cm³), toward the *n*-side, where they are concentrated only sparingly (2×10^4 holes/cm³.) Likewise, electrons will diffuse in the opposite direction. However, every hole diffusing across the metallurgical junction leaves behind a *negatively charged ion*, just as every diffusing electron leaves behind a *positively charged ion*. These ions are bound to their fixed positions in the crystal lattice and do not contribute to current. However, they form a *space charge layer* (SCL), also called *depletion layer* because it is essentially depleted of mobile charges

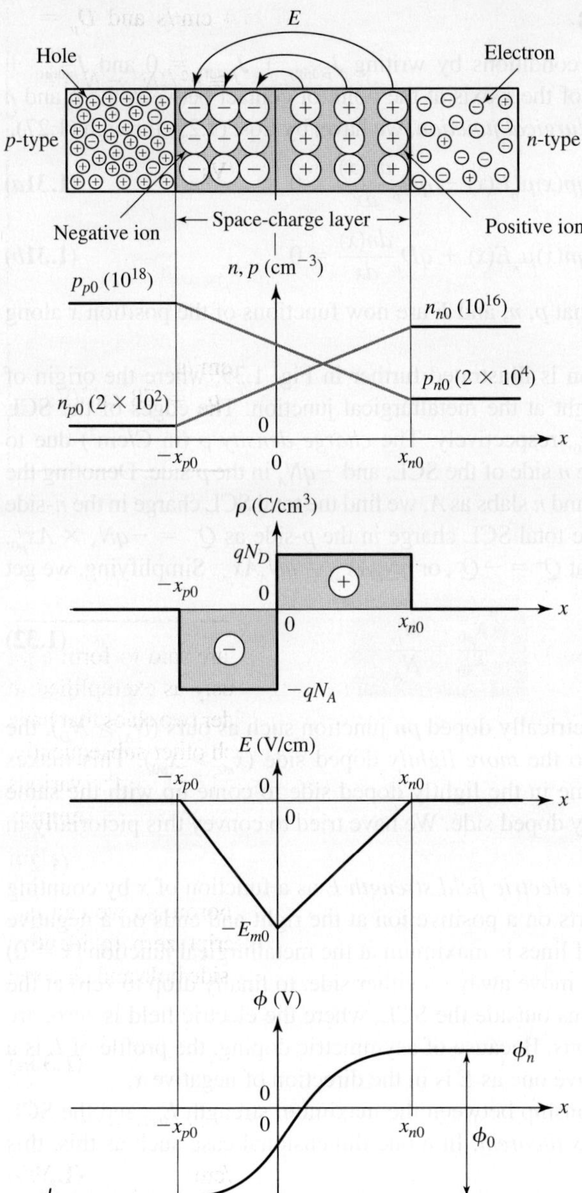

FIGURE 1.39 Equilibrium conditions in a *pn* slab.

due to their diffusion across the junction. The SCL, in turn, establishes an electric field E in the direction *opposing* diffusion. As holes and electrons keep diffusing, the SCL keeps building up, until an *equilibrium condition* is reached whereby the electric field E will *exactly* counterbalance the tendency of holes and electrons to diffuse further. Thereafter, the net current across the junction will be *zero*.

Equilibrium Conditions

We express the equilibrium conditions by writing $J_{p(\text{drift})} + J_{p(\text{diff})} = 0$ and $J_{n(\text{drift})} + J_{n(\text{diff})} = 0$. Taking the origin of the x-axis at the point of contact between the p and n regions, also called the *metallurgical junction*, we have, by Eqs. (1.25) through (1.27),

$$qp(x)\mu_p E(x) - qD_p \frac{dp(x)}{dx} = 0 \qquad \textbf{(1.31a)}$$

$$qn(x)\mu_n E(x) + qD_n \frac{dn(x)}{dx} = 0 \qquad \textbf{(1.31b)}$$

where we are emphasizing that p, n, and E are now functions of the position x along the *pn* slab.

The equilibrium situation is illustrated further in Fig. 1.39, where the origin of the x axis has been taken right at the metallurgical junction. The edges of the SCL are located at $-x_{p0}$ and $+x_{n0}$, respectively. The *charge density* ρ (in C/cm^3) due to immobile ions is $+qN_D$ in the n side of the SCL, and $-qN_A$ in the p side. Denoting the *cross-sectional area* of the p and n slabs as A, we find the total SCL charge in the n-side as $Q^+ = qN_D \times Ax_{n0}$, and the total SCL charge in the p-side as $Q^- = -qN_A \times Ax_{p0}$. Charge neutrality requires that $Q^+ = -Q^-$, or $qN_D Ax_{n0} = qN_A Ax_{p0}$. Simplifying, we get

$$\frac{x_{p0}}{x_{n0}} = \frac{N_D}{N_A} \qquad \textbf{(1.32)}$$

indicating that in an asymmetrically doped *pn* junction such as ours ($N_A \gg N_D$), the SCL will extend mostly into the *more lightly* doped side ($x_{n0} \gg x_{p0}$). This makes sense as it takes more volume in the lightly doped side to come up with the same number of ions as the heavily doped side. We have tried to convey this pictorially in Fig. 1.39, top.

We readily visualize the *electric field strength E* as a function of x by counting the field lines. Each line starts on a positive ion at the right and ends on a negative ion at the left. The number of lines is maximum at the metallurgical junction ($x = 0$) and decreases linearly as we move away on either side, to finally drop to zero at the edges of the SCL. The regions outside the SCL, where the electric field is zero, are aptly called the *neutral regions*. Because of asymmetric doping, the profile of E is a scalene triangle, and a negative one as E is in the direction of negative x.

We readily find a relationship between the maximum strength E_{m0} and the SCL edges x_{p0} and x_{n0} via *Gauss's theorem*. In a one-dimensional case such as this, this theorem is expressed as

$$\frac{dE}{dx} = \frac{\rho(x)}{\varepsilon_{si}} \qquad \textbf{(1.33)}$$

where ε_{si} is silicon's *permittivity* ($\varepsilon_{si} = 1.04$ pF/cm). In the right portion of the SCL we have $dE/dx = E_{m0}/x_n$ and $\rho/\varepsilon_{si} = qN_D/\varepsilon_{si}$, so $E_{m0}/x_{n0} = qN_D/\varepsilon_{si}$. Applying similar considerations to the left side of the SCL and solving for E_{m0} gives

$$E_{m0} = \frac{qN_A x_{p0}}{\varepsilon_{si}} = \frac{qN_D x_{n0}}{\varepsilon_{si}} \qquad \textbf{(1.34)}$$

The Built-in Potential ϕ_0

From basic electrostatics we know that an *electric field* is always accompanied by an *electric potential gradient*. For a one-dimensional situation such as ours, the relationship between field E and potential ϕ is, by Eq. (1.24), $E = -d\phi/dx$. Rewriting as $\phi = -\int E\,dx$, we visualize ϕ as the *negative* of the *area* enclosed by the E curve. Since E has a *linear* profile, ϕ will have a *quadratic* profile, as shown in Fig. 1.39, bottom. We observe that outside the SCL the profile of ϕ is flat because $E = 0$ there. The potentials there are denoted as ϕ_p and ϕ_n, respectively. We now wish to find expressions for ϕ_p and ϕ_n, as well as for the *built-in equilibrium potential* ϕ_0, defined as

$$\phi_0 = \phi_n - \phi_p$$

This potential acts as a *barrier* preventing holes and electrons from diffusing further, and is the result of the charge redistribution taking place automatically at either side of the metallurgical junction when we create it. Solving for $E(x)$ in Eq. (1.31) and using Einstein's relations of Eq. (1.28) gives

$$E(x)dx = V_T\frac{dp(x)}{p(x)} = -V_T\frac{dn(x)}{n(x)}$$

Using again $\phi(x) = -\int E(x)\,dx$ and integrating from $-x_{p0}$ to $+x_{n0}$ gives

$$\int_{\phi_p}^{\phi_n} d\phi(x) = -V_T\int_{p_{p0}}^{p_{n0}}\frac{dp(x)}{p(x)} = V_T\int_{n_{p0}}^{n_{n0}}\frac{dn(x)}{n(x)}$$

where the integration limits are, respectively, the values of ϕ, p, and n at $x = -x_{p0}$ and $x = +x_{n0}$. This gives

$$\phi_0 = \phi_n - \phi_p = V_T\ln\frac{p_{p0}}{p_{n0}} = V_T\ln\frac{n_{n0}}{n_{p0}} \tag{1.35}$$

Using Eq. (1.10), we can also write

$$\phi_0 = V_T\ln\frac{N_A N_D}{n_i^2} \tag{1.36a}$$

$$\phi_n = V_T\ln\frac{N_D}{n_i} \tag{1.36b}$$

$$\phi_p = V_T\ln\frac{n_i}{N_A} \tag{1.36c}$$

Note that since in practical junctions N_A and N_D are greater than n_i, we have $\phi_n > 0$ and $\phi_p < 0$. Moreover, since N_A and N_D appear in the argument of the *logarithmic* function, the ϕs of Eq. (1.36) are not that sensitive to variations in the doping doses.

EXAMPLE 1.6

(a) Find the room-temperature values of ϕ_n, ϕ_p, and ϕ_0 for a junction with the dopings of Eq. (1.29).

(b) Find ϕ_0 if both doping doses are increased by an order of magnitude.

Solution

(a) We have $\phi_n = (26\ \text{mV})\ln[10^{16}/(1.4 \times 10^{10})] = 0.350\ \text{V}$, $\phi_p = -0.470\ \text{V}$, and $\phi_0 = 0.350 - (-0.470) = 0.820\ \text{V}$.

(b) Now $\phi_0 = 0.940\ \text{V}$, not much of a change because of the logarithmic dependence. It pays to think of ϕ_0 as being close to 1 V, regardless of the particular doping values.

The Electric Field E_{m0}, the SCL Width X_{d0}, and the SCL Charge Q_{j0}

We now wish to derive an expression for all other pertinent junction parameters in equilibrium. The maximum field strength E_{m0} is found once again via $\int d\phi(x) = -\int E(x)\,dx$, where we integrate from $-x_{p0}$ to x_{n0}. The left-hand side integrates simply to ϕ_0, while the right-hand side represents the negative of the area of the electric-field triangle. Consequently, we have

$$\phi_0 = \frac{(x_{p0} + x_{n0})E_{m0}}{2} \tag{1.37}$$

But, according to Eq. (1.34),

$$x_{p0} = \frac{\varepsilon_{si}}{qN_A}E_{m0} \qquad x_{n0} = \frac{\varepsilon_{si}}{qN_D}E_{m0} \tag{1.38}$$

Substituting x_{p0} and x_{n0} into Eq. (1.37), expressing ϕ_0 via Eq. (1.36a), and solving for E_{m0}, we finally get

$$E_{m0} = \sqrt{\frac{2q\phi_0}{\varepsilon_{si}}\frac{N_A N_D}{N_A + N_D}} \tag{1.39}$$

If we insert Eq. (1.39) back into Eq. (1.38), we obtain the equilibrium edges of the SCL as

$$x_{p0} = \sqrt{\frac{2\varepsilon_{si}\phi_0}{qN_A}\frac{N_D}{N_A + N_D}} \qquad x_{n0} = \sqrt{\frac{2\varepsilon_{si}\phi_0}{qN_D}\frac{N_A}{N_A + N_D}} \tag{1.40}$$

The sum of the two is aptly called the equilibrium *SCL width*, $X_{d0} = x_{p0} + x_{n0}$. By Eq. (1.40),

$$X_{d0} = \sqrt{\frac{2\varepsilon_{si}\phi_0}{q}\left(\frac{1}{N_A} + \frac{1}{N_D}\right)} \tag{1.41}$$

The equilibrium *junction charge* is $Q_{j0} = Q^+ = qN_D A x_{n0}$, where A is the aforementioned junction's *cross-sectional area*. Using Eq. (1.40),

$$Q_{j0} = A\sqrt{2\varepsilon_{si}q\phi_0\frac{N_A N_D}{N_A + N_D}} \tag{1.42}$$

Assuming a cross-sectional area $A = (100\ \mu\text{m}) \times (100\ \mu\text{m})$ for a *pn* junction with   **EXAMPLE 1.7**
the doping doses of Eq. (1.29), find E_{m0}, X_{d0}, x_{p0}, x_{n0}, and Q_{j0}.

Solution

From Example 1.6, $\phi_0 = 0.820$ V. Also, since in our case $N_A \gg N_D$, we can approximate $N_A N_D / (N_A + N_D) \cong N_D = 10^{16}$ cm^{-3}. Then, Eq. (1.39) gives

$$E_{m0} \cong \sqrt{\frac{2 \times 1.602 \times 10^{-19} \times 0.820 \times 10^{16}}{1.04 \times 10^{-12}}} = 5.03 \times 10^4 \text{ V/cm}$$

and Eq. (1.41) gives

$$X_{d0} \cong \sqrt{\frac{2 \times 1.04 \times 10^{-12} \times 0.820}{1.602 \times 10^{-19}}\left(\frac{1}{10^{16}}\right)} = 3.26 \times 10^{-5} \text{ cm} = 0.326\ \mu\text{m}$$

Similarly, Eq. (1.40) gives $x_{p0} = 0.003\ \mu\text{m}$ and $x_{n0} = 0.323\ \mu\text{m}$, confirming that the SCL extends almost entirely into the more lightly doped side, which in our example is the *n* side. Finally, since the junction area is $A = (100 \times 10^{-4} \text{ cm})^2 = 10^{-4}$ cm^2, Eq. (1.42) gives $Q_{j0} = 5.23$ pC.

Exercise 1.5

Show that

$$\phi(0) = \phi_p + \frac{N_D}{N_A + N_D}\phi_0 \qquad\qquad \textbf{(1.43)}$$

Hence, verify that $\phi(0) = 0$ only in the case of symmetrically doped junctions ($N_D = N_A$). Otherwise, $\phi(0) > 0$ for $N_D > N_A$, and $\phi(0) < 0$ for $N_D < N_A$ (as in the case of Fig. 1.39).

1.6 EFFECT OF EXTERNAL BIAS ON THE SCL PARAMETERS

Let us now investigate the effect of applying a voltage v across our *pn* junction in the manner of Fig. 1.40, top. (Note that the polarity convention for v is *positive* at the *p* side and *negative* at the *n* side; for $v > 0$ the *pn* junction is said to be *forward biased*, and for $v < 0$ it is *reverse biased*.) By KVL, the potential barrier across the space-charge layer (SCL) becomes $\phi_0 - v$. With a changed profile for ϕ, the electric field strength E will also have to change. Since the field lines making up E come from the uncovered ions of the SCL, the SCL's *width* $X_d = x_n + x_p$ will have to change accordingly. Specifically, we can state that:

- *Forward biasing* the junction ($v > 0$) *lowers* the potential barrier as well as the electric field compared to the unbiased case, and thus *shrinks* X_d.
- Conversely, *reverse biasing* the junction ($v < 0$) *raises* the potential barrier and the electric field, and thus *widens* X_d. For a visual comparison, Fig. 1.40 uses gray lines to show the unbiased situation.

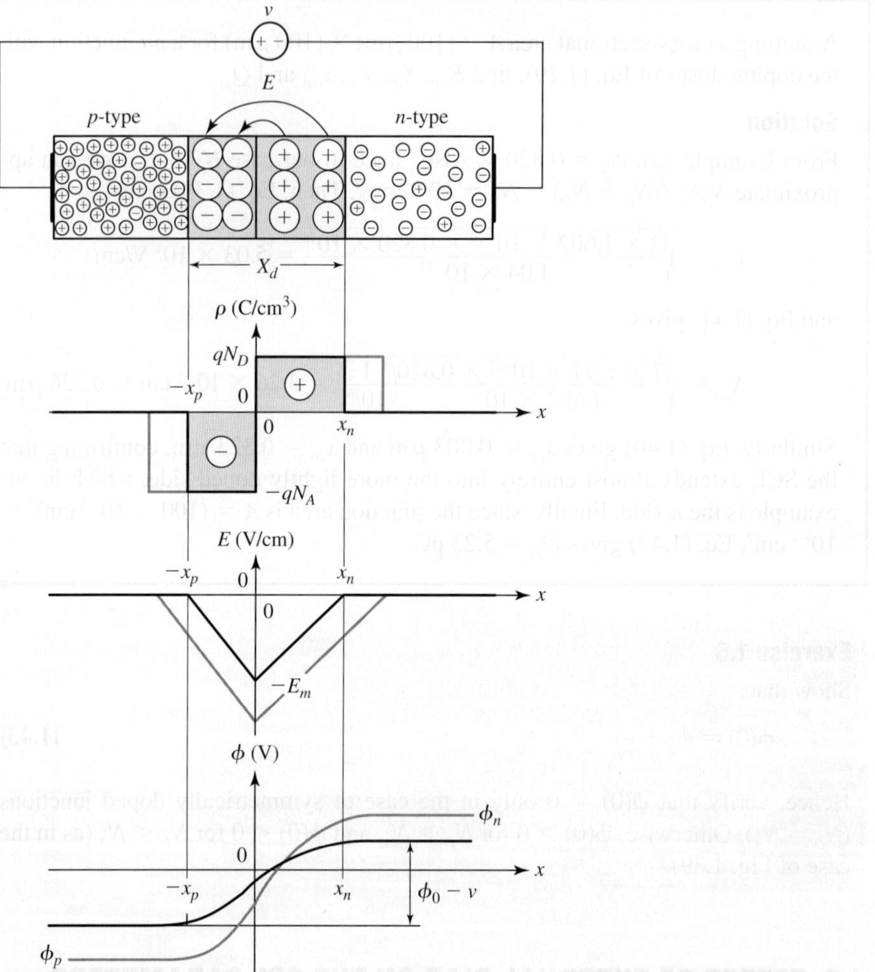

FIGURE 1.40 Effect of forward-biasing a *pn* junction.

To investigate the effect of the external bias v quantitatively, we simply replace ϕ_0 with $(\phi_0 - v)$ in Eqs. (1.39) through (1.42). Thus, rewriting Eq. (1.39) with $E_m(v)$ in place of E_{m0} and $(\phi_0 - v)$ in place of ϕ_0, we get the *maximum electric-field strength* as a function of v

$$E_m(v) = \sqrt{\frac{2q(\phi_0 - v)}{\varepsilon_{si}} \frac{N_A N_D}{N_A + N_D}} = \sqrt{\frac{2q\phi_0}{\varepsilon_{si}} \frac{N_A N_D}{N_A + N_D}} \sqrt{1 - \frac{v}{\phi_0}}$$

This is expressed more concisely as

$$E_m(v) = E_{m0}\sqrt{1 - \frac{v}{\phi_0}} \tag{1.44}$$

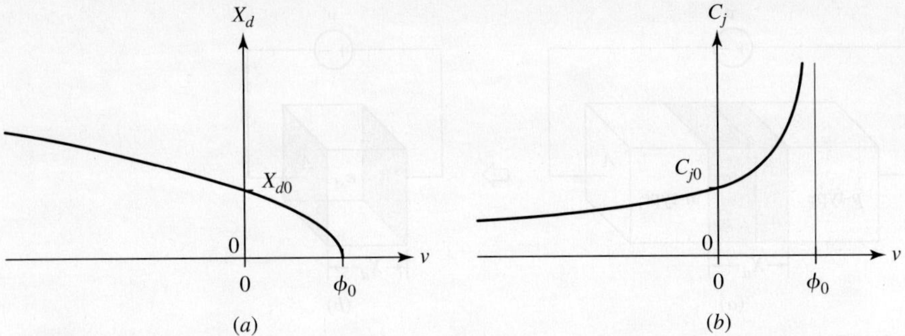

FIGURE 1.41 Voltage dependence of the (a) *SCL width* and (b) *junction capacitance* for $m = \frac{1}{2}$.

where E_{m0}, aptly called the *zero-bias* ($v = 0$) value of E_m, was derived in Eq. (1.39). Proceeding in similar manner for the *SCL's width*, we find

$$X_d(v) = X_{d0}\sqrt{1 - \frac{v}{\phi_0}}$$

(1.45)

where X_{d0} is the *zero-bias* ($v = 0$) value of X_d, which was derived in Eq. (1.41). The voltage dependence of X_d is depicted in Fig. 1.41a. Finally, the *junction charge* is

$$Q_j(v) = Q_{j0}\sqrt{1 - \frac{v}{\phi_0}}$$

(1.46)

where Q_{j0} is the *zero-bias* ($v = 0$) value of Q_j as given in Eq. (1.42).

The Junction Capacitance C_j

Since applying a *voltage* across a *pn* junction redistributes its SCL *charge*, the junction exhibits *capacitive behavior*. The *junction capacitance* is $C_j = dQ_j/dv$. Differentiating Eq. (1.46) and rearranging,

$$C_j(v) = \frac{C_{j0}}{\left(1 - v/\phi_0\right)^m}$$

(1.47a)

where

$$C_{j0} = A\sqrt{\frac{\varepsilon_{si}q}{2\phi_0}\frac{N_A N_D}{N_A + N_D}}$$

(1.47b)

is the *zero-bias* ($v = 0$) value of C_j and m, called the *grading coefficient*, is ½ in the present case, which assumes an *abrupt junction*. Practical junctions often have a *graded* doping profile, in which case it can be shown[2] that a more appropriate value is $m = \frac{1}{3}$. The actual value of m can be found experimentally by measuring C_j for different values of v and then using data-interpolation to indirectly find m. The voltage dependence of C_j is depicted in Fig. 1.41b.

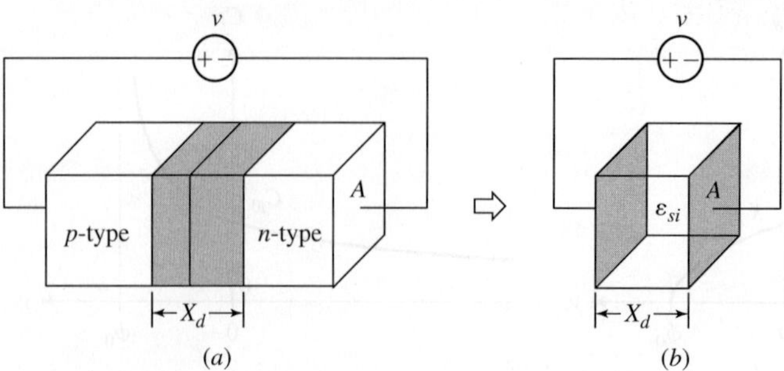

FIGURE 1.42 (a) The junction capacitance C_j, and (b) its parallel-plate equivalent.

Combining Eqs. (1.41), (1.45), and (1.47) with $m = \frac{1}{2}$, we put C_j in yet another insightful form

$$C_j(v) = \varepsilon_{si}\frac{A}{X_d(v)} \tag{1.48}$$

This is the same form as that of a parallel-plate capacitor consisting of two plates of area A separated by a dielectric material having permittivity ε_{si} and thickness X_d. This equivalence is illustrated in Fig. 1.42. However, unlike a fixed capacitor, the present one exhibits a voltage-dependent plate separation $X_d(v)$, indicating nonlinear capacitance behavior, as already seen in Fig. 1.41*b*. We also observe that Eq. (1.47*a*) predicts $C_j \to \infty$ for $v \to \phi_0$. This, of course, cannot happen in practice, indicating that Eq. (1.47*a*), whose derivation is based on a number of simplifying assumptions, no longer holds as v approaches ϕ_0.

EXAMPLE 1.8   Find C_{j0} for the junction of Example 1.7. Then, assuming $m = \frac{1}{2}$, calculate E_m, X_d, Q_j, and C_j for (a) $v = +0.65$ V, and (b) $v = -5$V.

Solution

(a) Using Eq. (1.47*b*) we readily find $C_{j0} = 3.19$ pF. Moreover

$$\sqrt{1 - v/\phi_0} = \sqrt{1 - 0.65/0.82} = 0.455$$

indicating a *decrease* in E_m, X_d, and Q_j, but an *increase* in C_j. Indeed, at $v = +0.65$ V we have $E_m = 5.03 \times 10^4 \times 0.455 = 2.29 \times 10^4$ V/cm, $X_d = 0.148$ μm, $Q_j = 2.38$ pC, and $C_j = 3.19/0.455 = 7.01$ pF.

(b) Now E_m, X_d, and Q_j will *increase* but C_j will *decrease* by $\sqrt{1 - (-5)/0.82} = 2.66$, so $E_m = 13.4 \times 10^4$ V/cm, $X_d = 0.869$ μm, $Q_j = 13.9$ pC, and $C_j = 1.20$ pF.

Remark: It pays to keep in mind the following orders of magnitude for low-power junctions:

$$\phi_0 \sim 1 \text{ V} \qquad E_m \sim 10^4 \text{ V/cm} \qquad X_d \sim 1 \text{ } \mu\text{m} \qquad Q_j \sim 1 \text{ pC} \qquad C_j \sim 1 \text{ pF}$$

1.7 THE *pn* DIODE EQUATION

Forward-biasing a *pn* junction affects not only the parameters of its space-charge layer (SCL), but also the concentration profiles of its charge carriers in the neutral regions, and quite dramatically so, as we are about to see. The starting point is offered by Eq. (1.35), which we turn around by taking the logarithms throughout and solving for the minority concentrations,

$$p_{n0} = p_{p0}e^{-\phi_0/V_T} \tag{1.49a}$$

$$n_{p0} = n_{n0}e^{-\phi_0/V_T} \tag{1.49b}$$

These equations relate the hole and the electron concentrations at either side of the SCL for the unbiased ($v = 0$) equilibrium situation. If we now forward bias the junction ($v > 0$), E will decrease and thus allow holes to diffuse from the p side, through the SCL, to the n side, and electrons from the n to the p side. We can still use Eq. (1.49) to relate the concentrations right at the edges of the SCL, also called *boundary concentrations*, provided we replace ϕ_0 with $\phi_0 - v$. The result is

$$p_n(x_n) = p_p(-x_p)e^{-(\phi_0-v)/V_T} \tag{1.50a}$$

$$n_p(-x_p) = n_n(x_n)e^{-(\phi_0-v)/V_T} \tag{1.50b}$$

The so-called *low-level injection* assumption stipulates that even after biasing, the minority concentrations at either side of the SCL continue to remain *much lower* than the majority concentrations there, and thus leave the latter essentially undisturbed compared to the unbiased case. This means that we can let $p_p(-x_p) = p_{p0}$ and $n_n(x_n) = n_{n0}$ in Eq. (1.50). Reusing Eq. (1.49), we can then write

$$p_n(x_n) = p_{n0}e^{v/V_T} \tag{1.51a}$$

$$n_p(-x_p) = n_{p0}e^{v/V_T} \tag{1.51b}$$

Referred to as the *law of the junction*, Eq. (1.51) relates the boundary values of the minority concentrations to the applied voltage v. Though the derivations were made for the forward-bias case ($v > 0$), this law holds also for the reverse-bias case ($v < 0$).

Assuming the doping doses of Eq. (1.29), find the minority and majority concentrations at either edge of the SCL if the junction is forward biased with $v = 0.65$ V. Comment on your results.

EXAMPLE 1.9

Solution

By Eq. (1.30), $p_p(-x_p) = p_{p0} \cong 10^{18}/cm^3$ and $n_n(x_n) \cong n_{n0} \cong 10^{16}/cm^3$. Moreover, $n_{p0} \cong 2 \times 10^2/cm^3$, and $p_{n0} \cong 2 \times 10^4/cm^3$. Since $\exp(0.650/0.026) \cong 7.2 \times 10^{10}$, Eq. (1.51) gives

$$p_n(x_n) \cong 2 \times 10^4 \times 7.2 \times 10^{10} = 1.44 \times 10^{15}/cm^3$$

$$n_p(-x_p) \cong 2 \times 10^2 \times 7.2 \times 10^{10} = 1.44 \times 10^{13}/cm^3$$

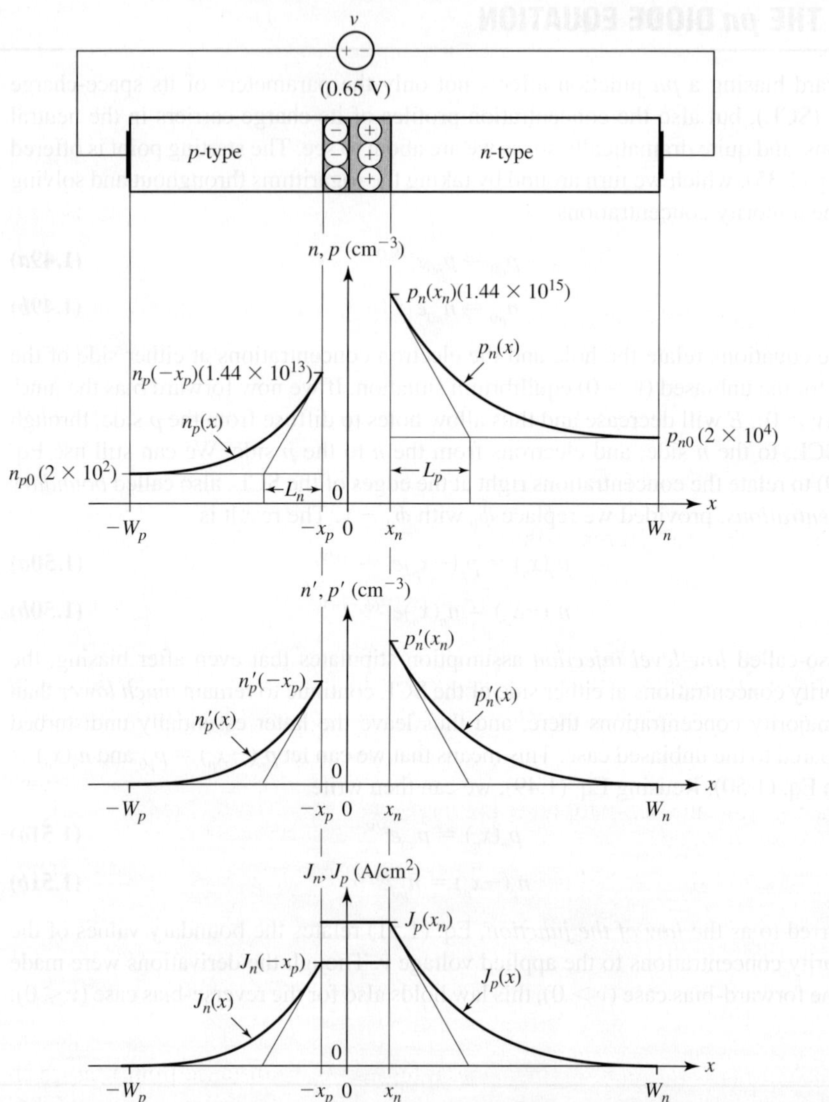

FIGURE 1.43 Forward biasing a *pn* junction creates an excess of minority charges in the neutral regions. These charges diffuse away from the SCL, giving raise to diffusion currents.

These boundary values are shown numerically in Fig. 1.43. We observe that a forward bias of only 0.65 V causes $p_n(x_n)$ to shoot up from $2 \times 10^4/\text{cm}^3$ to $1.44 \times 10^{15}/\text{cm}^3$! However, this is still less than the majority concentration there ($10^{16}/\text{cm}^3$), thus confirming low-level injection. Likewise, $n_p(-x_p)$ has jumped from $2 \times 10^2/\text{cm}^3$ to $1.44 \times 10^{13}/\text{cm}^3$. This too is quite a number, yet it is much less than the majority concentration there ($10^{18}/\text{cm}^3$), again indicating low-level injection.

Excess Minority Concentrations

It is apparent that forward biasing the *pn* junction establishes an *excess* of minority carriers at both edges of the SCL. The excess concentrations are $p'_n(x_n) = p_n(x_n) - p_{n0}$ at the right edge, and $n'_p(-x_p) = n_p(-x_p) - n_{p0}$ at the left edge. By Eq. (1.51), these excesses can be expressed as

$$p'_n(x_n) = p_{n0}(e^{v/V_T} - 1) \tag{1.52a}$$

$$n'_p(-x_p) = n_{p0}(e^{v/V_T} - 1) \tag{1.52b}$$

Once minority excesses have been established, the carriers will *diffuse* away from the SCL toward regions of lower excess concentrations (holes from x_n toward the right, electrons from $-x_p$ toward the left). In both cases, their diffusion takes place in a sea of oppositely-charged majority carriers, indicating a high chance of recombination. In fact, the further away we go from the SCL's edges, the less and less excess minority carriers we are likely to find.

This diffuse-and-recombine process is governed by the *diffusion equation*, which for excess holes takes on the form[2]

$$D_p \frac{d^2 p'_n(x)}{dx^2} = \frac{p'_n(x)}{\tau_p}$$

where τ_p is the *mean recombination time*, also called the *mean lifetime* of the excess holes. A similar equation holds for excess electrons, but with τ_n as the mean recombination time. The solution to the diffusion equation is an *exponential decay* with x for holes, and with $-x$ for electrons, in the manner illustrated in Fig. 1.43, middle. Mathematically, the decay for holes is expressed as

$$p'_n(x) = p_{n0}(e^{v/V_T} - 1)e^{-(x-x_n)/L_p} \tag{1.53}$$

where the quantity

$$L_p = \sqrt{D_p \tau_p} \tag{1.54a}$$

is called the *hole diffusion length*, in cm. It represents the distance from x_n at which $p'_n(x)$ drops to $1/e$ (36.8%) of its boundary value $p'_n(x_n)$. A similar expression holds for excess electrons in the *p* side, but with the *electron diffusion length*

$$L_n = \sqrt{D_n \tau_n} \tag{1.54b}$$

The lengths L_p and L_n are typically on the order of 10^1 μm, so we generally have $L_p \gg x_n$ and $L_n \gg x_p$.

Note that $p'_n \to 0$ at the right end of the *n*-type slab ($x = W_n$), as the excess holes undergo *total recombination* with the electrons of the metal electrode there. Likewise, $n'_p \to 0$ at the left end of the *p*-type slab ($x = -W_p$), as excess electrons undergo *total removal* by the metal electrode there.

By Eq. (1.27a) the diffusion of excess holes toward the right results in the current density $J_p(x) = -qD_p\,dp'_n(x)/dx$. Differentiating Eq. (1.53) and substituting gives

$$J_p(x) = qD_p\frac{p_{n0}}{L_p}(e^{v/V_T} - 1)e^{-(x-x_n)/L_p} \tag{1.55a}$$

A similar expression holds for the current density due to excess electrons diffusing toward the left, except that we need to replace x with $-x$ in the exponent,

$$J_n(x) = qD_n\frac{n_{p0}}{L_n}(e^{v/V_T} - 1)e^{(x+x_p)/L_n} \tag{1.55b}$$

We expect recombination within the (thin) depletion region to be negligible, so J_p and J_n will essentially be constant there. With reference to Fig. 1.43, bottom, we find the total current density within the SCL as $J_{\text{tot}} = J_p(x_n) + J_n(-x_p)$. Using Eq. (1.55), we readily get

$$J_{\text{tot}} = q\left(\frac{D_p p_{n0}}{L_p} + \frac{D_n n_{p0}}{L_n}\right)(e^{v/V_T} - 1) \tag{1.56}$$

Note that Fig. 1.43, bottom, shows only the *minority* diffusion currents. For a complete picture of conduction we need to show also the *majority* diffusion currents. Owing to the charge conservation principle, J_{tot} must be *constant* throughout the slab. We can thus obtain each majority current component by taking the graphical *difference* between the total current and the corresponding minority current component, or $J_p = J_{\text{tot}} - J_n$ to the left of $-x_p$, and $J_n = J_{\text{tot}} - J_p$ to the right of x_n. The result, shown in Fig. 1.44, shows how fascinating conduction is inside a *pn* slab.

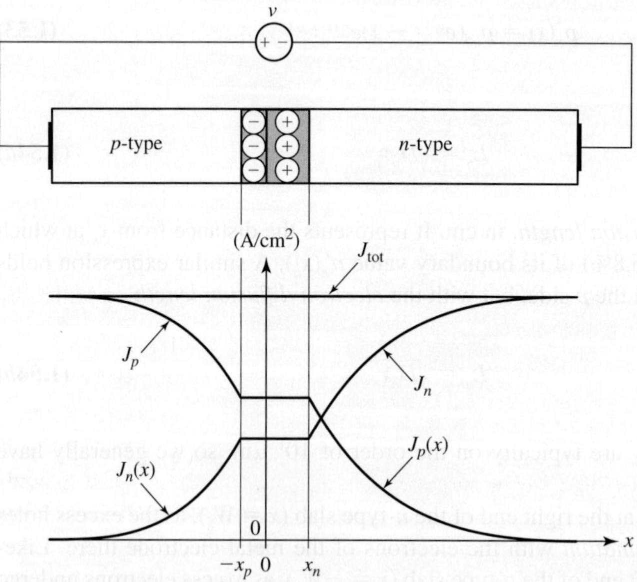

FIGURE 1.44 Minority and majority current densities inside a *pn* slab.

If we were to scan the slab from left to right, our description of conduction would be as follows

- At the *far left* we see a current of mainly holes diffusing toward the right on their way to recombine with electrons. Some of these holes disappear by recombination in the *p* side itself, others manage to make it all the way to the SCL, from where they emerge as minority carriers in the *n* side. As they progress toward the right, they are further annihilated by electrons.
- Moving *closer* to the SCL while still on the *p* side, we note that J_p has subsided somewhat, but at the expense of an increase in J_n, so as to ensure the constancy of J_{tot}. The fact that close to the SCL, J_p has subsided doesn't necessarily imply a reduction in hole concentration there. In fact, Example 1.9 has revealed that at $x = -x_p$ there are 10^{18} holes/cm^3 pressing against the SCL, quite a number compared to the 1.44×10^{13} electrons/cm^3 available there. Out of these 10^{18} holes/cm^3, only 1.44×10^{15} holes/cm^3 manage to emerge from the SCL, at the right.
- Moving now *inside* the SCL, we see a two-way traffic of holes and electrons, hardly recombining with each other because X_d is much shorter than the diffusion lengths L_p and L_n. Because of asymmetric doping as well as differences in the hole and electron diffusivities and diffusion lengths, J_p and J_n are generally different inside the SCL.
- As we move *past* the SCL into the *n* region, we note that excess holes fade away, annihilated by the majority of electrons present there. We instead observe a progressively larger current of electrons moving toward the left, on their way either to annihilate holes while still in the *n* side, or to be annihilated by holes once they cross the SCL to emerge in the *p* side.

The Diode Equation

The overall current i through a *pn* junction of cross-sectional area A is readily found as $i = AJ_{tot}$. Using Eqs. (1.56), along with Eq. (1.30), we get what is commonly referred to as the *diode equation*

$$i = I_s(e^{v/V_T} - 1) \tag{1.57}$$

where I_s is a *scaling factor* called the *saturation current*,

$$I_s = A \times n_i^2(T) \times q\left(\frac{D_p}{L_p N_D} + \frac{D_n}{L_n N_A}\right) \tag{1.58}$$

This factor gives an indication of how much current i we get for a given applied voltage v. For low-power junctions, I_s is typically on the order of femto-amperes (1 fA = 10^{-15} A). We observe that I_s depends on:

- The *cross-sectional area A* (the larger A, the higher the current—just as with ordinary resistors)
- *Temperature T*, especially via $n_i^2(T)$
- The *doping densities* N_A and N_D, the *diffusivities* D_p and D_n, and the *diffusion lengths* L_p and L_n.

As we proceed we shall find that most practical junctions are fabricated with one side much more heavily doped than the other. When this is the case, one of the terms within parentheses in Eq. (1.58) becomes negligible, and I_s is determined primarily by the term with the smaller doping concentration in its denominator. In our working *pn* junction example, for which $N_A \gg N_D$, the dominant term in Eq. (1.58) is the first one, stemming from the injection of holes into the more lightly doped *n*-side. As we know, this is also the side into which most of the SCL extends. For obvious reasons, asymmetrically-doped junctions are also referred to as *one-sided junctions*. An example will better illustrate.

EXAMPLE 1.10   (a) Estimate i if the *pn* junction of Example 1.7 is biased at $v = 0.65$ V. Assume $D_p = 10$ cm^2/s, $L_p = 5$ μm, $D_n = 7$ cm^2/s, and $L_n = 10$ μm. Comment on your results.

(b) Find the area A needed for the junction of part (*a*) to give $i = 0.15$ mA at $v = 0.65$ V.

Solution

(a) Inserting the given data into Eq. (1.58) gives

$$I_s = 10^{-4} \times 2 \times 10^{20} \times 1.602 \times 10^{-19}\left(\frac{10}{5 \times 10^{-4} \times 10^{16}} + \frac{7}{10 \times 10^{-4} \times 10^{18}}\right)$$

$$= 6.41 + 0.02 = 6.43 \text{ fA}$$

As expected of a one-sided junction such as the present one, I_s is determined primarily by one term, namely, the first term, representing hole injection. Electron injection into the heavily-doped *p*-side has little say in this case. Finally, we use Eq. (1.57) to find

$$i = 6.43 \times 10^{-15}(e^{650/26} - 1) = 463 \ \mu\text{A}$$

(b) To lower i from 0.463 mA to 0.15 mA, we need to lower A in proportion, that is, from 10^{-4} cm^2 to $(0.15/0.463) \times 10^{-4}$ cm^2, or to 0.324×10^{-4} cm^2. This requires a square area of about $(57 \ \mu\text{m}) \times (57 \ \mu\text{m})$.

Short-Base Diodes

In the diode example of Fig. 1.43 the neutral regions are long enough to provide sufficient distance for the minority charges to recombine with the majority charges as they diffuse away from the SCL. Aptly called a *long-base* diode, this structure occurs when the device is fabricated with dimensions $W_n \gg L_p$ and $W_p \gg L_n$. As we progress, we shall see that *pn* diodes are also fabricated with $W_n \ll L_p$, or $W_p \ll L_n$, or both. A popular example is the base-emitter junction of a bipolar junction transistor, this being the reason why such a structure is referred to as a *short-base diode*.

With $W_n \ll L_p$, the holes injected into the *n*-side don't have much of a chance to recombine while diffusing toward the right, indicating that J_p will essentially be

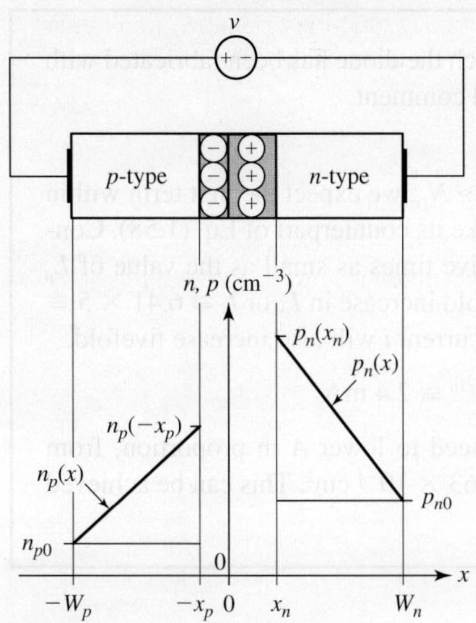

FIGURE 1.45 Minority carrier concentrations in a forward-biased *short-base* diode fabricated with $W_n \ll L_p$ and $W_p \ll L_n$.

constant in the *n*-side. By Eq. (1.27*a*), this implies, in turn, a constant slope for $p_n(x)$, as depicted in Fig. 1.45. If the condition $W_p \ll L_n$ holds, similar considerations apply to J_n and $n_p(x)$ in the *p*-side. To find the *i-v* characteristic of a short-base diode, we start with $J_p = -qD_p dp'_n(x)/dx$, where $p'_n(x)$ is the excess hole density in the *n*-side. The slope of the triangle in Fig. 1.45 is readily found as

$$\frac{dp'_n(x)}{dx} = -\frac{p_n(x_n) - p_{n0}}{W_n - x_n} \cong -\frac{p_{n0}(e^{v/V_T} - 1)}{W_n}$$

where we have exploited the fact that usually $x_n \ll W_n$. Similar considerations hold for the *p*-side slope $dn'_p(x)/dx$. Proceeding as in the derivation of Eq. (1.58), we readily find that a short-base diode still obeys Eq. (1.57), but with the following expression for the saturation current,

$$I_s \cong An_i^2 q\left(\frac{D_p}{W_n N_D} + \frac{D_n}{W_p N_A}\right) \tag{1.59}$$

This is identical to that of Eq. (1.58), except for the replacements $L_p \to W_n$ and $L_p \to W_n$. Considering that in a short-base diode the Ws are much smaller than the Ls, it is apparent that this structure requires a smaller cross-sectional area A to achieve the same value of I_s. Another advantage is that the amount of excess charge stored in a forward-biased short-base diode is much smaller than that of a long-base device operating at the same current. This results in much faster switching times, as we shall see in Chapter 6.

EXAMPLE 1.11   Repeat Example 1.10, but for the case in which the diode has been fabricated with $W_p = 0.5\ \mu$m and $W_n = 1\ \mu$m. Compare and comment.

Solution

Since we have a one-sided junction with $N_A \gg N_D$, we expect the first term within parentheses in Eq. (1.59) to dominate, just like its counterpart of Eq. (1.58). Considering that the value of W_n given here is five times as small as the value of L_p given in Example 1.10, we anticipate a fivefold increase in I_s, or $I_s \cong 6.41 \times 5 \cong$ 33 fA. With the same applied voltage v, the current i will also increase fivefold,

$$i \cong 33 \times 10^{-15} e^{650/26} \cong 2.4\ \text{mA}$$

To lower i from 2.4 mA to 0.15 mA, we need to lower A in proportion, from 10^{-4} cm^2 to $(0.15/2.4) \times 10^{-4}$ cm^2, or to 0.063×10^{-4} cm^2. This can be achieved with a square area of $(25\ \mu$m$) \times (25\ \mu$m$)$.

1.8 THE REVERSE-BIASED *pn* JUNCTION

Reverse biasing a *pn* junction further increases the existing potential barrier, thus inhibiting hole and electron diffusion across the metallurgical junction. Given this strong preference to conduct in the forward direction, the *pn* junction exhibits diode behavior, so from now on we shall use the words *pn junction* and *diode* interchangeably.

For sufficiently negative values of the applied voltage v (say, for $v < -4V_T \cong -0.1$ V), Eq. (1.57) predicts that i will saturate at $-I_s$ (hence, the name *saturation current*). As we know, for low-power diodes, I_s is typically in the fA range. However, the actual reverse current found in a *pn* junction, which we shall denote as I_R, is *orders of magnitude higher* than I_s, typically in the pA to nA range. This stems from the *thermal generation* of hole-electron pairs within the space-charge layer SCL, which we ignored in the course of our analysis. In fact, even though we have been referring to the SCL as the depletion region, thermal generation of hole-electron pairs does continue to take place there, and once generated, holes and electrons are swept in opposite directions by the strong local electric field E, resulting in a combined *drift current* from the n side, through the SCL, to the p side. Intuitively we expect I_R to be proportional to the SCL's volume AX_d, and since X_d increases with the amount of reverse bias, as per Eq. (1.45), I_R will also increase with the reverse voltage in *square-root fashion*. Depending on the quality of fabrication, *leakage current* may also flow across the surface of the *pn* junction, further contributing to I_R.

The overall reverse current I_R is a strong function of temperature, a behavior that engineers remember via the following important rule of thumb:

> The reverse current I_R of a *pn* junction *doubles* for about every 10°C of temperature increase

Once we know I_R at some reference temperature T_0, we can estimate it at any other temperature T using

$$I_R(T) \cong I_R(T_0) \times 2^{(T-T_0)/10} \qquad (1.60)$$

EXAMPLE 1.12

If at 25 °C a certain diode exhibits $I_R = 1$ pA, estimate I_R (*a*) at 125 °C, and (*b*) at −25 °C.

Solution

(*a*) By Eq. (1.60), $I_R(125\,°C) \cong 10^{-12} \times 2^{(125-25)/10} \cong 1$ nA. (*b*) Likewise, $I_R(-25\,°C) \cong 0.03$ pA.

Reverse Breakdown

If we gradually increase the reverse bias of a *pn* junction, a voltage is reached, called the *breakdown voltage* (*BV*), at which the reverse current shoots up in magnitude from the negligible value I_R discussed above to much higher values. The name stems from the fact that the *i-v* curve bends sharply, or breaks down. This does not necessarily imply a destructive process—in fact, one always limits the reverse current within safety levels by interposing a suitable resistor in *series* between the driving voltage source and the reverse-biased junction. Figure 1.46 shows the complete *i-v* characteristic of a typical *pn* junction.

The breaking down of the *i-v* curve evidently indicates the sudden availability of huge quantities of mobile charges to produce the much increased current levels. This sudden availability is the result of either of two separate mechanisms: *Zener*

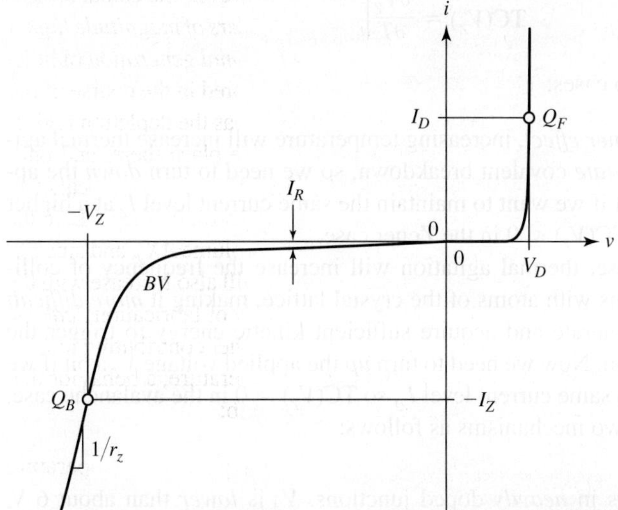

FIGURE 1.46 The complete *i-v* characteristic of a *pn* junction.

breakdown, and *avalanche breakdown*. The former occurs in heavily doped junctions, the latter in lightly doped junctions.

- In *heavily doped* junctions the electric field within the SCL is already fairly strong, and increasing it further with several volts of reverse bias will give it enough strength to strip electrons away from the covalent bonds and thus create electron-hole pairs. The field itself then sweeps these newly freed charges out of the SCL (holes into the *p* side, electrons into the *n* side), thus sustaining much higher currents than in the case of thermal generation alone. Called *Zener effect*, this phenomenon occurs for BV values on the order of 6 V or less.
- In *lightly doped* junctions the electric field is not strong enough to break covalent bonds directly. However, with the wider SCL widths now available, the field has more space to accelerate any free electrons that happen to be within the SCL. Given sufficient kinetic energy, these electrons will free new electron-hole pairs as they collide with the atoms of the crystal lattice. These secondary electrons can in turn free additional electrons, in an effect aptly called *avalanche effect*. This effect occurs for BV values on the order of 6 V or higher. In the neighborhood of 6 V the Zener and avalanche effects may coexist.

When designed to operate deliberately in the breakdown region, a diode is commonly referred to as a *Zener diode*, regardless of whether the actual breakdown mechanism is of the Zener or avalanche type. The coordinates of an operating point Q_B in the breakdown region are conveniently relabeled as $-I_Z$ and $-V_Z$, respectively. The *slope* of the diode curve in the breakdown region is denoted as $1/r_z$, and sufficiently to the left of the breakdown knee it is approximately constant. Depending on fabrication details, r_z is on the order of 10^1 to 10^3 Ω.

The *temperature coefficient* at a given breakdown-region operating point $Q_B(I_Z, V_Z)$ is defined as

$$\text{TC}(V_Z) = \left. \frac{\partial V_Z}{\partial T} \right|_{I_z}$$

We again distinguish two cases:

- In the case of the *Zener effect*, increasing temperature will increase thermal agitation and thus *facilitate* covalent breakdown, so we need to turn *down* the applied voltage V_Z a bit if we want to maintain the same current level I_Z at a higher temperature. Thus, $\text{TC}(V_Z) < 0$ in the Zener case.
- In the *avalanche* case, thermal agitation will increase the frequency of collisions of free electrons with atoms of the crystal lattice, making it *more difficult* for electrons to accelerate and acquire sufficient kinetic energy to trigger the avalanche mechanism. Now we need to turn *up* the applied voltage V_Z a bit if we want to maintain the same current level I_Z, so $\text{TC}(V_Z) > 0$ in the avalanche case. We summarize the two mechanisms as follows:

The *Zener effect* occurs in *heavily* doped junctions, V_Z is *lower* than about 6 V, and $\text{TC}(V_Z) < 0$

The *avalanche effect* occurs in *lightly* doped junctions, V_Z is *higher* than about 6 V, and $\text{TC}(V_Z) > 0$

By playing with the impurity concentrations during processing, the manufacturer can control the BV of a junction to a particular value. Two familiar representatives are the base-emitter (BE) and the base-collector (BC) junctions forming the bipolar junction transistor (BJT). The BE junction is heavily doped, and thus breaks down by the Zener effect in the neighborhood of 6 V. This low BV value poses no problem when the BJT is operated in the forward-active (FA) region, where the BE junction is forward biased. However, in the FA region the BC junction is reverse biased. To prevent it from going into breakdown, the collector region is doped lightly, indicating that BC breakdown is of avalanche type.

1.9 FORWARD-BIASED DIODE CHARACTERISTICS

For sufficiently high forward voltages (in practice, for $v > 4V_T \cong 0.1$ V), we can ignore unity in Eq. (1.57), and write

$$i_D = I_s e^{v_D/V_T} \tag{1.61}$$

where we are now using subscript D to signify operation well into the forward region. This equation represents a perfectly *exponential* i-v characteristic. Also called the *ideal diode equation*, it is satisfied by practical *pn* junctions quite well over a wide range of currents, typically on the order of six decades, making it one of the most predictable laws of electronics. The exponential law enjoys some fascinating properties, as we are about to see.

Equation (1.61) is readily turned around as

$$v_D = V_T \ln\!\left(\frac{i_D}{I_s}\right) \tag{1.62}$$

In this form, it allows us to find the voltage drop v_D needed to sustain a given current i_D.

Properties of the Exponential Characteristic

The *slope* of the diode curve at a given operating current I_D in the forward-bias region is defined as $g_d = di_D/dv_D|_{I_D}$, and is called the *dynamic conductance* of the diode. Differentiating Eq. (1.61), we get

$$g_d = \frac{I_D}{V_T} \tag{1.63}$$

indicating that *slope* is *linearly proportional to the operating current* I_D. The reciprocal of g_d is called the *dynamic resistance* of the diode, or $r_d = 1/g_d = V_T/I_D$. Both g_d and r_d span quite a range of values, depending on the operating current. Note the following significant values:

$$r_d\,(1\text{ mA}) = 26\ \Omega \qquad r_d\,(1\ \mu\text{A}) = 26\text{ k}\Omega \qquad r_d\,(1\text{ nA}) = 26\text{ M}\Omega$$

The parameters r_d and g_d form the basis of *small-signal* diode circuit analysis, to be studied later.

Given a diode carrying a certain current I_D, we wish to find the *voltage change* ΔV_D required to change its current from I_D to mI_D, where m is some multiplicative factor. Using Eq. (1.62), we find such a voltage change as $\Delta V_D = V_T \ln (mI_D/I_s) - V_T \ln(I_D/I_s) = V_T \ln[(mI_D/I_s)/(I_D/I_s)]$, or

$$\Delta V_D = V_T \ln m$$

Two popular cases are a change in current by an *octave* ($m = 2^{\pm 1}$), or by a *decade* ($m = 10^{\pm 1}$), which give, respectively, $\Delta V_{D(\text{oct})} = (26\text{ mV}) \times (\pm \ln 2) \cong \pm 18$ mV, and $\Delta V_{D(\text{dec})} = (26\text{ mV}) \times (\pm \ln 10) \cong \pm 60$ mV. These findings form the basis of the following important rules of thumb, illustrated pictorially in Fig. 1.47*a*:

> To effect an *octave* change in I_D we need to change V_D by 18 mV

> To effect a *decade* change in I_D we need to change V_D by 60 mV

A convenient feature of the above rules is that they are *independent* of the particular quiescent point Q_F on the diode curve where the changes are made. For instance, consider a diode initially operating at a quiescent current of 10 μA. If we wish to *double* its current to 20 μA, we need an increase $\Delta V_D = 18$ mV. To effect the *tenfold* change 10 μA $\rightarrow$ 100 μA we need an increase $\Delta V_D = 60$ mV. Likewise, the change 10 μA $\rightarrow$ 1 μA requires a decrease $\Delta V_D = -60$ mV. If you wish to change I_D from

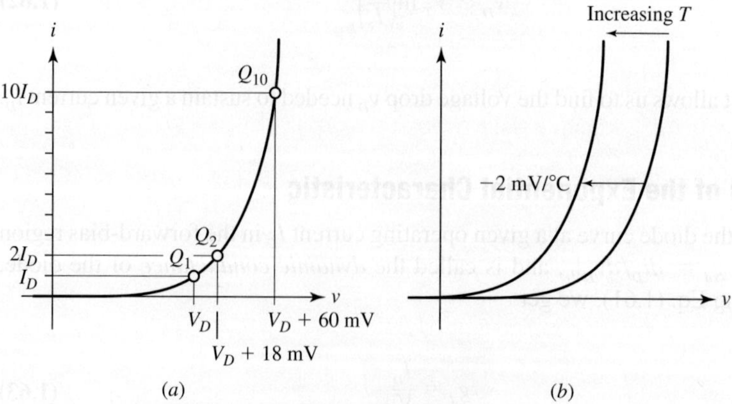

FIGURE 1.47 Illustrating some important rules of thumb for *pn* junctions.

$10\ \mu A$ to $50\ \mu A$, pretend first to raise it from $10\ \mu A$ to $100\ \mu A$ $(\Delta V_D = +60$ mV$)$, and then to lower it from $100\ \mu A$ to $50\ \mu A$ $(\Delta V_D = -18$ mV$)$, for a net change $V_D = 60 - 18 = 42$ mV.

Temperature Dependence

Whether we use Eq. (1.61) or (1.62), it is apparent that the diode characteristic depends on temperature via V_T as well as I_s. A convenient way to characterize the overall thermal dependence is via the *temperature coefficient* of the forward voltage drop at a given operating current I_D, defined as

$$\text{TC}(V_D) = \left. \frac{\partial V_D}{\partial T} \right|_{I_D} \tag{1.64}$$

Differentiating Eq. (1.62), we obtain

$$\text{TC}(V_D) = \left. \frac{\partial [V_T \ln(I_D/I_s)]}{\partial T} \right|_{I_D} = \frac{V_D}{T} - V_T \left(\frac{dI_s/dT}{I_s} \right) \tag{1.65}$$

According to Eqs. (1.58) and (1.59), $I_s \propto n_i^2(T)D(T)$, where n_i is the intrinsic concentration and D is the diffusivity, both of which are functions of temperature T. By Eq. (1.21), $n_i^2(T) \propto T^3 e^{-V_{G0}/V_T}$, where $V_{G0} = 1.205$ V is the *bandgap voltage* for silicon. By Eqs. (1.20) and (1.28), $D(T) = (kT/q)\mu(T)$, where $\mu(T)$ is the mobility, in turn a function of temperature of the type $\mu(T) \propto T^m$, $m \cong -1.5$. Combining all this we have

$$I_s \propto T^{4+m} e^{-V_{G0}(q/kT)}$$

$$\frac{dI_s/dT}{I_s} = \frac{(4+m)T^{4+m-1}e^{-V_{G0}(q/kT)} + T^{4+m}e^{-V_{G0}(q/kT)} \times V_{G0}(q/kT^2)}{T^{4+m}e^{-V_{G0}(q/kT)}} = \frac{4+m}{T} + \frac{V_{G0}}{TV_T}$$

Substituting into Eq. (1.65) and simplifying, we finally get

$$\text{TC}(V_D) = \frac{V_D - (4+m)V_T - V_{G0}}{T} \tag{1.66}$$

Assuming $V_D = 0.65$ V at $T = 300$ K, Eq. (1.66) gives $\text{TC}(V_D) \cong -2.1$ mV/°C. Engineers remember *pn* junction thermal behavior via the following rule of thumb, pictorially illustrated in Fig. 1.47*b*:

At room-temperature the forward voltage-drop of a *pn* diode drifts by about -2 mV/°C

Once we know V_D at some reference temperature T_0, we can estimate it at any other temperature T using

$$V_D(T) \cong V_D(T_0) - (2\ \text{mV}) \times (T - T_0) \tag{1.67}$$

EXAMPLE 1.13   If at 25 °C a certain diode exhibits $V_D = 650$ mV at $I_D = 0.1$ mA, estimate V_D at:

(a) $T = 70$ °C and $I_D = 0.1$ mA.
(b) $T = 0$ °C and $I_D = 0.1$ mA.
(c) $T = 50$ °C and $I_D = 0.02$ mA.
(d) $T = 40$ °C and $I_D = 4$ mA.

Solution

Using the rules of thumb learned thus far we get:

(a) $V_D \cong 650 - 2 \times (70 - 25) = 650 - 90 = 560$ mV.
(b) $V_D \cong 650 - 2 \times (0 - 25) = 650 + 50 = 700$ mV.
(c) First, pretend to lower $I_D = $ from 0.1 mA to 0.01 mA ($\Delta V_D = -60$ mV), and then double it ($\Delta V_D = +18$ mV) to end up at 0.02 mA with the net value $V_D \cong 650 - 60 + 18 = 608$ mV. Finally, pretend to increase temperature from 25 °C to 50 °C, for an additional change $\Delta V_D = -2 \times (50 - 25) = -50$ mV. The final value is thus $V_D \cong 608 - 50 = 558$ mV.
(d) Proceeding as in part (*c*) we find $V_D = 650 + 60 + 18 + 18 - 2 \times (40 - 25) = 716$ mV.

If a forward-biased diode is placed in *series* with a reverse-biased Zener diode of the *opposite* temperature coefficient, or $\text{TC}(V_Z) = -\text{TC}(V_D) \cong +2$ mV, the thermal variations of the two devices will *cancel* each other out to yield a composite voltage drop $V_{REF} = V_Z + V_D$ approaching *zero temperature coefficient*. This technique is used in the implementation of thermally stable voltage references. The best stability is achieved for $V_Z \cong 6.2$ V, thus yielding $V_{REF} \cong 6.2 + 0.7 = 6.9$ V with $\text{TC}(V_{REF}) \to 0$.

Deviations from Ideality

Rewriting Eq. (1.62) as $v_D = V_T(\ln i_D - \ln I_s)$, or

$$\ln i_D = \left(\frac{1}{V_T}\right)v_D + \ln I_s$$

indicates that if we plot i_D versus v_D on *semi-logarithmic* paper (v_D on the *linear* axis, i_D on the *logarithmic* axis), we get a curve of the type

$$y = (1/V_T)x + y(0)$$

This is a *straight line* with slope $(1/V_T)$ and 0-V intercept at $i_D = I_s$. This feature proves very convenient in the characterization of a diode. Given a set of measured data, we can easily find the *best-fit straight line*; then, we find its *slope* to obtain the experimental value of V_T, and we *extrapolate* the line in the limit $v_D \to 0$ to find its *intercept* on the i_D axis and thus obtain the experimental value of I_s.

The semi-log-plot of an actual low-power *pn* diode characteristic is more likely to appear as in Fig. 1.48. The curve is fairly straight over a wide range of currents, typically from 1 nA to 1 mA, but deviates from ideality both at the high and

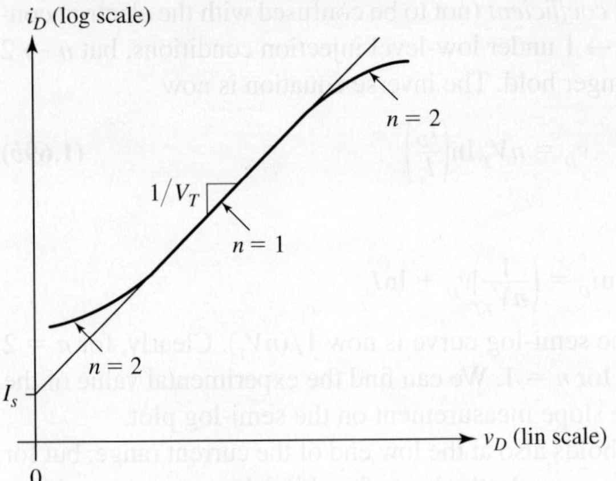

FIGURE 1.48 Illustrating the effect of the emission coefficient n at low and at high current levels.

low ends of the range. These deviations stem from various approximations that were made in the course of the derivations leading to the ideal diode equation. Specifically, deviations at the high end of the current range are caused by the presence of *bulk resistance* in the neutral regions as well as *high-level injection* effects, while deviation at the low end is caused by loss of mobile charges to *recombination* within the space-charge layer (SCL). We now wish to examine these approximations in greater detail.

- In our derivations we have assumed zero electric field inside the regions at either side of the SCL, so that the voltage that we apply across the terminals is transmitted *entirely* to the edges of the SCL itself. However, like any conductor, each of these regions exhibits a nonzero—if small—ohmic resistance called the *bulk resistance*. Denoting the overall resistance (sum of the p-side and n-side bulk resistances) as r_S, we observe that in response to the externally applied voltage v_D, the actual voltage v_J reaching the junction is, by KVL,

$$v_J = v_D - r_S i_D \tag{1.68}$$

  When the diode is operated at low current levels, the voltage drop across the bulk material is negligible and $v_J \cong v_D$. Not so at the upper end of the current range, where this drop may become significant and cause a deviation of the actual i-v characteristic from exponential ideality. On semi-log paper, this appears as a curvature for values of i_D on the order of 1 mA or higher.

- If the applied voltage across the junction is increased to the point of making the minority densities at the SCL edges comparable to the majority densities there, the low-level injection assumption no longer holds, and the diode equation now takes on the form

$$i_D = I_s e^{v_D/n V_T} \tag{1.69a}$$

where n, called the *emission coefficient* (not to be confused with the electron concentration n!) is such that $n \rightarrow 1$ under low-level injection conditions, but $n \rightarrow 2$ when these conditions no longer hold. The inverse equation is now

$$v_D = nV_T \ln\left(\frac{i_D}{I_s}\right)$$ **(1.69b)**

Rewriting as

$$\ln i_D = \left(\frac{1}{nV_T}\right)v_D + \ln I_s$$

indicates that the slope of the semi-log curve is now $1/(nV_T)$. Clearly, for $n = 2$ the slope is only *half* of that for $n = 1$. We can find the experimental value of the quantity nV_T through a mere slope measurement on the semi-log plot.

- Equation (1.69) with $n \rightarrow 2$ holds also at the low end of the current range, but for a completely different reason, namely the loss of mobile charges to recombination (trapping) within the SCL. Evidently, our assumption of constant current densities within the SCL is not exactly valid. It must be said that loss to recombination occurs regardless of the operating point on the *i-v* curve, but its effect is noticeable only at the low-end, where the forward current is comparable to, or even smaller than that due to loss. On semi-log paper, this appears as a curvature for values of i_D on the order of 1 pA or lower.

EXAMPLE 1.14   If a *pn* junction with $I_s = 1$ fA gives $I_D = 1$ mA at $V_D = 725$ mV, find r_S.

Solution

By Eq. (1.62) the actual junction voltage is $V_J = (26 \text{ mV}) \ln (10^{-3}/10^{-15}) = 718.4 \text{ mV}$. By Eq. (1.68), $r_S = (725 - 718.4)/1 = 6.6 \, \Omega$.

1.10 Dc ANALYSIS OF *pn* DIODE CIRCUITS

A task that arises all the time in the analysis of diode circuits is finding a diode's quiescent point $Q = Q(I_D, V_D)$. As we know, if the diode is embedded in an otherwise linear circuit, this task simplifies considerably if we replace the surrounding circuitry with its *Thévenin equivalent* and thus end up with the situation of Fig. 1.49a. Here, V_{OC} is the open-circuit voltage that the external circuit would produce between the nodes corresponding to the *anode* and the *cathode*, but with the *diode removed*, and R_{eq} is the external circuit's equivalent resistance, that is, the resistance as seen by the diode.

Load-Line Analysis

The operating point can be visualized *graphically* as the intercept of the *diode curve* and the *load line*, as already seen in Fig. 1.5b for the case of the ideal diode. The situation is depicted in Fig. 1.49b for the case of a forward-biased *pn* junction diode. Though graphical analysis helps us develop a *visual* feel for a circuit's workings, we usually need to find the operating point Q *numerically*, the issue that we are going to address next.

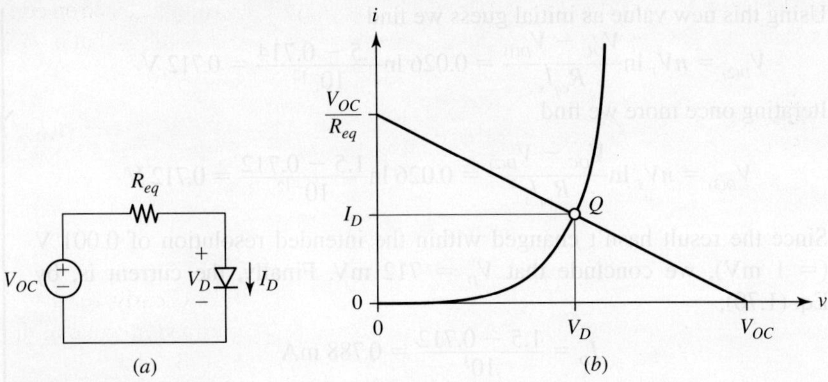

FIGURE 1.49 Finding the quiescent point Q of a *pn*-junction diode embedded in a linear circuit.

Iterative Analysis

With reference to Fig. 1.49a we observe that the diode current is, by Ohm's law,

$$I_D = \frac{V_{OC} - V_D}{R_{eq}} \tag{1.70}$$

Also, the diode voltage is, by Eq. (1.69b),

$$V_D = nV_T \ln\frac{I_D}{I_s} \tag{1.71}$$

where n is the emission coefficient, $1 \le n \le 2$. In the following we shall assume $n = 1$ for simplicity, but the ensuing analysis can readily be generalized to the case $n \ne 1$ by replacing V_T with nV_T. Substituting Eq. (1.70) into (1.71) gives

$$V_D = nV_T \ln\frac{V_{OC} - V_D}{R_{eq}I_s} \tag{1.72}$$

This is a transcendental equation in that it does not provide us with a closed-form expression for the unknown V_D. However, we can solve it by iterations. To this end, we start out with a reasonable initial estimate for V_D, we insert it in the right-hand side of Eq. (1.72) to come up with a better estimate, and then we insert this new estimate back in the right-hand side to come up with yet a better estimate, repeating the procedure if necessary, until the result settles within a predetermined resolution.

EXAMPLE 1.15

In the circuit of Fig. 1.49a let $R_{eq} = 1\ \text{k}\Omega$, and let the diode have $n = 1$, $V_T = 26\ \text{mV}$, and $I_s = 1\ \text{fA}$. Estimate V_D down to the mV, as well as I_D, if (*a*) $V_{OC} = 1.5\ \text{V}$, (*b*) $V_{OC} = 3\ \text{V}$, and (*c*) $V_{OC} = 0.75\ \text{V}$.

Solution

(**a**) In the various *pn*-junction examples encountered so far we have found that V_D is typically in the range of 0.6–0.7 V, so let us arbitrarily start out with the initial guess $V_{D(0)} = 0.65\ \text{V}$. Plugging into Eq. (1.72) gives the new estimate

$$V_{D(1)} = nV_T \ln\frac{V_{OC} - V_{D(0)}}{R_{eq}I_s} = 0.026 \ln\frac{1.5 - 0.65}{10^3 \times 10^{-15}} = 0.714\ \text{V}$$

Using this new value as initial guess we find

$$V_{D(2)} = nV_T \ln\frac{V_{OC} - V_{D(1)}}{R_{eq}I_s} = 0.026 \ln\frac{1.5 - 0.714}{10^{-12}} = 0.712 \text{ V}$$

Iterating once more we find

$$V_{D(3)} = nV_T \ln\frac{V_{OC} - V_{D(2)}}{R_{eq}I_s} = 0.026 \ln\frac{1.5 - 0.712}{10^{-12}} = 0.712 \text{ V}$$

Since the result hasn't changed within the intended resolution of 0.001 V ($= 1$ mV), we conclude that $V_D = 712$ mV. Finally, the current is, by Eq. (1.70),

$$I_D = \frac{1.5 - 0.712}{10^3} = 0.788 \text{ mA}$$

(b) This time let us start out with the initial guess $V_{D(0)} = 0.7$ V. Then,

$$V_{D(1)} = 0.026 \ln\frac{3 - 0.7}{10^{-12}} = 0.740 \text{ V}$$

One more iteration confirms that this is actually the desired value, or $V_D = 740$ mV within a 1-mV resolution. Then, $I_D = (3 - 0.74)/1 = 2.26$ mA.

(c) Starting again with $V_{D(0)} = 0.65$ V, we find

$$V_{D(1)} = 0.026 \ln\frac{0.75 - 0.65}{10^{-12}} = 0.659 \text{ V}$$

Three more iterations confirm the final value $V_D = 657$ mV. Consequently, $I_D = (0.75 - 0.657)/1 = 93 \ \mu\text{A}$.

Remark 1: In going from (*a*) to (*b*) we have *doubled* V_{OC}, and in going from (*a*) to (*c*) we have *halved* V_{OC}; however, I_D has neither doubled nor halved, as the presence of the diode renders the circuit *nonlinear*.

Remark 2: We could have found I_D also using the diode equation. For instance, in part (*a*) we could have calculated

$$I_D = I_s e^{V_D/nV_T} = 10^{-15}e^{712/26} = 0.782 \text{ mA}$$

This disagrees with the value of 0.788 mA found via Eq. (1.70). True, both calculations are afflicted by roundoff errors, but even so, which result is the more dependable of the two? Here's an important point to keep in mind: because of the high sensitivity of the exponential function to even small variations in its exponent, *the value of V_D in the exponent must be known very accurately* for the result to be also accurate. By contrast, the value of the V_D in the linear Eq. (1.70) need not be as accurate, and yet it will still yield a dependable I_D value. Consequently, 0.788 mA is the more accurate value in our example.

Remark 3: Of course, if we use the exponential form with a *more accurate* value of V_D the resulting I_D will also be more accurate. In fact, using just one more significant digit from the third-iteration result, which is $V_{D(3)} = 0.7122$ V (rather than the truncated value of 0.712 V) we get

$$I_D = 10^{-15}e^{712.2/26} = 0.788 \text{ mA}$$

This matches that found via Eq. (1.70). If, for some reason, you need to use the diode equation in its exponential form, *make sure that you know the value of V_D to an adequate degree of accuracy!*

Piecewise-Linear Approximation and Large-Signal Diode Models

Load-line analysis provides a graphical means for visualizing a diode's role in a circuit, and iterative analysis provides a numerical means for calculating a diode's operating point. In practice, whether we analyze an existing diode circuit or we design a new one, we need *faster*, if only *approximate*, analysis techniques. Moreover, as we apply these techniques, we wish to draw from the wealth of knowledge acquired in prerequisite linear-circuits courses. Both requirements are met by effecting upon the actual *pn*-junction characteristic of Fig. 1.46 a *piecewise-linear approximation* as depicted in Fig. 1.50. Specifically, the actual curve, shown in shaded form, is approximated with *three straight segments*, each corresponding to a different region of operation, namely, the *forward* (ON), the *cutoff* (CO), and the *breakdown* (BD) regions. We make the following observations:

* When a *pn* junction is forward biased, its characteristic is *exponential*, that is, a curve that after a brief knee, shoots up quite rapidly, no matter which current range we choose to display. The curves of Figs. 1.47 and 1.49*b* were doctored a bit to allow for a better visualization of the details under discussion, but the actual curve is indeed as in Fig. 1.46—quite steep. We know from one of the rules of thumb that increasing V_D by a mere 60 mV raises I_D tenfold, while decreasing it by 60 mV lowers I_D tenfold, indicating that within the

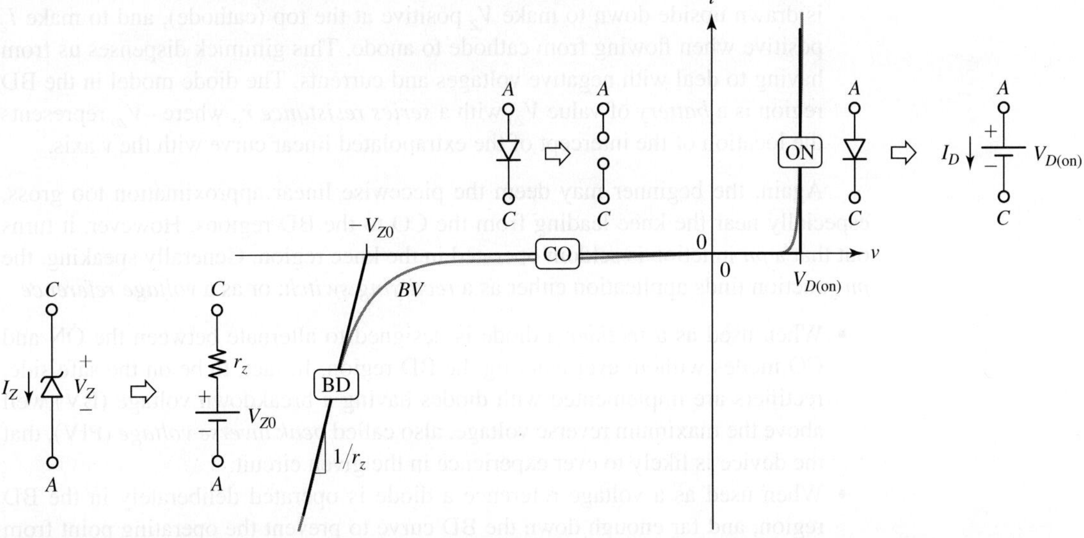

FIGURE 1.50 Piecewise-linear approximations and large signal models of the *pn* junction diode in its three regions of operation: *forward* (ON), *cutoff* (CO), and *breakdown* (BD).

range $V_D \pm 60$ mV, I_D undergoes a 100-to-1 change! For practical purpose, when a junction conducts only $1/100$ of its normal current, it can be regarded as being cut off (in digital applications, even a 10-to-1 ratio will do). Considering that a low-power silicon junction operating in the mA range typically exhibits a room-temperature voltage drop of about 0.7 V, we approximate the exponential curve with a *vertical segment* located at $v = V_{D(\text{on})} = 0.7$ V. The corresponding diode model is thus a *battery* of value $V_{D(\text{on})}$, as shown. For obvious reasons, this model is referred to as a *constant voltage-drop model*. The reader, who may find this approximation unconvincing, will soon realize that it is actually quite effective in providing us with a quick feel for the workings of diode circuits. During a second and more detailed pass, one can always use iterative techniques or computer simulation such as PSpice to refine the analysis and come up with more accurate results.

- From the knee leading to the exponential rise to the knee leading to the break-down drop, the current is negligibly small, indicating that for practical purposes we can regard the junction as *cut off* (CO). The corresponding diode model is an *open circuit*, as shown. Note that the cutoff region extends to the right of the origin, all the way to the onset of the exponential raise. Even though for $v > 0$ the junction is, strictly speaking, forward biased, the actual current between 0 V and the knee leading to the exponential raise is simply too small for the junction to be considered convincingly on. So, we take a weakly forward-biased junction as being effectively cut off.

- Well into the breakdown (BD) region, the characteristic is close to linear, so we approximate it with a straight segment, as shown. As seen in Fig. 1.46, the slope of the curve there is denoted as $1/r_z$, and a BD operating point is denoted as $Q_B = Q_B(-I_Z, -V_Z)$. To signify operation in the BD region, the circuit symbol for the diode is modified as shown at the lower left of Fig. 1.50. Also, the device is drawn upside down to make V_Z positive at the top (cathode), and to make I_Z positive when flowing from cathode to anode. This gimmick dispenses us from having to deal with negative voltages and currents. The diode model in the BD region is a *battery* of value V_{Z0} with a *series resistance* r_z, where $-V_{Z0}$ represents the location of the intercept of the extrapolated linear curve with the v axis.

Again, the beginner may deem the piecewise linear approximation too gross, especially near the knee leading from the CO to the BD regions. However, it turns out that a *pn* junction is seldom operated in the knee region. Generally speaking, the *pn*-junction finds application either as a *rectifying switch*, or as a *voltage reference*.

- When used as a *rectifier* a diode is designed to alternate between the ON and CO modes without ever entering the BD region. In fact, to be on the safe side, rectifiers are implemented with diodes having a breakdown voltage (BV) well above the maximum reverse voltage, also called *peak inverse voltage* (PIV), that the device is likely to ever experience in the given circuit.

- When used as a voltage reference a diode is operated deliberately in the BD region, and far enough down the BD curve to prevent the operating point from ever getting too close to the knee. This issue will be addressed in greater detail in Section 1.12.

EXAMPLE 1.16

Repeat Example 1.15, but using the constant voltage-drop diode model. Comment on your findings.

Solution

Equation (1.70) now approximates to

$$I_D = \frac{V_{OC} - V_{D(on)}}{R_{eq}} \tag{1.73}$$

or $I_D = (V_{OC} - 0.7)/1$.

(a) We now get $I_D = (1.5 - 0.7)/1 = 0.8$ mA, which is very close to 0.788 mA found in Example 1.15.

(b) Now $I_D = (3 - 0.7)/1 = 2.3$ mA, which is even closer to 2.26 mA found in Example 1.15.

(c) If we try $I_D = (0.75 - 0.7)/1 = 50$ μA, we obtain a result in significant disagreement with 93 μA found previously. It is apparent that Eq. (1.73) will give dependable results only if $V_{OC} \gg V_{D(on)}$. If this condition is not met, then the iterative method is the only reasonable alternative for hand calculations.

Remark: Once again we want to stress the *difference* between V_D and $V_{D(on)}$, repeated as follows:

- V_D is the *actual* voltage drop across the diode, and it can be used in critical calculations such as $I_D = I_s \exp(V_D/V_T)$, provided it is known to an adequate degree of accuracy, e.g. within millivolts.
- $V_{D(on)}$ is an *assumed* (~ 0.7 V) and thus only approximate value that we use in less critical calculations such as $I_D = [V_{OC} - V_{D(on)}]/R_{eq}$, provided $V_{OC} \gg V_{D(on)}$.

A potentially catastrophic mistake is to write $I_D = I_s \exp(V_{D(on)}/V_T)$. Make sure you never do this!

Circuit Analysis Using the Piecewise Linear Approximation

Following is a procedure for determining the operating point of a *pn* diode for the case in which the device is embedded in an otherwise *linear circuit*:

- First, find the open-circuit voltage V_{OC} produced by the external circuit with the diode removed (recall that the polarity of V_{OC} is defined *positive* at the node connected to the *anode*.)
- Next, determine the region of diode operation as follows:
  - If $V_{OC} > V_{D(on)}$ ($\cong 0.7$ V for a silicon diode), the diode is operating in the ON region
  - If $-V_{Z0} < V_{OC} < V_{D(on)}$, the diode is cut off (CO)
  - If $V_{OC} < -V_{Z0}$, the diode is operating in the BD region
- Finally, replace the diode with the model pertaining to that particular region, as depicted in Fig. 1.50, and proceed with the analysis of the resulting linear circuit using familiar analysis techniques.

Let us illustrate with practical examples.

The *pn* Junction Diode as a Rectifier

To investigate the operation of the *pn* junction diode as a rectifier we perform a PSpice simulation of a half-wave rectifier using the popular 1N4148 low-power *pn* diode, whose model is available in PSpice's analog library. With reference to the circuit of Fig. 1.51*a* we make the following observations:

- As long as $v_I < V_{D(on)}$, the diode is off and the circuit behaves as in Fig. 1.51*b*, top, giving

$$v_O = 0 \qquad \text{for } v_I < V_{D(on)} \tag{1.74a}$$

- For $v_I > V_{D(on)}$ the diode is conductive and acts as a battery $V_{D(on)} = 0.7$ V, as shown in Fig. 1.51*b*, bottom. The output now follows the input, but with an offset of -0.7 V, by KVL. Thus,

$$v_O = v_I - V_{D(on)} \qquad \text{for } v_I > V_{D(on)} \tag{1.74b}$$

Circuit behavior is illustrated further via the input and output waveforms of Fig. 1.52*a*, and the voltage transfer curve (VTC) of Fig. 1.52*b*. Compared to the ideal diode, for which $V_{D(on)} = 0$, a silicon diode requires about 0.7 V to turn on, and once on, it introduces an error in the form of a -0.7-V offset at the output. This offset may or may not be a problem, depending on the performance requirements of the application at hand. Also, a closer look at both plots reveals that the diode's transition from on to off (or vice versa) is not abrupt, as implied by our piecewise-linear approximation, but *gradual*, owing to the knee of the exponential curve. This feature is actually quite beneficial in the case of piecewise-linear function generators as it ensures a smoother transition from one segment to the next of the VTC.

Now that we understand how a silicon diode behaves as a half-wave rectifier, we find it easier to reexamine all other ideal-diode circuits investigated in Section 1.2, such as full-wave rectifiers, voltage clamps, piecewise-linear function generators, clamped

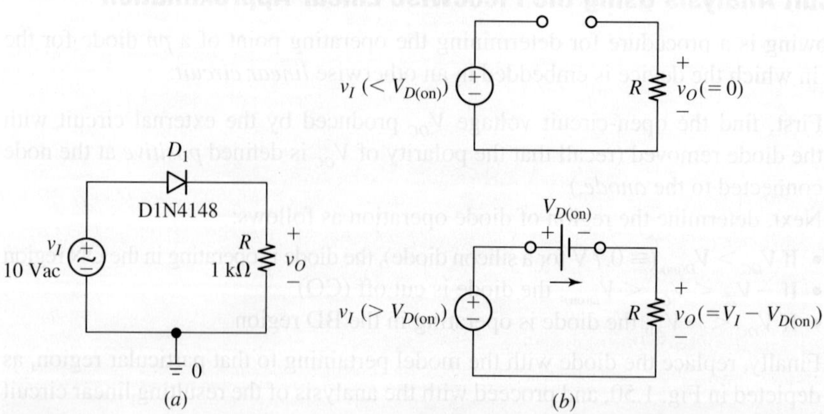

FIGURE 1.51 Investigating a half-wave rectifier using the 1N4148 *pn* diode: (*a*) PSpice circuit, and (*b*) its equivalent circuits when the diode is *off* (top) and *on* (bottom).

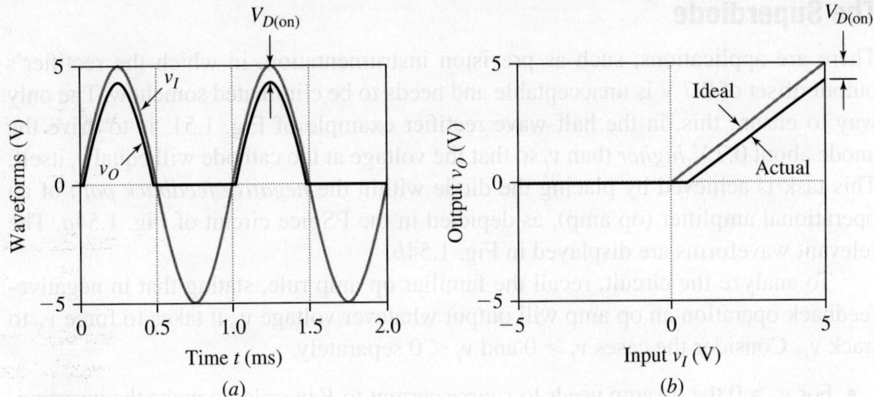

FIGURE 1.52 PSpice plots for the circuit of Fig. 1.51a. (*a*) Input and output waveforms, and (*b*) VTC.

capacitors, and voltage multipliers, but using real-life junction diodes instead. As an example, Fig. 1.53 depicts the full-wave rectifier. Due to the fact that we now have *two* diodes *in series* with the load, the output offset error is $2V_{D(on)}$ ($\cong 1.4$ V).

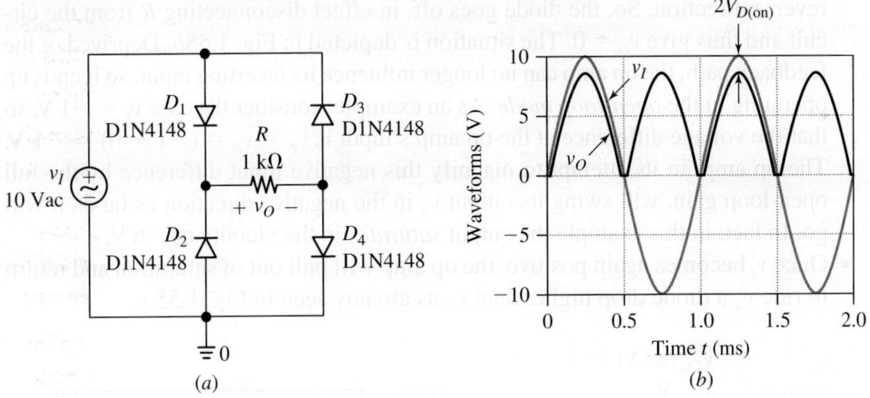

FIGURE 1.53 (*a*) PSpice circuit to simulate a *full-wave rectifier* using 1N4148 diodes, and (*b*) the input and output waveforms.

Exercise 1.6

a. In the following, let the diodes be silicon types with $V_{D(on)} = 0.7$ V. Assuming input logic levels of 0 V and 5 V in the OR gate of Fig. 1.15, find the output logic levels.

b. Repeat, but for the AND gate of Fig. 1.16.

c. Assuming $V_S = 5$ V in the IC circuit of Fig. 1.17, find the range of values for v_{IC}.

Ans. (*a*) 0 V and 4.3 V. (*b*) 0.7 V and 5 V. (*c*) $-0.7 \text{ V} \leq v_{IC} \leq 5.7 \text{ V}$.

The Superdiode

There are applications, such as precision instrumentation, in which the rectifier's output offset of 0.7 V is unacceptable and needs to be eliminated somehow. The only way to ensure this, in the half-wave rectifier example of Fig. 1.51, is to drive the anode about 0.7 V *higher* than v_I so that the voltage at the cathode will equal v_I itself. This task is achieved by placing the diode within the *negative feedback path* of an operational amplifier (op amp), as depicted in the PSpice circuit of Fig. 1.54*a*. The relevant waveforms are displayed in Fig. 1.54*b*.

To analyze the circuit, recall the familiar op amp rule, stating that in negative-feedback operation an op amp will output whatever voltage v_O it takes to force v_N to track v_P. Consider the cases $v_I > 0$ and $v_I < 0$ separately.

- For $v_I > 0$ the op amp needs to *source* current to R in order to make the inverting-input voltage (v_O) follow the non-inverting-input voltage (v_I). This the op amp can readily do via the diode, whose direction conforms to that of the needed current. To turn on the diode, the op amp must swing its output (v_A) a diode drop higher than v_O. For instance, with $v_I = 1$ V, the op amp achieves $v_O = 1$ V by outputting $v_A \cong 1.7$ V. The situation during the positive alternations in v_I is depicted in Fig. 1.55*a*.
- For $v_I < 0$ the op amp would have to *sink* current from R in order to make v_O follow v_I. But, this is impossible due to the diode's inability to conduct in the reverse direction. So, the diode goes off, in effect disconnecting R from the circuit and thus give $v_O = 0$. The situation is depicted in Fig. 1.55*b*. Deprived of the feedback path, the op amp can no longer influence its inverting input, so it ends up operating in the *open-loop mode*. As an example, consider the case $v_I = -1$ V, so that the voltage difference at the op amp's input is $v_P - v_N = (-1 - 0) = -1$ V. The op amp, in its attempt to magnify this negative input difference by the full open-loop gain, will swing its output v_A in the negative direction as far as it will go. In fact, in this example the output *saturates* in the vicinity of -6 V.
- Once v_I becomes again positive, the op amp will pull out of saturation and return to ride v_A a diode drop higher than v_I, as already seen in Fig. 1.55*a*.

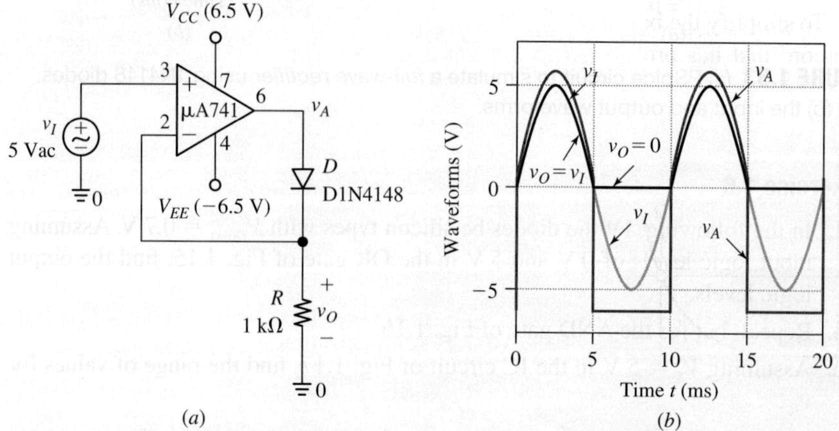

FIGURE 1.54 (a) Superdiode PSpice circuit, and (b) its waveforms

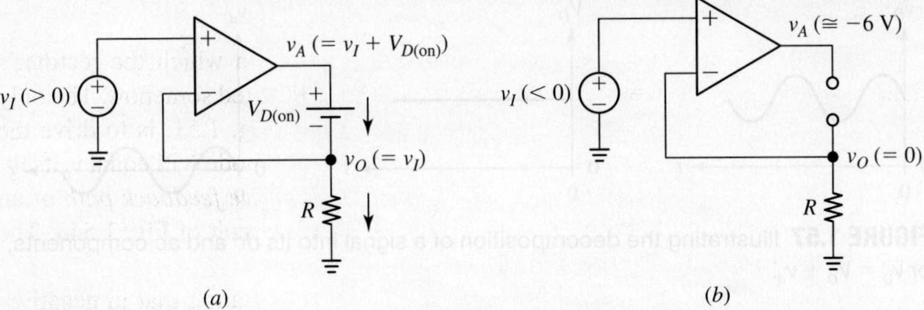

FIGURE 1.55 Equivalent circuits of the superdiode circuit when the diode is (a) *on* and (b) *off*.

1.11 Ac ANALYSIS OF *pn* DIODE CIRCUITS

The piecewise-linear approximation of Fig. 1.50 encompasses the *entire i-v* characteristic of the *pn* diode. There is another important form of linearization in use, but involving only a *limited portion* of the forward-region characteristic. As we know, this characteristic is exponential and thus a highly nonlinear curve. However, if we restrict the diode's operation within a sufficiently *small portion* of this curve, as highlighted in Fig. 1.56a, then we can approximate this portion with a *straight segment* and thus apply familiar linear-circuit analysis techniques. This alternative form of linearization, aptly called *small-signal approximation* and depicted in expanded form in Fig. 1.56b, relies on two premises:

- First, we *bias* the diode at a suitable operating point $Q_0 = Q_0(I_D, V_D)$ up the diode curve, in effect establishing the origin of a new system of *i-v* axes for *signal variations* about this point.
- Then, we *vary* the diode's operating point up and down the curve by amounts denoted as i_d and v_d, keeping i_d and v_d sufficiently small so as to ensure that i_d is *linearly proportional* to v_d, just as in the case of ordinary resistance.

To simplify the bookkeeping, engineers have developed a special form of signal notation[7] that has proved quite convenient not only for diodes, but also for other

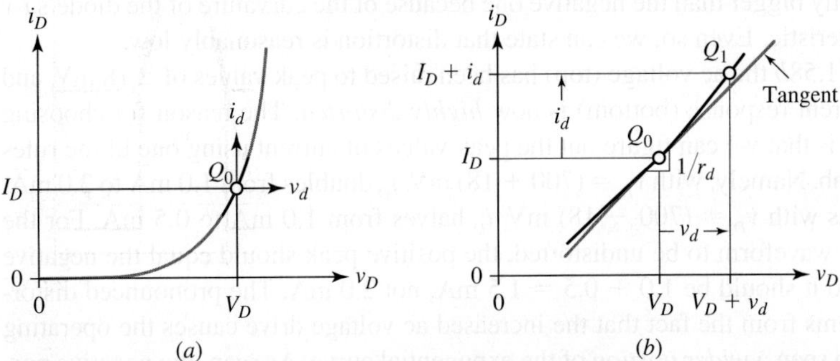

FIGURE 1.56 (a) Illustrating the small-signal operation of a *pn* diode. (b) Expanded view.

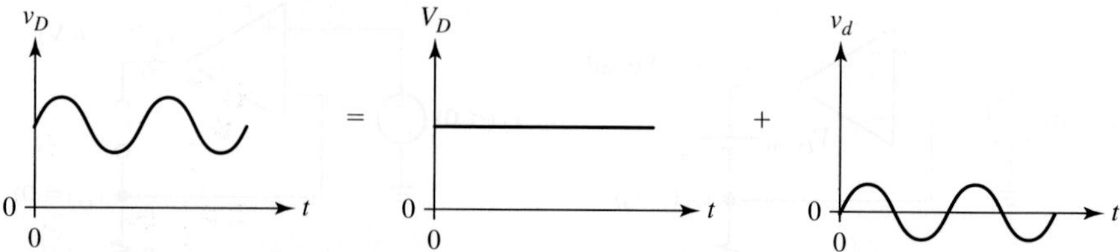

FIGURE 1.57 Illustrating the decomposition of a signal into its *dc* and *ac* components, or $v_D = V_D + v_d$.

nonlinear devices like transistors, as we shall see. By this notation, the voltage and current of the *pn* diode are expressed in the form

$$v_D = V_D + v_d \qquad (1.75a)$$

$$i_D = I_D + i_d \qquad (1.75b)$$

where:

- v_D and i_D are referred to as the *total signals* (lower-case symbols with upper-case subscripts)
- V_D and I_D are their *dc components* (upper-case symbols with upper-case subscripts)
- v_d and i_d are their *ac components* (lower-case symbols with lower-case subscripts)

This form of signal decomposition is illustrated in Fig. 1.57 for the case of the voltage v_D, but a similar picture holds also the current i_D.

Figure 1.58 shows two sets of waveforms obtained via PSpice simulation. The diode used is such that with $V_D = 700$ mV it gives exactly $I_D = 1.0$ mA. Moreover, to make it easier for the naked eye to observe any distortion, we have chosen a triangular waveform for the ac voltage component. We make the following observations:

- In Fig. 1.58*a* the diode is subjected to an ac voltage component (top) having peak values of ± 5 mV, and it responds with a current waveform (bottom) that is only *slightly distorted*. In a truly undistorted waveform, the positive and negative portions are mirror images of each other. In the example shown the positive portion is slightly bigger than the negative one because of the curvature of the diode's *i-v* characteristic. Even so, we can state that distortion is reasonably low.
- In Fig. 1.58*b* the ac voltage (top) has been raised to peak values of ± 18 mV, and the current response (bottom) is now *highly distorted*. The reason for choosing 18 mV is that we can figure out the peak values of current using one of the rules of thumb. Namely, with $v_D = (700 + 18)$ mV, i_D doubles from 1.0 mA to 2.0 mA, whereas with $v_D = (700 - 18)$ mV, i_D halves from 1.0 mA to 0.5 mA. For the current waveform to be undistorted, the positive peak should equal the negative peak, so it should be $1.0 + 0.5 = 1.5$ mA, not 2.0 mA. The pronounced distortion stems from the fact that the increased ac voltage drive causes the operating point to span a *wider portion* of the exponential curve. As seen, the negative portions of the waveforms get compressed, and the positive portions get expanded.

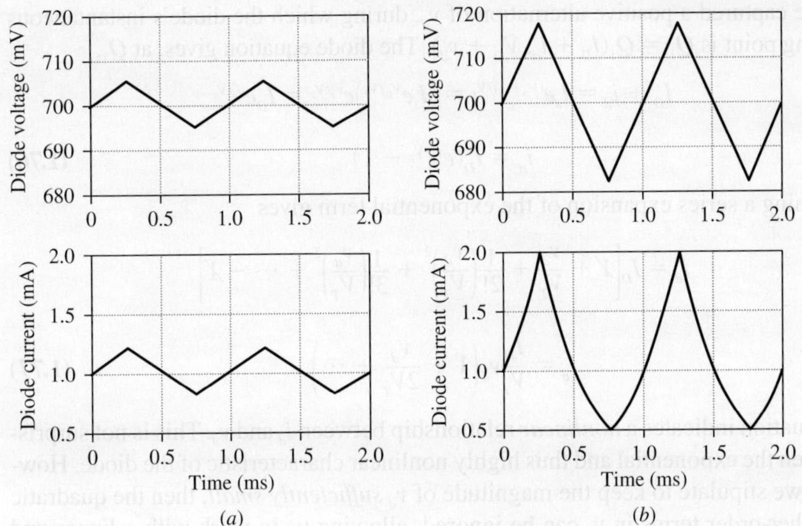

FIGURE 1.58 Voltage waveforms (top) and current waveforms (bottom) for two different ac voltage drives.

It is apparent that under the voltage drive conditions of Fig. 1.58*a* diode behavior can be regarded fairly close to linear, but that under the conditions of Fig. 1.58*b* it can't. Next, we wish to investigate quantitatively the range of validity of the small-signal approximation.

Small-Signal Operation

As we know, the function of the dc source V_D in Fig. 1.59*a* is to bias the diode at a specific quiescent point $Q_0 = Q_0(I_D, V_D)$ up the diode curve. Assuming an emission coefficient of unity ($n = 1$) for simplicity, the corresponding dc current is

$$I_D = I_s e^{V_D/V_T}$$

If we now turn on the *ac source* v_d as in Fig. 1.59*b*, the operating point will move up and down the diode curve, yielding the *ac current* i_d. In the expanded view of Fig. 1.56*b*

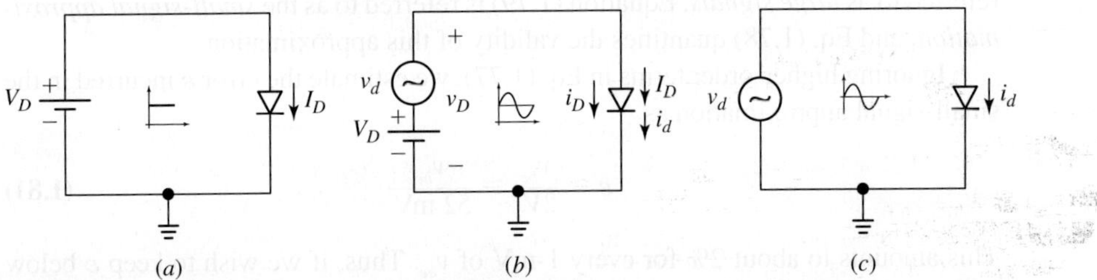

FIGURE 1.59 Systematic analysis of small-signal diode operation. The actual circuit is shown in the center (*b*), while (*a*) shows its *large-signal* or *dc* version, and (*c*) shows its *small-signal* or *ac* version.

we have captured a positive alternation of v_d, during which the diode's instantaneous operating point is $Q_1 = Q_1(I_D + i_d, V_D + v_d)$. The diode equation gives, at Q_1,

$$I_D + i_d = I_s e^{(V_D + v_d)/V_T} = \left(I_s e^{V_D/V_T}\right) e^{v_d/V_T} = I_D e^{v_d/V_T}$$

or

$$i_d = I_D(e^{v_d/V_T} - 1) \tag{1.76}$$

Performing a series expansion of the exponential term gives

$$i_d = I_D\left[\cancel{1} + \frac{v_d}{V_T} + \frac{1}{2!}\left(\frac{v_d}{V_T}\right)^2 + \frac{1}{3!}\left(\frac{v_d}{V_T}\right)^3 + \cdots - \cancel{1}\right]$$

or

$$i_d = \frac{I_D}{V_T} v_d\left(1 + \frac{v_d}{2V_T} + \cdots\right) \tag{1.77}$$

This equation indicates a *nonlinear* relationship between i_d and v_d. This is not surprising, given the exponential and thus highly nonlinear characteristic of the diode. However, if we stipulate to keep the magnitude of v_d *sufficiently small*, then the quadratic and higher-order terms in v_d can be ignored, allowing us to work with a *linear* and thus simpler relationship. Specifically, if we stipulate to keep

$$|v_d| \ll 2V_T \,(\cong 52 \text{ mV}) \tag{1.78}$$

then Eq. (1.77) simplifies as $i_d = (I_D/V_T)v_d$. This can be put in the form of Ohm's law,

$$i_d = \frac{v_d}{r_d} \tag{1.79}$$

where

$$r_d = \frac{V_T}{I_D} \tag{1.80}$$

is the diode's *dynamic resistance*. Its reciprocal $1/r_d$ is simply the *slope* of the diode curve calculated at the quiescent point Q_0. With reference to Fig. 1.56*b* we see that if v_d is sufficiently *small*, the portion of the curve from Q_0 to Q_1 can be approximated with the *straight segment* tangent to the diode curve right at Q_0. For obvious reasons the ac variations v_d and i_d are referred to as *small signals*. By contrast, V_D and I_D are referred to as *large signals*. Equation (1.79) is referred to as the *small-signal approximation*, and Eq. (1.78) quantifies the validity of this approximation.

Ignoring higher-order terms in Eq. (1.77), we estimate the error ε incurred in the small-signal approximation as

$$\varepsilon \cong \frac{v_{be}}{2V_T} \cong \frac{v_{be}}{52 \text{ mV}} \tag{1.81}$$

This amounts to about 2% for every 1-mV of v_{be}. Thus, if we wish to keep ε below 10% (an acceptable error in most practical situations), then we need to ensure that

$$|v_{be}| \leq 5 \text{ mV} \tag{1.82}$$

This shall be our working condition, as we move along.

EXAMPLE 1.17

(a) If $I_D = 1$ mA, find the peak values of i_d in Fig. 1.58a, where v_d has peak values of ± 5 mV. Calculate the peaks *approximately*, via Eq. (1.79), and *exactly*, via Eq. (1.76). What is the percentage error incurred in the small-signal approximation?

(b) Repeat, but for the case of Fig. 1.58b, where v_d has peak values of ± 18 mV.

Solution

(a) By Eq. (1.80), $r_d = 26/1 = 26\ \Omega$. By Eq. (1.79), i_d peaks at the values

$$i_{d(\text{pk})} = \frac{\pm 5 \times 10^{-3}}{26} \cong \pm 192\ \mu\text{A}$$

By Eq. (1.76), the exact values of the positive and negative peaks of i_d are, respectively,

$$i_{d(\text{pos pk})} = 10^{-3}(e^{5/26} - 1) = 212\ \mu\text{A} \qquad i_{d(\text{neg pk})} = 10^{-3}(e^{-5/26} - 1) = -175\ \mu\text{A}$$

We see that the small-signal approximation *underestimates* the positive peak by $(212 - 192)/192$, or 10.3%, and *overestimates* the negative peak by $(192 - 175)/192$, or 8.9%. Both errors are consistent with Eq. (1.81) which predicts errors of approximately $\pm 5/52 = \pm 9.6\%$.

(b) Now Eq. (1.79) predicts

$$i_{d(\text{pk})} = \frac{\pm 18 \times 10^{-3}}{26} \cong \pm 692\ \mu\text{A}$$

Using Eq. (1.76), or more simply the rule of thumb, we find the exact peak values to be $i_{d(\text{pos pk})} = 1000\ \mu\text{A}$ and $i_{d(\text{neg pk})} = -500\ \mu\text{A}$. The underestimation error is now 31% and the overestimation error is 38%, both of which are generally unacceptable. The far more pronounced distortion of Fig. 1.58b confirms this.

Remark: The above results, derived for the case of an emission coefficient $n = 1$, are readily generalized if we let $V_T \to nV_T$ in Eqs. (1.78), (1.80), and (1.81), and let 5 mV $\to n5$ mV in Eq. (1.82). For example, for $n = 1.5$, Eq. (1.78) becomes $|v_d| \ll 2 \times 1.5\ V_T\ (\cong 78$ mV), and Eq. (1.82) becomes $|v_d| \le 7.5$ mV.

The Small-Signal Diode Model

Equations (1.79) and (1.80) indicate that, under the condition of Eq. (1.78), a *pn* diode behaves with respect to the small signals v_d and i_d as a mere *resistor* r_d. The *small-signal model* for the diode, also called *incremental model*, is shown in Fig. 1.60 (*right*).

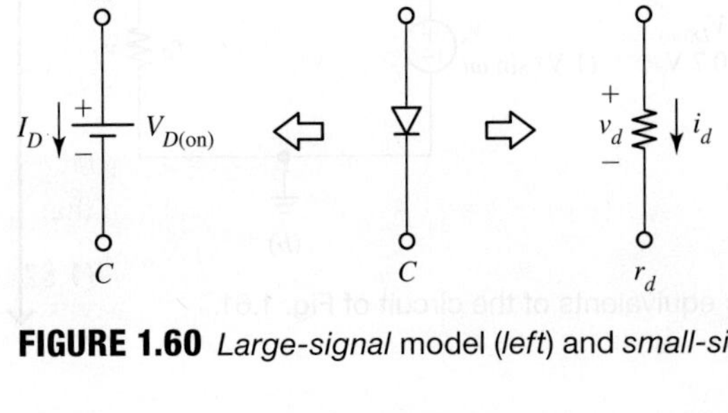

FIGURE 1.60 *Large-signal* model (*left*) and *small-signal* model (*right*) of the *pn* diode.

For convenience, also shown is the *large-signal model (left)*. The beginner should be careful not to confuse the two! We use the large-signal model to investigate *dc biasing*, for instance to find the quiescent current I_D. We use the small-signal model to investigate the diode's response to *ac signals* of suitably small magnitude.

The *decomposition* of the diode voltage and current into separate dc and ac components, along with the fact that both the large-signal and the small-signal diode models are *linear*, allows us to perform the dc and ac analyses *separately*, as pictured in Fig. 1.59a and 1.59c. We then apply the *superposition principle* and add the *dc* and *ac* results to obtain the *total* result. An example will better illustrate.

EXAMPLE 1.18   Let the diode of Fig. 1.61 have $I_s = 1$ fA and $n = 1$. Find $v_D = V_D + v_d$ if $v_S = V_S + v_s = 8\ \text{V} + (1\ \text{V})\sin\omega t$. Is the condition of Eq. (1.82) satisfied?

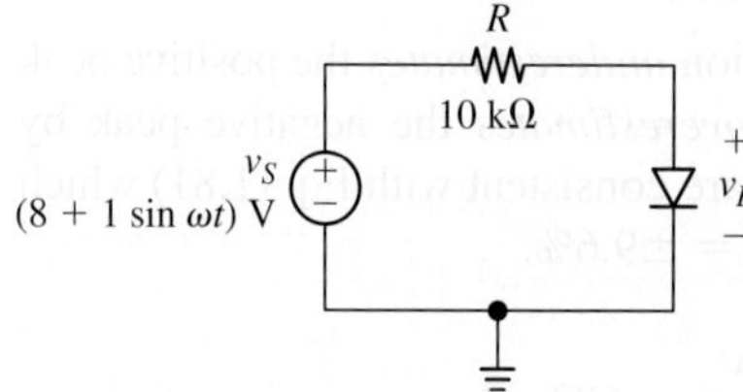

FIGURE 1.61 Circuit of Example 1.18.

Solution

Perform the dc and ac analyses *separately* to find, respectively, V_D and v_d. Then, apply the superposition principle to obtain $v_D = V_D + v_d$.

- For *dc analysis* refer to the dc version of Fig. 1.62a, showing only the *dc components* V_S and I_D. The *ac components* (v_s and v_d) have deliberately been set to *zero* as they don't intervene in dc analysis. Moreover, the diode has been replaced by its *large-signal model* (the battery $V_{D(\text{on})}$). We have

$$I_D = \frac{V_S - V_{D(\text{on})}}{R} = \frac{8 - 0.7}{10} = 0.73\ \text{mA}$$

$$V_D = V_T \ln\frac{I_D}{I_s} = 26 \ln\frac{0.73 \times 10^{-3}}{10^{-15}} = 710\ \text{mV}$$

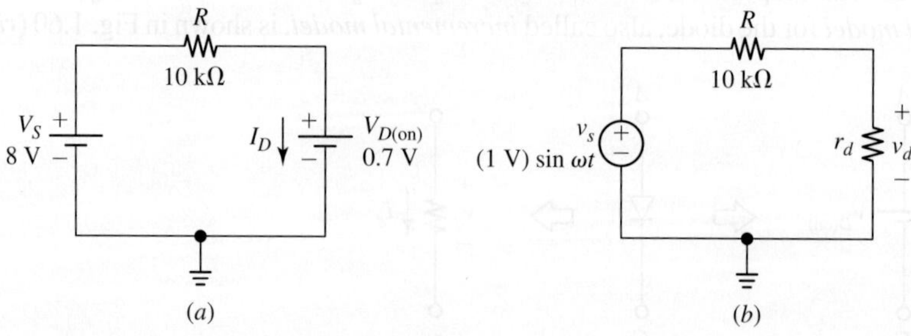

FIGURE 1.62 (a) *Dc* and (b) *ac* equivalents of the circuit of Fig. 1.61.

- For *ac analysis* refer to the ac version of Fig. 1.62*b*, showing only the *ac components* v_s and v_d. The *dc components* (V_S, I_D, V_D) have deliberately been set to *zero* in this case. Moreover, the diode has been replaced by its *small-signal model* (the incremental resistance r_d). We easily find

$$r_d = \frac{V_T}{I_D} = \frac{26}{0.73} \cong 36 \ \Omega$$

$$v_d = \frac{r_d}{R + r_d} v_s = \frac{36}{10,000 + 36} \ (1 \ \text{V}) \sin\omega t \cong (3.6 \ \text{mV}) \sin\omega t$$

Since 3.6 mV $<$ 5 mV $\ll 2V_T$ ($\cong$ 52 mV), the error of our approximate ac calculations is less than 10%.

- Finally, applying the *superposition principle*, we find the *total* diode voltage as

$$v_D = (710 + 3.6 \sin\omega t) \ \text{mV}$$

Exercise 1.7

Repeat Example 1.18 if a 6-kΩ resistance is connected in parallel with the diode.

Hint: Apply Thévenin Theorem to the circuit external to the diode.

Ans. $v_D = (706 + 4.2 \sin\omega t)$ mV.

The *pn* Diode as a Current-Controlled Resistance

The relation $r_d = V_T/I_D$ indicates that in small-signal operation a *pn* diode acts as a *variable resistance* whose value can be *programmed* via the bias current I_D. If we make r_d part of a voltage divider, then we can achieve *current-controlled attenuation*. Alternatively, making it part of the feedback network of an op amp, we can achieve *current-controlled amplification*. These concepts find application, among others, in *automatic gain control* (AGC), where one circuit controls the gain of another circuit.

Figure 1.63*b* shows a *current-controlled attenuator*. Current control is provided by the source I_D, which also causes the diode to develop the voltage drop $V_D = V_T \ln (I_D/I_s)$. To prevent the signal source v_i from disturbing the dc bias conditions of the diode, we interpose an *ac-coupling capacitor* C, as shown.

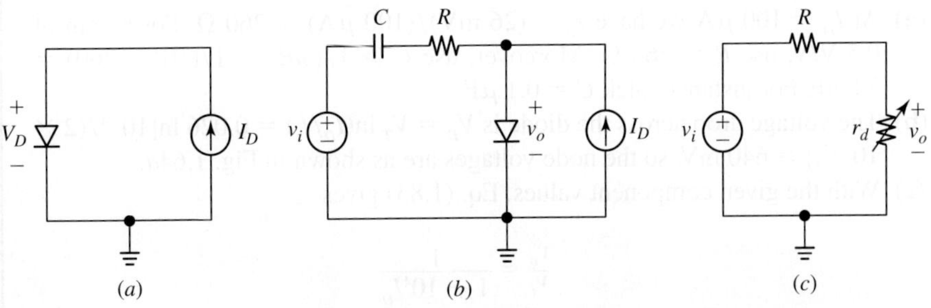

FIGURE 1.63 (*b*) Current-controlled attenuator and its (*a*) dc and (*c*) ac equivalents.

We make the following observations about the capacitor:

- At dc, C draws zero current and thus acts as an *open circuit*. In fact, in the dc equivalent of Fig. 1.63a, C has been omitted altogether.
- When power is applied to the circuit, C will charge up until its plates attain the open-circuit dc voltages of the corresponding nodes. So, the left plate remains at 0-V dc, the right plate charges to V_D.
- C is chosen sufficiently large to ensure that its impedance is negligible compared to R, in effect acting as an *ac short*. This requires that $1/(\omega C) \ll R$, or $C \gg 1/(\omega R)$, where ω is the input signal frequency.

To develop the ac equivalent of our circuit we replace C by a short circuit and the diode by the variable resistance r_d. The result is shown in Fig. 1.63c, where I_D has been omitted altogether as it is a dc quantity, thus having a zero ac component. Applying the voltage divider rule, we get

$$v_o = \frac{r_d}{R + r_d}v_i = \frac{V_T/I_D}{R + V_T/I_D}v_i = \frac{1}{1 + (R/V_T)I_D}v_i$$

indicating that the gain of the circuit is

$$\frac{v_o}{v_i} = \frac{1}{1 + (R/V_T)I_D} \tag{1.83}$$

EXAMPLE 1.19 In the circuit of Fig. 1.63b let

$$v_i = (5\ \text{mV}) \cos 10^6 t$$

and let I_D be variable over the range

$$(10\ \mu\text{A}) \le I_D \le (1\ \text{mA})$$

(a) Specify suitable values for R and C so that for $I_D = 100\ \mu\text{A}$ the gain is 0.5 V/V.
(b) If $I_s = 2$ fA, show all node voltages (dc as well as ac components) in the circuit of part (a).
(c) Plot the gain v_o/v_i versus I_D over the specified range of I_D values. What are the values of gain and V_D at the extremes of the range?

Solution

(a) At $I_D = 100\ \mu\text{A}$ we have $r_d = (26\ \text{mV})/(100\ \mu\text{A}) = 260\ \Omega$. For a gain of 0.5 V/V, use $R = 260\ \Omega$. Moreover, use $C \gg 1/(\omega R) = 1/(10^6 \times 260) = 3.8$ nF. For instance, pick $C = 0.1\ \mu\text{F}$.
(b) The voltage drop across the diode is $V_D = V_T \ln(I_D/I_s) = 0.026 \ln[10^{-4}/(2 \times 10^{-15})] \cong 640$ mV, so the node voltages are as shown in Fig. 1.64a.
(c) With the given component values, Eq. (1.83) gives

$$\frac{v_o}{v_i} = \frac{1}{1 + 10^4 I_D}$$

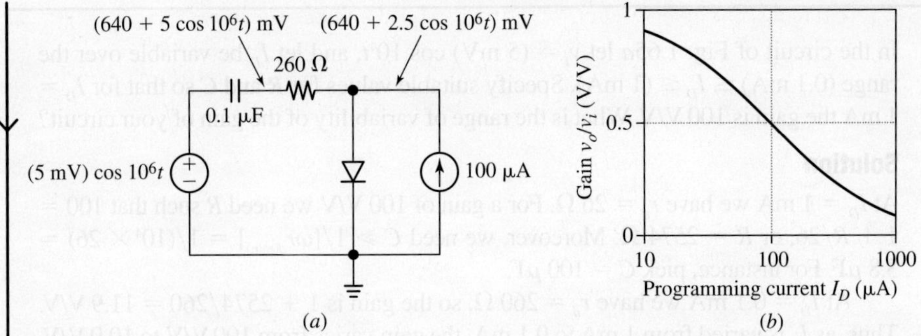

FIGURE 1.64 (a) Circuit of Example 10.3 at $I_D = 100$ μA, and (b) the range of variability of its gain.

whose plot is shown in Fig. 1.64b. At $I_D = 10$ μA this gives $v_o/v_i = 0.91$ V/V. Moreover, by the rule of thumb, $V_D = 640 - 60 = 580$ mV. Similarly, at $I_D = 1$ mA we get $v_o/v_i = 0.091$ V/V and $V_D = 700$ mV.

Figure 1.65a shows how a *pn* diode can be used in connection with an op amp to implement a *current-controlled amplifier*. As in the case of the controlled attenuator, the source I_D programs the value of r_d, and the capacitor is used to block the dc voltage component V_D developed by the diode. The ac equivalent, shown in Fig. 1.65b, reveals the familiar noninverting op amp configuration, whose gain is

$$\frac{v_o}{v_i} = 1 + \frac{R}{r_d} = 1 + \frac{R}{V_T}I_D \tag{1.84}$$

In this case the impedance provided by C must be negligible compared to r_d over the entire range of r_d values. This condition is hardest to satisfy when r_d is minimized, so we must have $C \gg 1/[\omega r_{d(\min)}]$.

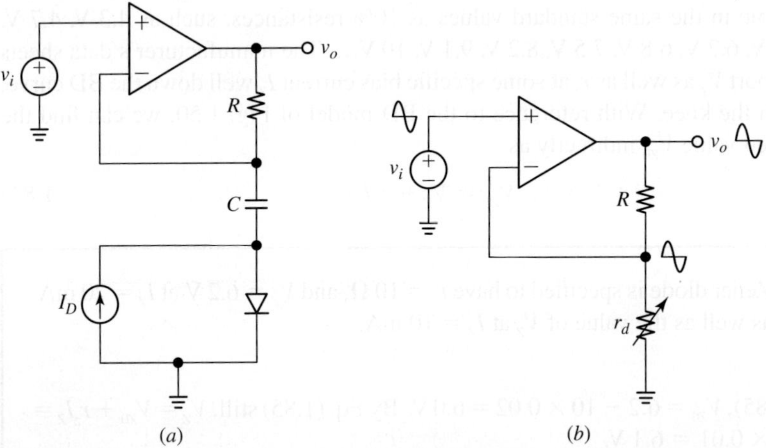

FIGURE 1.65 (a) Current-controlled amplifier, and (b) its small-signal ac equivalent.

EXAMPLE 1.20   In the circuit of Fig. 1.65*a* let $v_i = (5 \text{ mV}) \cos 10^4 t$, and let I_D be variable over the range $(0.1 \text{ mA}) \leq I_D \leq (1 \text{ mA})$. Specify suitable values for R and C so that for $I_D = 1$ mA the gain is 100 V/V. What is the range of variability of the gain of your circuit?

Solution

At $I_D = 1$ mA we have $r_d = 26\ \Omega$. For a gain of 100 V/V we need R such that $100 = 1 + R/26$, or $R = 2574\ \Omega$. Moreover, we need $C \gg 1/[\omega r_{d(min)}] = 1/(10^4 \times 26) = 3.8\ \mu$F. For instance, pick $C = 100\ \mu$F.

At $I_D = 0.1$ mA we have $r_d = 260\ \Omega$, so the gain is $1 + 2574/260 = 11.9$ V/V. Thus, as I_D is varied from 1 mA to 0.1 mA, the gain varies from 100 V/V to 10.9 V/V.

1.12 BREAKDOWN-REGION OPERATION

The *very steep* diode curve in the breakdown (BD) region suggests that we can use the *pn* diode as a *voltage reference*, if only an approximate one. A voltage reference is a circuit that provides a constant output voltage V_O in spite of variations in its own supply voltage V_I and in the load current I_L. Such a voltage is used as a reference by other circuits, like data converters, multimeters, and regulated power supplies, or also to power circuits with moderate power requirements, which we shall generally refer to as the *load* (LD). As we know, the reciprocal of slope of the BD-region characteristic is denoted as r_z and is called the *dynamic resistance* in the BD region. The smaller r_z, the steeper the characteristic. According to the BD diode model depicted in Fig. 1.50, in the limit $r_z \to 0$ the diode would act as a perfect voltage source V_{Z0}.

As we know, the breakdown effect is due to two different mechanisms: *Zener breakdown* for BD voltages of less than about 6 V, and *avalanche breakdown* for BD voltages of more than about 6 V. Diodes specifically intended for BD-region operation are referred to as *Zener diodes*, regardless of the BD mechanism, and exhibit r_z values on the order of a few ohms to a few tens of ohms. Commercial Zener diodes are available in the same standard values as 10% resistances, such as 4.3 V, 4.7 V, 5.1 V, 5.6 V, 6.2 V, 6.8 V, 7.5 V, 8.2 V, 9.1 V, 10 V…. The manufacturer's data sheets usually report V_Z as well as r_z at some specific bias current I_Z well down the BD curve, away from the knee. With reference to the BD model of Fig. 1.50, we can find the extrapolated value V_{Z0} indirectly as

$$V_{Z0} = V_Z - r_z I_Z \tag{1.85}$$

EXAMPLE 1.21   A certain Zener diode is specified to have $r_z = 10\ \Omega$, and $V_Z = 6.2$ V at $I_Z = 20$ mA. Find V_{Z0}, as well as the value of V_Z at $I_Z = 10$ mA.

Solution

By Eq. (1.85), $V_{Z0} = 6.2 - 10 \times 0.02 = 6.0$ V. By Eq. (1.85) still, $V_Z = V_{Z0} + r_z I_Z = 6.0 + 10 \times 0.01 = 6.1$ V.

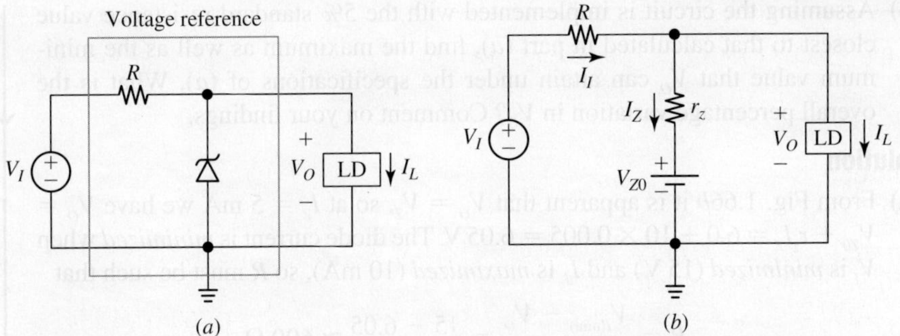

FIGURE 1.66 (a) The Zener diode as a *voltage reference*. (b) To investigate the reference's behavior, replace the Zener diode with its BD-region model.

The Zener Diode as a Voltage Reference

Figure 1.66 illustrates the application of the Zener diode as a simple voltage reference. In general, V_I is a poorly-defined voltage, whose value is only known to fall within a specified *range*

$$V_{I(min)} \le V_I \le V_{I(max)}$$

(If V_I were a stable and predictable voltage, then V_I itself would be a voltage reference!) Moreover, the circuit is designed to operate for any load current up to a specified *full-scale value* $I_{L(fs)}$, or

$$0 \le I_L \le I_{L(fs)}$$

The function of the *series resistance R* is twofold: to *drop* the *voltage difference* between the input source and the diode, and to *supply* the *current* needed to feed the load as well as to ensure that at all times the diode operates sufficiently down the BD curve, where r_z is small. The diode's operating point must be prevented from ever approaching the curve's knee (not to mention spilling into the cutoff region!) where the diode would cease behaving as a voltage source V_{Z0}.

The function of R can be better appreciated with the help of Fig. 1.66*b*, where the circuit has been redrawn with the diode replaced by its BD-region equivalent. It is apparent that V_O is a function of V_I, V_{Z0}, and I_L. In fact, using the superposition principle, we readily find

$$V_O = \frac{r_z}{R + r_z}V_I + \frac{R}{R + r_z}V_{Z0} - (R//r_z)I_L \qquad (1.86)$$

EXAMPLE 1.22

(a) Let the circuit of Fig. 1.66*a* be implemented with the diode of Example 1.21, for which $r_z = 10\ \Omega$ and $V_{Z0} = 6.0$ V. If $V_I = (20 \pm 5)$ V and $I_{L(fs)} = 10$ mA, find the resistance R needed to ensure that the diode current I_Z never drops below 5 mA.

(b) Assuming the circuit is implemented with the 5% standard resistance value closest to that calculated in part (*a*), find the maximum as well as the minimum value that V_O can attain under the specifications of (*a*). What is the overall percentage variation in V_O? Comment on your findings.

Solution

(a) From Fig. 1.66*b* it is apparent that $V_O = V_Z$, so at $I_Z = 5$ mA we have $V_O = V_{Z0} + r_z I_z = 6.0 + 10 \times 0.005 = 6.05$ V. The diode current is *minimized* when V_I is *minimized* (15 V) and I_L is *maximized* (10 mA), so R must be such that

$$R = \frac{V_{I(min)} - V_O}{I_{Z(min)} + I_{L(fs)}} = \frac{15 - 6.05}{5 + 10} \cong 600 \; \Omega$$

The closest standard values are 620 Ω and 560 Ω. To be on the safe side, use 560 Ω.

(b) Again from Fig. 1.66*b* we note that V_O is *maximized* when V_I is *maximized* (25 V) and I_L is *minimized* (0 mA), so using Eq. (1.86) we find

$$V_{O(max)} = \frac{10}{560 + 10} 25 + \frac{560}{560 + 10} 6.0 - 0 = 6.333 \text{ V}$$

Conversely, V_O is *minimized* when V_I is *minimized* (15 V) and I_L is *maximized* (10 mA), so

$$V_{O(min)} = \frac{10}{560 + 10} 15 + \frac{560}{560 + 10} 6.0 - (560//10)0.01 = 6.060 \text{ V}$$

The overall variation in V_O is $6.333 - 6.060 = 0.273$ V, or $0.273/6.2 = 4.4\%$. Given the much wider range of variability of V_I as well as the full-scale variability of I_L, this is a remarkable result!

Line and Load Regulation

The output of a voltage reference should be independent of V_I and I_L, so only the second term in the right-hand side of Eq. (1.86) is a desirable one. The other two terms should be zero, a condition that the present circuit could achieve only in the limit $r_z \rightarrow 0$. The quality of a voltage reference is specified in terms of two parameters borrowed from voltage-regulator terminology, namely, (*a*) the *line regulation*, representing the mV change in V_O for every 1-V change in V_I, and (*b*) the *load regulation*, representing the mV change in V_O for every 1-mA change in I_L. From Eq. (1.86) we find, for the present circuit,

$$\text{Line regulation} = \frac{\Delta V_O}{\Delta V_I} = \frac{r_z}{R + r_z} \tag{1.87a}$$

$$\text{Load regulation} = \frac{\Delta V_O}{\Delta I_L} = -(R//r_z) \tag{1.87b}$$

Find the line and load regulation of the circuit of Example 1.22. Express your results in mV/V and mV/mA, as well as in percentage form.   **EXAMPLE 1.23**

Solution

Line Regulation $= 10/570 = 17.5$ mV/V. Since 17.5 mV represents 0.27% of 6.2 V, we can also say that *Line Regulation* $= 0.27\%/V$. Likewise, *Load Regulation* $= -(560//10) = -9.8$ mV/mA, or $-0.16\%/$mA.

Using an Op Amp to Improve Voltage-Reference Performance

The performance of a Zener diode reference can be improved dramatically with the help of an operational amplifier (op amp). As a first step, we can *isolate* the diode from the load by interposing a non-inverting amplifier between the two, in the manner of Fig. 1.67. Since *no current* flows into the input pin of the op amp, the voltage across the Zener diode is now

$$V_Z = \frac{r_z}{R + r_z}V_I + \frac{R}{R + r_z}V_{Z0} - 0 \tag{1.88}$$

indicating that the load regulation is zero—a highly desirable feature indeed! The presence of the op amp offers the additional advantage of amplifying V_Z to yield a voltage V_O that can be adjusted by varying the amplifier's gain. Indeed, by the *non-inverting amplifier rule*, the op amp gives

$$V_O = \left(1 + \frac{R_2}{R_1}\right)V_Z \tag{1.89}$$

so we can adjust V_O to any value we wish ($V_O > V_Z$) by varying one of the two resistances, say R_2.

The circuit of Fig. 1.67 still suffers from poor line regulation in that any variation ΔV_I at the input will cause the variation $\Delta V_Z = [r_z/(R + r_z)]\Delta V_I$ across the diode, which the op amp then passes on to the output as $\Delta V_O = (1 + R_2/R_1)\Delta V_Z$. This last source of error is admirably eliminated by powering the Zener diode from the very

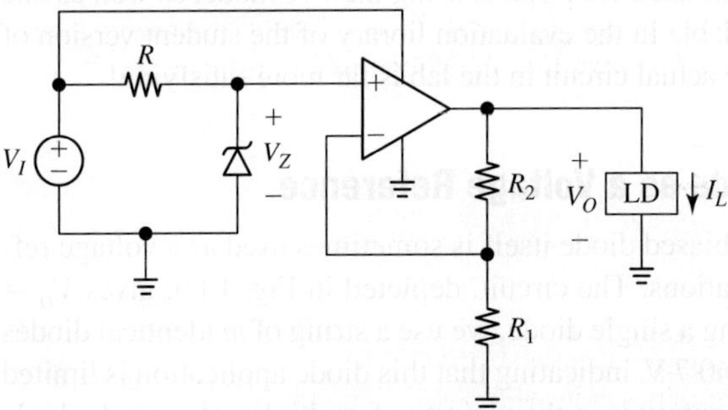

FIGURE 1.67 Using an op amp to improve load regulation.

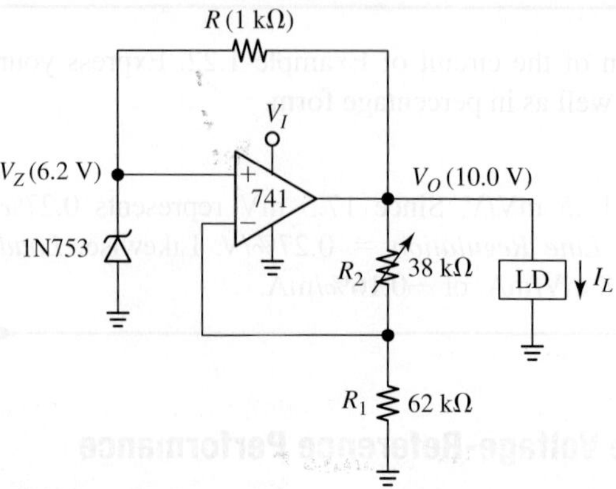

FIGURE 1.68 Self-regulated 10.0-V voltage reference.

voltage V_O that we are trying to regulate in the first place! This results in the circuit of Fig. 1.68, aptly called *self-regulated voltage reference*. With this modification, Eq. (1.88) becomes

$$V_Z = \frac{r_z}{R + r_z}V_O + \frac{R}{R + r_z}V_{Z0}$$

Substituting V_Z into Eq. (1.89) and solving for V_Z we obtain

$$V_O = \frac{(1 + R_2/R_1)}{1 - (R_2 r_z)/(R_1 R)}V_{Z0} \tag{1.90}$$

that is, V_O is now *independent* of both V_I and I_L! Clearly, both the line and load regulation have been driven to zero. If implemented with the popular 741 op amp as shown, the circuit can operate with V_I ranging from about 12 V to about 36 V, and it offers an output current drive of up to about 25 mA. The student who has access to a laboratory is encouraged to try out this circuit experimentally to appreciate first-hand how stable V_O is in the face of such wide variations in V_I and I_L. If any deviations are observed, they are astoundingly small and are due primarily to departures of op amp behavior from the ideal. The circuit can also be simulated via PSpice, using the 741 model as well as one of the Zener diode models available in the evaluation library of the student version of PSpice. However, trying out the actual circuit in the lab is far more satisfying!

The Forward-Biased Diode as a Voltage Reference

The voltage drop of a forward-biased diode itself is sometimes used as a voltage reference, especially in IC applications. The circuit, depicted in Fig. 1.69, gives $V_O = V_{D(on)} \cong 0.7$ V. If, instead of using a single diode, we use a string of m identical diodes in series, then $V_O = mV_{D(on)} \cong m0.7$ V, indicating that this diode application is limited to situations where the desired output is in the vicinity of multiples of a single diode voltage-drop (e.g. 0.7 V, 1.4 V, 2.1 V, and so forth).

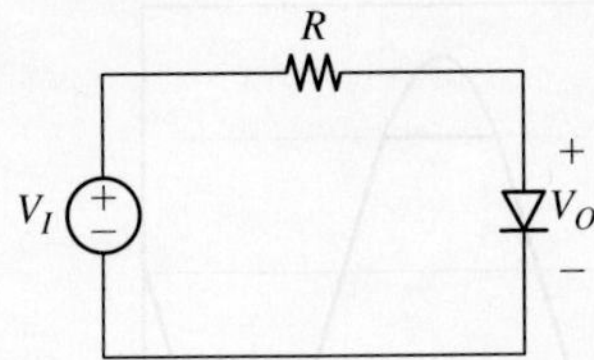

FIGURE 1.69 Using the voltage drop of a forward-biased diode as a voltage reference of about 0.7 V.

EXAMPLE 1.24

In the circuit of Fig. 1.69, let $V_I = 5$ V, and let the diode have $I_s = 1$ fA and $nV_T = 26$ mV.

(a) Specify R for a 1-mA diode current. Hence, assuming small enough line and load variations to justify the small-signal diode approximation, find the line and load regulation of this voltage reference.

(b) Compare with the case in which a resistor is used instead of the diode, and comment.

Solution

(a) We have $R = (V_I - V_{D(\text{on})})/I_D \cong (5 - 0.7)/1.0 = 4.3$ kΩ. Moreover, $V_O = (26 \text{ mV}) \ln(10^{-3}/10^{-15}) = 0.718$ V. At $I_D = 1$ mA the diode has $r_d = 26/1 = 26$ Ω, so *Line Regulation* $= r_d/(R + r_d) = 26/(4300 + 26) = 6$ mV/V, and *Load Regulation* $= -R//r_d \cong -26$ mV/mA.

(b) To achieve the same output voltage with a voltage divider we need a second resistor of value $0.718/1.0 = 0.718$ kΩ. We now have *Line Regulation* $= 718/(4300 + 718) = 143$ mV/V, and *Load Regulation* $= -4300//718 \cong -615$ mV/mA. Clearly, a diode reference is much better than a voltage divider.

Zener Diodes as Voltage Clamps

Its voltage-source behavior in the BD region makes the Zener diode suited to stable voltage-clamp applications. The example of Fig. 1.70 is based on a pair of Zener diodes connected back to back. Since they are in series, they are either both off, or both on (one in the forward region and the other in the BD region). We have three possibilities:

- For v_I sufficiently *positive* to turn on both diodes, D_1 will operate in the BD region and D_2 in the forward region, as depicted in Fig. 1.71*a*. The output is clamped at $V_{OH} = V_{Z1} + V_{D2(\text{on})}$, where we are ignoring the Zener-diode resistance r_{z1} compared to R. Clearly, this situation occurs for $v_I > V_{OH}$.
- For v_I sufficiently *negative*, the situation reverses, with D_1 now operating in the forward region and D_2 in the BD region. As depicted in Fig. 1.71*b*, the output is now clamped at $V_{OL} = -(V_{D1(\text{on})} + V_{Z2})$. This situation occurs for $v_I < V_{OL}$.

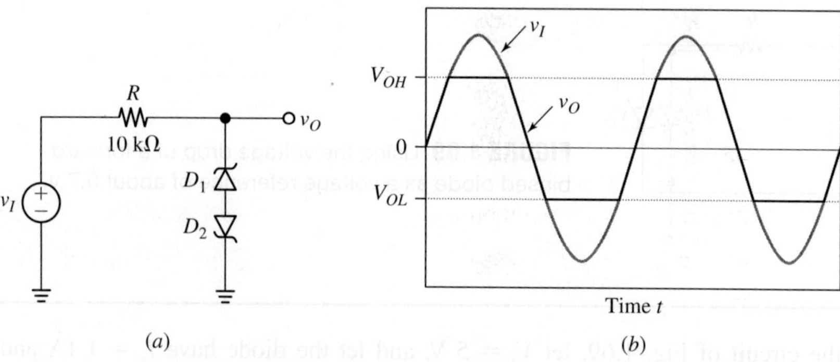

FIGURE 1.70   A Zener-diode clamp, and its effect upon the input waveform.

- For $V_{OL} < v_I < V_{OH}$ there isn't sufficient voltage drive to turn on the diodes, so they both act as open circuits. The situation is depicted in Fig. 1.71c, where the lack of current causes R to drop 0 V, thus giving $v_O = v_I$. We say that now R, being unencumbered, *pulls* v_O to v_I.

The clamping effect is depicted in Fig. 1.70b. For instance, if D_1 is a 4.3-V Zener diode and D_2 is a 6.8-V Zener diode, v_O will be clamped in the positive direction at $V_{OH} = 4.3 + 0.7 = 5.0$ V, and in the negative direction at $V_{OL} = -(0.7 + 6.8) = -7.5$ V.

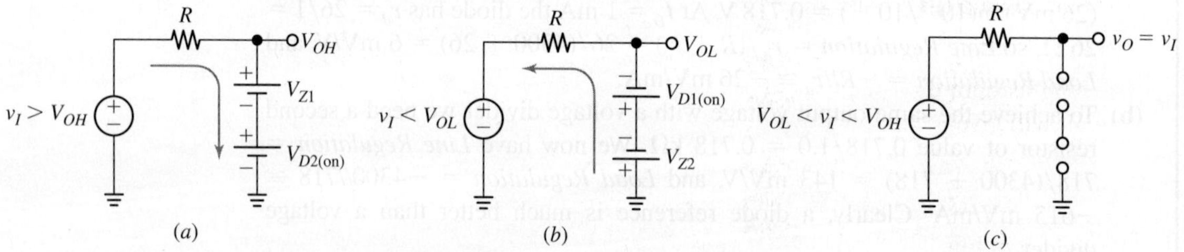

FIGURE 1.71   Illustrating the three possible conditions for the circuit of Fig. 1.70a.

There are situations in which it is desirable that clamping be *symmetric* about 0 V, or $V_{OL} = -V_{OH}$. This requires that the two Zener diodes be *matched*. Alternatively, we can use only one Zener diode, but along with a diode bridge to make it perform the same clamping action both in the positive and negative directions. The circuit is shown in Fig. 1.72, where we have again three possibilities:

- For v_I sufficiently *positive* to turn on the Zener diode via the diode bridge, current flows down the path

$$\text{source} \to R \to D_1 \to D_Z \to D_4 \to \text{ground}$$

The output is clamped at $V_{OH} = V_{D1(on)} + V_Z + V_{D4(on)} = V_Z + 2V_{D(on)}$. Clearly, this situation arises for $v_I > (V_Z + 2V_{D(on)})$.
- For v_I sufficiently *negative*, current flows down the path

$$\text{ground} \to D_2 \to D_Z \to D_3 \to R \to \text{source}$$

and this occurs for $v_I < -(V_Z + 2V_{D(on)})$. The output is now clamped at $V_{OL} = -(V_Z + 2V_{D(on)}) = -V_{OH}$.

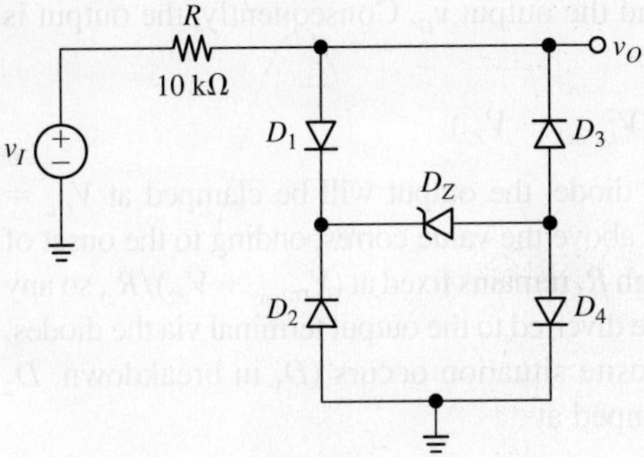

FIGURE 1.72 Symmetric voltage clamp.

- For $|v_I| < (V_Z + 2V_{D(\text{on})})$ there isn't sufficient voltage drive to turn on any of the diodes, so they all act as open circuits. The lack of current allows R to pull v_O to v_I, thus giving $v_O = v_I$.

For instance, using a 5.1-V Zener diode will cause the output to be clamped at $\pm(5.1 + 1.4) = \pm 6.5$ V.

Figure 1.73a illustrates the use of a Zener-diode clamp to limit the output swing of an op amp (in this example an inverting-type amplifier). We identify three possibilities:

- As long as both diodes are off, the *Inverting Amplifier Rule* indicates that the circuit will give

$$v_O = -\frac{R_2}{R_1}v_I$$

  As shown in Fig. 1.73b, the voltage transfer curve (VTC) is a straight line with the *gain* $(-R_2/R_1)$ as its *slope*. With both diodes in cutoff, any current through R_1 flows right through R_2.
- For v_I sufficiently *positive* both diodes will go on (D_1 forward, D_2 in breakdown), thus establishing a *fixed voltage drop* between the op amp's inverting

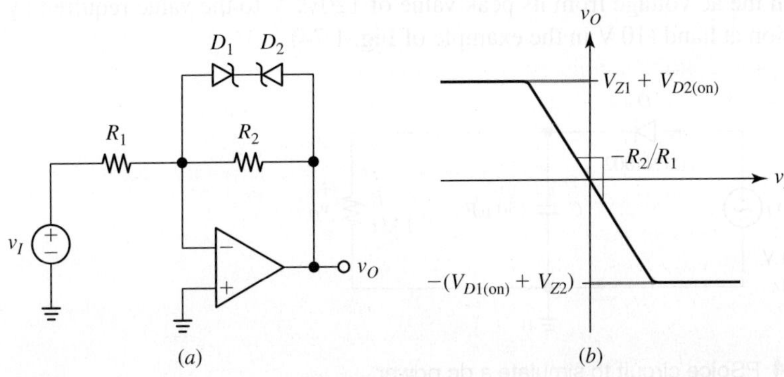

FIGURE 1.73 Using Zener diodes to limit the output swing of an op amp.

input, which is at virtual ground, and the output v_O. Consequently, the output is now clamped at

$$V_{OL} = -(V_{D1(on)} + V_{Z2})$$

For instance, if D_2 is a 4.3-V Zener diode, the output will be clamped at $V_{OL} = -(0.7 + 4.3) = -5$ V. As we raise v_I above the value corresponding to the onset of the clamping action, the current through R_2 remains fixed at $(V_{D1(on)} + V_{Z2})/R_2$, so any excess current coming in via R_1 will be diverted to the output terminal via the diodes.

- For v_I sufficiently *negative* the opposite situation occurs (D_1 in breakdown, D_2 forward), and the output is now clamped at

$$V_{OH} = +(V_{Z1} + V_{D2(on)})$$

For instance, if D_1 is a 6.8-V Zener diode, the output will be clamped at $V_{OH} = +(6.8 + 0.7) = +7.5$ V.

If *symmetric* clamping is desired, the Zener diodes need to be *matched*. Alternatively, we replace the back-to-back Zener pair with a single Zener diode and a diode bridge, in the manner discussed in connection with Fig. 1.72.

1.13 Dc POWER SUPPLIES

As implied by its name, a *dc power supply* is a circuit that provides a specified dc voltage to power other circuits. The supply is powered in turn by another power source, which in the following we shall assume to be the household ac power, which in the United States is 120 V rms, 60 Hz. Since the household ac voltage is sinusoidal, we need to convert it to a dc voltage. As a first step we rectify it to eliminate its negative alternations and thus ensure that the voltage is always positive. However, the result is a *pulsating* voltage, that is, a voltage that periodically returns to zero. We need an *energy-storage* device to hold the voltage above some specified level during the times when the rectifier's output would drop to zero. Such a device is the *capacitor*, and the result is shown in Fig. 1.74 for the simpler case of a *half-wave rectifier* (the full-wave rectifier will be discussed later), and a load that we model with resistor R. Not shown in the figure for simplicity is a transformer, which we use to step down the ac voltage from its peak value of $120\sqrt{2}$ V to the value required by the application at hand (10 V in the example of Fig. 1.74).

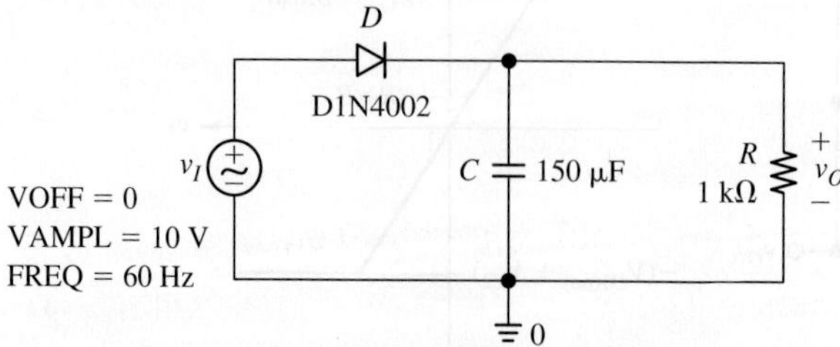

FIGURE 1.74 PSpice circuit to simulate a dc power supply with a load R.

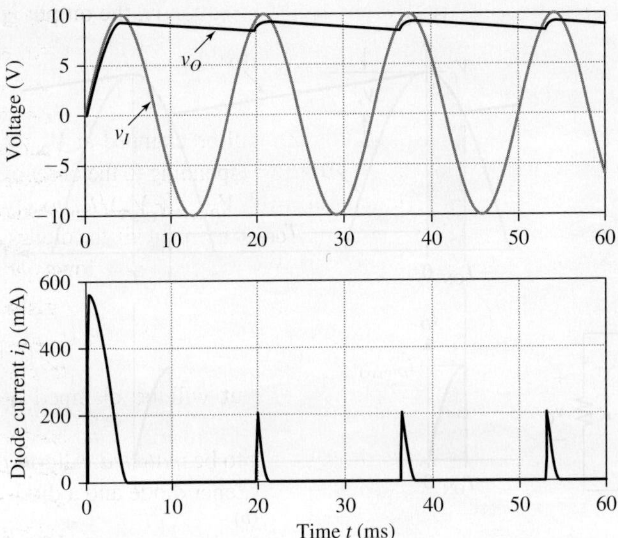

FIGURE 1.75 Voltage (*top*) and current (*bottom*) waveforms for the dc power supply of Fig. 1.74.

The relevant waveforms of the circuit of Fig. 1.74 are shown in Fig. 1.75, with respect to which we make the following observations:

- At power turn-on the diode D and the capacitor C act as a peak detector, and we witness a large *initial current surge* to charge C, initially assumed discharged, to the peak value of the input, short of the diode voltage drop ($10 - 0.7 = 9.3$ V in our example.)
- Once v_I peaks out, the diode goes off, leaving the capacitor as the sole source of power for the load. As R draws current, C will discharge. However, in a well-designed dc supply, C is chosen *large enough* to keep discharge fairly small during the time intervals when the diode is off.
- As the next positive input alternation comes along and v_I rises above the present capacitor voltage by a diode voltage drop, the diode becomes again conductive, recharging the capacitor to the peak value of the input, short of the diode voltage drop.
- Henceforth all the waveforms repeat, with the output voltage exhibiting *ripple*, and the diode current consisting of *spikes* during the diode conduction intervals.

Ripple, Conduction Interval, and Peak Diode Current

We now wish to investigate the interdependence of the various parameters intervening in the design of a dc power supply. Dc supplies of the type under discussion here are hardly precise systems, so we make a number of simplifying assumptions to speed up our estimations. As mentioned, in a well-designed dc supply the capacitor

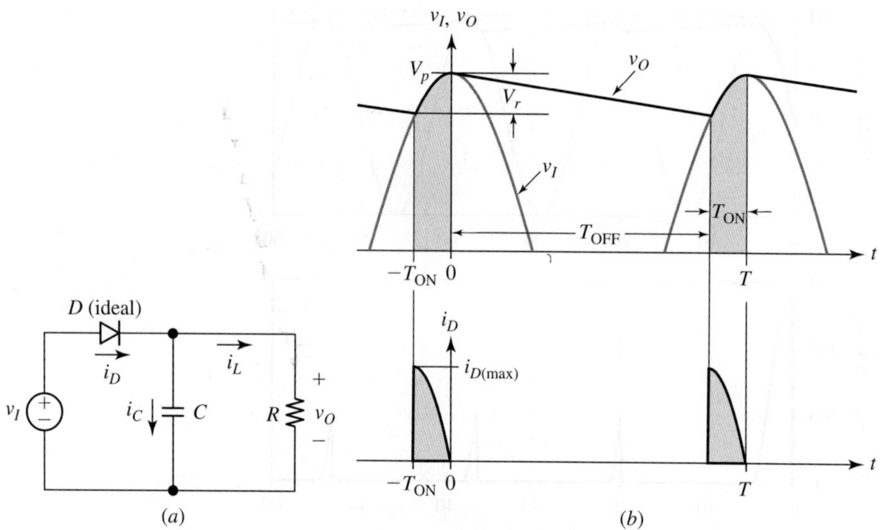

FIGURE 1.76 (a) Dc power supply, and (b) voltage and current waveforms.

is chosen sufficiently large to ensure a small ripple relative to the peak value V_p of the ac input, or

$$V_r \ll V_p$$

where V_r denotes the peak-to-peak amplitude of the output ripple. Also, it is convenient to assume the diode to be ideal, or $V_{D(on)} \cong 0$. With these approximations in mind, we redraw the circuit and the repetitive portion of its waveforms as in Fig. 1.76. As seen, capacitor discharge is approximately linear, and thus governed by the rule that engineers express in the form $C\Delta v = I\Delta t$ (*cee delta vee equals aye delta tee*), where presently Δv is the *ripple* V_r, I is the *average load current* I_L, and Δt is the *time interval* during which the diode is *off*, or T_{OFF}. Thus, $CV_r \cong I_L T_{OFF}$. Solving for the ripple, we obtain

$$V_r \cong \frac{I_L T_{OFF}}{C} \tag{1.91}$$

With reference to Fig. 1.76b, top, we approximate $I_L \cong (V_p - 0.5V_r)/R \cong V_p/R$ and $T_{OFF} \cong T = 1/f$, where f is the input frequency. Consequently, Eq. (1.91) becomes

$$V_r \cong \frac{V_p}{fRC} \tag{1.92}$$

If we wish to know to a better degree of accuracy the *average value of the output*, also called the *output dc component* and aptly denoted as V_O, then the diode's voltage drop needs to be taken into consideration. By inspection,

$$V_O = V_p - V_{D(on)} - 0.5V_r \tag{1.93}$$

EXAMPLE 1.25

(a) Estimate the ripple in the circuit of Fig. 1.74.

(b) If $V_{D(on)} = 0.7$ V, what is the dc component of the output? What value must V_p be changed to if we want $V_O = 10$ V?

(c) If V_p is changed to 50 V and R to 10 kΩ, find C for a ripple of not more than 2 V.

Solution

(a) By Eq. (1.92),

$$V_r \cong \frac{10}{60 \times 10^3 \times 150 \times 10^{-6}} = 1.1 \text{ V}$$

which is in fairly good agreement with Fig. 1.75, top.

(b) By Eq. (1.93), $V_O = 10 - 0.7 - 0.5 \times 1.1 = 8.75$ V, also in fairly good agreement with Fig. 1.75, top. For $V_O = 10$ V, we need to raise V_p to $10 + 0.7 + 0.5 \times 1.1$, or to $V_p = 11.25$ V.

(c) Using again Eq. (1.91), we need

$$C \geq \frac{V_p}{fRV_r} = \frac{50}{60 \times 10^4 \times 2} \cong 42 \ \mu\text{F}$$

We now wish to develop expressions for the diode *conduction interval* T_{ON} and the *peak diode current* $i_{D(max)}$. To simplify our estimations, we again assume $V_{D(on)} = 0$. With reference to Fig. 1.76b, top, we express the input as $v_I(t) = V_p \cos(2\pi ft)$. Imposing $v_O(-T_{ON}) = v_I(-T_{ON}) = V_p - V_r$ yields

$$V_p \cos[2\pi f(-T_{ON})] = V_p - V_r$$

Rearranging and expanding the cosine function, we write

$$1 - \frac{V_r}{V_p} = \cos[2\pi f(-T_{ON})] = \cos(2\pi f T_{ON}) = 1 - \frac{1}{2}(2\pi f T_{ON})^2 + \cdots$$

It is apparent from the figure that the condition $V_r \ll V_p$ implies $T_{ON} \ll T \,(= 1/f)$, so we can ignore higher-order terms in the series expansion. Solving for T_{ON} we get

$$T_{ON} \cong \frac{1}{2\pi f}\sqrt{\frac{2V_r}{V_p}} \tag{1.94}$$

With reference to Fig. 1.76a, we observe that during diode conduction we have, by KCL

$$i_D = C\frac{dv_O}{dt} + i_L = C\frac{d[V_p \cos(2\pi ft)]}{dt} + i_L \tag{1.95}$$

The diode current is maximum at the onset of the conduction interval, or $t = -T_{ON}$. At this instant we have

$$\frac{d\cos(2\pi ft)}{dt}\bigg|_{t=-T_{ON}} = -2\pi f \sin[2\pi f(-T_{ON})] \cong (2\pi f)^2 T_{ON} = 2\pi f\sqrt{\frac{2V_r}{V_p}}$$

where we have approximated $\sin x \cong x$ and have also used Eq. (1.94). Substituting into Eq. (1.95), along with the approximation $i_L \cong I_L \cong V_p/R$, we finally get

$$i_{D(\text{max})} \cong I_L\left(1 + 2\pi\sqrt{\frac{2V_p}{V_r}}\right) \qquad (1.96)$$

EXAMPLE 1.26

(a) Estimate the conduction interval in the circuit of Fig. 1.74, and express it as a percentage of the period T. Comment on your result.

(b) Find $i_{D(\text{max})}$ for the same circuit.

(c) Estimate the average diode current $i_{D(\text{avg})}$ during the conduction interval T_{ON}, and again comment.

Solution

(a) Using $V_r = 1.1$ V from Example 1.25, we have, by Eq. (1.94),

$$T_{\text{ON}} = \frac{1}{2\pi60}\sqrt{\frac{2 \times 1.1}{10}} = 1.24 \text{ ms}$$

Since the period is $T = 1/60 = 16.7$ ms, the diode conducts during $1.24/16.7 = 0.075$, or only 7.5% of each cycle, thus confirming the validity of the approximation $T_{\text{OFF}} \cong T$.

(b) We have $I_L \cong V_p/R = 10/1 = 10$ mA. By Eq. (1.96),

$$i_{D(\text{max})} = 10\left(1 + 2\pi\sqrt{\frac{2 \times 10}{1.1}}\right) \cong 280 \text{ mA}$$

(c) Given the approximately *triangular* shape of the current spikes, we estimate $i_{D(\text{avg})} \cong i_{D(\text{max})}/2 = 140$ mA. Note that $i_{D(\text{avg})} \gg I_L$, an obvious consequence of the fact that the charge delivered by the capacitor to the load during the time interval T_{OFF} must be replenished by the diode during the much shorter conduction interval T_{ON}.

When designing a dc power supply we must select a diode type with adequate current-handling capability during the brief spikes of conduction. Also, we must ensure that the breakdown voltage is sufficiently higher than the *peak inverse voltage* (PIV), which is the maximum reverse voltage that the diode ever experiences in a given circuit. In the circuit of Fig. 1.74 this condition occurs when v_I reaches its negative peak, at which point the voltage across the diode is, by KVL, $v_D = -V_p - v_O \cong -V_p - V_p = -2V_p$. Consequently,

$$\text{PIV} \cong 2V_p \qquad (1.97)$$

With the data of Fig. 1.74, PIV $= 20$ V. To be on the safe side, specify a PIV at least 50% higher than the calculated value. In our case, let PIV $\geq 1.5 \times 20 = 30$ V.

Dc Supplies with Full-Wave Rectifiers

It is apparent that if we use a *full-wave rectifier* instead of a half-wave, the ripple will be approximately reduced in half, everything else remaining the same. Alternatively, we can guarantee the same ripple but with a capacitance half as large, which is quite desirable because large-valued capacitors are bulky. As mentioned, a dc supply that is powered from the household ac power is likely to be preceded by a transformer to scale down the 120-V ac voltage to a more manageable value as required by the application at hand. We can take advantage of this and use a *center-tapped* transformer, along with a pair of diodes, to achieve full-wave rectification in the manner depicted in Fig. 1.77. A center tapped winding can be viewed as two identical windings connected in series but in *anti-phase*, so that the voltages at its terminals, relative to the common node, are, respectively, $+v_S$ and $-v_S$. Once we add two diodes as shown, we end up with two separate half-wave rectifiers, but operating on alternate half cycles, in effect resulting in full-wave rectification. Equation (1.91) still holds. However, in Eq. (1.92) we must replace f with $2f$ to reflect the fact that the period of the full-wave rectified ac voltage is *half* that of the original voltage and its frequency is thus $2f$. Consequently, Eqs. (1.92) and (1.96) become

$$V_r \cong \frac{V_p}{2fRC} \tag{1.98}$$

and

$$i_{D(max)} \cong I_L\left(1 + 2\pi\sqrt{\frac{V_p}{2V_r}}\right) \tag{1.99}$$

indicating that, everything else remaining the same, both V_r and $i_{D(max)}$ are approximately reduced in half compared to the half-wave case. However, we still have $\text{PIV} \cong 2V_p$. The conduction interval of each diode is still governed by Eq. (1.94), if with the new value of V_r.

A notorious disadvantage of the center-tapped transformer is that its secondary coil requires *twice* as many turns. The alternative implementation of Fig. 1.78 utilizes a *single* secondary, but with a *diode bridge* to achieve full-wave rectification. Its

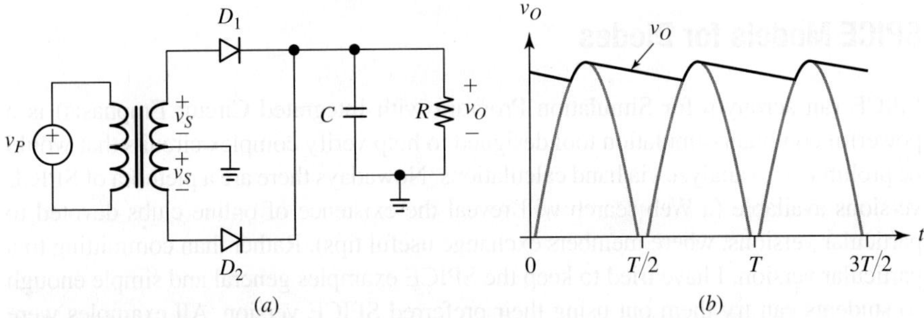

FIGURE 1.77 (a) Dc power supply with a *full-wave rectifier* using a center-tapped transformer, and (b) voltage waveforms.

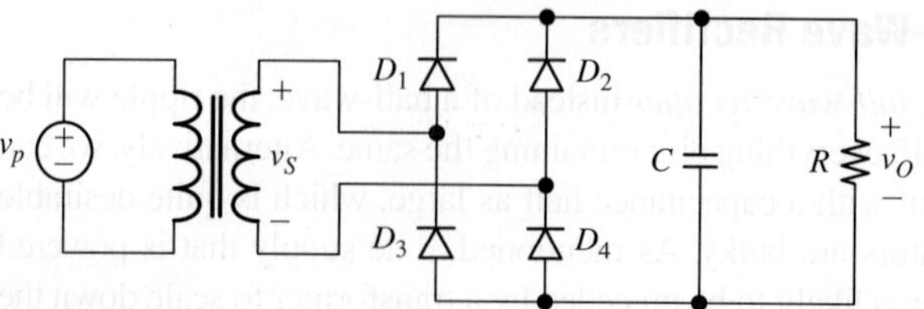

FIGURE 1.78 Dc power supply using a diode bridge rectifier.

waveforms are similar to those of Fig. 1.77*b*, and Eqs. (1.98) and (1.99) hold also in the present case. However, with two diode drops, Eq. (1.93) changes to

$$V_O = V_p - 2V_{D(\text{on})} - 0.5V_r \qquad (1.100)$$

Moreover, we now have

$$\text{PIV} \cong V_p \qquad (1.101)$$

Given the various options available—half wave, full-wave with center-tapped transformer, or full-wave with diode bridge—a designer will evaluate the advantages and disadvantages of each and settle for the one that best meets the cost and complexity constraints of the application at hand.

Concluding Observation

It is important to realize that the expressions of Eqs. (1.94), (1.96), (1.98), and (1.99) have been derived for the case of *sinusoidal inputs*, as most dc power supplies are powered from the household ac line. However, there are specialized cases in which the input waveform is not necessarily sinusoidal. It is left as an exercise, in the end-of-chapter problems, to apply the same line of reasoning to non-sinusoidal cases and develop ad-hoc expressions for T_{ON}, $i_{D(\text{max})}$, and $i_{D(\text{avg})}$. Also, for simplicity, the load is usually simulated with a mere resistance. A more realistic load model is a Norton equivalent, consisting of a fixed load-current in parallel with a resistance.

APPENDIX 1A

SPICE Models for Diodes

SPICE (an acronym for Simulation Program with Integrated Circuit Emphasis) is a powerful computer-simulation tool designed to help verify complex circuits that would be prohibitive to analyze via hand calculations. Nowadays there are a plethora of SPICE versions available (a Web search will reveal the existence of online clubs devoted to particular versions, where members exchange useful tips). Rather than committing to a particular version, I have tried to keep the SPICE examples general and simple enough so students can try them out using their preferred SPICE version. All examples were created using the Student Versions of Cadence's PSpice. This version, available free of charge, is a powerful pedagogical tool especially for the beginner, as you can glean from the various PSpice examples scattered throughout the current and subsequent chapters.

However powerful, SPICE is *no substitute* for true understanding—the kind attainable through diligent reasoning and patient paper-and-pencil calculations. Even when circuit complexity mandates the use of SPICE, one must still be able to examine the results provided by the computer and spot-check them by a combination of hand calculations and intuitive insight. Consequently, SPICE is only an analytical aid—if a most powerful one.

It is assumed that the student is already familiar with SPICE basics from prerequisite courses such as circuits courses and labs (how to create a circuit schematic using parts from SPICE's internal libraries, how to direct SPICE to perform a specific type of analysis among the various possible, and how to display traces using Probe, the oscilloscope-like feature provided by PSpice). The focus of this Appendix is SPICE models for diodes.

When we use a semiconductor device in a SPICE circuit schematic, we need to specify the characteristics of that device. The characteristics are expressed in terms of a *list of parameters* that SPICE then uses to create an internal *model* of the device. Shown in Table 1A.1 are the parameters intervening in the creation of the model for *pn* diodes. PSpice comes with a library of models for some of the most popular semiconductor devices. Moreover, the user can create additional models by suitably editing one of the already available models.

As an example, consider the PSpice circuit of Fig. 1.51*a*, utilizing the popular 1N4148 *pn* diode. By PSpice convention, diode names must start with the letter D, so the part number has been designated as `D1N4148`. To create the circuit we use the **Place → Part** commands to lay out the various components, and the **Place → Wire** commands to interconnect them. When it comes to placing the diode, we import it from PSpice's library by going down its list of entries and left-clicking on the `D1N4148` part. To visualize its model, first left-click on the diode to select it, then right-click to activate a pull-down menu of possible actions. Left-click on **Edit PSpice Model**, and the following list will appear:

```
.model  D1N4148    D(Is=2.682n N=1.836 Rs=.5664 Ikf=44.17m Xti=3
+       Eg=1.11 Cjo=4p M=.3333 Vj=.5 Fc=.5 Isr=1.565n Nr=2
+       Bv=100 Ibv=100u Tt=11.54n)
```

TABLE 1A.1 **Partial parameter list of the PSpice diode model.**

Symbol	Name	Parameter description	Units	Default	Example
I_s	Is	Saturation current	A	10 fA	2 fA
n	N	Emission coefficient		1	1.5
r_S	Rs	Bulk resistance	Ω	0	2
C_{j0}	Cjo	Zero-bias junction capacitance	F	0	1.0 pF
m	M	Grading coefficient		0.5	0.33
ϕ_0	Vj	Built-in potential	V	1 V	0.8 V
τ_T	Tt	Transit time	s	0	0.2 ns
V_Z	Bv	Voltage at onset of breakdown	V	∞	100 V
I_Z	Ibv	Current at onset of breakdown	A	0.1 nA	100 μA

The parameter values shown are designed to match as closely as possible those given in the manufacturer's data sheets. We easily see that this particular diode type has $I_s = 2.682$ nA, $n = 1.836$, $r_S = 0.5664$ Ω, $C_{j0} = 4$ pF, $m = 0.3333$, $\phi_0 = 0.5$ V, and $\tau_T = 11.54$ ns. We also note that the onset of the breakdown region (knee of the *i-v* curve in reverse bias) is specified to be 100 V at 100 μA ($V_Z = 100$ V at $I_Z = 100$ μA). The complete list contains additional parameters representing higher-order effects that are beyond our present scope. Also, note the usage of a space to separate individual parameters, and the usage of the $+$ symbol to denote continuity when the list is too long to fit in a single line. For more details, see Ref. [11].

If you wish to create your own diode model to experiment with, you can do so simply by overwriting (editing) the parameter values of an existing diode model, such as the `D1N4148` considered above. However, to avoid losing the original `D1N4148` model, a *new name* must be given to the newly formed model. This is what was done to create a model for the pseudo-ideal diode of Figs. 2.14 and 2.16. The diode model was renamed as `Dideal`, and the parameter list was edited as follows:

```
.model Dideal    D(Is=1n  N=0.001)
```

Even though we know that a practical *pn* junction has $1 \leq n \leq 2$, here we have specified an artificially small value of n to ensure that the exponential *i-v* curve shoots up for very small values of v, thus approximating an ideal diode. (To get an idea, use the logarithmic diode law to verify that to sustain $i = 1$ mA this diode requires only $v = 0.36$ mV $-$ almost 0 V, and certainly much less than the typical voltage drop of 0.7 V!) This is all we need to simulate an almost ideal diode, so all other parameters have been omitted from the list. All omitted parameters are automatically assigned *default values* according to Table 1A.1.

PSpice's library contains also a model for the 1N750 4.7-V Zener diode:

```
.model  D1N750    D(Is=880.5E-18 Rs=.25 Ikf=0 N=1 Xti=3 Eg=1.11
+       Cjo=175p M=.5516 Vj=.75 Fc=.5 Isr=1.859n Nr=2 Bv=4.7
+       Ibv=20.245m Nbv=1.6989 Ibvl=1.9556m Nbvl=14.976
+       Tbvl=-21.277u)
*       Vz = 4.7 @ 20mA,  Rz = 300 @ 1mA,  Rz = 12.5 @ 5mA,  Rz = 2.6 @ 20mA
```

The last line, starting with an asterisk, is by convention a *comment line* summarizing salient characteristics. It states that this diode is rated to provide $V_Z = 4.7$ V at $I_Z = 20$ mA. Moreover, the *reciprocal of the slope* of the *i-v* curve in the breakdown region, denoted as r_z in the text, is specified at different points as $r_z = 300$ Ω at $I_Z = 1$ mA, $r_z = 12.5$ Ω at $I_Z = 5$ mA, and $r_z = 2.6$ Ω at $I_Z = 20$ mA. Clearly, the further down the curve past the knee, the steeper the slope.

The user can readily edit the above model to create a Zener diode with a different rating. For instance, changing `Bv=4.7` to `Bv=6.2` will turn the model into that of a 6.2-V Zener.

REFERENCES

1. R. F. Pierret, *Semiconductor Fundamentals*, Modular Series on Solid State Devices, 2/E, Vol. I, and G. W. Neudeck, *The PN Junction Diode*, Modular Series on Solid State Devices, 2/E, Vol. II, G. W. Neudeck and R. F. Pierret, eds., Addison-Wesley, 1989.

2. R. S. Muller and T. I. Kamins, *Device Electronics for Integrated Circuits*, 2/E, J. Wiley and Sons, 1986.

3. R. T. Howe and C. G. Sodini, *Microelectronics: An Integrated Approach*, Prentice Hall, 1997.

4. A. S. Sedra and K. C. Smith, *Microelectronic Circuits*, 6/E, Oxford University Press, 2010.

5. R. C. Jaeger and T. N. Blalock, *Microelectronic Circuit Design*, 3/E, McGraw-Hill, 2008.

6. G. W. Gordon and A. S. Sedra, *SPICE for Microelectronic Circuits*, 2/E, Oxford University Press, 1996.

7. IRE Symbols Committee et al., "IEEE Standard Letter Symbols for Semiconductor Devices," IEEE Transactions on Electron Devices, Vol. 11, No. 8, pp. 392–397, August 1964.

PROBLEMS

1.1 The Ideal Diode

1.1 (*a*) In the circuit of Fig. P1.1 let $v_S = 25$ V, $R_1 = 2$ kΩ, $R_2 = 3$ kΩ, $R_3 = 1$ kΩ, and $i_S = 4$ mA. Find the voltage across and the current through the diode.

(*b*) Repeat, but with $v_S = 10$ V and $i_S = 10$ mA.

(*c*) Repeat, but with $v_S = 5$ V and $i_S = 3$ mA.

(*d*) Repeat parts (*a*) through (*c*), but with a fourth resistance $R_4 = 1.8$ kΩ connected in parallel with the diode. List the cases in which this additional resistance make a difference, and those in which it doesn't.

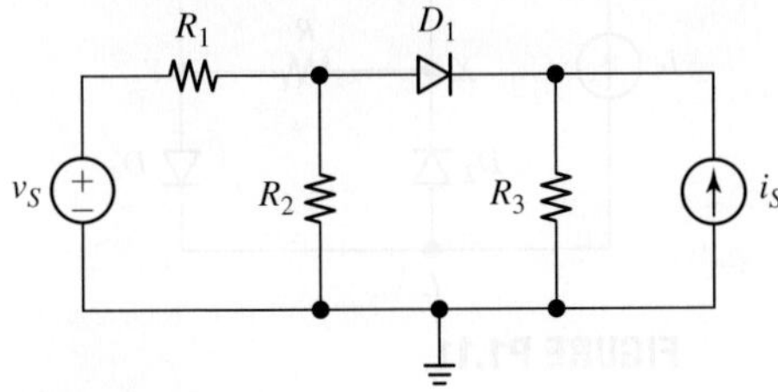

FIGURE P1.1

1.2 (*a*) In the circuit of Fig. P1.1 let $v_S = 24$ V, $R_1 = 300$ Ω, $R_2 = 200$ Ω, and $R_3 = 400$ Ω. Find i_S for a 20-mA diode current.

(*b*) If $i_S = 15$ mA, find v_S for a 4-V drop across the diode.

(*c*) If $i_S = 10$ mA, find v_S so that the diode voltage and current are both zero.

(*d*) Find v_S and i_S for a 30-mA current through the 200-Ω resistance, and a 20-mA current through the 300-Ω resistance. Is the solution unique?

1.3 Repeat Problem 1.2, but with the diode reversed, so that the anode is now at the right and the cathode at the left.

1.4 (*a*) In the circuit of Fig. P1.1, let $R_1 = 200$ Ω, $R_2 = 300$ Ω, $R_3 = 100$ Ω, and $i_S = 30$ mA. If v_S is swept from 0 to 10 V, sketch and label, as a function of v_S, the current i supplied by v_S and the voltage v across i_S. Show all breakpoints and slopes.

(*b*) If $v_S = 5$ V and i_S is swept from 0 to 60 mA, sketch and label, versus i_S, the current i supplied by v_S and the voltage v across i_S.

1.5 (*a*) Find the current and voltage of each diode in the circuit of Fig. P1.5.

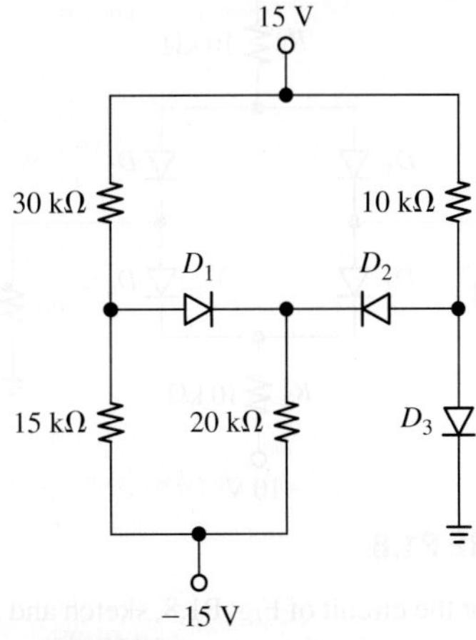

FIGURE P1.5

(b) Repeat part *(a)* if the 10-kΩ resistance is changed to 30 kΩ.

(c) Repeat part *(a)* if the 10-kΩ resistance is changed to 60 kΩ.

Hint: apply Thévenin's theorem to the network seen by D_1's anode.

(d) Repeat part *(a)* if the 15-kΩ resistance is removed from the circuit.

1.6 Find the current and voltage of each diode in the circuit of Fig P1.5 if:

(a) D_1 is reversed, so that its anode is now at the right and its cathode at the left;

(b) D_2 is reversed, so its anode is now at the left and its cathode at the right;

(c) D_3 is reversed, so its anode is now at the bottom and its cathode at the top;

(d) all three diodes are reversed simultaneously. Use the hint of Problem 1.5.

1.7 In the circuit of Fig. P1.5, find the value to which

(a) we must change the 10-kΩ resistance if we want D_3 to drop 5 V;

(b) we must change the 20-kΩ resistance if we want D_1 to carry a current of 0.25 mA;

(c) we must change the 15-kΩ resistance if we want D_1 to drop 0 V and carry 0 mA.

1.8 (a) In the circuit of Fig. P1.8 find v_o for the following cases: $v_I = 0$ V, $v_I = \pm3$ V, and $v_I = \pm6$ V.

(b) Repeat part *(a)* if R_3 is changed to 30 kΩ.

(c) Repeat part *(a)* if R_1 is changed to 30 kΩ.

(d) Repeat part *(a)* if R_2 is removed from the circuit.

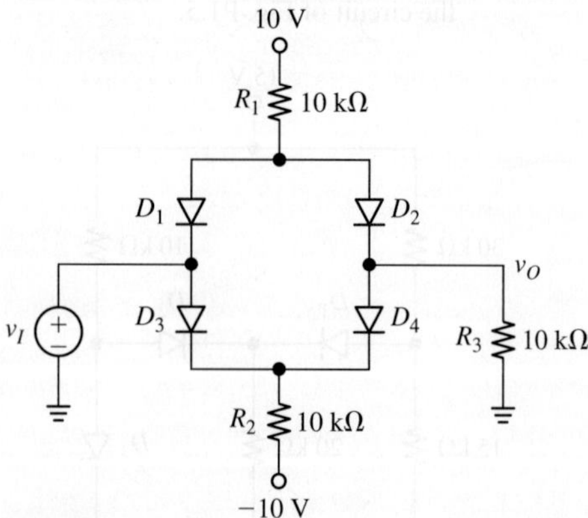

FIGURE P1.8

1.9 (a) For the circuit of Fig. P1.8, sketch and label v_o versus v_I over the range -10 V $\le v_I \le +10$ V. Show all breakpoints and slopes.

Hint: if, starting at 0 V, you gradually increase v_I, at what point will D_1 go off? If instead you gradually decrease v_I, at what point will D_3 go off?

(b) Repeat *(a)* if R_3 is changed to 30 kΩ.

(c) Repeat *(a)* if another 10-kΩ resistance is connected between the input and output nodes.

1.10 In the circuit of Fig. P1.8, let i represent the current out of the positive terminal of the source v_I.

(a) Sketch and label i versus v_I over the range -10 V $\le v_I \le +10$ V. Show all breakpoints and slopes.

(b) Repeat if an additional resistance $R_4 = 5$ kΩ is connected between the node labeled v_I and the node labeled v_o.

Hint: exploit the fact that circuit's behavior for $v_I < 0$ is *symmetric* of that for $v_I > 0$.

1.2 Basic Diode Applications

1.11 In Fig. P1.11 let i_S be a triangular wave with peak values of ±1 mA and let $R = 5$ kΩ.

(a) Sketch and label v_H, v_L, and v_X if node Y is grounded.

(b) Sketch and label v_H, v_L, and v_Y if node X is grounded.

Hint: consider the cases $i_S > 0$ and $i_S < 0$ separately.

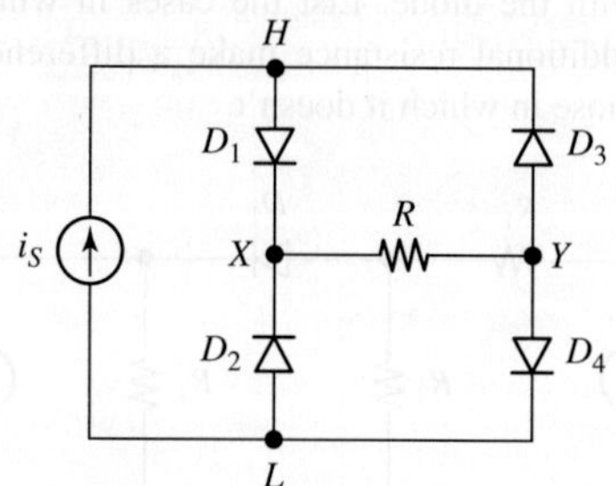

FIGURE P1.11

1.12 (a) Show that the circuit of Fig. P1.12 gives $v_o = v_I$ for $|v_I| < R_L I_S$, $v_o = R_L I_S$ for $v_I > R_L I_S$, and $v_o = -R_L I_S$ for $v_I < -R_L I_S$.

(b) Assuming $R_L = 1$ kΩ and v_I is a 1-kHz sine wave with peak values of ±5 V, sketch and label v_I and v_o versus time for the following cases: $I_S = 5$ mA, $I_S = 2.5$ mA, and $I_S = 0$ mA.

(c) Can you suggest possible applications for this circuit?

Hint: exploit the fact that circuit's behavior for $v_I < 0$ is symmetric of that for $v_I > 0$.

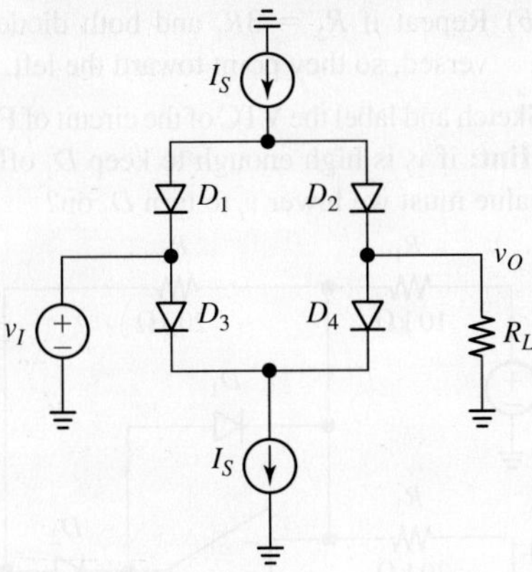

FIGURE P1.12

1.13 In the circuit of Fig. P1.12 let $R_L = 1\text{ k}\Omega$ and $I_S = 1$ mA, and let the input be swept over the range $-2\text{ V} \le v_I \le +2\text{ V}$. Sketch and label, versus v_I, the current i supplied by the input source as well as the current through each diode. Show all breakpoints and slopes.

Hint: exploit the fact that at all times we have $i_{D_1} + i_{D_2} = i_{D_3} + i_{D_4} = I_S$.

1.14 (*a*) In the circuit of Fig. P1.14 let $R_1 = 25\text{ k}\Omega$, $R_2 = R_3 = 30\text{ k}\Omega$, $R_4 = 120\text{ k}\Omega$, and $V_S = 3$ V. Sketch and label v_I and v_O versus time if v_I is a 500-Hz triangular wave with peak values of ± 5 V. Show all breakpoints and slopes. Can you suggest a possible application for this circuit?

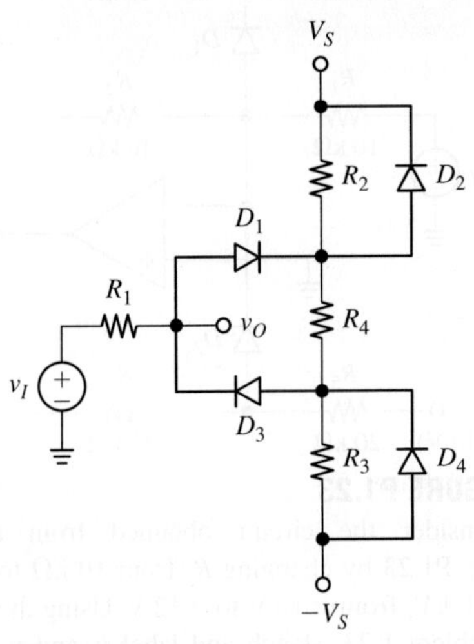

FIGURE P1.14

Hint: consider first the case $v_I \ge 0$, and show that D_3 and D_4 are off, and that if v_I is gradually increased from 0 V, a point is reached at which D_1 goes on, and then another at which D_2 goes on. Then, exploit the fact that circuit behavior for $v_I < 0$ is symmetric of that for $v_I > 0$.

(*b*) What happens if D_2 and D_4 are removed?

1.15 Sketch and label the VTC of the circuit of Problem 1.14 over the range $-6\text{ V} \le v_I \le 6\text{ V}$ if:

(*a*) the directions of D_2 and the D_4 are reversed, so that their anodes are now at the top and the cathodes at the bottom;

(*b*) D_2 and D_4 are removed from the circuit;

(*c*) D_2, D_4, and R_4 are removed from the circuit.

1.16 Redraw the circuit of Fig. P1.14, but with an additional 20-kΩ resistance between the node labeled v_O and ground.

(*a*) Specify suitable values for V_S and R_1 through R_4 to implement a symmetric VTC that goes through the origin and has a slope of 1/2 V/V for $|v_I| \le 6$ V, a slope of 1/3 V/V for $6\text{ V} \le |v_I| \le 12$ V, and a slope of 0 V/V for $|v_I| \ge 12$ V.

Hint: exploit the symmetry of the VTC, draw it for the case for $v_I \ge 0$, and show the subcircuit corresponding to each of its three segments.

(*b*) What happens to the VTC if D_2 and D_4 are removed?

(*c*) What happens to the VTC if the value of V_S is doubled?

1.17 In a clamped capacitor circuit of the type of Fig. 1.22a let v_I be a 1-kHz triangular wave having peak values of $+10$ V and -5 V, and let $C = 1\ \mu\text{F}$.

(*a*) Sketch and label, versus time, v_I, v_O, and the diode current i_D. Assume C is initially discharged, and let $t = 0$ be the instant when v_I starts rising from 0 V.

(*b*) Repeat, but with the direction of the diode reversed, so that the anode is now at the top and the cathode at the bottom.

(*c*) What happens if the frequency is doubled? Halved?

1.18 Let $C = 1\ \mu\text{F}$ be initially discharged in the circuit of Fig. P1.18. Sketch and label $v_O(t)$ if $v_I(t)$ is the square-wave shown. Do it first with $R = \infty$, then with $R = 5\text{ k}\Omega$. Compare the two cases, comment.

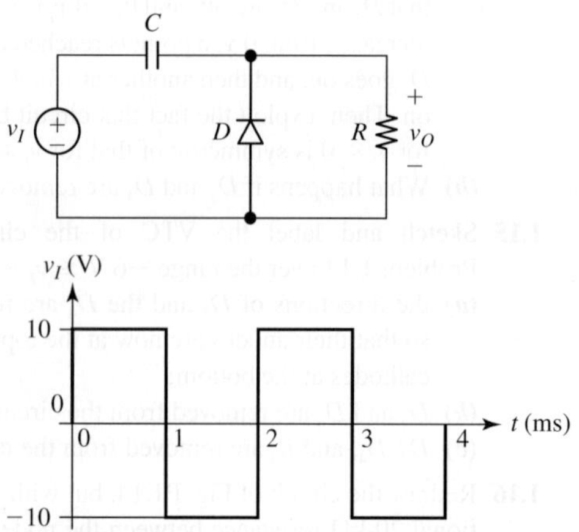

FIGURE P1.18

1.19 (*a*) Redraw the circuit of Fig. P1.18, but with R replaced by a 1-mA current source flowing downward. Assuming C is initially discharged, sketch and label $v_O(t)$ if $v_I(t)$ is the square-wave shown, and $C = 1\ \mu\text{F}$.

(*b*) Repeat, but with the 1-mA source now flowing upward. Compare the two cases, comment.

1.3 Operational Amplifiers and Diode Applications

1.20 (*a*) Obtain relationships between v_O and v_I for the circuit of Fig. P1.20 if $R_2 = R_1$ and $V_P = 0$.
Hint: consider the cases $v_I > 0$ and $v_I < 0$ separately.

(*b*) Repeat, if $R_2 = 2R_1$.

(*c*) Repeat if $R_2 = 4R_1$ and both diodes are reversed, so they point toward the left.

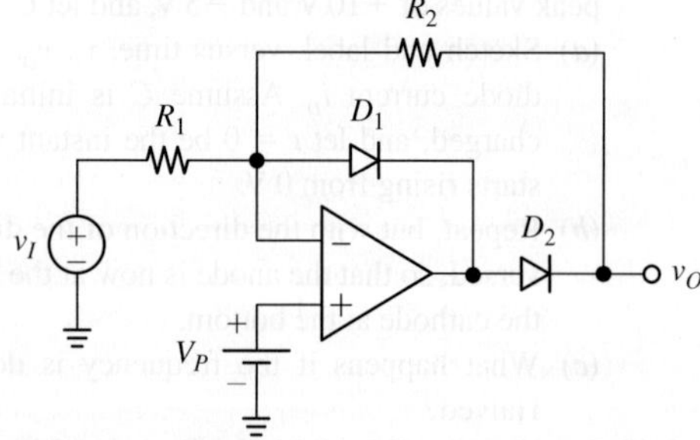

FIGURE P1.20

1.21 (*a*) Sketch and label the VTC of the circuit of Fig. P1.20 if $R_2 = 2R_1$ and $V_P = 1.0$ V.
Hint: if v_I is high enough to keep D_1 on, to what value must we lower v_I to turn D_1 off?

(*b*) Repeat if $R_2 = 3R_1$ and both diodes are reversed, so they point toward the left.

1.22 Sketch and label the VTC of the circuit of Fig. P1.22.
Hint: if v_I is high enough to keep D_1 off, to what value must we lower v_I to turn D_1 on?

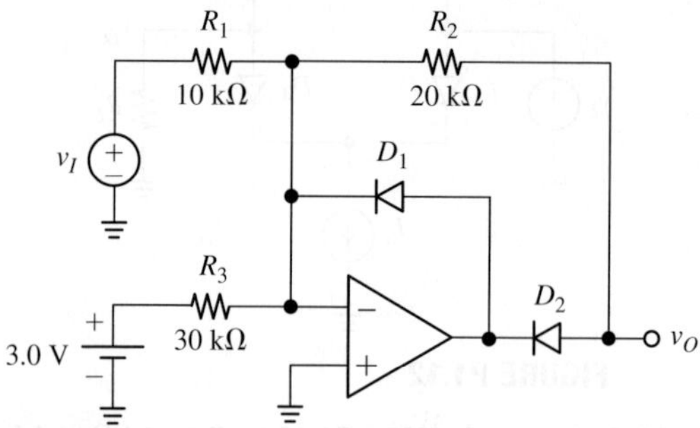

FIGURE P1.22

1.23 As long as both diodes are off, the circuit of Fig. P1.23 is an inverting amplifier giving $v_O = -(R_2/R_1)v_I$. But what if either diode conducts? Analyze the circuit and then sketch and label its VTC over the range $-10\ \text{V} \le v_I \le +10\ \text{V}$. Show all breakpoints and slopes.
Hint: examine first the case $v_I \ge 0$ and show that with $v_I = 0$ V, D_1 is off, but increasing v_I will eventually turn D_1 on. Then, exploit the fact that circuit behavior for $v_I < 0$ is symmetric of that for $v_I > 0$.

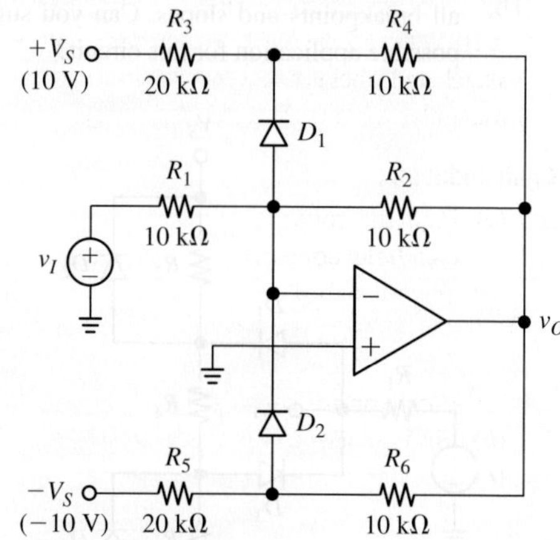

FIGURE P1.23

1.24 Consider the circuit obtained from that of Fig. P1.23 by changing R_2 from 10 kΩ to 20 kΩ and $\pm V_S$ from ± 10 V to ± 12 V. Using the hint of Problem 1.23, sketch and label v_I and v_O versus time if v_I is a 500-Hz triangular wave with peak values of ± 9 V.

1.25 As long as both diodes are off, the circuit of Fig. P1.25 is an inverting amplifier giving $v_O = -R_2/R_1)v_I$. But what if either diode conducts? Analyze the circuit and then sketch and label its VTC over the range $-7.5\ \text{V} \le v_I \le +7.5\ \text{V}$. Show all breakpoints and slopes.

Hint: examine first the case $v_I \ge 0$, and show that with $v_I = 0\ \text{V}$, D_2 is off, but increasing v_I will eventually turn D_2 on. Then, exploit the fact that circuit behavior for $v_I < 0$ is symmetric of that for $v_I > 0$.

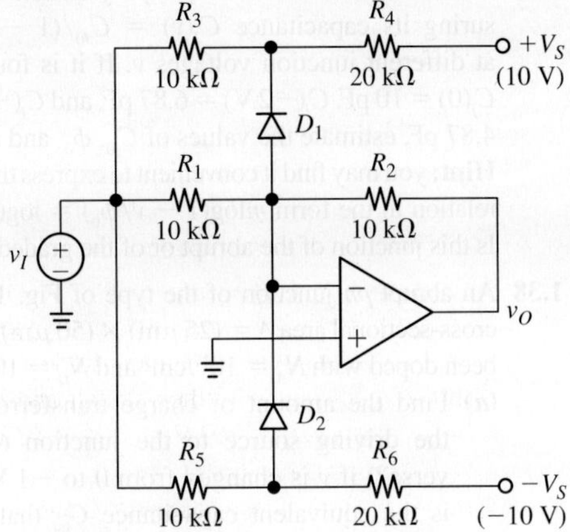

FIGURE P1.25

1.26 Consider the circuit obtained from that of Fig. P1.25 by changing R_2 from 10 kΩ to 30 kΩ and $\pm V_S$ from ± 10 V to ± 2 V. Using the hint of Problem 1.25, sketch and label v_I and v_O versus time if v_I is a 500-Hz triangular wave with peak values of ± 2 V.

1.4 Semiconductors

1.27 (*a*) Find the room-temperature (300 K) electron and hole concentrations n and p for a bulk silicon slab that has been doped first with $4 \times 10^{14}/\text{cm}^3$ atoms of boron, and subsequently with $10^{15}/\text{cm}^3$ atoms of arsenic. Is the slab p-type or n-type?

(*b*) Find n and p if T is raised to 400 K.

(*c*) If a slab of n-type silicon with $N_D = 10^{16}/\text{cm}^3$ is to be turned into a p-type slab with a hole concentration $p = 5 \times 10^{15}/\text{cm}^3$, what acceptor concentration N_A is needed?

(*d*) Find the mobilities μ_n and μ_p of the p-type slab of (*c*).

1.28 To develop a feel for the conductive properties of different materials available in IC technology, let us estimate the resistance R of a bar of the type

of Fig. 1.36*a* if it is 10-μm long, 1-μm thick, and 2-μm wide (1 μm = 10^{-4} cm), and is made up of the following materials:

(*a*) pure silicon;

(*b*) n^--type silicon with $N_D = 10^{16}/\text{cm}^3$ atoms of phosphorous;

(*c*) p-type silicon with $N_A = 10^{18}/\text{cm}^3$ atoms of boron;

(*d*) n^+-type silicon with $N_D = 10^{20}/\text{cm}^3$ atoms of phosphorous;

(*e*) p^+-type silicon with $N_A = 10^{20}/\text{cm}^3$ atoms of boron;

(*f*) aluminum, for which $\rho = 2.7\ \mu\Omega$ cm.

Hint: recall from basic physics that $R = \rho L/A$, where L is the bar's length, A its cross-sectional area, and $\rho = 1/[q(n\mu_n + p\mu_p)]$ is its resistivity; use the empirical formulas of Fig. 1.37 to find the mobilities.

1.29 A slab of the type of Fig. 1.36*a* is 20-μm long, 2-μm thick, and 5-μm wide (1 μm = 10^{-4} cm), and has been uniformly doped with $10^{14}/\text{cm}^3$ atoms of phosphorous. If a 1-V battery is connected across the slab, find:

(*a*) the electric field E inside the slab;

(*b*) the drift velocities of electrons and holes;

(*c*) the electron and hole drift currents (compare, comment);

(*d*) the average time taken by an electron and a hole to drift along the entire length of the slab;

(*e*) the resistance R of the slab. Use the hint of Problem 1.28.

1.30 A slab of p-type silicon of the type of Fig. 1.36*b* has been doped with $N_A = 10^{16}/\text{cm}^3$, has a length $L = 1\ \mu$m, and a cross-sectional area $A = (20\ \mu\text{m}) \times (50\ \mu\text{m})$. Electrons are being injected into the slab at the left ($x = 0$) and are removed from the slab at the right ($x = L$) in such a way as to maintain a *linear* profile $n(x)$ with the following values: $n(L) = n_i^2/N_A$ and $n(0) = 10^{10}n(L)$. Find the magnitude and direction of the current i. Use the empirical formulas of Fig. 1.37 to find the mobilities.

1.31 A 5-μm long slab of p-type silicon of the type of Fig. 1.36*b* has been doped non-uniformly along the x-axis according to the profile $N_A(x) = 10^{14}[1 + 10^3\exp(-x/(1\ \mu\text{m})]/\text{cm}^3$. Sketch $N_A(x)$ vs. x, and show that even if not part of any circuit, the slab possesses a nonzero internal electric field $E(x)$. Calculate $E(0)$ and $E(5\ \mu\text{m})$.

Hint: since the slab is not part of any circuit, in equilibrium it must have $J_{(\text{drift})} + J_{(\text{diff})} = 0$.

1.32 A slab of the type of Fig. 1.36*b* has been doped with $N_D = 10^{16}/\text{cm}^3$ atoms of phosphorous. Holes are being injected into the slab at the right ($x = L$) and are removed from the slab at the left ($x = 0$) in such a way as to maintain the profile $p(x) = 10^{14}[1 - \exp(-x/(10\ \mu\text{m})]/\text{cm}^3$.

   (*a*) Sketch and label the hole concentration $p(x)$ and current density $J_p(x)$ for $0 \le x \le 50\ \mu\text{m}$.

   (*b*) You will find that the current density is highest where the concentration is lowest, and vice versa. Do you see a contradiction here? Explain!

1.5 The *pn* Junction in Equilibrium

1.33 Sketch and label the charge density $\rho(x)$, the electric field $E(x)$, and the electrostatic potential $\phi(x)$ for a *pn* junction of the type of Fig. 1.39 if the *p*-side doping is kept fixed at $N_A = 10^{16}/\text{cm}^3$ but the *n*-side doping is progressively increased as follows:

   (*a*) $N_D = 10^{16}/\text{cm}^3$ (symmetric doping),

   (*b*) $N_D = 4 \times 10^{16}/\text{cm}^3$ (asymmetric doping), and

   (*c*) $N_D = 10^{17}/\text{cm}^3$ (even more asymmetry). Compare the three cases, comment.

1.34 Given that a certain slab of silicon material exhibits the charge-density distribution of Fig. P1.34, sketch and label the electric field $E(x)$ as well as the potential $\phi(x)$. Assume $\phi(0) = 0$.

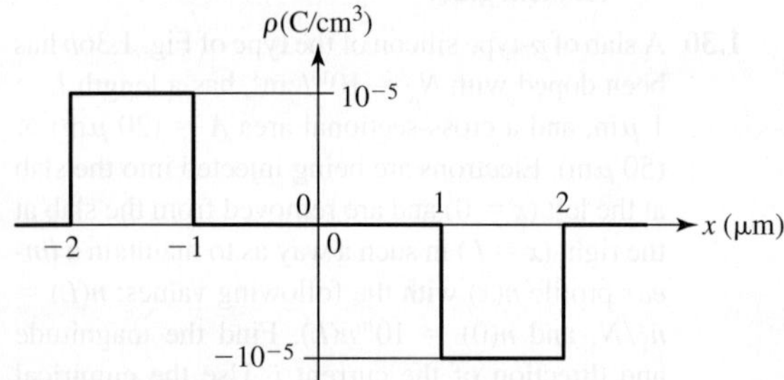

FIGURE P1.34

1.35 (*a*) For a *pn* junction with a cross-sectional area $A = (10\ \mu\text{m}) \times (20\ \mu\text{m})$, find the doping concentrations needed to ensure $\phi_0 = 0.7$ V, as well as the resulting values of E_{m0}, x_{p0}, x_{n0}, and Q_{j0}, under the constraint $N_D = N_A$.

   (*b*) Repeat, but under the constraint $N_D = 10N_A$.

   (*c*) Repeat, but under the constraint $N_D = 0.1N_A$. Comment on your various results.

1.6 Effect of External Bias on the SCL Parameters

1.36 An abrupt *pn* junction of the type of Fig. 1.40 has been doped with $N_A = 10^{16}/\text{cm}^3$ and $N_D = 10^{18}/\text{cm}^3$.

   (*a*) Find the voltage v needed to make $X_d = 1\ \mu\text{m}$. **Hint:** exploit the fact that $N_D \gg N_A$.

   (*b*) Sketch and label, versus x, the resulting potential $\phi(x)$; assume $\phi(x \ge x_n) = \phi_n$.

   (*c*) Find the cross-sectional square area needed to achieve $C_j = 1$ pF at the voltage of part (*a*).

1.37 A student is characterizing a *pn* junction by measuring its capacitance $C_j(v) = C_{j0}/(1 - v/\phi_0)^m$ at different junction voltages v. If it is found that $C_j(0) = 10$ pF, $C_j(-2\,\text{V}) = 6.87$ pF, and $C_j(-8\,\text{V}) = 4.87$ pF, estimate the values of C_{j0}, ϕ_0, and m. **Hint:** you may find it convenient to express the above relation in the form $m\log(1 - v/\phi_0) = \log(C_{j0}/C_j)$. Is this junction of the abrupt or of the graded type?

1.38 An abrupt *pn* junction of the type of Fig. 1.40 has cross-sectional area $A = (25\ \mu\text{m}) \times (50\ \mu\text{m})$ and has been doped with $N_A = 10^{17}/\text{cm}^3$ and $N_D = 10^{19}/\text{cm}^3$.

   (*a*) Find the amount of charge transferred from the driving source to the junction (or vice versa?) if v is changed from 0 to -1 V. What is the equivalent capacitance C_{eq} that would cause the same amount of charge transfer?

   (*b*) Repeat if v is changed from -1 V to -2 V.

   (*c*) Repeat if v is changed from -3 V to -2 V. Compare, and comment.

1.7 The *pn* Diode Equation

1.39 (*a*) A *pn* junction has been doped with $N_D = 10^{18}/\text{cm}^3$ atoms of phosphorous and $N_A = 10^{16}/\text{cm}^3$ atoms of boron, and the voltage across its terminals has been adjusted for a forward current of 1.0 mA. If you were to observe this current right inside the SCL, what fraction of it would be due to electron flow, and what fraction to hole flow? **Hint:** use the formulas of Fig. 1.37, and assume $\tau_n/\tau_p = 1$ for simplicity.

   (*b*) Repeat if $N_D = 10^{16}/\text{cm}^3$ and $N_A = 10^{18}/\text{cm}^3$.

   (*c*) Assuming $N_A = 10^{17}/\text{cm}^3$ and $\mu_n = 1.6\mu_p$, find the value of N_D needed to make electron flow and hole flow *equal* inside the SCL.

1.40 (*a*) Two *pn* junctions having $I_{s1} = 1$ fA and $I_{s2} = 5$ fA are connected in series and forward biased via a common 1-V source. Assuming $V_T = 26$ mV, find the individual voltage drops as well as the common current. **Hint:** use $v = V_T \ln(i/I_s)$.

(b) If the diodes are connected in parallel and forward biased via a common 1-mA source, find their individual currents as well as the common voltage drop.

1.41 Consider a short-base diode that has been doped with $N_D = 10^{17}/cm^3$ atoms of phosphorous and $N_A = 10^{16}/cm^3$ atoms of boron, and has been fabricated with $W_p = W_n = 1\ \mu m$ and a cross-sectional area of $(25\ \mu m) \times (50\ \mu m)$.

(a) Find i if the device is forward-biased with $v = 700$ mV.

(b) Estimate the error incurred in ignoring x_p and x_n compared to W_p and W_n.

1.8 The Reverse-Biased *pn* Junction

1.42 (a) Assuming a silicon junction breaks down when its maximum internal electric field E_m reaches 300 kV/cm, estimate the breakdown voltage BV of the *pn* junction of Example 1.7.

(b) Repeat if N_D is doubled to $2 \times 10^{16}/cm^3$.

1.43 When studying BJTs we will see that the base-collector junction of an *npn* BJT is a *pn* junction with the *n*-side doped *lightly* to ensure a *high* breakdown voltage BV. Assuming a silicon junction breaks down when the maximum internal electric field E_m reaches 300 kV/cm, what doping N_D is needed to result in

(a) $BV = 100$ V?

(b) $BV = 1$ kV?

Hint: exploit the fact that $N_D \ll N_A$ and that $BV \gg \phi_0$.

1.44 A student is characterizing a *pn* junction by performing a series of *I-V* measurements. The final data, tabulated in increasing voltage-magnitude order, are: $(V, I) = (600$ mV, 4.8 μA$)$, $(700$ mV, 100 pA$)$, $(800$ mV, 0.81 mA$)$, $(9.900$ V, 10 mA$)$, $(10.100$ V, 30 mA$)$. The student did not bother to record voltage polarities or current directions, claiming that one can figure them out via educated reasoning. Assuming a forward-region characteristic of the type $I = I_s \exp[V/(nV_T)]$, $V_T = 26$ mV, and a breakdown-region characteristic of the type $I_z = (V - V_{Z0})/r_z$, use the above data to find n, I_s, I_R, V_{Z0}, and r_z.

1.9 Forward-Biased Diode Characteristics

1.45 (a) A student is using a constant current source $I_S = 1$ mA to forward bias a diode D_1 having $nV_T = 50$ mV. What is the diode's dynamic resistance r_d?

(b) If a second identical diode D_2 is connected in *parallel* with D_1, what are the dynamic resistances of the individual diodes as well as the overall dynamic resistance of the two-diode structure?

(c) If a third identical diode D_3 is connected in *series* with the parallel structure of part (*b*), what are the individual dynamic resistances as well as the overall dynamic resistance of the three-diode structure? Again, comment.

1.46 A certain power diode with $n = 2$ is driven with a forward current of 10 A. Immediately after current turn-on, the voltage drop across the diode is found to be 900 mV. However, as temperature rises because of internal power dissipation, the voltage decreases and eventually settles to 750 mV.

(a) Using the rule of thumb, estimate the junction's temperature rise.

(b) Given that the temperature rise per watt (°C/W) represents *thermal resistance*, what is the thermal resistance of the present diode in its final state? What happens if another identical diode is connected

(c) in *series* with the existing one?

(d) In parallel?

1.47 (a) In the circuit of Fig. P1.47, D_2 is known to have, at room temperature, $I_s = 2$ fA and $nV_T = 26$ mV. If it found that $V = 340$ mV, what is the reverse current I_R of D_1?

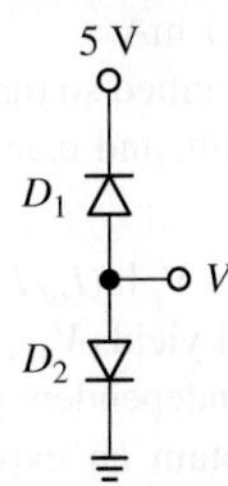

FIGURE P1.47

(b) If in the circuit of (*a*) a diode D_3 identical to D_1 is connected in parallel with D_1 itself (anode with anode and cathode with cathode), what is the new value of V?

(c) If in the circuit of (*a*) a diode D_4 identical to D_2 is connected in parallel with D_2 itself (anode with anode and cathode with cathode), what is the new value of V?

(*d*) Going back to the original circuit of (*a*), if temperature is increased by 50 °C, what is the new value of *V*?

Hint: like a good practicing engineer, solve this problem using the rules of thumb.

1.48 The diodes in the circuit of Fig. P1.48 are such that at 1 mA they drop 700 mV and also have $nV_T = 26$ mV. If $V_S = 3$ V, find suitable resistance values to bias the diodes at $I_{D_1} = 0.5$ mA, $I_{D_2} = 0.2$ mA, and $I_{D_3} = 0.1$ mA.

Hint: This problem can be solved using the rules of thumb.

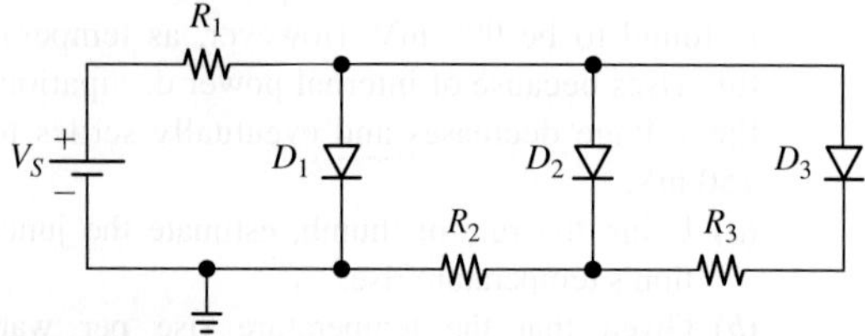

FIGURE P1.48

1.49 The diodes in the circuit of Fig. P1.48 are such that at 1 mA they drop 700 mV and also have $nV_T = 26$ mV. If $V_S = 2.4$ V, $R_2 = 84$ Ω, and $R_3 = 360$ Ω, find R_1 so that $I_{D_3} = 0.1$ mA.

Hint: start out at the right and progress toward the left using the rules of thumb.

1.50 The diodes in the circuit of Fig. P1.48 are such that at 1 mA they drop 700 mV and also have $nV_T = 26$ m. If, $R_1 = 2.0$ kΩ, $R_2 = 336$ Ω, and $R_3 = 1440$ Ω, find V_S so that $I_{D_2} = 0.1$ mA.

Hint: the problem has been specified so that it can be solved via the rules of thumb, and using iterations to find I_{D_3}.

1.51 (*a*) Using the relationship $V_D = V_T \ln(I_D/I_s)$, show that the circuit of Fig. P1.51 yields $V_{PTAT} = KT$, where K is a temperature-independent proportionality constant, and obtain an expression for K in terms of the diodes' saturation currents I_{s1} and I_{s2} and bias currents I_1 and I_2. Assume $n = 1$. The voltage V_{PTAT} is *proportional to absolute temperature* (PTAT), and as such it is at the basis of digital thermometers and other temperature-related instrumentation.

(*b*) Find the value of the proportionality constant K if $I_1 = I_2 = 100$ μA, and D_2 has a junction area 10 times as large as that of D_1 so that $I_{s2} = 10I_{s1}$.

(*c*) What happens if the bias currents of part (*b*) are inadvertently reduced to $I_1 = I_2 = 50$ μA? Will the value of V_{PTAT} change?

(*d*) Find the gain by which the voltage V_{PTAT} of (*b*) needs to be amplified if we want to synthesize the voltage $V(T) = (10 \text{ mV/°C})T$.

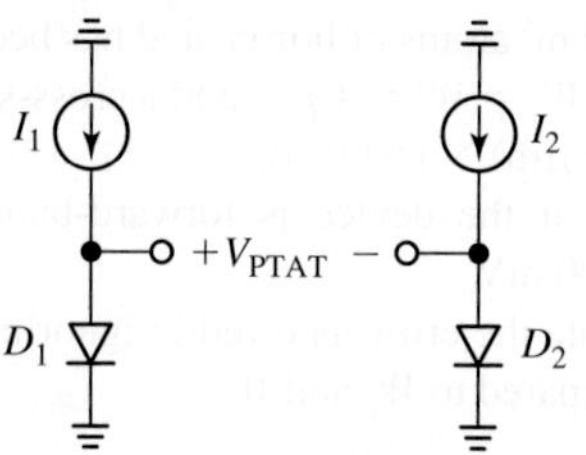

FIGURE P1.51

1.52 Figure P1.52 shows an alternative realization of the V_{PTAT} concept of Problem 1.51, and its advantage is that V_{PTAT} is now referenced to ground instead of being the difference between two floating node voltages. For the present circuit to function, we must have $I_1 > I_2$.

(*a*) Show that the circuit of Fig. P1.52 yields $V_{PTAT} = KT$, where K is a temperature-independent proportionality constant, and obtain an expression for K in terms of the diodes' saturation currents I_{s1} and I_{s2} and the bias currents I_1 and I_2. Assume $n = 1$.

(*b*) Find the value of the proportionality constant K if $I_1 = 100$ μA, $I_2 = 20$ μA, and D_1 and D_2 are matched.

(*c*) Find the value of the proportionality constant K if $I_1 = 100$ μA, $I_2 = 50$ μA, and D_2 has a junction area twice as large as that of D_1.

(*d*) For the conditions of part (*b*), use an op amp to design a circuit that synthesizes the voltage $V_o(T) = (10 \text{ mV/°C})T$.

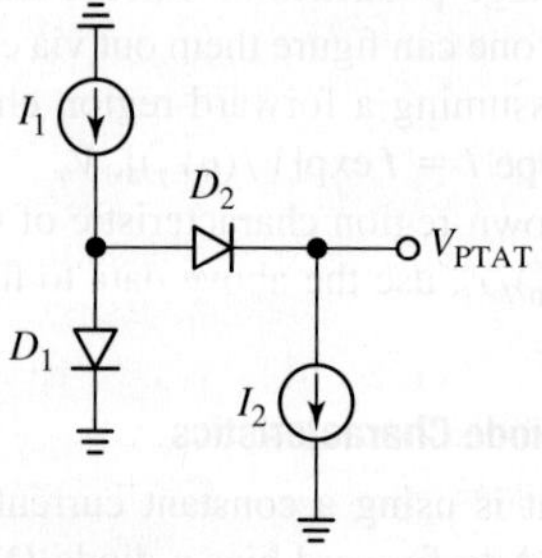

FIGURE P1.52

1.10 Dc Analysis of *pn* Diode Circuits

1.53 In the circuit of Fig. P1.1 let $v_S = 5$ V, $R_1 = 2$ kΩ, $R_2 = 3$ kΩ, $R_3 = 1$ kΩ, and $i_S = 2$ mA. Assuming the diode has $I_s = 5$ fA and $nV_T = 26$ mV, find the diode voltage (in mV) and the diode current (in μA), as well as the voltage v across the current source and the current i through the voltage source. Use both the iterative method and the 0.7-V diode model, compare the results, and comment.

1.54 Consider the circuit obtained from that of Fig. P1.1 by reversing the diode direction so that the anode is now at the right and the cathode at the left. Let $v_S = 4$ V, $R_1 = 2$ kΩ, $R_2 = 3$ kΩ, $R_3 = 1$ kΩ, and $i_S = 4$ mA. Assuming the diode has $I_s = 20$ fA and $nV_T = 35$ mV, find the diode voltage (in mV) and the diode current (in μA), as well as the voltage v across the current source and the current i through the voltage source. Use both the iterative method and the 0.7-V diode model, compare the results, comment.

1.55 In the half-wave rectifier of Fig. 1.10a, v_I is a 500-Hz saw-tooth wave with peak values of ± 5 V. Assuming $V_{D(\text{on})} = 0.7$ V, find the average value of v_O as well as the percentage of the period during which the diode conducts. Compare with the ideal-diode case of $V_{D(\text{on})} = 0$, and comment.

1.56 In the full-wave rectifier of Fig. 1.12a, v_I is a 250-Hz triangular wave with peak values of ± 8 V. Assuming $V_{D(\text{on})} = 0.7$ V, find the average value of v_O as well as the percentage of the period during which the diodes conduct. Compare with the ideal-diode case of $V_{D(\text{on})} = 0$, and comment.

1.57 In the design of signal generators, the need arises to convert a triangular wave v_T to a sine wave v_S. The VTC of a triangle-to-sine converter is shown in Fig. P1.57a for the case in which v_T and v_S have the same slope at the origin. In this case one can readily prove that the peak values of the two waveforms are related as $V_{tm} = (\pi/2)V_{sm}$. The circuit of Fig. P1.57b uses D_1 to clip the top and D_2 to clip the bottom of v_T. Thanks to the rounded knee of the diode characteristic, clipping occurs gradually, thus providing a crude approximation to the VTC of Fig. P1.57a. Let us arbitrarily impose $V_{sm} = 700$ mV at a diode current of 1 mA. Then, assuming $nV_T = 26$ mV, we need diodes with $I_s = 2$ fA. Moreover, we have $V_{tm} = (\pi/2)0.7 = 1.1$ V, so $R = (1.1 - 0.7)/1 = 0.4$ kΩ.

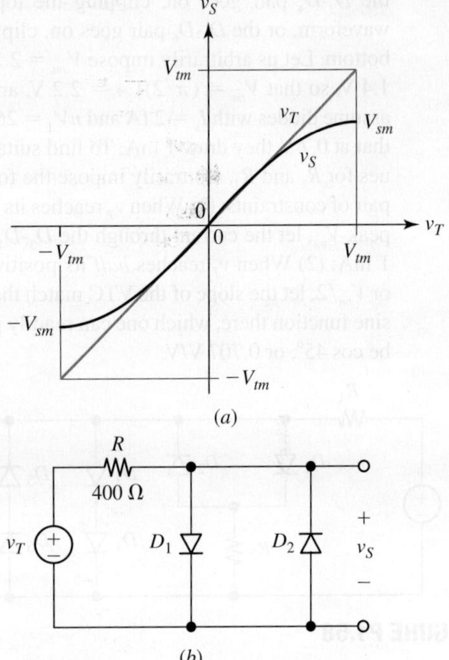

FIGURE P1.57

(a) Use the rules of thumb (18 mV/octave and 60 mV/decade) to calculate v_S as well as $v_T = Ri_D + v_S$ at the following diode currents: $i_D = $ 1 μA, 4 μA, 10 μA, 20 μA, 40 μA, 100μA, 0.2 mA, 0.4 mA, 0.5 mA, 0.8 mA, and 1.0 mA.

(b) Using the above data, plot v_S versus v_T and verify that the circuit of Fig. P1.57b provides an approximation, if crude, to the VTC of Fig. P1.57a.

(c) Simulate the circuit via PSpice. Use a 1-kHz triangular wave with peak values of ± 1.1 V, and display both v_T and v_S versus time.

1.58 The crude triangle-to-sine converter of Fig. P1.57b can be improved considerably by *rounding* the sides of the triangular wave input, besides *clipping* it at the top and bottom. The circuit of Fig. P1.58 provides a VTC with a slope of 1 V/V near the origin, where all diodes are off. As the magnitude of v_T rises and approaches a diode drop, either D_1 or D_2 goes on, in effect switching R_2 into the circuit. At this point, the slope of the VTC decreases to about $R_1/(R_1 + R_2)$. As the magnitude of v_T rises further and v_S approaches two diode drops, either the D_3-D_4 pair goes on, clipping the top of the waveform, or the D_5-D_6 pair goes on, clipping the bottom. Let us arbitrarily impose $V_{sm} = 2 \times 0.7 = 1.4$ V, so that $V_{tm} = (\pi/2)1.4 = 2.2$ V, and let us assume diodes with $I_s = 2$ fA and $nV_T = 26$ mV, so that at 0.7 V they draw 1 mA. To find suitable values for R_1 and R_2, arbitrarily impose the following pair of constraints: (1) When v_T reaches its positive peak V_{tm}, let the current through the D_3-D_4 pair be 1 mA; (2) When v_T reaches *half* its positive peak, or $V_{tm}/2$, let the slope of the VTC match that of the sine function there, which one can readily prove to be cos 45°, or 0.707 V/V.

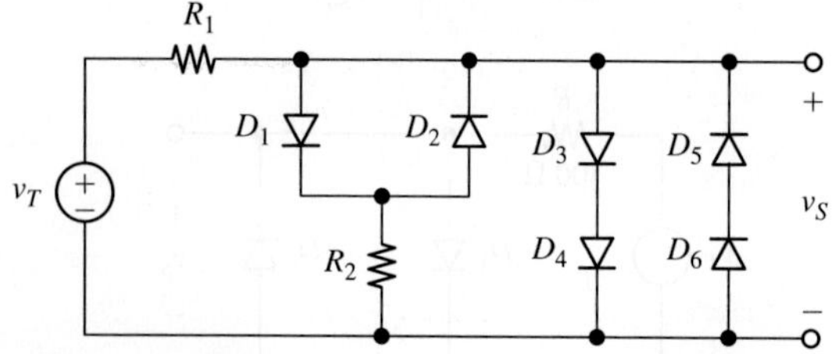

FIGURE P1.58

(*a*) Guided by the above constraints, find suitable values for R_1 and R_2.

(*b*) Simulate the circuit via PSpice. Use a 1-kHz triangular wave with peak values of ± 2.2 V, and display both v_T and v_S versus time. Make multiple runs, each time changing the values of R_1 and R_2 a bit until you come up with a set that gives what you think is the best sine wave.

(*c*) Using the optimized wave shaper of part (*b*) as basis, design a circuit that accepts a triangular wave with peak values of ± 5 V and yields a sine wave also with peak values of ± 5 V.

Hint: at the input, replace R_1 by a suitable voltage divider to accommodate the increased triangular wave while still meeting the aforementioned constraints. At the output, use a suitable amplifier implemented with a 741-type op amp.

1.59 (*a*) For the circuit of Fig. P1.59, sketch and label v_O versus v_I over the range -5 V $\leq v_I \leq 5$ V.

(*b*) Assuming $V_{D(on)} = 0.7$ V, sketch and label v_I, v_O, and v_A versus time if $v_I = (5 \text{ V}) \sin \omega t$.

Hint: use the op amp rule, and convince yourself that at any given time one diode is on and the other is off.

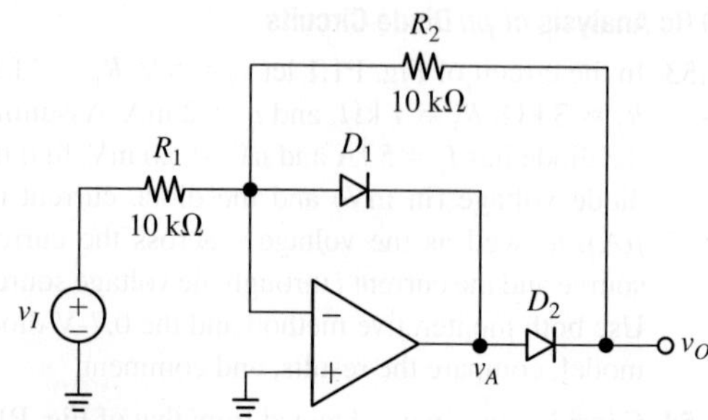

FIGURE P1.59

1.60 Repeat Problem 1.59 if the op amp's non-inverting input pin is lifted off ground and driven by a 2.5-V source.

Hint: use the op amp rule, and investigate separately the case in which the current through R_1 flows toward the right, and the case in which it flows toward the left.

1.61 Repeat Problem 1.59 if both diodes are reversed and R_2 is raised to 20 kΩ.

1.62 Consider the circuit obtained from that of Fig. P1.59 by raising R_2 to 20 kΩ, and by connecting an additional resistance $R_4 = 5$ kΩ between the positive terminal of the source v_I and the output node v_O. Draw the modified circuit, and then sketch and label v_O versus v_I over the range -5 V $\leq v_I \leq 5$ V.

1.63 Analyze the circuit of Fig. P1.63, and show that it gives $v_O = |v_I|$.

Hint: investigate the cases $v_I > 0$ and $v_I < 0$ separately.

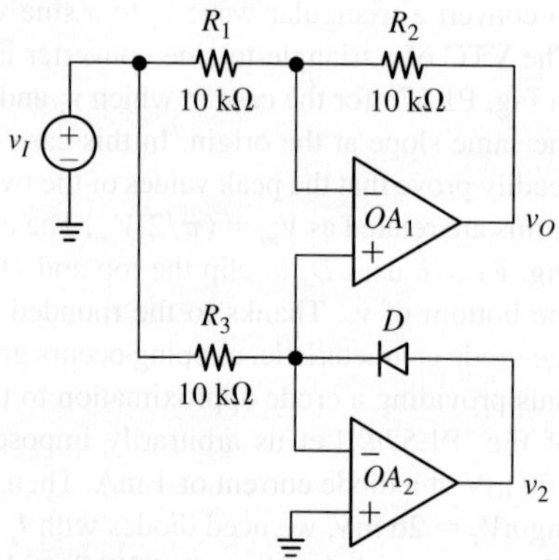

FIGURE P1.63

1.64 Consider the circuit obtained from that of Fig. P1.63 by connecting an additional resistance R_4 between the inverting-input node of OA_1 and ground. Analyze the circuit, and specify suitable values for R_2 and R_4 (while leaving $R_1 = R_3 = 10\ \text{k}\Omega$) so that the circuit gives $v_O = 2|v_I|$.
Hint: investigate the cases $v_I > 0$ and $v_I < 0$ separately.

1.11 Ac Analysis of *pn* Diode Circuits

1.65 The circuit of Fig. P1.65 utilizes a *pn* diode operating in the small-signal mode to implement an *RC low-pass filter* with a -3-dB frequency that is current controlled.
(*a*) Draw the small-signal equivalent circuit, derive its transfer function $H(j\omega)$, and show that its -3 dB frequency is $\omega_0 = I_D/(V_T C)$.
(*b*) Find C so that $\omega_0 = 100$ krad/s at $I_D = 0.1$ mA.
(*c*) Find I_D so that the magnitude of $H(j\omega)$ is -6 dB at $\omega = 50$ krad/s. (*d*) Find I_D so that the phase angle of $H(j\omega)$ is $-30°$ at $\omega = 500$ krad/s.

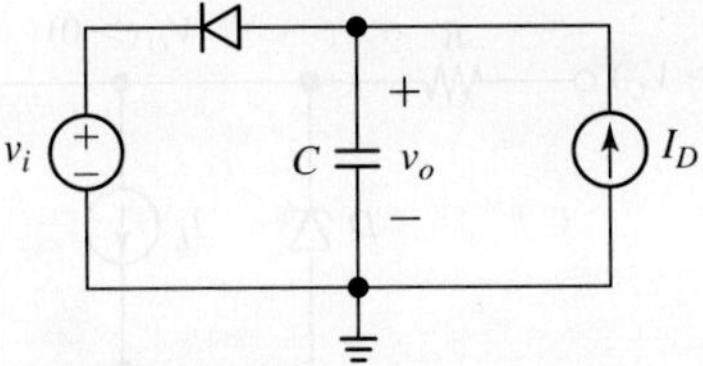

FIGURE P1.65

1.66 The circuit of Fig. P1.66 utilizes a *pn* diode operating in the small-signal mode to implement a *CR high-pass filter* with a -3-dB frequency that is current controlled.

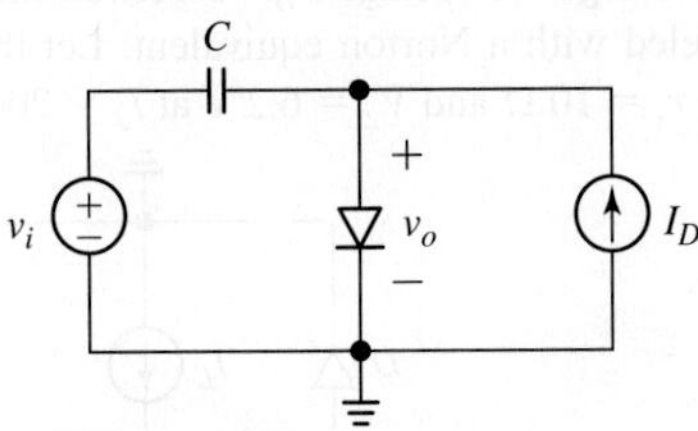

FIGURE P1.66

(*a*) Draw the equivalent small-signal circuit, derive its transfer function $H(j\omega)$, and show that its -3 dB frequency is $\omega_0 = I_D/(V_T C)$.
(*b*) If $C = 33$ nF, what is the value of ω_0 at $I_D = 0.1$ mA?
(*c*) Assuming $v_i = (5\ \text{mV})\cos 10^5 t$, find v_o (amplitude as well as phase) at $I_D = 0.2$ mA. If the

diode used is such at that at 700 mV it gives 1 mA, what is the DC component of the output?
(*d*) Repeat (*c*), but at $I_D = 50\ \mu\text{A}$.

1.67 Shown in Fig. P1.67 is an elegant current-controlled attenuator based on a pair of identical *pn* diodes operating in the small-signal mode.
(*a*) Assuming C is large enough to act as an ac short, draw the equivalent small-signal circuit, and show that its gain is $v_o/v_i = I/I_{REF}$.
(*b*) If $I_{REF} = 1$ mA, find the current I needed for the following gains: 1 V/V, 0.75 V/V, 0.5 V/V, 0.25 V/V, and 0 V/V. What are the corresponding values of the equivalent resistance R_{eq} seen by the capacitor?
(*c*) Specify a suitable value for C so that it acts as an ac short at $\omega = 1$ Mrad/s for *all* possible gain settings.

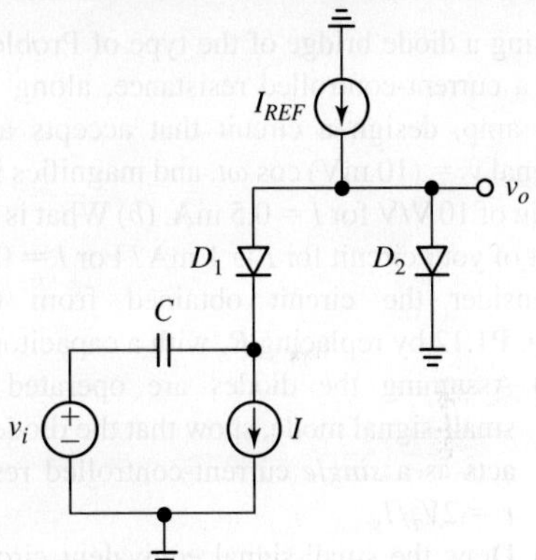

FIGURE P1.67

1.68 In small-signal operation the diode-bridge arrangement of Fig. P1.68 allows for *direct signal coupling* between v_i and v_o, without the need for any ac-coupling capacitor. It also relaxes the small-signal constraint. The increased circuit complexity is well worth these advantages, especially in integrated-circuit implementations, where large capacitances are impractical.
(*a*) Show that in small-signal operation, the diode bridge acts as a *single* current-controlled resistance $r = 2V_T/I$.
(*b*) Show that the small-signal constraint of Eq. (1.82) is now more relaxed and becomes $|v_o| \leq 10$ mV.
(*c*) If $R = 10\ \text{k}\Omega$ and $v_i = (1.0\ \text{V})\cos\omega t$, find I for $v_o = (10\ \text{mV})\cos\omega t$.

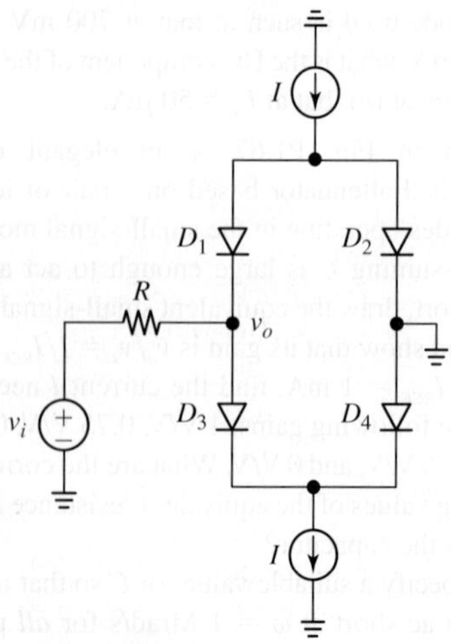

FIGURE P1.68

1.69 Using a diode bridge of the type of Problem 1.68 as a current-controlled resistance, along with an op amp, design a circuit that accepts an input signal $v_i = (10 \text{ mV}) \cos \omega t$, and magnifies it with a gain of 10 V/V for $I = 0.5$ mA. (*b*) What is the output of your circuit for $I = 1$ mA? For $I = 0.1$ mA?

1.70 Consider the circuit obtained from that of Fig. P1.12 by replacing R_L with a capacitor C.

(*a*) Assuming the diodes are operated in the small-signal mode, show that the diode bridge acts as a *single* current-controlled resistance $r = 2V_T/I_S$.

(*b*) Draw the small-signal equivalent circuit, derive its transfer function $H(j\omega)$, and show that its -3 dB frequency is $\omega_0 = I_S/(2V_TC)$.

(*c*) Specify C so that $\omega_0 = 1$ Mrad/s at $I_S = 0.1$ mA.

(*d*) If $v_i = (10 \text{ mV}) \cos 10^6 t$, find v_o (magnitude and phase) for $I_S = 1$ mA and $I_S = 10 \ \mu\text{A}$.

1.12 Breakdown-Region Operation

1.71 In the circuit of Fig. P1.1, let $R_1 = R_2 = 200 \ \Omega$, $R_3 = 300 \ \Omega$, and $i_S = 10$ mA, and let the diode be a Zener diode with $V_{D(on)} = 0.7$ V and $V_Z = 5.6$ V. Assuming r_z is negligible, sketch and label the open-circuit voltage $v_{oc}(t)$ seen by the diode, as well as the diode current waveform $i_D(t)$, if $v_S(t)$ is a 250-Hz triangular wave with peak values of ± 20 V.

1.72 In the circuit of Fig. P1.1 let $R_1 = R_3 = 12$ kΩ, $R_2 = 24$ kΩ, and $v_S = -6$ V, and let the diode be a Zener diode with $V_{D(on)} = 0.7$ V and $V_Z = 10$ V. Assuming r_z is negligible, sketch and label the anode voltage

$v_A(t)$ and the cathode voltage $v_C(t)$, if $i_S(t)$ is a 250-Hz triangular wave with peak values of ± 1 mA.

Hint: replace the circuit to the left of the diode and that to the right with their Thévenin equivalents.

1.73 In the voltage reference of Fig. 1.66*a* let $V_I = 18$ V, $R = 390 \ \Omega$, and $I_L = 10$ mA.

(*a*) Given that the diode exhibits $V_Z = 9.85$ V at $I_Z = 10$ mA and $V_Z = 10.0$ V at $I_Z = 20$ mA, find V_O.

(*b*) If all the above-listed values have a tolerance of $\pm 10\%$, estimate $V_{O(\text{max})}$ and $V_{O(\text{min})}$.

1.74 The voltage-reference circuit of Fig. P1.74 uses a Norton equivalent as a more realistic model for the load. Let the diode have $r_z = 12 \ \Omega$ and $V_Z = 12$ V at $I_Z = 25$ mA.

(*a*) Assuming $V_I = (24 \pm 6)$ V, 2 k$\Omega \leq R_L \leq 5$ kΩ, and $0 \leq I_L \leq 8$ mA, specify a 5% standard resistance R for a guaranteed diode current of 4 mA.

(*b*) What is the line regulation (in mV/V) and the load regulation (in mV/mA) of your circuit?

(*c*) Estimate $V_{O(\text{max})}$ and $V_{O(\text{min})}$.

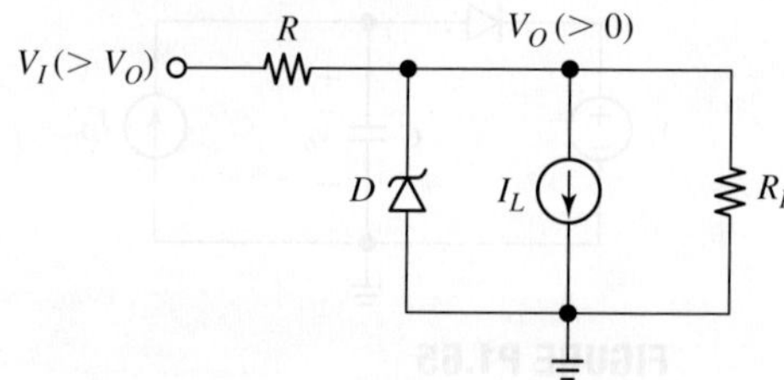

FIGURE P1.74

1.75 The circuit of Fig. P1.75 is a *negative* voltage reference because it accepts a negative voltage V_I to yield a negative voltage V_O. Moreover, the load is modeled with a Norton equivalent. Let the diode have $r_z = 10 \ \Omega$ and $V_Z = 6.2$ V at $I_Z = 20$ mA.

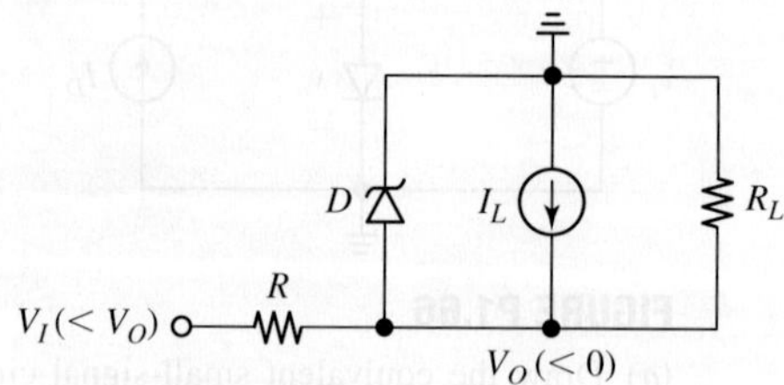

FIGURE P1.75

(*a*) Assuming $V_I = -(15 \pm 3)$ V, 2 k$\Omega \leq R_L \leq 3$ kΩ, and $0 \leq I_L \leq 4$ mA, specify a 5% standard resistance R for a guaranteed diode current of 3 mA.

(b) What is the line regulation (in mV/V) and the load regulation (in mV/mA) of your circuit? Be careful with the polarity of the load regulation!

(c) Estimate $V_{O(\max)}$ and $V_{O(\min)}$.

1.76 The breakdown voltage V_Z varies with temperature and this is undesirable especially in *precision voltage-reference* applications. A popular way to compensate for this variation is to use an avalanche diode, which has TC > 0, in *series* with an ordinary forward-biased diode, which has TC < 0, as depicted in Fig. P1.76. As we know, the latter has TC = −2 mV/°C, so if we use an avalanche diode with TC = +2 mV/°C, the opposing TCs will cancel each other out, resulting in a very stable combined voltage drop. Avalanche diodes with TC = +2 mV/°C fall in the neighborhood of 6.2 V, so the voltage drop of the series combination is about 6.2 + 0.7 = 6.9 V.

(a) Assuming $V_O \cong 6.9$ V, specify R for an avalanche-diode current of 3 mA at $V_I = 12$ V and $I_L = 2$ mA.

(b) Assuming sufficiently small line and load variations to justify the small-signal approximation for D_1, find the line and load regulation (in mV/V and mV/mA) if $r_z = 8$ Ω and $nV_T = 26$ mV.

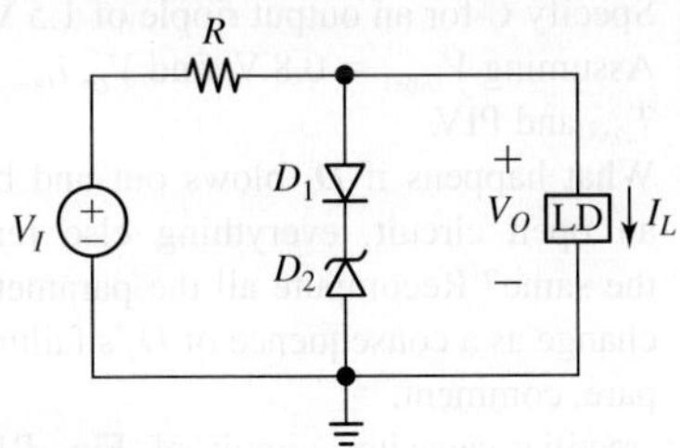

FIGURE P1.76

1.77 Using the series diode combination described in Problem 1.76, along with a pair of 741-type op amps, design a *self-regulated complementary-output voltage reference circuit* giving outputs of +10.0 V and −10.0 V. Impose a bias current of 5 mA for the series diode combination, and assume the availability of ±15 V poorly regulated power supplies to feed your op amps.

Hint: once you have synthesized +10.0 V, you can obtain −10.0 V via an ordinary inverting op amp.

1.78 The line regulation of a reference diode D_2 can be improved dramatically if we precede it by another reference diode D_1, as depicted in Fig. P1.78. Clearly, for the circuit to work properly, V_{Z1} must be sufficiently larger than V_{Z2}, say by at least 30% or so, and the resistances R_1 and R_2 must be small enough to guarantee that each diode remains in the BD region under all possible line and load conditions.

(a) If D_1 is a 10-V Zener diode with $r_{z1} = 15$ Ω, and D_2 is a 6.2-V Zener diode with $r_{z2} = 10$ Ω, find the line and load regulation (in mV/V and mV/mA) if $V_I = 18$ V, $R_1 = 1$ kΩ, $R_2 = 0.75$ kΩ, and $I_L = 2$ mA.

(b) To appreciate the benefits brought about by D_1, suppose we pull D_1 out of the circuit, while leaving everything else the same. How is the line regulation affected? How is the load regulation affected?

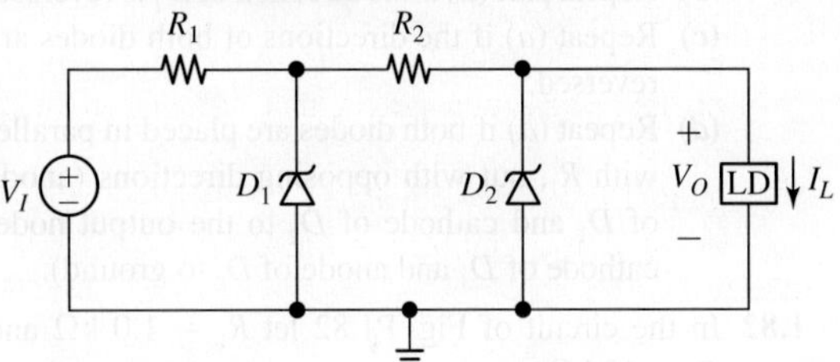

FIGURE P1.78

1.79 As mentioned in Problem 1.78, the function of D_1 in Fig. P1.78 is to improve the line-regulation capabilities of D_2. Moreover, R_1 and R_2 must be small enough to guarantee a predetermined minimum current through each diode under all possible line and load conditions.

(a) Assuming $V_I = (30 \pm 5)$ V, $V_{Z01} = 14.7$ V, $r_{z1} = 15$ Ω, $V_{Z02} = 9.8$ V, $r_{z2} = 10$ Ω, and $0 \le I_L \le 10$ mA, specify 5% standard values for R_1 and R_2 to ensure $I_{Z(\min)} = 5$ mA for each diode.

(b) Estimate the line and load regulation (in mV/V and mV/mA) of your circuit.

1.80 In the circuit of Fig. P1.80 let $R_1 = 20$ kΩ and $R_2 = 10$ kΩ, and let both diodes have $V_Z = 4.3$ V and $V_{D(\text{on})} = 0.7$ V.

(a) Sketch and label the VTC.

(b) Repeat if R_1 is removed from the circuit.

(c) Repeat parts **(a)** and **(b)** if D_1 is replaced with a 6.8-V Zener diode while D_2 remains the same.

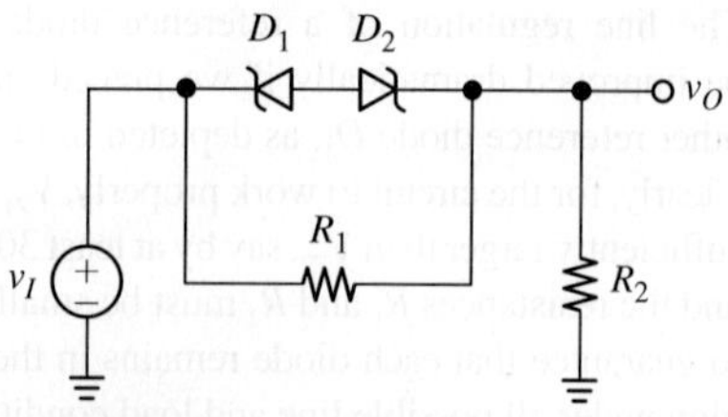

FIGURE P1.80

1.81 Consider the circuit obtained from that of Fig. P1.80 by swapping R_2 with the sub-network made up of D_1, D_2 and R_1, so that now R_2 is between v_I and v_O, and the sub-network is between v_O and ground. Let $R_1 = 30$ kΩ and $R_2 = 10$ kΩ, and let both diodes have $V_Z = 6.8$ V and $V_{D(on)} = 0.7$ V.

   (*a*) Draw the modified circuit and then sketch and label its VTC.

   (*b*) Repeat part (*a*) if the direction of D_1 is reversed.

   (*c*) Repeat (*a*) if the directions of both diodes are reversed.

   (*d*) Repeat (*a*) if both diodes are placed in parallel with R_1, but with opposing directions (anode of D_1 and cathode of D_2 to the output node, cathode of D_1 and anode of D_2 to ground).

1.82 In the circuit of Fig. P1.82 let $R_1 = 1.0$ kΩ and $R_2 = 13$ kΩ.

   (*a*) Assuming the Zener diode has $V_Z = 5.1$ V and all diodes have $V_{D(on)} = 0.7$ V, sketch and label the VTC of the circuit.

   (*b*) Repeat part (*a*) if the Zener diode direction is inadvertently reversed, so that the anode is at the top and the cathode at the bottom.

   (*c*) Repeat (*a*) if a resistance $R_3 = 39$ kΩ is connected in parallel with the Zener diode.

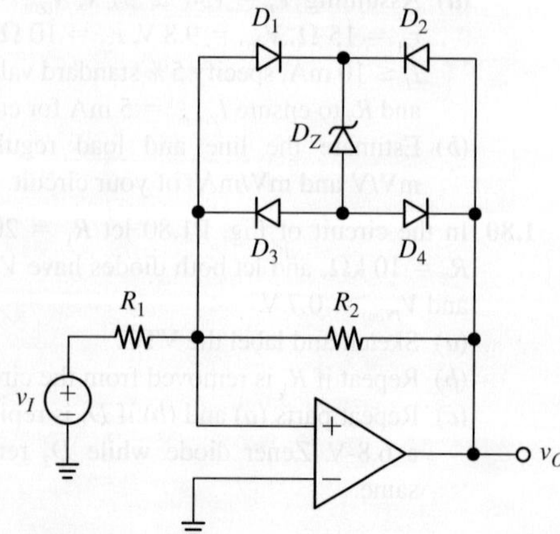

FIGURE P1.82

1.83 Consider the circuit obtained from that of Fig. P1.82 by lifting the op amp's non-inverting pin off ground and driving it with the source v_I, while the left terminal of R_1 is connected to ground, in effect transforming the op amp circuit from the inverting to the non-inverting configuration.

   (*a*) Draw the modified circuit. Then, assuming the Zener diode has $V_Z = 5.1$ V and all diodes have $V_{D(on)} = 0.7$ V, sketch and label the VTC if $R_1 = 1.0$ kΩ and $R_2 = 13$ kΩ.

   (*b*) Repeat if a resistance $R_3 = 39$ kΩ is connected in parallel with the Zener diode.

1.13 Dc Power Supplies

1.84 In the center-tapped full-wave rectifier of Fig. 1.77*a* let v_S be a 60-Hz, 24-V (rms) ac voltage, and let $R = 1$ kΩ.

   (*a*) Specify C for an output ripple of 2 V.

   (*b*) Assuming $V_{D(on)} = 0.8$ V, find V_O, as well as $i_{D(max)}$, $i_{D(avg)}$, T_{ON}, and PIV for each diode.

   (*c*) What happens if D_1 blows out and becomes an open circuit, everything else remaining the same? Recompute all the parameters that change as a consequence of D_1's failure, compare and comment.

1.85 In the diode-bridge full-wave rectifier of Fig. 1.78, let v_S be a 60-Hz, 18-V (rms) ac voltage, and let the load draw 10 mA of current.

   (*a*) Specify C for an output ripple of 1.5 V.

   (*b*) Assuming $V_{D(on)} = 0.8$ V, find V_O, $i_{D(max)}$, $i_{D(avg)}$, T_{ON}, and PIV.

   (*c*) What happens if D_4 blows out and becomes an open circuit, everything else remaining the same? Recompute all the parameters that change as a consequence of D_4's failure, compare, comment.

1.86 The rectifier-capacitor circuit of Fig. P1.86*b* is driven by the asymmetric triangular input of Fig. P1.86*a* and drives a load that has been modeled with a Norton equivalent. The input wave has $V_m = 10$ V, and the load has $I_L = 10$ mA and $R_L = 1$ kΩ.

   (*a*) Assuming $V_{D(on)} = 0.7$ V, specify C for a 0.5-V output ripple.

   (*b*) Sketch and label, versus time, both v_O and the diode current i_D (sketch v_O on the same time diagram as v_I).

   (*c*) Find V_O, $i_{D(max)}$, $i_{D(avg)}$, T_{ON}, and PIV.

Warning: the equations of the text, derived for the case of a *sinusoidal* input and a *purely resistive* load, do not apply to the present case. You need to develop your own through simple reasoning.

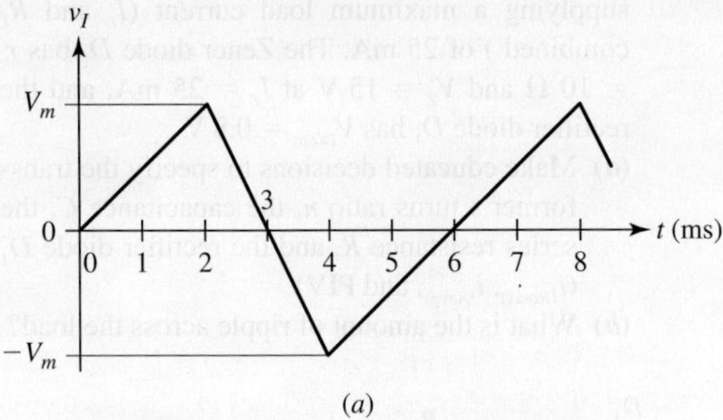

(a)

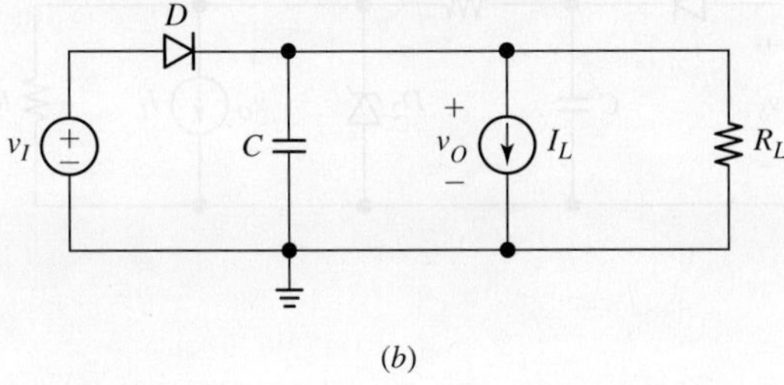

(b)

FIGURE P1.86

1.87 Shown in Fig. P1.87 is a *negative* dc power supply. Let v_I be the asymmetric triangular waveform of Fig. P1.86a with $V_m = 12$ V, and let the load be $I_L = 10$ mA.

(**a**) Assuming the diode has $V_{D(on)} = 0.85$ V, specify C for a 0.5-V output ripple.

(**b**) Sketch and label $v_o(t)$ on the same diagram as $v_I(t)$, and the diode current $i_D(t)$ on a separate diagram.

(**c**) Find V_O, $i_{D(max)}$, $i_{D(avg)}$, T_{ON}, and PIV. Read the warning of Problem 1.86.

1.88 In the circuit of Fig. P1.88, let v_I be a 1-kHz *sawtooth wave* with peak values of ± 24 V, and let $R = 2$ kΩ.

(**a**) Assuming diodes with $V_{D(on)} = 0.75$ V, specify C for a 1-V output ripple.

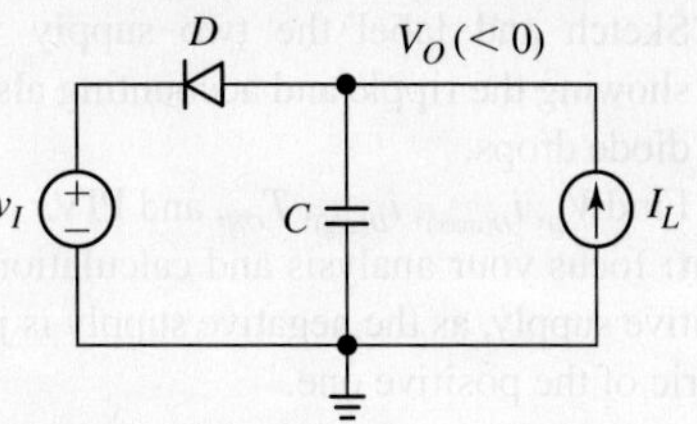

FIGURE P1.87

(**b**) Assuming v_I starts at -24 V at $t = 0$ and increases to reach $+24$ V at $t = 1$ ms, sketch and label, versus time, both v_o and the diode current i_D (sketch v_o on the same time diagram as v_I).

(**c**) Find $i_{D(max)}$, $i_{D(avg)}$, T_{ON}, and PIV. Read the warning of Problem 1.86.

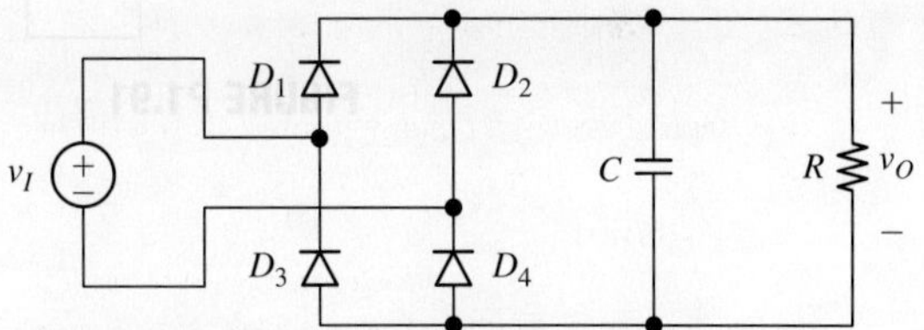

FIGURE P1.88

1.89 Repeat Problem 1.88, but for the case in which v_I is a *symmetric triangular wave* with $f = 1$ kHz and peak values of ± 24 V.

1.90 A student has decided to design and build the *complementary-output dc power supply* of Fig. P1.90 to power a bunch of op amp circuits. The supply is to provide dc outputs of ± 12 V to loads of up to 100 mA each, and with ripples not greater than 0.5 V.

(**a**) Assuming the diodes have $V_{D(on)} = 0.8$ V, find the required amplitude of the sine-wave across the entire secondary winding, as well as the required values of C_1 and C_2.

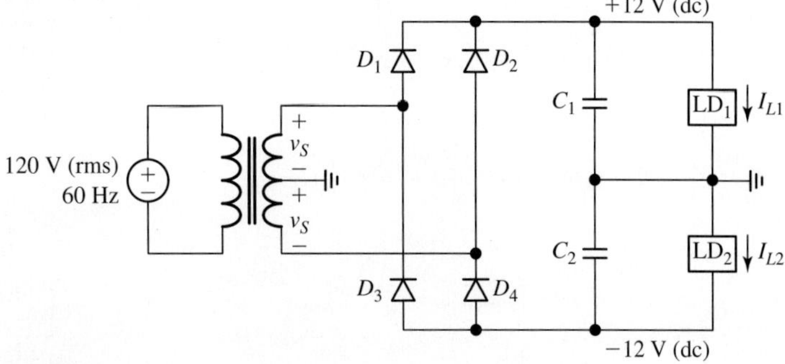

FIGURE P1.90

(*b*) Sketch and label the two supply voltages, showing the ripple and accounting also for the diode drops.

(*c*) Find V_O, $i_{D(\text{max})}$, $i_{D(\text{avg})}$, T_{ON}, and PIV.

Hint: focus your analysis and calculations on the positive supply, as the negative supply is just symmetric of the positive one.

1.91 A student wishes to use the circuit of Fig. P1.91 to design a 15-V voltage reference capable of supplying a maximum load current (I_L and R_L combined) of 25 mA. The Zener diode D_2 has r_z = 10 Ω and V_Z = 15 V at I_Z = 25 mA, and the rectifier diode D_1 has $V_{D(\text{on})}$ = 0.8 V.

(*a*) Make educated decisions to specify the transformer's turns ratio n, the capacitance C, the series resistance R, and the rectifier diode D_1 ($i_{D(\text{max})}$, $i_{D(\text{avg})}$, and PIV).

(*b*) What is the amount of ripple across the load?

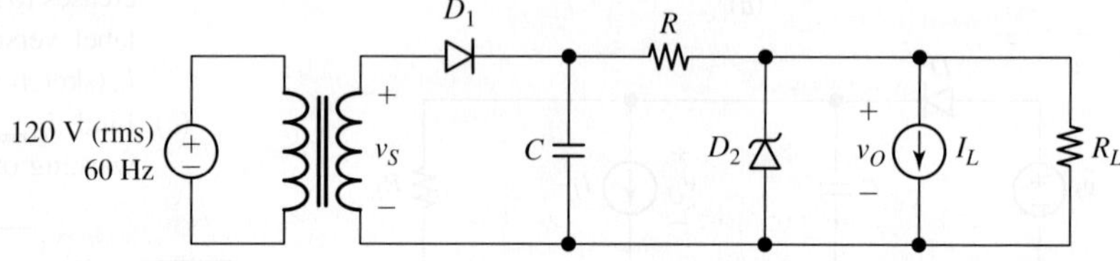

FIGURE P1.91

Chapter 2

Bipolar Junction Transistors

Chapter Outline

There is no doubt that the invention of the vacuum-tube diode paved the way for a number of useful applications that would have been impossible using only the linear devices that we study in introductory circuits courses. However, the horizons of electronics were expanded much further with the invention of a *three-element* vacuum tube called the *triode*. In 1906 Lee DeForest modified Fleming's diode valve, discussed at the beginning of Chapter 1, by inserting a third element between the cathode and the anode, called the *grid*, which he used to *modulate* the electron flow from cathode to anode. In fact, by driving the grid with an audio signal from a microphone, he could get a much stronger signal at the anode (which he renamed *plate*), and he then used this signal to drive a pair of earphones. Aptly called *audion*, this new valve was in effect an *amplifier*, which subsequently he applied to the design of radio signal detectors and oscillators. In fact, the electronics of the next half-century was dominated by vacuum tubes, which in the course of the years were further refined by incorporating additional grids to create first the *tetrode* and finally the *pentode*. During this period, a number of milestone innovations took place, such as the development of radio communications (AM and FM), television, radar, and finally, right after World War II, the first digital computers.

Central to the operation of the triode was the idea of using one of its elements (the grid) to *control* the current flow between the other two (cathode and plate). By a hydraulic analogy, the triode could be viewed as a *valve*. From a circuit viewpoint, it simply implemented the concept of *controlled* (or *dependent*) *source* that we study in introductory circuit courses. The next electronics-era milestone occurred when the triode function was realized on a piece of semiconductor material. Such a device, called the *transistor*, was invented in 1947 by John Bardeen, Walter Brattain, and William Shockley at Bell Labs. The closest counterpart of the triode is what today we refer to as the *npn bipolar junction transistor* (*npn* BJT). It consists of a slab of *n*-type semiconductor material with: (*a*) one side doped very heavily (n^+-type) to act as a copious source of free electrons, and thus called the *emitter*; (*b*) the opposite side, designed to act like the plate of a triode and thus called the *collector*; and (*c*) an extremely thin layer of *p*-type material sandwiched in between and called the *base*, designed to entice electrons from the emitter and direct them in *controlled* fashion to the collector—just like the grid in the triode. What made triodes and then transistors so useful is that *the power being controlled can be much higher than that expended in effecting the control itself*. For this reason, triodes and transistors are said to be *active devices*, and are also referred to as *amplifiers*. Of course, power cannot be created or magnified out of nothing—by active device we simply mean a device that uses little power to control the transfer of a much greater amount of power from an *external energy source* such as a power supply. A classic example is the car radio, where a low-power radio receiver controls, via an audio power amplifier, the *transfer* of a much higher amount of power from the car battery to the loudspeakers. In fact, the very name *transistor* was coined as a contraction of the words *transfer* and *resistor*. In summary, a transistor in itself is a passive device; however, when used in conjunction with an external source of energy, it can be made to act as an active device.

The newly invented transistor was soon put to use as a replacement for the much bulkier, power-hungry, and only moderately reliable vacuum tube. In fact, the first transistor circuits were virtual replicas of vacuum-tube circuit prototypes, if with suitable scaling of the power supplies and surrounding components. The 1950s saw the appearance of the first electronic consumer product using this new device, namely, the hand-held all-transistor radio. Toward the end of the decade it was also realized that the dramatic miniaturization and power savings brought about by the transistor could be exploited further by fabricating entire circuits (transistors, diodes, resistors, and small capacitors, along with their interconnections) *monolithically*, that is, on the same piece of semiconductor material, or *chip*. Called the *integrated circuit* (IC), it was first implemented in 1958 by Jack Kilby at Texas Instruments, and, independently, in 1959 by Robert Noyce at Fairchild Semiconductor. The 1960s saw a feverish activity that resulted in the development, among others, of the first IC *operational amplifier* (IC op amp) by Fairchild Semiconductor (μA Series), as well as the digital IC families known as *transistor-transistor logic* (TTL) by Texas Instruments (7400 Series), and *emitter-coupled logic* (ECL) by Motorola (10K Series).

Meanwhile, in the very early 1970s, another type of transistor known as the *metal-oxide-semiconductor* (MOS) *field-effect transistor* (FET), or MOSFET for short, become a commercial reality. Compared to its BJT predecessor, the MOSFET could be fabricated in an even *smaller size*, and digital MOSFET circuits could be designed to consume practically *zero standby power*. The first battery-powered electronic calculators and wristwatches made precise use of this new technology. Also, a new digital IC family known as *complementary* MOS (or CMOS for short) was introduced by RCA (4000 Series) as a low-power alternative to the bipolar families of the TTL and ECL types. Finally, Intel used MOS technology to develop the first *microprocessor*, in 1971. Since then, IC electronics has advanced *exponentially* and has penetrated virtually every aspect of modern life. This impressive growth has been governed by *Moore's law*, roughly stating that thanks to continued advances in IC fabrication, the number of devices that can be integrated on a given chip area *doubles* approximately every 18 months. Originally formulated in 1965, the law still holds to this day, though it has been pointed out that technology is bound to approach physical limits that will eventually lead to the demise of this law.

The BJT, after having been the dominant semiconductor device for nearly three decades, has been overtaken by the MOSFET especially in digital electronics, thanks to the latter's aforementioned advantages of smaller size and lower power consumption. Nonetheless, the BJT continues to be the device of choice in a number of specialized areas, such as high-performance analog electronics and radiofrequency electronics. It is also preferred in discrete designs, thanks to the availability of a wide selection of devices. BJTs and MOSFETs can also be fabricated simultaneously on the same chip, if at an increase in fabrication steps and thus production cost. The resulting technology, aptly called *biCMOS technology*, exploits the advantages of *both* transistor types to provide even more innovative design opportunities. Contemporary ICs often combine digital as well as analog functions on the same chip, this being the reason for the name *mixed-signal* or also *mixed-mode ICs*.

There is no question that microelectronics is one of the most exciting, challenging, and rapidly evolving fields of human endeavor. The beginner may feel overwhelmed by all this, and rightly so. But, as we embark upon the study of today's dominant processes and devices, we will try to focus on *general principles* that transcend the particular technological milieu of the moment and that we can apply in the future to understand new processes and devices as they become available and commercially mature. Focus on general principles, combined with continuing education, is a necessity for the young engineer determined to establish and maintain a satisfying career in a seemingly ever-changing field.

CHAPTER HIGHLIGHTS

This chapter begins with the physical structure of the BJT, basic underlying semi-conductor principles, device characteristics, operating regions and modeling. As in Chapter 1's treatment of the *pn* junction, emphasis is placed on practical aspects of relevance to today's industrial environment (rules of thumb).

Next, the BJT is investigated in its two most important class of applications, namely, as an *amplifier* for analog electronics, and as a *switch* for digital electronics. The *large-signal* and *small-signal models* developed for the *pn* junction are now expanded to accommodate the more complex BJT.

After a discussion of BJT biasing techniques, the remainder of the chapter investigates the three basic amplifier configurations, namely, the *common-emitter* (CE), *common-collector* (CC), and *common-base* (CB) configurations. The CE configuration is presented as the natural realization *of voltage amplification*, whereas the CC and CB configurations serve most naturally as *voltage* and *current buffers*, respectively. Also, proper emphasis is placed on the role of the BJT as a *resistance transformation device* (which actually provided the basis for its very name.) The transformation equations are conveniently tabulated for easy reference in later chapters.

The amplifiers investigated in this chapter are of the so-called *discrete* type because they can be built using off-the-shelf components (transistors, resistors, and capacitors), and as such they can easily be tried out by the student in the lab. Though nowadays electronic circuits comprise large numbers of transistors fabricated monolithically on the same semiconductor chip, we need to understand the workings of a *single-transistor* amplifier before tackling the complexity of *multi-transistor* ICs, a subject that will be undertaken in Chapter 4. In this respect, a discrete amplifier offers the pedagogical advantage that the transistor is surrounded by the circuit elements most familiar to the student (resistors, and when necessary, capacitors). Also, the separation between dc and ac components is usually more obvious than in more complex monolithic realizations. Lastly, it must be said that when an engineer tests or applies an IC, the need often arises to surround it with ancillary circuitry made up of discrete components (a buffer, a driver, a power booster, etc.). In summary, the objective of this chapter is to introduce the student to the basics of single-transistor amplifiers, and in so doing, to lay a solid foundation for the study of multi-transistor monolithic realizations in later chapters.

The chapter makes abundant use of PSpice both as a software oscilloscope to display BJT characteristics, transfer curves and waveforms, and as a verification tool for dc as well as ac calculations.

2.1 PHYSICAL STRUCTURE OF THE BJT

Figure 2.1 shows, in simplified form, the structure of an *npn bipolar junction transistor* (*npn* BJT) of the type encountered in more traditional integrated-circuit (IC) technology. The device is fabricated through a complex sequence of steps including *pattern definition*, *crystal growth*, *diffusion*, *material deposition* and *material removal*, on a wafer of *lightly-doped p*-type silicon (p^-) called the *substrate*. The wafer provides physical support for the device under consideration as well as all other devices of the same IC.

A BJT is equipped with three terminals, called respectively, the *emitter* (*E*), *base* (*B*), and *collector* (*C*). A fourth terminal (*S*) provides access to the substrate, but serves no active function except for ensuring electric isolation between adjacent devices. Starting out with a polished p^- wafer shown at the bottom, the fabrication of a *npn* BJT proceeds according to the following steps:

- First, some *heavily doped n*-type silicon (n^+) is deposited in the area to be occupied by the BJT, and then is diffused into the wafer to form a low-resistance path known as *buried layer*.

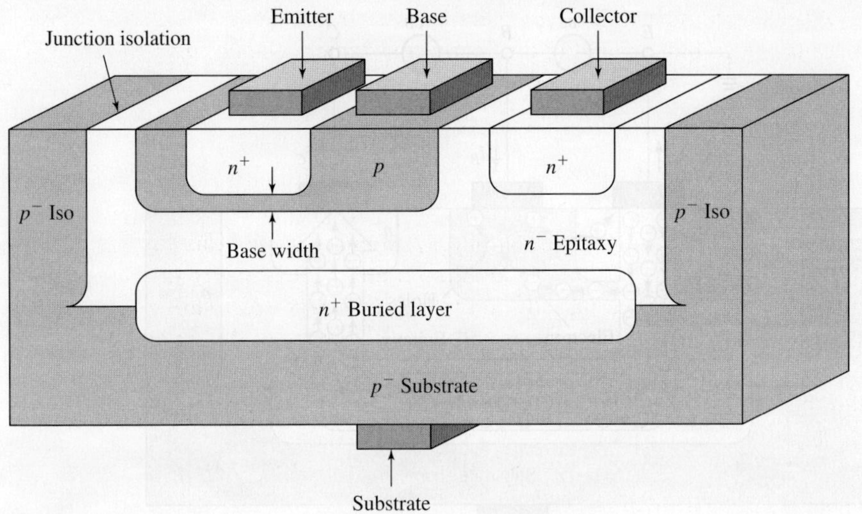

FIGURE 2.1 Basic physical structure of an IC *npn* BJT.

- Next, a crystal layer of *lightly doped n*-type silicon (n^-) is grown on top of the n^+ buried layer and the surrounding p^- wafer area. This layer, known as *epitaxial* (or epi) *layer*, is designed to form the *collector region*.
- A *p*-type diffusion is made into the epi layer along the perimeter of the region destined to be occupied by the BJT, until it joins the *p*-type substrate underneath. As we shall see, this diffusion, referred to as p^- *iso*, is designed to provide *junction isolation* between adjacent BJTs.
- Another but less deep *p*-type diffusion is made into the epi layer to create the *base region* of the BJT.
- Two heavily doped *n*-type diffusions are made simultaneously, one into the *p*-type base region to form the n^+ *emitter region*, and the other into the n^- epi layer to provide what is referred to as an *ohmic contact* between the n^- epi region and the collector electrode. (Nowadays the n^+ collector diffusion is made to extend all the way to the buried layer below.)
- Finally, three metal depositions form the *E*, *B*, and *C electrodes*. An additional connection is made to the substrate (*S*), which, in the case of the *npn* BJT shown, is always kept at the *most negative voltage* (MNV) in the circuit. (The interested student is encouraged to search the Web for videos and articles illustrating the fascinating subject of transistor fabrication.)

Viewing the collector regions as *n*-type material surrounded by *p*-type regions, we identify a distributed *pn* junction of sorts. Biasing the p^- substrate (and thus the outer p^- iso regions) at the MNV ensures that this distributed junction is at all times *reverse biased*. Aside for usually negligible leakage currents, the collectors of adjacent BJTs are thus kept *isolated* from each other, allowing for different BJTs to operate in electrically independent fashion.

We identify two basic ingredients in a BJT: (*a*) the *pn* junction formed by the *base-emitter* (B-E) regions, and (*b*) the *pn* junction formed by the *base-collector* (B-C)

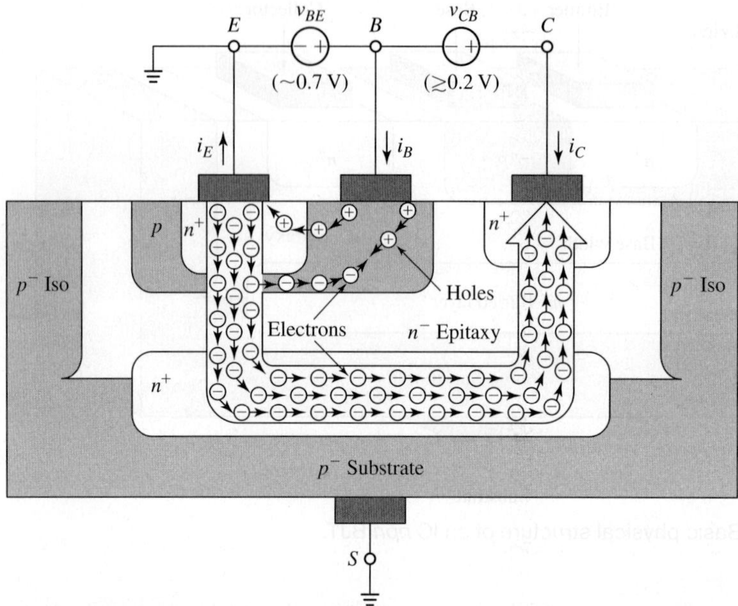

FIGURE 2.2 Currents in a monolithic *npn* BJT operating in the *forward active mode.*

regions. The most common mode of operation of a BJT is with the B-E junction *forward biased* and the B-C junction *reverse biased*. This mode, which is at the basis of BJT applications as an amplifier, is called the *forward active* (FA) *mode*. Briefly stated, the main event taking place in an *npn* BJT operating in the FA mode is

> A current of electrons flowing from the emitter, through the base, to the collector, under control by the base-emitter voltage drop v_{BE}.

The situation is illustrated in Fig. 2.2, where we note that since electrons are *negative*, the directions of the terminal currents i_E and i_C are *opposite* to the direction of electron flow. Additionally, we make the two following observations:

- Not all of the electrons emitted from the emitter are collected by the collector. As they transit through the *p*-type base region, a small fraction *recombines* with the many holes present there, indicating the existence of a base current component of holes to sustain this recombination process. By fabricating the **base region very thin** (fractions of 1 μm) the manufacturer ensures that the recombination component is suitably small.

- As is the case with all *pn* junctions, the diffusion of electrons from *E* to *B* is accompanied by a diffusion of holes from *B* to *E*. By **doping the emitter much more heavily than the base** (typically by two or more orders of magnitude), the manufacturer ensures that the inrush of electrons from the emitter literally dwarfs the diffusion of holes from the base, thus keeping this second base-current component also suitably small.

The reasons for desiring a small base current i_B for a given collector current i_C is that the ratio

$$\beta_F = \frac{i_C}{i_B} \tag{2.1}$$

which is aptly called the ***forward current gain***, represents a figure of merit of the BJT. It also provides us with additional flexibility in that we can operate the BJT either as a *voltage-controlled current source* (i_C controlled by v_{BE}) or as a *current-controlled current source* (i_C controlled by i_B). Integrated-circuit BJTs have typically $\beta_F \cong 250$, though it is possible to fabricate special BJTs, known as *superbeta* BJTs, with β_F as high as 10,000. As we proceed, we shall re-examine BJT behavior in much greater detail.

pnp BJTs

A *pnp* BJT is obtained by *negating* the doping type of each region of an *npn* BJT so that the emitter region is now p^+, the base is n, and the collector is p^-. This indeed is the case of *discrete pnp* BJTs, that is, devices that are fabricated and packaged individually. However, in conventional bipolar IC technology *pnp* BJTs must be fabricated using the same processing steps as *npn* BJTs. In this technology two *pnp* BJT structures are in common use: (*a*) the ***lateral pnp BJT*** and (*b*) the ***substrate*** or ***vertical pnp BJT*** (see Fig. 2.3). We make the following observations:

- In both *pnp* types the n^- *epi layer*, which served as the collector region of the *npn* BJT, is now put to use as the *base* region (the function of the n^+ diffusion is only to ensure an ohmic contact to the external terminal B).
- In both *pnp* types the p-type *diffusion*, which served as the base region of the *npn* BJT, is now put to use as the *emitter* region, whose function is to act as a source of holes.
- In the *lateral pnp* BJT the role of the collector is played by another p-type diffusion, fabricated in the guise of a ring surrounding the emitter (as viewed from above). As shown, holes flow *laterally* out of the centralized p-type emitter region and into the surrounding p-type collector ring.
- In the *substrate pnp* BJT the role of the collector is played by the p^- substrate, so holes now flow *vertically* from the emitter (fabricated in the guise of a ring surrounding the n^+ contact region), through the n^- epi layer, to the p^- substrate.
- Unlike lateral BJTs, vertical BJTs are constrained to have their collector (formed by the substrate) connected to the MNV in the circuit.
- In either *pnp* structure the base region is *not as thin* as that of the *npn* structure, indicating a higher base-current component of electrons recombining with the holes in transit through this wider base.
- Since the p-type emitters are *not as heavily doped* as the n^+ emitter of the *npn* structure, the injection of holes from emitter to base will not dwarf that of electrons from base to emitter as dramatically.

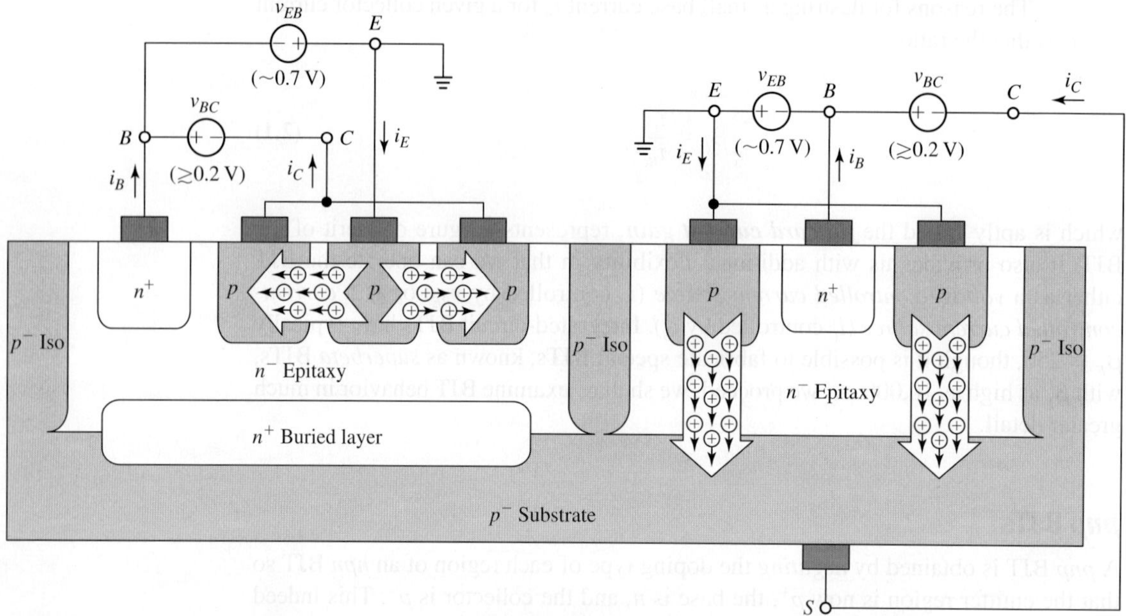

FIGURE 2.3 Physical structures and currents of planar-process *pnp* BJTs: the *lateral pnp* (left) and the *substrate* or *vertical pnp* (right).

- For these and other reasons, lateral *pnp* BJTs offer lower current gains than their *npn* counterparts (typically $\beta_F \approx 50$). The fabrication process has been optimized for the fabrication of *npn* devices, so *pnp* BJTs come out as second-rate devices. But, if no such constraints are imposed, *pnp* BJTs can indeed be fabricated with comparable performance characteristics as their *npn* counterparts, if at the price of additional processing steps and therefore higher cost.

The fabrication process just discussed, aptly referred to as *junction-isolated, double-diffusion bipolar planar process*, is just one of various processes in use today. In high-speed applications, processes using *dielectric isolation* are preferable as they are exempt from the parasitic capacitance provided by the isolating junction. Also, in BiCMOS technology, BJTs are fabricated via processes compatible with the fabrication of MOSFETs. Finally, it is possible to fabricate both *npn* and *pnp* BJTs with comparable high-quality performance, if at the price of increased processing complexity and cost.

Circuit Symbols for the BJTs

Figure 2.4 shows the circuit symbols used for the two BJT types, along with the current directions and voltage polarities. In either case, the terminal with the *arrow* is by definition the *emitter*, and the arrow itself indicates emitter-current direction. In the *pnp* BJT, where the main current is due to holes, the arrow coincides with the direction of hole flow. However, in the *npn* BJT, where the main current is due to

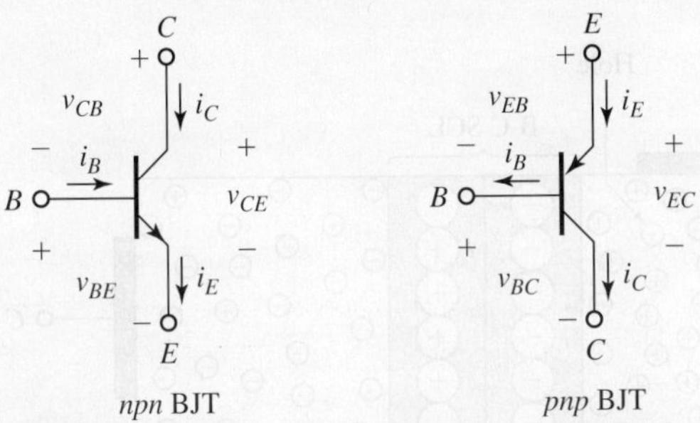

FIGURE 2.4 BJT symbols, showing the correct *current directions* and *voltage polarities*. Note that the two devices have *opposite current directions* as well as *opposite voltage polarities*. To signify opposite current directions, we simply reverse the current arrows (for instance, i_C flows *into* the *npn* BJT, but *out* of the *pnp* BJT). To signify opposite voltage polarities, we simply interchange the order of the subscripts in voltage drops. In so doing, we avoid having to work with negative voltage drops. For instance, we use $v_{BE} > 0$ to turn on the *npn* BJT, and $v_{EB} > 0$ (rather than $v_{BE} < 0$) to turn on the *pnp* BJT.

electrons, which are negative, the emitter arrow is *opposite* to the direction of internal electron flow.

Regardless of the device type and operating mode, a BJT must at all times satisfy KCL,

$$i_E = i_C + i_B \tag{2.2}$$

When analyzing a BJT circuit, it pays to draw a circle around the device and regard it as a *supernode*.

2.2 BASIC BJT OPERATION

Figure 2.5, top, shows a vertical slice of the emitter-base-collector structure of Fig. 2.1, but rotated counterclockwise by 90° to facilitate analysis. To develop a feel for the various physical quantities involved, we shall assume the following doping densities:

$$\text{Emitter } (n^+): \quad N_{DE} = 10^{20} \text{ donor atoms/cm}^3 \tag{2.3a}$$

$$\text{Base } (p): \quad N_{AB} = 10^{18} \text{ acceptor atoms/cm}^3 \tag{2.3b}$$

$$\text{Collector } (n^-): \quad N_{DC} = 10^{16} \text{ donor atoms/cm}^3 \tag{2.3c}$$

At room temperature, virtually all dopant atoms are ionized, so the electron concentrations in the emitter and collector regions are, respectively, $n_{E0} \cong N_{DE}$ and $n_{C0} \cong N_{DC}$, and the hole concentration in the base region is $p_{B0} \cong N_{AB}$. Also referred to as *majority-carrier concentrations*, they are thus

$$n_{E0} \cong 10^{20} \text{ electrons/cm}^3 \quad p_{B0} \cong 10^{18} \text{ holes/cm}^3 \quad n_{C0} \cong 10^{16} \text{ electrons/cm}^{-3} \tag{2.4}$$

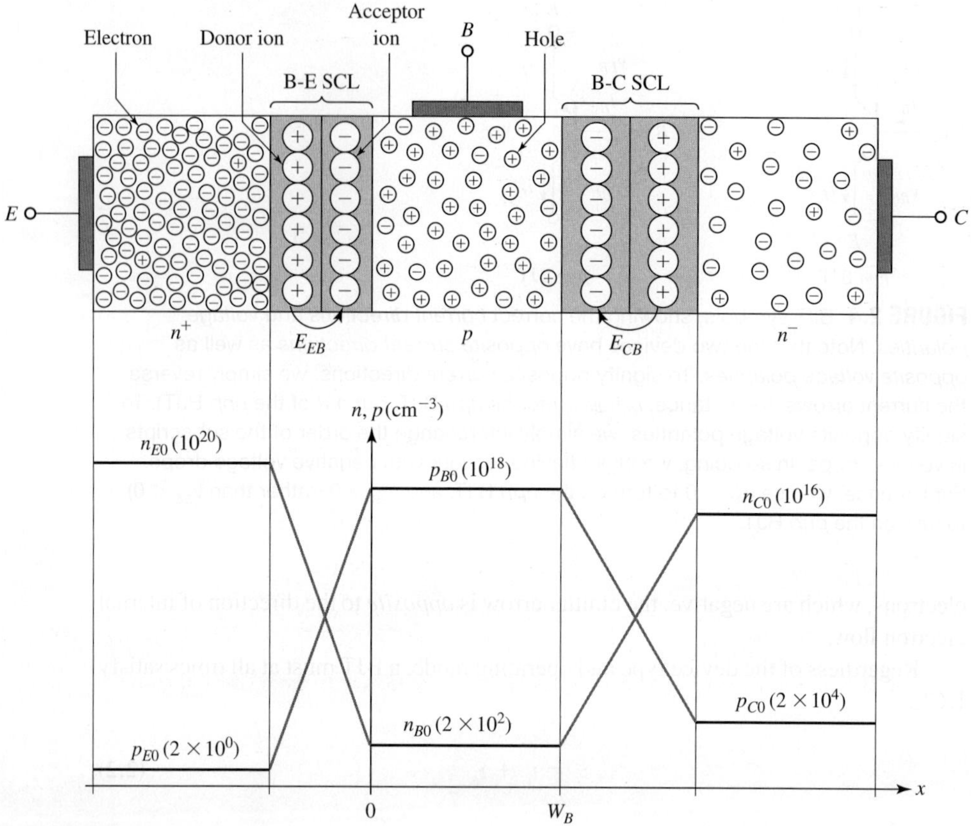

FIGURE 2.5 Equilibrium electron and hole concentrations inside an *npn* BJT.

In each region, the electron and hole concentrations n and p satisfy the *mass-action law*,

$$n \times p = n_i^2$$

where n_i is the *intrinsic concentration* of holes and electrons in silicon. This concentration is a strong function of temperature, and for silicon it takes on the form

$$n_i^2(T) = 1.5 \times 10^{33} T^3 e^{-14028/T}\ \mathrm{cm}^{-6} \tag{2.5}$$

At room temperature ($T = 300$ K), Eq. (2.5) gives $n_i^2 = 2 \times 10^{20}\,\mathrm{cm}^{-6}$. The *minority-carrier concentrations* are thus found as $p_{E0} = n_i^2/n_{E0}$, $n_{B0} = n_i^2/p_{B0}$, and $p_{C0} = n_i^2/n_{C0}$. For the above doping densities, they are

$$p_{E0} \cong 2 \times 10^{0}\ \text{holes/cm}^3 \quad n_{B0} \cong 2 \times 10^{2}\ \text{electrons/cm}^{-3} \quad p_{C0} \cong 2 \times 10^{4}\ \text{holes/cm}^{-3} \tag{2.6}$$

Both the majority and minority concentrations are diagrammed (not to scale) in Fig. 2.5, bottom.

During fabrication itself, electrons from the electron-rich emitter region will spontaneously diffuse into the adjacent electron-starved base region, leaving behind positively charged donor ions. Likewise, holes from the base region will diffuse to the adjacent emitter region, leaving behind negatively charged acceptor ions. Both ion types are bound to fixed positions of the crystal lattice, and end up forming a *space-charge layer* (SCL) at each side of the B-E metallurgical junction. Inside the SCL there is an electric field E_{EB} directed from the positive ions to the negative ions. A similar situation occurs at the B-C junction, where the electric field is denoted as E_{CB}. In *equilibrium*, the strengths of both fields are such as to exactly counterbalance the tendency by electrons to diffuse from the emitter and collector regions to the base region, and by holes to diffuse from the base to the emitter and collector regions.

Operation in the Forward Active (FA) Mode

The BJT reveals its unique abilities when we perturb the above equilibrium by applying suitable voltages across its junctions. As mentioned, the most useful mode of operation is the *forward active* (FA) *mode*, which occurs when we *forward-bias* the B-E junction and *reverse bias* the B-C junction. Forward-biasing the B-E junction with a voltage v_{BE} will reduce the potential barrier that keeps electrons and holes from diffusing across the junction. We are particularly interested in the situation right at the base edge of the B-E SCL, which has been taken as the origin of the x-axis in Fig. 2.5. By the *law of the junction* discussed in connection with Eq. (1.51), the effect of applying v_{BE} is to cause the electron concentration $n_B(0)$ at the SCL's edge to shoot up from its equilibrium value n_{B0} ($\cong 2 \times 10^2$ cm^{-3}) to the new value

$$n_B(0) = n_{B0}e^{v_{BE}/V_T} \tag{2.7}$$

where $V_T = kT/q$ is the *thermal voltage* ($V_T = 25.9$ mV $\cong 26$ mV at $T = 300$ K). To get an idea, with the typical value $v_{BE} = 700$ mV, $n_B(0)$ shoots up from $n_{B0} \cong 2 \times 10^2$ cm^{-3} to

$$n_B(0) \cong (2 \times 10^2)e^{700/26} \cong (2 \times 10^2) \times 5 \times 10^{11} = 10^{14} \text{ cm}^{-3}$$

This is quite an increase! However, since we still have $n_B(0) \ll p_{B0}$ (as $10^{14} \ll 10^{18}$), the majority of mobile charges there continue to be holes. We refer to this situation as *low-level injection* (in this case, injection of electrons from the emitter).

Reverse-biasing the B-C junction with a voltage $v_{BC} \le 0.2$ V or, equivalently, with a voltage $v_{CB} \ge 0.2$ V, will, once again by the law of the junction, change the existing electron concentration at the base-edge of the B-C SCL to the new value

$$n_B(W_B) = n_{B0}e^{v_{BC}/V_T} = n_{B0}e^{-v_{CB}/V_T} \tag{2.8}$$

where W_B denotes the *effective base width*, defined as the distance between the base edges of the two SCLs. With a reverse bias of as little as 0.2 V, Eq. (2.8) gives $n_B(W_B) \cong (2 \times 10^2)e^{-200/26} \cong (2 \times 10^2)/2191 \cong 0.1$ cm^{-3}, which for practical purposes can be regarded as 0 compared to other concentrations. Given that the base is fabricated deliberately very thin, the electron distribution in the base takes on a *straight-line* profile, as shown. Were the base made wide, the distribution profile would be an exponential decay, indicating that most electrons would recombine in the base and thus fail to make it successfully to the collector. We can see why it is critical that the base be fabricated very thin (typically a fraction of a μm).

The Collector Current

It is apparent that operating the BJT in the forward-active mode, as depicted in Fig. 2.6, establishes an *excess* of minority carriers (electrons, in the present case) in its base region. This excess electron distribution, defined as

$$n_B'(x) = n_B(x) - n_{B0}$$

is shown in Fig. 2.7, along with its values at $x = 0$ and $x = W_B$. Once injected into the base, the excess electrons diffuse towards the SCL of the B-C junction, where the electric field E_{CB} sweeps them away from the base, through the SCL, and into the collector region, from where they proceed toward the collector electrode (see Fig. 2.6). Electron diffusion within the base region is governed by the diffusion equation

$$J_n = qD_n \frac{dn_B'(x)}{dx} \tag{2.9}$$

where J_n is the electron *current density* (in A/cm^2), q is the electron charge, and D_n is the electron *diffusion constant*, also called *electron diffusivity* (in cm^2/s). Calculating the slope of the triangle, we get

$$J_n = qD_n\left[-\frac{n_{B0}(e^{v_{BE}/V_T} - 1) - (-n_{B0})}{W_B}\right] = -\frac{qD_n n_{B0}}{W_B}e^{v_{BE}/V_T}$$

where the negative sign indicates that the direction of J_n is *opposite* to x, that is, from right to left. This is not surprising, as J_n is due to the flow of electrons, which are *negative*.

The collector current i_C (in A) is obtained by multiplying the current density J_n by the *emitter area* A_E (in cm^2) shared by the solid-state emitter diffusion with the base region in Fig. 2.1. Choosing the direction of i_C from right to left as in Fig. 2.6 allows us to write $i_C = -A_E J_n$, and thus avoid the negative sign of J_n. Using also $n_{B0} = n_i^2/p_{B0} \cong n_i^2/N_{AB}$, we put i_C in the insightful form

$$i_C = I_s e^{v_{BE}/V_T} \tag{2.10}$$

where

$$I_s = A_E \times \frac{1}{W_B} \times n_i^2(T) \times \frac{qD_n}{N_{AB}} \tag{2.11}$$

Called the *collector saturation current*, I_s is just a *scale factor* giving a measure of how much collector current i_C a BJT will provide for a given voltage drive v_{BE}. For

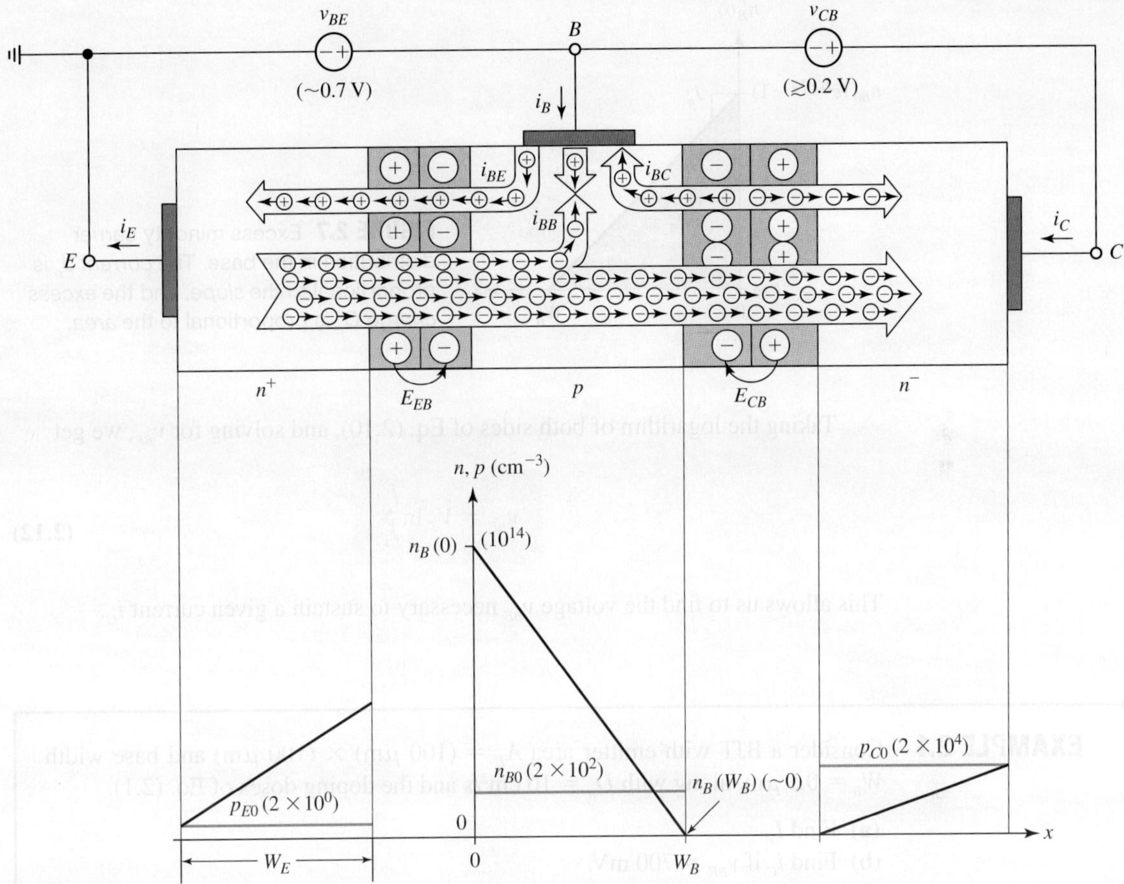

FIGURE 2.6 The *npn* BJT operating in the *forward active* mode: *top*: relevant currents; *bottom*: minority-carrier distributions.

low-power BJTs, I_s is typically in the range of femto-amperes (1 fA = 10^{-15} A.) We make the following important observations:

- I_s is directly proportional to the *emitter area*. The larger the emitter, the more current the collector will draw for a given drive v_{BE}. Indeed, power BJTs have suitably large emitters.
- I_s is inversely proportional to the *base width* W_B. The narrower the base, the steeper the distribution of Fig. 2.7 for a given drive v_{BE}. Also, this dependence upon W_B is at the basis of the Early effect, to be discussed in Section 2.3.
- I_s is a strong *function of temperature*, especially because of $n_i^2(T)$. Engineers quantify temperature dependence via the following useful rule of thumb:

The collector saturation current I_s doubles for about every 5°C of temperature increase

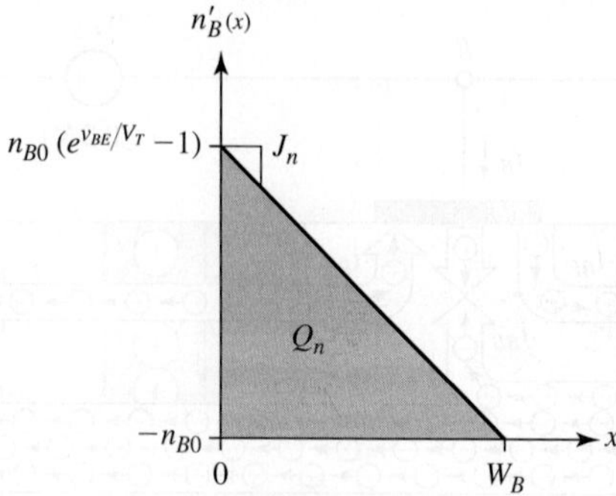

FIGURE 2.7 Excess minority carrier distribution in the base. The *current* J_n is proportional to the *slope*, and the *excess charge* Q_n is proportional to the *area*.

Taking the logarithm of both sides of Eq. (2.10), and solving for v_{BE}, we get

$$v_{BE} = V_T \ln\frac{i_C}{I_s} \tag{2.12}$$

This allows us to find the voltage v_{BE} necessary to sustain a given current i_C.

EXAMPLE 2.1   Consider a BJT with emitter area $A_E = (100\ \mu m) \times (100\ \mu m)$ and base width $W_B = 0.5\ \mu m$, along with $D_n = 10\ cm^2/s$ and the doping doses of Eq. (2.1).

(a) Find I_s.
(b) Find i_C if $v_{BE} = 700$ mV.
(c) Find v_{BE} for $i_C = 1.0$ mA.
(d) Find the square area A_E needed to achieve $i_C = 1$ mA with $v_{BE} = 700$ mV.

Solution

(a) By Eq. (2.11),

$$I_s = \frac{100 \times 10^{-4} \times 100 \times 10^{-4} \times 2 \times 10^{20} \times 1.602 \times 10^{-19} \times 10}{0.5 \times 10^{-4} \times 10^{18}}$$

$$= 0.64\ \text{fA}$$

(b) By Eq. (2.10),

$$i_C = 0.64 \times 10^{-15}e^{700/26} = 0.316\ \text{mA}$$

(c) By Eq. (2.12),

$$v_{BE} = 0.026 \ln\frac{1 \times 10^{-3}}{0.64 \times 10^{-15}} = 730\ \text{mV}$$

(d) From part (*b*) we find that A_E needs to be scaled in proportion as

$$A_E = \frac{1.0\ \text{mA}}{0.316\ \text{mA}}(100\ \mu m) \times (100\ \mu m) = (178\ \mu m) \times (178\ \mu m)$$

The Three Base-Current Components

With reference to Fig. 2.6, top, we observe that the base current consists of three components, denoted as i_{BE}, i_{BB}, and i_{BC}. Let us examine each one in detail.

- The component i_{BE} consists of **holes injected into the emitter.** This component is the *counterpart* of the electrons injected into the base, so we can adapt Eqs. (2.10) and (2.11) to the case of holes injected into the emitter and write

$$i_{BE} = \frac{A_E n_i^2(T) q D_p}{W_E N_{DE}} e^{v_{BE}/V_T} \tag{2.13}$$

where D_p is the *hole diffusivity* and W_E the *emitter width*, also referred to as the *emitter length.*

- The component i_{BB} consists of **holes recombining with the electrons transiting from emitter to collector.** To develop an expression for this component, we observe that in the forward active mode the BJT sustains a cloud of excess electrons in its base region. Call its total charge Q_n. If a transiting electron takes on average τ_n seconds to recombine with a hole in the base region, then the excess charge lost to recombination in one second is Q_n/τ_n, and its negative is precisely the hole current i_{BB} needed to replenish the holes lost to recombination.

  To obtain an expression for Q_n, refer to Fig. 2.8 and consider a vertical slice of thickness dx. The *volume* of this slice is $A_E dx$. To find the excess charge $dQ_n(x)$ inside this slice, first multiply its volume by $n_B'(x)$ to find the *number* of excess electrons, and then multiply by $-q$ to find the *charge* itself, or $dQ_n(x) = -n_B'(x) q A_E dx$. The *total excess charge* within the base region is then found by integrating over the length of this region,

$$Q_n = -qA_E \int_0^{W_B} n_B'(x)\,dx = -qA_E \frac{W_B[n_{B0}(e^{v_{BE}/V_T} - 1) - (-n_{B0})]}{2} = -\frac{qA_E n_i^2(T) W_B}{2 N_{AE}} e^{v_{BE}/V_T}$$

where we have used simple geometry to find the area of the triangle, and have also substituted $n_{B0} = n_i^2/N_{AE}$. Letting $i_{BB} = -Q_n/\tau_n$, we finally get

$$i_{BB} = \frac{qA_E n_i^2(T) W_B}{2 \tau_n N_{AE}} e^{v_{BE}/V_T} \tag{2.14}$$

where τ_n is called the *mean electron lifetime* in the base.

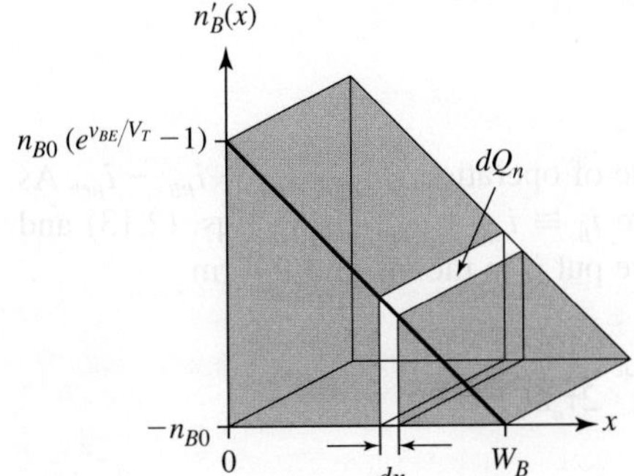

FIGURE 2.8 Calculating the *excess charge* Q_n in the base region.

- The component i_{BC} accounts for the **_thermal generation of electron-hole pairs within the SCL_** of the reverse-biased B-C junction. Once generated, holes and electrons are swept into *opposite directions* by the strong electric field E_{CB} present there. Depending on the quality of fabrication, surface leakage may also be present. BJT data sheets usually report I_{CB0}, the C-B *leakage current* with the *emitter open*. At room temperature, I_{CB0} is typically in the range of 1 nA to 1 pA, and since it is so small, it is usually ignored in the course of hand calculations. However, this current is a strong function of temperature, something that engineers quantify via the following rule of thumb:

> The leakage current I_{CB0} doubles for about every 10°C of temperature increase

Consequently, if sufficiently high operating temperatures are anticipated, it may prove necessary to take the leakage current into account even in the course of our hand analysis.

Looking back at Fig. 2.6 we observe that all base-current components consist of *holes*. Yet, the base electrode, usually made of metal, conducts only by *electron* flow. It follows that in order to ensure the continuity between the two current types, *free electron-hole pairs must be generated automatically right at the metal-base interface*. Once generated, holes progress into the base region to sustain $i_{BE} + i_{BB}$, while electrons drift up the base wire to sustain i_B.

A fraction of the holes needed to sustain $i_{BE} + i_{BB}$ comes also from the collector, in the form of i_{BC}. As mentioned, i_{BC} increases with temperature. It is worth mentioning that we can increase i_{BC} also by shining light upon the B-C SCL. As the quanta of light impinge upon the crystal, they impact enough energy to create free hole-electron pairs, which are then swept into opposite directions by the electric field present there. With sufficient light, the current of holes from collector to base can be raised to the point of turning the BJT convincingly on, with no need for any externally supplied current i_B! When used as a light-controlled device, the BJT is called a *phototransistor* and finds application as part of an *optocoupler*. An optocoupler consists of a light-emitting diode (LED) and a phototransistor mounted in the same package. Forcing current through the LED causes it to emit light, which then turns on the BJT. Since the LED and the BJT are coupled only via light, they can be part of separate circuits and thus provide *galvanic isolation* between the two. We can also couple a LED and a BJT via a fiber optic, thus allowing for the transmission of information over long distances and with minimal signal loss and interference.

The Forward Current Gain β_F

The base current in the forward active mode of operation is $i_B = i_{BE} + i_{BB} - i_{BC}$. As mentioned, we usually ignore i_{BC} and write $i_B \cong i_{BE} + i_{BB}$. Using Eqs. (2.13) and (2.14), along with Eqs. (2.10) and (2.11), we put i_B in the insightful form

$$i_B = \left(\frac{D_p}{D_n} \frac{N_{AB}}{N_{DE}} \frac{W_B}{W_E} + \frac{W_B^2}{2\tau_n D_n} \right) I_s e^{v_{BE}/V_T}$$

Comparing with Eq. (2.10), we observe that i_B is *linearly proportional* to i_C, a relationship expressed as

$$i_B = \frac{i_C}{\beta_F}$$

where β_F is the *forward current gain* mentioned earlier,

$$\beta_F = \frac{1}{\dfrac{D_p}{D_n}\dfrac{N_{AB}}{N_{DE}}\dfrac{W_B}{W_E} + \dfrac{W_B^2}{2\tau_n D_n}} \tag{2.15}$$

This expression confirms the already familiar criteria for the fabrication of high-beta BJTs:

- Make W_B small by fabricating the *base very thin*
- Make the ratio N_{DE}/N_{AB} large by doping the emitter much more *heavily* than the base.

Equation (2.15) reveals two additional features that the manufacturer can exploit to maximize β_F: (*a*) fabricate the emitter *long* ($W_E \gg W_B$) to further reduce i_{BE}; (*b*) create favorable conditions for a long *minority-carrier lifetime* in the base (long τ_n for *npn* BJTs, and long τ_p for *pnp* BJTs) to lower i_{BB} further.

EXAMPLE 2.2

Assuming $\beta_F = 100$ in the BJT of Fig. 2.9*a*, find all terminal currents (*a*) first for the case in which the collector terminal is left disconnected, and (*b*) then with the collector connected to the +5-V supply, as shown. Discuss the various current components as well as the main differences between the two cases.

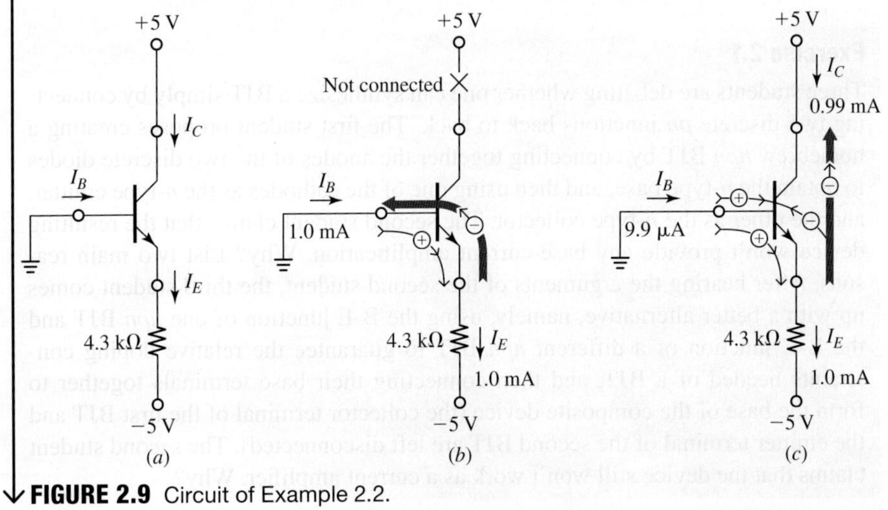

FIGURE 2.9 Circuit of Example 2.2.

Solution

(a) Leaving the collector disconnected results in $I_C = 0$. The only functioning part of the BJT is now its B-E junction, which acts as a plain diode with the base as the anode and the emitter as the cathode. Assuming a typical junction voltage-drop of 0.7 V, the emitter is at -0.7 V, so the emitter current is

$$I_E = \frac{-0.7 - (-5)}{4.3} = 1.0 \text{ mA}$$

Since $I_C = 0$ (collector disconnected), we must have $I_B = I_E = 1.0$ mA. The situation is depicted in Fig. 2.9b, where we observe that both I_E and I_B consist primarily of electrons injected from the emitter into the base region and thence progressing to the base terminal. Additionally, there is a current of holes injected from the base into the emitter, as it is the case with all *pn* junctions, but this component is much smaller because of the much heavier emitter doping. In this situation the BJT works as an ordinary *pn* diode.

(b) Connecting the collector to the $+5$-V supply reverse biases the B-C junction, making it now possible for the electrons injected into the *thin* base to proceed to the *positively* biased collector. A few will recombine with the many holes in the base, but the majority will now make it to the collector. We now have

$$I_B = \frac{I_C}{\beta_F} = \frac{I_E - I_B}{\beta_F}$$

or

$$I_B = \frac{I_E}{\beta_F + 1}$$

Presently, $I_B = 1.0/101 = 9.9\ \mu$A and $I_C = 100 \times 9.9 = 0.99$ mA. The situation is shown in Fig. 2.9c, where the main event now is electron flow from E to C.

Exercise 2.1

Three students are debating whether one can synthesize a BJT simply by connecting two discrete *pn* junctions back to back. The first student proposes creating a homebrew *npn* BJT by connecting together the anodes of the two discrete diodes to obtain the *p*-type base, and then using one of the cathodes as the *n*-type emitter, and the other as the *n*-type collector. The second student claims that the resulting device won't provide any base-current amplification. Why? List two main reasons. After hearing the arguments of the second student, the third student comes up with a better alternative, namely, using the B-E junction of one *npn* BJT and the B-C junction of a different *npn* BJT to guarantee the relative doping constraints needed of a BJT, and then connecting their base terminals together to form the base of the composite device (the collector terminal of the first BJT and the emitter terminal of the second BJT are left disconnected). The second student claims that the device still won't work as a current amplifier. Why?

EXAMPLE 2.3

(a) Assuming Eq. (2.15) gives $\beta_F = 1/(1/150 + 1/300) = 100$ for the BJT of Fig. 2.9c, find I_B as well as the base current components I_{BE} and I_{BB}.

(b) What happens if the manufacturer reduces W_B in half? What if the manufacturer doubles W_B?

Solution

(a) We easily find

$$I_B = I_{BE} + I_{BB} = \frac{I_C}{\beta_F} = \left(\frac{1}{150} + \frac{1}{300}\right)0.99 \text{ mA} = 6.6 \ \mu A + 3.3 \ \mu A = 9.9 \ \mu A$$

(b) Considering that the denominator terms in Eq. (2.15) involve both W_B and W_B^2, it is apparent that *halving* W_B gives

$$\beta_F = \frac{1}{\dfrac{1}{150}(0.5) + \dfrac{1}{300}(0.5)^2} = \frac{1}{\dfrac{1}{300} + \dfrac{1}{1200}} = 240$$

Retracing similar calculations, we still get $I_E = 1.0$ mA. However, we now have $I_B = 1.0/241 = 4.15 \ \mu A$, $I_C = 0.996$ mA, $I_{BE} = 3.32 \ \mu A$, and $I_{BB} = 0.83 \ \mu A$. Doubling W_B gives $\beta_F = 1/(1/75 + 1/75) = 37.5$, $I_B = 1.0/38.5 = 26 \ \mu A$, $I_C = 0.974$ mA, $I_{BE} = I_{BB} = 13 \ \mu A$.

A Practical Application: The BJT as a Current Booster

To start appreciating the usefulness of the BJT, let us examine what could easily be a hobbyist's project: the design of a 5-V, 200-mA regulated power supply starting from an unregulated voltage source such as a 12-V car battery. Based on the knowledge acquired so far, we could start out with the design of Fig. 2.10a. Here, the LM336-5.0 diode provides a high-quality 5-V voltage reference, which we then buffer to the load

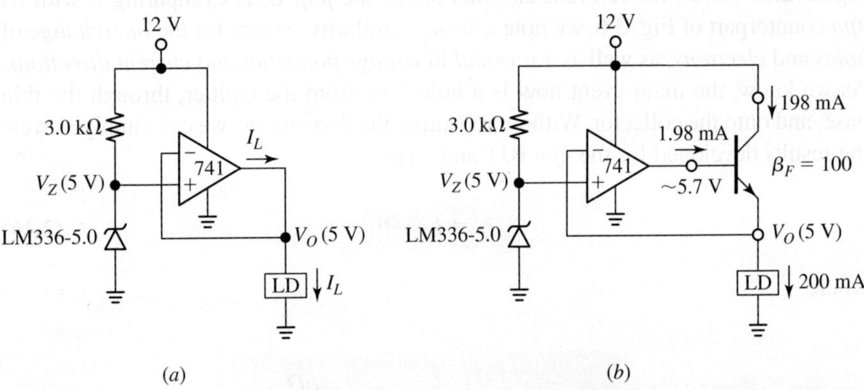

FIGURE 2.10 Showing the use of a BJT to increase the output current drive capability of an op amp.

via a 741 op amp connected as a voltage follower. This circuit will work properly but only for load currents of up to about 25 mA, which the data-sheets report as the *maximum* output-current capability of the 741 op amp. If the circuitry that we intend to power, here denoted as a load LD, attempts to draw a current I_L exceeding this value, the output voltage will simply drop, and regulation will be lost.

But this is precisely where the BJT comes to our rescue! If we interpose a BJT between the op amp and the load as in Fig. 2.10b, the op amp's output current will get boosted by $\beta_F + 1$, as seen in Example 2.2, thus causing a much heftier current to flow from the unregulated 12-V source, through the BJT, to the load. Circuit behavior is governed by the familiar *op amp rule*, stating that the op amp will provide whatever output voltage and current it takes to make its inverting input voltage (V_O) in this case) track its non-inverting input voltage (V_Z). In our case the op amp must source to the BJT the current

$$I_B = \frac{I_E}{\beta_F + 1} = \frac{200 \text{ mA}}{100 + 1} = 1.98 \text{ mA}$$

which is well within the 25-mA capability of the 741 op amp. Assuming a typical B-E junction voltage drop of about 0.7 V, we observe that the op amp must also supply the base voltage

$$V_B = V_{BE} + V_E \cong 0.7 + 5.0 = 5.7 \text{ V}$$

But this too is well within the 741's output voltage capability.

In the application just illustrated the BJT is put to use to *boost* the output current drive capability of a low-power device such as an op amp. In this function, the BJT finds application in a wide variety of power-related circuits, of which regulated power supplies and audio power amplifiers are two common examples. Even though voltage regulators are available in IC form, it is instructive for the beginner to assemble the circuits of Fig. 2.10 and try them out in the lab.

The *pnp* BJT Operating in the Forward-Active Mode

Figure 2.11 shows the relevant currents inside the *pnp* BJT. Comparing it with its *npn* counterpart of Fig. 2.6, we note a strong similarity, except for the *interchange* of *holes* and *electrons*, as well as a *reversal* in *voltage polarities* and *current directions*. As we know, the main event now is a hole flow from the emitter, through the thin base, and onto the collector. Without repeating the derivations, we can simply recycle the results developed for the *npn* BJT and write

$$i_C = I_s e^{v_{EB}/V_T} \tag{2.16}$$

where

$$I_s = A_E \times \frac{1}{W_B} \times n_i^2 (T) \times \frac{qD_p}{N_{DB}} \tag{2.17}$$

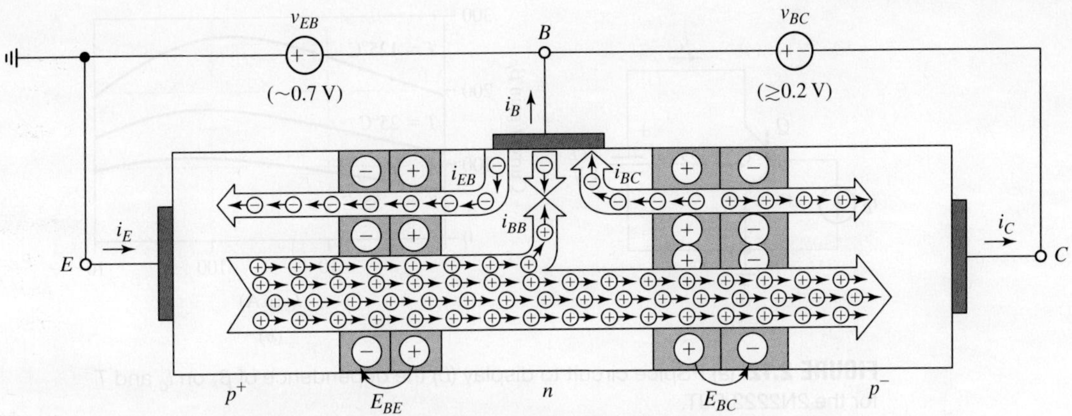

FIGURE 2.11 Relevant currents in a *pnp* BJT operating in the *forward active mode*.

Here, D_p is the hole diffusivity and N_{DB} the donor concentration in the base. More-over, we have

$$\beta_F = \cfrac{1}{\cfrac{D_n}{D_p}\cfrac{N_{DB}}{N_{AE}}\cfrac{W_B}{W_E} + \cfrac{W_B^2}{2\tau_p D_p}} \tag{2.18}$$

where D_n is the electron diffusivity, N_{DB} is the donor concentration in the base, N_{AE} is the acceptor concentration in the emitter, and τ_p is the *mean hole lifetime* in the base.

Equations (2.11) and (2.17) reveal an additional interesting feature, namely, that the scale factor I_s is proportional to the *diffusivity* (D_n or D_p) of the charges producing the main current in the BJT. This is not surprising, as the collector current is of the *diffusion* type. When studying MOSFETs, in the next chapter, we will encounter a similar scale factor, called the transconductance parameter k, which instead is pro-portional to the *mobility* (μ_n or μ_p) of the charges responsible for the main current in the device. This is so because MOSFET current is of the *drift* type. Figure 1.37 shows that for a given doping density, electron mobility and diffusivity are two to three times higher than hole mobility and diffusivity, respectively. For these reasons, *npn* BJTs are generally preferred over *pnp* types, and *n*-channel FETs are preferred over *p*-channel types.

Dependence of β_F Upon I_C and T

The current gain β_F is not constant but varies both with the operating current I_C and the temperature T. This is illustrated in Fig. 2.12 for the case of the popular 2N2222 BJT. For a fixed value of T, say $T = 25\ °C$, β_F is approximately constant only in the neighborhood of $I_C = 100\ \mu A$ for this particular device. At higher current lev-els β_F decreases due to current crowding as well as high-level injection effects.[1] At lower current levels it decreases due to carrier recombination inside the B-E SCL.[1]

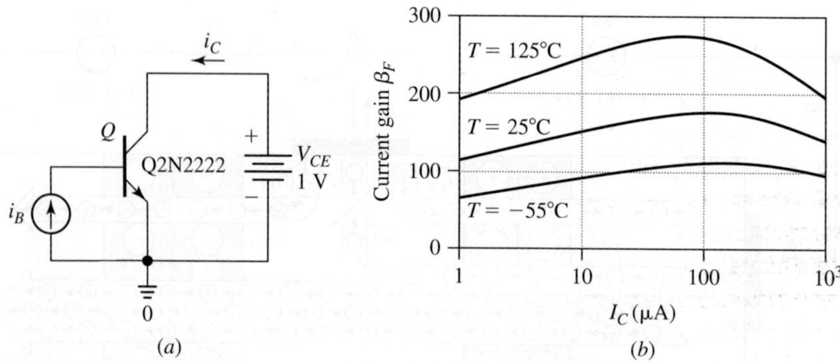

FIGURE 2.12 (a) PSpice circuit to display (b) the dependence of β_F on I_C and T for the 2N2222 BJT.

This results in an additional base-current component that, though always present, becomes relevant only at low current levels. Unless stated to the contrary, we shall assume constant β_F throughout for simplicity.

2.3 THE *i-v* CHARACTERISTICS OF BJTS

The most important characteristics of an *npn* BJT are the forward active plot of i_C versus v_{BE}, and the plot of i_C versus v_{CE} for different B-E driving conditions. Since the B-E junction can be either voltage driven or current driven, we have two families of curves, one obtained by plotting i_C versus v_{CE} for *different values* of V_{BE}, and the other by plotting i_C versus v_{CE} for or *different values* of I_B. Each family offers valuable insight into BJT operation. It is understood that the body of knowledge that we shall acquire for the *npn* BJT can readily be extended to its *pnp* counterpart by *interchanging holes* and *electrons* as well as *reversing voltage polarities* and *current directions*. The various *i-v* curves can be displayed either in the lab, via an oscilloscope equipped with a suitable curve-tracer module, or on a computer monitor via PSpice. In the following we shall use the popular 2N2222 *npn* BJT as a working example. The reader is encouraged to perform a web search for the data sheets of popular BJTs such as the 2N2222 type, and consult them frequently to develop a feel for the practical side of the theory we are going to study.

The Exponential Characteristic

Figure 2.13*a* shows a PSpice circuit to display the i_C-v_{BE} characteristic of the 2N2222 *npn* BJT using the model available in PSpice's library (See Appendix 2A). The result, shown in Fig. 2.13*b*, is the familiar exponential curve predicted by Eq. (2.10), albeit with a minor modification to be discussed below, in connection with Eq. (2.21). The PSpice model uses $I_s = 14.34$ fA. Recall that the B-E junction acts like an ordinary *pn* diode, except that almost all electrons injected into the base go to the collector

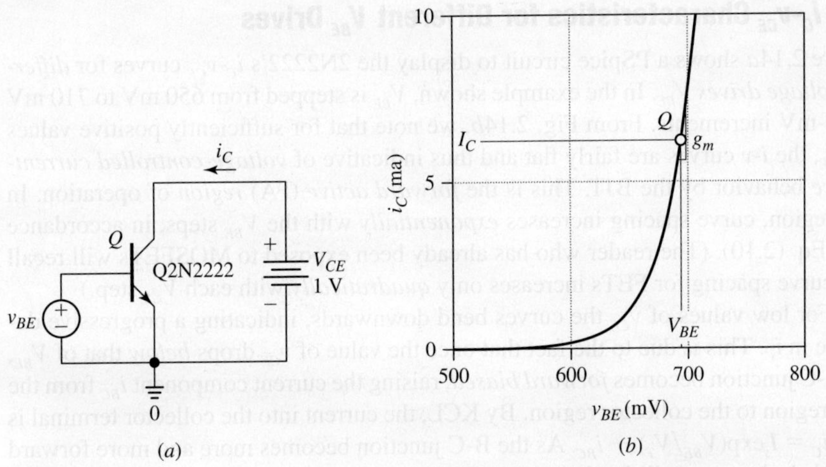

FIGURE 2.13 Using PSpice to display the exponential *i-v* curve of the 2N2222 BJT.

rather than to the base terminal. Consequently, all properties exhibited by the *pn* junction's *i-v* characteristic hold also for the BJT's i_C-v_{BE} characteristic. In particular, the following rules of thumb hold also for BJTs:

- To effect an *octave* change in i_C we need to change v_{BE} by 18 mV (**18-mV Rule**)
- To effect a *decade* change in i_C we need to change v_{BE} by 60 mV (**60-mV Rule**)
- The voltage drop V_{BE} exhibits a *temperature coefficient* of about -2 mV/°C (**−2-mV Rule**)

The *slope* of the i_C-v_{BE} curve at a particular operating point $Q = Q(I_C, V_{BE})$ is denoted as g_m and is called the *transconductance* (in A/V),

$$g_m = \left.\frac{\partial i_C}{\partial v_{BE}}\right|_Q \tag{2.19}$$

Differentiating Eq. (2.10), we readily find

$$g_m = \frac{I_C}{V_T} \tag{2.20}$$

which allows us to calculate the BJT's transconductance at any bias current I_C. To get a feel, at $I_C = 1$ mA we get $g_m = 38$ mA/V, which is often expressed in the alternative form $g_m = 1/(26\ \Omega)$.

(The reader who has already been exposed to MOSFETs may have noted a similarity with the MOSFET's transconductance expressed in the form $g_m = I_D/(0.5V_{OV})$, where V_{OV} is the overdrive voltage needed to sustain a given drain current I_D. Typically $0.5V_{OV} \gg V_T$, so for the same bias current a FET will generally exhibit much lower transconductance than a BJT. This is a notorious drawback of FETs compared to BJTs, especially in the design of high-gain amplifiers.)

The i_C-v_{CE} Characteristics for Different V_{BE} Drives

Figure 2.14a shows a PSpice circuit to display the 2N2222's i_C-v_{CE} curves for *different voltage drives* V_{BE}. In the example shown, V_{BE} is stepped from 650 mV to 710 mV in 10-mV increments. From Fig. 2.14b, we note that for sufficiently positive values of v_{CE}, the *i*-*v* curves are fairly flat and thus indicative of *voltage-controlled current-source* behavior by the BJT. This is the *forward active* (FA) *region* of operation. In this region, curve spacing increases *exponentially* with the V_{BE} steps, in accordance with Eq. (2.10). (The reader who has already been exposed to MOSFETs will recall that curve spacing for FETs increases only *quadratically* with each V_{GS} step.)

For low values of v_{CE} the curves bend downwards, indicating a progressive decrease in i_C. This is due to the fact that once the value of v_{CE} drops *below* that of V_{BE}, the B-C junction becomes *forward biased*, raising the current component i_{BC} from the base region to the collector region. By KCL, the current into the collector terminal is now $i_C = I_s\exp(V_{BE}/V_T) - i_{BC}$. As the B-C junction becomes more and more forward biased, i_C decreases till its drops to zero when the two terms cancel each other out. This occurs for v_{CE} near to 0 V, though not necessarily exactly 0 V, as the saturation currents of the two junctions are generally not identical in value. When *both* its junctions are forward biased, the BJT is said to be operating in the *saturation* (Sat) *mode*. The reason for this name will become apparent as we proceed.

The Early Effect

If we display the i_C-v_{CE} curves on a more compressed horizontal scale as in Fig. 2.15, we note that the *slope* of the curves in the active region increases progressively with v_{BE}. Moreover, the extrapolations of all curves meet at a common point[4] located at $v_{CE} = -V_A$. Called the *Early voltage* for James M. Early, who first investigated this phenomenon, V_A is typically in the range of 10 V to 100 V. The value used in the PSpice model of the 2N2222 BJT is $V_A > 75$ V.

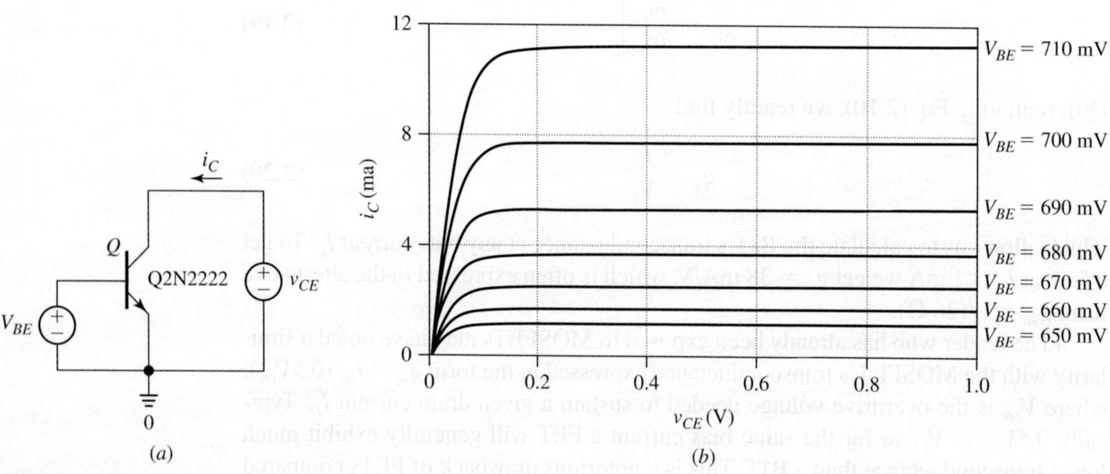

FIGURE 2.14   Using PSpice to display the 2N2222's i_C-v_{CE} curves for different V_{BE} values and $0 \leq v_{CE} \leq 1$ V.

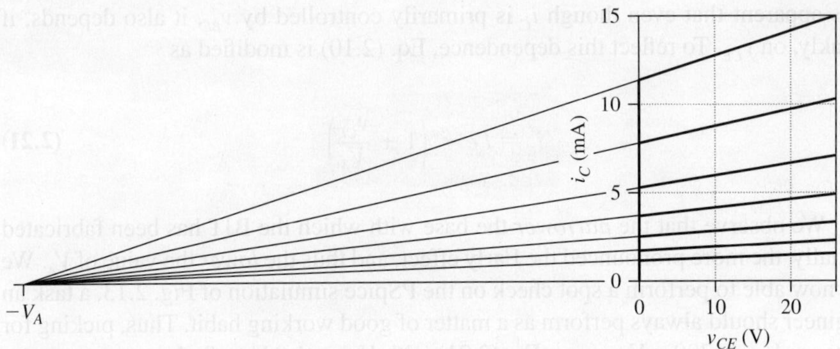

FIGURE 2.15 Illustrating the Early effect and the Early voltage V_A.

The slant of the i_C-v_{CE} curves is due to the fact that the effective base width W_B *shrinks* with *increasing* v_{CE}, a phenomenon aptly referred to as *base-width modulation* or also as the *Early effect*. To appreciate it, suppose the BJT is initially biased at some voltage $v_{CE} = V_{CE1}$ in the active region (see Fig. 2.16a). Let the corresponding base width be W_{B1}, the electron current density be J_{n1} (see Fig. 2.16b), and the collector current be I_{C1} ($=J_{n1}A_E$). If we now increase v_{CE} to the new value V_{CE2}, the amount of reverse bias across the B-C junction will also increase, indicating a corresponding increase in the electric field E_{CB} inside its space-charge layer (SCL). Referring back to Fig. 2.5, we can state that the increase in the number of field lines can only come at the price of a *widening* of the SCL, so as to uncover more ions. Consequently, the base width will *shrink* to a new value W_{B2} ($<W_{B1}$), as depicted in Fig. 2.16b. But, with a narrower base width, the slope of $n_B'(x)$ increases, ultimately raising the current density to the new value J_{n2} ($>J_{n1}$), and, hence, the collector current to I_{C2} ($>I_{C1}$). The chain of cause-effect relationships is summarized as follows:

$$(\text{increasing } v_{CE}) \Rightarrow (\text{widens the B-C SCL}) \Rightarrow (\text{shrinks } W_B)$$
$$\Rightarrow (\text{increases the slope of } n_B') \Rightarrow (\text{increases } J_n)$$

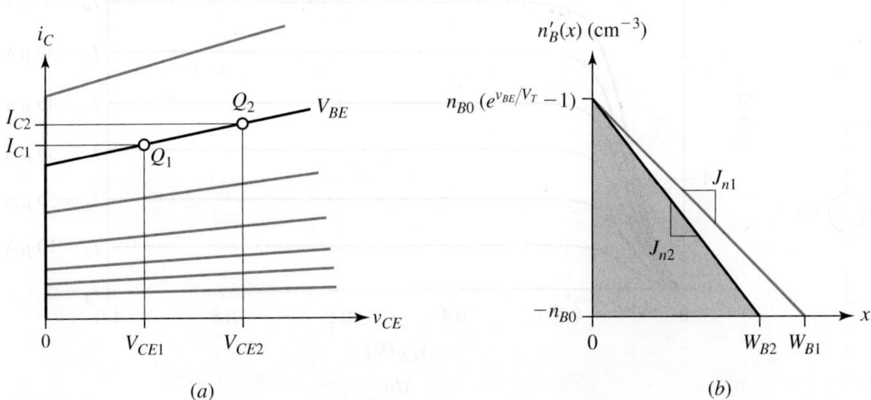

(a)   (b)

FIGURE 2.16 Illustrating the effects of base-width modulation.

It is apparent that even though i_C is primarily controlled by v_{BE}, it also depends, if weakly, on v_{CE}. To reflect this dependence, Eq. (2.10) is modified as

$$i_C = I_s e^{v_{BE}/V_T}\left(1 + \frac{v_{CE}}{V_A}\right)$$

$$(2.21)$$

We observe that the *narrower* the base with which the BJT has been fabricated initially, the more pronounced the Early effect, and thus the *lower* the value of V_A. We are now able to perform a spot check on the PSpice simulation of Fig. 2.13, a task an engineer should always perform as a matter of good working habit. Thus, picking for instance $V_{BE} = 700$ mV, we use Eq. (2.21) with $V_{CE} = 1.0$ V to find

$$I_C = 14.34 \times 10^{-15} e^{700/25.9}\left(1 + \frac{1}{75}\right) \cong 7.9 \text{ mA}$$

which agrees with the plotted value of Fig. 2.13b.

The i_C-v_{CE} Characteristics for Different I_B Drives

Given that the BJT lends itself to both voltage and current control, an alternative way of characterizing it is by displaying its i_C-v_{CE} curves for *different current drives I_B*. In the example of Fig. 2.17, I_B is stepped from 0 to 60 μA in 10-μA increments. For $I_B = 0$ we have $I_C = 0$, and the BJT is said to be *cut off* (CO). We note that curve spacing in the forward active region is now much more uniform. Also, as in the voltage-driven case, the curves are a bit slanted because of the Early effect. An advantage of this type of characteristics is that they allow us to find β_F at a given active-region operating point Q_F by mere inspection. For instance, consider the point Q_F lying right on the $I_B = 60$ μA curve at $V_{CE} = 0.8$ V. A visual readout of the collector current

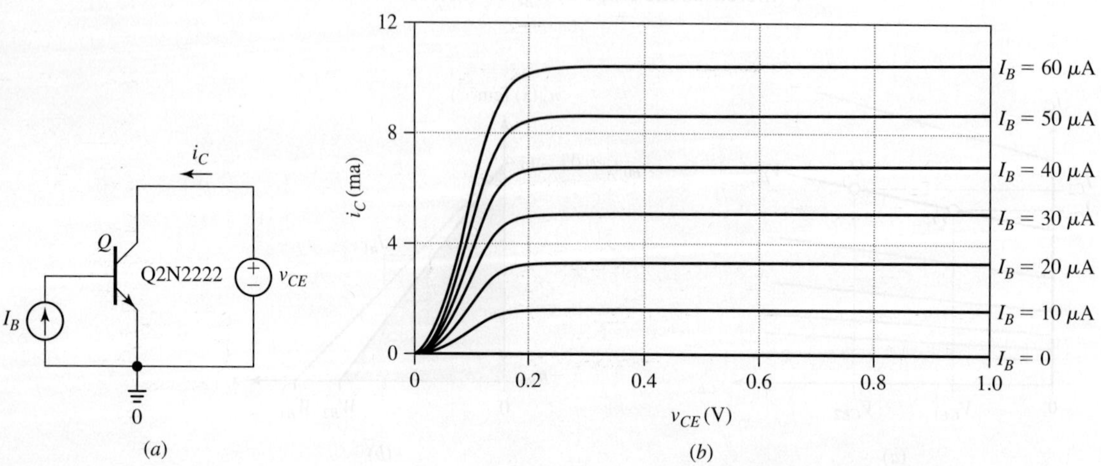

FIGURE 2.17 Using PSpice to display the 2N2222's i_C-v_{CE} curves for different I_B values and $0 \le v_{CE} \le 1$ V.

gives approximately $I_C = 10.5$ mA. Consequently, the current gain at the given operating point $Q_F(I_C, V_{CE}) = Q_F(10.5 \text{ mA}, 0.8 \text{ V})$ is simply $\beta_F = 10.5/0.060 = 174$.

As in the voltage-driven case, the curves bend downward at low values of v_{CE}. This bending is again due to the fact that once v_{CE} drops *below* V_{BE}, the B-C junction becomes *forward biased*, thus drawing the current i_{BC} from the base to the collector. This new current component *subtracts* from the existing currents both at the base and collector, so the net current into the collector terminal is now $i_C = \beta_F(I_B - i_{BC}) - i_{BC} = \beta_F I_B - (\beta_F + 1)i_{BC}$. As the B-C junction becomes more and more forward biased, i_C decreases till it drops to zero when the two terms cancel each other out. Since the effect of i_{BC} is now magnified by $\beta_F + 1$, it takes a *smaller* amount of forward bias to bring about the onset of saturation. Compared to Fig. 2.14b, the curves now start to bend "earlier", that is, slightly more to the right.

The Reverse Active (RA) Mode

If we allow v_{CE} to take on *negative* values in the circuit of Fig. 2.17a, then the roles of the emitter and collector will be *interchanged* (B-C junction forward biased, B-E junction reverse biased.) When operating in this mode, referred to as the *reverse-active* (RA) *mode*, the BJT exhibits a notoriously low current gain, now defined as

$$\beta_R = \frac{-I_E}{I_B}$$

This is depicted in the lower-left area of Fig. 2.18a for the popular 2N2222 BJT. The lower gain stems from the fact that electrons are now injected into the base from the *lightly doped collector region*, and holes are injected into the collector region from the *more heavily doped base region*. Clearly, the conditions that were critical to ensure a high current gain in the FA region turn out to be detrimental for RA operation. Actual transistors exhibit β_R values ranging from as high as 10 to as low as 0.1 or lower (see the end-of-chapter problems for examples of how to measure β_R).

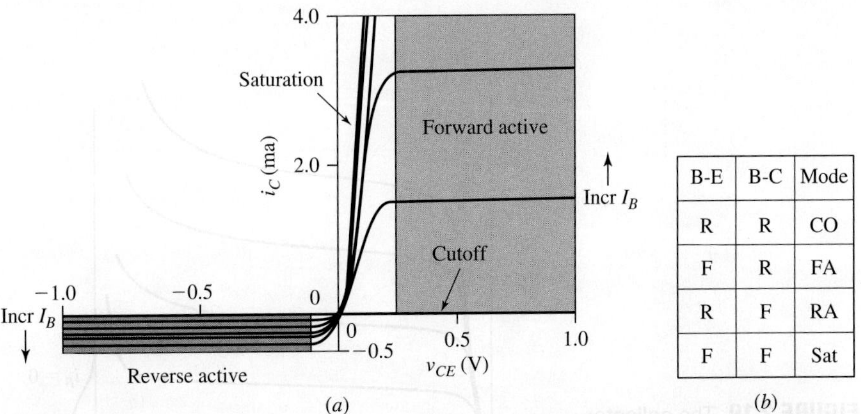

B-E	B-C	Mode
R	R	CO
F	R	FA
R	F	RA
F	F	Sat

(b)

FIGURE 2.18 (a) The four regions of operation of the popular 2N2222 *npn* BJT. (b) The four modes of BJT operation (F stands for forward biased, and R for reverse biased.)

Except for a few specialized cases, operation of the BJT in the RA region hardly offers any practical advantages. We will take up this mode again in Chapter 6, when studying the storage time of the BJT in switching applications. For convenience, the four operating modes of a BJT are tabulated in Fig. 2.18b. This table applies both to *npn* and *pnp* BJTs.

Transistor Breakdown Voltages

With sufficient reverse bias, each junction of a BJT can be driven in the breakdown region. Since the emitter side is heavily doped, the B-E junction (with the collector terminal) open breaks down by the Zener mechanism, typically at a breakdown voltage $BV_{EBO} \cong 6$ V. This mode, while not necessarily destructive so long as we limit power dissipation, is to be avoided as it tends to degrade in the value of β_F significantly.[1]

On the other hand, to permit operation over an adequately wide range of v_{CE} values, the collector side is doped much more lightly, so in this case breakdown occurs by the avalanche mechanism. The breakdown process of the i_C-v_{CE} characteristics, pictured in Fig. 2.19, is far more complex[4] than that of a basic *pn* junction because of the current amplification provided by the BJT itself. Any current entering the base, whether generated thermally in the B-C SCL or triggered by the onset of the avalanche process, gets magnified by β_F. Moreover, β_F starts out low but grows with the current itself, as per Fig. 2.12. Suffice it to say here that because of current amplification, the value of BV_{CEO} is appreciably *lower* than the breakdown voltage of the B-C junction alone, that is, with the emitter open. It is apparent that BJT operation must be restricted over a range of v_{CE} values *lower* than BV_{CEO} by an adequately *safe* margin. If a given BJT fails to meet the requirements of the application at hand, a different type with a higher BV_{CEO} rating must be selected. BJTs are available with a wide range of BV_{CEO} ratings, from a half a dozen volts for BJTs intended for digital applications to many hundreds of volts for BJTs intended for power-handling applications.

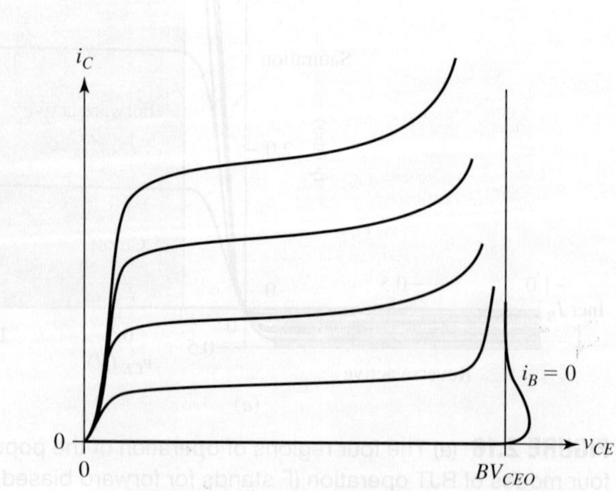

FIGURE 2.19 The collector-emitter breakdown characteristics.

2.4 OPERATING REGIONS AND BJT MODELS

We now re-examine the various BJT operating modes in greater detail, and develop suitable BJT models to expedite hand calculations in dc analysis. By analogy with *pn* diodes, we shall stipulate that to make a BJT fully conductive it takes a B-E voltage drop of approximately 0.7 V. We express this by writing

$$V_{BE(on)} = 0.7 \text{ V} \tag{2.22a}$$

for an *npn* BJT, and $V_{EB(on)} = 0.7$ V for a *pnp* BJT. For bookkeeping purposes, we shall stipulate that to bring a BJT to the *edge of conduction* (EOC) it takes a B-E drop of approximately 0.6 V, or

$$V_{BE(EOC)} = 0.6 \text{ V} \tag{2.22b}$$

for an *npn* BJT, and $V_{EB(EOC)} = 0.6$ V for a *pnp* BJT. As we move along we will find that when a BJT is meant to be in saturation, its base current is made on purpose considerably higher than in the forward active mode, to be on the safe side. Consequently, the B-E drop also tends to be slightly higher. We shall stipulate

$$V_{BE(sat)} = 0.8 \text{ V} \tag{2.22c}$$

for an *npn* BJT, and $V_{EB(sat)} = 0.8$ V for a *pnp* BJT. The above differences are minor, and many authors use the same value of 0.7 V throughout. But, as mentioned, we shall keep these distinctions primarily for bookkeeping purposes.

The Cutoff (CO) Region

The *cutoff* (CO) region is defined as

$$i_C = 0$$

in the i_C-v_{CE} characteristics (see Fig. 2.20a). A BJT operates in cutoff (CO) when neither junction is sufficiently forward-biased to conduct significant current. We can thus ignore all currents and say that for practical purposes the B-E and C-B ports act as *open circuits*, as modeled in Fig. 2.20b. But, as we know, a junction's reverse current doubles for about every 10°C of temperature increase, so if high operating

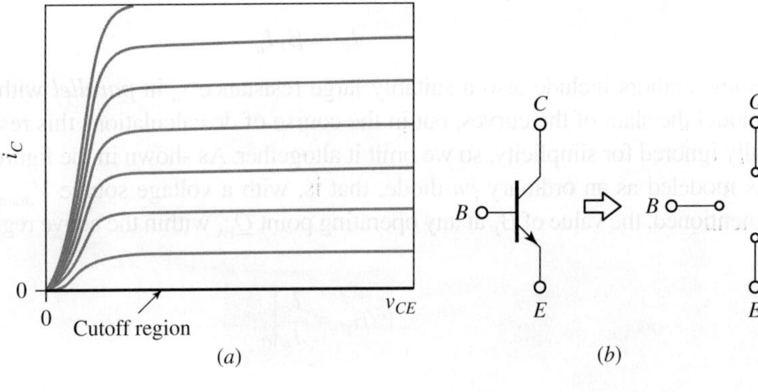

FIGURE 2.20 The *cutoff region*, and the corresponding large-signal model for the *npn* BJT.

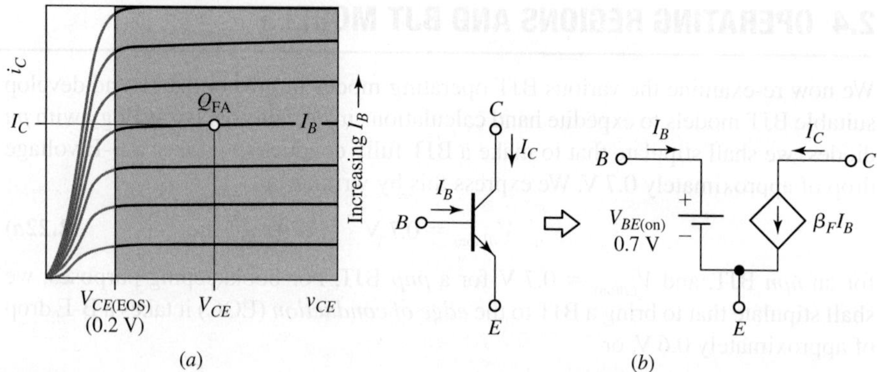

FIGURE 2.21 The *forward active* region, and the corresponding large-signal model of the *npn* BJT.

temperatures are anticipated, the designer must check whether the actual leakage currents affect circuit performance at the high end of the temperature range.

The Forward Active (FA) Region

The *forward active* (FA) region, highlighted in Fig. 2.21a, is defined as

$$i_C > 0 \qquad v_{CE} > V_{CE(EOS)} \;(\cong 0.2\ \text{V})$$

where the subscript EOS designates the *edge of saturation*. (For a *pnp* BJT, these conditions are $i_C > 0$ and $v_{EC} > V_{EC(EOS)} \cong 0.2$ V.) In order for an *npn* BJT to operate in the FA region, its B-E junction must be *forward biased* at $v_{BE} = V_{BE(on)} \cong 0.7$ V, and its B-C junction must be *reverse biased*, or at most it can be *slightly forward biased*, but not enough to conduct any significant forward current i_{BC}. Judging by the 2N2222's characteristics of Fig. 2.17b, which are fairly typical of low-power to medium-power BJTs, we see that this BJT tolerates a slight amount of B-C forward bias and still gives flat curves all the way down to $v_{CE} \cong 0.2$ V, where $v_{BC} = v_{BE} - v_{CE} \cong 0.7 - 0.2 = 0.5$ V.

The curves in the FA region are approximately horizontal, indicating current-source behavior there. So, we model the C-E port with a *current controlled current source* (CCCS) as in Fig. 2.21b,

$$I_C = \beta_F I_B \tag{2.23}$$

Some authors include also a suitably large resistance r_o in *parallel* with the CCCS to model the slant of the curves, but in the course of dc calculations this resistance is usually ignored for simplicity, so we omit it altogether. As shown in the figure, the B-E port is modeled as an ordinary *pn* diode, that is, with a voltage source $V_{BE(on)} \cong 0.7$ V. As mentioned, the value of β_F at any operating point Q_{FA} within the active region is found as

$$\beta_F = \left.\frac{I_C}{I_B}\right|_{Q_{FA}} \tag{2.24}$$

By KCL, we have $I_E = I_B + I_C = I_B + \beta_F I_B$ or

$$I_E = (\beta_F + 1)I_B \tag{2.25}$$

We also have $I_C = \beta_F I_B = \beta_F I_E / (\beta_F + 1)$, or

$$I_C = \alpha_F I_E \tag{2.26}$$

where α_F, referred to as the *common-base forward current gain*, is

$$\alpha_F = \frac{\beta_F}{\beta_F + 1} \tag{2.27}$$

By contrast, β_F is called the *common-emitter forward current gain*. If α_F is known, then β_F is found as

$$\beta_F = \frac{\alpha_F}{1 - \alpha_F} \tag{2.28}$$

As mentioned, a BJT may typically have $\beta_F = 100$, and thus $\alpha_F = 100/101 = 0.99$. Likewise, if $\beta_F = 250$, then $\alpha_F = 0.996$. We observe that α_F is *less* than unity, though quite close to it. A small variation in α_F generally results in a much greater variation in β_F, so care must be exercised in applying Eq. (2.28). Because of α_F's closeness to unity, Eq. (2.26) is often approximated as $I_C \cong I_E$, but *only* in the FA region!

The Saturation (Sat) Region

The *saturation* region, highlighted in Fig. 2.22*a*, is defined as

$$i_C > 0 \qquad 0 < v_{CE} < V_{CE(EOS)} \cong 0.2 \text{ V}$$

(For a *pnp* BJT, these conditions are $i_C > 0$ and $v_{EC} < V_{EC(EOS)} \cong 0.2$ V.) As we know, in saturation the B-C junction becomes forward biased, drawing current away from the base region and thus allowing only a fraction of I_B to undergo amplification by β_F. Consequently, for a given base drive I_B, the collector current at any operating point Q_{sat} in the saturation region will always be *less* than the collector current at an operating point Q_{FA} in the forward active region. In other words, we always have $I_{C(\text{sat})} < \beta_F I_B$. The ratio $I_{C(\text{sat})}/I_B$, aptly denoted as β_{sat}, is such that

$$\beta_{sat} = \frac{I_{C(sat)}}{I_B}\bigg|_{Q_{sat}} < \beta_F \tag{2.29}$$

This inequality provides an alternative test of determining whether a BJT is operating in saturation.

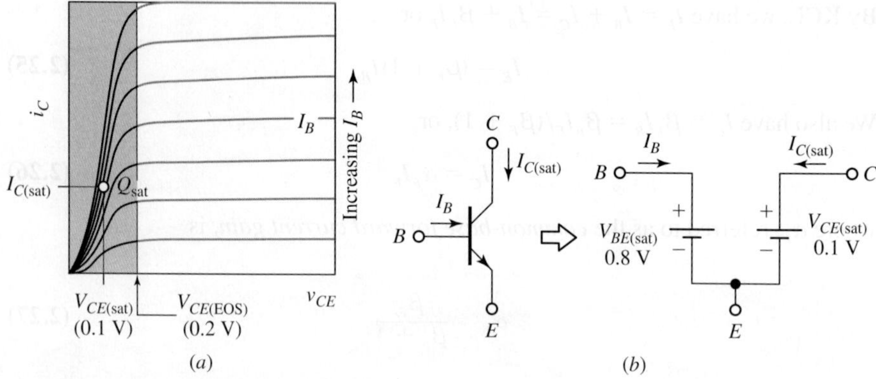

FIGURE 2.22 The *saturation* region, and the corresponding large-signal model of the *npn* BJT.

The curves in the saturation region are relatively steep, indicating *voltage-source* behavior there. Moreover, they are bundled together somewhere in the middle of the saturation region. Consequently, as depicted in Fig. 2.22*b*, we model the C-E port with a voltage source

$$V_{CE(\text{sat})} \cong 0.1 \text{ V} \tag{2.30}$$

Some authors include also a suitably small resistance $R_{CE(\text{sat})}$ in *series* with this source to model the slant of the curves, but in the course of dc calculations this resistance is usually ignored for simplicity, so we omit it altogether. As shown in the figure, the B-E port is modeled like an ordinary *pn* diode, but with a slightly higher voltage drop $V_{BE(\text{sat})} \cong 0.8 \text{ V}$ to account for the fact that in actual applications a BJT is driven in saturation with higher than usual base currents. It is important to realize that because of the inequality of Eq. (2.29), Eqs. (2.23) through (2.28) no longer hold once the BJT enters saturation. The region of their validity is only the FA region! However, we still have, by KCL, $I_E = I_B + I_C$.

EXAMPLE 2.4   Suppose a BJT is biased in the active region with $V_{BE} = 700$ mV and $V_{CE} = 1$ V, and it is found to have $\beta_F = 100$. If V_{CE} is gradually decreased until a 10% drop in β_F is observed, what is the value of V_{CE}? For simplicity assume the B-E and B-C junctions have identical values of I_s.

Note: when this drop occurs, the BJT is said to be in *soft saturation*, a condition also referred to as the *edge of saturation* (EOS).

Solution

In soft saturation the B-C junction becomes *weakly forward biased* and carries a current $I_{BC} \neq 0$. To cause a 10% drop in β_F ($=I_C/I_B$), I_B must increase by about 10%, and this increase is precisely I_{BC}. Imposing $I_{BC} \cong 0.1I_B = 0.1(I_C/100) = I_C/10^3$ indicates that I_{BC} is 3 decades lower than I_C. By the 60-mV Rule, $V_{BC} = V_{BE} - 3 \times 60$ mV $= 700 - 180 = 520$ mV. By KVL, $V_{CE} = V_{BE} - V_{CE} = 0.18$ V ($\cong 0.2$ V typically assumed).

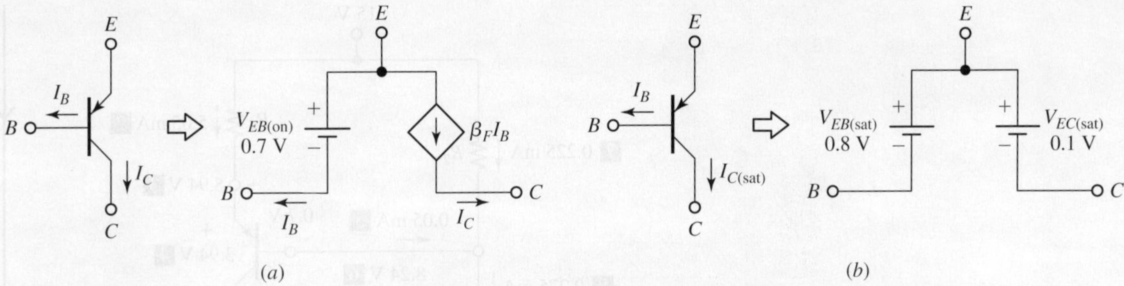

FIGURE 2.23 Large-signal models of the *pnp* BJT in the (a) *forward active* and (b) *saturation* region.

Large-Signal Models for the *pnp* BJT

As mentioned, the body of knowledge pertaining to the *npn* BJT can readily be extended to its *pnp* counterpart provided we *reverse* all *current directions* and *voltage polarities*. The *pnp* models are shown in Fig. 2.23. To point out similarities and differences between the *npn* and *pnp* devices, and also to initiate the reader to basic BJT circuit methodologies, let us consider a practical circuit example.

In the circuit of Fig. 2.24 find V_1 so that $V_2 = 5$ V. Show all voltages and currents in your final circuit.

EXAMPLE 2.5

Solution

Start out at V_2 and work your way back to V_1, one step at a time. The numerical result of each step is identified by the corresponding step number in Fig. 2.25.

1. By Ohm's law, Q_2's collector current is $I_{C2} = V_2/R_5 = 5/1 = 5.0$ mA, flowing *out* of the device.
2. *Assume* Q_2 is in the FA region. Then, its emitter current is $I_{E2} = I_{C2}/\alpha_{F2} = 5/(100/101) = 5.05$ mA, flowing *into* Q_2.

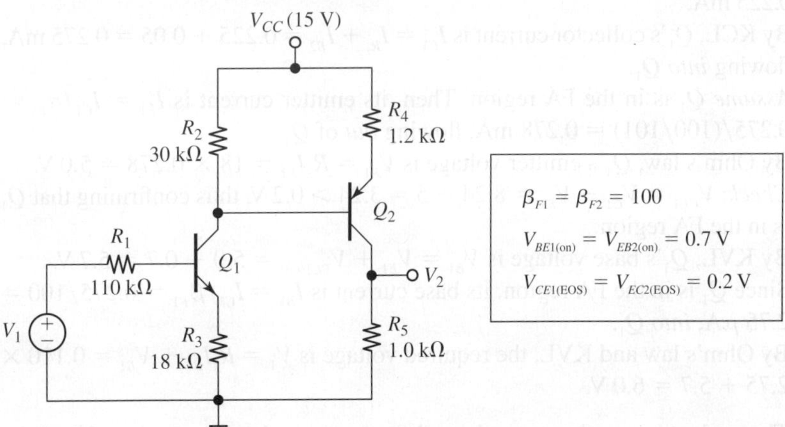

FIGURE 2.24 Circuit of Example 2.5.

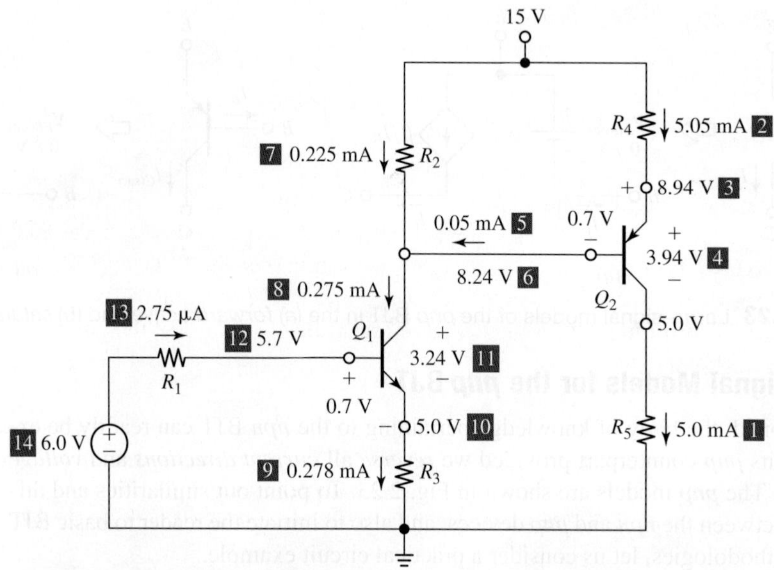

FIGURE 2.25 Circuit of Fig. 2.24 with each voltage and current identified by the corresponding computational step number in the text.

3. By Ohm's law and KVL, Q_2's emitter voltage is $V_{E2} = V_{CC} - R_4 I_{E2} = 15 - 1.2 \times 5.05 = 8.94$ V.

4. *Check*: $V_{EC2} = V_{E2} - V_{C2} = 8.94 - 5 = 3.94$ V > 0.2 V, thus confirming that Q_2 is in the FA region.

5. Since Q_2 is in the FA region, its base current is $I_{B2} = I_{C2}/\beta_{F2} = 5/100 = 0.05$ mA, flowing *out* of Q_2.

6. By KVL, Q_2's base voltage is $V_{B2} = V_{E2} - V_{EB2(on)} = 8.94 - 0.7 = 8.24$ V. This is also Q_1's collector voltage V_{C1}.

7. By Ohm's law, the current through R_2 is $I_{R_2} = (V_{CC} - V_{C1})/R_2 = (15 - 8.24)/30 = 0.225$ mA.

8. By KCL, Q_1's collector current is $I_{C1} = I_{R_2} + I_{B2} = 0.225 + 0.05 = 0.275$ mA, flowing *into* Q_1.

9. *Assume* Q_1 is in the FA region. Then, its emitter current is $I_{E1} = I_{C1}/\alpha_{F1} = 0.275/(100/101) = 0.278$ mA, flowing *out* of Q_1.

10. By Ohm's law, Q_1's emitter voltage is $V_{E1} = R_3 I_{E1} = 18 \times 0.278 = 5.0$ V.

11. *Check*: $V_{CE1} = V_{C1} - V_{E1} = 8.24 - 5 = 3.24 > 0.2$ V, thus confirming that Q_1 is in the FA region.

12. By KVL, Q_1's base voltage is $V_{B1} = V_{E1} + V_{BE1(on)} = 5.0 + 0.7 = 5.7$ V.

13. Since Q_1 is in the FA region, its base current is $I_{B1} = I_{C1}/\beta_{F1} = 0.275/100 = 2.75$ μA, *into* Q_1.

14. By Ohm's law and KVL, the required voltage is $V_1 = R_1 I_{B1} + V_{B1} = 0.110 \times 2.75 + 5.7 = 6.0$ V.

The student is urged to trace through each step in detail, referring also to the FA models of Figs. 2.21*b* and 2.23*a*.

Finding the Operating Mode of a BJT

A frequent task in the dc analysis of BJT circuits is to find a BJT's operating mode in a given circuit. Barring the seldom-used reverse-active mode, a BJT can be either in the cutoff (CO), or forward active (FA), or saturation (Sat) regions. Following is one possible way to proceed for the case of an *npn* BJT:

- Find the open-circuit voltage $V_{BE(oc)}$ produced by the external circuit across the B-E junction, that is, the voltage between the B and E nodes with the BJT *removed* from the circuit. If $V_{BE(oc)} < V_{BE(EOC)} \cong 0.6$ V, then the BJT is in CO, and we are finished. Otherwise, it is on, and it is either in FA or in Sat.
- Assume FA operation. Using the FA large-signal BJT model, find the operating point $Q = Q(I_C, V_{CE})$, and check whether the assumption was correct by examining V_{CE}. If $V_{CE} > V_{CE(EOS)} \cong 0.2$ V, then the BJT is indeed in FA, and we are finished.
- Conversely, if our calculation yields $V_{CE} < V_{CE(EOS)} \cong 0.2$ V, the BJT must be saturated, and we need to *recalculate* I_C, but via the saturation model, which uses $V_{CE} = V_{CE(sat)} \cong 0.1$ V. As a final check, take the ratio $\beta_{sat} = I_C/I_B$, and verify that indeed you get $\beta_{sat} < \beta_F$.

This procedure applies to the *pnp* BJT as well, provided we *reverse* all current directions and voltage polarities. Specifically, if $V_{EB(oc)} < V_{EB(EOC)} \cong 0.6$ V, the *pnp* device is in CO. Otherwise it is on, and we need to examine V_{EC} to tell its operating mode. If our calculation yields $V_{EC} > V_{EC(EOS)} \cong 0.2$ V, then the BJT is in FA. Otherwise, it is saturated, and we recalculate I_C via the saturation model, which uses $V_{EC} = V_{EC(sat)} \cong 0.1$ V.

EXAMPLE 2.6

Let the BJT of Fig. 2.26a have $\beta_F = 125$, as well as the voltages of Eqs. (2.22) and (2.30).

(a) Find all BJT voltages and currents.
(b) To what value must we increase R_C to bring the BJT to the edge of saturation (EOS)?
(c) What happens if R_C is raised to twice the value found in part (*b*)?

Solution

(a) Since the base is at 0 V and R_E is pulling the emitter toward -5 V, it is clear that the BJT is on. Assume it is in the forward active region, so that $V_E = -0.7$ V. Then, $I_E = [-0.7 - (-5)]/4.3 = 1.0$ mA, $I_C = (125/126) \times 1.0 = 0.992$ mA, $V_C = 5 - 2 \times 0.992 \cong 3.0$ V, and $V_{CE} = V_C - V_E = 3 - (-0.7) = 3.7$ V.

Check: Since $V_{CE} > V_{CE(EOS)}$ (3.7 V > 0.2 V), the BJT is indeed in the FA region, as assumed. The situation is summarized in Fig. 2.26b, which also shows that $I_B = 1.0/126 = 8 \ \mu A$.

Remark: For the purpose of finding V_C, we could have approximated $I_C \cong I_E = 1.0$ mA to speed up our calculations, obtaining results that are still fairly accurate.

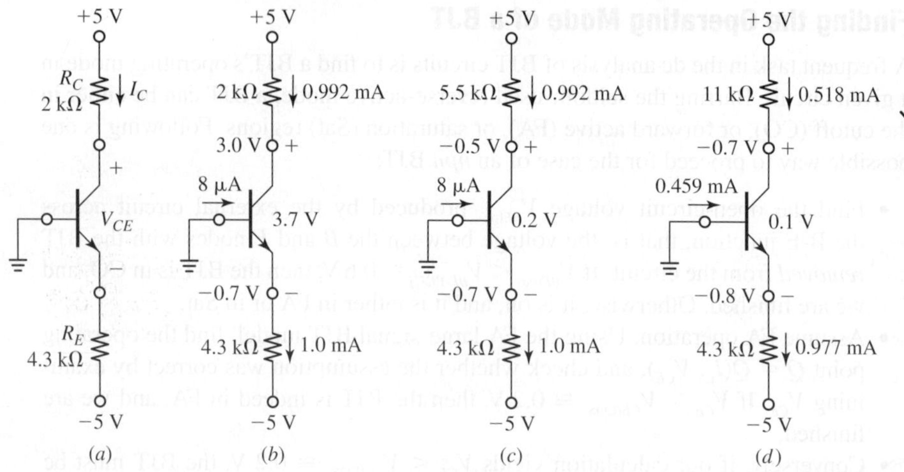

FIGURE 2.26 (a) Circuit of Example 2.6, and operation (b) in the *active region*, (c) at the EOS, and (d) in *saturation*.

(b) At the EOS all currents are still as in the FA case, the only difference being that now $V_{CE} = V_{CE(EOS)} = 0.2$ V. As depicted in Fig. 2.26c, we now have $V_C = V_E + V_{CE(EOS)} = -0.7 + 0.2 = -0.5$ V. Thus, the value of R_C that brings the BJT to the EOS is $R_C = [5 - (-0.5)]/0.992 \cong 5.5$ kΩ, as shown.

(c) We claim that with $R_C = 2 \times 5.5 = 11$ kΩ the BJT is saturated. To convince ourselves, let us pretend it is still in FA. Then, we'd have $V_C = 5 - 11 \times 0.992 = -5.9$ V, and $V_{CE} = -5.9 - (-0.7) = -5.2$ V, an impossible proposition! So, the BJT must be saturated at $V_{CE} = V_{CE(sat)} = 0.1$ V and $V_{BE(sat)} = -0.8$. V

This is depicted in Fig. 2.26d. We now have $I_E = (5 - 0.8)/4.3 = 0.977$ mA, $V_C = -0.8 + 0.1 = -0.7$ V, and $I_C = 5.7/11 = 0.518$ mA. By KCL, the base current is now

$$I_B = I_E - I_C = 0.977 - 0.518 = 0.459 \text{ mA}$$

quite an increase from the FA value of 0.008 mA!

Check: Just to make sure, calculate $\beta_{sat} = 0.518/0.459 \cong 1.1$. Since $\beta_{sat} < \beta_F$ (1.1 < 125), the BJT is indeed in deep saturation!

In the popular circuit of Fig. 2.27a, the voltage divider made up of R_1 and R_2 establishes a bias voltage for the base, while R_E and R_C set the BJT's operating point. To simplify analysis, it is convenient to replace the voltage divider with its Thévenin equivalent, consisting of the open-circuit voltage source

$$V_{BB} = \frac{R_2}{R_1 + R_2} V_{CC} \tag{2.31a}$$

and series resistance

$$R_B = R_1 /\!/ R_2 \tag{2.31b}$$

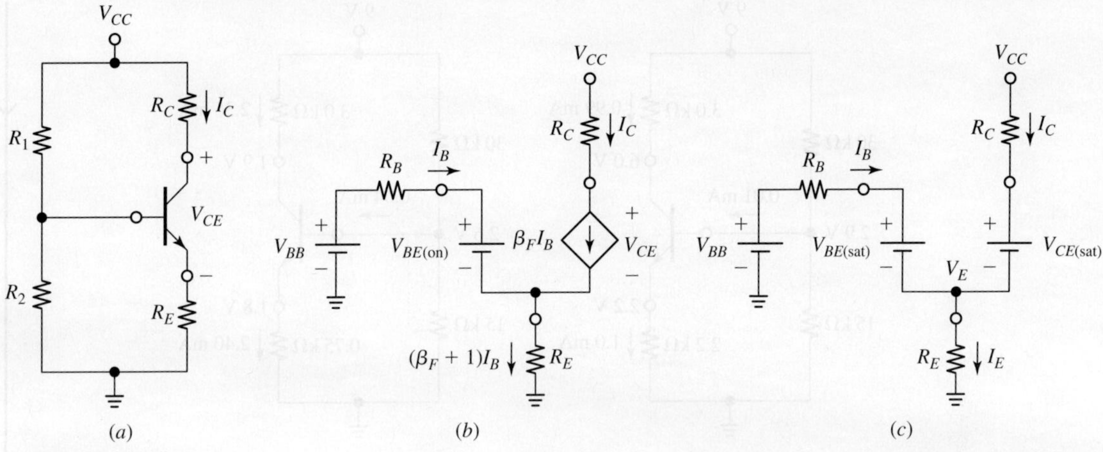

FIGURE 2.27 (a) A popular single-supply BJT circuit, and its equivalents for the case of the BJT operating in the (b) *forward active* and (c) *saturation* region.

Then, the circuit reduces to either equivalent of Fig. 2.27b or 2.27c, depending on whether the BJT is operating in the forward active or saturation mode. In this regard, we make the following observations:

- In the *forward active* equivalent of Fig. 2.27b we apply KVL and write

$$V_{BB} = R_B I_B + V_{BE\text{(on)}} + R_E(\beta_F + 1)I_B$$

Solving for I_B, and then multiplying by β_F, we get

$$I_C = \beta_F \frac{V_{BB} - V_{BE\text{(on)}}}{R_B + (\beta_F + 1)R_E} \tag{2.32}$$

- In the *saturation* equivalent of Fig. 2.27c we apply KCL and write

$$\frac{V_{BB} - (V_E + V_{BE\text{(sat)}})}{R_B} + \frac{V_{CC} - (V_E + V_{CE\text{(sat)}})}{R_C} = \frac{V_E}{R_E} \tag{2.33}$$

This equation is readily solved for V_E, after which we can find all other voltages and all currents in the circuit via repeated usage of KVL and Ohm's law.

In the circuit of Fig. 2.27a, let $V_{CC} = 9$ V, $R_1 = 30$ kΩ, $R_2 = 15$ kΩ, $R_C = 3.0$ kΩ, and $R_E = 2.2$ kΩ, and let the BJT have $\beta_F = 100$ as well as the voltages of Eqs. (2.22) and (2.30).

EXAMPLE 2.7

(a) Find all BJT voltages and currents, and show them in the circuit.
(b) Repeat part (a) if R_E is lowered to 0.75 kΩ.
(c) Repeat (a) if R_2 is lowered to 1.0 kΩ.

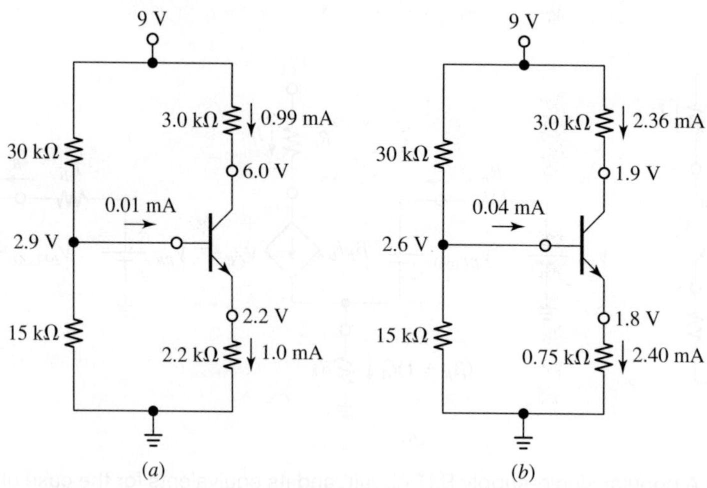

FIGURE 2.28 Circuits of Example 2.6, showing (a) *forward active* operation with $\beta_F = 100$, and (b) *saturation* operation with $\beta_{sat} = 59$.

Solution

(a) Assume the BJT is in the FA region. Applying Eq. (2.31), we get

$$V_{BB} = \frac{15}{30 + 15}9 = 3 \text{ V} \qquad\qquad R_B = \frac{30 \times 15}{30 + 15} = 10 \text{ k}\Omega$$

Using Eq. (2.32), we obtain

$$I_C = 100\frac{3 - 0.7}{10 + 101 \times 2.2} = 0.99 \text{ mA}$$

Consequently, we have $I_B = I_C/\beta_F = 0.99/100 \cong 0.01$ mA, $I_E = I_C/\alpha_F = 0.99/(100/101) = 1.0$ mA. Also,

$$V_B = V_{BB} - R_B I_B = 3 - 10 \times 0.01 = 2.9 \text{ V}$$
$$V_C = V_{CC} - R_C I_C = 9 - 3.0 \times 0.99 \cong 6.0 \text{ V}$$
$$V_E = V_B - V_{BE(\text{on})} = 2.9 - 0.7 = 2.2 \text{ V}$$

All voltages and currents are shown in Fig. 2.28a.

Check: We have $V_{CE} = V_C - V_E = 6.0 - 2.2 = 3.8$ V. Since $V_{CE} > V_{CE(\text{EOS})}$ (3.8 V > 0.2 V), the BJT is indeed in the FA region, as assumed.

Remark: The actual base voltage ($V_B = 2.9$ V) is a bit *less* than the open-circuit voltage ($V_{BB} = 3$ V) because of the tiny base current, which causes the BJT to *load* down the base-biasing network.

(b) We can start out again assuming forward active operation. Repeating the above calculations but with $R_E = 0.75$ kΩ we get $I_C \cong 2.7$ mA, so $V_{CE} \cong 9 - 3 \times 2.7 - 0.75 \times 2.7 = -1$ V. Since this would imply $V_{CE} < V_{CE(\text{EOS})}$ (-1 V < 0.2 V), we conclude that the BJT is now saturated, and we can *no longer* use the active-region relationships, such as $I_C = \beta_F I_B$ and the like. We must apply Eq. (2.33) instead, and write

$$\frac{3 - (V_E + 0.8)}{10} + \frac{9 - (V_E + 0.1)}{3} = \frac{V_E}{0.75}$$

Solving for V_E, we get

$$V_E = 95.6/53 = 1.8 \text{ V}$$
$$V_B = 1.8 + 0.8 = 2.6 \text{ V}$$
$$V_C = 1.8 + 0.1 = 1.9 \text{ V}$$

Moreover,

$$I_B = \frac{3 - 2.6}{10} = 0.04 \text{ mA} \qquad I_C = \frac{9 - 1.9}{3} = 2.36 \text{ mA} \qquad I_E = \frac{1.8}{0.75} = 2.40 \text{ mA}$$

(Note, as a check, that the currents satisfy KCL.) All voltages and currents are shown in Fig. 2.28b.

Check: Just to make sure, calculate $\beta_{sat} = 2.36/0.04 = 59$. Since $\beta_{sat} < \beta_F$ ($59 < 100$), the BJT is indeed in saturation!

Remark: In the present example we have driven the BJT in saturation by *lowering* R_E and thus increasing I_C. In Example 2.5 we drove it in saturation by *raising* R_C. Both changes *lowered* V_C to the point of trying to make $V_{CE} < V_{CE(EOS)}$. Any other change in the circuit that will decrease the value of V_{CE} will tend to drive the BJT in saturation. Possible examples are increasing V_{BB}, or replacing the BJT with another unit with higher β_F, or a combination of both.

(c) We now have $V_{BB} = [1/(30 + 1)]9 = 0.29$ V. This is insufficient to turn on the B-E junction convincingly. The BJT is now cut off, and all currents are for practical purposes zero. Moreover, $V_B \cong 0.29$ V, $V_E \cong 0$, and $V_C \cong 9$ V.

Assuming the BJTs in the circuit of Fig. 2.29a have $\beta_{F1} = \beta_{F2} = 100$ and base-emitter voltage drops of 0.7 V, find all BJT terminal voltages and currents, and show them explicitly.

EXAMPLE 2.8

Solution

Consider the two BJT circuits *separately*, one at a time. Turning first to Q_1 and its related resistors, and performing the usual Thévenin reduction of its base-biasing network, we end up with the equivalent of Fig. 2.30a. By KVL we have

$$V_{CC} = R_3(\beta_{F1} + 1)I_{B1} + V_{EB1(on)} + R_{B1}I_{B1} + V_{BB1}$$

or

$$12 = 18(100 + 1)I_{B1} + 0.7 + 75I_{B1} + 7.5$$

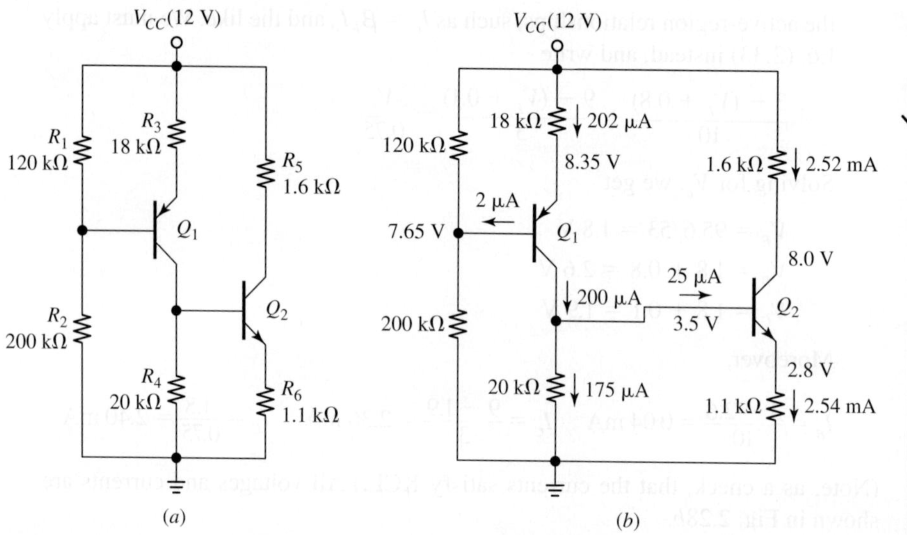

FIGURE 2.29 (a) Circuit of Example 2.7. (b) Showing all BJT voltages and currents explicitly.

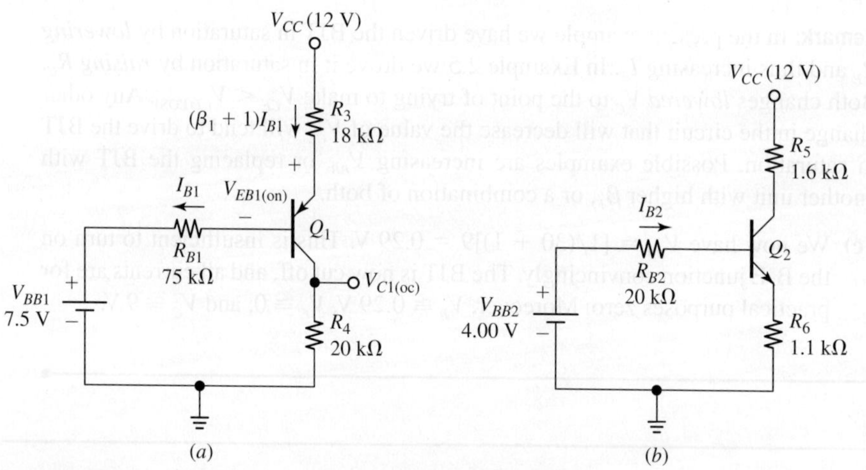

FIGURE 2.30 Intermediate steps in the analysis of the circuit of Fig. 2.29a.

This gives

$$I_{B1} = 2 \ \mu A \qquad I_{C1} = 100 \times 2 = 200 \ \mu A \qquad I_{E1} = 101 \times 2 = 202 \ \mu A$$

Moreover, we have

$$V_{B1} = 7.5 + 75 \times 0.002 = 7.65 \ V$$
$$V_{E1} = 7.65 + 0.7 = 8.35 \ V$$
$$V_{C1(oc)} = 20 \times 0.200 = 4.00 \ V$$

where $V_{C1(oc)}$ is the *open-circuit voltage* at Q_1's collector. Once we bring Q_2's back into the picture, this voltage will act as V_{BB2} to Q_2, and R_4 will act as R_{B2}. For the purpose of our calculations we can thus replace the entire circuit based on Q_1 with its Thévenin equivalent and work with the much simpler equivalent of Fig. 2.30b. This circuit is similar to that of Fig. 2.27b, so we can proceed in the manner of Example 2.7, and come up with the voltages and currents shown in Fig. 2.29b.

Remark: Because of I_{B1}, V_{B1} is a bit *higher* than V_{BB1}, and because of I_{B2}, V_{B2} is a bit *lower* than V_{BB2}. As we know, these effects are commonly referred to as *loading*.

The Diode Mode of Operation

Tying the base and collector together turns a BJT into a two-terminal device. To find its i-v characteristic we subject it to a test voltage v and examine its current response i, as shown in Fig. 2.31a. By KCL,

$$i = i_B + i_C = \left(\frac{1}{\beta_F} + 1\right)i_C = \frac{I_s}{\alpha_F}e^{v/V_T} \tag{2.34}$$

indicating that the two-terminal device acts as a diode having the B-C terminal as the anode (A), the emitter terminal as the cathode (C), and I_s/α_F ($\cong I_s$) as the saturation current. In fact, in integrated circuits this is how a diode is often created, namely, using a BJT with its B and C terminals tied together. The analysis of circuits incorporating a diode-connected BJT proceeds as in the case of ordinary diodes.

In the circuit of Fig. 2.31b let the BJT have $I_s = 10$ fA, $\beta_F = 100$, and $V_{BE(on)} = 0.7$ V.   **EXAMPLE 2.9**

(a) Find I if $V_{CC} = 5$ V and $R = 10$ kΩ.
(b) Repeat, but for $V_{CC} = 0.75$ V and $R = 1$ kΩ.

Solution

(a) Since $V_{CC} \gg V$, we don't need to know V that accurately, so we approximate $V \cong V_{BE(on)} = 0.7$ V, and write

$$I = \frac{V_{CC} - V}{R} \cong \frac{5 - 0.7}{10} = 0.43 \text{ mA}$$

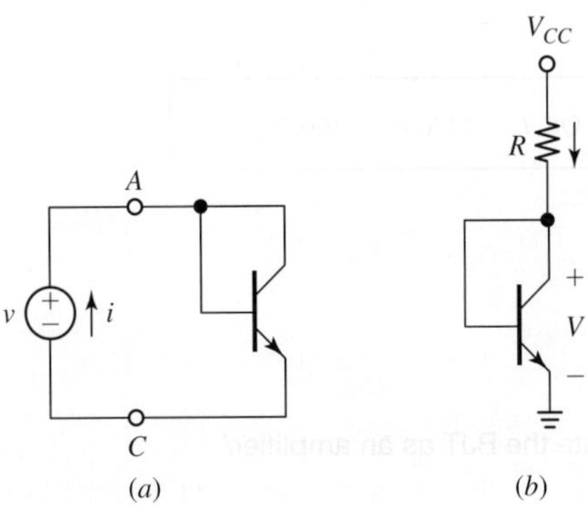

FIGURE 2.31 Diode-connected BJT. (a) Test circuit to find its i-v characteristic. (b) Circuit example.

(b) In this case V_{CC} is too close to $V_{BE(\text{on})}$ for us to proceed as in part (*a*). Rather, we need to apply the familiar iterative technique for diodes, and find V first,

$$V = V_T \ln \frac{I}{I_s/\alpha_F} \cong V_T \ln \frac{(V_{CC} - V)/R}{I_s} = 0.026 \ln \frac{0.75 - V}{10^{-11}}$$

Starting out with the initial estimate $V = 0.65$ V, we find, after a few iterations, $V = 0.6078$ V. Consequently, $I = (0.75 - 0.608)/1 = 0.142$ mA.

2.5 THE BJT AS AN AMPLIFIER/SWITCH

We shall now investigate the two most important BJT applications: *amplification* and *switching*. To this end, refer to the basic circuit of Fig. 2.32, where R_B serves the function of converting v_I to the base drive i_B, and R_C and Q can be viewed as forming a voltage divider of sorts: R_C tends to *pull v_O up* toward V_{CC}, and Q tends to *pull v_O down* toward ground. Depending on which pull action prevails, v_O will assume a value somewhere in between. The plot of v_O versus v_I, called the *voltage transfer curve* (VTC), provides much insight into the capabilities of this circuit. Figure 2.33 shows the VTC as well as other pertinent curves for the case in which v_I is swept from 0 V to 2 V and the BJT possesses the characteristics tabulated in Fig. 2.32. We make the following observations:

- For $v_I < V_{BE(\text{EOC})} \cong 0.6$ V, the B-E junction is insufficiently biased and the BJT is therefore in *cutoff*. With no current being drawn by the collector, the voltage across R_C is 0 V, indicating that R_C is *pulling v_O* all the way up to V_{CC}. We express this by writing $v_O = V_{OH}$, where

$$V_{OH} = V_{CC} = 5 \text{ V} \tag{2.35}$$

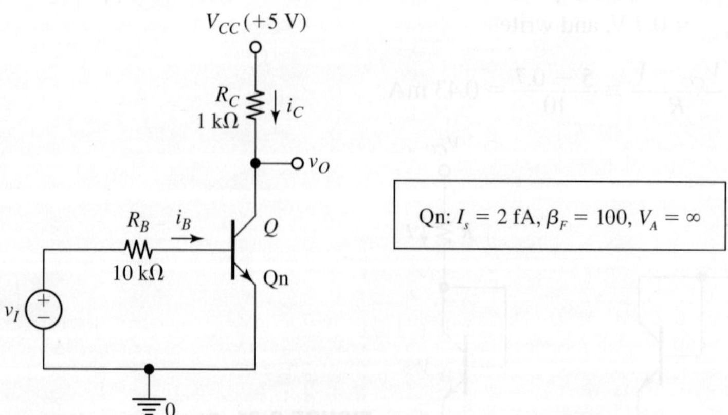

FIGURE 2.32 PSpice circuit to investigate the BJT as an amplifier/switch.

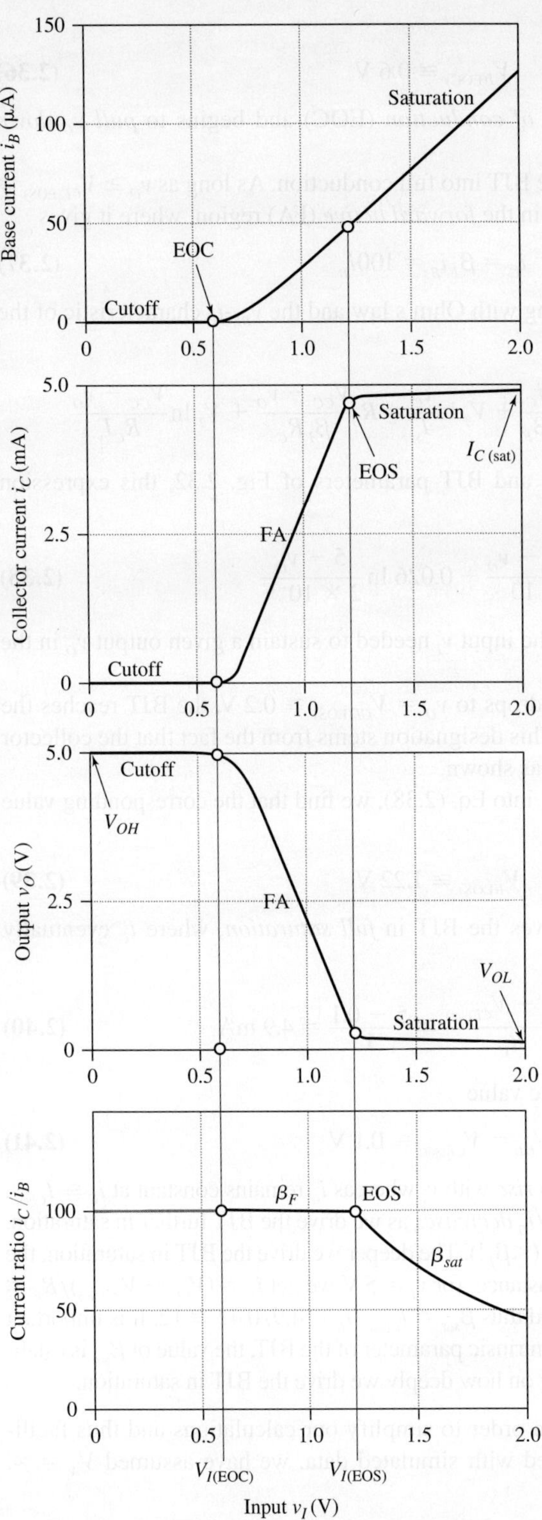

FIGURE 2.33 Plots for the circuit of Fig. 2.32, showing the variations of i_B, i_C, v_O, and beta ($=i_C/i_B$) as the BJT is swept from cutoff (CO), to the edge of conduction (EOC), through the forward active (FA) region, to the edge of saturation (EOS), to full saturation (sat).

- As we raise v_I to the value

$$V_{I(EOC)} \cong 0.6 \text{ V} \tag{2.36}$$

the BJT reaches the *edge of conduction* (EOC) and begins to *pull v_O down* from V_{CC}.

- Raising v_I further brings the BJT into full conduction. As long as $v_O \geq V_{CE(EOS)} \cong$ 0.2 V, the BJT is operating in the *forward active* (FA) region, where it gives

$$i_C = \beta_F i_B = 100 i_B \tag{2.37}$$

Moreover, using KVL, along with Ohm's law and the v_{BE}-i_C characteristic of the BJT, we write

$$v_I = R_B i_B + v_{BE} = R_B \frac{i_C}{\beta_F} + V_T \ln \frac{i_C}{I_s} = R_B \frac{V_{CC} - v_O}{\beta_F R_C} + V_T \ln \frac{V_{CC} - v_O}{R_C I_s}$$

With the resistance values and BJT parameters of Fig. 2.32, this expression becomes

$$v_I = \frac{5 - v_O}{10} + 0.026 \ln \frac{5 - v_O}{2 \times 10^{-12}} \tag{2.38}$$

which can be used to find the input v_I needed to sustain a given output v_O in the FA region.

- Once the collector voltage drops to $v_O = V_{CE(EOS)} \cong 0.2$ V, the BJT reaches the *edge of saturation* (EOS). This designation stems from the fact that the collector current i_C starts to saturate, as shown.

Substituting $v_O = 0.2$ V into Eq. (2.38), we find that the corresponding value of v_I is

$$V_{I(EOS)} \cong 1.22 \text{ V} \tag{2.39}$$

Raising v_I above $V_{I(EOS)}$ drives the BJT in *full saturation*, where i_C eventually settles to the value

$$I_{C(sat)} = \frac{V_{CC} - V_{CE(sat)}}{R_C} \cong \frac{5 - 0.1}{1} = 4.9 \text{ mA} \tag{2.40}$$

Accordingly, v_O settles to the value

$$V_{OL} = V_{CE(sat)} \cong 0.1 \text{ V} \tag{2.41}$$

- Past the EOS, i_B continues to *rise* with v_I whereas i_C remains constant at $i_C \cong I_{C(sat)}$. It is apparent that the ratio i_C / i_B *decreases* as we drive the BJT further in saturation, so we denote this ratio as β_{sat} ($<\beta_F$!). The deeper we drive the BJT in saturation, the lower the value of β_{sat}. For instance, for $v_I = 5$ V we get $i_B = (V_{CC} - V_{BE(sat)})/R_B \cong$ $(5 - 0.8)/10 = 0.42$ mA, and thus $\beta_{sat} = I_{C(sat)}/i_B = 4.9/0.42 \cong 12$. It is important to realize that while β_F is an intrinsic parameter of the BJT, the value of β_{sat} is established by the user, depending on how deeply we drive the BJT in saturation.

We wish to point out that in order to simplify our calculations and thus facilitate the comparison of calculated with simulated data, we have assumed $V_A = \infty$.

In practice, the effect of non-infinite V_A will alter the curves a little, but our general observations still stand.

The BJT as an Amplifier

The *slope* of the VTC represents *voltage gain*, denoted as a. Differentiating both sides of Eq. (2.38) with respect to v_I we get

$$\frac{dv_I}{dv_I} = -\frac{1}{10}\frac{dv_O}{dv_I} + 0.026\frac{2 \times 10^{-12}}{5 - v_O}\left(-\frac{1}{2 \times 10^{-12}}\frac{dv_O}{dv_I}\right)$$

After simplifying, we obtain, for the parameter values of Fig. 2.32,

$$a = \frac{dv_O}{dv_I} = -10\frac{5 - v_O}{5.26 - v_O} \tag{2.42}$$

Figure 2.34 shows the VTC as well as the slope a. In cutoff and in saturation we have $a = 0$. However, there are two points, denoted as V_{IL} and V_{IH}, such that for $V_{IL} \leq v_I \leq V_{IH}$ we have $|a| > 1$ V/V, indicating that the circuit can be used as an *amplifier*. As seen, the voltage gain peaks at about -9 V/V just before the EOS.

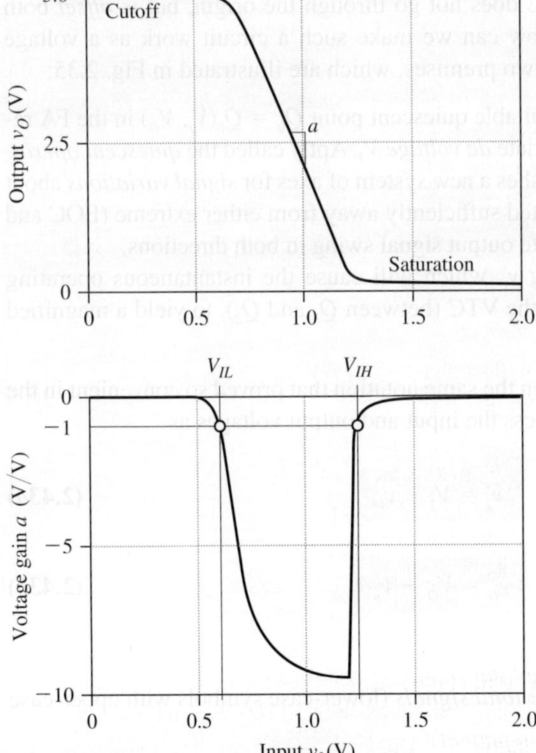

FIGURE 2.34 The voltage transfer curve (VTC) and
its *slope,* representing the voltage gain a.

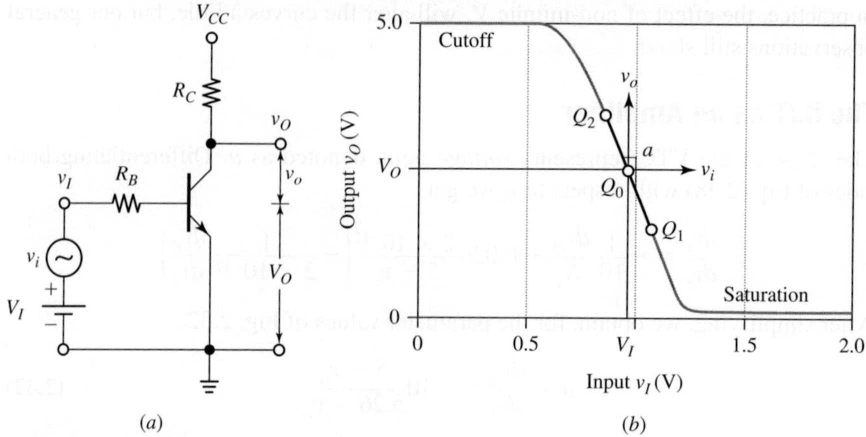

FIGURE 2.35 (a) The BJT of Fig. 2.32 as a *voltage amplifier,* and (b) variations about the operating point Q_0.

A circuit whose gain is not constant but varies with the value of the signal itself is *nonlinear.* Moreover, the VTC does not go through the origin, but is *offset* both along the v_I and the v_O axes. How can we make such a circuit work as a voltage amplifier? The answer relies on two premises, which are illustrated in Fig. 2.35:

- First, we bias the BJT at a suitable quiescent point $Q_0 = Q_0(V_I, V_O)$ in the FA region by applying the appropriate *dc voltage* V_I. Aptly called the *quiescent operating point,* Q_0 in effect establishes a new system of axes for *signal variations* about this point. Q_0 should be located sufficiently away from either extreme (EOC and EOS) to allow for an adequate output signal swing in both directions.
- Then, we apply an *ac input* v_i, which will cause the instantaneous operating point to move up and down the VTC (between Q_1 and Q_2), to yield a magnified *ac voltage* v_o at the output.

In our discussion we are relying on the same notation that proved so convenient in the study of diodes, namely, we express the input and output voltages as

$$v_I = V_I + v_i \tag{2.43a}$$

$$v_O = V_O + v_o \tag{2.43b}$$

where:

- v_I and v_O are referred to as the *total signals* (lower-case symbols with upper-case subscripts)
- V_I and V_O are their *dc components* (upper-case symbols with upper-case subscripts)
- v_i and v_o are their *ac components* (lower-case symbols with lower-case subscripts)

As depicted in Fig. 2.35*b* for the circuit with the parameters of Fig. 2.32, the bias point Q_0 has been chosen in the *middle* of the active portion of the VTC so that $V_O = 2.5$ V. The voltage gain there is $a(Q_0)$.

Find the voltage V_I needed to bias the BJT of Fig. 2.32 at $V_O = 2.5$ V. What is the voltage gain a there?   **EXAMPLE 2.10**

Solution

By Eq. (2.38),

$$V_I = \frac{5 - 2.5}{10} + 0.026 \ln\frac{5 - 2.5}{2 \times 10^{-12}} = 0.25 + 0.724 = 0.974 \text{ V}$$

indicating that we need 0.724 V to bias the B-E junction, and 0.25 V to supply the required base current via R_B. By Eq. (2.42),

$$a(Q_0) = -10\frac{5 - 2.5}{5.26 - 2.5} \cong -9 \text{ V/V}$$

PSpice simulations with a *triangular input* v_i of progressively increasing magnitude yield the waveforms of Fig. 2.36. We make the following observations:

- In Fig. 2.36*a* the ac input v_i has peak values of ± 0.1V, and the ac output v_o is an inverted and magnified version of v_i, or $v_o \cong -9v_i$. The amount of output distortion is imperceptible.
- Doubling v_i's peak values to ± 0.2 V still yields a fairly undistorted output, as seen in Fig. 2.36*b*. The operating point moves up and down a *wider* portion of the VTC, which however is still approximately straight.

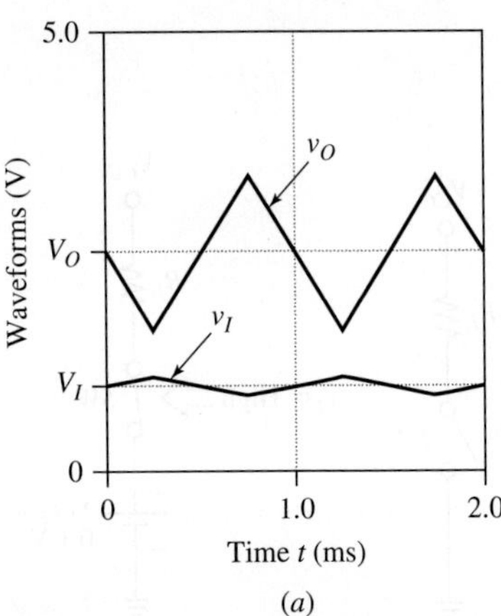
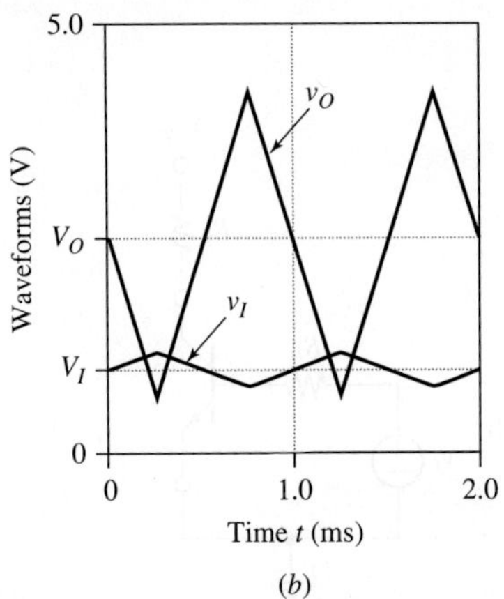
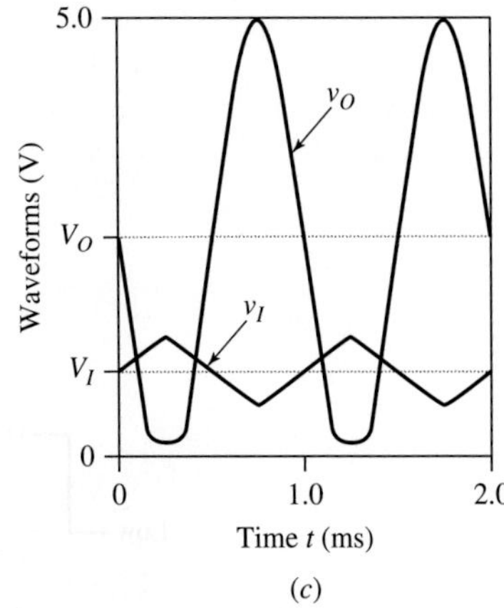

FIGURE 2.36 The responses of the circuit of Fig. 2.39 to a triangular wave v_i with peak values of: (a) ± 0.1 V, (b) ± 0.2 V, and (c) ± 0.4 V. The BJT is biased at $V_I = 0.94$ V and $V_O = 2.5$ V.

- Raising v_i's peak values to ± 0.4 V forces the operating point to *spill over* into the *cutoff* and the *saturation* regions, where gain drops dramatically. The result is the highly distorted output waveform of Fig. 2.36c. Distortion at the top is due to BJT cutoff, and distortion at the bottom to BJT saturation.

We can now better appreciate the reason for biasing the BJT somewhere in the middle of the FA region, sufficiently away from both cutoff and saturation, as well as the reason for keeping the magnitude of v_i and, hence, of v_o, sufficiently small. In fact, the smaller the signals the lesser the amount of output distortion. Viewed in this light, v_i and v_o are also referred to as *small signals*. A more rigorous treatment of this subject will be undertaken in the next section.

The BJT as a Switch/Logic Inverter

When BJT operation alternates between the *cutoff* and the *saturation* modes, the device acts as an *electronic switch SW*. In this capacity, illustrated in Fig. 2.37, the BJT can be used to turn on/off power to some arbitrary load R_L, such as a light-emitting device, a dc motor, or a heating element. With reference to Fig. 2.37, we make the following observations:

- When v_I is *low*, near 0 V, the BJT is in cutoff, and since it draws no current, it can be regarded as an *open* switch. This is pictured in Fig. 2.37b.
- When v_I is *high*, say near V_{CC}, the BJT is designed to saturate. Consequently, it can be regarded as a *closed* switch, but in series with a tiny battery $V_{CE(\text{sat})} \cong$ 0.1 V, as pictured in Fig. 2.37c. Consequently, the load R_L now receives the full power supplied by V_{CC}. To guarantee a reliably closed switch under all possible conditions, especially in view of fluctuations in the value of β_F, we must force the BJT in sufficiently deep saturation. We must thus ensure that $\beta_{\text{sat}} < \beta_{F(\text{min})}$ with a certain margin of safety.

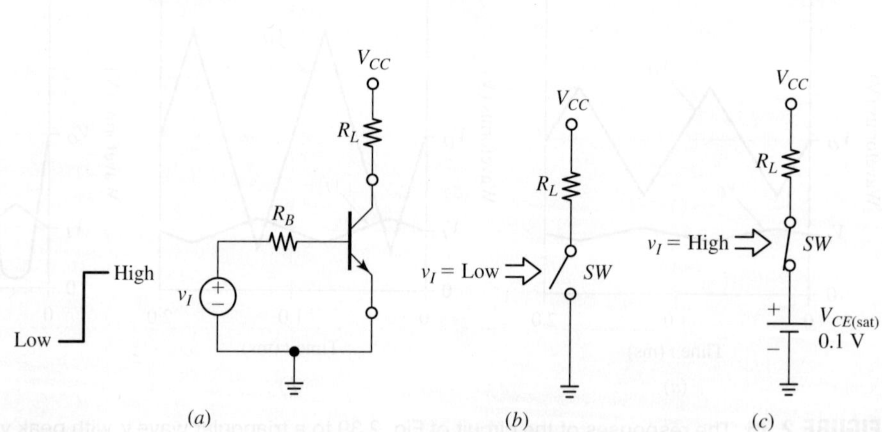

FIGURE 2.37 Operating the BJT as an *electronic switch*.

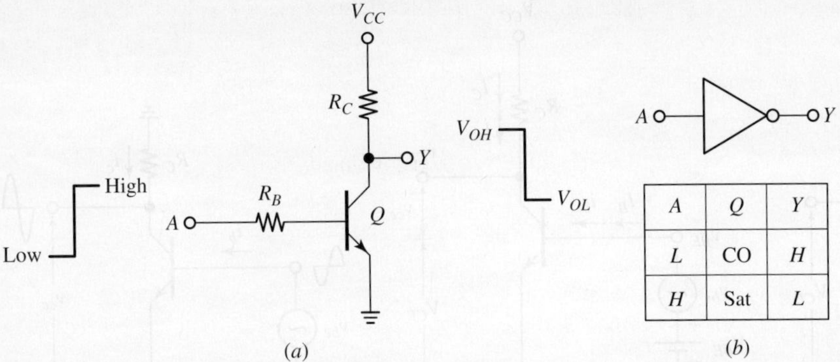

FIGURE 2.38 The BJT as a *logic inverter*. (a) circuit, (b) logic symbol and truth table.

A BJT with β_F specified to be anywhere within the range $50 \leq \beta_F \leq 200$ is to be **EXAMPLE 2.11**
used as a switch for a 100-mA load. If v_I is a logic signal with logic levels of 0 V
and 5 V, find a suitable value for R_B.

Solution
To guarantee a saturated BJT under all possible conditions, including the worst-
case scenario of $\beta_F = 50$, we need $I_B > 100/50 = 2$ mA. Impose $I_B = 3$ mA to be
on the safe side. Then, $R_B = (5 - 0.8)/3 = 1.4$ kΩ.

A popular application of the BJT switch is to provide *logic inversion* in computer
circuitry. As implied by its name, a logic inverter outputs a high voltage level (H) in
response to a low input level (L), and outputs a low level in response to a high input
level. For the BJT inverter of Fig. 2.38a, these levels are, respectively,

$$V_{OH} = V_{CC} \tag{2.44a}$$

$$V_{OL} = V_{CE(\text{sat})} \tag{2.44b}$$

Typically $V_{OH} = 5$ V, and $V_{OL} = 0.1$ V. Figure 2.38b shows the logic symbol for the
inverter, along with the *truth table* listing the BJT's operating mode as well as the
logic output for the two possible logic input combinations.

2.6 SMALL-SIGNAL OPERATION OF THE BJT

We now wish to pursue a more systematic investigation of the small-signal operation introduced in the previous section. Let us start with the circuit of Fig. 2.39a,
where we are using the dc source V_{BE} to bias the BJT at some *quiescent point*
$Q_0 = Q_0(I_C, V_{BE})$ up the exponential curve (see Fig. 2.40a), and the source V_{CC}, along

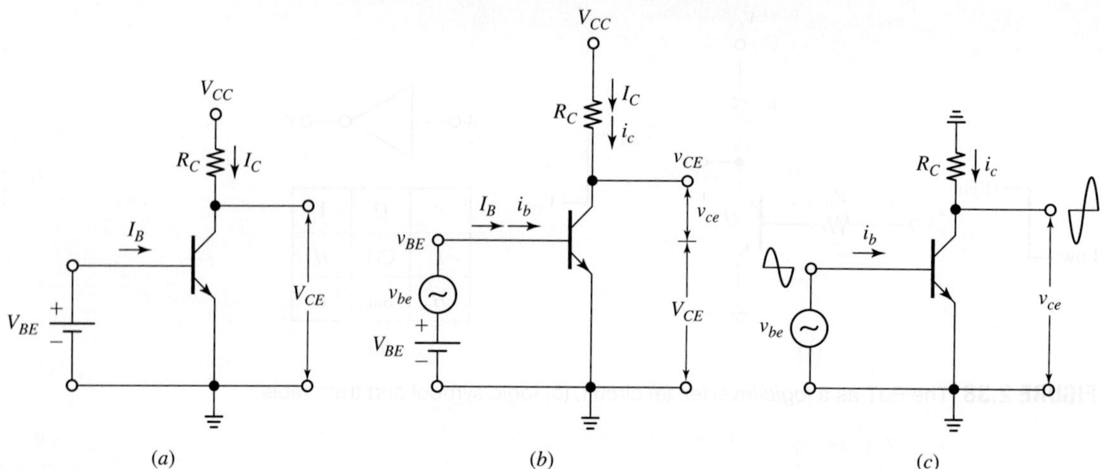

FIGURE 2.39 Systematic analysis of the BJT as a small-signal amplifier. The actual circuit is shown in (b), while (a) shows its large-signal or *dc* version, and (c) shows its small-signal or *ac* version.

with the resistance R_C, to bias the BJT at the corresponding operating point $Q_0 = Q_0(I_C, V_{CE})$ in the active region (see Fig. 2.40b). Applying Eq. (2.21) at Q_0 gives

$$I_C = I_s e^{V_{BE}/V_T}\left(1 + \frac{V_{CE}}{V_A}\right) \tag{2.45}$$

Along with the i_C-v_{CE} curves of the BJT, Fig. 2.40b also shows the curve of the circuit external to the collector, a curve known as the *load line*,

$$i_C = \frac{V_{CC} - v_{CE}}{R_C} \tag{2.46}$$

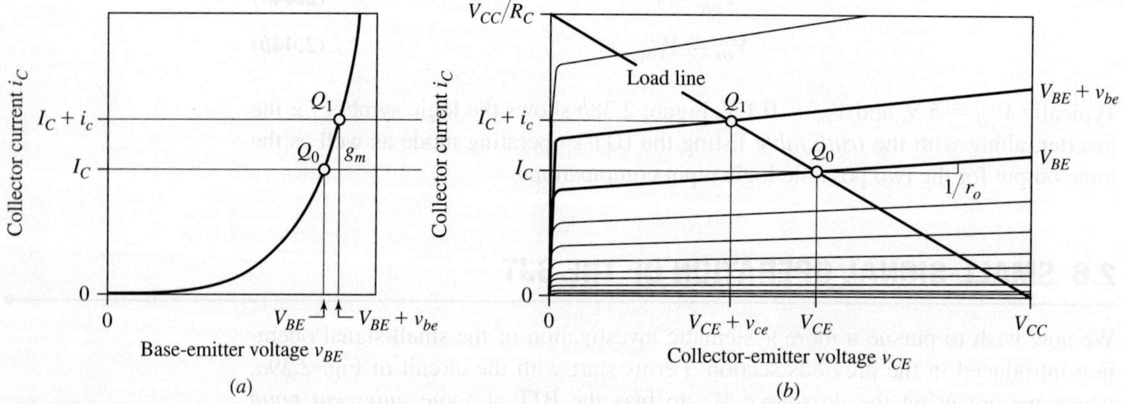

FIGURE 2.40 Graphical illustrations of the BJT amplifier of Fig. 2.39.

The quiescent point $Q_0 = Q_0(I_C, V_{CE})$ lies right where the BJT curve corresponding to the given value of V_{BE} intersects the load line.

If we now turn on the ac source v_{be} as in Fig. 2.39b, the operating point will move up and down the exponential curve of Fig. 2.40a as well as up and down the load line of Fig. 2.40b. In Fig. 2.40 we have captured a *positive* alternation of v_{be}, during which the instantaneous operating point in Fig. 2.40a is $Q_1 = Q_1(I_C + i_c, V_{BE} + v_{be})$, and in Fig. 2.40b is $Q_1 = Q_1(I_C + i_c, V_{CE} + v_{ce})$. We wish to find a relationship between the ac current i_c and the ac voltages v_{be} and v_{ce}. Applying Eq. (2.21) at the new operating point Q_1,

$$I_C + i_c = I_s e^{(V_{BE}+v_{be})/V_T}\left(1 + \frac{V_{CE} + v_{ce}}{V_A}\right) = I_s e^{V_{BE}/V_T}\left(1 + \frac{V_{CE}}{V_A} + \frac{v_{ce}}{V_A}\right)e^{v_{be}/V_T}$$

For $V_{CE}/V_A \ll 1$ we approximate $I_s \exp(V_{BE}/V_T) \cong I_C$ and rewrite as

$$I_C + i_c \cong I_C\left(1 + \frac{v_{ce}}{V_A/I_C}\right)e^{v_{be}/V_T}$$

Performing the series expansion of the exponential term gives

$$I_C + i_c \cong I_C\left(1 + \frac{v_{ce}}{V_A/I_C}\right)\left[1 + \frac{v_{be}}{V_T} + \frac{1}{2!}\left(\frac{v_{be}}{V_T}\right)^2 + \frac{1}{3!}\left(\frac{v_{be}}{V_T}\right)^3 + \cdots\right]$$

$$\cong I_C + \frac{I_C}{V_T}v_{be} + \frac{v_{ce}}{V_A/I_C} + \cdots \tag{2.47}$$

As long as we can ignore higher-order terms involving ac products and powers, Eq. (2.47) allows us to write

$$i_c = g_m v_{be} + \frac{v_{ce}}{r_o} \tag{2.48}$$

where

$$g_m = \frac{I_C}{V_T} \tag{2.49}$$

is the *transconductance* of the BJT, and

$$r_o = \frac{V_A}{I_C} \tag{2.50}$$

is the collector's *output resistance*. As shown in Fig. 2.40, g_m and $1/r_o$ represent, respectively, the *slopes* of the i_C-v_{BE} and the *slope* of the i_C-v_{CE} curve at Q_0. Both g_m and r_o depend on the operating current I_C. Moreover, the fact that $V_A \gg V_T$ indicates a much weaker dependence of i_c on v_{ce} than on v_{be}.

We wish to assess under what conditions we can ignore higher-order powers and products in Eq. (2.47). By inspection, we can stop at the second term within brackets in Eq. (2.47) provided we keep v_{be} small enough to satisfy the condition $\frac{1}{2}(v_{be}/V_T)^2 \ll |v_{be}|/V_T$, that is,

$$|v_{be}| \ll 2V_T \,(\cong 52 \text{ mV}) \tag{2.51}$$

For obvious reasons, Eq. (2.48) is referred to as the *small signal approximation*, and Eq. (2.51) quantifies the validity of such an approximation. The error ε incurred in the small-signal approximation is

$$\varepsilon \cong \frac{v_{be}}{2V_T} \cong \frac{v_{be}}{52 \text{ mV}} \tag{2.52}$$

or about 2% for every mV of v_{be}. Thus, if we wish to keep ε below 10% (an acceptable error in most practical situations), then we need to ensure that

$$|v_{be}| \le 5 \text{ mV} \tag{2.53}$$

This shall be our *working condition* as we move along.

EXAMPLE 2.12   (a) With reference to Fig. 2.40*a*, suppose v_{be} is an ac signal with peak values of ± 5 mV. Assuming $V_A = \infty$ for simplicity, use the small-signal approximation to estimate the peak values of i_c at $I_C = 1$ mA.

(b) Find the *exact* peak values of i_c, compare with the approximated values of part (*a*) and comment.

Solution

(a) By Eq. (2.49), $g_m = 1/26$ A/V. The small-signal approximation of Eq. (2.48) predicts, for i_c, peak values of $(1/26)(\pm 5 \times 10^{-3}) \cong \pm 192\ \mu$A.

(b) The exact peak values of i_c are

$$i_c = I_C(1 - e^{v_{be}/V_T}) = (1.0 \text{ mA}) \times (e^{\pm 5/26} - 1)$$

or $+212\ \mu$A and $-175\ \mu$A, respectively. Because of the curvature of the i_C-v_{BE} characteristic, the small-signal approximation *underestimates* the positive current peak by $(212 - 192)/212 \cong 9.4\%$, and *overestimates* the negative peak by $(192 - 175)/175 \cong 9.7\%$. These errors are consistent with Eq. (2.52).

Just like we use the *dc* equivalent of Fig. 2.39*a* to investigate the biasing conditions of our BJT, we use the *ac* equivalent of Fig. 2.39*c* to investigate its operation as an amplifier. Indeed, the latter gives, by KVL, Ohm's law, and Eq. (2.48),

$$v_{ce} = 0 - R_C i_c = -R_C\left(g_m v_{be} + \frac{v_{ce}}{r_o}\right)$$

Collecting, and solving for v_{ce}, we can write

$$v_{ce} = -g_m(R_C /\!/ r_o)v_{be}$$

indicating that our circuit magnifies v_{be} by the *gain* $-g_m(R_C /\!/ r_o)$.

Assuming $R_C = 10$ kΩ and $V_A = 100$ V in Fig. 2.39b, find the small-signal gain at $I_C = 1$ mA. **EXAMPLE 2.13**

Solution

We have $g_m = 1/26$ A/V, $r_o = 100/1 = 100$ kΩ, and $-gm(R_C//r_o) = -(1/26) \times (10//100)10^3 \cong -350$ V/V.

The Small-Signal BJT Model

Figure 2.41 shows the *small-signal model* for the BJT. The function of this model, also referred to as *incremental model* or simply as *ac model*, is to provide a circuit representation of the dependence of i_c upon v_{be} and v_{ce} as expressed by Eq. (2.48). As we know, the dependence on v_{ce} is much weaker than that on v_{be}, so the term v_{ce}/r_o is sometimes ignored to speed up calculations. This is tantamount to assuming $r_o = \infty$ in the ac model. The model includes also the resistance

$$r_\pi = \frac{v_{be}}{i_b} \tag{2.54}$$

to account for the fact that the BJT responds to v_{be} not only with the collector current i_c but also with the base current i_b. The ratio of the two is called the *common-emitter ac current gain*

$$\beta_0 = \frac{i_c}{i_b} \tag{2.55}$$

and it is customary to assume $\beta_0 \cong \beta_F$, even though there are subtle differences between the two betas.[4] By the above equations we have $r_\pi = v_{be}/i_b = (i_c/g_m)/i_b = (i_c/i_b)/g_m$, or

$$r_\pi = \frac{\beta_0}{g_m} = \beta_0 \frac{V_T}{I_C} \tag{2.56}$$

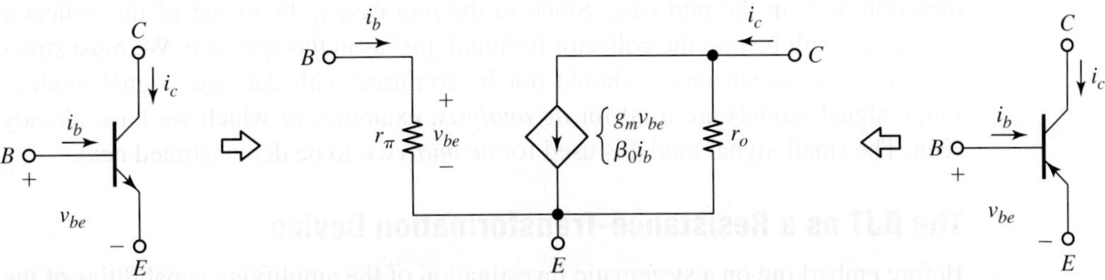

FIGURE 2.41 Small-signal BJT model. This model applies *both* to the *npn* and the *pnp* BJT.

TABLE 2.1   Small-signal parameter summary.

Definition	Calculation
$g_m = \left.\dfrac{\partial i_C}{\partial v_{BE}}\right|_{V_{CE}}$	$g_m = \dfrac{I_C}{V_T}$
$\dfrac{1}{r_\pi} = \left.\dfrac{\partial i_B}{\partial v_{BE}}\right|_{V_{CE}}$	$r_\pi = \dfrac{\beta_0}{g_m} = \beta_0 \dfrac{V_T}{I_C}$
$\dfrac{1}{r_o} = \left.\dfrac{\partial i_C}{\partial v_{CE}}\right|_{V_{BE}}$	$r_o = \dfrac{V_A}{I_C}$

indicating that r_π itself depends on the bias current I_C, just like g_m and r_o do. To develop a feel for the various parameters, consider a BJT with $\beta_0 = 100$ and $V_A = 100$ V that is operating at $I_C = 1$ mA. Then,

$$1/g_m = 26 \ \Omega \ \text{(small)} \qquad r_\pi = 2.6 \ \text{k}\Omega \ \text{(medium)} \qquad r_o = 100 \ \text{k}\Omega \ \text{(large)}$$

It pays to have an idea of the orders of magnitude of these parameters when dealing with a BJT amplifier. Their definitions (for the case of the *npn* BJT) as well as their calculations are summarized in Table 2.1.

As we know, in forward-active operation a BJT can be regarded either as a *voltage-controlled* (VC) or as a *current-controlled* (CC) device. This versatility carries over also to the small-signal domain, where we can express the value of the dependent source either as $g_m v_{be}$ (VCCS) or as $\beta_0 i_b$ (CCCS). This is shown explicitly in the ac model of Fig. 2.41. When we use this second alternative, Eq. (2.48) becomes

$$i_c = \beta_0 i_b + \frac{v_{ce}}{r_o} \tag{2.57}$$

As we proceed, we have the option of using whichever of the two expressions makes our analysis simpler.

We wish to point out that the model of Fig. 2.41 applies *both* to the *npn* BJT and the *pnp* BJT, with *no change* in voltage polarities or current directions. Indeed, consider the effect of *increasing* v_{BE} by v_{be} in both devices. In the *npn* BJT i_C will also *increase*, but in the *pnp* BJT i_C will *decrease*. Consequently, i_c will have the same direction as i_C in the *npn* case (i_c *into* the collector terminal), but the opposite direction as i_C in the *pnp* case. Since in the *pnp* case i_C flows out of the collector terminal, i_c will be *into* the collector terminal, just as in the *npn* case. We must stress that the small-signal model should not be confused with the large signal models. Large-signal models are used for *dc analysis*, examples of which we have already seen. The small-signal model is used for *ac analysis*, to be demonstrated next.

The BJT as a Resistance-Transformation Device

Before embarking on a systematic investigation of the amplifying capabilities of the BJT, we wish to explore some intriguing resistance-transformation properties that will prove quite useful as we proceed. Specifically, we wish to find the *small-signal*

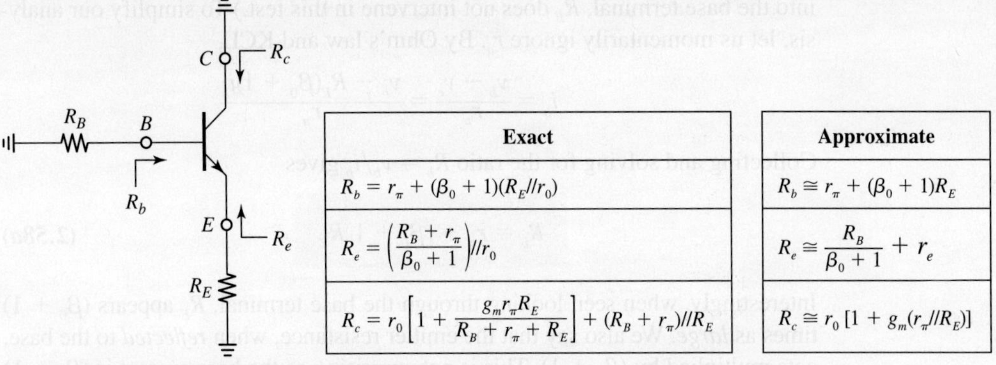

	Exact	**Approximate**
	$R_b = r_\pi + (\beta_0 + 1)(R_E//r_0)$	$R_b \cong r_\pi + (\beta_0 + 1)R_E$
	$R_e = \left(\dfrac{R_B + r_\pi}{\beta_0 + 1}\right)//r_0$	$R_e \cong \dfrac{R_B}{\beta_0 + 1} + r_e$
	$R_c = r_0\left[1 + \dfrac{g_m r_\pi R_E}{R_B + r_\pi + R_E}\right] + (R_B + r_\pi)//R_E$	$R_c \cong r_0\left[1 + g_m(r_\pi//R_E)\right]$

FIGURE 2.42 The small-signal resistances seen looking into the BJT's terminals.

resistances seen looking *into* the base, the emitter, and the collector terminals of the circuit of Fig. 2.42. We denote these resistances as R_b, R_e, and R_c (lower-case subscripts). Conversely, we denote the resistances *external* to the BJT as R_B and R_E (upper-case subscripts.) To find the small-signal resistance R_x seen looking into terminal X ($X = B, E, C$), proceed as follows:

- Replace the BJT with its small-signal model
- Apply a test voltage v_x to the terminal X under consideration
- Determine the resulting current i_x into X
- Find the resistance seen looking into that terminal as $R_x = v_x/i_x$

This procedure will give us an opportunity to put the small-signal BJT model to use. At times we shall find it more convenient to express the dependent source as $g_m v_{be}$, at others as $\beta_0 i_b$.

- **The Small-Signal Resistance R_b Seen Looking into the Base.** The circuit to find this resistance is shown in Fig. 2.43a (note that since we are looking right

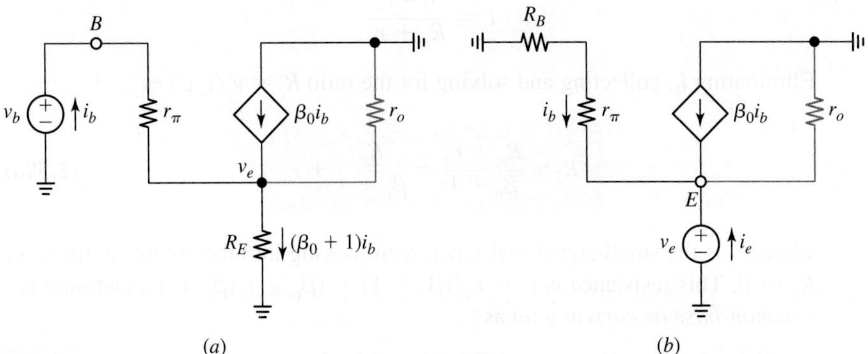

(a) (b)

FIGURE 2.43 Test circuits to find the small-signal resistance (a) R_b seen looking into the *base*, and (b) R_e seen looking into the *emitter*. In both cases we ignore r_o to simplify our initial analysis.

into the base terminal, R_B does not intervene in this test.) To simplify our analysis, let us momentarily ignore r_o. By Ohm's law and KCL,

$$i_b = \frac{v_b - v_e}{r_\pi} = \frac{v_b - R_E(\beta_0 + 1)i_b}{r_\pi}$$

Collecting and solving for the ratio $R_b = v_b/i_b$ gives

$$R_b = r_\pi + (\beta_0 + 1)R_E \tag{2.58a}$$

Interestingly, when seen looking through the base terminal, R_E appears $(\beta_0 + 1)$ times as *large*. We also say that the emitter resistance, when *reflected* to the base, gets multiplied by $(\beta_0 + 1)$. This is not surprising as the base current is $(\beta_0 + 1)$ times as *small* as that flowing through R_E. It pays to think of the combination made up of R_E and the dependent source $\beta_0 i_b$ as a *single resistance* of value $(\beta_0 + 1)R_E$.

In the above analysis we have deliberately ignored r_o, but we can readily take it into account by noting that it is in *parallel* with R_E (r_o shares the same node pair as R_E, namely, v_e and ac ground). We thus modify Eq. (2.58a) by replacing R_E with $R_E//r_o$ and write

$$R_b = r_\pi + (\beta_0 + 1)(R_E//r_o) \tag{2.58b}$$

(The reader should be observant of simple tricks, such as the present one, to simplify hand analysis.)

- **The Small-Signal Resistance R_e Seen Looking into the Emitter.** The circuit to find this resistance is shown in Fig. 2.43*b* (note that since we are looking right into the emitter terminal, R_E does not intervene in this test.) Again, to simplify our analysis, we momentarily ignore r_o. Summing currents into the emitter node, we get

$$i_b + \beta_0 i_b + i_e = 0$$

Also, by Ohm's law,

$$i_b = \frac{0 - v_e}{R_B + r_\pi}$$

Eliminating i_b, collecting and solving for the ratio $R_e = v_e/i_e$ gives

$$R_e = \frac{R_B + r_\pi}{\beta_0 + 1} = \frac{R_B}{\beta_0 + 1} + r_e \tag{2.59a}$$

where r_e is the small-signal resistance seen looking into the emitter in the limit $R_B \to 0$. This resistance is $r_e = r_\pi/(\beta_0 + 1) = (\beta_0/g_m)/(\beta_0 + 1)$. Defining the *common-base ac current gain* as

$$\alpha_0 = \frac{\beta_0}{\beta_0 + 1} \tag{2.60}$$

we get

$$r_e = \frac{\alpha_0}{g_m} \cong \frac{1}{g_m} \qquad\qquad (2.61)$$

In general, r_e is very small compared to r_π and r_o: for instance at $I_C = 1$ mA we have $r_e \cong 1/g_m = (26\text{ mV})/(1\text{ mA}) = 26\ \Omega$. Equation (2.59a) indicates that the presence of the BJT makes R_B appear $(\beta_0 + 1)$ times as *small* when seen looking through the emitter terminal. We also say that the base resistance R_B, *reflected* to the emitter, gets divided by $(\beta_0 + 1)$. This stems from the fact that the emitter current is $(\beta_0 + 1)$ times as *large* as that through R_B. Clearly, the effect of the BJT upon R_B is *inverse* of that upon R_E.

In the above analysis we have deliberately ignored r_o, but we can readily take it into account by noting that it is in parallel with the test source. We thus modify Eq. (2.59a) as

$$R_e = \left(\frac{r_\pi + R_B}{\beta_0 + 1}\right)//r_o \qquad\qquad (2.59b)$$

Let the BJT of Fig. 2.42 have $\beta_0 = 100$, $g_m = 1/26$ A/V, $r_\pi = 2.6$ kΩ, and $r_o = 100$ kΩ. If $R_B = 10$ kΩ and $R_E = 1.0$ kΩ, estimate R_b and R_e.   **EXAMPLE 2.14**

Solution

Applying the approximate expressions of Eqs. (2.58a) and (2.59a) we get

$$R_b \cong 2.6 + 101 \times 1.0 = 103.6 \text{ k}\Omega \text{ (large)}$$

$$R_e \cong \frac{10{,}000}{101} + 26 = 125\ \Omega \text{ (small)}$$

Using instead the exact Eqs. (2.58b) and (2.59b) we get $R_b = 102.6$ kΩ and $R_e = 124.6\ \Omega$. The difference is so small that ignoring r_o is quite acceptable, at least in this example.

- **The Small-Signal Resistance R_c Seen Looking into the Collector.** The circuit to find this resistance is shown in Fig. 2.44, with r_o now deliberately included. This time we are using the alternative form $g_m v_{be}$ for the dependent source. By KCL and Ohm's law,

$$i_c = g_m v_{be} + \frac{v_c - v_e}{r_o}$$

By the voltage divider formula,

$$v_{be} = -\frac{r_\pi}{R_B + r_\pi} v_e$$

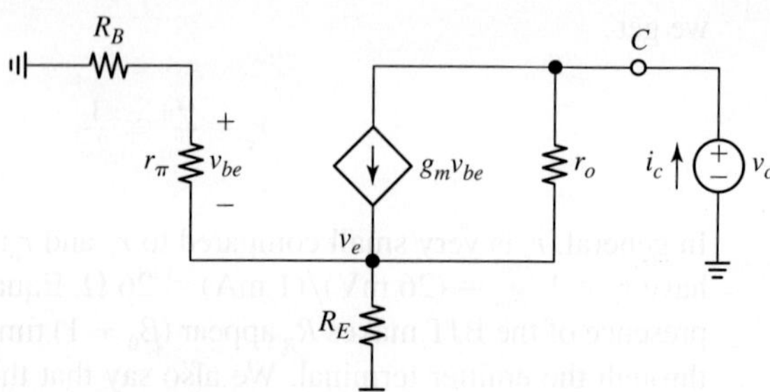

FIGURE 2.44 Test circuit to find the small-signal resistance R_c seen looking into the *collector.*

The test current i_c splits between r_o and the dependent source and then re-converges at node v_e, so we apply Ohm's law to write

$$v_e = [(R_B + r_\pi)/\!/R_E] \times i_c$$

Eliminating v_{be} and v_e, collecting, and solving for the ratio $R_c = v_c/i_c$ gives, after some algebra,

$$R_c = r_o\left[1 + \frac{g_m(r_\pi/\!/R_E)}{1 + R_B/(r_\pi + R_E)}\right] + [(R_B + r_\pi)/\!/R_E] \qquad (2.62a)$$

In most cases of practical interest the last term is negligible compared to the rest, so it will be ignored. Of particular interest is the limiting case $R_B \ll R_E + r_\pi$, for then Eq. (2.62a) simplifies to

$$R_c \cong r_o[1 + g_m(r_\pi/\!/R_E)] \qquad (2.62b)$$

Two additional sub-cases are of great interest. One is $R_E \ll r_\pi$, for then Eq. (2.62b) reduces to

$$R_c \cong r_o(1 + g_m R_E) \qquad (2.62c)$$

The other case is $R_E \gg r_\pi$, for then Eq. (2.61a) reduces to $R_c \cong (1 + g_m r_\pi)$, or

$$R_c \cong r_o(1 + \beta_0) \qquad (2.62d)$$

Regardless, we note that the presence of R_E *raises* the resistance seen looking into the collector. To justify this physically, consider first the case $R_E = 0$, which results in $v_e = 0$ in Fig. 2.44. In this case we have $v_{be} = 0$, indicating that the dependent source in Fig. 2.44 is off, thus giving $i_c = v_c/r_o$, or $R_c = v_c/i_c = r_o$. Now consider the case $R_E \neq 0$, which results in $v_e > 0$ because of the current coming from the test source via r_o. We now have $v_{be} < 0$, indicating that the dependent source is on and its direction is reversed, so that it pumps current *into* node C. But, this *reduces* the current i_c required of the test source, thus *raising* the ratio v_c/i_c. The fact that the test-source current is met by a counter-action that tends to reduce it indicates that R_E provides a *negative-feedback* action. In Chapter 7 we shall encounter alternative forms of negative feedback for the purpose of raising R_c.

EXAMPLE 2.15

(a) Find R_c for the BJT of Example 2.14.

(b) Repeat if R_E is raised from 1.0 kΩ to 100 kΩ. Comment on your findings.

EXAMPLE 2.15

Solution

(a) Applying Eq. (2.62a) we get

$$R_c = 100\left[1 + \frac{(2.6//1.0)/0.026}{1 + 10/(2.6 + 1)}\right] + (10 + 2.6)//1 = 835 + 0.93$$

$$= 836 \text{ k}\Omega \text{ (quite large)}$$

(b) By similar calculation we now find $R_c \cong 9$ MΩ (huge). Clearly, the larger R_E, the higher R_c. In the limit $R_E \to \infty$, Eq. (2.62d) predicts $R_c \cong 101r_o \cong 10$ MΩ, a truly huge value.

Equations (2.58) through (2.62) reveal an intriguing BJT feature, namely, the ability to alter the values of resistances seen looking into one of its terminals. This is not surprising, as the dependent source across the C-E port, in turn controlled by the B-E port, establishes interdependence among the three BJT terminals. The expressions for the three resistances are tabulated in Fig. 2.42. Also shown separately in Fig. 2.45 are simpler, if approximate, expressions for the student to keep handy as we embark upon the study of discrete BJT amplifiers in the rest of this chapter. We note that the last two examples reveal a tendency that holds in general, namely, a *crescendo* in resistance levels as we go from R_e (small), to R_b (medium), to R_c (large).

A Practical Application: The BJT as a Current Source

The BJT's ability to achieve truly high R_c values makes it suited to the implementation of current sources/sinks. Figure 2.46 offers an example of how a *pnp* BJT can be configured to *source* current to a load (LD). If we need to *sink*

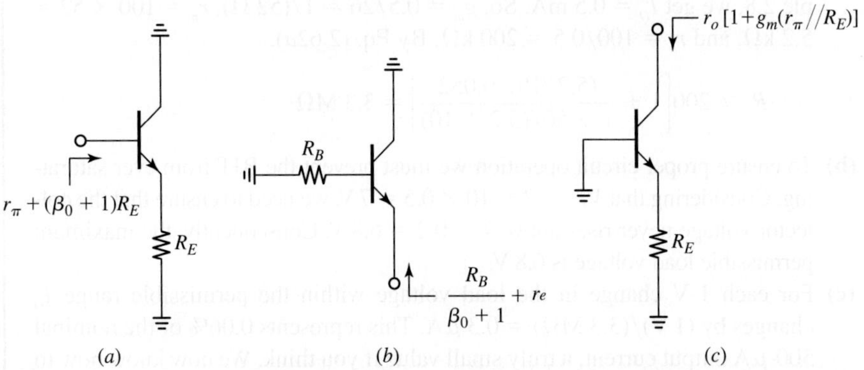

FIGURE 2.45 Visualizing the small-signal resistances seen looking into the base, emitter, and collector. The effect of r_o is negligible in (a) and (b), so it has been ignored.

current from the load, then we use an *npn* BJT with all voltages and currents reversed. As we shall see, a common application of current sources/sinks is to bias other circuits.

EXAMPLE 2.16   Let the BJT of Fig. 2.46 have $\beta_0 = \beta_F = 100$, $V_A = 100$ V, $V_{EB(on)} = 0.7$ V, and $V_{EC(EOS)} = 0.2$ V.

(a) Find I_O and R_o.
(b) What is the maximum load voltage for which the circuit will still operate properly?
(c) By how much does I_O change for each 1-V change in the voltage across the load? Express this change in percentage form.

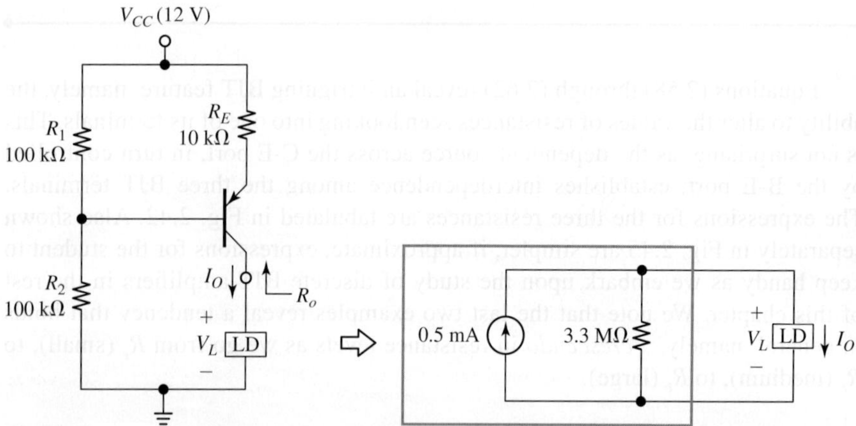

FIGURE 2.46 Using a *pnp* BJT to implement a *current source*, and its Norton equivalent.

Solution

(a) We have $V_{BB} = 6$ V and $R_B = 50$ kΩ. Proceeding as in the first part of Example 2.8, we get $I_O = 0.5$ mA. So, $g_m = 0.5/26 = 1/(52\ \Omega)$, $r_\pi = 100 \times 52 = 5.2$ kΩ, and $r_o = 100/0.5 = 200$ kΩ. By Eq. (2.62a),

$$R_o \cong 200\left[1 + \frac{(5.2//10)/0.052}{1 + 50/(5.2 + 10)}\right] = 3.3\ \text{M}\Omega$$

(b) To ensure proper circuit operation we must prevent the BJT from ever saturating. Considering that $V_E \cong 12 - 10 \times 0.5 = 7$ V, we need to ensure that the collector voltage never rises above $7 - 0.2 = 6.8$ V. Consequently, the maximum permissible load voltage is 6.8 V.

(c) For each 1-V change in the load voltage within the permissible range I_O changes by $(1\ \text{V})/(3.3\ \text{M}\Omega) = 0.3\ \mu\text{A}$. This represents 0.06% of the nominal 500-μA output current, a truly small value if you think. We now know how to apply a BJT to implement a good-quality current source/sink!

2.7 BJT BIASING FOR AMPLIFIER DESIGN

As we know, to operate a BJT as an amplifier we must *bias* it in the *forward active* (FA) region. Since the amplifier's characteristics are determined by the small-signal parameters g_m, r_π, and r_o, which in turn depend on the bias current I_C, it is apparent that to achieve predictable and stable amplifier characteristics we need to establish a predictable and stable bias current I_C. Among the factors conspiring against us is the spread in the parameter values of the BJTs that we use, particularly $V_{BE(on)}$ and β_F. When designing a BJT circuit, it is customary to assume the *typical values*

$$V_{BE(on)} = 0.7 \text{ V} \tag{2.63a}$$

$$\beta_F = 100 \tag{2.63b}$$

which we shall also refer to as *nominal values*. However, because of fabrication process variations, the *actual values* will vary from one BJT sample to another, even for samples of the same device type, such as the popular 2N2222 considered previously. For example, Eqs. (2.11) and (2.15) indicate that both I_s and β_F are inversely proportional to the base width W_B, which is fabricated very thin (a fraction of a micro-meter) to ensure high betas. With reference to Fig. 2.1, it is not difficult to imagine the impact of even a tiny variation in the depth of the n^+ emitter diffusion during fabrication. Moreover, both I_s and β_F drift with temperature as well as with time, so their actual values can lie anywhere within a range that can be quite wide. For the sake of discussion we shall assume the following parameter-value spreads:

$$0.4 \text{ V} \le V_{BE(on)} \le 0.8 \text{ V} \tag{2.64a}$$

$$50 \le \beta_F \le 200 \tag{2.64b}$$

As a rule, the performance of a transistor circuit should be independent of the particular transistor sample being used, so a good BJT amplifier requires a quiescent point $Q = Q(I_C, V_{CE})$ that is relatively independent of $V_{BE(on)}$ and β_F. As an added bonus, should a BJT fail, we can simply replace it with one of the same family and still expect the same overall performance level, even though the actual parameter values of the new unit may be quite different from those of the old unit.

Based on our studies so far, a given bias current I_C can be established in three different ways,

$$I_C = \beta_F I_B = \alpha_F I_E = I_s e^{V_{BE}/V_T}$$

that is, via I_B, via I_E, or via V_{BE}. In the following we shall investigate merits and drawbacks of each.

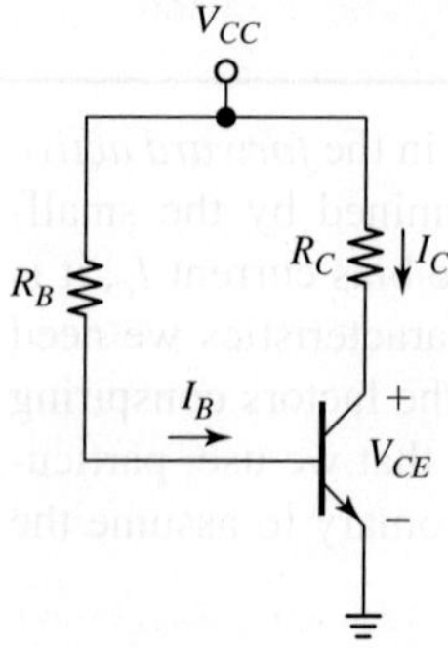

FIGURE 2.47 BJT biasing via I_B.

BJT Biasing via I_B

The biasing scheme of Fig. 2.47 uses R_B to establish the base drive I_B necessary to sustain the desired bias current as $I_C = \beta_F I_B$. Moreover, it uses R_C to achieve the desired voltage V_{CE}. We have

$$I_C = \beta_F \frac{V_{CC} - V_{BE(on)}}{R_B} \tag{2.65a}$$

$$V_{CE} = V_{CC} - R_C I_C \tag{2.65b}$$

We immediately note a serious drawback of this biasing scheme, namely, I_C being directly proportional to β_F, which is notoriously an ill-defined parameter. Consequently, the spread of Eq. (2.64b) will affect I_C as well. An actual example will better illustrate.

EXAMPLE 2.17
(a) Assuming $V_{CC} = 12$ V in the circuit of Fig. 2.47, along with the nominal specifications of Eq. (2.63), specify standard 5% values for R_B and R_C to bias the BJT at $I_C = 1$ mA and $V_{CE} = 5$ V.
(b) Find the range of variability of I_C and V_{CE} as a consequence of the parameter spread of Eq. (2.64). Comment on your results.

Solution

(a) By Eq. (2.65a),

$$R_B = 100 \frac{12 - 0.7}{1} = 1.13 \text{ M}\Omega$$

The closest standard value is 1.1 MΩ. Reinserting it into Eq. (2.65a) gives $I_{C(nom)} \cong 1.03$ mA. By Eq. (2.65b),

$$R_C = \frac{12 - 5}{1.03} = 6.8 \text{ k}\Omega$$

which is a standard value. In summary, using $R_B = 1.1$ MΩ and $R_C = 6.8$ kΩ yields $I_{C(nom)} \cong 1.03$ mA and $V_{CE(nom)} = 5$ V.

(b) Equation (2.65) indicates that I_C is minimized when β_F is minimum and $V_{BE(on)}$ is maximum. Moreover, when I_C is minimized, V_{CE} is maximized, and vice versa. We thus have

$$I_{C(min)} = 50\frac{12 - 0.8}{1100} = 0.51 \text{ mA} \qquad V_{CE(max)} = 12 - 6.8 \times 0.51 = 8.5 \text{ V}$$

Likewise, using the maximum value of β_F and the minimum value of $V_{BE(on)}$ we would get

$$I_C = 200\frac{12 - 0.4}{1100} = 2.1 \text{ mA} \qquad V_{CE} = 12 - 6.8 \times 2.1 = -2.3 \text{ V}$$

The last result is impossible, indicating a saturated BJT! Consequently, under the given conditions, we have

$$V_{CE(min)} = V_{CE(sat)} = 0.1 \text{ V} \qquad I_{C(max)} = \frac{12 - 0.1}{6.8} = 1.75 \text{ mA}$$

It is apparent that under the given conditions, this biasing scheme is unacceptable, for not only does it result in a wide spread of operating points, but it may even drive the BJT in saturation.

However tempting because of its simplicity, the biasing scheme of Fig. 2.47 is seldom used in actual BJT amplifier design. It is used, however, to bias the BJT in saturation in switching applications. As we have already seen in Example 2.11, the BJT is biased in *deep* saturation precisely to cope with the wide spread in the values of β_F. The reason for covering this scheme is to pave the way for the alternative scheme to be discussed next.

BJT Biasing via I_E

A much better alternative is to bias the BJT via its emitter current I_E to get $I_C = \alpha_F I_E$. The reason is that the spread in the values of α_F is far more limited than that of β_F. Indeed, since $\alpha_F = \beta_F/(\beta_F + 1)$, the spread of Eq. (2.64b) translates into the following spread for α_F,

$$0.980 \leq \alpha_F \leq 0.995$$

that is, a spread of less than 1.5%! To establish I_E we need to bias the emitter at some voltage $V_E > 0$, and then use an emitter resistance R_E to set $I_E = V_E/R_E$. This is achieved by biasing the base via the voltage divider R_1-R_2, as depicted in Fig. 2.48. This circuit is identical to that of Fig. 2.27a, so we simply recycle the results developed there and write

$$I_C = \beta_F \frac{V_{BB} - V_{BE(on)}}{R_B + (\beta_F + 1)R_E} \tag{2.66a}$$

$$V_{CE} \cong V_{CC} - (R_C + R_E)I_C \tag{2.66b}$$

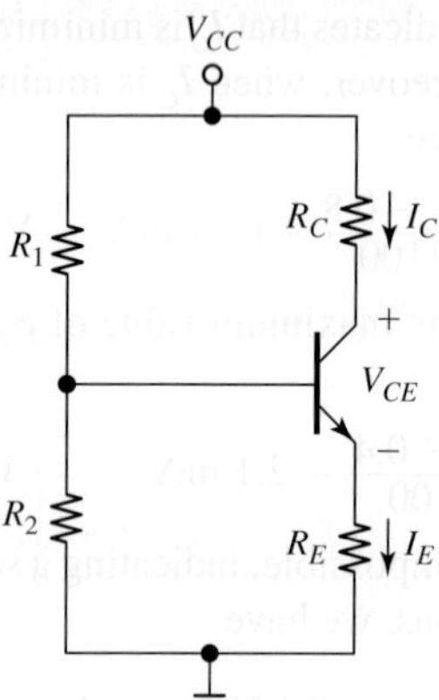

FIGURE 2.48 BJT biasing via I_E.

where

$$V_{BB} = \frac{R_2}{R_1 + R_2} V_{CC} \qquad\qquad R_B = R_1 /\!/ R_2 \qquad\qquad (2.67)$$

(Note that in Eq. (2.66*b*) we have approximated $I_E \cong I_C$.) Dividing numerator and denominator by β_F in the right-hand side of Eq. (2.66*a*), and letting $(\beta_F + 1)/\beta_F \cong 1$, yields the more insightful expression

$$I_C \cong \frac{V_{BB} - V_{BE(on)}}{R_B/\beta_F + R_E}$$

This instructs us on how to go to ensure a fairly stable and predictable bias current I_C:

- To render I_C relatively insensitive to variations in $V_{BE(on)}$ impose

$$V_{BB} \gg \Delta V_{BE(on)}$$

  where $\Delta V_{BE(on)}$ is the expected spread in $V_{BE(on)}$. (In our running example we have $\Delta V_{BE(on)} = 0.8 - 0.4 = 0.4$ V.) Indirectly this condition implies that the emitter voltage V_E be made *large enough* to swamp any variations in $V_{BE(on)}$. Clearly, the larger V_E the better. However, this erodes the signal swing of the collector. A reasonable compromise is to impose $V_E \cong 10\Delta V_{BE(on)}$ and then to bias the collector *halfway* between V_{CC} and V_E for a symmetric collector signal swing. Some authors simply propose the so-called 1/3-1/3-1/3 *Rule:* Let $V_E = \frac{1}{3}V_{CC}$ and $V_{CE} = \frac{1}{3}V_{CC}$, so $R_C I_C = \frac{1}{3}V_{CC}$.
- To render I_C relatively insensitive to variations in β_F impose

$$R_B/\beta_F \ll R_E$$

Indirectly, this condition implies that R_1 and R_2 be chosen *small enough* so that their standby current is *sufficiently large* to swamp the effect of any variations in the base current I_B stemming from the spread in β_F. Clearly, the smaller R_1 and R_2 the better. However, this increases the current drain from V_{CC} and, even more undesirable, it lowers the input resistance when the base is used as the amplifier's input. A reasonable compromise is to impose a standby current of about $10 I_{B(nom)}$.

EXAMPLE 2.18

(a) Assuming $V_{CC} = 12$ V in the circuit of Fig. 2.48, along with the nominal specifications of Eq. (2.63), specify standard 5% resistances to bias the BJT at $I_C = 1$ mA and the collector *halfway* between V_{CC} and V_E. Show your final circuit and calculate the nominal values of I_C and V_{CE}.

(b) Find the range of variability of I_C and V_{CE} as a consequence of the parameter spread of Eq. (2.64). Comment on your results.

Solution

(a) Let $V_E = 10\Delta V_{BE(on)} = 10 \times 0.4 = 4$ V. To bias the collector halfway we need $V_C = (V_{CC} + V_E)/2 = (12 + 4)/2 = 8$ V. Then, approximating $I_E \cong I_C$, we find

$$R_E = \frac{V_E}{I_E} \cong \frac{4}{1} = 4 \text{ k}\Omega \qquad R_C = \frac{V_{CC} - V_C}{I_C} = \frac{12 - 8}{1} = 4 \text{ k}\Omega$$

The closest standard values are $R_E = R_C = 3.9$ kΩ.

We have $I_{B(nom)} = 1/100 = 10 \ \mu$A. Impose a current through R_2 of $10 I_{B(nom)}$, or $10 \times 10 = 100 \ \mu$A. Considering that we have, by KVL, $V_B = V_E + V_{BE(on)} = 4 + 0.7 = 4.7$ V, then

$$R_2 = \frac{V_B}{I_{R_2}} = \frac{4.7}{0.1} = 47 \text{ k}\Omega$$

which is a standard value. The current through R_1 is, by KCL, $10 + 100 = 110 \ \mu$A, so

$$R_1 = \frac{V_{CC} - V_B}{I_{R_1}} = \frac{12 - 4.7}{0.11} = 66 \text{ k}\Omega$$

The closest standard value is 68 kΩ. The circuit is shown in Fig. 2.49.

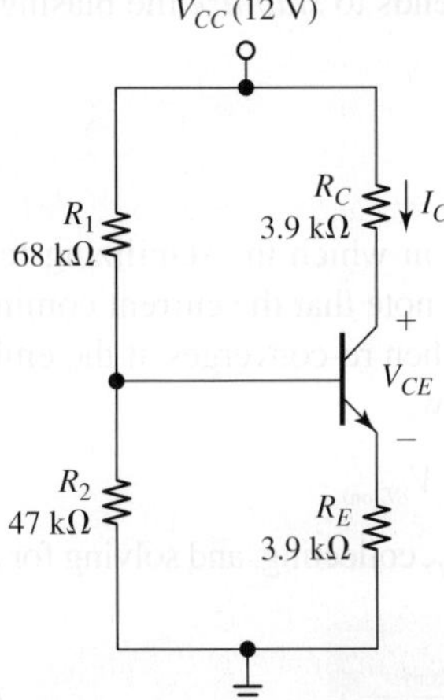

	I_C (mA)	V_{CE} (V)
Min	0.904	3.4
Nom	0.996	4.2
Max	1.109	4.9

FIGURE 2.49 Circuit of Example 2.18.

To find the nominal values of I_C and V_{CE}, insert $R_1 = 68$ kΩ and $R_2 = 47$ kΩ into Eq. (2.67) and find $V_{BB} = 4.9$ V and $R_B = 27.8$ kΩ. Then, use Eq. (2.66) to find

$$I_{C(nom)} = 100 \frac{4.9 - 0.7}{27.8 + (100 + 1)3.9} = 0.996 \text{ mA}$$

$$V_{CE(nom)} = 12 - (3.9 + 3.9)0.996 = 4.2 \text{ V}$$

(b) Equation (2.66) indicates that I_C is minimized when β_F is minimum and $V_{BE(on)}$ is maximum. Moreover, when I_C is minimized, V_{CE} is maximized, and vice versa. We thus have

$$I_{C(min)} = 50 \frac{4.9 - 0.8}{27.8 + (50 + 1)3.9} = 0.904 \text{ mA}$$

$$V_{CE(max)} = 12 - (3.9 + 3.9)0.904 = 4.9 \text{ V}$$

Likewise, using the maximum value of β_F and the minimum value of $V_{BE(on)}$ we get

$$I_{C(max)} = 200 \frac{4.9 - 0.4}{27.8 + (200 + 1)3.9} = 1.109 \text{ mA}$$

$$V_{CE(min)} = 12 - (3.9 + 3.9)1.109 = 3.4 \text{ V}$$

The data are tabulated in Fig. 2.49, where we observe a spread in I_C on the order of $\pm 10\%$. Not at all bad, considering the much wider spreads of Eq. (2.64)!

You may be wondering what makes the BJT maintain I_C at its prescribed value of 1 mA. Suppose the BJT attempted to draw *less* than 1 mA. Then, the voltage drop across R_E would decrease, causing a *decrease* also in V_E. But this, in turn, would *increase* V_{BE}, thus directing the BJT to draw *more* current. Conversely, any attempt to draw *more* than 1 mA would be met by a *decrease* in V_{BE}, and thus an invitation to draw *less* current. In either case, any attempt by I_C to deviate from its prescribed value of about 1 mA is met by a *counteraction* that tends to restore I_C to its prescribed value. This state of affairs is summarized by saying that R_E provides a *negative feedback* action around the BJT, and that this action tends to *stabilize* the biasing condition of the device (more on this in Chapter 7).

Feedback Bias

Figure 2.50 shows an alternative biasing scheme in which the stabilizing feedback action is now provided by R_C and R_F. To see how, note that the current coming from R_C splits between the base and the collector, and then re-converges at the emitter, so it must equal I_E, as shown. By KVL and Ohm's law,

$$V_{CC} = R_C I_E + R_F I_B + V_{BE(on)}$$

Substituting $I_E = [(\beta_F + 1)/\beta_F] I_C$ and $I_B = I_C/\beta_F$, collecting, and solving for I_C gives

$$I_C = \beta_F \frac{V_{CC} - V_{BE(on)}}{R_F + (\beta_F + 1)R_C} \tag{2.68a}$$

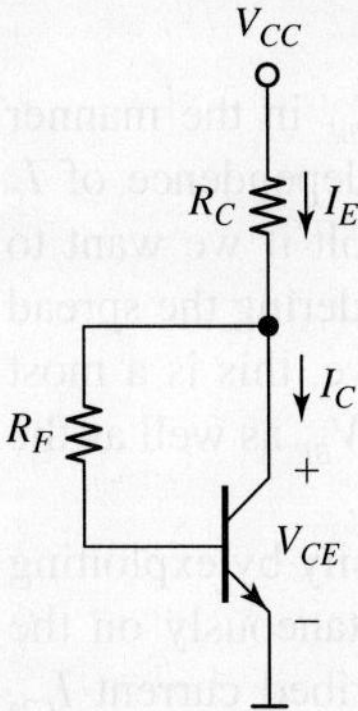

FIGURE 2.50 Feedback bias.

This expression is similar to that of Eq. (2.66a), but with R_C in place of R_E, R_F in place of R_B, and V_{CC} in place of V_{BB}. Consequently, the stabilizing advantages discussed above hold also in the present case. We also have, by KVL, $V_{CE} = V_{BE(on)} + R_F I_B$, or

$$V_{CE} = V_{BE(on)} + \frac{R_F}{\beta_F} I_C \qquad (2.68b)$$

indicating that this scheme does not offer much flexibility for the specification of V_{CE}. Typically, V_{CE} is just a bit higher than $V_{BE(on)}$, so the collector's downswing capability is much more limited than with the scheme of Fig. 2.48. The circuit of Fig. 2.50 finds application in preamplifier situations, where the signals are small enough to fit within the modest signal swing.

EXAMPLE 2.19

(a) Assuming $V_{CC} = 12$ V in the circuit of Fig. 2.50, along with the nominal specifications of Eq. (2.63), specify standard 5% resistances to bias the BJT at $I_C = 1$ mA.

(b) What are the nominal collector swing capabilities of your circuit?

Solution

(a) Impose $R_F \ll (\beta_F + 1)R_C = 101R_C$. Pick $R_F = 10R_C$. Then, substitution into Eq. (2.68a) gives

$$1 = 100 \frac{12 - 0.7}{10R_C + (100 + 1)R_C}$$

or $R_C = 10.2$ kΩ. Pick the standard value $R_C = 10$ kΩ. Then, $R_F = 100$ kΩ.

(b) Plugging the above resistance values into Eq. (2.68) gives $I_C = 1.02$ mA and $V_{CE} = 1.7$ V. The nominal downswing is $V_{CE} - V_{CE(EOS)} = 1.7 - 0.2 = 1.5$ V. The nominal upswing is $V_{CC} - V_{CE} = 12 - 1.7 = 10.3$ V, indicating a highly asymmetric situation.

BJT Biasing via V_{BE} (Current Mirror)

The third way of biasing a BJT is via a suitable voltage drive V_{BE} in the manner illustrated previously in Fig. 2.39a. Because of the exponential dependence of I_C upon V_{BE}, the V_{BE} drive needs to be accurate down to the milli-volt if we want to ensure a prescribed I_C with a good degree of reproducibility. Considering the spread of Eq. (2.64a), along with the fact that V_{BE} is temperature sensitive, this is a most arduous task, unless we have a means for anticipating the required V_{BE} as well as the ability to continuously adjust it to follow the whims of temperature.

In integrated circuits (ICs) this task is accomplished rather easily by exploiting the superior matching characteristics of devices fabricated simultaneously on the same substrate. Simply put, to bias a certain BJT Q_2 at a prescribed current I_{C2}, we use a twin BJT Q_1 connected as a diode, and we drive it with a current I_{C1} of the *same* magnitude as the desired current I_{C2}. Q_1 then develops a voltage V_{BE} that is fed also to Q_2. Since the two devices are matched and experience the *same* V_{BE} drop, I_{C2} will simply *mimic* I_{C1}, this being the reason why the two BJTs are said to form a *current mirror*. Moreover, if the BJTs lie next to each other on the chip, they will experience the same temperature variations, so their characteristics will drift identically, a feature known as *temperature tracking*. This ingenious technique is illustrated in Fig. 2.51. Ignoring the tiny base currents, and also ignoring the Early effect as is customary in dc analysis, we express the bias conditions of Q_2, the device being biased, as $I_{C2} = I_{C1}$, or

$$I_{C2} = \frac{V_{CC} - V_{BE}}{R_1} \tag{2.69a}$$

$$V_{CE2} = V_{CC} - R_2 I_{C2} \tag{2.69b}$$

Current mirrors find wide application in IC design, both as current-signal processing blocks also known as *current reversers*, and as dc biasing blocks for other circuits. For instance, Q_2 of this circuit could be used to provide the emitter current bias for yet another BJT for its use as an amplifier.

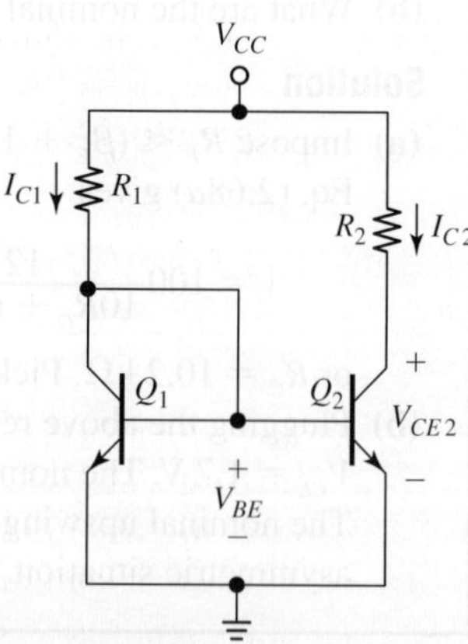

FIGURE 2.51 Current-mirror arrangement to bias Q_2.
Note: Q_1 and Q_2 are matched BJTs.

> Assuming $V_{CC} = 5$ V in the circuit of Fig. 2.51, along with the nominal specifica-
> tions of Eq. (2.63), specify standard 5% resistances to bias Q_2 at $I_C = 1$ mA and
> its collector right in the middle of the active region.
>
> **Solution**
>
> By Ohm's law, $R_1 = (5 - 0.7)/1 = 4.3$ kΩ, and $R_2 = (5 - 2.5)/1 = 2.5$ kΩ
> (use 2.4 kΩ).

EXAMPLE 2.20

It goes without saying that the biasing scheme of Fig. 2.51 works well only in IC situations. If we were to use discrete BJT samples, they would most likely be mismatched, and certainly would not track each other in temperature as closely as in the case of monolithic devices. If you use PSpice to simulate a circuit, be aware that all BJTs with the same device model are treated as identical, giving the beginner the false impression of matched devices throughout. If needed, one can use Monte Carlo techniques to simulate the parameter spreads of real-life discrete devices.[6]

2.8 BASIC BIPOLAR VOLTAGE AMPLIFIERS

Depending on which terminal we apply the input to, and which terminal we obtain the output from, a BJT can be used in any one of three amplifier configurations: the *common-emitter*, *common-collector*, and *common-base* configurations. Viewing an amplifier as a *two-port* block, it is apparent that one of the three terminals of the BJT will have to be *common* to both ports (hence the reason for the above designations). The circuit implementations to be discussed herewith are referred to as *discrete* because we can build them using discrete transistors, resistors, and capacitors. Though nowadays a great deal of BJT amplifiers are implemented in integrated-circuit (IC) form, the motivation for studying discrete BJT implementations is not only historical, but also pedagogical as discrete designs are somewhat easier to grasp, and yet reveal important aspects that apply to IC implementations as well. Moreover, students who have access to an electronics lab can try them out experimentally to reinforce their understanding while simultaneously developing experimental skills as part of a well-rounded academic formation. Lacking a lab, the student can simulate the circuits via PSpice (see Appendix 2A).

Unilateral Voltage Amplifiers

Figure 2.52 shows the block diagram of a *voltage* amplifier of the *unilateral* type, so called because the signal progresses only in the *forward* direction, from source to load, with no return signal paths. (In the next section we will encounter an example of a non-unilateral amplifier in the common-collector configuration. More examples will be found in the study of IC amplifiers.) The amplifier receives its input v_i from a signal source v_{sig} with *internal resistance* R_{sig}, and supplies its output v_o to a resistive load R_L. The amplifier is uniquely characterized in terms of its *input resistance* R_i,

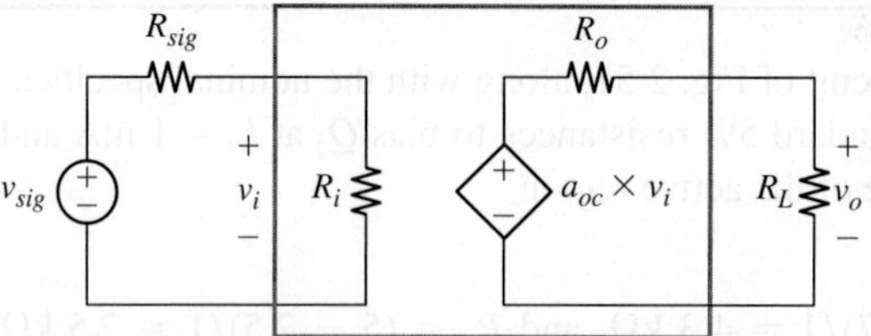

FIGURE 2.52 Block diagram of a voltage amplifier of the unilateral type.

output resistance R_o, and the *open-circuit voltage gain a_{oc}*. At the amplifier's input we have a voltage divider, resulting in *input loading*

$$v_i = \frac{R_i}{R_{sig} + R_i} v_{sig} \tag{2.70}$$

Likewise, at the amplifier's output we have another voltage divider, resulting in *output loading*

$$v_o = \frac{R_L}{R_o + R_L} a_{oc} \times v_i \tag{2.71}$$

We observe that

$$a_{oc} = \lim_{R_L \to \infty} \frac{v_o}{v_i} \tag{2.72a}$$

that is, a_{oc} represents the gain with which the amplifier would amplify its input v_i in the absence of any output load. Consequently, a_{oc} is called the *open-circuit voltage gain*, or also the *unloaded gain*. Eliminating v_i from the above equations we obtain the *signal-to-load voltage gain*

$$\frac{v_o}{v_{sig}} = \frac{R_i}{R_{sig} + R_i} \times a_{oc} \times \frac{R_L}{R_o + R_L} \tag{2.73}$$

As the signal progresses from the source to the load, first it undergoes some attenuation at the amplifier's input, then it gets magnified by a_{oc}, and finally it undergoes some additional attenuation at the output. In light of Eq. (2.73) we can also write

$$a_{oc} = \frac{v_o}{v_{sig}} \bigg|_{R_{sig} \to 0,\, R_L \to \infty} \tag{2.72b}$$

which provides an alternative way of finding the unloaded gain.

The Common-Emitter (CE) Configuration

In the circuit of Fig. 2.53 the amplifier proper is made up of the BJT and its surrounding components. To prevent the source and the load from disturbing the dc conditions of the amplifier, we use the *ac-coupling capacitors C_1 and C_2*. Moreover, to establish an *ac ground* at the emitter terminal, we use the *bypass capacitor C_3*.

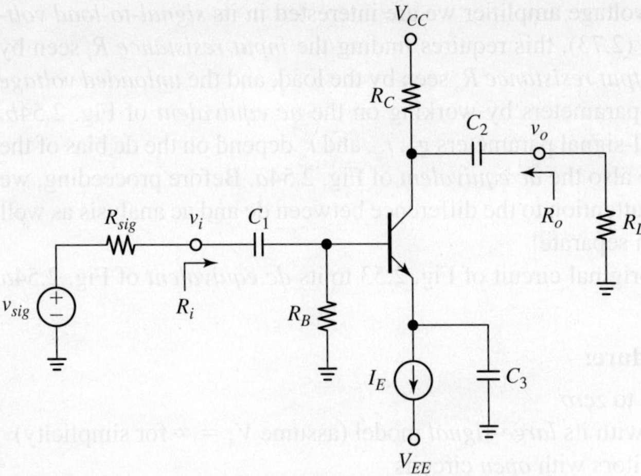

FIGURE 2.53 The *common-emitter* (CE) amplifier.

At dc, the capacitors draw zero current and thus act as *open circuits*. In fact, in the dc equivalent of Fig. 2.54a, the capacitors have been omitted altogether. Also, to simplify dc analysis, we assume $V_A = \infty$, and we use a dc current sink I_E to bias the BJT. Such a sink could be implemented with a current mirror of the type of Fig. 2.51 (let us not worry about these details here). The dc voltages are then

$$V_B = -R_B\frac{I_E}{\beta_F + 1} \qquad V_C = V_{CC} - \alpha_F R_C I_E \qquad V_E = V_B - V_{BE(on)} \qquad (2.74)$$

Moreover, we have $I_C = \alpha_F I_E \cong I_E$. When power is applied to the circuit, each capacitor will charge up until its plates attain the dc voltages of their corresponding nodes. For instance, while the bottom plate of C_3 remains at ground potential, the top plate will charge to V_E, which in this circuit is negative. Likewise, the left plate of C_2 will charge to V_C, while the right plate is pulled to 0 V by R_L.

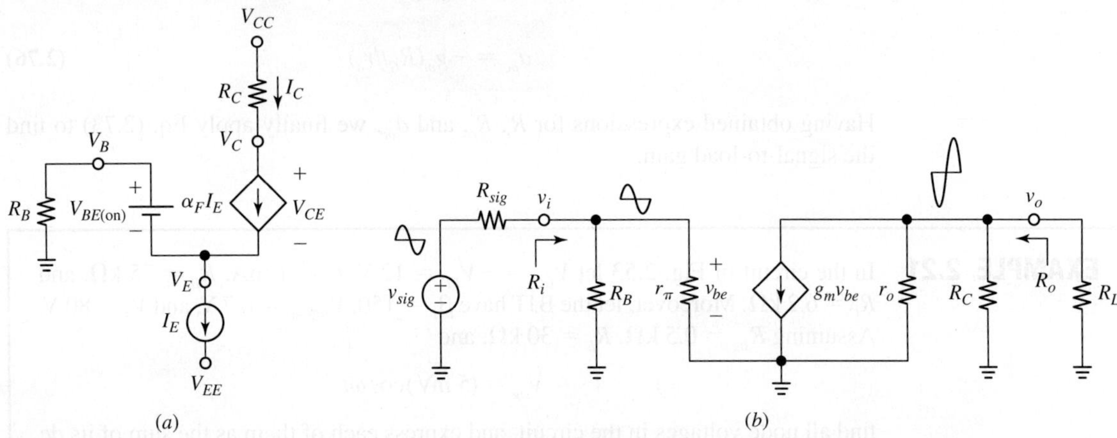

FIGURE 2.54 (a) Dc and (b) ac equivalents of the CE amplifier of Fig. 2.53.

When analyzing a voltage amplifier we are interested in its *signal-to-load volt-age gain* v_o/v_{sig}. By Eq. (2.73), this requires finding the *input resistance* R_i seen by the signal source, the *output resistance* R_o seen by the load, and the *unloaded voltage gain* a_{oc}. We find these parameters by working on the *ac equivalent* of Fig. 2.54b. However, since the small-signal parameters g_m, r_π, and r_o depend on the dc bias of the BJT, we need to analyze also the *dc equivalent* of Fig. 2.54a. Before proceeding, we wish to call the reader's attention to the difference between dc and ac analysis as well as the need to keep them separate!

In going from the original circuit of Fig. 2.53 to its *dc equivalent* of Fig. 2.54a we apply the following

- **Dc Analysis Procedure:**

  - Set all *ac* sources to *zero*
  - Replace the BJT with its *large-signal* model (assume $V_A = \infty$ for simplicity)
  - Replace all capacitors with *open* circuits

Conversely, in going from the original circuit of Fig. 2.53 to its *ac equivalent* of Fig. 2.54b we apply by applying the following

- **Ac Analysis Procedure:**

  - Set all *dc* sources to *zero*
  - Replace the BJT with its *small-signal* model, inclusive of r_o
  - Replace all capacitors with *short* circuits

  With reference to Fig. 2.54b, we note by inspection that

$$R_i = R_B /\!/ r_\pi \qquad\qquad R_o = R_C /\!/ r_o \qquad\qquad (2.75)$$

Moreover, by Ohm's law, we have

$$0 - v_o = (r_o /\!/ R_C /\!/ R_L)g_m v_{be} = (r_o /\!/ R_C /\!/ R_L)g_m v_i$$

Letting $R_L \to \infty$ gives $v_o = -(r_o /\!/ R_C)g_m v_i$. But, according to Eq. (2.72a), the ratio v_o/v_i in the limit $R_L \to \infty$ is the *unloaded voltage gain*, so

$$a_{oc} = -g_m(R_C /\!/ r_o) \qquad\qquad (2.76)$$

Having obtained expressions for R_i, R_o, and a_{oc}, we finally apply Eq. (2.73) to find the signal-to-load gain.

EXAMPLE 2.21   In the circuit of Fig. 2.53 let $V_{CC} = -V_{EE} = 12$ V, $I_E = 1$ mA, $R_B = 75$ kΩ, and $R_C = 6.2$ kΩ. Moreover, let the BJT have $\beta_F = 150$, $V_{BE(on)} = 0.7$ V, and $V_A = 80$ V. Assuming $R_{sig} = 0.5$ kΩ, $R_L = 30$ kΩ, and

$$v_{sig} = (5 \text{ mV})\cos\omega t$$

find all node voltages in the circuit, and express each of them as the sum of its dc and ac component, in the manner of Eq. (2.43).

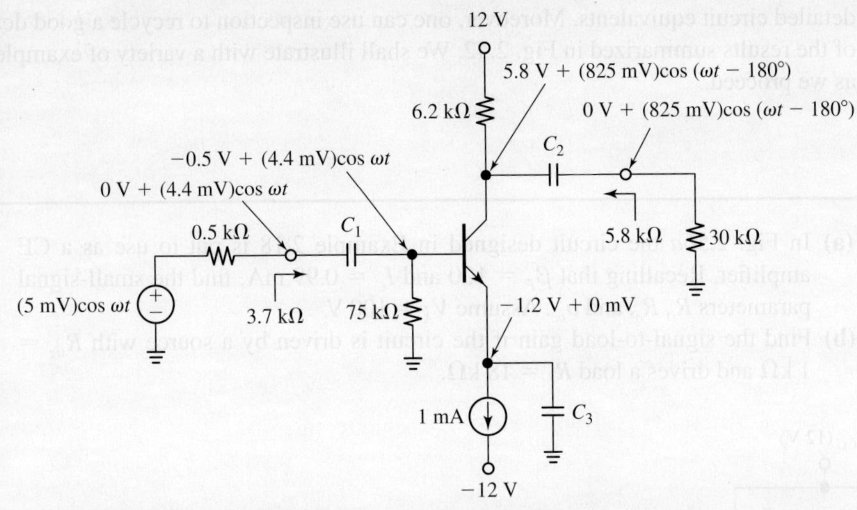

FIGURE 2.55 Circuit of Example 2.21, with each node voltage expressed as the sum of its *dc* and *ac* component.

Solution

We have $I_C \cong I_E = 1$ mA. By Eq. (2.74) we have

$$V_B = -75\frac{1}{151} = -0.5 \text{ V}$$

$$V_C \cong 12 - 6.2 \times 1 = 5.8 \text{ V}$$

$$V_E = -0.5 - 0.7 = -1.2 \text{ V}$$

Moreover, $g_m = 1/26$ A/V, $r_\pi = 150 \times 26 = 3.9$ kΩ, and $r_o = 80/1 = 80$ kΩ. Consequently, Eqs. (2.75) and (2.76) give

$$R_i = 75//3.9 = 3.7 \text{ k}\Omega$$

$$R_o = 6.2//80 = 5.8 \text{ k}\Omega$$

$$a_{oc} = -5{,}800/26 = -223 \text{ V/V}$$

Also, Eq. (2.70) gives $v_i = [3.7/(0.5 + 3.7)]v_{sig} = 0.88v_{sig} = (4.4 \text{ mV})\cos \omega t$. Finally, Eq. (2.73) gives

$$v_o = \left[\frac{3.7}{0.5 + 3.7}(-223)\frac{30}{5.8 + 30}\right]v_{sig} = -165v_{sig} = -165(5 \text{ mV})\cos \omega t$$

$$= (825 \text{ mV})\cos(\omega t - 180°)$$

The node voltages are shown in Fig. 2.55. The reader is encouraged to verify each of them in detail.

The systematic procedure of redrawing an amplifier both in dc and ac form, as exemplified in Fig. 2.54, though highly recommended for the beginner, may soon prove an overkill as one seeks to speed up the analysis process. With experience, some of the intermediate steps can be carried out mentally without having to draw detailed circuit equivalents. Moreover, one can use inspection to recycle a good deal of the results summarized in Fig. 2.42. We shall illustrate with a variety of examples as we proceed.

EXAMPLE 2.22 (a) In Fig. 2.56a the circuit designed in Example 2.18 is put to use as a CE amplifier. Recalling that $\beta_F = 100$ and $I_C = 0.99$ mA, find the small-signal parameters R_i, R_o, and a_{oc}. Assume $V_A = 100$ V.

(b) Find the signal-to-load gain if the circuit is driven by a source with $R_{sig} = 1$ kΩ and drives a load $R_L = 18$ kΩ.

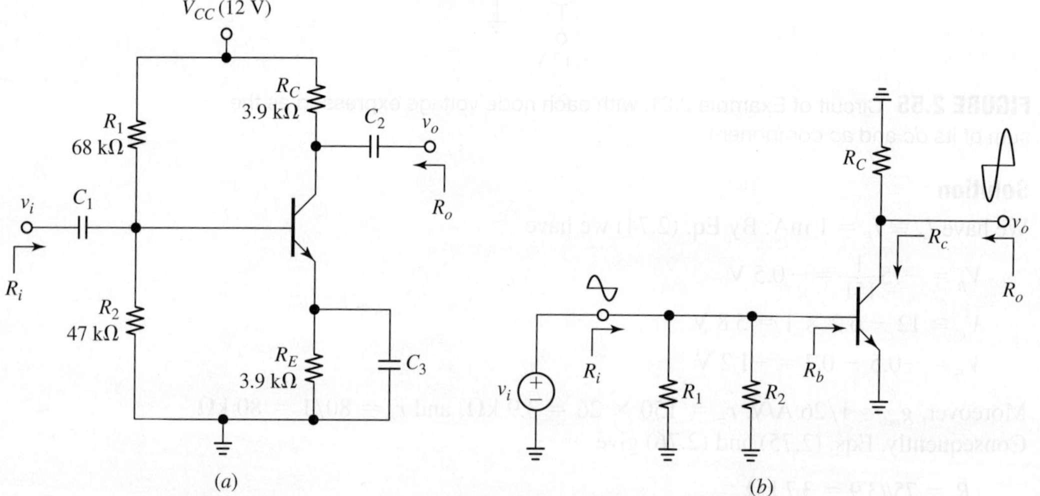

(a)

(b)

FIGURE 2.56 (a) Single-supply CE amplifier of Example 2.22 and (b) it ac equivalent.

Solution

(a) We have $g_m = 0.99/26 = 38$ mA/V, $r_\pi = 100 \times 26/0.99 = 2.6$ kΩ, and $r_o = 100/0.99 = 101$ kΩ. Next, refer to the ac equivalent of Fig. 2.56b, where we observe that because of the bypass action by C_3, the emitter is at *ac ground*. By inspection, $R_b = r_\pi$ and $R_c = r_o$. Consequently,

$$R_i = R_1//R_2//R_b = 68//47//2.6 = 2.4 \text{ k}\Omega$$

$$R_o = R_C//R_c = 3.9//101 = 3.8 \text{ k}\Omega$$

$$a_{oc} = -g_m R_o = -38 \times 3.8 = -144 \text{ V/V}$$

(b) Due to input and output loading, the gain drops to

$$\frac{v_o}{v_{sig}} = \frac{2.4}{1 + 2.4}(-144)\frac{18}{3.8 + 18} = -84 \text{ V/V}$$

Quick Estimates for the CE Configuration

In everyday practice the circuit designer often needs to come up with quick if rough estimates for the relevant parameters of a BJT amplifier. In light of the above derivations and examples, we draw the following conclusions for the CE configuration:

- R_i is dominated by r_π
- R_o is dominated by R_C
- a_{oc} is approximately $-g_m R_C$

Using $g_m = I_C/V_T$ we can also write

$$a_{oc} \cong -\frac{R_C I_C}{V_T} \qquad (2.77a)$$

indicating that the unloaded gain magnitude of a CE amplifier is approximately the ratio of the *voltage drop across* R_C to the *thermal voltage* V_T. For instance, in the split-supply amplifier of Example 2.21, R_C dropped approximately $V_{CC}/2$, so Eq. (2.77a) gives the estimate $a_{oc} \cong -(V_{CC}/2)/V_T = -(12/2)/0.026 = -230$ V/V. Similarly, in the single-supply design of Example 2.22, where the BJT was biased according to the 1/3-1/3-1/3 Rule, Eq. (2.77a) gives $a_{oc} \cong -(V_{CC}/3)/V_T = -(12/3)/0.026 = -160$ V/V. Both estimates are in reasonable agreement with the respective examples, so keep Eq. (2.77a) in mind!

Exercise 2.2

Show that if we take also r_o into account, then Eq. (2.77a) becomes

$$a_{oc} = -\frac{R_C I_C}{V_T(1 + R_C I_C/V_A)} \qquad (2.77b)$$

The Common-Emitter with Emitter-Degeneration (CE-ED) Configuration

The circuit of Fig. 2.57 is similar to that of Fig. 2.53, except for the presence of the unbypassed resistance R_E in series with the emitter. To find the small-signal parameters, we turn to the ac equivalent of Fig. 2.58. Equation (2.62b) indicates that the presence of R_E can raise R_c considerably to the point of making $R_c \gg R_C$. In discrete designs this is usually the case, so we approximate $R_o = R_c//R_c \cong R_C$. In fact, to simplify the analysis and also help the beginner develop a quick feel for the circuit, it is customary to ignore r_o altogether in discrete CE-ED circuits (though this is not necessarily the case with their IC counterparts, as we shall see in Chapter 4.) This allows us to make the approximations

$$R_i = R_B//R_b \cong R_B//[r_\pi + (\beta_0 + 1)R_E] \qquad R_o \cong R_C \qquad (2.78)$$

Next we wish to derive an expression for i_c. With r_o out of the way, Eq. (2.52) simplifies as $i_c = g_m v_{be}$. With reference to the ac equivalent of Fig. 2.58, we have

$$i_c = g_m v_{be} = g_m(v_b - v_e) = g_m(v_i - R_E i_e) \cong g_m(v_i - R_E i_c)$$

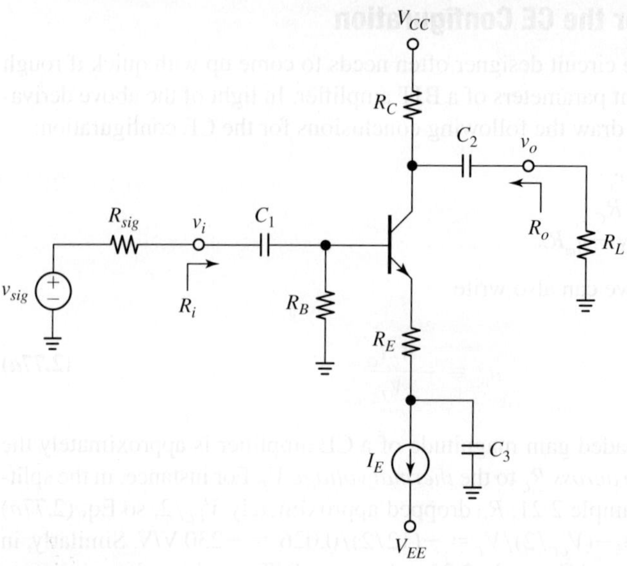

FIGURE 2.57 The *common-emitter* with *emitter-degeneration* (CE-ED) amplifier.

having used the fact that $i_e = i_c/\alpha_0 \cong i_c$. Collecting and solving for i_c gives

$$i_c = G_m v_i \qquad (2.79a)$$

where

$$G_m \cong \frac{g_m}{1 + g_m R_E} \qquad (2.79b)$$

We observe that with $R_E = 0$ the entire signal v_i appears across the B-E port, giving $G_m = g_m$. However, with $R_E \neq 0$, only a *portion* of v_i appears across the B-E port,

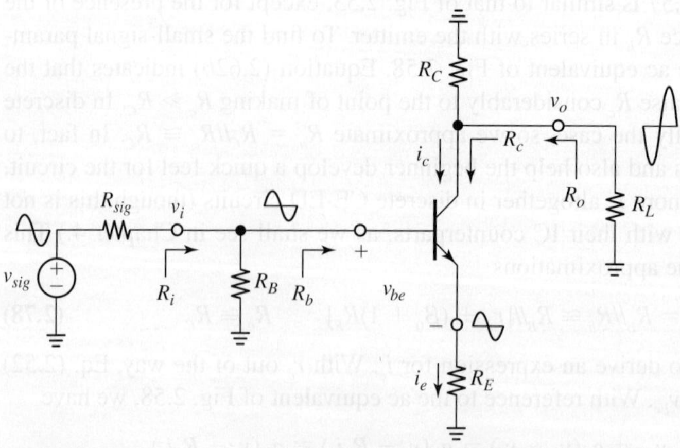

FIGURE 2.58 Ac equivalent of the CE-ED amplifier of Fig. 2.57.

the rest being dropped across R_E. Consequently we have $G_m < g_m$, indicating a smaller response i_c for the same v_i. This transconductance reduction is referred to as *degeneration* because R_E provides a *negative feedback* function, as mentioned previously. Though this subject will be investigated systematically in Chapter 7, suffice it to say here that the presence of R_E, aptly called *emitter-degeneration resistance*, not only reduces the transconductance, but also raises the value of R_b, and hence the value of R_i, by Eq. (2.78).

To find the voltage gain, refer again to the ac equivalent of Fig. 2.58 and write Ohm's law,

$$0 - v_o = (R_o /\!/ R_L)i_c \cong (R_C /\!/ R_L)i_c = (R_C /\!/ R_L)G_m v_i$$

Letting $R_L \to \infty$ gives $v_o \cong -R_C G_m v_i$. But, according to Eq. (2.72a), the ratio v_o/v_i in the limit $R_L \to \infty$ is the *unloaded voltage gain*, so

$$a_{oc} \cong -G_m R_C \cong -\frac{g_m R_C}{1 + g_m R_E} \qquad (2.80)$$

Comparing with Eq. (2.76) we note that the presence of the *degeneration resistance* R_E causes a_{oc} to drop by about $(1 + g_m R_E)$. Rewriting Eq. (2.80) in the alternative form

$$a_{oc} \cong -\frac{R_C}{1/g_m + R_E} \cong -\frac{R_C}{r_e + R_E} \qquad (2.81a)$$

provides us with a useful rule of thumb for a quick estimation of the gain of the CE-ED configuration:

The unloaded voltage gain from base to collector is the (negative of the) *ratio* of the *total collector resistance* to the *total emitter resistance*

We observe that if $g_m R_E \gg 1$ (or, equivalently, if $R_E \gg 1/g_m$), then

$$a_{oc} \cong -\frac{R_C}{R_E} \qquad (2.81b)$$

indicating that the gain becomes independent of g_m and thus of the biasing conditions of the BJT—an important advantage of emitter degeneration! Having obtained expressions for R_i, R_o, and a_{oc}, we finally apply Eq. (2.73) to find the signal-to-load gain.

EXAMPLE 2.23

(a) Investigate the effect of inserting an emitter-degeneration resistance $R_E = 220\ \Omega$ in the CE circuit of Example 2.21, and thus turning it into the CE-ED circuit of Fig. 2.57a.

(b) Find R_E for an unloaded gain of -10 V/V.

Solution

(a) All dc voltages and dc current remain the same, and so do g_m ($= 38$ mA/V) and r_π ($= 3.9$ kΩ). The insertion of $R_E = 220$ Ω in the circuit has the following effects:

- R_b increases from 3.9 kΩ to $[3.9 + (150 + 1)0.22] = 37$ kΩ (almost a tenfold increase!)
- R_i increases from 3.7 kΩ to $75//37 = 25$ kΩ (a desirable affect)
- a_{oc} drops (or degenerates) from -223 V/V to $-[6{,}200/(26 + 220)] = -25$ V/V

Using Eq. (2.73) we find that v_o drops to

$$v_o = \left[\frac{25}{0.5 + 25}(-25)\frac{30}{6.2 + 30}\right]v_{sig} = -20v_{sig} = -20(5\text{ mV})\cos\omega t$$

$$= (100\text{ mV})\cos(\omega t - 180°)$$

(b) Use Eq. (2.80) to impose $-6{,}200/(26 + R_E) = -10$. This yields $R_E = 594$ Ω.

Summary for the CE-ED Configuration

We summarize the effects of the emitter-degeneration resistance R_E as follows:

- The transconductance g_m as well as the unloaded gain a_{oc} are *reduced* by the amount $(1 + g_m R_E)$.
- R_b is raised from r_π to $r_\pi + (\beta_0 + 1)R_E \cong r_\pi(1 + g_m R_E)$, that is, it is *increased* by the amount $(1 + g_m R_E)$.
- The input signal range is raised to $v_i = v_{be} + v_e = v_{be} + R_E i_e \cong v_{be} + R_E i_c = v_{be}(1 + g_m R_E)$, that is, it is *increased* by the amount $(1 + g_m R_E)$, thus widening the range of applicability of the small-signal approximation.
- The signal-to-load gain is less dependent on β_0 and I_C, and is established by external resistance ratios.

Though gain reduction may be undesirable in certain situations, all other effects are generally welcome, and as such they are often exploited on purpose by the designer to optimize the circuit at hand (in Chapter 6 we will see another important effect of degeneration, namely, faster frequency/time responses).

Capacitance Selection

To complete our understanding of discrete BJT amplifiers we need to address the issue on how to go about selecting the various capacitances involved in discrete design. As the signal source is turned on, we want each capacitance C to act as an *ac short* at the source's frequency f_{sig}. Physically, this requires that we select C large enough so as to prevent it from charging/discharging appreciably in response to the ac alternations of v_{sig}.

As we know, the impedance presented by a capacitance C at the signal frequency f_{sig} is $Z_C(jf_{sig}) = 1/(j2\pi f_{sig})$. For this capacitance to act effectively as an ac short at f_{sig}, its impedance must be such that

$$|Z_C(jf_{sig})| \ll R_{eq}$$

where R_{eq} is the *equivalent resistance* seen by C. This condition is readily rephrased in terms of C as

$$C \gg \frac{1}{2\pi R_{eq} f_{sig}} \qquad (2.82)$$

If the circuit is designed to operate over a *range* of signal frequencies, then we must use the lowest frequency in the above condition, that is, $f_{sig(min)}$.

Specify suitable capacitances in the CE amplifier of Example 2.21 for operation over the *audio range*.   **EXAMPLE 2.24**

Solution

The audio range extends from 20 Hz to 20 kHz, so $f_{sig(min)} = 20$ Hz.

- For C_1 we have $R_{eq1} = R_{sig} + R_i = 0.5 + 3.7 = 4.2$ kΩ, so we need $C_1 \gg 1/[2\pi \times 4.2 \times 10^3 \times 20) \cong 2.0$ μF.
- For C_2 we have $R_{eq2} = R_o + R_L = 5.8 + 30 = 35.8$ kΩ, so impose $C_2 \gg 1/[2\pi \times 35.8 \times 10^3 \times 20) \cong 0.22$ μF.
- For C_3 we adapt Eq. (2.59a) to the present case and write

$$R_{eq3} \cong \frac{(R_{sig}//R_B) + r_\pi}{\beta_0 + 1} = \frac{(0.5//75) + 3.9}{150 + 1} = 29 \ \Omega$$

  so impose $C_3 \gg 1/[2\pi \times 29 \times 20) \cong 275$ μF.

A reasonable way to go is to use standard capacitance values that exceed the calculated lower limits by an order of magnitude or more. Thus, use $C_1 = 22$ μF, $C_2 = 2.2$ μF, and $C_3 = 2.7$ mF. Since it sees a very small equivalent resistance, C_3 comes out quite large. To save on size, it is common to compromise and use a smaller value, such as 330 μF in the present case.

PSpice Simulation

The circuits discussed herein can readily be verified via PSpice (see Appendix 2A). The student can either make up ad-hoc BJT models, or use the device models available in the library that comes with the Student Version of PSpice, such as the popular 2N2222 *npn* BJT. Figure 2.59 shows a PSpice circuit to simulate the CE amplifier of Example 2.22, but using a 2N2222 BJT. All relevant waveforms are displayed in Fig. 2.60.

As seen, the input signal v_{sig} is a 1-kHz sine wave with $\pm$5-mV peak values. The waveform v_B at the base is still a sine wave with peak values of almost $\pm$5-mV, but with a dc component $V_B = 4.717$ V as established by the dc biasing resistances R_1 and R_2. The waveform v_E at the emitter has a dc component $V_E = 4.072$ V, about 0.7-V lower than V_B. We note also a small ac component v_e. Ideally, v_e should be zero (perfect ac ground at the emitter), but this would require $C_2 \rightarrow \infty$. In practice we specify C_2 to be *large enough* to ensure $|v_e| \ll |v_b|$. Finally, we note that the waveform v_C at the collector has a dc component $V_C = 7.968$ and an ac component v_c which is a

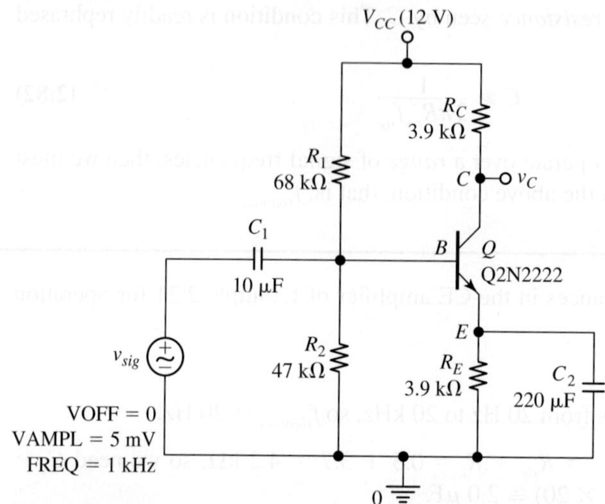

FIGURE 2.59 PSpice circuit to display the waveforms of a CE amplifier.

magnified version of the input v_{sig}, but shifted by 180°. Its peak-to-peak amplitude is $(8.659 - 7187) = 1.472$ V. Since the peak-to-peak amplitude of the input is 10 mV, the gain is $-1.472/0.010 = -147.2$ V/V, which is in reasonable agreement with the value of -144 V/V predicted in Example 2.22. In practice the output waveform is slightly distorted, though imperceptibly so in the present case, because of the nonlinear (exponential) BJT characteristic. In fact, one can verify that its average value is not exactly halfway between its peaks. Consequently, the calculations of Example 2.22, based on the small-signal approximation, are bound to be in slight disagreement with the more dependable results provided by computer simulation or actual lab measurements.

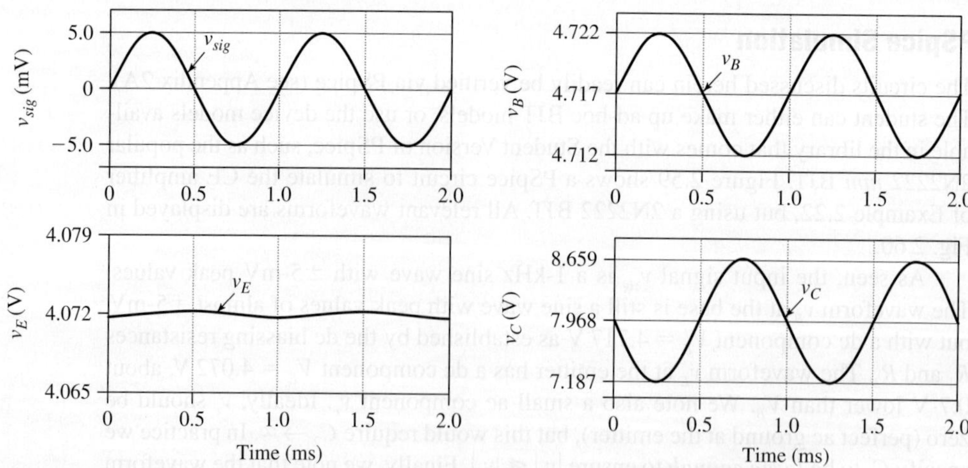

FIGURE 2.60 Probe plots of the waveforms of the CE amplifier of Fig. 2.59.

2.9 BIPOLAR VOLTAGE AND CURRENT BUFFERS

In this section we examine the two remaining single-transistor amplifier configurations of interest, namely, the *common collector* and the *common base* configurations. As we shall see, these configurations find application as *voltage buffers* and *current buffers*, respectively.

The Common Collector (CC) Configuration

The common-collector (CC) amplifier receives the input at the base and delivers the output from the emitter. In the circuit realization of Fig. 2.61*a* we have chosen to drive the base directly from the signal source so that we can focus on the bare essentials of the circuit. Turning to its ac equivalent of Fig. 2.61*b* we note that it is identical to that of Fig. 2.42, so we just recycle the results developed there, provided we suitably re-label the resistances.

The input resistance in Fig. 2.61*b* is simply R_b as given in Eq. (2.58*b*), but with $R_E \rightarrow R_L$,

$$R_i = r_\pi + (\beta_0 + 1)(R_L /\!/ r_o) \tag{2.83}$$

This resistance is usually *large*, thanks to the magnifying effect by β_0. The output resistance in Fig. 2.61*b* is simply R_e as given in Eq. (2.59*b*), but with $R_B \rightarrow R_{sig}$,

$$R_o = \left(\frac{R_{sig} + r_\pi}{\beta_0 + 1}\right) /\!/ r_o \tag{2.84}$$

This resistance is usually *small*, thanks to the presence of β_0 in the denominator. Just like Eq. (2.83) reveals that R_i is a function of R_L, Eq. (2.84) reveals that R_o is a function of R_{sig}. As mentioned, an amplifier exhibiting this interdependence is said to be a *non-unilateral* amplifier.

To find the voltage gain, we observe that if we replace the BJT with its small-signal model and look into its base, we see r_π followed by the trio $\beta_0 i_b$, R_L, and r_o. In light of

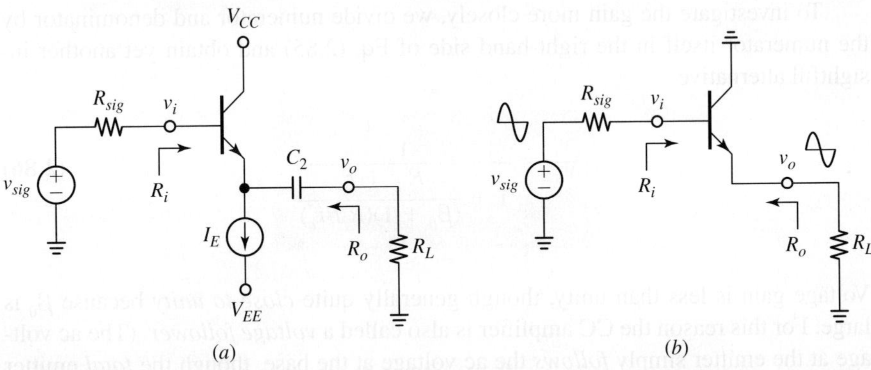

FIGURE 2.61 The *common-collector* (CC) amplifier, and it ac equivalent.

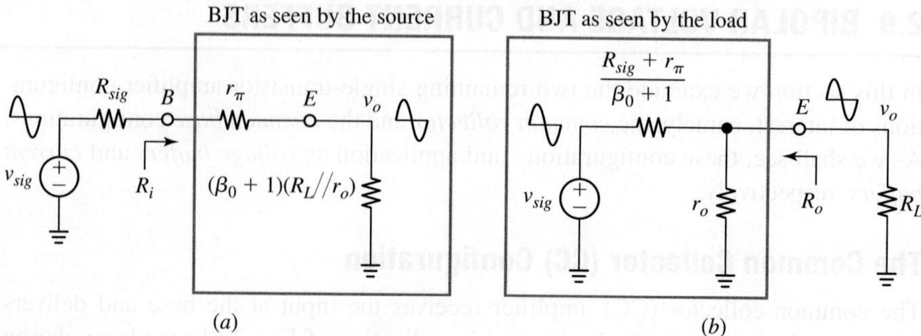

FIGURE 2.62 Small-signal equivalents of the CC amplifier of Fig. 2.61 as seen (a) by the *source* and (b) as seen by the *load*.

Eq. (2.83), it is comforting to realize that we can lump these three elements together and treat them as a *single equivalent resistance* of value $(\beta_0 + 1) \times (R_L//r_o)$. This is precisely what has been done in Fig. 2.62a, which shows the equivalent ac circuit as *seen by the signal source*. Using the voltage divider rule, we find the signal-to-load gain to be

$$\frac{v_o}{v_{sig}} = \frac{(\beta_0 + 1)(R_L//r_o)}{R_{sig} + r_\pi + (\beta_0 + 1)(R_L//r_o)} \tag{2.85a}$$

Dividing numerator and denominator by $(\beta_0 + 1)$ yields an insightful alternative expression for the signal-to-load gain,

$$\frac{v_o}{v_{sig}} = \frac{R_L//r_o}{\dfrac{R_{sig} + r_\pi}{\beta_0 + 1} + (R_L//r_o)} \tag{2.85b}$$

Viewing this expression as a voltage divider rule, we can readily draw the circuit originating it. Shown in Fig. 2.62b, this is the equivalent ac circuit as *seen by the load*. These alternative equivalents ought to help the reader develop a better feel for this intriguing circuit. As mentioned, a CC amplifier provides *high input resistance* and *low output resistance*.

To investigate the gain more closely, we divide numerator and denominator by the numerator itself in the right-hand side of Eq. (2.85) and obtain yet another insightful alternative

$$\frac{v_o}{v_{sig}} = \frac{1}{1 + \dfrac{R_{sig} + r_\pi}{(\beta_0 + 1)(R_L//r_o)}} \tag{2.86}$$

Voltage gain is less than unity, though generally quite *close to unity* because β_0 is large. For this reason the CC amplifier is also called a *voltage follower*. (The ac voltage at the emitter simply *follows* the ac voltage at the base, though the *total* emitter voltage is *offset* by about -0.7 V relative to that at the base.) In accordance with

Eq. (2.72*b*), we readily find the unloaded gain a_{oc} by letting $R_{sig} \to 0$ and $R_L \to \infty$ in Eq (2.86). The result is

$$a_{oc} \cong \frac{1}{1 + 1/(g_m r_o)} = \frac{1}{1 + V_T/V_A} \qquad (2.87)$$

a value extremely close to unity. We conclude with the following observations:

- Even though not much of a winner as a voltage amplifier, the CC configuration offers the advantages of potentially *high input resistance* and *low output resistance*, which make it a good *voltage buffer*. Ideally, a voltage buffer has,

$$R_i \to \infty \qquad R_o \to 0 \qquad \frac{v_o}{v_{sig}} \to 1 \text{ V/V} \qquad (2.88)$$

Even though the CC configuration is only an approximation of the ideal voltage buffer, it is widely used to reduce inter-stage loading. For instance, preceding a CE amplifier by a voltage buffer can provide a much higher input resistance at the price of a negligible reduction in the overall gain, just like following it by a buffer can result in a much lower output resistance.
- The CC amplifier can also be viewed as a *current* amplifier that accepts the ac current i_b at the base and delivers the ac current i_e at the emitter with the gain

$$\frac{i_e}{i_b} = \beta_0 + 1 \qquad (2.89)$$

When used in this capacity, the CC configuration provides *power amplification* and finds application as the output stage of power-handling circuits such as dc power supplies and audio power amplifiers.

EXAMPLE 2.25

(a) In the circuit of Fig. 2.61*a*, let $V_{CC} = -V_{EE} = 12$ V and $I_E = 1$ mA, and let the BJT have $\beta_0 = \beta_F = 150$, $V_{BE(on)} = 0.7$ V, and $V_A = 80$ V. If $R_{sig} = 47$ kΩ, $R_L = 10$ kΩ, and

$$v_{sig} = 0 \text{ V} + (2.0 \text{ V}) \cos \omega t$$

find all node voltages in the circuit, and express each of them as the sum of the dc and ac components, in the manner of Eq. (2.43). Show them explicitly in the circuit.

(b) Check that the BJT satisfies the small-signal approximation condition of Eq. (2.51).

Solution

(a) The dc conditions of the circuit are

$$V_B = 0 - 47\frac{1}{151} \cong -0.3 \text{ V} \qquad V_E \cong -0.3 - 0.7 = -1.0 \text{ V}$$

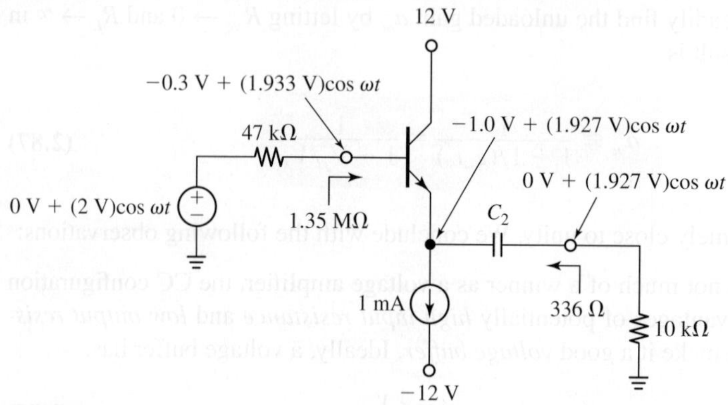

FIGURE 2.63 Circuit of Example 2.25, with each node voltage expressed as the sum of its *dc* and *ac* components.

Moreover, $I_C \cong I_E = 1$ mA, $r_\pi \cong 150 \times 26/1 = 3.9$ kΩ and $r_o \cong 80/1 = 80$ kΩ. So, Eqs. (2.83) and (2.84) give

$$R_i = 3.9 + (150 + 1)(10//80) \cong 1.35 \text{ M}\Omega \text{ (high)}$$

$$R_o = \frac{47 + 3.9}{150 + 1}//80 = 0.336 \text{ k}\Omega \text{ (low)}$$

By Eq. (2.86),

$$v_o = \cfrac{1}{1 + \cfrac{47 + 3.9}{(150 + 1)(10//80)}} v_{sig} = 0.963 v_{sig} = (1.927 \text{ V})\cos\omega t$$

Also,

$$v_i = \frac{R_i}{R_{sig} + R_i} v_{sig} = \frac{1.35}{0.047 + 1.35} v_{sig} = (1.933 \text{ V})\cos\omega t$$

The node voltages are shown in Fig. 2.63. The reader is encouraged to verify each of them in detail.

(b) For the BJT we have $v_{be} = v_i - v_o \cong (6 \text{ mV})\cos\omega t$, indicating a peak value close to the 5-mV limit agreed upon in Eq. (2.51). The CC amplifier's ability to handle in fairly linear fashion external signals that are not strictly of the small-signal type stems from the *negative feedback action* provided by the emitter resistance, in this case R_L. As previously noted in the CE-ED case, the presence of R_L increases the input signal range by the factor $(1 + g_m R_L) = (1 + 10,000/26) = 386$!

EXAMPLE 2.26 Figure 2.64 shows a single-supply emitter follower. As we know, R_1 and R_2 bias the base, and R_E sets the emitter current bias. Assuming $\beta_0 = \beta_F = 100$, $V_{BE(on)} = 0.7$ V, and $V_A = 75$ V, find the small-signal resistances R_i and R_o, as well as the *signal-to-load gain*.

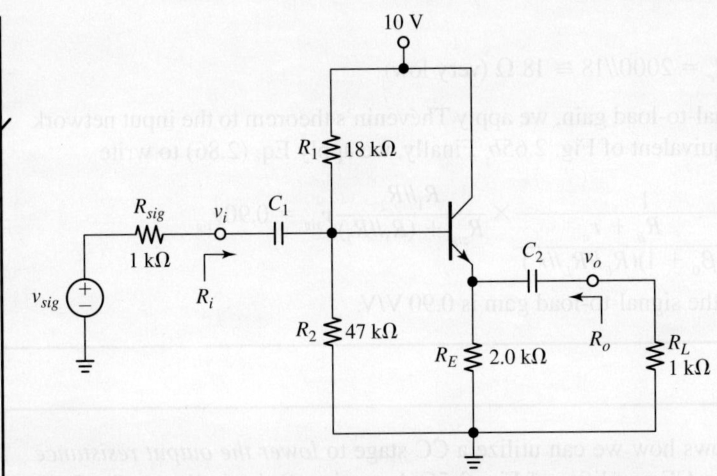

FIGURE 2.64 Single-supply emitter follower of Example 2.26.

Solution

Proceeding in the now familiar manner we find $I_C = 3$ mA. Consequently, $r_\pi = 100(26/3) = 0.87$ kΩ and $r_o = 75/3 = 25$ kΩ. With reference to the ac equivalent of Fig. 2.65a, we adapt Eq. (2.58) to write

$$R_b = r_\pi + (\beta_0 + 1)(R_E/\!/r_o/\!/R_L) = 0.87 + 101(2/\!/25/\!/1) = 66 \text{ k}\Omega$$

and

$$R_i = R_1/\!/R_2/\!/R_b = 18/\!/47/\!/66 = 11 \text{ k}\Omega$$

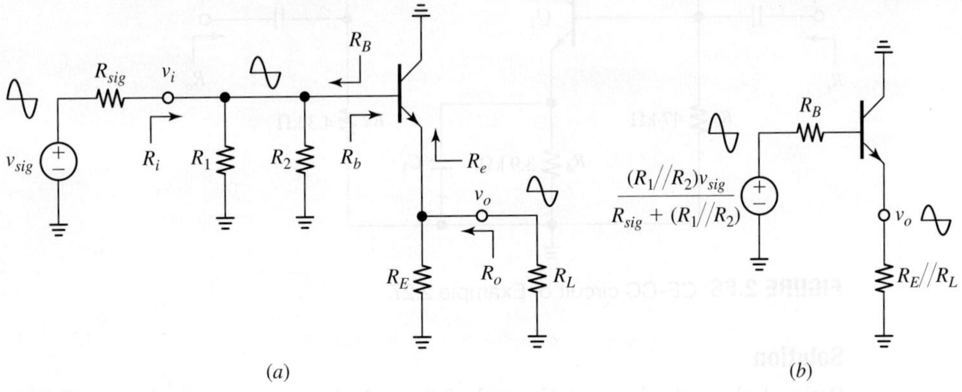

FIGURE 2.65 (a) Ac equivalent of the circuit of Fig. 2.64, and (b) the same after further reduction.

The equivalent resistance seen looking towards the left of the base is

$$R_B = R_{sig}/\!/R_1/\!/R_2 = 1/\!/18/\!/47 = 0.93 \text{ k}\Omega$$

so, we adapt Eq. (2.59b) to write

$$R_e = \frac{R_B + r_\pi}{\beta_0 + 1}/\!/r_o = \frac{0.93 + 0.87}{101}/\!/25 \cong 0.018 \text{ k}\Omega = 18 \ \Omega$$

and

$$R_o = R_E /\!/ R_e = 2000 /\!/ 18 \cong 18 \ \Omega \ \text{(very low)}$$

To find the signal-to-load gain, we apply Thévenin's theorem to the input network to obtain the equivalent of Fig. 2.65b. Finally, we apply Eq. (2.86) to write

$$v_o = \cfrac{1}{1 + \cfrac{R_B + r_\pi}{(\beta_0 + 1)(R_E /\!/ R_L /\!/ r_o)}} \times \frac{R_1 /\!/ R_2}{R_{sig} + (R_1 /\!/ R_2)} v_{sig} = 0.90 v_{sig}$$

indicating that the signal-to-load gain is 0.90 V/V.

EXAMPLE 2.27 Figure 2.66 shows how we can utilize a CC stage to *lower the output resistance* presented by the CE amplifier of Fig. 2.56. Assuming Q_2 is similar to Q_1 ($\beta_0 = \beta_F = 100$, $V_{BE(\text{on})} = 0.7$ V, and $V_A = 100$ V), find the small-signal parameters R_i, R_o, and $a_{oc} = v_o/v_i$ for the overall circuit. Comment on your results.

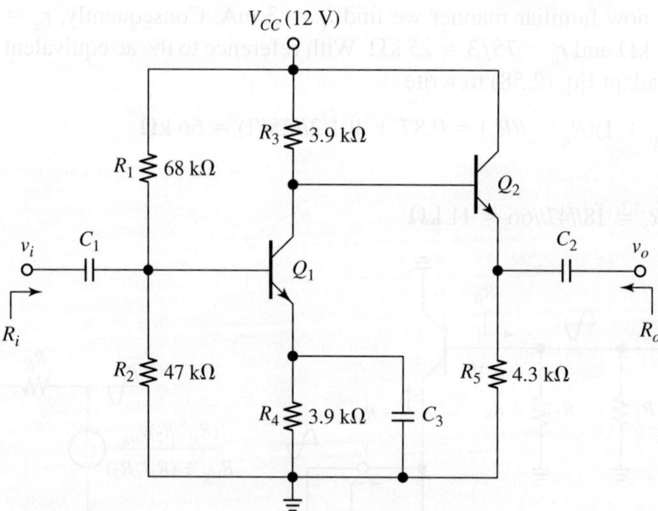

FIGURE 2.66 CE-CC circuit of Example 2.27.

Solution

Proceed along the lines of Example 2.8, analyzing one stage at a time. For dc analysis, replace the CE stage with its *dc* Thévenin equivalent as in Fig. 2.67a. Recall from previous analysis that $I_{C1} = 0.99$ mA, so the CE stage provides the open-circuit dc voltage $V_{C1} = V_{CC} - R_3 I_{C1} = 12 - 3.9 \times 0.99 = 8.1$ V with the equivalent series resistance $R_{eq} = R_3 = 3.9$ kΩ. Adapting Eq. (2.32) with $V_{BB} = V_{C1} = 8.1$ V we get

$$I_{C2} = 100 \frac{8.1 - 0.7}{3.9 + 101 \times 4.3} = 1.7 \text{ mA}$$

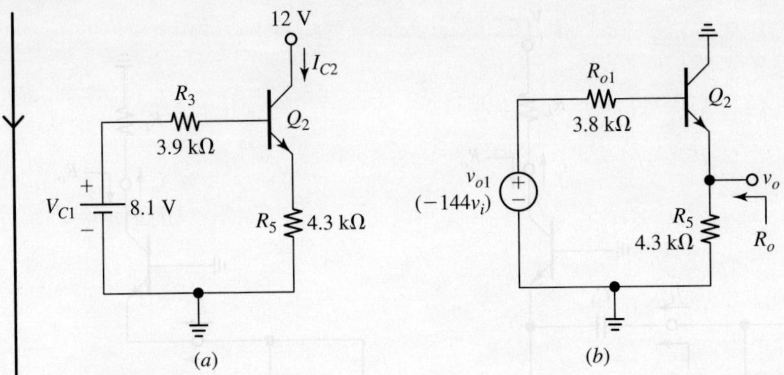

FIGURE 2.67 (a) Dc and (b) ac equivalents of the CE-CC circuit of Fig. 2.66.

Consequently, $r_{\pi 2} = 100(26/1.7) = 1.5\ \text{k}\Omega$ and $r_{o2} = 100/1.7 = 59\ \text{k}\Omega$.

For ac analysis, replace the CE stage with its Thévenin ac equivalent as in Fig. 2.67b. Recall from Example 2.22 that the CE stage provides the open-circuit ac voltage $v_{o1} = -144v_i$ with the equivalent output resistance $R_{o1} = 3.8\ \text{k}\Omega$. We can thus adapt Eq. (2.86) and write

$$\frac{v_o}{v_{o1}} = \frac{v_o}{-144v_i} = \frac{1}{1 + \dfrac{3.8 + 1.5}{(100 + 1)(4.3//59)}} = 0.987$$

so the overall voltage gain is $a_{oc} = v_o/v_i = -144 \times 0.987 = -142\ \text{V/V}$. Finally, we adapt Eq. (2.59a) (the experience gained with Example 2.25 indicates that we can ignore r_o) and obtain

$$R_{e2} = \frac{R_{o1} + r_{\pi 2}}{\beta_{02} + 1} = \frac{3.8 + 1.5}{101} = 0.052\ \text{k}\Omega = 52\ \Omega$$

Consequently,

$$R_o = R_5//R_{e2} = 4{,}300//52 \cong 52\ \Omega \text{ (very low)}$$

Since a_{oc} is reduced from -144 V/V to -142 V/V, loading of the CE stage by the CC stage is minimal. However, the reduction in R_o is truly dramatic, from $3{,}800\ \Omega$ to $52\ \Omega$!

The Common-Base (CB) Configuration

The common-base (CB) amplifier receives the input at the emitter and delivers the output from the collector. Since the resistance seen looking into the emitter is generally low ($R_e \cong 1/g_m$), the natural input signal for this configuration is a *current*, i_{sig}. Also, since the resistance seen looking into the collector is generally high ($R_c \cong r_o$, or even higher if there is emitter degeneration), the natural output signal is also a *current*, i_o. Just like the CC configuration approximates a *voltage buffer*, which ideally has $R_i \to \infty$, $R_o \to 0$, and $v_o/v_{sig} \to 1$ V/V, the CB configuration approximates a *current buffer*, which ideally has

$$R_i \to 0 \qquad R_o \to \infty \qquad \frac{i_o}{i_{sig}} \to 1\ \text{A/A} \tag{2.90}$$

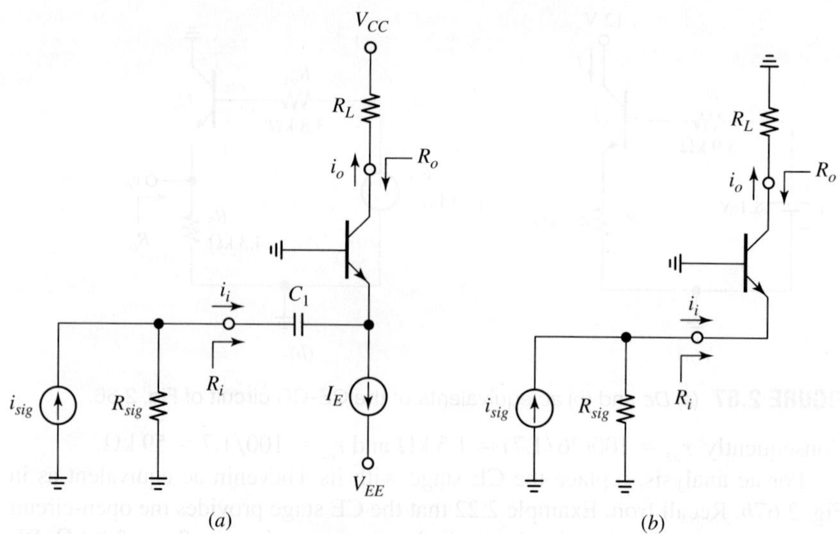

FIGURE 2.68 The *common-base* (CB) amplifier and its ac equivalent.

The CB configuration is shown in Fig. 2.68*a*, where we observe that the signal source is now modeled with a Norton equivalent. To find the small-signal parameters, refer to the ac equivalent of Fig. 2.68*b*. The output resistance R_o is readily found by recycling Eq. (2.62*a*), but with $R_B \to 0$ and $R_E \to R_{sig}$,

$$R_o \cong r_o\left[1 + g_m(r_\pi /\!/ R_{sig})\right] \tag{2.91}$$

If $R_{sig} \gg r_\pi$, then $R_o \cong r_o(1 + g_m r_\pi) = r_o(1 + \beta_0)$, a truly large value. In the calculation of the (low) input resistance R_i it is customary to ignore the presence of the (high) resistance r_o, at least in discrete designs, where the coupling from emitter to collector via r_o is generally negligible (this, however, may not necessarily be the case in IC designs, as we shall see in Chapter 4.) We thus adapt Eq. (2.59) and write

$$R_i \cong r_e \tag{2.92}$$

To find the signal-to-load current gain, we note that the source resistance R_{sig} forms a *current divider* with the input resistance R_i, giving

$$i_i = \frac{R_{sig}}{R_{sig} + R_i}i_{sig}$$

But, $i_o = \alpha_0 i_i$, so, combining with Eq. (2.92) we get

$$\frac{i_o}{i_{sig}} = \frac{\alpha_0}{1 + r_e/R_{sig}} \tag{2.93}$$

It is apparent that this gain is less than unity, but it can be very close to unity for $R_{sig} \gg r_e$. The CB configuration is particularly useful when its signal input is supplied by the collector of another BJT. The resulting two-transistor configuration, known as the *cascode configuration*, enjoys advantages of speed and flexibility that make it particularly suited to IC implementations, as we shall see in Chapters 4 and 6.

In the circuit of Fig. 2.68a let $V_{CC} = -V_{EE} = 5$ V and $I_E = 1$ mA, and let the BJT have $\beta_0 = \beta_F = 100$ and $V_A = 100$ V. Find R_i, R_o, and i_o/i_{sig} if $R_{sig} = 10$ kΩ, and comment on your findings.

EXAMPLE 2.28

Solution

We have $\alpha_0 = 0.99$, $g_m = 38.5$ mA/V, $r_e = 26$ Ω, $r_\pi = 2.6$ kΩ, and $r_o = 100$ kΩ. By Eqs. (2.91) through (2.93),

$$R_i \cong 26\ \Omega$$

$$R_o = 100[1 + 38.5(2.6//10)] \cong 8\ \text{M}\Omega$$

$$\frac{i_o}{i_{sig}} = \frac{0.99}{1 + 0.026/10} = 0.987\ \text{A/A}$$

As expected of a current buffer, R_i is truly *low*, R_o truly *high*, and the gain is fairly close to *unity*.

The CB Configuration as a Voltage Amplifier

Even though the most appropriate application of the CB configuration is as a current buffer, there are situations in which it is used as a *voltage amplifier* with gain v_c/v_e. Considering that $v_c = -g_m(R_L//r_o)v_{be} = -g_m(R_L//r_o)(v_b - v_e) = -g_m(R_L//r_o)(0 - v_e)$, it follows that

$$\frac{v_c}{v_e} = +g_m(R_L//r_o) \tag{2.94}$$

In words, the voltage gain of the CB configuration has the *same magnitude* but *opposite polarity* as that of the CE configuration. The other major difference is in the input resistance, which is r_π in the CE case, but r_e (which is $\beta_0 + 1$ times as small) in the CB case. For instance, with $R_L = 10$ kΩ the circuit of Example 2.28 gives the voltage gain $v_c/v_e = 38 \times (10//100) \cong +350$ V/V, a fairly large value.

The *T* Model of the BJT

Even though the small-signal model of Fig. 2.41 is adequate for the ac analysis of all three BJT configurations, an alternative model is available, which offers additional insight especially into the CB configuration. To illustrate, consider the ac model of Fig. 2.41, repeated in Fig. 2.69a but with r_o omitted for simplicity. Since it is reminiscent of the Greek letter π, it is also referred to as the π *model*.

If we now replace the dependent source with two identical sources connected in series as in Fig. 2.69b, circuit behavior won't change. Indeed, since the node common

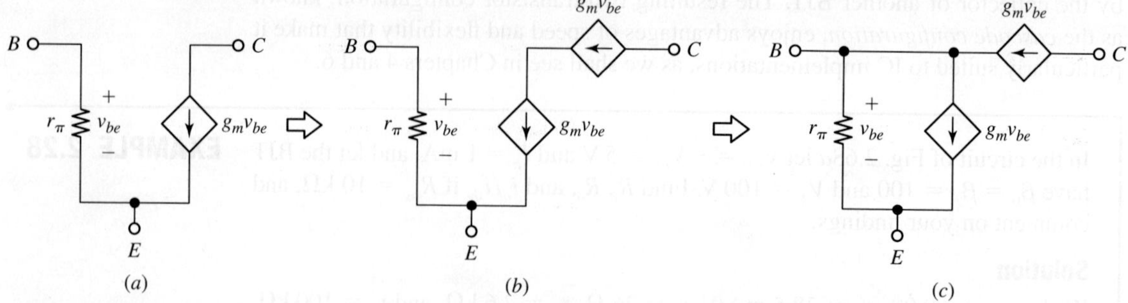

FIGURE 2.69 Steps for transforming the π model into the T model of the BJT.

to the two sources is such that the current entering it equals that exiting it, we can tie it to the base terminal B as in Fig. 2.69c, and circuit behavior still won't change. However, by this artifice we avoid having a dependent source directly between the collector and emitter, a definite advantage in the ac analysis of the CB configuration, as we shall see. Now, we can simplify the model further because the bottom dependent source is controlled by the voltage v_{be} across its *own* terminals and thus acts as a *resistance* of value $v_{be}/(g_m v_{be}) = 1/g_m$. Its parallel combination with r_π is

$$r_\pi /\!/ \frac{1}{g_m} = \frac{\beta_0}{g_m} /\!/ \frac{1}{g_m} = \frac{\beta_0}{\beta_0 + 1} \frac{1}{g_m} = \frac{\alpha_0}{g_m} = r_e$$

that is, it is simply the dynamic resistance of the B-E junction as encountered in Eq. (2.61). (We should have anticipated this intuitively!) As for the top dependent source, we can regard it either as controlled by v_{be}, as shown, or as controlled by the current $i_e = v_{be}/r_e$, in which case we express it as

$$g_m v_{be} = g_m r_e i_e = \alpha_0 i_e$$

We can finally draw the alternative small-signal BJT model as in Fig. 2.70, where we have included also the small-signal resistance r_o to account for the Early effect.

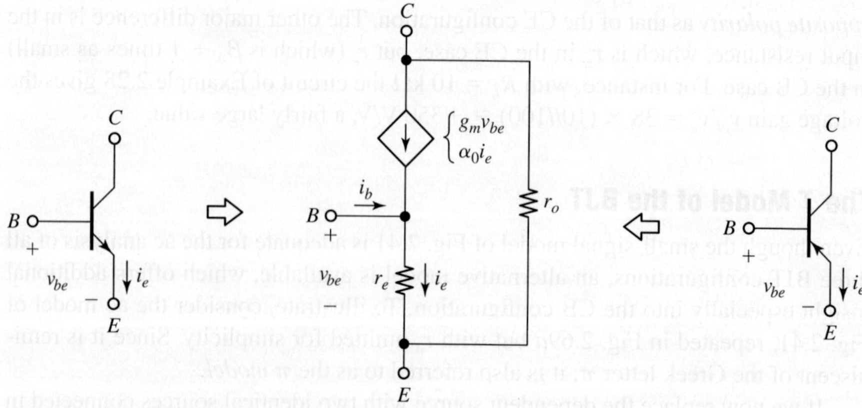

FIGURE 2.70 The small-signal BJT model known as the T model. Note that it applies both to the *npn* and the *pnp* BJT.

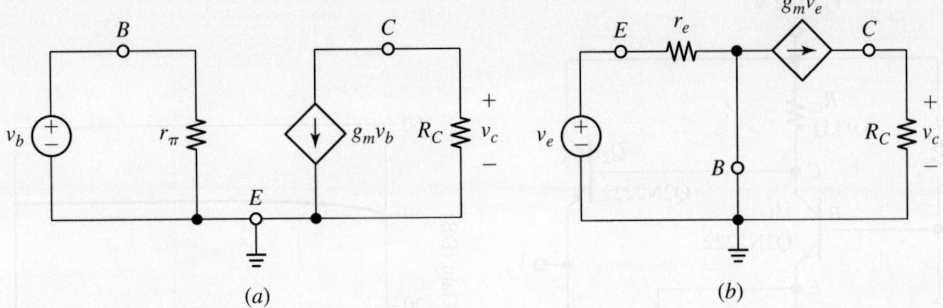

FIGURE 2.71 The CE and CB configurations are compared most effectively if we use, respectively, (a) the π *model*, and (b) and *T model* for the BJT.

Since this model, with r_o omitted, looks like a rotated letter T, it is aptly called the *T model*.

Figure 2.71 utilizes both BJT models (r_o is again omitted for simplicity) for a quick comparison of the CE and CB configurations. By inspection, the CE amplifier of Fig. 2.71a presents an input resistance of r_π, and yields $v_c = -R_C g_m v_b$, indicating a gain of $v_c/v_b = -g_m R_C$. Turning to Fig. 2.71b, we note that the direction of the dependent source has been reversed compared to Fig. 2.70, owing to the fact that $g_m v_{be} = g_m(v_b - v_e) = g_m(0 - v_e) = -g_m v_e$. By inspection, the CB amplifier presents an input resistance of r_e and yields $v_c = +R_C g_m v_e$, indicating a gain of $v_c/v_e = +g_m R_C$. These quick analyses confirm that the input resistances differ by a factor of $(\beta_0 + 1)$, and the gains have *equal magnitudes* but *opposite polarities*. Also, in both cases, applying a voltage across the input resistance (r_π or r_e) results in the transfer of current to another resistance (R_C at the output.) Not surprisingly, the name *transistor* was coined as a contraction of the words *transfer* and *resistor*.

PSpice Simulation

As we study BJT circuits, it is always good practice to corroborate the predictions of hand calculations with computer simulation or laboratory measurements. Figure 2.72a shows a PSpice circuit to simulate the CE-CC amplifier of Example 2.27, but using 2N2222 BJTs.

Shown in Fig. 2.72b is the Bode plot of the gain magnitude $|V_o/V_i|$. At 1 kHz, PSpice predicts a gain of 43.3 dB, or a gain magnitude of $10^{43.3/20} = 146$ V/V. In Example 2.27 we predicted a gain magnitude of 142 V/V. Considering that the present simulation uses 2N2222 BJTs, whose characteristics differ a bit from those of the BJTs used in the example, some discrepancy is expected. We also note that gain drops at the low end of the frequency range. This too is expected, as the capacitances cease to act as ac shorts at low frequencies. In the present case, the gain decrease is dominated by C_2.

For a complete picture, we need to plot also the *input* and *output impedances* $|Z_i|$ and $|Z_o|$. Recall from basic circuits courses that the impedance Z seen looking into a

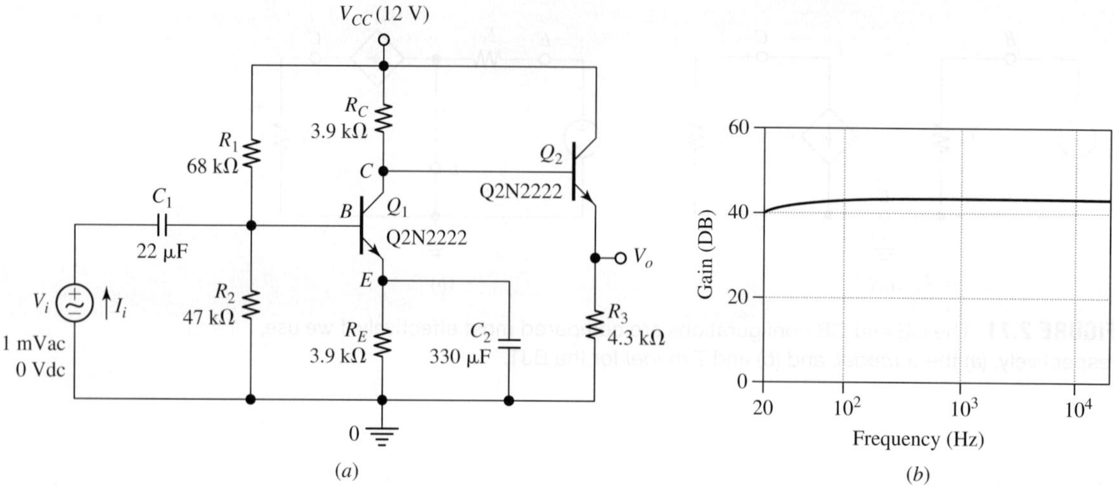

(a)

(b)

FIGURE 2.72 PSpice circuit to display the Bode plot of gain for the CE-EC amplifier of Example 2.27, but using 2N2222 BJTs.

given terminal is obtained by applying an ac test voltage V (or an ac test current I), obtaining the resulting ac current I (or ac voltage V), and then taking the ratio $Z = V/I$. Here, V and I represent the *phasors* associated with the given ac test signals.

To find Z_i, we still use the PSpice circuit of Fig. 2.72a but with the input source V_i now acting as the *test voltage*. Then, $Z_i = V_i/I_i$, where I_i is the ac current *out* of the source V_i. The result, plotted in Fig. 2.73b, top, yields $|Z_i| = 3.7$ kΩ at 1 kHz. Note that $|Z_i|$ increases at low frequencies, again because of the fact that the capacitors cease to act as ac shorts as frequency is lowered.

To find Z_o, use the PSpice circuit of Fig. 2.73a, where the input signal has been set to zero and the output terminal is subjected to a *test current* I_o. Then, $Z_o = V_o/I_o$,

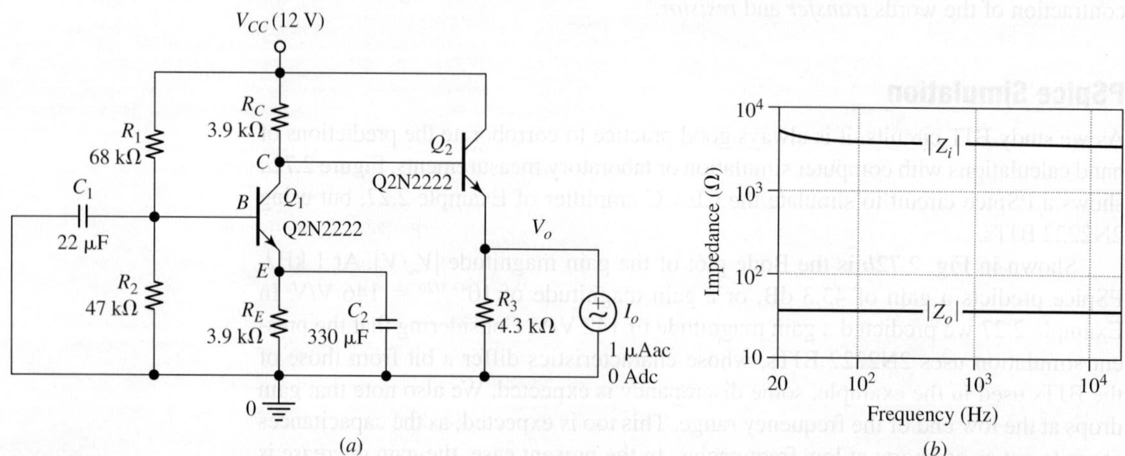

(a)

(b)

FIGURE 2.73 (a) PSpice circuit to find the output impedance Z_o. (b) Frequency plots of the input and output impedances.

where V_o is the ac voltage appearing at the output. The result, plotted in Fig. 2.73b, bottom, yields $|Z_o| = 36\ \Omega$ at 1 kHz. Again, these results are fairly consistent with those predicted in Examples 2.22 and 2.27. The main cause for the discrepancies are the differences in the betas of the transistors used in the examples and those used in the simulation. The motivated student may wish to create ad hoc PSpice models for the BJTs of the examples, and verify a much closer agreement between predictions and simulations.

APPENDIX 2A

SPICE Models for BJTs

As in the case of the *pn* diode, the characteristics of a BJT are expressed in terms of a *list of parameters* that SPICE then uses to create an internal *model* of the device. Such a list is shown in Table 2A.1. PSpice's library comes with the models of several popular BJTs, such as the *npn* 2N2222 and the *pnp* 2N2907. The user can create additional models by editing any one of the models already provided.

As an example, consider the PSpice circuit of Fig. 2.59, utilizing the popular 2N2222 *npn* BJT. By PSpice convention, BJT names must start with the letter Q, so the part number has been designated as Q2N2222. To create a PSpice circuit schematic we use the **Place → Part** commands to lay out the various components, and the **Place → Wire** commands to interconnect them. When it comes to placing the BJT, we import it from the library by going down the list of entries and selecting the Q2N2222 part by left-clicking on it. Once the BJT has been placed in the circuit schematic, we can visualize its model by left-clicking on the BJT itself to select it, and then right-clicking to activate a pull-down menu of possible actions. If we left-click on **Edit PSpice Model**, the following list will appear:

```
.model    Q2N2222 NPN(Is=14.34f Xti=3 Eg=1.11 Vaf=74.03 Bf=255.9
+         Ne=1.307 Ise=14.34f Ikf=.2847 Xtb=1.5 Br=6.092 Nc=2
+         Isc=0 Ikr=0 Rc=1 Cjc=7.306p Mjc=.3416 Vjc=.75 Fc=.5
+         Cje=22.01p Mje=.377 Vje=.75 Tr=46.91n Tf=411.1p Itf=.6
+         Vtf=1.7 Xtf=3 Rb=10)
```

The parameter values shown are designed to match as closely as possible those given in the manufacturer's data sheets. Scanning the list we easily see that this particular BJT type has $I_s = 14.34$ fA, $V_A = 74.03$ V, $\beta_F = 255.9$, $\beta_R = 6.092$, $r_c = 1\ \Omega$, $C_{ic0} = 7.306$ pF, $m_c = 0.3416$, $\phi_c = 0.75$ V, $C_{je0} = 22.01$ pF, $m_e = 0.377$, $\phi_e = 0.75$ V, $\tau_R = 46.91$ ns, $\tau_F = 411.1$ ps, and $r_b = 10\ \Omega$. The list contains additional parameters representing higher-order effects that are beyond our present scope. One such effect is the dependence of β_F upon I_C, as illustrated in Fig. 2.12b. For more details, see Ref. [12]. Also shown in the list are the parameters intervening in the calculation of the *junction capacitances* associated with the base-emitter (C_{je}), base-collector (C_{jc}), and collector-substrate junctions (C_s). This subject will be taken up in great detail in Chapter 6, when we will investigate the frequency response of integrated circuits.

TABLE 2A.1   Partial parameter list of the PSpice model for BJTs.

Symbol	Name	Parameter description	Units	Default	Example
I_s	Is	Saturation current	A	0.1 fA	2 fA
β_F	Bf	Forward current gain		100	250
V_A	Vaf	Forward Early voltage	V	∞	75 V
β_R	Br	Reverse current gain		1	2.5
r_b	Rb	Bulk resistance of the base	Ω	0	200 Ω
r_c	Rc	Bulk resistance of the collector	Ω	0	50 Ω
r_{ex}	Re	Bulk resistance of the emitter	Ω	0	1 Ω
C_{je0}	Cje	Zero-bias B-E junction capacitance	F	0	1.0 pF
ϕ_e	Vje	B-E built-in potential	V	0.75 V	0.8 V
m_e	Mje	B-E grading coefficient		0.33	0.5
C_{jc0}	Cjc	Zero-bias B-C junction capacitance	F	0	0.5 pF
ϕ_c	Vjc	B-C built-in potential	V	0.75 V	0.7 V
m_c	Mjc	B-C grading coefficient		0.33	0.5
C_{s0}	Cjs	Zero-bias collector-substrate junction capacitance	F	0	1.0 pF
ϕ_s	Vjs	Collector-substrate built-in potential	V	0.75 V	0.6 V
m_s	Mjs	Collector-substrate grading coefficient		0	0.5
τ_F	Tf	Forward transit time	s	0	0.2 ns
τ_R	Tr	Reverse transit times	s	0	15 ns

As another example, consider the 2N2907A *pnp* BJT, whose model is also available in PSpice's library. Its parameter list is as follows:

```
.model   Q2N2907A PNP(Is=650.6E-18 Xti=3 Eg=1.11 Vaf=115.7
+        Bf=231.7 Ne=1.829 Ise=54.81f Ikf=1.079 Xtb=1.5
+        Br=3.563 Nc=2 Isc=0 Ikr=0 Rc=.715 Cjc=14.76p
+        Mjc=.5383 Vjc=.75 Fc=.5 Cje=19.82p Mje=.3357
+        Vje=.75 Tr=111.3n Tf=603.7p Itf=.65 Vtf=5
+        Xtf=1.7 Rb=10)
```

As in the 2N2222 case, the name has been changed to Q2N2907A, and the (parenthesized) list is now preceded by the designation PNP specifying the device type. The parameter list is very similar to that of the Q2N2222 model, except for the different parameter values reflecting differences in the data-sheet characteristics.

If you wish to create your own BJT model, you can do so simply by overwriting (editing) the parameter values of an existing BJT model, such as the Q2N2222 or the Q2N2907A models seen above. However, to avoid losing the original parameter list, a *new name* must be given to the newly formed model before saving it. This is what was done to create the model for the simplified *npn* BJT of Fig. 2.32. The BJT model was renamed as Qn, and the parameter list was edited as follows:

```
.model Qn NPN(Is=2fA Bf=100)
```

Likewise, the model of a homebrew *pnp* BJT with, say, $I_s = 0.5$ fA, $\beta_F = 75$, and $V_A = 50$ V would be

```
.model Qp PNP(Is=0.5fA Bf=75 Vaf=50V)
```

All omitted parameters are automatically assigned *default values* according to Table 2A.1.

REFERENCES

1. G. W. Neudeck, *The Bipolar Junction Transistor,* Modular Series on Solid State Devices, 2/E, Vol. II, G. W. Neudeck and R. F. Pierret, eds., Addison-Wesley, 1989.
2. R. S. Muller and T. I. Kamins, *Device Electronics for Integrated Circuits,* 2/E, J. Wiley and Sons, 1986.
3. R. T. Howe and C. G. Sodini, *Microelectronics: An Integrated Approach,* Prentice Hall, 1997.
4. P. R. Gray, P. J. Hurst, S. H. Lewis, and R. G. Meyer, *Analysis and Design of Analog Integrated Circuits,* 4/E, Wiley and Sons, 2001.
5. A. S. Sedra and K. C. Smith, *Microelectronic Circuits,* 5/E, Oxford University Press, 2004.
6. R. C. Jaeger and T. N. Blalock, *Microelectronic Circuit Design,* 3/E, McGraw-Hill, 2007.
7. G. W. Gordon and A. S. Sedra, *SPICE for Microelectronic Circuits,* 2/E, Oxford University Press, 1996.

PROBLEMS

2.1 Physical Structure of the BJT

2.1 A student is trying to create a homebrew *pnp* BJT by connecting together the cathode terminals of two separate *pn* diodes to obtain the *n*-type base, and by designating one of the anodes as the *p*-type emitter, and the other as the *p*-type collector. Soon, the student finds that the resulting device doesn't provide any base-current amplification. Why? List at least two reasons. (*b*) Undaunted, the student decides to use the B-E junction of one *pnp* BJT sample and the B-C junction of another *pnp* BJT sample to meet the relative doping constraints needed of a BJT. He then connects together their base terminals to form the base terminal of the composite device. The student still finds that the device doesn't provide any current gain. Why?

2.2 Basic BJT Operation

2.2 (*a*) A certain *npn* BJT has an emitter area of $(10 \ \mu m) \times (20 \ \mu m)$, and its base and emitter doping concentrations are $N_{AB} = 10^{17}/cm^3$ and $N_{DE} = 10^{19}/cm^3$. Assuming $D_p = 1.8 \ cm^2/s$, $D_n = 18 \ cm^2/s$, $W_B = W_E = 1 \ \mu m$, and $\tau_n = 150$ ns, find I_s and β_F.

(*b*) If $V_{BE} = 700$ mV, find I_C and I_B.

(*c*) What portion of I_B is due to holes diffusing from base to emitter, and what portion is due to electrons recombining inside the base?

2.3 A certain *npn* BJT has $N_{DE} = 10^{19}/cm^3$, $N_{AB} = 10^{17}/cm^3$, and $N_{DC} = 10^{15}/cm^3$. Moreover, $D_p = 1.8 \ cm^2/s$, $D_n = 18 \ cm^2/s$, and $\tau_n = 150$ ns.

(a) Find W_B and W_E so that $\beta_F = 250$ under the constraint $I_{BE} = I_{BB}$.

(b) With reference to Fig. 2.6, find the distance between the B-E and the B-C metallurgical junctions, as well as the distance between the B-E metallurgical junction and the E electrode, given that $V_{BE} = 700$ mV and $V_{BC} = -2.0$ V. Assume grading coefficients of 0.5 for both junctions.

2.4 A certain *pnp* BJT has an emitter area of $(25 \ \mu m) \times (50 \ \mu m)$, and its base and emitter doping concentrations are $N_{DB} = 10^{17}/cm^3$ and $N_{AE} = 10^{19}/cm^3$.

(a) If $D_n = 3 \ cm^2/s$, $D_p = 8 \ cm^2/s$, $W_E = W_B = 1 \ \mu m$, and $\tau_p = 100$ ns, find I_s and β_F.

(b) Repeat part (a) if W_B is lowered to 0.5 μm.

(c) Repeat (a) if W_E is raised to 2 μm. In both cases, comment on your findings.

2.5 Suppose Eqs. (2.17) and (2.18) give, for the *pnp* BJT of Fig. P2.5, $I_s = 4$ fA and $\beta_F = 1/(0.002 + 0.004)$.

(a) Find I_E, I_B, I_C, and V_{EB}.

(b) What portion of I_B is due to holes recombining inside the base region, and what portion is due to electrons diffusing from base to emitter?

(c) Recalculate I_E, I_B, and I_C if the collector is disconnected from the -4-V supply and is left floating.

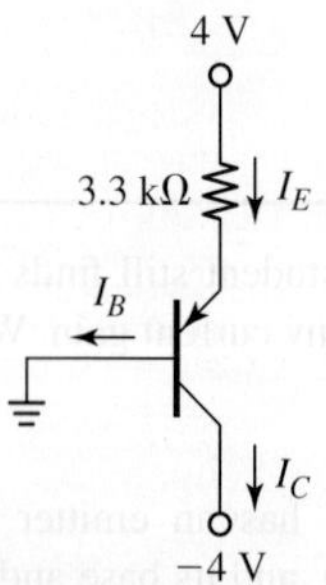

FIGURE P2.5

2.6 Assuming the BJT of Fig. P2.6 has $V_{BE(on)} = 0.7$ V and $\beta_F = 100$, estimate I_B, I_C, and I_E for each of the three cases shown (collector terminal left floating, connected to the base, and connected to the power supply), and give an indication of the composition of each current.

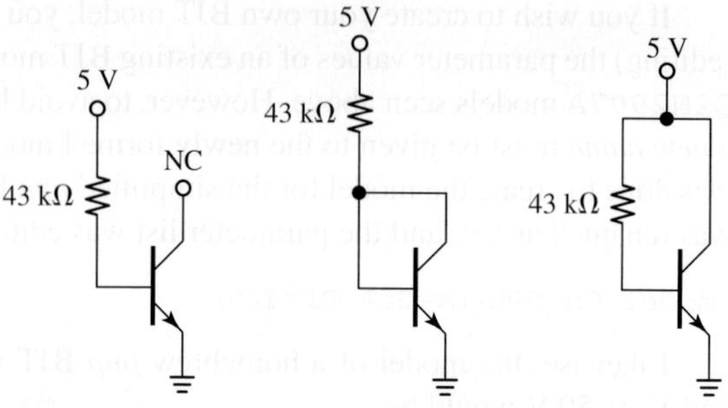

FIGURE P2.6

2.7 Figure P2.7 shows how a *pnp* BJT can boost the output current capability of an op amp when it comes to *sinking* current from the load (as opposed to *sourcing* current to the load, as discussed in the text). Assuming the BJT has $\beta_F = 80$ and an E-B junction voltage drop of 0.8 V, find the current and voltage at each of the BJT terminals if $I_L = 500$ mA. What is the voltage and current required of the op amp's output?

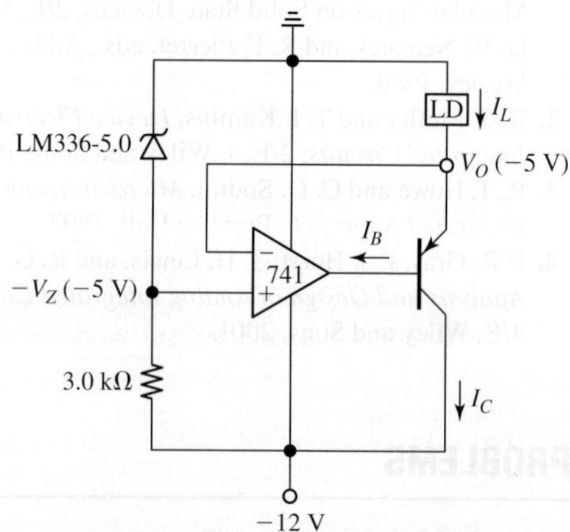

FIGURE P2.7

2.8 When operated as a current amplifier, an *npn* BJT will only *source* emitter current, while a *pnp* BJT will only *sink* emitter current. In ac situations, current must be sourced during the positive alternations of the load voltage, but must be sunk during the negative alternations. To serve these situations, the configuration of Fig. P2.8 is used. During positive alternations Q_1 conducts and Q_2 is off; during negative alternations Q_2 conducts and Q_1 is off. This popular transistor team, known as a *push-pull* stage, is at the basis of a variety of output stages, including those found in op amps and audio power amplifiers.

Assume a 100-Ω load and ±12-V power supplies, and let the BJTs have, respectively, $I_{s1} = 10$ fA and $\beta_{F1} = 150$, $I_{s2} = 20$ fA and $\beta_{F2} = 100$. Using $v_{BE1} = V_T \ln(i_{C1}/I_{s1})$ for Q_1 and $v_{EB2} = V_T \ln(i_{C2}/I_{s2})$ for Q_2, with $i_C \cong i_E$ and $i_B = i_E/(\beta_F + 1)$ for both BJTs, find the input voltage v_I as well as the input current (either i_{I1} or i_{I2}) needed to achieve:

(a) $v_O = +1$ V,
(b) $v_O = -1$ V,
(c) $v_O = +5$ V, and
(d) $v_O = -8$ V.

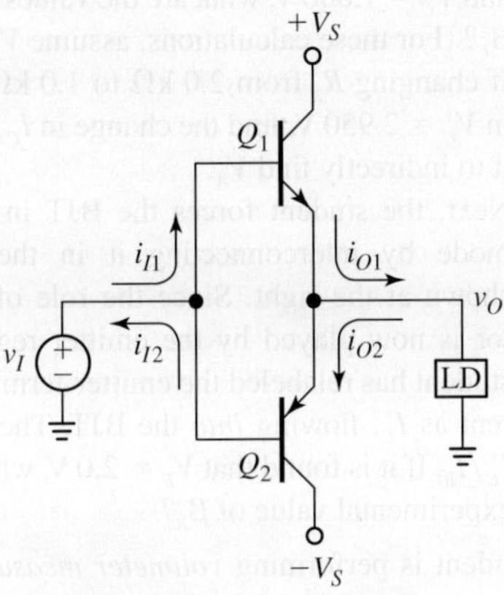

FIGURE P2.8

2.9 If the current gain provided by a given BJT is not sufficient for the application at hand, we can follow it by another BJT to further amplify the current already amplified by the first BJT. This is similar to following the lens of a telescope by a second lens to increase the overall magnifying power. The resulting two-transistor team, known as the *Darlington configuration*, finds application especially in power-handling circuits. Figure P2.9 illustrates the Darlington concept for the case of *npn* BJTs (the *pnp* case is shown in Fig. P2.10). The current i_I enters Q_1's base and exits Q_1's emitter amplified by $(\beta_{F1} + 1)$. This current is then fed to Q_2's base, and exits Q_2's emitter amplified by $(\beta_{F2} + 1)$. The overall current gain is thus $i_O/i_I = (\beta_{F1} + 1) \times (\beta_{F2} + 1) \cong \beta_{F1} \times \beta_{F2}$, or β_F^2 when the BJTs have identical betas. Assume $R_L = 8$ Ω and $V_S = 24$ V, and let the BJTs have, respectively, $I_{s1} = 10$ fA and $\beta_{F1} = 100$, $I_{s2} = 1$ pA and $\beta_{F2} = 50$. Using $v_{BE1} = V_T \ln(i_{C1}/I_{s1})$ for Q_1, and $v_{BE2} = V_T \ln(i_{C2}/I_{s2})$ for Q_2, with $i_C \cong i_E$ and $i_B = i_E/(\beta_F + 1)$ for both BJTs, find the input current i_I as well as

the input voltage v_I (accurate within 1 mV) needed to achieve:

(a) $v_O = 1$ V,
(b) $v_O = 4$ V,
(c) $v_O = 16$ V.

Hint: use the rules of thumb to speed up your calculations.

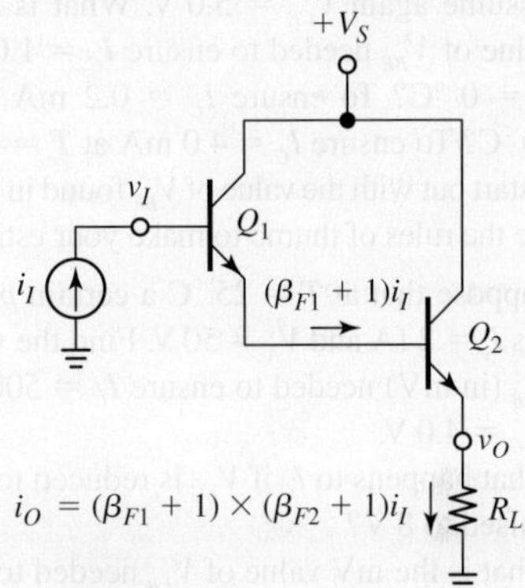

FIGURE P2.9

2.10 Read the Darlington configuration concept described in Problem 2.9, and then turn to its *pnp* realization of Fig. P2.10. Assume $R_L = 4$ Ω and $-V_S = -12$ V, and let the BJTs have, respectively, $I_{s1} = 5$ fA and $\beta_{F1} = 100$, $I_{s2} = 100$ pA and $\beta_{F2} = 40$. Using $v_{EB1} = V_T \ln(i_{C1}/I_{s1})$ for Q_1, and $v_{EB2} = 1.5V_T \ln(i_{C2}/I_{s2})$ for Q_2, find the input current i_I as well as the input voltage v_I (accurate within 1 mV) needed to achieve:

(a) $v_O = -1$ V,
(b) $v_O = -5$ V.

Hint: you can speed up your calculations using the rules of thumb.

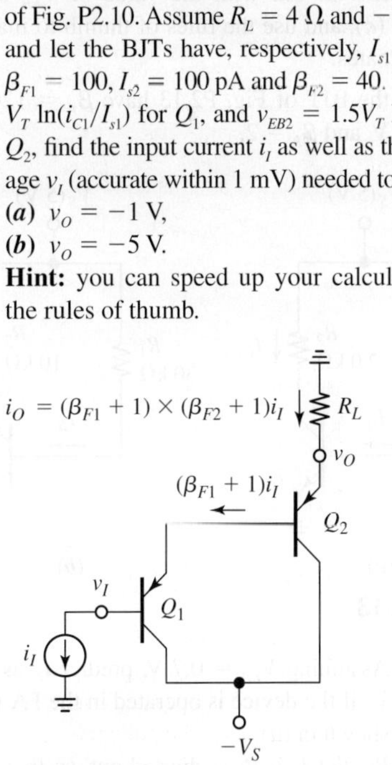

FIGURE P2.10

2.3 The *i-v* Characteristics of BJTs

2.11 (*a*) Suppose that at $T = 25\ °C$ a certain *npn* BJT has $I_s = 1$ fA and $V_A = 75$ V. Find the value of V_{BE} (in mV) needed to ensure $I_C = 1.0$ mA at $V_{CE} = 5.0$ V.

(*b*) What happens to I_C if V_{CE} is raised to 10 V? Lowered to 1 V?

(*c*) Assume again $V_{CE} = 5.0$ V. What is the mV value of V_{BE} needed to ensure $I_C = 1.0$ mA at $T = 0\ °C$? To ensure $I_C = 0.2$ mA at $T = 50\ °C$? To ensure $I_C = 4.0$ mA at $T = 40\ °C$?

Hint: start out with the value of V_{BE} found in part (*a*), and use the rules of thumb to make your estimates.

2.12 (*a*) Suppose that at $T = 25\ °C$ a certain *pnp* BJT has $I_s = 2$ fA and $V_A = 50$ V. Find the value of V_{EB} (in mV) needed to ensure $I_C = 500\ \mu A$ at $V_{EC} = 4.0$ V.

(*b*) What happens to I_C if V_{EC} is reduced to 1.0 V? Raised to 8 V?

(*c*) What is the mV value of V_{EB} needed to ensure $I_C = 0.2$ mA at $T = 75\ °C$ and $V_{EC} = 4.0$ V?

(*d*) Suppose V_{EB} and V_{EC} are kept at the values of (*a*), while T is raised from 25 °C to 55 °C. What is the new value of I_C?

Hint: start out with the value of V_{EB} found in part (*a*), and use the rules of thumb to make your estimates.

2.13 Let the BJT of Fig. P2.13 have $\beta_F = 120$, $V_A = 100$ V, and $\beta_R = 2$.

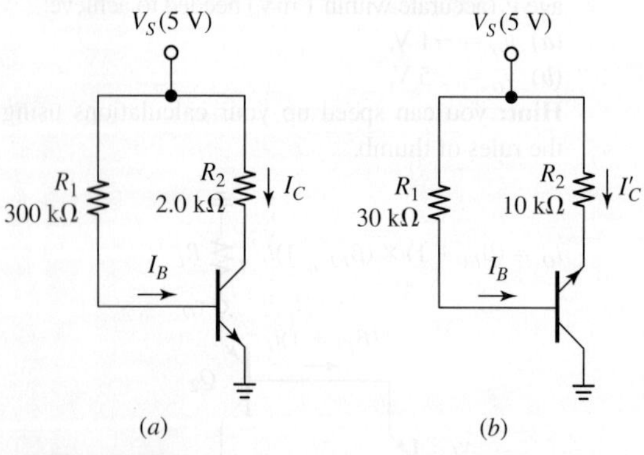

FIGURE P2.13

(*a*) Assuming $V_{BE} = 0.7$ V, predict I_C as well as V_C if the device is operated in the FA mode as shown in (*a*).

(*b*) Predict I_C if R_2 is shorted out, so that the collector voltage is forced to 5 V.

(*c*) Assuming $V_{BC} = 0.7$ V, predict the emitter voltage if the device is operated in the RA mode as shown in (*b*). (Since the role of collector is now being played by the emitter region, the current at the emitter terminal has been relabeled as I'_C.)

2.14 A student is performing *voltmeter measurements* on the circuits of Fig. P2.13 in order to extract the principal BJT parameters.

(*a*) In the circuit at the left, the BJT operates in the FA mode. If it is found that $V_B = 710$ mV and $V_C = 1.000$ V, what are the values of I_s and β_F? (For these calculations, assume $V_A = \infty$.)

(*b*) If changing R_2 from 2.0 kΩ to 1.0 kΩ results in $V_C = 2.950$ V, find the change in I_C, and use it to indirectly find V_A.

(*c*) Next, the student forces the BJT in the RA mode by interconnecting it in the circuit shown at the right. Since the role of collector is now played by the emitter region, the student has relabeled the emitter-terminal current as I'_C, flowing *into* the BJT. Then, $\beta_R = I'_C/I_B$. If it is found that $V_E = 2.0$ V, what is the experimental value of β_R?

2.15 A student is performing *voltmeter measurements* on the circuits of Fig. P2.15 in order to extract the principal BJT parameters.

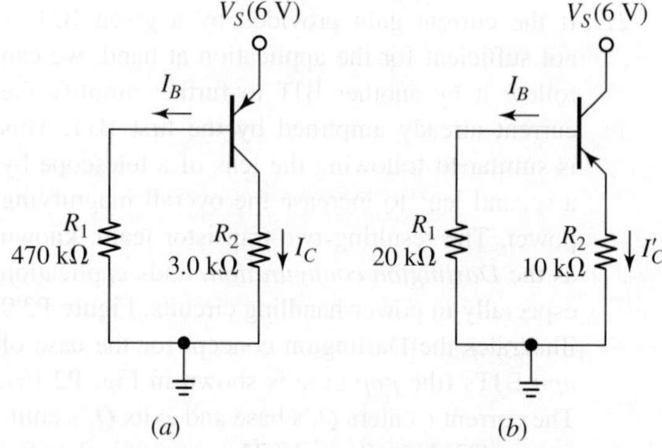

FIGURE P2.15

(*a*) In the circuit of (*a*) the BJT operates in the FA mode. If it is found that $V_{EB} = 690$ mV and $V_{EC} = 1.0$ V, what are the values of I_s and β_F? (For these calculations, assume $V_A = \infty$.)

(*b*) If letting $R_2 \rightarrow 0$ causes a 10% increase in I_C, what is the value of V_A?

(*c*) Next, the student forces the BJT in the RA mode by interconnecting it in the circuit of (*b*).

Since the role of collector is now played by the emitter region, the student has relabeled the emitter-terminal current as I'_C, flowing *out* of the BJT. Then, $\beta_R = I'_C/I_B$. If it is found that $V_E = 3.5$ V, what is the experimental value of β_R?

2.16 *(a)* In the circuit of Fig. P2.16, the power supply has been adjusted so as to result in $V_{BC} = 0$ with the switch open. In this state, the effective base width is $W_B = 500$ nm, and the portion of the B-C SCL extending onto the base region has the width $x_p = 20$ nm. What is the value of β_F?

(b) If the switch is closed, the B-C junction becomes reverse biased by 10 V, and x_p widens, shrinking W_B, and increasing I_C (Early effect). Assuming the B-C junction has the built-in potential $\phi_c = 0.8$ V and the grading coefficient $m_c = 0.5$, estimate the changes in x_p, W_B, and I_C as a consequence of switch closure.

(c) Use the change in I_C to indirectly estimate the Early voltage V_A.

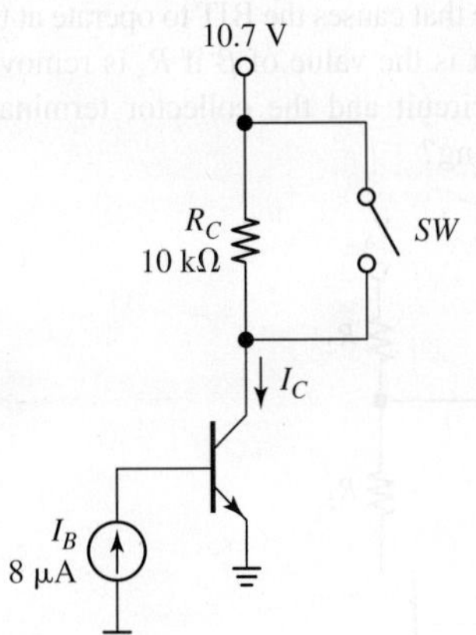

FIGURE P2.16

2.4 Operating Regions and BJT Models

2.17 Assuming the BJT of the circuit of Fig. P2.17 has $V_{BE(EOC)} = 0.6$ V, $V_{BE(on)} = 0.7$ V, $V_{BE(sat)} = 0.8$ V, $V_{CE(EOS)} = 0.2$ V, $V_{CE(sat)} = 0.1$ V, and β_F large, sketch and label the plots of v_B, v_C, and v_E (all on the same graph) versus v_I as v_I is swept from 0 V to 7 V. On a separate graph, sketch and label the plot of i_C. Identify the various regions of BJT operation, and show all relevant breakpoints and slopes. Comment on anything that you may find puzzling.

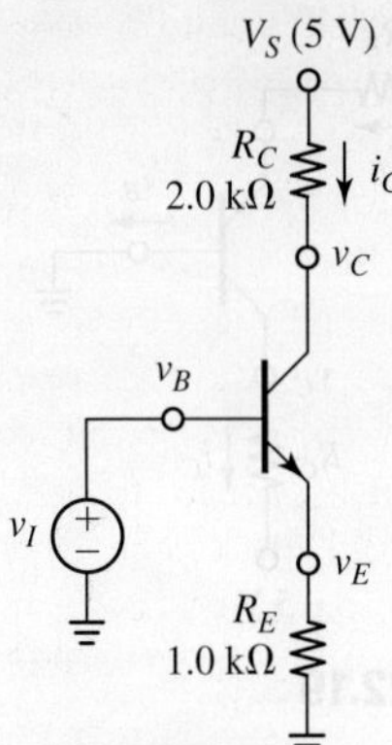

FIGURE P2.17

2.18 Assuming the BJT of the circuit of Fig. P2.18 has $V_{EB(EOC)} = V_{EB(on)} = V_{EB(EOS)} = 0.8$ V, $V_{EC(EOS)} = V_{EC(sat)} = 0.1$ V, and β_F large, sketch and label the plots of v_B, v_C, and v_E (all on the same graph) versus v_I as v_I is swept from 5 V to -2 V. On a separate graph, sketch and label the plot of i_C. Identify the various regions of BJT operation, and show all relevant breakpoints and slopes. Comment on anything that you may find surprising.

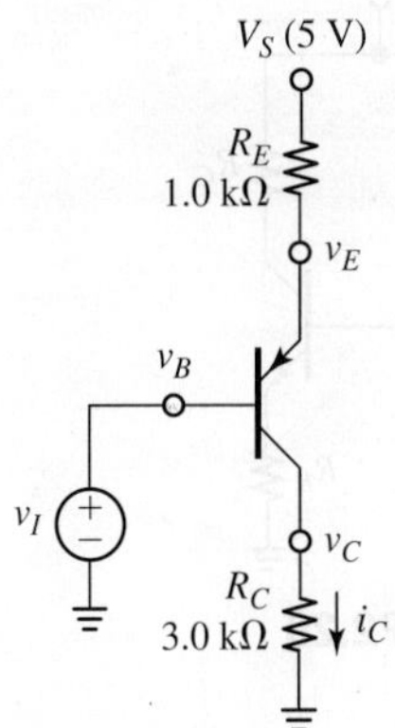

FIGURE P2.18

2.19 In the circuit of Fig. P2.19 let $R_E = 1.0$ kΩ and $R_C = 2.0$ kΩ. Assuming the BJT has $V_{EB(EOC)} = V_{EB(on)} = V_{EB(sat)} = 0.7$ V, $V_{EC(EOS)} = V_{EC(sat)} = 0.1$ V, and $\beta_F = 100$, sketch and label the plots of i_E, i_C, i_B, v_C, and the ratio i_C/i_B as v_S is swept from 0 to 5 V. Comment on your results.

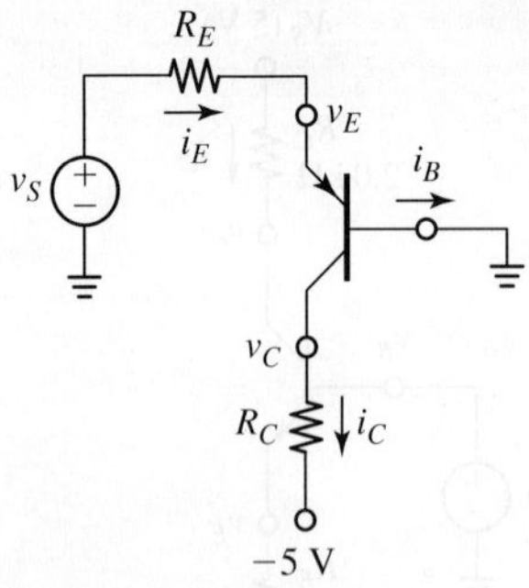

FIGURE P2.19

2.20 In the circuit of Fig. P2.20 let $R_B = 300\ \text{k}\Omega$, $R_C = 1.0\ \text{k}\Omega$, $R_E = 2.0\ \text{k}\Omega$, and $V_{CC} = 5$ V.

(a) If it is found that $V_{CE} = 2.0$ V, what is the value of β_F? Assume $V_{BE(\text{on})} = 0.7$ V.

(b) What value must R_B be changed to if we wish to bring the BJT to the EOS, where $V_{CE(\text{EOS})} = 0.2$ V?

(c) What value of R_B will cause the BJT to saturate with $\beta_{sat} = \beta_F/5$? Assume $V_{CE(\text{sat})} = 0.1$ V and $V_{BE(\text{sat})} = 0.8$ V.

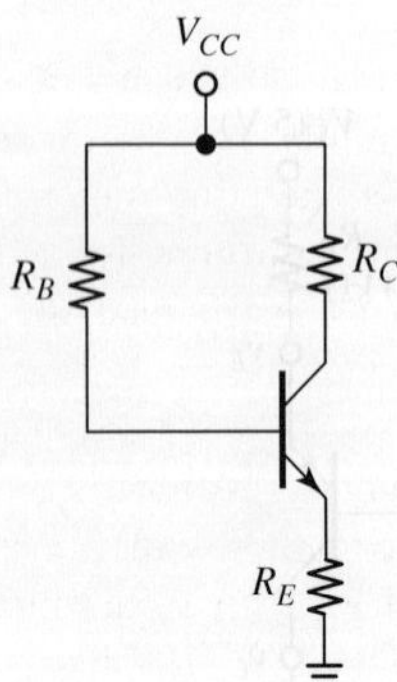

FIGURE P2.20

2.21 In the circuit of Fig. P2.21 let $R_B = 180\ \text{k}\Omega$, $R_C = 1.0\ \text{k}\Omega$, $R_E = 3.0\ \text{k}\Omega$, and $V_{CC} = -5$ V. Moreover, let $V_{EB(\text{on})} = 0.7$ V, $V_{EC(\text{EOS})} = 0.2$ V, $V_{EC(\text{sat})} = 0.1$ V, and $V_{EB(\text{sat})} = 0.8$ V.

(a) If it is found that $V_{EC} = 1.0$ V, what is the value of β_F?

(b) Find all voltages and currents at the BJT terminals if R_C and R_E are swapped with each other.

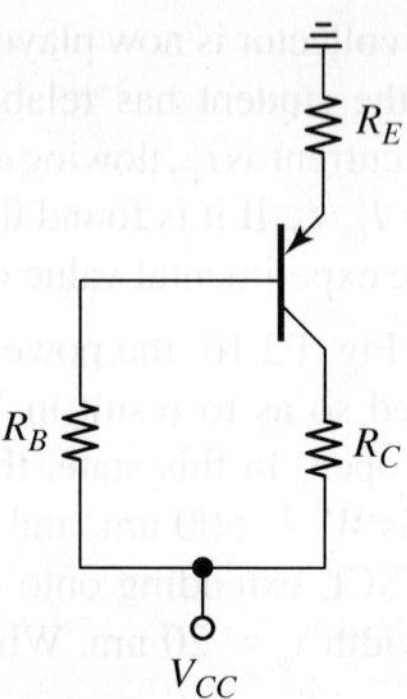

FIGURE P2.21

2.22 In the circuit of Fig. P2.22 let $V_S = 5$ V, $R_1 = 20\ \text{k}\Omega$ and $R_2 = 3.0\ \text{k}\Omega$. Moreover, let the BJT have $\beta_F = 150$, $V_{BE(\text{on})} = 0.7$ V, $V_{CE(\text{EOS})} = 0.2$ V, $V_{CE(\text{sat})} = 0.1$ V, and $V_{BE(\text{sat})} = 0.8$ V.

(a) Find the range of values for R_3 for which the BJT will operate in the FA region, and calculate the BJT's terminal currents.

(b) Find the BJT's beta if R_3 is raised to *twice* the value that causes the BJT to operate at the EOS.

(c) What is the value of β if R_3 is removed from the circuit and the collector terminal is left floating?

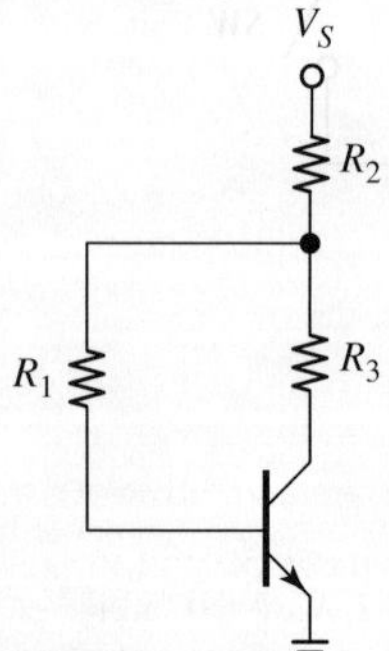

FIGURE P2.22

2.23 In the circuit of Fig. P2.22 let $V_S = 9$ V, $R_1 = 100\ \text{k}\Omega$, $R_2 = 20\ \text{k}\Omega$, and $R_3 = 1.0\ \text{k}\Omega$.

(a) Assuming the BJT parameters of Problem 2.22, find all voltages and currents in the circuit.

(b) Repeat, if R_2 and R_3 are swapped with each other.

2.24 In the circuit of Fig. P2.24 let $R_B = R_C = R_E = 10\ \text{k}\Omega$, and let the BJT have $\beta_F = 100$ and $\beta_R = 5$. Moreover, to simplify calculations somewhat, assume that when either junction is fully forward biased, it drops 0.7 V regardless of the BJT's operating mode. Identify the BJT's operating region, and estimate its terminal voltages and currents (magnitudes and directions!) if:

(a) $V_{CC} = 10$ V, $V_{BB} = 5$ V, and $V_{EE} = 0$ (ground).
(b) $V_{BB} = V_{CC} = 5$ V, and $V_{EE} = 0$.
(c) $V_{EE} = 10$ V, $V_{BB} = 5$ V, and $V_{CC} = 0$.
(d) $V_{BB} = V_{EE} = 5$ V, and $V_{CC} = 0$.
(e) $V_{BB} = 0$ V, $V_{CC} = 10$ V, and $V_{EE} = 5$ V.

Hint: for each individual case, you may wish to reorient the circuit so that positive voltages appear at the top of the figure, and ground at the bottom.

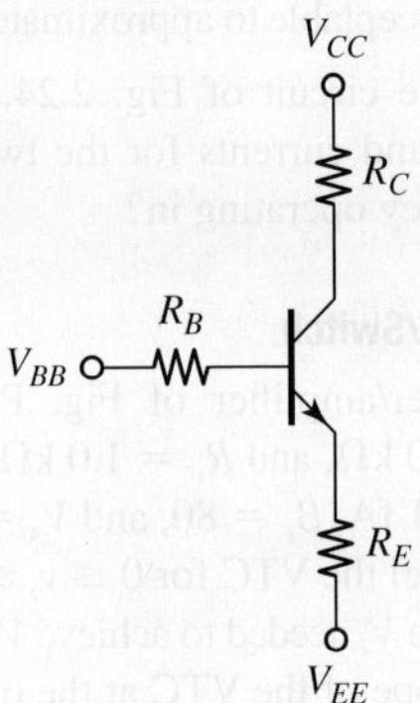

FIGURE P2.24

2.25 Let the BJT of Fig. P2.24 have $\beta_F = 100$ and $\beta_R = 5$. Moreover, assume that either junction drops 0.7 V when fully forward biased, regardless of the BJT's operating mode, and that its forward current is still zero for a forward bias of 0.65 V or less. Also, read the hint of Problem 2.24.

(a) If $R_C = R_E = 10$ kΩ, $R_B = 100$ kΩ, $V_{CC} = 10$ V, and $V_{EE} = 0$ V, find V_{BB} so that $V_{CE} = 1.0$ V.
(b) Reconsider the circuit of part (a), but with $V_{EE} = 10$ V and $V_{CC} = 0$ V, and find V_{BB} so that $V_{EC} = 1.0$ V.
(c) If $R_B = R_C = R_E = 10$ kΩ, $V_{BB} = 10$ V, and $V_{EE} = 0$ V, estimate V_{CC} so that $I_C \cong 0$.
(d) If $R_B = R_C = R_E = 10$ kΩ, $V_{CC} = 5$ V and $V_{EE} = 0$ V, find V_{BB} so that $I_B = 1.5I_E$. What is the value of beta in this case?

2.26 In the circuit of Fig. P2.26, assume that the BJT has $\beta_F = 150$ and $\beta_R = 4$, and assume that when either junction is fully forward biased, it drops

0.7 V regardless of the BJT's operating mode. Also, read the hint of Problem 2.24. If $R_C = 1$ kΩ, $R_E = 2$ kΩ, and $R_B = 3$ kΩ, identify the BJT's operating mode, and estimate its terminal voltages and currents (magnitudes and directions!) if:

(a) $V_{EE} = 12$ V, $V_{BB} = 6$ V, and $V_{CC} = 0$ (ground).
(b) $V_{CC} = 10$ V, $V_{BB} = 5$ V, and $V_{EE} = 0$.
(c) $V_{EE} = 6$ V, and $V_{BB} = V_{CC} = 0$.

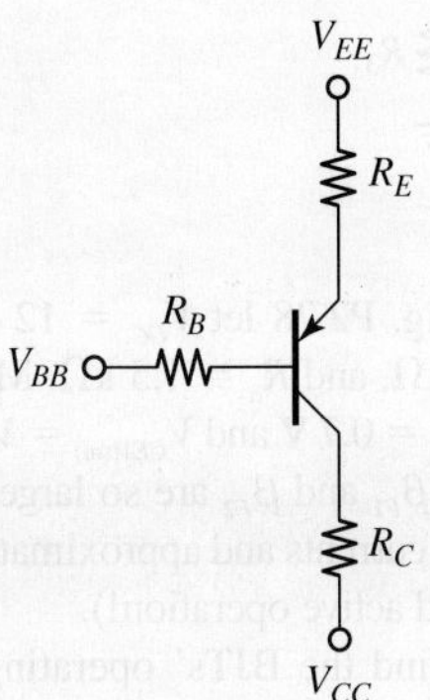

FIGURE P2.26

2.27 Let the BJT of Fig. P2.26 have $\beta_F = 150$ and $\beta_R = 4$. Moreover, assume that either junction drops 0.7 V when fully forward biased, regardless of the BJT's operating mode, and that its forward current is still zero for a forward bias of 0.65 V or less. Also, read the hint of Problem 2.24.

(a) If $R_C = R_E = 10$ kΩ, $R_B = 100$ kΩ, $V_{EE} = 5$ V, and $V_{CC} = -5$ V, find V_{BB} so that $V_{EC} = 1.0$ V.
(b) Reconsider the circuit of part (a), but with $V_{EE} = -5$ V and $V_{CC} = 5$ V, and find V_{BB} so that $V_{CE} = 1.0$ V.
(c) If $R_B = R_C = R_E = 10$ kΩ, $V_{EE} = 9$ V, and $V_{BB} = 0$ V, estimate V_{CC} so that $I_C \cong 0$.
(d) If $R_B = R_C = R_E = 10$ kΩ, $V_{EE} = 6$ V and $V_{CC} = 0$ V, find V_{BB} so that $I_B = 0.75I_E$. What is the value of beta in this case?
(e) Repeat part (d), but for $I_B = 1.25I_E$.

2.28 In the circuit of Fig. P2.28 let $V_{CC} = 10$ V, $V_1 = 4$ V, $R_2 = 10$ kΩ, and $R_3 = 3.0$ kΩ. Draw and label the circuit and then, assuming $\beta_{F1} = 55$, $\beta_{F2} = 50$, $V_{BE1(on)} = V_{EB2(on)} = 0.7$ V, and $V_{CE1(sat)} = V_{EC2(sat)} = 0.1$ V, find R_1 so that $I_{C2} = 0.5$ mA. What is the current supplied by V_1?

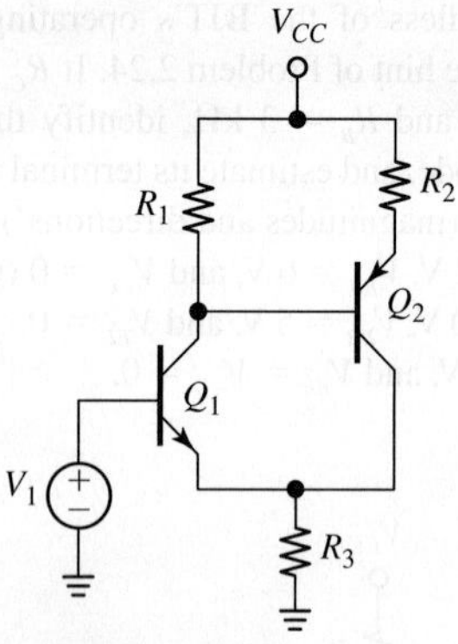

FIGURE P2.28

2.29 In the circuit of Fig. P2.28 let $V_{CC} = 12$ V, $R_1 = 3.0$ kΩ, $R_2 = 2.0$ kΩ, and $R_3 = 1.5$ kΩ. Moreover, let $V_{BE1(on)} = V_{EB2(on)} = 0.7$ V and $V_{CE1(sat)} = V_{EC2(sat)} = 0.1$ V, and assume β_{F1} and β_{F2} are so large that we can ignore the base currents and approximate $I_C \cong I_E$ (but only in forward active operation!).
(*a*) If $V_1 = 5$ V, find the BJTs' operating points $Q_1 = Q_1(I_{C1}, V_{CE1})$ and $Q_2 = Q_2(I_{C2}, V_{EC2})$.
(*b*) Repeat if V_1 is raised to 8 V. What is the current supplied by V_1 in this case?

2.30 In the circuit of Fig. P2.30 let $V_{CC} = 12$ V, $V_1 = 6$ V, $R_2 = 33$ kΩ, $R_3 = 3.0$ kΩ, and $R_4 = 2.0$ kΩ. Moreover, let $\beta_{F1} = \beta_{F2} = 50$, $V_{EB1(on)} = V_{BE2(on)} = 0.7$ V, and $V_{EC1(sat)} = V_{CE2(sat)} = 0.1$ V.
(*a*) Find R_1 so that $I_{C1} = 1.5$ mA.
(*b*) Look at your calculations again, and identify the cases in which base currents could have been ignored to speed up calculations, and those in which they couldn't. Recalculate using these approximations, compare, and comment.

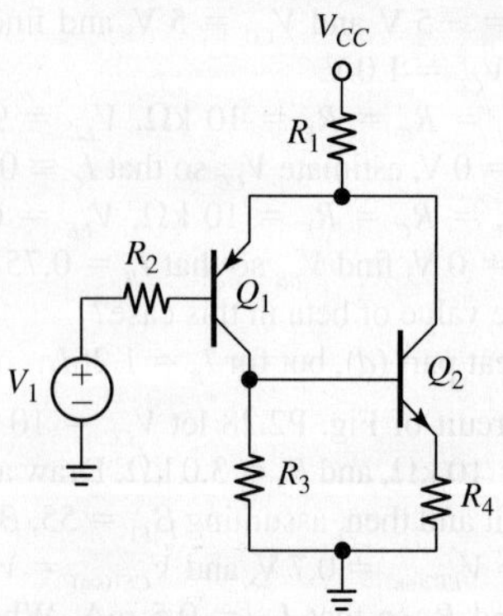

FIGURE P2.30

2.31 In the circuit of Fig. P2.30 let $V_{CC} = 12$ V, $V_1 = 5$ V, and $R_1 = R_2 = R_3 = R_4 = 10$ kΩ. Moreover, let $V_{EB1(on)} = V_{BE2(on)} = 0.7$ V, and $V_{EC1(sat)} = V_{CE2(sat)} = 0.1$ V, and assume β_{F1} and β_{F2} are so large that we can ignore the base currents as well as the voltage drop across R_2 (but only in forward active operation!).
(*a*) Find the BJTs' operating points $Q_1 = Q_1(I_{C1}, V_{EC1})$ and $Q_2 = Q_2(I_{C2}, V_{CE2})$.
(*b*) Repeat if R_3 is removed altogether from the circuit.

2.32 If $V_1 = 5$ V in the circuit of Fig. 2.24, what value does R_3 need to be changed to if we want Q_2 to operate right at the edge of saturation (EOS)?
Hint: as long as a BJT is known to be in the FA region, it is quite acceptable to approximate $I_C \cong I_E$.

2.33 If $V_1 = 5$ V in the circuit of Fig. 2.24, find all terminal voltages and currents for the two BJTs. What modes are they operating in?

2.5 The BJT as an Amplifier/Switch

2.34 In the *pnp* inverter/amplifier of Fig. P2.34 let $V_{CC} = 5$ V, $R_B = 10$ kΩ, and $R_C = 1.0$ kΩ, and let the BJT have $I_s = 1$ fA, $\beta_F = 80$, and $V_A = \infty$.
(*a*) Sketch and label the VTC for $0 \le v_I \le 5$ V.
(*b*) Find the voltage V_I needed to achieve $V_O = 4$ V.
(*c*) What is the slope of the VTC at the operating point of part (*b*)?

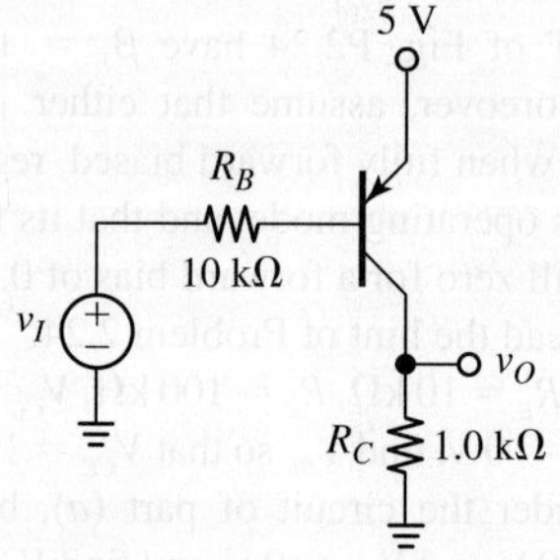

FIGURE P2.34

2.35 In logic applications of the BJT inverter, it is desirable that the VTC be centered near the middle of the v_I range. The VTC of the simple inverter of Fig. 2.32, shown in Fig. 2.34, top, is centered about 1 V, but it can easily be shifted to the right by connecting an additional resistance R_{BE} across the B-E junction, as depicted in Fig. P2.35.

(**a**) Show that we now have $v_I = R_B i_B + (1 + R_B/R_{BE})v_{BE}$.

(**b**) Assuming the same BJT parameters as in Fig. 2.32 ($\beta_F = 100$ and $I_s = 2$ fA), obtain an input-output relationship of the type of Eq. (2.38), of course suitably modified to reflect the presence of R_{BE}.

(**c**) Find the voltage V_I needed to bias the BJT at $V_O = 2.5$ V.

(**d**) What is the voltage gain a there? How does it compare with that of Example 2.10? Justify any difference.

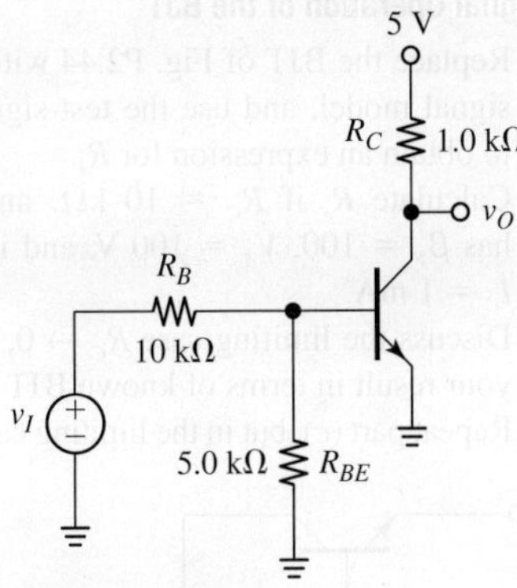

FIGURE P2.35

2.36 (**a**) Find a resistance R_{EB} that, when connected in parallel with the E-B junction of the *pnp* BJT of Problem 2.34, will yield $V_O = 2.5$ V for $V_I = 2.5$ V.

Hint: read Problem 2.35 to understand the effect of R_{EB} upon the VTC, and show that the modified circuit of Fig. P2.34 gives $v_I = 5\,V - R_B i_B - (1 + R_B/R_{EB})v_{EB}$, where i_B is the base current, which for a *pnp* BJT flows *out* of the base, and v_{EB} is the voltage drop across the E-B junction.

(**b**) Obtain a numerical relationship between v_I and v_O, and differentiate it with respect to v_I to estimate the gain a at $V_O = 2.5$ V.

2.37 Consider the circuit of Fig. P2.34 with $V_{CC} = 5$ V, $R_B = 10$ kΩ, and $R_C = 2.0$ kΩ, plus an additional resistance $R_{EB} = 10$ kΩ connected in parallel with the E-B junction of the BJT. Moreover, let the BJT have $I_s = 5$ fA, $\beta_F = 125$, and $V_A = \infty$.

(**a**) Show that we have $v_I = 5 - 10^4 i_B - 2v_{EB}$, where i_B is the current *out* of the base, and v_{EB}

is the voltage drop across the E-B junction (v_I and v_{EB} are in V, i_B is in A).

(**b**) Find v_{I1} needed to achieve $v_{O1} = 2$ V, and v_{I2} needed to achieve $v_{O2} = 3$ V. Hence, estimate the voltage gain of your circuit as $a \cong (v_{O2} - v_{O1})/(v_{I2} - v_{I1})$.

(**c**) Sketch and label v_I and v_O versus time if v_I is a 1-kHz sine wave alternating between v_{I1} and v_{I2}.

2.38 Consider a push-pull BJT pair of the type of Fig. P2.8, with the common base node being driven by a source v_s having a series resistance $R_s = 10$ kΩ, and the common emitter node driving a load $R_L = 1$ kΩ. The BJTs have $\beta_F = 100$, $V_{BE1(EOC)} = V_{EB2(EOC)} = 0.6$ V, $V_{BE1(on)} = V_{EB2(on)} = 0.7$ V, $V_{CE1(EOS)} = V_{EC2(EOS)} = 0.2$ V, and $V_{CE1(sat)} = V_{EC2(sat)} = 0.1$ V.

(**a**) Assuming ±5-V power supplies, sketch and label v_O versus v_S over the range $(-7\,V) \le v_S \le (+7V)$.

(**b**) What are the values of v_O at $v_S = \pm 2.5$ V?

2.39 Shown in Fig. P2.39 is a BJT realization of the logic function known as NOR.

(**a**) Assuming $L = 0$ V and $H = 5$ V, prepare the truth table, showing the state of each BJT (CO or Sat) as well as the logic level (H or L) of Y for each of the four possible input combinations: $(A, B) = (L, L), (L, H), (H, L), (H, H)$.

(**b**) If $V_{BE(sat)} = 0.8$ V and $V_{CE(sat)} = 0.1$ V, what is the minimum β_F required of each BJT for proper operation?

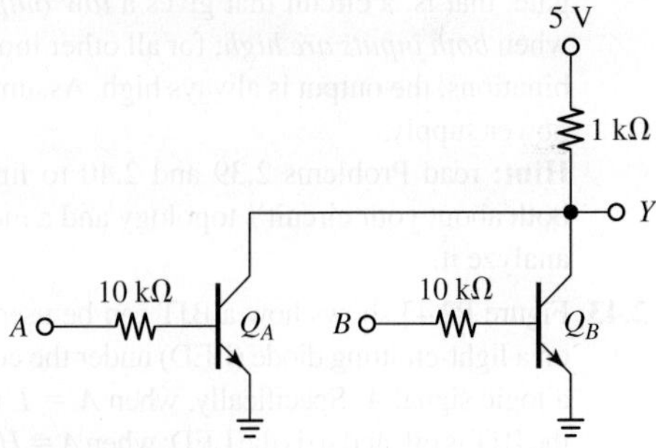

FIGURE P2.39

2.40 Shown in Fig. P2.40 is a BJT realization of the logic function known as NAND.

(**a**) Assuming $L = 0$ V and $H = 5$ V, prepare the truth table, showing the state of each BJT (CO or Sat) as well as the logic level (H or L) of Y for each of the four possible input combinations: $(A, B) = (L, L), (L, H), (H, L), (H, H)$.

(**b**) If $V_{BE(sat)} = 0.8$ V and $V_{CE(sat)} = 0.1$ V, what is the minimum β_F required of each BJT for proper operation?

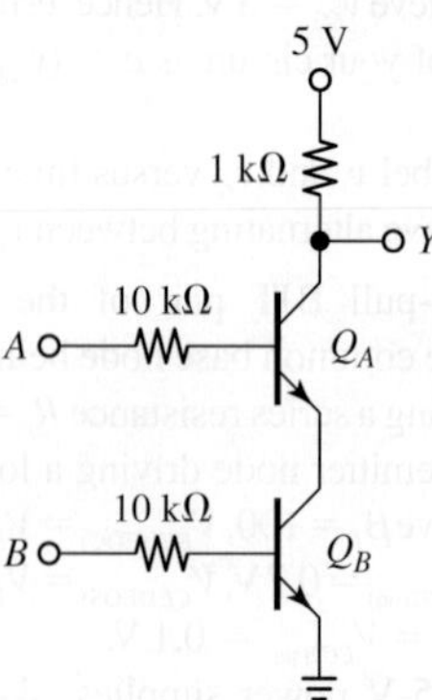

FIGURE P2.40

2.41 Using two *pnp* BJTs and resistors as needed, design a circuit to implement a two-input NOR gate, that is, a circuit that gives a *high output* only when *both inputs are low*; for all other input combinations, the output is always low. Assume a 5-V power supply.

Hint: read Problems 2.39 and 2.40 to come up with ideas both about your circuit's topology and a method to analyze it.

2.42 Using two *pnp* BJTs and resistors as needed, design a circuit to implement a two-input NAND gate, that is, a circuit that gives a *low output* only when *both inputs are high*; for all other input combinations, the output is always high. Assume a 5-V power supply.

Hint: read Problems 2.39 and 2.40 to find ideas both about your circuit's topology and a method to analyze it.

2.43 Figure P2.43 shows how a BJT can be used to turn on a light-emitting diode (LED) under the control of a logic signal A. Specifically, when $A = L\,(= 0$ V) the BJT is off, and so is the LED; when $A = H\,(= 5$ V) the BJT saturates and the LED glows.

(**a**) Assuming the BJT has $\beta_{F(min)} = 50$, $V_{BE(sat)} = 0.8$ V, and $V_{CE(sat)} = 0.1$ V, specify standard 5% values for R_1 and R_2 to make the LED glow with a current $I_D = 10$ mA and a voltage drop $V_D = 1.5$ V.

(**b**) Design a circuit that accepts a logic input B and causes the LED to glow when $B = L\,(= 0$ V), but causes it to be off when $B = H\,(= 5$ V).

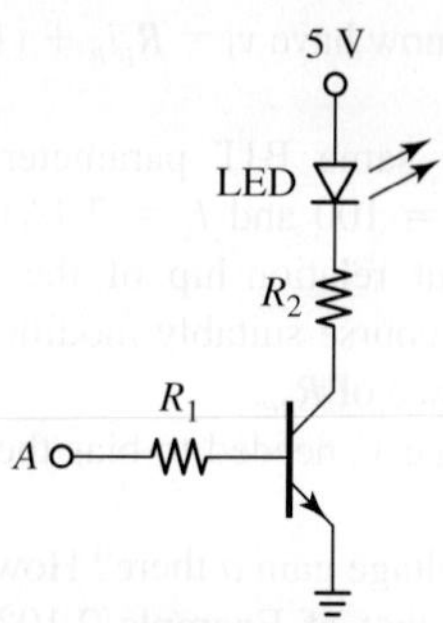

FIGURE P2.43

2.6 Small-Signal Operation of the BJT

2.44 (**a**) Replace the BJT of Fig. P2.44 with its small-signal model, and use the test-signal method to obtain an expression for R_e.

(**b**) Calculate R_e if $R_C = 10$ kΩ, and the BJT has $\beta_F = 100$, $V_A = 100$ V, and is biased at $I_C = 1$ mA.

(**c**) Discuss the limiting case $R_C \to 0$, and justify your result in terms of known BJT behavior.

(**d**) Repeat part (**c**), but in the limiting case $R_C \to \infty$.

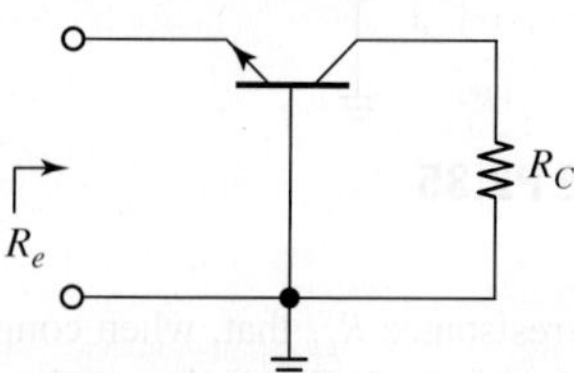

FIGURE P2.44

2.45 (**a**) Replace the BJT of Fig. P2.45 with its small-signal model, and use the test-signal method to show that the resistance seen looking into the input port is

$$R_i = r_\pi /\!/ \left[\frac{R_F + (R_C /\!/ r_o)}{1 + g_m(R_C /\!/ r_o)} \right]$$

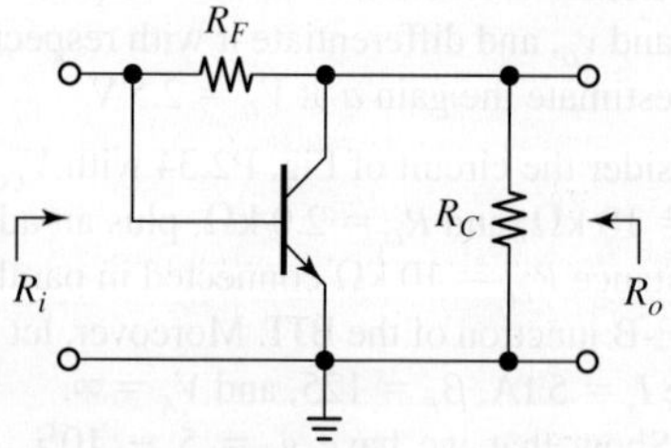

FIGURE P2.45

   (b) Calculate R_i if $R_F = 10\ \text{k}\Omega$, $R_C = 1\ \text{k}\Omega$, and the BJT has $\beta_F = 100$, $V_A = 100\ \text{V}$, and is biased at $I_C = 1\ \text{mA}$.

   (c) Discuss the limiting case $R_F \to 0$ and $R_C \to \infty$, and justify your result in terms of previously known BJT behavior.

   (d) Repeat part (c), but in the limiting case $R_F \to \infty$.

2.46 (a) Replace the BJT of Fig. P2.45 with its small-signal model, and use the test-signal method to prove that the resistance seen looking into the output port with the input port open-circuited, as shown, is

$$R_o = R_C /\!/ r_o /\!/ \left(r_e + \frac{R_F}{\beta_0 + 1} \right)$$

   (b) Calculate R_o if $R_F = 10\ \text{k}\Omega$, $R_C = 1\ \text{k}\Omega$, and the BJT has $\beta_F = 100$, $V_A = 100\ \text{V}$, and is biased at $I_C = 1\ \text{mA}$.

   (c) Discuss the limiting case $R_F \to 0$ and $R_C \to \infty$, and justify your result in terms of already familiar BJT behavior.

   (d) Repeat part (c), but with an ac short-circuit across the input port.

2.47 Assuming the small-signal model for the BJT of Fig. P2.47 has $r_\pi = 10\ \text{k}\Omega$, $g_m = 1/(50\ \Omega)$, and $r_o = \infty$, find R via the test-signal method.

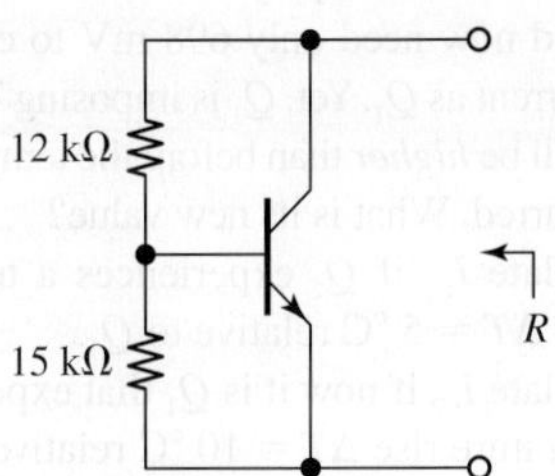

FIGURE P2.47

2.48 (a) Use simple inspection to come up with an expression for R_i in the circuit of Fig. P2.48 (switch open).

   (b) Next, investigate the effect of closing the switch. Replace the BJT with its small-signal model, and use the test-signal method to show that now R_i is *increased* to

$$R_i = R_x + (1 + g_m R_x)R_y$$
$$R_x = R_1 /\!/ r_\pi$$
$$R_y = R_2 /\!/ R_E$$

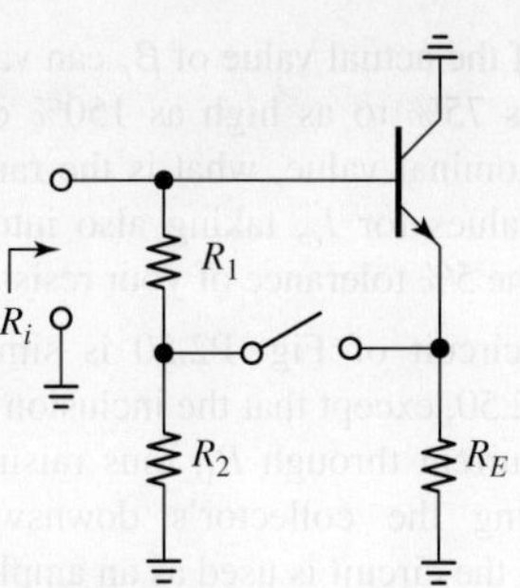

FIGURE P2.48

   (c) Calculate R_i for each of the two above cases if $R_1 = R_2 = R_E = 10\ \text{k}\Omega$, and the BJT has $\beta_F = 100$ and $V_A = \infty$, and is biased at $I_C = 1\ \text{mA}$. Compare the two results, and comment!

Note: this problem illustrates the so called *boot-strapping technique*, which is used to *raise the input resistance* of an emitter follower. In a practical situation, what is shown here as a switch is a capacitor, which acts as an open circuit at dc, but is designed to act as a short circuit at ac.

2.7 BJT Biasing for Amplifier Design

2.49 (a) Assuming $V_{EE} = 9\ \text{V}$ and $V_{CC} = 0\ \text{V}$, along with typical BJT parameters in the circuit of Fig. P2.49, specify standard 5% resistances to bias the BJT at $I_C = 2\ \text{mA}$ using the 1/3-1/3-1/3 Rule.

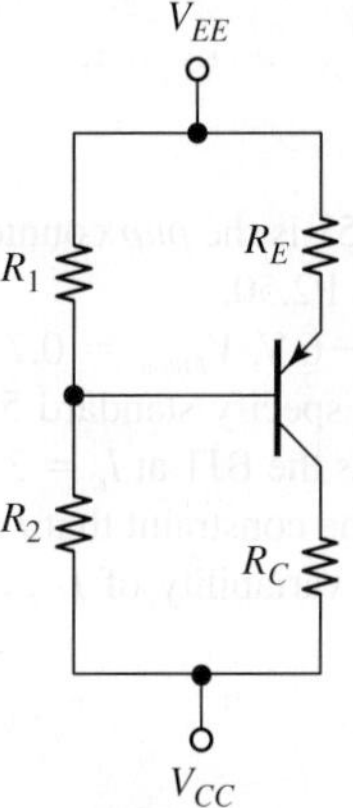

FIGURE P2.49

(b) If the actual value of β_F can vary from as low as 75% to as high as 150% of the assumed nominal value, what is the range of possible values for I_C, taking also into consideration the 5% tolerance of your resistances?

2.50 The circuit of Fig. P2.50 is similar to that of Fig. 2.50, except that the inclusion of R_2 increases the current through R_1, thus raising V_{CE} and expanding the collector's downswing capability when the circuit is used as an amplifier.

(a) Assuming $V_{CC} = 5$ V, $V_{BE(on)} = 0.7$ V, $\beta_F = 100$, and $V_A = \infty$, specify standard 5% resistance values to bias the BJT at $I_C = 2$ mA and $V_{CE} = 2$ V under the constraint that $I_{R_2} = I_B$. What is the resulting value of V_{CE}?

(b) Find the range of variability of I_C and V_{CE} if $50 \leq \beta_F \leq 200$.

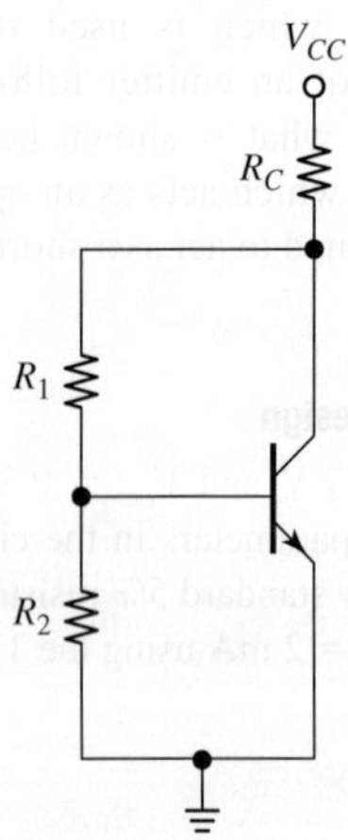

FIGURE P2.50

2.51 The circuit of Fig. P2.51 is the *pnp* counterpart of the *npn* version of Fig. P2.50.

(a) Assuming $V_{CC} = -6$ V, $V_{EB(on)} = 0.7$ V, $\beta_F = 150$, and $V_A = \infty$, specify standard 5% resistance values to bias the BJT at $I_C = 3$ mA and $V_{EC} = 3$ V under the constraint that $I_{R_1} = 2\,I_B$.

(b) Find the range of variability of I_C and V_{EC} if $75 < \beta_F < 250$.

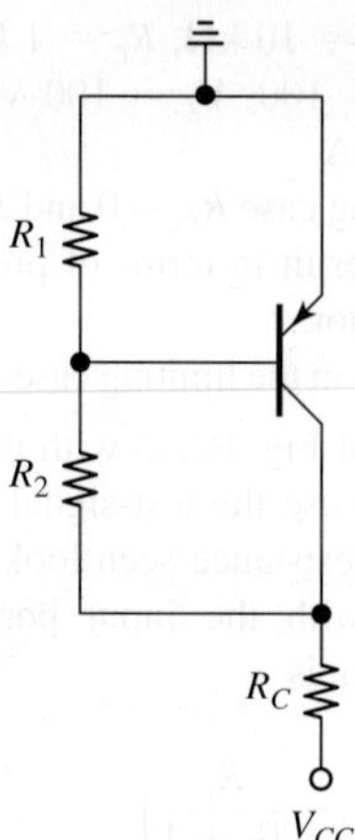

FIGURE P2.51

2.52 Assume the current mirror of Fig. 2.51 is implemented with perfectly matched BJTs having negligible base currents and $V_A = \infty$. Moreover, let $V_{CC} = 5$ V and $R_1 = 4.3$ kΩ.

(a) Assuming $V_{BE} = 700$ mV and R_2 is small enough to ensure that Q_2 is always in the FA region, find I_{C2}.

(b) Suppose now Q_2 experiences a temperature rise $\Delta T = 1\ ^\circ$C relative to Q_1. Given that the temperature coefficient of a *pn* junction is -2 mV/$^\circ$C, Q_2 would now need only 698 mV to carry the same current as Q_1. Yet, Q_1 is imposing 700 mV, so I_{C2} will be *higher* than before the temperature rise occurred. What is its new value?

(c) Recalculate I_{C2} if Q_2 experiences a temperature rise $\Delta T = 5\ ^\circ$C relative to Q_1.

(d) Recalculate I_{C2} if now it is Q_1 that experiences a temperature rise $\Delta T = 10\ ^\circ$C relative to Q_2.

(e) If it is found that $I_{C2} = 0.75$ mA, what do you conclude?

2.53 There are situations in which it is desired to have $I_{C2} < I_{C1}$ in a current mirror. A popular way to lower I_{C2} relative to I_{C1} is to reduce V_{BE2} by lifting Q_2's emitter off ground and inserting a suitable series resistance R to drop the required voltage difference, as shown in Fig. P2.53. For instance, if we want to make $I_{C2} = I_{C1}/2$, R will have to drop 18 mV, by the familiar rule of thumb. The circuit is known as the *Widlar current source* for its inventor Bob Widlar, the designer of the first monolithic op amp.

In Fig. P2.53 let $V_{CC} = 5$ V and $R_1 = 4.3$ kΩ. Assuming $V_{BE1} = 700$ mV, negligible base currents, and a small enough voltage drop across the load to ensure that Q_2 is always in the FA region, find R so that:

(a) $I_{C2} = 0.4$ mA.
(b) $I_{C2} = 50$ μA.
(c) $I_{C2} = 123$ μA.
Hint: use the rules of thumb when possible.

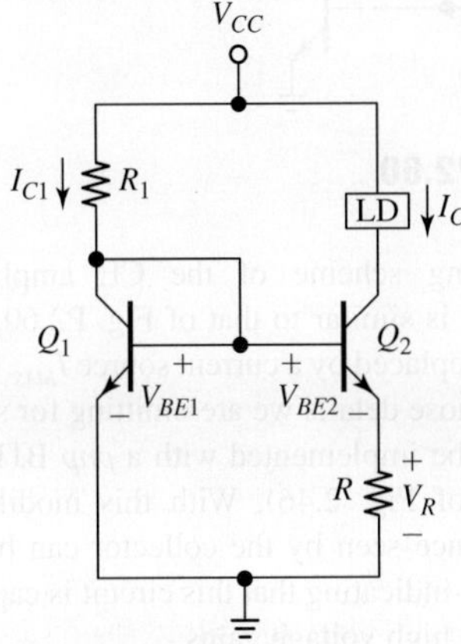

FIGURE P2.53

2.54 In Fig. P2.53 let $V_{CC} = 6$ V, $R_1 = 10$ kΩ, and $R = 1.0$ kΩ. Moreover, let the BJTs be matched devices with $I_s = 2$ fA, $V_A = \infty$, and negligible base currents.

(a) Assuming a small enough voltage drop across the load to ensure that Q_2 is always in the FA region, find V_{BE1}, in mV.

(b) Using the relations $V_{BE2} = V_T \ln(I_{C2}/I_s)$ and $I_{C2} = (V_{BE1} - V_{BE2})/R$, iterate until you find the value of I_{C2}.

(c) What value must V_{CC} be lowered to if we want I_{C1} to drop to 50% of its initial value? What is the resulting value of I_{C2}? Does I_{C2} also drop to 50% of its initial value? Explain!

2.55 A current mirror consisting of two matched BJTs can be *imbalanced* on purpose by inserting a resistance R in series with the emitter of one of its two BJTs. In Fig. P2.53 we have inserted R in series with the emitter of Q_2 to *lower* I_{C2} relative to I_{C1}. Let us now investigate the opposite case, namely, the insertion of R in series with the emitter of Q_1 to *increase* I_{C2} relative to I_{C1}. Redraw the circuit of Fig. P2.53, but with Q_2's emitter at ground and with R in series between Q_1's emitter and ground.

(a) If $V_{CC} = 5$ V and the BJTs are matched and have $I_s = 2$ fA, $V_A = \infty$, and negligible base currents, specify suitable values for R_1 and R to ensure $I_{C1} = 0.5$ mA and $I_{C2} = 2$ mA.

(b) If V_{CC} is increased until $I_{C1} = 1$ mA, what is the new value of I_{C2}?
Hint: use the rules of thumb.

2.56 Shown in Fig. P2.56 is a *current sink* that could be used to provide the emitter bias for the *npn* BJT amplifiers discussed in Sections 8 and 9, here referred to simply as load (LD). The circuit is the *npn* counterpart of the *pnp* version of Fig. 2.46a, except for the use of a Zener diode to *stabilize* the circuit against possible variations in the supply voltage V_{EE}. Its output current is $I_O = \alpha_F(V_Z - V_{BE(on)})/R_2$.

(a) Let $V_{EE} = -12$ V, and let the diode be a 5.6-V Zener diode with $r_z = 15$ Ω. Assuming the BJT has $\beta_F = 100$ and $V_A = 75$ V, specify standard 5% values for R_1 and R_2 to give $I_O = 2$ mA and $I_Z = 3$ mA.

(b) Hence, find the *Load Regulation* $\Delta I_O/\Delta V_L$ and the *Line Regulation* $\Delta I_O/\Delta V_{EE}$ of your circuit.

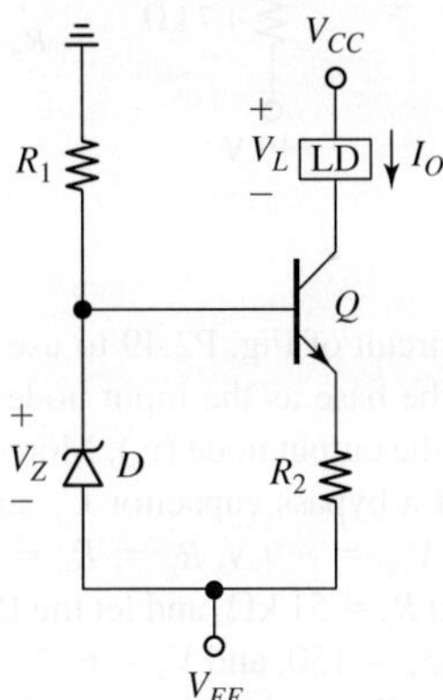

FIGURE P2.56

2.57 Redraw the *current source* of Fig. 2.46a, but with R_1 replaced by a 6.2-V Zener diode (cathode at the top) having $r_z = 20$ Ω, and by changing R_2 to 2.0 kΩ. The function of the Zener diode is to *stabilize* the source against possible variations in the supply voltage V_{CC}. This circuit could be used to provide the emitter bias for the *pnp* BJT amplifiers discussed in Sections 8 and 9, here referred to simply as load (LD). Its output current is $I_O = \alpha_F(V_Z - V_{EB(on)})/R_E \cong (6.2 - 0.7)/10 = 0.55$ mA.

Assuming the BJT has $\beta_F = 100$ and $V_A = 100$ V, find the *Load Regulation* $\Delta I_O / \Delta V_L$ and the *LineRegulation* $\Delta I_O / \Delta V_{CC}$ of the circuit.

2.8 Basic Bipolar Voltage Amplifiers

2.58 The CE amplifier of Fig. P2.58 uses a *pnp* BJT with $\beta_F = 150$, $V_{EB(on)} = 0.7$ V, and $V_A = 50$ V.

(a) Find I_C as well as the small-signal parameters R_i, R_o, and v_o / v_{sig}.

(b) Assuming $v_{sig} = (5\ \text{mV}) \cos \omega t$, find all node voltages in the circuit, and express each one as the sum of its dc and ac component.

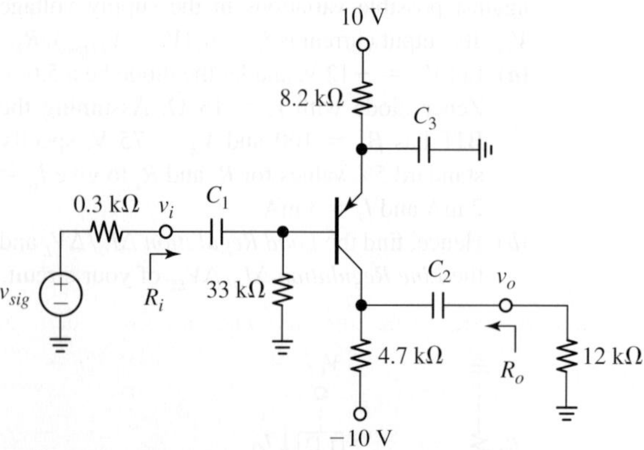

FIGURE P2.58

2.59 Let us put the circuit of Fig. P2.49 to use as a CE amplifier with the base as the input node (v_i) and the collector as the output node (v_o). Moreover, we need to connect a bypass capacitor C_E across R_E. Let $V_{EE} = 0$ V, $V_{CC} = -9$ V, $R_C = R_E = 2.0$ kΩ, $R_1 = 36$ kΩ, and $R_2 = 51$ kΩ, and let the BJT have $V_{EB(on)} = 0.7$ V, $\beta_F = 150$, and $V_A = 60$ V.

(a) Find the small-signal resistances R_i and R_o seen looking into the input and output nodes, and the small-signal voltage gain v_o / v_i.

(b) Specify C_E for operation at $f \geq 1$ kHz.

2.60 The CE amplifier of Fig. P2.60 is based on the feedback-bias scheme of Fig. 2.50. Since the emitter is already at ground, we don't need to use any emitter-bypass capacitor, a highly desirable feature. Assuming the BJT has $V_{BE(on)} = 0.7$ V, $\beta_F = 120$, and $V_A = 80$ V, find the small-signal parameters R_i, R_o, and v_o / v_i.

Hint: you may wish to take a look at Problems 2.45 and 2.46.

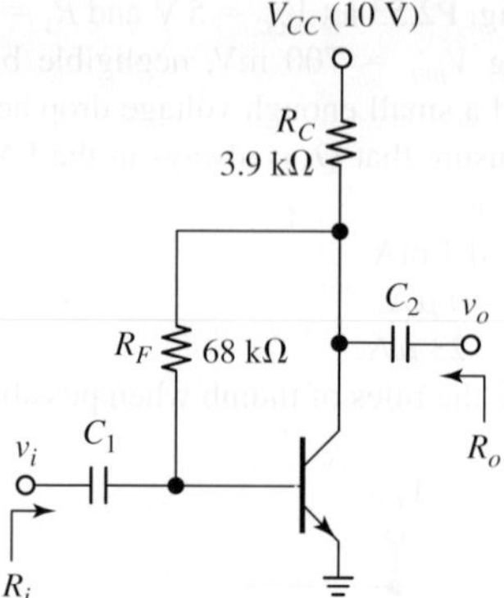

FIGURE P2.60

2.61 The biasing scheme of the CE amplifier of Fig. P2.61 is similar to that of Fig. P2.60, except that R_C is replaced by a current source I_{BIAS}. (Such a source, whose details we are omitting for simplicity, could be implemented with a *pnp* BJT, along the lines of Fig. 2.46). With this modification, the resistance seen by the collector can be made quite high, indicating that this circuit is capable of potentially high voltage gains.

(a) Assuming $\beta_F = 100$, $V_{BE(on)} = 0.7$ V, and $V_A = 100$ V, use the BJT large-signal and small-signal models to find the dc collector voltage V_C and the unloaded ac voltage gain v_o / v_i.

(b) Repeat if I_{BIAS} is doubled to 2 mA. Comment on your findings.

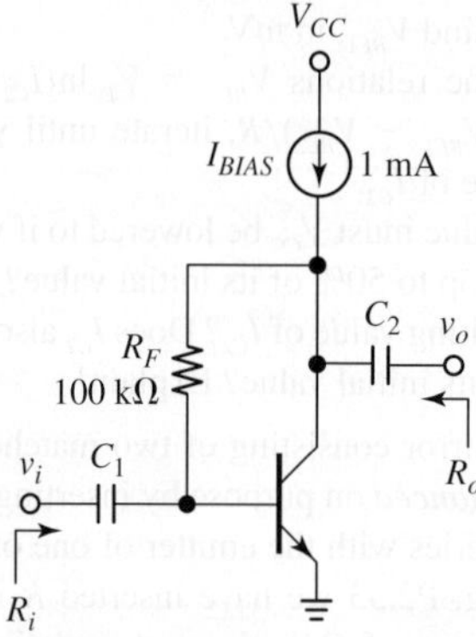

FIGURE P2.61

2.62 (a) Use the small-signal BJT model to find the input resistance R_i of the CE amplifier of Problem 2.61, both for the unloaded case, as shown, and the case in which the output is ac coupled to a 100-kΩ load. Comment on your results, and justify the claim that this amplifier is of the non-unilateral type.

(**b**) Find the output resistance R_o both for the case in which the amplifier is driven by an ideal source v_{sig}, and the case of a real-life source v_{sig} having output resistance $R_{sig} = 1$ kΩ. Again, comment.

2.63 Consider the circuit obtained from that of Fig. P2.58 by lifting the right plate of C_3 off ground and inserting a 220-Ω series resistance between that plate and ground. Draw the modified circuit, and convince yourself that this modification turns the circuit into a CE-ED amplifier with a net emitter-degeneration resistance of $(8200//220)$ Ω. Then, assuming $\beta_F = 125$, $V_{EB(on)} = 0.7$ V, and $V_A = \infty$, find the small-signal parameters R_i, R_o, and v_o/v_{sig}.

2.64 The CE-ED amplifier of Fig. P2.64 has an emitter-degeneration resistance of $(15//0.1)$ kΩ.

(**a**) If the BJT has $\beta_F = 125$, $V_{BE(on)} = 0.7$ V, and $V_A = \infty$, find I_C as well as the small-signal parameters R_i, R_o, and v_o/v_i.

(**b**) Find the equivalent resistance R_{eq} seen by C, and then specify C for operation at 100 Hz.

(**c**) What happens if C is omitted altogether?

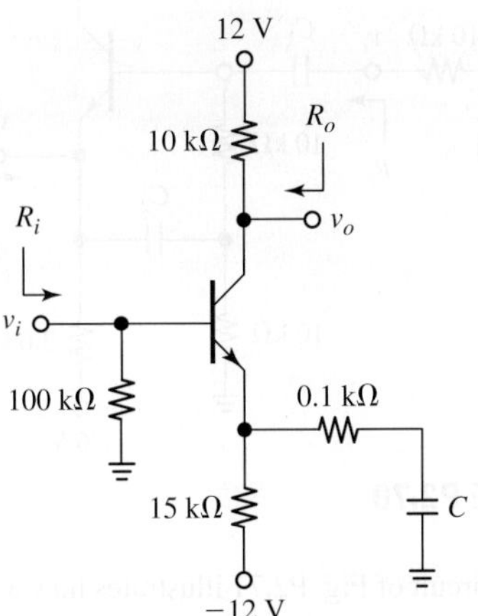

FIGURE P2.64

2.65 The CE-ED amplifier of Fig. P2.65 has an emitter-degeneration ac resistance of 0.2 kΩ.

(**a**) If the BJT has $\beta_F = 100$, $V_{BE(on)} = 0.7$ V, and $V_A = \infty$, find I_C as well as the small-signal parameters R_i, R_o, and v_o/v_i.

(**b**) Find the equivalent resistance R_{eq3} seen by C_3, and then specify C_3 for operation at 1 kHz.

(**c**) What happens if C_3 is omitted altogether?

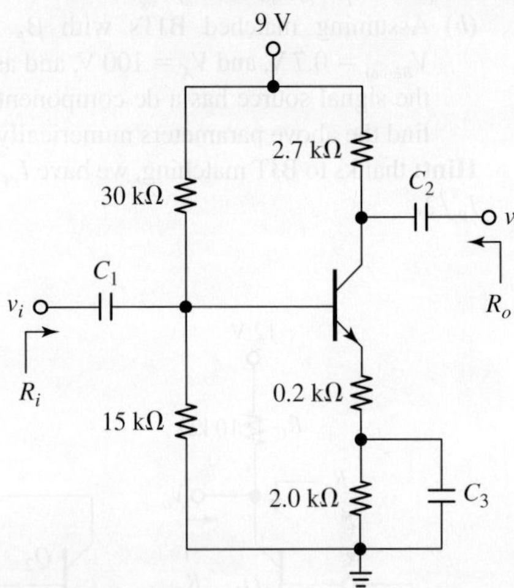

FIGURE P2.65

2.66 Consider the circuit obtained from that of Fig. P2.61 by lifting the emitter off ground, and inserting a resistance $R_E = 1.0$ k in series between the emitter and ground to turn it into a CE-ED amplifier. Also, let the output node v_o be ac coupled to a load $R_L = 100$ kΩ. Assuming $\beta_F = 150$ and $V_A = \infty$, use the BJT small-signal model to find the ac voltage gain v_o/v_i.

2.67 As we know, the function of the emitter-bypass capacitor in the CE configuration is to establish an *ac ground at the emitter*. The fact that the resistance seen looking into the emitter is usually low requires a fairly large capacitance, which is undesirable. The circuit of Fig. P2.67 eliminates altogether the need for such a capacitance by using the diode-connected BJT Q_2 instead. The ac resistance r_{e2} presented by Q_2, though not zero, is *small* (26 Ω at $I_C = 1$ mA), so the small amount of emitter degeneration that it introduces for Q_1 is a price well worth the elimination of the bulky bypass capacitor. This technique is widely used in IC implementations, where Q_1 and Q_2 are matched devices. With a signal source having a dc component of 0 V, the two BJTs experience the same V_{BE} drop and thus carry the same current I_C. Consequently, R_E has to be specified so as to carry twice as much current.

(*a*) Regarding Q_1 as a CE-ED amplifier with a total emitter-degeneration resistance of $R_E//r_{e2}$, find expressions for the small-signal parameters R_i, R_c, R_o, and v_o/v_{sig}.

(*b*) Assuming matched BJTs with $\beta_F = 200$, $V_{BE(on)} = 0.7$ V, and $V_A = 100$ V, and assuming the signal source has a dc component of 0 V, find the above parameters numerically.

Hint: thanks to BJT matching, we have $I_{E1} = I_{E2} = I_{R_E}/2$.

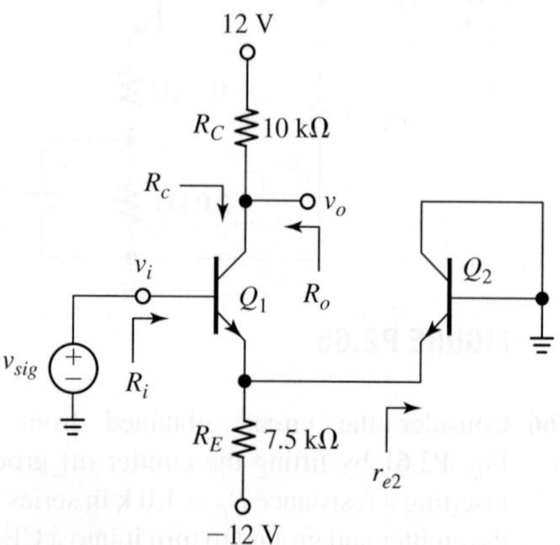

FIGURE P2.67

2.9 Bipolar Voltage and Current Buffers

2.68 A student is measuring the gain of the voltage follower of Fig. 2.61*a* for different source and load conditions. First, with $R_{sig} = 0$ and $R_L = 300$ Ω, the gain is found to be $v_o/v_{sig} = 0.853$ V/V. Next, inserting a resistance $R_{sig} = 10$ kΩ while leaving $R_L = 300$ Ω, causes the gain to drop to 0.718 V/V. Based on the above measurements, can the student predict the gain with $R_{sig} = 20$ kΩ and $R_L = 1.2$ kΩ?

Hint: in fact, the student can even tell the values of β_0 and I_C!

2.69 In the *pnp* emitter follower of Fig. P2.69 let the BJT have $\beta_F = 125$, $V_{EB(on)}$ 0.7 V, and $V_A = 80$ V. Find the small-signal resistances R_i and R_o and the signal-to-load gain v_o/v_{sig}.

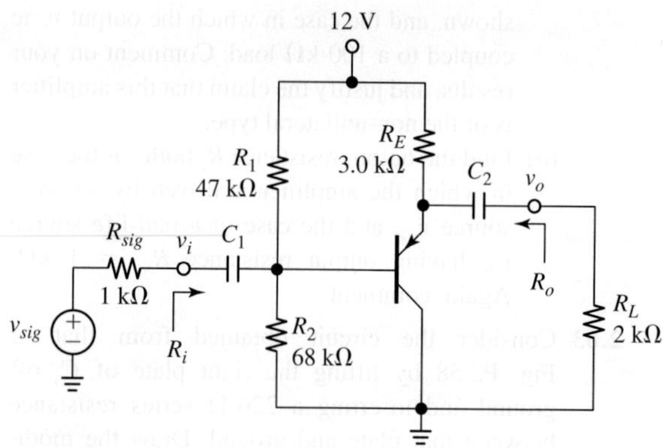

FIGURE P2.69

2.70 The CC amplifier of Fig. P2.70 is said to be of the *bootstrapped* type because it uses the feedback capacitance C_2 to *raise* its own ac input resistance and thus reduce input loading. Let the BJT have $\beta_F = 100$, $V_{BE(on)} = 0.7$ V, and $V_A = \infty$.

(*a*) Find R_i, R_o, and v_o/v_{sig}, but without C_2.

(*b*) Repeat, but with C_2 in place. Compare with part (*a*) and comment.

Hint: take a look at Problem 2.48.

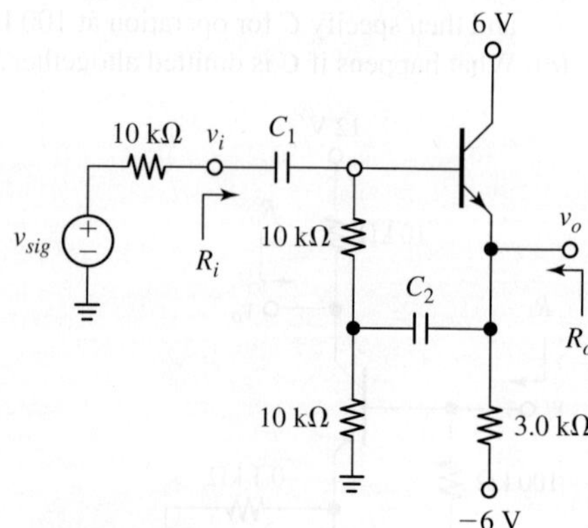

FIGURE P2.70

2.71 The circuit of Fig. P2.71 illustrates how a BJT can be put to use as a *voltage-to-current* (V-I) converter. Assuming $\beta_F = 150$, $V_{BE(on)} = 0.7$ V, $V_A = 75$ V, and $v_i = (0.5\ \text{V})\cos \omega t$, find the small-signal element values of the Norton equivalent seen by the load LD.

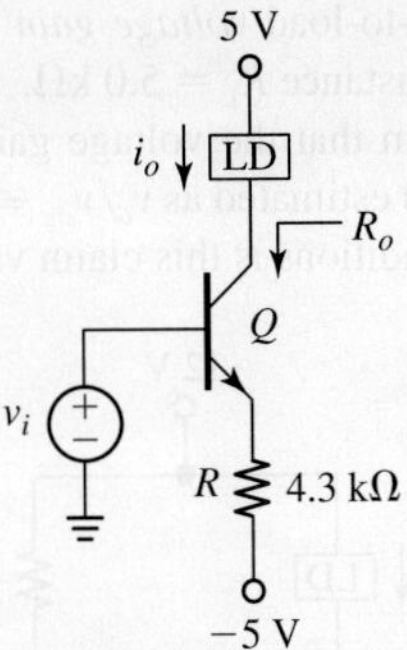

FIGURE P2.71

2.72 As we know, the emitter voltage of the CC config-uration follows the base voltage, but with an *offset* of about 0.7 V. This offset is often undesirable, as one would rather have the dc level of the output be the same as that of the input. The circuit of Fig. P2.725 uses a *pnp* CC stage, whose offset is $+0.7$ V, followed by an *npn* CC stage, whose off-set is -0.7 V. The two offsets tend to cancel each other out, making the output dc level identical to that at the input. In particular, if the signal source has a zero dc component, so will the output. For this cancellation to be effective, we must have $V_{BE2} = V_{EB1}$. This can be achieved, for instance, if the BJTs have $I_{s2} = I_{s1}$ and we bias them identi-cally by letting $R_{E2} = R_{E1}$. In the circuit shown, let the BJTs have $\beta_{F1} = \beta_{F2} = 100$, $V_{A1} = V_{A2} = \infty$, and $I_{s1} = I_{s2}$.

Hint: to find R_i, suitably adapt Eq. (2.58) and apply it twice; to find R_o, suitably adapt Eq. (2.59) and apply it twice.

(b) Assuming $v_{sig} = (5 \text{ V})\cos\omega t$, find all node voltages in the circuit, and express each one as the sum of its dc and ac component.

2.73 Using a typical *npn* BJT and 5% standard resis-tance values, design a circuit that accepts a sig-nal at the base having a 0-V dc component and ac component v_b, and gives two ac outputs, v_c at the collector and v_e at the emitter, such that $v_c = -v_e$. Assume the availability of ±12 V power supplies. If you use any capacitors, specify them for opera-tion at a signal frequency of 10 kHz. This circuit is called a *phase splitter*.

2.74 Consider the circuit obtained from that of Fig. P2.64 by grounding the base terminal, lifting the bottom plate of C off ground, and driving this plate with an ideal signal source v_{sig}. This turns the circuit into a CB voltage amplifier. Find the small-signal voltage gain v_o/v_{sig}.

Hint: first find the gain from v_{sig} to v_e, the emitter signal, and then from v_e to v_o.

2.75 Assuming ±10-V power supplies as well as typi-cal BJT parameters in the circuit of Fig. P2.75, specify suitable standard 5% resistances to make the BJT operate at $Q = Q(2 \text{ mA}, 5 \text{ V})$ and provide a gain of $v_o/v_{sig} = 10$ V/V. Finally, specify C for operation at 1 kHz.

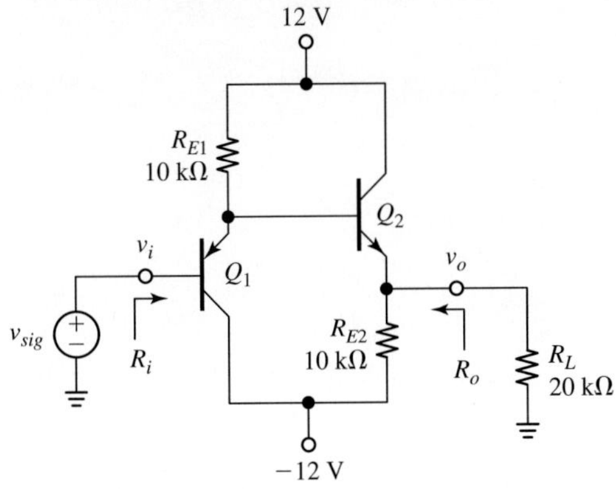

FIGURE P2.72

(a) Assuming the signal source has a dc compo-nent of 0 V, find the small-signal parameters R_i, R_o, and v_o/v_i.

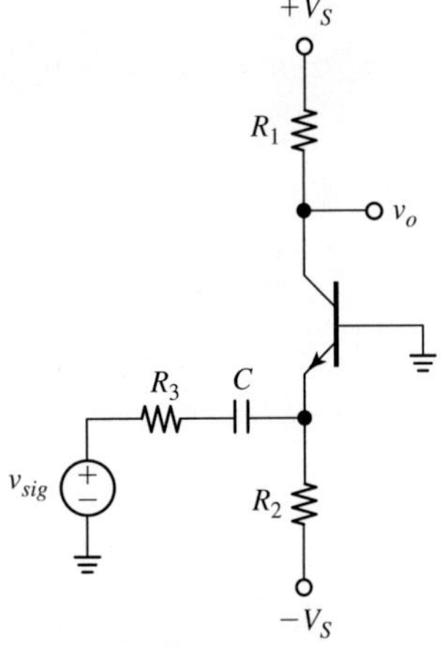

FIGURE P2.75

2.76 In Fig. P2.76, Q_1 and Q_2 are matched BJTs. Q_1 is operated in the CB mode, and Q_2 in the diode mode. The function of Q_2 is to bias the base of Q_1 at about $+0.7$ V, and thus ensure a dc voltage of 0 V at the emitter of Q_1, which in the CB mode represents the input node. An input dc level of 0 V is highly desirable as it allows us to couple the signal source to the amplifier *directly*, without the need for any ac-coupling, dc-blocking capacitors. Moreover, the circuit works all the way down to low frequencies, including zero, or dc. For this scheme to work, we must have $V_{BE2} = V_{EB1}$. This can be achieved, for instance, if we use BJTs with matched values of I_s and we bias them identically by letting $R_2 = R_1$. In the circuit shown, the CB stage is used as a *voltage-to-current* (V-I) converter. Let the BJTs be matched with $\beta_F = 150$ and $V_A = 80$ V.

(*a*) Assuming the signal source has a dc component of 0 V, find the small-signal parameters R_i, R_o, and the *transconductance gain* i_o/v_{sig}. **Hint:** after finding R_i, find the voltage gain v_i/v_{sig}, and obtain the transconductance gain as $i_o/v_{sig} = (v_i/v_{sig}) \times (i_o/v_i)$.

(*b*) Find the signal-to-load *voltage gain* v_o/v_{sig} if the load is a resistance $R_L = 5.0$ kΩ.

(*c*) Justify the claim that the voltage gain of (*b*) could have been estimated as $v_o/v_{sig} \cong R_L/R_{sig}$. Under what conditions is this claim valid?

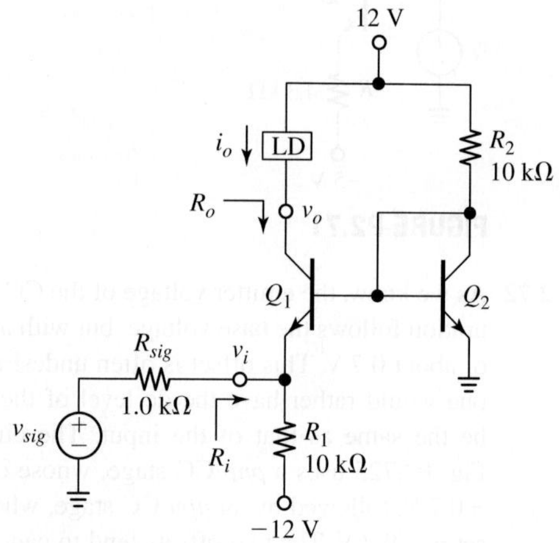

FIGURE P2.76

MOS Field-Effect Transistors

Chapter Outline

The age of semiconductor electronics began when the triode function (a controlled current source—see the introduction to Chapter 2 for a discussion of the vacuum-tube triode) was implemented on a piece of semiconductor material. This occurred in 1947 with the invention of the *bipolar junction transistor* (BJT), the first working realization of the semiconductor triode concept. However, neither is the BJT the only transistor type possible, nor was it the first transistor to be conceived. In fact, as early as 1925, Julius Lilienfeld patented a device of the type today known as the *field-effect transistor* (FET). However, because of fabrication difficulties at the time, he could never get it to work. It took another 35 years or so before Dawon Kahng and John Atalla of Bell Laboratories demonstrated, in 1960, the first FET of the so-called *metal-oxide-semiconductor* (MOS) type, or MOSFET for short.

The closest MOSFET counterpart of the vacuum-tube triode is what is today known as the *n-channel depletion-type MOSFET* (*n*-channel DMOSFET), which is one of four possible MOSFET types. Briefly stated, a DMOSFET consists of a thin layer of *n*-type material called the *channel*, which forms a parallel-plate capacitor with an electrode called the *gate*. One end of the channel, called the *source*, acts as a copious source of free electrons, which are designed to flow to the opposite end of the channel, aptly called the *drain*. The roles of source and drain are similar to those of cathode and plate in the triode (or emitter and collector in the BJT). The role of the gate, similar to that of the grid in the triode (or the base in the BJT), is to *modulate* the channel's conductivity and thus *control* the electron flow from source to drain. Specifically, driving the gate voltage negative will induce a positive charge in the channel, at the expense of a reduction in the concentration of free electrons there. For a sufficiently negative gate voltage, the channel will be depleted of free electrons and current flow will cease altogether. By a hydraulic analogy, FET behavior can be likened to that of a garden hose being squeezed for the purpose of controlling water flow, or even being shut off completely.

Following the successful demonstration of the first MOSFET, the new technology was put to use especially in those applications in which the advantages of smaller size and lower power consumption of the MOSFET made it competitive with its BJT counterpart. The first battery-powered electronic calculators and wristwatches made use precisely of this new technology. Also, a new digital integrated circuit (IC) family known as *complementary* MOS (or CMOS for short) was introduced by RCA as a low-power alternative to the then prevalent bipolar logic family known as TTL. In 1971 Intel used MOS technology to develop the first *microprocessor*. Since then, IC electronics has advanced *exponentially* and has penetrated virtually every aspect of modern life. This impressive growth has been governed by *Moore's law*, roughly stating that thanks to continued advances in IC fabrication, the number of devices that can be integrated on a given chip area *doubles* every approximately 18 months. Originally formulated in 1965, this law still holds to this day, though it has been pointed out that technology is bound to approach physical limits that will eventually lead to the demise of this law.

Over the years, the MOSFET has overtaken its BJT predecessor especially in high-density IC electronics, owing to its aforementioned advantages of smaller size and lower power consumption. Nonetheless, there are applications such as high-performance analog electronics, in which the BJT continues to be the preferred transistor type. To exploit the advantages of both BJTs and MOSFETs, the two device types are sometimes fabricated simultaneously on the same chip. The resulting technology, aptly called *biCMOS technology*, provides even greater design opportunities than the all-BJT or all-MOSFET technologies individually. Also, contemporary ICs often combine digital as well as analog functions on the same chip, this being the reason for the name *mixed-signal* or also *mixed-mode ICs*.

There is no question that microelectronics is a most exciting, challenging, and rapidly evolving field. The beginner may feel overwhelmed by all this, and rightly so. But, as we embark upon the study of today's dominant processes and devices, we will try to focus on *general principles* that transcend the particular technological milieu of the moment and that we can apply to understand new processes and devices as they become available and commercially mature. Focus on general principles, combined with continuing education, is a necessity for the young engineer aiming at establishing and maintaining a satisfying career in a seemingly ever-changing field.

CHAPTER HIGHLIGHTS

The chapter begins with a study of the physical structure of the MOSFET, basic underlying semiconductor principles, device characteristics, operating regions and modeling. Much emphasis is placed on practical aspects of relevance to today's industrial environment (rules of thumb). The FETs under scrutiny are of the so-called *long-channel* type (channel lengths in the range of several micrometers or longer) because their behavior conforms reasonably well to theoretical prediction, so they are easier to model as well as easier to grasp by the beginner. However, the devices available in today's ever shrinking IC processes are of the *short-channel* type (channel lengths of a fraction of 1 μm). At such small sizes, a number of higher-order effects arise, particularly carrier velocity saturation, which may cause significant departure from long-channel behavior. Aptly called *short-channel effects*, they require more complex formalism and more sophisticated models as the price for providing more realistic results. These advanced models, while realizable in computer simulators, are too complicated for hand analysis. We shall nevertheless continue to rely on the formalism and models of long-channel devices to develop an *intuitive understanding* of MOSFETs, and then turn to computer simulation for more accurate results.

After examining a variety of resistive MOSFET circuits so as to develop a basic feel for MOSFET circuit operation, we investigate the MOSFET in its two most important class of applications, namely, as an *amplifier* in analog electronics, and as a *switch* in digital electronics. Next, we develop suitable *large-signal* models as well as *small-signal models* for the FETs, so we can turn to the *three basic amplifier configurations*, namely, the *common-source* (CS), *common-drain* (CD), and *common-gate* (CG) configurations. The CS configuration is presented as the natural realization *of voltage amplification*, whereas the CD and CG configurations serve most naturally as *voltage* and *current buffers*, respectively. Suitable emphasis is placed on the role of the MOSFET as a *resistance transformation device*, which actually provided the basis for the name *transistor*. The transformation equations are conveniently tabulated for easy reference in later chapters.

The amplifiers studied in this chapter are of the *discrete* type because they can be built using individual transistors, resistors, and capacitors. (In this respect, a handy device to experiment with in the lab is the CD4007 CMOS Transistor Array, comprising three *n*MOSFETs and three *p*MOSFETs.) Though nowadays MOSFET amplifiers are implemented mostly in IC form, the motivation for studying discrete designs is primarily pedagogical, as discrete circuits are somewhat easier to grasp, and yet they reveal important aspects that apply to IC implementations as well. Once we master the basics of discrete design involving a *single-transistor* amplifier, we will be in a better position to tackle the complexity of *multi-transistor* ICs, a subject that will be undertaken in Chapter 4.

For the benefit especially of computer engineering majors, the chapter concludes with a detailed analysis of the CMOS *inverter/amplifier* as a simple yet important IC building block that demonstrates the flexibility of CMOS technology both in the analog and digital domains. Basic CMOS logic gates are also addressed in some detail.

The chapter makes abundant use of PSpice both as a software oscilloscope to display MOSFET characteristics, transfer curves and waveforms, and as a verification tool for dc as well as ac calculations.

3.1 PHYSICAL STRUCTURE OF THE MOSFET

Figure 3.1 shows, in simplified form, the structure of the *n-channel, metal-oxide-semiconductor* (MOS) *field-effect transistor* (FET), or *n*MOSFET for short. The device is fabricated through a complex sequence of steps involving *pattern definition, oxidation, diffusion, ion implantation, material deposition* and *material removal*, on a wafer of *lightly-doped p*-type silicon (p^-) called the *body* or also the *bulk* of the *n*MOSFET. The wafer is also called the *substrate* because it provides physical support for the device under consideration as well as all other devices of the same integrated circuit (IC). Starting out with a polished wafer, the fabrication of an *n*MOSFET consists of the following principal steps:

- First, a thin (t_{ox}) insulating layer of *silicon oxide* (SiO_2) is thermally grown on the surface of the substrate.
- Next, the *gate* electrode is created by growing over the oxide a layer of *heavily-doped n*-type silicon (n^+). Being extremely rich in free electrons, this electrode acts for all practical purposes like a metal. The resulting *metal-oxide-semiconductor* (MOS) structure is the reason for the name of the device.
- Next, the oxide is removed from each side of the gate, and ion implantation is used to create two *heavily-doped n*-type regions (n^+) extending into the substrate, called the *source* and the *drain* regions.
- Finally, two metal depositions form the *source* and *drain* electrodes. (The interested student is encouraged to search the Web for videos and articles illustrating the fascinating subject of MOSFET fabrication.)

The region of the body just below the oxide is called the *channel region*. Its *length* and *width* are denoted as L and W, respectively. In current VLSI technology, L and W can be as small as fractions of a *micrometer* (1 μm = 10^{-6} m = 10^{-4} cm), while the oxide thickness t_{ox} can be as low as ten *nanometers* (1 nm = 10^{-9} m = 10^{-7} cm = 10 Å). We identify two basic ingredients in a MOSFET:

- the *channel region* extending between the source and drain regions, and
- the *parallel-plate capacitor* formed by the gate and the channel region.

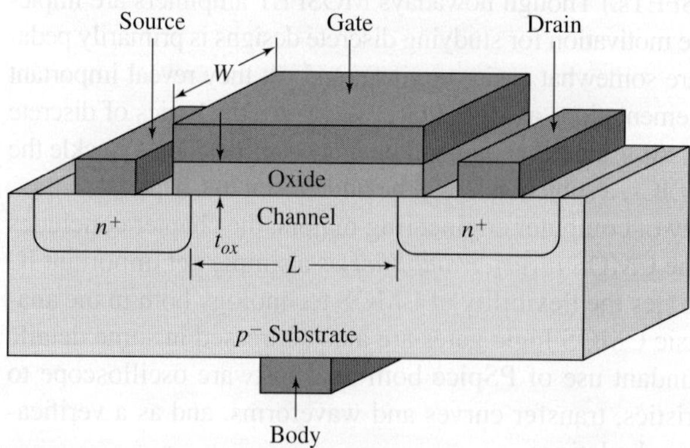

FIGURE 3.1 Basic physical structure of the *n*MOSFET.

Briefly stated, the principle at the basis of the MOSFET is to utilize the gate-body *capacitance* to control the channel-region *conductance*. The path extending from the source to the drain regions includes two back-to-back *pn* junctions (the body-source and the body-drain junctions), so it normally exhibits very high resistance to current flow (typically $\sim 10^{12}$ Ω). However, by raising the potential of the gate electrode to a suitable value, we can create *favorable conditions for free electrons to exist in the channel region*, and thus form a continuous conductive path, or *channel*, from source to drain, along which electrons can flow and produce current. To investigate device behavior, we will have to address two basic questions:

- What is the *threshold voltage* V_t to which we need to raise the gate potential relative to the bulk to form a channel and thus *turn on* the device?
- Once the device has been turned on, what are the *i-v characteristics* of its channel?

Both issues will be addressed in the following sections.

Complementary MOSFETs

The dominant IC technology today utilizes the *n*MOSFET as well as its *dual* (or *complementary*) device, the *p*MOSFET. Aptly called *complementary* MOS (or CMOS) technology, it requires that both device types be fabricated on the *same* substrate. A *p*MOSFET is obtained by *negating* the doping types of the body, source, and drain regions of an *n*MOSFET, so that the body is now n^-, and the source and drain regions are p^+. To allow for the coexistence of the two devices on a common substrate, the *p*MOSFET is placed inside a local lightly doped *n*-type (n^-) substrate, also called *well* or *tub*, which is formed by a separate diffusion into the existing p^- wafer prior to the fabrication of the transistors themselves.

The structure is depicted in cross-sectional form in Fig. 3.2, where subscripts *n* and *p* identify the terminals of the *n*MOSFET and the *p*MOSFET, respectively. The *n*MOSFET, located left of center, is similar to that of Fig. 3.1, except that the connection to its body (B_n) is not at the bottom, as shown in the simplified rendition of Fig. 3.1, but at the top, left. This is mandated by the planar-IC requirement that all interconnections be made at the top of the wafer. The *p*MOSFET, located right of center, is placed inside its own well (n^-), and connection to the well (B_p) is at the top, right. To ensure good-quality ohmic contacts, the metal connections to the bulks are implemented through heavily doped regions, as shown.

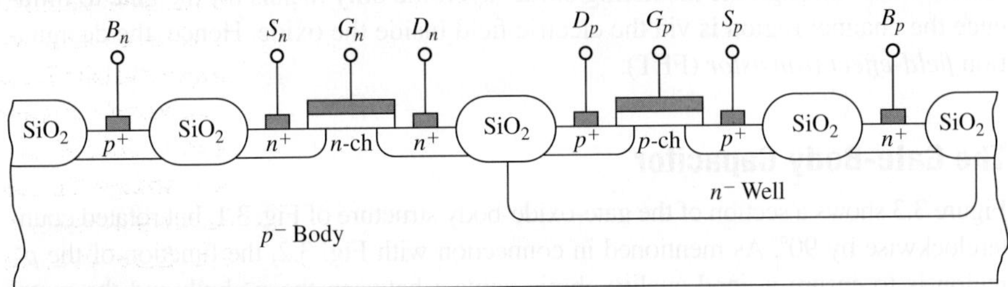

FIGURE 3.2 Cross-sectional view of CMOS transistors.

Figure 3.2 illustrates also another important aspect of ICs, namely, the need for adjacent devices to be *electrically isolated* from each other. This constraint is met by growing, prior to fabrication of the actual transistors, a ring of SiO_2 insulator material, also called *field oxide*, around each transistor's intended site. Indeed, each transistor must be kept electrically isolated not only from its neighboring devices, but also from its own body! The *p*-type body of the *n*MOSFET forms *pn* junctions with the *n*-type source and drain regions, so in this case body isolation is achieved by anchoring B_n to the *most negative voltage* (MNV) in the circuit. This will keep both junctions reverse biased, and therefore in cutoff, under all possible circuit conditions. Likewise, the *n*-type well of the *p*MOSFET forms *np* junctions with the *p*-type source and drain regions, so anchoring B_p to the *most positive voltage* (MPV) in the circuit will keep both junctions in cutoff under all possible circuit conditions. For instance, in the case of a digital CMOS circuit powered between 5 V and ground, B_n is connected to ground, and B_p is connected to +5 V. The manufacturer makes these connections internally to the IC.

3.2 THE THRESHOLD VOLTAGE V_t

To investigate the mechanism of channel formation in a *n*MOSFET, we focus on its *gate-oxide-bulk* structure, which forms a parallel-plate capacitor, albeit one with plates of different materials. Though in earlier MOSFETs the gate electrode was made of metal, such as aluminum, nowadays it is fabricated using n^+ silicon, this being the reason why modern processes are also referred to as *silicon-gate* processes. Since the n^+ silicon film is grown over amorphous oxide, it consists of sub-micrometer-sized crystallites, rather than a single crystal, and it is thus referred to a *polysilicon*. Regardless, n^+ polysilicon is very rich in free electrons, just like a metal, and it is used not only to create the gate electrode, but also to interconnect different devices in an IC. The reason for making the gate electrode of polysilicon is that the subsequent ion implantation to create the source and drain regions will inherently guarantee a high degree of *alignment* among the different regions. In particular, as the ions diffuse downward into the body, they also diffuse sideways a bit, resulting in a small amount of *overlap* between the edges of the gate and those of the source/drain regions. As we proceed, we shall appreciate how this slight overlap, clearly shown in both pictures above, is critical for the proper functioning of the MOSFET.

We now wish to investigate the effect of an external bias upon the type of charges as well as their distributions in the bulk region just below the oxide layer. Since no current flows through the insulating oxide layer, the only means for the gate to influence the channel region is via the electric field inside the oxide. Hence, the designation *field-effect transistor* (FET).

The Gate-Body Capacitor

Figure 3.3 shows a section of the gate-oxide-body structure of Fig. 3.1, but rotated counterclockwise by 90°. As mentioned in connection with Fig. 3.2, the function of the p^+ region is to ensure a good-quality ohmic contact between the p^- bulk and the metal connection, so it will play no role in our analysis. The well-known formula for parallel-plate capacitance gives, in the present case, $C = \varepsilon_{ox}(W \times L)/t_{ox}$, where W and L are the

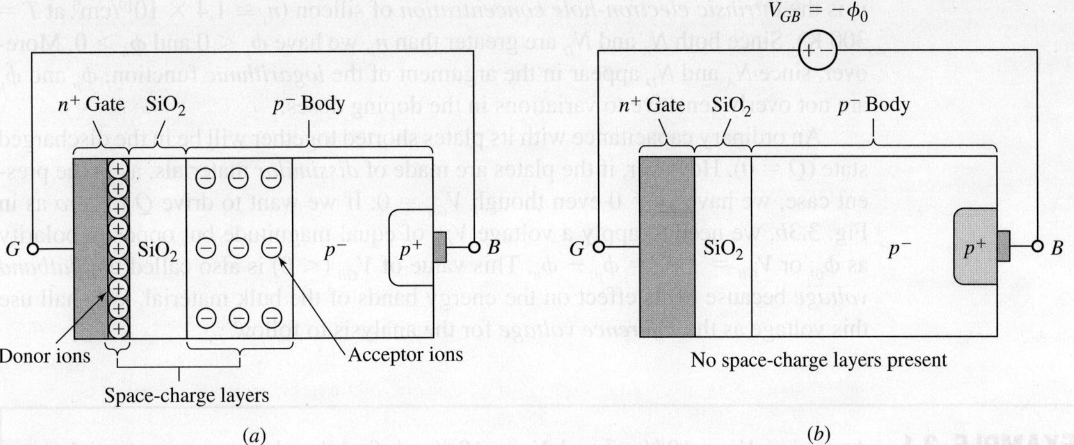

FIGURE 3.3 Gate-body capacitor (a) at 0-V bias, and (b) biased at $-\phi_0$ to eliminate the space-charge layers.

channel-region's width and length depicted in Fig. 3.1, ε_{ox} is the *permittivity* of the oxide layer, and t_{ox} is its thickness. To make analysis independent of the particular device size, it is convenient to work with the *capacitance per unit area*, defined as $C_{ox} = C/(W \times L)$, or

$$C_{ox} = \frac{\varepsilon_{ox}}{t_{ox}} \tag{3.1}$$

In today's technology W and L are typically in the *sub-micron range* ($1\ \mu$m $= 10^{-9}$ m), and t_{ox} can be in the range of 10 nm (1 nm $= 10^{-9}$ m) or less. Silicon oxide has $\varepsilon_{ox} = 345$ fF/cm, so a fabrication process with, say, $t_{ox} = 10$ nm gives $C_{ox} = 3.45$ fF/μm^2. (As a rule, $C_{ox} = 34.5/t_{ox}$, C_{ox} in fF/μm^2 and t_{ox} in nm.)

The bulk-oxide-gate structure is reminiscent of the familiar *pn* junction, except that the present p^- and n^+ materials are separated by an *insulator layer*, which prevents direct current flow. However, if we connect G and B externally as in Fig. 3.3a, electrons will diffuse from the electron-rich n^+ gate, through the connection, to the electron-starved p^- bulk, leaving behind a layer of immobile *positive* donor ions in the gate. Once in the bulk, these excess electrons recombine with holes there in order to satisfy the mass-action law, leading in turn to the formation of a layer of immobile *negative* acceptor ions in the bulk. The two layers concentrate near the gate-oxide and the bulk-oxide interfaces in order to minimize the electrostatic energy of the system. Just as in the case of the *pn* junction, these space-charge layers yield an electric field E from the gate, through the oxide, to the bulk, and an equilibrium condition is reached whereby this field opposes further diffusion of electrons through the external connection. Associated with this field is a *built-in potential* $\phi_0 = \phi_n - \phi_p$ across the gate-bulk structure, where

$$\phi_p = V_T \ln\frac{n_i}{N_A} \qquad\qquad \phi_n = V_T \ln\frac{N_D}{n_i} \tag{3.2}$$

are, respectively, the *equilibrium electrostatic potentials* (also called *Fermi potentials*) of the bulk and of the gate. Here, $V_T = kT/q$ is the *thermal voltage* ($V_T \cong 26$ mV at $T = 300$ K), N_A and N_D are the *doping densities* in the bulk and gate materials, and

n_i is the *intrinsic electron-hole concentration* of silicon ($n_i \cong 1.4 \times 10^{10}$/cm³ at $T =$ 300 K). Since both N_A and N_D are greater than n_i, we have $\phi_p < 0$ and $\phi_n > 0$. Moreover, since N_A and N_D appear in the argument of the *logarithmic* function, ϕ_p and ϕ_n are not overly sensitive to variations in the doping doses.

An ordinary capacitance with its plates shorted together will be in the discharged state ($Q = 0$). However, if the plates are made of *dissimilar* materials, as in the present case, we have $Q \neq 0$ even though $V_{GB} = 0$. If we want to drive Q to *zero* as in Fig. 3.3*b*, we need to apply a voltage V_{GB} of equal magnitude but opposite polarity as ϕ_0, or $V_{GB} = -\phi_0 = \phi_p - \phi_n$. This value of V_{GB} (< 0) is also called the *flatband voltage* because of its effect on the energy bands of the bulk material. We shall use this voltage as the *reference voltage* for the analysis to follow.

EXAMPLE 3.1   Assuming $N_A = 10^{16}$/cm³ and $N_D = 10^{20}$/cm³, find the electrostatic potentials as well as the value of V_{GB} needed to eliminate the space-charge layers.

Solution

From Eq. (3.2),

$$\phi_p = 0.026 \ln\frac{1.4 \times 10^{10}}{10^{16}} = -0.35 \text{ V}$$

$$\phi_n = 0.026 \ln\frac{10^{20}}{1.4 \times 10^{10}} = +0.59 \text{ V}$$

To achieve charge neutrality in the gate and bulk, the gate must be biased more *negatively* than the body such that $V_{GB} = -\phi_0 = -0.35 - 0.59 = -0.94$ V.

Remark: Were the gate-bulk an ordinary *np* junction, with $V_{GB} = -0.94$ V it would be forward biased quite heavily and conduct a large forward current from the *p*-bulk (anode) to the *n*-gate (cathode). In the present case, however, no current flows because of the oxide insulator separating the two.

Inversion

Let us now gradually increase V_{GB}, starting at $V_{GB} = -\phi_0$ (or $V_{GB} = -0.94$ V in our example). The effect of this increase is to re-establish space-charge layers on both sides of the oxide, uncovering *positive* charge in the gate and *negative* charge in the bulk. We are particularly interested in the situation in the bulk, so we will ignore that in the gate, keeping in mind that the charge in the gate is always of *equal magnitude* but *opposite polarity* as the charge in the bulk. The situation in the bulk is depicted in Fig. 3.4, where we have chosen the origin of the *x*-axis to coincide with the *oxide-bulk interface*, a surface that will play an important role in our analysis. Initially, the negative charge in the bulk consists of the *negative* acceptor ions there (the holes are simply pushed away from the oxide-bulk interface, leaving behind the bound ions). The resulting space-charge layer is also referred to as a *depletion layer* because it is devoid of holes. However, as we increase V_{GB} and thus widen the depletion layer of the bulk, the *surface potential* $\phi(0)$

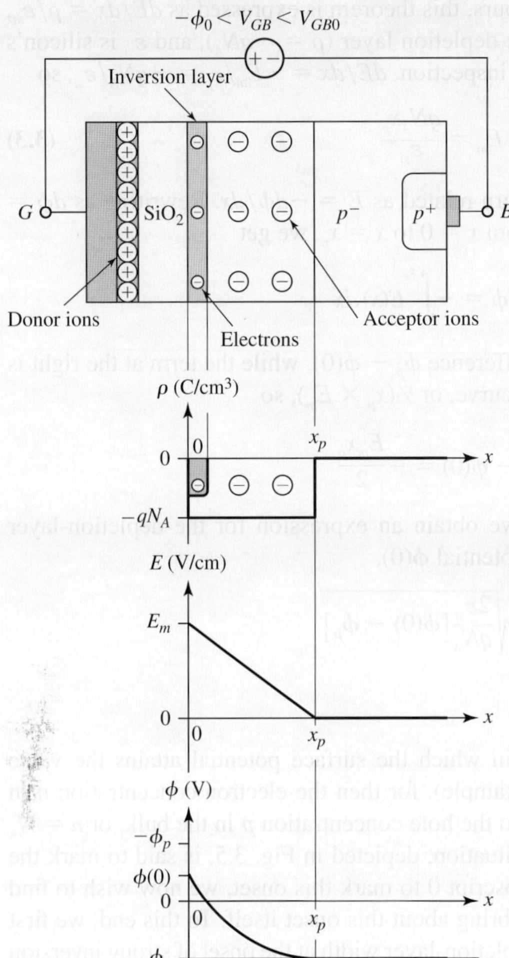

FIGURE 3.4 The situation in a nMOSFET *before* the onset of strong inversion.

also *increases* because of the *quadratic* dependence ϕ versus x (recall Fig. 1.39). When $\phi(0)$ changes *from negative to positive*, the bulk near the surface is said to undergo *inversion* because it turns from p-type to n-type, at least electrostatically speaking. For this reason, the bulk region near the surface is called *inversion layer.*

With reference to Fig. 3.4, we observe that the space-charge layers yield in turn an electric field $E(x)$. The field strength as a function of x is readily visualized by counting the field lines, each of which starts on a positive ion in the gate and ends on a negative ion in the bulk. We are interested in the lines in the bulk, whose number is maximum at the oxide-bulk interface ($x = 0$), and decreases linearly with x to finally drop to zero at the edge of the depletion layer ($x = x_p$). We readily find a relationship between the maximum strength E_m and the layer's width x_p using Gauss's theorem.

In a one-dimensional case such as ours, this theorem is expressed as $dE/dx = \rho/\varepsilon_{si}$, where ρ is the charge density in the depletion layer ($\rho = -qN_A$), and ε_{si} is silicon's permittivity ($\varepsilon_{si} = 1.04$ pF/cm). By inspection, $dE/dx = -E_m/x_p = -qN_A/\varepsilon_{si}$, so

$$E_m = \frac{qN_A x_p}{\varepsilon_{si}} \tag{3.3}$$

Electric field and potential are in turn related as $E = -d\phi/dx$. Rewriting as $d\phi = -Edx$ and integrating both sides from $x = 0$ to $x = x_p$, we get

$$\int_0^{x_p} d\phi = -\int_0^{x_p} E(x)\,dx$$

The term at the left is simply the difference $\phi_p - \phi(0)$, while the term at the right is the area of the triangle under the E curve, or $\frac{1}{2}(x_p \times E_m)$, so

$$\phi_p - \phi(0) = -\frac{E_m x_p}{2}$$

Using Eq. (3.3) to eliminate E_m, we obtain an expression for the depletion-layer width as a function of the surface potential $\phi(0)$,

$$x_p = \sqrt{\frac{2\varepsilon_{si}}{qN_A}[\phi(0) - \phi_p]}$$

Onset of Strong Inversion

We are interested in the situation in which the surface potential attains the value $\phi(0) = -\phi_p$ (or $+0.35$ V in our example), for then the electron concentration n in the inversion layer becomes *equal* to the hole concentration p in the bulk, or $n = N_A$ ($=10^{16}$/cm^3 in our example). This situation, depicted in Fig. 3.5, is said to mark the *onset of strong inversion*. Using subscript 0 to mark this onset, we now wish to find the gate-body bias V_{GB0} required to bring about this onset itself. To this end, we first substitute $\phi(0) = -\phi_p$ to find the depletion-layer width at the onset of strong inversion

$$x_{p0} = \sqrt{\frac{2\varepsilon_{si}}{qN_A}2(-\phi_p)} \tag{3.4}$$

Next, we observe that the unit-area charge in the bulk depletion-layer is $Q_{b0} = -qN_A x_{p0}$. Using Eq. (3.4),

$$Q_{b0} = -\sqrt{2qN_A\varepsilon_{si}2(-\phi_p)} \tag{3.5}$$

This *negative* charge in the bulk is matched by a *positive* charge in the gate. By the capacitance law, the voltage required to sustain this charge redistribution is $V_{ox0} = -Q_{b0}/C_{ox}$. Finally, the gate-to-body voltage drop required to bring about the onset of strong inversion is, by KVL, $V_{GB0} = -\phi_0 + 2(-\phi_p) + V_{ox0}$, or

$$V_{GB0} = -\phi_0 - 2\phi_p - \frac{Q_{b0}}{C_{ox}} \tag{3.6}$$

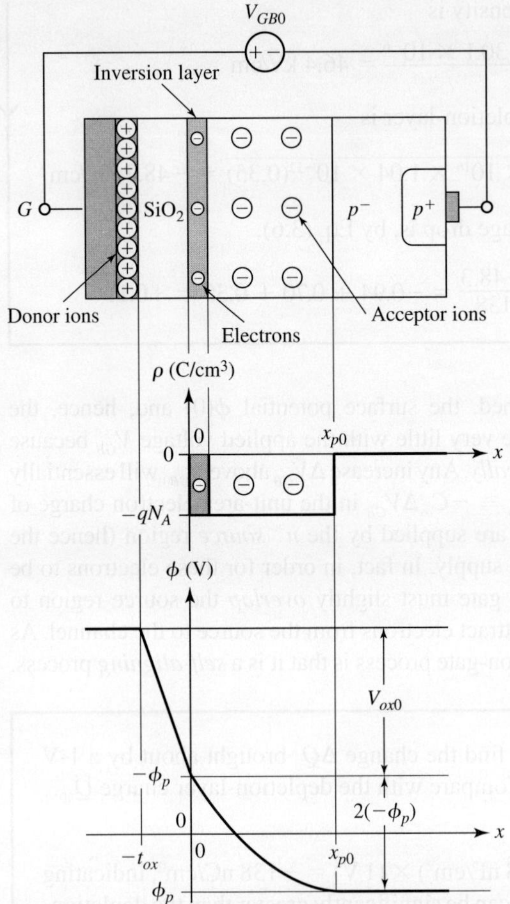

FIGURE 3.5 Situation at the onset of strong inversion.

In other words, to bring about the onset of strong inversion, we need to increase V_{GB}, starting from the reference level $-\phi_0$, (a) by the term $2(-\phi_p)$ to raise the surface potential $\phi(0)$ from ϕ_p, through zero, to $-\phi_p$, and (b) by the term $-Q_{b0}/C_{ox}$ to sustain the unit-area charge Q_{b0} in the bulk depletion layer.

EXAMPLE 3.2

Assuming the doping densities of Example 3.1, along with $t_{ox} = 25$ nm, find the values of all relevant physical quantities right at the onset of strong inversion.

Solution

The unit-area capacitance is

$$C_{ox} = \frac{345 \times 10^{-15}}{2.5 \times 10^{-6}} = 138 \text{ nF/cm}^2$$

At the onset of strong inversion, the depletion layer width is

$$x_{p0} = \sqrt{\frac{2 \times 1.04 \times 10^{-12}}{1.602 \times 10^{-19} \times 10^{16}} 2(0.35)} = 301 \text{ nm}$$

The corresponding electric field intensity is

$$E_{m0} = \frac{1.602 \times 10^{-19} \times 10^{16} \times 30.1 \times 10^{-6}}{1.04 \times 10^{-12}} = 46.4 \text{ kV/cm}$$

The unit-area charge in the bulk depletion-layer is

$$Q_{b0} = -\sqrt{4 \times 1.602 \times 10^{-19} \times 10^{16} \times 1.04 \times 10^{-12}(0.35)} = -48.3 \text{ nC/cm}^2$$

Finally, the required gate-body voltage drop is, by Eq. (3.6),

$$V_{GB0} = -0.94 - 2(-0.35) - \frac{-48.3}{138} = -0.94 + 0.70 + 0.35 = +0.11 \text{ V}$$

Once strong inversion is reached, the surface potential $\phi(0)$ and, hence, the depletion-layer width x_p, will change very little with the applied voltage V_{GB} because $\phi(0)$ depends on V_{GB} only *logarithmically*. Any increase ΔV_{GB} above V_{GB0} will essentially be accompanied by an increase $\Delta Q_n \cong -C_{ox}\Delta V_{GB}$ in the unit-area electron charge of the inversion layer. These electrons are supplied by the n^+ *source* region (hence the name), where they exist in abundant supply. In fact, in order for these electrons to be enticed into the inversion layer, the gate must slightly *overlap* the source region to allow for the fringe electric field to attract electrons from the source to the channel. As mentioned, the advantage of the silicon-gate process is that it is a *self-aligning* process.

EXAMPLE 3.3   Assuming the data of Example 3.2, find the change ΔQ_n brought about by a 1-V increase ΔV_{GB} in strong inversion. Compare with the depletion-layer charge Q_{b0}.

Solution

We have $\Delta Q_n \cong -C_{ox}\Delta V_{GB} = -(138 \text{ nF/cm}^2) \times (1 \text{ V}) = -138 \text{ nC/cm}^2$, indicating that the inversion-layer charge $|\Delta Q_n|$ can be significantly greater than the depletion-layer charge $|Q_{b0}|$ ($= 48.3 \text{ nC/cm}^2$ in our example), even though the inversion layer is much thinner than the depletion layer.

The Threshold Voltage V_{t0}

We now wish to apply the above findings to the full-fledged MOSFET, starting from the situation at the onset of strong inversion depicted in Fig. 3.6 for both MOSFET types. Note the presence of the inversion layer immediately below the oxide-bulk surface, along with the depletion layer extending not only below the inversion layer, but also around the source and drain regions, as they form *pn* junctions with the body. The *threshold voltage* V_t is defined as the *gate-source voltage* v_{GS} needed to bring about the *onset of strong inversion* in the channel region. When body and source are at the *same potential* (ground in Fig. 3.6), the threshold is denoted as V_{t0}. For the case of an *n*MOSFET it takes on the general form

$$V_{t0} = -\phi_0 - 2\phi_p - \frac{Q_{b0}}{C_{ox}} - \frac{Q_{ox}}{C_{ox}} - \frac{Q_i}{C_{ox}} \tag{3.7}$$

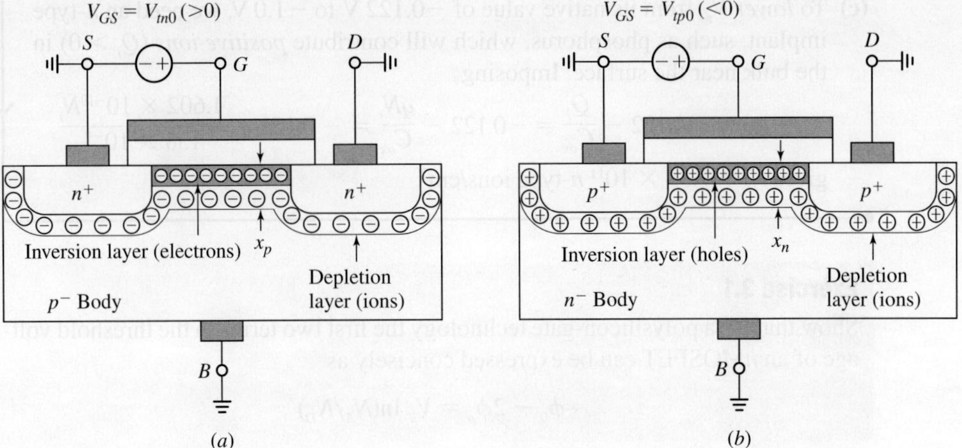

FIGURE 3.6 The *onset of strong inversion* in (a) the *n*MOSFET and (b) the *p*MOSFET.

The first *three* terms are simply those of Eq. (3.6). The next term, involving the unit-area charge Q_{ox}, accounts for the presence of dangling bonds in the bulk right at the interface, as well as positive ions that during fabrication get trapped in the oxide right near the oxide-bulk interface.[2] The first *four* terms form what is known as the *native threshold* of the *n*MOSFET. The last term, involving the unit-area charge Q_i, accounts for impurities that the manufacturer introduces deliberately in the bulk, right at the oxide-bulk interface, to adjust V_{t0} to the prescribed value. For *p*-type impurities we have $Q_i < 0$, and for *n*-type impurities we have $Q_i > 0$. For obvious reasons, the native threshold is also called the *undoped threshold*.

EXAMPLE 3.4

Assuming the data of Example 3.2, along with a surface state density $N_{ox} = 2 \times 10^{11}$ positive ions/cm^2,

(a) Find the native threshold of the *n*MOSFET.
(b) Find the implant type and dosage N_i needed for $V_{t0} = +1.0$ V.
(c) Find the implant type and dosage N_i needed for $V_{t0} = -1.0$ V.

Solution

(a) We have $Q_{ox} = qN_{ox} = 1.602 \times 10^{-19} \times 2 \times 10^{11} = 32$ nC/cm^2. So, using the result of Example 3.2,

$$V_{t0} = 0.11 - \frac{32}{138} = -0.122 \text{ V}$$

(b) To *raise* $V_{t0} =$ from its native value of -0.122 V to $+1.0$ V, we need a *p*-type implant, such as boron, which will contribute *negative ions* ($Q_i < 0$) in the bulk near the surface. Imposing

$$+1.0 = -0.122 - \frac{Q_i}{C_{ox}} = -0.122 - \frac{-qN_i}{C_{ox}} = -0.122 + \frac{1.602 \times 10^{-19}N_i}{138 \times 10^{-9}}$$

gives $N_i = 9.66 \times 10^{11}$ *p*-type ions/cm^2.

(c) To *lower* V_{t0} from its native value of -0.122 V to -1.0 V, we need an *n*-type implant, such as phosphorus, which will contribute *positive ions* $(Q_i > 0)$ in the bulk near the surface. Imposing

$$-1.0 = -0.122 - \frac{Q_i}{C_{ox}} = -0.122 - \frac{qN_i}{C_{ox}} = -0.122 - \frac{1.602 \times 10^{-19} N_i}{138 \times 10^{-9}}$$

gives $N_i = 7.56 \times 10^{11}$ *n*-type ions/cm^2.

Exercise 3.1

Show that for a polysilicon-gate technology the first two terms in the threshold voltage of an *n*MOSFET can be expressed concisely as

$$-\phi_0 - 2\phi_p = V_T \ln(N_A/N_D)$$

The Four MOSFET Types and Their Circuit Symbols

Depending on the body type (p^- or n^-) and threshold-voltage polarity ($V_t > 0$ or $V_t < 0$), we can have *four* different types of MOSFET. We now make some important observations:

- An *n*MOSFET with $V_{t0} > 0$ is said to be *normally off* because with $v_{GS} = 0$ there is no channel present. We need to *raise* V_{GS} above V_{t0} (>0) in order to create a channel or, equivalently, to *enhance* the channel-region conductivity. This type of device is aptly referred to as *enhancement n*MOSFET. The higher the (p-type) implant dosage, the more positive the value of V_{t0}. The circuit symbol for this device, shown in Fig. 3.7a, uses a *broken* line to signify a normally nonconductive channel.

- An *n*MOSFET with $V_{t0} < 0$ is said to be *normally on* because with $v_{GS} = 0$ there is already a channel present. In this case we need to lower v_{GS} *below* V_{t0} (<0) in order to eliminate the channel or, equivalently, to *deplete* the channel region of its free electrons. This device type is aptly referred to as *depletion n*MOSFET. The higher the (n-type) implant dosage, the more negative the value of V_{t0}. The circuit symbol for this device, shown in Fig. 3.7b, uses a *continuous* line to signify a normally conductive channel.

- A *p*MOSFET with $V_{t0} < 0$ is said to be *normally off* because with $v_{GS} = 0$ there is no channel present. We need to lower v_{GS} *below* V_{t0} (<0) in order to create a

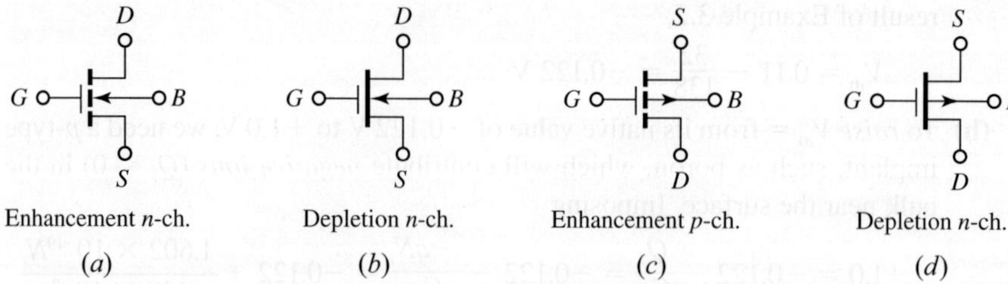

FIGURE 3.7 Full-fledged circuit symbols for the four MOSFET types.

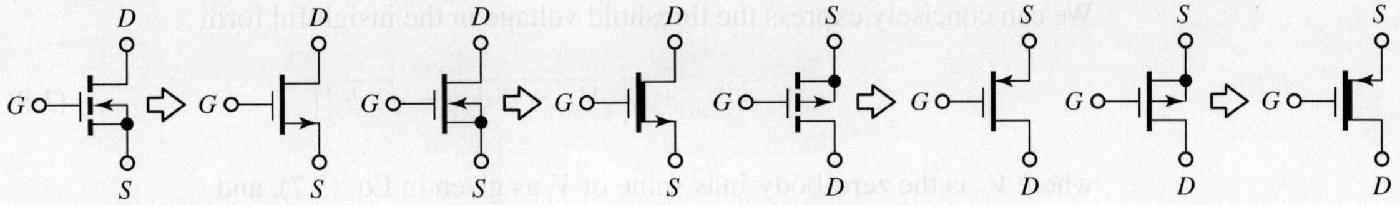

Enhancement n-ch.   Depletion n-ch.   Enhancement p-ch.   Depletion p-ch.

FIGURE 3.8 Simplified circuit symbols for the four MOSFET types.

channel, or, equivalently, to *enhance* its channel conductivity. Aptly called *enhancement p*MOSFET, this device is shown in Fig. 3.7*c*.

- A *p*MOSFET with $V_{t0} > 0$ is said to be *normally on* because with $v_{GS} = 0$ there is already a channel. If we want to *deplete* it of its free holes, we need to raise v_{GS} above V_{t0} (>0). The circuit symbol of this device, aptly called *depletion p*MOSFET, is shown in Fig. 3.7*d*.

The preferred mode of operation of a MOSFET is with the *body tied to the source*, resulting in a *three-terminal* device. This is the case, for instance, of discrete devices. Figure 3.8 shows the simplified MOSFET symbols most commonly used for this type of connection. To avoid the awkward broken lines, the enhancement types are given solid lines. To signify that the channels of the depletion types are already present, thicker lines are used.

The Body Effect and the Threshold Voltage V_t

When multiple devices share the same substrate, the latter must be tied to the most negative voltage (MNV) to avoid inadvertently turning on any of the body-source or body-drain *pn* junctions. Likewise, the common substrate of *p*MOSFETs must be tied to the most positive voltage (MPV). It is thus possible for the source of an *n*MOSFET to find itself at a *higher* voltage than the body, or $V_S > V_B$ (likewise, we can have $V_S < V_B$ for a *p*MOSFET). We wish to investigate the effect of body bias on the threshold voltage of an *n*MOSFET.

Denoting the source-body voltage of an *n*MOSFET as V_{SB} ($V_{SB} \geq 0$), we can simply recycle our previous findings by replacing $[2(-\phi_p)]$ with $[2(-\phi_p) + V_{SB}]$ in Eq. (3.5). The result is

$$Q_b = -\sqrt{2qN_A\varepsilon_{si}(V_{SB} + 2|\phi_p|)}$$

where we are using the absolute value of ϕ_p ($\phi_p < 0$) to reduce the possibility of confusion. Clearly, the increase in the depletion-region charge Q_b ($Q_b > 0$) comes at the expense of a simultaneous decrease in the inversion-layer charge Q_n ($Q_n < 0$). To return the channel to its former state we need to suitably increase v_{GS}. To find out by how much, we rewrite Eq. (3.7) as

$$V_t = -\phi_0 - 2\phi_p - \frac{Q_b}{C_{ox}} - \frac{Q_{ox}}{C_{ox}} - \frac{Q_i}{C_{ox}} = -\phi_0 - 2\phi_p - \frac{Q_{b0}}{C_{ox}} - \frac{Q_{ox}}{C_{ox}} - \frac{Q_i}{C_{ox}} - \frac{Q_b - Q_{b0}}{C_{ox}}$$

$$= V_{t0} - \frac{Q_b - Q_{b0}}{C_{ox}}$$

We can concisely express the threshold voltage in the insightful form

$$V_t = V_{t0} + \gamma\left[\sqrt{V_{SB} + 2|\phi_p|} - \sqrt{2|\phi_p|}\right] \tag{3.8}$$

where V_{t0} is the zero-body-bias value of V_t as given in Eq. (3.7), and

$$\gamma = \frac{\sqrt{2qN_A\varepsilon_{si}}}{C_{ox}} \tag{3.9}$$

is called the *body-effect parameter*. Its value, in $V^{1/2}$, is typically on the order of a fraction of 1 $V^{1/2}$.

EXAMPLE 3.5

(a) For the *enhancement nMOSFET* of Example 3.4b, which has $V_{t0} = 1.0$ V, find V_t at $V_{SB} = 1$ V, as well as at $V_{SB} = 5$ V.

(b) For the *depletion nMOSFET* of Example 3.4c, which has $V_{t0} = -1.0$ V, find V_t at $V_{SB} = 1$ V, as well as at $V_{SB} = 5$ V. For what value of V_{SB} do we get $V_t = -0.5$ V?

Solution

$$\gamma = \frac{\sqrt{2 \times 1.602 \times 10^{-19} \times 10^{16} \times 1.04 \times 10^{-12}}}{138 \times 10^{-9}} = 0.418 \ V^{1/2}$$

(a) For the *enhancement nMOSFET* we have

$$V_t(V_{SB} = 1 \ V) = 1.0 + 0.418(\sqrt{1 + 0.7} - \sqrt{0.7}) = 1.0 + 0.195 = 1.195 \ V.$$

$$V_t(V_{SB} = 5 \ V) = 1.0 + 0.418(\sqrt{5 + 0.7} - \sqrt{0.7}) = 1.0 + 0.648 = 1.648 \ V.$$

(b) For the *depletion nMOSFET* we have

$$V_t(V_{SB} = 1 \ V) = -1.0 + 0.418(\sqrt{1 + 0.7} - \sqrt{0.7}) = -1.0 + 0.195 = -0.805 \ V.$$

$$V_t(V_{SB} = 5 \ V) = -1.0 + 0.418(\sqrt{5 + 0.7} - \sqrt{0.7}) = -1.0 + 0.648 = -0.352 \ V.$$

Imposing

$$-0.5 = -1.0 + 0.418(\sqrt{V_{SB} + 0.7} - \sqrt{0.7})$$

gives $V_{SB} = 3.43$ V.

The example indicates that the effect of body bias is to *shift* the threshold voltage of an *nMOSFET* in the *positive* direction, regardless of whether it is a depletion or enhancement type. By contrast, for a *pMOSFET* the shift is in the *negative* direction. Adapted to the *pMOSFET* case, Eq. (3.8) becomes

$$V_t = V_{t0} - \gamma\left[\sqrt{V_{BS} + 2\phi_n} - \sqrt{2\phi_n}\right] \tag{3.10}$$

where γ is still given by Eq. (3.9), but with N_A replaced by N_D. The dependence of V_t upon the body bias is referred to as the *body effect*, and the body itself is sometimes referred to as *back gate* because it influences the inversion layer like the gate, though in the *opposite* direction and also in *square-root* fashion.

A certain *enhancement* pMOSFET has $V_{t0} = -1.5$ V and $\gamma = 0.5$ V$^{1/2}$. If $\phi_n = +0.3$ V, find V_t at $V_{BS} = 3$ V.   **EXAMPLE 3.6**

Solution

$$V_t(V_{BS} = 3 \text{ V}) = -1.5 - 0.5(\sqrt{3 + 2 \times 0.3} - \sqrt{2 \times 0.3}) = -1.5 - 0.56 = -2.06 \text{ V}$$

As mentioned, body bias causes a *negative* shift in the threshold voltage of a pMOSFET, regardless of the polarity of V_{t0}.

3.3 THE *n*-CHANNEL CHARACTERISTIC

We are now ready to investigate the *i-v* characteristics of the *n*-channel, anticipating that our understanding of the *p*-channel will follow easily once we have mastered the *n*-channel. Figure 3.9 shows the sequence of situations an *n*-channel goes through as we gradually increase v_{DS} starting out with $v_{DS} \cong 0$. Once the MOSFET is biased in strong inversion, its channel can be viewed as a *resistor* of length *L*, width *W*, and thickness proportional to the *overdrive voltage*, which is defined as the amount by which the gate-source voltage *exceeds* the threshold voltage V_t,

$$V_{OV} = V_{GS} - V_t \tag{3.11}$$

For instance, in the device of Example 3.3, every volt of V_{OV} induces an electron charge of -138 nC/cm^2 in the channel, so the greater V_{OV}, the more conductive the channel will be. If we now apply a voltage $v_{DS} > 0$ to the drain, electrons will drift from the source, through the channel, to the drain, like in an ordinary resistor (hence the designation *ohmic* for this region of operation), thus producing current. But, electrons are negative, so the current i_D at the drain terminal will flow *into* the device, as shown in Fig. 3.9*a*. The source and drain designations reflect the fact that mobile charges (electrons in *n*MOSFETS, holes in *p*MOSFETs) are *sourced* to the channel at one end, and *drained* from the channel at the other.

The Triode Region

If we now increase v_{DS} further, an interesting effect occurs, namely, the channel becomes *tapered*, as depicted in Fig. 3.9*b*. This stems from the fact that while at the source end we have $V_{OV} = V_{GS} - V_t$, at the drain end we only have $V_{OV} = (V_{GS} - v_{DS}) - V_t$, indicating a thinner channel there. For instance, let $V_t = 1$ V, $V_{GS} = 5$ V, and $v_{DS} = 2$ V. Then, the overdrive at the source end is $V_{OV(source)} = 5 - 1 = 4$ V, but that at the drain end is only $V_{OV(drain)} = (5 - 2) - 1 = 2$ V. In this example, the channel at the drain end is only half as thick as at the source end.

To investigate quantitatively, refer to Fig. 3.10, where we imagine that we have sliced the channel like a loaf of bread, and we focus on the slice of width *dy* located

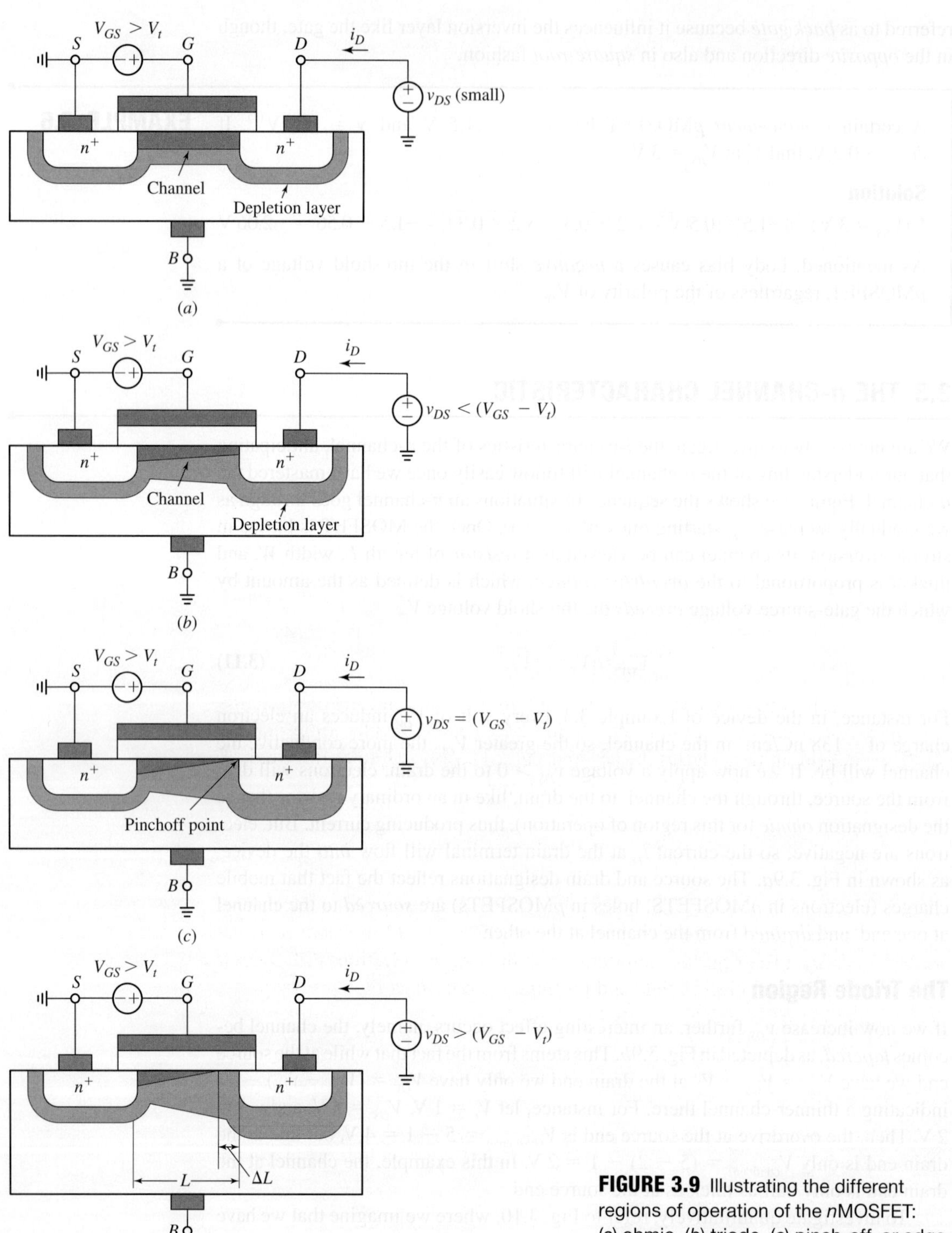

FIGURE 3.9 Illustrating the different regions of operation of the *n*MOSFET: (*a*) ohmic, (*b*) triode, (*c*) pinch-off, or edge of saturation (EOS), and (*d*) saturation, or active region.

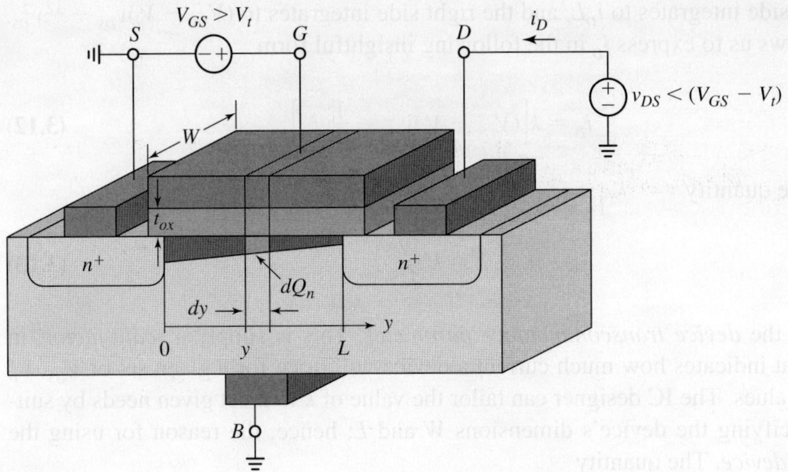

FIGURE 3.10 The *n*MOSFET in the triode region.

at some distance y from the source. The voltage at each slice varies from 0 V at the leftmost slice to v_{DS} at the rightmost slice, so the voltage $v(y)$ at our particular slice will lie somewhere in between, or $0 \leq v(y) \leq v_{DS}$. Now, the gate strip immediately above our slice forms a capacitance $dC = C_{ox} \times W \times dy$ with the channel itself, so the charge packet dQ_n induced in the channel is, by the capacitance law,

$$dQ_n = -dC\{[V_{GS} - v(y)] - V_t\} = -C_{ox}Wdy[V_{GS} - V_t - v(y)]$$

(This charge is negative because it consists of electrons.) The voltage drop v_{DS} across the channel produces an electric field E inside the channel, oriented from drain to source. This field, in turn, causes the negative charge packet dQ_n to drift toward the drain, thus producing the current i_D. By definition,

$$i_D = -\frac{dQ_n}{dt} = C_{ox}W[V_{GS} - V_t - v(y)]\frac{dy}{dt}$$

where dy/dt represents the velocity with which dQ_n drifts toward the drain. This velocity is proportional to the electric field, or $dy/dt = -\mu_n E(y)$, where μ_n is the *electron mobility*. (The negative sign stems from the fact that electrons drift *against* the electric field.) But, electric field and potential are related as $E(y) = -dv(y)/dy$, so $dy/dt = \mu_n dv(y)/dy$. Substituting in the above equation gives

$$i_D = \mu_n C_{ox}W[V_{GS} - V_t - v(y)]\frac{dv(y)}{dy}$$

Multiplying both sides by dy and integrating from $y = 0$, where $v(y) = 0$, to $y = L$, where $v(y) = v_{DS}$, we get

$$\int_0^L i_D\,dy = \mu_n C_{ox}W\int_0^{v_{DS}} [V_{GS} - V_t - v(y)]\,dv(y)$$

The left side integrates to $i_D L$, and the right side integrates to $(V_{GS} - V_t)v_{DS} - \frac{1}{2}v_{DS}^2$. This allows us to express i_D in the following insightful form

$$i_D = k\left[(V_{GS} - V_t)v_{DS} - \frac{1}{2}v_{DS}^2\right] \tag{3.12}$$

where the quantity

$$k = k'\frac{W}{L} \tag{3.13}$$

is called the *device transconductance parameter*. This is simply a *scale factor*, in A/V^2, that indicates how much current a device will draw for a given set of V_{GS}, V_t, and v_{DS} values. The IC designer can tailor the value of k to meet given needs by suitably specifying the device's dimensions W and L; hence, the reason for using the qualifier *device*. The quantity

$$k' = \mu_n C_{ox} = \frac{\mu_n \varepsilon_{ox}}{t_{ox}} \tag{3.14}$$

is called the *process transconductance parameter*, in A/V^2. Being common to all devices, it is unique of the particular fabrication process; hence, the reason for the qualifier *process*. Figure 3.11 shows the plot of i_D versus v_{DS} for a given overdrive voltage V_{OV}.

We observe that near the origin, where v_{DS} is small enough to render the quadratic term negligible in Eq. (3.12), the i_D-v_{DS} characteristic approaches, for a given gate-source drive V_{GS}, a *straight line*, or

$$i_D \cong k(V_{GS} - V_t)v_{DS} \tag{3.15}$$

For this reason, the region corresponding to small values of v_{DS} is called the *linear region*. Rewriting Eq. (3.15) in the form of Ohm's law as

$$i_D = \frac{1}{r_{DS}}v_{DS} \tag{3.16}$$

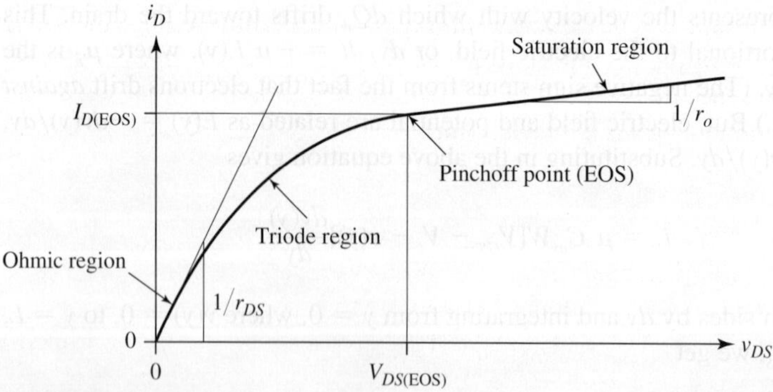

FIGURE 3.11 The complete i_D-v_{DS} characteristic for a given overdrive voltage $V_{OV} = V_{GS} - V_t > 0$. Note that $V_{DS(EOS)} = V_{OV}$.

confirms that the channel acts as a *resistor*, this being the reason why this region is also referred to as the *ohmic region*. The channel resistance r_{DS} is *controlled* by the overdrive V_{OV} as

$$r_{DS} = \frac{1}{k(V_{GS} - V_t)} = \frac{1}{k'\dfrac{W}{L}V_{OV}} \qquad (3.17)$$

This resistance depends also on the W/L ratio, also called the *aspect ratio*, indicating that by proper choice of this ratio, the IC designer can set this resistance to virtually any value for a given overdrive V_{OV}.

EXAMPLE 3.7

(a) Assuming $\mu_n = 600$ cm²/Vs, $C_{ox} = 83$ nF/cm², and $V_t = 1.0$ V, specify the W/L ratio so that $r_{DS} = 1$ kΩ for $V_{GS} = 5$ V.

(b) Calculate r_{DS} for $V_{GS} = 4$ V, 3 V, 2 V, 1 V, 0 V.

Solution

(a) By Eq. (3.14), $k' = 600 \times 83 \times 10^{-9} \cong 50$ μA/V². Using Eq. (3.17) to impose

$$10^3 = \frac{1}{50 \times 10^{-6}(W/L)(5 - 1)}$$

we get $W/L = 5$. Consequently, $k = (50\ \mu\text{A/V}^2)5 = 250\ \mu\text{A/V}^2$.

(b) By Eq. (3.17), for $V_{GS} = 4$ V we have

$$r_{DS} = \frac{1}{250 \times 10^{-6}(4 - 1)} = 1.333 \text{ k}\Omega$$

Likewise, for $V_{GS} = 3$ V we find $r_{DS} = 2$ kΩ, and for $V_{GS} = 2$ V we find $r_{DS} = 4$ kΩ. For $V_{GS} \leq 1$ V the MOSFET is in cutoff and $r_{DS} = \infty$.

As we increase v_{DS} further, the channel becomes progressively thinner at the drain end, and the quadratic term of Eq. (3.12) becomes more and more pronounced. Consequently, the slope of the curve decreases, indicating a corresponding increase in the dynamic resistance of the channel. This region of operation is called the *triode region* by analogy with vacuum tubes, which exhibit similar characteristics. We also observe in the sequence depicted in Fig. 3.9 that the depletion layer associated with the body-drain junction *widens* as we keep *increasing* v_{DS}.

The Pinchoff Point

Once v_{DS} achieves the critical value $V_{DS(EOS)} = V_{GS} - V_t$, or

$$V_{DS(EOS)} = V_{OV} \qquad (3.18)$$

the channel thickness at the drain end reduces to zero, as depicted in Fig. 3.9c, and the corresponding point on the i_D-v_{DS} curve is referred to as the *pinchoff point*. As we shall see shortly, this point marks the beginning, or *edge of saturation* (EOS) condition. The current at this point is readily found by substituting $v_{DS} = V_{GS} - V_t$ in Eq. (3.12). The result is

$$I_{D(EOS)} = \frac{k}{2}(V_{GS} - V_t)^2 \qquad (3.19)$$

with k as given in Eq. (3.13). This is also expressed as $I_{D(EOS)} = (k/2)V^2_{DS(EOS)}$, or better yet as

$$I_{D(EOS)} = \frac{k}{2}V^2_{OV} \qquad (3.20)$$

The Saturation Region

If we raise v_{DS} *above* the critical value $V_{DS(EOS)}$, the voltage at the pinchoff point continues to remain at $V_{DS(EOS)}$, and the *excess difference* $v_{DS} - V_{DS(EOS)}$ is dropped across a narrow depletion layer of width ΔL between the pinchoff point and the edge of the drain. As depicted in Fig. 3.9d, the pinchoff point moves leftwards away from the drain, in effect *shortening* the channel by the amount ΔL. This situation, aptly referred to as *channel-length modulation*, results in the *actual channel length*

$$L_{actual} = L - \Delta L = L\left(1 - \frac{\Delta L}{L}\right)$$

To find the i_D-v_{DS} characteristic past the pinchoff point we adapt Eq. (3.19) and write

$$i_D = \frac{1}{2}\left(k'\frac{W}{L_{actual}}\right)(V_{GS} - V_t)^2 = \frac{1}{2}k'\frac{W}{L\left(1 - \frac{\Delta L}{L}\right)}(V_{GS} - V_t)^2$$

$$\cong \frac{1}{2}k'\frac{W}{L}\left(1 + \frac{\Delta L}{L}\right)(V_{GS} - V_t)^2$$

where we have exploited the fact that usually $\Delta L/L \ll 1$. It is an established practice in the literature to assume that the fractional change $\Delta L/L$ be *linearly proportional* to v_{DS}, or $\Delta L/L = \lambda v_{DS}$. Consequently, the i_D-v_{DS} characteristic past the pinchoff point is expressed as

$$i_D = \frac{k}{2}(V_{GS} - V_t)^2(1 + \lambda v_{DS}) \qquad (3.21)$$

with k as given in Eq. (3.13). The proportionality constant λ (in V^{-1}) is called the *channel-length modulation parameter*. Typically, λ is on the order of 0.01 to 0.1 V^{-1}, and for simplicity it is usually ignored ($\lambda \to 0$) in the course of hand dc calculations. The region past the pinchoff point is aptly referred to as the *saturation region* because i_D increases with v_{DS} only slightly there, in effect saturating.

The slope of the saturation-region characteristic is the reciprocal of a resistance r_o called the *output resistance* of the MOSFET in saturation. Differentiating Eq. (3.21) and calculating at the EOS gives

$$\frac{1}{r_o} = \frac{\partial i_D}{\partial v_{DS}} = \lambda I_{D(EOS)}$$

The output resistance is usually expressed in the form

$$r_o = \frac{1}{\lambda I_D} \qquad (3.22)$$

where I_D is the current at the actual saturation-region operating point ($I_D \cong I_{D(EOS)}$). In general, r_o is fairly large relative to other resistances in a MOSFET circuit. Indeed, the smaller the value of λ, the higher the value of r_o. In the limit $\lambda \to 0$, a saturated MOSFET would approach *ideal current-source* behavior, or, more precisely, it would act as an ideal voltage-controlled current-source (VCCS), with V_{GS} as the control voltage. As such, the MOSFET finds application as an amplifier.

Remark: To ensure continuity between Eqs. (3.12) and (3.21) at the EOS, the right-hand side of Eq. (3.12) should also be multiplied by the term $(1 + \lambda v_{DS})$. In practice, to simplify the triode-region calculations, the term λv_{DS} is usually ignored as v_{DS} is small in that region.

EXAMPLE 3.8

A certain *n*MOSFET has $V_{t0} = 1.0$ V, $k = 0.5$ mA/V^2, $\lambda = 0.02$ V^{-1}, $\gamma = 0.6$ V$^{1/2}$, and $\phi_p = -0.3$ V.

(a) If $V_{GS} = 3$ V and $V_{SB} = 0$, find $V_{DS(EOS)}$, $I_{D(EOS)}$, and r_o.
(b) What is the value of I_D at $V_{DS} = 0.5V_{DS(EOS)}$? At $V_{DS} = 2V_{DS(EOS)}$? At $V_{DS} = 4V_{DS(EOS)}$?
(c) Repeat parts (a) and (b), but with $V_{SB} = 2$ V. Comment on your findings.
(d) Find V_{SB} so that $V_{DS(EOS)} = 1$ V with $V_{GS} = 3$ V. What is the corresponding value of $I_{D(EOS)}$?

Solution

(a) We have $V_{DS(EOS)} = V_{OV} = V_{GS} - V_{t0} = 3 - 1 = 2$ V, so

$$I_{D(EOS)} = \frac{k}{2}V_{OV}^2(1 + \lambda V_{DS(EOS)}) = \frac{0.5}{2}2^2(1 + 0.02 \times 2) = 1.04 \text{ mA}$$

Moreover, $r_o = 1/(0.02 \times 1.04) = 48$ kΩ.

(b) With reference to Fig. 3.11, we observe that for $V_{DS} = 0.5V_{DS(EOS)} = 1$ V ($<V_{OV}$) the FET is operating in the triode region, while for $V_{DS} = 2V_{DS(EOS)} = 4$ V ($>V_{OV}$) the FET is operating in saturation. Consequently, we use Eqs. (3.12) and (3.21) to find

$$I_D(V_{DS} = 1 \text{ V}) = 0.5(2 \times 1 - 1^2/2) = 0.75 \text{ mA}$$

$$I_D(V_{DS} = 4 \text{ V}) = (0.25)2^2(1 + 0.02 \times 4) = 1.08 \text{ mA}$$

Similarly, $I_D(V_{DS} = 8 \text{ V}) = 1.16$ mA

(c) By Eq. (3.8), we now have

$$V_t(V_{SB} = 2 \text{ V}) = 1.0 + 0.6(\sqrt{2 + 2 \times 0.3} - \sqrt{2 \times 0.3}) = 1.5 \text{ V}$$

Consequently, $V_{DS(EOS)} = V_{OV} = 3 - 1.5 = 1.5$ V. By analogous calculations, we now get

$$I_{D(EOS)} = 0.25 \times 1.5^2(1 + 0.02 \times 1.5) = 0.58 \text{ mA}$$

and $r_o = 1/(0.02 \times 0.58) = 86$ kΩ. Moreover,

$$I_D(V_{DS} = 0.75 \text{ V}) = 0.5(1.5 \times 0.75 - 0.75^2/2) = 0.42 \text{ mA}$$

$$I_D(V_{DS} = 3 \text{ V}) = 0.25 \times 1.5^2(1 + 0.02 \times 3) = 0.60 \text{ mA}$$

Likewise, $I_D(V_{DS} = 6 \text{ V}) = 0.63$ mA. The increase in V_t due to the body affect has resulted in a less conductive channel, thus causing a decrease in the drain current values as well as an increase in r_o.

(d) We now have $V_t = V_{GS} - V_{DS(EOS)} = 3 - 1 = 2$ V. Using Eq. (3.8) to impose

$$2 = 1.0 + 0.6\left(\sqrt{V_{SB} + 0.6} - \sqrt{0.6}\right)$$

yields $V_{SB} = 5.36$ V. Finally, $I_{D(EOS)} = 0.25 \times 1^2(1 + 0.02 \times 1) = 0.255$ mA.

Determining the Operating Region of a *n*MOSFET

As we progress, we will often face the need to identify the operating region of a FET from a set of incomplete data. For the case of an *n*MOSFET, we will proceed as follows:

- If $V_{GS} \leq V_t$, the FET is operating in the cut-off (CO) region, where $i_D = 0$.
- If $V_{GS} > V_t$, the FET is on, but is it operating in the triode or in the saturation region? This depends on whether $V_{DS} < V_{OV}$ or $V_{DS} > V_{OV}$, respectively. To find out, proceed as follows:
- Assume the FET is *saturated*, and use Eq. (3.21) to find the missing data until you have both V_{OV} and V_{DS} in hand. If it turns out that $V_{DS} > V_{OV}$, the assumption was correct, and no further steps are needed.
- Otherwise you get a contradiction, indicating that the FET is in the *triode region*, and you must recalculate the missing data via Eq. (3.12) instead. As a final check, verify that indeed $V_{DS} < V_{OV}$.
- Alternatively, we could start out with the assumption that the FET be in the *triode region*, and then check that indeed $V_{DS} < V_{OV}$ to confirm our assumption. Otherwise, we get a contradiction signifying that the FET is instead *saturated*. An example will better illustrate the above procedure.

EXAMPLE 3.9   A certain *n*MOSFET has $V_t = 1.5$ V, $k = 1.0$ mA/V^2, and $\lambda = 0.02$ V^{-1}, and is operated at $V_{SB} = 0$.

(a) Find V_{GS} so that the FET gives $I_D = 2.2$ mA at $V_{DS} = 5$ V.

(b) Find V_{GS} for $I_D = 2$ mA at $V_{DS} = 1$ V.

(c) Find V_{DS} so that the FET gives $I_D = 4$ mA with $V_{GS} = 4.5$ V.

(d) Find V_{DS} for $I_D = 0.52$ mA with $V_{GS} = 2.5$ V.

Solution

(a) Assume the FET is saturated, and then check. By Eq. (3.21), the overdrive V_{OV} needed to sustain 2.2 mA in saturation is such that

$$2.2 = \tfrac{1}{2}V_{OV}^2(1 + 0.02 \times 5)$$

This gives $V_{OV} = 2$ V. Since $V_{DS} > V_{OV}$ $(5 > 2)$, the FET is indeed in saturation, confirming that our assumption was correct. Clearly, $V_{GS} = V_t + V_{OV} = 1.5 + 2 = 3.5$ V.

(b) Assume again saturation. Imposing

$$2 = 0.5V_{OV}^2(1 + 0.02 \times 1)$$

gives $V_{OV} = 1.98$ V, that is, $V_{DS} < V_{OV}$ $(1 < 1.98)$. This contradicts our assumption of a saturated FET, so the device must be operating in the triode region, where Eq. (3.12) holds. The overdrive V_{OV} needed to sustain 2 mA in the triode region is such that

$$2 = 1\left(V_{OV} \times 1 - \frac{1^2}{2}\right)$$

This gives $V_{OV} = 2.5$ V. The fact that $V_{DS} < V_{OV}$ $(1 < 2.5)$ confirms that the FET is indeed in the triode region. Moreover, $V_{GS} = 1.5 + 2.5 = 4$ V.

(c) As an alternative, assume this time the FET to be in the triode region, and then check as usual. Now, $V_{OV} = V_{GS} - V_t = 4.5 - 1.5 = 3$ V. By Eq. (3.12) we must have

$$4 = 1\left(3 \times V_{DS} - \frac{V_{DS}^2}{2}\right)$$

or $0.5V_{DS}^2 - 3V_{DS} + 4 = 0$. This quadratic equation admits two solutions, $V_{DS} = 2$ V and $V_{DS} = 4$ V. The second one is unacceptable as it would imply a saturated FET $(V_{DS} > V_{OV})$, for which Eq. (3.21) would predict $I_D = 4.86$ mA, in blatant contradiction with the desired value of I_D. Consequently, our FET is indeed in the triode region, and $V_{DS} = 2$ V.

(d) Assume again the triode region. We have $V_{OV} = 2.5 - 1.5 = 1$ V, and

$$0.52 = 1\left(1 \times V_{DS} - \frac{V_{DS}^2}{2}\right)$$

or $0.5V_{DS}^2 - V_{DS} + 0.52 = 0$. This quadratic equation admits the solutions $V_{DS} = 1 \pm 0.2j$, which are complex numbers and thus physically unacceptable. Evidently, our triode-region assumption was wrong. We must thus use Eq. (3.21) and impose

$$0.52 = \tfrac{1}{2}1^2(1 + 0.02V_{DS})$$

which yields $V_{DS} = 2$ V. The fact that $V_{DS} > V_{OV}$ $(2 > 1)$ confirms that the FET is indeed saturated.

Series/Parallel MOSFET Combinations

As we proceed we shall often encounter FETs connected in series or parallel. The channels combine just like resistors do, namely, in *series* combinations the channel *resistances* add up, while in *parallel* combinations the channel *conductances* add up.

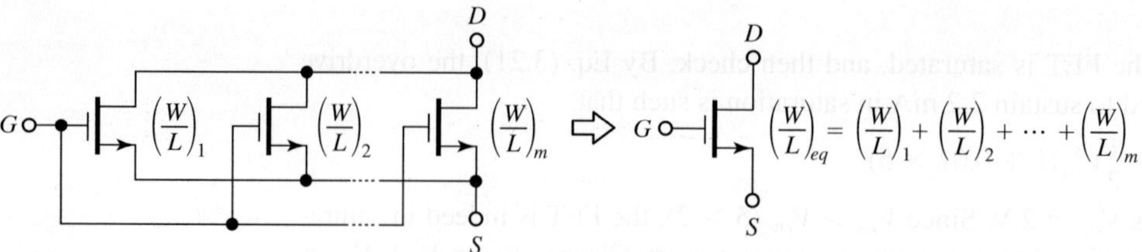

FIGURE 3.12 When *m* FETs are connected in *parallel*, they act like a single equivalent FET whose *W/L* ratio is the *sum* of the individual *W/L* ratios.

Consider *m* MOSFETs having the same V_t and k' but individual ratios $(W/L)_1$, $(W/L)_2, \ldots (W/L)_m$, and suppose they are connected in *parallel*, as in Fig. 3.12. The current drawn by each FET is linearly proportional to its *W/L* ratio, and since all FETs are subjected to the same input drive, the proportionality constant is the same for all FETs. But, the total current drawn from the drain terminal is the sum of the individual currents, so the *parallel* combination of *m* FETs acts like a a *single* equivalent FET having

$$\left(\frac{W}{L}\right)_{eq} = \left(\frac{W}{L}\right)_1 + \left(\frac{W}{L}\right)_2 + \cdots + \left(\frac{W}{L}\right)_m \tag{3.23}$$

A common occurrence is when two identical FETs are in parallel, which can intuitively be regarded as a single FET but with *W* twice as long.

Figure 3.13 shows the case of *m* FETs connected in series. Assuming for simplicity $\gamma = 0$ and $\lambda = 0$ throughout, one can prove that if all FETs have the same V_t and k' but individual ratios $(W/L)_1$, $(W/L)_2, \ldots (W/L)_m$, they act like *single* equivalent MOSFET such that

$$\left(\frac{W}{L}\right)_{eq}^{-1} = \left(\frac{W}{L}\right)_1^{-1} + \left(\frac{W}{L}\right)_2^{-1} + \cdots + \left(\frac{W}{L}\right)_m^{-1} \tag{3.24}$$

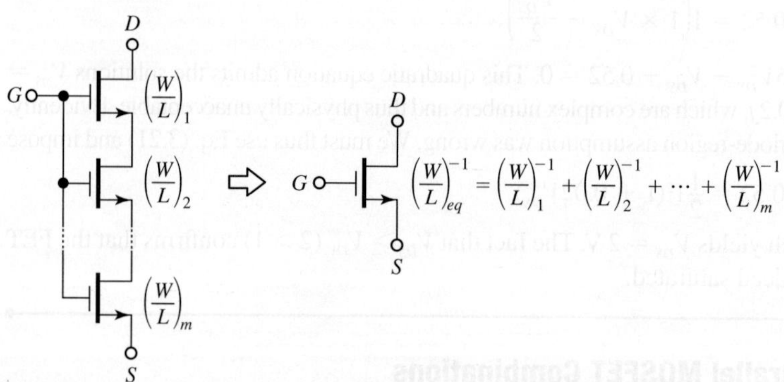

FIGURE 3.13 When *m* FETs are connected in *series*, they act like a single equivalent FET whose *L/W* ratio is the *sum* of the individual *L/W* ratios.

A common occurrence is one of two identical FETs in series, which can intuitively be regarded as a single FET but with L twice as large.

EXAMPLE 3.10

Suppose two FETs are fabricated in a 0.5-μm process with $(W/L)_1 = (1.0\ \mu\text{m})/(0.5\ \mu\text{m})$ and $(W/L)_2 = (0.5\ \mu\text{m})/(0.5\ \mu\text{m})$. Find $(W/L)_{eq}$ if the FETs are connected (*a*) in *parallel*, and (*b*) in *series*. In either case, express $(W/L)_{eq}$ in such a way that the smaller of W and L is 0.5 μm.

Solution

(a) By Eq. (3.23), $(W/L)_{eq} = (1.0/0.5) + (0.5/0.5) = (1.5\ \mu\text{m})/(0.5\ \mu\text{m})$.
(b) By Eq. (3.24), $(L/W)_{eq} = (0.5/1.0) + (0.5/0.5) = 1.5$, so $(W/L)_{eq} = 1/1.5 = (0.5\ \mu\text{m})/(0.75\ \mu\text{m})$.

3.4 THE *i-v* CHARACTERISTICS OF MOSFETS

The two most important MOSFET characteristics are the plot of i_D versus v_{GS} in saturation, and the plot of i_D versus v_{DS} for different values of V_{GS}. These curves can be displayed either in the lab, via an oscilloscope equipped with a suitable curve-tracer module, or on a computer monitor via PSpice (see Appendix 3A for PSpice models for MOSFETs).

Diode-Mode Operation

In the PSpice circuit of Fig. 3.14 (see Appendix 3A) the gate and drain terminals are tied together, turning the *n*MOSFET into a *two-terminal* device with $v_{DS} = v_{GS}$. For $v_{GS} < V_t$ the device is in cutoff. For $v_{GS} > V_t$ the device is on and in saturation because $v_{DS} = v_{GS}$ implies $v_{DS} > V_{GS} - V_t$, the condition for a saturated *n*MOSFET. Consequently, when on, the device is governed by Eq. (3.21), but with $v_{DS} = v_{GS}$. The result is the curve of Fig. 3.15, which reveals a tendency by a diode-connected MOSFET to favor current flow in one direction (drain-to-source for an *n*MOSFET, source-to-drain for a *p*MOSFET) while inhibiting it in the opposite direction. Hence the name *diode* for this mode of operation.

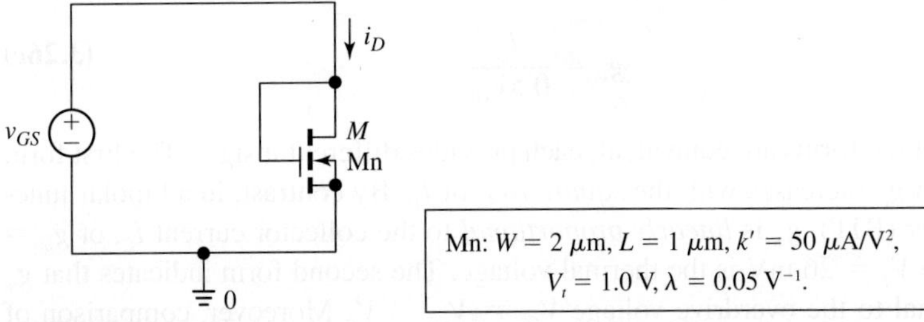

FIGURE 3.14 Diode-connected enhancement *n*MOSFET.

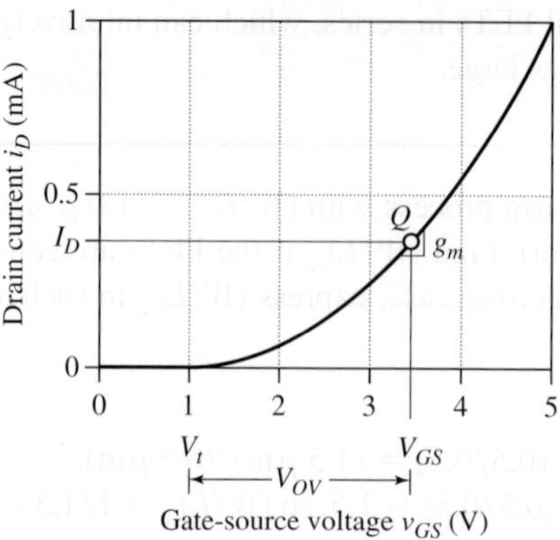

FIGURE 3.15 The *i-v* characteristic of the diode-connected *enhancement nMOSFET* of Fig. 3.14.

At this juncture it must be pointed out that the MOSFET's transition from off to on in the vicinity of V_t is not abrupt, but rather a *gradual* process. In fact, the channel already starts to conduct for a range of v_{GS} values less than, if close to, V_t. Over this range, aptly called the *subthreshold region,* i_D increases *exponentially*—rather than quadratically—with v_{GS}. The choice of V_t as the value of v_{GS} responsible for the onset of strong inversion is primarily a matter of mathematical convenience and mental bookkeeping.

The *slope* of the curve at a given point V_{GS} is denoted as g_m and is called the *transconductance*

$$g_m = \left.\frac{\partial i_D}{\partial v_{GS}}\right|_{V_{GS}} \tag{3.25}$$

Its units are A/V, or more likely μA/V for micropower devices. Differentiating Eq. (3.21) but with $\lambda = 0$ for simplicity, and suitably manipulating, we find three different expressions for the transconductance,

$$g_m = \sqrt{2kI_D} \tag{3.26a}$$

$$g_m = kV_{OV} \tag{3.26b}$$

$$g_m = \frac{I_D}{0.5V_{OV}} \tag{3.26c}$$

Though the three forms are equivalent, each provides different insight. The first form indicates that g_m increases with the *square root* of I_D. By contrast, in a bipolar junction transistor (BJT), g_m is *linearly proportional* to the collector current I_C, or $g_m = I_C/V_T$, where $V_T = 26$ mV is the thermal voltage. The second form indicates that g_m is proportional to the overdrive voltage $V_{OV} = V_{GS} - V_t$. Moreover, comparison of Eq. (3.26b) with Eq. (3.17) reveals the additional interesting relation $g_m = 1/r_{DS}$.

(a) Assuming the *n*MOSFET data of Fig. 3.14, but with $\lambda = 0$ to simplify the calculations, find V_{GS} for $I_D = 1$ mA. Compare with Fig. 3.15, and comment.

(b) Find g_m at that point, and compare with the g_m of a bipolar junction transistor (BJT) operating at the same current level.

(c) Find W/L to raise the g_m of the FET to the same value as that of the BJT.

EXAMPLE 3.11

Solution

(a) By Eq. (3.13), $k = 50 \times 10^{-6}(2/1) = 100\ \mu\text{A/V}^2$. Using Eq. (3.21) but with $\lambda = 0$ we get $1 \times 10^{-3} = \frac{1}{2}(100 \times 10^{-6}) \times (V_{GS} - 1.0)^2$, or $V_{GS} = 5.472$ V. This is a bit higher than the value (5 V) predicted by Fig. 3.15 because we have assumed $\lambda = 0$. This gives an idea of the error incurred by ignoring λ.

(b) By Eq. (3.26*a*), $g_m = \sqrt{2kI_D} = \sqrt{2 \times 100 \times 10^{-6} \times 10^{-3}} = 0.447$ mA/V. By contrast, at 1 mA a BJT gives $g_m = 1/26 = 38.5$ mA/V, almost two orders of magnitude higher.

(c) Since g_m is linearly proportional to $\sqrt{k}$, and thus to $\sqrt{W/L}$, we impose a simple proportion,

$$\frac{\sqrt{(W/L)_{\text{new}}}}{\sqrt{(W/L)_{\text{old}}}} = \frac{38.5}{0.447}$$

which gives $(W/L)_{\text{new}} \cong 14{,}800$, an outlandish number. This example illustrates a notorious drawback of FETs compared to BJTs, namely, their generally much lower g_ms. Indeed, Eq. (3.26*c*) gives $g_m = I_D/[0.5(5.472 - 1.0)\ \text{V}] = I_D/(2{,}236\ \text{mV})$, which compares quite unfavorably with the BJT relation $g_m = I_C/(26\ \text{mV})$.

The i_D-v_{DS} Characteristics

In Fig. 3.11 we have illustrated the behavior of the channel as we walk it through the different situations of Fig. 3.9, but for only a *single fixed value* of V_{GS} ($V_{GS} > V_t$). To get the complete picture, we need to display the characteristics for *different values* of V_{GS}. The PSpice circuit of Fig. 3.16 displays the i_D-v_{DS} characteristics of the FET of Fig. 3.14, but with V_{GS} stepped in 0.5-V increments. The result is the family of curves of Fig. 3.17, with respect to which we make the following observations:

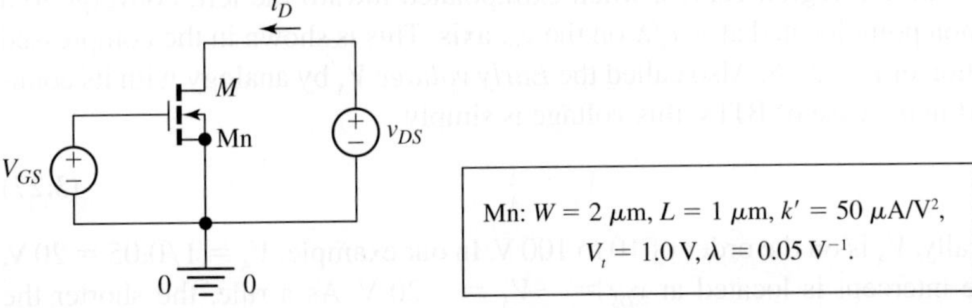

FIGURE 3.16 PSpice circuit to display the complete i_D-v_{DS} characteristics of the *n*MOSFET of Fig. 3.14.

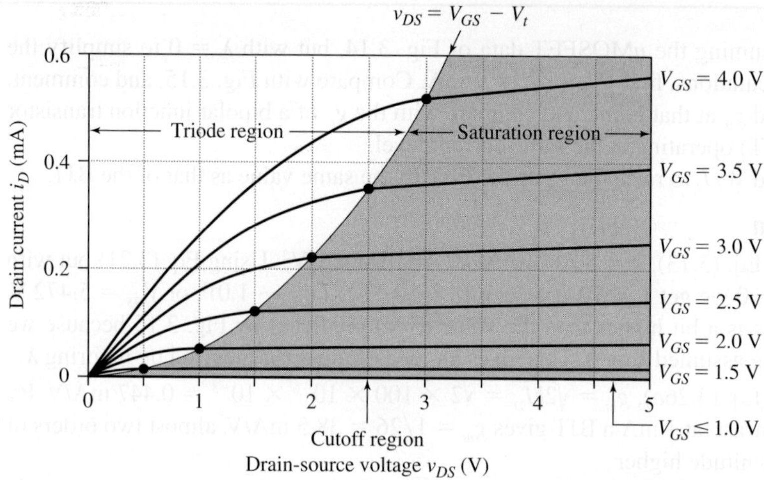

FIGURE 3.17 Complete i_D-v_{DS} characteristics of the *enhancement nMOSFET* of Fig. 3.16, and its regions of operation.

- For $V_{GS} < V_t$ (or $V_{GS} < 1.0$ V in our example), the device gives $i_D = 0$ and is thus in cutoff (CO). Its terminals draw only leakage currents, which are negligible in most practical situations.
- For $V_{GS} > V_t$, the device is on, either in the *triode* region if $v_{DS} < (V_{GS} - V_t)$, or in the *saturation* region if $v_{DS} > (V_{GS} - V_t)$. Either region requires a separate equation for finding i_D, namely,

$$v_{DS} < (V_{GS} - V_t) \Rightarrow \text{Triode region} \Rightarrow i_D = k\left[(V_{GS} - V_t)v_{DS} - \frac{1}{2}v_{DS}^2\right]$$

$$v_{DS} > (V_{GS} - V_t) \Rightarrow \text{Satur.n region} \Rightarrow i_D = \frac{k}{2}(V_{GS} - V_t)^2(1 + \lambda v_{DS})$$

- The locus of points for which $v_{DS} = V_{GS} - V_t = V_{OV}$ provides the *borderline* between the two regions (the borderline is aptly referred to as edge of saturation, or EOS for short). Since the abscissas are spaced *evenly* while the ordinates are spaced *quadratically*, this locus is a *parabola*. In fact, one can readily see that this locus is simply the *i-v* curve of Fig. 3.15, but shifted to the left by V_t.
- The saturation-region curves, when extrapolated toward the left, converge to a common point located at $-1/\lambda$ on the v_{DS} axis. This is shown in the compressed rendition of Fig. 3.18. Also called the *Early voltage* V_A by analogy with its counterpart in the case of BJTs, this voltage is simply

$$V_A = \frac{1}{\lambda} \tag{3.27}$$

Typically, V_A is on the order of 10 to 100 V. In our example, $V_A = 1/0.05 = 20$ V, so the intercept is located at $v_{DS} = -V_A = -20$ V. As a rule, the shorter the channel, the lower the value of V_A, indicating that V_A *scales* with L. For long-channel devices, this is sometimes expressed via the empirical form $V_A \cong L/(0.1\,\mu\text{m})$ V, or, equivalently, as $\lambda \cong (0.1\,\mu\text{m})/L$ V^{-1}.

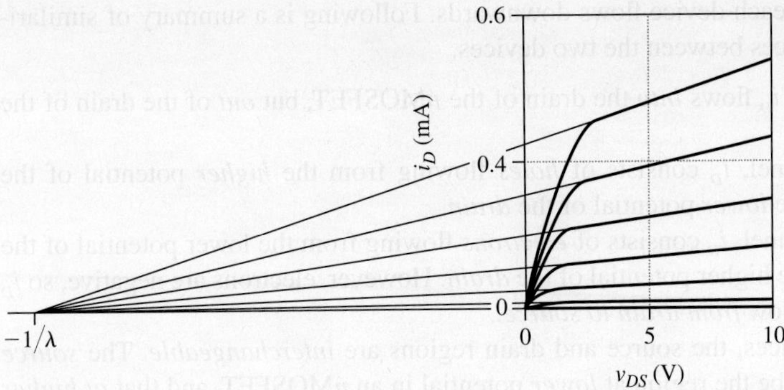

FIGURE 3.18 Effect of channel-length modulation on the *i-v* characteristics.

The *p*MOSFET and Comparison with the *n*MOSFET

The voltage-current relationships developed for the *n*MOSFET apply also to the *p*MOSFET, provided we (a) *reverse all current directions* and (b) *reverse all voltage polarities*. The two devices are compared in Fig. 3.19, where voltage is likened to height, so higher voltages are at the top and lower voltages at the bottom. Moreover,

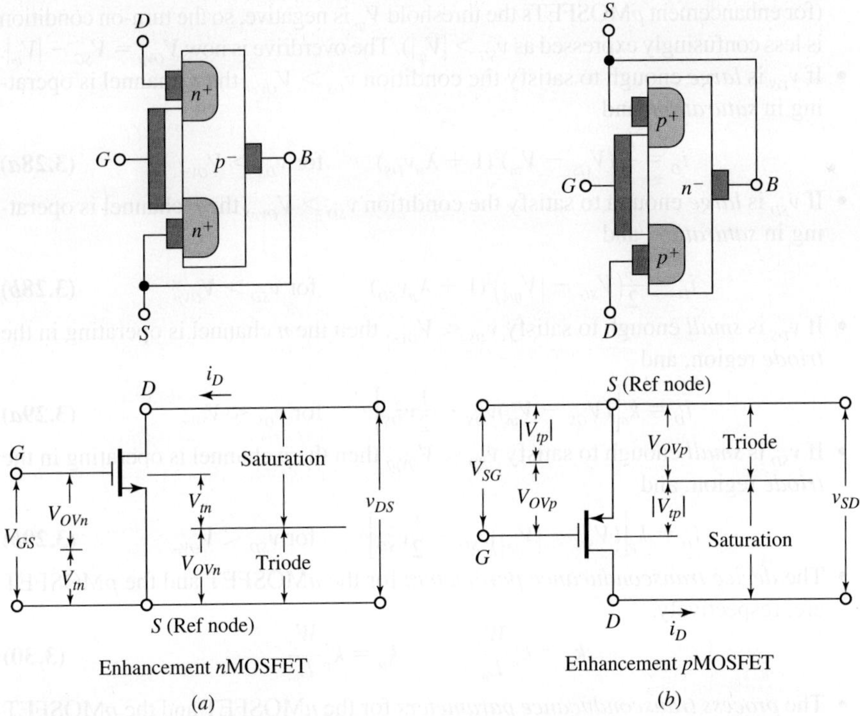

Enhancement *n*MOSFET

(a)

Enhancement *p*MOSFET

(b)

FIGURE 3.19 Comparing voltage *polarities*, current *directions*, voltage *ranges*, and *operating regions* for the *enhancement n*MOSFETS and *p*MOSFETs. All voltages are positive at the *top*.

current through each device flows downwards. Following is a summary of similarities and differences between the two devices.

- The current i_D flows *into* the drain of the *n*MOSFET, but *out* of the drain of the *p*MOSFET.
- In a *p*-channel, i_D consists of *holes* flowing from the *higher* potential of the *source* to the *lower* potential of the *drain*.
- In an *n*-channel, i_D consists of *electrons* flowing from the lower potential of the *source* to the higher potential of the *drain*. However, electrons are negative, so i_D is taken to flow *from drain to source*.
- In both devices, the source and drain regions are *interchangeable*. The *source* will always be the region at *lower* potential in an *n*MOSFET, and that *at higher* potential in a *p*MOSFET.
- An *enhancement n*MOSFET is *normally off*. To turn it on we need to create favorable conditions for *electrons* to exist in its channel region. Electrons are negative, so we need to induce an opposing positive charge on the gate. This requires that we *raise* the gate voltage v_G *above* the source voltage v_S by at least V_{tn}, the threshold voltage. The amount $V_{OVn} = V_{GS} - V_{tn}$, $V_{OVn} > 0$ is called the *overdrive*.
- An *enhancement p*MOSFET is *normally off*. To turn it on we need to create favorable conditions for *holes* to exist in its channel region. Holes are positive, so we need to induce an opposing negative charge on the gate. This requires that we *lower* the gate voltage v_G *below* the source voltage v_S by at least V_{tp}, the threshold voltage (for enhancement *p*MOSFETs the threshold V_{tp} is negative, so the turn-on condition is less confusingly expressed as $v_{SG} > |V_{tp}|$). The overdrive is now $V_{OVp} = V_{SG} - |V_{tp}|$.
- If v_{DS} is *large* enough to satisfy the condition $v_{DS} > V_{OVn}$, the *n* channel is operating in *saturation*, and

$$i_D = \frac{k_n}{2}(V_{GS} - V_{tn})^2(1 + \lambda_n v_{DS}) \qquad \text{for } v_{DS} > V_{OVn} \qquad \textbf{(3.28a)}$$

- If v_{SD} is *large* enough to satisfy the condition $v_{SD} > V_{OVp}$, the *p* channel is operating in *saturation*, and

$$i_D = \frac{k_p}{2}(V_{SG} - |V_{tp}|)^2(1 + \lambda_p v_{SD}) \qquad \text{for } v_{SD} > V_{OVp} \qquad \textbf{(3.28b)}$$

- If v_{DS} is *small* enough to satisfy $v_{DS} < V_{OVn}$, then the *n* channel is operating in the *triode* region, and

$$i_D = k_n\left[(V_{GS} - V_{tn})v_{DS} - \frac{1}{2}v_{DS}^2\right] \qquad \text{for } v_{DS} < V_{OVn} \qquad \textbf{(3.29a)}$$

- If v_{SD} is *small* enough to satisfy $v_{SD} < V_{OVp}$, then the *p* channel is operating in the *triode* region, and

$$i_D = k_p\left[(V_{SG} - |V_{tp}|)v_{SD} - \frac{1}{2}v_{SD}^2\right] \qquad \text{for } v_{SD} < V_{OVp} \qquad \textbf{(3.29b)}$$

- The *device transconductance parameters* for the *n*MOSFET and the *p*MOSFET are, respectively,

$$k_n = k_n'\frac{W_n}{L_n} \qquad\qquad k_p = k_p'\frac{W_p}{L_p} \qquad \textbf{(3.30)}$$

- The *process transconductance parameters* for the *n*MOSFET and the *p*MOSFET are, respectively,

$$k_n' = \frac{\mu_n\varepsilon_{ox}}{t_{ox}} \qquad\qquad k_p' = \frac{\mu_p\varepsilon_{ox}}{t_{ox}} \qquad \textbf{(3.31)}$$

- To avoid the inadvertent turn-on of the internal *pn* junctions, the bodies must be biased so that $V_{SB} \geq 0$ for the *n*MOSFET, and $V_{SB} \leq 0$ (that is, $V_{BS} \geq 0$) for the *p*MOSFET.
- Body bias in the *n*MOSFET shifts V_{tn} in the *positive* direction as

$$V_{tn} = V_{tn0} + \gamma_n \left[\sqrt{V_{SB} + 2|\phi_p|} - \sqrt{2|\phi_p|} \right] \qquad (3.32a)$$

- Body bias in the *p*MOSFET shifts V_{tp} in the *negative* direction as

$$V_{tp} = V_{tp0} - \gamma_p \left[\sqrt{V_{BS} + 2\phi_n} - \sqrt{2\phi_n} \right] \qquad (3.32b)$$

In general, V_{tn0}, V_{tp0}, λ_n, λ_p, γ_n, and γ_p are found experimentally via suitable measurements.

As we know, the *transconductance parameters* of Eq. (3.31) are proportional to the *mobilities* (μ_n or μ_p) of the charges intervening in the main current in the device. This is not surprising, as MOSFET current is of the *drift* type. When studying BJTs, in the previous chapter, we have encountered a similar parameter, namely, the *saturation current* I_s, which instead is proportional to the *diffusivity* (D_n or D_p) of the charges responsible for the main current in the device; this is so because BJT current is of the *diffusion* type. For a given doping density, electron mobility and diffusivity are two to three times higher than hole mobility and diffusivity, respectively. For these reasons, *n*-channel FETs are generally preferred over *p*-channel types, just like *npn* BJTs are generally preferred over *pnp* types.

A certain *p*MOSFET has $V_{t0} = -1.0$ V, $k = 0.4$ mA/V^2, $\lambda = 0.05$ V^{-1}, $\gamma = 0.73$ V$^{1/2}$, and $\phi_n = 0.3$ V. Unless otherwise specified, the device is operated at $V_{BS} = 0$.   **EXAMPLE 3.12**

(a) If $V_{SG} = 3$ V, find I_D at $V_{SD} = 1.5$ V.
(b) Repeat part (*a*) if $V_{SD} = 3$ V.
(c) Find V_{SD} so that the FET gives $I_D = 1.6$ mA with $V_{SG} = 4$ V.
(d) Find V_{SG} so that the FET gives $I_D = 0.24$ mA with $V_{SD} = 4$ V.
(e) Repeat part (*a*), but for $V_{BS} = 4$ V.
(f) Find V_{BS} so that the FET, with $V_{SG} = 3.5$ V, gives $I_D = 1$ mA at $V_{SD} = 5$ V.

Solution

(a) We have $V_{OV} = V_{SG} - |V_{t0}| = 3 - 1 = 2$ V. Since $V_{SD} < V_{OV}$ (1.5 < 2), the FET is operating in the triode region. By Eq. (3.29b),

$$I_D = 0.4 \left(2 \times 1.5 - \frac{1.5^2}{2} \right) = 0.75 \text{ mA}$$

(b) The FET is now saturated because $V_{SD} > V_{OV}$ (3 > 2). By Eq. (3.28b), we have

$$I_D = \frac{0.4}{2} 2^2 (1 + 0.05 \times 3) = 0.92 \text{ mA}$$

(c) We now have $V_{OV} = 4 - 1 = 3$ V, but we don't know whether the FET is in the triode region or in saturation. Assume saturation, and check. Using Eq. (3.28b), impose

$$1.6 = 0.2 \times 3^2 (1 + 0.05 V_{SD})$$

whose solution is $V_{SD} = -2.2$ V. This is physically unacceptable, indicating that our assumption was wrong. Evidently the FET is in the triode region, so we use Eq. (3.29b) to impose

$$1.6 = 0.4\left(3V_{SD} - \frac{V_{SD}^2}{2}\right)$$

or $0.5V_{SD}^2 - 3V_{SD} + 4 = 0$. The solutions to this quadratic equation are $V_{SD} = 4$ V and $V_{SD} = 2$ V. The first solution would imply a saturated FET ($V_{SD} > V_{OV}$ because $4 > 3$) which we have just proved not to be the case. So, the physically acceptable solution is $V_{SD} = 2$ V, which corroborates triode-region operation because $V_{SD} < V_{OV}$ ($2 < 3$).

(**d**) Assume triode-region operation and use Eq. (3.29b) to impose

$$0.24 = 0.4\left(V_{OV} \times 4 - \frac{4^2}{2}\right)$$

This yields $V_{OV} = 2.15$ V, or $V_{SD} > V_{OV}$ ($4 > 2.15$), contradicting our assumption. Evidently the FET is saturated, so impose

$$0.24 = 0.2 \times V_{OV}^2(1 + 0.05 \times 4)$$

This gives $V_{OV} = 1$ V. The fact that $V_{SD} > V_{OV}$ ($4 > 1$) corroborates saturation operation.

(**e**) By Eq. (3.32b), the threshold voltage is now

$$V_t = -1.0 - 0.73\left[\sqrt{4 + 2 \times 0.3} - \sqrt{2 \times 0.3}\right] = -2 \text{ V}$$

so the overdrive is $V_{OV} = 3 - |-2| = 1$ V. Since $V_{SD} > V_{OV}$ ($1.5 > 1$), the FET is now saturated, and

$$I_D = 0.2 \times 1^2(1 + 0.05 \times 1.5) = 0.215 \text{ mA}$$

(**f**) Assume operation in saturation, and then check. The required overdrive is such that

$$1 = 0.2 \times V_{OV}^2(1 + 0.05 \times 5)$$

or $V_{OV} = 2$ V. Since $V_{SD} > V_{OV}$ ($5 > 2$), the FET is indeed saturated. The required threshold voltage V_t must be such that $V_{OV} = V_{SG} - |V_t|$, or $2 = 3.5 - |V_t|$, or $|V_t| = 1.5$ V. As we know, V_t is negative, so we use Eq. (3.32b) to impose

$$-1.5 = -1.0 - 0.73\left[\sqrt{V_{BS} + 2 \times 0.3} - \sqrt{2 \times 0.3}\right]$$

This finally gives $V_{BS} = 1.53$ V.

Large-Signal Models for Saturated FETs

Figures 3.20 and 3.21 show the circuit models of the *n*MOSFET and the *p*MOSFET operating in saturation.

Also called *large-signal models* (to distinguish them from the small-signal models to be introduced later), they are used primarily in *dc calculations*. Since the gate is the plate of a capacitor, the G-S (or *input*) port appears as an *open circuit*, at least at

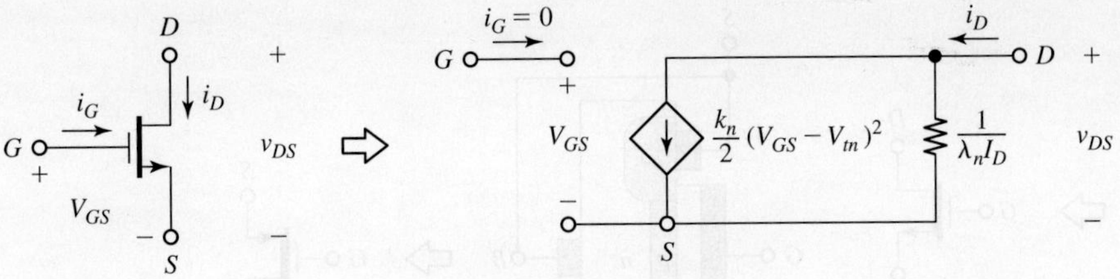

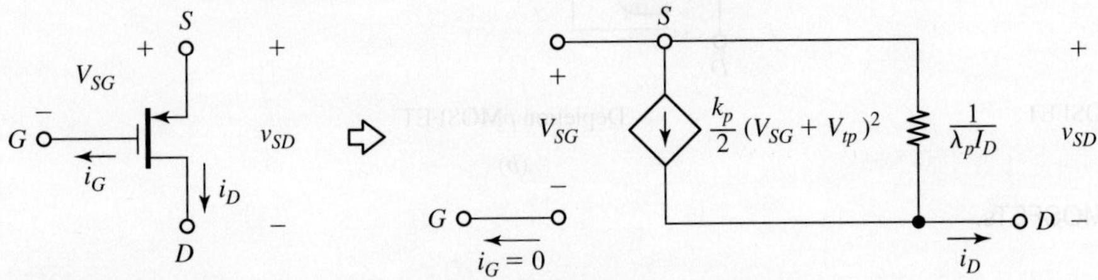

FIGURE 3.20 Large-signal model for the saturated *n*MOSFET.

FIGURE 3.21 Large-signal model for the saturated *p*MOSFET.

dc, so $i_G = 0$, and $i_S = i_D$. The D-S (or *output*) port is modeled with its *Norton equivalent* consisting of a dependent current source calculated at the edge of saturation, and an output resistance to model the slight increase of i_D with v_{DS} in *n*MOSFETs and v_{SD} in *p*MOSFETs.

To simplify dc calculations it is customary to ignore the output resistance, which is equivalent to assuming $\lambda = 0$. Also, for the *enhancement p*MOSFET, we shall express the dependent-source value in the form $(k_p/2)(V_{SG} - |V_{tp}|)^2$, more closely resembling that of the *n*MOSFET.

Depletion MOSFETs

As we know, enhancement-type FETs are normally off devices. To turn them on, we need to apply a suitable gate-source voltage exceeding the device's threshold voltage V_t. By contrast, depletion-type MOSFETs (or DFETs) are deliberately fabricated with a channel already present. As depicted in Fig. 3.22, an *n*DFET is created via a suitable *n*-type channel implant, and a *p*FET via a suitable *p*-type channel implant. For obvious reasons, DFETs are said to be *normally on* devices. In this case we ask ourselves what needs to be done to turn a DFET off. Keeping in mind that the gate-channel structure forms a parallel-plate capacitor, we make the following statements:

- To turn off the *n*DFET of Fig. 3.22*a* we need to induce *positive charges* in its channel region so as to neutralize the free electrons already present there. This requires inducing *negative* charge on the gate electrode. Consequently, we need to *lower* the gate voltage v_G *below* the source voltage v_S by a suitable amount, a condition expressed as $v_{GS} \leq V_{tn}$, where the threshold voltage V_{tn} is now *negative*. Thus, for $v_{GS} \leq V_{tn}$, $V_{tn} < 0$, the *n*DFET is off, while for $v_{GS} > V_{tn}$ it is on, and its conductivity is controlled by the overdrive voltage $V_{OVn} = V_{GS} - V_{tn}$, as usual.

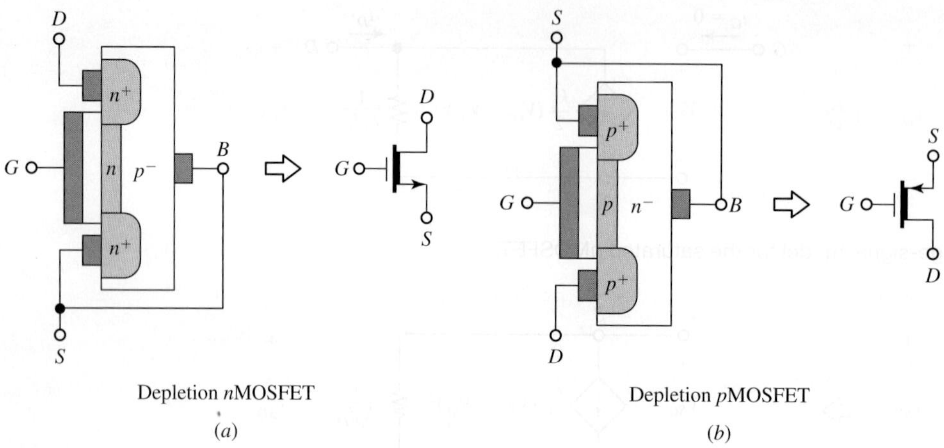

FIGURE 3.22 Depletion MOSFETs.

To reduce the possibility of confusion stemming from the negative threshold, the overdrive for an *n*DMOSFET is often expressed as $V_{OVn} = V_{GS} + |V_{tn}|$. Note, in particular, that with $V_{GS} = 0$ the device is conductive with $V_{OVn} = |V_{tn}|$.

- To turn off the *p*DFET of Fig. 3.22*b* we need to induce *negative charges* in its channel region so as to neutralize the free holes already present there. This requires inducing *positive* charge on the gate electrode. Consequently, we need to *raise* the gate voltage v_G *above* the source voltage v_S by a suitable amount, a condition expressed as $v_{GS} \geq V_{tp}$, where the threshold voltage V_{tp} is now *positive*. Thus, for $v_{GS} \geq V_{tp}$, $V_{tp} > 0$, the *p*DFET is off, while for $v_{GS} < V_{tp}$ it is on, and its conductivity is controlled by the overdrive voltage, which now is $V_{OVp} = V_{SG} + V_{tp}$. Note, in particular, that with $V_{SG} = 0$, the device is conductive with $V_{OVp} = V_{tp}$.

The beginner may feel confused by the different FET types and corresponding threshold-voltage polarites, and trying to memorize them may provide even more confusion. The best approach is to refer to the physical structures of Figs. 3.19 and 3.22, and ask oneself what kind of gate-to-source voltage is needed to turn the device on if it is a normally off type, or to turn it off if it is a normally on type.

The *i-v* characteristics of the depletion FET are similar to those of its enhancement counterpart, except for a shift along the v_{GS} axis. This is depicted in Fig. 3.23 for the case of an *n*DFET with $V_t = -1.5$ V, $k = 1$ mA/V², and $\lambda = 0.05$ V$^{1/2}$. Note that the curve of Fig. 3.23*a* is similar to that of Fig. 3.15, except that it is shifted to the *left* because now $V_t < 0$. Starting at $v_{GS} = 0$ V, we can either make the device *more conductive* by raising v_{GS} above 0 V, or make it *less conductive* by lowering v_{GS} *below* 0 V, until it shuts off completely once v_{GS} reaches V_t (= -1.5 V in our example). The effect of sweeping v_{GS} can be appreciated also from Fig. 3.23*b*, where we note that the curve corresponding to $v_{GS} = 0$ V is now somewhere in the middle. However, the locus of the pinchoff points is still $v_{DS} = V_{GS} - V_t$ (=$V_{GS} + 1.5$ V in our example.)

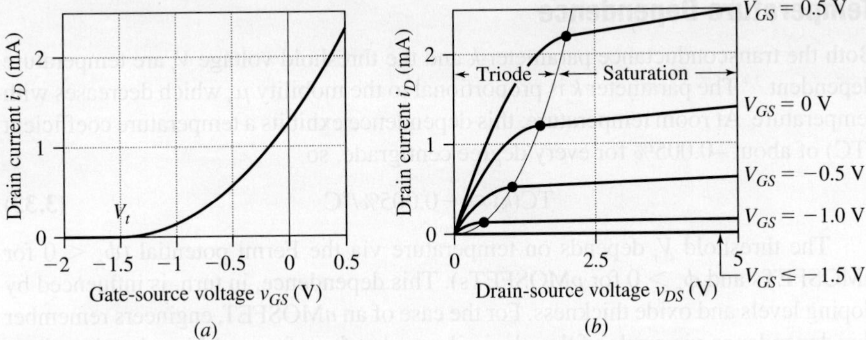

FIGURE 3.23 The *i-v* curves of a *depletion n*MOSFET with $V_t = -1.5$ V, $k = 1$ mA/V^2, and $\lambda = 0.05$ V$^{1/2}$.

A common DMOSFET application is illustrated in Fig. 3.24*a*, where the gate and source are tied together ($V_{GS} = 0$) to form a two-terminal, normally on device. Its *i-v* characteristic, shown in Fig. 3.24*b*, indicates that the device can be used either as a resistor, if operated near the origin, or as a current source or sink, if operated to the right of the edge of saturation (EOS). When operated as a current source or sink, the DFET finds application in the biasing of other FETs, such as amplifiers.

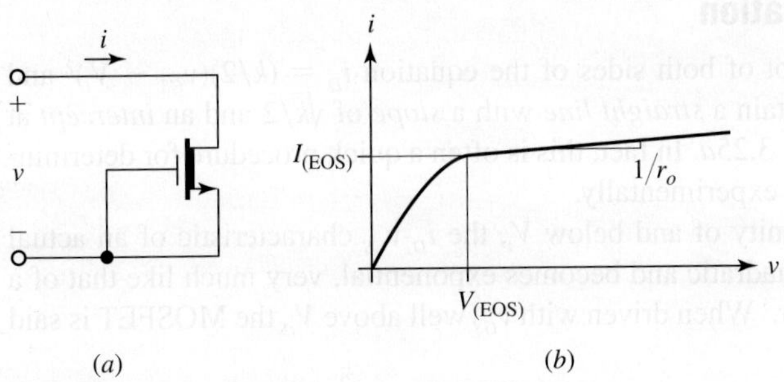

FIGURE 3.24 The depletion *n*MOSFET as a current source.

(a) Assuming the data of Fig. 3.23 for the DFET of Fig. 3.24, find the channel resistance r_{DS} in the ohmic region.   **EXAMPLE 3.13**

(b) What are the edge-of-saturation values $V_{(EOS)}$ and $I_{(EOS)}$?

(c) What is the output resistance r_o in the saturation region?

Solution

(a) With $V_{GS} = 0$ we have $V_{OV} = 0 - V_t = 0 - (-1.5) = 1.5$ V. So, $r_{DS} = 1/(kV_{OV}) = 1/(1 \times 1.5) = 667$ Ω.

(b) $V_{(EOS)} = V_{OV} = 1.5$ V. $I_{(EOS)} = (k/2)V^2_{(EOS)}(1 + \lambda V_{(EOS)}) = (1/2)(1.5)^2(1 + 0.05 \times 1.5) = 1.21$ mA.

(c) $r_o = 1/(\lambda I_{(EOS)}) = 1/(0.05 \times 1.21 \times 10^{-3}) = 16.5$ kΩ.

Temperature Dependence

Both the transconductance parameter k and the threshold voltage V_t are temperature dependent.[3,4] The parameter k is proportional to the mobility μ, which decreases with temperature. At room temperature, this dependence exhibits a temperature coefficient (TC) of about -0.005% for every degree centigrade, so

$$\text{TC}(k) \cong -0.005\%/°\text{C} \tag{3.33}$$

The threshold V_t depends on temperature via the Fermi potential ($\phi_p < 0$ for nMOSFETs and $\phi_n > 0$ for pMOSFETs). This dependence, in turn, is influenced by doping levels and oxide thickness. For the case of an nMOSFET, engineers remember this dependence via a rule of thumb similar to that for a forward-biased pn junction,

$$\text{TC}(V_{tn}) \cong -2 \text{ mV/°C} \tag{3.34}$$

For a pMOSFET, $\text{TC}(V_{tp}) \cong +2 \text{ mV/°C}$.

The two TCs have opposing effects, as a temperature decrease in k tends to reduce I_D, while a temperature decrease in V_t increases V_{OV} and thus tends to increase I_D. These opposing tendencies can be exploited on purpose to bias a FET at a point where they cancel each other out, resulting in a temperature-independent current I_D (see Problem 3.33)

Sub-Threshold Operation

If we take the square root of both sides of the equation $i_D = (k/2)(v_{GS} - V_t)^2$ and plot $\sqrt{i_D}$ versus v_{GS}, we obtain a *straight line* with a *slope* of $\sqrt{k/2}$ and an *intercept* at $v_{GS} = V_t$, as shown in Fig. 3.25a. In fact, this is often a quick procedure for determining the values of k and V_t experimentally.

However, in the vicinity of and below V_t, the i_D-v_{GS} characteristic of an actual MOSFET ceases to be quadratic and becomes exponential, very much like that of a bipolar junction transistor.[5] When driven with v_{GS} well above V_t, the MOSFET is said

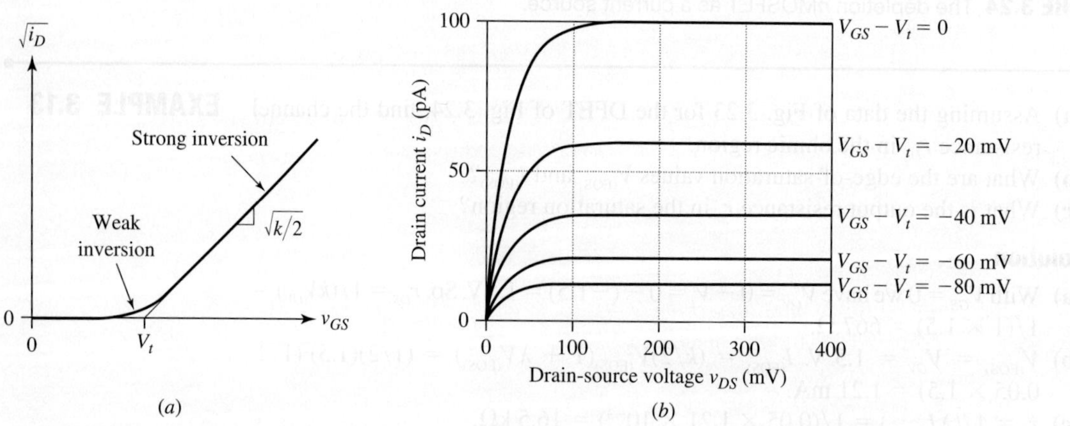

FIGURE 3.25 (a) Weak inversion, and (b) the i_D-v_{DS} characteristics in the sub-threshold region.

to be operating in *strong inversion*, but when confined to v_{GS} near or below V_t it is said to be operating in *weak inversion*, or also in the *sub-threshold* region. In weak inversion, the characteristic is expressed as[4,5]

$$i_D = \frac{W}{L} I_0 \exp\left(\frac{v_{GS} - V_t}{nV_T}\right)\left[1 - \exp\left(-\frac{v_{DS}}{V_T}\right)\right] \tag{3.35}$$

where W and L are the channel width and length; I_0 is a suitable scaling factor, typically on the order of 1 μA or less; V_T is the familiar thermal voltage; and n is a suitable factor, typically $1 < n < 3$. Figure 3.25b shows the i_D-v_{DS} characteristics for different values of $v_{GS} - V_t$. Sub-threshold operation finds application in very low power circuits with limited frequency bandwidths. As we proceed, we shall assume operation in strong inversion.

3.5 MOSFETS IN RESISTIVE Dc CIRCUITS

We now wish to familiarize ourselves with the behavior of a MOSFET when embedded in a resistive circuit. We find it convenient to start out with simple dc circuits, and to assume $\lambda = 0$ both to speed up our calculations and to focus on the essentials of MOSFET behavior.

As we know, to sustain a conductive channel, a FET requires a suitable overdrive voltage V_{OV}. For an nMOSFET we have $V_{OVn} = V_{GS} - V_{tn}$, where $V_{tn} > 0$ for an enhancement type and $V_{tn} < 0$ for a depletion type. For a pMOSFET we have $V_{OVp} = V_{SG} + V_{tp}$, where $V_{tp} < 0$ for an enhancement type and $V_{tp} > 0$ for a depletion type. Once a channel is conductive, the question arises as to whether operation is in saturation or in the triode region. In certain cases, such as the diode mode, the operating region is evident by inspection. When it isn't, we need to determine (and verify!) it via suitable calculations. As already outlined in Section 3 for the nMOSFET case, one way to proceed is as follows:

- Assume the FET is saturated, and utilize the expression $I_D = (k/2)V_{OV}^2$ to carry out the necessary calculations until the values of both V_{OV} and V_{DS} (or V_{SD} for the pMOSFET case) are known.
- If it turns out that $V_{DS} > V_{OV}$ (or $V_{SD} > V_{OV}$ for the pMOSFET case), then the FET is indeed saturated.
- Otherwise you get some kind of contradiction, such as a physically unacceptable result, indicating that our assumption was wrong, and that the FET is in the triode region.

Diode-Connected MOSFETs

Figure 3.26a shows the resistive biasing of a diode-connected nMOSFET. As we know, for $v \leq V_{tn}$ the FET is in *cutoff*, giving $i = 0$. For $v > V_{tn}$, the FET is *saturated*, giving the quadratic i-v characteristic

$$i = \frac{k_n}{2}(v - V_{tn})^2 \tag{3.36a}$$

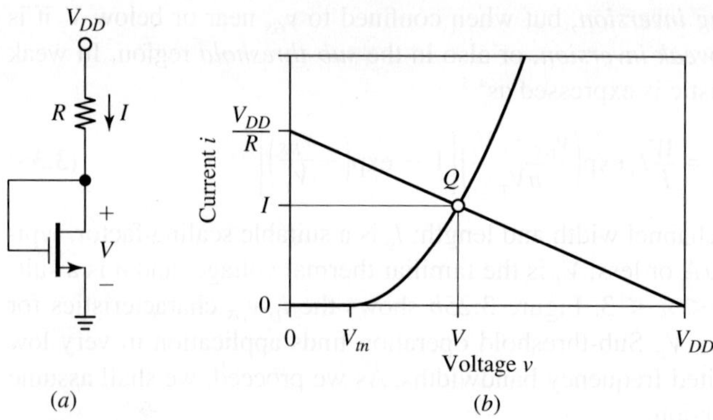

FIGURE 3.26 Investigating the dc operation of a diode-connected *n*MOSFET.

The *i-v* characteristic of the circuit external to the FET is the straight line

$$i = \frac{V_{DD} - v}{R} \tag{3.36b}$$

The two curves intercept at a common point called the *operating point*, or also the *quiescent point Q*. To find its abscissa V, we simply equate the two currents,

$$\frac{V_{DD} - V}{R} = \frac{k_n}{2}(V - V_{tn})^2 \tag{3.37}$$

The resulting quadratic equation is readily solved for V. Of the two solutions, only one makes physical sense, so the other must be discarded. We then substitute the physically correct value of V into either of Eq. (3.36) to find I. This is a common situation arising when we deal with dc circuits incorporating FETs.

EXAMPLE 3.14 In the circuit of Fig. 3.26a let $V_{DD} = 7$ V and $R = 10$ kΩ, and let the FET have $V_{tn} = 1.0$ V and $k_n = 0.2$ mA/V^2. Assuming $\lambda_n = 0$, find V and I.

Solution

With resistance in kΩ and current in mA, Eq. (3.37) gives

$$\frac{7 - V}{10} = \frac{0.2}{2}(V - 1)^2$$

which results in the following quadratic equation

$$V^2 - V - 6 = 0.$$

Its solutions are

$$V = \frac{1 \pm \sqrt{1 + 24}}{2}$$

or $V = +3$ V and $V = -2$ V. The second value makes no physical sense, as it would imply $V_{GS} < V_{tn}$, and thus a FET in cutoff, when in fact we know that the FET is on. The first value implies the overdrive voltage $V_{OV} = 3 - 1 = 2$ V. The resulting current is $(7 - 3)/10 = 0.4$ mA.

(a) In the circuit of Example 3.14, find the value of R needed to give $I = 0.9$ mA.   **EXAMPLE 3.15**
(b) What happens if R is shorted out ($R \to 0$)?

Solution

(a) We use Eq. (3.36a) to impose

$$0.9 = \frac{0.2}{2}(V - 1)^2$$

The solutions are $V = +4$ V and -2 V, but only $V = 4$ V is acceptable, so $R = (7 - 4)/0.9 = 3.33$ kΩ.

(b) With $R = 0$, we have $V = 7$ V, and $I = (0.2/2) \times (7-1)^2 = 3.6$ mA.

Figure 3.27 shows the dual situation of Fig. 3.26, involving a pMOSFET but with the source referenced to V_{DD} instead of ground. Considering that $v_{SG} = V_{DD} - v$, the i-v characteristic of the combination consisting of the FET and the V_{DD} source is $i = 0$ for $v \geq V_{DD} - |V_{tp}|$, and the quadratic curve

$$i = \frac{k_p}{2}(V_{DD} - v - |V_{tp}|)^2 \qquad (3.38a)$$

for $v < V_{DD} - |V_{tp}|$. Compared to Fig. 3.26b, the quadratic curve is now *folded* about the vertical axis, and *shifted* along the v-axis so that conduction starts at $v = V_{DD} - |V_{tp}|$. The i-v characteristic of the resistor is the straight line

$$i = \frac{v}{R} \qquad (3.38b)$$

The abscissa of the *operating point Q* is readily found by solving the quadratic equation in V

$$\frac{V}{R} = \frac{k_p}{2}(V_{DD} - V - |V_{tp}|)^2 \qquad (3.39)$$

and retaining only the physically acceptable solution, as illustrated in the following example.

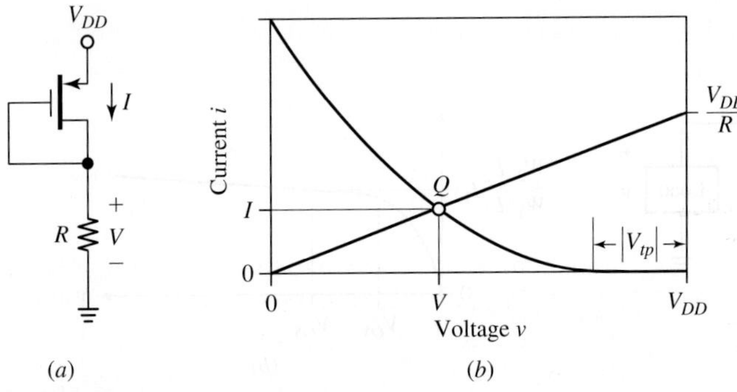

FIGURE 3.27 Investigating the dc operation of a diode-connected pMOSFET.

EXAMPLE 3.16

(a) In the circuit of Fig. 3.27a let $V_{DD} = 5$ V, $V_{tp} = -1.0$ V, and $k_p = 0.5$ mA/V². Assuming $\lambda_p = 0$, find V and I if $R = 2.0$ kΩ.

(b) Repeat, but with $R = 0$ (short).

Solution

(a) With resistance in kΩ and current in mA, Eq. (3.39) gives

$$\frac{V}{2.0} = \frac{0.5}{2}(5 - V - 1)^2$$

whose solutions are $V = 2$ V, and $V = 8$ V (physically impossible). Then, $I = 2/2.0 = 1.0$ mA.

(b) With $R = 0$ we get $V = 0$, so $I = (0.5/2)(5 - 1)^2 = 4$ mA.

Current Mirrors

A common application of a diode-connected FET, especially in integrated-circuit (IC) design, is the generation of a suitable gate-source voltage drop to bias another similar FET (or FETs). Owing to the fact that when fabricated in IC form, different FETs of the same type enjoy a high degree of matching in the values of V_t, k', and λ, both analysis and design are simplified considerably.

Figure 3.28a shows an example involving matched nMOSFETs. Denoting their common gate-source drive as V_{GS}, the current drawn by M_1 is

$$I_1 = \frac{1}{2}k'\frac{W_1}{L_1}(V_{GS} - V_t)^2(1 + \lambda V_{GS})$$

For $v = V_{GS}$, M_2 operates under the *same* voltage conditions as M_1, giving

$$I_2 = \frac{1}{2}k'\frac{W_2}{L_2}(V_{GS} - V_t)^2(1 + \lambda V_{GS})$$

Combining the two equations, we easily obtain, for $v = V_{GS}$,

$$I_2 = \frac{W_2/L_2}{W_1/L_1}I_1 \tag{3.40}$$

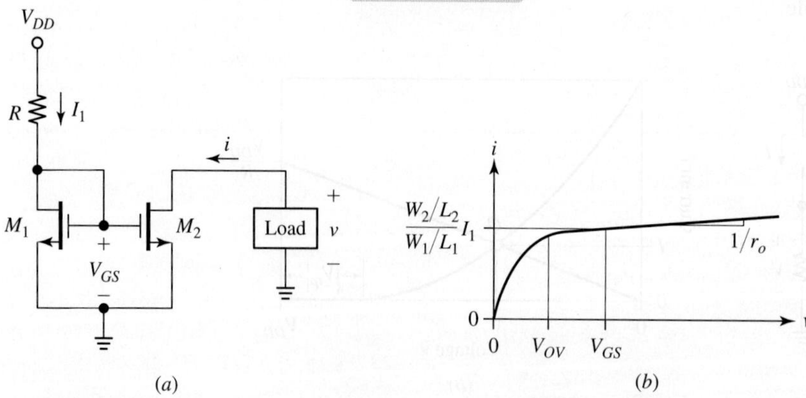

FIGURE 3.28 nMOSFET current mirror, and its i-v characteristic.

Of particular interest is the case of devices with identical W/L ratios, for then Eq. (3.40) gives, for $v = V_{GS}$, $I_2 = I_1$, indicating that the current of M_2 will *mirror* that of M_1, this being the reason for the circuit's name.

In the example shown, I_1 is established by R in a manner similar to that of Example 3.14. However, there are also other techniques for forcing current through M_1, but M_2 will just mirror the behavior of M_1 regardless of the technique. In IC design we have the flexibility of establishing virtually any ratio between the two currents by suitably specifying the W/L ratios of the two devices. For instance, if $W_1/L_1 = (1\ \mu\text{m})/(1\ \mu\text{m})$, then, specifying $W_2/L_2 = (2\ \mu\text{m})/(1\ \mu\text{m})$ we get, for $v = V_{GS}$, $I_2 = 2I_1$. Likewise, with $W_2/L_2 = (1\ \mu\text{m})/(2\ \mu\text{m})$ we get $I_2 = 0.5I_1$, whereas with $W_2/L_2 = (1\ \mu\text{m})/(1\ \mu\text{m})$ we get $I_2 = I_1$.

The *i-v* characteristic of the current mirror is shown in Fig. 3.28*b*. As we know, the saturation portion of the curve exhibits a slope of $1/r_o$. Moreover, the saturation region extends all the way *down* to $v = V_{OV} = V_{GS} - V_t$.

Assume $V_{DD} = 5$ V and identical devices with $V_t = 1.0$ V, $k = 0.8$ mA/V², and $\lambda = 0.02$ V^{-1} in Fig. 3.28*a*.   **EXAMPLE 3.17**

(a) Find R so that the circuit gives $i = 100\ \mu$A for $v = V_{GS}$. Here, assume $\lambda = 0$ for simplicity.
(b) Find the lower voltage limit for operation in saturation, as well as r_o and the per-volt change of i in the saturation region.

Solution

(a) The overdrive voltage required to sustain the given current is

$$V_{OV} = \sqrt{\frac{2I_1}{k}} = \sqrt{\frac{2 \times 100 \times 10^{-6}}{0.8 \times 10^{-3}}} = 0.5 \text{ V}$$

Consequently, $V_{GS} = V_t + V_{OV} = 1 + 0.5 = 1.5$ V, and $R = (V_{DD} - V_{GS})/I_1 = (5 - 1.5)/0.1 = 35$ kΩ.

(b) The saturation region extends over the range $v \geq 0.5$ V, where $r_o \cong 1/(\lambda I) = 1/(0.02 \times 100 \times 10^{-6}) = 500$ kΩ, and where i increases with v at the rate of $1/(500$ k$\Omega) = 2\ \mu$A/V.

The current mirror of Fig. 3.28*a* is also referred to as a *current sink* because M_2 sinks current from the load. By contrast, its *p*MOSFET counterpart of Fig. 3.29*a* is referred to as a *current source*, as M_2 in this case sources current to the load. Using similar reasoning we conclude that for $v = V_{SS} - V_{SG}$ the two transistors operate under identical voltage conditions, so M_2 gives $I_2 = [(W_2/L_2)/(W_1/L_1)]I_1$. As depicted in Fig. 3.29*b*, the saturation region extends all the way up to $V_{DD} - V_{OV}$, and the slope of the saturation-region curve is now $-1/r_o$.

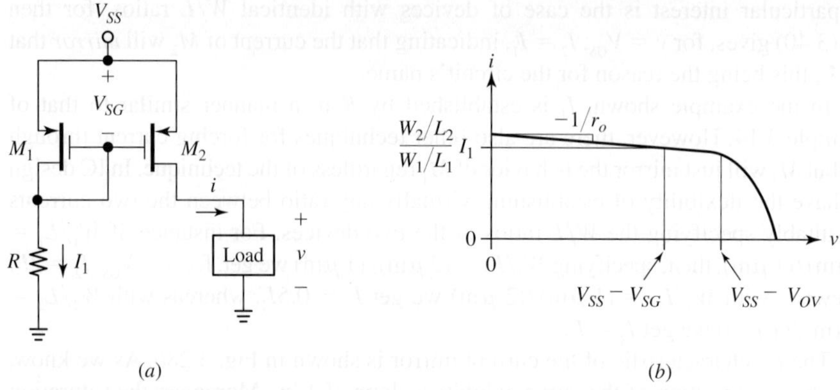

FIGURE 3.29 *p*MOSFET current mirror, and its *i-v* characteristic.

EXAMPLE 3.18

(a) In the circuit of Fig. 3.29*a* let $V_{SS} = 6$ V, and let the FETs have $V_{t2} = V_{t1} = -1.5$ V, $k_2 = 2k_1 = 0.5$ A/V^2, and $\lambda_2 = \lambda_1 = 0.04$ V^{-1}. Find R for $V_{OV} = 2$ V. What is the corresponding value of I_1?

(b) Find i at $v = V_{SS} - V_{SG}$, $v = 0$, and $v = V_{SS} - V_{OV}$.

Solution

(a) With $V_{OV} = 2$ V we have $V_{SD1} = V_{SG} = V_{OV} + |V_t| = 2 + 1.5 = 3.5$ V. Since $k_1 = 0.25$ mA/V^2,

$$I_1 = \frac{0.25}{2}2^2(1 + 0.04 \times 3.5) = 0.57 \text{ mA}$$

Moreover, $R = (V_{SS} - V_{SG})/I_1 = (6 - 3.5)/0.57 = 4.39$ kΩ.

(b) For $v = 6 - 3.5 = 2.5$ V we get $i = (k_2/k_1)I_1 = 2 \times 0.57 = 1.14$ mA. For $v = 0$ we have $V_{SD2} = 6$ V, so

$$i = \frac{0.5}{2}2^2(1 + 0.04 \times 6) = 1.24 \text{ mA}$$

Likewise, for $v = 6 - 2 = 4$ V we have $V_{SD2} = 2$ V, so $i = 1(1 + 0.04 \times 2) = 1.08$ mA.

Resistive Biasing of MOSFETs—Dual-Supply Schemes

When studying FET amplifiers, we shall find it necessary to bias a FET at a specific dc operating point $Q = Q(I_D, V_{DS})$ in the *saturation region*, a region also called the *active region*. FET biasing can be dealt with from either a *design* or an *analysis* viewpoint. The objective of design is to devise a suitable external circuit to bias the FET at the specified operating point Q. Conversely, the objective of analysis is to find the operating point Q, given the circuit in which the FET is embedded. The diode-connected FETs examined above have already provided examples of dc biasing in saturation.

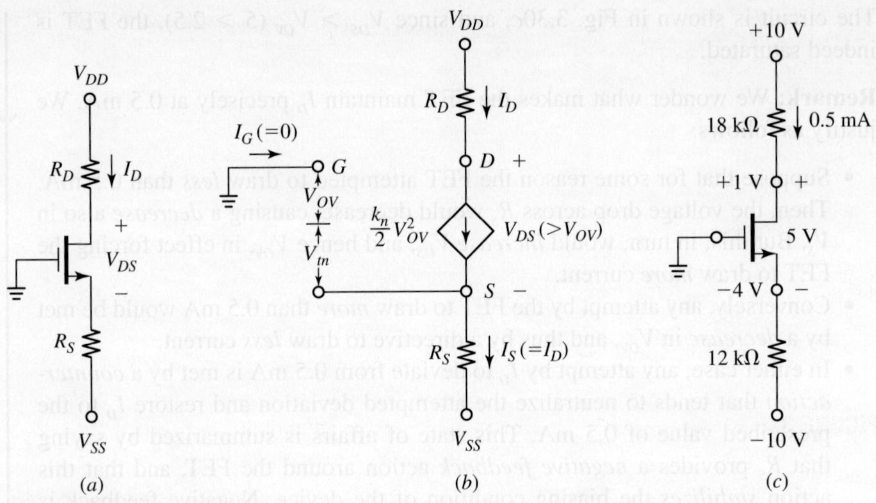

FIGURE 3.30 *Dual-supply biasing* of an *n*MOSFET in the active region. (a) General circuit, (b) its dc equivalent, and (c) an actual example, showing all voltages and currents.

The biasing scheme of Fig. 3.30a, based on a dual power-supply system, utilizes the resistance R_S to establish the value of I_D, and the resistance R_D to establish the value of V_{DS}. To better understand circuit operation, replace the *n*FET with its dc equivalent as in Fig. 3.30b, where for simplicity we are assuming $\lambda_n = 0$ and, hence, $r_o = \infty$. As we know, to draw a given current I_D, the *n*FET requires the overdrive voltage $V_{OV} = \sqrt{2I_D/k_n}$, indicating that in the present biasing scheme the source must be held at $V_S = -(V_{tn} + V_{OV})$. This task is performed by R_S, which is chosen on the basis of dropping the voltage difference $(V_S - V_{SS})$ at the given current I_D. An example will better illustrate.

EXAMPLE 3.19

In the circuit of Fig. 3.30a let $V_{DD} = 10$ V and $V_{SS} = -10$ V, and let the FET have $V_{tn} = 1.5$ V and $k_n = 0.16$ mA/V². Assuming $\lambda_n = 0$, specify values for R_S and R_D to bias the FET at $I_D = 0.5$ mA and $V_{DS} = 5$ V.

Solution

The required overdrive voltage is $V_{OV} = \sqrt{2I_D/k_n} = \sqrt{2 \times 0.5/16} = 2.5$ V, indicating that the source must be held at $V_S = -(V_{tn} + V_{OV}) = -(1.5 + 2.5) = -4$ V. This requires

$$R_S = \frac{V_S - V_{SS}}{I_D} = \frac{-4 - (-10)}{0.5} = 12 \text{ k}\Omega$$

To ensure $V_{DS} = 5$ V, the drain voltage must be held at $V_D = V_S + V_{DS} = -4 + 5 = +1$ V. Consequently,

$$R_D = \frac{V_{DD} - V_D}{I_D} = \frac{10 - 1}{0.5} = 18 \text{ k}\Omega$$

The circuit is shown in Fig. 3.30c, and since $V_{DS} > V_{OV}$ (5 > 2.5), the FET is indeed saturated.

Remark: We wonder what makes the FET maintain I_D precisely at 0.5 mA. We justify as follows:

- Suppose that for some reason the FET attempted to draw *less* than 0.5 mA. Then, the voltage drop across R_S would decrease, causing a *decrease* also in V_S. But this, in turn, would *increase* V_{GS}, and hence V_{OV}, in effect forcing the FET to draw *more* current.
- Conversely, any attempt by the FET to draw *more* than 0.5 mA would be met by a *decrease* in V_{OV}, and thus by a directive to draw *less* current.
- In either case, any attempt by I_D to deviate from 0.5 mA is met by a *counteraction* that tends to neutralize the attempted deviation and restore I_D to the prescribed value of 0.5 mA. This state of affairs is summarized by saying that R_S provides a *negative feedback* action around the FET, and that this action *stabilizes* the biasing condition of the device. Negative feedback is also referred to as degenerative feedback, to distinguish it from positive feedback, which may become regenerative. Consequently, R_S is also referred to as *degeneration resistance* (more on this in Chapter 7).

EXAMPLE 3.20

(a) In the circuit of Fig. 3.31a let the FET have $V_{tn} = 1.0$ V, $k_n = 0.5$ mA/V^2. Assuming $\lambda_n = 0$, find the FET's operating point Q.

(b) To what value must we increase R_D to bring the FET to operate at the edge of saturation (EOS)?

(c) What happens if R_D is raised to twice the value found in part (b)?

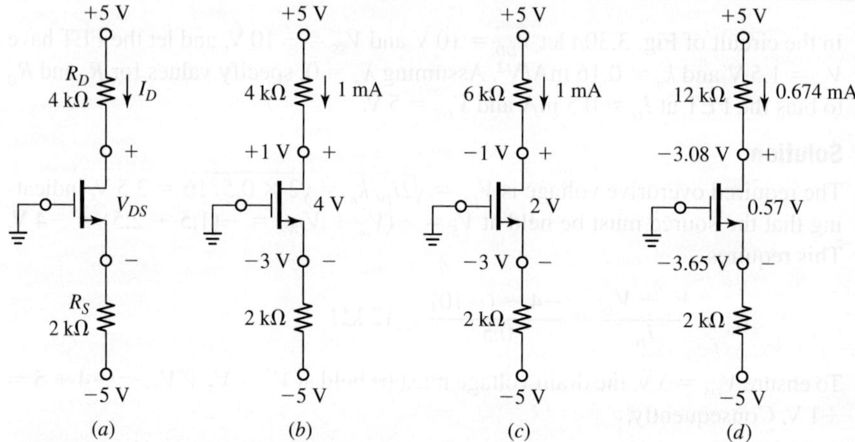

FIGURE 3.31 (a) Circuit of Example 3.20, and operation (b) in *saturation*, (c) at the EOS, and (d) in the *triode* region.

Solution

(a) Following the directive at the beginning of this section, we start out by assuming a saturated FET, and we check afterwards whether the assumption was correct. If the assumption proves wrong, it simply means that the FET is in the triode region, and that we must redo our calculations using the triode-region expression for I_D. In saturation we have

$$I_D = \frac{k_n}{2}(V_{GS} - V_{tn})^2 = \frac{k_n}{2}(V_G - V_S - V_{tn})^2 = \frac{k_n}{2}(0 - V_S - V_{tn})^2$$

$$= \frac{0.5}{2}[-(-5 + 2 \times I_D) - 1]^2 = \frac{1}{4}(4 - 2I_D)^2$$

After expanding, collecting, and simplifying, we end up with the quadratic equation

$$I_D^2 - 5I_D + 4 = 0$$

whose solutions are $I_D = 1$ mA and $I_D = 4$ mA (physically impossible). We finally get $V_D = 5 - 4 \times 1 = 1$ V and $V_S = -5 + 2 \times 1 = -3$ V. Consequently, $V_{DS} = 1 - (-3) = 4$ V, and $V_{OV} = -(-3) - 1 = 2$ V. Since $V_{DS} > V_{OV}$ (4 > 2), the FET is indeed in saturation. Summarizing, the operating point is $Q = Q(1$ mA, 4 V). The situation is shown in Fig. 3.31b.

(b) Raising R_D lowers V_D, in turn reducing V_{DS}. The EOS is reached when $V_{DS} = V_{OV} = 2$ V, or $V_D = V_S + V_{OV} = -3 + 2 = -1$ V. The corresponding resistance value is $R_D = [5 - (-1)]/1 = 6$ kΩ. The situation is shown in Fig. 3.31c, where the operating point is now $Q = Q(1$ mA, 2 V).

(c) With $R_D = 2 \times 6 = 12$ kΩ the FET will definitely be in the triode region. We can convince ourselves by noting that if it were still in saturation, we would have $V_D = 5 - 12 \times 1 = -7$ V, thus implying $V_{DS} = V_D - V_S = -7 - (-3) = -4$ V, an absurdity! Clearly, we must re-compute I_D, but using the triode expression. In part (a) we have found that

$$V_{GS} - V_{tn} = 4 - 2I_D$$

Moreover, we have $V_{DS} = V_D - V_S = (5 - 12 \times I_D) - (-5 + 2 \times I_D)$ or

$$V_{DS} = 10 - 14I_D$$

Consequently,

$$I_D = k_n\left[(V_{GS} - V_{tn})V_{DS} - \frac{V_{DS}^2}{2}\right] = 0.5\left[(4 - 2I_D)(10 - 14I_D) - \frac{(10 - 14I_D)^2}{2}\right]$$

Collecting terms and solving the ensuing quadratic equation we get

$$I_D = \frac{31 \pm \sqrt{261}}{70}$$

or $I_D = 0.674$ mA, and $I_D = 0.212$ mA (physically impossible—can you tell why?). The various voltages are easily found as $V_D = -3.08$ V, $V_S = -3.65$ V, and $V_{DS} = 0.57$ V. The situation is summarized in Fig. 3.31d, where the operating point is now $Q = Q(0.674$ mA, 0.57 V). Finally, we find $V_{OV} = 2.65$ V. The fact that $V_{DS} < V_{OV}$ (0.57 < 2.65) confirms that the FET is indeed operating in the triode region.

Figure 3.32 shows the *p*MOSFET counterpart of the *n*MOSFET circuit of Fig. 3.31. Again, this circuit can be investigated from either a design or an analysis standpoint. The procedure is similar to that of the *n*MOSFET, as long as we handle voltages and currents in dual fashion.

EXAMPLE 3.21

(a) In the circuit of Fig. 3.32*a* let $V_{SS} = 6$ V and $V_{DD} = -6$ V, and let the FET have $V_{tp} = -1.5$ V and $k_p = 0.5$ mA/V^2. Assuming $\lambda_p = 0$, specify suitable values for R_S and R_D to bias the FET at $I_D = 0.25$ mA and $V_{SD} = 4$ V. What region is the FET operating in?

(b) Find the FET's operating point and region if $R_S = 6$ kΩ and $R_D = 16$ kΩ.

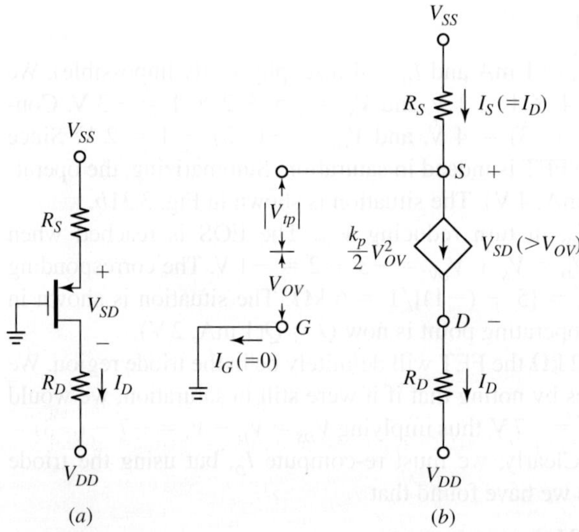

FIGURE 3.32 Dual-supply biasing of a *p*MOSFET in the saturation (or active) region.

Solution

(a) Assuming saturation, and retracing dual steps of those of Example 3.19, we find

$$V_{OV} = \sqrt{2I_D/k_p} = \sqrt{2 \times 0.25/0.5} = 1 \text{ V}$$

$$V_S = V_{OV} + |V_{tp}| = 1 + 1.5 = 2.5 \text{ V}$$

$$R_S = (V_{SS} - V_S)/I_D = (6 - 2.5)/0.25 = 14 \text{ k}\Omega$$

$$V_D = V_S - V_{SD} = 2.5 - 4 = -1.5 \text{ V}$$

$$R_D = (V_D - V_{DD})/I_D = [-1.5 - (-6)]/0.25 = 18 \text{ k}\Omega$$

The circuit is shown in Fig. 3.33*a*. The fact that $V_{SD} > V_{OV}$ (4 > 1) confirms saturation-region operation.

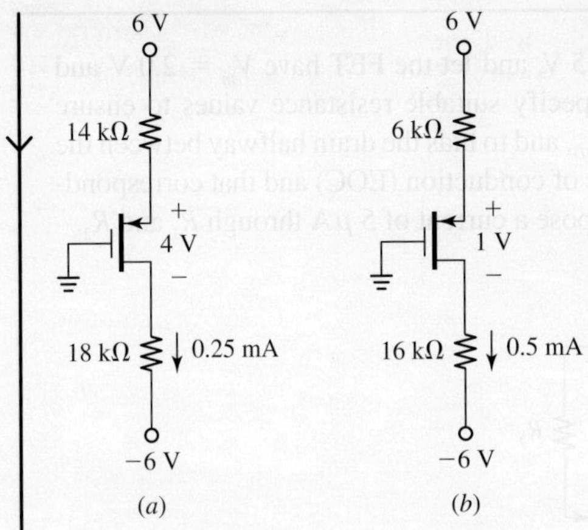

FIGURE 3.33 Circuits of Example 3.21.

(b) Assume saturation again, and check. We have

$$V_{SG} - |V_{tp}| = V_{SS} - R_S I_D - |V_{tp}| = 6 - 6I_D - 1.5 = 4.5 - 6I_D$$

Consequently, in saturation we would have

$$I_D = \frac{0.5}{2}(4.5 - 6I_D)^2$$

This quadratic equation admits the solutions $I_D = 1.1$ mA and $I_D = 0.51$ mA, both of which are untenable because the first would imply $V_{SG} < 0$ (cutoff), while the second would imply $V_{SD} < V_{OV}$ (triode), both of which contradict the saturation assumption. Evidently the FET is in the triode region. Considering that

$$V_{SD} = (V_{SS} - V_{DD}) - (R_D I_D - R_S I_D) = 12 - 22I_D$$

we have, in the triode region,

$$I_D = 0.5[(4.5 - 6I_D) \times (12 - 22I_D) - (12 - 22I_D)^2/2]$$

This quadratic equation admits two solutions, with the physically acceptable one being $I_D = 0.5$ mA. Consequently, $V_{SD} = 12 - 22 \times 0.5 = 1$ V, indicating the operating point $Q = Q(0.5$ mA, 1 V). The circuit is shown in Fig. 3.33b. As a check, we calculate $V_{OV} = 4.5 - 6 \times 0.5 = 1.5$ V, and confirm that $V_{SD} < V_{OV}$ (1 < 1.5), that is, triode-region operation.

Resistive Biasing of MOSFETs—Single-Supply Schemes

Figure 3.34 shows single-supply MOSFET biasing alternatives. The function of the voltage divider R_1-R_2 is to establish an intermediate bias voltage for the gate. Since the gate draws zero dc current, these resistances can be chosen to be fairly large if desired, say, in the MΩ-range.

EXAMPLE 3.22   In the circuit of Fig. 3.34*a* let $V_{DD} = 15$ V, and let the FET have $V_{tn} = 2.0$ V and $k_n = 0.5$ mA/V^2. Assuming $\lambda_n = 0$, specify suitable resistance values to ensure $I_D = 1$ mA, to bias the source at $(1/3)V_{DD}$, and to bias the drain halfway between the drain voltage corresponding to the edge of conduction (EOC) and that corresponding to the edge of saturation (EOS). Impose a current of 5 μA through R_1 and R_2.

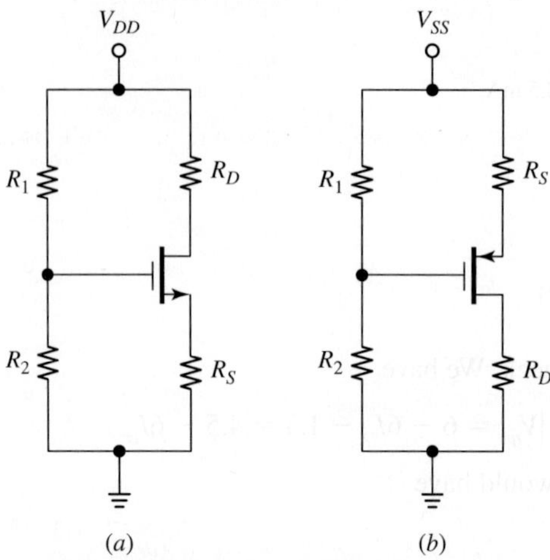

FIGURE 3.34 Single-supply biasing for (a) the *n*MOSFET and (b) the *p*MOSFET.

Solution

The required overdrive is $V_{OV} = \sqrt{2I_D/k_n} = \sqrt{2 \times 1/0.5} = 2.0$ V, so $V_{GS} = V_{tn} + V_{OV} = 2 + 2 = 4$ V. We thus have

$$V_S = (1/3)V_{DD} = (1/3)15 = 5 \text{ V}$$

$$V_G = V_S + V_{GS} = 5 + 4 = 9 \text{ V}$$

Consequently, $R_1 = (15 - 9)/5 = 1.2$ MΩ, $R_2 = 9/5 = 1.8$ MΩ, and $R_S = 5/1 = 5$ kΩ. Now, the value of V_D corresponding to the EOC is $V_{D(EOC)} = V_{DD} = 15$ V, and that corresponding to the EOS is $V_{D(EOS)} = V_S + V_{OV} = 5 + 2 = 7$ V. The halfway value is thus

$$V_D = (15 + 7)/2 = 11 \text{ V}$$

Finally, $R_D = (15 - 11)/1 = 4$ kΩ.

EXAMPLE 3.23   In the circuit of Fig. 3.34*a* let $V_{DD} = 12$ V, $R_1 = 1$ MΩ, $R_2 = 1.4$ MΩ, $R_D = 8$ kΩ, and $R_S = 10$ kΩ, and let the FET have $V_{tn} = 1.0$ V and $k_n = 0.4$ mA/V^2. Assuming $\lambda_n = 0$, find the operating point Q.

Solution

Start out by assuming the FET is in saturation, and check later that the assumption was correct. We have

$$V_{GS} = V_G - V_S = \frac{R_2}{R_1 + R_2}V_{DD} - R_S I_S = \frac{1.4}{1 + 1.4}12 - 10I_S = 7 - 10I_D$$

and

$$I_D = \frac{k_n}{2}(V_{GS} - V_{tn})^2 = \frac{0.4}{2}(7 - 10I_D - 1)^2$$

This quadratic equation admits the solutions $I_D = 0.45$ mA and $I_D = 0.8$ mA (physically impossible—can you tell why?). Consequently,

$$V_S = R_S I_S = R_S I_D = 10 \times 0.45 = 4.5 \text{ V}$$

$$V_D = V_{DD} - R_D I_D = 12 - 8 \times 0.45 = 8.4 \text{ V}$$

$$V_{DS} = V_D - V_S = 8.4 - 4.5 = 3.9 \text{ V}$$

Considering that the overdrive voltage is $V_{OV} = V_{GS} - V_t = 7 - 4.5 - 1 = 1.5$ V, it follows that $V_{DS} > V_{OV}$ (3.9 > 1.5), confirming that our initial assumption of a saturated FET was indeed correct.

In the circuit of Fig. 3.34b let $V_{SS} = 10$ V, $R_1 = 1.8$ MΩ, $R_2 = 2.2$ MΩ, $R_D = 10$ kΩ, and $R_S = 7.5$ kΩ, and let the FET have $V_{tp} = -0.5$ V and $k_p = 0.8$ mA/V^2. Assuming $\lambda_p = 0$, find the operating point Q.

EXAMPLE 3.24

Solution

Assume a saturated FET, and then check whether the assumption is correct. We have

$$V_{SG} = V_S - V_G = (V_{DD} - R_S I_S) - \frac{R_2}{R_1 + R_2}V_{DD} = (10 - 7.5I_S) - 5.5$$

$$= 4.5 - 7.5I_D$$

and

$$I_D = \frac{k_p}{2}(V_{SG} - |V_{tp}|)^2 = \frac{0.8}{2}(4.5 - 7.5I_D - 0.5)^2$$

This quadratic equation admits the solutions $I_D = 0.4$ mA and $I_D = 0.711$ mA (physically impossible). Consequently, $V_S = V_{DD} - R_S I_S = 10 - 7.5 \times 0.4 = 7$ V, $V_D = R_D I_D = 10 \times 0.4 = 4$ V, and $V_{SD} = V_S - V_D = 7 - 4 = 3$ V. Considering that the overdrive voltage is $V_{OV} = V_{SG} - |V_{tp}| = 1$ V, it follows that $V_{SD} > V_{OV}$ (3 > 1), thus confirming a saturated FET. The operating point is $Q(I_D, V_{SD}) = Q(0.4 \text{ mA}, 3 \text{ V})$.

EXAMPLE 3.25   In the circuit of Fig. 3.35 specify R_1 and R_2 so that $V_{DSn} = 4$ V under the constraint that the smaller of R_1 and R_2 be 1 MΩ. Show all voltages and currents in your circuit.

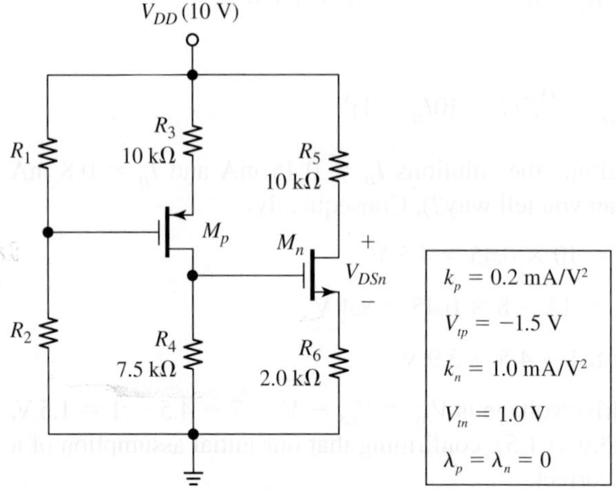

FIGURE 3.35 Circuit of Example 3.25.

Solution

Start out at V_{DSn} and work your way back to R_1 and R_2, one step at a time. The numerical result of each step is identified by the corresponding step number in Fig. 3.36.

1. By KVL and Ohm's law, M_n's current I_{Dn} is such that $V_{DD} = R_5 I_{Dn} + V_{DSn} + R_6 I_{Dn}$, or $10 = 10 I_{Dn} + 4 + 2 I_{Dn}$. This gives $I_{Dn} = 0.5$ mA.
2. By Ohm's law, M_n's source voltage is $V_{Sn} = R_6 I_{Dn} = 2 \times 0.5 = 1$ V.
3. Assume M_n is saturated. Then, the overdrive needed to sustain 0.5 mA is $V_{OVn} = \sqrt{2 \times 0.5/1} = 1$ V. This is less than V_{DSn} ($= 4$ V), so the FET is saturated. We have $V_{GSn} = V_{tn} + V_{OVn} = 1 + 1 = 2$ V.
4. By KVL, M_n's gate voltage is $V_{Gn} = V_{Sn} + V_{GSn} = 1 + 2 = 3$ V. This is also M_p's drain voltage V_{Dp}.
5. M_p's drain current is $I_{Dp} = V_{Dp}/R_4 = 3/7.5 = 0.4$ mA.
6. By KVL and Ohm's law, M_p's source voltage is $V_{Sp} = V_{DD} - R_3 I_{Dp} = 10 - 10 \times 0.4 = 6$ V.
7. By KVL, M_p's source-drain voltage drop is $V_{SDp} = V_{Sp} - V_{Dp} = 6 - 3 = 3$ V.
8. Assume M_p is saturated. Then, the overdrive needed to sustain 0.4 mA is $V_{OVp} = \sqrt{2 \times 0.4/0.2} = 2$ V. This is less than V_{SDp} ($= 3$ V), so the FET is saturated. We have $V_{SGp} = |V_{tp}| + V_{OVp} = 1.5 + 2 = 3.5$ V.
9. By KVL, M_p's gate voltage is $V_{Gp} = V_{Sp} - V_{SGp} = 6 - 3.5 = 2.5$ V.
10. We note that R_2 drops 2.5 V and R_1 drops 7.5 V, so $R_1 = 3R_2$. Use $R_2 = 1$ MΩ and $R_1 = 3$ MΩ.

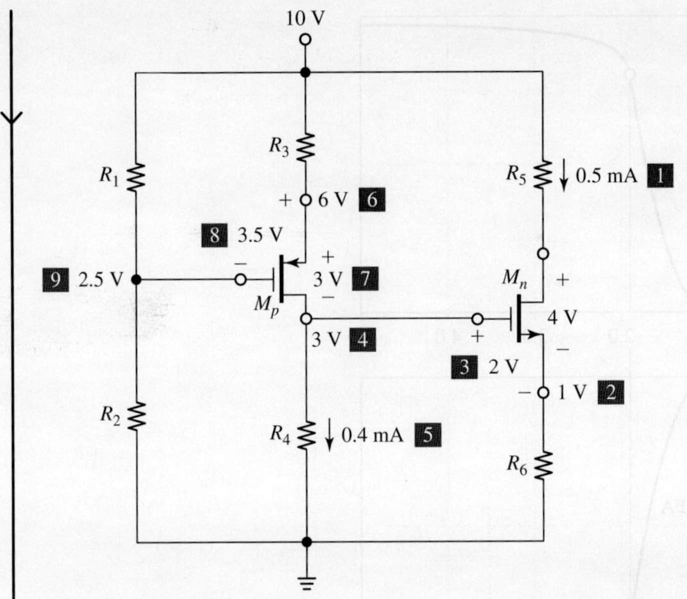

FIGURE 3.36 Circuit of Fig. 3.35 with each voltage and current identified by the corresponding computational step number in the text.

3.6 THE MOSFET AS AN AMPLIFIER/SWITCH

Let us now investigate the two fundamental MOSFET applications, *amplification* and *switching*. To this end, refer to the basic circuit of Fig. 3.37, where R_D and M can be viewed as forming a voltage divider of sorts: R_D tends to *pull v_O up* toward V_{DD}, and M tends to *pull v_O down* toward ground. Depending on which pull action prevails, v_O will assume a value somewhere in between. The plot of v_O versus v_I, called the *voltage transfer curve* (VTC), provides much insight into the capabilities of this

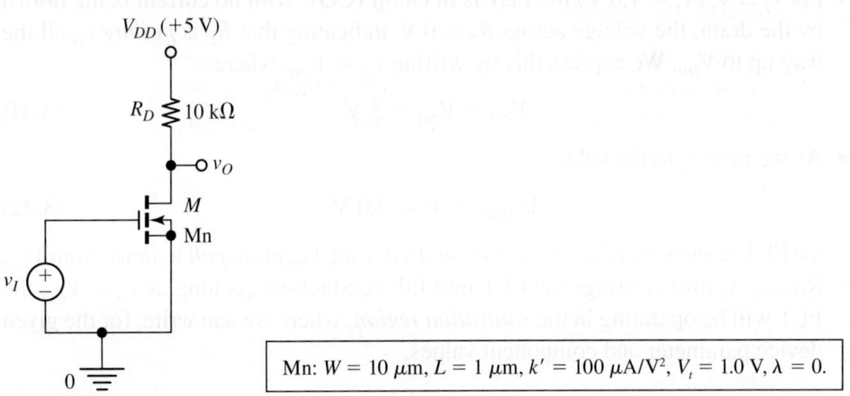

FIGURE 3.37 PSpice circuit to investigate the MOSFET as an amplifier/switch.

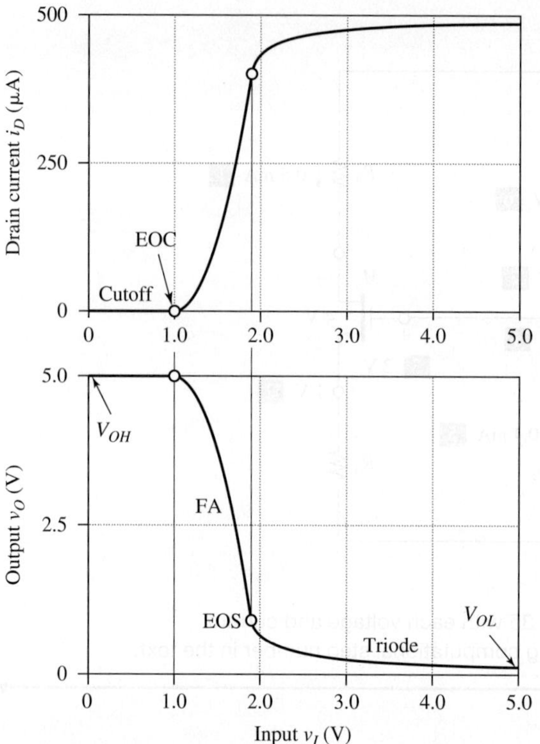

FIGURE 3.38 Sweeping the MOSFET of Fig. 3.37 from cutoff (CO), to the edge of conduction (EOC), through the forward-active (FA) region, to the edge of saturation (EOS), into the triode region.

circuit. Figure 3.38 shows the drain current as well as the VTC for the case in which v_I is swept from 0 V to 5 V and the FET possesses the characteristics tabulated in Fig. 3.37. We make the following observations:

- For $v_I \leq V_t$ ($V_t = 1.0$ V) the FET is in *cutoff* (CO). With no current being drawn by the drain, the voltage across R_D is 0 V, indicating that R_D is *pulling v_O* all the way up to V_{DD}. We express this by writing $v_O = V_{OH}$, where

$$V_{OH} = V_{DD} = 5 \text{ V} \tag{3.41}$$

- As we raise v_I to the value

$$V_{I(EOC)} = V_t = 1.0 \text{ V} \tag{3.42}$$

the FET reaches the *edge of conduction* (EOC) and begins to *pull v_O down* from V_{DD}.
- Raising v_I further brings the FET into full conduction. As long as $v_O \geq V_{OV}$, the FET will be operating in the *saturation region*, where we can write, for the given device parameter and component values,

$$v_O = V_{DD} - R_D I_D = 5 - 10\tfrac{1}{2}(v_I - 1)^2 = 10v_I - 5v_I^2 \tag{3.43}$$

- As v_O drops to the value $v_O = V_{OV}$, the FET reaches the *edge of saturation* (EOS), where we have $v_I = V_t + V_{OV} = 1 + V_{OV}$. Letting $v_O = V_{OV}$ and $v_I = 1 + V_{OV}$ in Eq. (3.43), solving the ensuing quadratic equation, and keeping only the physically acceptable solution, we get $V_{OV} = 0.905$ V. Denoting the corresponding value of v_I as $V_{I(EOS)}$, we have, for the given device and components,

$$V_{I(EOS)} \cong 1.9 \text{ V} \qquad (3.44)$$

- For $v_I \geq V_{I(EOS)}$, v_O drops below V_{OV}, so the FET enters the *triode region*, where we now have, for the given device and components,

$$v_O = V_{DD} - R_D I_D = 5 - 10 \times 1\left[(v_I - 1)v_O - \frac{1}{2}v_O^2\right]$$
$$= 5 - 10(v_I - 1)v_O + 5v_O^2 \qquad (3.45)$$

- For $v_I = V_{DD} = 5$ V, Eq. (3.45) gives, for the given device and component values, $v_O = V_{OL}$, where

$$V_{OL} = 0.124 \text{ V} \qquad (3.46)$$

We wish to point out that in order to simplify our calculations and thus facilitate the comparison with simulated data, we have assumed $\lambda = 0$ in Fig. 3.37. In practice, a nonzero λ will alter the curves a little, but our general observations still stand. The interested reader can easily visualize the differences by running the above PSpice circuit with, say, $\lambda = 0.05$ V^{-1}.

The MOSFET as an Amplifier

The *slope* of the VTC represents *voltage gain*. Denoted as a, it is readily found by differentiating Eq. (3.43). The result is, for the given device and component values,

$$a = \frac{dv_O}{dv_I} = 10(1 - v_I) \qquad (3.47)$$

Figure 3.39 shows the VTC as well as the slope a. In cutoff and in the triode region a is small or even zero. However, there are two points, denoted as V_{IL} and V_{IH}, such that for $V_{IL} \leq v_I \leq V_{IH}$ we have $|a| > 1$ V/V, indicating that the circuit can be used as an *amplifier* (if of the inverting type). As seen, the voltage gain peaks at about -9 V/V just before the EOS.

A circuit whose gain is not constant but varies with the value of the signal itself is *nonlinear*. Moreover, the VTC does not go through the origin, but is *offset* both along the v_I and the v_O axes. How can we make such a circuit work as a voltage amplifier? The answer relies on two premises, which are illustrated in Fig. 3.40:

- First, we bias the FET at a suitable operating point $Q_0 = Q_0(V_I, V_O)$ in the FA region by applying the appropriate *dc voltage* V_I. Aptly called the *quiescent operating point*, Q_0 in effect establishes a new system of axes for *signal variations* about this point. Q_0 should be located sufficiently away from either extreme (EOC and EOS) to allow for an adequate output signal swing in both directions.

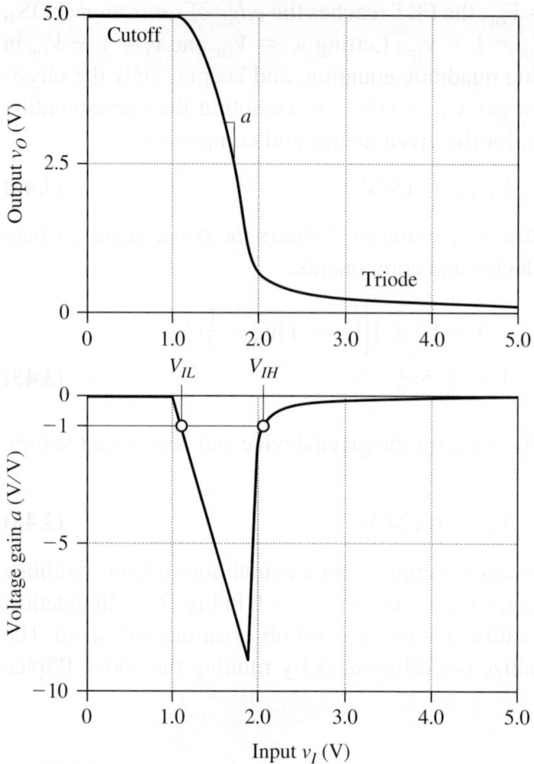

FIGURE 3.39 The voltage transfer curve (VTC) and its *slope*, representing the voltage gain a.

- Then, we apply an *ac input* v_i, which will cause the instantaneous operating point to move up and down the VTC (between Q_1 and Q_2), to yield a magnified *ac voltage* v_o at the output.

In our discussion we are relying on the same notation that proved so convenient in the study of diodes, namely, we express the input and output voltages as

$$v_I = V_I + v_i \qquad (3.48a)$$

$$v_O = V_O + v_o \qquad (3.48b)$$

where:

- v_I and v_O are referred to as the *total signals* (lower-case symbols with upper-case subscripts)
- V_I and V_O are their *dc components* (upper-case symbols with upper-case subscripts)
- v_i and v_o are their *ac components* (lower-case symbols with lower-case subscripts)

As depicted in Fig. 3.40b for the circuit of Fig. 3.37, the bias point Q_0 has been chosen half-way between the value of v_O corresponding to the EOC and that corresponding to the EOS, or $V_O = (5 + 0.905)/2 \cong 3$ V. The voltage gain there is denoted as $a(Q_0)$.

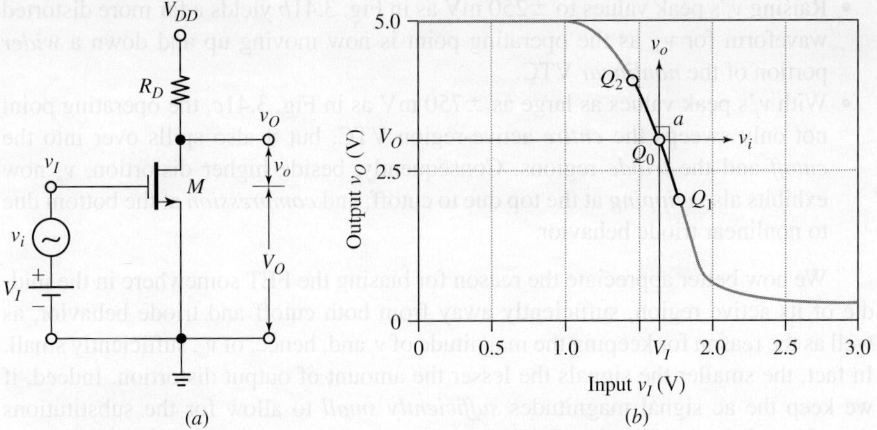

FIGURE 3.40 (a) The MOSFET of Fig. 3.37 as a *voltage amplifier*, and (b) variations about the operating point Q_0.

Exercise 3.2

Find the value of V_I needed to bias the drain at $V_O = 3.0$ V in the circuit of Fig. 3.37. What is the corresponding voltage gain a there?

Ans. $V_I = 1.632$ V, $a(Q_0) = -6.324$ V/V.

PSpice simulations with a *triangular input* v_i of progressively increasing magnitude give the waveforms of Fig. 3.41, where we make the following observations:

- In Fig. 3.41a the ac input v_i has peak values of ± 100 mV and the ac output v_o is an inverted and magnified version of v_i. The slight distortion exhibited by v_o is due to the quadratic (rather than linear) active-region VTC.

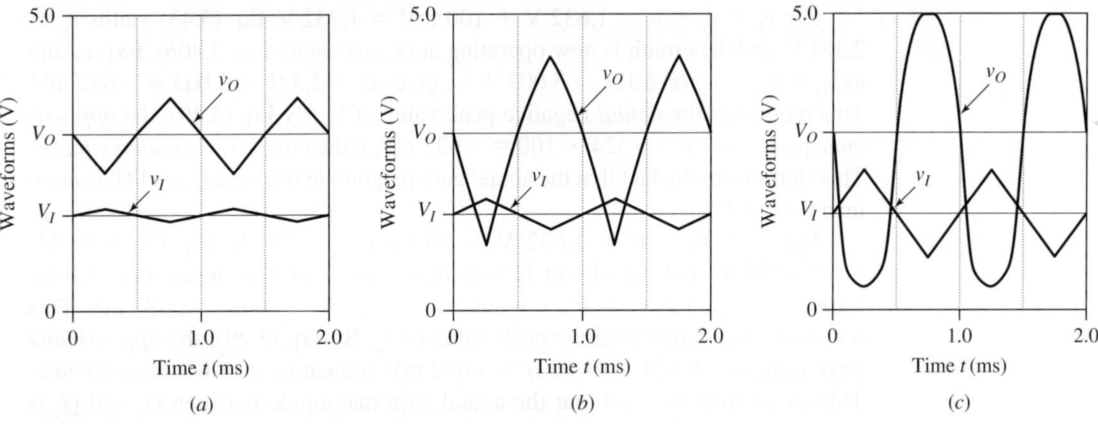

FIGURE 3.41 The responses of the circuit of Fig. 3.37 to a triangular wave v_i with peak values of: (a) ± 100 mV, (b) ± 250 mV, and (c) ± 750 mV. The FET is biased at $V_I = 1.63$ V.

- Raising v_i's peak values to ± 250 mV as in Fig. 3.41b yields a far more distorted waveform for v_o, as the operating point is now moving up and down a *wider* portion of the *nonlinear* VTC.
- With v_i's peak values as large as ± 750 mV as in Fig. 3.41c, the operating point not only sweeps the *entire* active-region VTC, but it also spills over into the *cutoff* and the *triode* regions. Consequently, beside higher distortion, v_o now exhibits also *clipping* at the top due to cutoff, and *compression* at the bottom due to nonlinear triode behavior.

We now better appreciate the reason for biasing the FET somewhere in the middle of its active region, sufficiently away from both cutoff and triode behavior, as well as the reason for keeping the magnitude of v_i and, hence, of v_o, sufficiently small. In fact, the smaller the signals the lesser the amount of output distortion. Indeed, if we keep the ac signal magnitudes *sufficiently small* to allow for the substitutions $dv_I \rightarrow v_i$ and $dv_O \rightarrow v_o$ in Eq. (3.47), then we can approximate $a = v_o/v_i$, or

$$v_o = a(Q_0) \times v_i \tag{3.49}$$

where $a(Q_0)$ is the gain at the operating point Q_0. Viewed in this light, v_i and v_o are also referred to as *small signals*. A more rigorous treatment of this subject will be undertaken in the next section.

EXAMPLE 3.26   With reference to Fig. 3.41a, investigate the ac output v_o using both *exact* analysis via Eq. (3.43), and *approximate* analysis via Eq. (3.49), and compare the two cases.

Solution

For $v_I = V_I + v_i = 1.632 + 0$ V, Eq. (3.43) yields $v_o = V_O + v_o = 3.003 + 0 = 3.003$ V, and the circuit is operating at Q_0 (see Fig. 3.40b), where the gain is $a = -6.324$ V/V.

For $v_I = V_I + v_i = 1.632$ V $+ 100$ mV $= 1.732$ V, Eq. (3.43) yields $v_o = 2.321$ V, and the circuit is now operating at Q_1 (see again Fig. 3.40b). Expressing as $v_O = V_O + v_o$, or $2.321 = 3.003 + v_o$, gives $v_o = 2.321 - 3.003 = -682$ mV. This represents the *actual* negative peak value of v_o. By Eq. (3.49), the *approximate* peak value is $-6.324 \times 100 = -632$ mV, indicating a 7.3% *underestimate*. This stems from the fact that the actual gain magnitude between Q_0 and Q_1 is *more* than 6.324 V/V.

For $v_I = V_I - v_i = 1.632$ V $- 100$ mV $= 1.532$ V, Eq. (3.43) yields $v_o = 3.585$ V, and the circuit is now operating at Q_2 (see again Fig. 3.40b). Expressing again as $v_O = V_O + v_o$ gives $v_o = 3.585 - 3.003 = +582$ mV. This represents the *actual* positive peak value of v_o. By Eq. (3.49), the *approximate* peak value is $-6.324 \times (-100) = +632$ mV, indicating an 8.6% *overestimate*. This stems from the fact that the actual gain magnitude between Q_0 and Q_2 is *less* than 6.324 V/V.

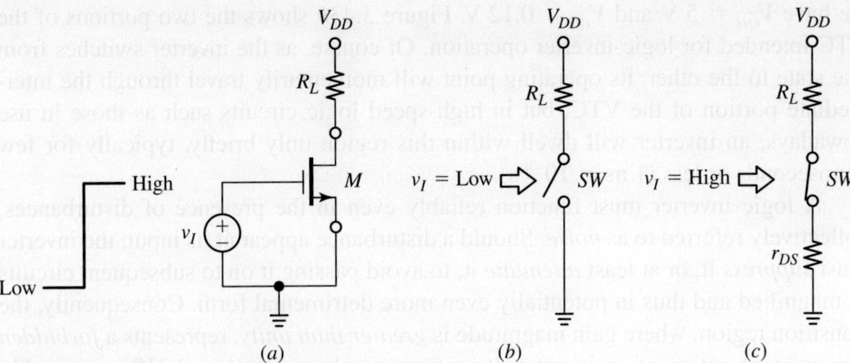

FIGURE 3.42 Operating the MOSFET as an *electronic switch*.

The MOSFET as an Electronic Switch

When FET operation alternates between the *cutoff* and the *ohmic* regions, the device acts as an *electronically controlled switch SW*. This function is illustrated in Fig. 3.42, where we observe the following:

- When the input voltage level in the circuit of Fig. 3.42*a* is *low*, such as near 0 V, the FET is in cutoff, and since it draws no current, it can be regarded as an *open* switch, as pictured in Fig. 3.42*b*.
- When the input voltage level is *high*, such as near V_{DD}, or 5 V, the FET is in the ohmic region, and acts like a *closed* switch with a resistance r_{DS} as pictured in Fig. 3.42*c*. In a well designed circuit $r_{DS} \ll R_L$. With $v_I = V_{DD} = 5$ V the circuit of Fig. 3.37 gives, by Eq. (3.17), $r_{DS} = 1/(5 - 1) = 250\ \Omega$.

The MOSFET as a Logic Inverter

One of the most important applications of the FET switch is *logic inversion* in computer circuitry. As depicted in Fig. 3.43*a*, a logic inverter outputs a high level V_{OH} in response to a low input level ($v_I \cong 0$ V), and a low level V_{OL} in response to a high input level ($v_I \cong 5$ V). As already mentioned, with the component values of Fig. 3.37,

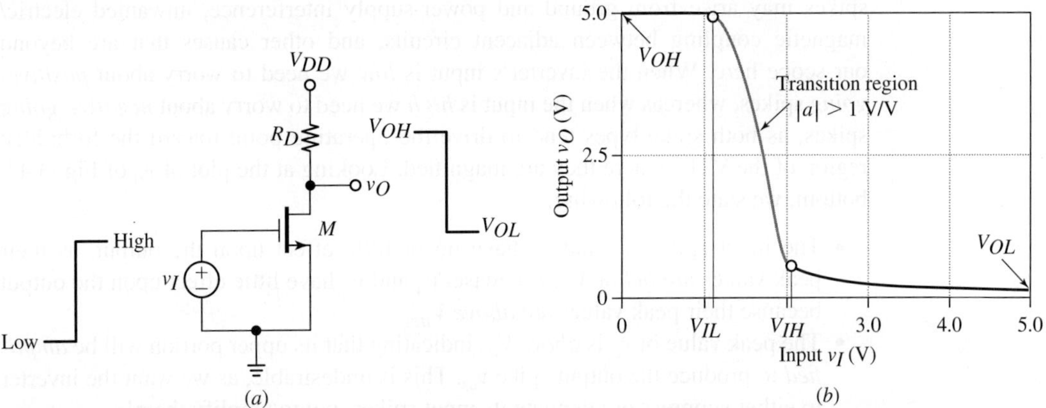

FIGURE 3.43 (*a*) The MOSFET of Fig. 3.37 as a *logic inverter*, and (*b*) the relevant portions of its VTC.

we have $V_{OH} = 5$ V and $V_{OL} = 0.12$ V. Figure 3.43*b* shows the two portions of the VTC intended for logic-inverter operation. Of course, as the inverter switches from one state to the other, its operating point will momentarily travel through the intermediate portion of the VTC, but in high-speed logic circuits such as those in use nowadays, an inverter will dwell within this region only briefly, typically for few nanoseconds or less (1 ns $= 10^{-9}$ s).

A logic inverter must function reliably even in the presence of disturbances, collectively referred to as *noise*. Should a disturbance appear at its input, the inverter must *suppress* it, or at least *attenuate* it, to avoid passing it on to subsequent circuits in magnified and thus in potentially even more detrimental form. Consequently, the transition region, where gain magnitude is *greater than unity*, represents a *forbidden region* of operation of an inverter. Its extremes, where gain is -1 V/V, are readily visualized with the help of Fig. 3.38, bottom. The values of v_I at these extremes are denoted as V_{IL} and V_{IH}.

To find V_{IL}, impose $-1 = 10(1 - V_{IL})$ in Eq. (3.47). This gives

$$V_{IL} = 1.1 \text{ V} \tag{3.50a}$$

To find V_{IH}, we differentiate both sides of Eq. (3.45) with respect to v_I

$$\frac{dv_O}{dv_I} = -10\left[v_O + (v_I - 1)\frac{dv_O}{dv_I} - v_O\frac{dv_O}{dv_I}\right]$$

Imposing $dv_O/dv_I = -1$ yields a relationship between v_O and v_I right at the point of negative unity slope,

$$v_O = 0.5v_I - 0.45$$

Substituting back into Eq. (3.45), solving the resulting quadratic equation in v_I, and retaining only the physically acceptable solution, which is obviously V_{IH}, we finally get

$$V_{IH} = 2.055 \text{ V} \tag{3.50b}$$

To better appreciate the inverter's ability to reject noise, we run a PSpice simulation of the circuit of Fig. 3.37 for the case in which the input levels are contaminated by *noise spikes*, as exemplified in the plot of v_I of Fig. 3.44, top. In practice such spikes may arise from ground and power-supply interference, unwanted electric/magnetic coupling between adjacent circuits, and other causes that are beyond our scope here. When the inverter's input is *low* we need to worry about *positive-going* spikes, whereas when the input is *high* we need to worry about *negative-going* spikes, as both spike types tend to drive the operating point toward the forbidden region of the VTC, where they are magnified. Looking at the plot of v_O of Fig. 3.44, bottom, we state the following:

- The input spikes v_{i1} and v_{i2} have no or little effect upon the output, as their peak values are *below* V_{IL}. Likewise, v_{i4} and v_{i5} have little effect upon the output because their peak values are *above* V_{IH}.
- The peak value of v_{i3} is *above* V_{IL}, indicating that its upper portion will be *amplified* to produce the output spike v_{o3}. This is undesirable, as we want the inverter to either suppress or attenuate its input spikes, not to amplify them!

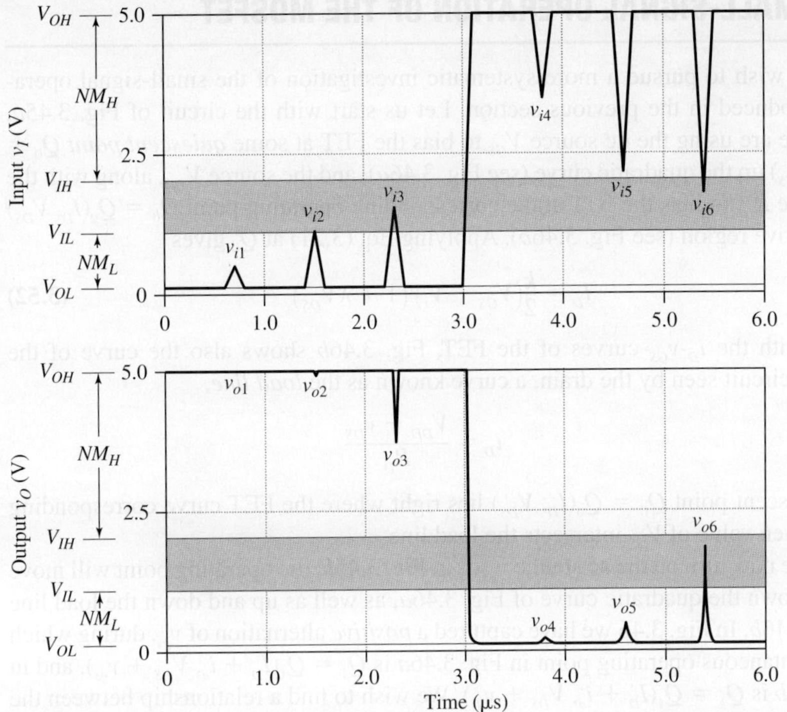

FIGURE 3.44 Logic-inverter behavior in the presence of noise, showing the noise margins NM_L and NM_H (the shaded area represents the *transition* or *forbidden region*, where noise gets magnified).

- The peak value of v_{i6} is *below* V_{IH}, indicating that its lower portion will be *amplified* to produce the output spike v_{o6}. Again, this is undesirable.

An inverter's ability to function properly in spite of input disturbances is quantified in terms of its *noise margins*, defined as

$$NM_L = V_{IL} - V_{OL} \qquad (3.51a)$$

$$NM_H = V_{OH} - V_{IH} \qquad (3.51b)$$

As visualized in the plot of v_I of Fig. 3.44, NM_L represents the maximum tolerable noise at the input in the *low* state. In our example, $NM_L = 1.1 - 0.124 = 0.98$ V, indicating that positive-going input spikes of magnitudes not exceeding 0.98 V will be either suppressed or attenuated. Likewise, NM_H represents the maximum tolerable noise at the input in the *high* state. In our example, $NM_H = 5 - 2.055 = 2.95$ V, indicating our circuit's ability to attenuate negative-going input spikes of magnitudes not exceeding 2.95 V. The circuit of our particular example tolerates noise more easily in the high than in the low input state. Using a FET with a higher value of V_t, such as $V_t = 2$ V, will shift the VTC toward the right, thus increasing NM_L, if at the expense of a decrease in NM_H, and resulting in more balanced noise margins.

3.7 SMALL-SIGNAL OPERATION OF THE MOSFET

We now wish to pursue a more systematic investigation of the small-signal operation introduced in the previous section. Let us start with the circuit of Fig. 3.45*a*,
where we are using the dc source V_{GS} to bias the FET at some *quiescent point* $Q_0 =
Q_0(I_D, V_{GS})$ up the quadratic curve (see Fig. 3.46*a*), and the source V_{DD}, along with the
resistance R_D, to bias the BJT at the corresponding operating point $Q_0 = Q_0(I_D, V_{DS})$
in the active region (see Fig. 3.46*b*). Applying Eq. (3.21) at Q_0 gives

$$I_D = \frac{k}{2}(V_{GS} - V_t)^2(1 + \lambda V_{DS}) \tag{3.52}$$

Along with the i_D-v_{GS} curves of the FET, Fig. 3.46*b* shows also the curve of the
external circuit seen by the drain, a curve known as the *load line*,

$$i_D = \frac{V_{DD} - v_{DS}}{R_D}$$

The quiescent point $Q_0 = Q_0(I_D, V_{DS})$ lies right where the FET curve corresponding
to the given value of V_{GS} intersects the load line.

If we now turn on the ac source v_{gs} as in Fig. 3.45*b*, the operating point will move
up and down the quadratic curve of Fig. 3.46*a*, as well as up and down the load line
of Fig. 3.46*b*. In Fig. 3.46 we have captured a *positive* alternation of v_{gs}, during which
the instantaneous operating point in Fig. 3.46*a* is $Q_1 = Q_1(I_D + i_d, V_{GS} + v_{gs})$, and in
Fig. 3.46*b* is $Q_1 = Q_1(I_D + i_d, V_{DS} + v_{ds})$. We wish to find a relationship between the
ac current i_d and the ac voltages v_{gs} and v_{ds}. Applying Eq. (3.21) at Q_1 gives

$$I_D + i_d = \frac{k}{2}[V_{GS} + v_{gs} - V_t]^2[1 + \lambda(V_{DS} + v_{ds})]$$

Regrouping and multiplying out gives

$$I_D + i_d = \frac{k}{2}[(V_{GS} - V_t)^2 + 2(V_{GS} - V_t)v_{gs} + v_{gs}^2][(1 + \lambda V_{DS}) + \lambda v_{ds}] \tag{3.53}$$

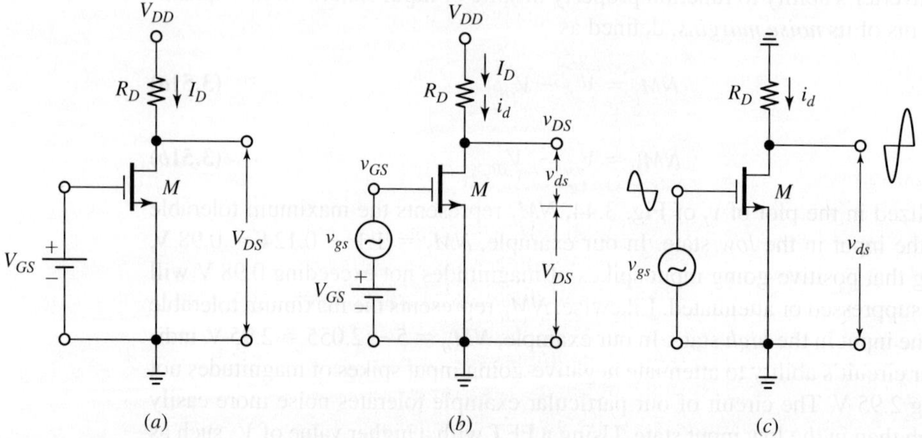

FIGURE 3.45 Systematic analysis of the MOSFET as a small-signal amplifier. The actual circuit is
shown in (*b*), while (*a*) shows its large-signal or *dc* version, and (*c*) shows its small-signal or *ac* version.

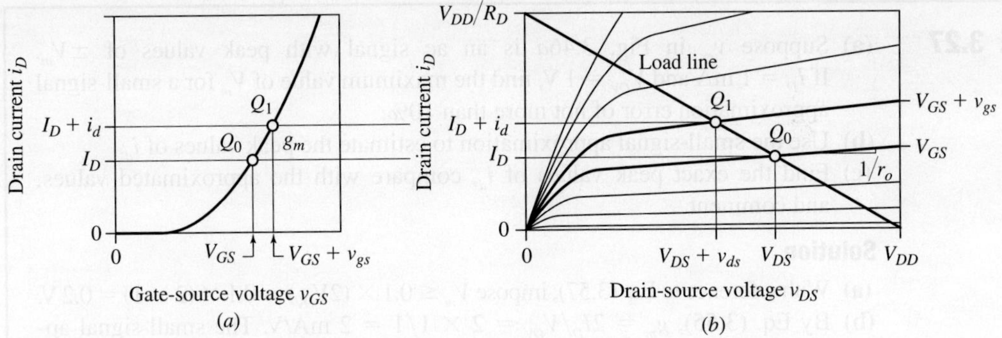

FIGURE 3.46 Graphical illustrations of the FET amplifier of Fig. 3.45.

For $\lambda V_{DS} \ll 1$ we can approximate $(k/2)(V_{GS} - V_t)^2 \cong I_D$ and rewrite as

$$I_D + i_d = I_D + kV_{ov}v_{gs} + \lambda I_D v_{ds} + \cdots$$

where $V_{OV} = V_{GS} - V_t$, as usual. So long as we can ignore higher-order terms involving ac products and powers, the above expression allows us to write

$$i_d = g_m v_{gs} + \frac{v_{ds}}{r_o} \tag{3.54}$$

where g_m is the already familiar transconductance of the FET introduced in Eq. (3.25) and repeated here in its three equivalent forms

$$g_m = kV_{OV} = \sqrt{2kI_D} = 2\frac{I_D}{V_{OV}} \tag{3.55}$$

and

$$r_o = \frac{1}{\lambda I_D} \tag{3.56}$$

is the familiar drain's *output resistance*. As shown in Fig. 3.46, g_m and $1/r_o$ represent, respectively, the slope of the i_D-v_{GS} and the slope of the i_D-v_{DS} curve at Q_0. Both parameters depend on the operating current I_D. Moreover, the fact that $1/r_o \ll g_m$ indicates a much weaker dependence of i_d on v_{ds} than on v_{gs}.

We wish to assess under what conditions we can ignore higher-order terms in Eq. (3.53). By inspection, the quadratic term can be ignored as long as we keep v_{gs} small enough to satisfy the condition $v_{gs}^2 \ll 2(V_{GS} - V_t)|v_{gs}|$, that is,

$$|v_{gs}| \ll 2V_{OV} \tag{3.57}$$

For obvious reasons Eq. (3.54) is referred to as the *small signal approximation*, and Eq. (3.57) quantifies the validity of such an approximation.

EXAMPLE 3.27

(a) Suppose v_{gs} in Fig. 3.46a is an ac signal with peak values of $\pm V_m$. If $I_D = 1$ mA and $V_{OV} = 1$ V, find the maximum value of V_m for a small-signal approximation error of not more than 10%.

(b) Use the small-signal approximation to estimate the peak values of i_d.

(c) Find the exact peak values of i_d, compare with the approximated values, and comment.

Solution

(a) With reference to Eq. (3.57), impose $V_m \leq 0.1 \times (2V_{OV}) = 0.1 \times (2 \times 1) = 0.2$ V.

(b) By Eq. (3.55), $g_m = 2I_D/V_{OV} = 2 \times 1/1 = 2$ mA/V. The small-signal approximation of Eq. (3.54) predicts, for i_d, peak values of $\pm I_m = g_m(\pm V_m) = 2(\pm 0.2) = \pm 0.4$ mA.

(c) With the given data, $k = 2I_D/V_{OV}^2 = 2 \times 1/1^2 = 2$ mA/V². By Eq. (3.53), the exact peak values of i_d are, respectively, $\pm g_m V_m + (k/2)V_m^2 = [\pm 0.4 + (2/2)0.2^2]$ mA, or 0.44 mA and -0.36 mA, respectively. Because of the curvature of the i_D-v_{DS} characteristic, the small-signal approximation *underestimates* the positive current peak by 10%, and *overestimates* the negative current peak by 10%.

Just like we use the *dc* equivalent of Fig. 3.45a to investigate the biasing conditions of our FET, we shall use the *ac* equivalent of Fig. 3.45c to investigate its operation as an amplifier. Indeed, the latter gives, by KVL, Ohm's law, and Eq. (3.54),

$$v_{ds} = 0 - R_D i_d = -R_D\left(g_m v_{gs} + \frac{v_{ds}}{r_o}\right)$$

Collecting, and solving for v_{ds}, we can write

$$v_{ds} = -g_m(R_D//r_o)v_{gs}$$

indicating that our circuit magnifies v_{gs} by the *gain* $-g_m(R_D//r_o)$. To give an idea, let $g_m = 1$ mA/V, $R_D = 10$ kΩ, and $r_o = 100$ kΩ. Then, $v_{ds} = -9.1v_{gs}$.

The need to perform dc and ac analysis *separately* will be illustrated further as we proceed. Figures 3.20 and 3.21 presented large-signal MOSFET models for the purpose of facilitating *dc analysis*, as exemplified in Figs. 3.30 and 3.32. We now need to develop a small-signal MOSFET model to facilitate *ac analysis*.

The Small-Signal Model

Figure 3.47 shows the *small-signal model* of the MOSFET. The function of this model, also called the *incremental model* or simply the *ac equivalent*, is to provide a circuit representation of the dependence of i_d upon v_{gs} and v_{ds} as expressed by Eq. (3.54). As we know, the dependence on v_{ds} is much weaker than that on v_{gs}, this being the reason why the second term in Eq. (3.54) is sometimes ignored in order to speed up calculations. In this chapter we are limiting our study to MOSFET applications in which body and source are *tied together*, and the model of Fig. 3.47 is adequate for this kind of investigation. If the source is allowed to float independently of the body, the ensuing body effect requires further refinements in the small-signal model presented here (this will be the subject of the next chapter). The small-signal parameter definitions (for the case of the *n*MOSFET) and their calculations are summarized in Table 3.1.

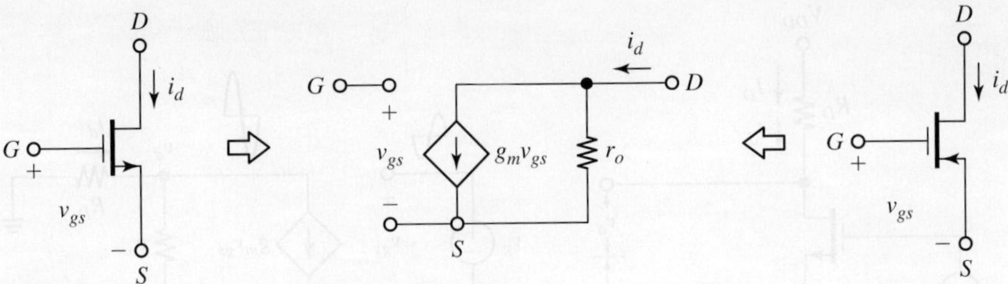

FIGURE 3.47 Small-signal MOSFET model. This model applies *both* to the *n*MOSFET and the *p*MOSFET.

We wish to point out that the model of Fig. 3.47 applies *both* to the *n*MOSFET and the *p*MOSFET, with no change in voltage polarities or current directions. To see why, consider the effect of *increasing* v_{GS} by v_{gs} in both devices. In the *n*MOSFET i_D will also *increase*, but in the *p*MOSFET i_D will *decrease*. Consequently, i_d will have the same direction as i_D in the *n*MOSFET (i_d *into* the drain terminal), but the opposite direction as i_D in the *p*MOSFET: since i_D flows out of the drain terminal, i_d will be *into* the drain terminal, just as in the *n*MOSFET case. We must stress that the small-signal model should not be confused with the large signal models. The latter are used in *dc analysis*, examples of which we have already seen. The former is used in *ac analysis*, to be demonstrated next.

TABLE 3.1 Small-signal parameter summary.

Definition	Calculation
$g_m = \left.\dfrac{\partial i_D}{\partial v_{GS}}\right|_{V_{DS}}$	$g_m = \sqrt{2kI_D} = 2\dfrac{I_D}{V_{OV}} = kV_{OV}$
$\dfrac{1}{r_o} = \left.\dfrac{\partial i_D}{\partial v_{DS}}\right|_{V_{GS}}$	$r_o = \dfrac{1}{\lambda I_D}$

A Generalized MOSFET Circuit

As our first application of the small-signal model we investigate the generalized MOSFET circuit of Fig. 3.48*a*, similar to that of Fig. 3.45*b*, except for the inclusion of the additional resistance R_S. Our analysis will reveal a number of interesting relationships stemming from the interaction between the FET and the surrounding circuit elements. Moreover, it will provide some general expressions that will help us expedite the ac analysis of future MOSFET amplifiers.

The function of the dc sources V_G and V_{DD} is to bias the FET at some dc current I_D in the saturation region. Turning on the input ac source v_g will result in the drain ac current i_d as well as the output ac voltages v_d and v_s. We wish to investigate how the small signals i_d, v_d, and v_s are related to v_g. We also wish to find the *small-signal resistances* seen looking into the gate, source, and drain terminals, which we shall denote as R_g, R_s, and R_d. (Note the use of lower-case subscripts to distinguish them from the resistances external to the FET, identified by upper-case subscripts.)

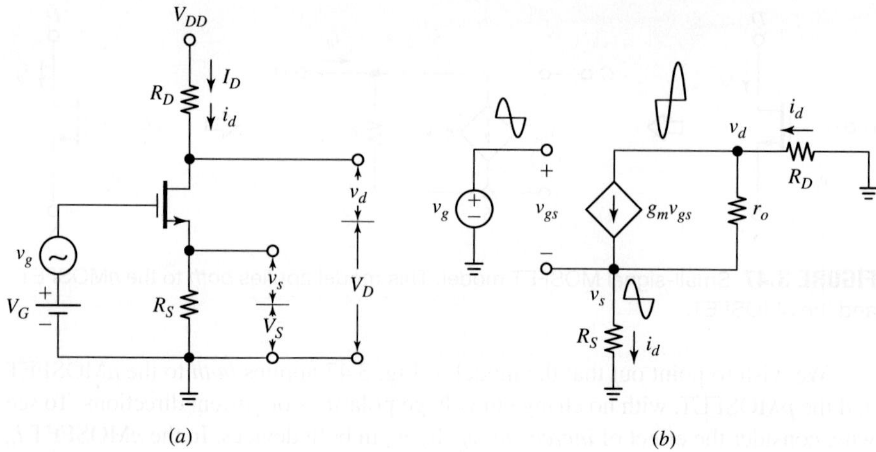

FIGURE 3.48 (a) A generalized MOSFET circuit, and (b) its small-signal equivalent.

As a rule, in order to perform the small-signal analysis of a MOS circuit,

- We replace the FET with its *small-signal model*
- We show only the *ac signals*, as depicted in the equivalent circuit of Fig. 3.48b. The bias voltages (V_G and V_{DD}) as well as all other dc components (V_S, V_D, and I_D) do not appear in this equivalent because they are constant and thus possess *no* ac components. Consequently, as we go from the actual circuit to its small-signal equivalent, we set all *dc voltages* and all *dc currents* to *zero*.

Let us now proceed to find the aforementioned small-signal relationships and resistances.

- **The Circuit's Small-Signal Transconductance $G_m = i_d/v_g$.** With reference to Fig. 3.48b, KCL gives,

$$i_d = g_m v_{gs} + \frac{v_d - v_s}{r_o}$$

Now, by Ohm's law, $0 - v_d = R_D i_d$, or

$$v_d = -R_D i_d \qquad (3.58)$$

By Ohm's law again,

$$v_s = R_S i_d \qquad (3.59)$$

Also, by definition, $v_{gs} = v_g - v_s$, or

$$v_{gs} = v_g - R_S i_d \qquad (3.60)$$

Substituting v_{gs}, v_d, and v_s into the expression for i_d and collecting, we get

$$i_d = G_m v_g \qquad (3.61)$$

where G_m, referred to as the *circuit's transconductance* (not to be confused with the individual FET's transconductance g_m) is given by

$$G_m = \frac{g_m}{1 + g_m R_S + (R_D + R_S)/r_o} \qquad (3.62a)$$

As we proceed, we will find that in discrete MOSFET amplifiers we usually have $(R_D + R_S)/r_o \ll 1$, so Eq. (3.62a) simplifies to

$$G_m \cong \frac{g_m}{1 + g_m R_S} \qquad (3.62b)$$

Note that for $R_S = 0$ we have $G_m = g_m$, but for $R_S \neq 0$ we have $G_m < g_m$. This transconductance reduction, referred to as *degeneration*, stems from the fact that the voltage drop $R_S i_d$ *subtracts* from the input v_g to yield a *reduced* control signal v_{gs} for the dependent source, as per Eq. (3.60). Hence, $i_d\ (=g_m v_{gs})$ will also get reduced. A more systematic investigation of degeneration reveals that R_S provides a *negative feedback* function, as mentioned in connection with Example 3.19. Though this subject will be investigated systematically in Chapter 7, suffice it to say here that the presence of R_S, aptly called *source-degeneration resistance*, not only reduces the transconductance, but also affects the small-signal resistance R_d seen looking into the drain, as we are about to find out.

- **The Small-Signal Voltage Gain v_d/v_g from Gate to Drain.** By Eqs. (3.58), (3.61), and (3.62a), this gain is

$$\frac{v_d}{v_g} = \frac{-g_m R_D}{1 + g_m R_S + (R_D + R_S)/r_o} \qquad (3.63a)$$

The negative sign indicates *inverting amplification* from gate to drain: positive (negative) alternations in v_g result in negative (positive) alternations in v_d. If the condition $(R_D + R_S) \ll r_o$ holds, Eq. (3.63a) simplifies to

$$\frac{v_d}{v_g} \cong -\frac{g_m R_D}{1 + g_m R_S} \qquad (3.63b)$$

- **The Small-Signal Voltage Gain v_s/v_g from Gate to Source.** By Eqs. (3.59), (3.61), and (3.62), this gain is

$$\frac{v_s}{v_g} = \frac{g_m R_S}{1 + g_m R_S + (R_D + R_S)/r_o} \qquad (3.64a)$$

indicating that v_s is *in phase* with v_g. If the condition $(R_D + R_S) \ll r_o$ holds, Eq. (3.64a) simplifies to

$$\frac{v_s}{v_g} \cong \frac{1}{1 + 1/(g_m R_S)} \qquad (3.64b)$$

The gain from gate to source is a bit less than unity, depending on how large the product $g_m R_S$ is compared to unity. We summarize this by saying that the source *follows* the gate.

- **The Small-Signal Resistance R_g Seen Looking into the Gate.** As we know, the gate electrode is the plate of a tiny capacitor, which draws negligible current, at

EXAMPLE 3.28   A certain FET with $k = 0.5$ mA/V^2 and $\lambda = 0.01$ V^{-1} is biased at $I_D = 1$ mA. If $R_D = 10$ kΩ and $R_S = 2.0$ kΩ, estimate the gains v_d/v_g and v_s/v_g.

Solution

By Eqs. (3.55) and (3.56) we have $g_m = 1/(1 \text{ k}\Omega)$ and $r_o = 100$ kΩ. Exploiting the fact that $(R_D + R_S) \ll r_o$ $(12 \ll 100)$, we use the approximate expressions of Eqs. (3.63b) and (3.64b) to find

$$\frac{v_d}{v_g} \cong -\frac{1 \times 10}{1 + 1 \times 2} = -3.33 \text{ V/V} \qquad \frac{v_s}{v_g} \cong \frac{1}{1 + 1/(1 \times 2)} = 0.67 \text{ V/V}$$

least up to moderate frequencies. For practical purposes we can thus say that the resistance seen looking into the gate of a MOSFET is

$$R_g = \infty \tag{3.65}$$

- **The Small-Signal Resistance R_s Seen Looking into the Source.** To find this resistance, set $v_g \to 0$ in the ac equivalent of Fig. 3.48b, apply a test voltage v_s right to the source terminal as in Fig. 3.49a (note that the external resistance R_S does not intervene in this test), find the resulting current i_s, and finally let $R_s = v_s/i_s$. Thus, summing currents into the source node we get

$$i_s + g_m v_{gs} + \frac{v_d - v_s}{r_o} = 0$$

But, $v_{gs} = v_g - v_s = 0 - v_s = -v_s$, and $v_d = R_D i_s$. Substituting, collecting, and solving for the ratio v_s/i_s we get, after suitable algebraic manipulation,

$$R_s = \left(\frac{1}{g_m}//r_o\right) + \frac{R_D}{1 + g_m r_o} \tag{3.66a}$$

Because of the coupling action by r_o, R_s depends also on the external drain resistance R_D, and we say that reflected to the source, R_D gets *divided* by $(1 + g_m r_o)$. MOSFETs usually have $g_m r_o \gg 1$. Moreover, discrete MOSFET amplifiers usually have $R_D \ll r_o$. Under these conditions, Eq. (3.66a) simplifies to

$$R_s \cong \frac{1}{g_m}//r_o \cong \frac{1}{g_m} \tag{3.66b}$$

Note the similarity with BJTs.

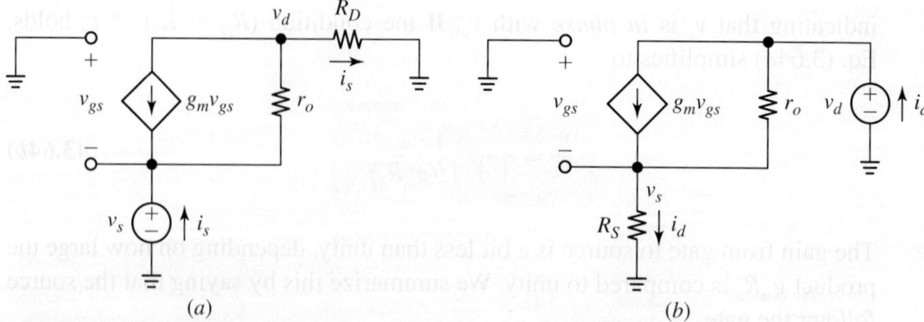

FIGURE 3.49 Test circuits to find (a) the small-signal resistance R_s seen looking into the *source*, and (b) the small-signal resistance R_d seen looking into the *drain*.

- **The Small-Signal Resistance R_d Seen Looking into the Drain.** To find this resistance, let $v_g \to 0$ in the ac equivalent of Fig. 3.48b, apply a test voltage v_d right to the drain terminal as in Fig. 3.49b (note that R_D does not intervene in this test), find the resulting current i_d, and finally let $R_d = v_d/i_d$. Thus, KCL at the source node gives

$$i_d = g_m v_{gs} + \frac{v_d - v_s}{r_o} = g_m(0 - v_s) + \frac{v_d}{r_o} - \frac{v_s}{r_o} = \frac{v_d}{r_o} - \left(g_m + \frac{1}{r_o}\right)v_s$$

$$= \frac{v_d}{r_o} - \left(\frac{g_m r_o + 1}{r_o}\right)R_s i_d$$

Collecting and solving for the ratio v_d/i_d gives

$$R_d = r_o + (1 + g_m r_o)R_S \qquad (3.67a)$$

Because of the coupling action by r_o, R_d depends also on the external source resistance R_S, and we say that reflected to the drain, R_S gets *multiplied* by $(1 + g_m r_o)$. Again exploiting the fact that $g_m r_o \gg 1$ and that discrete MOSFET amplifiers usually have $R_S \ll r_o$, we simplify Eq. (3.67a) as

$$R_d \cong r_o(1 + g_m R_S) \qquad (3.67b)$$

We observe that with $R_S = 0$ we get $R_d = r_o$, but with $R_S \neq 0$ we get $R_d > r_o$. This increase in R_d is the result of the aforementioned *negative-feedback action* by the source-degeneration resistance R_S. (Note also the similarity with BJTs.)

It is interesting to contrast the effects that the FET has upon its external resistances: while R_D reflected to the source gets *divided* by $(1 + g_m r_o)$, R_S reflected to the drain gets *multiplied* by $(1 + g_m r_o)$. For convenience the small-signal characteristics are tabulated in Fig. 3.50 for the ac equivalent shown.

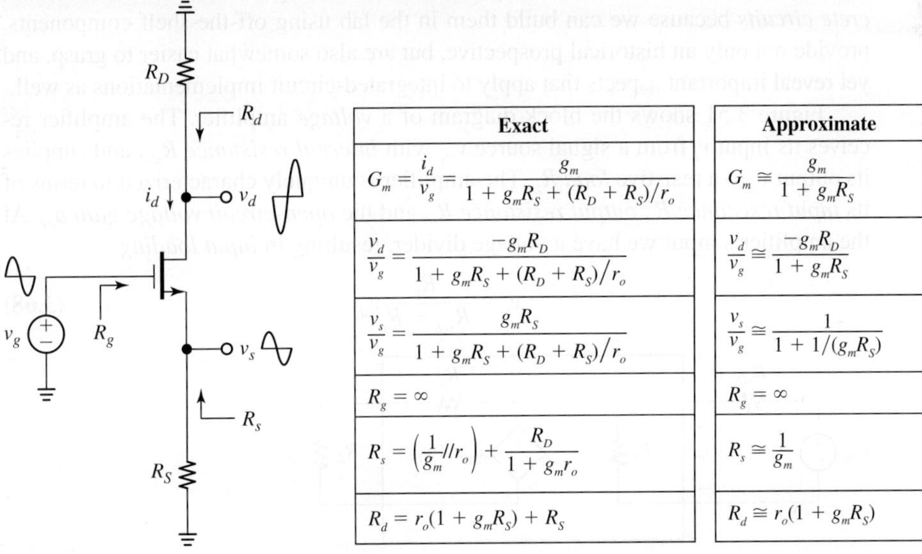

Exact	Approximate
$G_m = \dfrac{i_d}{v_g} = \dfrac{g_m}{1 + g_m R_S + (R_D + R_S)/r_o}$	$G_m \cong \dfrac{g_m}{1 + g_m R_S}$
$\dfrac{v_d}{v_g} = \dfrac{-g_m R_D}{1 + g_m R_S + (R_D + R_S)/r_o}$	$\dfrac{v_d}{v_g} \cong \dfrac{-g_m R_D}{1 + g_m R_S}$
$\dfrac{v_s}{v_g} = \dfrac{g_m R_S}{1 + g_m R_S + (R_D + R_S)/r_o}$	$\dfrac{v_s}{v_g} \cong \dfrac{1}{1 + 1/(g_m R_S)}$
$R_g = \infty$	$R_g = \infty$
$R_s = \left(\dfrac{1}{g_m}//r_o\right) + \dfrac{R_D}{1 + g_m r_o}$	$R_s \cong \dfrac{1}{g_m}$
$R_d = r_o(1 + g_m R_S) + R_S$	$R_d \cong r_o(1 + g_m R_S)$

FIGURE 3.50 Summary of small-signal gains and terminal resistances, and approximations for discrete designs.

EXAMPLE 3.29   (a) Using the small-signal FET parameters of Example 3.28, estimate R_s if $R_D = 0$.

(b) Estimate R_d if $R_S = 5\ \text{k}\Omega$.

Solution

(a) Exploiting the fact that $1/g_m \ll r_o$, we use Eq. (3.66b) to find

$$R_s \cong 1/g_m = 1\ \text{k}\Omega$$

(b) Since $R_S \ll r_o$, we use Eq. (3.67b) to find

$$R_d \cong 100(1 + 1 \times 5) = 600\ \text{k}\Omega$$

3.8 BASIC MOSFET VOLTAGE AMPLIFIERS

Depending on which terminal we apply the input to and which terminal we obtain the output from, a MOSFET can be used in any one of three amplifier configurations: the *common-source*, *common-drain*, and *common-gate* configurations. In the following we shall assume that body and source are *tied* together so that the MOSFET is operated as a *three-terminal* device. Viewing an amplifier as a *two-port* block, it is apparent that one of the terminals of the three-terminal FET will have to be *common* to both ports; hence, the reason for the above designations.

Nowadays the majority of MOSFET circuits are implemented in *integrated-circuit* form using MOSFETs to provide both active functions (such as amplification) and passive functions (such as dc biasing and active loading). However, before turning to multiple-transistor circuitry, we need to master single-transistor stages, and this is best done if we focus on a single MOSFET surrounded by already familiar components such as resistors and capacitors. The resulting circuits, referred to as *discrete circuits* because we can build them in the lab using off-the-shelf components, provide not only an historical prospective, but are also somewhat easier to grasp, and yet reveal important aspects that apply to integrated-circuit implementations as well.

Figure 3.51 shows the block diagram of a *voltage* amplifier. The amplifier receives its input v_i from a signal source v_{sig} with *internal resistance* R_{sig}, and supplies its output v_o to a resistive *load* R_L. The amplifier is uniquely characterized in terms of its *input resistance* R_i, output resistance R_o, and the *open-circuit voltage gain* a_{oc}. At the amplifier's input we have a voltage divider, resulting in *input loading*

$$v_i = \frac{R_i}{R_{sig} + R_i} v_{sig} \tag{3.68}$$

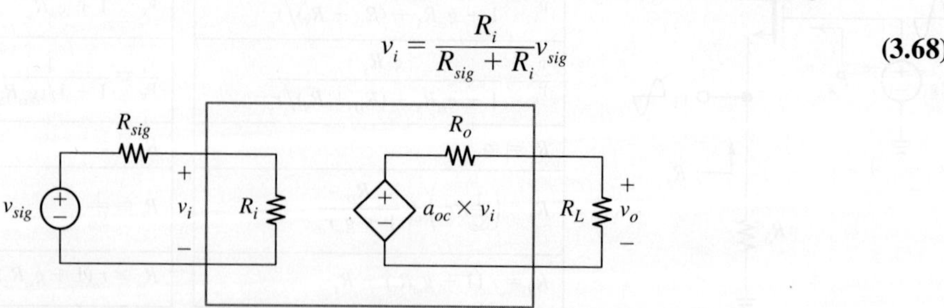

FIGURE 3.51 Block diagram of a voltage amplifier.

Likewise, at the amplifier's output we have another voltage divider, resulting in *output loading*

$$v_o = \frac{R_L}{R_o + R_L} a_{oc} \times v_i \tag{3.69}$$

We observe that

$$a_{oc} = \frac{v_o}{v_i}\Big|_{R_L \to \infty} \tag{3.70}$$

that is, a_{oc} represents the gain with which the amplifier would amplify its input v_i in the absence of any output load. Consequently, a_{oc} is called the *open-circuit voltage gain*, or also the *unloaded voltage gain*. Eliminating v_i from the above equations, we obtain the *signal-to-load voltage gain*

$$\frac{v_o}{v_{sig}} = \frac{R_i}{R_{sig} + R_i} \times a_{oc} \times \frac{R_L}{R_o + R_L} \tag{3.71}$$

As the signal progresses from the source to the load, first it undergoes some attenuation at the amplifier's input, then it gets magnified by a_{oc}, and finally it undergoes some additional attenuation at the output.

The Common-Source (CS) Configuration

In the circuit of Fig. 3.52 the amplifier proper is made up of the FET and surrounding components. To prevent the source and the load from disturbing the dc conditions of the FET we use the *ac-coupling capacitors* C_1 and C_2. Moreover, to establish an *ac ground* at the source terminal, we use the *bypass capacitor* C_3.

At dc the capacitors draw zero current and thus act as *open circuits*. In fact, in the dc equivalent of Fig. 3.53a, the capacitors have been omitted altogether. Also, to simplify dc analysis, we assume $\lambda = 0$ and we use a dc current sink I_D to bias the

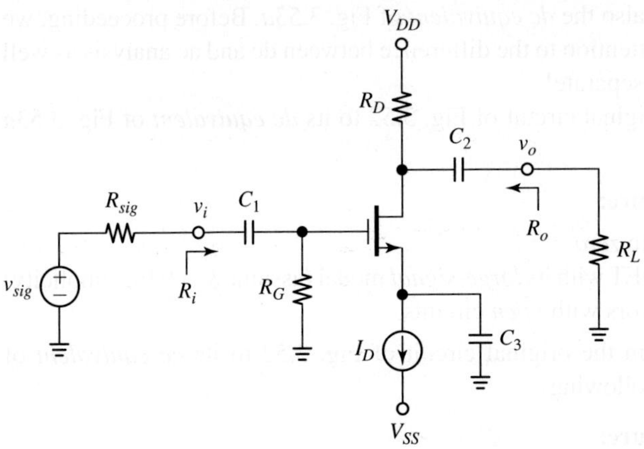

FIGURE 3.52 The *common-source* (CS) amplifier.

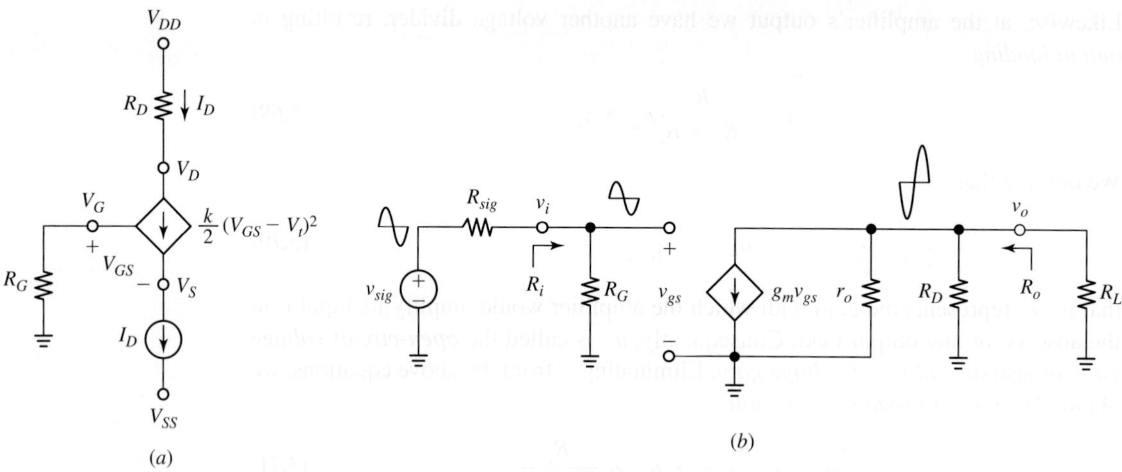

FIGURE 3.53 (a) Dc and (b) ac equivalents of the CS amplifier of Fig. 3.52.

FET. Such a sink could be implemented with another FET, such as a current mirror (let us not worry about these details here). The dc voltages are then

$$V_G = 0 \qquad V_D = V_{DD} - R_D I_D \qquad V_S = -(V_t + V_{OV}) \tag{3.72}$$

where $V_{OV} = \sqrt{2I_D/k}$. When power is applied to the circuit, each capacitor will charge up until its plates attain the dc voltages of their corresponding nodes. For instance, while the bottom plate of C_3 remains at ground potential, the top plate will charge to V_S, which in this circuit is negative. Likewise, the left plate of C_2 will charge to V_D, while the right plate is pulled to 0 V by R_L.

When analyzing a voltage amplifier we are interested in its *signal-to-load voltage gain* v_o/v_{sig}. By Eq. (3.71) this requires finding the *input resistance* R_i seen by the signal source, the *output resistance* R_o seen by the load, and the *unloaded voltage gain* a_{oc}. We find these parameters by working with the *ac equivalent* of Fig. 3.53b. However, since the small-signal parameters g_m and r_o depend on the dc bias of the FET, we need to analyze also the *dc equivalent* of Fig. 3.53a. Before proceeding, we wish to call the reader's attention to the difference between dc and ac analysis as well as the need to keep them separate!

In going from the original circuit of Fig. 3.52 to its *dc equivalent* of Fig. 3.53a we apply the following

- **Dc Analysis Procedure:**
  - Set all *ac* sources to *zero*
  - Replace the MOSFET with its *large-signal* model (assume $\lambda = 0$ for simplicity)
  - Replace all capacitors with *open* circuits

Conversely, in going from the original circuit of Fig. 3.52 to its *ac equivalent* of Fig. 3.53b we apply the following

- **Ac Analysis Procedure:**
  - Set all *dc* sources to *zero*

- Replace the MOSFET with its *small-signal* model, inclusive of r_o
- Replace all capacitors with *short* circuits

With reference to Fig. 3.53b, we note by inspection that

$$R_i = R_G \qquad\qquad R_o = R_D//r_o \qquad\qquad (3.73)$$

Moreover, by Ohm's law, we have

$$0 - v_o = (r_o//R_D//R_L)g_m v_{gs} = (r_o//R_D//R_L)g_m v_i$$

Letting $R_L \to \infty$ gives $v_o = -(r_o//R_D)g_m v_i$. But, according to Eq. (3.70), the ratio v_o/v_i in the limit $R_L \to \infty$ is the *unloaded voltage gain*. Consequently,

$$a_{oc} = -g_m(R_D//r_o) \qquad\qquad (3.74)$$

Having obtained expressions for R_i, R_o, and a_{oc}, we now apply Eq. (3.71) to find the signal-to-load gain.

EXAMPLE 3.30

In the circuit of Fig. 3.52 let $V_{DD} = -V_{SS} = 12$ V, $I_D = 1$ mA, $R_G = 1$ MΩ, and $R_D = 10$ kΩ, and let the FET have $V_t = 1.0$ V, $k = 0.5$ mA/V^2, and $\lambda = 0.01$ V^{-1}. Assuming $R_{sig} = 0.1$ MΩ, $R_L = 30$ kΩ, and

$$v_{sig} = (100 \text{ mV})\cos\omega t$$

find all node voltages in the circuit, express each of them as the sum of the dc and ac component, in the manner of Eq. (3.48), and show them explicitly in the circuit.

Solution

We have $V_{OV} = \sqrt{2I_D/k} = \sqrt{2 \times 1/0.5} = 2$ V, so Eq. (3.72) gives

$$V_G = 0 \qquad V_D = 12 - 10 \times 1 = 2 \text{ V} \qquad V_S = -(1 + 2) = -3 \text{ V}$$

Moreover, $g_m = \sqrt{2kI_D} = \sqrt{2 \times 0.5 \times 1} = 1$ mA/V and $r_o = 1/\lambda I_D = 1/(0.01 \times 1) = 100$ kΩ. Consequently, Eqs. (3.73) and (3.74) give

$$R_i = 1 \text{ M}\Omega \qquad R_o = 10//100 = 9.09 \text{ k}\Omega \qquad a_{oc} = -1 \times 9.09 = -9.09 \text{ V/V}$$

Also, Eq. (3.68) gives $v_i = [1/(0.1 + 1)]v_{sig} = (90.9 \text{ mV})\cos\omega t$. Finally, using Eq. (3.71) we find

$$v_o = \left[\frac{1}{0.1 + 1}(-9.09)\frac{30}{9.09 + 30}\right]v_{sig} = -6.34 \times (100 \text{ mV})\cos\omega t$$

$$= (634 \text{ mV})\cos(\omega t - 180°)$$

The node voltages are shown in Fig. 3.54. The reader is encouraged to verify each of them in detail.

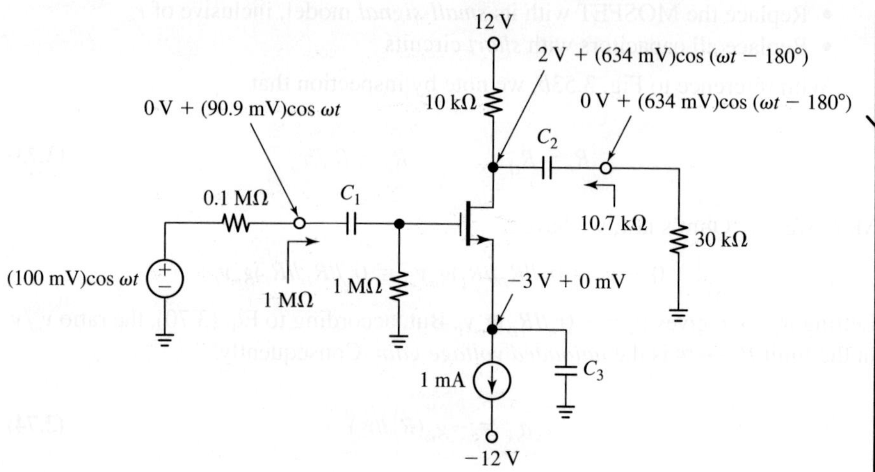

FIGURE 3.54 Circuit of Example 3.30, showing the *dc* and *ac* voltage component at each node.

The systematic procedure of redrawing an amplifier both in dc and ac form, as exemplified in Fig. 3.53, though highly recommended for the beginner, may soon prove an overkill as one seeks to speed up the analysis process. With experience, some of the intermediate steps can be carried out mentally without having to draw detailed circuit equivalents. Moreover, one can use inspection to recycle a good deal of the results developed in connection with the generalized circuit of Fig. 3.48, and summarized in Fig. 3.50. We shall illustrate with a variety of examples as we proceed.

EXAMPLE 3.31   Shown in Fig. 3.55a is a popular single-supply realization of the CS amplifier. The function of R_1 and R_2 is to bias the gate at some intermediate voltage between the 10-V supply and ground, and that of R_S is to set the value of the bias current I_D.

(a) Assuming $V_t = 1.5$ V, $k = 0.8$ mA/V^2, and $\lambda = 0.02$ V^{-1}, find the small-signal parameters R_i, R_o, and $a_{oc} = v_o/v_i$.

(b) Find the signal-to-load gain if the circuit is driven by a source with $R_{sig} = 100$ kΩ, and it drives a load $R_L = 75$ kΩ.

Solution

(a) First we need to find the bias current I_D. Considering that at dc all capacitors act as open circuits, it is apparent that the dc version of our amplifier is of the type of Fig. 3.34a. Proceeding as in Example 3.23 with $\lambda = 0$ to simplify dc analysis, we readily find $I_D = 0.4$ mA. We thus have

$$g_m = \sqrt{2kI_D} = \sqrt{2 \times 0.8 \times 0.4} = 0.8 \text{ mA/V}$$

$$r_o = 1/\lambda I_D = 1/(0.02 \times 0.4) = 125 \text{ k}\Omega$$

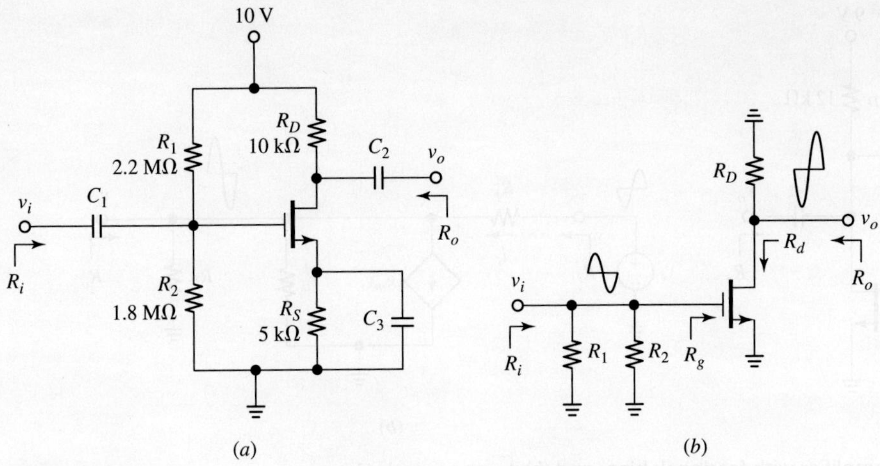

FIGURE 3.55 (a) Single-supply CS amplifier of Example 3.31, and (b) its ac equivalent.

Next, refer to the ac equivalent of Fig. 3.55b, where we note that the by-pass action by C_3 puts the source at *ac ground*. Consequently, we recycle the results tabulated in Fig. 3.50, but with $R_S = 0$. By inspection,

$$R_i = R_1//R_2//R_g = 2.2//1.8//\infty = 990 \text{ k}\Omega$$

$$R_o = R_D//R_d = R_D//r_o = 10//125 = 9.26 \text{ k}\Omega$$

$$a_{oc} = -g_m R_o = -0.8 \times 9.26 = -7.4 \text{ V/V}$$

(b) Owing to input and output loading, the gain drops to

$$\frac{v_o}{v_{sig}} = \frac{990}{100 + 990}(-7.4)\frac{75}{9.26 + 75} = -5.98 \text{ V/V}$$

EXAMPLE 3.32

The single-supply CS amplifier of Fig. 3.56a utilizes a popular dc biasing alterna-tive known as *feedback bias*. The name stems from the presence of resistance R_F, which feeds the dc voltage of the drain back to the gate. Since $I_G = 0$, the voltage drop across R_F is 0 V, indicating that the FET operates with $V_G = V_D$, that is, in the diode mode, and therefore in saturation.

(a) Assuming $V_t = 2.0$ V, $k = 1.0$ mA/V^2, and $\lambda = 0.033$ V^{-1}, find the dc operat-ing point of the FET, as well as $a_{oc} = v_o/v_i$, R_i (the input resistance with the output port open-circuited), and R_o (the output resistance with the input port short-circuited).

(b) Assuming v_i is an ac signal with amplitude V_{im}, what is the maximum value of V_{im} if we wish to contain the small-signal approximation error within 10%? What is the corresponding amplitude V_{om} of the output? Is the FET operating at all times in the active region?

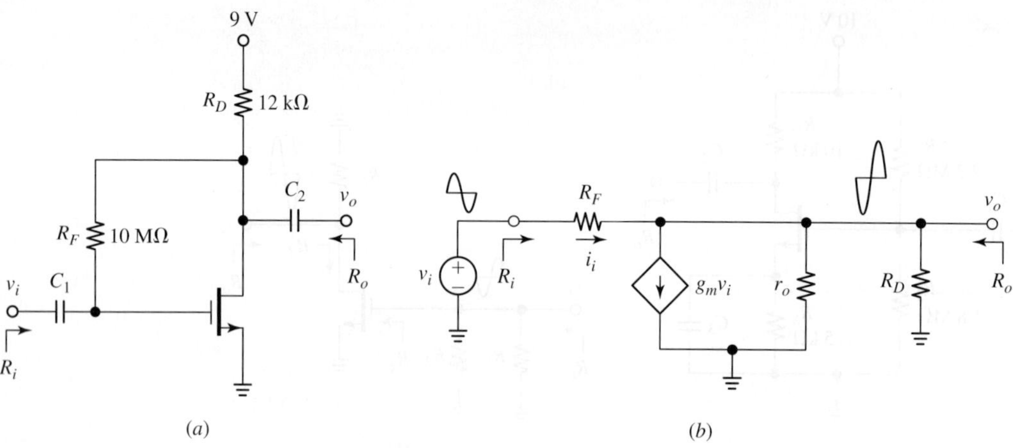

FIGURE 3.56 (a) CS amplifier with feedback bias, and (b) its ac equivalent.

Solution

(a) We have $V_{GS} = V_{DS} = V_{DD} - R_D I_D = 9 - 12 I_D$, so $V_{OV} = V_{GS} - V_t = 7 - 12 I_D$. Letting

$$I_D = \frac{1}{2}(7 - 12 I_D)^2$$

and solving as usual, we get $I_D = 0.5$ mA and $V_{DS} = 3.0$ V. Moreover, $g_m = 1.0$ mA/V and $r_o = 60.6$ kΩ.

To find the small-signal parameters we need to work explicitly with the small-signal equivalent of Fig. 3.56*b*. (Note that the presence of the feedback resistance R_F precludes us from recycling the results of Fig. 3.50!) Given that R_o is the output-node resistance in the limit $v_i \rightarrow 0$, we find, by inspection,

$$R_o = R_F /\!/ r_o /\!/ R_D = 10{,}000 /\!/ 60.6 /\!/ 12 = 10 \text{ k}\Omega$$

To find a_{oc} we apply KCL at the output node, and get

$$\frac{v_i - v_o}{R_F} = g_m v_i + \frac{v_o}{r_o} + \frac{v_o}{R_D}$$

Collecting terms and solving for the ratio v_o/v_i gives

$$a_{oc} = \frac{v_o}{v_i} = -\left(g_m - \frac{1}{R_F}\right)R_o \cong -g_m R_o = -1 \times 10 = -10 \text{ V/V}$$

where we have exploited the fact that $1/R_F \ll g_m$. The input resistance is found as $R_i = v_i/i_i$, where $i_i = (v_i - v_o)/R_F$. Substituting $v_o = a_{oc} \times v_i$ and collecting gives $i_i = v_i(1 - a_{oc})/R_F$. Consequently,

$$R_i = \frac{v_i}{i_i} = \frac{R_F}{1 - a_{oc}} = \frac{10 \times 10^6}{1 - (-10)} = \frac{10 \times 10^6}{11} = 0.91 \text{ M}\Omega$$

Interestingly enough, when reflected to the input, the feedback resistance R_F gets *divided* by the factor $(1 - a_{oc})$, a phenomenon known as the *Miller effect* (more on this in Chapters 6 and 7).

(b) For an error of not more than 10%, Eq. (3.57) requires that we keep $V_{im} \leq (2V_{OV})/10 = (2 \times 1)/10 = 0.2\,\text{V}$. The corresponding output amplitude is $V_{om} \cong |a_{oc}| \times V_{im} = 10 \times 0.2 = 2\,\text{V}$. The dc voltage at the drain is $V_D = 3\,\text{V}$, and operation in the active region, or saturation, is maintained all the way down to $V_D = V_{OV} = 1\,\text{V}$. The maximum allowed downswing for the drain is thus 2 V, indicating that our amplifier will just barely accommodate an input with 0.2-V of peak amplitude.

Quick Estimate for the Gain of the CS Configuration

In daily practice an engineer often needs to come up with a quick if rough estimate for the gain of a MOSFET amplifier. The above examples reveal that the CS configuration tends to give $a_{oc} = -g_m R_o$. In discrete designs R_o is usually dominated by R_D, indicating that we can approximate $a_{oc} \cong -g_m R_D$. Letting $g_m = 2I_D/V_{OV}$, we can estimate the unloaded gain of a discrete-type CS amplifier as

$$a_{oc} \cong -g_m R_D = -\frac{R_D I_D}{0.5V_{OV}} \tag{3.75a}$$

In words, the magnitude of the unloaded gain of a CS amplifier is the ratio of the *voltage-drop across R_D* to half the *overdrive V_{OV}*. For the circuit of Example 3.30, Eq. (3.75a) gives the estimate $a_{oc} \cong -2(10 \times 1)/2 = -10\,\text{V/V}$, in fair agreement with the calculated value of $-9.09\,\text{V/V}$. Likewise, the estimates for Examples 3.31 and 3.32 are $a_{oc} \cong -2(10 \times 0.4)/1 = -8\,\text{V/V}$, and $a_{oc} \cong -2(12 \times 0.5)/1 = -12\,\text{V/V}$. Both compare reasonably well with the calculated values of $-7.4\,\text{V/V}$ and $-10\,\text{V/V}$.

The student already familiar with bipolar junction transistors (BJTs) will note that the expression $a_{oc} \cong -R_D I_D/(0.5V_{OV})$ for the CS amplifier is similar to the expression $a_{oc} \cong -R_C I_C/V_T$ for the common-emitter (CE) amplifier. However, considering that $V_T = 26\,\text{mV}$, while $0.5V_{OV}$ is typically on the order of a volt or so, it is apparent that under similar biasing conditions the gain available from a FET tends to be *much lower* than that available from a BJT. This stems from the notoriously lower transconductance of FETs. In IC design the gain of the CS configuration is boosted by using a current source in place of R_D. As we shall see in Chapter 4, such a source is implemented with a pMOSFET.

Exercise 3.3

Show that if we take also r_o into account, then Eq. (3.75a) becomes

$$a_{oc} = -\frac{R_D I_D}{0.5V_{OV}} \times \frac{1}{1 + \lambda R_D I_D} \tag{3.75b}$$

The Common-Source with Source-Degeneration (CS-SD) Configuration

The circuit of Fig. 3.57 is similar to the CS amplifier of Fig. 3.52, except for the presence of the unbypassed resistance R_S in series with the FET's source. Turning to its ac equivalent of Fig. 3.58 we note that, aside from the presence of the input voltage

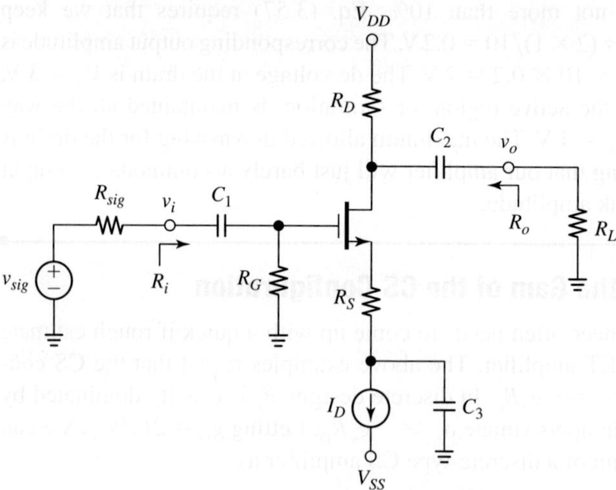

FIGURE 3.57 The *common-source with source-degeneration* (CS-SD) amplifier.

divider and output load, the circuit is identical to that of Fig. 3.50. We can again reuse the relationships developed there, provided we let $v_s/v_g \rightarrow v_o/v_i$. By inspection,

$$R_i = R_G \qquad (3.76a)$$

As we know, one of the effects of the source-degeneration resistance R_S is to raise the resistance seen looking into the drain from r_o to $R_d \cong r_o(1 + g_m R_S)$. Consequently, we now have

$$R_o = R_D /\!/ R_d \cong R_D /\!/ [r_o(1 + g_m R_S)] \cong R_D \qquad (3.76b)$$

The unloaded voltage gain, defined as the ratio v_o/v_i in the limit $R_L \rightarrow \infty$, is now

$$a_{oc} = -\frac{g_m R_D}{1 + g_m R_S + (R_D + R_S)/r_o} \cong -\frac{g_m R_D}{1 + g_m R_S} \qquad (3.77)$$

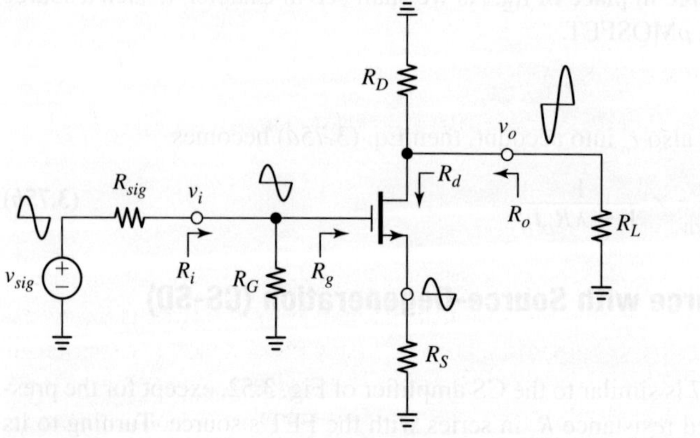

FIGURE 3.58 Ac equivalent of the CS-SD amplifier of Fig. 3.57.

where we have exploited the fact that in discrete designs we have $(R_D + R_S) \ll r_o$. Comparing Eq. (3.77) with Eq. (3.74) we observe that the presence of R_S causes a_{oc} to drop from approximately $-g_m R_D$ to approximately $-g_m R_D/(1 + g_m R_S)$. This magnitude reduction stems from the negative-feedback action, or *degenerative* action, provided by R_S. Rewriting in the alternative form

$$a_{oc} \cong -\frac{R_D}{1/g_m + R_S} \qquad (3.78)$$

provides us with a useful rule of thumb for a quick estimation of the gain of the CS-SD configuration:

> The unloaded voltage gain from gate to drain is the (negative of the) *ratio* of the *drain resistance* to the *total source resistance*

Having obtained expressions for R_i, R_o, and a_{oc}, we can apply Eq. (3.71) to find the signal-to-load gain.

EXAMPLE 3.33

(a) Investigate the effect of inserting a source-degeneration resistance $R_S = 2\ \text{k}\Omega$ in the CS circuit of Example 3.30, and thus turning it into a CS-SD circuit of the type of Fig. 3.57.

(b) Estimate R_S for an unloaded gain of -2 V/V.

Solution

(a) All dc voltages and dc current remain the same, and so do g_m ($=1$ mA/V) and r_o ($=100$ kΩ). The insertion of $R_S = 2\ \text{k}\Omega$ in the circuit has the following effects:

- R_d increases from 100 kΩ to $100(1 + 1 \times 2) = 300$ kΩ.
- R_o increases from 9.09 kΩ to $10//300 = 9.68$ kΩ.
- a_{oc} drops (or degenerates) from -9.09 V/V to $-(1 \times 9.68)/(1/1 + 1 \times 2) = -3.23$ V/V.

Using Eq. (3.71), we find that v_o changes to

$$v_o = \left[\frac{1}{0.1 + 1}(-3.23)\frac{30}{9.68 + 30}\right]v_{sig} = -2.22 \times (100\ \text{mV})\cos\omega t$$

$$= (222\ \text{mV})\cos(\omega t - 180°)$$

(b) Use Eq. (3.78) to impose $-2 \cong -10/(1/1 + R_S)$. This yields $R_S = 4\ \text{k}\Omega$.

Capacitor Selection

Before leaving the subject of discrete MOSFET amplifiers, we wish to address the issue of how to go about selecting the various capacitances in the circuits discussed above. As the signal source is turned on, we want each capacitance C to act as an *ac short* at the source's frequency f_{sig}. Physically, this requires that we select C large enough to prevent it from charging/discharging appreciably in response to the ac signal alternations.

As we know, the impedance presented by a capacitance C at the signal frequency f_{sig} is $Z_C(jf_{sig}) = 1/(j2\pi f_{sig})$. For this capacitance to act effectively as an ac short at f_{sig}, its impedance must be such that

$$|Z_C(jf_{sig})| \ll R_{eq}$$

where R_{eq} is the *equivalent resistance* seen by C itself. This condition is rephrased in terms of C as

$$C \gg \frac{1}{2\pi R_{eq} f_{sig}} \tag{3.79}$$

If the circuit is designed to operate over a *range* of frequencies, then we must use the lowest frequency $f_{sig(min)}$ in the above condition. It is good practice to use $C \cong 10/(2\pi R_{eq} f_{sig(min)})$.

EXAMPLE 3.34   Specify suitable capacitances in the CS amplifier of Fig. 3.54 for operation over the *audio range*.

Solution

The audio range extends from 20 Hz to 20 kHz, so $f_{sig(min)} = 20$ Hz.

- For C_1 we have $R_{eq1} = R_{sig} + R_i = 0.1 + 1 = 1.1$ MΩ, so $C_1 \gg 1/[2\pi \times 1.1 \times 10^6 \times 20) \cong 7$ nF (use 100 nF).
- For C_2 we have $R_{eq2} = R_o + R_L = 10.7 + 30 = 40.7$ kΩ, so $C_2 \gg 1/[2\pi \times 40.7 \times 10^3 \times 20) \cong 0.2$ μF (2 μF).
- For C_3 we have $R_{eq3} = R_s \cong 1/g_m = 1$ kΩ, so $C_3 \gg 1/[2\pi \times 10^3 \times 20) \cong 8$ μF (100μF).

3.9 MOSFET VOLTAGE AND CURRENT BUFFERS

In this section we examine the two remaining single-FET amplifier configurations of interest, namely, the *common-drain* and the *common-gate* configurations. We shall see that these configurations find application as *voltage buffers* and *current buffers*, respectively.

The Common-Drain (CD) Configuration

The common-drain (CD) amplifier receives the input at the gate and delivers the output at the source. The circuit realization of Fig. 3.59a utilizes the same biasing scheme as the CS amplifier of Fig. 3.52. Turning to its ac equivalent of Fig. 3.59b, we observe that it is similar to that of Fig. 3.50, but with $R_D = 0$. Instead of repeating the small-signal analysis routine all over again, we simply reuse the results tabulated there, but after letting $R_D \to 0$ and re-labeling resistors and signals as $R_s \to R_L$, $R_s \to R_o$, and $v_s/v_g \to v_o/v_i$. This yields the following results:

$$R_i = R_G \qquad R_o = \frac{r_o}{1 + g_m r_o} = \frac{1}{g_m}//r_o \tag{3.80}$$

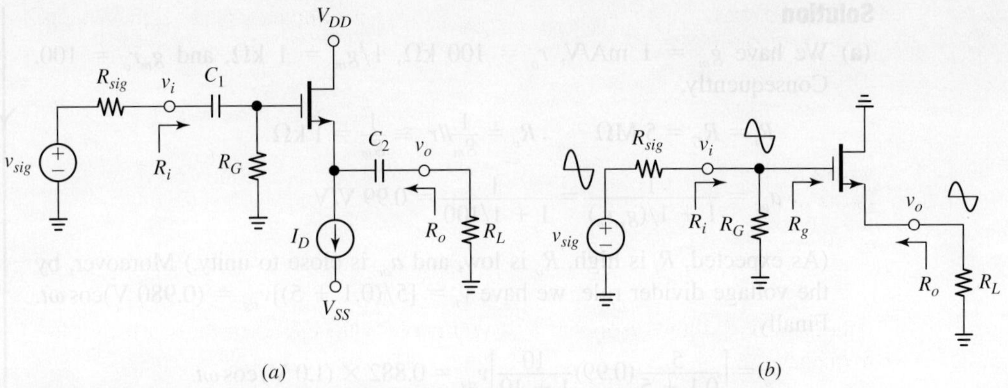

FIGURE 3.59 (a) The *common-drain* (CD) configuration, and (b) its ac equivalent.

$$v_o = \frac{g_m R_L}{1 + g_m R_L + R_L/r_o} v_i$$

In the limit $R_L \to \infty$ we get

$$v_o = \frac{g_m}{g_m + 1/r_o} v_i = \frac{1}{1 + 1/(g_m r_o)} v_i$$

But, according to Eq. (3.70), the ratio v_o/v_i in the limit $R_L \to \infty$ is the *unloaded voltage gain*, so

$$a_{oc} = \frac{1}{1 + 1/(g_m r_o)} \tag{3.81}$$

Since in general $g_m r_o \gg 1$, the unloaded gain from gate to source is fairly close to unity. Physically, the source voltage v_s *follows* the gate voltage v_g closely, this being the reason why the CD amplifier is also called *source follower*. Even though not much of a winner as a voltage amplifier, the CD configuration offers the advantages of potentially *high input resistance* and *low output resistance*, which makes it suited to applications as a *voltage buffer*, either to reduce inter-stage loading, or to equip a CS amplifier with low output resistance. Having obtained expressions for R_i, R_o, and a_{oc}, we can now apply Eq. (3.71) to find the signal-to-load gain.

EXAMPLE 3.35

(a) In the circuit of Fig. 3.59, let $V_{DD} = -V_{SS} = 10$ V, $I_D = 1$ mA, and $R_G = 5$ MΩ, and let the FET have $V_t = 1.0$ V, $k = 0.5$ mA/V^2, and $\lambda = 0.01$ V^{-1}. Assuming $R_{sig} = 0.1$ MΩ, $R_L = 10$ kΩ, and

$$v_{sig} = 5.0 \text{ V} + (1.0 \text{ V})\cos \omega t$$

find all node voltages in the circuit, express each of them as the sum of its dc and ac component, in the manner of Eq. (3.48), and show them explicitly in the circuit.

(b) Check that the FET satisfies the small-signal approximation condition of Eq. (3.57).

Solution

(a) We have $g_m = 1$ mA/V, $r_o = 100$ kΩ, $1/g_m = 1$ kΩ, and $g_m r_o = 100$. Consequently,

$$R_i = R_G = 5 \text{ M}\Omega \qquad R_o = \frac{1}{g_m}//r_o \cong \frac{1}{g_m} = 1 \text{ k}\Omega$$

$$a_{oc} = \frac{1}{1 + 1/(g_m r_o)} = \frac{1}{1 + 1/100} = 0.99 \text{ V/V}$$

(As expected, R_i is high, R_o is low, and a_{oc} is close to unity.) Moreover, by the voltage divider rule, we have $v_i = [5/(0.1 + 5)]v_{sig} = (0.980 \text{ V})\cos\omega t$. Finally,

$$v_o = \left[\frac{5}{0.1 + 5}(0.99)\frac{10}{1 + 10}\right]v_{sig} = 0.882 \times (1.0 \text{ V})\cos\omega t$$

$$= (0.882 \text{ V})\cos\omega t$$

The node voltages are shown in Fig. 3.60. The reader is encouraged to verify each of them in detail.

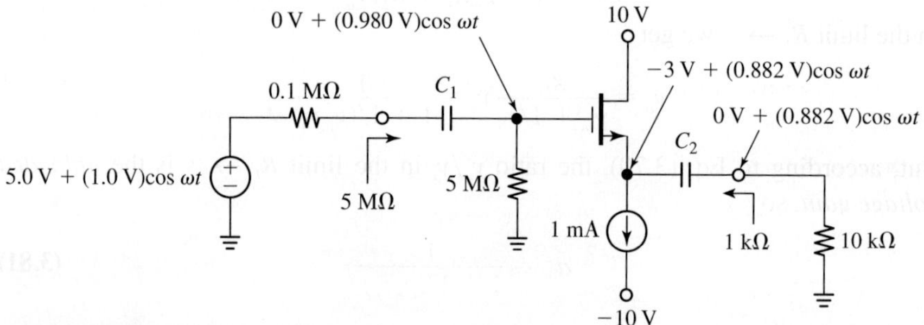

FIGURE 3.60 Circuit of Example 3.35, showing the *dc* and *ac* component of each node voltage.

(b) For the FET we have $v_{gs} = v_i - v_o \cong (98 \text{ mV})\cos\omega t$. Considering that $V_{OV} = 2$ V, so that $2V_{OV} = 4$ V, we have $0.098 \ll 4$, thus confirming the validity of the small-signal approximation.

Remark: Even though neither v_i nor v_o can be regarded as small signals in this circuit, v_{gs} nevertheless is! The reason for the CD amplifier's ability to handle linearly even signals that are not strictly of the small-signal type stems from the *negative feedback action* (more in Chapter 7) provided by the source resistance (in this case R_L). Indeed, R_L develops a voltage v_o close to v_i to yield the small-signal difference $v_{gs} = v_i - v_o$.

EXAMPLE 3.36

Figure 3.61 shows a single-supply source follower. To prevent the follower from loading the signal source appreciably, R_1 and R_2 have been chosen so that $R_1//R_2 \gg R_{sig}$. As usual, the function of R_S is to establish the bias current I_D. Assuming $V_t = 1.0$ V, $k = 0.625$ mA/V^2, and $\lambda = 0.025$ V^{-1}, find the small-signal resistances R_i and R_o, and estimate the signal-to-load gain.

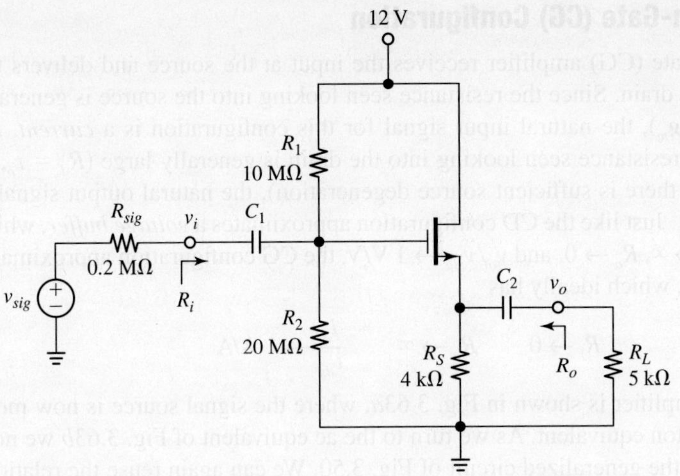

FIGURE 3.61 Single-supply CD amplifier of Example 3.36.

Solution

Proceeding as usual, but with $\lambda = 0$ to facilitate our dc calculations, we find $I_D = 1.25$ mA, $1/g_m = 0.8$ kΩ, $r_o = 32$ kΩ, and $g_m r_o = 40$. Referring to the ac equivalent of Fig. 3.62 we find, by inspection,

$$R_i = R_1 /\!/ R_2 /\!/ R_g = 10/\!/20/\!/\infty = 6.7 \text{ M}\Omega \text{ (large)}$$

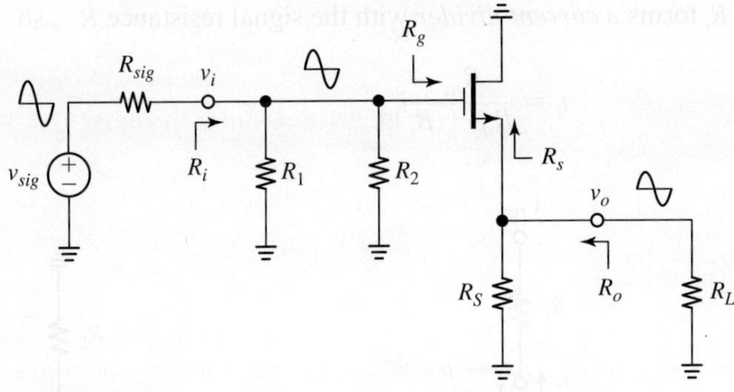

FIGURE 3.62 Ac equivalent of the circuit of Fig. 3.61.

We also have $R_s = (1/g_m)/\!/r_o = 0.8/\!/32 = 0.78$ kΩ, so we can write, by inspection,

$$R_o = R_s /\!/ R_S = 0.78/\!/4 = 0.65 \text{ k}\Omega$$

Finally, the source-to-load gain is

$$\frac{v_o}{v_s} = \frac{6.7}{0.2 + 6.7} \times \frac{1}{1 + 1/40} \times \frac{5}{0.65 + 5} = 0.838 \text{ V/V}$$

The Common-Gate (CG) Configuration

The common-gate (CG) amplifier receives the input at the source and delivers the output from the drain. Since the resistance seen looking into the source is generally small ($R_s \cong 1/g_m$), the natural input signal for this configuration is a *current*, i_{sig}. Also, since the resistance seen looking into the drain is generally large ($R_d = r_o$, or even $R_d \gg r_o$ if there is sufficient source degeneration), the natural output signal is also a *current*, i_o. Just like the CD configuration approximates a *voltage buffer*, which ideally has $R_i \to \infty$, $R_o \to 0$, and $v_o/v_{sig} \to 1$ V/V, the CG configuration approximates a *current buffer*, which ideally has

$$R_i \to 0 \qquad R_o \to \infty \qquad \frac{i_o}{i_{sig}} \to 1 \text{ A/A}$$

The CG amplifier is shown in Fig. 3.63a, where the signal source is now modeled with a Norton equivalent. As we turn to the ac equivalent of Fig. 3.63b we note its similarity to the generalized circuit of Fig. 3.50. We can again reuse the relationships developed there, provided we let $R_S \to R_{sig}$, $R_D \to R_L$, $R_s \to R_i$, and $R_d \to R_o$. The results are

$$R_i = \left(\frac{1}{g_m}//r_o\right) + \frac{R_L}{1 + g_m r_o} \tag{3.82}$$

and

$$R_o = r_o(1 + g_m R_{sig}) + R_{sig} \tag{3.83}$$

The input resistance R_i forms a *current divider* with the signal resistance R_{sig}, so

$$i_i = \frac{R_{sig}}{R_{sig} + R_i} i_{sig}$$

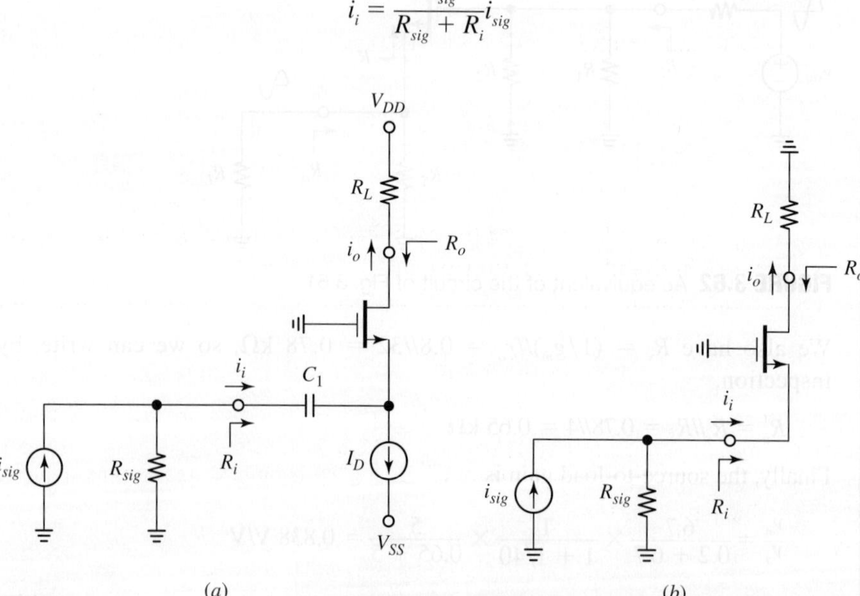

FIGURE 3.63 (a) The *common-gate* (CG) amplifier, and (b) its ac equivalent.

Since the gate current is zero, KCL gives $i_o = i_i$. Combining with Eq. (3.82) we obtain, after suitable algebraic manipulation, the *signal-to-load current gain*

$$\frac{i_o}{i_{sig}} = \frac{1}{1 + \dfrac{r_o + R_L}{(1 + g_m r_o)R_{sig}}} \tag{3.84}$$

It is apparent that this gain is *less* than (though close to) unity. The CG configuration is particularly useful when its signal input is supplied by the drain of another FET. The resulting two-transistor configuration, known as the *cascode configuration*, enjoys advantages of speed and flexibility that make it particularly suited to integrated-circuit realizations, as we shall see in Chapters 4 and 6.

EXAMPLE   3.37

(a) In the circuit of Fig. 3.63a let $V_{DD} = -V_{SS} = 12$ V, $I_D = 1$ mA, and $R_{sig} = 50$ kΩ, and let the FET have the same parameters as in Example 3.35 ($V_t = 1.0$ V, $k = 0.5$ mA/V^2, and $\lambda = 0.01$ V^{-1}). Estimate R_i, R_o, and i_o/i_{sig} if $R_L = 0$. Comment on your findings.

(b) Repeat if $R_L = 10$ kΩ, and comment.

Solution

(a) By Eq. (3.82) through (3.84) we have

$$R_i \cong 1//100 = 0.99 \text{ kΩ (low)}$$

$$R_o = 100(1 + 1 \times 50) + 50 = 5.15 \text{ MΩ (high)}$$

$$\frac{i_o}{i_{sig}} = \frac{1}{1 + \dfrac{100}{(1 + 100)50}} = 0.980 \text{ A/A (close to unity)}$$

(b) Recalculating with $R_L = 10$ kΩ we get

$$R_i \cong (1//100) + 10/101 = 1.09 \text{ kΩ} \qquad R_o = 5.15 \text{ MΩ}$$

$$\frac{i_o}{i_{sig}} = \frac{1}{1 + \dfrac{100 + 10}{101 \times 50}} = 0.979 \text{ A/A}$$

The presence of $R_L = 10$ kΩ has no effect on R_o. However, it causes a slight increase in R_i and a slight drop in gain.

The CG Configuration as a Voltage Amplifier

Even though the most common application of the CG configuration is as a current buffer, there are situations in which it is used as a *voltage amplifier* with gain v_d/v_s. Considering that $v_d = -g_m(R_L//r_o)v_{gs} = -g_m(R_L//r_o)(v_g - v_s)$, and that the CG configuration has $v_g = 0$, we get

$$\frac{v_d}{v_s} = +g_m(R_L//r_o) \tag{3.85}$$

In words, the voltage gain of the CG configuration has the *same magnitude* but *opposite polarity* as that of the CS configuration. The other major difference is in the input resistance, which is ∞ in the CS case, but generally low in the CG case. For the circuit of Example 3.37*b* this voltage gain is $v_d/v_s = 1 \times (10//100) \cong +9.1$ V/V. In the limit $R_L \to \infty$ this gain tends to $g_m r_o = 1 \times 100 = 100$ V/V (more on this in Chapter 4).

3.10 THE CMOS INVERTER/AMPLIFIER

A simple yet most elegant and useful circuit configuration, the CMOS inverter/amplifier is at the basis of a wide variety of contemporary circuitry, both digital and analog. As shown in Fig. 3.64, it consists of an *n*MOSFET and a *p*MOSFET with their gates tied together to form the input node, and their drains tied together to form the output node. For each device, body and source are tied together, so there are no body effects. Moreover, the source of the *n* channel is connected to the lowest potential, and that of the *p* channel to the highest potential. In our example, these potentials are *ground* and V_{DD}, but other arrangements are possible, such as split power supplies. The circuit is usually implemented with *matched* FETs, whose parameters we shall concisely express as

$$k_p = k_n = k \qquad V_{tn} = -V_{tp} = V_t \qquad \lambda_n = \lambda_p = \lambda \qquad (3.86)$$

Considering that the process transconductance parameter k'_p is typically 2 to 3 times smaller than its counterpart k'_n, the manufacturer compensates for this imbalance by fabricating the *p*MOSFET with a W/L ratio 2 to 3 times greater than that of the *n*MOSFET, thus ensuring matched device transconductance parameters, or $k_p = k_n$. Moreover, the manufacturer effects suitable doping implants to ensure $|V_{tp}| = V_{tn}$. Typically, the implant dosages are chosen so that $V_t \cong 0.2V_{DD}$, or $V_t = 1$ V for $V_{DD} = 5$ V. Also shown in Fig. 3.64 is the logic symbol used for the inverter, with the power-supply details omitted to reduce cluttering the circuit schematics.

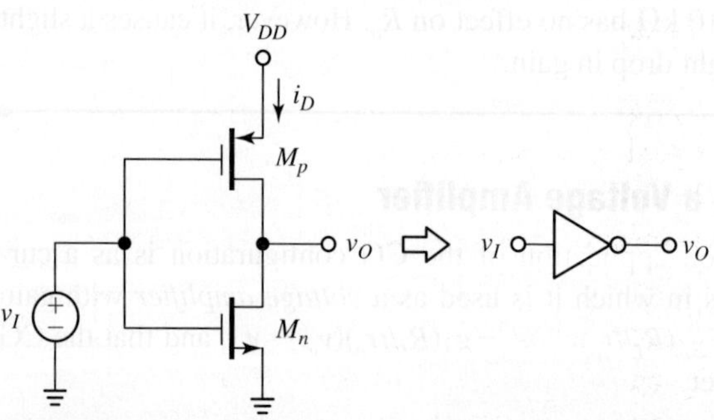

FIGURE 3.64 Circuit schematic and logic symbol of the CMOS inverter/amplifier.

The Voltage Transfer Curve (VTC)

To investigate circuit operation we sweep v_I from 0 to V_{DD} and examine the ensuing response v_O. Keeping in mind that $V_{GSn} = v_I$ and $V_{SGp} = V_{DD} - v_I$, we make two general observations:

- As v_I is swept from 0 to V_{DD}, M_n goes from the cutoff state to a highly conductive state, while M_p goes from a highly conductive state to the cutoff state, indicating *complementary* behavior by the FETs.
- With respect to the output v_O, M_p exerts a *pull-up* action toward V_{DD}, while M_n exerts a *pull-down* action toward ground. As a consequence, v_O will assume a value somewhere in between, depending on which pull action prevails.

The plot of v_O versus v_I, called the *voltage transfer curve* (VTC), is readily visualized via PSpice. Figure 3.65 shows one such example, and Fig. 3.66, top, displays the VTC for the component and device parameter values shown. We make the following considerations:

- For $v_I \leq V_{tn}$, or $V_{GSn} \leq 1.0$ V, M_n is in *cutoff* and acts as an *open* switch. On the other hand, M_p is highly conductive because $V_{SGp} \geq 4$ V. But, owing to the cutoff state of M_n, no current can flow through M_p, forcing it to operate right at the origin of its i_D-v_{SD} characteristics, that is, in the *ohmic* region. The situation is illustrated further in Fig. 3.67a, top, which shows the curves corresponding to $V_{GSn} = 0$ and $V_{SGp} = 5$ V. The operating point Q_H lies right at the intersection of the two curves, that is, at $v_O = V_{OH} = V_{DD} = 5$ V and $i_D = 0$. The FETs thus act as in Fig. 3.67a, bottom. The p-channel resistance pulling v_O to V_{DD} is $r_{SDp} = 1/[k_p(V_{SGp} - |V_{tp}|)] = 1/[1(5 - 1)] = 0.25$ kΩ.
- For $v_I \geq V_{DD} - |V_{tp}|$, or $v_I \geq 5 - 1 = 4$ V, we have the *opposite* situation to the one just described, namely, M_n is highly conductive while M_p is in cutoff. As depicted in Fig. 3.67c, top, the operating point Q_L lies right at the intersection of the curves corresponding to $V_{GSn} = 5$ V and $V_{SGp} = 0$ V, that is, at $v_O = V_{OL} = 0$ V and $i_D = 0$. The FETs now act as in Fig. 3.67c, bottom. The n-channel resistance pulling v_O to ground is $r_{DSn} = 1/[k_n(V_{GSn} - V_{tn})] = 0.25$ kΩ.

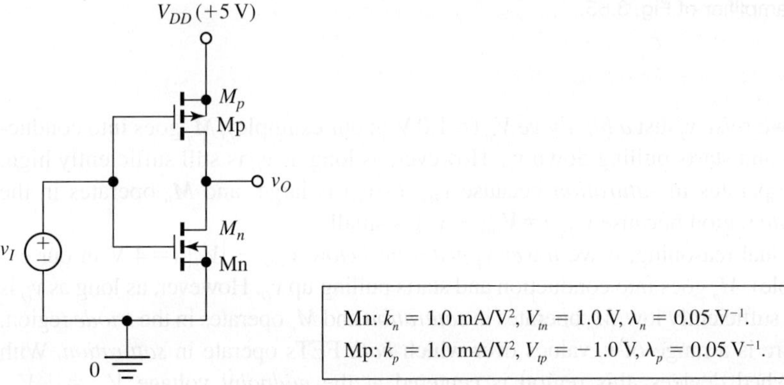

FIGURE 3.65 PSpice circuit for the simulation of the CMOS inverter/amplifier.

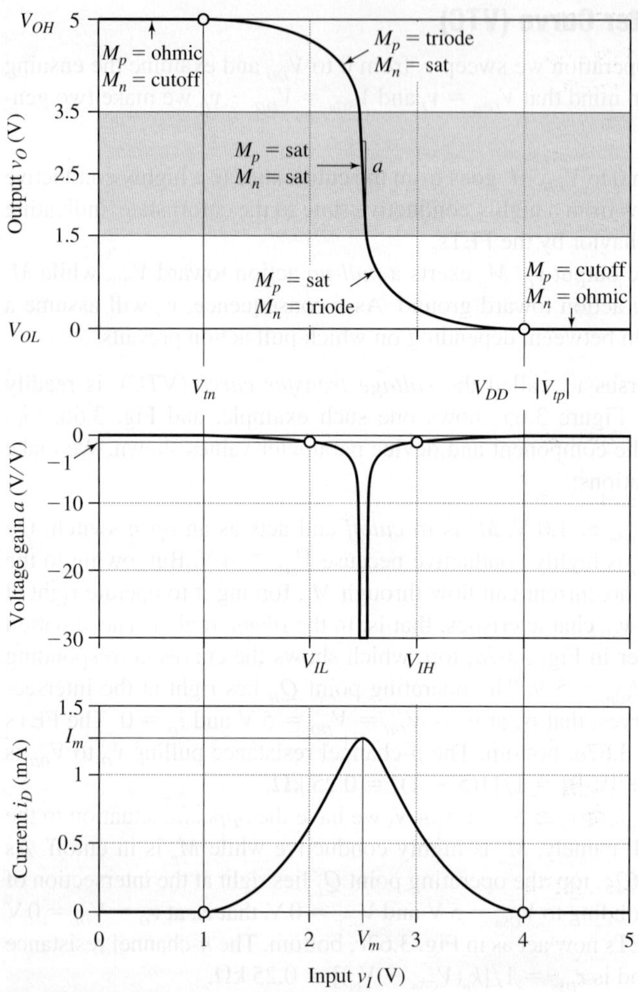

FIGURE 3.66 Plots of v_O, a, and i_D versus v_I for the CMOS inverter/amplifier of Fig. 3.65.

- As we *raise v_I just a bit above* V_{tn} ($=1.0$ V in our example), M_n goes into conduction and starts pulling down v_O. However, as long as v_O is still sufficiently high, M_n operates in *saturation* because v_{DSn} ($=v_O$) is large, and M_p operates in the *triode* region because v_{SDp} ($=V_{DD} - v_O$) is small.
- By dual reasoning, if we *lower v_I just a bit below* $V_{DD} - |V_{tp}|$ ($=4$ V in our example), M_p goes into conduction and starts pulling up v_O. However, as long as v_O is still sufficiently low, M_p operates in *saturation* and M_n operates in the *triode* region.
- There is a range of v_I values over which *both* FETs operate in *saturation*. With matched devices, this region is centered at the *midpoint* voltage $V_m = \frac{1}{2}V_{DD}$ ($=2.5$ V in our example). As depicted in Fig. 3.67b, top, the operating point Q_m lies right at the intersection of the curves corresponding to $V_{GSn} = V_m$ ($=2.5$ V) and

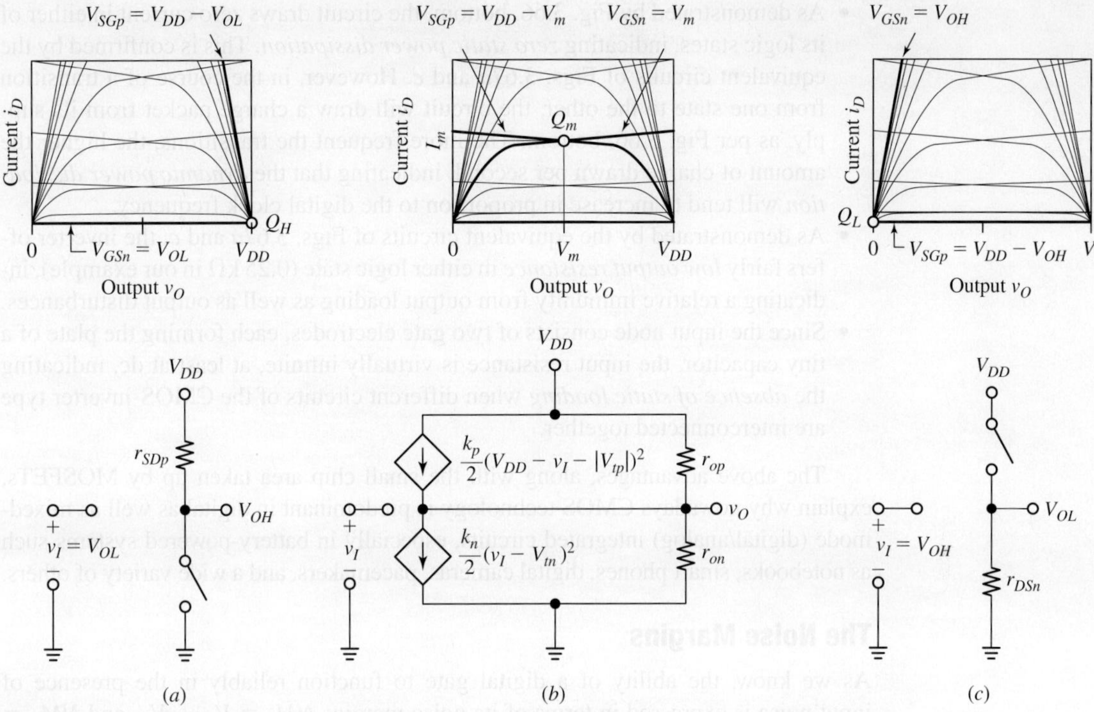

FIGURE 3.67 Operating points and large-signal models of the CMOS inverter/amplifier for the cases (a) $v_I = V_{OL}$, (b) v_I near V_m, and (c) $v_I = V_{OH}$.

$V_{SGp} = V_{DD} - V_m (= 2.5\text{ V})$. The coordinates of Q_m are thus $v_O = V_m$ and $i_D = I_m$, where I_m is readily found via the nMOSFET expression

$$I_m = \frac{k}{2}(V_m - V_t)^2(1 + \lambda V_m) \tag{3.87}$$

In our example,

$$I_m = \frac{1}{2}(2.5 - 1)^2(1 + 0.05 \times 2.5) = 1.27\text{ mA}$$

The FETs now act as in Fig. 3.67b, bottom.

As we know, the slope of the VTC represents *voltage gain*. Figure 3.66, top, indicates that slope is steepest in the region where both FETs are saturated. Figure 3.66, center, shows the gain a as a function of v_I. Also shown in Fig. 3.66, bottom, is the current i_D drawn from the power supply.

The CMOS Inverter as a Logic Element

When used as a logic element, the CMOS inverter of Fig. 3.64 offers a number of unique advantages:

- The output swings from rail to rail, or

$$V_{OL} = 0\text{ V} \qquad V_{OH} = V_{DD} \tag{3.88}$$

thus providing *maximum signal swing* and, hence, the widest noise margins.

- As demonstrated by Fig. 3.66, bottom, the circuit draws zero current in either of its logic states, indicating *zero static power dissipation*. This is confirmed by the equivalent circuits of Figs. 3.67a and c. However, in the course of a transition from one state to the other, the circuit will draw a charge packet from its supply, as per Fig. 3.66, bottom. The more frequent the transitions, the higher the amount of charge drawn per second, indicating that the *dynamic power dissipation* will tend to increase in proportion to the digital clock frequency.
- As demonstrated by the equivalent circuits of Figs. 3.67a and c, the inverter offers fairly *low output resistance* in either logic state (0.25 kΩ in our example), indicating a relative immunity from output loading as well as output disturbances.
- Since the input node consists of two gate electrodes, each forming the plate of a tiny capacitor, the input resistance is virtually infinite, at least at dc, indicating the *absence of static loading* when different circuits of the CMOS-inverter type are interconnected together.

The above advantages, along with the small chip area taken up by MOSFETs, explain why nowadays CMOS technology is predominant in digital as well as mixed-mode (digital/analog) integrated circuits, especially in battery-powered systems such as notebooks, smart phones, digital cameras, pacemakers, and a wide variety of others.

The Noise Margins

As we know, the ability of a digital gate to function reliably in the presence of input noise is expressed in terms of its noise margins $NM_L = V_{IL} - V_{OL}$ and $NM_H = V_{OH} - V_{IH}$, where V_{IL} and V_{IH} are the values of v_I at which $a = -1$ V/V, and V_{OL} and V_{OH} are given by Eq. (3.88). Considering the symmetry of the CMOS inverter, we only need to find one of the two, say, V_{IH}. Then, we obtain the other as $V_{IL} = V_{DD} - V_{IH}$.

Turning again to Fig. 3.66, top, we observe that V_{IH} is located in the region where M_p operates in saturation and M_n in the triode region. Imposing $i_{Dp(\text{sat})} = i_{Dn(\text{triode})}$ yields an equation relating v_I and v_O. For the case of matched FETs this is expressed as

$$\frac{k}{2}(V_{DD} - v_I - V_t)^2 = k\left[(v_I - V_t)v_O - \frac{1}{2}v_O^2\right] \tag{3.89}$$

Differentiating both sides with respect to v_I and simplifying gives

$$-(V_{DD} - v_I - V_t) = (v_I - V_t)\frac{dv_O}{dv_I} + v_O - v_O\frac{dv_O}{dv_I}$$

Imposing $dv_O/dv_I = -1$ yields a relationship between v_O and v_I right at the point of negative unity slope,

$$v_O = v_I - \frac{V_{DD}}{2}$$

Substituting back into Eq. (3.89), letting $v_I = V_{IH}$, and solving for V_{IH}, we finally obtain

$$V_{IH} = \frac{5V_{DD} - 2V_t}{8} \tag{3.90a}$$

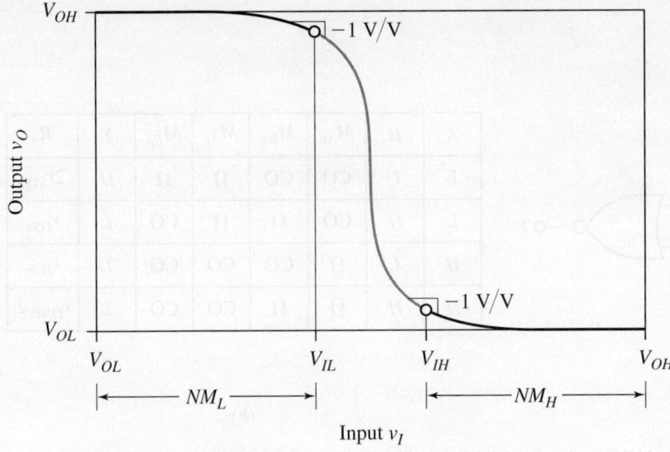

FIGURE 3.68 Visualizing the noise margins of the CMOS inverter.

By symmetric reasoning, $V_{IL} = V_{DD} - V_{IH}$, or

$$V_{IL} = \frac{3V_{DD} + 2V_t}{8} \qquad (3.90b)$$

Combining with Eq. (3.88), we readily find the noise margins for the case of matched FETs as

$$NM_L = NM_H = \frac{3V_{DD} + 2V_t}{8} \qquad (3.91)$$

The noise margins are illustrated further in Fig. 3.68.

EXAMPLE 3.38

Find V_{IL} and V_{IH} as well as the noise margins of the inverter of Fig. 3.65.

Solution

Using Eqs. (3.90) and (3.91), we find $V_{IL} = (3 \times 5 + 2 \times 1)/8 = 2.1\,\text{V}$, $V_{IH} = 2.9\,\text{V}$, and $NM_L = NM_H = 2.1\,\text{V}$.

Basic NOR and NAND Gates

Figures 3.69a and 3.70a show how the CMOS inverter topology can serve as basis for the implementation of the basic logic functions known as NOR and NAND. Here, A and B are the inputs, Y is the output, and their logic levels are L for 0 V and H for V_{DD} (such as 5 V). In either case, circuit behavior is best understood by tracing through the different table entries, one row at a time.

- With reference to the 1st row in the table of Fig. 3.69b we observe that with $AB = LL$ both M_{An} and M_{Bn} are cut off (CO) while both M_{Ap} and M_{Bp} are in the ohmic region (Ω),

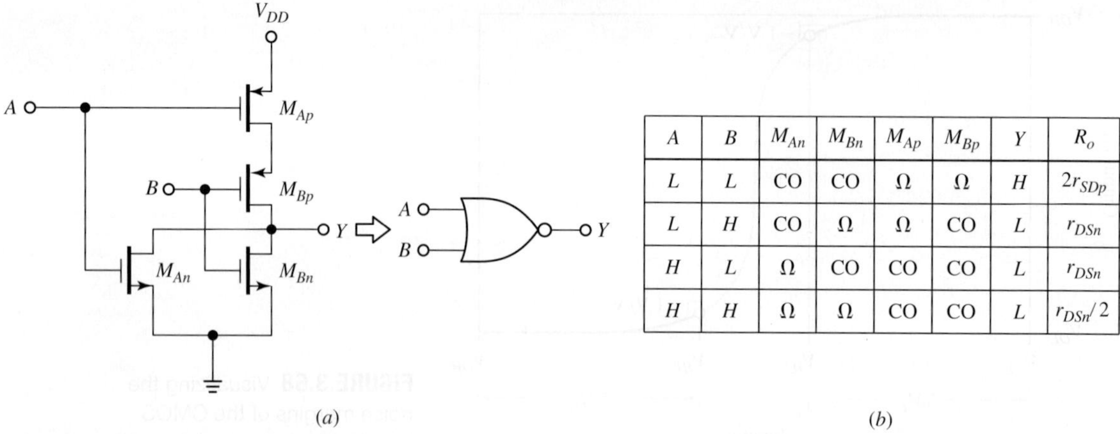

A	B	M_{An}	M_{Bn}	M_{Ap}	M_{Bp}	Y	R_o
L	L	CO	CO	Ω	Ω	H	$2r_{SDp}$
L	H	CO	Ω	Ω	CO	L	r_{DSn}
H	L	Ω	CO	CO	CO	L	r_{DSn}
H	H	Ω	Ω	CO	CO	L	$r_{DSn}/2$

(a)   (b)

FIGURE 3.69 (a) CMOS implementation of the NOR gate, and (b) table illustrating the various circuit conditions for each of the four possible input combinations.

thus pulling Y high. The resistance R_o seen looking into the Y node is simply the *series* combination of the two p-channel resistances, $2r_{SDp}$, pulling Y to V_{DD}.

- Proceeding to the 2nd row of Fig. 3.69b, where $AB = LH$, we observe that now M_{Bn} becomes conductive (Ω) and M_{Bp} goes in cutoff (CO), while the states of M_{An} and M_{Ap} remain unchanged relative to the first row. Consequently, M_{Bn} will now pull Y low, and R_o is M_{Bn}'s resistance r_{DSn} pulling Y to ground.
- The 3rd row of Fig. 3.69b is similar to the second row, but with the roles of A and B interchanged, so the output conditions are similar to those of the second row.
- Finally, in the 4th row of Fig. 3.69b, where $AB = HH$, both n-channels are in the ohmic region (Ω), while both p-channels are in cutoff (CO). Consequently, Y is

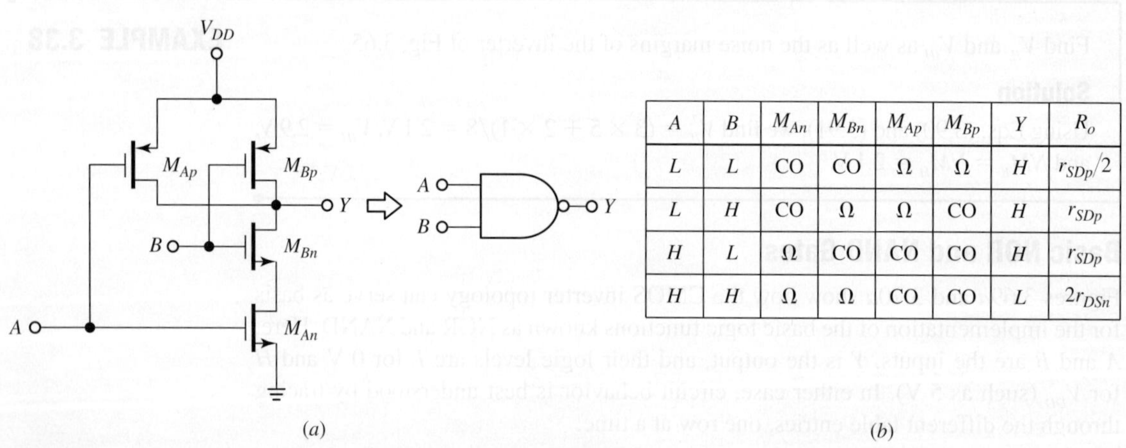

A	B	M_{An}	M_{Bn}	M_{Ap}	M_{Bp}	Y	R_o
L	L	CO	CO	Ω	Ω	H	$r_{SDp}/2$
L	H	CO	Ω	Ω	CO	H	r_{SDp}
H	L	Ω	CO	CO	CO	H	r_{SDp}
H	H	Ω	Ω	CO	CO	L	$2r_{DSn}$

(a)   (b)

FIGURE 3.70 (a) CMOS implementation of the NAND gate, and (b) table illustrating the various circuit conditions for each of the four possible input combinations.

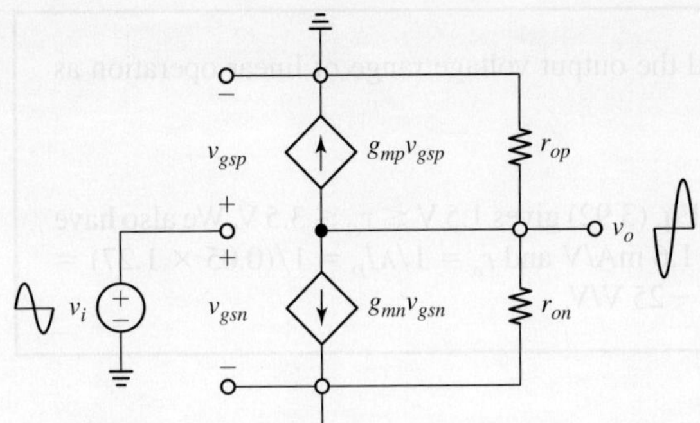

FIGURE 3.71 Small-signal model of the CMOS inverter in the active region.

pulled low by the action of two channel resistances in *parallel*, and such that R_o is now $r_{DSn}/2$. It is apparent that the NOR gate differentiates the state $AB = LL$ from all others.

The reader is encouraged to repeat a similar analysis for the circuit of Fig. 3.70a, and to trace through each row in detail. It is apparent that the NAND gate differentiates the state $AB = HH$ from all others.

The CMOS Inverter as an Amplifier

As mentioned, in the vicinity of the midpoint $V_m = \frac{1}{2}V_{DD}$ ($= 2.5$ in our example) both FETs are saturated and thus provide linear amplification. The corresponding output voltage range is

$$(V_m - V_{tn}) \le v_O \le (V_m + |V_{tp}|) \tag{3.92}$$

We wish to find the *small-signal gain a* over this range. To this end we replace the FETs with their respective small-signal models, and obtain the ac equivalent of Fig. 3.71. By Ohm's law,

$$v_o = -(g_{mn}v_{gsn} + g_{mp}v_{gsp}) \times (r_{on}//r_{op})$$

Considering that $v_{gsn} = v_{gsp} = v_i$, we get

$$a = \frac{v_o}{v_i} = -(g_{mn} + g_{mp}) \times (r_{on}//r_{op}) \tag{3.93}$$

With matched devices ($g_{mn} = g_{mp} = g_m$, $r_{on} = r_{op} = r_o$), this simplifies as $a = -(2g_m) \times (r_o/2)$, or

$$a = -g_m r_o \tag{3.94}$$

Note that the two FETs *reinforce* each other in pulling ac current out of the output node. Moreover, each FET can be viewed as a CS amplifier with the output resistance r_o of the other FET as its load resistance. There's no doubt that the CMOS inverter/amplifier, despite its simplicity, is a clever circuit!

EXAMPLE 3.39   For the inverter of Fig. 3.65, find the output voltage range of linear operation as well as the gain there.

Solution

Since $V_m = 2.5$ V and $V_t = 1.0$ V, Eq. (3.92) gives $1.5\text{ V} \le v_O \le 3.5$ V. We also have $g_m = \sqrt{2kI_D} = \sqrt{2 \times 1 \times 1.27} = 1.6$ mA/V and $r_o = 1/\lambda I_D = 1/(0.05 \times 1.27) = 15.7$ kΩ, so $a = -1.6 \times 15.7 \cong -25$ V/V.

Exercise 3.4

(a) Show that the voltage gain of a CMOS inverter/amplifier with matched FETs can be estimated as

$$a = -\frac{2V_A}{0.5V_{DD} - V_t} \tag{3.95}$$

where $V_A = 1/\lambda$.

(b) Use Eq. (3.95) to verify the result of Example 3.39.

(c) What happens if V_{DD} is increased from 5 V to 10 V? Comment.

Ans. (b) $a = -26.7$ V/V. (c) $a = -10$ V/V; raising V_{DD} lowers a.

APPENDIX 3A

SPICE Models for MOSFETs

As in the case of the BJT, the characteristics of a MOSFET are expressed in terms of a *list of parameters* that PSpice then uses to create an internal *model* of the device. Over the last several decades CMOS technology has evolved dramatically along the lines of Moore's law, as mentioned at the beginning of this chapter. As channel lengths keep shrinking to the nanometer range, various higher-order effects come into play, which make the task of modeling a MOSFET for computer simulation an increasingly complex and challenging task. Presently, three different model levels are available. **Level 1**, also known as the *Shichman-Hodges* model, works well for devices with channel lengths in the micrometer range, where the *i-v* characteristics are governed by the *square law* espoused in this chapter. **Level 2** is a more advanced model utilizing *analytical techniques* to calculate the higher-order effects that arise at the submicron level. **Level 3** calculates higher-order effects utilizing a combination of analytical and empirical tools.

For our scope here we shall limit ourselves to Level 1, whose parameter list is shown in Table 3A.1. Going down the list, we easily recognize a number of already familiar parameters in the first half or so. The second half contains parameters intervening in the calculation of the various internal capacitances of the MOSFET, a

TABLE 3A.1 **Partial parameter list of the PSpice model for Level-1 MOSFETs.**

Symbol	Name	Parameter description	Units	Default	Example
	Level	Model level number		1	3
V_{t0}	Vto	Zero-bias threshold voltage	V	0	1.0
k'	Kp	Process transconductance parameter	A/V^2	$20\,\mu$	$50\,\mu$
γ	Gamma	Body-effect parameter	V$^{1/2}$	0	0.5
$2\phi_f$	Phi	Surface potential	V	0.6 V	0.65
λ	Lambda	Channel-length modulation parameter	V^{-1}	0	0.05
r_d	Rd	Bulk resistance of the drain	Ω	0	1
r_s	Rs	Bulk resistance of the source	Ω	0	1
μ	Uo	Surface mobility	cm^2/Vs	600	500
t_{ox}	Tox	Oxide thickness	m	100 n	10 n
N_A or N_D	Nsub	Substrate doping	cm^{-3}	0	10^{15}
C_{db0}	Cbd	Zero-bias B-D junction capacitance	F	0	10 fF
C_{sb0}	Cbs	Zero-bias B-S junction capacitance	F	0	10 fF
ϕ_0	Pb	B-D and B-S junction built-in potential	V	0.8	0.75
C_{ov}/W	Cgso	G-S overlap capacitance per unit W	F/m	0	100 p
C_{ov}/W	Cgdo	G-D overlap capacitance per unit W	F/m	0	100 p
C_{gb}/L	Cgbo	G-B overlap capacitance per unit L	F/m	0	250 p
$C_{j0(btm)}$	Cj	Unit-area zero-bias bulk junction bottom capacitance	F/m^2	0	$250\,\mu$
m_{btm}	Mj	Bulk junction bottom grading coefficient		0.5	0.5
$C_{j0(sw)}$	Cjsw	Unit-perimeter zero-bias bulk junction sidewall capacitance	F/m	0	0.5 n
m_{sw}	Mjsw	Bulk junction sidewall grading coefficient		0.33	0.33
X_j	Xj	Metallurgical S-B and D-B junction depth	m	0	$0.5\,\mu$
L_{ov}	LD	Lateral diffusion	m	0	100 n

subject that will be taken up in great detail in Chapter 6, when we will investigate the frequency and transient responses of integrated circuits.

The library of the PSpice version used in this book comes with the Level 3 models of two power MOSFETs, the n-channel IRF150 and the p-channel IRF9140. The user can create additional models by editing either of these models. As an example, consider the PSpice circuits of Figs. 3.14 and 3.16 that were created to plot the i-v curves of a homebrew MOSFET having $k' = 50\ \mu\text{A/V}^2$, $V_{t0} = 1.0$ V, $\lambda = 0.05$ V^{-1}, $W = 2\ \mu$m, and $L = 1\ \mu$m. As usual, we create a PSpice circuit schematic via the **Place** $\rightarrow$ **Part** commands to lay out the various components, and the **Place** $\rightarrow$ **Wire** commands to interconnect them. When it comes to placing the FET, we import it from PSpice's library by going down the list of entries and selecting the IRF150 part by left-clicking on it. Once the FET has been placed in the circuit schematic, we can visualize its model by left-clicking on the FET itself to select it, and then right-clicking to activate a pull-down menu of possible actions. If we left-click on **Edit PSpice Model**, the following list will appear:

```
.model   IRF150  NMOS(Level=3 Gamma=0 Delta=0 Eta=0 Theta=0
+        Kappa=0 Vmax=0 Xj=0 Tox=100n Uo=600 Phi=.6 Rs=1.624m
+        Kp=20.53u W=.3 L=2u Vto=2.831 Rd=1.031m Rds=444.4K
+        Cbd=3.229n Pb=.8 Mj=.5 Fc=.5 Cgso=9.027n Cgdo=1.679n
+        Rg=13.89 Is=194E-18 N=1 Tt=288n)
```

To create our own homebrew nMOSFET model we simply edit (overwrite) the above list, making sure to give our new model a different name before saving it, such as Mn, in order to avoid destroying the existing one. The result is

```
.model Mn NMOS(W=2u L=1u Kp=50u Vto=1.0V Lambda=0.05)
```

Likewise, a homebrew pMOSFET model called Mp and having $k' = 20\ \mu\text{A/V}^2$, $V_{t0} = -0.75\ \text{V}$, $\lambda = 0.1\ \text{V}^{-1}$, $W = 5$, and $L = 1\ \mu\text{m}$ would be

```
.model Mp PMOS(W=5u L=1u Kp=20u Vto=-0.75V lambda=0.1)
```

All omitted parameters are automatically assigned *default values* according to Table 3A.1.

REFERENCES

1. R. F. Pierret, *Field Effect Devices*, Modular Series on Solid State Devices, 2/E, Vol. IV, G. W. Neudeck and R. F. Pierret, eds., Addison-Wesley, 1989.
2. R. S. Muller and T. I. Kamins, *Device Electronics for Integrated Circuits*, 2/E, J. Wiley and Sons, 1986.
3. R. T. Howe and C. G. Sodini, *Microelectronics: An Integrated Approach*, Prentice Hall, 1997.
4. P. E. Allen and D. R. Holberg, *CMOS Analog Circuit Design*, 2/E, Oxford University Press, 2002.
5. P. R. Gray, P. J. Hurst, S. H. Lewis, and R. G. Meyer, *Analysis and Design of Analog Integrated Circuits*, 4/E, Wiley and Sons, 2001.
6. A. S. Sedra and K. C. Smith, *Microelectronic Circuits*, 5/E, Oxford University Press, 2004.
7. R. C. Jaeger and T. N. Blalock, *Microelectronic Circuit Design*, 2/E, McGraw-Hill, 2004.
8. G. W. Gordon and A. S. Sedra, *SPICE for Microelectronic Circuits*, 2/E, Oxford University Press, 1996.

PROBLEMS

3.1 Physical Structure of the MOSFET

3.1 (*a*) Figure 3.2 reveals the presence of a parasitic npn BJT whose base region is the p^- body (with B_n as its terminal), and whose emitter and collector regions are the n^+ source and drain regions (with S_n and D_n as their terminals). Would it be possible to make this BJT operate with a reasonably high current gain? Could it serve any useful function?

(*b*) The same figure indicates the presence of another parasitic BJT, namely, a pnp type whose base region is the n^- well (with B_p as its terminal), and whose emitter and collector regions are the p^+ source and drain regions (with S_p and D_p as their terminals). Would it be possible to make this BJT operate with a reasonably high current gain? Could it serve any useful function?

(c) Identify two additional parasitic BJTs (you may wish to search the web for "CMOS latchup").

3.2 The unorthodox interconnection of Fig. P3.2 violates the requirement that the p^- body be held at the MNV. It is nevertheless instructive to investigate it because it will help us better appreciate the structure's behavior when configured for proper operation.

(a) Assuming typical pn junction parameter values, predict all node voltages and all terminal currents in the circuit of Fig. P3.2.

(b) What might happen if the left terminal of the 10-kΩ resistor is lifted off ground and connected to a 10-V source?

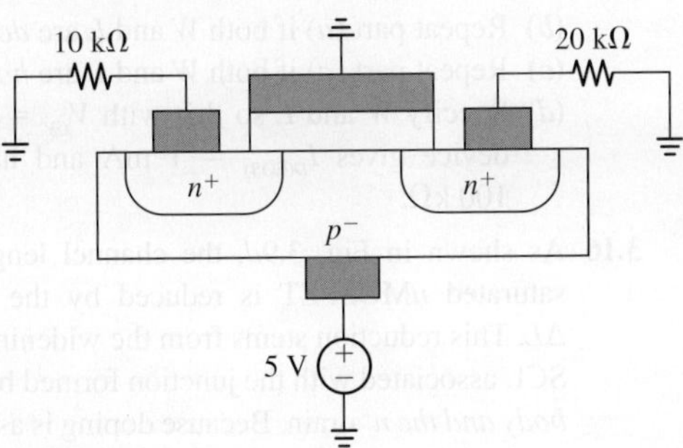

FIGURE P3.2

3.2 The Threshold Voltage V_t

3.3 Let the pMOSFET of Fig. 3.6b have an n^+-type polysilicon gate with $N_D = 2 \times 10^{19}$ cm^{-3}, an n^--type bulk with $N_D = 10^{16}$ cm^{-3}, and $t_{ox} = 30$ nm.

(a) Sketch and label the equilibrium potential $\phi(x)$.

(b) Find the gate-to-body voltage needed to eliminate the space-charge layers.

(c) Find the gate-to-body voltage V_{GB0} needed to bring about the onset of strong inversion.

(d) Sketch and label $\phi(x)$ for $V_{GB} = V_{GB0}$.

3.4 *(a)* Assuming an nMOS process with bulk doping $N_A = 10^{16}$ cm^{-3}, $t_{ox} = 50$ nm, and a native threshold of -0.1 V, find the implant type and dosage N_i needed to create an *enhancement-type device* with $V_{t0} = 1.0$ V.

(b) Sketch and label the plot of V_t versus V_{SB} for $0 \le V_{SB} \le 5$ V.

(c) Repeat parts *(a)* and *(b)*, but for a depletion-type device with $V_{t0} = -0.5$ V. What value of V_{SB} yields $V_t = 0$ V? $V_t = +0.5$ V?

3.5 *(a)* Assuming a pMOS process with bulk doping $N_D = 2 \times 10^{16}$ cm^{-3}, $t_{ox} = 40$ nm, and a native threshold of -1.5 V, find the implant type and dosage N_i needed to create an *enhancement-type device* with $V_{t0} = -1$ V.

(b) Sketch and label the plot of V_t versus V_{BS} over the range $0 \le V_{BS} \le 5$ V.

(c) Repeat parts *(a)* and *(b)*, but for a depletion-type device with $V_{t0} = 1.0$ V.

3.6 *(a)* How is the native threshold of Example 3.4a affected by a 10% increase in N_A? In t_{ox}? In N_{ox}?

(b) How is the threshold of Example 3.4b affected by a 10% increase in t_{ox}? By a 10% increase in N_i?

3.7 *(a)* If an nMOS process has $t_{ox} = 25$ nm, find N_A so that $\gamma = 0.5$ V$^{1/2}$.

(b) If t_{ox} is doubled, what is the new value of γ?

(c) If it is desired to restore the original value $\gamma = 0.5$ V$^{1/2}$, what must the new value of N_A be?

(d) If $t_{ox} = 100$ nm, find N_A so that V_t increases by 1 V as V_{SB} is changed from 0 V to 5 V.

Hint: you may need to iterate.

3.3 The n-Channel Characteristic

3.8 Figure 3.10 indicates that the charge/unit length in a tapered channel *decreases* as we move from source to drain. Yet, in the course of the integration leading to Eq. (3.12), it was argued that the current i_D must be *constant* throughout the entire channel. How can a constant current exist in spite of a decreasing charge/unit length? Do you sense any contradiction here? Explain!

3.9 Suppose a nMOS process has $\mu_n = 500$ cm^2/Vs, $t_{ox} = 25$ nm, and a native threshold of -0.1 V.

(a) Assuming $\lambda = 0$, specify the implant type and dosage N_i as well as the W/L ratio needed to create an nMOSFET that gives $I_{D(EOS)} = 8$ μA with $V_{GS} = 1.0$ V, and $I_{D(EOS)} = 98$ μA with $V_{GS} = 2.0$ V.

(b) Repeat, but for an nMOSFET that gives $I_{D(EOS)} = 25$ μA with $V_{GS} = 0$ V, and $I_{D(EOS)} = 225$ μA with $V_{GS} = 1.0$ V.

3.10 *(a)* A certain nMOSFET is operated in the ohmic region with $v_{DS} = 0.1$ V, and gives $I_D = 90$ μA for $V_{GS} = 2$ V, and $I_D = 165$ μA for $V_{GS} = 3$ V. Assuming $k' = 50$ μA/V^2, find W/L and V_t for this device.

(b) Find i_D for $V_{GS} = 4$ V and $v_{DS} = 0.2$ V.

(c) If $V_{GS} = 2.5$ V, find $V_{DS(EOS)}$ as well as $I_{D(EOS)}$.

3.11 A certain nMOSFET is to be operated at low values of v_{DS} as a *voltage-controlled resistance r_{DS}*. The FET has $k' = 50$ μA/V^2, $W = 10$ μm, $L = 1$ μm, and $V_t = 1.0$ V.

(*a*) Find the range of values of V_{GS} that will cause r_{DS} to vary from 500 Ω to 20 kΩ.

(*b*) For the range of V_{GS} values of part (*a*), what is the range of r_{DS} values if W is halved? If L is halved? If both W and L are halved? If both W and L are doubled? If W is raised to 50 μm while L is kept constant at 1 μm?

3.12 An nMOSFET with $k = 1$ mA/V^2 is to be used as a *voltage-controlled resistance*, but only over a limited range of v_{DS} values in order to keep nonlinearities low. As we know, the channel resistance in the limit $v_{DS} \to 0$ is $r_{DS} = 1/(k \times V_{OV})$. However, as v_{DS} is increased, the slope of the i_D-v_{DS} curve *decreases*, so its reciprocal, representing channel resistance, *increases* from the above value.

(*a*) If $V_{OV} = 1$ V, what is the maximum value of $|v_{DS}|$ for which the actual channel resistance deviates by 5% from its value in the limit $v_{DS} \to 0$? What is this limiting value?

(*b*) Repeat, but for $V_{OV} = 2$ V, 5 V, 0.5 V, and 0.2 V.

3.13 Consider three MOSFETs having the same values of V_t and k' but individual ratios $(W/L)_1 = 1/1$, $(W/L)_2 = 2/1$, and $(W/L)_3 = 4/1$, all values being in μm. For simplicity, assume $\gamma = 0$ and $\lambda = 0$ for all devices.

(*a*) Given that if connected in *parallel* as in Fig. 3.12, the three FETs act as a single equivalent FET, find $(W/L)_{eq}$; let the smaller of W and L in $(W/L)_{eq}$ be 1 μm.

(*b*) Repeat, if the FETs are connected in *series* as in Fig. 3.13.

(*c*) Repeat, but if device #3 is in series with the parallel combination of devices #1 and #2.

(*d*) Repeat, but if device #1 is in parallel with the series combination of devices #2 and #3.

(*e*) Repeat, but if device #3 is in parallel with the series combination of devices #1 and #2.

3.14 A certain nMOSFET is operated in saturation with a 1-V overdrive, and gives $I_D = 110$ μA at $V_{DS} = 2$ V, and $I_D = 120$ μA at $V_{DS} = 4$ V.

(*a*) Find λ and r_o.

(*b*) Find $I_{D(EOS)}$.

(*c*) If the FET is known to have $L = 1$ μm in the limit $V_{DS} \to 0$, what is its actual channel length L_{actual} at $V_{DS} = 5$ V?

Hint: exploit the fact that $L_{actual} = L - \Delta L = L/(1 + \lambda V_{DS})$.

(*d*) For what value of V_{DS} does the device have $L_{actual} = (2/3)L$?

3.15 Assume an nMOSFET process in which the modulation parameter λ scales with the channel length L according to $\lambda = (0.1\mu\text{m})/L$ V^{-1}, with L in μm. Also, let $k' = 50$ μA/V^2.

(*a*) If a particular device has $W/L = (10$ μm$)/(2$ μm$)$ and is driven with $V_{OV} = 2$ V, find I_D at $V_{DS} = 5$ V and at $V_{DS} = 10$ V. What is the value of r_o?

(*b*) Repeat part (*a*) if both W and L are *doubled*.

(*c*) Repeat part (*a*) if both W and L are *halved*.

(*d*) Specify W and L so that with $V_{OV} = 1$ V the device gives $I_{D(EOS)} = 1$ mA and has $r_o = 100$ kΩ.

3.16 As shown in Fig. 3.9*d*, the channel length of a saturated nMOSFET is reduced by the amount ΔL. This reduction stems from the widening of the SCL associated with the junction formed by the p^- *body and the n^+ drain*. Because doping is asymmetrical ($N_D \gg N_A$), the SCL extends mostly into the lightly-doped body region. Adapting Eq. (1.45) to the present case gives $\Delta L \cong \sqrt{2\varepsilon_{si}(\phi_0 - V)/(qN_A)}$, where ϕ_0 is the junction's built-in potential, and V is the junction's voltage drop right at the drain end of the channel, or $V = V_{OV} - V_{DS}$.

(*a*) Given that with the overdrive voltage $V_{OV} = 1$ V a certain nMOSFET yields $I_D = 210$ μA at $V_{DS} = 2$ V, and $I_D = 220$ μA at $V_{DS} = 4$ V, what are the values of λ and k?

Hint: consider the *ratio* of the two currents.

(*b*) Assuming a process with bulk doping $N_A = 10^{16}$ cm^{-3}, polysilicon gate doping $N_D = 10^{20}$ cm^{-3}, and $k' = 50$ μA/V^2, what are the values of ΔL in the two cases?

(*c*) Exploiting the fact that I_D is inversely proportional to $L - \Delta L$, estimate L. What is the value of W?

3.17 Let an nMOSFET have $V_{t0} = 1.0$ V, $k = 200$ μA/V^2, $\lambda = 0.02$ V^{-1}, and $\gamma = 0.46$ V$^{1/2}$. Assuming $\phi_p = -0.35$ V, fill in the blanks in Table P3.17.

TABLE P3.17

#	V_{GS} (V)	V_{DS} (V)	V_{SB} (V)	I_D (μA)
1	2	2	0	
2	2	5	0	
3	2		0	106
4	3	4	3	
5	1.5	3	5	
6		6	3	252
7	2	5		55
8	3	1	0	

3.18 A student is using the test setup of Fig. P3.18 to perform measurements on an *n*MOSFET in order to extract its most significant parameters, and the results are shown in the first 4 rows of Table P3.18.

(a) Assuming $\phi_p = -0.35$ V, find V_{t0}, k, λ, and γ. **Hint:** use rows 2 and 3 to find λ, rows 1 and 2 to find V_{t0} and k, and rows 3 and 4 to find γ.

(b) Use the results of part (a) to fill in the blanks in the remaining rows.

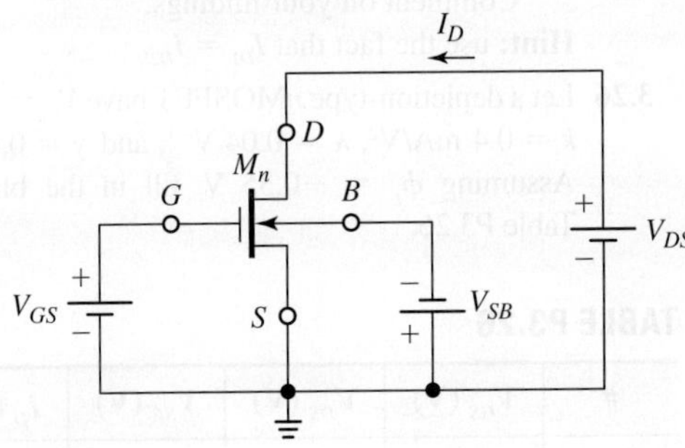

FIGURE P3.18

TABLE P3.18

#	V_{GS} (V)	V_{DS} (V)	V_{SB} (V)	I_D (μA)
1	2	4	0	174
2	3	3	0	672
3	3	5	0	720
4	3	5	3	405
5	3	3	4	
6	4	1	3	

3.19 In the test circuit of Fig. P3.18, let $V_{SB} = 0$ and $V_{DS} = 2$ V, and let v_{GS} be variable from 0 to 5 V.

(a) Assuming $V_t = 1.0$ V, $k = 1$ mA/V^2, and $\lambda = 0$, sketch and label the plot of i_D versus v_{GS}. **Hint:** as v_{GS} is varied from 0 to 5 V, MOSFET operation goes from cutoff, to saturation, to triode.

(b) Find the device's dynamic resistance r at $v_{GS} = 2$ V and at $v_{GS} = 4$ V. (Recall that r is the reciprocal of the slope.)

3.4 The *i-v* Characteristics of MOSFETs

3.20 Consider two *diode-connected n*MOSFETs having, respectively, $V_{t1} = 1.0$ V and $k_1 = 100$ μA/V^2, and $V_{t2} = 2.0$ V and $k_2 = 400$ μA/V^2. Moreover, assume $\gamma = 0$ and $\lambda = 0$ for both devices.

(a) Show that if the two diodes are connected in *series*, they still behave like a diode-connected MOSFET. What are the values of V_t and k for this equivalent device? **Hint:** derive the *i-v* characteristic of the series combination, exploiting the fact that $i = i_{D1} = i_{D2}$, and $v = v_{DS1} + v_{DS2}$.

(b) If the two diodes are connected in *parallel*, will the resulting structure still exhibit diode behavior? Sketch and label its *i-v* characteristic, calculate it at a few significant points, and discuss.

3.21 In the circuit of Fig. P3.21 let $R = 1$ kΩ and let the FET have $V_t = 1.0$ V and $k = 2$ mA/V^2. Assuming $\lambda = 0$, sketch and label the *i-v* characteristic for $0 \le v \le 5$ V. Calculate it at a few significant points, and use physical insight to predict the ultimate slope of this characteristic for large values of v. **Hint:** as v is raised from 0 V, the FET operates first in cutoff, then in saturation, and finally in the triode region.

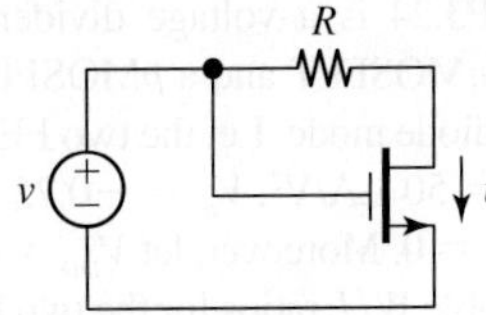

FIGURE P3.21

3.22 (a) The depletion *n*MOSFET of Fig. P3.22 is operated with $v \ge 0$. Show that the device is always in the triode region.

(b) Assuming $V_t = -1.0$ V, $k = 100$ μA/V^2, and $\lambda = 0$, sketch and label the *i-v* curve for $0 \leq v \leq 5$ V.

(b) Find the device's dynamic resistance r at $v = 0$ V, 3 V, and 5 V. (Recall that r is the reciprocal of the slope of the *i-v* curve.)

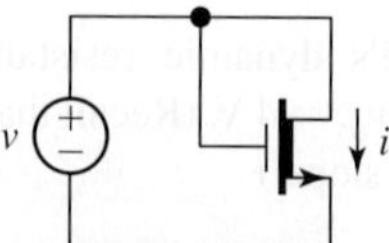

FIGURE P3.22

3.23 Shown in Fig. P3.23 is a *voltage divider* implemented with two diode-connected *n*MOSFETs. Let both FETs have $V_t = 0.5$ V, $k' = 50$ μA/V^2, $\gamma = 0$, and $\lambda = 0$.

(a) If $V_{DD} = 5$ V, specify suitable W/L ratios for M_1 and M_2 so that the circuit gives $V = 2.5$ V while drawing $I = 25$ μA from V_{DD}.

(b) Repeat, but for $V = 1.5$ V and $I = 10$ μA.

(c) If $(W/L)_1 = (1\ \mu\text{m})/(1\ \mu\text{m})$, specify $(W/L)_2$ so that $V = 3$ V; let the smaller of W and L in $(W/L)_2$ be 1 μm. What is the power dissipated by the circuit?

Hint: use the fact that $I_{D1} = I_{D2}$.

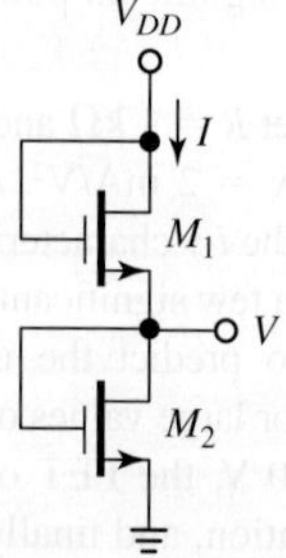

FIGURE P3.23

3.24 Shown in Fig. P3.24 is a voltage divider implemented with an *n*MOSFET and a *p*MOSFET, both operating in the diode mode. Let the two FETs have $V_{tn} = 0.5$ V, $k'_n = 50$ μA/V^2, $V_{tp} = -0.75$ V, $k'_p = 20$ μA/V^2, and $\lambda = 0$. Moreover, let $V_{DD} = 4$ V.

(a) Specify suitable W/L ratios for the two FETs so that the circuit gives $V = 1.5$ V while drawing $I = 20$ μA; let the smaller of W and L be 1 μm.

(b) Repeat, but for $V = 1.0$ V and $I = 10$ μA.

Hint: use the fact that $I_{D1} = I_{D2}$.

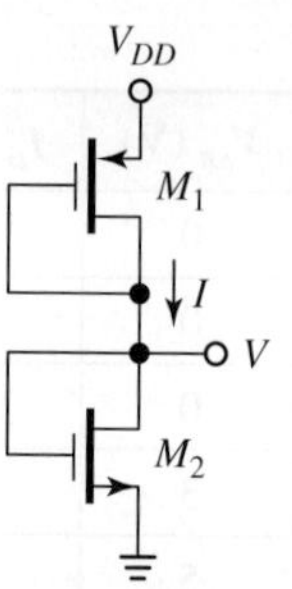

FIGURE P3.24

3.25 In the voltage divider of Fig. P3.23 both FETs have $V_t = 0.5$ V, $k' = 50$ μA/V^2, $\gamma = 0$, and $\lambda = 0$.

(a) If $V_{DD} = 3$ V, and the W/L ratio for each device is constrained within the range $(1\ \mu\text{m})/(10\ \mu\text{m}) \leq W/L \leq (10\ \mu\text{m})/(1\ \mu\text{m})$, what is the possible range of values for V? What is the current drawn by the circuit from V_{DD}?

(b) Specify W/L ratios for the two devices such that the circuit yields $V = V_{DD}/2$ while drawing the same current as it does at the extremes of part (*a*); let the smaller of W and L be 1 μm.

(c) What happens if V_{DD} is doubled to 6 V? Comment on your findings.

Hint: use the fact that $I_{D1} = I_{D2}$.

3.26 Let a depletion-type *n*MOSFET have $V_{t0} = -1.5$ V, $k = 0.4$ mA/V^2, $\lambda = 0.04$ V^{-1}, and $\gamma = 0.62$ V$^{1/2}$. Assuming $\phi_p = -0.35$ V, fill in the blanks in Table P3.26.

TABLE P3.26

#	V_{GS} (V)	V_{DS} (V)	V_{SB} (V)	I_D (μA)
1	0	2	0	
2	-0.5	1	0	
3		3	0	56
4	0	2	2	
5	-1	1	5	
6		0.5	2	250
7	0	5		160
8	0		0	522

3.27 A certain enhancement-type pMOSFET has $V_{t0} = -1.5$ V, $k = 0.25$ mA/V^2, $\lambda = 0.04$ V^{-1}, and $\gamma = 0.62$ V$^{1/2}$. Assuming $\phi_n = 0.35$ V, fill in the blanks in Table P3.27.

TABLE P3.27

#	V_{SG} (V)	V_{SD} (V)	V_{BS} (V)	I_D (μA)
1	3	3	0	
2	3	1	0	
3	3		0	130
4	4		2	550
5		3	2	35
6	4		1	250

3.28 A certain depletion-type pMOSFET has $V_{t0} = +0.5$ V, $k = 240$ μA/V^2, $\lambda = 0.05$ V^{-1}, and $\gamma = 0.65$ V$^{1/2}$. Assuming $\phi_n = 0.35$ V, fill in the blanks in Table P3.28.

TABLE P3.28

#	V_{SG} (V)	V_{SD} (V)	V_{BS} (V)	I_D (μA)
1	0	4	0	
2	0.5	0.5	0	
3	1.5		0	360
4	0	2	5	
5	1.5		5	90
6		1	5	360

3.29 (*a*) In the test circuit of Fig. P3.18, let $V_{GS} = 0$ and $V_{SB} = 0$. If it is found that $I_D = 112$ μA at $V_{DS} = 3$ V, and $I_D = 120$ μA at $V_{DS} = 5$ V, identify the device type (enhancement or depletion?) and find λ.

(*b*) Now let $V_{GS} = 1.5$ V and $V_{DS} = 5$ V. If it is found that $I_D = 750$ μA with $V_{SB} = 0$, and $I_D = 480$ μA with $V_{SB} = 2$ V, find k, V_{t0}, and γ, assuming $\phi_p = -0.35$ V.

(*c*) Predict I_D if $V_{GS} = 0$, $V_{SB} = 5.3$ V, and $V_{DS} = 1$ V, and comment on your findings.

(*d*) Repeat part (*c*) if V_{GS} is raised to 2 V.

3.30 A student is using the test setup of Fig. P3.30 to perform measurements on a pMOSFET in order to extract its most significant parameters, and the results are shown in the first 4 rows of Table P3.30.

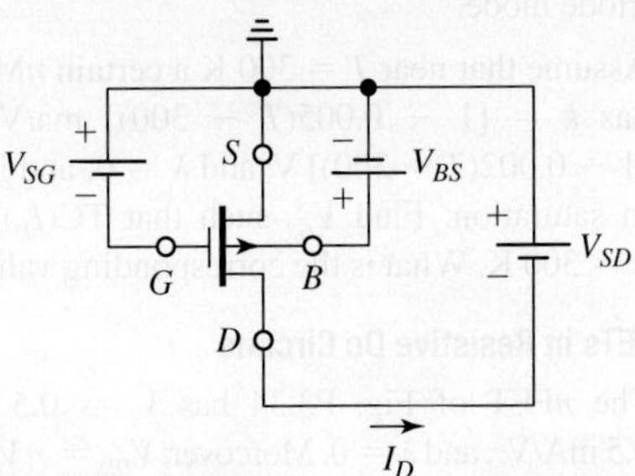

FIGURE P3.30

TABLE P3.30

#	V_{SG} (V)	V_{SD} (V)	V_{BS} (V)	I_D (μA)
1	2	2	0	110
2	2	4	0	120
3	3	5	0	500
4	4	5	4	720
5	5		4	580
6	4	1		400

(*a*) Identify the device type (enhancement or depletion?) and find V_{t0}, k, λ, and γ, assuming $\phi_n = 0.35$ V.

(*b*) Use the results of part (*a*) to fill in the blanks in the remaining rows.

3.31 (*a*) The test circuit of Fig. P3.30 is adjusted so that $V_{SG} = V_{BS} = 0$. Given that $I_D = 240$ μA for $V_{SD} = 8$ V, $I_D = 210$ μA for $V_{SD} = 2$ V, and $I_D = 150$ μA for $V_{SD} = 0.5$ V, and given that I_D drops to zero if V_{BS} is raised to 6 V, identify the device type (enhancement or depletion?) and find V_{t0}, k, γ, and λ. Assume $\phi_n = 0.35$ V.

(*b*) Predict I_D with $V_{SG} = V_{BS} = V_{SD} = 2$ V.

3.32 In the test circuit of Fig. P3.30 let $V_{SD} = 2$ V, and let v_{SG} be variable from 0 to 5 V.

(*a*) Assuming the FET has $V_t = -1.0$ V, $k = 2.0$ mA/V^2, $\gamma = 0.46$ V$^{1/2}$, $\phi_n = 0.35$ V, and $\lambda = 0$, sketch and label the plot of i_D versus v_{SG} if $V_{BS} = 0$.

(**b**) Repeat, but with $V_{BS} = 3$ V. Compare the two curves, and comment.

Hint: as v_{SG} is varied from 0 to 5 V, the MOSFET goes from cutoff, to the saturation mode, to the triode mode.

3.33 Assume that near $T = 300$ K a certain *n*MOSFET has $k = [1 - 0.005(T - 300)]$ ma/V², $V_t = [1 - 0.002(T - 300)]$ V, and $\lambda = 0$, and is biased in saturation. Find V_{GS} such that $\text{TC}(I_D) \cong 0$ at $T = 300$ K. What is the corresponding value of I_D?

3.5 MOSFETs in Resistive Dc Circuits

3.34 The *n*FET of Fig. P3.34 has $V_t = 0.5$ V, $k = 0.5$ mA/V², and $\lambda = 0$. Moreover, $V_{DD} = -V_{SS} = 5$ V.
(**a**) Specify R_D and R_S to bias the FET at $I_D = 1$ mA and $V_{DS} = 3$ V.
(**b**) Repeat, but for $V_{DS} = 1$ V.
(**c**) What happens if R_D is set to 0 (shorted out) in part (*b*)? What is the new operating point of the FET?

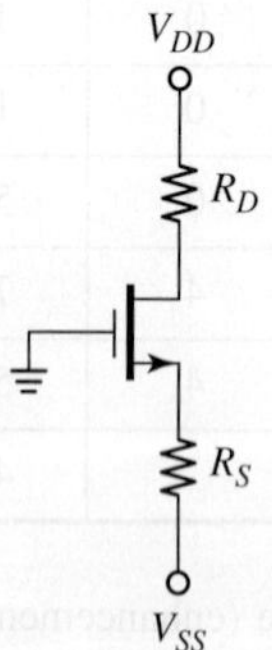

FIGURE P3.34

3.35 The *p*FET of Fig. P3.35 has $V_t = -1.0$ V, $k = 0.25$ mA/V², and $\lambda = 0$. Also, let $V_{SS} = -V_{DD} = 6$ V.
(**a**) Specify R_D and R_S to bias the FET at the edge of saturation (EOS) with $I_D = 0.5$ mA.

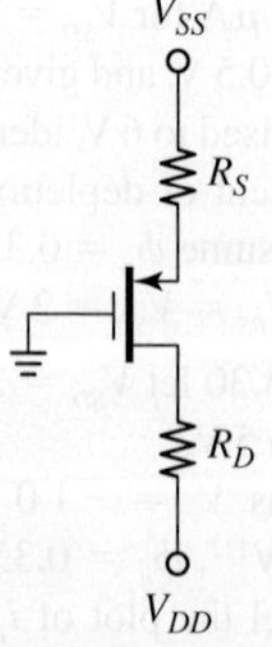

FIGURE P3.35

(**b**) Repeat, but to bias the FET at $I_D = 2$ mA and $V_{SD} = 2$ V.
(**c**) What happens if R_D in part (*b*) is doubled in value? What is the new operating point of the FET?

3.36 In the circuit of Fig. P3.34 let $V_{DD} = -V_{SS} = 6$ V, $R_D = 36$ kΩ, and $R_S = 10$ kΩ.
(**a**) If the FET has $V_t = 2.0$ V, $k = 0.4$ mA/V², and $\lambda = 0$, find all voltages and currents in the circuit.
(**b**) Repeat, but with R_S increased to 15 kΩ.

3.37 In the circuit of Fig. P3.35 let $V_{SS} = -V_{DD} = 12$ V, $R_S = 16$ kΩ, and $R_D = 30$ kΩ.
(**a**) If the FET has $V_t = -1.5$ V, $k = 0.25$ mA/V², and $\lambda = 0$, find all voltages and currents in the circuit.
(**b**) Repeat, if the gate terminal is lifted off ground and connected to the drain terminal.

3.38 Let the *n*FET of Fig. P3.38 have $V_t = 0.5$ V, $k = 0.8$ mA/V², and $\lambda = 0$.
(**a**) If $V_{DD} = 5$ V, $R_1 = 2$ MΩ, and $R_S = 5$ kΩ, find suitable values for R_2 and R_D to bias the FET at $I_D = 0.3$ mA and $V_{DS} = 0.5$ V.
(**b**) How is the FET's operating point affected if R_1 is increased from 2 MΩ to 3 MΩ? Comment!

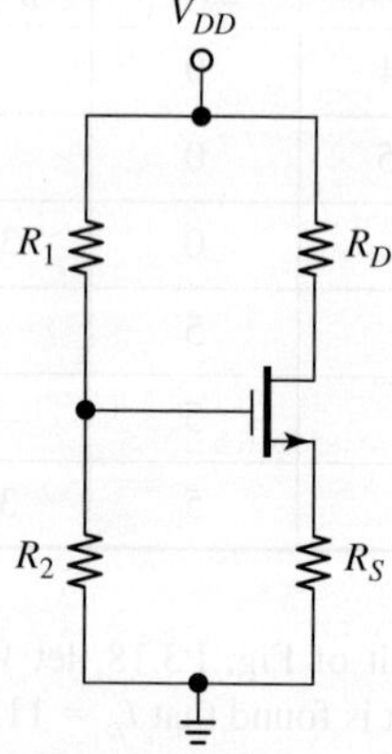

FIGURE P3.38

3.39. Let the *p*FET of Fig. P3.39 have $V_t = -2$ V, $k = 0.25$ mA/V², and $\lambda = 0$.
(**a**) If $V_{SS} = 12$ V, specify suitable resistances to bias the FET at $I_D = 0.5$ mA under the following constraints: V_S is to be biased at $(2/3)V_{SS}$; V_D is to be biased in the middle of the saturation region, and $R_1 + R_2 \geq 3$ MΩ.

(b) What happens if in the circuit designed in part (*a*) we change V_{SS} to 15 V? How is the FET's operating point affected?

(c) Repeat, but with $V_{SS} = 6$ V.

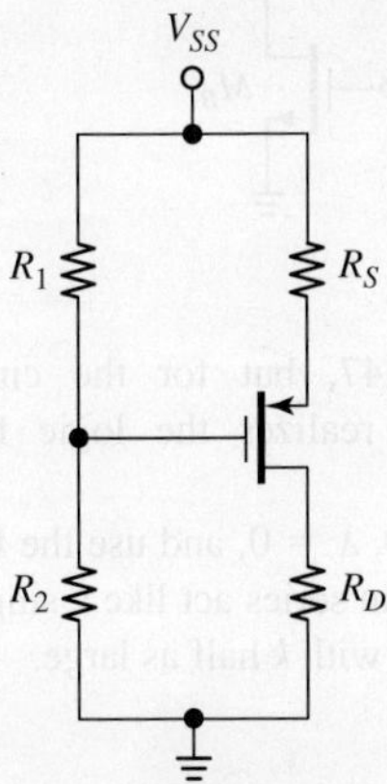

FIGURE P3.39

3.40. In the circuit of Fig. P3.38 let $V_{DD} = 15$ V, $R_1 = 1.2$ MΩ, $R_2 = 1.8$ MΩ, $R_D = 15$ kΩ, and $R_S = 5$ kΩ. Moreover, let the FET have $V_t = 1.5$ V, $k = 0.2$ mA/V², and $\lambda = 0$.

(a) Find all voltages and currents in the circuit.

(b) Repeat part (*a*), but with $R_D = 0$.

(c) Repeat (*a*), but with $R_S = 0$.

3.41 In the circuit of Fig. P3.39 let $V_{SS} = 10$ V, $R_1 = R_2 = 10$ MΩ, $R_S = 2$ kΩ, and $R_D = 10$ kΩ. Moreover, let the FET have $V_t = -1.5$ V, $k = 0.5$ mA/V², and $\lambda = 0$.

(a) Find all voltages and currents in the circuit.

(b) Repeat, if the gate terminal is disconnected from the voltage divider and connected instead to the drain terminal.

3.42 In the circuit of Fig. P3.42 let $V_{DD} = 10$ V, $R_1 = R_2 = 10$ MΩ, and $R_D = 10$ kΩ.

(a) Assuming the FET has $V_t = 1.0$ V, $k = 0.2$ mA/V², and $\lambda = 0$, estimate the FET's operating point.

Hint: note that 10 kΩ ≪ 10 MΩ.

(b) How is the operating point Q affected if R_1 is increased from 10 MΩ to 20 MΩ?

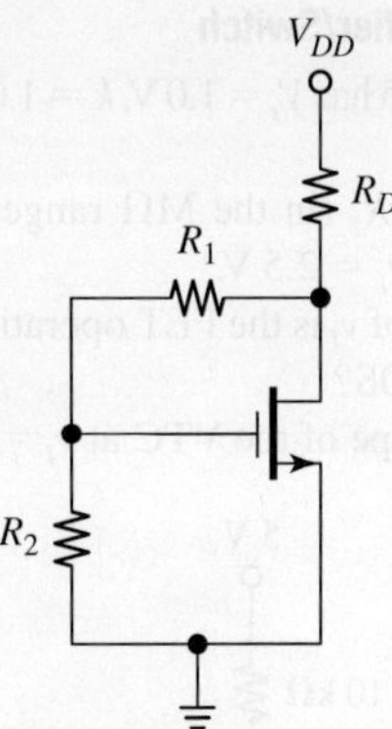

FIGURE P3.42

3.43 The FET of Fig. P3.43 has $V_t = -1.0$ V, $k = 0.75$ mA/V², and $\lambda = 0$.

(a) Assuming $V_{DD} = -8$ V, specify suitable resistance values to bias the device at $I_D = 1.5$ mA with V_D half-way between the values corresponding to the edge of conduction and the edge of saturation; specify R_1 and R_2 in the MΩ range.

(b) How is the FET's operating point Q affected if V_{DD} is changed to -5 V?

(c) To -1.5 V?

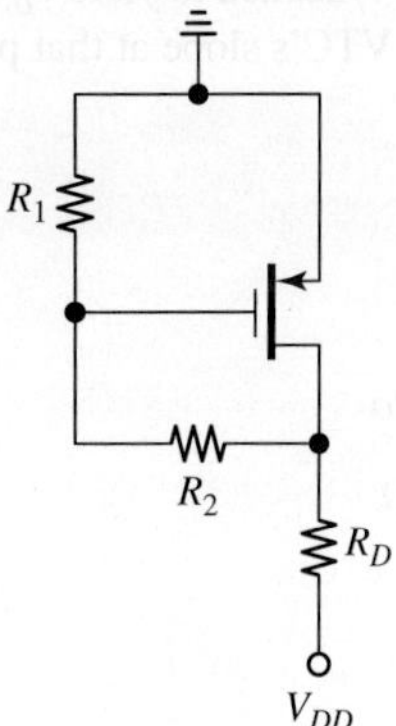

FIGURE P3.43

3.44 **(a)** In the circuit of Fig. 3.35 specify R_1 and R_2 (in the MΩ range) to bias M_p at the edge of saturation (EOS).

(b) What is the resulting drain current of M_n?

3.6 The MOSFET as an Amplifier/Switch

3.45 The FET of Fig. P3.45 has $V_t = 1.0$ V, $k = 1.0$ mA/V^2, and $\lambda = 0$.

(*a*) Specify R_1 and R_2 (in the MΩ range) so that $v_O = 2.5$ V for $v_I = 2.5$ V.

(*b*) For what value of v_I is the FET operating at the EOC? At the EOS?

(*c*) Estimate the slope of the VTC at $v_I = 2.5$ V.

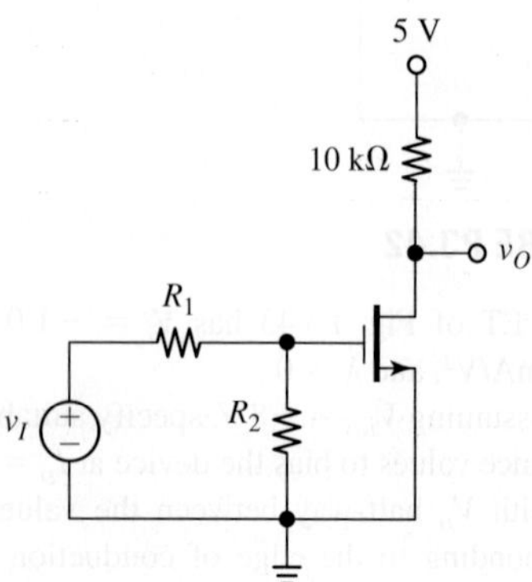

FIGURE P3.45

3.46 The FET of Fig. P3.46 has $V_t = -1.5$ V, $k = 1.0$ mA/V^2, and $\lambda = 0$.

(*a*) Sketch and label the VTC.

(*b*) Find the value of v_I needed to yield $v_O = 3.0$ V, and estimate the VTC's slope at that point.

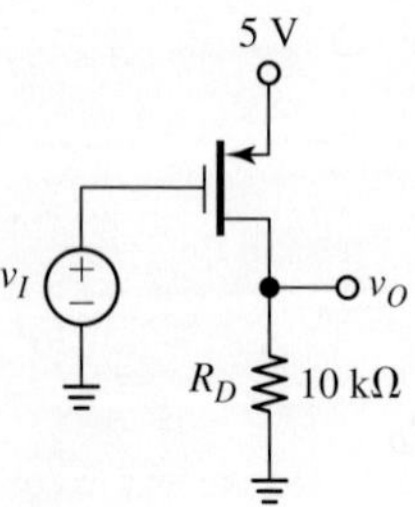

FIGURE P3.46

3.47 Shown in Fig. P3.47 is an *n*MOSFET realization of the logic function known as NOR. Assuming identical FETs with $V_t = 1.0$ V, $k = 0.25$ mA/V^2, and $\lambda = 0$, prepare the truth table, identifying the operating region of each FET (CO or Ohmic), and calculate the output node voltage for the following voltage combinations at the input nodes A and B: $(A, B) = (0\text{ V}, 0\text{ V}), (0\text{ V}, 5\text{ V}), (5\text{ V}, 0\text{ V}), (5\text{ V}, 5\text{ V})$.

Hint: two identical FETS in parallel act like a single FET with the same V_t but with k twice as large.

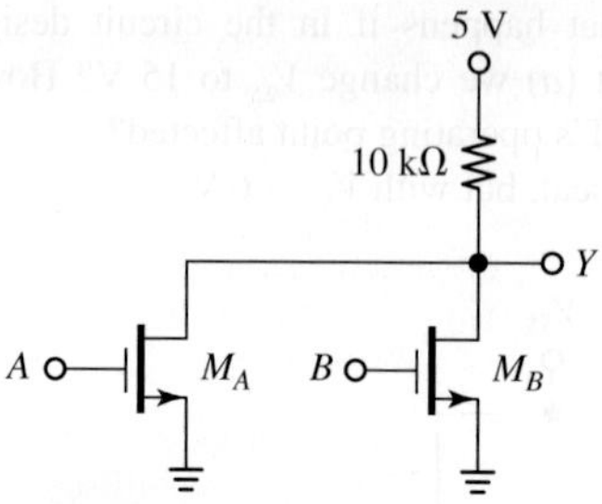

FIGURE P3.47

3.48 Repeat Problem 3.47, but for the circuit of Fig. P3.48, which realizes the logic function known as NAND.

Hint: assume $\gamma = 0$, $\lambda = 0$, and use the fact that two identical FETS in series act like a single FET with the same V_t but with k half as large.

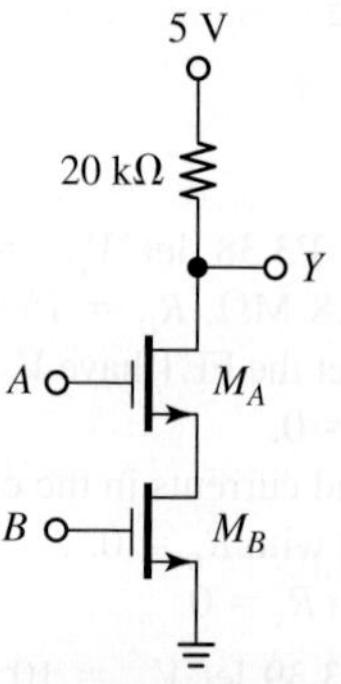

FIGURE P3.48

3.7 Small-Signal Operation of the MOSFET

3.49 (*a*) Replace the FET of Fig. P3.49 with its small-signal model, and use the test-signal method to find R_{gs}.

(*b*) Calculate R_{gs} if $R_F = 1$ MΩ and $R_D = 10$ kΩ, and the FET has $g_m = 1$ mA/V and $r_o = 100$ kΩ.

(*c*) Investigate the limiting case $R_F \rightarrow 0$ and $R_D \rightarrow \infty$, and justify in terms of known FET properties.

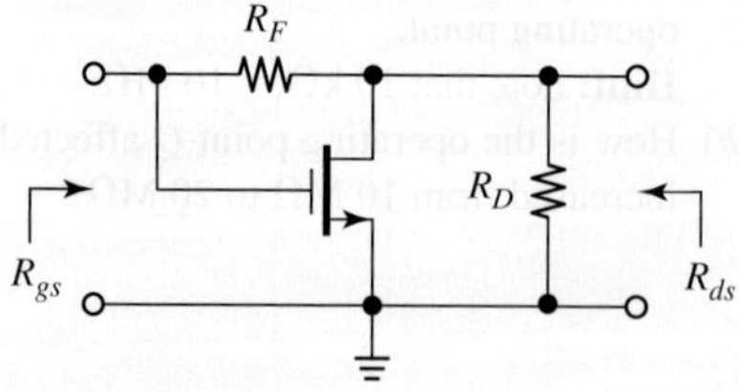

FIGURE P3.49

(d) Repeat, but for the case $R_F \gg R_D$, R_D now being finite.

(e) What happens in the limit $R_D \to 0$?

3.50 (a) Replace the FET of Fig. P3.49 with its small-signal model, and use the test-signal method to obtain an expression for R_{ds}. What is the function of R_F in this circuit?

(b) Calculate R_{ds} if $R_F = 1.0\,\text{M}\Omega$ and $R_D = 10\,\text{k}\Omega$, and the FET has $g_m = 1.0$ mA/V and $r_o = 100\,\text{k}\Omega$.

(c) What happens if the gate and source terminals are ac shorted together?

3.51 (a) Replace the FET of Fig. P3.51 with its small-signal model, and use the test-signal method to obtain an expression for R_{ds}.

(b) Calculate R_{ds} if $R_1 = R_2 = 1.0$ MΩ, and the FET has $g_m = 1.0$ mA/V and $r_o = 100$ kΩ.

(c) What is the limiting value of R_{ds} if R_1, R_2, and r_o are very large?

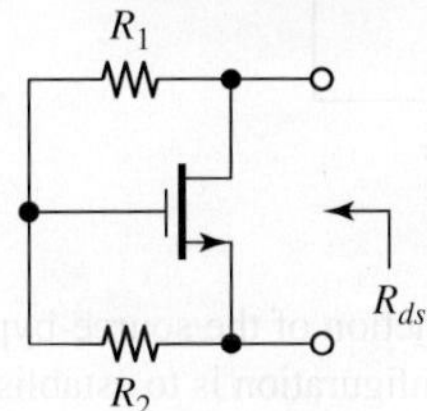

FIGURE P3.51

3.52 The depletion nMOSFET of Fig P3.52 has $V_t = -2.0$ V, $k = 2.0$ mA/V^2, and $\lambda = 0.01$ V^{-1}.

(a) Find $V_{DS(EOS)}$ and $I_{D(EOS)}$ for case $R = 0$, and then sketch and label the i-v curve for $v \geq 0$. What is its slope in the saturation region? What is the per-volt change if i in the saturation region? Express this change as a percentage of $I_{D(EOS)}$.

(b) Repeat, but for the case $R = 1.0$ kΩ. Compare the percentage changes, and comment.

Hint: exploit the fact that R introduces source degeneration.

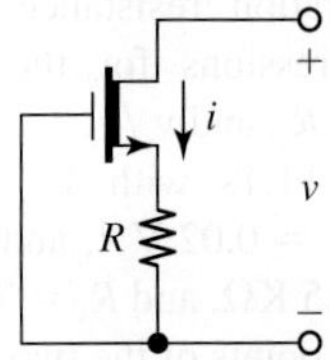

FIGURE P3.52

3.53 (a) Replace both FETs of Fig. P3.53 with their small-signal models, and derive an expression for R_d.

(b) If both devices have $g_m = 1$ mA/V and $r_o = 100$ kΩ, calculate R_d and comment on your result.

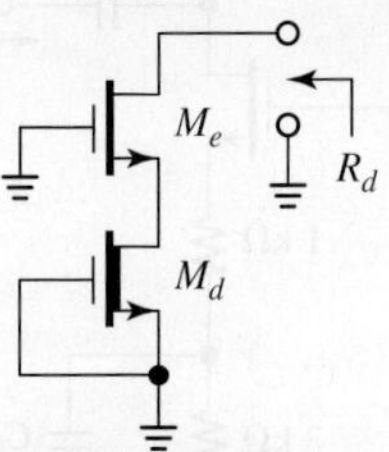

FIGURE P3.53

3.8 Basic MOSFET Voltage Amplifiers

3.54 In the circuit of Fig. P3.54 the FET has $V_t = 1.5$ V, $k = 1$ mA/V^2, and $\lambda = 0.02$ V^{-1}.

(a) Find the dc operating point of the FET, assuming $\lambda = 0$ to simplify your dc calculations.

(b) If $R = 0$, find R_i, R_o, and v_o/v_i, and specify C for operation at a signal frequency of 10 kHz.

(c) Find R to lower the gain to about half the value of part (b). How does this affect R_i, R_o, and the choice of C?

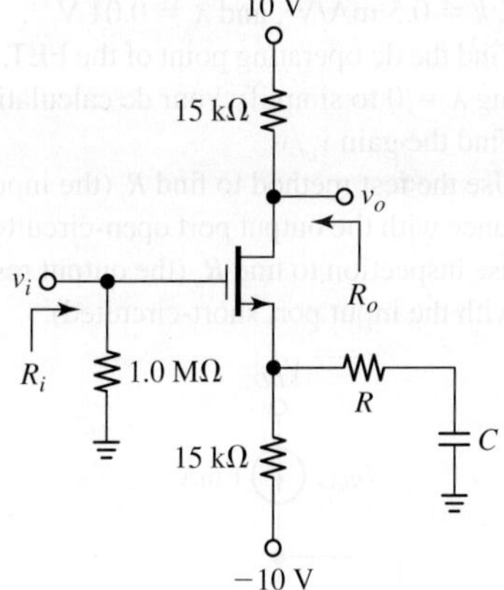

FIGURE P3.54

3.55 In the circuit of Fig. P3.55 the FET has $V_t = 1.0$ V, $k = 2$ mA/V^2, and $\lambda = 0.02$ V^{-1}.

(a) Find the dc operating point of the FET, assuming $\lambda = 0$ to simplify your dc calculations.

(b) Find R_i, R_o, and v_o/v_i.

(c) Repeat part (a) and (b) if the 1.0-kΩ resistance is set to 0 (shorted out).

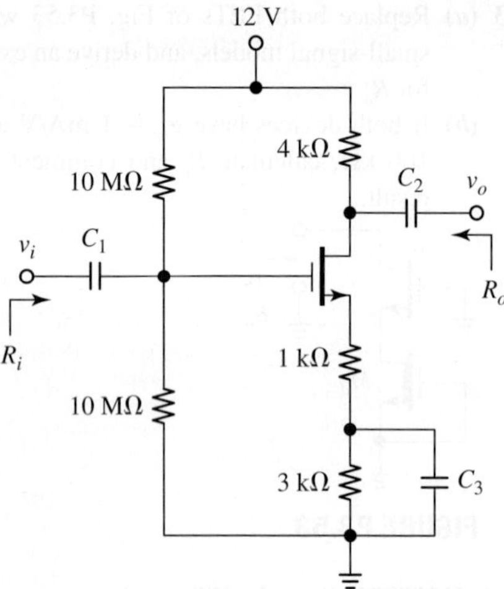

FIGURE P3.55

3.56 Repeat Example 3.32a if the source terminal of the FET of Fig. 3.57a is lifted off ground to allow for the insertion of a series resistance $R_S = 2.0\,\text{k}\Omega$ between source and ground, and thus introduce source degeneration. Discuss the effect of R_S upon the dc operating point and the small-signal parameters.

3.57 In the circuit of Fig. P3.57 let the FET have $V_t = 1.5\,\text{V}$, $k = 0.5\,\text{mA/V}^2$, and $\lambda = 0.01\,\text{V}^{-1}$.
  (a) Find the dc operating point of the FET, assuming $\lambda = 0$ to simplify your dc calculations.
  (b) Find the gain v_o/v_i.
  (c) Use the test method to find R_i (the input resistance with the output port open-circuited), and use inspection to find R_o (the output resistance with the input port short-circuited).

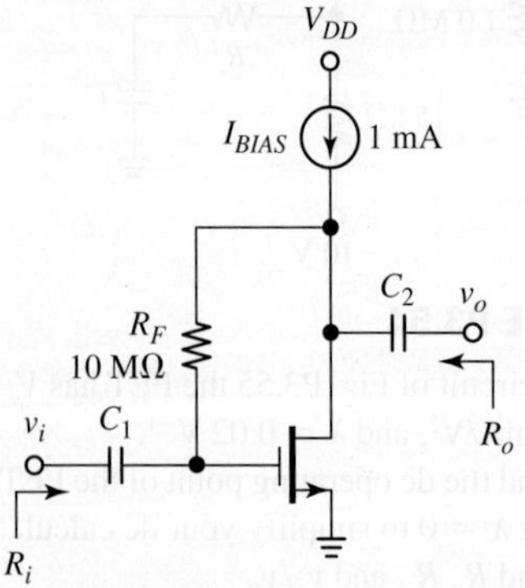

FIGURE P3.57

3.58 In the circuit of Fig. P3.58 let the $V_{DD} = 12$ V, $R_D = 2.0\,\text{k}\Omega$, $R_1 = 1.0\,\text{M}\Omega$, and $R_2 = 2.0\,\text{M}\Omega$.
  (a) If the FET has $V_t = 2.0\,\text{V}$, $k = 1.5\,\text{mA/V}^2$, and $\lambda = 0.015\,\text{V}^{-1}$, find its dc operating point; assume $\lambda = 0$ to simplify your dc calculations.
  (b) Find the gain v_o/v_i.
  (c) Use the test method to find R_i (the input resistance with the output port open-circuited), and use inspection to find R_o (the output resistance with the input port short-circuited).

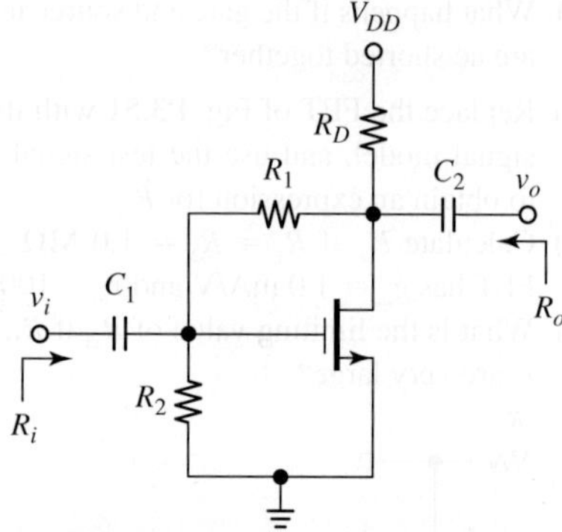

FIGURE P3.58

3.59 As we know, the function of the source-bypass capacitor in the CS configuration is to establish an *ac ground at the source*. The circuit of Fig. P3.59 eliminates the need for such a capacitance by utilizing the diode-connected FET M_2 instead. Even though M_2 does not provide a true ac ground at M_1's source, its ac resistance (R_{s2}) is *relatively low*, and the small amount of source degeneration that it introduces for M_1 is a price well worth the elimination of the by-pass capacitor. This technique is widely used in IC implementations, where M_1 and M_2 are matched devices. With a signal source having a dc component of 0 V, the two FETs experience the same V_{GS} drop and thus carry the same current I_D. Consequently, R_S has to be specified to carry a current of $2I_D$.
  (a) Regarding M_1 as a CS-SD amplifier with a total source-degeneration resistance $R_{S1} = R_S/\!/R_{s2}$), derive expressions for the small-signal parameters R_i, R_o, and v_o/v_i.
  (b) Assuming matched FETs with $V_t = 1$ V, $k = 1\,\text{mA/V}^2$, and $\lambda = 0.02\,\text{V}^{-1}$, and $V_{DD} = -V_{SS} = 12$ V, $R_D = 15\,\text{K}\Omega$, and $R_S = 7.5\,\text{K}\Omega$, find the dc operating points of the two FETs.
  (c) Calculate R_i, R_o, and v_o/v_i numerically.

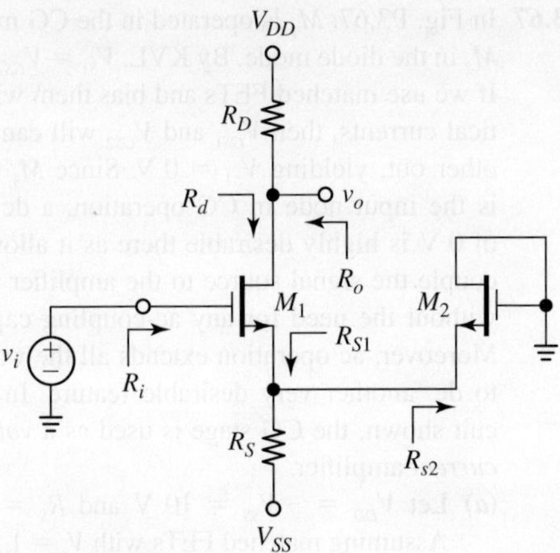

FIGURE P3.59

3.60 Using the circuit topology of Fig. P3.54 but with $R = 0$, design an amplifier that accepts a 1-kHz signal source v_{sig} having $R_{sig} = 100\ \text{k}\Omega$, and outputs $v_o = -4 \times v_{sig}$ to a load $R_L = 20\ \text{k}\Omega$. Your circuit is to be powered from ± 6-V supplies, and is to use an nMOSFET with $V_t = 1.0\ \text{V}$, $k = 1.0\ \text{mA/V}^2$, and $\lambda = 0.02\ \text{V}^{-1}$.

(a) Draw the circuit, and specify standard 5% resistance and capacitance values to achieve your goal.

(b) Verify that your circuit operates properly by showing all node voltages (dc as well as ac component) if $v_{sig} = (50\ \text{mV})\cos(2\pi 10^3 t)$.

Hint: the design is not unique, and it may take some iterations for you to come up with a circuit that meets the given specifications.

3.61 Using the circuit topology of Fig. P3.39, design a CS amplifier with $a_{oc} = -5\ \text{V/V}$ and $R_i \geq 1\ \text{M}\Omega$, and specify the capacitances for operation at 100 kHz. Your amplifier is to be powered from a 9-V supply, and is to use a pMOSFET with $V_t = -1.5\ \text{V}$, $k = 1.25\ \text{mA/V}^2$, and $\lambda = 0.02\ \text{V}^{-1}$.

Hint: start out by imposing $V_D = (1/3)V_{SS}$ and $V_S = (2/3)V_{SS}$ (1/3-1/3-1/3 Rule).

3.62 In the CS amplifier of Fig. 3.56a the FET is biased in the diode mode, giving $V_D = V_{D(EOS)} + V_t$. To allow more headroom for the output signal, it may be desirable to bias the drain somewhat higher, at $V_D = V_{D(EOS)} + mV_t$, $m > 1$. The CS configuration of Fig. P3.58 achieves this goal by utilizing the additional resistance R_2, which forces R_1 to drop some voltage, thus raising V_D.

(a) Show that if R_1 and R_2 are sufficiently large to draw negligible current compared to I_D, the CS amplifier of Fig. P3.58 gives, for $\lambda = 0$,

$$a_{oc} = 2\left(\frac{V_{DD} - mV_t}{V_{OV}} - 1\right)$$

(b) If $V_{DD} = 5\ \text{V}$ and the FET has $V_t = 0.5\ \text{V}$, $k = 2.0\ \text{mA/V}^2$, and $\lambda = 0$, specify suitable resistances for $a_{oc} = -10\ \text{V/V}$ with $m = 2$.

(c) Recalculate a_{oc} if $\lambda = 0.02\ \text{V}^{-1}$, and comment.

(d) Verify that your circuit operates properly by showing all node voltages (dc as well as ac component) if $v_i = (100\ \text{mV})\cos\omega t$.

3.9 MOSFET Voltage and Current Buffers

3.63 In the circuit of Fig. P3.63 the enhancement FET M_e acts as a voltage follower, and the depletion FET M_d as a current source to bias M_e. Let M_e have $V_t = 1.0\ \text{V}$, $k = 2.0\ \text{mA/V}^2$, and $\lambda = 0.025\ \text{V}^{-1}$, and let M_d have $V_t = -1.0\ \text{V}$, $k = 2.0\ \text{mA/V}^2$, and $\lambda = 0.02\ \text{V}^{-1}$.

(a) Find the dc operating points of the two FETs for $v_I = 0$. To simplify your dc calculations, assume $\lambda = 0$.

(b) Find R_i, R_o, and v_o/v_i.

(c) What is the range of values of v_I over which the circuit will operate properly, with each FET in the active region?

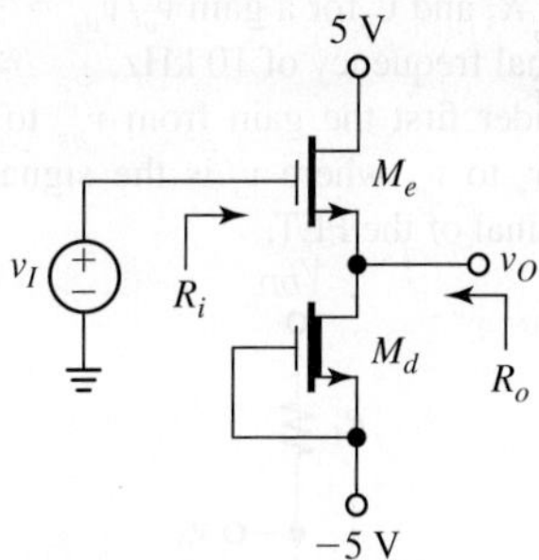

FIGURE P3.63

3.64 Repeat Problem 3.63, but for the case in which the enhancement FET M_e is replaced with a *depletion* FET having the same characteristics as M_d.

3.65 In the circuit of Fig. P3.65, utilizing two depletion-type FETs, M_1 acts as a *voltage follower*, M_2 as a *current sink* to bias M_1, and R controls the bias current.

(*a*) Assuming ± 5-V power supplies and matched FETs with $V_t = -1.5$ V and $k = 2.0$ mA/V^2, specify R to bias the FETs at 1 mA. What is the value of v_O when $v_I = 0$?

(*b*) Assuming $\lambda = 0.025$ V^{-1}, find R_i, R_o, and v_o/v_i.

(*c*) What is the range of values of v_I over which the circuit will operate properly, with each FET in the active region?

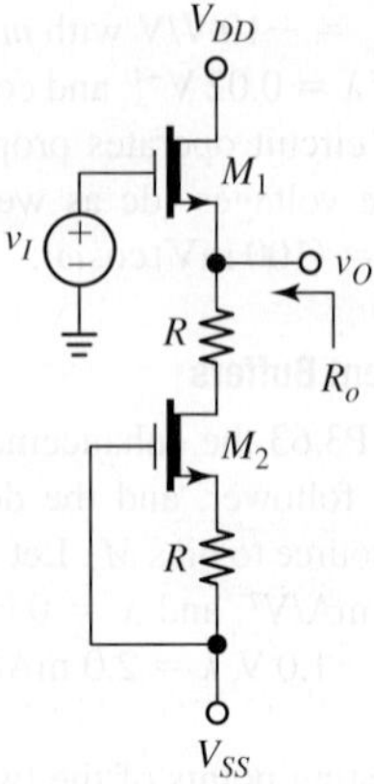

FIGURE P3.65

3.66 The CG circuit of Fig. P3.66 has $V_{DD} = -V_{SS} = 10$ V, and the FET has $k = 1.0$ mA/V^2, $V_t = 1.0$ V, and $\lambda = 0$.

(*a*) Specify R_1 and R_2 to ensure $I_D = 2$ mA and to bias the drain halfway between $(V_S + V_{OV})$ and V_{DD}.

(*b*) Specify R_3 and C for a gain $v_o/v_{sig} = +2$ V/V at a signal frequency of 10 kHz.

Hint: consider first the gain from v_{sig} to v_s, and then from v_s to v_o, where v_s is the signal at the source terminal of the FET.

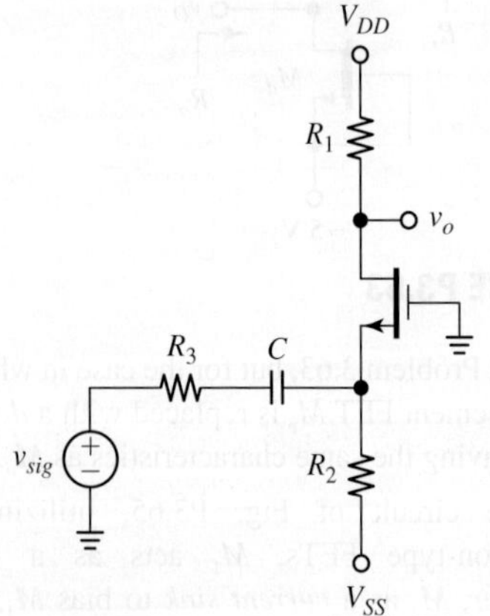

FIGURE P3.66

3.67 In Fig. P3.67, M_1 is operated in the CG mode and M_2 in the diode mode. By KVL, $V_{S1} = V_{GS2} - V_{GS1}$. If we use matched FETs and bias them with identical currents, then V_{GS1} and V_{GS2} will cancel each other out, yielding $V_{S1} = 0$ V. Since M_1's source is the input node in CG operation, a dc voltage of 0 V is highly desirable there as it allows us to couple the signal source to the amplifier *directly*, without the need for any ac-coupling capacitors. Moreover, ac operation extends all the way down to dc, another very desirable feature. In the circuit shown, the CG stage is used as a *voltage-to-current* amplifier.

(*a*) Let $V_{DD} = -V_{SS} = 10$ V and $R_s = 10$ kΩ. Assuming matched FETs with $V_t = 1.5$ V, $k = 2.0$ mA/V^2, and $\lambda = 0.02$ V^{-1}, and assuming the signal source has $R_{sig} = 10$ kΩ and a dc component of 0 V, specify R_B to ensure $V_{S1} = 0$.

(*b*) Find the small-signal parameters R_i, R_o, and i_o/v_{sig}. How does the gain i_o/v_{sig} compare with the ideal case $R_i \to 0$?

Hint: after finding R_i, find the intermediate voltage gain v_i/v_{sig}.

(*c*) What is the maximum voltage that the load can develop and still ensure active-mode operation for M_1?

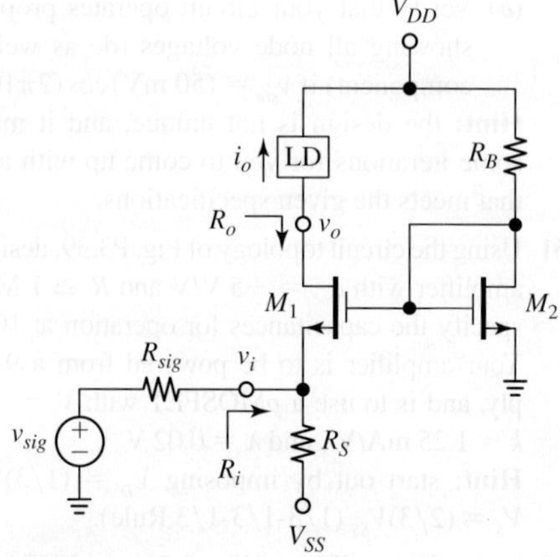

FIGURE P3.67

3.10 The CMOS Inverter/Amplifier

3.68 (*a*) Given that the inverter of Fig. P3.68 has the current transfer curve shown, find k_n and k_p assuming $\lambda_n = \lambda_p = 0$.

(*b*) Repeat, but with $\lambda_n = \lambda_p = 0.08$ V^{-1}.

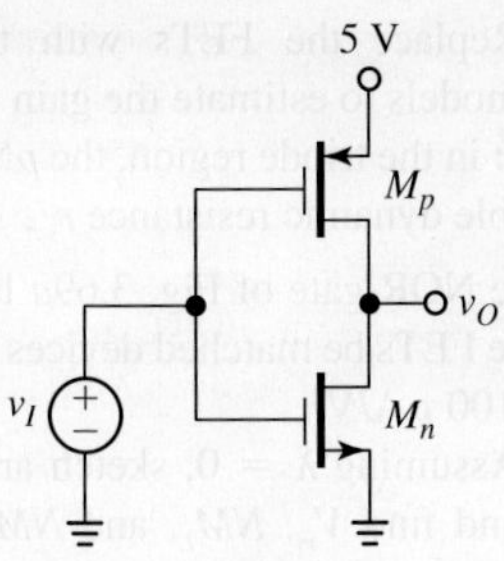

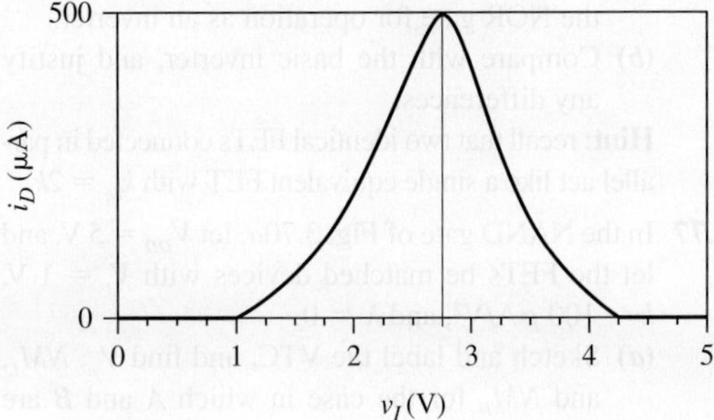

FIGURE P3.68

3.69 A CMOS inverter is implemented with matched FETs having $k = 200\ \mu\text{A/V}^2$ and $V_t = 0.6$ V, and is powered from $V_{DD} = 3$ V.

(a) Assuming $\lambda = 0$, find V_{IL}, V_{IH}, NM_L, NM_H, V_m, and I_m.

(b) Find the output resistance for $v_O = V_{OL}$ and for $v_O = V_{OH}$.

(c) Find the maximum output current that the inverter can *sink* from and external load with its output within 0.1 V of ground, and can *source* to an external load with its output within 0.1 V of V_{DD}.

(d) Find the maximum output current that the inverter can source/sink while retaining noise margins of 1 V.

3.70 A CMOS inverter is implemented with matched FETs having $V_t = V_{tn} = -V_{tp} = 0.2V_{DD}$.

(a) Show that the maximum output current that the inverter can sink from an external load with its output within $0.1V_{DD}$ of ground, and can source to an external load with its output within $0.1V_{DD}$ of V_{DD}, is $I_{O(\max)} = 0.075kV_{DD}^2$, where $k = k_n = k_p$.

(b) Assuming $k'_n = 100\ \mu\text{A/V}^2$ and $k'_p = 40\ \mu\text{A/V}^2$, specify $(W/L)_n$ and $(W/L)_p$ for $I_{O(\max)} = 0.5$ mA if $V_{DD} = 3$ V. What is the resulting value of I_m?

(c) Find the output resistance for $v_O = V_{OL}$ and for $v_O = V_{OH}$.

3.71 Even though today's digital electronics is dominated by CMOS technology, there are systems still in operation that were implemented in TTL technology, the dominant BJT technology before the advent of CMOS technology. Figure P3.71 depicts the interfacing of a CMOS system to a TTL system, using two ordinary inverters as an example. The CMOS inverter is of the type of Fig. 3.64. The TTL inverter, whose internal details need not concern us here, is specified in terms of its terminal characteristics in the manufacturer's data sheets. In the following, assume a CMOS process with $V_{tn} = -V_{tp} = 1$ V, $k'_n = 50\ \mu\text{A/V}^2$, and $k'_p = 20\ \mu\text{A/V}^2$.

(a) The TTL data sheets specify that for proper operation, a TTL gate of the so-called 7400 Series requires that when $v_I = 0$ V, the CMOS inverter *source* to the TTL gate a current $i_O \geq 40\ \mu\text{A}$ at $v_O \geq 2.4$ V. Specify a lower limit on $(W/L)_p$ to meet this requirement.

(b) Likewise, when $v_I = 5$ V, the TTL gate requires that the CMOS inverter *sink* from the TTL gate a current $i_O \geq 1.6$ mA at $v_O \leq 0.4$ V. Specify a lower limit on $(W/L)_n$ to meet this requirement.

(c) Repeat, but for a gate of the so-called 74LS00 Series (low-power Schottky TTL), which requires that the CMOS inverter source $i_O \geq 4\ \mu\text{A}$ at $v_O \geq 2.7$ V when $v_I = 0$ V, and sink $i_O \geq 0.4$ mA at $v_O \leq 0.4$ V. Comment on the differences.

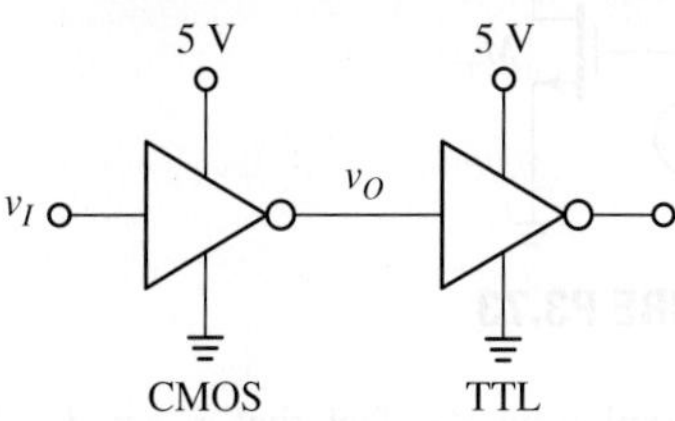

FIGURE P3.71

3.72 As we know, with matched MOSFETs the VTC of a CMOS inverter is centered at the midpoint voltage $v_I = V_m = V_{DD}/2$. What if the devices are mismatched due to fabrication process variations?

(a) Exploiting the fact that $i_{Dp} = i_{Dn}$, show for the case of devices not necessarily matched, the VTC is centered at

$$V_m = \frac{V_{DD} - |V_{tp}| + V_{tn}\sqrt{k_n/k_p}}{1 + \sqrt{k_n/k_p}}$$

(b) Assuming $V_{DD} = 5$ V, find the range of possible values for V_m if, for each FET, k may lie anywhere within the range of 100 μA/V^2 ± 20%, and V_t may lie anywhere within the range of 1.0 V ± 20%. What is the maximum percentage range of variability of V_m from its ideal value of $V_{DD}/2 = 2.5$ V?

(c) What is the possible range of values for the peak current I_m?

3.73 The inverter of Fig. P3.73 is known as *pseudo nMOS*, and finds application as an alternative to CMOS in special situations that are beyond the scope of the present chapter.

(a) Assuming $V_{DD} = 5$ V, $V_{tn} = -V_{tp} = 1$ V, $k_p = 40$ μA/V^2, $k_n = 200$ μA/V^2, and $\lambda_n = \lambda_p = 0$, find V_{OH}, V_{OL}, NM_L, and NM_H.

(b) How much current does this gate draw from the supply when $v_I = 0$ V? When $v_I = 5$ V? Compare with the CMOS inverter of Fig. 3.64, and comment.

Hint: exploit the fact that $i_{Dp} = i_{Dn}$.

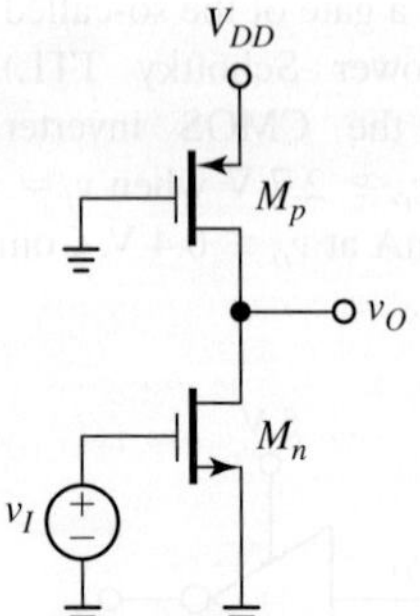

FIGURE P3.73

3.74 (a) Exploiting the fact that $i_{Dp} = i_{Dn}$, find the value of v_I needed to bias the inverter of Problem 3.73 at $v_O = 2.5$ V.

(a) Estimate the small-signal gain (slope of the VTC) at $v_O = 2.5$ V.

3.75 In the circuit of Fig. P3.73, let $V_{DD} = 5$ V, $V_{tn} = -V_{tp} = 1$ V, $k_p = 40$ μA/V^2, $k_n = 400$ μA/V^2, and assume $\lambda_n = \lambda_p = 0$ for simplicity.

(a) Find the point Q_0 on the VTC at which $v_O = v_I$.
Hint: exploit the fact that $i_{Dp} = i_{Dn}$.

(b) Replace the FETs with their small-signal models to estimate the gain v_o/v_i at Q_0.
Hint: in the triode region, the *p*MOSFET acts as a suitable dynamic resistance r_{sd}.

3.76 In the NOR gate of Fig. 3.69a let $V_{DD} = 5$ V, and let the FETs be matched devices with $V_t = 1$ V and $k = 100$ μA/V^2.

(a) Assuming $\lambda = 0$, sketch and label the VTC, and find V_m, NM_L, and NM_H for the case in which A and B are tied together to configure the NOR gate for operation as an inverter.

(b) Compare with the basic inverter, and justify any differences.

Hint: recall that two identical FETs connected in parallel act like a single equivalent FET with $k_{eq} = 2k$.

3.77 In the NAND gate of Fig. 3.70a, let $V_{DD} = 5$ V, and let the FETs be matched devices with $V_t = 1$ V, $k = 100$ μA/V^2, and $\lambda = 0$.

(a) Sketch and label the VTC, and find V_m, NM_L, and NM_H for the case in which A and B are tied together to configure the NAND gate for operation as an inverter.

(b) Compare with the basic inverter, and justify the differences.

Hint: recall that two identical FETs connected in series act like a single equivalent FET with $k_{eq} = k/2$.

3.78 In the circuit of Fig. P3.78, R biases the FETs at $V_O = V_I$, and since the FETs are assumed to be matched, we have $V_O = V_I = V_{DD}/2$.

(a) Replace the FETs with their small-signal models, and use the test method to obtain expressions for R_i (the input resistance with the output port open-circuited), and R_o (the output resistance with the input port ac short-circuited) in terms of g_m, r_o, and R.

(b) If $R = 1$ MΩ, $V_t = 1$ V, $k = 1$ mA/V^2, and $\lambda = 0.01$ V^{-1}, calculate R_i and R_o for $V_{DD} = 5$ V.

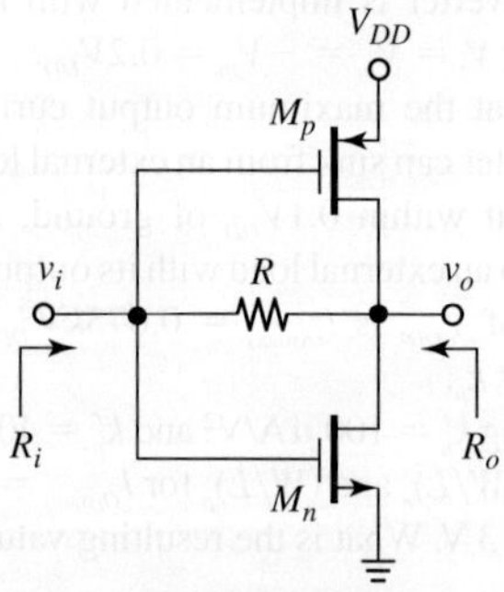

FIGURE P3.78

3.79 (*a*) In the circuit of Problem 3.78, replace the FETs with their small-signal models and obtain an expression for the small-signal gain $a = v_o/v_i$ in terms of g_m, r_o, and R. Hence, calculate this gain for

(*b*) $V_{DD} = 5$ V,

(*c*) $V_{DD} = 10$ V, and

(*d*) $V_{DD} = 3$ V.

(*e*) Develop an expression for a in terms of V_{DD} and λ in the limit $R \gg r_o$, compare with the calculated values, and comment.

3.80 The circuit of Fig. P3.80 illustrates the application of a CMOS inverter as an *inverting amplifier*. The function of the negative-feedback resistance R_2 is to bias the FETs at $V_O = V_I$, that of R_1 is to set the closed-loop gain ideally at $-R_2/R_1$, and that of C is to ac-couple the input source to the amplifier without disturbing its dc operating point. Assume matched FETs, each having $g_m = 1$ mA/V and $r_o = 100$ kΩ. If $R_1 = 200$ kΩ, $R_2 = 1.0$ MΩ, and $v_{sig} = (100 \text{ mV})\cos \omega t$, estimate v_o and v_i as well as the *open-loop gain* v_o/v_i and the *closed-loop gain* v_o/v_{sig}. How does the latter compare with its ideal value?

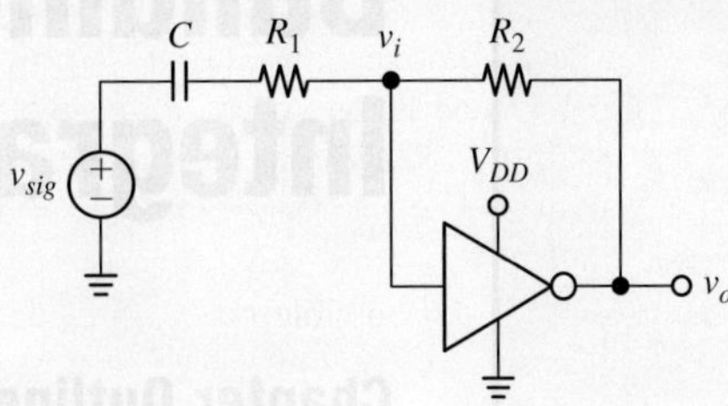

FIGURE P3.80

Chapter

Building Blocks for Analog Integrated Circuits

Chapter Outline

s mentioned in Chapter 2, following its invention in 1947, the bipolar junction transistor (BJT) was first put to use as a replacement for the much bulkier, power hungry and far less reliable vacuum tube. In fact, the first transistor circuits were virtual replicas of vacuum-tube circuit prototypes, if with suitable scaling of the power supplies and surrounding components. Broadly speaking, they belong to the class of circuits that we have investigated so far and that are generally referred to as *discrete circuits*.

Toward the end of the 1950s it was realized that the dramatic miniaturization and reduction in power consumption brought about by the transistor could be exploited further by fabricating entire circuits (transistors, diodes, resistors, and small capacitors, along with their interconnections) in *monolithic* form, that is, on the same piece of semiconductor material, or *chip*. Also called an *integrated circuit* (IC), it was first implemented in 1958 by Jack Kilby at Texas Instruments, and, independently, in 1959 by Robert Noyce at Fairchild Semiconductor. The 1960s saw a feverish activity that resulted in the development of the first monolithic *operational amplifier* by Fairchild Semiconductor (mA Series) along with the digital IC families known as *transistor-transistor logic* (TTL) by Texas Instruments (7400 Series) and *emitter-coupled logic* (ECL) by Motorola (10K Series).

Meanwhile, in the early 1960s the field-effect transistor of the metal-oxide-semiconductor type (MOSFET) became a commercial reality. Compared to the BJT, the MOSFET offered the advantages of smaller size and lower power consumption. This alternative technology led to the development of the first battery-powered electronic calculators and wristwatches, as well as a low-power alternative to the prevailing TTL bipolar logic family, namely, the CMOS digital family of the 4000 Series, by RCA. These products were followed by the first microprocessor, by Intel, in 1971. Since then, IC electronics has progressed *exponentially* and has penetrated virtually every aspect of modern life. This impressive growth has been governed by *Moore's law*, roughly stating that thanks to continued advances in IC fabrication, the number of devices that can be integrated on a given chip area *doubles* every approximately 18 months. Originally formulated in 1965, this law still holds to this day, though it has been pointed out that technology is bound to approach physical limits that will eventually lead to the demise of this law.

The BJT, after having been the dominant semiconductor device for nearly three decades, has been overtaken by the MOSFET, especially in high-density ICs, owing to the MOSFET's aforementioned advantages of smaller size and lower power consumption. Nonetheless, the BJT continues to be the device of choice in high-performance general-purpose analog ICs. It is also preferred in specialized discrete designs, thanks to the availability of a wide selection of devices. It is also possible to fabricate BJTs and MOSFETs simultaneously on the same chip, if at an increase in the number of fabrication steps and cost. The resulting technology, aptly known as BiCMOS technology, exploits the advantages of both transistor types to provide even more innovative design solutions. Contemporary ICs tend to combine digital as well as analog functions on the same chip, this being the reason for the name *mixed-signal* or also *mixed-mode ICs*. As a rule, one would try to implement as many functions as possible in digital form, and use analog circuitry only when interfacing to the external physical world, which is analog by nature.

CHAPTER HIGHLIGHTS

The chapter begins with a comparison between discreet and monolithic design. This is followed by a review of BJT and FET characteristics emphasizing second-order effects that were deliberately omitted in the earlier introductory chapters, but that are relevant to monolithic circuit realizations. Also reviewed are the basic single-transistor configurations of Chapters 2 and 3, but from a monolithic perspective, where the functions of dc biasing and output loading are no longer provided by resistors, but by other transistors operating as current sources or sinks.

Next, the chapter investigates a variety of multiple-transistor circuit configurations that have become standard building blocks for analog ICs. These include the *Darlington* and *cascode* configurations, *differential pairs*, *current mirrors*, *dc current sources and sinks*, *active-loaded gain stages*, and *push-pull output stages*. Particular attention is devoted to the differential pair because it forms the core of most analog ICs. First, we investigate the pair for the idealized case of perfectly matched components, then we examine the effect of fabrication mismatches upon input offset errors and the common-mode rejection ratio. Whenever possible, BJT and FET realizations are covered in parallel both to void duplications and to compare the two technologies in their similarities and differences.

This chapter exposes the student to a variety of clever design solutions that contribute to making analog electronics a most fascinating discipline. But the best is yet to come, in the next chapter, when these building blocks will be combined together in the design of some representative analog ICs.

It is reassuring to know that no matter how sophisticated a given circuit, the tools we use to comprehend it are still those mastered in a basic electric circuits course, namely, Ohm's law, Kirchhoff's laws (KVL and KCL), Thévenin's/Norton's reductions, and the test-signal technique to find terminal resistances in the presence of dependent sources.

The chapter makes abundant use of PSpice both as a software oscilloscope to display transfer curves and waveforms, and as a verification tool for dc and ac hand calculations.

4.1 DESIGN CONSIDERATIONS IN MONOLITHIC CIRCUITS

The widespread usage of *monolithic circuits*, also called *integrated circuits* (ICs), stems from our ability to fabricate large numbers of interconnected devices on the same semiconductor chip while keeping power consumption suitably low. Compared to their discrete predecessors, ICs present unique constraints as well as advantages for the designer. The most relevant ones are as follows:

- Capacitors are highly undesirable in IC technology as they tend to take up inordinate amounts of chip area. While small capacitors (on the order of a few picofarads or less) are still acceptable, large-valued ones such as those used for ac coupling and ac bypassing in the discrete CE and CS amplifier examples of Chapters 2 and 3 must definitely be ruled out. Consequently, inter-stage *coupling must be of the dc type*, and suitable techniques must be devised to *avoid ac bypass capacitors*. All considered, these constraints turn out to be a blessing because once they are met, a monolithic circuit will function all the way down to dc. By contrast, the discrete designs of Chapters 2 and 3 will function properly only above a certain frequency, as at low frequencies capacitors will start acting as open circuits, no longer providing the intended functions of coupling and bypass.

- Resistors too tend to take up precious chip area, though not as severely as capacitors, so they too must be avoided whenever possible. When implemented with the same materials that are utilized to form the regions of a transistor, monolithic resistances also suffer from gross tolerances and can be quite sensitive to temperature. On the other hand, transistors are the most natural devices in IC technology, both in terms of size and ease of fabrication, so in monolithic implementations, resistive functions such as *dc biasing* must be rephrased in terms of transistors.

- Devices fabricated simultaneously on the same chip tend to exhibit highly *matched characteristics*. True, these characteristics are sensitive to temperature and also drift with time, but if two or more devices are fabricated in close proximity to each other so that they share the same environment, their characteristics will track over a range of environmental variations. *Matching* and *tracking* are exploited on purpose in IC design to implement clever functions that would be far more difficult to achieve in discrete form. Two popular examples are *current mirrors* and *differential pairs*, to be investigated in great detail as we proceed.

An Illustrative Example

To illustrate similarities as well as differences between discrete and monolithic design, let us reconsider the familiar CE configuration of Fig. 4.1, which in the days of discrete designs preceding ICs was the workhorse of voltage amplification. As long as each capacitor acts as an *ac short* compared to the equivalent resistance presented by the surrounding circuitry, the small-signal gain takes on the form

$$a = \frac{v_o}{v_i} = -g_m(r_o /\!/ R_C) \qquad (4.1a)$$

To convert this design to a form suited to monolithic fabrication we need to get rid of the capacitors and replace the resistors with suitably biased transistors whenever possible. Following are the relevant steps:

- The function of C_1 in Fig. 4.1 is to resolve the dc difference between the input source (typically with a dc component of 0 V) and the base (typically biased at $\frac{1}{3}V_{CC}$) while providing an ac short between source and base ($v_b \rightarrow v_i$). The best way to get rid of C_1 is to *couple the input source to the base directly*, as in Fig. 4.2a. Dc coupling also eliminates the need for the biasing resistors R_1 and R_2.
- The function of R_E in Fig. 4.1 is to establish the bias current I_E, whereas that of C_2 is to ensure an ac ground at the emitter ($v_e \rightarrow 0$). Both functions can be met, at least in principle, by biasing the emitter negatively via a suitable voltage source $V_{BE1} = V_T \ln(I_{C1}/I_{s1})$, as also shown in Fig. 4.2a.

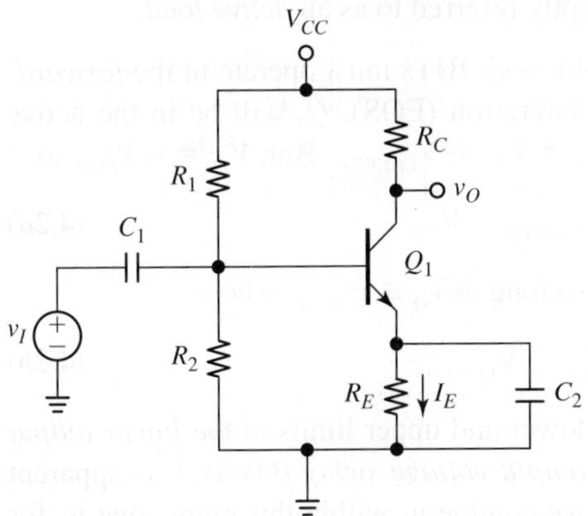

FIGURE 4.1 A discrete voltage amplifier.

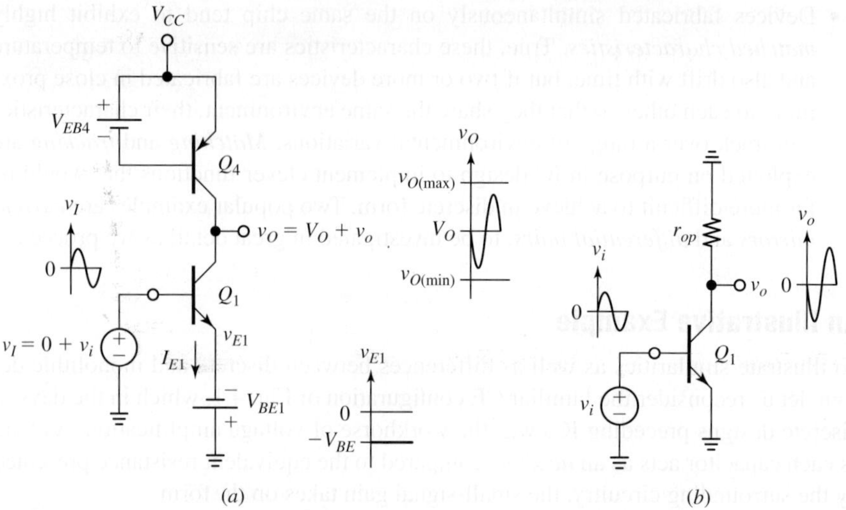

FIGURE 4.2 (a) Conceptual circuit showing how to turn the discrete amplifier of Fig. 4.1 into a form suitable to monolithic fabrication. (b) Ac equivalent of the circuit of (a).

- The function of R_C in Fig. 4.1 is to bias the collector, typically at $V_O \cong \frac{2}{3}V_{CC}$, while also setting the gain, as per Eq. 4.1a. The best way to get rid of R_C is to replace it with a *pnp* BJT operating as a current source, as also shown in Fig. 4.2a. Once this is done, the ac equivalent becomes as in Fig. 4.2b, where r_{op} is the ac resistance seen looking into Q_4's collector, now playing the role of R_C in the expression for the gain. Consequently, Eq. (4.1a) becomes

$$a = \frac{v_o}{v_i} = -g_{mn}(r_{on}//r_{op}) \tag{4.1b}$$

where we are using subscripts n and p to distinguish between the parameters of the *npn* and *pnp* BJTs. Given that R_C (a passive element) has been replaced by a transistor (an active element), Q_4 is aptly referred to as an *active load*.

We observe that for Eq. (4.1b) to hold, both BJTs must operate in the *forward-active* region or at most at the edge of saturation (EOS). Q_1 will be in the active region so long as $v_O \geq v_{O(min)}$, where $v_{O(min)} = V_{E1} + V_{CE1(EOS)}$. But, $V_{E1} = -V_{BE1}$, so

$$v_{O(min)} = V_{CE1(EOS)} - V_{BE1} \tag{4.2a}$$

Similarly, Q_4 will be in the active region so long as $v_O \leq v_{O(max)}$, where

$$v_{O(max)} = V_{CC} - V_{EC4(EOS)} \tag{4.2b}$$

The voltages $v_{O(min)}$ and $v_{O(max)}$ define the lower and upper limits of the *linear output voltage range*, commonly known as the *output voltage swing* (OVS). It is apparent that Eq. (4.1b) will hold only so long as we confine v_O within this range, that is, for $v_{O(min)} \leq v_O \leq v_{O(max)}$.

(a) In the circuit of Fig. 4.2a let Q_1 have $I_s = 10\text{ fA}$, $V_A = 100\text{ V}$, and $V_{CE(EOS)} = 0.2\text{ V}$, and let Q_4 have $I_s = 5\text{ fA}$, $V_A = 75\text{ V}$, and $V_{EC(EOS)} = 0.2\text{ V}$. If $V_{CC} = 10\text{ V}$, specify suitable values for V_{BE} and V_{EB} so that with $v_I = 0$ the BJTs draw 1 mA and V_O lies in the middle of the OVS.

(b) Find the ac gain a as well as $v_O(t)$ if $v_I(t) = V_I + v_i(t) = 0\text{ V} + (2.5\text{ mV})\cos\omega t$. Does $v_O(t)$ fall within the linear output range at all times?

EXAMPLE 4.1

Solution

(a) By Eq. (4.2) we have

$$v_{O(min)} \cong 0.2 - 0.7 = -0.5\text{ V} \qquad v_{O(max)} \cong 10 - 0.2 = 9.8\text{ V}$$

To bias the output in the middle of the OVS we need

$$V_O = \frac{1}{2}[v_{O(max)} + v_{O(min)}] = \frac{1}{2}(9.8 - 0.5) = 4.65\text{ V}$$

For Q_1 we have, by Eq. (2.21),

$$I_{C1} = I_s e^{V_{BE1}/V_T}\left(1 + \frac{V_{CE1}}{V_{A1}}\right)$$

or

$$10^{-3} = 10 \times 10^{-15} \times e^{V_{BE1}/(26\text{ mV})}\left(1 + \frac{4.65 - (-0.7)}{100}\right)$$

Solving, we get $V_{BE1} = 657.2\text{ mV}$. Adapting the above expression to Q_4's collector current yields

$$10^{-3} = 5 \times 10^{-15} \times e^{V_{EB4}/(26\text{ mV})}\left(1 + \frac{10 - 4.65}{75}\right)$$

which gives $V_{EB4} = 674.8\text{ mV}$.

(b) We have $g_{mn} = 1/(26\ \Omega)$, $r_{on} = 100\text{ k}\Omega$, and $r_{op} = 75\text{ k}\Omega$, so Eq. (4.1$b$) gives

$$a = -\frac{100//75}{0.026} = -1{,}648\text{ V/V}$$

Finally, since $(2.5\text{ mV}) \times 1648 = 4.12\text{ V}$, we have

$$v_O(t) = V_O + v_o(t) = 4.65\text{ V} + (4.12\text{ V})\cos(\omega t - 180°)$$

Note that v_O alternates between $4.65 + 4.12 = 8.77\text{ V}$ and $4.65 - 4.12 = 0.53\text{ V}$, indicating that the condition $v_{O(min)} < v_O < v_{O(max)}$ is satisfied at all times.

We observe that the gain achievable with an active load can be much higher than with a discrete load, thanks to the fact that usually $r_{op} \gg R_C$. High gains are especially welcome in negative-feedback systems such as op amp circuits, where the higher the gain the more pronounced the benefits of negative feedback tend to be (this will become clear when we study negative feedback in Chapter 7.)

As we move along we shall see that we can increase gain further by suitably raising the effective resistance seen looking into Q_4's collector, for instance, by introducing emitter degeneration for Q_4. If this resistance is made much higher than r_{on}, then the ac gain of Eq. (4.1b) simplifies as $a = a_{\text{intrinsic}}$, where

$$a_{\text{intrinsic}} = -g_m r_o = -\frac{V_A}{V_T} \tag{4.3}$$

Here, V_A is the *Early voltage* of the amplifying BJT (Q_1 in the present case), V_T is the *thermal voltage* (26 mV at $T = 300$ K), and $a_{\text{intrinsic}}$ is the *maximum gain* achievable with a single BJT, a gain aptly called the *intrinsic gain*. In Example 4.1 we have $a_{\text{intrinsic}} = -100/0.026 = -3846$ V/V. Compare this with the case of a typical passive load of the types investigated in Chapter 2, say $R_C = 10$ kΩ, which would give $a = -(100//10)/0.026 = -350$ V/V, a whole order of magnitude lower than the intrinsic gain!

At this point we wonder how to implement the sources V_{BE1} and V_{EB4} of Fig. 4.2a. Example 4.1 indicates that their values must be accurate within millivolts. We also know that these values drift with temperature by about -2 mV/°C, so we need biasing schemes capable of assuring stable collector currents against a range of fabrication and environmental variations. In IC technology these schemes are made possible by the aforementioned availability of *matching* and *tracking*, as it will be demonstrated next.

Emitter-Coupled Pairs

The scheme shown in Fig. 4.3a utilizes a current sink ($2I_E$) to provide dc biasing, and a second BJT (Q_2) to provide a low ac-resistance path between Q_1's emitter and ground. (Viewed this way, Q_2 in effect replaces the ac bypass capacitance C_2 of Fig. 4.1.) If Q_1 and Q_2 are matched, then for $v_i = 0$ the sink current $2I_{EE}$ will split *equally* between the two BJTs as they experience the same V_{BE} drop. Moreover, the two currents I_E will *track* each other over a range of thermal variations in V_{BE}.

The ac resistance between Q_1's emitter and ground is the resistance seen looking into Q_2's emitter, or $r_{e2} = \alpha_{02}/g_{m2} \cong 1/g_{m2}\,(=1/g_{m1})$. This resistance, though not zero, is small (26 Ω at 1 mA) and introduces emitter degeneration for Q_1, as depicted in the ac equivalent of Fig. 4.3b. The degenerated transconductance is

$$G_m = \frac{g_{m1}}{1 + g_{m1}r_{e2}} \cong \frac{g_{m1}}{1 + g_{m1}/g_{m2}} = \frac{g_m}{2} \tag{4.4}$$

where the n and p subscripts have been dropped as the BJTs have identical g_ms. Because of degeneration the voltage gain will also be reduced, but this price is well worth the elimination of C_2. In fact, even with degeneration, gain continues to remain fairly high, and it does so all the way down to dc!

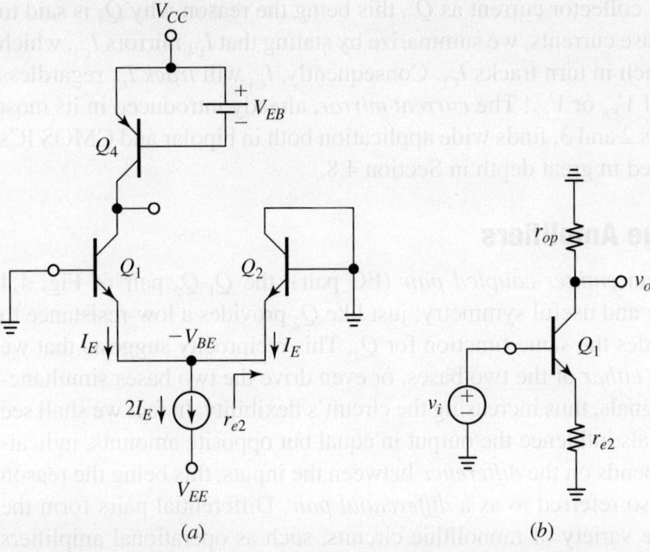

FIGURE 4.3 (a) Biasing scheme for the amplifier Q_1, and (b) ac equivalent for the entire circuit.

Current Mirrors

Next, we wish to investigate a suitable IC scheme to bias Q_4. Since the current I_{C4} sourced by Q_4 must at all times *equal* the current I_{C1} sunk by Q_1, it is apparent that V_{EB} must be *adjusted continuously* against any thermal variations. The scheme of Fig. 4.4 does this automatically via Q_2 and the diode-connected transistor Q_3 as follows: as we know, I_{C2} matches I_{C1}; moreover, in response to I_{C2}, Q_3 develops a voltage drop V_{EB} that is in turn transmitted to Q_4. Being matched and subject to the *same* V_{EB} drive,

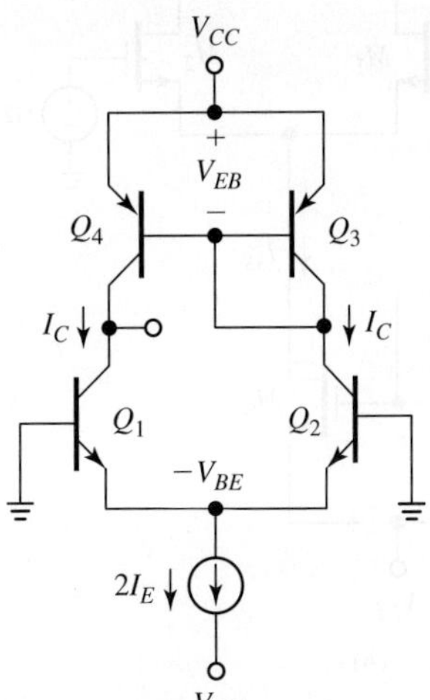

FIGURE 4.4 Biasing scheme for the active load Q_4.

Q_4 will draw as much collector current as Q_3, this being the reason why Q_4 is said to *mirror* Q_3. Ignoring base currents, we summarize by stating that I_{C4} mirrors I_{C3}, which is the same as I_{C2}, which in turn tracks I_{C1}. Consequently, I_{C4} will *track* I_{C1} regardless of any thermal drift of V_{EB} or V_{BE}! The *current mirror*, already introduced in its most basic form in Chapters 2 and 3, finds wide application both in bipolar and CMOS ICs and will be investigated in great depth in Section 4.8.

Monolithic Voltage Amplifiers

Aptly referred to as an *emitter-coupled pair* (EC pair), the Q_1-Q_2 pair of Fig. 4.4 exhibits an interesting and useful symmetry: just like Q_2 provides a low-resistance to Q_1's emitter, Q_1 provides the same function for Q_2. This reciprocity suggests that we can apply the input to *either* of the two bases, or even drive the two bases simultaneously with separate signals, thus increasing the circuit's flexibility. In fact we shall see that the two base signals influence the output in equal but opposite amounts, indicating that the output depends on the *difference* between the inputs, this being the reason why the EC pair is also referred to as a *differential pair*. Differential pairs form the input stages of a wide variety of monolithic circuits, such as operational amplifiers and voltage comparators, and they will be investigated in great depth in Section 4.5.

Figure 4.5a shows a popular monolithic counterpart of the discrete design of Fig. 4.1. Its basic ingredients are as follows:

- The EC pair Q_1-Q_2 provides signal amplification, and it does so in *differential* form.

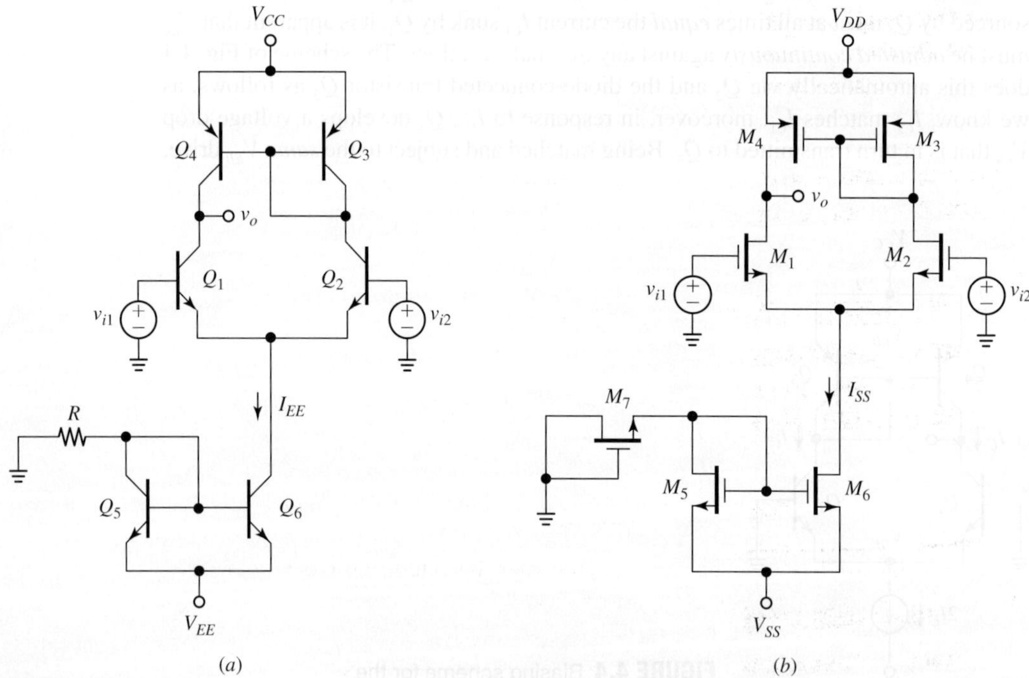

(a)   (b)

FIGURE 4.5 High-gain monolithic amplifiers: (a) bipolar and (b) CMOS.

- The current mirror Q_3-Q_4 forms an *active load* for the EC pair, assuring high gain and also converting the input signal difference to a single-ended output. As we move along we shall see that the gain is

$$a = \frac{v_o}{v_{i1} - v_{i2}} = -g_{mn}(r_{on}/\!/r_{op}) \tag{4.5}$$

- Another current mirror (Q_5-Q_6) is used to provide the dc bias for the EC pair. Here, R establishes the current into the diode-connected BJT Q_5, and Q_6 then mirrors this current (now re-labeled as I_{EE}) to the EC pair, but at a higher output-resistance level (r_{o6} in the present case).

Compared to the discrete predecessor of Fig. 4.1, the monolithic version of Fig. 4.5a might seem far more complex and expensive to fabricate. However, considering that it uses no capacitors and just one resistor, and that transistors are the preferred devices in IC technology, the monolithic version is highly desirable indeed. It is also more flexible as it responds to the *difference* between a pair of input signals, not to mention its ability to provide much *higher voltage gains*.

By the time MOS technology became commercially viable, the canons of IC design were already well established in bipolar technology, so it was a straightforward matter to transfer them directly to the new technology. Shown in Fig. 4.5b is the CMOS counterpart of the bipolar version of Fig. 4.5a. In this example even the current-setting resistor R has been replaced by a transistor, namely, the diode-connected MOSFET M_7, whose W/L ratio is chosen so as to achieve the desired bias current I_{SS}. As we move along, we shall see that Eq. (4.5) applies to the CMOS amplifier version as well, the only difference being in the lower transconductances of FETs. (The circuits of Fig. 4.5 will be investigated in great depth in Section 4.9.)

Figure 4.6 shows a PSpice circuit to display the VTC of the bipolar amplifier of Fig. 4.5a. The steepness of the curve confirms the high gain of the circuit. Also shown are the saturation limits of the linear region of operation.

What to Expect

The discussion leading to the amplifiers of Fig. 4.5 provides a nutshell overview of the most relevant considerations in monolithic design:

- Avoid the use of on-chip resistors and capacitors. If capacitors are required, they should be on the order of a few picofarads or less.
- Resistances will continue to appear in our circuits and calculations, but for the most part they will be the resistances of small-signal transistor models, such as r_{op} and r_{e2} of Fig. 4.3b.
- Exploit the availability of matched and tracking characteristics to devise creative design solutions, such as current mirrors and differential pairs.
- Even though IC design follows a different set of rules than discrete design, the study of ICs still relies very heavily on the foundations provided by Chapters 1 through 3.

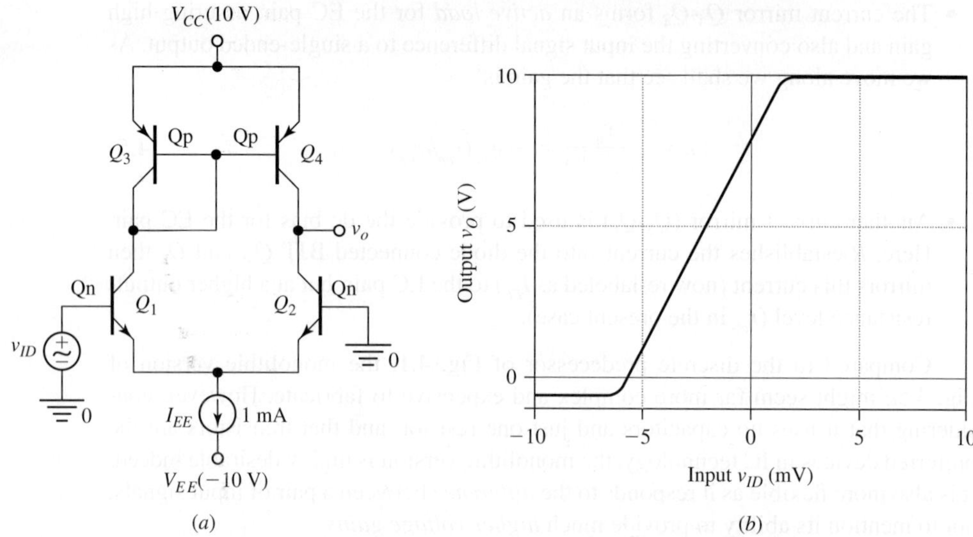

FIGURE 4.6 (a) PSpice circuit to plot (b) the VTC of the monolithic amplifier of Fig. 4.5a. The following parameters are assumed: Qn: $I_s = 2$ fA, $\beta_F = 200$, $V_A = 100$ V. Qp: $I_s = 1$ fA, $\beta_F = 50$, $V_A = 50$ V.

- We shall make frequent use of PSpice simulation to corroborate the result of hand analysis as well as to investigate higher-order nuances that often escape paper-and-pencil calculations.

In order to proceed, we need to investigate in more systematic detail the current-mirror and differential-pair concepts introduced above, as well as a variety of other canonical blocks that are at the basis of contemporary monolithic design. However, before embarking on these tasks, we need to reexamine the characteristics and models for active devices (both BJTs and MOSFETs) in finer detail.

4.2 BJT CHARACTERISTICS AND MODELS REVISITED

The BJT characteristics and models of Chapter 2 were deliberately kept as simple as possible to allow the beginner to focus on the essentials of discrete circuit design and develop a basic feel for circuit operation. As we embark upon the study of monolithic circuits, we need to suitably refine our BJT models to include second-order effects that tend to play more prominent roles in these types of circuits.

Base Width Modulation Revisited

In our study of *npn* BJT operation we found that *increasing* v_{CE} *expands* the width of the base-collector depletion region, thus *shrinking* the effective base width W_B. This phenomenon, known as the Early effect, yields a *steeper* profile of excess electrons within the base, ultimately *raising* the collector current i_C because it is proportional to

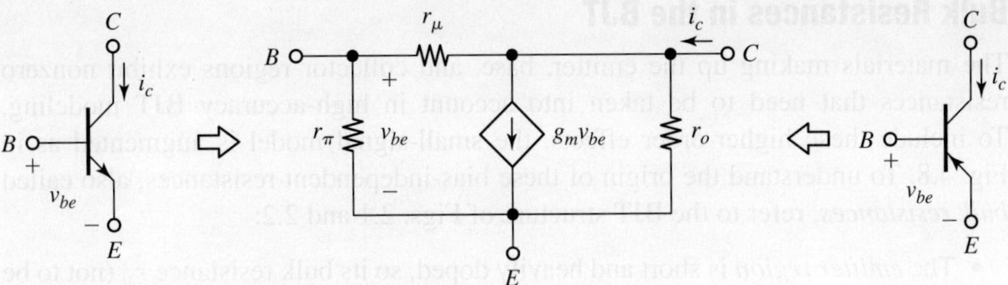

FIGURE 4.7 Small-signal BJT model including r_μ.

this profile's slope. There is yet another more subtle effect of base-width modulation, for a reduction in W_B reduces also the recombination component i_{BB} of the total base current i_B. We recall that $i_B = i_{BE} + i_{BB}$, where i_{BE} is the *diffusion* component from base to emitter, and i_{BB} is the *recombination* component within the base. According to Eq. (2.14), i_{BB} is proportional to the excess charge in the base region, in turn proportional to the *volume* $A_E \times W_B$, so a reduction in W_B will also reduce i_{BB}, and hence, i_B. In summary, if we *increase* v_{CE} while keeping v_{BE} *constant*, we witness (*a*) an *increase* in the current i_C drawn from the v_{CE} source, and (*b*) a *decrease* in the current i_B drawn from the v_{BE} source. In small-signal operation we model the former by means of the familiar resistance r_o between collector and emitter, and the latter by means of an additional resistance r_μ between collector and base. The resulting more refined model is depicted in Fig. 4.7.

Assuming v_{BE} is held constant, we find r_μ as

$$\frac{1}{r_\mu} = \frac{di_{BB}}{dv_{CE}} = \frac{di_C}{dv_{CE}}\frac{di_{BB}}{di_C} = \frac{1}{r_o}\frac{di_{BB}}{di_C}$$

or

$$r_\mu = \frac{i_c}{i_{bb}}r_o \qquad (4.6)$$

where i_c and i_{bb} are small-signal variations of i_C and i_{BB}. If i_B consisted entirely of i_{BB}, then we would have $i_c/i_{bb} = i_c/i_b = \beta_0$, and Eq. (4.6) would give $r_\mu = r_{\mu(min)}$, where

$$r_{\mu(min)} = \beta_0 r_o \qquad (4.7)$$

However, i_{BB} is only a *fraction* of i_B, so if we write $i_{bb} = (1/m)i_b$, $m \geq 1$, then Eq. (4.6) becomes

$$r_\mu = m\beta_0 r_o \qquad (4.8)$$

In monolithic *npn* BJTs, i_{BE} usually dominates and i_{BB} is on the order of only 10% of i_B ($m = 10$), so it is common practice to assume $r_\mu \cong 10\beta_0 r_o$. Because of its sheer size, r_μ is usually ignored except for a few special cases, as we shall see. In lateral *pnp* BJTs (see Fig 2.3) the recombination component is far more significant, so in this case r_μ is closer to the lower limit of Eq. (4.7).

Bulk Resistances in the BJT

The materials making up the emitter, base, and collector regions exhibit nonzero resistances that need to be taken into account in high-accuracy BJT modeling. To include these higher-order effects, the small-signal model is augmented as in Fig. 4.8. To understand the origin of these bias-independent resistances, also called *bulk resistances*, refer to the BJT structure of Figs. 2.1 and 2.2:

- The *emitter region* is short and heavily doped, so its bulk resistance r_{ex} (not to be confused with $r_e = \alpha_0/g_m$) is rather small and fixed. In monolithic BJTs r_{ex} is on the order of just a few Ohms.
- The *base region* is doped more lightly than the emitter and it also forms a longer conductive path, especially transversally, so r_b is one to two orders of magnitude higher than r_{ex}. In monolithic BJTs r_b typically ranges from 50 to 500 Ω. Note that the voltage controlling the dependent source is the internal voltage v_π, not the terminal voltage v_{be}! For $r_{ex} \cong 0$, the two are related as $v_\pi = v_{be} \times r_\pi/(r_b + r_\pi)$.
- The *collector region* is doped even more lightly in order to assure an adequately high value of BV_{CEO}, and it forms an even longer conductive path. This would result in an unacceptably high bulk resistance. In the planar process this drawback is reduced drastically by fabricating a highly conductive *buried layer*, as shown in Fig. 2.2. This heavily doped layer provides a low-resistance path for electrons once they have crossed from the p base into the n^- collector region and thence to the buried layer itself. By this artifice, the overall resistance r_c is kept in the range of 20 to 500 Ω. (Nowadays the n^+ collector diffusion is extended all the way to the buried layer so as to minimize r_c.)

The model of Fig. 4.8 is referred to as a *low-frequency* model. When studying the frequency response of BJT circuits, in Chapter 6, we will need to augment this model with appropriate parasitic capacitances. At low operating frequencies these capacitances act as open circuits and will thus be ignored. Not so at high frequencies, where they become significant and tend to alter circuit behavior dramatically. To keep hand calculations manageable we will ignore r_b, r_{ex}, r_c, and r_μ whenever possible. Only when their presence has an appreciable impact shall we take them into proper consideration.

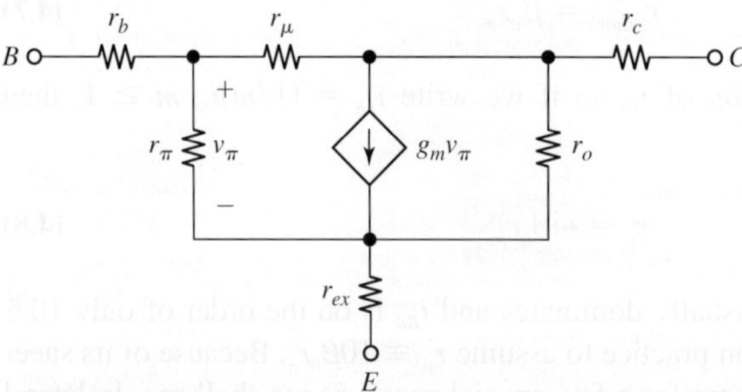

FIGURE 4.8 Complete small-signal model of the BJT at low frequencies.

The Small-Signal Resistances Seen Looking into the Terminals of a BJT

While investigating the resistance-transformation abilities of the BJT in connection with Fig. 2.42, we stipulated an ac grounded collector in order to keep our derivations simple. The results obtained there provide good approximations for discrete designs, where the net equivalent resistance external to the collector is not very large. However, in monolithic circuits the collector is often terminated on an active load whose equivalent resistance can be quite high, rendering the results of Chapter 2 no longer adequate. We need to reexamine the resistance transformation ability of the BJT but using the more general ac circuit of Fig. 4.9. The main novelty now is the coupling established by r_o between R_C and the internal circuitry of the BJT model. As usual, we utilize lower-case subscripts to distinguish the small-signal resistances R_b, R_e, and R_c from the external resistances R_B, R_E, and R_C.

- **The Resistance R_b Seen Looking into the Base**. The circuit to find this resistance is shown in Fig. 4.10a. Ignoring r_μ because of its sheer size, we use Ohm's law to write

$$i_b = \frac{v_b - v_e}{r_\pi}$$

We need another equation to eliminate v_e, so we apply KCL at node v_e and write

$$\frac{v_b - v_e}{r_\pi} + \beta_0 i_b + \frac{v_c - v_e}{r_o} = \frac{v_e}{R_E}$$

Now, we need yet another equation to eliminate v_c, so we again apply KCL at v_c and write

$$\frac{0 - v_c}{R_C} = \beta_0 i_b + \frac{v_c - v_e}{r_o}$$

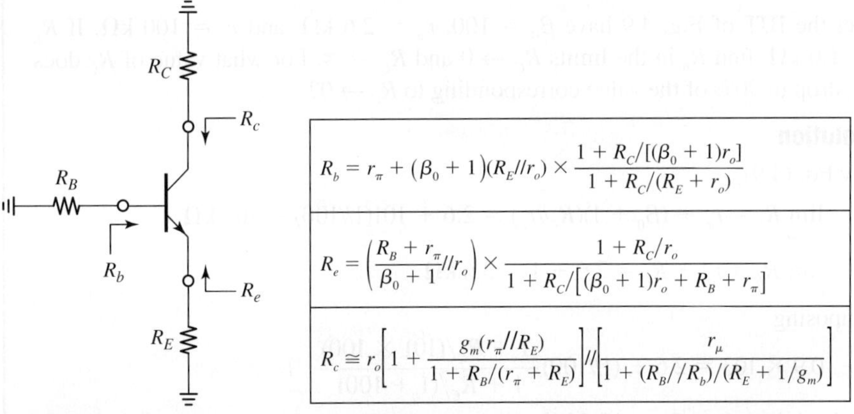

$$R_b = r_\pi + (\beta_0 + 1)(R_E//r_o) \times \frac{1 + R_C/[(\beta_0 + 1)r_o]}{1 + R_C/(R_E + r_o)}$$

$$R_e = \left(\frac{R_B + r_\pi}{\beta_0 + 1}//r_o\right) \times \frac{1 + R_C/r_o}{1 + R_C/[(\beta_0 + 1)r_o + R_B + r_\pi]}$$

$$R_c \cong r_o\left[1 + \frac{g_m(r_\pi//R_E)}{1 + R_B/(r_\pi + R_E)}\right]//\left[\frac{r_\mu}{1 + (R_B//R_b)/(R_E + 1/g_m)}\right]$$

FIGURE 4.9 The small-signal resistances seen looking into the BJT's terminals.

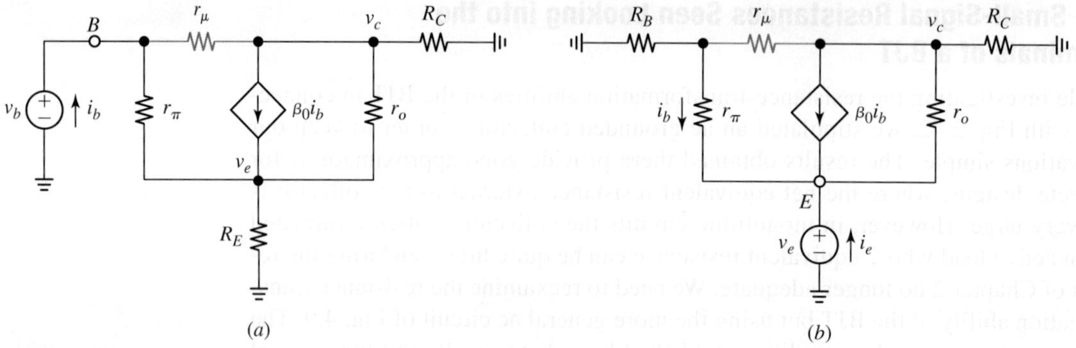

FIGURE 4.10 Test circuits to find (a) R_b and (b) R_e.

Eliminating v_e and v_c, collecting, and taking the ratio $R_b = v_b/i_b$ gives, after some algebra,

$$R_b = r_\pi + (\beta_0 + 1)(R_E//r_o) \times \frac{1 + R_C/[(\beta_0 + 1)r_o]}{1 + R_C/(R_E + r_o)} \tag{4.9}$$

It is apparent that as long as $R_C \ll (\beta_0 + 1)r_o$ and $R_C \ll (R_E + r_o)$, Eq. (4.9) reduces to the familiar form $R_b = r_\pi + (\beta_0 + 1)(R_E//r_o)$. In the discrete designs of Chapter 2, R_C was always *small enough* to meet these conditions. However, this is not necessarily the case in monolithic designs, where the collector may be terminated on an active load and therefore R_C can be quite large. In fact, you can verify that in the limit $R_C \to \infty$, Eq. (4.9) gives $R_b \to r_\pi + R_E$! To justify this surprising result, note that letting $R_C = \infty$ in Fig. 4.10a will force the current $\beta_0 i_b$ to circulate entirely in r_o, making the two elements act as a self-contained subcircuit. Consequently, the source v_i sees just r_π in series with R_E!

EXAMPLE 4.2   Let the BJT of Fig. 4.9 have $\beta_0 = 100$, $r_\pi = 2.6$ kΩ, and $r_o = 100$ kΩ. If $R_E = 1.0$ kΩ, find R_b in the limits $R_C \to 0$ and $R_C \to \infty$. For what value of R_C does R_b drop to 90% of the value corresponding to $R_C \to 0$?

Solution

By Eq. (4.9),

$$\lim_{R_C \to 0} R_b = r_\pi + (\beta_0 + 1)(R_E//r_o) = 2.6 + 101(1//100) \cong 103 \text{ k}\Omega$$

$$\lim_{R_C \to \infty} R_b = r_\pi + R_E = 2.6 + 1 = 3.6 \text{ k}\Omega$$

Imposing

$$0.9 \times 103 = 2.6 + (1//100)\frac{1 + R_C/(101 \times 100)}{1 + R_C/(1 + 100)}$$

and solving gives $R_C = 11.5$ kΩ.

- **The Resistance R_e Seen Looking into the Emitter**. Summing currents into the emitter node of Fig. 4.10*b* gives

$$(\beta_0 + 1)i_b + i_e + \frac{v_c - v_e}{r_o} = 0$$

We need two additional equations to eliminate i_b and v_c. Again ignoring r_μ, we use Ohm's law to write

$$i_b = \frac{0 - v_e}{R_B + r_\pi}$$

Moreover, KCL at the collector node gives

$$\frac{0 - v_c}{R_C} = \beta_0 i_b + \frac{v_c - v_e}{r_o}$$

Eliminating, collecting and solving for the ratio $R_e = v_e/i_e$ gives, after some algebra,

$$R_e = \left(\frac{R_B + r_\pi}{\beta_0 + 1}//r_o\right) \times \frac{1 + R_C/r_o}{1 + R_C/[(\beta_0 + 1)r_o + R_B + r_\pi]} \tag{4.10}$$

As long as $R_C \ll r_o$ this equation reduces to the familiar form of discrete designs, namely, $R_e \cong [(R_B + r_\pi)/(\beta_0 + 1)]//r_o$, which is small because of the term $(\beta_0 + 1)$ in the denominator. However, if the collector is terminated on a current source, R_C can be sufficiently large to raise R_e dramatically. In fact, one can easily verify that in the limit $R_C \to \infty$ Eq. (4.10) gives $R_e \to R_B + r_\pi$! Again, with the collector ac open circuited, the current $\beta_0 i_b$ will circulate exclusively in r_o, making the two elements act as a self-contained subcircuit. Consequently, the source v_e sees just r_π in series with R_B.
- **The Resistance R_c Seen Looking into the Collector**. With sufficient emitter degeneration the resistance seen looking into the collector can be quite large, so ignoring r_μ may no longer be acceptable. Unfortunately, the inclusion of r_μ in the calculations complicates the algebra significantly and without providing much insight. A quicker, if approximate, approach is to investigate the effects of r_o and r_μ separately and then combine them together. For the test circuit of Fig. 4.11*a* we ignore r_μ and recycle familiar results to write $i_{c1} = v_c/R_{c1}$, where

$$R_{c1} \cong r_o\left[1 + \frac{g_m(r_\pi//R_E)}{1 + R_B/(r_\pi + R_E)}\right] \tag{4.11}$$

In the circuit of Fig. 4.11*b* we ignore r_o and apply KCL,

$$i_{c2} = \frac{v_c - v_b}{r_\mu} + \frac{g_m}{1 + g_m R_E}v_b$$

By the voltage divider formula,

$$v_b = \frac{R_B//R_b}{(R_B//R_b) + r_\mu}v_c \cong \frac{R_B//R_b}{r_\mu}v_c$$

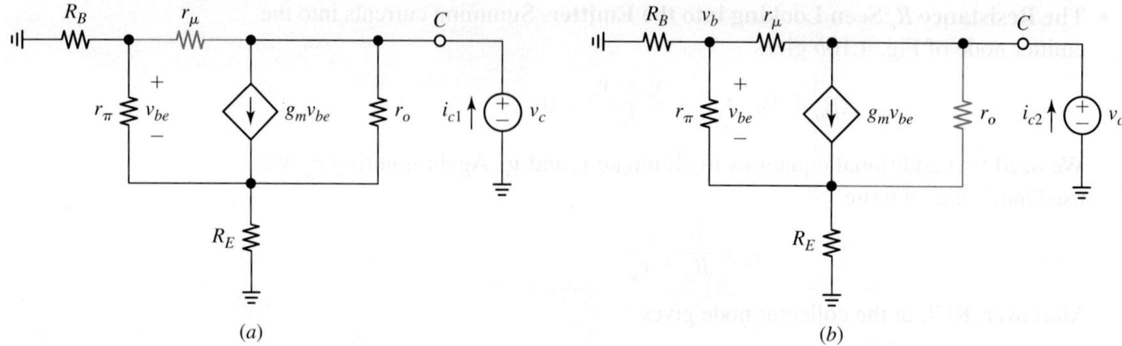

FIGURE 4.11 Test circuits to find the contributions to R_c due to (a) r_o and (b) r_μ.

where we have used the fact that in circuits of practical interest $r_\mu \gg R_B//R_b$, $R_b = r_\pi + (\beta_0 + 1)R_E$. Clearly v_b is much smaller than v_c, so we can simplify further, as

$$i_{c2} = \frac{v_c}{r_\mu} + \frac{g_m}{1 + g_m R_E}\frac{R_B//R_b}{r_\mu}v_c \cong \frac{1}{r_\mu}\left(1 + \frac{R_B//R_b}{R_E + 1/g_m}\right)v_c = \frac{v_c}{R_{c2}}$$

where

$$R_{c2} = \frac{r_\mu}{1 + (R_B//R_b)/(R_E + 1/g_m)} \tag{4.12}$$

With r_o and r_μ present simultaneously, the overall current drawn from the v_c source is $i_c \cong i_{c1} + i_{c2} = v_c/R_{c1} + v_c/R_{c2}$. Letting $R_c = v_c/i_c$, we find

$$R_c \cong R_{c1}//R_{c2} \cong r_o\left[1 + \frac{g_m(r_\pi//R_E)}{1 + R_B/(r_\pi + R_E)}\right]//\left[\frac{r_\mu}{1 + (R_B//R_b)/(R_E + 1/g_m)}\right] \tag{4.13}$$

We observe that the presence of R_B reduces both R_{c1} and R_{c2}. In applications where it is desired to maximize R_c, it would be best to keep R_B as small as the design will allow.

EXAMPLE 4.3   Let the BJT of Fig. 4.9 have $\beta_0 = 100$, $r_\pi = 2.6$ kΩ, $r_o = 100$ kΩ, and $r_\mu = 2\beta_0 r_o$. If $R_B = 10$ kΩ and $R_E = 1.0$ kΩ, estimate R_c. What happens if $R_B = 0$? Compare with PSpice.

Solution

We have $g_m = 100/2600 = 1/(26\ \Omega)$ and $r_\mu = 2 \times 100 \times 100 = 20$ MΩ. Equations (4.11) through (4.13) give

$$R_{c1} \cong 100\left(1 + \frac{(2.6//1)/0.026}{1 + 10/(2.6 + 1)}\right) = 835\ \text{k}\Omega$$

$$R_{c2} = \frac{20}{1 + (10//103.6)/(1 + 0.026)} \cong 2\ \text{M}\Omega$$

$$R_c \cong 835//2{,}000 = 591\ \text{k}\Omega$$

With $R_B = 0$ we get $R_{c1} = 2.88$ MΩ, $R_{c2} = r_\mu = 20$ MΩ, and $R_c = 2.88//20{,}000 = 2.52$ MΩ, in excellent agreement with PSpice, which gives $R_c = 594$ kΩ for $R_B = 10$ kΩ, and $R_c = 2.52$ MΩ for $R_B = 0$.

As mentioned, a monolithic transistor amplifier is likely to be biased and/or loaded by another transistor configured as a current source or sink, so it is instructive to reexamine the basic CE, CC, and CB configurations from the perspective of monolithic design. To focus on the distinguishing aspects of this type of design, we first investigate a circuit in the *absence of loading* and for the case *of ideal current sources/ sinks* (to keep things simple we also assume $r_\mu = \infty$). Then, we adapt our results to the case of non-ideal sources/sinks, in the presence of output loading, and with $r_\mu \neq \infty$.

The CE Configuration with an Idealized Active Load

Recall that the common-emitter (CE) configuration (with or without emitter degeneration) is best suited to voltage amplification. The monolithic rendition of Fig. 4.12 uses the current source I_{LOAD} as an active load, and the voltage source V_{BIAS} to bias the BJT. We assume that V_{BIAS} has been adjusted so as to bias the collector well within the linear region of operation. We readily find the input and output resistances by adapting Eqs. (4.9) and (4.13) in the limits $R_B \to 0$, $R_C \to \infty$, and $r_\mu \to \infty$. The results are

$$R_i = r_\pi + R_E \tag{4.14}$$

$$R_o = r_o[1 + g_m(r_\pi /\!/R_E)] \tag{4.15}$$

Comparing to the case $R_C \to 0$, for which the familiar expression $R_i = r_\pi + (\beta_0 + 1)(R_E/\!/r_o)$ holds, one may find Eq. (4.14) surprising. However, we readily justify it by turning to the ac equivalent of Fig. 4.13, where we observe that in the limit $R_C \to \infty$ the collector acts as an open circuit, at least on an ac basis. This means that the $g_m v_\pi$ current must circulate entirely in r_o, so these two elements act as a self-contained subcircuit. In this case the source v_i sees just r_π in series with R_E! Here is one significant difference between monolithic (high R_C) and discrete (low R_C) design!

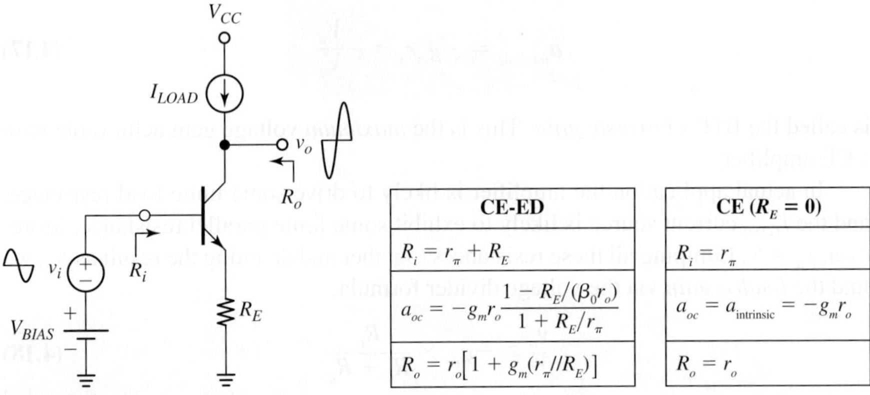

CE-ED	CE ($R_E = 0$)
$R_i = r_\pi + R_E$	$R_i = r_\pi$
$a_{oc} = -g_m r_o \dfrac{1 - R_E/(\beta_0 r_o)}{1 + R_E/r_\pi}$	$a_{oc} = a_{\text{intrinsic}} = -g_m r_o$
$R_o = r_o[1 + g_m(r_\pi /\!/R_E)]$	$R_o = r_o$

FIGURE 4.12 Voltage amplifier with an ideal active load, and its small-signal characteristics for $r_\mu = \infty$.

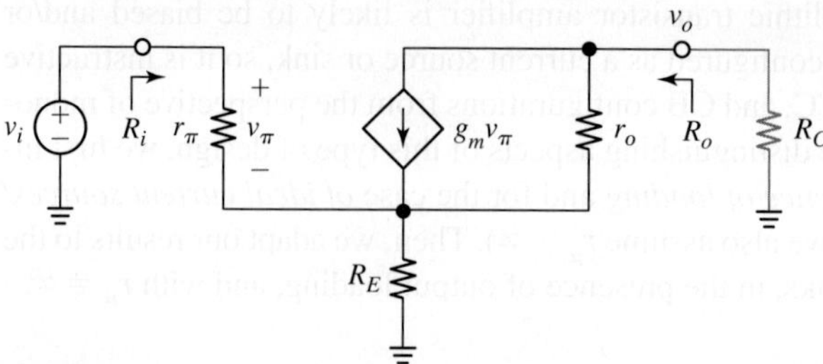

FIGURE 4.13 Ac equivalent of the voltage amplifier of Fig. 4.12.

To find the voltage gain, consider again Fig. 4.13 in the limit $R_C \to \infty$. By KVL and Ohm's Law,

$$v_o = v_i - v_\pi - r_o g_m v_\pi = v_i - (1 + g_m r_o) v_\pi$$

By the voltage divider formula,

$$v_\pi = \frac{r_\pi}{r_\pi + R_E} v_i$$

Substituting v_π, collecting, and solving for the ratio v_o/v_i we get, after some algebra,

$$a_{oc} = \lim_{R_C \to \infty} \frac{v_o}{v_i} = -g_m r_o \frac{1 - R_E/(\beta_0 r_o)}{1 + R_E/r_\pi} \tag{4.16a}$$

where a_{oc} is the *open-circuit* voltage gain, also called the *unloaded* voltage gain. A practical circuit has $R_E/(\beta_0 r_o) \ll 1$, so this gain simplifies to

$$a_{oc} \cong \frac{-g_m r_o}{1 + R_E/r_\pi} \tag{4.16b}$$

The above expressions were derived in the presence of emitter degeneration, but we readily adapt them to the nondegenerated case by letting $R_E = 0$. This gives $R_i = r_\pi$, $R_o = r_o$, and $a_{oc} = a_{\text{intrinsic}}$, where

$$a_{\text{intrinsic}} = -g_m r_o = -\frac{V_A}{V_T} \tag{4.17}$$

is called the BJT's *intrinsic gain*. This is the *maximum* voltage gain achievable with a CE amplifier.

In actual application the amplifier is likely to drive some finite load resistance, and the I_{LOAD} current source is likely to exhibit some finite parallel resistance. Moreover, $r_\mu \neq \infty$. Lumping all these resistances together and denoting the result as R_C, we find the *loaded gain* via the voltage divider formula,

$$\frac{v_o}{v_i} \cong -a_{oc} \times \frac{R_C}{R_o + R_C} \tag{4.18}$$

Note that the presence of R_C will also alter the input resistance R_i, as per Eq. (4.9).

Exercise 4.1

Show that for $R_C \ll r_o$, Eqs. (4.9) and (4.18) can be put in the more familiar forms of discrete design

$$R_i \cong r_\pi + (\beta_0 + 1)R_E \qquad \frac{v_o}{v_i} \cong -\frac{R_C}{r_e + R_E} \qquad (4.19)$$

EXAMPLE 4.4

Let the BJT of Fig. 4.12 have $g_m = 1/25$ A/V, $r_\pi = 4.0$ kΩ, and $r_o = 50$ kΩ. Assuming $r_\mu = \infty$, find R_i, R_o, and v_o/v_i if:

(a) $R_E = 0$ and $R_C = \infty$.
(b) $R_E = 1.0$ kΩ and $R_C = \infty$.
(c) $R_E = 1.0$ kΩ and R_C is adjusted so that $R_C = R_o$.
(d) $R_E = 1.0$ kΩ and $R_C = 10$ kΩ. Comment on your results at each step.
(e) Repeat parts (a) and (b) assuming $r_\mu = 5\beta_0 r_o$.

Solution

(a) With $R_E = 0$ and $R_C = \infty$ the BJT amplifies by its intrinsic gain, and we have

$$R_i = r_\pi = 4.0 \text{ k}\Omega$$
$$R_o = r_o = 50 \text{ k}\Omega$$
$$a_{\text{intrinsic}} = -g_m r_o = -50/0.025 = -2000 \text{ V/V}$$

(b) With emitter degeneration we expect an increase in R_i and R_o and a drop in a,

$$R_i = r_\pi + R_E = 4 + 1 = 5 \text{ k}\Omega$$
$$R_o = r_o[1 + g_m(r_\pi // R_E)] = 50[1 + (4//1)/0.025] = 1.65 \text{ M}\Omega$$
$$\frac{v_o}{v_i} \cong \frac{-g_m r_o}{1 + R_E/r_\pi} = \frac{-2000}{1 + 1/4} = -1600 \text{ V/V}$$

(c) We still have $R_o = 1.65$ MΩ. However, loading the collector with $R_C = R_o = 1.65$ MΩ will reduce gain to one-half the unloaded value, by Eq. (4.18). Thus

$$\frac{v_o}{v_i} = -1600\frac{1.65}{1.65 + 1.65} = -800 \text{ V/V}$$

Moreover, R_i will increase because of the coupling action by r_o. Using Eq. (4.9) with $\beta_0 = g_m r_\pi = 160$,

$$R_i = 4 + 161(1//50) \times \frac{1 + 1650/(161 \times 50)}{1 + 1650/(1 + 50)} = 9.7 \text{ k}\Omega$$

(d) Using again Eqs (4.18) and (4.9), but with $R_C = 10$ kΩ, we now get

$$\frac{v_o}{v_i} = -9.63 \text{ V/V} \qquad R_i = 136 \text{ k}\Omega$$

With R_C this low, we could have used the familiar expressions of Eq. (4.19) to estimate

$$\frac{v_o}{v_i} \cong -\frac{10}{0.025 + 1} = -9.76 \text{ V/V} \qquad R_i \cong 4 + 161 \times 1 = 165 \text{ k}\Omega$$

which are in fair agreement with the above. It is apparent that reducing R_C lowers a while raising R_i. The difference between monolithic and discrete design is evidenced further in Exercise 4.3, p. 357.

(e) We have $r_\mu = 5 \times 160 \times 50 = 40$ MΩ. To assess the impact of r_μ we adapt Eq. (4.18). Thus, in (a) gain drops to $-2000 \times 40/(0.05 + 40) = -1997$ V/V and in (b) gain drops to $-1600 \times 40/(1.65 + 40) = -1537$ V/V. In both cases the change is fairly small, indicating that we can ignore r_μ, at least in this example.

Exercise 4.2

Use Eq. (4.16a) to predict the gain of the circuit of Fig. 4.12 if (a) $R_E = \beta_0 r_o$, and (b) $R_E = \infty$. Hence, justify your results via physical reasoning.

The CC Configuration with an Idealized Active Load

Recall that the common-collector (CC) configuration is best suited to voltage buffering. The IC rendition of Fig. 4.14 admits the ac equivalent of Fig. 4.15, which we wish to investigate in the limit $R_E \to \infty$. In the absence of r_μ the resistance seen looking into the base would be $r_\pi + (\beta_0 + 1)r_o \cong \beta_0 r_o$. This is large and comparable to r_μ ($= m\beta_0 r_o$), so taking the latter into account, we write, for $R_E \to \infty$,

$$R_i \cong (\beta_0 r_o)//(m\beta_0 r_o) = \frac{m}{m+1}\beta_0 r_o \tag{4.20}$$

Moreover, adapting familiar expressions in the limit $R_E \to \infty$, we get

$$R_o = r_e//r_o \cong r_e \tag{4.21}$$

and

$$a_{oc} = \lim_{R_E \to \infty}\frac{v_o}{v_i} = \frac{1}{1 + \dfrac{r_\pi}{(\beta_0 + 1)r_o}} = \frac{1}{1 + r_e/r_o} \cong \frac{1}{1 + 1/(g_m r_o)} \tag{4.22}$$

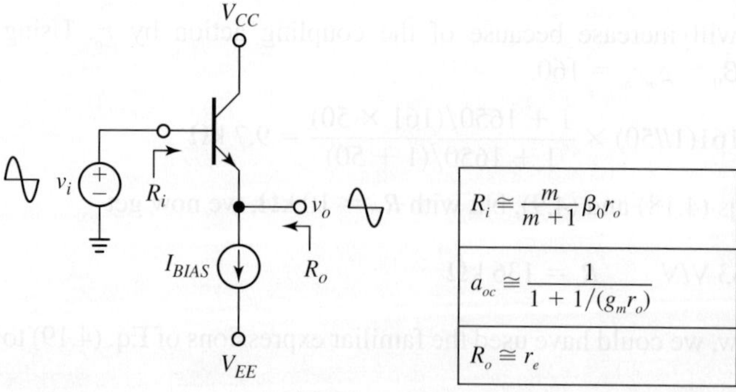

FIGURE 4.14 Voltage follower with ideal current sink bias, and its small-signal characteristics.

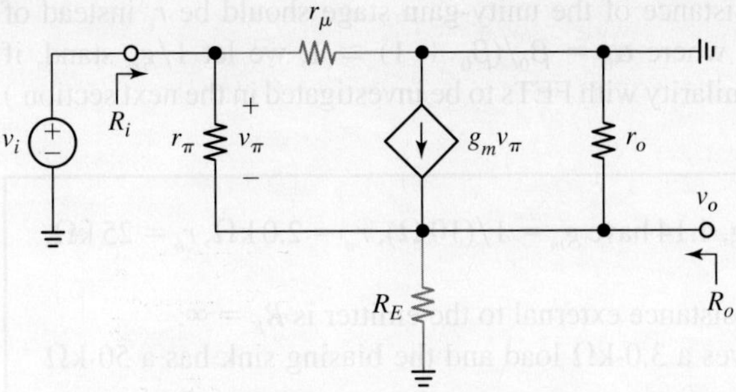

FIGURE 4.15 Ac equivalent of the voltage follower of Fig. 4.14.

where a_{oc} is the *unloaded* or *open-circuit voltage gain*. This gain, slightly less than unity, is the *maximum gain* achievable with the CC configuration, and it can also be expressed in the bias-independent form

$$a_{oc} \cong \frac{1}{1 + V_T/V_A} \tag{4.23}$$

In actual application the circuit is likely to drive some finite load, and the I_{BIAS} current sink is likely to exhibit some finite parallel resistance. Lumping these resistances together and denoting the result as R_E, we observe that R_E appears in *parallel* with r_o in Fig. 4.15, so we recycle Eq. (4.22) but with $r_o//R_E$ in place of r_o to get the loaded gain

$$\frac{v_o}{v_i} \cong \frac{1}{1 + 1/[g_m(r_o//R_E)]} \tag{4.24}$$

You can readily verify that the above expression can also be manipulated into a voltage-divider form as

$$v_o \cong \frac{(r_o//R_E)}{(1/g_m) + (r_o//R_E)} v_i \tag{4.25}$$

The reason for doing this is that we can visualize the voltage buffer according to Fig. 4.16. This insightful form will be used extensively as we move along.

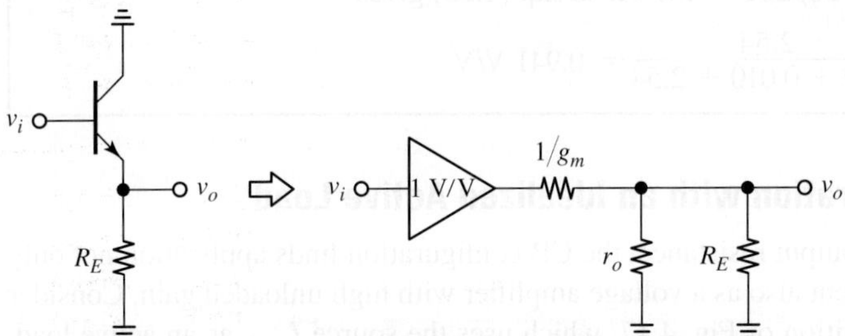

FIGURE 4.16 A BJT voltage follower acts as a *unity-gain* buffer with *output resistance* $1/g_m$, in turn forming a *voltage divider* with the *rest* of the resistances associated with the emitter ($r_o//R_E$ in the present case).

(Truth be told, the output resistance of the unity-gain stage should be r_e instead of $1/g_m$; but, since $r_e = \alpha_0/g_m$, where $\alpha_0 = \beta_0/(\beta_0 + 1) \cong 1$, we let $1/g_m$ stand, if nothing else because of the similarity with FETs to be investigated in the next section.)

EXAMPLE 4.5   Let the voltage follower of Fig. 4.14 have $g_m = 1/(10\ \Omega)$, $r_\pi = 2.0\ \text{k}\Omega$, $r_o = 25\ \text{k}\Omega$, and $m = 5$.

(a) Find R_i and v_o/v_i if net resistance external to the emitter is $R_E = \infty$.

(b) Repeat, if the circuit drives a 3.0-kΩ load and the biasing sink has a 50-kΩ parallel resistance.

(c) Repeat part (b) if the v_i source has a series resistance $R_B = 30\ \text{k}\Omega$.

Solution

(a) The BJT has $\beta_0 = g_m r_\pi = (1/10)2000 = 200$. For $R_E = \infty$ we have, by Eqs. (4.20) and (4.22),

$$R_i \cong \frac{5}{5+1} 200 \times 25 \cong 4.2\ \text{M}\Omega$$

$$a_{oc} = \frac{1}{1 + 1/(25/0.01)} = \frac{1}{1 + 1/2500} = 0.9996\ \text{V/V}$$

(b) We now have $R_E = 50//3 = 2.83\ \text{k}\Omega$ and $r_o//R_E = 25//2.83 = 2.54\ \text{k}\Omega$. By inspection,

$$R_i = r_\mu//[r_\pi + (\beta_0 + 1)(r_o//R_E)] = (5 \times 200 \times 25)//(2 + 201 \times 2.54)$$

$$= (25,000//512) = 502\ \text{k}\Omega$$

indicating that r_μ now has a negligible effect and we could have ignored it. Finally, Eq. (4.24) gives

$$\frac{v_o}{v_i} = \frac{1}{1 + 1/(2.54/0.010)} = \frac{1}{1 + 1/254} \cong 0.996\ \text{V/V}$$

(c) Here is an example where the viewpoint of Fig. 4.16 comes in handy. We can still apply the voltage-divider formula, but after reflecting R_B to the emitter, where it will appear in series with the $1/g_m$ resistance. The reflected value is $R_B/(\beta_0 + 1) = 30/201 = 149\ \Omega$, so Eq. (4.25) gives

$$\frac{v_o}{v_i} = \frac{2.54}{0.149 + 0.010 + 2.54} = 0.941\ \text{V/V}$$

The CB Configuration with an Idealized Active Load

Thanks to its high output resistance, the CB configuration finds application not only as a current buffer, but also as a voltage amplifier with high unloaded gain. Consider the monolithic rendition of Fig. 4.17, which uses the source I_{LOAD} as an active load, and the source V_{BIAS} to bias the BJT (as usual, we assume that V_{BIAS} has been adjusted so as to bias the collector well within the linear region of operation). We wish to find the *unloaded* or *open-circuit* voltage gain a_{oc} as well as R_i and R_o. To this end, refer

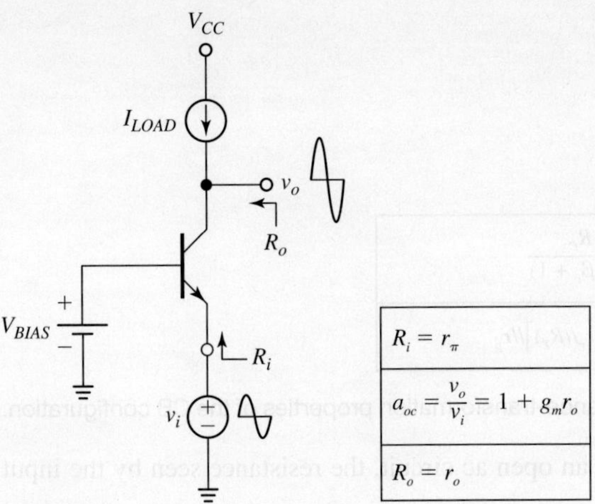

$R_i = r_\pi$
$a_{oc} = \dfrac{v_o}{v_i} = 1 + g_m r_o$
$R_o = r_o$

FIGURE 4.17 The CB configuration as a *high-gain voltage amplifier*, and its ac characteristics for $r_\mu = \infty$.

to the ac equivalent of Fig. 4.18 in the limit $R_C \to \infty$ (for the time being assume also $r_\mu = \infty$). Noting that $v_\pi = -v_i$, we relabel and reorient the dependent source to effect the circuit transformation shown. It is apparent that the $g_m v_\pi$ current circulates entirely in r_o, so we apply KVL and Ohm's law to write $v_o = v_i + r_o g_m v_i = (1 + g_m r_o)v_i$. Taking the ratio v_o/v_i we get

$$a_{oc} = \lim_{R_C \to \infty} \frac{v_o}{v_i} = 1 + g_m r_o \qquad (4.26)$$

It is interesting to observe that the product $g_m r_o$ appears in the expressions for a_{oc} for *each* of the three basic BJT configurations, as per Eqs. (4.17), (4.22), and (4.26). As we know, $g_m r_o \ (=V_A/V_T)$ is a fairly large number that depends only on physical properties of the BJT, regardless of the biasing conditions. We can easily say that this number represents a *figure of merit* of a BJT in monolithic design.

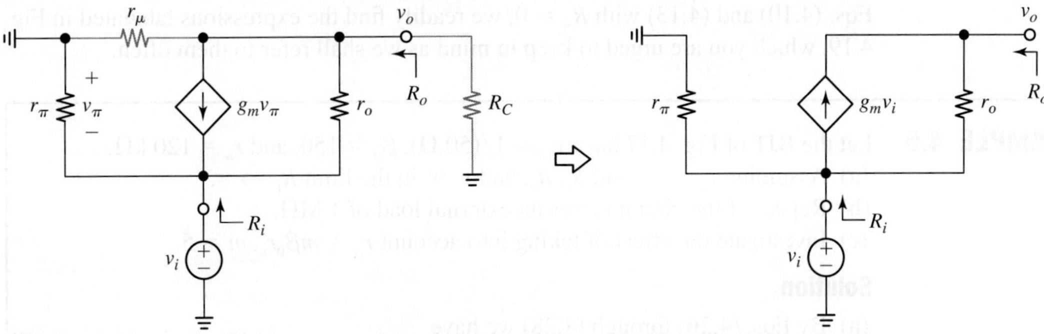

FIGURE 4.18 Ac equivalent of the CB voltage amplifier of Fig. 4.17.

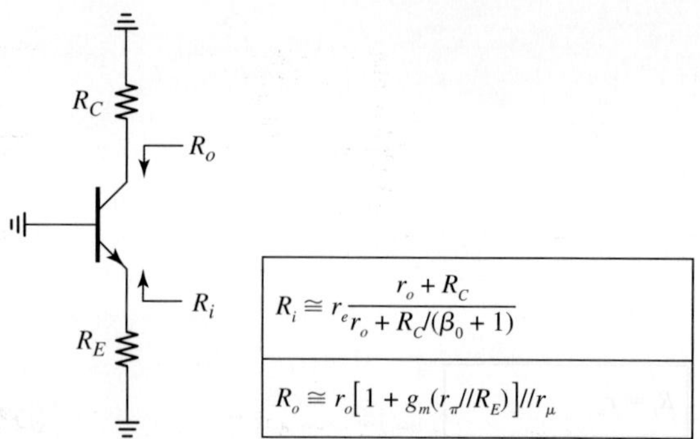

FIGURE 4.19 Illustrating the resistance-transformation properties of the CB configuration.

With the collector acting as an open ac circuit, the resistance seen by the input source is simply

$$R_i = r_\pi \tag{4.27}$$

Comparing to the case $R_C \to 0$, for which the familiar expression $R_i = r_\pi//r_o \cong r_e$ holds, one may find Eq. (4.27) surprising. Again, because of the ac open-circuited collector, the current supplied by the v_i source flows entirely in r_π. This is another significant difference between monolithic (high R_C) and discrete (low R_C) design! Finally, to find the output resistance, we let $v_i \to 0$ and write, by inspection

$$R_o = r_o \tag{4.28}$$

In actual application the amplifier is likely to drive some finite load resistance, and the I_{LOAD} source is likely to exhibit some finite parallel resistance. Moreover, $r_\mu \neq \infty$. Lumping all these resistances together and denoting the result as R_C, we find the *loaded gain* via the voltage divider formula,

$$\frac{v_o}{v_i} \cong a_{oc} \times \frac{R_C}{r_o + R_C} \tag{4.29}$$

The CB configuration is widely used in monolithic design both because of its potentially *high unloaded voltage gain* and because of its *resistance-transformation abilities*. (We will soon see an example in the cascode configuration.) Adapting Eqs. (4.10) and (4.13) with $R_B = 0$, we readily find the expressions tabulated in Fig. 4.19, which you are urged to keep in mind as we shall refer to them often.

EXAMPLE 4.6   Let the BJT of Fig. 4.17 have $g_m = 1/(50\ \Omega)$, $\beta_0 = 150$, and $r_o = 120\ \text{k}\Omega$.

(a) Assuming $r_\mu = \infty$, find R_i, R_o, and v_o/v_i in the limit $R_C \to \infty$.

(b) Repeat, if the circuit drives an external load of 1 MΩ.

(c) Investigate the effect of taking into account $r_\mu = m\beta_0 r_o$, $m = 5$.

Solution

(a) By Eqs. (4.26) through (4.28) we have

$$R_i = 150 \times 50 = 7.5\ \text{k}\Omega \qquad R_o = 120\ \text{k}\Omega \qquad a_{oc} = 1 + \frac{120}{0.05} = 2401\ \text{V/V}$$

(b) R_o remains the same. However, the input resistance and gain drop. Using the expression for R_i tabulated in Fig. 4.19, along with Eq. (4.29), we find

$$R_i \cong 50\frac{0.12 + 1}{0.12 + 1/151} \cong 442\ \Omega \qquad \frac{v_o}{v_i} = 2401\frac{1}{0.12 + 1} \cong 2144\ \text{V/V}$$

(c) We have $r_\mu = 5 \times 150 \times 120 = 90\ \text{M}\Omega$, so the parameters of part (a) change to

$$R_i \cong 50\frac{0.12 + 90}{0.12 + 90/151} \cong 6.29\ \text{k}\Omega \qquad \frac{v_o}{v_i} = 2401\frac{90}{0.12 + 90} \cong 2398\ \text{V/V}$$

Similarly, we recalculate the parameters of part (b), but using (90//1) $\text{M}\Omega$ instead of 1 $\text{M}\Omega$. The difference is on the order of 1%, indicating that r_μ plays a negligible role in this example.

Exercise 4.3

The voltage gain $a = v_c/v_e$ of the CB configuration may lie anywhere over the range $g_m R_C \leq a \leq 1 + g_m r_o$, depending on how R_C compares with r_o. The lower extreme is reached in the limit $R_C \ll r_o$, typical of discrete designs, whereas the upper extreme is reached in the limit $R_C \gg r_o$, typical of monolithic designs. Let us arbitrarily define the *discrete* design range as the range $R_C \leq R_{C(\max)}$ over which a departs from its *lower* extreme by no more than 10%, and the *monolithic* design range as the range $R_C \geq R_{C(\min)}$ over which a departs from its *upper* extreme by no more than 10%.

  a. Estimate $R_{C(\max)}$ and $R_{C(\min)}$ for a BJT with $\beta_0 = 100$ and $V_A = 100$ V at $I_C = 1$ mA.

  b. What are the values of R_e at the lower and upper extremes?

  c. What are the values of R_e at $R_C = R_{C(\max)}$ and at $R_C = R_{C(\min)}$?

Ans: (a) $R_{C(\max)} \cong 11\ \text{k}\Omega$, $R_{C(\min)} \cong 900\ \text{k}\Omega$. (b) 26 Ω, 2.6 kΩ. (c) 28.8 Ω, 239 Ω

4.3 MOSFET CHARACTERISTICS AND MODELS REVISITED

The MOSFET models and analytical methodologies of Chapter 3 were deliberately kept as simple as possible so we could focus on the very basics. As we embark upon the study of monolithic circuits, we need to suitably refine models and techniques to include higher-order effects that tend to play a more prominent role in monolithic design.

The Role of λ in Dc Calculations

Recall from Chapter 3 that the dc current of a saturated nMOSFET is

$$I_D = \frac{1}{2}k'\frac{W}{L}V_{ov}^2(1 + \lambda V_{DS}) \tag{4.30}$$

where λ is the *channel-length modulation parameter*, a device parameter typically on the order of 0.01 to 0.1 V^{-1}. As we know, this type of modulation stems from the portion of the drain-body SCL extending into the channel. Its width (x_p for the *n*MOSFET, x_n for the *p*MOSFET) depends on doping levels as well as on junction bias, as per Eqs. (1.40) and (1.45). As we know, the shorter the channel the more pronounced the effect of channel-length modulation in Eq. (4.30). IC designers relate λ to L as

$$\lambda = \frac{\lambda'}{L}$$

(4.31)

where λ' (in μm/V) is a *process parameter*, and L (in μm) is the *channel length*. Typically, λ' ranges from 0.02 to 0.2 μm/V. An IC designer will specify L to achieve a given λ, and W to achieve a given k, $k = k'(W/L)$. Note that this flexibility is not available in bipolar IC design!

As long as $\lambda V_{DS} \ll 1$ we ignore the term λV_{DS} in Eq. (4.30) to simplify dc calculations. However, in monolithic circuits this may no longer be acceptable. An example will give a better idea.

EXAMPLE 4.7   Let the diode-connected FET of Fig. 4.20 have $V_t = 1.0$ V, $k' = 100\ \mu\text{A/V}^2$, and $\lambda' = 0.2\ \mu$m/V.

(a) Find W and L to achieve $\lambda = 0.1\ \text{V}^{-1}$ and $k = 0.5\ \text{mA/V}^2$.
(b) If $I = 1.0$ mA, find V assuming $\lambda = 0$ to simplify dc calculations.
(c) Refine your calculation of part (b), but with $\lambda = 0.1\ \text{V}^{-1}$.
(d) How would you adjust I to retain the value of V found in part (b)?
(e) How would you alter the W/L ratio to retain the values of I and V of part (b) without changing λ?

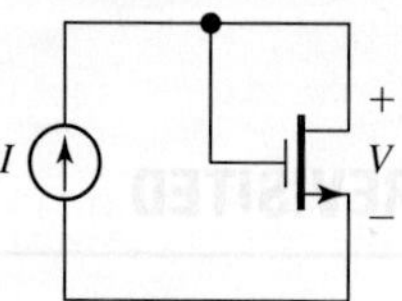

FIGURE 4.20 Circuit of Example 4.7.

Solution

(a) We need $L = \lambda'/\lambda = 0.2/0.1 = 2\ \mu m$, and $W/L = k/k' = 0.5/0.1 = 5$. So, $W = 5L = 10\ \mu$m.
(b) We have $V_{OV} = \sqrt{2I_D/k} = \sqrt{2 \times 1/0.5} = 2$ V, and $V = V_t + V_{OV} = 1.0 + 2.0 = 3.0$ V.
(c) We now have

$$1\ \text{mA} = \frac{0.5\ \text{mA}}{2}(V - 1.0)^2 \times (1 + 0.1 \times V)$$

Taking the square root of both sides and solving for the linear term in V we get

$$V = 1.0 + \frac{2}{\sqrt{1 + 0.1 \times V}}$$

Starting out with the estimate $V = 3.0$ V and iterating, we find $V = 2.77$ V (<3.0 V because of λ).

(d) $I = (1 \text{ mA}) \times (1 + 0.1 \times 3.0) = 1.3$ mA (>1 mA because of λ).

(e) To lower I from 1.3 mA to 1.0 mA while retaining $V = 3.0$ V we need to reduce the W/L ratio in proportion. So, $(W/L)_{\text{new}} = (W/L)_{\text{old}} \times (1.0/1.3) = 5 \times 0.77 = 3.85$. To retain the same λ, keep $L_{\text{new}} = L_{\text{old}}$. So, we need $W_{\text{new}} = W_{\text{old}} \times 0.77 = 7.7 \ \mu\text{m}$.

The Body Transconductance

In Chapter 3 we assumed MOSFETs with the body and source terminals tied together. This freed us from having to worry about the body effect so we could focus on the very basics of MOSFET behavior. In monolithic circuits different FETs share the same tub or body. The body common to all nMOSFETs is tied to the most negative voltage (MNV) in the circuit in order to avoid the inadvertent turn-on of the diode formed by the p-type body and the n-type source. Likewise, the body common to all pMOSFETs is tied to the most positive voltage (MPV). In the case of a saturated nMOSFET whose source is allowed to attain a different potential than the body, the drain current takes on the more general form

$$i_D = \frac{k}{2}[v_{GS} - v_t(v_{SB})]^2(1 + \lambda v_{DS}) \tag{4.32}$$

showing explicitly that the threshold voltage v_t is a function of the source-to-body voltage v_{SB} (≥ 0). This dependence takes on the form of Eq. (3.8), repeated here for convenience

$$v_t(v_{SB}) = V_{t0} + \gamma\left[\sqrt{v_{SB} + 2|\phi_p|} - \sqrt{2|\phi_p|}\right] \tag{4.33}$$

where

$$\gamma = \frac{\sqrt{2qN_A\varepsilon_{si}}}{C_{ox}} \tag{4.34}$$

For a pMOSFET Eq. (4.33) takes on the form $v_t(v_{BS}) = V_{t0} - \gamma\left[\sqrt{v_{BS} + 2\phi_n} - \sqrt{2\phi_n}\right]$, where γ is still found via Eq. (4.34), but with N_A replaced by N_D. It is apparent that FETs operating with $v_B = v_S$ will also have $v_t = V_{t0}$. However, FETs with $v_S \neq v_B$ will have $v_t \neq V_{t0}$, a phenomenon referred to as *the body effect*. Since v_{SB} (or v_{SB} for a pMOSFET) affects i_D, the body acts as a second gate, though with a lesser impact upon i_D than the gate proper because of the presence of the square root function in Eq. (4.33). For this reason the body is also referred to as the *back gate*.

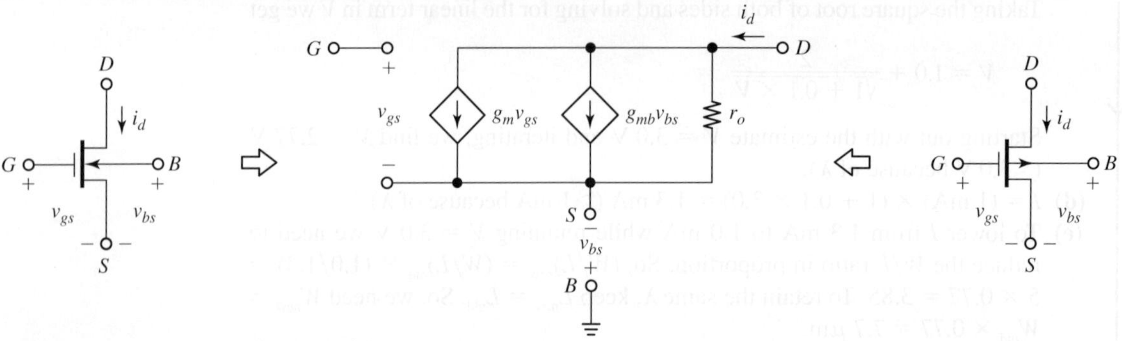

FIGURE 4.21 Small-signal MOSFET model with the source $g_{mb}v_{bs}$ stemming from the body effect.

Just like we use the dependent source $g_m v_{gs}$ to model the effect of a small-signal change in the gate voltage relative to that of the source, we use an additional dependent source $g_{mb}v_{bs}$ to model the effect of a small-signal change in the back-gate voltage relative to that of the source. This results in the small-signal model of Fig. 4.21, where g_{mb}, aptly called the *body transconductance*, is defined as

$$g_{mb} = \left. \frac{\partial i_D}{\partial v_{BS}} \right|_{V_{GS}, V_{DS}} \tag{4.35}$$

with v_{GS} and v_{DS} held constant. (Conversely, the ordinary transconductance is defined as $g_m = \partial i_D / \partial v_{GS}$ with v_{BS} and v_{DS} held constant.) Differentiation of Eq. (4.32) with respect to v_{BS} $(=-v_{SB})$ gives

$$g_{mb} = \frac{\partial i_D}{\partial (-v_{SB})} = -\frac{\partial i_D}{\partial v_{SB}} = -k[v_{GS} - v_t(v_{SB})](1 + \lambda v_{DS})\left(-\frac{\partial v_t}{\partial v_{SB}}\right)$$

$$= g_m(1 + \lambda v_{DS})\frac{\gamma}{2\sqrt{v_{SB} + 2|\phi_p|}}$$

where we have used the fact that $k[v_{GS} - v_t(v_{SB})] = g_m$. For $\lambda v_{DS} \ll 1$ we can write

$$g_{mb} = \chi g_m \tag{4.36}$$

where

$$\chi = \frac{\gamma}{2\sqrt{v_{SB} + 2|\phi_p|}} \tag{4.37}$$

The proportionality constant χ depends on process parameters as well as on v_{SB} (or v_{BS} for the *p*MOSFET case), and is typically on the order of 0.1 to 0.3. Note that since the MNV (or MPV for the *p*MOSFET) is usually of the dc type, the body is shown at ac ground in Fig. 4.21.

Let the FET of Fig. 4.22 have $V_{t0} = 1.0$ V, $k = 0.5$ mA/V^2, $\lambda = 1/(40$ V), $|2\phi_p| = 0.6$ V, and $\gamma = 0.445$ V$^{1/2}$. If $V_{DD} = 9.0$ V, specify suitable resistances to bias the FET at $I_D = 1$ mA according to the 1/3-1/3-1/3 Rule (to simplify dc calculations, assume $\lambda = 0$). Hence, calculate the small-signal parameters.

EXAMPLE 4.8

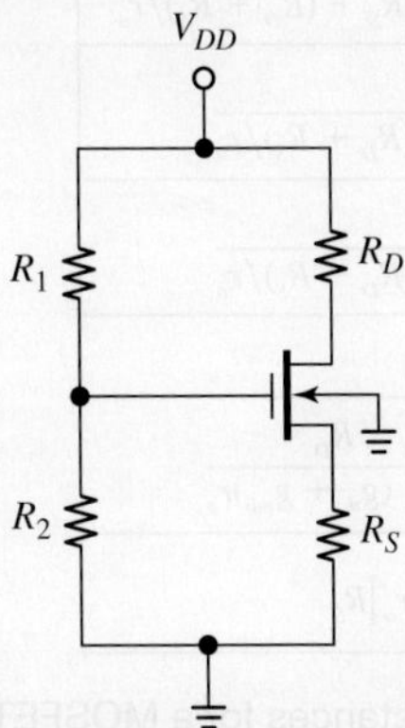

FIGURE 4.22 Circuit of Example 4.8

Solution

We have $V_S = (1/3)9 = 3$ V, so $R_D = R_S = 3/1 = 3$ kΩ. Clearly, $V_{SB} = 3$ V, so Eqs. (4.33) and (4.37) give

$$V_t = 1.0 + 0.445[\sqrt{3 + 0.6} - \sqrt{0.6}] = 1.5 \text{ V} \qquad \chi = \frac{0.445}{2\sqrt{3 + 0.6}} = 0.117$$

Also, $V_{OV} = (2I_D/k)^{1/2} = (2 \times 1/0.5)^{1/2} = 2$ V, so $V_G = V_S + V_t + V_{OV} = 3 + 1.5 + 2 = 6.5$ V. We need a resistance pair such that $R_1/R_2 = 2.5/6.5 = 1/2.6$. Use $R_1 = 1.0$ MΩ and $R_2 = 2.6$ MΩ. Finally, the FET's small-signal parameters are

$$g_m = (2 \times 0.5 \times 1)^{1/2} = 1.0 \text{ mA/V}$$

$$g_{mb} = 0.117 \times 1 = 0.117 \text{ mA/V}$$

$$r_o = 40/1 = 40 \text{ kΩ}$$

A Generalized Ac Circuit

We now wish to investigate how the presence of the $g_{mb}v_{bs}$ source affects the various parameters in the generalized ac circuit of Fig. 4.23. We anticipate that except for the CS configuration, for which $v_{bs} = 0$, all other configurations will be sensitive to the body effect.

- **The Transconductance $G_m = i_d/v_g$.** With reference to the small-signal equivalent of Fig. 4.24a we observe that $v_{bs} = -v_s$, so we can reverse the direction of the $g_{mb}v_{bs}$ source and relabel it as $g_{mb}v_s$. This is depicted in Fig. 4.24b. Applying KCL at the drain node we get

$$i_d + g_{mb}v_s = g_m(v_g - v_s) + \frac{v_d - v_s}{r_o}$$

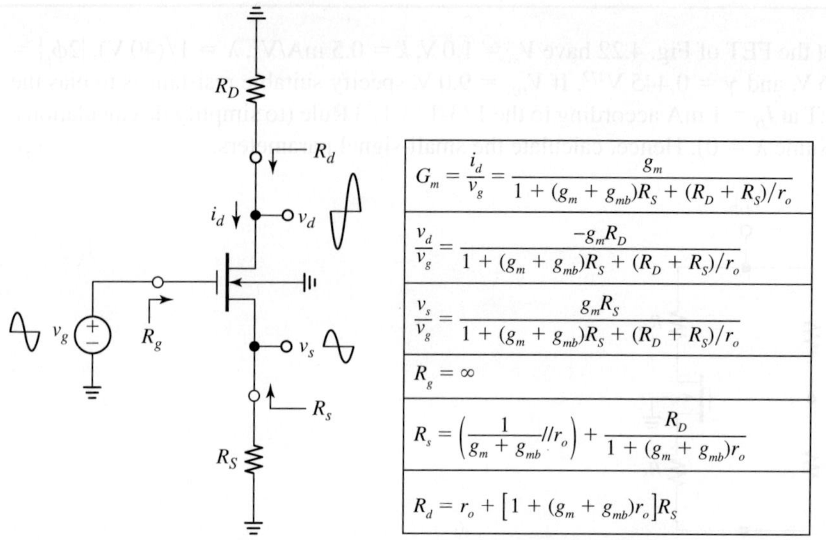

$$G_m = \frac{i_d}{v_g} = \frac{g_m}{1 + (g_m + g_{mb})R_S + (R_D + R_S)/r_o}$$

$$\frac{v_d}{v_g} = \frac{-g_m R_D}{1 + (g_m + g_{mb})R_S + (R_D + R_S)/r_o}$$

$$\frac{v_s}{v_g} = \frac{g_m R_S}{1 + (g_m + g_{mb})R_S + (R_D + R_S)/r_o}$$

$$R_g = \infty$$

$$R_s = \left(\frac{1}{g_m + g_{mb}}//r_o\right) + \frac{R_D}{1 + (g_m + g_{mb})r_o}$$

$$R_d = r_o + \left[1 + (g_m + g_{mb})r_o\right]R_S$$

FIGURE 4.23 Summary of small-signal gains and terminal resistances for a MOSFET.

Since i_d enters the drain and exits the source, we have, by Ohm's law, $v_s = R_S i_d$ and $v_d = -R_D i_d$. Substituting in the above expression and collecting we get the *circuit's transconductance*

$$G_m = \frac{i_d}{v_g} = \frac{g_m}{1 + (g_m + g_{mb})R_S + (R_D + R_S)/r_o} \tag{4.38}$$

This is similar to the expression tabulated in Fig. 3.50, except for the denominator term increase from $g_m R_S$ to $(g_m + g_{mb})R_S$. By Eq. (4.36) this represents a percentage increase of 100χ.

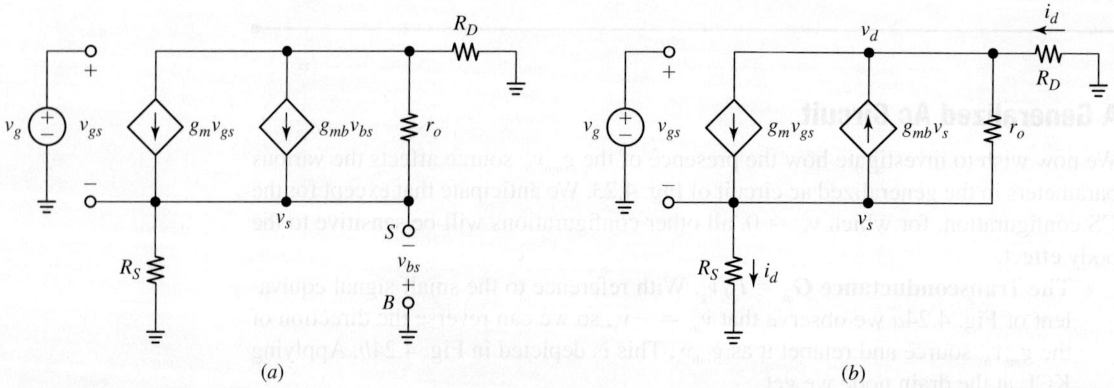

FIGURE 4.24 (a) Small-signal equivalent of the generalized ac circuit of Fig. 4.22. (b) The circuit of (a) after suitable transformation.

- **The Voltage Gains v_s/v_g and v_d/v_g.** We also have $v_s = R_S i_d = R_S G_m v_g$ and $v_d = -R_D i_d = -R_D G_m v_g$, so the small-signal voltage gains from v_g to v_s and from v_g to v_d are, respectively

$$\frac{v_d}{v_g} = \frac{-g_m R_D}{1 + (g_m + g_{mb})R_S + (R_D + R_S)/r_o} \tag{4.39a}$$

$$\frac{v_s}{v_g} = \frac{g_m R_S}{1 + (g_m + g_{mb})R_S + (R_D + R_S)/r_o} \tag{4.39b}$$

Note again the denominator term increase from $g_m R_S$ to $(g_m + g_{mb})R_S$.

- **The Resistance R_s Seen Looking into the Source**. To find this resistance we let $v_g \to 0$ and we subject the source terminal to a test voltage v_s. But, with $v_g = 0$ we get $v_{gs} = v_g - v_s = 0 - v_s = -v_s$, so now we can reverse and relabel also the $g_m v_{gs}$ source, as in Fig. 4.25a. By KCL we have

$$i_s = g_m v_s + g_{mb} v_s + \frac{v_s - v_d}{r_o} = (g_m + g_{mb})v_s + \frac{v_s - R_D i_s}{r_o}$$

Collecting and taking the ratio $R_s = v_s/i_s$ gives, after some algebra,

$$R_s = \left(\frac{1}{g_m + g_{mb}}//r_o\right) + \frac{R_D}{1 + (g_m + g_{mb})r_o} \tag{4.40a}$$

Typically $1/(g_m + g_{mb}) \ll r_o$, so the above expression simplifies to

$$R_s \cong \frac{1}{g_m + g_{mb}} + \frac{R_D}{1 + (g_m + g_{mb})r_o} \tag{4.40b}$$

The first term is similar to that of discrete designs, except for the change from g_m to $g_m + g_{mb}$. The second term represents the effect of the external drain resistance R_D reflected to the source. This term is usually ignored in discrete designs, but not necessarily in monolithic designs, where the drain may be terminated on a current source and therefore R_D may be quite large.

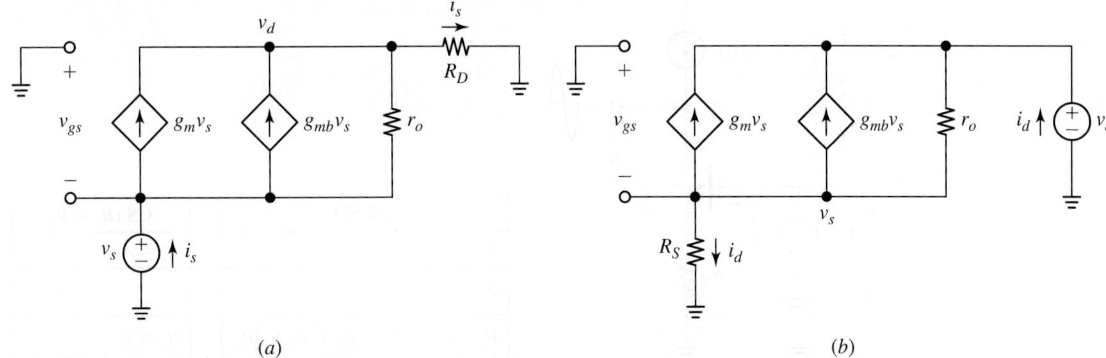

FIGURE 4.25 Test circuits to find (a) the small-signal resistance R_s seen looking into the *source* and (b) the small-signal resistance R_d seen looking into the *drain*.

- **The Resistance R_d Seen Looking into the Drain**. To find this resistance we let $v_g \to 0$ and we subject the drain to a test voltage v_d. But, with $v_g = 0$ we again get $v_{gs} = -v_s$, so we again reverse and relabel both dependent sources as in Fig. 4.25b. By KCL,

$$i_d + g_m v_s + g_{mb} v_s = \frac{v_d - v_s}{r_o}$$

Substituting $v_s = R_S i_d$, collecting, and taking the ratio $R_d = v_d/i_d$, we get

$$R_d = r_o + [1 + (g_m + g_{mb})r_o]R_S \qquad (4.41)$$

The first component is just the FET's internal resistance r_o, while the second component represents the effect of the external source resistance R_S reflected to the drain. It is interesting to contrast the effects of the MOSFET upon its external resistances: while R_D reflected to the source gets *divided* by the amount $1 + (g_m + g_{mb})r_o$, R_S reflected to the drain gets *multiplied* by the same amount. We shall have more to say about these symmetric actions when studying cascode configurations.

As mentioned, a monolithic MOSFET amplifier is likely to be biased or to be loaded by another FET operating as a current source or sink, so it is instructive to reexamine the basic FET configurations from the IC design viewpoint. We first investigate the limiting case of ideal current sources/sinks, then we adapt our results to the case of non-ideal sources/sinks, and in the presence of an output load.

The CS Configuration with an Idealized Active Load

Recall that the common-source (CS) configuration (with or without source degeneration) is best suited to voltage amplification. The monolithic rendition of Fig. 4.26 uses the current source I_{LOAD} as an active load, and the voltage source V_{BIAS} to bias the FET. We assume that V_{BIAS} has been adjusted so as to ensure that the drain is biased at a voltage well within the linear region of operation. The amplifier's characteristics are readily obtained by adapting the expressions of Fig. 4.23 in the limit $R_D \to \infty$. The results are tabulated in Fig. 4.26 both for the CS-SD case ($R_S \neq 0$), and the plain CS case ($R_S = 0$).

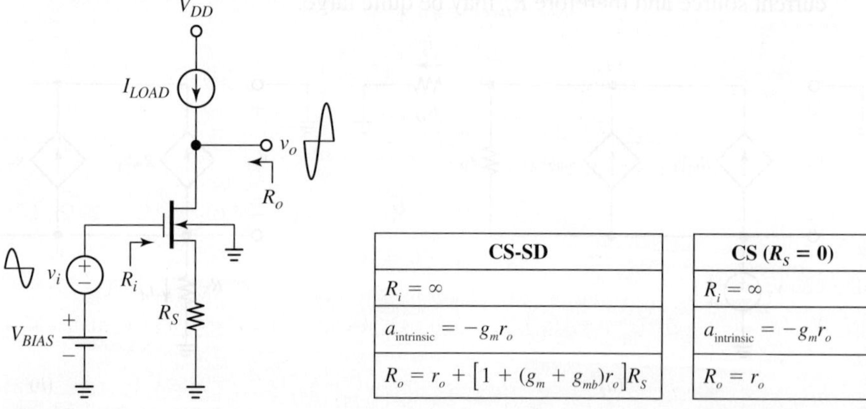

	CS-SD		CS ($R_S = 0$)
	$R_i = \infty$		$R_i = \infty$
	$a_{\text{intrinsic}} = -g_m r_o$		$a_{\text{intrinsic}} = -g_m r_o$
	$R_o = r_o + \left[1 + (g_m + g_{mb})r_o\right]R_S$		$R_o = r_o$

FIGURE 4.26 Voltage amplifier with a current-source load and summary of its small-signal characteristics.

The voltage gain achieved in the limit $R_D \to \infty$ is $v_o/v_i = -g_m r_o$. Aptly called the *unloaded* or also the *open-circuit* voltage gain, this is the maximum voltage gain attainable with the CS configuration and for this reason it is also called the *intrinsic gain* $a_{intrinsic}$. It is interesting that this gain is unaffected by the presence of R_S. This is so because with an ideal current-source load the drain is an *ac open circuit* giving $i_d = 0$. So, $i_s = 0$ and $v_s = R_S i_s = 0$, indicating that neither R_S nor g_{mb} intervene in the expression for the gain in this case. (However, they do intervene in the expression for R_o in the CS-SD case.)

In actual application a CS amplifier is likely to drive some finite load resistance, and the I_{LOAD} source is likely to exhibit some finite parallel resistance. Lumping these resistances together and denoting their combination as R_D, we adapt Eq. (4.39) to find the *loaded gain* as

$$\frac{v_o}{v_i} = \frac{-g_m R_D}{1 + (g_m + g_{mb})R_S + (R_D + R_S)/r_o} \tag{4.42}$$

EXAMPLE 4.9

Suppose the FET of Fig. 4.26 has $g_m = 1$ mA/V^2, $r_o = 50$ kΩ, and $\chi = 0.15$. Find R_o and v_o/v_i if:

(a) $R_S = 0$ and $R_D = \infty$.
(b) $R_S = 0$ and $R_D = 100$ kΩ.
(c) $R_S = 2.0$ kΩ and $R_D = \infty$.
(d) $R_S = 2.0$ kΩ and $R_D = 100$ kΩ. Comment on your results at each step.

Solution

(a) With $R_S = 0$ and $R_D = \infty$ the FET amplifies by its intrinsic gain and we have

$$R_o = r_o = 50 \text{ k}\Omega \qquad a_{intrinsic} = -g_m r_o = -1 \times 50 = -50 \text{ V/V}$$

(b) With $R_S = 0$ we still have $R_o = 50$ kΩ. With $R_D = 100$ kΩ, the loaded gain is, by Eq. (4.42),

$$\frac{v_o}{v_i} = \frac{-1 \times 100}{1 + 100/50} = -33.3 \text{ V/V}$$

(c) With source degeneration we expect R_o to increase. However, so long as $R_D = \infty$, gain is unaffected by R_S.

$$R_o = r_o + [1 + (g_m + g_{mb})r_o]R_S = 50 + [1 + (1 + 0.15 \times 1)50]2 = 167 \text{ k}\Omega$$

$$\frac{v_o}{v_i} = -50 \text{ V/V}$$

(d) The resistance seen by R_D is still 167 kΩ. However, by Eq. (4.42), gain drops to

$$\frac{v_o}{v_i} = \frac{-1 \times 100}{1 + 1 \times 1.15 \times 2 + (100 + 2)/50} = 18.7 \text{ V/V}$$

Combining familiar expressions for g_m and r_o we can put the intrinsic gain in the alternative form

$$a_{\text{intrinsic}} = -\frac{1}{\lambda}\sqrt{\frac{2k}{I_D}} \tag{4.43}$$

Unlike the BJT's intrinsic gain of Eq. (4.17), which is bias-independent, that of the FET is inversely proportional to $\sqrt{I_D}$, indicating that lowering I_D increases $a_{\text{intrinsic}}$. This is certainly desirable in low-power IC design. However, two observations are in order. First, Eq. (4.43) is a result of the quadratic characteristic of the MOSFET. At sufficiently low operating currents the FET enters the *sub-threshold region* where its characteristic changes from quadratic to *exponential*. Consequently, at low currents the intrinsic gain of the MOSFET becomes bias-independent, just like in the case of the BJT. Second, low operating currents reduce a circuit's ability to drive capacitive loads, resulting in slower operation and thus reduced frequency bandwidth (more on this in Chapter 6).

EXAMPLE 4.10

(a) If a MOSFET gives $a_{\text{intrinsic}} = -50$ V/V at $I_D = 1$ mA, find $a_{\text{intrinsic}}$ at $I_D = 100$ μA. What value of I_D is needed for $a_{\text{intrinsic}} = -100$ V/V?

(b) If it is found that this FET enters the sub-threshold region in the vicinity of $I_D = 10$ μA, estimate its maximum intrinsic gain.

Solution

(a) Lowering I_D by a factor of 10 will change $a_{\text{intrinsic}}$ from -50 V/V to $-50\sqrt{10} = -158$ V/V. In order to *double* $a_{\text{intrinsic}}$ from -50 V/V to -100 V/V we must lower I_D by a factor of *four*, so $I_D = 1/4 = 250$ μA.

(b) For $I_D \leq 10$ μA gain will settle at $a_{\text{intrinsic(max)}} = -50\sqrt{1000/10} = -500$ V/V.

Writing $g_m r_o = (2I_D/V_{OV})/(\lambda I_D) = 2/(\lambda V_{OV})$ we get yet another insightful form for the intrinsic gain

$$a_{\text{intrinsic}} = -\frac{2L}{\lambda' V_{OV}} \tag{4.44}$$

where V_{OV} is the overdrive voltage, L the channel length, and λ' is the process parameter characterizing channel-length modulation, as per Eq. (4.31). This form indicates that the *longer the channel* for a given V_{OV}, the *higher the intrinsic gain*. This flexibility is quite convenient for the MOS IC designer!

EXAMPLE 4.11

Assuming a process with $k' = 75$ μA/V^2 and $\lambda' = 0.1$ μm/V, specify W and L so that at $I_D = 100$ μA and $V_{DS} = 2$ V a MOSFET provides $a_{\text{intrinsic}} = -50$ V/V with $V_{OV} = 0.5$ V.

Solution

By Eqs. (4.44) and (4.31) we have

$$L = -\lambda' \times V_{OV} \times a_{\text{intrinsic}}/2 = -0.1 \times 0.5 \times (-50)/2 = 2.5 \ \mu m$$

$$100 = \frac{1}{2}75\frac{W}{2.5}0.5^2\!\left(1 + \frac{0.1}{2.5} \times 2\right)\!.$$

Solving we get $W = 24.7 \ \mu m$.

The CD Configuration with an Idealized Active Load

Recall that the common-drain (CD) configuration is best suited to *voltage buffering*. In monolithic realizations a FET buffer is usually biased via a current sink, so let us investigate the voltage transfer curve (VTC) of such a realization. Figure 4.27 shows a PSpice circuit to serve this purpose, along with its voltage transfer curve (VTC). The latter is shown both for the already familiar case of body and source tied together ($\gamma = 0$), and the henceforth more common case of the body tied to the most negative voltage (MNV) in the circuit, so that ($\gamma \neq 0$). We observe that both curves are shifted downward by v_{GS}, namely, $v_O = v_I - v_{GS} = v_I - (V_t + V_{OV})$. In the case $\gamma = 0$ we have $V_t = V_{t0} = $ constant, but in the case $\gamma \neq 0$, the body effect makes V_t itself a function of v_O and, hence, of v_I. We also note that the *pn* junction formed by the body and source clamps v_O a diode drop *below* V_{SS} (or at about -5.6 V in the present example). Also, v_O saturates at V_{DD} ($=5$ V in the present example) for v_I high enough to drive the FET into the ohmic region.

We now wish to find the unloaded voltage gain and output resistance, as depicted in Fig. 4.28. We do this by adapting the expressions tabulated in Fig. 4.23 in the

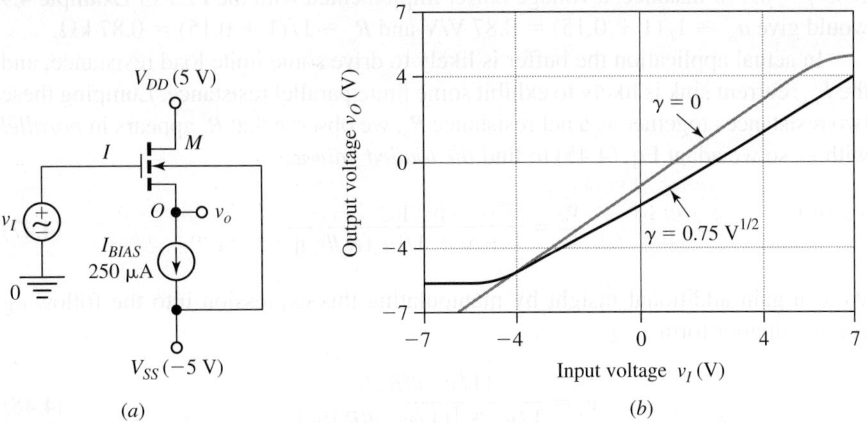

FIGURE 4.27 (a) PSpice circuit of a monolithic MOS buffer, and (b) its VTC. The MOSFET parameters are $k' = 100 \ \mu A/V^2$, $W = 10 \ \mu m$, $L = 1 \ \mu m$, $V_{t0} = 0.5$ V, $\gamma = 0.7$ V$^{1/2}$, $|2\phi_p| = 0.6$ V, and $\lambda = 0.05$ V^{-1}.

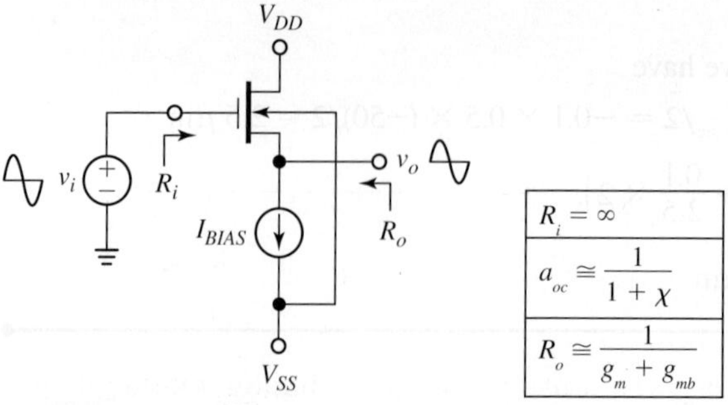

FIGURE 4.28 Voltage buffer with current-sink bias, and summary of its small-signal characteristics.

limits $R_D \rightarrow 0$ and $R_S \rightarrow \infty$. With a bit of algebra, we can put these expressions in the insightful forms

$$a_{oc} = \lim_{R_S \rightarrow \infty} \frac{v_o}{v_i} = \frac{1}{1 + \chi + 1/(g_m r_o)} \qquad R_o = \left(\frac{1}{g_m + g_{mb}}\right)//r_o \qquad (4.45)$$

with χ as given in Eq. (4.37). In view of the fact that $(g_m + g_{mb})r_o > g_m r_o \gg 1$, we can simplify as

$$a_{oc} \cong \frac{1}{1 + \chi} \qquad R_o \cong \frac{1}{g_m + g_{mb}} \qquad (4.46)$$

Clearly, the effect of g_{mb} is to *lower* both a_{oc} and R_o by about $1 + \chi$ compared to the case $\gamma = 0$. For instance, a voltage buffer implemented with the FET of Example 4.9 would give $a_{oc} \cong 1/(1 + 0.15) = 0.87$ V/V and $R_o \cong 1/(1 + 0.15) = 0.87$ kΩ.

In actual application the buffer is likely to drive some finite load resistance, and the I_{BIAS} current sink is likely to exhibit some finite parallel resistance. Lumping these two resistances together as a net resistance R_S, we observe that R_S appears in *parallel* with r_o, so we adapt Eq. (4.45) to find *the loaded gain* as

$$\frac{v_o}{v_i} = \frac{1}{1 + \chi + 1/[g_m(R_S//r_o)]} \qquad (4.47)$$

We can gain additional insight by manipulating this expression into the following voltage-divider form

$$v_o = \frac{(1/g_{mb})//R_S//r_o}{1/g_m + [(1/g_{mb})//R_S//r_o]} v_i \qquad (4.48)$$

which allows us to visualize the voltage buffer more intuitively as in Fig. 4.29. Note the similarity to the bipolar counterpart of Fig. 4.16, aside from the $1/g_{mb}$ resistance, which is absent in the bipolar version.

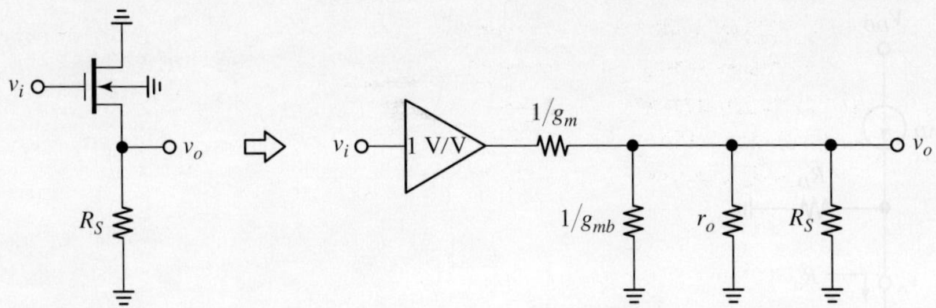

FIGURE 4.29 A FET voltage buffer acts as a *unity-gain* buffer with *output resistance* $1/g_m$, in turn forming a *voltage divider* with the *rest* of the resistances associated with the source terminal, or $(1/g_{mb})//r_o//R_S$.

EXAMPLE 4.12

(a) Let the FET of Fig. 4.28 have $V_{t0} = 0.5$ V, $k = 2$ mA/V^2, $\lambda = 0.025$ V^{-1}, $|2\phi_p| = 0.6$ V, and $\gamma = 0.4$ V$^{1/2}$. (*a*) If $V_{DD} = -V_{SS} = 5.0$ V and $I_{BIAS} = 1$ mA, find the output voltage V_O corresponding to the input voltage $V_I = 0$.

(b) Estimate a_{oc} and R_o.

(c) Find the gain if the circuit drives a 10-kΩ load.

Solution

(a) To sustain $I_D = 1$ mA the FET needs $V_{OV} = \sqrt{2 \times 1/2} = 1$ V. As a first estimate, let $V_{O(0)} = -(V_{t0} + V_{OV}) = -(0.5 + 1) = -1.5$ V, so $V_{SB(0)} = -1.5 - (-5) = 3.5$ V. Then, $V_{t(1)} = 0.5 + 0.4(\sqrt{3.5 + 0.6} - \sqrt{0.6}) = 1.0$ V, by Eq. (4.33). So a better estimate is $V_{O(1)} = -(V_{t(1)} + V_{OV}) = -(1 + 1) = -2$ V. This corresponds to $V_{SB(1)} = 3$ V, so we iterate to find $V_{t(2)} = 0.95$ V and get yet a better estimate as $V_{O(2)} = -1.95$ V. One more iteration with $V_{SB(2)} = 3.05$ V yields a negligible change, so we conclude that with $V_I = 0$ the circuit gives

$$V_O \cong -1.95 \text{ V}$$

(b) We have $g_m = \sqrt{2 \times 2 \times 1} = 2$ mA/V and $r_o = 1/(\lambda I_D) = 1/(0.025 \times 1) = 40$ kΩ. Also, Eq. (4.37) gives $\chi = 0.4/(2\sqrt{3.05 + 0.6}) = 0.105$. Consequently, use Eq. (4.46) to get

$$a_{oc} \cong \frac{1}{1 + 0.105} = 0.905 \text{ V/V} \qquad R_o \cong \frac{1}{2(1 + 0.105)} = 0.452 \text{ k}\Omega$$

(Using the exact expressions of Eq. (4.45) we find $a_{oc} = 0.895$ V/V and $R_o = 0.447$ kΩ, hardly worth the extra computational effort.)

(c) We have $1/g_m = 1/2 = 0.5$ kΩ and $1/g_{mb} = 1/(0.105 \times 2) = 4.76$ kΩ. Using Eq. (4.48) we get

$$\frac{v_o}{v_i} = \frac{4.76//10//40}{0.5 + 4.76//10//40} = 0.835 \text{ V/V}$$

The Active-Loaded CG Configuration as a Voltage Amplifier

The inherently high output resistance of the common-gate (CG) configuration makes it suited not only to *current buffering*, as we already know, but also to *high-gain voltage amplification*. To investigate this alternative aspect, refer to the monolithic

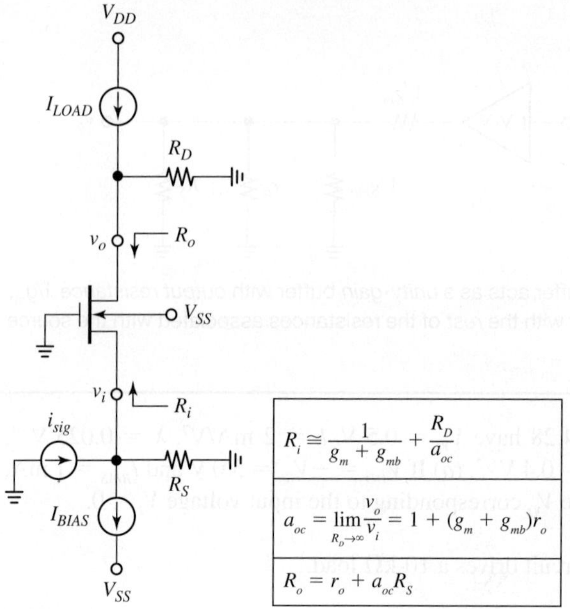

$$R_i \cong \frac{1}{g_m + g_{mb}} + \frac{R_D}{a_{oc}}$$

$$a_{oc} = \lim_{R_D \to \infty} \frac{v_o}{v_i} = 1 + (g_m + g_{mb})r_o$$

$$R_o = r_o + a_{oc}R_s$$

FIGURE 4.30 The CG configuration as a high-gain voltage amplifier, and summary of its ac characteristrics.

rendition of Fig. 4.30, where the I_{BIAS} current sink provides source bias, and the I_{LOAD} current source acts as an active load. Also shown are the net resistance R_D external to the drain, combining the parallel resistance of the I_{LOAD} source as well as that of a possible output load, and the net resistance R_S external to the source, combining the parallel resistances of the i_{sig} source and the I_{BIAS} sink.

We wish to find the *unloaded voltage gain* from v_i to v_o, that is, the gain v_o/v_i in the limit $R_D \to \infty$, for this reason also called the *open-circuit voltage gain*. To this end, refer to the ac equivalent of Fig. 4.31. Since $v_{bs} = v_{gs} = -v_i$, we invert the direction of the dependent sources, and since they are in parallel, we lump them together

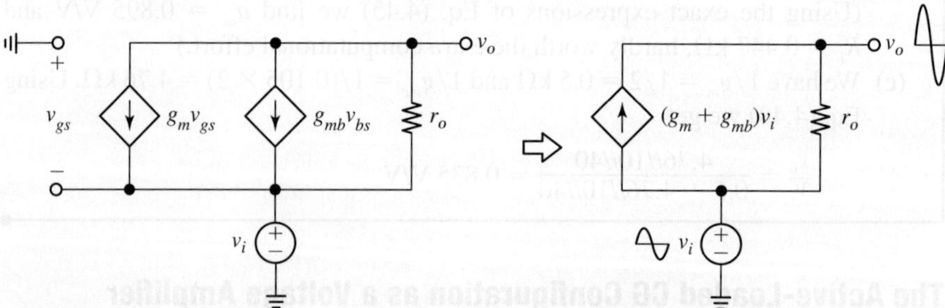

FIGURE 4.31 Ac equivalent to find the open-circuit voltage gain of the CG configuration.

as a single dependent source of value $(g_m + g_{mb})v_i$, as shown. By KVL and Ohm's Law, $v_o = v_i + r_o(g_m + g_{mb})v_i$, so

$$a_{oc} = \lim_{R_D \to \infty} \frac{v_o}{v_i} = 1 + (g_m + g_{mb})r_o \tag{4.49}$$

To find the resistance R_i seen looking into the source and the resistance R_o seen looking into the drain, we simply recycle the results tabulated in Fig. 4.23 and write

$$R_i \cong \frac{1}{g_m + g_{mb}} + \frac{R_D}{a_{oc}} \qquad R_o = r_o + a_{oc}R_S \tag{4.50}$$

As already seen, the drain resistance R_D reflected to the source gets *divided* by a_{oc}, whereas the source resistance R_S reflected to the drain gets *multiplied* by a_{oc}. These symmetric resistance-transformation properties are worth remembering, as they will appear frequently as we move along.

4.4 DARLINGTON, CASCODE, AND CASCADE CONFIGURATIONS

Up to now we have focused on single-transistor configurations, namely, the CE/CS *voltage amplifier*, the CC/CD *voltage buffer*, and the CB/CG *current buffer*. Each configuration implements its intended function but only as an approximation to the ideal. For instance, to avoid loading both at its input and output ports, a voltage amplifier/buffer should have $R_i \to \infty$ and $R_o \to 0$, while a current amplifier/buffer should have $R_i \to 0$ and $R_o \to \infty$. The CE amplifier, while capable of relatively high gain, has $R_i = r_\pi$, which is often not high enough. On the other hand, the CC configuration is renowned for its high input resistance, but at the price of a voltage gain a bit less than unity. It makes sense to combine the two configurations so that as a *team* they exploit the advantages of one to reduce the shortcomings of the other. The resulting CC-CE pair, along with its CC-CC sibling, belongs to a class of two-transistor circuits broadly referred to as the *Darlington configuration* and characterized by high input resistance.

On another subject, the design of high-gain amplifiers such as operational amplifiers and voltage comparators requires that high gains be achieved with as few amplifier stages as possible. The CE/CS configurations offer $a_{\text{instrinsic}} = -g_m r_o$, which may not be sufficiently high for this purpose, especially in the CS case. On the other hand, the CB/CG current buffers are renowned for providing high output resistance, so if we combine a CE/CS amplifier with a current buffer we can raise the unloaded voltage gain way above the intrinsic gain of a single transistor. The resulting CE-CB and CS-CG pairs are known as the *cascode configuration*, a term coined in the days of vacuum tubes.

On yet another note, the advent of monolithic circuits with matched devices has ushered in other transistor pairings, such as the CC-CB and CD-CG configurations. Also known as *emitter-coupled* (EC) and *source-coupled* (SC) *pairs*, or collectively as *differential pairs*, they are used as the input stage of popular ICs such as the operational amplifier and the voltage comparator. These pairs play such an important role that they merit detailed attention separately. In the present section we focus on Darlington-type and cascode-type transistor pairs.

The Darlington Configuration

Named after Sidney Darlington who patented it in 1952, this configuration is based on the idea of feeding the emitter of one BJT to the base of another to achieve an overall current gain approaching the *product* of the individual current gains. Figure 4.32*a* shows the *npn* realization of the Darlington concept, whereas Fig. 4.32*b* shows its *complementary* version using *pnp* transistors. The function of the current source *I* is to establish the operating point of Q_1 and also to help remove the base charge of Q_2 during turnoff (more on this in Chapter 6). This source is usually implemented with a plain resistor, or it may even be absent.

When it is absent, Q_1 is biased by Q_2's base current. This current may be too low for Q_1 to provide a reasonably high beta, hence the need to bias Q_1 more convincingly. We shall now demonstrate that a Darlington pair can be regarded as a *single composite transistor*. Though the discussion will focus on the *npn* pair, the results are readily adapted to the *pnp* pair, provided we *reverse* all voltage polarities and current directions.

In response to the applied current I_B, the configuration of Fig. 4.33*a* develops the composite base-emitter voltage drop

$$V_{BE} = V_{BE1} + V_{BE2} \tag{4.51}$$

where we are using subscripts 1 and 2 to denote parameters pertaining to Q_1 and Q_2, and no numerical subscripts for the composite transistor. To signify the need for two B-E voltage drops to turn it on, the composite device is sometimes drawn with two emitter arrows, as shown in Fig. 4.32. Assuming both transistors are operating in the forward-active region, the collector current of the composite device in Fig. 4.33*a* is

$$I_C = I_{C1} + I_{C2} = \beta_1 I_B + \beta_2 (I_{E1} - I) = \beta_1 I_B + \beta_2 [(\beta_1 + 1)I_B - I]$$
$$= (\beta_1 + \beta_1\beta_2 + \beta_2)I_B - \beta_2 I$$

If we rewrite as

$$I_C = \beta I_B - \beta_2 I$$

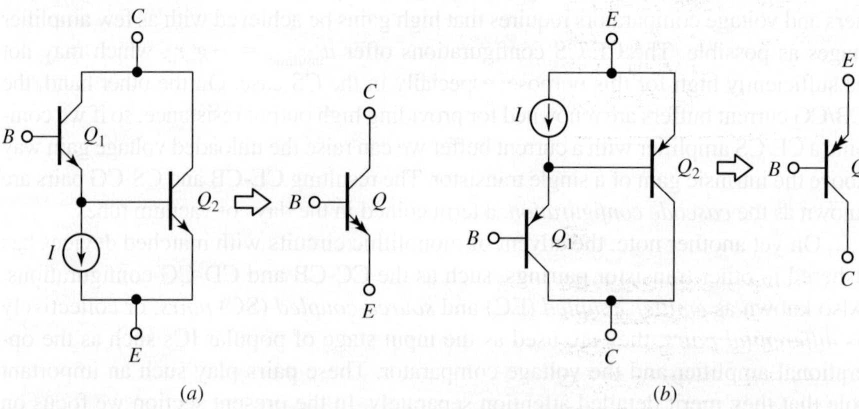

FIGURE 4.32 The Darlington configuration: (a) *npn* and (b) *pnp*.

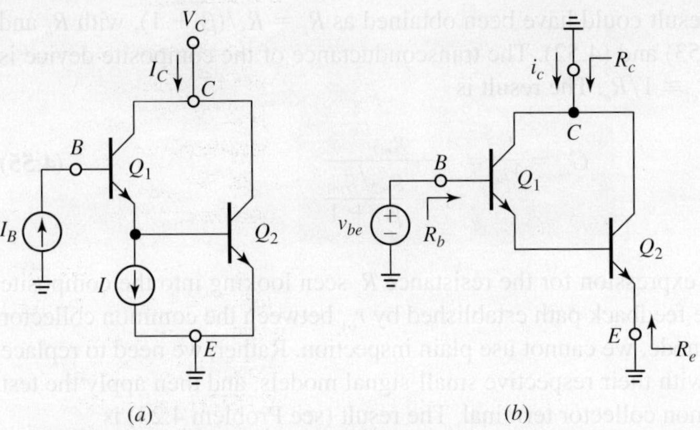

FIGURE 4.33 Circuits to investigate the (a) dc and (b) ac characteristics of the *npn* Darlington configuration.

it is apparent that

$$\beta = \beta_1 + \beta_1\beta_2 + \beta_2 \cong \beta_1\beta_2 \qquad (4.52)$$

that is, the current gain of the composite device is approximately the *product* of the individual gains. We can look at the Darlington configuration from two alternative viewpoints (assume $I = 0$ for simplicity). If we focus on Q_2, we can view Q_1 as providing a means to *lower* the required input current by a factor of β_1. For instance, to sustain $I_{C2} = 1$ mA with $\beta_1 = \beta_2 = 100$, we need $I_{B2} \cong 10$ μA, but only $I_{B1} \cong 0.1$ μA. Alternatively, if we focus on Q_1, we can view Q_2 as providing a means to *boost* the output current capability by a factor of β_2. For instance, with $I_{E1} = 10$ mA and $\beta_2 = 50$, the Darlington configuration yields $I_{E2} \cong 5$ A. In the former capacity, the Darlington configuration is often used in the design of low input bias current amplifiers. In the latter capacity, it is used in the design of the output stage of power amplifiers.

Turning next to the ac equivalent of Fig. 4.33b, we note that Q_1 acts as an emitter follower with $r_{\pi2}$ as its emitter load, so the resistance seen looking into the base of the composite device is

$$R_b = r_{\pi1} + (\beta_1 + 1)r_{\pi2} \qquad (4.53)$$

We likewise observe that the base of Q_2 is terminated on the resistance r_{e1} seen looking into Q_1's emitter, so the resistance seen looking into the emitter of the composite device is, by inspection,

$$R_e = r_{e2} + \frac{r_{e1}}{\beta_2 + 1} \qquad (4.54)$$

Alternatively, this result could have been obtained as $R_e = R_b/(\beta + 1)$, with R_b and β given by Eqs. (4.53) and (4.52). The transconductance of the composite device is found as $G_m = \alpha/R_e \cong 1/R_e$. The result is

$$G_m = \frac{i_c}{v_{be}} = \frac{g_{m2}}{1 + \dfrac{g_{m2}/g_{m1}}{\beta_2 + 1}} \tag{4.55}$$

Finally, we seek an expression for the resistance R_c seen looking into the composite collector. Due to the feedback path established by r_{o1} between the common collector node and Q_2's base node, we cannot use plain inspection. Rather, we need to replace the two transistors with their respective small-signal models, and then apply the test method to the common collector terminal. The result (see Problem 4.23) is

$$R_c \cong r_{o2}//\left(r_{o1}\frac{1 + \beta_2 g_{m1}/g_{m2}}{1 + \beta_2} \right) \tag{4.56}$$

To get a better feel for the various parameters, it is instructive to consider the special case $I = 0$, $\beta_1 = \beta_2$, and $V_{A1} = V_{A2}$, for then the above expressions simplify as

$$\beta = \beta_1^2 \qquad R_b = 2r_{\pi1} \qquad G_m = \frac{g_{m2}}{2} \qquad R_e = 2r_{e2} \qquad R_c = \frac{2}{3}r_{o2} \tag{4.57}$$

EXAMPLE 4.13   Assuming a Darlington configuration with $\beta_1 = \beta_2 = 100$ and $V_{A1} = V_{A2} = 100$ V, find its small-signal parameters if $I_{C2} = 1$ mA and $I = 90$ μA.

Solution

We have $I_{C1} \cong I_{E1} = I_{B2} + I = I_{C2}/\beta_2 + I \cong 1000/100 + 90 = 100$ μA. Consequently, $g_{m2} = 1/(26 \ \Omega)$, $r_{\pi2} = 2.6$ kΩ, $r_{o2} = 100$ kΩ, $g_{m1} = 1/(260 \ \Omega)$, $r_{\pi1} = 26$ kΩ, and $r_{o1} = 1$ MΩ. Using Eqs. (4.52) through (4.56),

$$\beta \cong 100 \times 100 = 10{,}000 \qquad G_m = \frac{1/26}{1 + (260/26)/101} = \frac{1}{28.6 \ \Omega}$$

$$R_b \cong 26 + 101 \times 2.6 \cong 290 \text{ k}\Omega \qquad R_e \cong 26 + \frac{260}{100 + 1} = 28.3 \ \Omega$$

$$R_c \cong 100//\left(1000\frac{1 + 100 \times 26/260}{101} \right) = 100//109 = 52 \text{ k}\Omega$$

Recall that *pnp* BJTs fabricated in the standard planar process exhibit poorer characteristics than their *npn* counterparts, so *pnp* BJTs should be avoided whenever possible. Shown in Fig. 4.34a is a popular alternative to the complementary pair of Fig. 4.32b. Named for G. C. Sziklai who patented it in the 1950s, this configuration acts like a composite *pnp* BJT, except that it uses an *npn* device for Q_2, which usually provides a much higher current gain than a *pnp* type. Also referred to as *quasi-complementary* Darlington configuration, it enjoys the additional advantage of requiring only a single junction

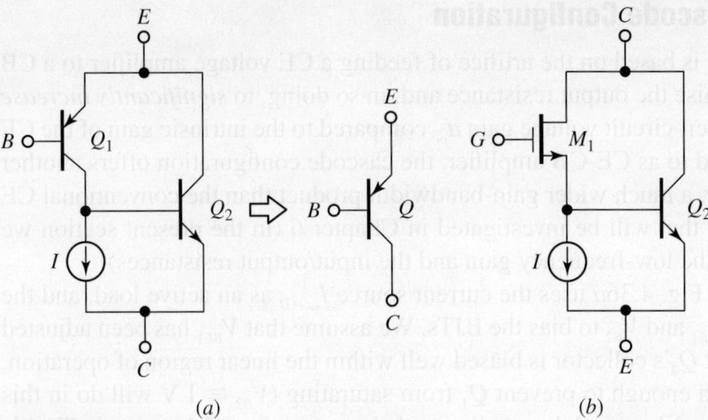

FIGURE 4.34 Alternative Darlington realizations: (a) quasi-complementary and (b) BiMOS.

voltage drop between its composite emitter and base, or $V_{EB} = V_{EB1}$. The small-signal characteristics of the Sziklai pair are investigated further in Problem 4.24.

BiCMOS technology exploits the advantages of both BJTs and MOSFETs to further improve circuit performance. The Darlington rendition of Fig. 4.34b uses a MOSFET to provide $R_i = \infty$ and a BJT to provide a higher current-drive capability as well as a higher transconductance.

The CC-CE and CC-CC Configurations

IC design often utilizes slight variants of the Darlington configuration to meet special needs. Two examples are shown in Fig. 4.35. The CC-CE version of Fig. 4.35a avoids the reduction in r_o due to r_{o1}'s feedback action by *separating* the collectors, as shown. Even more important, Q_1 provides a low-impedance drive to Q_2 to alleviate the impact of the Miller effect and thus maximize the frequency bandwidth, a subject we'll address in great detail in Chapter 6 when studying frequency/transient responses.

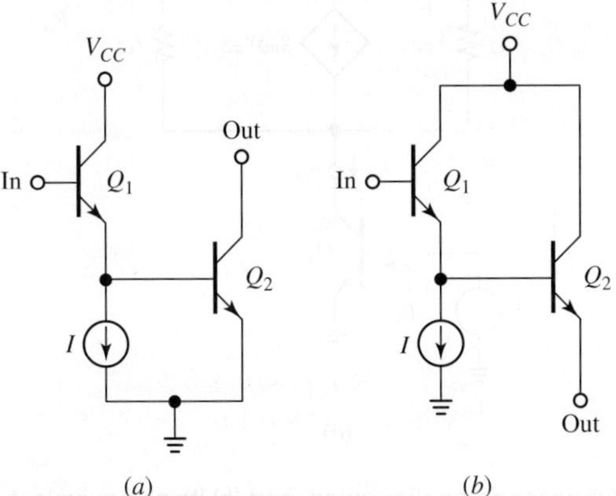

FIGURE 4.35 The (a) CC-CE and (b) CC-CC configurations.

The Bipolar Cascode Configuration

This configuration is based on the artifice of feeding a CE voltage amplifier to a CB current buffer to raise the output resistance and, in so doing, to *significantly increase* the unloaded or open-circuit voltage gain a_{oc} compared to the intrinsic gain of the CE stage. Also referred to as CE-CB amplifier, the cascode configuration offers another advantage, namely, a much wider gain-bandwidth product than the conventional CE amplifier—a topic that will be investigated in Chapter 6 (in the present section we limit ourselves to the low-frequency gain and the input/output resistances).

The circuit of Fig. 4.36a uses the current source I_{LOAD} as an active load, and the voltage sources V_{BE1} and V_{B2} to bias the BJTs. We assume that V_{BE1} has been adjusted so as to ensure that Q_2's collector is biased well within the linear region of operation, and that V_{B2} is high enough to prevent Q_1 from saturating ($V_{B2} \cong 1$ V will do in this example). We now wish to find the small-signal characteristics of this circuit. To this end, refer to its ac equivalent of Fig. 4.36b where we have, by inspection,

$$R_i = r_{\pi1} \tag{4.58}$$

To find R_o we observe that Q_2 is subject to a good deal of emitter degeneration, with Q_1's collector resistance r_{o1} acting as Q_2's degeneration resistance. We thus write

$$R_o = r_{o2}[1 + g_{m2}(r_{\pi2}//r_{o1})] \tag{4.59a}$$

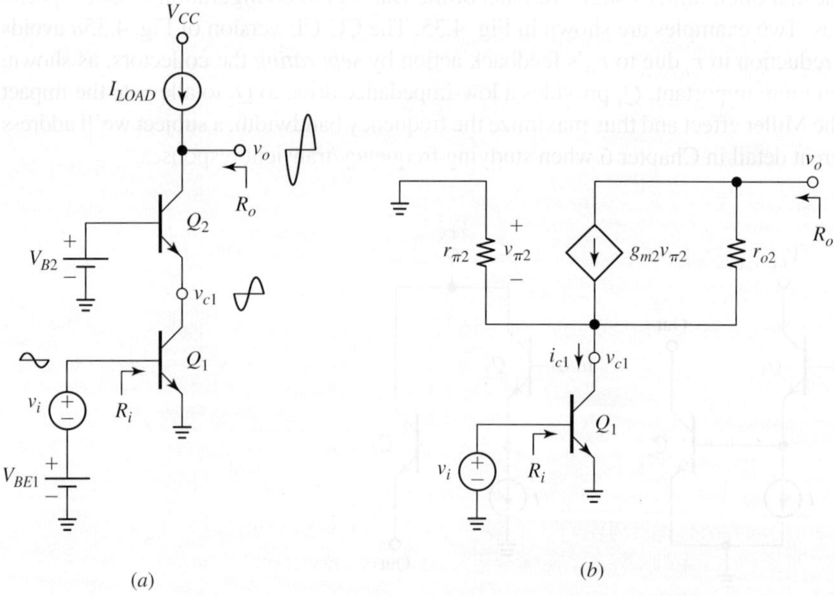

FIGURE 4.36 (a) The CE-CB, or bipolar cascode configuration, and (b) its ac equivalent.

For $r_{\pi2} \ll r_{o1}$ we can approximate $g_{m2}(r_{\pi2}//r_{o1}) \cong g_{m2}r_{\pi2} = \beta_{02}$, so the above expression simplifies as

$$R_o \cong (\beta_{02} + 1)r_{o2} \qquad (4.59b)$$

indicating that cascoding *raises* the output resistance by about $\beta_{02} + 1$. To find the voltage gain, recall that in the absence of any external ac load the current $g_{m2}v_{\pi2}$ circulates entirely in the resistance r_{o2}. Consequently, Q_1 acts as a CE amplifier with $r_{\pi2}$ as its collector resistance, providing the voltage gain

$$a_1 = \frac{v_{c1}}{v_i} = -g_{m1}(r_{\pi2}//r_{o1}) = -g_m\frac{r_\pi r_o}{r_\pi + r_o} = -\beta_0\frac{r_o}{r_\pi + r_o}$$

where the numerical subscripts have been dropped to reflect the fact that matched and equally-biased BJTs exhibit identical small-signal parameters. We also know from Eq. (4.26) that the unloaded voltage gain of the CB stage Q_2 is

$$a_2 = \frac{v_o}{v_{c1}} = 1 + g_{m2}r_{o2}$$

Consequently, the overall unloaded voltage gain is $a_{oc} = v_o/v_i = (v_o/v_{c1}) \times (v_{c1}/v_i) = a_2 \times a_1 = a_1 \times a_2$, or

$$a_{oc} = -\beta_0\frac{r_o}{r_\pi + r_o}(1 + g_mr_o) \qquad (4.60a)$$

where the numerical subscripts have again been omitted. For $r_\pi \ll r_o$ and $g_mr_o \gg 1$ we can approximate

$$a_{oc} \cong -\beta_0 g_m r_o \qquad (4.60b)$$

Since $-g_mr_o$ represents a BJT's intrinsic gain, it is apparent that the artifice of cascoding raises this gain by almost β_0! This is not surprising once we realize that Q_1's *collector* current flows through $r_{\pi2}$ to become Q_2's *base* current. Hence, the amplification factor β_0.

We observe that R_o of Eq. (4.59b) is comparable to r_μ, which has been deliberately omitted for the sake of simplicity. A more accurate expression is thus $R_o \cong (\beta_{02}r_{o2})//r_\mu$. Using Eq. (4.8), we write

$$R_o \cong \frac{m}{m+1}\beta_0 r_o \qquad\qquad a_{oc} \cong -\beta_0 g_m R_o = -\frac{m}{m+1}\beta_0 g_m r_o \qquad (4.61)$$

Note that in order to fully realize the high-gain potential of the bipolar cascode we must avoid loading its output appreciably. We could buffer the cascode's output with a high input-resistance stage such as a Darlington-type voltage follower, or, in the case of BiMOS technology, with a MOSFET follower.

EXAMPLE 4.14

(a) Assuming a cascode configuration with $I_{LAD} = 1$ mA, $\beta_1 = \beta_2 = 100$, $V_{A1} = V_{A2} = 100$ V, and $m = 5$, find quick estimates for R_i, R_o, and v_o/v_i.

(b) Repeat if the circuit drives a 1-MΩ external load, and comment.

Solution

(a) Applying Eqs. (4.58) and (4.61) we get

$$R_i = 2.6 \text{ k}\Omega \qquad R_o \cong \frac{5}{5+1}100 \times 10^5 = 8.33 \text{ M}\Omega$$

$$a_{oc} = -g_m R_o = -\frac{8.33 \times 10^6}{26} \cong -320 \times 10^3 \text{ V/V}$$

This is quite a gain!

(b) With an external load $R_L = 1$ MΩ the gain drops to

$$\frac{v_o}{v_i} = a_{oc}\frac{R_L}{R_o + R_L} \cong -320 \times 10^3 \frac{1}{8.33 + 1} = -34.3 \times 10^3 \text{ V/V}$$

Note almost an order of magnitude drop due to heavy output loading, even though a 1-MΩ load might seem large by common standards. Clearly, in order to fully realize the high-gain capabilities of the cascode, we must ensure that output loading is negligible.

EXAMPLE 4.15

To gain deeper insight it is instructive to follow the signal as it progresses from the input to the output node. In so doing, we will also get a chance to refine the quick estimate provided by Eq. (4.61).

(a) With reference to the cascode of Example 4.14a, find the individual gains a_1 and a_2 as well as the gain a_{oc}. Verify with PSpice and compare with Example 4.14a.

(b) Assuming a maximum output swing of $|v_{o(max)}| = 2.5$ V, find the corresponding input swing $|v_{i(max)}|$. Do both BJTs meet the small-signal constraint $|v_{be}| \leq 5$ mV?

Solution

(a) For the CE stage Q_1 we have $a_1 = -g_{m1}R_{C1}$, where R_{C1} is the total collector resistance of Q_1. By inspection, $R_{C1} \cong r_{\mu 1}//r_{o1}//R_{e2}$, where R_{e2} is the resistance seen looking into the emitter of Q_2. Adapting the expression tabulated in Fig. 4.19 we get

$$R_{e2} \cong r_{e2}\frac{r_{o2} + r_{\mu 2}}{r_{o2} + r_{\mu 2}/(\beta_{02} + 1)} = \frac{100}{101}26\frac{0.1 + 50}{0.1 + 50/101} = 2.17 \text{ k}\Omega$$

so

$$a_1 = -g_{m1}R_{C1} = -\frac{50{,}000//100//2.17}{0.026} = -81.7 \text{ V/V}$$

in excellent agreement with PSpice's prediction $a_1 = -81.6$ V/V. A quick estimate would have given $a_1 \cong -\beta_{01} = -100$ V/V and $R_{e2} = r_{\pi 2} = 2.6$ kΩ.

Both estimates are higher than the actual values because they ignore the presence of r_{o2} and $r_{\mu 2}$. The gain of the CB stage Q_2 is

$$a_2 = 1 + g_{m2}r_{o2} = 1 + \frac{100}{0.026} = 3846 \text{ V/V}$$

so $a_{oc} = -81.7 \times 3846 = -314 \times 10^3$ V/V, in excellent agreement with PSpice's $a_{oc} = -313 \times 10^3$ V/V.

(b) We have $|v_{i(\max)}| = |v_{o(\max)}|/|a_{oc}| = 2.5/(314,000) \cong 8 \ \mu\text{V}$. Moreover, $|v_{be2}| = |v_{o(\max)}|/a_2 = 2.5/3846 = 0.65$ mV, indicating that both BJTs amply meet the small-signal constraint.

The MOS Cascode Configuration

The cascode concept, originally developed for vacuum tubes and subsequently adapted to BJTs, is widely used in MOS technology as well. Here, we feed a CS amplifier to a CG current buffer to raise the output resistance and, in so doing, we raise also the unloaded voltage gain. This is especially desirable in the case of FETs, which are notorious for their lower g_ms compared to BJTs. Also referred to as the CS-CG amplifier, the cascode configuration is also an inherently faster circuit than an ordinary CS amplifier (this subject will be investigated in Chapter 6; in this section we limit ourselves to the low-frequency gain and the input/output resistances).

The circuit of Fig. 4.37*a* uses the current source I_{LOAD} as active load, and the voltage sources V_{GS1} and V_{G2} to bias the FETs. We assume that V_{GS1} has been adjusted so

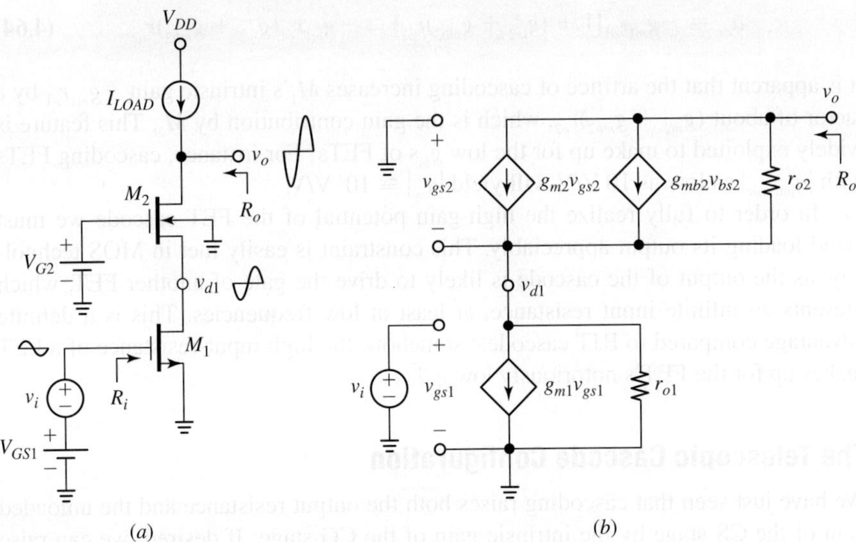

(a) (b)

FIGURE 4.37 (a) The CS-CG, or MOSFET cascode configuration, and (b) its small-signal equivalent circuit.

as to ensure that M_2's drain is biased well within the linear region of operation, and that V_{G2} is high enough to bias M_1 at or above its edge of saturation. We now wish to find the small-signal characteristics of the overall circuit. To this end, refer to the ac equivalent of Fig. 4.37b. By inspection, the input resistance is

$$R_i = \infty \tag{4.62}$$

To find the output resistance, let $v_i \to 0$, thereby deactivating the $g_{m1}v_{gs1}$ source and leaving only r_{o1} in place. This resistance causes a great deal of source degeneration for M_2, so we adapt Eq. (4.41) to write

$$R_o = r_{o2} + [1 + (g_{m2} + g_{mb2})r_{o2}]r_{o1} \cong (g_{m2} + g_{mb2})r_{o1}r_{o2} \tag{4.63}$$

indicating that cascoding two FETs raises the output resistance dramatically. We expect the unloaded gain to increase accordingly. Referring again to Fig. 4.37b, we observe that in the absence of any external ac load, M_2's drain is an ac open circuit, so the entire ac equivalent of M_2 forms a self-contained subcircuit. The source $g_{m1}v_{gs1}$ sinks current out of r_{o1} to give, by Ohm's law, $v_{d1} = -g_{m1}v_ir_{o1}$, so the gain of the CS stage M_1 is

$$a_1 = \frac{v_{d1}}{v_i} = -g_{m1}r_{o1}$$

As we know, this is M_1's intrinsic gain. By Eq. (4.49) the unloaded voltage gain of the CG stage M_2 is

$$a_2 = 1 + (g_{m2} + g_{mb2})r_{o2}$$

Consequently, the overall voltage gain is $a_{oc} = v_o/v_i = (v_o/v_{d1}) \times (v_{d1}/v_i) = a_2 \times a_1 = a_1 \times a_2$, or

$$a_{oc} = -g_{m1}r_{o1}[1 + (g_{m2} + g_{mb2})r_{o2}] \cong -g_{m1}r_{o1}(g_{m2} + g_{mb2})r_{o2} \tag{4.64}$$

It is apparent that the artifice of cascoding increases M_1's intrinsic gain $-g_{m1}r_{o1}$ by a factor of about $(g_{m2} + g_{mb2})r_{o2}$, which is the gain contribution by M_2. This feature is widely exploited to make up for the low g_ms of FETs! For instance, cascoding FETs with $|a_{\text{intrinsic}}|$ as low as 10 V/V will yield $|a_{oc}| \cong 10^2$ V/V.

In order to fully realize the high-gain potential of the FET cascode we must avoid loading its output appreciably. This constraint is easily met in MOS technology, as the output of the cascode is likely to drive the gate of another FET, which presents an infinite input resistance, at least at low frequencies. This is a definite advantage compared to BJT cascodes: somehow, the high input-resistance of a FET makes up for the FET's notoriously low g_m!

The Telescopic Cascode Configuration

We have just seen that cascoding raises both the output resistance and the unloaded gain of the CS stage by the intrinsic gain of the CG stage. If desired, we can raise a_{oc} and R_o further by adding another level of cascoding, in the manner exemplified in Fig. 4.38 for the case of double cascoding. The stacking of transistors is reminiscent

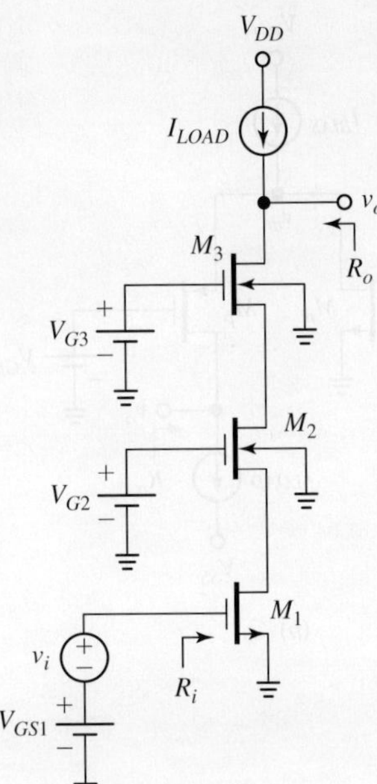

FIGURE 4.38 The telescopic cascode configuration (double cascode shown).

of a telescope extension, so this arrangement is aptly referred to as *telescopic cascode*. For instance, a three-transistor cascode using FETs with intrinsic gains as low as 10 V/V each will yield $|a_{oc}| \cong 10^3$ V/V! This is a definite advantage of MOSFETs compared to BJTs, as a bipolar cascode, after one level of cascoding, already approaches its maximum possible gain of about $\beta_0 a_{\text{instrinsic}}$.

The advantages of telescopic cascoding come at a price: a reduction in the output voltage swing (OVS). Assuming M_1 and M_2 are biased at the edge of saturation (EOS), the output must swing above

$$v_{O(\text{min})} = V_{OV1} + V_{OV2} + V_{OV3} \tag{4.65}$$

where V_{OV1} through V_{OV3} are the overdrive voltages of the stacked FETs. This limit may be unacceptably high in low power-supply systems. Fortunately, this drawback can be avoided by using the folded cascode technique discussed next.

The Folded Cascode Configuration

A notorious drawback of the cascode configurations of Figs. 4.36 and 4.37 is limited voltage headroom at the output. With proper choice of the bias voltages V_{B2} and V_{G2}, the best we can achieve is $v_{O(\text{min})} = 2V_{CE(\text{EOS})}$ in the bipolar case, and $v_{O(\text{min})} = 2V_{DS(\text{EOS})} = 2V_{OV}$ in the MOS case. In both cases $v_{O(\text{min})} > 0$ V. Yet, many applications call for v_O's ability to swing both *above* and *below* 0 V. We can meet this requirement by using a *complementary* transistor as the CB/CG stage, in the manner of Fig. 4.39. Aptly called *folded cascode*, this arrangement requires an additional current source I_{BIAS} (typically $I_{BIAS} = 2I_{LOAD}$) to bias the transistor pair, but this is a price well worth paying for expanded output headroom. In both cases $v_{O(\text{min})}$ is now established by the

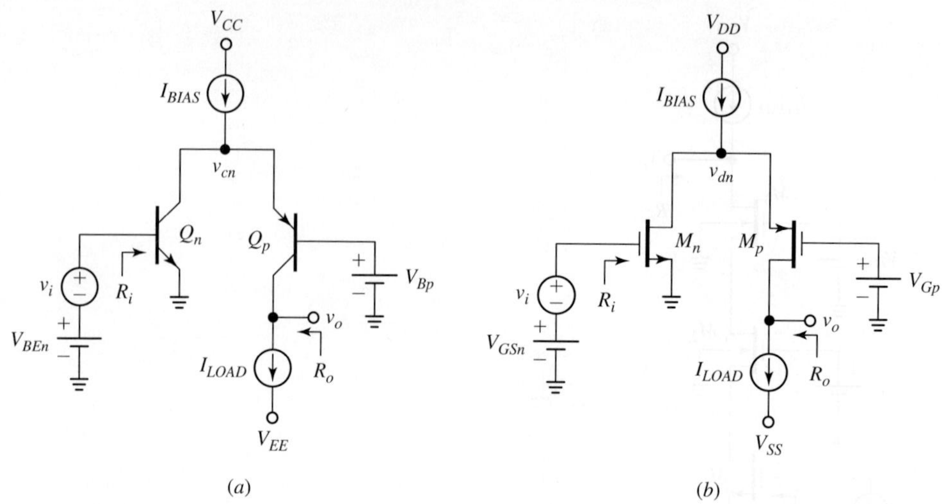

FIGURE 4.39 Folded cascode: (a) bipolar, and (b) CMOS.

negative supply, along with the minimum permissible voltage drop $V(I_{LOAD})_{min}$ across the I_{LOAD} current sink. We thus have for the BJT and CMOS circuits, respectively,

$$v_{O(min)} = V_{EE} + V(I_{LOAD})_{min} \qquad v_{O(min)} = V_{SS} + V(I_{LOAD})_{min} \qquad \textbf{(4.66a)}$$

The upper bound on the output swing is established by the bias voltage V_{Bp} and V_{Gp}, along with the minimum permissible voltage $V_{BIAS(min)}$ across the I_{BIAS} source. Selecting V_{Bp} and V_{Gp} so that this source is operated right at its minimum voltage drop, we have for the BJT and CMOS circuits, respectively,

$$v_{O(max)} = V_{CC} - V(I_{BIAS})_{min} - V_{ECp(sat)} \qquad v_{O(max)} = V_{DD} - V(I_{BIAS})_{min} - V_{OVp} \qquad \textbf{(4.66b)}$$

For instance, assuming ±5-V power supplies, along with $V(I_{BIAS})_{min} = V(I_{LOAD})_{min} = V_{ECp(sat)} = 0.2$ V in the BJT case, and $V(I_{BIAS})_{min} = V(I_{LOAD})_{min} = V_{OVp} = 0.5$ V in the MOS case, the output voltage swings are

$$-4.8 \text{ V} \le v_{O(BJT)} \le 4.6 \text{ V} \qquad -4.5 \text{ V} \le v_{O(MOS)} \le 4.0 \text{ V}$$

These compare quite favorably with the swings of the conventional cascodes of Figs. 4.36 and 4.37, which, for the same parameter values, are $0.4 \text{ V} \le v_{O(BJT)} \le 4.8$ V and $1.0 \text{ V} \le v_{O(MOS)} \le 4.5$ V.

Cascoded Current Sources/Sinks

In Sections 4.2 and 4.3 we have made extensive use of current sources/sinks to provide dc bias (I_{BIAS}) as well as active loading (I_{LOAD}). Current sources/sinks are themselves implemented with transistors. A good starting point is the collector or drain terminal, owing to the relative flatness of the active-region i_C-v_{CE} and i_D-v_{DS} characteristics, whose slope is $1/r_o$. There are many situations in which the curves are not

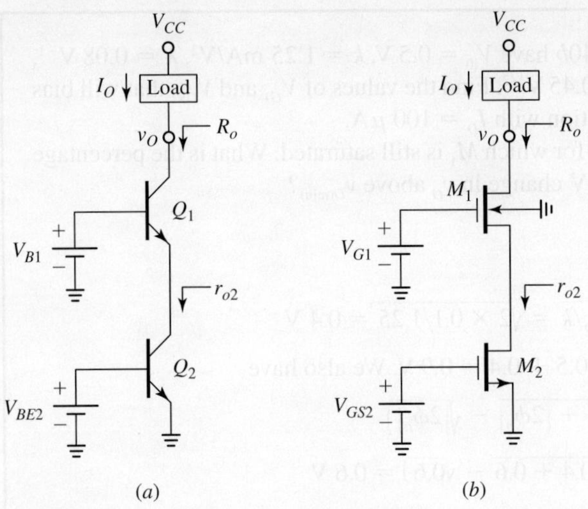

FIGURE 4.40 Using the (a) bipolar cascode and (b) MOS cascode to achieve high output resistance.

flat enough, so we must devise ways to raise the effective value of r_o. Already familiar examples are emitter and source *degeneration*, where we use a suitable resistance R_E or R_S in series with the emitter or source to provide a biasing function as well as raise the effective resistance seen looking into the collector or drain. However, it has already been pointed out that physical resistances are undesirable in monolithic circuits. Another serious drawback is that if a high degree of degeneration is needed, the degeneration resistance would have to be large, and its voltage drop could drastically reduce the available voltage headroom at the output of the source/sink.

An ingenious artifice is to replace the emitter/source degeneration resistance with a second transistor Q_2/M_2, and to let its resistance r_{o2} play the role of the degeneration resistance for the transistor Q_1/M_1. The result is the two-transistor circuits of Fig. 4.40a and b. These are the already familiar cascode circuits, except that we are now taking an alternative viewpoint: while in Figs. 4.36 and 4.37 we focused on the bottom transistor and added the top transistor to raise the bottom transistor's intrinsic gain, in Fig. 4.40 we focus on the top transistor and add the bottom transistor to raise the top transistor's output resistance.

Regardless of the viewpoint, the formulas developed earlier still stand, so we adapt them to the present case to write

$$R_{o(\text{BJT})} = r_{o1}[1 + g_{m1}(r_{\pi 1}//r_{o2})] \tag{4.67}$$

and

$$R_{o(\text{MOS})} = r_{o1} + [1 + (g_{m1} + g_{mb1})r_{o1}]r_{o2} \tag{4.68}$$

As we know, for $r_{o2} \gg r_{\pi 1}$ Eq. (4.67) simplifies as $R_o \cong (1 + \beta_{01})r_{o1}$. Similarly, for $(g_{m1} + g_{mb1})r_{o1} \gg 1$ Eq. (4.68) simplifies as $R_o \cong (1 + \chi)g_{m1}r_{o1}r_{o2}$.

EXAMPLE 4.16   (a) Let the FETs of Fig. 4.40b have $V_{t0} = 0.5$ V, $k = 1.25$ mA/V^2, $\lambda = 0.08$ V^{-1}, $|2\phi_p| = 0.6$ V, and $\gamma = 0.45$ V$^{1/2}$. Find the values of V_{G1} and V_{GS2} that will bias M_2 at the edge of saturation with $I_D = 100$ μA.

(b) Find R_o as well as $v_{O(\min)}$ for which M_1 is still saturated. What is the percentage change in I_O for each 1-V change in v_O above $v_{O(\min)}$?

Solution

(a) We have

$$V_{OV1} = V_{OV2} = \sqrt{2I_D/k} = \sqrt{2 \times 0.1/1.25} = 0.4 \text{ V}$$

so $V_{GS2} = V_{t0} + V_{OV2} = 0.5 + 0.4 = 0.9$ V. We also have

$$V_{t1} = V_{t0} + \gamma\left(\sqrt{V_{SB1} + |2\phi_p|} - \sqrt{|2\phi_p|}\right)$$

$$= 0.5 + 0.45(\sqrt{0.4 + 0.6} - \sqrt{0.6}) = 0.6 \text{ V}$$

so $V_{G1} = V_{GS1} + V_{DS2} = V_{OV1} + V_{t1} + V_{OV2} = 0.4 + 0.6 + 0.4 = 1.4$ V.

(b) We have

$$g_{m1} = k_1 V_{OV1} = 1.25 \times 0.4 = 0.5 \text{ mA/V}$$

$$r_{o1} = r_{o2} = \frac{1}{\lambda I_D} = \frac{1}{0.08 \times 0.1} = 125 \text{ k}\Omega$$

$$\chi = \frac{\gamma}{2\sqrt{V_{SB1} + 2|\phi_p|}} = \frac{0.45}{2\sqrt{0.4 + 0.6}} = 0.225$$

$$R_o = r_o + [1 + (1 + \chi)g_{m1}r_o]r_o = 125 + [1 + 1.225 \times 0.5 \times 125]125$$

$$= 9.82 \text{ M}\Omega$$

Also, $v_{O(\min)} = V_{OV2} + V_{OV1} = 0.8$ V. For a 1-V change in v_O within the linear region of operation we get $\Delta I_O = \Delta v_O/R_o = (1 \text{ V})/(9.82 \text{ M}\Omega) = 102$ nA, representing a change of only $(102 \times 10^{-9})/(100 \times 10^{-6}) \cong 0.1\%$.

Cascaded Stages

When its gain specifications cannot be met with a single transistor, a circuit is implemented as a *cascade* of two or more individual stages. Already familiar examples are Darlington amplifiers, which use two stages to raise the input resistance as well as the current gain, and cascode amplifiers, which use CE-CB or CS-CG pairs to raise the output resistance as well as the unloaded voltage gain. (Incidentally, the word *cascode*, which denotes a particular example of a *cascade*, has not made it into mainstream dictionaries. So, chances are that a spell-checker will annoyingly change typed instances of cascode to cascade.)

When analyzing a cascade of two or more individual stages, an engineer often needs to come up with a quick estimate for the overall signal-to-load gain, with interstage loading automatically taken into account. To accomplish this, we label each interstage node separately, and we find the *loaded gain* from one node to the next using the familiar rules of thumb:

> The loaded gain of a CE/CS voltage amplifier stage is the (negative of the) ratio of the *total* resistance associated with the collector/drain to the *total* resistance associated with emitter/source.

> A voltage follower acts as a unity-gain buffer with output resistance $1/g_m$, in turn forming a voltage divider with the rest of the resistances associated with the emitter/source.

Then, we find the gain of the entire cascade as the *product* of the individual gains. Study the following example carefully, as we will have plenty of occasions to retrace this procedure as we move along.

EXAMPLE 4.17

Shown in Fig. 4.41 is the ac equivalent of a three-stage amplifier consisting of CE stage Q_1, followed by CE-ED stage Q_2, followed in turn by CC stage Q_3. Assuming for simplicity that all three BJTs have $g_m = 1/(25\ \Omega)$, $r_\pi = 5\ \text{k}\Omega$, and $r_o = 50\ \text{k}\Omega$, estimate the overall gain v_o/v_i.

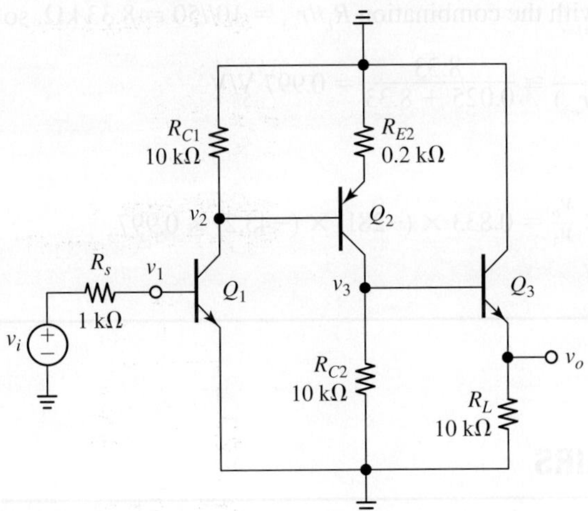

FIGURE 4.41 Circuit of Example 4.17.

Solution

Even though the node voltages of interest to us are v_i and v_o, we label also the intermediate nodes v_1, v_2, and v_3, as they will appear in our intermediate calculations. We make the following considerations:

- In progressing from node v_i to node v_1 the signal encounters a voltage divider formed by R_s and the resistance R_1 presented by node v_1. By inspection, $R_1 = r_{\pi 1}$, so

$$\frac{v_1}{v_i} = \frac{R_1}{R_s + R_1} = \frac{r_{\pi 1}}{R_s + r_{\pi 1}} = \frac{5}{1 + 5} = 0.833\ \text{V/V}$$

- In progressing from node v_1 to node v_2 the signal undergoes amplification by the CE stage Q_1, whose total emitter resistance is r_{e1} $(=\alpha_{01}/g_{m1} \cong 1/g_{m1})$, and whose total collector resistance R_2 consists of three components: R_{C1}, r_{o1}, and the ac resistance R_{b2} seen looking into Q_2's base. By inspection, $R_{b2} = r_{\pi2} + (\beta_2 + 1)R_{E2} = 5 + [(5/0.025) + 1] \times 0.2 = 45.2$ kΩ, so we write

$$\frac{v_2}{v_1} = -\frac{R_{C1}//r_{o1}//R_{b2}}{r_{e1}} \cong -\frac{10//50//45.2}{0.025} = -281 \text{ V/V}$$

- In progressing from v_2 to v_3 the signal undergoes amplification by the CE-ED stage Q_2. The total emitter resistance is $\alpha_{02}/g_{m2} + R_{E2} \cong 25 + 200 = 225$ Ω. The total collector resistance R_3 consists of three components: R_{C2}, the ac resistance R_{c2} seen looking into Q_2's collector, and the ac resistance R_{b3} seen looking into Q_3's base. By inspection, $R_{c2} = r_{o2}[1 + g_{m2}(r_{\pi2}//R_{E2})] = 50[1 + (5//0.2)/0.025] = 435$ kΩ, and $R_{b3} = r_{\pi3} + (\beta_3 + 1)R_{E3} = 5 + 201 \times 10 = 2.015$ MΩ. Consequently,

$$\frac{v_3}{v_2} = -\frac{R_{C2}//R_{c2}//R_{b3}}{r_{e2} + R_{E2}} \cong -\frac{10//435//2015}{0.025 + 0.200} = -43.2 \text{ V/V}$$

- In progressing from v_3 to v_o the signal is buffered by the CC stage Q_3, acting as a unity-gain amplifier with output resistance $1/g_{m3} = 25$ Ω. This resistance forms a voltage divider with the combination $R_L//r_{o3} = 10//50 = 8.33$ kΩ, so

$$\frac{v_o}{v_3} = 1 \times \frac{R_L//r_{o3}}{r_{e3} + (R_L//r_{o3})} \cong \frac{8.33}{0.025 + 8.33} = 0.997 \text{ V/V}$$

We finally have

$$\frac{v_o}{v_i} = \frac{v_1}{v_i} \times \frac{v_2}{v_1} \times \frac{v_3}{v_2} \times \frac{v_o}{v_3} = 0.833 \times (-281) \times (-43.2) \times 0.997$$
$$= 10.1 \times 10^3 \text{ V/V}$$

4.5 DIFFERENTIAL PAIRS

We now turn our attention to another important transistor pair, namely, the *differential pair*. Already introduced informally in Section 4.1, this is probably the most widely used subcircuit in analog ICs. Even though the concept was initially developed for vacuum tubes and subsequently adapted to discrete BJTs, it was the advent of monolithic circuits that made it possible to realize its full potential. This is because performance is critically dependent upon matching between the two transistors, and the monolithic process offers *matched* devices also capable of *tracking* each other with temperature and time. Compared to single-transistor amplifiers, differential amplifiers are much more immune to a type of interference noise known as common-mode noise, and lend themselves to dc coupling, thus avoiding the bulky capacitors of discrete designs. In fact, the differential amplifier functions over a frequency range extending all the way down to dc. All this comes at the price of increased circuit complexity, but it is not a problem in monolithic circuits, where additional transistors are easily available at insignificant extra cost.

The monolithic differential pair was first realized in bipolar form, taking on the name of *emitter-coupled* (EC) *pair*. Once MOS technology became commercially viable, the concept was adapted to MOSFETs as the *source-coupled* (SC) *pair*. Let us investigate both implementations in proper detail.

The Emitter-Coupled Pair

The simplest form of the emitter-coupled (EC) pair is shown in Fig. 4.42. Emitter bias for the pair is provided by a current source, here modeled by the Norton equivalent consisting of I_{EE} and R_{EE}. (In a well-designed circuit, R_{EE} is large and is often ignored, especially in the course of dc analysis.) Since one side is the exact *mirror image* of the other, the circuit is also said to be *balanced*, and this symmetry is exploited to facilitate circuit analysis, as we shall see. To develop a basic feel for the circuit, let us investigate its *large-signal* behavior (to keep things simple, in this section we assume $V_A = \infty$).

Applying KVL around the input loop gives

$$v_{I1} - v_{BE1} + v_{BE2} - v_{I2} = 0$$

that is, $v_{BE1} - v_{BE2} = v_{I1} - v_{I2}$. Assuming forward-active operation for the BJTs, we use the familiar exponential relationships and write $i_{C1} = I_{s1}\exp(v_{BE1}/V_T)$ and $i_{C2} = I_{s2}\exp(v_{BE2}/V_T)$. Consequently,

$$\frac{i_{C1}}{i_{C2}} = \frac{I_{s1}}{I_{s2}} \exp\left(\frac{v_{BE1} - v_{BE2}}{V_T}\right) = \frac{I_{s1}}{I_{s2}} \exp\left(\frac{v_{I1} - v_{I2}}{V_T}\right)$$

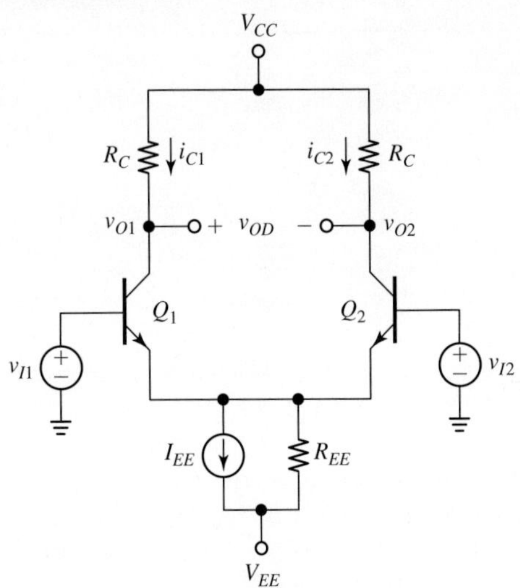

FIGURE 4.42 The emitter-coupled (EC) pair.

Matched transistors have $I_{s1} = I_{s2}$, so the above relation simplifies as

$$\frac{i_{C1}}{i_{C2}} = \exp\left(\frac{v_{ID}}{V_T}\right) \tag{4.69}$$

where

$$v_{ID} = v_{I1} - v_{I2} \tag{4.70}$$

is the *difference* between the input voltages. By KCL, the emitter currents are such that $i_{E1} + i_{E2} = I_{EE}$, where we are ignoring R_{EE}, which in a well-designed circuit is very large. This allows us to write

$$i_{C1} + i_{C2} = \alpha_F I_{EE} \tag{4.71}$$

Combining Eqs. (4.69) and (4.71) we readily get

$$i_{C1} = \frac{\alpha_F I_{EE}}{1 + \exp(-v_{iD}/V_T)} \qquad i_{C2} = \frac{\alpha_F I_{EE}}{1 + \exp(v_{iD}/V_T)} \tag{4.72}$$

We also have, by KVL and Ohm's law,

$$v_{O1} = V_{CC} - R_C i_{C1} \qquad v_{O2} = V_{CC} - R_C i_{C2}$$

Substituting the expressions of Eq. (4.72) and simplifying we obtain an expression for the voltage *difference* at the output,

$$v_{OD} = v_{O1} - v_{O2} = \alpha_F I_{EE} R_C \tanh\left(-\frac{v_{ID}}{2V_T}\right) \tag{4.73}$$

We can readily plot Eqs. (4.72) and (4.73) using PSpice. In the circuit of Fig. 4.43 we have set $v_{I2} = 0$, so $v_{ID} = v_{I1}$ in this case. After directing PSpice to sweep

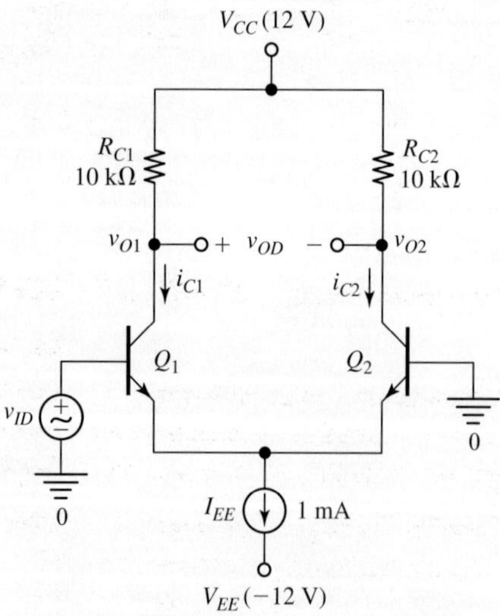

FIGURE 4.43 PSpice circuit to plot the transfer curves of an EC pair using BJTs with $\beta_F = 100$ and $I_s = 2$ fA.

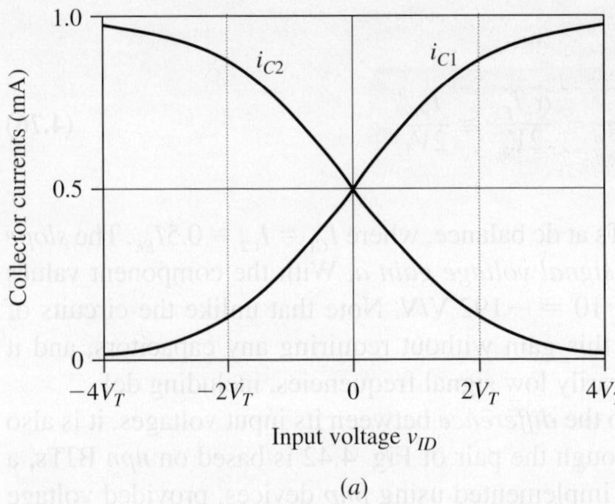

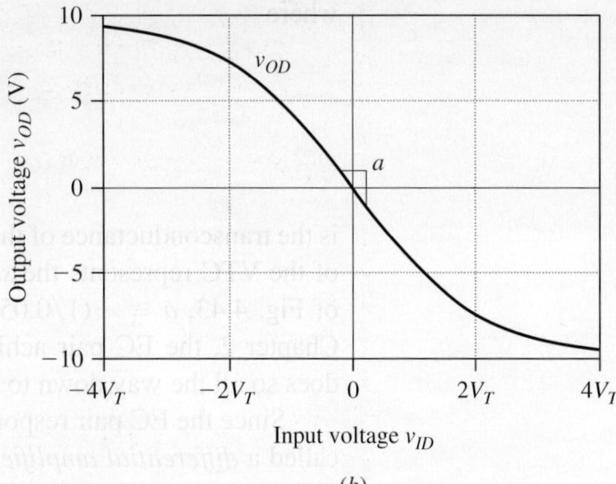

FIGURE 4.44 Plots of (a) the collector currents i_{C1} and i_{C2}, and (b) plot of the difference v_{OD} between the collector voltages of the PSpice circuit of Fig. 4.43

v_{ID} from $-4V_T (\cong -100\text{ mV})$ to $4V_T (\cong +100\text{ mV})$, we obtain the traces of Fig. 4.44. We make the following observations:

- For $v_{ID} = 0$, I_{EE} splits *equally* between Q_1 and Q_2, giving $i_{C1} = i_{C2} = \alpha_F I_{EE}/2 \cong 0.5$ mA. With equal currents, R_{C1} and R_{C2} drop equal voltages, giving $v_{O1} = v_{O2} = 12 - 10 \times 0.5 = 7$ V. Consequently, $v_{OD} = v_{O1} - v_{O2} = 0$, and we say that with $v_{ID} = 0$ the EC pair is in *dc balance*.
- Raising v_{ID} *above* 0 V makes Q_1 more conductive at the expense of Q_2 becoming less conductive (i_{C1} increases while i_{C2} decreases). This current imbalance causes v_{O1} to drop while v_{O2} rises, causing a two-fold *decrease* in v_{OD}.
- As v_{ID} reaches about $4V_T (\cong 100\ \mu\text{V})$, we can say that virtually all of I_{EE} comes from Q_1 while Q_2 is essentially off. Consequently, $v_{O1} \cong 12 - 10 \times 1 = 2$ V, $v_{O2} \cong 12$ V, and $v_{OD} \cong 2 - 12 = -10$ V. Increasing v_{ID} further will simply cause v_{OD} to saturate at -10 V.
- Lowering v_{ID} *below* 0 V makes Q_2 more conductive and Q_1 less conductive. The roles of Q_1 and Q_2 are now interchanged, and the curves are therefore *symmetric* with respect to the origin. Lowering v_{ID} below about -100 mV causes v_{OD} to saturate at $+10$ V.

We are particularly interested in the voltage transfer curve (VTC) of Fig. 4.44b. Since BJTs are nonlinear devices, it is not surprising that this curve is also nonlinear. However, if we restrict operation in the vicinity of the origin, the VTC can be approximated with its tangent there, so we can write

$$v_{od} = av_{id} \tag{4.74}$$

where v_{od} and v_{id} are *small variations* of v_{OD} and v_{ID} about 0 V, and a is the *slope* of the VTC at the origin, as indicated in the figure. Differentiating Eq. (4.73) and calculating the derivative at the origin gives

$$a = -\frac{\alpha_F I_{EE} R_C}{2V_T} = -g_m R_C \tag{4.75}$$

where

$$g_m = \frac{\alpha_F I_{EE}}{2V_T} \cong \frac{I_{EE}}{2V_T}$$

(4.76)

is the transconductance of the BJTs at dc balance, where $I_{C1} = I_{C2} \cong 0.5I_{EE}$. The *slope* of the VTC represents the *small-signal voltage gain a*. With the component values of Fig. 4.43, $a \cong -(1/0.052) \times 10 = -192$ V/V. Note that unlike the circuits of Chapter 2, the EC pair achieves this gain without requiring any capacitors, and it does so all the way down to arbitrarily low signal frequencies, including dc!

Since the EC pair responds to the *difference* between its input voltages, it is also called a *differential amplifier*. Though the pair of Fig. 4.42 is based on *npn* BJTs, a differential amplifier can also be implemented using *pnp* devices, provided voltage and current polarities are suitably reversed (see Problem 4.40). As we move along, we will work with both *npn* and *pnp* pairs.

The Source-Coupled Pair

Figure 4.45 shows the MOS counterpart of the bipolar differential amplifier of Fig. 4.42. Its analysis proceeds pretty much as in the BJT case, except that the exponential characteristics of BJTs are now replaced by the quadratic characteristics of MOSFETs. (To keep things simple, in this section we assume $\lambda = 0$ and $\gamma = 0$.) Applying KVL around the input loop gives

$$v_{I1} - v_{GS1} + v_{GS2} - v_{I2} = 0$$

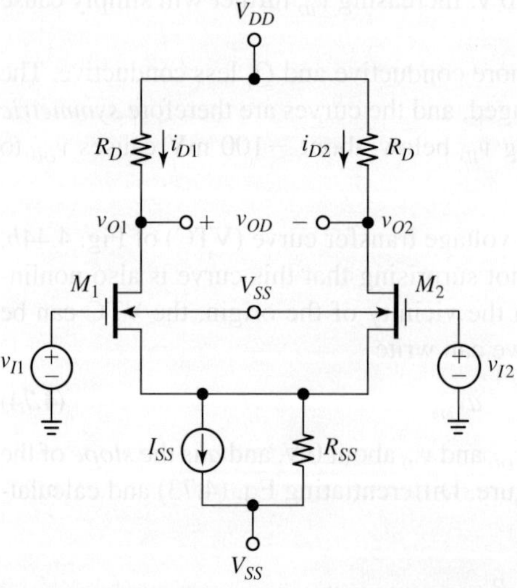

FIGURE 4.45 The source-coupled (SC) pair.

that is, $v_{GS1} - v_{GS2} = v_{I1} - v_{I2}$. Assuming saturated FETs, we use the familiar quadratic relationships to write

$$v_{GS1} = V_{t1} + \sqrt{2i_{D1}/k_1} \qquad v_{GS2} = V_{t2} + \sqrt{2i_{D2}/k_2}$$

Matched FETs have $V_{t1} = V_{t2} = V_t$ and $k_1 = k_2 = k$, so the above relationships yield

$$v_{ID} = v_{I1} - v_{I2} = v_{GS1} - v_{GS2} = \frac{\sqrt{i_{D1}} - \sqrt{i_{D2}}}{\sqrt{k/2}} \qquad (4.77)$$

By KCL, the transistor currents satisfy the condition

$$i_{D1} + i_{D2} = I_{SS} \qquad (4.78)$$

where we are ignoring R_{SS}, which in a well-designed circuit is very large. We have two equations in the two unknowns i_{D1} and i_{D2}. Letting $i_{D2} = I_{SS} - i_{D1}$ in Eq. (4.77), squaring both sides, and rearranging, we get

$$\sqrt{i_{D1}(I_{SS} - i_{D1})} = \frac{1}{2}\left(I_{SS} - \frac{k}{2}v_{ID}^2\right)$$

Squaring once again and rearranging,

$$i_{D1}^2 - I_{SS}i_{D1} + \frac{1}{4}\left(I_{SS} - \frac{k}{2}v_{ID}^2\right)^2 = 0$$

Solving this quadratic equation and retaining only the physically acceptable solution we get

$$i_{D1} = \frac{I_{SS}}{2}\left[1 + \frac{v_{ID}}{\sqrt{I_{SS}/k}}\sqrt{1 - \frac{v_{ID}^2}{4I_{SS}/k}}\right] \qquad (4.79a)$$

Using again $i_{D2} = I_{SS} - i_{D1}$, we likewise find

$$i_{D2} = \frac{I_{SS}}{2}\left[1 - \frac{v_{ID}}{\sqrt{I_{SS}/k}}\sqrt{1 - \frac{v_{ID}^2}{4I_{SS}/k}}\right] \qquad (4.79b)$$

We also have, by KVL and Ohm's law,

$$v_{O1} = V_{DD} - R_D i_{D1} \qquad v_{O2} = V_{DD} - R_D i_{D2}$$

Using the expressions of Eq. (4.79) we get, after suitable simplifications, an expression for the voltage *difference* at the output,

$$v_{OD} = v_{O1} - v_{O2} = -R_D \sqrt{kI_{SS}}\, v_{ID}\sqrt{1 - \frac{v_{ID}^2}{4I_{SS}/k}} \qquad (4.80)$$

We can readily plot Eqs. (4.79) and (4.80) using PSpice. In the circuit of Fig. 4.46 we have set $v_{I2} = 0$, so $v_{ID} = v_{I1}$ in this case. After directing PSpice to sweep v_{ID} from

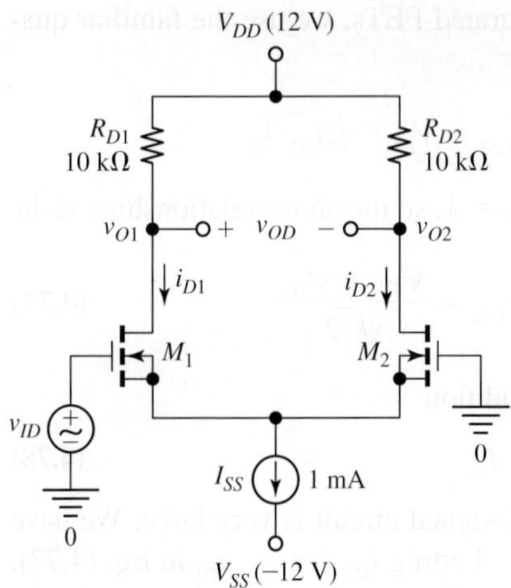

FIGURE 4.46 PSpice circuit to plot the transfer curves of an SC pair using FETs with $k = 1.0$ mA/V² and $V_t = 1.0$ V.

-2 V to $+2$ V, we obtain the traces of Fig. 4.47, in connection with which we make the following observations:

- For $v_{ID} = 0$, I_{SS} splits *equally* between M_1 and M_2, giving $i_{D1} = i_{D2} = I_{SS}/2 = 0.5$ mA. With equal currents, R_{D1} and R_{D2} drop equal voltages, giving $v_{O1} = v_{O2} = 12 - 10 \times 0.5 = 7$ V. Consequently, we have $v_{OD} = v_{O1} - v_{O2} = 0$, and we say that the SC pair is in *dc balance*. The overdrive voltage V_{OV} required of *each* FET to sustain $I_{SS}/2$ is such that $I_{SS}/2 = (k/2)V_{OV}^2$, or

$$V_{OV} = \sqrt{\frac{I_{SS}}{k}} \qquad (4.81)$$

In our example, $V_{OV} = \sqrt{1/1} = 1$ V.

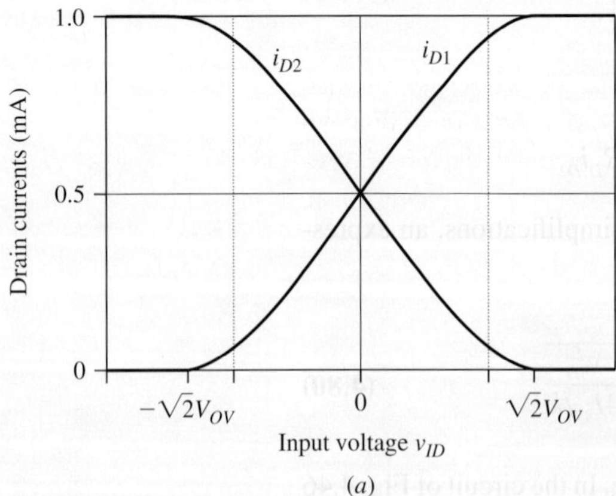

(a)

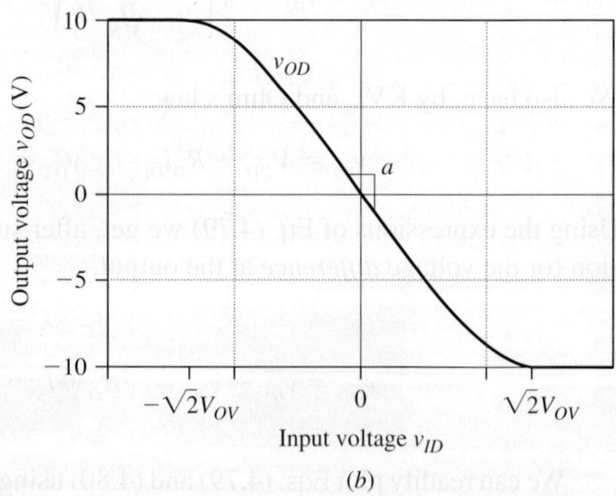

(b)

FIGURE 4.47 Plots of (a) the drain currents i_{D1} and i_{D2}, and (b) plot of the difference v_{OD} in the drain voltages for the PSpice circuit of Fig. 4.46.

- Raising v_{ID} *above* 0 V makes M_1 more conductive at the expense of M_2 becoming less conductive (i_{D1} increases while i_{D2} decreases). This imbalance causes v_{O1} to drop while v_{O2} rises, causing a two-fold *decrease* in v_{OD}.
- As we raise v_{ID} further, we reach a point at which all of I_{SS} will come from M_1 and none from M_2. For this to occur, M_1's overdrive must be $\sqrt{2}$ times as large as in Eq. (4.81), owing to the quadratic dependence of current on overdrive. At this point we have $v_{GS2} = V_t$ and $v_{GS1} = V_t + \sqrt{2}V_{OV}$, that is, $v_{ID} = \sqrt{2}V_{OV}$, with V_{OV} as given in Eq. (4.81). Beyond this point, the voltages saturate at $v_{O1} \cong 12 - 10 \times 1 = 2$ V, $v_{O2} \cong 12$ V, and $v_{OD} \cong 2 - 12 = -10$ V.
- Lowering v_{ID} *below* 0 V makes M_2 more conductive and M_1 less conductive. The roles of M_1 and M_2 are now interchanged, thereby yielding curves that are *symmetric* about the origin. Saturation now occurs for $v_{ID} \leq -\sqrt{2}V_{OV}$, where $v_{OD} = +10$ V.

We are especially interested in the voltage transfer curve (VTC) of Fig. 4.47*b*. Since FETs are nonlinear devices, it is not surprising that this curve is also nonlinear. However, if we restrict operation in the vicinity of the origin, the VTC can be approximated with its tangent there, so we can write

$$v_{od} = av_{id} \tag{4.82}$$

where v_{od} and v_{id} represent *small variations* of v_{OD} and v_{ID} about 0 V, and a is the *slope* of the VTC at the origin, as indicated in the figure. Differentiating Eq. (4.80) and calculating it at the origin gives

$$a = -R_D\sqrt{kI_{SS}} = -g_mR_D \tag{4.83}$$

where

$$g_m = \sqrt{kI_{SS}} \tag{4.84a}$$

is the transconductance of the FETs at dc balance, where $I_{D1} = I_{D2} = 0.5I_{SS}$. The *slope* of the VTC represents the *small-signal voltage gain a*. With the component values of Fig. 4.46 we have $a \cong -10 \times \sqrt{1/1} = -10$ V/V. Unlike the circuits of Chapter 3, the SC pair achieves this gain without requiring any capacitors and it functions all the way down to arbitrarily low signal frequencies, including dc!

Since the SC pair responds to the *difference* between its input voltages, it is also called a *differential amplifier*. Though the pair of Fig. 4.45 is based on *n*-channel FETs, a differential amplifier can also be implemented using *p*-channel devices, provided voltage and current polarities are suitably reversed (see Problem 4.48). As we move along, we will work with both types of pairs.

Rewriting Eq. (4.84*a*) as

$$g_m = \frac{I_{SS}}{V_{OV}} \tag{4.84b}$$

allows for the direct comparison with the BJT expression of Eq. (4.76): if an SC pair and an EC pair are biased equally, the ratio of the transconductances is $g_{m(SC)}/g_{m(EC)} = 2V_T/V_{OV}$. Typically, $g_{m(SC)}/g_{m(EC)} \ll 1$ because usually $V_{OV} \gg 2V_T$.

Intuitive Ac Analysis of Differential Pairs

Even though VTC analysis has already provided us with expressions for the small-signal gains, in daily practice one needs to come up with quick ac parameter estimates by performing ac analysis directly on the circuit itself. The ability to investigate a circuit from different perspectives is highly desirable because it reinforces our understanding, not to mention that we can use one method to find results and another to check them.

Let us turn first to the **EC pair** of Fig. 4.48a, representing the ac equivalent of the circuit of Fig. 4.42. In small-signal operation the BJTs of Fig. 4.42 draw identical dc currents,

$$I_{C1} = I_{C2} = I_C = \frac{\alpha_F I_{EE}}{2} \cong \frac{I_{EE}}{2} \tag{4.85}$$

so their transconductances are also identical,

$$g_{m1} = g_{m2} = g_m = \frac{I_C}{V_T} \cong \frac{I_{EE}}{2V_T} \tag{4.86}$$

Focusing on the lower part of the circuit, we can regard Q_1 as a unity-gain emitter follower with output resistance R_{e1} in turn loaded by the resistance R_{e2} provided by Q_2. Since $R_{e1} = R_{e2}$ at dc balance, the voltage divider rule indicates that the ac voltage at the shared emitter terminal is $v_i/2$, as shown. The ac current *into* Q_1's collector is thus $i_c = g_{m1}v_{be1} = g_{m1}(v_{b1} - v_{e1}) = g_{m1}(v_i - v_i/2) = g_m v_i/2$. This current emerges *out* of Q_1's emitter as $i_e (= i_c/\alpha_0)$, it flows *into* Q_2's emitter, and emerges once again *out* of Q_2's collector as $\alpha_0 i_e$, that is, it emerges again as i_c (the fact that $v_{be1} = v_i/2$ and $v_{be2} = -v_i/2$ confirms equal magnitudes but opposite polarities for the two i_c currents). By KVL and Ohm's Law we have $v_o = v_{o1} - v_{o2} = -R_C i_c - (R_C i_c) = -2R_C i_c = -2R_C g_m v_i/2 = -g_m R_C v_i$, so

$$a = \frac{v_o}{v_i} = -g_m R_C \tag{4.87}$$

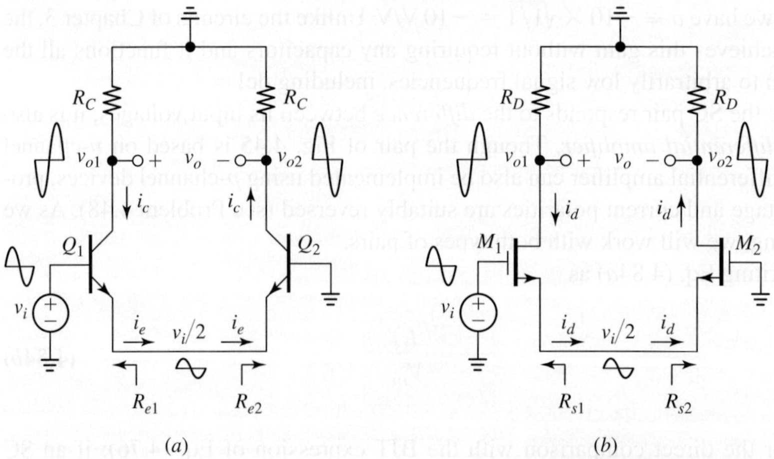

FIGURE 4.48 Small-signal equivalents of (a) the EC pair, and (b) the SC pair.

where a is the already familiar differential gain of the EC pair. Expressing it in the alternative form

$$a = -\frac{R_C I_{EE}/2}{V_T} \qquad (4.88)$$

allows us to estimate the gain of the EC pair by taking the ratio of the voltage $R_C \times (I_{EE}/2)$ dropped at dc balance by the collector resistors to the thermal voltage V_T. Recall from Chapter 2 that the small-signal approximation holds so long as $|v_{be}| \ll 2V_T$. Considering that presently we have $v_{be1} = -v_{be2} = v_i/2$, the small-signal condition is now

$$|v_i| \ll 4V_T \qquad (4.89)$$

Thus, for the small-signal approximation to hold with an error of no more than about 10%, we approximately need $|v_i| \leq 10$ mV.

Turning next to the **SC pair** of Fig. 4.48*b*, representing the ac equivalent of the circuit of Fig. 4.45, we note its formal similarity to its bipolar counterpart of Fig. 4.48*a*, and conclude that we can recycle a good deal of previous derivations. Thus, at dc balance the FETs draw identical dc currents,

$$I_{D1} = I_{D2} = I_D = \frac{I_{SS}}{2} \qquad (4.90)$$

so their transconductances are also identical,

$$g_{m1} = g_{m2} = g_m = \sqrt{k I_{SS}} \qquad (4.91)$$

We can view M_1 as a unity-gain source follower with output resistance R_{s1}, in turn loaded by the resistance R_{s2} provided by M_2's source. Since $R_{s1} = R_{s2}$ at dc balance, the voltage divider rule indicates that the ac voltage at the shared source terminal is $v_i/2$, as shown. The ac current into M_1's drain is thus $i_d = g_{m1}v_{gs1} = g_{m1}(v_{g1} - v_{s1}) = g_m v_i/2$. This current exits M_1's source, enters M_2's source, and finally exits M_2's drain (this is confirmed by the fact that with $v_{gs1} = v_i/2$ and $v_{gs2} = -v_i/2$ the two i_d currents have equal magnitudes but opposite polarities). Finally, $v_o = v_{o1} - v_{o2} = -R_D i_d - (R_D i_d) = -2R_D g_m v_i/2 = R_D g_m v_i$, so

$$a = \frac{v_o}{v_i} = -g_m R_D \qquad (4.92)$$

This is the already familiar differential gain of the SC pair. Expressing it in the alternative form

$$a = -\frac{R_D I_{SS}/2}{0.5V_{OV}} \qquad (4.93)$$

allows us to estimate the gain achievable with this circuit as the ratio of the voltage $R_D \times (I_{SS}/2)$ dropped by the drain resistors at dc balance to half the overdrive voltage. Recall from Chapter 3 that the small-signal approximation holds provided we keep

$|v_{gs}| \ll 2V_{OV}$. Considering that presently we have $v_{gs1} = -v_{gs2} = v_i/2$, the small-signal condition is now

$$|v_i| \ll 4V_{OV} \qquad (4.94)$$

Looking at Fig. 4.47b, where $V_{OV} = 1$ V, we see that a reasonable choice is to restrict the input within the range $|v_i| \leq 0.5V_{OV}$ (or 0.5 V in our example).

4.6 COMMON-MODE REJECTION RATIO IN DIFFERENTIAL PAIRS

As implied by its name, a differential amplifier responds only to the difference $v_{id} = v_{i1} - v_{i2}$, regardless of the individual values of v_{i1} and v_{i2}. For instance, when subjected to any one of the following input pairs

$$(v_{i1}, v_{i2}) = (0.005 \text{ V}, 0.000 \text{ V}), (1.005 \text{ V}, 1.000 \text{ V}), (-2.000 \text{ V}, -2.005 \text{ V})$$

a differential amplifier with a gain of, say, -100 V/V will respond only to their difference ($v_{id} = 5$ mV in each of the three cases) to give $v_o = -100 \times 5 = -500$ mV, even though the individual inputs are in the vicinity of 0 V in the first case, 1 V in the second case, and -2 V in the third case. For a more systematic analysis, it is convenient to define the *differential-mode input* as

$$v_{id} = v_{i1} - v_{i2} \qquad (4.95a)$$

and the average, or *common-mode input* as

$$v_{ic} = \frac{v_{i1} + v_{i2}}{2} \qquad (4.95b)$$

Turning these equations around allows us to express the original signals in the insightful forms

$$v_{i1} = v_{ic} + \frac{v_{id}}{2} \qquad v_{i2} = v_{ic} - \frac{v_{id}}{2} \qquad (4.96)$$

The process is illustrated pictorially in Fig. 4.49. Based in this decomposition, we can now state succinctly that a true differential amplifier responds *only* to v_{id}, *regardless* of v_{ic}.

In practice, because of unavoidable imbalances in the two circuit halves that are processing v_{i1} and v_{i2}, a real-life amplifier will be somewhat sensitive also to v_{ic}. The small-signal output takes on the more general form

$$v_o = a_{dm}v_{id} + a_{cm}v_{ic} \qquad (4.97)$$

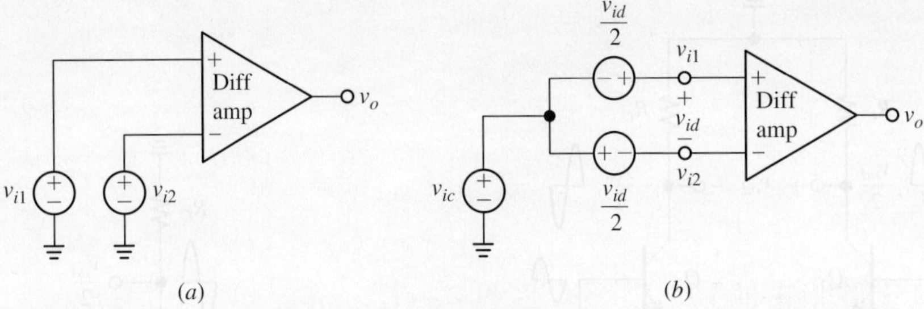

FIGURE 4.49 Expressing (*a*) the input signals v_{i1} and v_{i2} of a differential amplifier in terms of (*b*) their common-mode and differential-mode components v_{ic} and v_{id}.

where a_{dm} and a_{cm} are the *differential-mode* and *common-mode* gains. Ideally, a_{cm} should be zero, but in practice it will differ from zero, though presumably by very little. To tell how close a practical differential amplifier is to ideal, we use a figure of merit known as the *common-mode rejection ratio* (CMRR),

$$\text{CMRR} = \left| \frac{a_{dm}}{a_{cm}} \right| \tag{4.98}$$

In a high-quality differential amplifier this ratio can be rather high, such as 10^5, so CMRR is usually expressed in decibels ($20 \log 10^5 = 100$ dB).

Being referenced to ground, both v_{i1} and v_{i2} in Fig. 4.49 are said to be *single-ended signals*. As a signal propagates down a wire or a printed-circuit trace, it tends to pick up all sorts of interference noise from nearby circuits via capacitive and inductive coupling, as well as via the imperfect common ground wire or trace. Consequently, by the time it reaches its destination, the signal may have experienced significant corruption, making recovery of the useful information a formidable task. A clever trick is to work with *double-ended signals*, where the useful signal is now the *voltage difference* v_{id} between a *pair* of dedicated wires, rather than between a single wire and ground. If these dedicated wires are run in *parallel* and *close* to each other, interference with respect to ground will affect them *equally*, thus appearing as a *common-mode* component v_{ic}. Then, processing a double-ended signal with a high-CMRR difference amplifier will magnify v_{id} while rejecting v_{ic}, thus assuring a high degree of signal integrity.

CMRR and Input Resistances of the EC Pair

We wish to find the CMRR of the EC pair of Fig. 4.42. With reference to Fig. 4.49*b* and Eq. (4.97), we proceed in two steps. First, we let $v_{ic} = 0$ and find $a_{dm} = v_o/v_{id}$. Then, we let $v_{id} = 0$ and find $a_{cm} = v_o/v_{ic}$. Finally, we plug our results into Eq. (4.98).

With $v_{ic} = 0$, the circuit of Fig. 4.42 reduces to the ac equivalent of Fig. 4.50*a*. Looking at the lower portion of the circuit, we can regard Q_1 and Q_2 as emitter followers with Q_1 trying to pull the common emitter voltage v_e up, and Q_2 trying to pull v_e down by the same amount. Consequently, v_e will remain at ac *ground*. Also, v_{od} splits symmetrically between the two halves of the circuit, with $v_{od}/2$ appearing

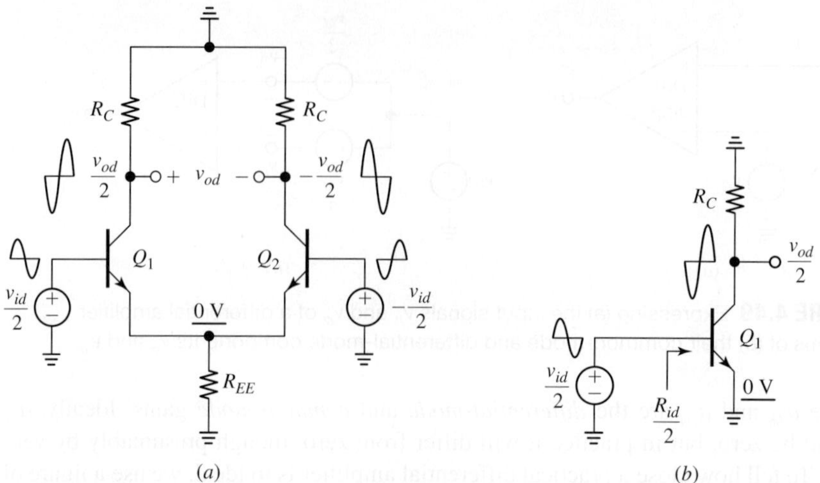

FIGURE 4.50 (a) Ac circuit to find the differential-mode parameters of the EC pair, and (b) differential-mode half circuit.

at Q_1's collector, and $-v_{od}/2$ at Q_2's collector. We can thus focus on just one of the two halves, say, the left half, as shown in Fig. 4.50b. (Note that because the common emitter node is at ac ground, R_{EE} plays no role in this case). This is the familiar CE configuration, for which $v_{od}/2 = -g_m(R_C//r_o)(v_{id}/2)$, so

$$a_{dm} = \frac{v_{od}}{v_{id}} = -g_m(R_C//r_o) \tag{4.99}$$

This is an already familiar result, but the ability to confirm it via the half-circuit analysis technique offers further insight into balanced operation.

Next, we let $v_{id} = 0$, reducing the circuit of Fig. 4.42 to that of Fig. 4.51a, where the reason for splitting R_{EE} into a pair of $2R_{EE}$s in parallel will be obvious in a moment. Since the two identical halves are driven by the same voltage v_{ic}, the collector voltages will be identical, so we label them with the same symbol v_{oc}. Moreover, no current will flow through the wire connecting the two emitters, so we can remove this wire altogether without altering circuit operation, and focus on just one half of the circuit, say, the left half, as shown in Fig. 4.51b. This is the familiar CE-ED configuration for which we readily find

$$a_{cm} = \frac{v_{oc}}{v_{ic}} = -\frac{g_m R_C}{1 + 2g_m R_{EE}} \tag{4.100}$$

(In this case r_o has been ignored altogether as emitter degeneration raises its effective value dramatically.)

Depending on the intended use of the differential pair outputs, we have two possibilities:

- The output is taken *single-endedly* from either one of the two collectors. In this case the CMRR is the ratio of *one-half* the gain of Eq. (4.99) to that of Eq. (4.100). If we ignore r_o, this ratio simplifies as

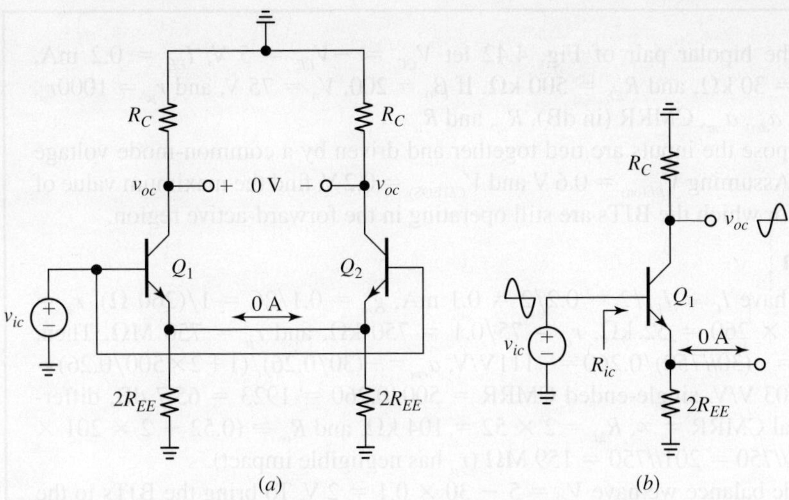

FIGURE 4.51 (a) Ac circuit to find the *common-mode* parameters of the EC pair, and (b) common-mode half circuit.

$$\text{CMRR} \cong \frac{1}{2} + g_m R_{EE} \cong g_m R_{EE} \qquad (4.101)$$

- The output is taken *differentially*, that is, between the two collectors. Since their voltage difference is zero regardless of v_{ic}, then $a_{cm} = (v_{o1} - v_{o2})/v_{ic} = 0$ in this case, so Eq. (4.98) gives

$$\text{CMRR} = \infty \qquad (4.102)$$

This last result predicates two perfectly balanced circuit halves. In practice, unavoidable mismatches between the two halves will result in a non-infinite CMRR, as we shall see shortly.

An EC pair is fully characterized once we know also its differential-mode and common-mode input resistances R_{id} and R_{ic}. With reference to Fig. 4.50b, we have, by inspection, $R_{id}/2 = r_\pi$, so the overall resistance *between* the inputs of the EC pair is the sum of the two halves, or

$$R_{id} = 2r_\pi \qquad (4.103)$$

On the other hand, the resistance between either input and ground is, from Fig. 4.51b,

$$R_{ic} = r_\pi + 2(\beta_0 + 1)R_{EE} \qquad (4.104)$$

This may be truly large, in which case it may be advisable to take r_μ into account.

EXAMPLE 4.18

(a) In the bipolar pair of Fig. 4.42 let $V_{CC} = -V_{EE} = 5$ V, $I_{EE} = 0.2$ mA, $R_C = 30$ kΩ, and $R_{EE} = 500$ kΩ. If $\beta_0 = 200$, $V_A = 75$ V, and $r_\mu = 1000r_o$, find a_{dm}, a_{cm}, CMRR (in dB), R_{id}, and R_{ic}.

(b) Suppose the inputs are tied together and driven by a common-mode voltage v_{IC}. Assuming $V_{BE(on)} = 0.6$ V and $V_{CE(EOS)} = 0.2$ V, find the maximum value of v_{IC} for which the BJTs are still operating in the forward-active region.

Solution

(a) We have $I_C \cong I_{EE}/2 = 0.2/2 = 0.1$ mA, $g_m = 0.1/26 = 1/(260\ \Omega)$, $r_\pi = 200 \times 260 = 52$ kΩ, $r_o = 75/0.1 = 750$ kΩ, and $r_\mu = 750$ MΩ. Then, $a_{dm} = -(30//750)/0.260 = -111$ V/V, $a_{cm} = -(30/0.26)/(1+2\times500/0.26) = -0.03$ V/V, single-ended CMRR $= 500/0.260 = 1923 = 65.7$ dB, differential CMRR $= \infty$, $R_{id} = 2 \times 52 = 104$ kΩ, and $R_{ic} \cong (0.52 + 2 \times 201 \times 0.5)//750 = 201//750 = 159$ MΩ (r_μ has negligible impact).

(b) At dc balance we have $V_C = 5 - 30 \times 0.1 = 2$ V. To bring the BJTs to the edge of saturation we need to raise their common emitter voltage to $V_E = V_C - V_{CE(EOS)} = 2 - 0.2 = 1.8$ V. Consequently, $v_{IC(max)} = V_E + V_{BE(on)} = 1.8 + 0.6$ V $= 2.4$ V.

CMRR of the SC Pair

The half-circuit analysis technique applied to bipolar pairs holds for MOS pairs just as well. These circuits are shown in Fig. 4.52, and their similarity to their bipolar counterparts indicates that we can reuse a good deal of already familiar insight and equations. Thus, the half circuit of Fig. 4.52a gives

$$a_{dm} = \frac{v_{od}}{v_{id}} = -g_m(R_D//r_o) \tag{4.105}$$

Likewise, the half circuit of Fig. 4.52b gives, by Eq. (4.39a),

$$a_{cm} = \frac{v_{oc}}{v_{ic}} = -\frac{g_m R_D}{1 + 2(g_m + g_{mb})R_{SS} + (R_D + 2R_{SS})/r_o} \tag{4.106}$$

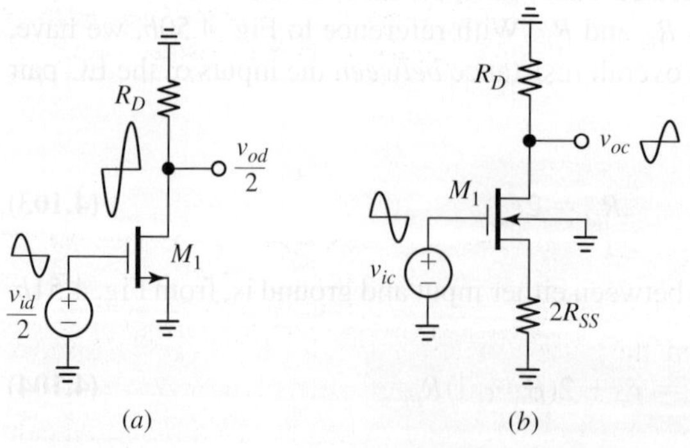

FIGURE 4.52 Half circuits to find the (a) differential-mode gain and (b) common-mode gain of an SC pair.

where g_{mb} is the body transconductance. For *single-ended* utilization, the CMRR is the ratio of *one-half* the gain of Eq. (4.105) to that of Eq. (4.106). If r_o is sufficiently large, we can approximate

$$\text{CMRR} \cong \frac{1}{2} + (g_m + g_{mb})R_{SS} \cong (g_m + g_{mb})R_{SS} \qquad (4.107)$$

For *differential* output utilization we have

$$\text{CMRR} = \infty \qquad (4.108)$$

An advantage of FETs compared to BJTs is that

$$R_{id} = R_{ic} = \infty, \qquad (4.109)$$

(a) In the MOS pair of Fig. 4.45 let $V_{DD} = -V_{EE} = 2.5$ V, $I_{SS} = 0.2$ mA, $R_D = 10$ kΩ, and $R_{SS} = 1$ MΩ. If the FETs have $k = 1.25$ mA/V^2, $V_t = 0.4$ V, $\chi = 0.2$, and $\lambda = 1/(10$ V$)$, find a_{dm}, a_{cm}, and CMRR.

(b) If the inputs are tied together and driven by a common-mode voltage v_{IC}, what is the maximum value of v_{IC} for which the FETs will still operate in saturation?

EXAMPLE 4.19

Solution

(a) We have $I_D \cong I_{SS}/2 = 0.2/2 = 0.1$ mA, $g_m = (2kI_D)^{1/2} = (2 \times 1.25 \times 0.1)^{1/2} = 0.5$ mA/V, $r_o = 1/(\lambda I_D) = 10/0.1 = 100$ kΩ; $a_{dm} = -0.5 \times (10//100) = -4.55$ V/V, $a_{cm} = -0.5 \times 10/[1 + 2(1.2 \times 0.5 \times 1000) + (10 + 2000)/100] = -4.1 \times 10^{-3}$ V/V; single-ended CMRR $= 0.5 \times 1.2 \times 1000 = 600 = 55.6$ dB, differential CMRR $= \infty$.

(b) At dc balance we have $V_D = 2.5 - 10 \times 0.1 = 1.5$ V. To bring the FETs to the edge of saturation we need to raise v_{IC} till $V_{DS} = V_{OV}$, or $V_S = V_D - V_{OV}$. The corresponding input, denoted as $v_{IC(max)}$, is such that $v_{IC(max)} = V_S + (V_t + V_{OV}) = V_D - V_{OV} + V_t + V_{OV} = V_D + V_t = 1.5$ V $+ 0.4 = 1.9$ V.

Effect of Mismatches on the CMRR

With perfectly matched halves, the collector resistors drop identical voltages in Fig. 4.53a, so the collector signals cancel each other out *exactly* to give $a_{cm} = (v_{o1} - v_{o2})/v_{ic} = 0/v_{ic} = 0$, and therefore CMRR $= \infty$, by Eq. (4.98). (Similar considerations hold in the MOS circuit of Fig. 4.53b, so the following analysis applies both to the EC and SC pairs.) In practice, the two halves of a differential pair are likely to be mismatched, even if only slightly, so it is of interest to investigate the impact on the CMRR. The two main factors to be considered are (a) mismatched

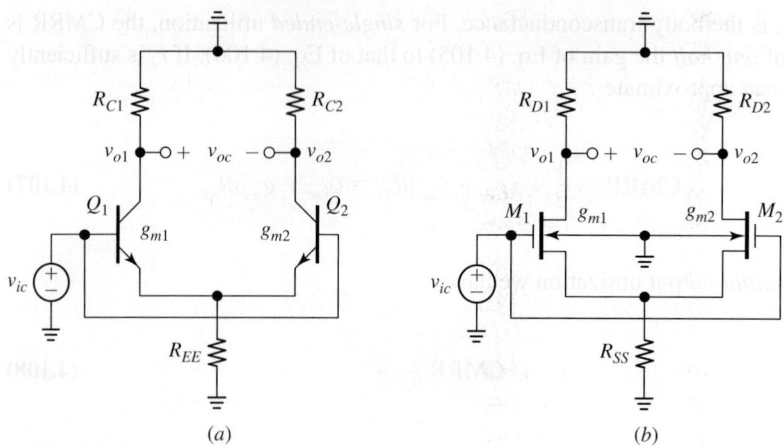

FIGURE 4.53 Ac circuits to investigate the effect of R and g_m mismatches upon the CMRR.

collector resistances R_{C1} and R_{C2}, and (*b*) mismatched BJT transconductances g_{m1} and g_{m2}. We assume that these mismatches are small enough that we can still approximate

$$a_{dm} = \frac{v_{od}}{v_{id}} \cong -g_m R_C \qquad (4.110)$$

with g_m and R_C representing the *average values* of the transconductances and resistances,

$$g_m = \frac{g_{m1} + g_{m2}}{2} \qquad R_C = \frac{R_{C1} + R_{C2}}{2}$$

Moreover, with sufficiently small mismatches in Fig. 4.53a, we can still approximate

$$a_{cm} = \frac{v_{o1} - v_{o2}}{v_{ic}} \cong -\frac{g_{m1} R_{C1} - g_{m2} R_{C2}}{1 + 2g_m R_{EE}} \qquad (4.111)$$

If we introduce the *differences*

$$\Delta g_m = g_{m1} - g_{m2} \qquad \Delta R_C = R_{C1} - R_{C2}$$

then it is readily seen that the individual transconductances and resistances can be expressed as

$$g_{m1} = g_m + \frac{\Delta g_m}{2} \qquad g_{m2} = g_m - \frac{\Delta g_m}{2}$$

$$R_{C1} = R_C + \frac{\Delta R_C}{2} \qquad R_{C2} = R_C - \frac{\Delta R_C}{2}$$

Substituting into Eq. (4.111) gives

$$a_{cm} \cong -\frac{\left(g_m + \dfrac{\Delta g_m}{2}\right)\left(R_C + \dfrac{\Delta R_C}{2}\right) - \left(g_m - \dfrac{\Delta g_m}{2}\right)\left(R_C - \dfrac{\Delta R_C}{2}\right)}{1 + 2g_m R_{EE}}$$

Expanding and ignoring higher-order terms (Δ-term products or squares), we get, after simplifying,

$$a_{cm} \cong -\frac{g_m \Delta R_C + \Delta g_m R_C}{1 + 2g_m R_{EE}} \tag{4.112}$$

Finally, substituting Eqs. (4.112) and (4.110) into Eq. (4.98) gives

$$\text{CMRR} \cong \frac{1 + 2g_m R_{EE}}{\Delta R_C / R_C + \Delta g_m / g_m} \tag{4.113}$$

The above equations refer to the worst-case scenario in which the mismatches conspire to reinforce each other. However, the two causes of mismatch are usually uncorrelated, so a more realistic estimate for the CMRR of the EC pair is found as[1]

$$\text{CMRR}_{(\text{BJT})} \cong \frac{1 + 2g_m R_{EE}}{\sqrt{(\Delta R_C / R_C)^2 + (\Delta g_m / g_m)^2}} \tag{4.114a}$$

We can readily adapt this expression to the SC pair of Fig. 4.53b by writing

$$\text{CMRR}_{(\text{MOS})} \cong \frac{1 + 2(g_m + g_{mb})R_{SS}}{\sqrt{(\Delta R_D / R_D)^2 + (\Delta g_m / g_m)^2}} \tag{4.114b}$$

It is apparent that for given amounts of mismatch, the CMRR is approximately proportional to the equivalent resistance R_{EE} or R_{SS} presented by the biasing current sink. To ensure high CMRR values, this source is typically a high output-resistance source, such as a cascode source or other source types to be investigated in Section 4.8.

EXAMPLE 4.20

(a) Suppose the g_m's in an SC pair are in the range of $100 \pm 10 \ \mu\text{A/V}$, and the drain resistances have tolerances of $\pm 5\%$. If $\chi = 0.15$ and $R_{SS} = 500 \ \text{k}\Omega$, estimate the worst-case scenario CMRR. What if the tolerances are uncorrelated?

(b) Find the value of R_{SS} needed to ensure CMRR $\geq$ 60 dB.

Solution

(a) By inspection, $\Delta R_D / R_D = 0.1$. We also have $g_m + g_{mb} = (1 + \chi)g_m = 115 \ \mu\text{A/V}$, $\Delta g_m = 20 \ \mu\text{A/V}$, and $\Delta g_m / g_m = 20/100 = 0.2$. In the worst-case scenario,

$$\text{CMRR} \cong \frac{1 + 2(115)10^{-6} \times 500 \times 10^3}{0.1 + 0.2} = \frac{116}{0.3} = 387 = 51.7 \ \text{dB}$$

If the mismatches are uncorrelated, then use Eq. (4.114b) and write

$$\text{CMRR} \cong \frac{116}{\sqrt{0.1^2 + 0.2^2}} = \frac{116}{0.224} = 519 = 54.3 \text{ dB}$$

(b) To raise CMRR to 60 dB, or 1,000, we need to increase R_{SS} in proportion, that is, $R_{SS} = (500 \text{ k}\Omega) \times (1000/519) = 964 \text{ k}\Omega$.

4.7 INPUT OFFSET VOLTAGE/CURRENT IN DIFFERENTIAL PAIRS

If we ground both inputs of a differential pair, as depicted in Fig. 4.54 for a bipolar pair, we expect $V_O = 0$. However, because of fabrication process variations, the two halves of the circuit are likely to be somewhat mismatched, resulting in an output error $E_O \neq 0$. We can visualize the effect of mismatches as a shift in the VTC of Fig. 4.44b either to the right or to the left, depending on the direction of the mismatch. If we want to drive the output to zero we need to apply a corrective input voltage that will offset the VTC in the *opposite* direction until it goes through the origin. This corrective voltage is the *input offset voltage V_{OS}* depicted in Fig. 4.54b. By inspection, we find V_{OS} by reflecting the negative of E_O to the input, or

$$V_{OS} = \frac{E_O}{-a_{dm}} \tag{4.115}$$

where a_{dm} is the differential-mode gain of the pair under consideration.

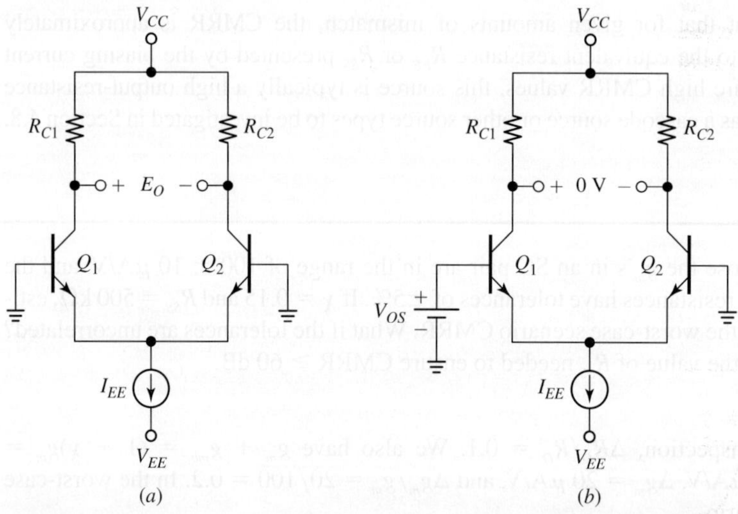

FIGURE 4.54 (a) A practical EC pair with grounded inputs generally gives an output error $E_O \neq 0$. (b) V_{OS} is defined as the input voltage required to null E_O.

Input Offset Voltage of the EC Pair

Two main factors contribute to the V_{OS} of an EC pair: (*a*) mismatched collector resistances R_{C1} and R_{C2}, and (*b*) mismatched saturation currents I_{s1} and I_{s2} of the BJTs. By Eq. (2.11), mismatches between I_{s1} and I_{s2} stem, in turn, from mismatches between the emitter areas A_{E1} and A_{E2} and between the base widths W_{B1} and W_{B2}, as well as differences in the base-region doping densities and in the individual temperatures of the two devices. Moreover, because of the Early effect, W_{B1} and W_{B2} depend on the voltages V_{CE1} and V_{CE2}, so even two BJTs fabricated perfectly identical will exhibit saturation-current mismatches if operated at different V_{CE} values.

- To investigate the effect of mismatched R_{C1} and R_{C2}, assume the BJTs are perfectly matched so that in the circuit of Fig. 4.54*a* we have $I_{C1} = I_{C2} \cong I_{EE}/2$. The contribution to E_O is

$$E_{O1} = V_{CC} - R_{C1}I_{C1} - (V_{CC} - R_{C2}I_{C2}) = -\Delta R_C \frac{I_{EE}}{2}$$

where $\Delta R_C = R_{C1} - R_{C2}$. Dividing by $-a_{dm} \cong g_m R_C$, with R_C representing the mean value of the two resistances, or $R_C = (R_{C1} + R_{C2})/2$, and $g_m = (I_{EE}/2)/V_T$, we get

$$V_{OS1} \cong -V_T \frac{\Delta R_C}{R_C} \qquad (4.116)$$

- To investigate the effect of mismatched I_{s1} and I_{s2}, assume the resistances are now perfectly matched. Since the BJTs of Fig. 4.54*a* experience the same drive V_{BE}, the current I_{EE} must split between Q_1 and Q_2 in proportion to I_{s1} and I_{s2},

$$I_{C1} \cong \frac{I_{EE}}{2}\left(1 + \frac{\Delta I_s}{2I_s}\right) \qquad I_{C2} \cong \frac{I_{EE}}{2}\left(1 - \frac{\Delta I_s}{2I_s}\right)$$

where $\Delta I_s = I_{s1} - I_{s2}$ and $I_s = (I_{s1} + I_{s2})/2$, as usual. The contribution to E_O is now

$$E_{O2} = V_{CC} - R_C I_{C1} - (V_{CC} - R_C I_{C2}) = -R_C \frac{I_{EE}}{2}\frac{\Delta I_s}{I_s}$$

Again dividing by $-a_{dm}$ we get

$$V_{OS2} \cong -V_T \frac{\Delta I_s}{I_s} \qquad (4.117)$$

Note that the negative signs in Eqs. (4.116) and (4.117) are not relevant because a mismatch can occur in either direction, depending on the random variation of the fabrication process. It is thus customary to drop the sign and to express V_{OS} always as a *positive* number. In general the two causes of offset voltage are uncorrelated, so the total offset voltage is usually estimated as[1]

$$V_{OS} \cong V_T \sqrt{\left(\frac{\Delta R_C}{R_C}\right)^2 + \left(\frac{\Delta I_s}{I_s}\right)^2} \qquad (4.118)$$

EXAMPLE 4.21   Suppose the EC pair of Example 4.18 is implemented with resistances afflicted by $\pm 5\%$ tolerances and with BJTs whose saturation currents are afflicted by $\pm 10\%$ tolerances. Estimate V_{OS} and E_O.

Solution

We have $\Delta R_C / R_C = 0.1$ and $\Delta I_s / I_s = 0.2$. Applying Eq. (4.118) gives

$$V_{OS} \cong (26 \text{ mV})\sqrt{(0.1)^2 + (0.2)^2} = 5.8 \text{ mV}$$

Note that V_{OS} may be positive or negative, depending on the direction of the mismatch. $E_O = |a_{dm}|V_{OS} = 111 \times 5.8 = 644 \text{ mV}$.

Input Bias Current and Offset Current of the EC Pair

When an EC pair is driven by sources having nonzero series resistances it is of interest to know the currents into the bases of the BJTs because these currents flow through the source resistances, causing voltage drops that may upset the dc balance of the pair appreciably. With perfectly matched devices, the base currents are $I_{B1} = I_{B2} = I_B$, where

$$I_B = \frac{I_{EE}}{2(\beta_F + 1)} \tag{4.119}$$

However, any mismatches in the β_Fs of the two BJTs will cause mismatches in the base currents. If we define the *input bias current* I_B and the *input offset current* I_{OS} of an EC pair as

$$I_B = \frac{I_{B1} + I_{B2}}{2} \qquad I_{OS} = I_{B1} - I_{B2} \tag{4.120}$$

then a beta mismatch $\Delta\beta_F$ will result in $I_{OS} \neq 0$. The two are related as

$$I_{OS} \cong -I_B \frac{\Delta\beta_F}{\beta_F} \tag{4.121}$$

where $\beta_F = (\beta_{F1} + \beta_{F2})/2$, as usual. For instance, with a 10% beta mismatch, I_{OS} is 10% of I_B.

Input Offset Voltage of the SC Pair

Three main factors contribute to V_{OS} in the case of an SC pair: (*a*) mismatched drain resistances R_{D1} and R_{D2}, (*b*) mismatched device transconductance parameters k_1 and k_2, and (*c*) mismatched threshold voltages V_{t1} and V_{t2}. By Eqs. (3.13) and (3.14), mismatches between k_1 and k_2 stem, in turn, from mismatches between the ratios W_1/L_1 and W_2/L_2 as well as variations in the oxide thickness t_{ox} and temperature gradients from one device to the other. By Eqs. (3.7) through (3.9), mismatches between V_{t1} and V_{t2} are the result of differences in the implant densities, oxide thickness, and temperature of the two devices. Finally, because of the channel-length modulation effect, L_1 and L_2 depend on the voltages V_{DS1} and V_{DS2}, so even two FETs fabricated perfectly identical will exhibit transconductance-parameter mismatches if operated at different V_{DS} values.

- To investigate the effect of mismatched R_{D1} and R_{D2}, proceed as in the bipolar case and write

$$E_{O1} \cong -\Delta R_D \frac{I_{SS}}{2}$$

where $\Delta R_D = R_{D1} - R_{D2}$. Dividing by $-a_{dm} \cong g_m R_D$, with R_D representing the mean value of the two resistances, or $R_D = (R_{D1} + R_{D2})/2$, and with $g_m = 2(I_{SS}/2)/V_{OV} = I_{SS}/V_{OV}$, we get

$$V_{OS1} \cong -\frac{V_{OV}}{2}\frac{\Delta R_D}{R_D} \qquad (4.122)$$

where V_{OV} is the dc-balance overdrive voltage of Eq. (4.81).

- To investigate the effect of mismatched k_1 and k_2 assume the resistances as now perfectly matched and the FETs have identical threshold voltages. Since the FETs of Fig. 4.55a are subject to the same overdrive voltage V_{OV}, the current I_{SS} must split between M_1 and M_2 in proportion to k_1 and k_2,

$$I_{D1} = \frac{I_{SS}}{2}\left(1 + \frac{\Delta k}{2k}\right) \qquad I_{D2} = \frac{I_{SS}}{2}\left(1 - \frac{\Delta k}{2k}\right)$$

where $\Delta k = k_1 - k_2$ and $k = (k_1 + k_2)/2$. Consequently,

$$E_{O2} = V_{DD} - R_D I_{D1} - (V_{DD} - R_D I_{D2}) = -R_D\frac{I_{SS}}{2}\frac{\Delta k}{k}$$

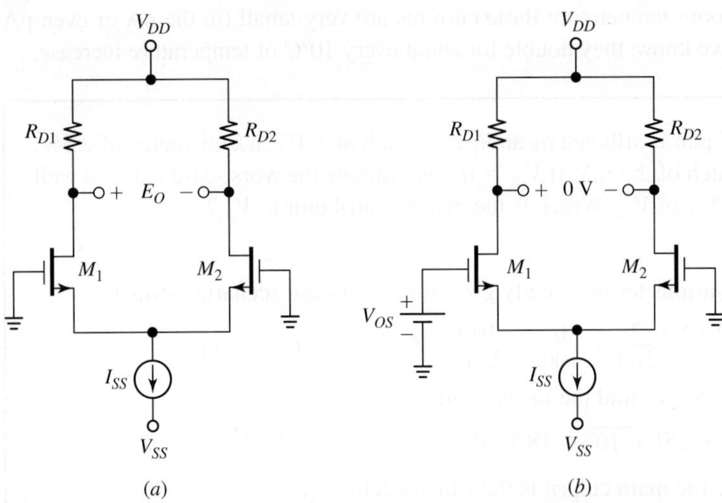

FIGURE 4.55 (a) A practical SC pair with grounded inputs generally gives an output error $E_O \neq 0$. (b) V_{OS} is defined as the input voltage required to null E_O.

Again dividing by $-a_{dm} \cong g_m R_D$, we get

$$V_{OS2} \cong -\frac{V_{ov}}{2}\frac{\Delta k}{k} \tag{4.123}$$

- To investigate the effect of mismatched V_{t1} and V_{t2}, assume all other parameters are matched. Then, to achieve the dc balance of Fig. 4.55b, we must ensure that the FETs experience identical overdrive voltages. This occurs for

$$V_{OS3} = \Delta V_t = V_{t1} - V_{t2} \tag{4.124}$$

Note the similarity of the MOS Eqs. (4.122) and (4.123) to the BJT Eqs. (4.116) and (4.117), except that we now have $V_{ov}/2$ in place of V_T. Since $V_T = 26$ mV whereas $V_{ov}/2$ is typically at least an order of magnitude higher, it is apparent that SC pairs tend to exhibit larger offsets than BJT pairs.

As in the bipolar case, the signs of the various offset components are not relevant. Moreover, the three causes of offset voltage are not correlated, so the total offset voltage is usually estimated as[1]

$$V_{OS} \cong \frac{V_{ov}}{2}\sqrt{\left(\frac{\Delta R_D}{R_D}\right)^2 + \left(\frac{\Delta k}{k}\right)^2 + \left(\frac{\Delta V_t}{0.5V_{ov}}\right)^2} \tag{4.125}$$

Finally, since the gate current of a MOSFET is zero at dc, the input bias current and input offset current are non-issues in SC pairs. When the input terminals are made externally accessible, they are equipped with internal diode clamps to prevent electric discharge that might damage the dielectric of the FETs. In normal operation these diodes are reverse biased, so the gate-terminal currents are those of reverse-biased pn junctions. At room temperature these currents are very small (in the nA or even pA range), but as we know, they double for about every 10°C of temperature increase.

EXAMPLE 4.22   Suppose an SC pair is afflicted by an R_D mismatch of $\pm 1\%$, a k mismatch of $\pm 3\%$, and a V_t mismatch of ± 5 mV. If $V_{OV} = 0.5$ V, estimate the worst-case value as well as the likely value of V_{OS}. Which is the major contributor to V_{OS}?

Solution

Summing the various terms directly gives the worst-case scenario estimate,

$$V_{OS} \cong \frac{500\ \text{mV}}{2}\left(\frac{2}{100} + \frac{6}{100} + \frac{10}{250}\right) = 5 + 15 + 10 = 30\ \text{mV}$$

Using Eq. (4.125), we find the likely estimate

$$V_{OS} \cong \sqrt{5^2 + 15^2 + 10^2} = 18.7\ \text{mV}$$

In this example the main culprit is the k mismatch.

Input Offset Voltage Drift

Like virtually all device and circuit parameters, V_{OS} drifts with temperature, and in low-level signal processing, such as in instrumentation and measurement, it is necessary to know both V_{OS} and its thermal drift.

In the case of **EC pairs**, Eq. (4.118) indicates that V_{OS} is proportional to the thermal voltage V_T, which in turn is proportional to temperature T. Consequently,

$$\frac{dV_{OS}}{dT} = \frac{V_{OS}}{T} \tag{4.126}$$

At room temperature ($T = 300$ K), V_{OS} drifts by $1 \times 10^{-3}/300 = 3.3$ μV/°C for every mV of offset. So, for an EC pair with $V_{OS} = 1.5$ mV, V_{OS} drifts by 5 μV/°C.

The offset drift of **SC pairs** is more complex[1,3] than in the bipolar case. Suffice it to say here that CMOS IC designers continuously strive to reduce both V_{OS} and its drift using clever compensation techniques such as autozero techniques and floating-gate programming.

4.8 CURRENT MIRRORS

Along with differential pairs, current mirrors are the workhorses of analog ICs. A common current mirror application is the generation of stable and predictable dc currents to *bias* other circuits. When used in this capacity, a current mirror is also referred to as a *current reference*. Mirrors are also used to *steer* current signals. As such they find application as active loads in a variety of analog signal processing ICs such as operational amplifiers (op amps), current-feedback amplifiers (CFAs), and operational transconductance amplifiers (OTAs). Current mirroring is made possible by the high degree of matching and thermal tracking of transistors fabricated in close proximity of each other on the same chip.

The function of a current mirror is to accept a current i_I at a *low* (ideally zero) *input-resistance* terminal, and deliver a current i_O ($=i_I$) at a *high* (ideally infinite) *output-resistance* terminal. The current mirror is similar to the current buffer, except that both currents flow into (or out of) the circuit. For this reason, a current mirror is also said to provide *current reversal*. We have already been exposed to current mirrors in the introductory chapters on BJTs and MOSFETs, as well as in Section 4.1. We now wish to examine them in systematic detail.

Basic Bipolar Current Mirrors

Shown in Fig. 4.56a is the basic bipolar current mirror. As the input source i_I is turned on, the diode-connected transistor Q_1 develops a voltage drop v_{BE} related to i_I by the well-known logarithmic law. But Q_2 is subjected to the *same* drive v_{BE} as Q_1, so Q_2 will just *mirror* the current of Q_1. We wish to find a precise relationship between i_O and i_I, as well as the small-signal input and output resistances R_i and R_o. We also wish to display the *i-v* characteristic of the output port.

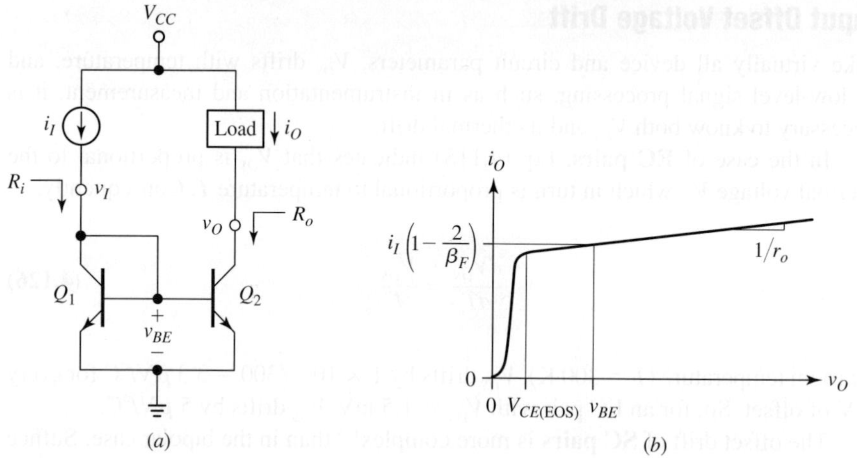

FIGURE 4.56 (a) Basic BJT current mirror, and (b) output-port i-v characteristic.

For a detailed analysis refer to Fig. 4.57a and assume $V_A = \infty$ for simplicity. Since the BJTs are matched and experience the same v_{BE} drop, they must draw identical currents, here denoted as i_C. Also, together they draw a total base current of $2i_C/\beta_F$. Consequently, KCL gives $i_I = i_C + 2i_C/\beta_F = i_C(1 + 2/\beta_F)$. Substituting $i_C = i_O$ and solving for i_O gives

$$i_O = \frac{i_I}{1 + 2/\beta_F} \cong i_I\left(1 - \frac{2}{\beta_F}\right) \tag{4.127}$$

Because of the finite current gain β_F, i_O does not mirror i_I *exactly*, but is afflicted by a small systematic error $\varepsilon = -2/\beta_F$. For instance, with $\beta_F = 100$, we have $\varepsilon = -2\%$.

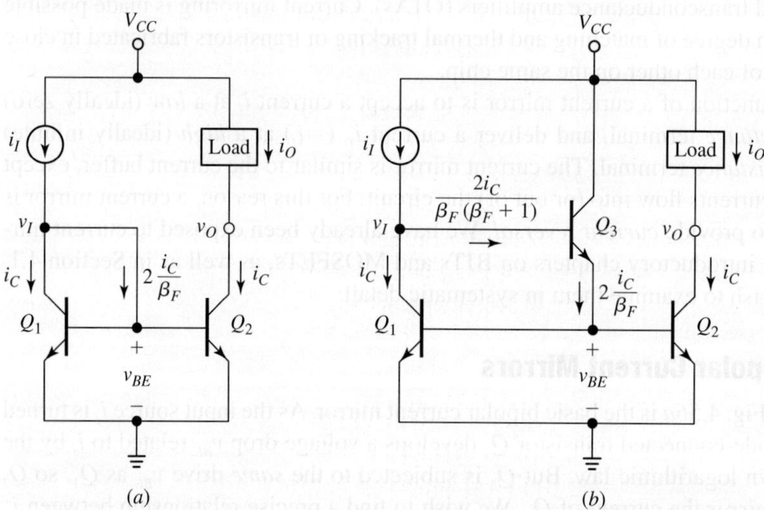

FIGURE 4.57 (a) Currents in (a) the basic BJT current mirror, and (b) basic mirror with beta helper.

Note that Eq. (4.127) holds in dc as well as in ac form, so we have made no small-signal approximations.

The above equation holds only as long as $v_{CE2} = v_{CE1}$, a condition that we shall refer to as *dc balance* for the two BJTs. In the present case, this occurs for $v_O = v_I = v_{BE}$. If v_O is raised above this value, i_O will increase due to the Early effect. To account for this, we need to modify Eq. (4.127) as

$$i_O \cong i_I\left(1 - \frac{2}{\beta_F}\right) \times \left(1 + \frac{v_O - v_{BE}}{V_A}\right) \qquad (4.128)$$

The i_O-v_O characteristic is shown in Fig. 4.56b, along with the value of i_O at dc balance and the corresponding value of v_O. Note that Eq. (4.128) holds so long as $v_O \geq v_{O(\min)}$, where

$$v_{O(\min)} = V_{CE(EOS)} \; (\cong 0.2 \text{ V}) \qquad (4.129)$$

If v_O drops below this limit, Q_2 will enter the saturation region, where we witness a rapid decrease in i_O. By inspection we also have $R_i = r_{o1} // r_{e1} // r_{\pi 2} \cong r_{e1} \cong 1/g_{m1}$, and $R_o = r_{o2}$. Dropping subscripts 1 and 2 as the BJTs are assumed matched and are also biased identically, we thus have

$$R_i \cong \frac{1}{g_m} \qquad R_o = r_o \qquad (4.130)$$

As we know, the i_O-v_O curve has a slope of $1/r_o$, and its extrapolation intercepts the horizontal axis at $v_O = -V_A$, where V_A is the Early voltage.

The above analysis stipulates perfect matching between Q_1 and Q_2. At times the transistors are deliberately fabricated with unequal emitter areas in order to provide current amplification or attenuation, depending on the case. For instance, if Q_2's emitter area is made *twice* as large as Q_1's, then Q_2 will draw twice as much current as Q_1, giving $i_O \cong 2i_I$. Denoting the saturation currents of the two BJTs as I_{s1} and I_{s2}, respectively, we can readily generalize Eq. (4.128) as

$$i_O \cong i_I\left(\frac{I_{s2}}{I_{s1}}\right) \times \left(1 - \frac{1 + I_{s2}/I_{s1}}{\beta_F}\right) \times \left(1 + \frac{v_O - v_{BE}}{V_A}\right) \qquad (4.131)$$

In order to reduce the error stemming from the fact that β_F is not infinite, a third transistor Q_3 is often added as in Fig. 4.57b. Aptly referred to as *beta helper*, Q_3 reduces the current component being subtracted from i_I by a factor of $\beta_F + 1$, in effect changing the dc balance value of Eq. (4.127) to

$$i_O \cong i_I\left(1 - \frac{2}{\beta_F^2}\right) \qquad (4.132)$$

Note that with this modification v_I is raised to $2v_{BE}$, so the dc balance condition is now $v_O = 2v_{BE}$. The ac resistance seen by the input source also doubles to $R_i \cong 2/g_m$ (see Problem 4.71). Beta helpers find application especially in multiple-output current mirrors.

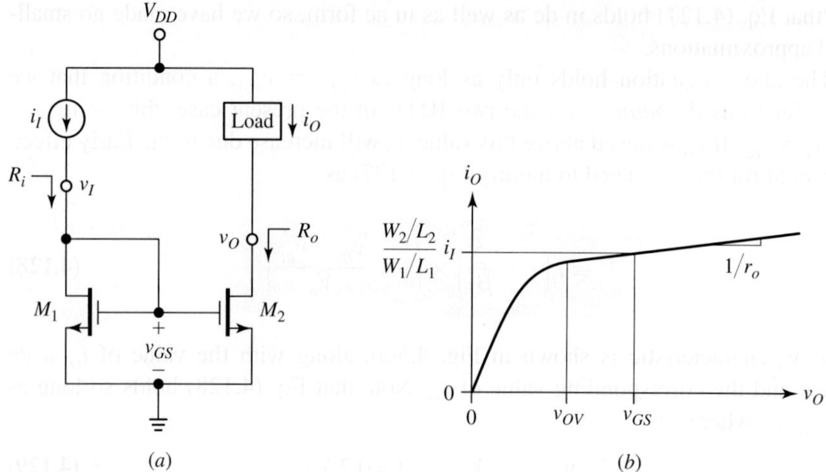

FIGURE 4.58 (a) Basic MOSFET current mirror, and (b) output-port *i-v* characteristic.

The above discussion has focused on current mirrors using *npn*-type BJTs, whose collectors *sink* current. If the application calls for the current mirror to *source* current, then the circuit is implemented with *pnp*-type BJTs, as already discussed in Chapter 2. Since IC *pnp* BJTs exhibit notoriously low betas, *pnp* mirrors often use beta helpers to reduce their inherently higher systematic error.

Basic MOSFET Current Mirror

The MOS counterpart of the BJT mirror is shown in Fig. 4.58*a*. Thanks to the fact that gate currents are zero, a MOSFET current mirror does not suffer from the systematic error due to β_F. As the input source i_I is turned on, the diode-connected transistor M_1 responds with the voltage drop $v_{GS} = V_t + v_{OV}$, where v_{OV} is the overdrive voltage necessary to sustain i_I. The incoming current and v_{GS} are related as

$$i_I = \frac{k_1}{2}(v_{GS} - V_t)^2(1 + \lambda v_{GS})$$

But, M_2 is subjected to the same drive v_{GS} as M_1, so M_2 will draw the current

$$i_O = \frac{k_2}{2}(v_{GS} - V_t)^2(1 + \lambda v_O)$$

As we know, the device trasconductance parameter of a MOSFET is $k = k'(W/L)$, where W and L are the channel width and length of the particular FET, and k' is the process transconductance parameter, common to all same-type FETs on the chip. Taking the ratio i_O/i_I and simplifying, we get, under the assumption $\lambda v_{GS} \ll 1$,

$$i_O \cong i_I \frac{W_2/L_2}{W_1/L_1} \times \left[1 + \lambda(v_O - v_{GS})\right] \tag{4.133}$$

The i_O-v_O characteristic is shown in Fig. 4.58*b*, along with the value of i_O at dc balance, a condition now expressed as $v_O = v_I = v_{GS}$. Note that Eq. (4.133) holds so long as $v_O \geq v_{O(min)}$, where

$$v_{O(min)} = v_{OV} \qquad (4.134)$$

If v_O drops below this limit, M_2 will enter the triode region where i_O eventually drops to zero. By inspection we also have $R_i = r_{o1}//(1/g_{m1}) \cong 1/g_{m1}$ and $R_o = r_{o2} = 1/(\lambda i_O)$. Dropping subscripts 1 and 2 as the FETs are assumed matched and are also biased identically, we thus have

$$R_i \cong \frac{1}{g_m} \qquad R_o = r_o \qquad (4.135)$$

As we know, the slope of the *i-v* curve in Fig. 4.58*b* is $1/r_o$. Moreover, its extrapolation intercepts the horizontal axis at $v_O = -1/\lambda$. If the W/L ratios of the two FETs are equal, then $i_O = i_I$ at dc balance.

Cascode Current Mirrors

According to Eqs. (4.130) and (4.135), the output resistance of basic current mirrors is r_o. There are many situations requiring a much higher output resistance, and the cascoding techniques of Section 4.4 provide a popular way to raise output resistance significantly.

Figure 4.59*a* shows a **bipolar cascode mirror**. The matched BJT pair Q_3-Q_4 provides the mirror action proper, whereas the CB BJT Q_2 raises the output resistance way above r_o. The function of diode-connected Q_1 is to bias Q_2's base a v_{BE} drop above Q_4's

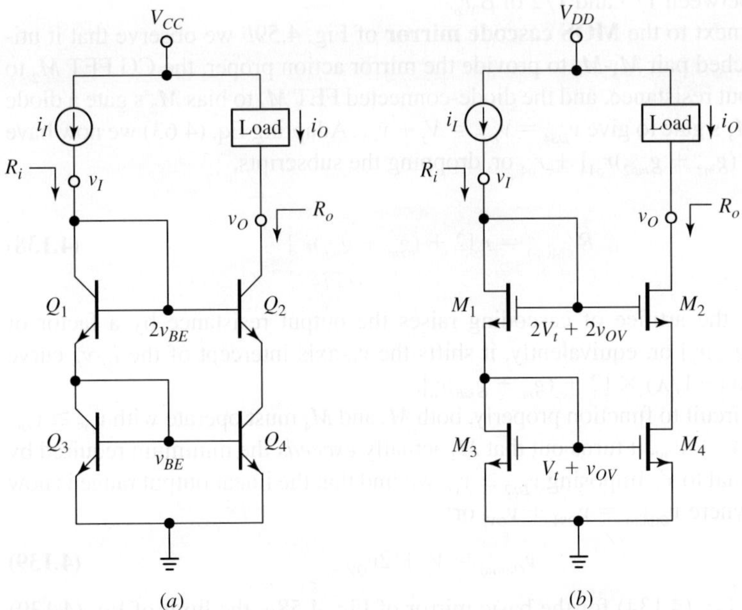

FIGURE 4.59 Cascode current mirrors: (*a*) bipolar and (*b*) MOS.

base to give $v_{CE4} = v_{CE3} = v_{BE}$. Compared to the basic mirror, the input parameters are doubled to $v_I = 2v_{BE}$ and $R_i \cong 2/g_m$. Also, the lower limit of the linear range of operation is now raised by one v_{BE} drop to $v_{O(min)} = v_{CE4} + V_{CE2(EOS)} = v_{BE} + V_{CE2(EOS)}$. It is left as an exercise for the student (see Problem 4.72) to prove that $R_i \cong 2/g_m$, $R_o \cong (\beta_0/2)r_o$, and

$$i_O \cong i_I\left(1 - \frac{4}{\beta_F}\right) \times \left(1 + \frac{v_O - 2v_{BE}}{(\beta_0/2)V_A}\right) \qquad (4.136)$$

We can visualize the effect of cascoding as *shifting* the extrapolated intercept of the i_O-v_O curve with the horizontal axis from $-V_A$ to about $-(\beta_0/2)V_A$, thus making the i_O-v_O curve *much flatter*.

When the output resistance of a collector terminal is raised significantly above r_o as in the present case, the base-collector resistance r_μ may no longer be negligible. As we know, r_μ models the effect of base-width modulation by v_{CE} upon the recombination current in the base, and is expressed as $r_\mu = m\beta_0 r_o$, where $1/m$ $(m \geq 1)$ represents the fraction of the total base current due to recombination. A more accurate expression for the output resistance is thus

$$R_{o(BJT)} \cong \left(\frac{\beta_0}{2}r_o\right)//r_\mu = \frac{m}{1 + 2m}\beta_0 r_o \qquad (4.137)$$

In the worst-case scenario of the base current being predominantly of the recombination type $(m \to 1)$, we get $R_o \to (\beta_0/3)r_o$. In a practical cascode mirror R_o will lie somewhere between $1/3$ and $1/2$ of $\beta_0 r_o$.

Turning next to the **MOS cascode mirror** of Fig. 4.59*b* we observe that it utilizes the matched pair M_3-M_4 to provide the mirror action proper, the CG FET M_2 to raise the output resistance, and the diode-connected FET M_1 to bias M_2's gate a diode drop above M_4's gate to give $v_{DS4} = v_{DS3} = V_t + v_{OV}$. Adapting Eq. (4.63) we now have $R_o = r_{o2}[1 + (g_{m2} + g_{mb2})r_{o4}] + r_{o4}$ or, dropping the subscripts,

$$R_{o(MOS)} = r_o[2 + (g_m + g_{mb})r_o] \qquad (4.138)$$

As expected, the artifice of cascoding raises the output resistance by a factor of $[2 + (g_m + g_{mb})r_o]$ or, equivalently, it shifts the v_O-axis intercept of the i_O-v_O curve from $-1/\lambda$ to $(-1/\lambda) \times [2 + (g_m + g_{mb})r_o]$.

For the circuit to function properly, both M_2 and M_4 must operate with $v_{DS} \geq v_{OV}$. Since $v_{DS4} = V_t + v_{OV}$, it turns out that M_4 actually *exceeds* the minimum required by an amount equal to V_t. Imposing $v_{DS2} \geq v_{OV}$ we find that the linear output range is now $v_O \geq v_{O(min)}$, where $v_{O(min)} = v_{DS4} + v_{OV}$, or

$$v_{O(min)} = V_t + 2v_{OV} \qquad (4.139)$$

Compared to Eq. (4.134) for the basic mirror of Fig. 4.58*a*, the limit of Eq. (4.139) may prove too high in low-voltage applications where even fractions of a volt matter.

While we can make the $2v_{OV}$ term as small as needed by fabricating the FETs with suitably large W/L ratios, the V_t term poses the ultimate limit on $v_{O(min)}$.

Wide-Swing MOS Cascode Mirrors

Wide-swing cascode mirrors eliminate the V_t term from Eq. (4.139) by *downshifting* M_2's bias from $v_{G2} = 2V_t + 2v_{OV}$ of Fig. 4.59b to $v_{G2} = 1V_t + 2v_{OV}$ in order to bring M_4 right to the edge of saturation, where $v_{DS4} = v_{OV}$. This results in

$$v_{O(min)} = 2v_{OV} \qquad\qquad (4.140)$$

In the modified cascode mirror of Fig. 4.60a this downshift is provided by the source follower M_5. (M_5 is biased by M_6, in turn mirroring M_3.) To yield $v_{S5} = V_t + 2v_{OV}$, M_5 requires $v_{G5} = v_{S5} + v_{GS5} = (V_t + 2v_{OV}) + (V_t + v_{OV}) = 2V_t + 3v_{OV}$. Compared to Fig. 4.59$b$, where $v_I = 2V_t + 2v_{OV}$, we now need $v_I = 2V_t + 3v_{OV}$, or $1v_{OV}$ higher. We achieve this by fabricating M_1 with a W/L ratio that is $1/4$ that of all other FETs so that, by virtue of the relation $v_{OV} = \sqrt{(2/k)i_D}$, M_1 will require an overdrive voltage of $2v_{OV}$ to sustain the same current i_D that all other FETs are sustaining with only $1v_{OV}$.

A drawback of the circuit of Fig. 4.60a is that it requires an additional branch (M_5-M_6) to provide level shifting. The two branches are cleverly combined into one in the circuit of Fig. 4.60b, called the *Sooch cascode current mirror* for its

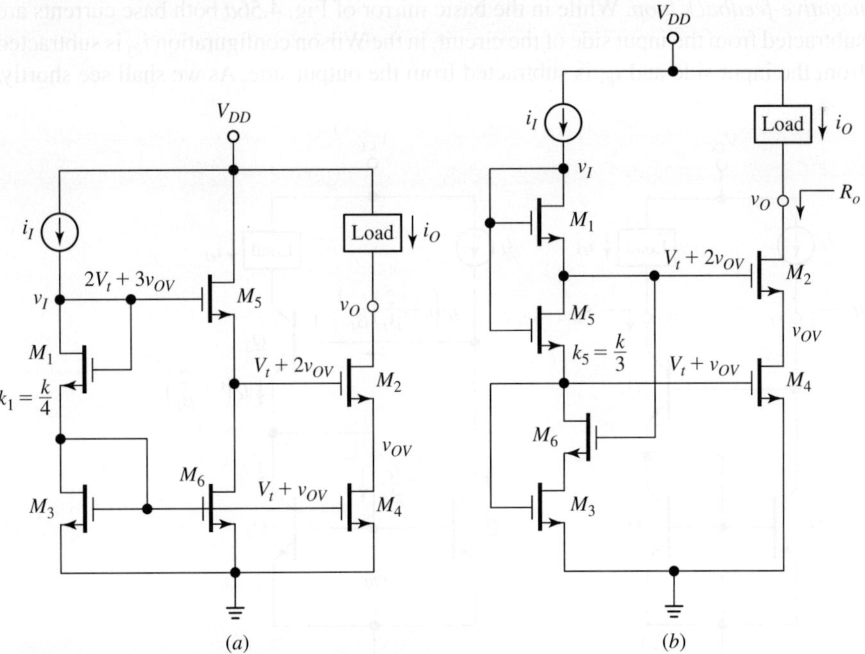

FIGURE 4.60 Wide-swing cascode mirrors. (a) Using source follower M_5 to downshift M_2's bias by V_t (note that the W/L ratio of M_1 is ¼ that of all other FETs). (b) The Sooch cascode current mirror.

inventor N. S. Sooch. Though the details of its analysis are left as an exercise for the student (see Problem 4.73), suffice it to list here its principal features, which are:

- The M_6-M_3 pair synthesizes the voltage $v_{G4} = V_t + v_{OV}$ needed to bias M_4.
- The M_1-M_5 pair synthesizes the voltage drop $v_{DS5} = v_{OV}$ needed to bias M_2 at $v_{G2} = V_t + 2v_{OV}$, that is, $1v_{OV}$ higher than v_{G4}. As discussed in Problem 4.73, we achieve this by fabricating M_5 with a W/L ratio that is $1/3$ that of the other FETs.
- M_6 is designed to drop $v_{DS6} = V_t$ and thus force M_3 to operate at $v_{DS3} = v_{OV} = v_{DS4}$. This eliminates any channel-length modulation differences between M_3 and M_4 and thus results in perfectly matched currents (so long as the W/L ratios of M_3 and M_4 are matched).

In the above analyses we have neglected the body effect for simplicity. In practice, all FETs with source voltages different from the body voltage will exhibit slightly higher values of V_t. The IC designer can compensate for threshold shifts by suitably adjusting the W/L ratios when needed.

Wilson Current Mirror

The Wilson current mirror, shown in Fig. 4.61a, was developed to improve the characteristics of the basic bipolar current mirror. As the input source is turned on, i_I will initially flow into Q_3's base, turning on Q_3 as well as the diode-connected transistor Q_2. The current through Q_2 is then mirrored by Q_1 back to the input node, thus closing a *negative-feedback loop*. While in the basic mirror of Fig. 4.56a both base currents are subtracted from the input side of the circuit, in the Wilson configuration i_{B3} is subtracted from the input side and i_{B1} is subtracted from the output side. As we shall see shortly,

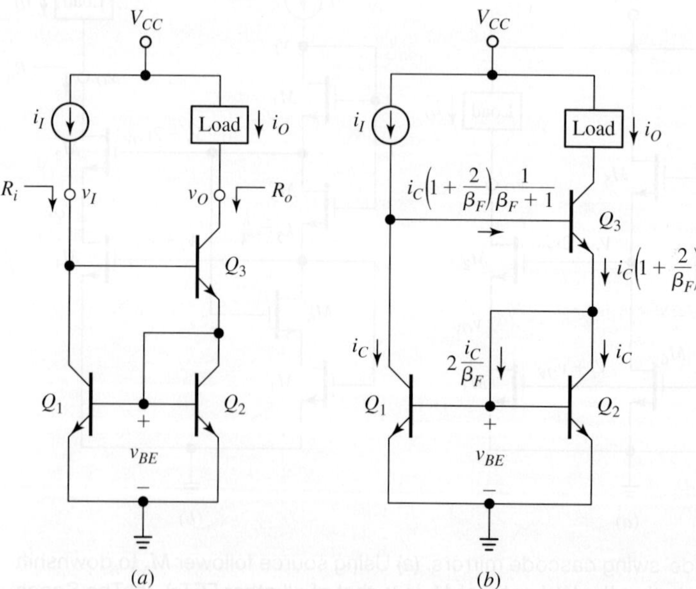

(a)   (b)

FIGURE 4.61 (a) Wilson current mirror, and (b) its various current components.

this form of cancellation reduces the output current error to a level comparable to that of the beta helper, provided the BJTs have matched betas. Moreover, the presence of negative feedback raises the output resistance dramatically.

To find a relationship between i_o and i_I, refer to Fig. 4.61b. Assuming $V_A = \infty$ for simplicity and starting out at the bottom, we observe that Q_1 and Q_2 are subject to the same drive v_{BE}, so they draw identical currents, here denoted as i_C. Moving upward, we repeatedly apply KCL as well as the forward-region current relationships of BJTs to end up with the relationships

$$i_I = i_C\left[1 + \left(1 + \frac{2}{\beta_F}\right)\frac{1}{\beta_F + 1}\right] \qquad i_O = \beta_F i_C\left(1 + \frac{2}{\beta_F}\right)\frac{1}{\beta_F + 1}$$

Eliminating i_C we find, after minor algebra,

$$i_o = i_I\frac{1}{1 + \dfrac{2}{\beta_F(\beta_F + 2)}} \cong i_I\frac{1}{1 + \dfrac{2}{\beta_F^2}} \cong i_I\left(1 - \frac{2}{\beta_F^2}\right) \qquad (4.141)$$

For instance, with $\beta_F = 100$ the error is $\varepsilon = -0.02\%$, which is truly negligible. We observe that the voltage at the input node is now $2v_{BE}$ and that the circuit will work properly so long as $v_O \geq v_{O(\min)}$, where

$$v_{O(\min)} = v_{BE} + V_{CE3(EOS)} \; (\cong 0.9 \text{ V}) \qquad (4.142)$$

To find the output resistance R_o we replace the circuit with its small-signal equivalent and use the test-voltage method of Fig. 4.62a. Here, $r_{\pi1}$ and r_{o2} have been lumped together with r_{e2}, the dynamic resistance of diode-connected transistor Q_2. Moreover, since Q_1 mirrors the current of Q_2, we model it with a unity-gain controlled source $1i_2$. Since $r_{o2} \gg r_{\pi1} \gg r_{e2}$, we approximate $r_{\pi1}//r_{e2}//r_{o2} \cong r_{e2} = \alpha_{02}/g_{m2} \cong 1/g_m$, as shown in Fig. 4.62b. In fact, it turns out that we can also ignore r_{o1} in the course of our calculations. To see why, we apply KCL at the upper-left node, along with KVL and Ohm's law and get

$$i_{b3} + 1i_2 + \frac{r_{e2}i_2 + r_{\pi3}i_{b3}}{r_{o1}} = 0 \Rightarrow i_{b3}\left(1 + \frac{r_{\pi3}}{r_{o1}}\right) + i_2\left(1 + \frac{r_{e2}}{r_{o1}}\right) = 0$$

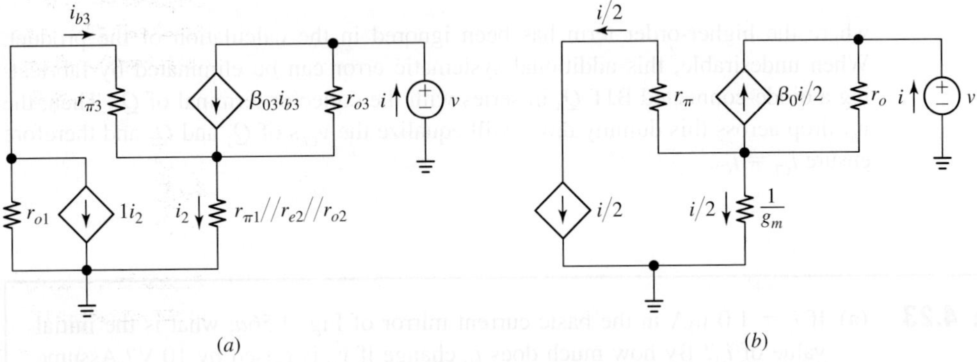

FIGURE 4.62 (a) Small-signal model of the Wilson current mirror, and (b) its simplified version.

Given that $r_{e2} \ll r_{\pi3} \ll r_{o1}$, we can ignore r_{o1} and write $i_{b3} + i_2 \cong 0$, or $i_{b3} = -i_2$. This means that the ac current i_{b3} is actually flowing *out* of Q_3's base, and that i_{b3} coincides with the current $1i_2$ drawn by the dependent source modeling Q_1. The ac equivalent of Fig. 4.62a simplifies as in Fig. 4.62b, where we have exploited the fact that $i = 1i_2 + i_2$, or $i_2 = i/2$. Applying Kirchoff's laws and Ohm's law gives

$$i + \beta_0 \frac{i}{2} = \frac{v - (1/g_m)i/2}{r_o} \Rightarrow i\left(1 + \frac{\beta_0}{2} + \frac{1}{2g_m r_o}\right) = \frac{v}{r_o}$$

But, $1/(2g_m r_o) \ll 1$, so we finally get

$$R_o = \frac{v}{i} \cong \left(1 + \frac{\beta_0}{2}\right)r_o \cong \frac{\beta_0}{2}r_o \tag{4.143a}$$

It is left as an exercise for the student (see Problem 4.71) to prove that $R_i \cong 2/g_m$ in Fig. 4.61a. Compared to the basic-mirror characteristic of Fig. 4.56b, the Wilson mirror yields a much flatter curve, but only down to $v_O = v_{BE} + V_{CE3(EOS)}$. Alternatively, we can say that the intercept of the Wilson's *i-v* curve with the horizontal axis is shifted from $-V_A$ to $-(\beta_0/2)V_A$. This great improvement is the result of the negative-feedback action provided by Q_1, a subject we shall return to in Chapter 7. As in the cascode realization, the output resistance is raised significantly above r_o, so r_μ may no longer be negligible. As in the cascode case, a better estimate for R_o is then

$$R_o \cong \left(\frac{\beta_0}{2}r_o\right)//r_\mu = \frac{m}{1 + 2m}\beta_0 r_o \tag{4.143b}$$

Finally, it must be said that the calculation of the various currents in Fig. 4.62b postulates identical currents for Q_1 and Q_2, when in fact the two BJTs are operating at *different* values of v_{CE}, namely, $v_{CE2} = v_{BE}$ and $v_{CE1} = 2v_{BE}$. Consequently, $i_{C1} \cong i_{C2}(1 + v_{BE}/V_A)$. This difference results in a *systematic error* for i_O. To account for this error, typically around 1%, we need to refine the initial value of Eq. (4.141) as

$$i_O \cong i_I\left(1 - \frac{2}{\beta_F^2}\right) \times \left(1 - \frac{v_{BE}}{V_A}\right) \cong i_I\left(1 - \frac{2}{\beta_F^2} - \frac{v_{BE}}{V_A}\right) \tag{4.144}$$

where the higher-order term has been ignored in the calculation of the product. When undesirable, this additional systematic error can be eliminated by fabricating a diode-connected BJT Q_4 in series with the collector terminal of Q_1. Then, the v_{BE} drop across this dummy diode will equalize the v_{CE}s of Q_1 and Q_2 and therefore ensure $i_{C1} = i_{C2}$.

EXAMPLE 4.23   (a) If $i_I = 1.0$ mA in the basic current mirror of Fig. 4.56a, what is the initial value of i_O? By how much does i_O change if v_O is raised by 10 V? Assume matched BJTs with $\beta_0 = 100$ and $V_A = 80$ V.

(b) Repeat, but for the Wilson mirror of Fig. 4.61a. Compare and comment.

Solution

(a) By Eq. (4.127), $i_O = 1.0(1 - 2/100) = 0.98$ mA. Also, $R_o = r_o = 80/0.98 = 81.6$ kΩ. Consequently, $\Delta i_O = \Delta v_O/R_o = 10/81.6 = 0.1225$ mA, indicating that i_O will increase to $0.98 + 0.1225 = 1.1025$ mA.

(b) By Eq. (4.144), $i_O = 1.0(1 - 2/100^2 - 0.7/80) = 0.9910$ mA. Moreover, $R_o = (\beta_0/2)r_o = (100/2)80.7 = 4.04$ MΩ, so $\Delta i_O = \Delta v_O/R_o = 10/4.04 = 2.5$ μA, indicating that i_O will increase to $0.9910 + 0.0025 = 0.9935$ mA. The Wilson source is superior both in terms of the initial error and the current variation with voltage.

Widlar Current Source/Sink

In low-current dc biasing the need often arises for a current mirror capable of giving $I_O \ll I_I$. Though this can, in principle, be achieved by fabricating Q_2 in the basic mirror of Fig. 4.56a with a much smaller emitter area than Q_1, a more viable alternative is to place a resistance R in series with Q_2's emitter terminal to suitably reduce its V_{BE} drop and hence lower the output current I_O. The result is the modified circuit of Fig. 4.63a, known as the *Widlar current source* for its inventor B. Widlar. (More properly, the circuit shown ought to be called Widlar current *sink* as it utilizes *npn* BJTs, with the designation Widlar current *source* reserved for its *pnp* version.) A side benefit of this circuit is that R introduces emitter degeneration and therefore raises the output resistance to $R_o \cong r_{o2}[1 + g_{m2}(r_{\pi2}//R)]$.

To investigate circuit behavior, neglect base currents and apply Ohm's law and KVL to write $RI_O = V_{BE1} - V_{BE2} = V_T \ln(I_I/I_s) - V_T \ln(I_O/I_s)$, or

$$RI_O = V_T \ln\frac{I_I}{I_O} \tag{4.145}$$

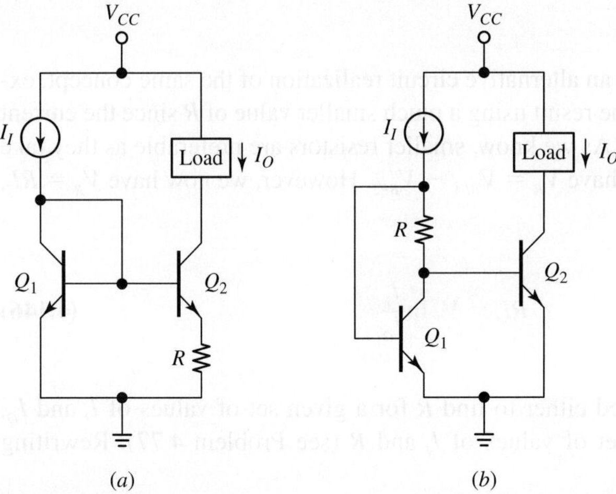

FIGURE 4.63 (a) Widlar current sink. (b) An alternative implementation of the same concept.

where V_T is the familiar thermal voltage ($V_T = 26$ mV at room temperature). Two issues arise in connection with the Widlar circuit: find R to achieve a specified I_O/I_I ratio; or, given R, find I_O/I_I.

EXAMPLE 4.24   (a) Find R so that the Widlar circuit of Fig. 4.63a gives $I_O = 30\ \mu$A for $I_I = 0.5$ mA. Assuming $\beta_0 = 100$ and $V_A = 60$ V, find the source's output resistance as seen by the load.

(b) Find I_O if $I_I = 1.0$ mA and $R = 5$ kΩ.

Solution

(a) By Eq. (4.145),

$$R = \frac{V_T}{I_O}\ln\frac{I_I}{I_O} = \frac{26 \times 10^{-3}}{30 \times 10^{-6}}\ln\frac{0.5 \times 10^{-3}}{30 \times 10^{-6}} = 2.44\ \text{k}\Omega$$

We have $g_m = 1/(0.87\ \text{k}\Omega)$, $r_\pi = 87$ kΩ, and $r_o = 2$ MΩ. Due to the degeneration introduced by R, we get

$$R_o = r_{o2}[1 + g_{m2}(r_{\pi2}/\!/R)] = 2[1 + (87/\!/2.44)/0.87] = 7.5\ \text{M}\Omega$$

(b) Using again Eq. (4.145) we get

$$I_O = \frac{V_T}{R}\ln\frac{I_I}{I_O} = \frac{26 \times 10^{-3}}{5 \times 10^3}\ln\frac{10^{-3}}{I_O} = 5.2 \times 10^{-6}\ln\frac{10^{-3}}{I_O}$$

This transcendental equation is solved via iterations. We expect $I_O \ll I_I$, so start out with an educated guess, say, $I_{O(0)} = 10\ \mu$A, and plug it in the right-hand side to obtain the new estimate $I_{O(1)} = 24\ \mu$A. Iterate by plugging this new estimate in the right-hand side and get $I_{O(2)} = 19.4\ \mu$A. After a few more iterations the result settles at $I_O = 20.3\ \mu$A.

Shown in Fig. 4.63b is an alternative circuit realization of the same concept, except that it achieves the same result using a much smaller value of R since the current through R is now I_I ($\gg I_O$). (As we know, smaller resistors are preferable as they take up less chip area.) We still have $V_R = V_{BE1} - V_{BE2}$. However, we now have $V_R = RI_I$, so Eq. (4.145) becomes

$$RI_I = V_T\ln\frac{I_I}{I_O} \tag{4.146}$$

This expression can be used either to find R for a given set of values of I_I and I_O, or to find I_O for a given set of values of I_I and R (see Problem 4.77). Rewriting Eq. (4.146) as

$$I_O = I_I e^{-RI_I/V_T}$$

we observe that for *small* values of I_I the exponential term tends to unity, indicating that I_O *increases* approximately in proportion to I_I. On the other hand, for *large* values of I_I, the exponential term dominates, causing I_O to *decrease* with I_I. It is apparent that I_O must peak at some intermediate value of I_I (see Problem 4.78), this being the reason why the circuit of Fig. 4.63b is also called a *peaking current source* (strictly speaking, the designation peaking current *sink* would be more appropriate for the present case of *npn* BJTs, and peaking current *source* for the case of *pnp* BJTs).

4.9 DIFFERENTIAL PAIRS WITH ACTIVE LOADS

The most common application of the differential pair is as input stage to operational amplifiers and voltage comparators, where the two most critical requirements are (*a*) a *high differential gain* a_{dm} and (*b*) a *high common-mode rejection ratio* (CMRR). The circuits of Fig. 4.64 maximize both parameters by taking advantage of current mirrors.

Let us address the **common-mode rejection ratio** (CMRR) first. Previous analysis has shown that to ensure a high CMRR, the biasing circuitry must present a high resistance to the differential pair (high R_{EE} for EC pairs, high R_{SS} for SC pairs). In both circuits of Fig. 4.64 this constraint is met by using a current mirror that accepts

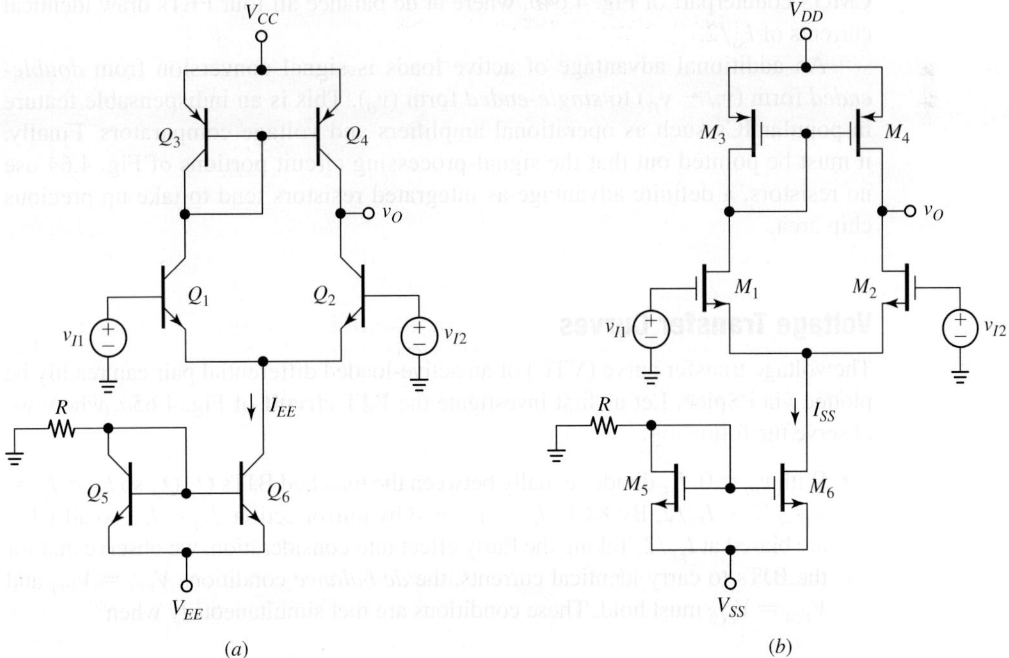

(a)   (b)

FIGURE 4.64 Differential pairs with active loads and current-source biasing: (*a*) BJT, and (*b*) CMOS.

the reference current established by R and mirrors it to the differential pair at a *high output* resistance (in this case the resistance r_{o6} of the mirror transistor). If desired, we can raise this resistance further by using a Wilson current mirror or a cascode current mirror. When used to provide a biasing function as in the present case, a current mirror is called a *current reference*.

Next, let us address the **differential-mode gain** a_{dm}. Equations (4.88) and (4.93) provide estimates for the gains achievable with *resistive-loaded* differential pairs, namely,

$$a_{dm(\text{BJT})} \cong -\frac{R_C(I_{EE}/2)}{V_T} \qquad a_{dm(\text{FET})} \cong -\frac{R_D(I_{SS}/2)}{0.5V_{OV}}$$

In both cases gain is proportional to the dc voltage dropped across the load resistance R_C/R_D. If a higher gain is desired for given biasing conditions, R_C/R_D will have to be increased. This, however, risks driving the transistors in saturation. We resolve this impasse by replacing resistive loads with *active loads*, as already mentioned in Section 4.1. In Fig. 4.64, Q_4/M_4 acts as the load to Q_2/M_2, so the role of R_C/R_D in the above gain expressions is now played by the usually *much larger* output resistance r_{o4} of Q_4/M_4.

For proper operation, the load Q_4/M_4 must be biased at the same current as Q_2/M_2. The circuit of Fig. 4.64a uses Q_1 to *mimic* the dc current $\alpha_F I_{EE}/2$ ($\cong I_{EE}/2$ drawn by its matched companion Q_2. This current is then fed to Q_3, which in turn forces its matched companion Q_4 to *mirror* it back to Q_2. So, at dc balance, all four BJTs draw identical currents, namely, $I_{EE}/2$! Similar considerations hold for the CMOS counterpart of Fig. 4.64b, where at dc balance all four FETs draw identical currents of $I_{SS}/2$.

An additional advantage of active loads is signal conversion from *double-ended* form ($v_{I1} - v_{I2}$) to *single-ended* form (v_O). This is an indispensable feature in popular ICs such as operational amplifiers and voltage comparators. Finally, it must be pointed out that the signal-processing circuit portions of Fig. 4.64 use no resistors, a definite advantage as integrated resistors tend to take up precious chip area.

Voltage Transfer Curves

The voltage transfer curve (VTC) of an active-loaded differential pair can readily be plotted via PSpice. Let us first investigate the **BJT circuit** of Fig. 4.65a, where we observe the following:

- With $v_{ID} = 0$, I_{EE} divides equally between the matched BJTs Q_1-Q_2, so $I_{C1} = I_{C2} = \alpha_F I_{EE}/2 \cong I_{EE}/2$. By KCL, $I_{C3} = I_{C1}$, and by mirror action, $I_{C4} = I_{C3}$, so all BJTs are biased at $I_{EE}/2$. Taking the Early effect into consideration, we observe that for the BJTs to carry identical currents, the *dc balance* conditions $V_{CE2} = V_{CE1}$ and $V_{EC4} = V_{EC3}$ must hold. These conditions are met simultaneously when

$$V_O = V_{CC} - V_{EBp} \qquad\qquad (4.147)$$

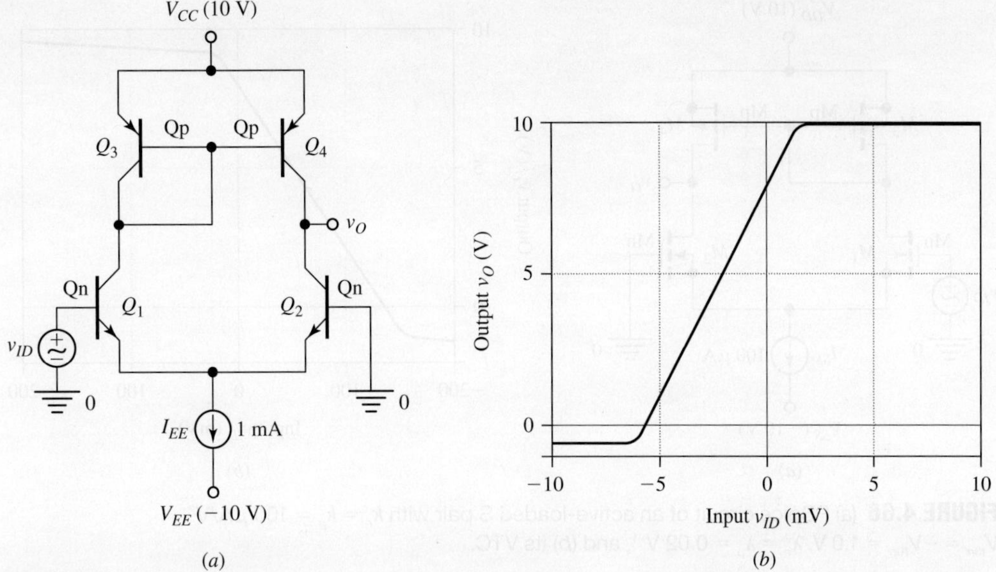

FIGURE 4.65 (a) PSpice circuit of an active-loaded EC pair with $I_{sn} = 2I_{sp} = 2$ fA, $\beta_{Fn} = 4\beta_{Fp} = 200$, $V_{An} = 2V_{Ap} = 100$ V, and (b) its VTC.

where V_{EBp} is the emitter-base voltage drop of the *pnp* BJTs. In the example given, $V_O \cong 10 - 0.7 = 9.3$ V. However, a closer examination of the VTC of Fig. 4.65*b* reveals that the actual value of V_O is *lower* than the above estimate. This is due to the beta error of the current-mirror load, which gives $I_{C4} < I_{EE}/2$ at the value of V_O of Eq. (4.147). Consequently, Q_2 will pull down V_O until $I_{C4} = I_{C2}$ *exactly*. From the plot, this occurs at $V_O \cong 8.0$ V (more on this in Example 4.27).

- Raising v_{ID} *above* 0 V makes Q_1 more conductive at the expense of Q_2 becoming less conductive. By mirror action, Q_4 also becomes more conductive, so the pull-up action by Q_4 will prevail over the pull-down action by Q_2. We thus witness a *rise* in v_O until Q_4 reaches the edge of saturation (EOS). Beyond this point Q_4 saturates, in turn causing the VTC to saturate at $v_O = V_{CC} - V_{EC4(\text{sat})} \cong 10 - 0.1 = 9.9$ V.

- Lowering v_{ID} *below* 0 V makes Q_1 less conductive and Q_2 more conductive, causing the pull-down action by Q_2 to prevail over the pull-up action by Q_4. We now witness a *drop* in v_O, until Q_2 reaches the EOS. Below this point, the VTC saturates at $v_O = V_{E2} + V_{CE2(\text{sat})} \cong -0.7 + 0.1 = -0.6$ V.

Next, let us turn to the **CMOS circuit** of Fig. 4.66*a*, where we observe the following:

- With $v_{ID} = 0$, I_{SS} divides equally between the matched FETs M_1-M_2, giving $I_{D1} = I_{D2} = I_{SS}/2$. By KCL, $I_{D3} = I_{D1}$, and by mirror action, $I_{D4} = I_{D3}$, so all FETs are biased at $I_{SS}/2$. Taking channel modulation into consideration, we observe that

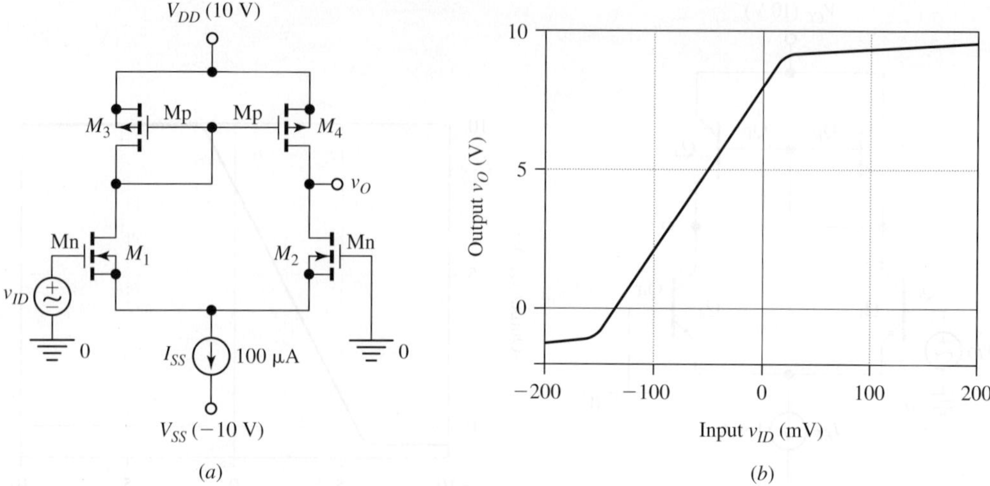

FIGURE 4.66 (a) PSpice circuit of an active-loaded S pair with $k_n = k_p = 100\ \mu\text{A/V}^2$, $V_{t0n} = -V_{t0p} = 1.0\ \text{V}$, $\lambda_n = \lambda_p = 0.02\ \text{V}^{-1}$, and (b) its VTC.

for the FETs to carry identical currents, the *dc balance* conditions $V_{DS2} = V_{DS1}$ and $V_{SD4} = V_{SD3}$ must hold. These conditions are met simultaneously for

$$V_O = V_{DD} - V_{SGp} \tag{4.148}$$

where V_{SGp} is the source-gate voltage drop of the pMOSFETs. In the example given, $V_{OV} = 1.0\ \text{V}$, so $V_{SGp} = |V_{tp}| + V_{OV} = 1 + 1 = 2\ \text{V}$ and $V_O = 10 - 2 = 8\ \text{V}$, in agreement with Fig. 4.66*b*.

- Raising v_{ID} *above* 0 V makes M_1 more conductive at the expense of M_2 becoming less conductive. By mirror action, M_4 also becomes more conductive, indicating that the pull-up action by M_4 will prevail over the pull-down action by M_2. We thus witness a *rise* in v_O until M_4 leaves the saturation region to enter the triode region. Beyond this point M_4 ceases to provide the mirror function and the VTC saturates, as shown.
- Lowering v_{ID} *below* 0 V makes M_1 less conductive and M_2 more conductive, causing the pull-down action by M_2 to prevail over the pull-up action by M_4. We now witness a *drop* in v_O until M_2 leaves the saturation region to enter the triode region. Below this point the VTC saturates, as shown.

The Differential-Mode Gain

An instructive method for finding the differential-mode gain of an active-loaded differential pair is via its *Norton equivalent*, consisting of a dependent source $i_{o(sc)}$ and a parallel resistance R_o.

To find the *short-circuit output current* $i_{o(sc)}$, refer to the ac equivalents of Fig. 4.67, whose similarity indicates that their analysis can be carried out in parallel. We observe that since the collectors of Q_1 and Q_2 are terminated *differently*, the shared emitter terminal is not, strictly speaking, at ac ground, as shown. The EC pair

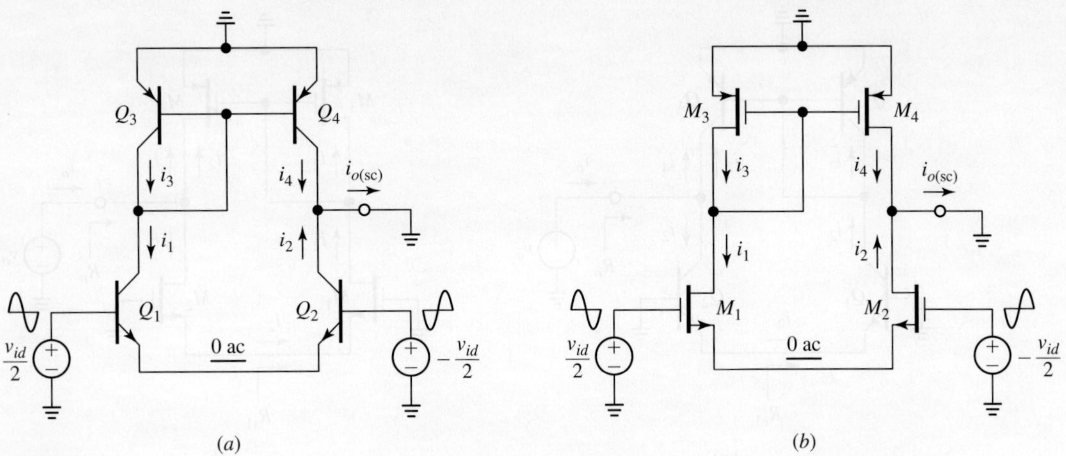

FIGURE 4.67 Half-circuits to find the short-circuit output current i_o: (a) BJT, and (b) CMOS.

will be slightly imbalanced because of the Early effect and so will be the SC pair due to the channel-length modulation effect. Yet, to expedite our estimations, let us continue to assume ac grounds, as shown. For both pairs we can thus approximate

$$i_2 = i_1 \cong g_{mn}\frac{v_{id}}{2}$$

where g_{mn} is the transconductance of the transistors in the differential pair. By KCL we have $i_3 = i_1$, and by current-mirror action we have $i_4 = i_3$. Consequently, we also have $i_4 = i_1$, so $i_{o(sc)} = i_4 + i_2 = 2i_1$, that is,

$$i_o(\text{sc}) \cong g_{mn}v_{id} \tag{4.149}$$

In the bipolar case all four BJTs have identical g_ms, so we can drop the subscript n and write in this case $i_{o(sc)} = g_m v_{id}$, $g_m = 0.5I_{EE}/V_T$. However, in the MOS case we need to keep the distinction as $g_{mn}\left(= \sqrt{k_n I_{SS}}\right)$ and $g_{mp}\left(= \sqrt{k_p I_{SS}}\right)$ may differ because k_n and k_p are not necessarily identical.

Next, let us turn to the task of finding the *small-signal output resistance R_o*. To this end, set the input sources to zero, apply a test voltage v, find the current i out of the test source, and let $R_o = v_o/i_o$. With reference to Fig. 4.68 we observe that in both circuits the test current i consists of three components:

- The component i_4 into Q_4's collector or into M_4's drain. By Ohm's law this component is simply

$$i_4 = \frac{v}{r_{o4}}$$

- The component i_2 into Q_2's collector or into M_2's drain. Due to the presence of the degeneration resistance $R_{e1} = r_{e1} = \alpha_{01}/g_{m1} \cong 1/g_{m1}$, the resistance seen looking into Q_2's collector is $r_{o2}(1 + g_{m2}R_{e1}) = r_{o2}(1 + g_{m2}/g_{m1}) = 2r_{o2}$. Likewise, because

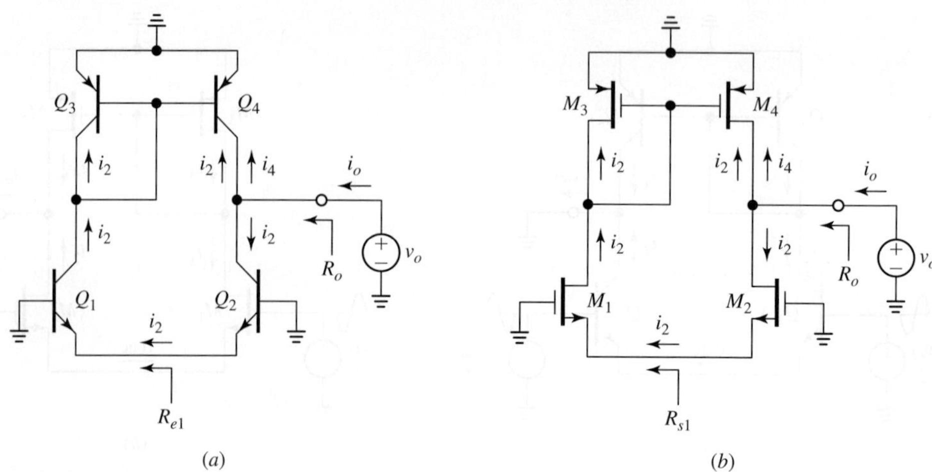

FIGURE 4.68 Test circuits to find the output resistance R_o of the (a) BJT and (b) CMOS differential amplifier.

of the degeneration resistance $R_{s1} \cong 1/(g_{m1} + g_{mb1})$, the resistance seen looking into M_2's drain is approximately $r_{o2}[1 + (g_{m2} + g_{mb2})/(g_{m1} + g_{mb1})] = 2r_{o2}$. Thus

$$i_2 = \frac{v}{2r_{o2}}$$

- By KCL, the component i_2 must leave Q_2's emitter or M_2's source, flow through Q_1 or M_1 and into Q_3 or M_3, from where it is finally *mirrored* by Q_4 or M_4, as shown.

We now apply KCL to write

$$i_o = i_4 + i_2 + i_2 = \frac{v}{r_{o4}} + \frac{v}{2r_{o2}} + \frac{v}{2r_{o2}} = \frac{v}{r_{o4}} + \frac{v}{r_{o2}} = \frac{v}{r_{o4}//r_{o2}}$$

Letting $R_o = v_o/i_o$ we finally get

$$R_o = r_{op}//r_{on} \tag{4.150}$$

where, as usual, we use subscripts n and p to denote the collector/drain resistances of the transistors in the differential pair and in the current mirror, respectively.

As we know, the differential input resistance of the bipolar circuit is $R_{id} = 2r_\pi$, whereas that of its CMOS counterpart is $R_{id} = \infty$. We visualize our findings via the Norton equivalents depicted in Fig. 4.69. Finally, we use Ohm's law to obtain $v_{od} = R_o i_{o(sc)}$, or $v_{od} = R_o g_m v_{id}$ in the bipolar case, and $v_{od} = R_o g_{mn} v_{id}$ in the CMNOS case. The unloaded voltage gain is $a_{dm} = v_{od}/v_{id}$, so

$$a_{dm(\text{BJT})} = g_m(r_{op}//r_{on}) \qquad\qquad a_{dm(\text{MOS})} = g_{mn}(r_{op}//r_{on}) \tag{4.151}$$

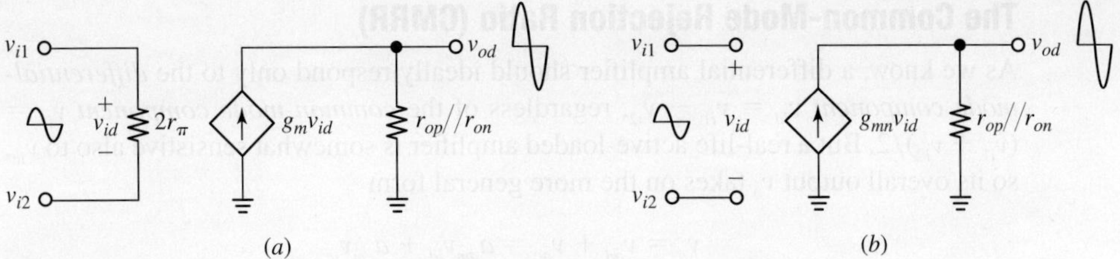

FIGURE 4.69 Norton equivalents of the active-loaded differential amplifier: (a) BJT, and (b) CMOS.

In Problem 4.84 it is shown that $a_{dm(MOS)}$ can be put in the insightful form

$$a_{dm(MOS)} = -\frac{2}{V_{OVn}(\lambda_n + \lambda_p)} \qquad (4.152a)$$

where V_{OVn} is the overdrive voltage of the FETs of the differential pair, and λ_n and λ_p are the channel-length modulation parameters of the nFETs and the pFETs, respectively. If all FETs are fabricated with the *same* channel length L, then Eq. (4.31) provides yet another insightful form for the differential gain

$$a_{dm(MOS)} = -\frac{2L}{V_{OVn}(\lambda'_n + \lambda'_p)} \qquad (4.152b)$$

where λ'_n and λ'_p are the process parameters characterizing channel-length modulation in the two FET types. It is apparent that the longer the channel for a given V_{OVn}, the higher the gain.

EXAMPLE 4.25

(a) Estimate the element values of the Norton equivalent of the bipolar circuit of Fig. 4.65a, as well as its voltage gain a_{dm}.
(b) Repeat, but for the CMOS circuit of Fig. 4.66a. Compare with part (a) and comment.

Solution

(a) We have $g_m = 0.5I_{EE}/V_T = 0.5/26 = 1/(52\ \Omega)$, $r_{op} = 50/0.5 = 100\ \text{k}\Omega$, $r_{on} = 100/0.5 = 200\ \text{k}\Omega$, $r_{op}//r_{on} = 100//200 = 67\ \text{k}\Omega$, $2r_\pi = 2 \times 200 \times 52 = 20.8\ \text{k}\Omega$, and $a_{dm} = 67/0.052 = 1282\ \text{V/V}$.
(b) We have $g_{mn} = \sqrt{kI_{SS}} = \sqrt{100 \times 100} = 100\ \mu\text{A/V}$, $r_{op} = r_{on} = 1/(0.02 \times 50 \times 10^{-6}) = 1\ \text{M}\Omega$, $r_{op}//r_{on} = 0.5\ \text{M}\Omega$, and $a_{dm} = 100 \times 0.5 = 50\ \text{V/V}$, a much lower gain than in the bipolar case due to lower g_{mn}.

The Common-Mode Rejection Ratio (CMRR)

As we know, a differential amplifier should ideally respond only to the *differential-mode component* $v_{id} = v_{i1} - v_{i2}$, regardless of the *common-mode component* $v_{ic} = (v_{i1} + v_{i2})/2$. But a real-life active-loaded amplifier is somewhat sensitive also to v_{ic}, so its overall output v_o takes on the more general form

$$v_o = v_{od} + v_{oc} = a_{dm}v_{id} + a_{cm}v_{ic}$$

where v_{od} and v_{oc} are the differential-mode and common-mode output components, and a_{dm} and a_{cm} are the corresponding gains. As we know, a figure of merit is the *common-mode rejection ratio* (CMRR),

$$\text{CMRR} = \left| \frac{a_{dm}}{a_{cm}} \right| \tag{4.153}$$

which should be as large as possible (ideally, a_{cm} should be zero, so CMRR $= \infty$). We already know a_{dm} from Eq. (4.151), so we only need to find a_{cm}, a task that we shall carry out with the help of the equivalent circuits of Fig. 4.70. In accordance with Section 4.6, the differential pairs Q_1-Q_2 and M_1-M_2 have been split into two common-mode halves. Moreover, the diode-connected transistors Q_3 and M_3 have been replaced by an equivalent resistance r_3, and the mirror transistors Q_4 and M_4 have been replaced by their small-signal equivalents (note that $r_{\pi 4}$ has been included in r_3). The obvious similarity between the two circuits suggests that we can analyze them simultaneously. (As usual, the following analysis assumes matched differential pairs as well as matched active-load pairs.)

In Fig. 4.70a we have, by inspection,

$$i_{1(\text{BJT})} = i_{2(\text{BJT})} = \frac{g_m}{1 + g_m 2R_{EE}} v_{ic} \tag{4.154a}$$

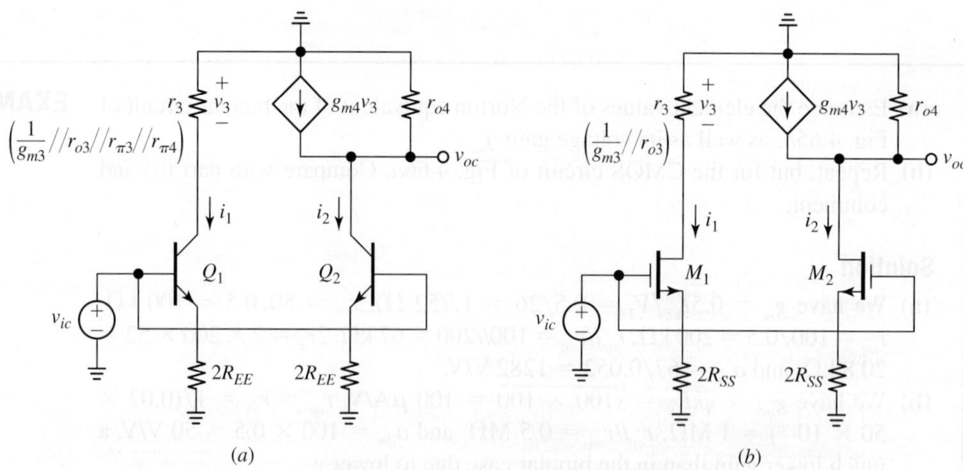

FIGURE 4.70 (a) BJT and (b) CMOS ac equivalents to find the common-mode gain $a_{cm} = v_{oc}/v_{ic}$.

Likewise, in Fig. 4.70*b* we have

$$i_{1(\text{MOS})} = i_{2(\text{MOS})} = \frac{g_{mn}}{1 + (g_{mn} + g_{mbn})2R_{SS}} v_{ic} \tag{4.154b}$$

Also in Fig. 4.70*a* we have, by inspection,

$$r_{3(\text{BJT})} = \frac{1}{g_{m3}} // r_{o3} // r_{\pi3} // r_{\pi4} \tag{4.155a}$$

and in Fig. 4.70*b* we have

$$r_{3(\text{MOS})} = \frac{1}{g_{m3}} // r_{o3} \tag{4.155b}$$

The voltage drop $v_3 = r_3 i_1$ causes Q_4 and M_4 to source the current $g_{m4}v_3$ to the output node, where KCL gives, for both circuits,

$$g_{m4}v_3 = \frac{v_{oc}}{r_{o4}} + i_2$$

that is, $v_{oc} = g_{m4}r_{o4}r_3 i_1 - r_{o4}i_2$. Exploiting the fact that $i_2 = i_1$ and $g_{m4} = g_{m3}$ we get, for both circuits,

$$v_{oc} = (g_{m3}r_3 - 1)r_{o4}i_1 \tag{4.156}$$

Note that r_3 is *slightly less* than $1/g_{m3}$, so the product $g_{m3}r_3$ will be *slightly less than unity*. Clearly, there is a *slight imbalance* between the current sourced by Q_4/M_4 and that sunk by Q_2/M_2, giving $v_{oc} \neq 0$. It is precisely this *inherent imbalance* that makes $a_{cm} \neq 0$ and therefore CMRR $< \infty$. Indeed, substituting Eq. (4.154) into Eq. (4.156) and letting $a_{cm} = v_{oc}/v_{ic}$ gives (see Problem 4.85)

$$a_{cm(\text{BJT})} = \frac{-g_m r_{op}}{(1 + 0.5\beta_{0p})(1 + 2g_m R_{EE})} \tag{4.157a}$$

$$a_{cm(\text{MOS})} = \frac{-g_{mn} r_{op}}{(1 + g_{mp} r_{op})[1 + 2(g_{mn} + g_{mbn})R_{SS}]} \tag{4.157b}$$

where, as usual, the numerical subscripts have been replaced by subscripts p and n when needed. So long as the various $g_m \times r$ products are much greater than unity, the above expressions simplify as

$$a_{cm(\text{BJT})} \cong \frac{-r_{op}}{\beta_{0p} R_{EE}} \qquad a_{cm(\text{MOS})} = \frac{-1}{2(1 + \chi_n)g_{mp} R_{SS}} \tag{4.157c}$$

EXAMPLE 4.26   (a) Assuming $R_{EE} = 100$ kΩ, find a_{cm} and CMRR for the active-loaded EC pair of Fig. 4.65a.

(b) Repeat, but for the CMOS circuit of Fig. 4.66a. Assume $R_{SS} = 0.5$ MΩ and $g_{mbn} = 0.1g_{mn}$.

Solution

(a) By Eq. (4.157a) we have

$$a_{cm(\text{BJT})} = \frac{-100/0.052}{(1 + 0.5 \times 50)(1 + 2 \times 100/0.052)} = -19.2 \text{ mV/V}$$

Example 4.25a gave $a_{dm} = 1282$ V/V, so CMRR$_{\text{BJT}} = 1282/(19.2 \times 10^{-3}) = 66,681$ ($= 96.5$ dB).

(b) In this particular example we have $g_{mn} = g_{mp}$ ($= 0.1$ mA/V). Using Eq. (4.157b),

$$a_{cm(\text{MOS})} = \frac{-0.1 \times 1000}{(1 + 0.1 \times 1000)[1 + 2 \times 0.1(1 + 0.1)500]} = -8.92 \text{ mV/V}$$

Example 4.25b gave $a_{dm} = 50$ V/V, so CMRR$_{\text{MOS}} = 50/(8.92 \times 10^{-3}) = 5,605$ ($=75$ dB).

It is instructive to develop direct expressions for the CMRRs. Substituting Eqs. (4.151) and (4.157) into Eq. (4.153) we get (see Problem 4.85) the BJT expression

$$\text{CMRR}_{\text{BJT}} = \frac{1 + 0.5\beta_{0p}}{1 + r_{op}/r_{on}}(1 + 2g_m R_{EE}) \rightarrow \frac{\beta_{0p}g_m R_{EE}}{1 + r_{op}/r_{on}} \tag{4.158a}$$

which shows an active-load improvement on the order of $(1 + 0.5\beta_{0p})/(1 + r_{op}/r_{on})$ compared to the passive-load case. Likewise, for the MOSFET case we get

$$\text{CMRR}_{\text{MOS}} \cong \frac{1 + g_{mp}r_{op}}{1 + r_{op}/r_{on}}[1 + 2(g_{mn} + g_{mbn})R_{SS}] \rightarrow 2g_{mp}(r_{on}//r_{op})g_{mn}(1 + \chi_n)R_{SS} \tag{4.158b}$$

indicating an improvement on the order of $(1 + g_{mp}r_{op})/(1 + r_{op}/r_{on})$ compared to the passive-load case.

Input Offset Voltage of Active-Loaded Differential Pairs

In an active-loaded differential pair the input offset voltage V_{OS} is the result of mismatches in the transistors of the differential pair as well as in those of the current mirror. Turning first to the **BJT circuit** of Fig. 4.64a, we adapt Eq. (4.118) and write

$$V_{OS(BJT)} \cong V_T \sqrt{\left(\frac{\Delta I_{sn}}{I_{sn}}\right)^2 + \left(\frac{\Delta I_{sp}}{I_{sp}}\right)^2} \qquad (4.159)$$

If the two mismatches are of equal magnitude, the effect of the active-load mismatch is to make the offset typically $\sqrt{2}$ times as large as that due to the EC pair alone. In the bipolar case we have an additional offset term due to the beta error of the *pnp* mirror. As we know, this error is

$$\Delta I_{C4} = -\frac{2}{\beta_{Fp}} I_{C4} \cong -\frac{2}{\beta_{Fp}} I_C$$

where β_{Fp} is the average beta of the *pnp* BJTs. Dividing this term by $-g_m \, (= -I_C/V_T)$ gives

$$V_{OS(systematic)} \cong \frac{\Delta I_{C4}}{-g_m} = V_T \frac{2}{\beta_{Fp}} \qquad (4.160)$$

This is the corrective voltage that we need to apply at the input in order to compensate for the beta error of the mirror, even if the *npn* and *pnp* pairs are *perfectly* matched. Unlike the offset terms resulting from random transistor mismatches, the term of Eq. (4.160) occurs always in the *same* direction and is therefore referred to as a *systematic* offset term. If necessary, it can be reduced by implementing the active load with a mirror equipped with a beta helper, or with a cascode-type or a Wilson-type mirror.

Using the data of Example 4.25a, discuss the effect of β_{Fp} in the bipolar circuit of Fig. 4.65a.

EXAMPLE 4.27

Solution

With $v_{ID} = 0$ we have $I_C \cong 500 \ \mu A$ and $V_{EBp} \cong 0.7$ V. If the *pnp* BJTs had infinite betas, the circuit would yield $V_O \cong 10 - 0.7 = 9.3$ V. However, because of non-infinite *pnp* betas, the mirror exhibits an error $\Delta I_{C4} \cong -(2/50)500 = -20 \ \mu A$. To achieve $I_{C4} = I_{C2}$, the output will change automatically by $\Delta V_O = R_o \Delta I_{C4} = 67 \times 10^3 \times (-20 \times 10^{-6}) \cong -1.3$ V and settle at $V_O \cong 9.3 - 1.3 = 8.0$ V, in agreement with the VTC of Fig. 4.65b. If we wish to ensure ideal dc balance we must drive V_O back to 9.3 V. This is achieved by applying a corrective input voltage $V_{ID} = -\Delta V_O/a_{dm} = 1.3/1282 \cong 1$ mV. Sure enough, this is the offset term predicted by Eq. (4.160), that is, $V_{OS(systematic)} = 26(2/50) = 1$ mV!

Turning next to the **CMOS circuit** of Fig. 4.64b, we note the absence of any systematic offset because the gate currents are zero. The offset now stems from k and V_t mismatches in each transistor pair. It is left as an exercise for the student (see Problem 4.96) to show that

$$V_{OS(MOS)} \cong \frac{V_{OVn}}{2} \sqrt{\left(\frac{\Delta k_n}{k_n}\right)^2 + \left(\frac{\Delta k_p}{k_p}\right)^2 + \left(\frac{\Delta V_{tn}}{0.5 V_{OVn}}\right)^2 + \left(\frac{\Delta V_{tp}}{0.5 V_{OVn}}\right)^2} \qquad (4.161)$$

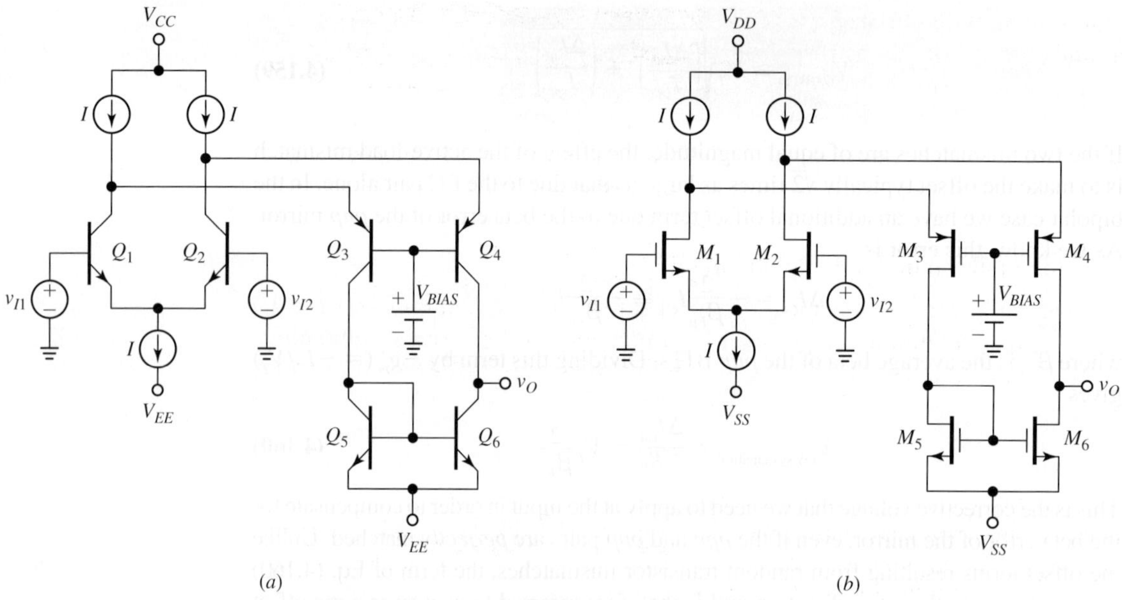

FIGURE 4.71 (a) Bipolar and (b) CMOS folded-cascoded differential pairs.

If the active-load mismatches have equal magnitudes as those of the differential pair, the effect of the active load is to make the offset typically $\sqrt{2}$ times as large as that due to the differential pair alone.

Folded-Cascoded Differential Pairs

A notorious drawback in both circuits of Fig. 4.64 is limited voltage headroom at the output. The *upper* limit of the output voltage swing (OVS) is reached when Q_4/M_4 is driven to the edge of saturation (EOS), so the bipolar circuit has $v_{O(max)} = V_{CC} - V_{EC4(EOS)}$ and the MOS version has $v_{O(max)} = V_{DD} - V_{OV4}$. The *lower* limit of the OVS is reached when Q_2/M_2 is driven to the EOS, so the bipolar circuit has $v_{O(min)} = v_{I2} - V_{BE2} + V_{CE2(EOS)}$ and the MOS circuit has $v_{O(min)} = v_{I2} - V_{GS2} + V_{OV2} = v_{I2} - (V_{t2} + V_{OV2}) + V_{OV2} = v_{I2} - V_{t2}$.

It is the *lower limit* that poses problems because it depends on v_{I2}. In fact, the higher v_{I2}, the less voltage headroom is available at the output.

The above drawback is ingeniously avoided via the *folded-cascode* arrangements of Fig. 4.71, where it is apparent that the lower limit is now reached when Q_6/M_6 is driven to the edge of saturation. In fact, the bipolar circuit has now $v_{O(min)} = V_{EE} + V_{CE6(EOS)}$ and the MOS version has $v_{O(min)} = V_{SS} + V_{OV6}$. In both cases $v_{O(min)}$ is independent of v_{I2} and is fairly close to the negative supply (more in Problems 4.97 and 4.98).

4.10 BIPOLAR OUTPUT STAGES

The primary function of the output stage of a voltage-output circuit is to *provide low output resistance* in order to reduce output loading. In general-purpose ICs such as op amps, the output stage must be capable of supplying enough current (and thus power) to meet the needs of a variety of loads, and it should do so while consuming a minimum of standby power. Finally, these functions should be provided over a suitably wide frequency bandwidth, with a minimum of distortion, and with a wide output voltage swing—ideally, from rail to rail. Among the various output-stage configurations available in bipolar technology, the ones that have gained most prominence are the so-called *push-pull output stages*.

The Class B Push-Pull Output Stage

A good candidate for the role of output stage is the common-collector (CC) configuration because it accepts a small base current to deliver an emitter current $\beta + 1$ times as large. Moreover, the driving source's resistance, reflected to the emitter, is $\beta + 1$ times as small. The *npn* BJT only *sources* emitter current and the *pnp* BJT only *sinks* emitter current, so we need both device types in order to accommodate both polarities. The *npn* BJT will handle *positive* voltage alternations, when current is to be *pushed* into the load; the *pnp* BJT will handle *negative* voltage alternations, when current is to be *pulled* out of the load. Aptly called *push-pull stage*, the circuit is shown in its basic form in Fig. 4.72a. With reference to its voltage transfer curve (VTC) of Fig. 4.72b we observe the following:

- As long as v_I falls within the range $-V_{EB2(on)} < v_I < V_{BE1(on)}$, both BJTs are off, giving $v_O = 0$.
- As we raise v_I above $V_{BE1(on)}$, Q_1 goes on while Q_2 continues to remain off. In forward-active operation, Q_1 acts as an emitter follower with a voltage gain of slightly less than 1 V/V.
- Raising v_I above the positive supply voltage will eventually bring Q_1 to the edge of saturation (EOS), thus estasblishing the upper limit of the *output voltage swing* (OVS) as $v_{O(max)} = V_{CC} - V_{CE1(EOS)}$.
- As we lower v_I below $-V_{EB2(on)}$, the roles of Q_1 and Q_2 are interchanged, yielding a symmetric VTC with respect to the origin, and such that $v_{O(min)} = V_{EE} + V_{EC2(EOS)}$.

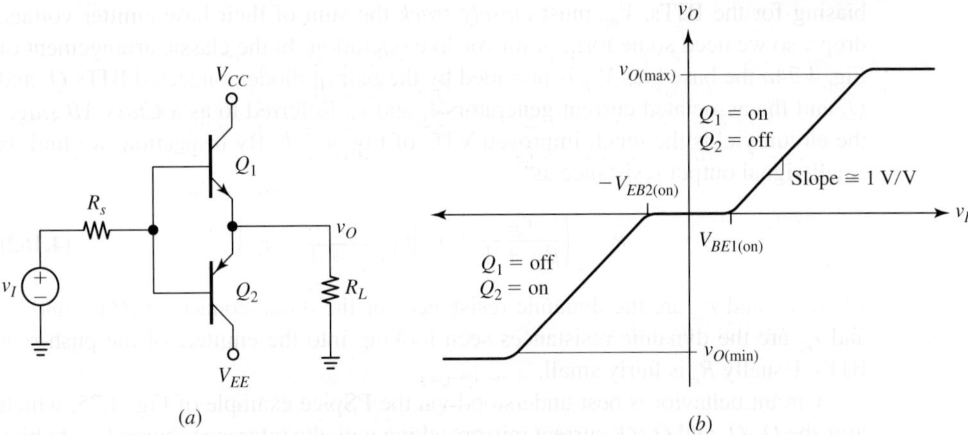

FIGURE 4.72 (a) Class B push-pull circuit and (b) its VTC.

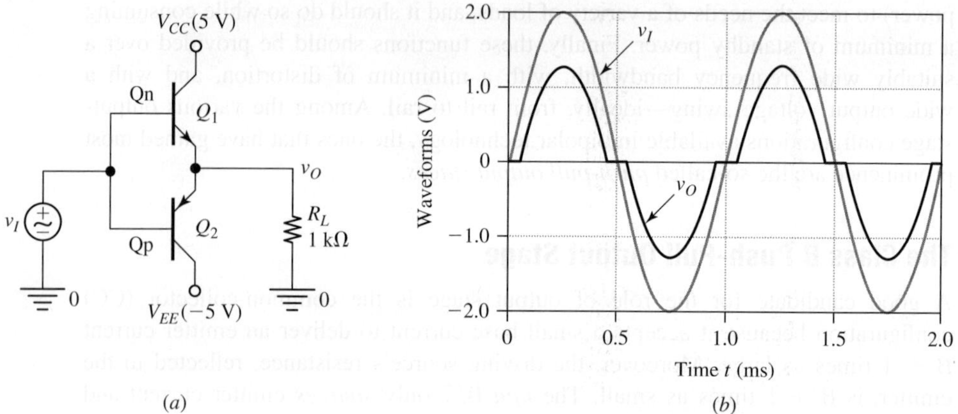

FIGURE 4.73 (*a*) PSpice Class B push-pull circuit and (*b*) its input and output waveforms.

We can gain additional insight by simulating the circuit via PSpice. As shown in Fig. 4.73, the circuit introduces considerable distortion due to the presence of the dead band $-V_{EB2(on)} < v_I < V_{BE1(on)}$, over which neither BJT conducts. Called *crossover distortion*, it is generally intolerable, so the present circuit is relegated primarily to handling square-wave signals, where crossover distortion is not an issue.

This circuit is said to be of the *Class B* type because each BJT conducts only during half a cycle (actually, because of the nonzero base-emitter voltage drops, the conduction angle is less than 180° for each BJT).

The Class AB Push-Pull Output Stage

If we could arrange for both BJTs to be *already conductive* for $v_O = 0$, as opposed to having to wait for v_I to raise above $V_{BE1(on)}$ or to drop below $-V_{EB2(on)}$, then crossover distortion would be eliminated altogether. This requires establishing between the two bases a voltage drop $V_{BB} = V_{BE1(on)} + V_{EB2(on)} \cong 2 \times 0.7 = 1.4$ V. To ensure predictable biasing for the BJTs, V_{BB} must *closely track* the sum of their base-emitter voltage drops, so we need some form of mirror-like operation. In the classic arrangement of Fig. 4.74*a* the base bias V_{BB} is provided by the pair of diode-connected BJTs Q_3 and Q_4 and the associated current generators I_1 and I_2. Referred to as a *Class AB stage*, the circuit yields the much-improved VTC of Fig. 4.74*b*. By inspection, we find its small-signal output resistance as

$$R_o = \left(\frac{r_{d3}}{\beta_{01} + 1} + r_{e1}\right) /\!/ \left(\frac{r_{d4}}{\beta_{02} + 1} + r_{e2}\right) \tag{4.162}$$

where r_{d3} and r_{d4} are the dynamic resistances of the diode-connected BJTs, and r_{e1} and r_{e2} are the dynamic resistances seen looking into the emitters of the push-pull BJTs. Usually R_o is fairly small.

Circuit behavior is best understood via the PSpice example of Fig. 4.75, which uses the Q_5-Q_6 and Q_7-Q_8 current mirrors, along with the reference source I_{REF}, to bias

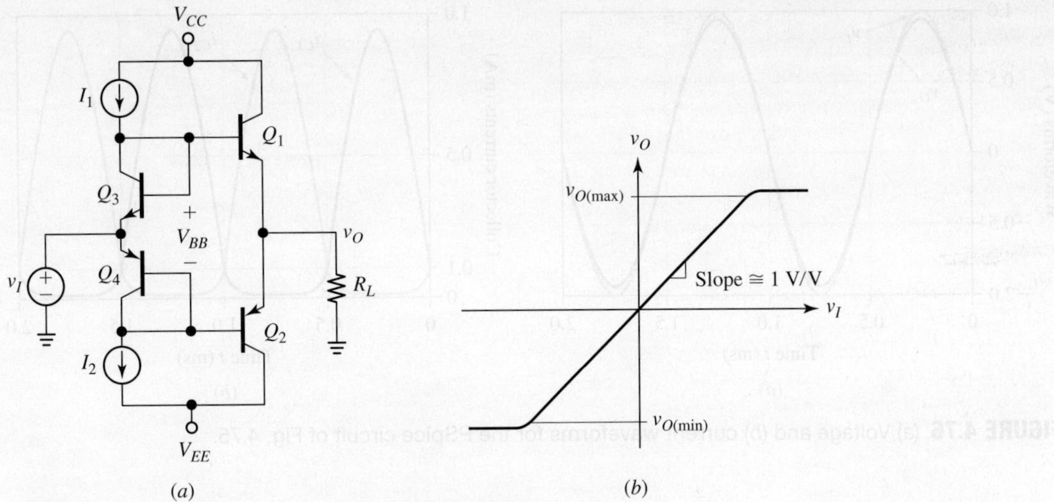

FIGURE 4.74 (a) Class AB push-pull circuit and (b) its VTC.

the diode-connected Q_3-Q_4 pair at $I_{C3} = I_{C4} = I_{REF} = 0.1$ mA. Referring also to the waveforms of Fig. 4.76, we make the following considerations:

- For $v_O = 0$ the current through R_L is zero, so Q_1 and Q_2 must carry identical currents, or $i_{C1} = i_{C2}$. Assuming the Q_1-Q_3 and Q_2-Q_4 pairs are matched, Q_1 will mirror Q_3 and Q_2 will mirror Q_4 to give $i_{C1} = i_{C2} = 0.1$ mA. This is called

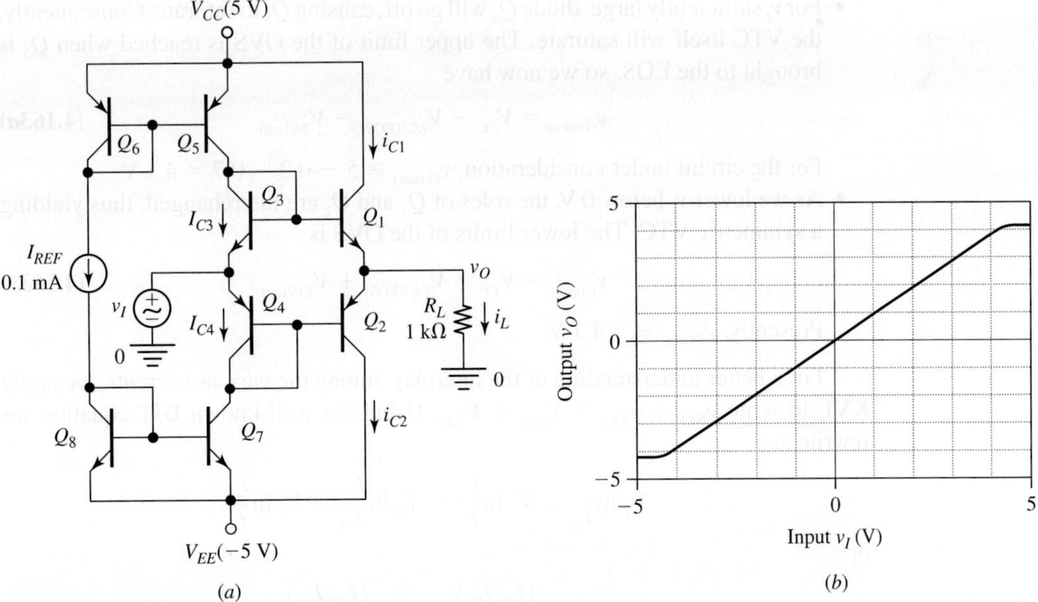

FIGURE 4.75 (a) PSpice circuit of a Class AB bipolar push-pull stage, and (b) its VTC.

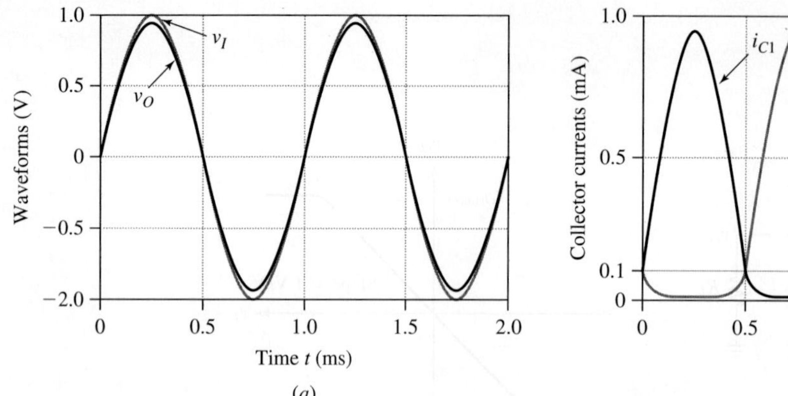
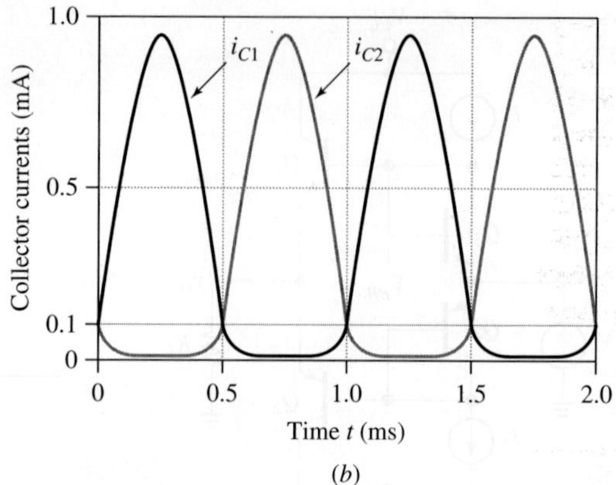

FIGURE 4.76 (a) Voltage and (b) current waveforms for the PSpice circuit of Fig. 4.75.

the *quiescent current I_Q* of the Q_1-Q_2 pair. For the sake of efficiency, in a well-designed circuit I_Q is kept at the minimum necessary to avoid distortion. It is apparent that $v_O = 0$ for $v_I = 0$. When $i_L = 0$, the circuit is said to be in *standby*.

- As we increase v_I, v_O will also increase due to the emitter-follower action by Q_1. So, the load current $i_L = v_O/R_L$ will also increase, in turn raising i_{C1}. For instance, as i_{C1} *doubles*, v_{BE1} will increase by 18 mV, by the well-known rule of thumb. But, since $v_{BE1} + v_{EB2} = V_{BB} \cong$ constant, the 18-mV *increase* in v_{BE1} will cause an 18-mV *decrease* in v_{EB2}, indicating that i_{C2} will *halve*. It is apparent from Fig. 4.76b that as v_I is raised further, we eventually get $i_{C2} \rightarrow 0$ and $i_{C1} \rightarrow i_L$.

- For v_I sufficiently large, diode Q_3 will go off, causing Q_5 to saturate. Consequently, the VTC itself will saturate. The upper limit of the OVS is reached when Q_5 is brought to the EOS, so we now have

$$v_{O(\max)} = V_{CC} - V_{EC5(EOS)} - V_{BE1(on)} \tag{4.163a}$$

For the circuit under consideration, $v_{O(\max)} \cong 5 - 0.2 - 0.7 = 4.1$ V.

- As we lower v_I below 0 V, the roles of Q_1 and Q_2 are interchanged, thus yielding a symmetric VTC. The lower limits of the OVS is

$$v_{O(\min)} = V_{EE} + V_{CE7(EOS)} + V_{EB2(on)} \tag{4.163b}$$

Presently, $v_{O(\min)} \cong -4.1$ V.

For a better understanding of the interplay among the various currents, we apply KVL to write $v_{BE1} + v_{EB2} = V_{BE3} + V_{EB4}$. Using the well-known BJT equation we rewrite as

$$V_T \ln\frac{i_{C1}}{I_{s1}} + V_T \ln\frac{i_{C2}}{I_{s2}} = V_T \ln\frac{I_{C3}}{I_{s3}} + V_T \ln\frac{I_{C4}}{I_{s4}}$$

or

$$V_T \ln\left(\frac{i_{C1}}{I_{s1}}\frac{i_{C2}}{I_{s2}}\right) = V_T \ln\left(\frac{I_{C3}}{I_{s3}}\frac{I_{C4}}{I_{s4}}\right)$$

For this equality to hold, the arguments of the logarithms must be identical, so we get

$$i_{C1}i_{C2} = \left(\frac{I_{s1}I_{s2}}{I_{s3}I_{s4}}\right)I_{C3}I_{C4} \tag{4.164}$$

indicating that the *product* $i_{C1}i_{C2}$ remains *constant*. In the example of Fig. 4.75 we have $i_{C1}i_{C2} = I_{C3}I_{C4} = (0.1\ \text{mA})^2 = 10^{-8}\ \text{A}^2$. This means that if one of the two currents *increases* because of i_L by a given number of octaves or decades, the other current must *decrease* by the same number of octaves or decades. As we know, for $i_L = 0$ the circuit is in standby with $i_{C1} = i_{C2}$ and it draws the quiescent current

$$I_Q = \sqrt{\frac{I_{s1}I_{s2}}{I_{s3}I_{s4}}I_{C3}I_{C4}} \tag{4.165}$$

Since all its BJTs are assumed identical, the present circuit has $I_Q = 0.1\ \text{mA}$.

EXAMPLE 4.28

(a) For the circuit of Fig. 4.75a, find v_I for $i_{C1} = 0.4$ mA. What is the corresponding value of v_O?

(b) Find v_I for $v_O = -0.25$ V. What are the values of i_{C1} and i_{C2}?

(c) Estimate v_O if $v_I = 1.0$ V.

Solution

(a) Increasing i_{C1} from 0.1 mA to 0.4 mA (two octaves) will lower i_{C2} from 0.1 mA to $0.1/(2 \times 2) = 0.025$ mA. By KCL, $i_L = i_{C1} - i_{C2} = 0.4 - 0.025 = 0.375$ mA. By Ohm's law, $v_O = R_L i_L = 1 \times 0.375 = 0.375$ V. By the rule of thumb, a two-octave increase in i_{C1} requires an increase in v_{BE1} of $2 \times 18 = 36$ mV. Thus, to raise v_O from 0 V to 0.375 V, we must raise v_I from 0 V to $0.375 + 0.036 = 0.411$ V.

(b) We now have $i_L = v_O/R_L = -0.25/1 = -0.25$ mA, that is, a 0.25-mA load current flowing into Q_2's emitter. KCL now gives $i_{C2} = i_L + i_{C1} = 0.25 \times 10^{-3} + 10^{-8}/i_{C2}$, which we solve to get $i_{C2} = 0.285$ mA. Moreover, $i_{C1} = 10^{-8}/(0.285 \times 10^{-3}) = 0.035$ mA. Raising i_{C2} from 0.1 mA to 0.285 mA requires that v_{EB2} be increased by $(26\ \text{mV})\ln(0.285/0.1) \cong 27$ mV. Consequently, $v_I = -0.25 - 0.027 = -0.277$ V.

(c) We anticipate v_O to be slightly less than 1.0 V. Start out with the initial estimate $v_{O(0)} \cong 1$ V and then iterate. We have $i_{L(0)} \cong 1/1 = 1$ mA and $i_{C2} \ll i_{C1}$, so $i_{C1(0)} \cong i_{L(0)} \cong 1$ mA. To increase i_{C1} from 0.1 mA to 1 mA (one decade), we must increase v_{BE1} by 60 mV, by the well-known rule of thumb. So, a better estimate is $v_{O(1)} = 1.0 - 0.06 = 0.94$ V. Then, $i_{L(1)} = 0.94/1 = 0.94$ mA and $i_{C2(1)} = 10^{-8}/(0.94 \times 10^{-3}) = 0.016$ mA. The reader can do one more iteration to verify that the last results are adequate enough.

Overload Protection

Bipolar push-pull stages are notoriously vulnerable to overload conditions such as the inadvertent shorting of the output terminal to ground. When an overload condition arises, the stage upstream will step up the base drive of the overloaded BJT, causing the latter to draw sufficient current to overheat and possibly destroy itself. We need watchdog circuitry to sense the current in each of the push-pull BJTs, and should this current try to exceed a prescribed safety limit, intervene in such a way as to prevent any further current increase. The output stage will no longer perform its intended function, but at least it will be saved from possible destruction.

Shown in Fig. 4.77 is the overload protection circuitry for Q_1, but a similar concept can be used to protect Q_2. This circuitry consists of a small series resistance R_{SC} to sense the emitter current of Q_1, and a watchdog BJT Q_5 designed to be off under normal operation, but to go on as soon as the load attempts to draw excessive current from Q_1. Once on, Q_5 will divert to the load any excess current from the I_1 source, letting into Q_1's base only the amount necessary to sustain Q_1's conduction at the safety limit. The current sensing resistance is chosen as

$$R_{SC} = \frac{V_{BE5(on)}}{I_{SC}} \tag{4.166}$$

where $V_{BE5(on)}$ is the voltage needed to turn on Q_5 and I_{SC} is the maximum allowed current for Q_1.

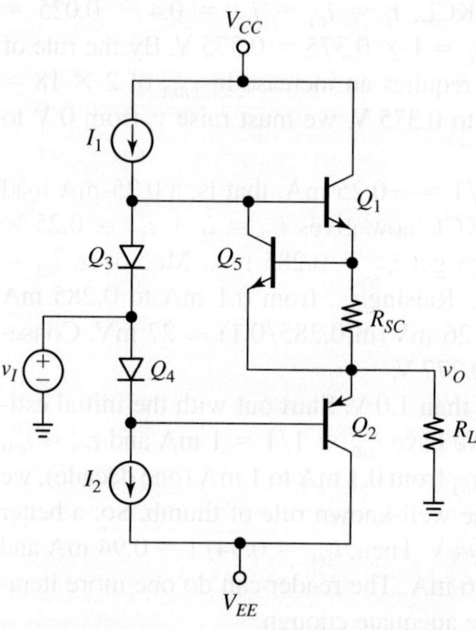

FIGURE 4.77 Overload protection for Q_1.

In the circuit of Fig. 4.77 let $\beta_{F1} = \beta_{F5} = 250$, $V_{BE5\text{(on)}} = 0.7$ V, $I_1 = I_2 = 300\ \mu$A, and $V_{CC} = 15$ V. Suppose the circuit is part of a negative-feedback system designed to regulate the output at $V_O = 10$ V.

EXAMPLE 4.29

(a) Specify R_{SC} for $I_{SC} = 20$ mA.
(b) Show all relevant voltages and currents if a student tries the circuit in the lab with $R_L = 2.0$ kΩ.
(c) Repeat if the student, because of a resistance color-code misreading, loads the circuit with $R_L = 20$ Ω instead of $R_L = 2.0$ kΩ.
(d) What is likely to happen if Q_1 does not have short-circuit protection? Comment on your findings.

Solution

(a) $R_{SC} = 0.7/0.020 = 35$ Ω.
(b) With $R_L = 2.0$ kΩ we have $I_L = V_O/R_L = 10/2.0 = 5$ mA. As it flows through R_{SC}, this current creates the voltage drop $V_{BE5} = R_{SC}I_L = 0.035 \times 5 = 0.175$ V. This is insufficient to turn on Q_5, so the latter will remain in a dormant state, like a good watchdog should do. To supply 5 mA, Q_1 draws the base current $I_{B1} = I_{E1}/(\beta_{F1} + 1) = 5/251 \cong 20\ \mu$A. This current comes from the I_1 source, so the remaining 280 μA go to the diode Q_3. The situation is illustrated in Fig. 4.78a.
(c) Installing a voracious load such as $R_L = 20$ Ω will pull V_O down toward ground. Sensing this drop in V_O via the feedback network, the circuitry upstream will try adjusting v_I in such a way as to boost the base drive of Q_1 in order to raise V_O. In fact, *all* of I_1 will now be diverted toward Q_1, thereby awakening Q_5 and leading to the overload situation of Fig. 4.78b. Here, Q_1's current is limited

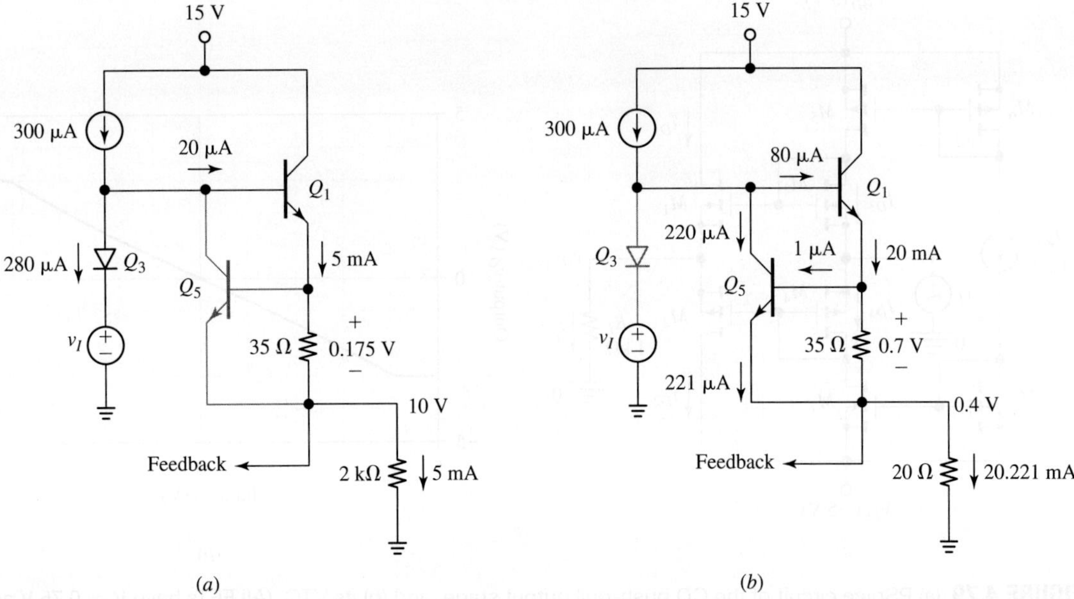

FIGURE 4.78 Circuit of Example 4.29 operating under (a) normal and (b) overload conditions.

to about $I_{SC} = 20$ mA. To sustain this current, Q_1 draws the base current $I_{B1} = I_{SC}/(\beta_F + 1) = 20/251 \cong 80$ μA. The remaining current of $300 - 80 = 220$ μA is passed on by Q_5 directly to the load. Allowing for about 1 μA of base current for Q_5, we have $I_{E5} = 220 + 1 = 221$ μA, $I_L = 20 + 0.221 = 20.221$ mA, and $V_O = 20 \times 20.221 \times 10^{-3} \cong 0.4$ V. This is a far cry from the intended value $V_O = 10$ V, but at least Q_1 is spared destruction.

(d) Without protection, Q_1 would try to draw $I_{E1} = (\beta_F + 1)I_1 = 251 \times 0.3 \cong 75$ mA, raising the output only to $V_O = 0.020 \times 75 = 1.5$ V. The power dissipated by Q_1 would be $P_1 \cong V_{CE1} \times I_{C1} = (15 - 1.5)75 > 1$ W, high enough to cause the monolithic BJT Q_1 most likely to blow out.

4.11 CMOS OUTPUT STAGES

As in the bipolar case, a CMOS output stage should provide low output impedance over a wide frequency bandwidth and with a wide OVS—ideally from rail to rail—while consuming a minimum of standby power. As we are about to see, the design of CMOS output stages presents some marked differences compared to the bipolar case.

The CD Push-Pull Output Stage

In principle, the bipolar Class AB configuration of Fig. 4.75 could be replicated in MOS form as shown in Fig. 4.79. Here, the common-drain (CD) M_1-M_2 pair forms the push-pull stage proper; the diode-connected M_3-M_4 pair biases the M_1-M_2 pair

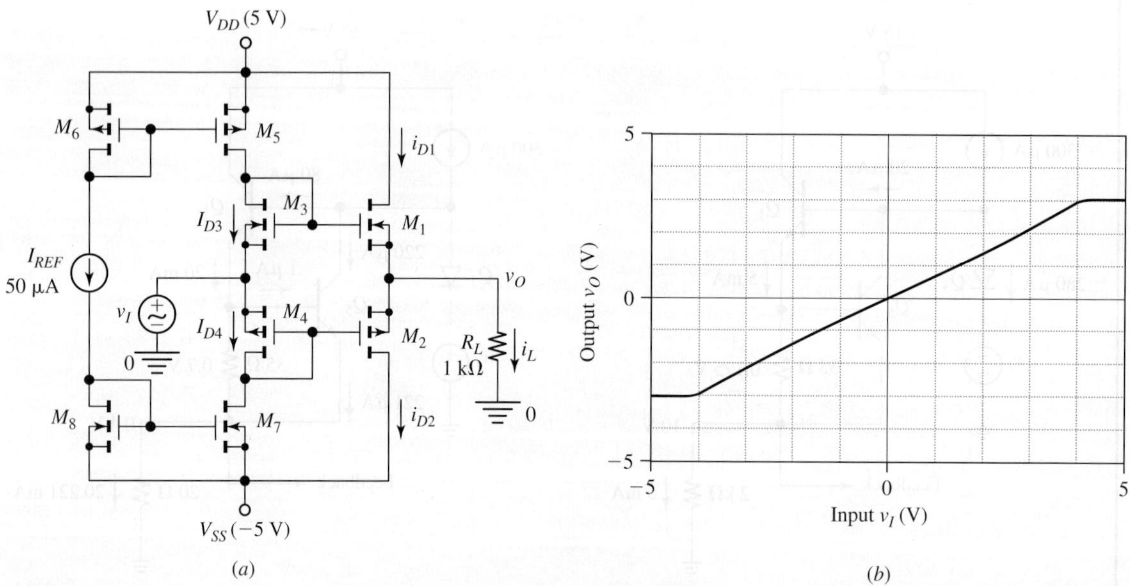

FIGURE 4.79 (a) PSpice circuit of the CD push-pull output stage, and (b) its VTC. (All FETs have $V_t = 0.75$ V and $\lambda = 0.02$ V^{-1}; moreover, $k_1 = k_2 = 4$ mA/V^2, and $k_3 = k_4 = k_5 = k_6 = k_7 = k_8 = 1.6$ mA/V^2.)

for Class AB operation, and the current mirrors M_5-M_6 and M_7-M_8, along with the reference current I_{REF}, provide the necessary current to bias the diode pair. By inspection, the output resistance is

$$R_o = \frac{1}{(g_{m1} + g_{m2})} // r_{o1} // r_{o2} \tag{4.167}$$

which is usually low. Using KVL we readily find the limits of the output voltage swing (OVS) as

$$v_{O(\max)} = V_{DD} - V_{OV5} - V_{t1} - v_{OV1} \tag{4.168a}$$

$$v_{O(\min)} = V_{SS} + V_{OV7} + |V_{t2}| + v_{OV2} \tag{4.168b}$$

We immediately note a glaring difference compared to the bipolar case. While the voltage drops $V_{EC5(EOS)} + V_{BE1(on)}$ and $V_{CE7(EOS)} + V_{EB2(on)}$ of Eq. (4.163) are relatively constant ($\sim$0.9 V), their counterpart drops $V_{OV5} + V_{t1} + v_{OV1}$ and $V_{OV7} + |V_{t2}| + v_{OV2}$ of Eq. (4.168) depend on i_L via v_{OV1} and v_{OV2}. Moreover, V_{t1} and V_{t2} are subject to the body effect, further reducing the OVS, at whose extremes V_{t1} and V_{t2} get maximized. In the CMOS example of Fig. 4.79 the edges of the OVS are within a couple of volts of either supply rail, whereas in the bipolar example of Fig. 4.75 they are approximately fixed and within less than a volt. The limited OVS of the Class AB CD stage can be a serious drawback in low-voltage power supply systems, where better design alternatives are necessary.

The CMOS Inverter as Output Stage

The OVS can be improved considerably if the FETs of the push-pull pair are operated in the common-source (CS) mode instead of the CD mode. This is the case of the already familiar CMOS inverter, shown again in Fig. 4.80. The ensuing VTC, obtained using the same transistor parameters and load resistance as in Fig. 4.79,

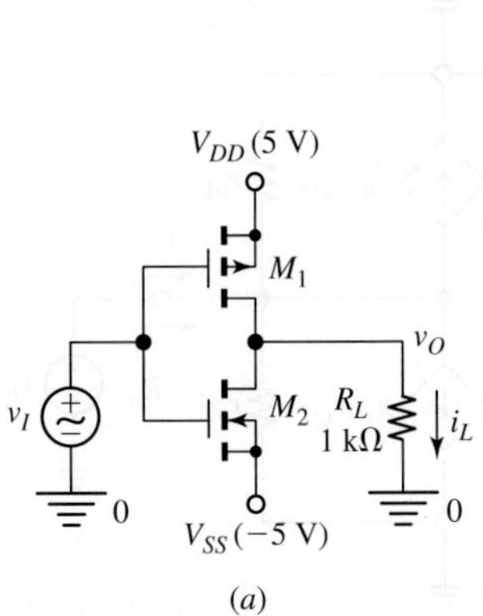

(a)

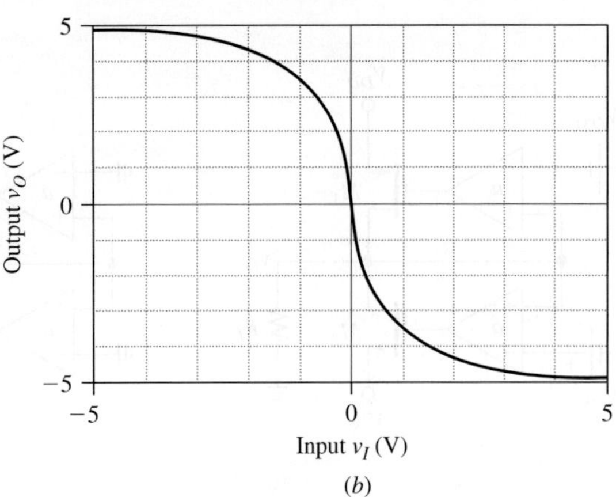

(b)

FIGURE 4.80 (a) PSpice circuit of the CMOS inverter, and (b) its VTC. (Both FETs have $k = 4$ mA/V², $V_t = 0.75$ V, and $\lambda = 0.02$ V⁻¹.)

clearly shows the circuit's ability to swing its output fairly close to the power-supply rails. However, as an output stage, the inverter suffers from a number of drawbacks: its VTC is highly nonlinear, its output resistance $R_o = r_{o1}//r_{o2}$ is usually too high, and and its quiescent current can be intolerably large. (The circuit provides also polarity inversion, but this is not a serious issue as we can deliberately invert signal polarity somewhere else in the system.) For these reasons, the CMOS inverter is relegated to logic-type output circuits such as voltage comparators, where v_O sits mostly at either logic level, undergoing only rapid transitions between one logic level and the other.

The CS Push-Pull Output Stage with Feedback Amplifiers

The shortcomings of the CMOS-inverter output stage are cleverly avoided by separately *predistorting* the gate drives in such a way as to ensure a *fairly linear* VTC with a comparatively wide OVS. This task is accomplished by means of negative feedback, according to the principle depicted in Fig. 4.81a. We make the following considerations:

- The circuit is made up of two complementary subcircuits, each consisting of a low-gain op amp and a CS FET connected for *negative-feedback operation* (note that since the CS configuration provides signal *inversion*, the ouput is fed back to the op amp's *noninverting* input, rather than to the more usual inverting input). In this mode of operation each op amp will provide the corresponding FET with *whatever gate drive* it takes to force v_O to *track* v_I. Consequently, as long as at least one of the op amps can exercise its negative-feedback control, we expect a fairly linear VTC.
- The op amps, themselves made up of MOSFETs, are deliberately unbalanced for the purpose of setting the quiescent current of the push-pull pair at a specified

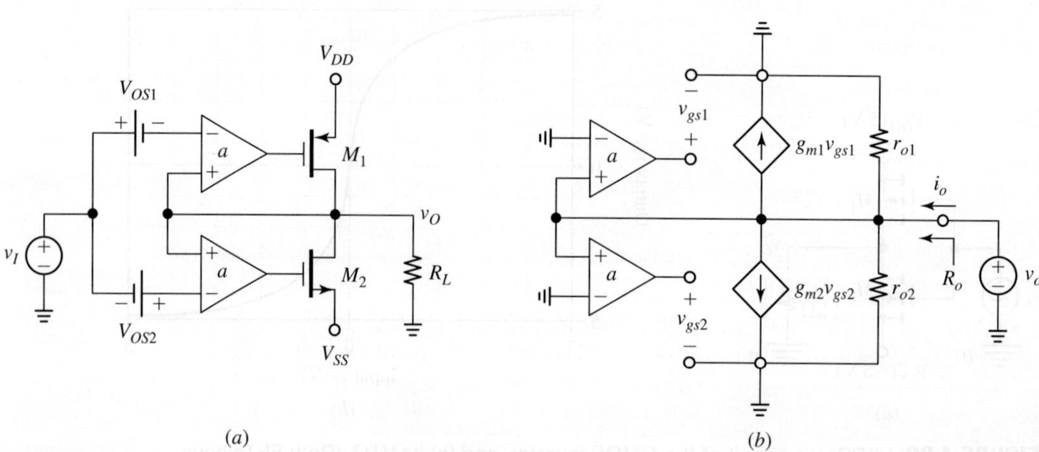

FIGURE 4.81 (a) CS push-pull output stage, and (b) ac equivalent to find its output resistance R_o.

value, as we shall see in Example 4.30. This imbalance, created by fabricating the two halves of each amplifier with differing W/L ratios, is modeled via the offset voltages V_{OS1} and V_{OS2}.

- Negative feedback, besides linearizing the VTC, reduces the output resistance R_o dramatically. To substantiate, set all independent sources to zero and apply a test source v_o as in Fig. 4.81b. By KCL,

$$i_o = \frac{v_o}{r_{o1}} + \frac{v_o}{r_{o2}} + g_{m1}av_o + g_{m2}av_o$$

Collecting, we get

$$R_o = \frac{v_o}{i_o} = \frac{1}{a(g_{m1} + g_{m2})}//r_{o1}//r_{o2} \qquad (4.169)$$

which shows a two-fold beneficial effect of negative feedback upon the output resistance: (a) it brings the g_ms into the picture ($1/g_m \ll r_o$), and (b) it multiplies them by the gain a to further lower R_o.

EXAMPLE 4.30

Let the MOSFETs of Fig. 4.81a be matched devices with $k = 4$ mA/V^2, $V_t = 0.75$ V, and $\lambda = 0.02$ V^{-1}. Moreover, let $V_{DD} = -V_{SS} = 5.0$ V and $a = 10$ V/V.

(a) Specify V_{OS1} and V_{OS2} for a standby current $I_Q = 125$ μA.
(b) Find the value of R_o in standby.
(c) Find v_O if $v_I = 4.0$ V and $R_L = 1$ kΩ.
(d) Use PSpice to display v_O, i_L, i_{D1}, i_{D2}, v_{G1} and v_{G2} versus v_I for -5 V $< v_I <$ $+5$ V. Hence, comment on your findings.

Solution

(a) In standby ($v_O = v_I = 0$) both FETs are saturated, so we impose

$$125 \ \mu A = \frac{1}{2}(4 \ \text{mA/V}^2)V_{OV(SBY)}^2$$

to obtain $V_{OV(SBY)} = 0.25$ V. The required gate voltage for M_1 in standby is $V_{G1(SBY)} = V_{DD} - V_t - V_{OV(SBY)} = 5 - 0.75 - 0.25 = 4$ V. But, $V_{G1(SBY)}$ is generated by the upper op amp as $V_{G1(SBY)} = a(v_P - v_N)$, so imposing $4 = 10[0 - (-V_{OS1})]$ gives $V_{OS1} = 0.4$ V. With matched FETs we have, by symmetry, $V_{OS2} = V_{OS1} = 0.4$ V.

(b) In standby we have $g_{m1} = g_{m2} = 2I_Q/V_{OV(SBY)} = 2 \times 125 \times 10^{-6}/0.25 = 1/(1 \ \text{k}\Omega)$ and $r_{o1} = r_{o2} = 1/(\lambda I_D) = 1/(0.02 \times 125 \times 10^{-6}) = 400$ kΩ, so

$$R_o = \frac{1}{10(10^{-3} + 10^{-3})}//(400 \times 10^3)//(400 \times 10^3) \cong 50 \ \Omega$$

which is quite low.

(c) For $v_I = 4.0$ V we expect v_O to approach 4 V, indicating that with a relatively small v_{SD1}, M_1 is likely to be in the triode region, where

$$i_{D1} = k\left[v_{OV1}v_{SD1} - \frac{1}{2}v_{SD1}^2\right] = 4\left[(5 - 0.75 - v_{G1})(5 - v_O) - \frac{1}{2}(5 - v_O)^2\right]$$

We also have, by inspection,

$$v_{G1} = a(v_P - v_N) = a[v_O - (v_I - V_{OS1})] = 10(v_O - (4 - 0.4)) = 10(v_O - 3.6)$$

Moreover, we expect M_2 to be in cutoff, so we can write

$$i_{D1} = i_L = \frac{v_O}{R_L} = \frac{v_O}{1}$$

Eliminating v_{G1} and i_{D1} and solving gives the physically acceptable solution $v_O = 3.88$ V (fairly close to 4 V, as expected). Back substituting gives $v_{G1} = 2.82$ V. Since $v_{SD1} < v_{OV1}$ (1.12 V $<$ 1.43 V), the FET is indeed in the triode region.

(d) Using the PSpice circuit of Fig. 4.82a we obtain the VTC of Fig. 4.82b, which is fairly linear over a wide OVS. The plots of Fig. 4.83a confirm that for $v_I = 0$ the op amps provide $v_{G1} = 4$ V and $v_{G2} = -4$ V in order to ensure the voltage overdrives of 0.25 V needed to bias both FETs at $I_Q = 125$ μA. As v_I moves away from 0 V, one of the FETs goes off while the other takes on the task of feeding the load. Even more revealing are the plots of Fig. 4.83b, showing how the op amps predistort the gate drives v_{G1} and v_{G2} in order to ensure a fairly linear voltage transfer characteristic, especially toward the extremes of the OVS. The predistorting action by negative feedback will be investigated more systematically in Chapter 7.

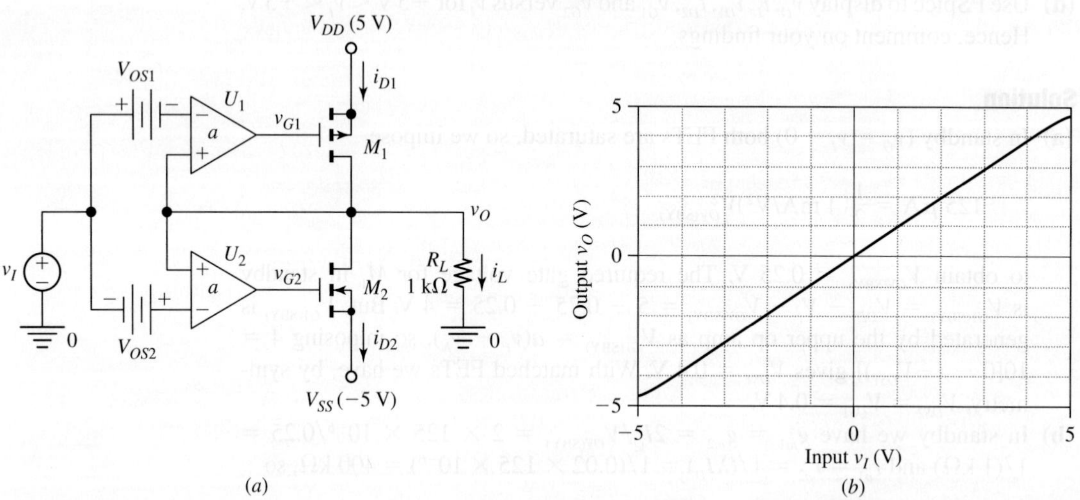

FIGURE 4.82 (a) PSpice circuit of the CS push-pull stage of Example 4.30, and (b) its VTC. (Both FETs have $k = 4$ mA/V², $V_t = 0.75$ V, and $\lambda = 0.02$ V⁻¹; both op amps have $a = 10$ V/V and $V_{os} = 0.4$ V.)

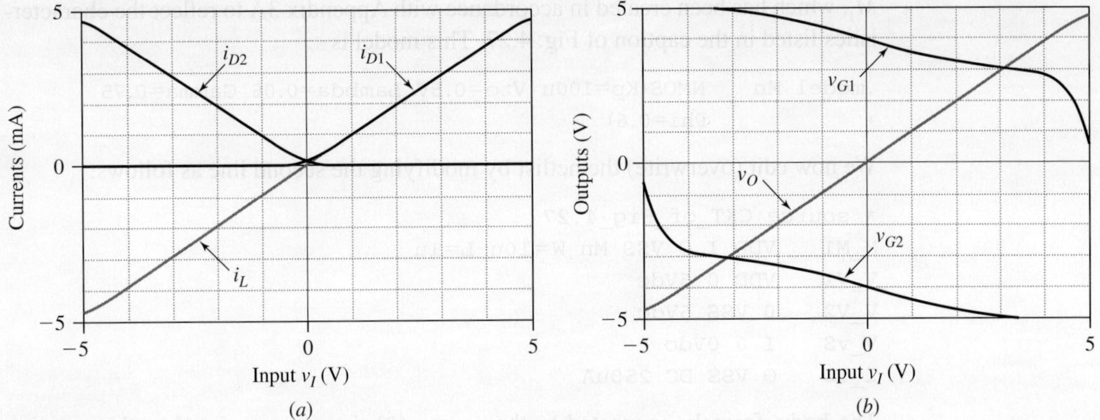

FIGURE 4.83 (a) Current and (b) voltage transfer curves for the PSpice circuit of Fig. 4.82a.

APPENDIX 4A

Editing SPICE Netlists

The MOSFET models available in the `Eval` library of Version 9.2 of PSpice refer to devices with body and source tied together. However, if we want to investigate the body effect, we can readily untie the two terminals and connect the body to the MNV in the case of *n*MOSFETs or to the MPV in the case of *p*MOSFETs by suitably editing the circuit's netlist. (A netlist is an internal code to which PSpice automatically converts the circuit entered via the Schematic Capture facility, before performing the actual simulation.)

To illustrate, refer to the PSpice example of Fig. 4.27. Once we have created the circuit schematic via the **Place → Part** and **Place → Wire** commands, we first use **PSpice → Create Netlist** to direct PSpice to generate the netlist, and then we use **PSpice → View Netlist** to visualize it. The result is the following lines of code:

```
* source CKT_of_Fig_4.27
M_M1     VDD I O O Mn
V_V1     VDD 0 5Vdc
V_V2     0 VSS 5Vdc
V_vS     I 0 0Vdc
I_ID     O VSS DC 250uA
```

We are interested in the second line, which refers to the MOSFET M_1 of Fig. 4.27, renamed by PSpice as M_M1. The remaining entries in this line refer to the circuit's nodes to which the *Drain*, *Gate*, *Source*, and *Body* (in that order) are connected. These are, respectively, the power supply (VDD), the input node (I), the output node (O), and again the output node (O). The last entry (Mn) refers to the PSpice model for

M_1, which has been created in accordance with Appendix 3A to reflect the characteristics listed in the caption of Fig. 4.27. This model is

```
.model Mn     NMOS(Kp=100u Vto=0.5V Lambda=0.05 Gamma=0.75
+             Phi=0.6)
```

We now edit (overwrite) the netlist by modifying the second line as follows:

```
* source CKT_of_Fig_4.27
M_M1     VDD I O VSS Mn W=10u L=1u
V_V1     VDD 0 5Vdc
V_V2     0 VSS 5Vdc
V_vS     I 0 0Vdc
I_ID     O VSS DC 250uA
```

The body, formely connected to the source (O), is now connected to the negative power supply (VSS). Moreover, the model name (Mn) is followed by the specifications of the channel width and length (W=10u L=1u). This is particularly convenient in multi-transistor circuits, where all FETs share the same process parameters as specified in a common model (Mn in this case), but each device is assigned its individual W and L values in the netlist line where the device appears.

Once the netlist has been edited, we must *save* it via the **File → Save** commands. We finally launch the simulation of the circuit pertaining to the modified netlist via the **PSpice → Run** commands, as usual.

REFERENCES

1. P. R. Gray, P. J. Hurst, S. H. Lewis, and R. G. Meyer, *Analysis and Design of Analog Integrated Circuits*, 5/E, Wiley and Sons, 2009.
2. R. S. Muller and T. I. Kamins, *Device Electronics for Integrated Circuits*, 2/E, J. Wiley and Sons, 1986.
3. P. E. Allen and D. R. Holberg, *CMOS Analog Circuit Design*, 2/E, Oxford University Press, 2002.
4. R. T. Howe and C. G. Sodini, *Microelectronics: An Integrated Approach*, Prentice Hall, 1997.
5. A. S. Sedra and K. C. Smith, *Microelectronic Circuits*, 6/E, Oxford University Press, 2010.
6. R. C. Jaeger and T. N. Blalock, *Microelectronic Circuit Design*, 2/E, McGraw-Hill, 2004.
7. D. A. Johns and K. Martin, *Analog Integrated Circuit Design*, Wiley and Sons, 1997.
8. H. Camenzind, *Designing Analog Chips*, www.designinganalogchips.com, 2005.

PROBLEMS

4.1 Design Considerations in Monolithic Circuits

4.1 Reconsider the circuit of Example 4.1, dealing with the circuit of Fig. 4.2a.

(a) How is v_O affected if V_{BE1} is 1 mV higher than the value calculated in the example?

(b) What if V_{EB4} is 1 mV higher than the calculated value?

(c) If $v_i = V_{im}\cos\omega t$, estimate the maximum value of V_{im} for which the output is still a relatively undistorted sine wave. Justify any approximations you may be making.

(d) What happens if V_{im} is raised to twice the value found in part (c)? Illustrate by sketching and labeling $v_O(t)$.

4.2 (a) If the BJTs of Fig. P4.2 have $I_{s1} = I_{s2} = 1.0$ fA, $V_A = \infty$, and negligible base currents, find I_{C1}, I_{C2}, and V_E.

(b) Repeat part (*a*) if $I_{s1} = I_{s2} = 6$ fA.

(c) Repeat (*a*) if the BJTs are mismatched with $I_{s1} = 4$ fA and $I_{s2} = 3$ fA.

(d) Repeat (*a*) if the BJTs are perfectly matched but Q_1 is 1°C warmer than Q_2.
   Hint: use a well-known rule of thumb to ask yourself what voltage V_{B1} would be needed at Q_1's base to ensure identical collector currents; then, what happens if V_{B1} is returned to 0 V?

(e) Repeat part (*a*) if the BJTs have $V_A = 50$ V.

(f) Repeat part (*e*) if Q_1's collector is lifted off ground and tied to a +10-V supply.

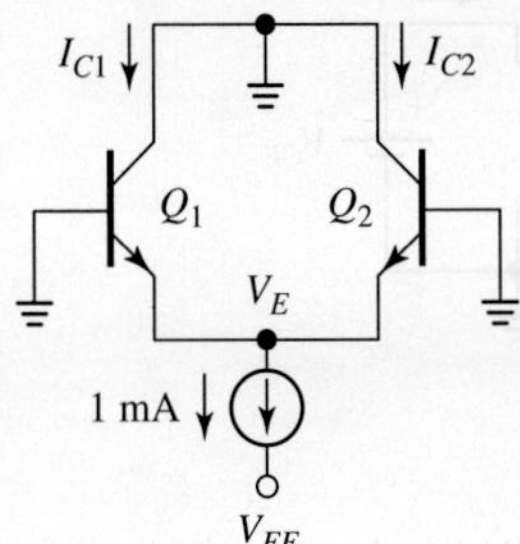

FIGURE P4.2

4.3 In MOS IC technology a voltage divider is usually implemented via a series combination of diode-connected FETs in order to avoid using resistors, which are undesirable in IC technology. An example is shown in Fig. P4.3.

(a) Assuming $k' = 50$ μA/V^2, $V_{t0} = 0.5$ V, $\gamma = 0.4$ V$^{1/2}$, $\lambda = 1/(25$ V$)$, and $|2\phi_p| = 0.6$ V, specify suitable W/L ratios for M_1 and M_2 so that with $V_{DD} = 3$ V the circuit gives $V = V_{DD}/2$ while dissipating $P_D = 100$ μW.

(b) Find V and P_D if V_{DD} is lowered to 2 V.

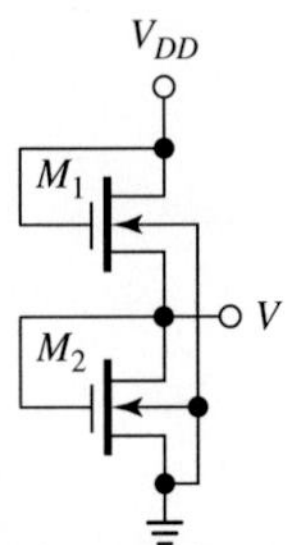

FIGURE P4.3

4.4 In Chapter 3 we found that connecting a feedback resistance between the output and input terminals

of a CMOS inverter will bias it right in the middle of its linear region of operation. In IC technology resistors are undesirable, so the circuit or Fig. P4.4 utilizes the FET M_3 to achieve the same function. Since M_1 and M_2 draw zero gate currents, M_3 operates at the origin of its i_D-v_{DS} characteristics, where it acts as a resistance r_{DS}.

(a) Assuming $k'_n = 2.5k'_p = 100$ μA/V^2 and $V_{tn0} = -V_{tp0} = 0.5$ V, specify W/L ratios for the three devices so that with $V_{DD} = 3$ V the inverter is biased at $V_I = V_O = V_{DD}/2$ with $r_{DS} = 1$ MΩ, and it dissipates $P_D = 150$ μW. Since M_3 is subject to the body effect, assume $\gamma = 0.4$ V$^{1/2}$ and $|2\phi_p| = 0.6$ V to find V_{t3}.

(b) What happens if due to a wiring error the gate and body terminals of M_3 are interchanged with each other, so that the gate goes to ground and the body to V_{DD}? How does P_D change?
   Hint: refer to the *n*MOSFET structure of Fig. 3.1

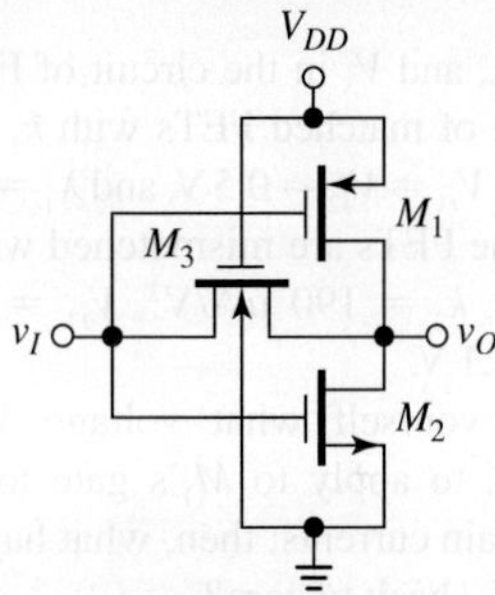

FIGURE P4.4

4.5 The circuit of Fig. P4.5 is the CMOS counterpart of the bipolar version of Fig. 4.2, and its analysis follows a line of reasoning similar to Example 4.1. Let M_1 have $k_n = 400$ μA/V^2, $V_{tn} = 1.0$ V, and $\lambda_n = 1/(25$ V$)$, and let M_2 have $k_p = 175$ μA/V^2, $V_{tp} = -0.75$ V, and $\lambda_p = 1/(20$ V$)$.

(a) If $V_{DD} = 5$ V and the FETs are biased at 200 μA, estimate $v_{O(\text{min})}$ and $v_{O(\text{max})}$, the lower and upper limits of the linear output swing (to simplify your calculations, assume $\lambda_p = \lambda_n = 0$ in this step).

(b) Find V_{GS} and V_{SG} so that the output node is biased right in the middle of the linear range.

(c) Find the gain $a = v_o/v_i$.

(d) How is V_O affected if V_{GS} is 10 mV higher than the value calculated in the example?

(e) What if V_{SG} is 10 mV higher than the calculated value?

(f) If $v_i = V_{im} \cos \omega t$, estimate the maximum value of V_{im} for which the output is still a relatively undistorted sinewave. Justify any approximations you may be making.

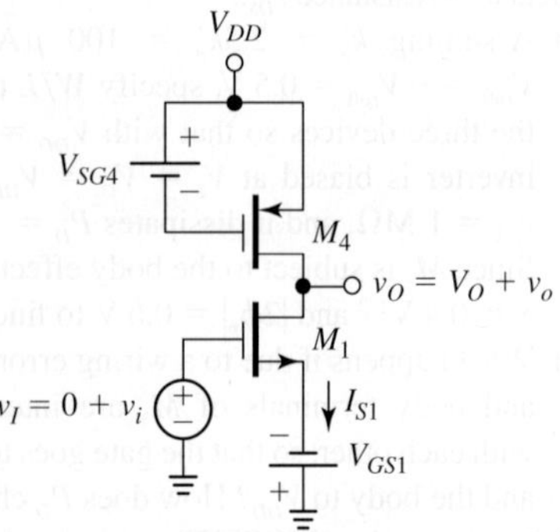

FIGURE P4.5

4.6 (a) Find I_{D1}, I_{D2}, and V_S in the circuit of Fig. P4.6 for the case of matched FETs with $k_1 = k_2 = 200\ \mu\text{A/V}^2$, $V_{t1} = V_{t2} = 0.5$ V, and $\lambda_1 = \lambda_2 = 0$.

(b) Repeat if the FETs are mismatched with $k_1 = 205\ \mu\text{A/V}^2$, $k_2 = 190\ \mu\text{A/V}^2$, $V_{t1} = 0.48$ V, and $V_{t2} = 5.1$ V.

Hint: ask yourself what voltage V_{G1} you would need to apply to M_1's gate to ensure identical drain currents; then, what happens if you drive V_{B1} back to zero?

(c) Repeat part (a) if the FETs have $\lambda = 1/(20\text{ V})$.

(d) Repeat part (c) if M_1's drain is lifted off ground and tied to a 5-V supply.

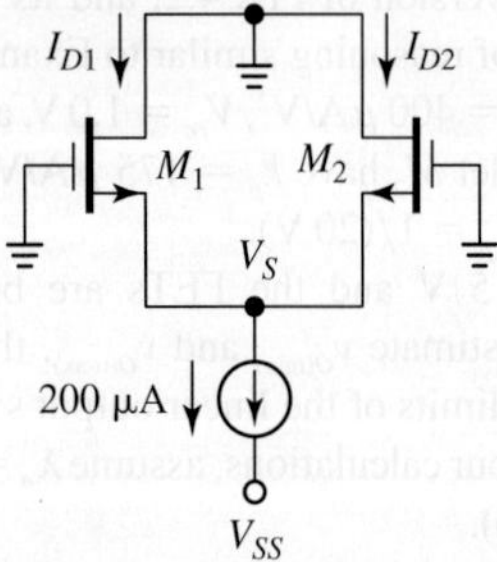

FIGURE P4.6

4.2 BJT Characteristics and Models Revisited

4.7 (a) Suppose V_{BE} in Fig. P4.7 has been adjusted for $I_C = 1.0$ mA at $V_{CE} = 1.0$ V. If Eq. (2.15)

predicts $\beta_F = 1/(1/120 + 1/600)$, find the base current components I_{BE} and I_{BB} as well as m.

(b) If at $V_{CE} = 1.0$ V the effective base width is 250 nm and the portion of the B-C SCL extending into the base region has a width of 20 nm, predict the values of I_C and I_B at $V_{CE} = 6.0$ V (assume the B-C junction has a built-in potential of 0.75 V and a grading coefficient of 0.4). What are the values of r_o and r_μ? What is the value of V_A?

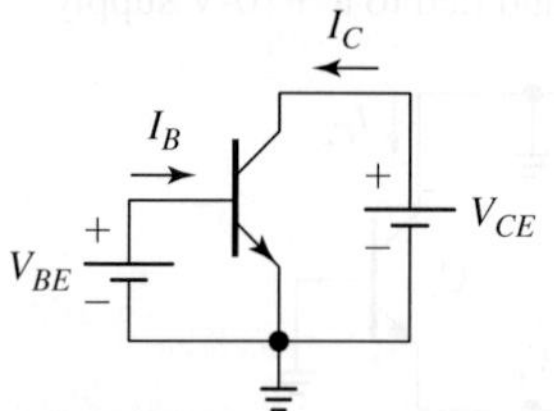

FIGURE P4.7

4.8 Shown in Fig. P4.8 is the ac equivalent of a current-driven CE amplifier. Let the BJT have $g_m = 1/(10\ \Omega)$, $r_\pi = 1.5$ kΩ, $r_o = 30$ kΩ, and $r_\mu = 18$ MΩ, and let i_b be a 1-μA ac current.

(a) Find the ac voltages v_b, v_e, and v_c, if $R_E = 0$ and $R_L = \infty$. To see the effect of r_μ, calculate first assuming $r_\mu = \infty$, then using $r_\mu = 18$ MΩ. Note that r_μ is subject to the Miller effect, so reflected to the base r_μ gets divided by $1 - a$, where $a = v_c/v_b$.

(b) Repeat part (a), but with an output load $R_L = r_o = 30$ kΩ.

(c) Repeat parts (a) and (b), but with an emitter-degeneration resistance $R_E = 0.5$ kΩ.

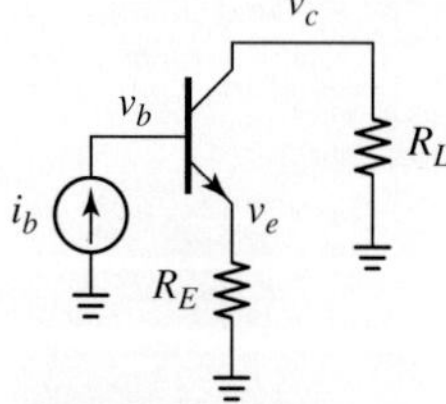

FIGURE P4.8

4.9 The BJT in the ac circuit of Fig. P4.9 has $g_m = 1/(25\ \Omega)$, $r_\pi = 5$ kΩ, $r_o = 50$ kΩ, and $r_\mu = 50$ MΩ. Moreover, i_i is a 1-μA ac source.

(a) Find the ac voltages v_i and v_o for the idealized case $R_E = R_L = r_\mu = \infty$.

(b) Repeat (a), but with $r_\mu = 50\ \text{M}\Omega$. Repeat (a), but with $r_\mu = 50\ \text{M}\Omega$ and for the following significant cases:

(c) $R_L = r_\mu$,

(d) $R_L = (\beta_0 + 1)r_o$,

(e) $R_L = r_o$,

(f) $R_L = r_\pi$. Comment on your results, and identify the two extremes of *monolithic* (large R_C) and *discrete* (small R_C) design.

(g) Repeat (a), but with $R_E = R_e$.

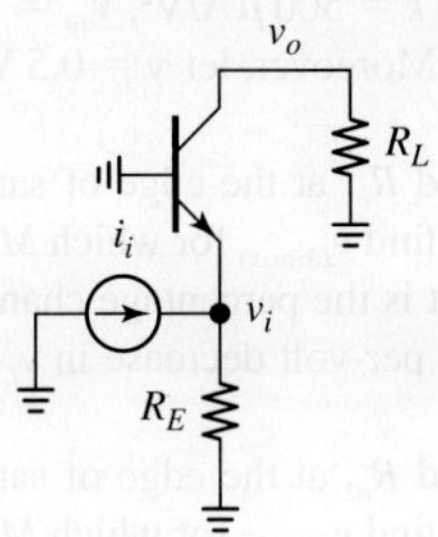

FIGURE P4.9

4.10 (a) Assuming the current drawn by R_E in Fig. P4.10 splits equally between the two BJTs, find R_i, R_o, and v_o/v_i if $\beta_0 = 200$, $V_A = 50\ \text{V}$, and $V_{BE(on)} = 0.7\ \text{V}$.

Hint: recall that for ac purposes Q_2 acts as a diode with ac resistance r_e.

(b) What function does Q_2 serve in this circuit? What happens if we remove Q_2 altogether from the circuit?

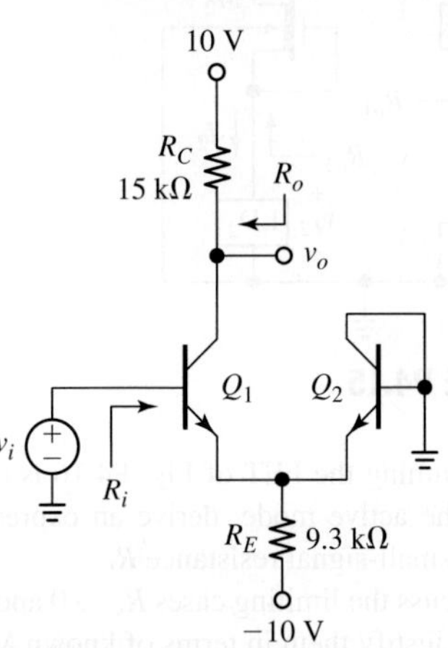

FIGURE P4.10

4.11 In the BiMOS circuit of Fig. P4.11 the emitter follower Q is biased by the depletion FET M, here operated as a current sink. Assume the BJT has $\beta_F = 200$, $V_A = 50\ \text{V}$, $V_{BE(on)} = 0.7\ \text{V}$, and $V_{CE(EOS)} = 0.2\ \text{V}$, and the FET has $V_t = -1.0\ \text{V}$, $k' = 100\ \mu\text{A/V}^2$, and $\lambda = 1/(25\ \text{V})$.

(a) Specify the W/L ratio to bias the BJT at 5 mA.

(b) Find R_i, R_o, and the gain $a = v_o/v_i$.

(c) Estimate $v_{O(max)}$ and $v_{O(min)}$, the upper and lower limits of the linear output swing, as well as the corresponding values of v_I.

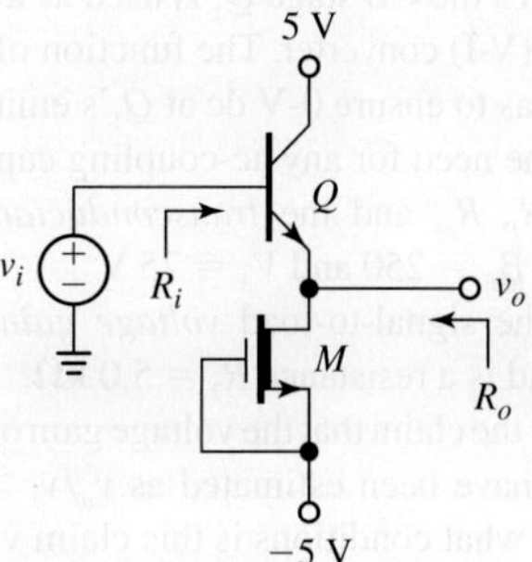

FIGURE P4.11

4.12 The characteristics of an emitter follower can be made much closer to ideal through the use of negative feedback, a subject that will be explored in greater detail in Chapter 7. The circuit of Fig. P4.12, known as *super emitter follower*, uses Q_1 as the emitter follower proper, and Q_2 to provide negative feedback around Q_1.

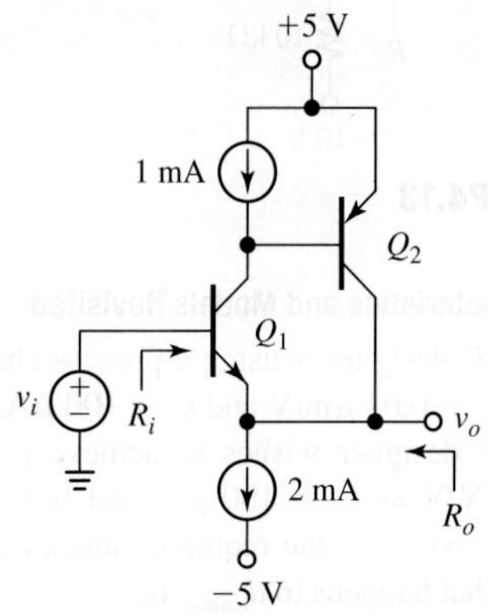

FIGURE P4.12

(a) Draw the ac equivalent of the circuit by replacing each BJT with its small-signal equivalent consisting of g_m, r_π, and r_o (ignore r_μ for simplicity). Hence, use the test-signal method to find an expression for the output resistance R_o.

(b) Assuming both BJTs have $\beta_0 = 100$ and $V_A = 50$ V, calculate R_o, compare with the value provided by a single-BJT emitter follower operating at $I_C = 1$ mA, and comment on your result.

4.13 In Fig. P4.13 the CB stage Q_1 is used as a *voltage-to-current* (V-I) converter. The function of Q_2 is to bias Q_1 so as to ensure 0-V dc at Q_1's emitter, thus avoiding the need for any ac-coupling capacitor.

(a) Find R_i, R_o, and the *transconductance gain* i_o/v_i if $\beta_F = 250$ and $V_A = 75$ V.

(b) Find the signal-to-load *voltage gain* v_o/v_i if the load is a resistance $R_2 = 5.0$ kΩ.

(c) Justify the claim that the voltage gain of part (b) could have been estimated as $v_o/v_i > R_2/R_1$. Under what conditions is this claim valid?

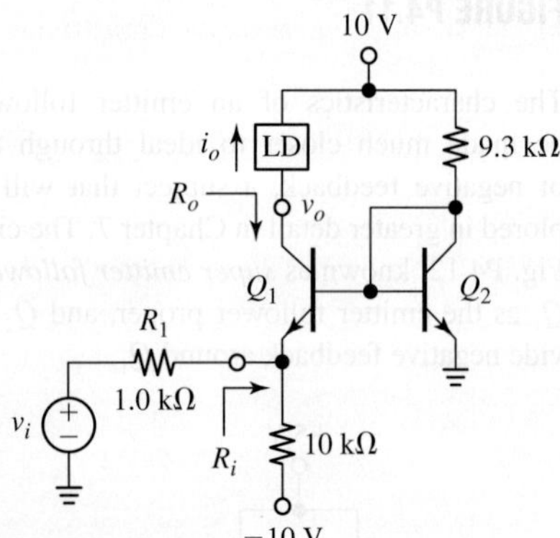

FIGURE P4.13

4.3 MOSFET Characteristics and Models Revisited

4.14 A MOS IC designer is using a process characterized by $\lambda' = 0.04$ μm/V and $k' = 100$ μA/V².

(a) If the designer wishes to achieve $a_{\text{intrinsic}} = -50$ V/V at $I_D = 100$ μA and with $V_{OV} = 0.4$ V, what are the required values of W and L? What happens to $a_{\text{intrinsic}}$ if:

(b) W is doubled?

(c) L is doubled?

(d) V_{OV} is doubled?

(e) If the value of L found in part (a) is doubled and the value of W found in part (a) is halved, find V_{OV} for $a_{\text{intrinsic}} = -80$ V/V. What is the corresponding value of I_D?

4.15 In Fig. P4.15, M_1 is designed to operate as a current sink and M_2 as a current source. If M_2 had the source and body tied together, there would be no difference in operation of the two FETs. However, M_2 is subject to the body effect, so its operation will differ from that of M_1. We wish to investigate this difference and see which of the two devices comes closer to ideal current source/sink behavior. Let both FETs have $k = 500$ μA/V², $V_{t0} = -1.0$ V, and $\lambda = 1/(25$ V$)$. Moreover, let $\gamma = 0.5$ V$^{1/2}$ and $|2\phi_p| = 0.65$ V.

(a) Calculate I_1 and R_{o1} at the edge of saturation for M_1. Hence, find $v_{L1(\text{max})}$ for which M_1 is still saturated. What is the percentage change in I_1 for a 1-V for a per-volt decrease in v_{L1} below $v_{L1(\text{max})}$?

(b) Calculate I_2 and R_{o2} at the edge of saturation for M_2. Hence, find $v_{L2(\text{max})}$ for which M_2 is still saturated. What is the percentage change in I_2 for a per-volt decrease in v_{L2} below $v_{L2(\text{max})}$?

(c) How must W_2 be changed if we want I_2 at M_2's edge of saturation to equal I_1 at M_1's edge of saturation? Does this affect the percentage change in I_2 for a 1-V decrease in v_{L2}? Comment on your results.

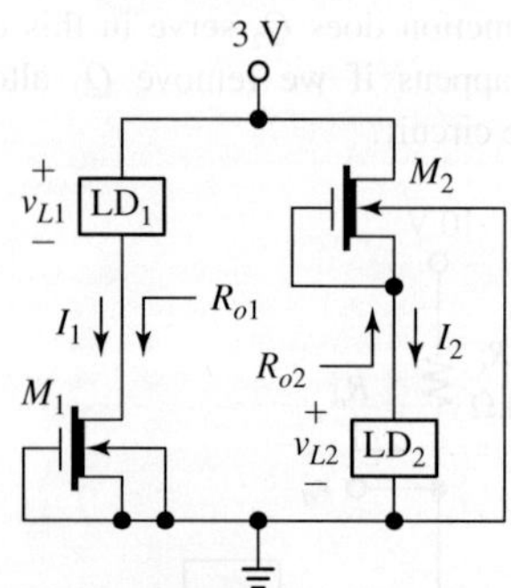

FIGURE P4.15

4.16 (a) Assuming the FET of Fig. P4.16 is operating in the active mode, derive an expression for the small-signal resistance R.

(b) Discuss the limiting cases $R_S \to 0$ and $r_o \to \infty$, and justify them in terms of known MOSFET properties.

Hint: replace the FET with its small-signal model, and use the test method.

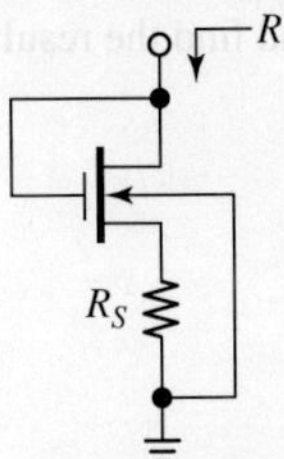

FIGURE P4.16

4.17 The depletion nMOSFET M_1 of Fig. P4.17 is a CS amplifier requiring no dc biasing source, and the pMOSFET M_2 is its active load.
(a) If $k_p = k_n = 200\ \mu\text{A/V}^2$, $V_{tp} = -V_{tn} = 1.0$ V, $\lambda_n = 1/(50\text{ V})$, and $\lambda_p = 1/(30\text{ V})$, find the dc voltage V_O at the output and the dc power P_D drawn by the circuit.
(b) Find R_o and v_o/v_i.
(c) Estimate $v_{O(\text{min})}$ and $v_{O(\text{max})}$, the lower and upper limits of the linear region of operation.

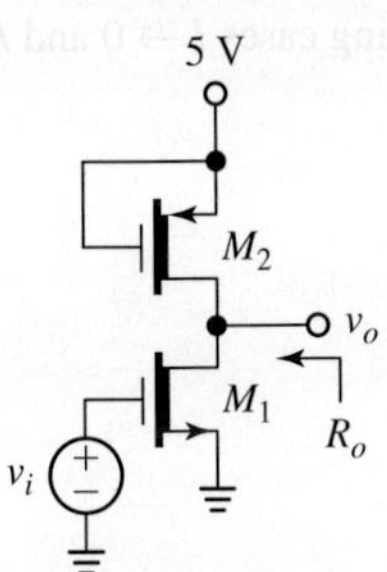

FIGURE P4.17

4.18 The depletion nMOSFET M_1 of Fig. P4.18 operates as a CS amplifier, and the nMOSFET M_2 as an active load.
(a) If $k' = 100\ \mu\text{A/V}^2$, $V_{t0} = -1.0$ V, $\lambda = 1/(25\text{ V})$, $\gamma = 0.4\ \text{V}^{1/2}$, and $|2\phi_p| = 0.6$ V. specify W/L ratios for the two devices so that with $V_{DD} = 5$ V the output node is biased at $V_O = 3.0$ V and the FETs draw $I_D = 50\ \mu\text{A}$.
(b) Find R_o and $a = v_o/v_i$.
(c) Estimate $v_{O(\text{min})}$ and $v_{O(\text{max})}$, the lower and upper limits of the linear region of operation.
(d) Suppose M_2's body is lifted off ground and tied to the source, and W_2/L_2 is made equal to W_1/L_1. If V_{DD} is adjusted so as to leave V_O

unchanged at 3.0 V, what is the new value of V_{DD}, and how do the values of R_o and a change? Comment on your findings.

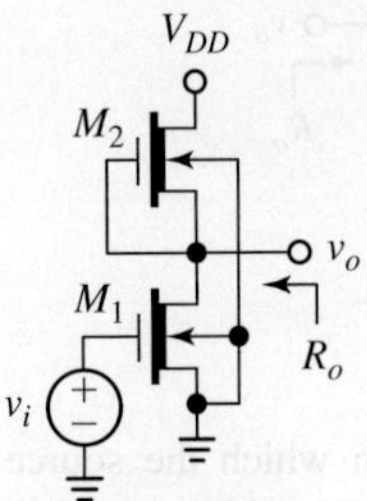

FIGURE P4.18

4.19 Let M_1 and M_2 in Fig. P4.19 be matched devices with $k = 200\ \mu\text{A/V}^2$, $V_t = 1.0$ V, and $\lambda = 1/(50\text{ V})$. Moreover, let M_3 have $k' = 30\ \mu\text{A/V}^2$, $V_t = -1.0$ V, and $\lambda = 1/(30\text{ V})$.
(a) Specify the ratio W_3/L_3 so as to bias the output node at $V_O = 0$ V.
(b) Find R_o and v_o/v_i.
(c) Estimate $v_{O(\text{min})}$ and $v_{O(\text{max})}$, the lower and upper limits of the linear region of operation.

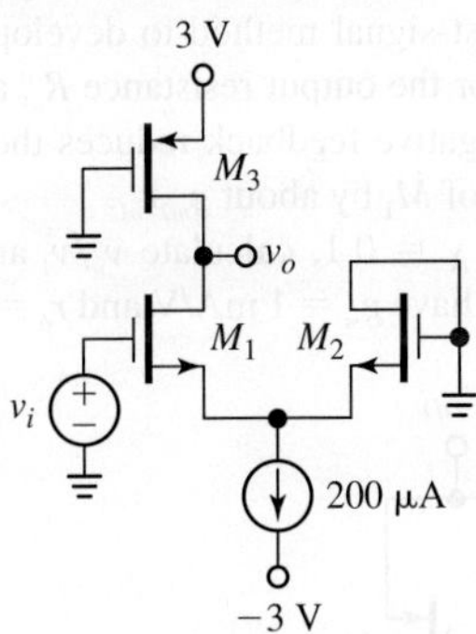

FIGURE P4.19

4.20 In Fig. P4.20 the source follower M_1 is biased by the current sink M_2. Let $k' = 100\ \mu\text{A/V}^2$, $\lambda = 1/(15\text{ V})$, $\gamma = 0.4\ \text{V}^{1/2}$, and $|2\phi_p| = 0.6$ V. Moreover, let M_1 have $V_{t0} = 0.5$ V and let M_2 have $V_{t0} = -1.0$ V.
(a) Specify W/L ratios for the two devices to bias the output at $V_O = -2$ V and the FETs at $I_D = 300\ \mu\text{A}$.
(b) Find R_o and v_o/v_i.

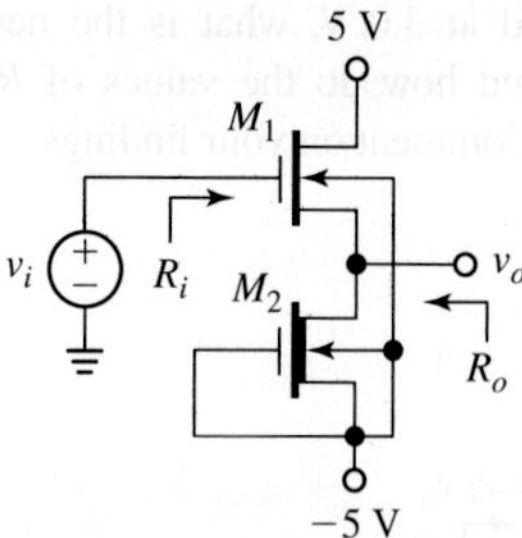

FIGURE P4.20

4.21 There are situations in which the source follower's output resistance R_o, which is dominated by $1/(g_m + g_{mb})$, is not sufficiently low due to the notoriously poor FET transconductance. A clever way to lower R_o is via negative feedback, a subject that will be explored in greater detail in Chapter 7. The circuit of Fig. P4.21, known as a *super source follower*, uses M_1 as the source follower proper, and M_2 to provide negative feedback around M_1. (Note that for the circuit to function we must have $I_2 > I_1$.)

(a) Draw the ac equivalent of the circuit by replacing each FET with its small-signal model, consisting of g_m, r_o, and g_{mb} (when appropriate). Hence, develop an expression for the gain v_o/v_i.

(b) Use the test-signal method to develop an expression for the output resistance R_o, and verify that negative feedback reduces the output resistance of M_1 by about $g_{m2}r_{o1}$.

(c) Assuming $\chi = 0.1$, calculate v_o/v_i and R_o if both FETs have $g_m = 1$ mA/V and $r_o = 20$ kΩ.

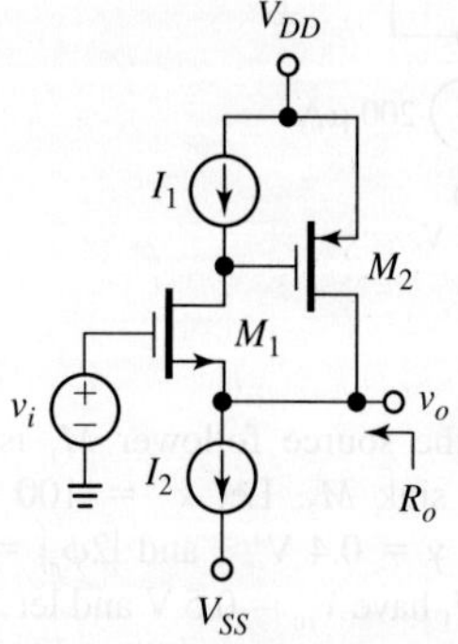

FIGURE P4.21

4.22 The circuit of Fig. P4.22 is required to sink $I_O = 1$ mA at $R_o \geq 100$ kΩ. Since the FET has $r_o = 20$ kΩ, the circuit uses source degeneration to raise the resistance seen looking into the drain.

(a) If $k = 2$ mA/V² and $V_{t0} = 0.5$ V, find the required values of R_S and V_G, assuming $\gamma = 0$.

(b) If $\gamma = 0.48$ V$^{1/2}$ and $|2\phi_p| = 0.6$ V, recalculate the required and V_G, and find the resulting R_o.

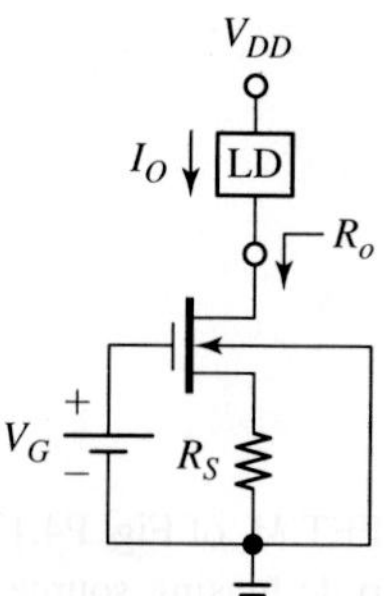

FIGURE P4.22

4.4 Darlington, Cascode, and Cascade Configurations

4.23 (a) In the Darlington configuration of Fig. P4.23 replace each BJT with its small-signal model, and use the test-signal method to obtain an expression for the resistance R_c seen looking into the collector of the composite device.

(b) Discuss the limiting cases $I \to 0$ and $I \to I_{C2}$.

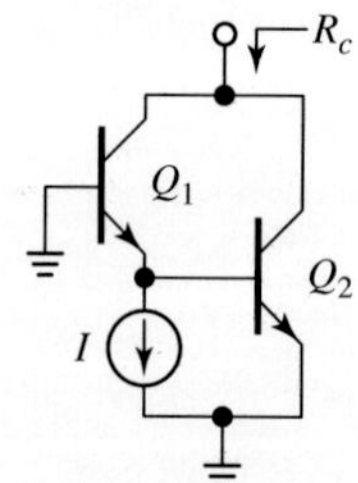

FIGURE P4.23

4.24 (a) In the Sziklai circuit of Fig. P4.24 replace each BJT with its small-signal model and use the test-signal method to obtain an expression for the ac resistance seen looking into each terminal of the composite device if the other two terminals are at ac ground.

(b) Assuming $\beta_1 = \beta_2 = 100$ and $V_{A1} = V_{A2} = 100$ V, calculate the above resistances if $I_{C2} = 1$ mA and $I = 90$ μA. Comment on your results and compare with the conventional Darlington circuit of Fig. 4.32b.

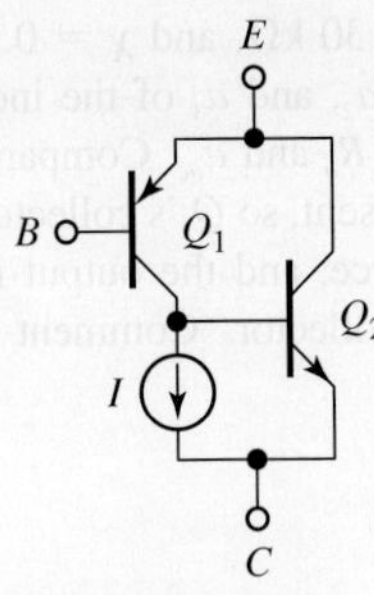

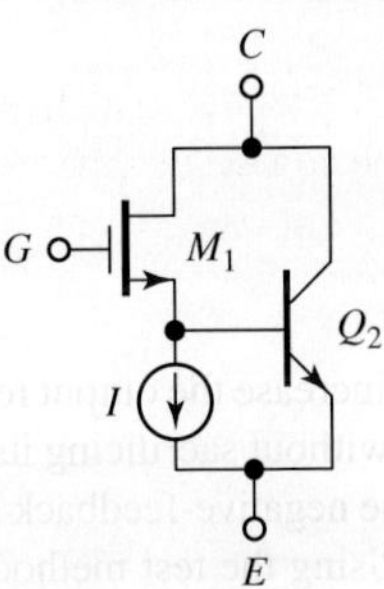

FIGURE P4.24

4.25 (a) In the BiMOS Darlington circuit of Fig. P4.25 replace each transistor with its small-signal model and use the test-signal method to obtain an expression for the ac resistance seen looking into each terminal of the composite device if the other two terminals are at ac ground.

(b) Calculate the above resistances if $g_{m1} = 0.5$ mA/V, $g_{m2} = 50$ mA/V, $r_{o1} = r_{o2} = 50$ kΩ, and $r_{\pi 2} = 2.0$ kΩ.

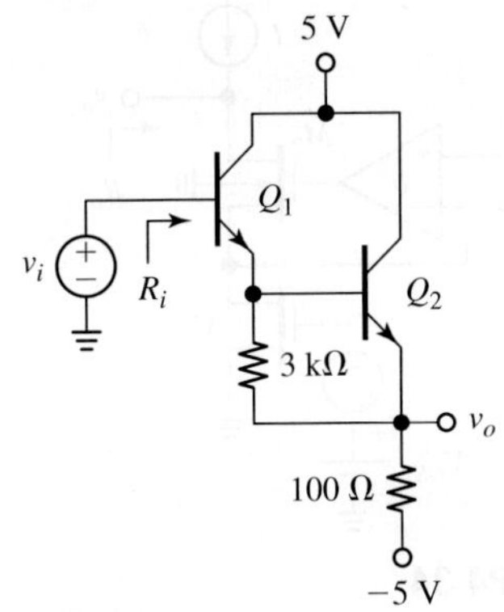

FIGURE P4.25

4.26 Assuming $\beta_1 = 150$, $\beta_2 = 100$, $V_{BE1(on)} = 0.7$ V, $V_{BE2(on)} = 0.8$ V, and $V_{A1} = V_{A2} = \infty$, find R_i and $a = v_o/v_i$ in the Darlington buffer of Fig. P4.26.

FIGURE P4.26

4.27 In the biMOS Darlington amplifier of Fig. P4.27 let M have $k' = 100$ μA/V^2, $V_t = 0.5$ V, $\chi = 0.2$, and let Q have $\beta = 100$ and $V_{BE} = 0.75$ V.

(a) Assuming $\lambda = 0$ and $V_A = \infty$, find W/L as well as V_G such that the lower limit of the linear output range is $v_{O(min)} = 1.5$ V, and the output node is biased in the middle of the range.

(b) Find $a = v_o/v_i$.

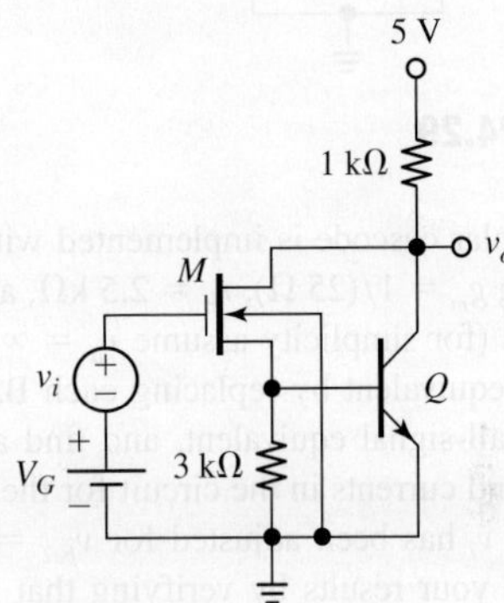

FIGURE P4.27

4.28 (a) For the CC-CB pair of Fig. P4.28 find R_i, R_o, and the gain v_o/v_i for the case $R_{E1} = R_{E2} = 0$. Assume matched BJTs with $\beta_0 = 200$ and $V_A = \infty$.

(b) Repeat, but for the case $R_{E1} = R_{E2} = 100$ Ω.

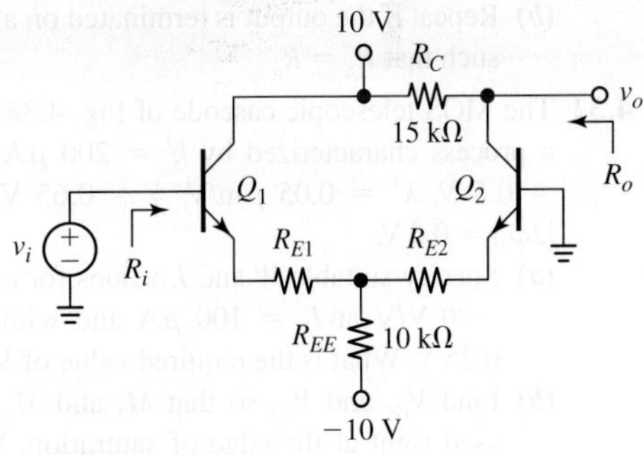

FIGURE P4.28

4.29 (a) For the CC-CE pair of Fig. P4.29 find the ratio I_{s2}/I_{s1} between the saturation currents of the two BJTs that will bias the output node at $V_O = 5$ V (assume negligible base currents).

(b) If $\beta_{npn} = 4\beta_{pnp} = 200$ and $V_{An} = 2V_{Ap} = 80$ V, find R_i, R_o, and $a = v_o/v_i$.

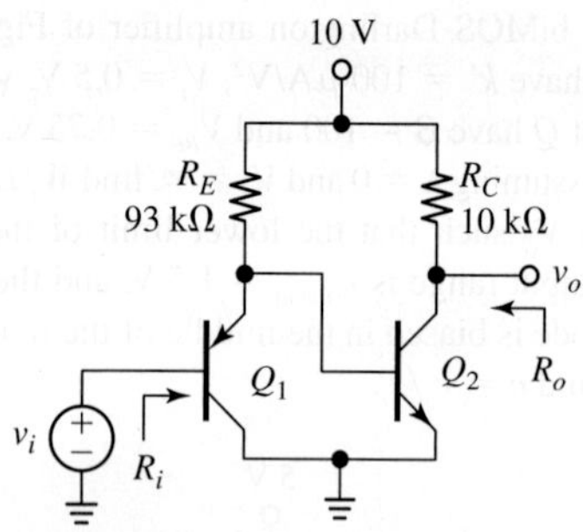

FIGURE P4.29

4.30 (*a*) A bipolar cascode is implemented with BJTs having $g_m = 1/(25\ \Omega)$, $r_\pi = 2.5\ \text{k}\Omega$, and $r_o = 50\ \text{k}\Omega$ (for simplicity assume $r_\mu = \infty$). Draw its ac equivalent by replacing each BJT with its small-signal equivalent, and find all voltages and currents in the circuit for the case in which v_i has been adjusted for $v_{be2} = 1$ mV. Check your results by verifying that KCL is satisfied at v_{c1} and v_o.

(*b*) Find the output load R_L that will reduce v_o to ½ the value of part (*a*) for the same value of v_i, and check again via KCL.

4.31 (*a*) Suppose the FETs in the cascode of Fig. 4.37*a* have $g_m = 1/(2\ \text{k}\Omega)$, $r_o = 25\ \text{k}\Omega$, and $\chi = 0.2$. Find all the node voltages in the circuit if v_i is a 1-mV ac signal.

(*b*) Repeat if the output is terminated on a load R_L such that $R_L = R_o$.

4.32 The MOS telescopic cascode of Fig. 4.38 utilizes a process characterized by $k' = 200\ \mu\text{A/V}^2$, $V_{t0} = 0.5$ V, $\lambda' = 0.05\ \mu\text{m/V}$, $\gamma = 0.65\ \text{V}^{1/2}$, and $|2\phi_p| = 0.6$ V.

(*a*) Specify suitable W and L values for $a_{\text{intrinsic}} = -20$ V/V at $I_D = 100\ \mu\text{A}$ and with $V_{OV} = 0.25$ V. What is the required value of V_{GS1}?

(*b*) Find V_{G2} and V_{G3} so that M_2 and M_3 are biased right at the edge of saturation. What is the lower limit $v_{O(\text{min})}$ of the linear region of operation?

(*c*) Find R_o and a_{oc}.

4.33 The BiMOS telescopic cascode of Fig. P4.33 uses M_3 to further raise the values of R_o and a_{oc} provided by the bipolar cascode. Assuming the BJTs have $g_m = 1/(50\ \Omega)$, $r_\pi = 5\ \text{k}\Omega$, and $r_o = 100\ \text{k}\Omega$ (for simplicity assume $r_\mu = \infty$), and the FET has

$g_m = 1/(2\ \text{k}\Omega)$, $r_o = 30\ \text{k}\Omega$, and $\chi = 0.15$, find the voltage gains a_1, a_2, and a_3 of the individual transistors, as well as R_o and a_{oc}. Compare to the case in which M_3 is absent, so Q_2's collector is terminated on the I source, and the output is taken directly from Q_2's collector. Comment on the differences.

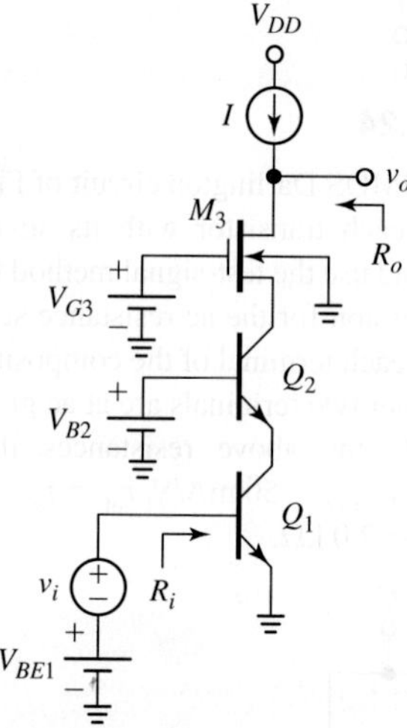

FIGURE P4.33

4.34 An ingenious way to increase the output resistance of the MOS cascode without sacrificing its OVS is to use an op amp in the negative-feedback arrangement of Fig. P4.34. Using the test method, obtain an expression for R_o, and show that for a large gain a, $R_o \cong a g_{m2} r_{o1} r_{22}$. What is the OVS?

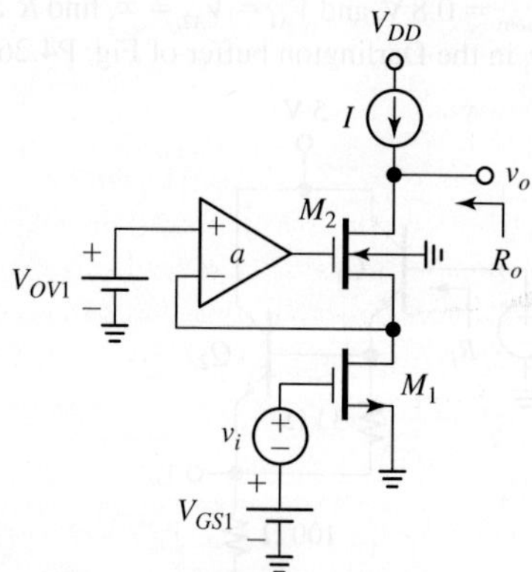

FIGURE P4.34

4.35 Suppose the BJTs of the folded cascode of Fig. 4.39a have $\beta_{01} = 150$, $V_{A1} = 75$ V, $\beta_{02} = 50$, and $V_{A2} = 30$ V.

   (a) If $I_{BIAS} = 2$ mA, $I_{LOAD} = 1$ mA, and V_{BE1} has been adjusted so as to bias the output node at 0 V dc, find R_i, R_o, a_1, a_2, and a_{oc}.

   (b) Assuming $V_{EC2(on)} = 0.7$ V and $V_{EC2(sat)} = 0.2$ V, specify V_{B2} so that $v_{O(max)} = 2.5$ V. Hence, find the maximum input-signal amplitude such that both constraints $|v_{be2}| \leq 5$ mV and $|v_o| \leq 2.5$ V are met.

   (c) Repeat part (a) if $I_{BIAS} = 1.5$ mA, $I_{LOAD} = 1$ mA, and V_{BE1} is readjusted so as to bias the output node still at 0 V dc.

   (d) Repeat (a) if $I_{BIAS} = 1.5$ mA, $I_{LOAD} = 0.5$ mA, and V_{BE1} is once again readjusted so as to bias the output node still at 0 V dc. Compare and comment.

4.36 Let the FETs of the folded cascode of Fig. 4.39b have $V_{t1} = -V_{t2} = 0.5$ V, $k_1 = k_2 = 0.8$ mA/V^2, $\lambda_1 = 1/(15$ V$)$, and $\lambda_2 = 1/(10$ V$)$.

   (a) If $I_{BIAS} = 200$ μA and $I_{LOAD} = 100$ μA, find the required V_{GS1}. Hence, find V_{G2} such that $v_{O(max)} = 1.0$ V.

   (b) If the I_{BIAS} source has an equivalent parallel resistance of 250 kΩ, and the I_{LOAD} sink has an equivalent parallel resistance of 5 MΩ, find a_1, a_2, and v_o/v_i.

   (c) What is the maximum amplitude of v_i that will still yield an undistorted v_o?

4.37 An IC designer, seeking to combine the advantages of BJTs and MOSFETs in a BiMOS cascode, is evaluating the two circuits of Fig. P4.37. Both BJTs have $g_m = 1/(25$ $\Omega)$, $r_\pi = 4$ kΩ, and $r_o = 50$ kΩ (for simplicity assume $r_\mu = \infty$), and both FET have $g_m = 1/(1$ k$\Omega)$, $r_o = 20$ kΩ, and $\chi = 0.2$.

   (a) Find R_i, R_o, and a_{oc} for each circuit.

   (b) Compare the two cascodes in terms of the above three parameters, and state which one, if any, is better. How do they compare against an all-bipolar and an all-MOS realization?

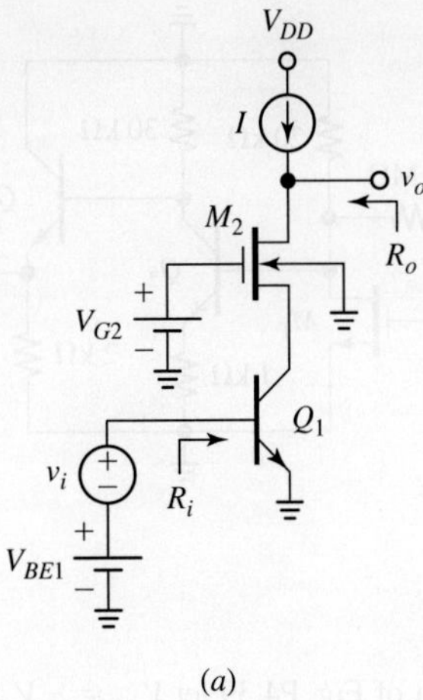

(a)

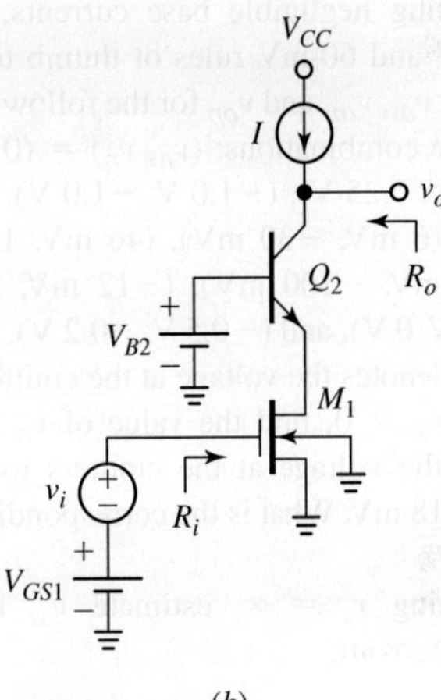

(b)

FIGURE P4.37

4.38 Suppose the transistors of the three-stage amplifier of Fig. P4.38 have, respectively, $g_{m1} = 1/(1.25$ k$\Omega)$, $g_{m2} = 1/(50$ $\Omega)$, $g_{m3} = 1/(10$ $\Omega)$, and $\beta_{02} = \beta_{03} = 150$. For simplicity, assume $r_{o1} = r_{o2} = r_{o3} = \infty$. What is the overall voltage gain v_o/v_i?

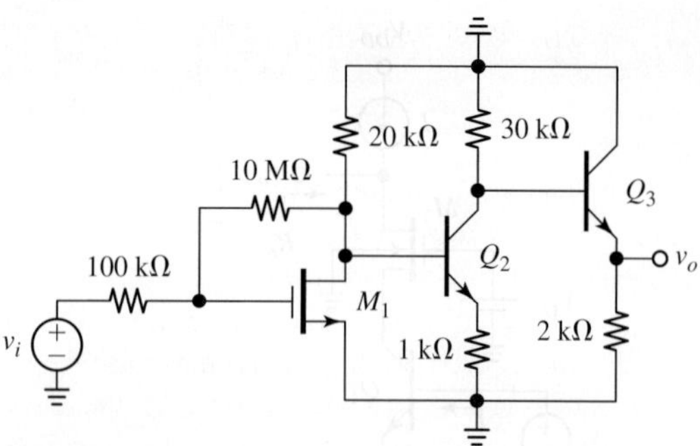

FIGURE P4.38

4.5 Differential Pairs

4.39 In the circuit of Fig. P4.39 let $V_{CC} = -V_{EE} = 5$ V, $R_{C1} = R_{C2} = 5$ kΩ, $I_{EE} = 1$ mA, and $R_{EE} = \infty$.

(a) Assuming negligible base currents, use the 18-mV and 60-mV rules of thumb to predict i_{C1}, i_{C2}, v_{O1}, v_{O2}, and v_{OD} for the following input voltage combinations: $(v_{I1}, v_{I2}) = (0$ V, 0 V$)$, $(0.25$ V, 0.25 V$)$, $(-1.0$ V, -1.0 V$)$, $(18$ mV, 0 V$)$, $(6$ mV, -30 mV$)$, $(46$ mV, 100 mV$)$, $(-40$ mV, -100 mV$)$, $(-12$ mV, 30 mV$)$, $(0.12$ V, 0 V$)$, and $(-0.5$ V, -0.2 V$)$.

(b) If V_{E0} denotes the voltage at the emitters when $v_{I1} = v_{I2} = 0$, find the value of v_{I1} that will cause the voltage at the emitters to raise to $V_{E0} + 18$ mV. What is the corresponding value of v_{OD}?

(c) Assuming $r_o = \infty$, estimate v_{od} if $v_{id} = (5$ mV$) \cos \omega t$.

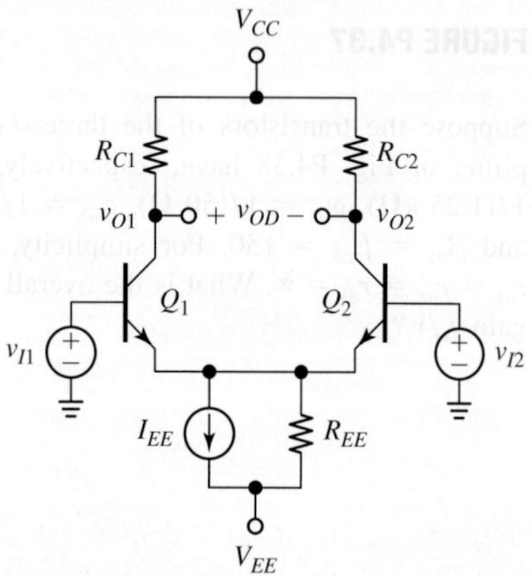

FIGURE P4.39

4.40 Shown in Fig. P4.40 is the *pnp* realization of the EC pair concept. Let $V_{EE} = -V_{CC} = 6$ V, $R_{C1} = R_{C2} = 5$ kΩ, $I_{EE} = 1$ mA, and $R_{EE} = \infty$.

(a) Find v_{O1}, v_{O2}, and v_{OD} if $v_{I1} = v_{I2} = 0$.

(b) What happens if $v_{I1} = v_{I2} = 1.0$ V? If $v_{I1} = v_{I2} = -1.0$ V?

(c) Find v_{O1}, v_{O2}, and v_{OD} if $v_{I1} = 30$ mV and $v_{I2} = 0$.

(d) If $v_{I1} = 0$, find v_{I2} so that $v_{OD} = 4$ V.

(e) If V_{E0} denotes the voltage at the emitters in part (a), what is the voltage at the emitters in part (b)? In part (c)?

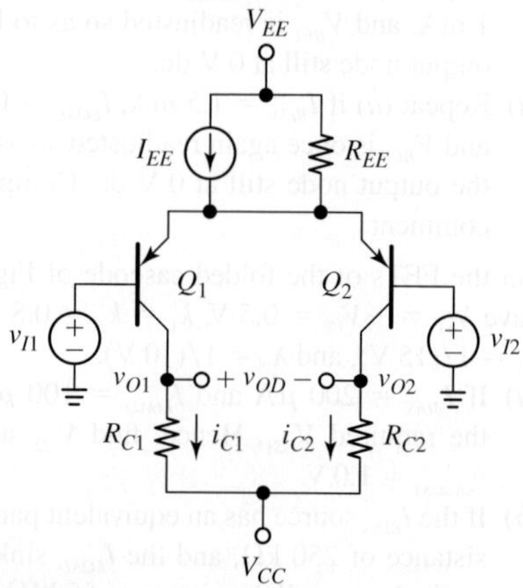

FIGURE P4.40

4.41 In the circuit of Fig. P4.39 let $V_{CC} = -V_{EE} = 6$ V, $R_{C1} = R_{C2} = 10$ kΩ, $I_{EE} = 0.45$ mA, and $R_{EE} = \infty$, and let the BJTs be mismatched, with Q_1's emitter area being 25% larger than Q_2's.

(a) If $v_{I1} = v_{I2} = 0$, find i_{C1}, i_{C2}, and v_{OD} (assume negligible base currents).

(b) If $v_{I2} = 0$, find the value of v_{I1} that will drive v_{OD} to zero.

(c) If $v_{I1} = v_{I2} = 0$, find ΔR such that raising one of the collector resistances (which one?) to $(10$ k$\Omega + \Delta R)$ while simultaneously lowering the other to $(10$ k$\Omega - \Delta R)$ will result in $v_{OD} = 0$. What are the collector voltages now?

4.42 We wish to design an EC pair of the type of Fig. P4.39 so that when driven with $v_{I1} = (10$ mV$) \cos \omega t$ and $v_{I2} = 0$ it produces the *largest possible* output signal under the constraint that neither collector voltage ever drops below 0 V, thus avoiding forward-biasing the B-C junctions of the BJTs.

(**a**) If $V_{CC} = -V_{EE} = 12$ V, $R_{C1} = R_{C2} = 10$ kΩ, and $R_{EE} = \infty$, what is the required I_{EE}, assuming negligible base currents?

Hint: increasing I_{EE} to raise a will also lower V_{O1} and V_{O2} and thus bring the BJTs closer to saturation. The most critical instants are when v_{o1} and v_{o2} reach their negative peaks.

(**b**) Assuming $r_o = \infty$, what is the ensuing gain a? What are the total signals v_{O1}, v_{O2}, and v_{OD}?

4.43 In the circuit of Fig. P4.40 let $V_{EE} = -V_{CC} = 5$ V, $R_{C1} = R_{C2} = 12$ kΩ, $I_{EE} = 0.6$ mA, and $R_{EE} = \infty$.

(**a**) Assuming $r_o = \infty$, find the gain a. Find v_{od} if $v_{I1} = (8$ mV$)\cos\omega t$ and $v_{I2} = 0$.

(**b**) Assuming negligible base currents, find the total voltages v_{O1} and v_{O2} (express each voltage as the sum of its dc and ac components).

(**c**) Find v_{od} if a 30-kΩ load is connected between the two collectors.

(**d**) Find the total signals v_{O1}, v_{O2}, and v_{OD} if R_{C2} is accidentally lowered to 10 kΩ. Comment on your results.

4.44 Figure P4.44 shows a differential amplifier variant that uses emitter degeneration. Let $V_{CC} = -V_{EE} = 5$ V, $R_{C1} = R_{C2} = 3.0$ kΩ, $R_{E1} = R_{E2} = 120$ Ω, and $I_{EE} = 1.0$ mA.

(**a**) If $v_{I2} = 0$, find the value of v_{I1} for which $i_{C1} = 2i_{C2}$ (assume negligible base currents).

(**b**) If $v_{I1} = 25$ mV, find v_{I2} so that $v_{OD} = 1.5$ V.

(**c**) Using iterations, find v_{OD} if $v_{ID} = 100$ mV.

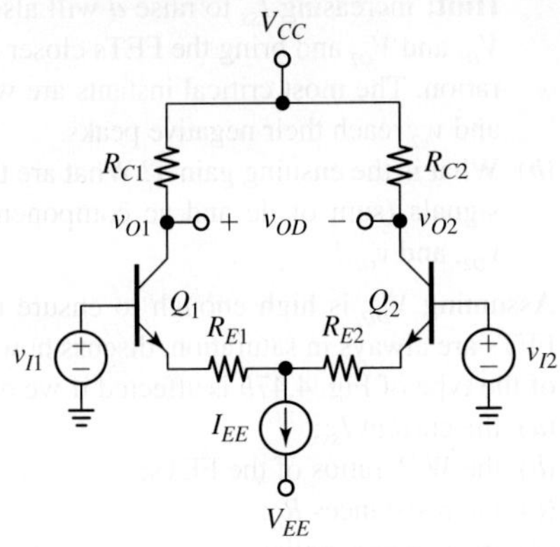

FIGURE P4.44

4.45 In the circuit of Fig. P4.44 let $V_{CC} = -V_{EE} = 6$ V, $R_{C1} = R_{C2} = 100$ kΩ, and $I_{EE} = 0.5$ mA. Moreover, let $v_{I1} = (100$ mV$)\cos\omega t$ and $v_{I2} = 0$.

(**a**) Specify suitable values for $R_{E1} = R_{E2}$ so as to ensure that both BJTs meet the small-signal constraint $|v_{be}| \le 5$ mV. What are the ensuing emitter ac voltages v_{e1} and v_{e2}?

(**b**) Find v_{od}. Find the small-signal resistance R_{i1} seen looking into Q_1's base if $\beta_0 = 250$.

4.46 The EC pair Q_1–Q_2 of Fig. P4.46 utilizes the emitter followers Q_3–Q_4 to lower the currents drawn from the sources v_{I1} and v_{I2} and thus raise the ac input resistances R_{i1} and R_{i2} seen by the same sources. Let $V_{CC} = -V_{EE} = 6$ V, $R_{C1} = R_{C2} = 10$ kΩ, $R_{E3} = R_{E4} = 15$ kΩ, and $I_{EE} = 1$ mA. Moreover, assume $V_{BE1} = V_{BE2} = 0.7$ V and $\beta_F = 100$ for all BJTs.

(**a**) Find the base currents I_{B3} and I_{B4} and the collector voltages V_{O1} and V_{O2} at dc balance.

(**b**) If $v_{I1} = v_i$ and $v_{I2} = 0$, obtain expressions for v_{e1}, v_{e2}, v_{e3}, and v_{e4} in terms of v_i.

Hint: exploit the symmetry of the circuit.

(**c**) Obtain expressions for v_{o1} and v_{o2} in terms of v_i. Hence, assuming $r_o = \infty$, find the gain v_{od}/v_i.

(**d**) Again exploiting circuit symmetry, find the input resistances R_{i1} and R_{i2}.

(**e**) What is the upper limit on $|v_i|$ for which the condition $|v_{be}| \le 5$ mV is met by all BJTs?

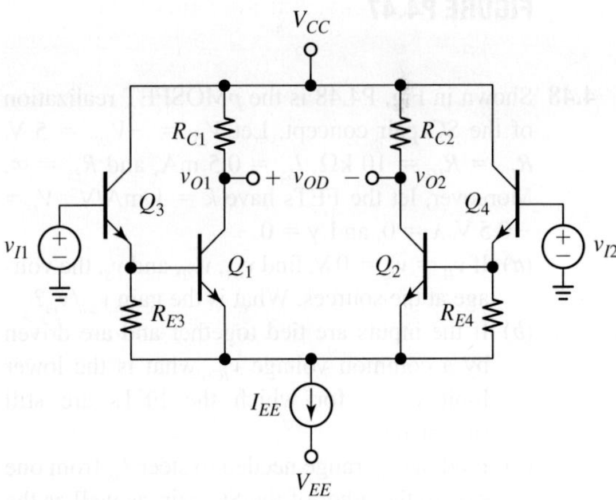

FIGURE P4.46

4.47 In the circuit of Fig. P4.47 let $V_{DD} = -V_{SS} = 3.5$ V, $R_{D1} = R_{D2} = 10$ kΩ, $I_{SS} = 0.4$ mA, and $R_{SS} = \infty$. Moreover, let the FETs have $k' = 100$ μA/V^2, $V_t = 0.6$ V, $\lambda = 0$, and $\gamma = 0$.

(*a*) Specify the W/L ratio for the FETs that will yield $v_{od}/v_{id} = -10$ V/V.

(*b*) If $v_{I1} = v_{I2} = 0$ V, find v_{O1}, v_{O2}, and v_S, the voltage at the sources.

(*c*) If the inputs are tied together and are driven by a common voltage v_{IC}, what is the upper limit on v_{IC} for which the FETs are still saturated?

(*d*) What is the v_{ID} range needed to steer I_{SS} from one side to the other of the SC pair? Find v_{O1}, v_{O2}, and v_S at the extremes of this range.

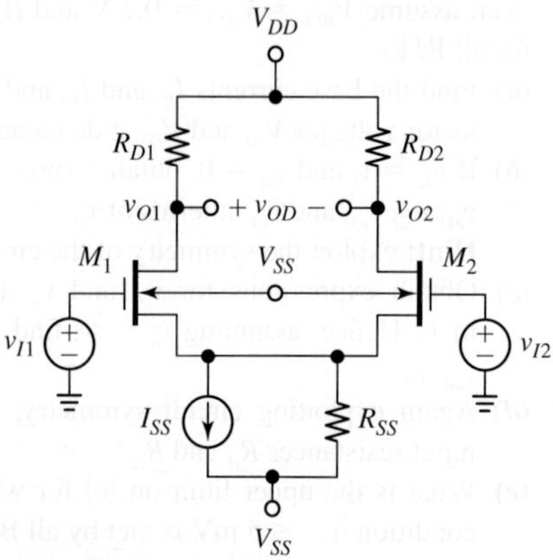

FIGURE P4.47

4.48 Shown in Fig. P4.48 is the *p*MOSFET realization of the SC pair concept. Let $V_{SS} = -V_{DD} = 5$ V, $R_{D1} = R_{D2} = 10$ kΩ, $I_{SS} = 0.5$ mA, and $R_{SS} = \infty$. Moreover, let the FETs have $k = 1$ mA/V^2, $V_t = -0.5$ V, $\lambda = 0$, and $\gamma = 0$.

(*a*) If $v_{I1} = v_{I2} = 0$ V, find v_{O1}, v_{O2}, and v_S, the voltage at the sources. What is the gain v_{od}/v_{id}?

(*b*) If the inputs are tied together and are driven by a common voltage v_{IC}, what is the lower limit $v_{IC(\text{min})}$ for which the FETs are still saturated?

(*c*) Find the v_{ID} range needed to steer I_{SS} from one side to the other of the SC pair, as well as the values of v_{O1}, v_{O2}, and v_S at the extremes of this range.

(*d*) Find v_{ID} so that $v_{OD} = 4$ V.

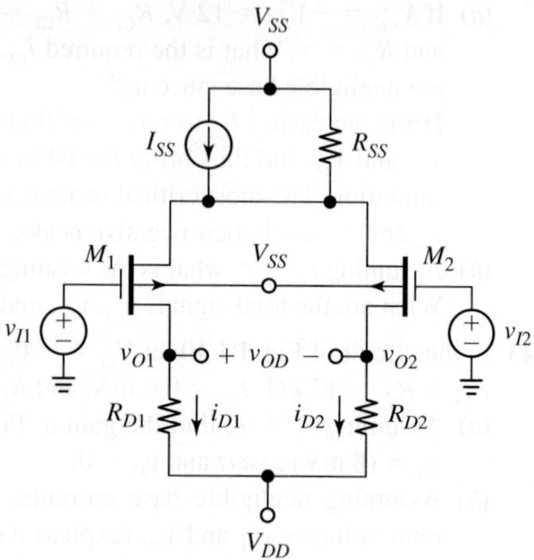

FIGURE P4.48

4.49 We wish to design a SC pair of the type of Fig. P4.47 so that when driven with $v_{I1} = (0.2$ V$)\cos\omega t$ and $v_{I2} = 0$ it produces the *largest possible* output signal under the constraint that neither FET ever leave the saturation region. The available components are $V_{DD} = -V_{SS} = 5$ V and $R_{D1} = R_{D2} = 10$ kΩ, and the FETs have $k' = 100$ μA/V^2, $V_t = 0.4$ V, $\lambda = 0$, and $\gamma = 0$.

(*a*) Assuming $R_{SS} = \infty$, and supposing we wish to satisfy Eq. (4.94) by an order of magnitude, or $4V_{OV} = 10 \times (0.2$ V$)$, what is the required I_{SS}? What are the required W/L ratios for the FETs? **Hint:** increasing I_{SS} to raise a will also lower V_{O1} and V_{O2} and bring the FETs closer to saturation. The most critical instants are when v_{o1} and v_{o2} reach their negative peaks.

(*b*) What is the ensuing gain a? What are the total signals (sum of dc and ac components) v_{O1}, v_{O2}, and v_{OD}?

4.50 Assuming V_{DD} is high enough to ensure that the FETs are always in saturation, discuss how a VTC of the type of Fig. 4.47*b* is affected if we double:

(*a*) the current I_{SS};

(*b*) the W/L ratios of the FETs;

(*c*) the resistances R_D;

(*d*) the power supplies.

(*e*) What happens if a load $R_L = 2R_D$ is connected between the drains? In each case, what is the effect (expansion, compression, by how much) upon the horizontal scale? The vertical scale? The gain a?

4.51 In the circuit of Fig. P4.48 let $V_{SS} = -V_{DD} = 5$ V, $R_{D1} = R_{D2} = 15$ kΩ, $I_{SS} = 0.3$ mA, and $R_{SS} = \infty$. The FETs have $k' = 0.1$ mA/V^2, $V_t = -0.5$ V, $\lambda = 0$, and $\gamma = 0$, but are mismatched in their W/L ratios, having $W_1/L_1 = 10$ and $W_2/L_2 = 15$.

   (a) If $v_{I1} = v_{I2} = 0$, find v_{OD} and v_S, the voltage at the sources.

   (b) If $v_{I2} = 0$, find the value of v_{I1} that will equalize the drain currents and thus drive v_{OD} to zero.

4.52 As we know, resistors are undesirable in MOS IC technology, so the circuit of Fig. P4.52 utilizes the diode-connected pair M_3-M_4 in lieu of the resistance pair R_{D1}-R_{D2}. Let $V_{SS} = -V_{DD} = 3$ V, $I_{SS} = 200$ μA, and $R_{SS} = \infty$. Moreover, let the FETs have $V_{tn} = -V_{tp} = 0.4$ V, $k'_n = 2.5k'_p = 100$ μA/V^2, $\lambda = 0$, and $\gamma = 0$.

   (a) Obtain an expression for the gain $a = v_{od}/v_{id}$.

   (b) Find a relationship between the nMOS ratio W_n/L_n and the pMOS ratio W_p/L_p that will result in $a = -4$ V/V.

   (c) Specify suitable values for W_n/L_n and W_p/L_p if the input is $v_{id} = (0.1\ \text{V})\cos\omega t$. Impose $4V_{OVn} = 10 \times (0.1$ V). What is v_{od}?

   (d) Find the total (sum of the dc and ac components) signals v_{O1}, v_{O2}.

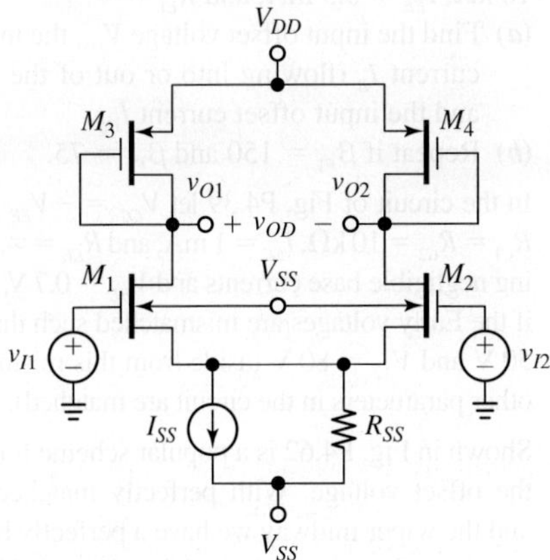

FIGURE P4.52

4.6 Common-Mode Rejection Ratio in Differential Pairs

4.53 We wish to investigate the performance of the EC amplifier of Fig. P4.39 at two extremes, one being when emitter bias is provided entirely by the I_{EE} source and thus $R_{EE} = \infty$, and the other being when it is provided entirely by the R_{EE} resistance and

thus $I_{EE} = 0$. Let $V_{CC} = -V_{EE} = 12$ V and $R_{C1} = R_{C2} = 10$ kΩ, and let the BJTs have $\beta_0 = 160$ and $V_A = 50$ V (assume $r_\mu = \infty$).

   (a) Find a_{dm}, a_{cm}, R_{id}, R_{ic}, and the CMRR (both for single-ended and double-ended utilization) if $I_{EE} = 1$ mA and $R_{EE} = \infty$.

   (b) Repeat, but with $I_{EE} = 0$ and $R_{EE} = 11.3$ kΩ. Compare the two cases and comment.

4.54 In the SC circuit of Fig. P4.47 let $V_{DD} = -V_{SS} = 2.5$ V, $R_{D1} = R_{D2} = 10$ kΩ, $I_{SS} = 0.2$ mA, and $R_{SS} = 200$ kΩ. Moreover, let the FETs have $k = 1.25 \pm 2\%$ mA/V^2, $V_t = 0.6$ V $\pm\ 1\%$ V, $\lambda = 1/(15$ V), and $\chi = 0.2$.

   (a) Using nominal parameter values, find a_{dm} and a_{cm}.

   (b) Investigate how the above tolerances affect the CMRR for the case of double-ended utilization. Find its value, in dB, both for the worst-case scenario and the case in which the tolerances are uncorrelated.

   Hint: develop an expression for $\Delta g_m/g_m$ in terms of $\Delta k/k$ and $\Delta V_{OV}/V_{OV}$.

4.55 In the EC circuit of Fig. P4.44 let $V_{CC} = -V_{EE} = 6$ V, $R_{C1} = R_{C2} = 12$ kΩ, $R_{E1} = R_{E2} = 250$ Ω, and $I_{EE} = 0.5$ mA. Moreover, assume the I_{EE} source has a parallel resistance $R_{EE} = 150$ kΩ, and the BJTs have $\beta_0 = 125$ and $r_o = 200$ kΩ.

   (a) Using half-circuit analysis, find a_{dm}, a_{cm}, R_{id}, R_{ic}, and the CMRR (both for single-ended and double-ended utilization).

   (b) If the BJTs suffer from $\pm 3\%$ mismatches in their emitter areas A_{E1} and A_{E2} and $\pm 5\%$ mismatches in their effective base widths W_{B1} and W_{B2}, how is the CMRR (double-ended utilization) affected? Consider both the worst-case scenario and the case in which the mismatches are uncorrelated.

   Hint: develop an expression for $\Delta g_m/g_m$ in terms of $\Delta A_E/A_E$ and $\Delta W_B/W_B$.

4.56 Let the BJTs in the differential amplifier of Fig. P4.56 have $\beta_0 = 150$ and $V_A = 75$ V.

   (a) Using half-circuit analysis, find a_{dm}, a_{cm}, R_{id}, R_{ic}, and the CMRR, both for single-ended and double-ended utilization.

   (b) If the current sink biasing Q_2 is 10% higher than the nominal value shown, how is the CMRR (double-ended utilization) affected?

   Hint: develop an expression for $\Delta g_m/g_m$ in terms of $\Delta g_{m1}/g_{m1}$ and $\Delta g_{m2}/g_{m2}$.

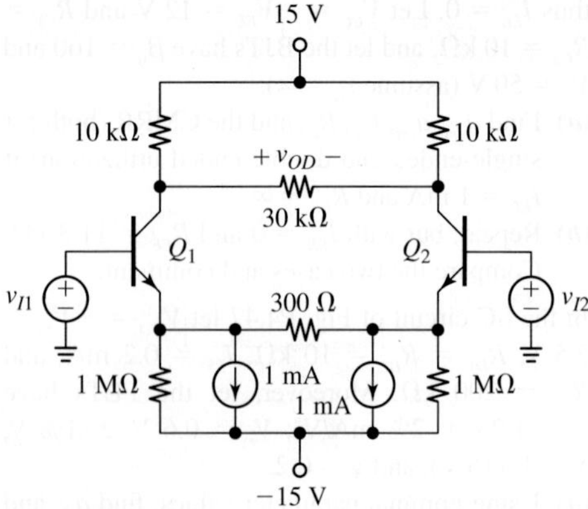

FIGURE P4.56

4.57 As we know, resistors are undesirable in MOS IC technology, so the circuit of Fig. P4.57 utilizes the pair M_3-M_4 in lieu of the resistance pair R_{D1}-R_{D2}. Let $V_{DD} = -V_{SS} = 2.5$ V and $R_{SS} = 100$ kΩ, and suppose I_{SS} has been adjusted so that at dc balance each FET draws 100 μA. Moreover, let $V_{tn} = -V_{tp} = 1.0$ V, $k'_n = 2.5\, k'_p = 100\, \mu$A/V^2, $\lambda_n = 1/(10\,\text{V})$, $\lambda_p = 1/(20\,\text{V})$, and $\chi_n = 0.2$.

(a) Specify the W_n/L_n ratio for the M_1-M_2 pair so as to achieve $g_{mn} = 1.25$ mA/V, and the W_p/L_p ratio for the M_3-M_4 pair so that at dc balance the circuit gives $V_{O1} = V_{O2} = 0$ V.

(b) Find $v_{O(\max)}$ and $v_{O(\min)}$, the upper and lower limits of the linear region of operation.

(c) Find a_{dm}, a_{cm}, and the CMRR both for single-ended and double-ended utilization.

(d) How does a $\pm 10\%$ mismatch in the λ_p's affect the CMRR for double-ended utilization?

Hint: use $a_{cm} = (v_{o1} - v_{o2})/v_{ic}$.

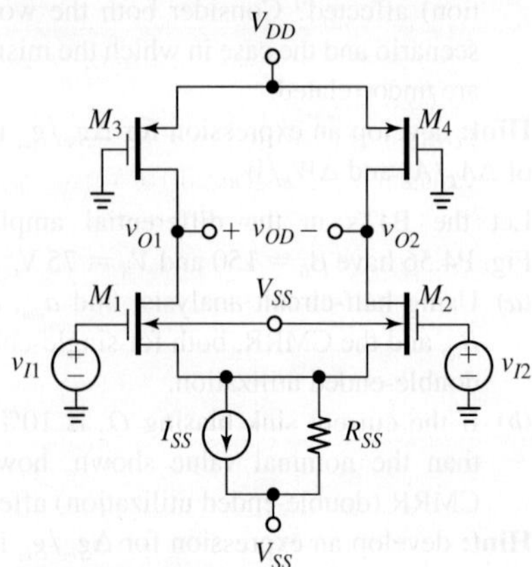

FIGURE P4.57

4.58 In the circuit of Fig. P4.52 let $I_{SS} = 250\,\mu$A and $R_{SS} = 250$ kΩ, and let the FETs have $k_n = 1$ mA/V^2, $\lambda_n = 1/(25\,\text{V})$, $\chi_n = 0.2$, $k_p = 0.1$ mA/V^2, and $\lambda_p = 1/(20\,\text{V})$.

(a) Find a_{dm}, a_{cm}, and the CMRR both for single-ended and double-ended utilization.

(b) How does a $\pm 10\%$ mismatch in the g_{mn}'s affect the CMRR for double-ended utilization?

(c) Repeat part (b), but for the case a $\pm 10\%$ mismatch in the g_{mp}'s.

Hint: adapt Eq. (4.111) to the present circuit.

4.59 Suppose that because of a production error the FETs of Fig. P4.47 are grossly mismatched such that $k_2 = 2k_1 = 1.6$ mA/V^2. Otherwise, the V_t's are matched, and let us also assume $\lambda = 0$ and $\gamma = 0$ for simplicity. If $R_{D1} = R_{D2} = 10$ kΩ, $I_{SS} = 0.3$ mA, and $R_{SS} = 100$ kΩ, find $a_{dm} = (v_{o1} - v_{o2})/v_{id}$ and $a_{cm} = (v_{o1} - v_{o2})/v_{ic}$. Hence, calculate the CMRR for double-ended utilization.

4.7 Input Offset Voltage/Current in Differential Pairs

4.60 Suppose the emitter areas of the *pnp* BJTs of Fig. P4.40 are mismatched such that $A_{E1} = 1.15 A_{E2}$. Otherwise, the BJTs have $\beta_{F1} = \beta_{F2} = 100$. Moreover, $V_{CC} = -V_{EE} = 5$ V, $R_{C1} = R_{C2} = 10$ kΩ, $I_{EE} = 0.5$ mA, and $R_{EE} = \infty$.

(a) Find the input offset voltage V_{OS}, the input bias current I_B (flowing into or out of the BJTs?), and the input offset current I_{OS}.

(b) Repeat if $\beta_{F1} = 150$ and $\beta_{F2} = 75$.

4.61 In the circuit of Fig. P4.39 let $V_{CC} = -V_{EE} = 10$ V, $R_{C1} = R_{C2} = 10$ kΩ, $I_{EE} = 1$ mA, and $R_{EE} = \infty$. Assuming negligible base currents and $V_{BE} = 0.7$ V, find V_{OS} if the Early voltages are mismatched such that $V_{A1} = 50$ V and $V_{A2} = 80$ V (aside from this mismatch, all other parameters in the circuit are matched).

4.62 Shown in Fig. P4.62 is a popular scheme for nulling the offset voltage. With perfectly matched halves and the wiper midway we have a perfectly balanced circuit, giving $V_{C1} = V_{C2}$ and thus $V_O = 0$. However, moving the wiper to the right or the left will imbalance the circuit, so by proper choice of the direction and extent of this deliberate imbalance we can cancel out the circuit's inherent imbalance and thus null V_O, creating the appearance of an *offset-less* circuit.

(a) Let $V_{CC} = -V_{EE} = 12$ V, $I_{EE} = 1$ mA, and $R_{EE} = \infty$. Moreover, suppose R_{C1} is 8% higher and R_{C2} is 5% lower than the intended nominal value of 10 kΩ, and suppose the BJTs are mismatched such that I_{s1} is 10.5% higher than I_{s2}. If R is a 3-kΩ potentiometer, find the wiper setting that will yield $V_O = 0$ (specify the wiper setting in terms of the portion of the 3-kΩ resistance assigned to the left of the wiper, and the portion to the right—for instance, 2.5 kΩ to the left and 0.5 kΩ to the right).

(b) Repeat if the R_C's are interchanged with each other, so that the smaller one is now at the left and the larger at the right.

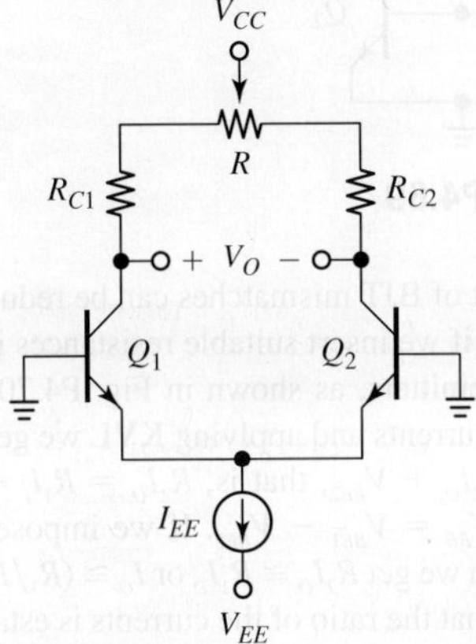

FIGURE P4.62

4.63 A student is performing a series of simple measurements on the circuit of Fig. P4.63 to determine mismatches in the BJT parameters. At each step the student adjusts V_I so as to drive V_O to 0 V and thus balance the circuit, and then measures V_I with the voltmeter. The component values are $V_{CC} = -V_{EE} = 6$ V and $R_{C1} = R_{C2} = 10$ kΩ.

(a) With $R_{B1} = R_{B2} = 0$ the student first adjusts V_I to balance the circuit, and then adjusts I_{EE} to place the collector voltages exactly at 1.0 V. If it is found that $V_I = -2.5$ mV, what is the saturation current ratio I_{s1}/I_{s2}?

(b) With $R_{B1} = 0$ and $R_{B2} = 1.0$ kΩ the student finds that balancing now requires $V_I = -4.5$ mV. What are I_{B2} and β_{F2}?

(c) With $R_{B1} = 1.0$ kΩ and $R_{B2} = 0$ the student finds that balancing now requires $V_I = 0$ V. What are I_{B1} and β_{F1}?

(d) Predict V_I with $R_{B1} = R_{B2} = 1.0$ kΩ.

(e) What is the input offset voltage V_{OS}, the input bias current I_B, and the input offset current I_{OS}?

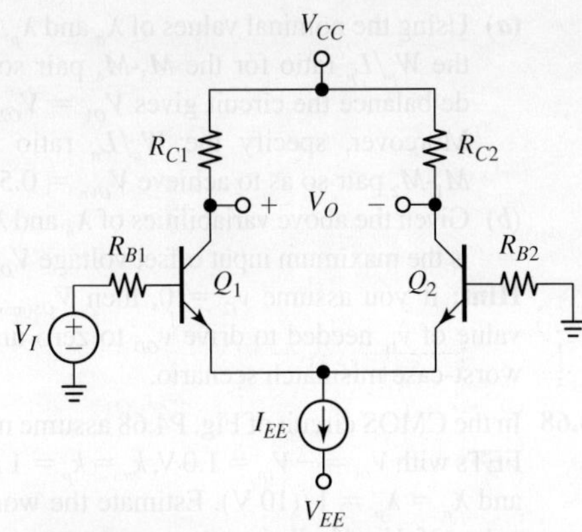

FIGURE P4.63

4.64 In the circuit of Fig. P4.44 let $R_{C1} = R_{C2} = 10$ kΩ, $R_{E1} = R_{E2} = 100$ Ω, and $I_{EE} = 1$ mA. If the resistances have $\pm2.5\%$ tolerances, and the saturation currents I_{s1} and I_{s2} exhibit a $\pm5\%$ mismatch, find the maximum input offset voltage $V_{OS(max)}$.
Hint: if you assume $v_{i2} = 0$, then $V_{OS(max)}$ is the value of v_{i1} needed to drive v_{OD} to zero under the worst-case mismatch scenario.

4.65 In the pMOS circuit of Fig. P4.48 let $I_{SS} = 200$ μA, $R_{SS} = \infty$, $R_{D1} = 16$ kΩ, and $R_{D2} = 14$ kΩ. Moreover, let the FETs have $k_1 = 0.9$ mA/V^2, $k_2 = 1.2$ mA/V^2, $V_{t1} = -495$ mV, $V_{t2} = -503$ mV. Assuming $\lambda_1 = \lambda_2 = 0$, estimate V_{OS}. Which is the major contribution to V_{OS}?

4.66 In the CMOS circuit of Fig. P4.52 let $I_{SS} = 200$ μA and $R_{SS} = \infty$, and let the FETs have $k_n = 1$ mA/V^2 and $k_p = 0.1$ mA/V^2 (for simplicity assume $\lambda_n = \lambda_p = 0$ and $\gamma_n = 0$.)

(a) If both k_n and k_p are afflicted by $\pm5\%$ tolerances, what is the maximum input offset voltage $V_{OS(max)}$?
Hint: if you assume $v_{i2} = 0$, then $V_{OS(max)}$ is the value of v_{i1} needed to drive v_{OD} to zero under the worst-case mismatch scenario.

(b) How are $V_{OS(max)}$ and a_{dm} affected if k_n is quadrupled, still with the same tolerance as in part (a)?

(c) Repeat part (b) but if k_p (rather than k_n) is now quadrupled, still with $\pm5\%$ tolerances.

4.67 In the CMOS circuit of Fig. P4.57 let $V_{DD} = -V_{SS} = 2.5$ V, $I_{SS} = 300$ μA, and $R_{SS} = \infty$. Also, let the FETs have $V_{tn} = -V_{tp} = 1.0$ V, $k'_n = 2.5 k'_p = 100$ μA/V^2, $\lambda_n = 1/[(7.5 \pm 2.5)$ V], $\lambda_p = 1/[(22.5 \pm 4.5)$ V]), and $\gamma_n = 0$.

(*a*) Using the nominal values of λ_n and λ_p, specify the W_p/L_p ratio for the M_3-M_4 pair so that at dc balance the circuit gives $V_{O1} = V_{O2} = 0$ V. Moreover, specify the W_n/L_n ratio for the M_1-M_2 pair so as to achieve $V_{OVn} = 0.5$ V.

(*b*) Given the above variabilities of λ_n and λ_p, what is the maximum input offset voltage $V_{OS(max)}$?

Hint: if you assume $v_{I2} = 0$, then $V_{OS(max)}$ is the value of v_{I1} needed to drive v_{OD} to zero under the worst-case mismatch scenario.

4.68 In the CMOS circuit of Fig. P4.68 assume matched FETs with $V_{tn} = -V_{tp} = 1.0$ V, $k_n = k_p = 1$ mA/V^2, and $\lambda_n = \lambda_p = 1/(10$ V$)$. Estimate the worst-case value of V_{OS} if all three parameter pairs are afflicted by a $\pm 5\%$ tolerance.

Hint: investigate one half at a time, and then exploit the circuit's symmetry to generalize.

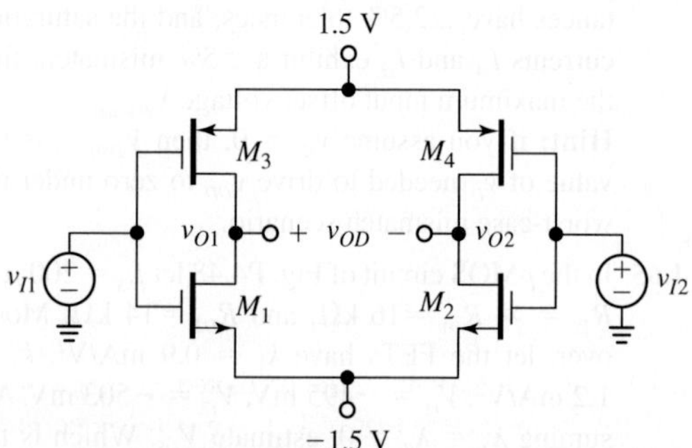

FIGURE P4.68

4.8 Current Mirrors

4.69 In the circuit of Fig. P4.69 let $V_{CC} = 10$ V and $R = 8.6$ kΩ. Moreover, let the BJTs have $I_{s3} = I_{s4}$, $I_{s1} = I_{s2}$, and $V_{EBp} = V_{BEn} = 0.7$ V, where subscripts p and n refer to the *pnp* and *npn* pairs, respectively.

(*a*) Assuming negligible base currents, find I_C and V_C if $V_{Ap} = V_{An} = 100$ V.

(*b*) Repeat if $V_{Ap} = 30$ V and $V_{An} = 100$ V.

(*c*) Repeat parts (*a*) and (*b*) if $\beta_{Fp} = 50$ and $\beta_{Fn} = 250$. Justify your findings at each step.

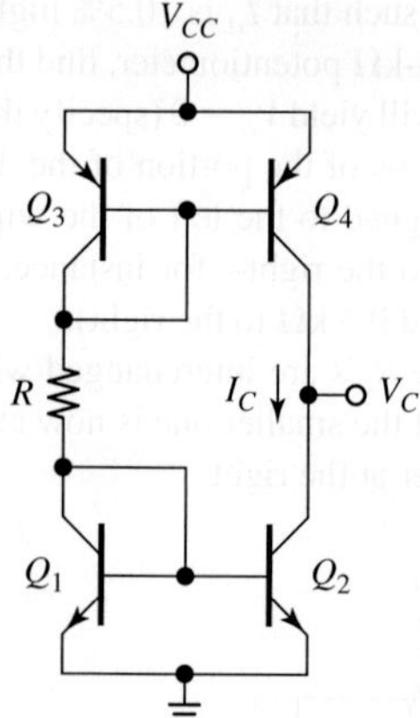

FIGURE P4.69

4.70 The effect of BJT mismatches can be reduced significantly if we insert suitable resistances in series with the emitters, as shown in Fig. P4.70. Ignoring base currents and applying KVL we get $R_1 I_1 + V_{BE1} = R_2 I_O + V_{BE2}$, that is, $R_2 I_O = R_1 I_1 + \Delta V_{BE}$, where $\Delta V_{BE} = V_{BE1} - V_{BE2}$. If we impose $R_1 I_1 \gg \Delta V_{BE}$, then we get $R_2 I_O \cong R_1 I_1$, or $I_O \cong (R_1/R_2)I_1$, indicating that the ratio of the currents is established by the resistors. In particular, with equal resistors, we get accurate mirroring regardless of any mismatches between the two BJTs. This technique, particularly popular in the days of unmatched discrete devices, is still used in current ICs so long as the resistors don't take up too much chip area. As an additional benefit, emitter degeneration raises the output resistance, resulting in a much flatter output-port *i-v* characteristic. If desired, one can add a beta helper to reduce the error even further.

(*a*) Let the BJTs be grossly mismatched such that $I_{s2} = 2I_{s1}$. Assuming $V_A = \infty$, $V_{CE2(EOS)} = 0.3$ V, and negligible base currents, find I_O if $I_I = 1.0$ mA and $R_1 = R_2 = 0$. What is the value of $v_{C2(min)}$, the lower limit of the linear region of operation?

(*b*) Repeat part (*a*) if $R_1 = R_2 = 100$ Ω.

(*c*) Repeat (*a*) if $R_1 = R_2 = 1.0$ kΩ.

(*d*) Find the values of R_1 and R_2 (= R_1) that will result in an output current error of 1% or less.

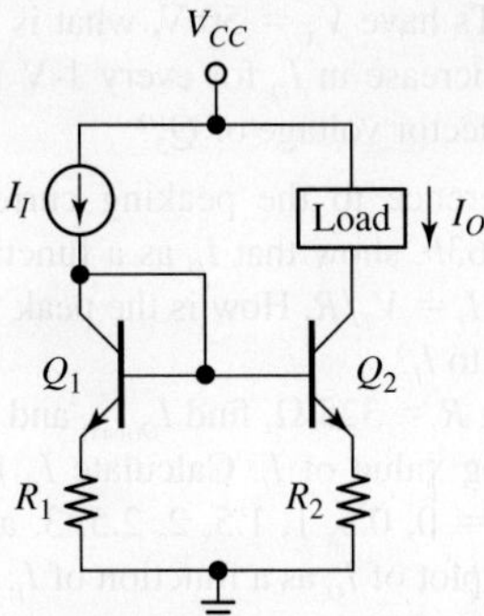

FIGURE P4.70

4.71 **(a)** Use the test method to prove that the small-signal resistance seen by the input current source in the beta-helper mirror of Fig. 4.57b is $R_i \cong 2V_T/I_I$.

   (b) Repeat, but for the Wilson mirror of Fig. 4.61a.

4.72 **(a)** Assuming $r_\mu = \infty$, show that the bipolar cascoded mirror of Fig. 4.59 yields $R_o \cong (\beta_0/2)r_o$. **Hint:** replace the BJTs with their small-signal equivalents, and use the test method, noting that because of the mirror action provided by Q_3 and Q_4, the small-signal current out of Q_2's emitter must mirror that out of Q_2's base. Show also that $R_i \cong 2V_T/I_I$.

   (b) Exploiting the fact that cascoding shifts the point where the extrapolated i_O-v_O curve intercepts the v_O axis from $-V_A$ to $-(\beta_0/2)V_A$, prove Eq. (4.136).

4.73 The Sooch current mirror of Fig. 4.60b utilizes the blocks of Fig. P4.73 to provide the proper voltage drives for the output FETs M_2 and M_4.

   (a) The function of M_3 is to synthesize the v_{GS} drive needed to sink i_I at $v_{DS} = v_{OV}$. This requires $v_{D3} = v_{G3} - V_t$, a task performed by M_6. Assuming $v_{OV} \le V_t$, find v_{G6} so that $v_{DS6} = V_t$.

   (b) For a diode-connected FET, the voltage at any point of its channel will lie somewhere between v_S and $v_S + V_t + v_{OV}$. In particular, there must be a point at which this voltage is $v_S + v_{OV}$. In order to access it, we split the FET into two FETs M_1 and M_5 in series, as shown in Fig 4.73b (recall that two FETs in series still act as a single FET!) Prove that in order to drop $v_{DS5} = v_{OV}$, M_5 must operate in the *triode region* (while M_1 is saturated) and must have $k_5 = k_1/3$ (a condition that is achieved

by fabricating M_5 with a W/L ratio $1/3$ that of M_1). What is the resulting voltage v_{D1}?

   (c) Verify that once the circuit of (b) is stacked on top of that of (a) as per the Sooch circuit of Fig. 4.60b, M_6's drain will provide the required drive for M_4's gate, and M_1's drain will provide the required drives for M_2's gate as well as M_6's gate. Find v_I.

   (d) Suppose a certain application requires the creation of the voltage $2v_{OV}$ (instead of $1v_{OV}$) using the topology of Fig. P4.73b. What is the required relationship between k_5 and k_1 in this case?

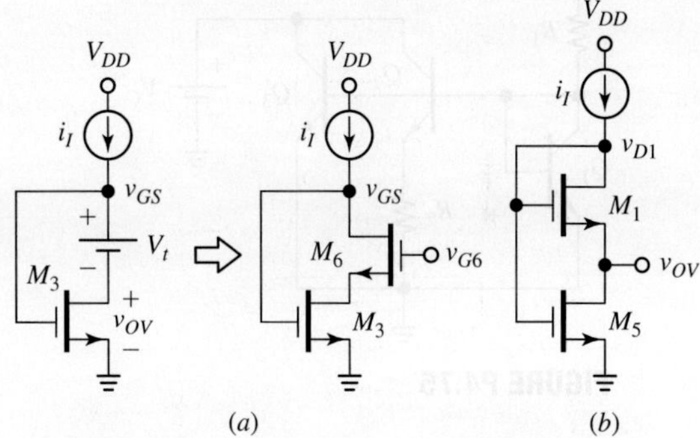

(a) (b)

FIGURE P4.73

4.74 The error term $\varepsilon = -2/\beta_F^2$ predicted by Eq. (4.141) for the Wilson mirror of Fig. 4.61a relies on perfectly matched betas. This matching causes first-order terms in $1/\beta_F$ to cancel each other out, leaving only the quadratic term.

   (a) To investigate what happens if the betas are mismatched, assume $\beta_{F2} = \beta_{F1}$ for simplicity, but arbitrary β_{F3}. Proceeding in a manner similar to that depicted in Fig. 4.61b, show that

$$\frac{i_O}{i_I} = 1 - \frac{2(\beta_{F1} - \beta_{F3}) + 2}{\beta_{F1}\beta_{F3} + 2\beta_{F1} + 2} \cong 1 - 2\frac{\beta_{F1} - \beta_{F3} + 1}{\beta_{F1}\beta_{F3}}$$

   (b) Assuming $\beta_{F2} = \beta_{F1} = 100$, examine the particular cases $\beta_{F3} = 100, 90, 110,$ and 101, and comment on your findings.

   (c) A student, worried about the error due to finite betas, is thinking of replacing Q_3 in Fig. 4.61a with a Darlington pair to exploit its high equivalent beta, or even with an nMOSFET, which draws zero gate current. What advice would you offer to that student?

4.75 Suppose the emitter areas of the BJTs of Fig. P4.75 are such that $A_{E1} = A_{E2} = 0.5A_{E3}$, and all three BJTs have $V_A = 75$ V.

   (a) Assuming negligible base currents, specify suitable values for R_1 and R_2 so that, with $V_C = 0.7$ V the circuit gives $I_{C3} = 1$ mA and $I_{C2} = 20$ μA.

   (b) If we increase V_C above 0.7 V, what is the percentage change in I_{C3} and I_{C2} for every 1-V increase in V_C?

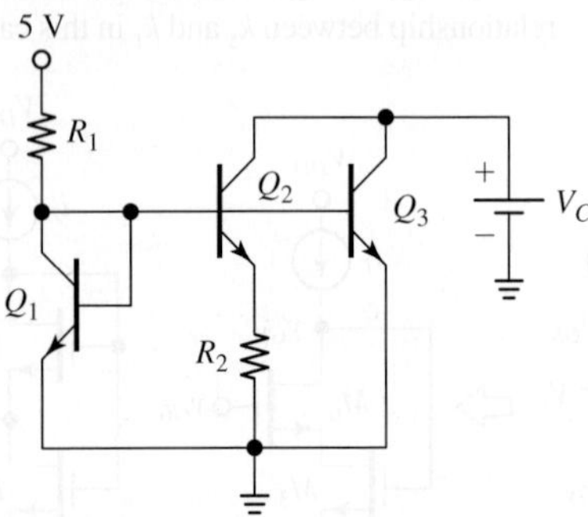

FIGURE P4.75

4.76 Let the BJTs of Fig. P4.76 have $I_s = 0.1$ fA.

   (a) Assuming $V_A = \infty$ and negligible base currents, specify suitable values for R_1 and R_2 so that $I_{C1} = 1.0$ mA and $I_{C2} = 0.2$ mA.

   (b) If $R_1 = 1.0$ kΩ and $R_2 = 3.0$ kΩ, find I_{C1} and I_{C2}. **Hint:** use iterations for both BJTs.

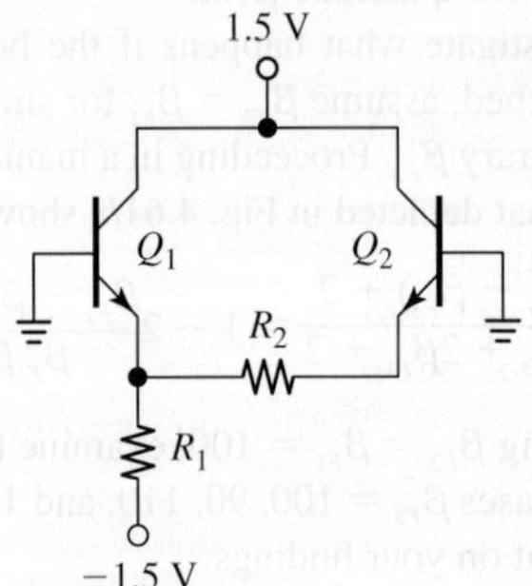

FIGURE P4.76

4.77 In the circuit of Fig. 4.63b let $I_I = 0.25$ mA.

   (a) Assuming negligible base currents, find I_O if $R = 100$ Ω.

   (b) Find R for $I_O = 10$ μA.

   (c) If the BJTs have $V_A = 50$ V, what is the percentage increase in I_O for every 1-V increase in the collector voltage of Q_2?

4.78 (a) With reference to the peaking current sink of Fig. 4.63b, show that I_O as a function of I_I peaks for $I_I = V_T/R$. How is the peak value of I_O related to I_I?

   (b) Assuming $R = 325$ Ω, find $I_{O(max)}$ and the corresponding value of I_I. Calculate I_O for $I_I = kV_T/R$, $k = 0, 0.5, 1, 1.5, 2, 2.5, 3$, and construct the plot of I_O as a function of I_I.

4.79 The circuit of Fig. P4.79, called *current conveyor*, finds application in fast current-signal processing. The circuit has two input terminals X and Y and one output terminal Z, and it is made up of the ordinary *pnp* current mirror Q_1-Q_2 and the dual-output *npn* current mirror Q_3-Q_4-Q_5.

   (a) Describe circuit behavior by asking yourself what happens when the source i_X is turned on. Hence, assuming negligible base currents, find v_X as a function of v_Y, and i_Y and i_Z as functions of i_X.

   (b) Obtain expressions for the small-signal resistances R_x, R_y, and R_z seen looking into the corresponding terminals.

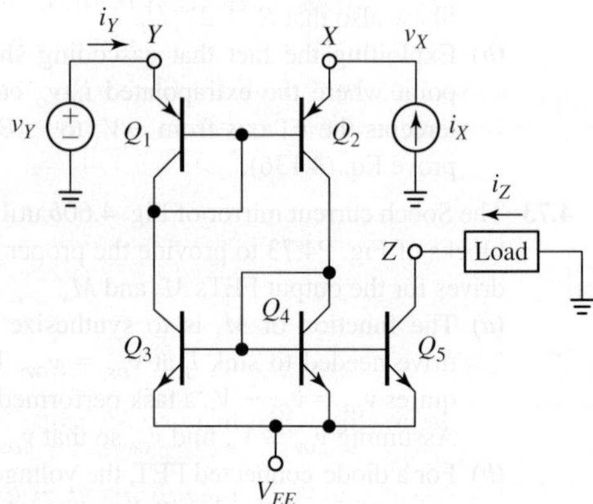

FIGURE P4.79

4.80 Let the FETs in the *p*MOS mirror of Fig. P4.80 be matched devices with $V_t = -0.5$ V, $k = 0.5$ mA/V^2, and $\lambda = 1/(10\ \text{V})$.

   (a) If $V_{DD} = 3$ V, find R so that $I_O = 0.1$ mA at dc balance. What is R_o? Find $v_{O(max)}$ for which Q_2 is still saturated.

(b) If $R = 10$ kΩ, find I_o, R_o, and $v_{O(\max)}$.

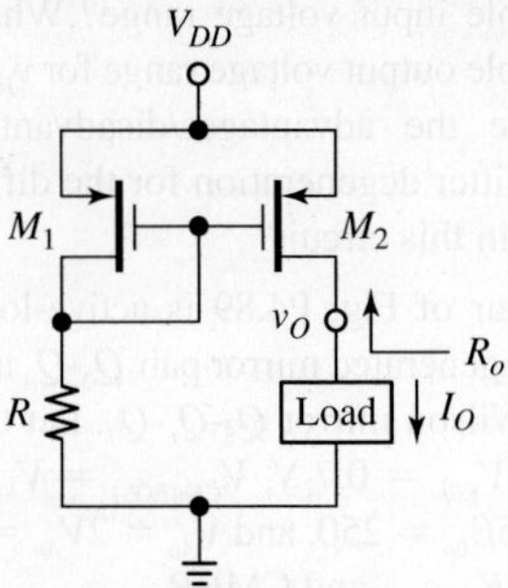

FIGURE P4.80

4.81 Let the FETs of the basic mirror of Fig. 4.58a have $k_2 = 4k_1 = 2$ mA/V^2, $V_{t0} = 0.4$ V and $\lambda = 1/(15$ V$)$.
(a) If $i_I = 50$ μA, obtain an expression for i_O as a function of v_O. Hence, find v_I, R_i, R_o, and $v_{O(\min)}$, the lower limit of the linear output range. What is the percentage change in i_O for each 1-V change in v_O above $v_{O(\min)}$?
(b) Repeat, but after adding a second FET pair M_3-M_4 to turn the circuit into a cascode mirror of type of Fig. 4.59b. Assume $k_4 = k_2 = 4k_3 = 4k_1 = 2$ mA/V^2, $\gamma = 0.5$ V$^{1/2}$, and $|2\phi_p| = 0.6$ V. Compare with part (a), and comment.

4.82 Consider the lower portion of Fig. 4.5b, comprising the current mirror M_5-M_6 and the diode-connected transistor M_7 establishing the bias current for M_5. Let the process parameters be $k' = 100$ μA/V^2, $\lambda' = 0.08$ μm/V and $V_t = 0.5$ V (for simplicity, ignore the body effect for M_7).
(a) Specify W_6 and L_6 so that M_6 draws 100 μA with a 0.4-V overdrive voltage, and exhibits an output resistance of 250 kΩ.
(b) Specify the W_5/L_5 ratio so that M_5 provides the required gate bias for M_6 while drawing 25 μA.
(c) Specify the W_7/L_7 ratio so that M_7 biases M_5 at 25 μA with $V_{SS} = -2.5$ V.

4.83 The circuit of Fig. P4.83 uses the nMOS mirror M_1-M_2 to *sink* the current I_1 from the load LD$_1$, and the pMOS mirror M_4-M_5 to *source* the current I_2 to the load LD$_2$. The nMOS mirror is biased by the input source I_I, and it comprises an additional FET, M_3, to establish the current I_3 needed to bias the pMOS mirror. Assume the process parameters to

be $V_{tn} = -V_{tp} = 0.5$ V, $k'_n = 2.5k'_p = 100$ μA/V^2, and $\lambda'_p = \lambda'_n = 0.05$ V^{-1}.
(a) If $I_I = 50$ μA, specify W/L ratios for M_1, M_3, and M_4 so that $I_3 = I_I$ and $V_{OV4} = V_{OV1} = 0.25$ V.
(b) Specify W_2 and L_2 so that at dc balance M_2 sinks $I_1 = 250$ μA with $r_{o2} = 100$ kΩ.
(c) Specify W_5 and L_5 so that at dc balance M_5 sources $I_5 = 100$ μA with $r_{o5} = 100$ kΩ.

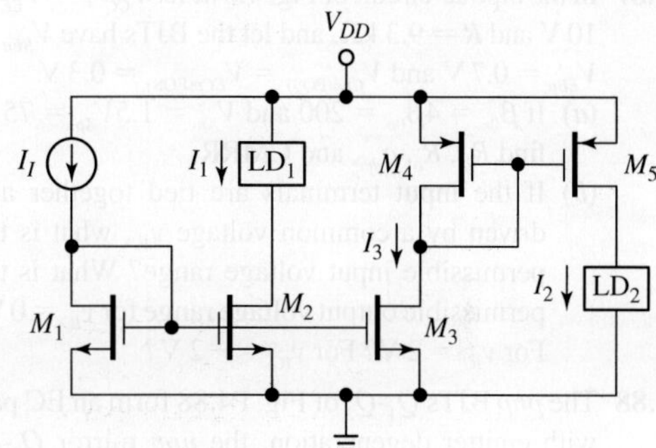

FIGURE P4.83

4.9 Differential Pairs with Active Loads

4.84 **(a)** Prove Eq. (4.152).
(b) If an active-loaded CMOS differential pair is powered from ± 2.5-V power supplies and has been fabricated in a process with $k'_n = 2.5k'_p = 100$ μA/V^2, $V_{tn} = -V_{tp} = 0.6$ V, and $\lambda'_n = 2\lambda'_p = 0.1$ μm/V, specify suitable values for W and L for the four FETs so that at dc balance the circuit gives $a_{dm} = -50$ V/V with $I_{SS} = 200$ μA and $V_{OVn} = V_{OVp} = 0.4$ V. Ignore the body effect.

4.85 **(a)** Prove Eq. (4.157).
(b) Prove Eq. (4.158).
(c) Use the simplified expressions of Eq. (4.158) to estimate the CMRRs, compare with Example 4.26, and comment.

4.86 Consider the active-loaded CMOS amplifier of Fig. 4.67b, but with both inputs at ac ground. Assume the M_1-M_2 and M_3-M_4 pairs are separately matched, and the body effect of M_1 and M_2 can be ignored.
(a) Replace each transistor with its small-signal model consisting of a $g_m v_{gs}$ source and parallel resistance r_o. Hence, apply a test voltage v to

the node common to M_1's and M_3's drains, find the resulting current i out of the test source, and show that

$$\frac{v}{i} = \frac{r_{op}//(2r_{on} + r_{op})}{1 + g_{mp}r_{op}}$$

(b) How does the ratio v/i simplify if $r_{on} = r_{op} = r_o$? What is its value if $g_{mp} = 1$ mA/V and $r_o = 20$ kΩ?

4.87 In the bipolar circuit of Fig. 4.64a let $V_{CC} = -V_{EE} = 10$ V and $R = 9.3$ kΩ, and let the BJTs have $V_{BEn} = V_{EBp} = 0.7$ V and $V_{CEn(EOS)} = V_{ECp(EOS)} = 0.3$ V.

(a) If $\beta_{0n} = 4\beta_{0p} = 200$ and $V_{An} = 1.5V_{Ap} = 75$ V, find R_{id}, R_o, a_{dm}, and CMRR.

(b) If the input terminals are tied together and driven by a common voltage v_{IC}, what is the permissible input voltage range? What is the permissible output voltage range for $v_{IC} = 0$ V? For $v_{IC} = 2$ V? For $v_{IC} = -2$ V?

4.88 The *pnp* BJTs Q_1-Q_2 of Fig. P4.88 form an EC pair with emitter degeneration, the *npn* mirror Q_3-Q_4 forms the active load, and the *pnp* mirror Q_5-Q_6 provides emitter bias for the EC pair. Let the BJTs have $V_{BEn} = V_{EBp} = 0.7$ V, $V_{CEn(EOS)} = V_{ECp(EOS)} = 0.3$ V, $\beta_{0n} = 4\beta_{0p}, = 200$, and $V_{An} = 1.5V_{Ap} = 75$ V.

(a) Find R_{id}, R_o, a_{dm}, and CMRR.

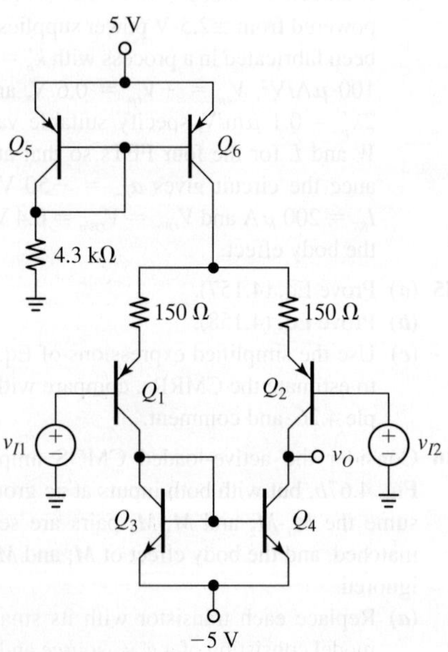

FIGURE P4.88

(b) If the input terminals are tied together and driven by a common voltage v_{IC}, what is the permissible input voltage range? What is the permissible output voltage range for $v_{IC} = 0$ V?

(c) What are the advantages/disadvantages of using emitter degeneration for the differential *pnp* pair in this circuit?

4.89 The Q_1-Q_2-pair of Fig. P4.89 is active-loaded by the emitter-degenerated mirror pair Q_3-Q_4 and is biased by the Wilson mirror Q_5-Q_6-Q_7. Let the BJTs have $V_{BEn} = V_{EBp} = 0.7$ V, $V_{CEn(EOS)} = V_{ECp(EOS)} = 0.3$ V, $\beta_{0n} = 5\beta_{0p} = 250$, and $V_{An} = 2V_{Ap} = 80$ V.

(a) Find R_{id}, R_o, a_{dm}, and CMRR.

(b) If the input terminals are tied together and driven by a common voltage v_{IC}, find the permissible range for v_{IC}. What is the permissible output voltage range if $v_{IC} = 0$ V?

(c) What are the advantages/disadvantages of using emitter degeneration in the active load? Of biasing the circuit via a Wilson mirror instead of a basic mirror?

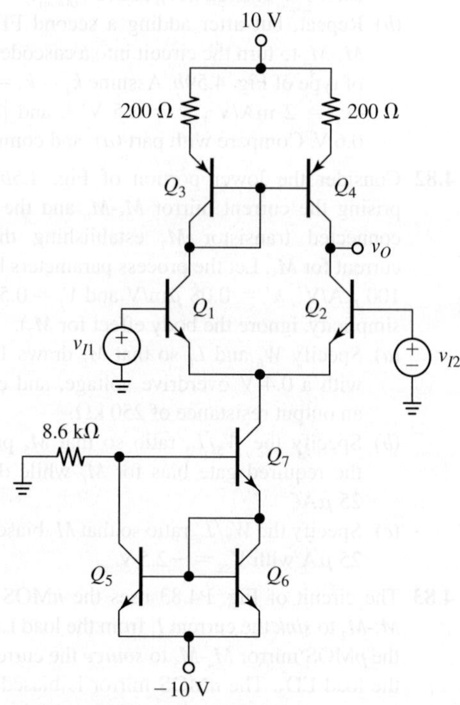

FIGURE P4.89

4.90 In Fig. P4.90 both the differential pair and the active load have been *cascoded* to raise the output resistance R_o and, hence, the unloaded gain a_{dm}. Specifically, the EC pair Q_1-Q_2 is cascoded by the CB pair Q_5-Q_6, and the mirror pair Q_3-Q_4 is cascoded by the CB pair Q_7-Q_8. (The function of the 1.4-V source, whose details have been omitted for simplicity, is to bias the Q_5-Q_6 pair so as to keep Q_1 and Q_2 in the active region over the entire common-mode input range.) (*a*) Obtain an expression for a_{dm} in terms of the betas and the Early voltages of the BJTs (assume $r_\mu = \infty$ for simplicity). Hence, show that if all BJTs have the same beta and Early voltage, then gain simplifies as $a_{dm} = (\beta_0/3) \times (V_A/V_T)$.
(*b*) Calculate R_o and a_{dm} if $I_{EE} = 1.0$ mA, $\beta_{0n} = 3\beta_{0p} = 150$, and $V_{An} = (4/3)V_{Ap} = 100$ V.
(*c*) Estimate the systematic input offset voltage due to non-infinite β_{Fp} (assume $\beta_{Fp} = \beta_{0p}$).
(*d*) If $V_{CC} = 10$ V, $V_{BEn} = V_{EBp} = 0.7$ V, and $V_{CEn(EOS)} = V_{ECp(EOS)} = 0.3$ V, what is the permissible output voltage range?

inherently high output resistance, so we expect a dramatic increase in both R_o and a_{dm}. (The function of the 1.4-V source, whose details have been omitted for simplicity, is to bias the Q_5-Q_6 pair so as to keep Q_1 and Q_2 in the active region over the entire common-mode input range.)
(*a*) Obtain an expression for a_{dm} in terms of the betas and the Early voltages of the BJTs (assume $r_\mu = \infty$ for simplicity). Hence, show that if all BJTs have the same beta and Early voltage, then the gain simplifies as $a_{dm} = (\beta_0/3)(V_A/V_T)$.
(*b*) Calculate R_o and a_{dm} if $I_{EE} = 0.4$ mA, $\beta_{0n} = 100$, $\beta_{0p} = 50$, $V_{An} = 75$ V, and $V_{Ap} = 30$ V.
(*c*) Estimate the systematic input offset voltage due to non-finite β_{Fp} (assume $\beta_{Fp} = \beta_{0p}$).
(*d*) If $V_{CC} = 9$ V, $V_{BEn} = V_{EBp} = 0.7$ V, and $V_{CEn(EOS)} = V_{ECp(EOS)} = 0.3$ V, what is the permissible output voltage swing?

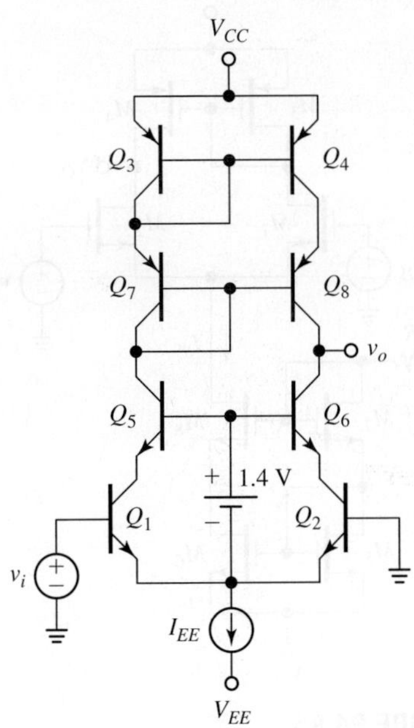

FIGURE P4.90

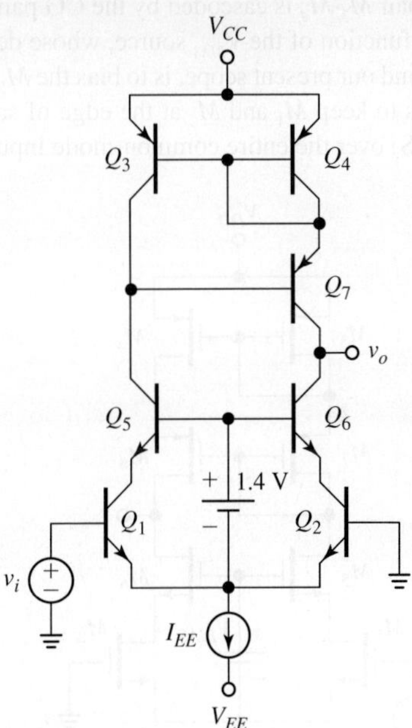

FIGURE P4.91

4.91 In Fig. P4.91 the differential pair Q_1-Q_2 is *cascoded* via the CB pair Q_5-Q_6, and the active load is a Wilson-type mirror. As we know, both the cascode and the Wilson configurations enjoy

4.92 (*a*) Assuming I_{sn} mismatches of $\pm4\%$ and I_{sp} mismatches of $\pm5\%$, along with $\beta_{0p} = 50$, find the worst-case input offset voltage for the circuit of Fig. P4.90. Which is the major contribution to V_{OS}?
Hint: not all transistors intervene in the offset error. Which ones do and which ones don't?
(*b*) Repeat, but for the circuit of Fig. P4.91.

4.93 In the CMOS circuit of Fig. 4.64b let $V_{DD} = -V_{SS} = 2.5$ V and let the FETs have $V_{tn} = 0.5$ V, $V_{tp} = -0.4$ V, $k_1 = k_2 = 2$ mA/V^2, $k_3 = k_4 = k_5 = k_6 = 1$ mA/V^2, $\lambda_n = 1/(30$ V$)$, $\lambda_p = 1/(25$ V$)$, and $\chi_n = 0.1$.

(a) Specify R so that $a_{dm} = 80$ V/V. What is the value of the CMRR?

   Hint: develop an expression for a_{dm} as a function of I_{SS}.

(b) If the input terminals are tied together and driven by a common voltage v_{IC}, what is the permissible input voltage range? What is the permissible output voltage range for $v_{IC} = 0$ V?

4.94 In the circuit of Fig. P4.94 both the differential pair and the active load have been cascoded in order to raise the output resistance R_o and thus the unloaded gain a_{dm}. Specifically, the SC pair M_1-M_2 is cascoded by the CG pair M_5-M_6, and the mirror pair M_3-M_4 is cascoded by the CG pair M_7-M_8. The function of the V_{BIAS} source, whose details are beyond our present scope, is to bias the M_5-M_6 pair so as to keep M_1 and M_2 at the edge of saturation (EOS) over the entire common-mode input range.

holds, the circuit gives, assuming $\chi = 0$, $a_{dm} \cong 2/(\lambda V_{OV})^2$, where V_{OV} is the overdrive voltage required to sustain the current $I_{SS}/2$.

(b) Calculate a_{dm} and CMRR if $k = 1$ mA/V^2 and $\lambda = 1/(20$ V$)$, and the biasing network sinks $I_{SS} = 0.25$ mA with $R_{SS} = 80$ kΩ.

(c) Specify a suitable value for V_{BIAS} if $V_t = 0.4$ V.

4.95 (a) Show that if all FETs in the circuit of Fig. 4.64b have the same λ and k, then, so long as the condition $g_m r_o \gg 1$ holds, the circuit gives CMRR $\cong 2/(\lambda V_{OV})^2$, where V_{OV} is the overdrive voltage required to sustain the current $I_{SS}/2$, and $\chi = 0$ has been assumed for simplicity.

(b) Show that if the mirror supplying I_{SS} is changed to a cascode type as in Fig. P4.95, then CMRR $\cong \sqrt{8}/(\lambda V_{OV})^3$.

(c) Compare the two circuits if $k = 2$ mA/V^2, $\lambda = 1/(15$ V$)$, and $I_{SS} = 0.18$ mA.

(d) What happens if I_{SS} is doubled? Halved?

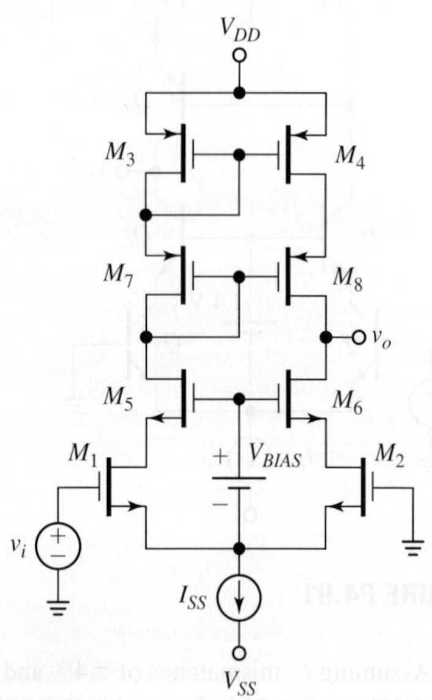

FIGURE P4.94

(a) Obtain an expression for a_{dm} in terms of the g_ms and the lambdas of the FETs. Hence, show that if all FETs have the same values of k and λ, then, so long as the condition $g_m r_o \gg 1$

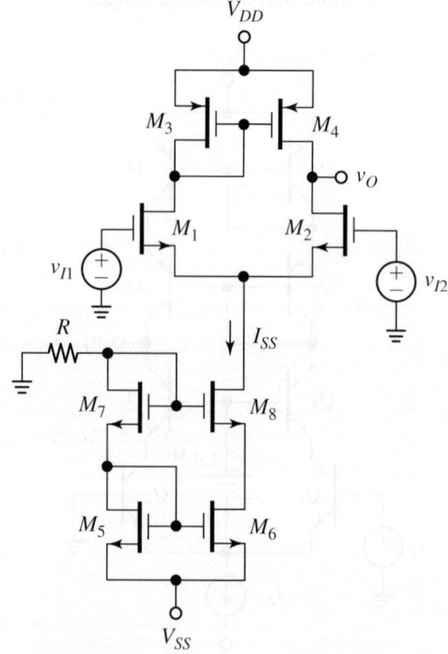

FIGURE P4.95

4.96 (a) Retracing similar steps to those leading to Eq. (4.125), prove Eq. (4.161).

   Hint: consider one mismatch at a time, and then add them up in rms fashion.

(b) Estimate the likely value of V_{OS} if $I_{SS} = 0.25$ mA, $k_n = (4 \pm 10\%)$ mA/V², $k_p = (2 \pm 10\%)$ mA/V², and both V_{tn} and V_{tp} have ± 10-mV tolerances. Which is the major contributor to V_{OS}?

4.97 **(a)** Assuming the I sources at the top of the folded-cascoded EC pair of Fig. 4.71*a* are basic *pnp* current mirrors, find V_{BIAS} so that $v_{O(\max)}$ is as high as possible, given that $V_{CC} = -V_{EE} = 10$ V, $V_{EBp} = V_{BEn} = 0.7$ V, and $V_{ECp(EOS)} = V_{CEn(EOS)} = 0.3$ V. What are $v_{O(\min)}$ and $v_{O(\max)}$?

(b) Find R_{id}, R_o, and a_{dm} if $I = 1$ mA and the BJTs have $\beta_{0n} = 4\beta_{0p} = 200$ and $V_{An} = 2V_{Ap} = 80$ V.

4.98 **(a)** Assuming the I sources at the top of the folded-cascoded SC pair of Fig. 4.71*b* are basic *p*MOS current mirrors, find V_{BIAS} so that $v_{O(\max)}$ is as high as possible, given that $V_{DD} = -V_{SS} = 3$ V, $I = 0.2$ mA, and the FETs have $k_n = 5k_p = 5$ mA/V², $V_{tn} = -V_{tp} = 0.3$ V, $\chi_p = 0.15$, $\lambda_n = 1/(15$ V$)$, and $\lambda_p = 1/(10$ V$)$. What are the values of $v_{O(\min)}$ and $v_{O(\max)}$?

(b) Find R_o and a_{dm}.

4.10 Bipolar Output Stages

4.99 **(a)** With reference to the Class AB circuit of Fig. 4.74*a*, suppose the currents I_1 and I_2 are perfectly matched, whereas the saturation currents I_{s1}, I_{s2}, I_{s3}, and I_{s4} of the BJTs are afflicted by a $\pm 10\%$ tolerance each. Find the worst-case voltage V_{OS} that we need to apply at the input in order to drive v_O to zero.

(b) Repeat, but for the case in which the saturation currents are perfectly matched, whereas I_1 and I_2 are afflicted by a $\pm 10\%$ tolerance.

4.100 The circuit of Fig. P4.100, known as a V_{BE} *multiplier*, provides an alternative method for the generation of the voltage drop V_{BB} necessary to bias the push-pull BJTs. Its advantage is that V_{BB} can be adjusted by playing with the ratio R_1/R_2.

(a) Show that $V_{BB} = (1 + R_1/R_2) \times [V_{BE} + (R_1//R_2)I_B]$. Usually the base current is small enough to meet the condition $(R_1//R_2)I_B \ll V_{BE}$, in which case the above expression simplifies as $V_{BB} = MV_{BE}$, with $M = 1 + R_1/R_2$ as a multiplier term.

(b) If $I_1 = I_2 = 200$ μA and the BJT has $\beta_F = 250$ and $I_s = 1$ fA, specify suitable values for R_1 and R_2 to achieve $V_{BB} = 1250$ mV under the constraint $I_C = 150$ μA.

(c) Replace the BJT with its small-signal model and find the dynamic resistance R_{bb} between the collector and emitter nodes.

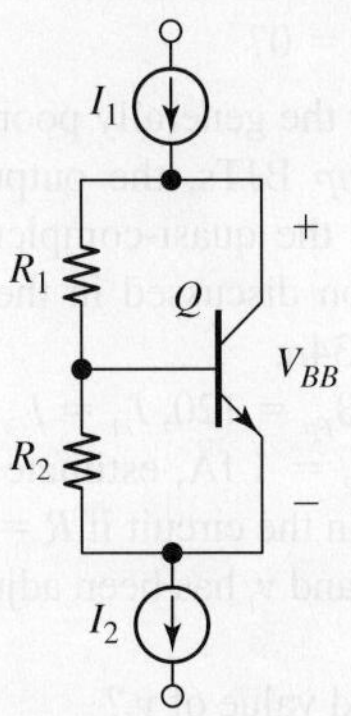

FIGURE P4.100

4.101 The V_{BE} *multiplier* of Problem 4.100 is put to use in the push-pull circuit of Fig. P4.101.

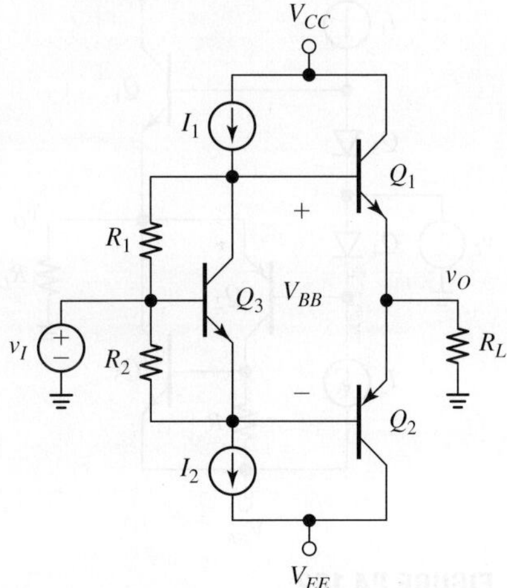

FIGURE P4.101

(a) If $I_{s1} = I_{s2} = 4I_{s3}$ and $I_1 = I_2 = I$, obtain an expression for the quiescent current I_Q of the Q_1-Q_2 pair as a function of I, I_{s3}, and the ratio R_1/R_2 (to simplify your analysis, neglect all base currents as well as the portion of I flowing through the R_1-R_2 pair).

(b) If $I = 200$ μA and $I_{s3} = 1$ fA, specify the R_1/R_2 ratio for $I_Q = 50$ μA.

(c) Find v_I so that $v_O = 0$.

(d) Given that I_{s3} is a strong function of temperature, find the R_1/R_2 ratio that will render I_Q independent of I_{s3}. What is the new value of I_Q for this R_1/R_2 ratio? What is the new value of v_I that will yield $v_O = 0$?

4.102 In order to make up for the generally poorer characteristics of planar *pnp* BJTs, the output stage of Fig. P4.102 utilizes the quasi-complementary Darlington configuration discussed in the text in connection with Fig. 4.34*a*.

(a) Assuming $\beta_{Fn} = 4\beta_{Fp} = 120$, $I_{s1} = I_{s3} = I_{s4} = I_{s5} = 10$ fA, and $I_{s2} = 1$ fA, estimate all currents and voltages in the circuit if $R = 20$ kΩ, $I_1 = I_2 = 200$ μA, and v_I has been adjusted so that $v_O = 0$.

(b) What is the required value of v_I?

(c) Find R so that $v_O = 0$ for $v_I = 0$.

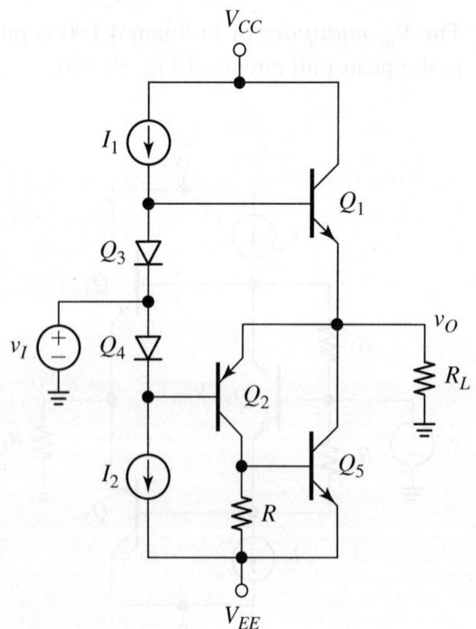

FIGURE P4.102

4.103 The circuit of Fig. P4.103 utilizes overload protection both for Q_1 and Q_2. Let $V_{CC} = -V_{EE} = 10$ V, $I_1 = I_2 = 0.25$ mA, and $R_1 = R_2 = 50$ Ω.

(a) Assuming negligible base currents and $I_{s3} = I_{s4} = 10$ fA, find I_{s1} and I_{s2} so that $I_Q = 0.1$ mA under the constraint $I_{s1} = I_{s2}$. Assuming $\beta_{Fn} = 2\beta_{Fp} = 200$ and $V_{BEn(on)} = V_{EBp(on)} = 0.7$ V, give estimates for all collector currents as well as v_O and the current in or out of the v_I source for the following cases:

(b) $v_I = 4$ V and $R_L = 1$ kΩ;

(c) $v_I = -5$ V and $R_L = 2$ kΩ;

(d) $v_I = 5$ V and $R_L = 100$ Ω;

(e) $v_I = -6$ V and $R_L = 75$ Ω;

(f) $v_I = -2$ V and $R_L = 0$ Ω.

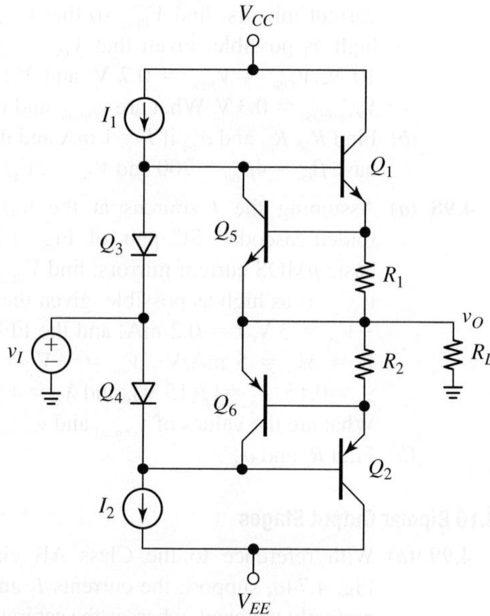

FIGURE P4.103

4.11 CMOS Output Stages

4.104 **(a)** For the CD push-pull stage of Fig. 4.79*a* find an expression for the quiescent current I_Q in terms of the I_{D3}, I_{D4}, and the device transconductance parameters k_1 through k_4 (assume $\lambda = 0$).

(b) How does the expression for I_Q simplify for the case of ideal current mirrors?

(c) Calculate I_Q for the component values shown in the figure.

4.105 **(a)** Estimate $v_{O(max)}$ for the CD push-pull stage of Fig. 4.79*a* (assume $\lambda = 0$). What is the corresponding value of v_I? Compare with the VTC of Fig. 4.79*b*, and comment.

(b) Repeat, but taking the body effect into consideration. Assume the bodies of all *p*FETs are tied to V_{DD}, those of the *n*FETs to V_{SS}, and use $\gamma = 0.4$ V$^{1/2}$ and $2|\phi| = 0.6$ V.

4.106 (*a*) Find the quiescent current I_Q of the CMOS inverter of Fig. 4.80*a* (assume $\lambda = 0$). What is its output resistance in standby? What is the power provided by V_{DD} and by V_{SS}?

(*b*) Find v_O for $v_I = 5$ V. What is the output resistance, and what is the power provided by the supplies? Compare with part (*a*), and comment.

4.107 The circuit made up for transistors M_3 through M_8 in Fig. P4.107 is designed to establish the quiescent current of the CS pair M_1-M_2 at an acceptable level while retaining the inherently wide OVS of the CMOS inverter. Assume all FETs have $V_t = 0.75$ V, and let $V_{DD} = -V_{SS} = 5$ V.

(*a*) Assuming M_3 through M_8 have $k = 1.6$ mA/V^2, find the values of the biasing voltages V_{G6} and V_{G8} that will cause M_3 through M_8 to draw a current of 50 μA each for $v_I = 0$.

(*b*) Find k_1 and k_2 so that in standby $I_{D1} = I_{D2} = 0.25$ mA.

(*c*) Find the limits $v_{O(max)}$ and $v_{O(min)}$ of the OVS if $R_L = 2$ kΩ. What are the corresponding values of v_I?

(*d*) Find v_O for $v_I = \pm 1$ V.

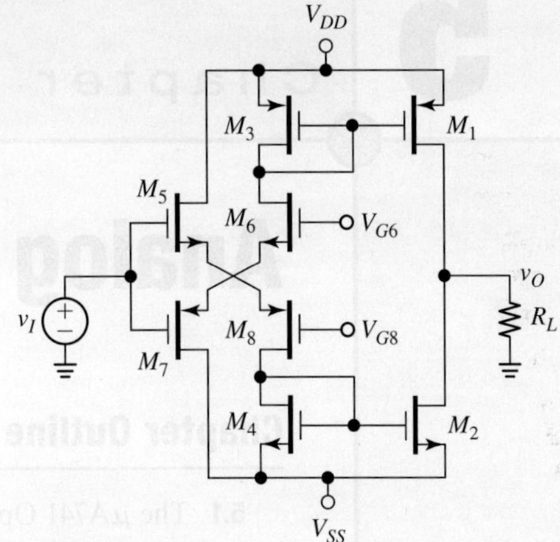

FIGURE P4.107

4.108 (*a*) For the CS push-pull stage of Fig. 4.82*a* find the voltage v_I that causes M_2 to operate right at the edge of conduction (assume $\lambda = 0$). What is the corresponding value of v_O? The values of v_{G1} and v_{G2}?

(*b*) Find the voltage v_I that causes M_1 to operate right at the edge of saturation. What are the corresponding values of v_O, v_{G1} and v_{G2}?

5

Chapter

Analog Integrated Circuits

Chapter Outline

The implementation of the integrated-circuit concept by Jack Kilby at Texas Instruments in 1958, and, independently by Robert Noyce at Fairchild Semiconductor in 1959, triggered a feverish activity that led to the development of the first analog integrated circuits (ICs), namely, the μA702 and μA709 *operational amplifiers*, the μA710 *voltage comparator*, and the μA723 *voltage regulator*. These devices were designed in the early 1960s by Robert J. Widlar (1937–1991) while at Fairchild. (He subsequently moved to National Semiconductor, where he continued to develop groundbreaking analog products.) Widlar devised a number of building blocks, such as the *Widlar current source* and *the bandgap voltage reference*, that became industry standards and are in widespread use to this very day.

In May 1968 Fairchild introduced the μA741, the first internally compensated op amp. Previous op amps required external components for frequency compensation. By incorporating this function on chip, the 741 relieved the user from having to contend with the arcane issue of frequency compensation, opening up a broad market for experts and neophytes alike. Though a great many other op amp families have been developed since then, the 741 is still the most widely documented op amp and it contains fundamental building blocks that are common to a variety of contemporary analog ICs. It pays, therefore, to start the chapter by studying this device, both from a historical perspective and a pedagogical standpoint.

CHAPTER HIGHLIGHTS

Following a detailed analysis of the classic 741 *op amp*, with an eye on the most common op amp nonidealities and limitations, the chapter turns to the two CMOS op amp topologies in widest use today, namely, the *two-stage* and the *folded-cascode* op amp types. (We return to these three classes of op amps in Chapter 6, where we investigate their frequency and time responses, and again in Chapter 7, where we investigate their stability in negative-feedback operation.)

We then turn to *voltage comparators*, another popular class of analog ICs. Both bipolar and CMOS types are addressed, including comparators with hysteresis. (Their transient responses are investigated in Chapter 6.)

Virtually every IC requires suitable circuitry to bias its transistors internally. Also, applications such as instrumentation and measurement require stable and predictable *current* and *voltage references*. This class of circuits is addressed next, including *bandgap-reference* types, both for bipolar and CMOS ICs.

It is interesting that the engineering community has traditionally favored the manipulation of voltages while in fact the manipulation of currents is an inherently faster physical process. Though their development has been delayed also because of technological factors, *current-mode ICs* are now in much use along with their voltage-mode counterparts. The circuits addressed in this chapter include *transconductors, current conveyors* (CCs), *operational transconductance amplifiers* (OTAs), *current feedback amplifiers* (CFAs), and *Gilbert Cells*.

In today's mixed-mode ICs, where sensitive analog circuitry is forced to coexist with digital circuitry in highly noisy environments and with low-valued power supplies, much signal processing and transmission is done in fully differential form. The chapter investigates some of the most common *fully differential op amp*s in use today.

The desire to implement analog and digital functions on the same chip mandates the rephrasing of some traditional analog functions in terms of CMOS switches and capacitors, the devices most easily available in today's prevalent digital technology. The chapter concludes with switched-capacitor techniques in two representative applications, autozeroing and filtering. A brief discussion of the discrete-time nature of switched-capacitor integrators illustrates their departure from their continuous-time counterparts.

The chapter makes abundant use of PSpice both as a software oscilloscope to display transfer curves and waveforms, and as a verification tool for dc and ac hand calculations.

5.1 THE μA741 OPERATIONAL AMPLIFIER

The 741 op amp incorporates a number of clever design ideas that are in widespread use to this very day. Being also the most widely documented analog IC, it offers the student a variety of sources that can be consulted and used as a vehicle to learn otherwise general analog-design techniques.

Shown in Fig. 5.1 is the circuit schematic of the μA741 op amp as provided by Fairchild, its original manufacturer (the 741 has been second-sourced by virtually every analog IC manufacturer, so you are likely to find slight variants in the literature). With two dozen BJTs, a dozen resistors, and a capacitor, the beginner has

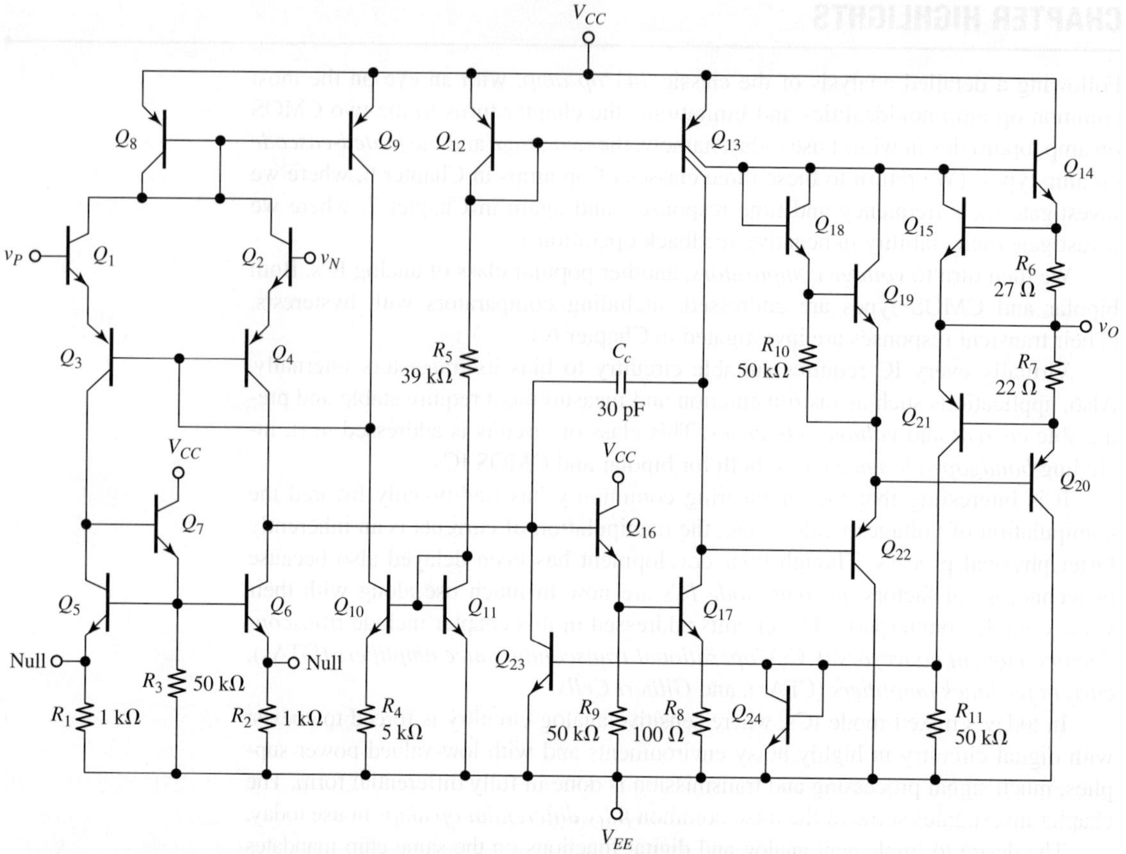

FIGURE 5.1 Circuit schematic of the 741 op amp. (Courtesy of Fairchild Semiconductor Corporation.)

legitimate reasons to feel intimidated. However, the proper approach is not to try to understand the whole circuit at once, but to identify already familiar subcircuits and analyze them individually. Only after understanding its parts can we expect to tackle the circuit as a whole. To facilitate this process, it is convenient to redraw the circuit in a simplified form highlighting only the essentials while leaving the finer details for later. This leads us to the schematic of Fig. 5.2, which the reader is encouraged to compare to the original of Fig. 5.1 before moving along.

A General Overview of the 741 Op Amp

With reference to Fig. 5.2 we make the following observations:

- Starting out at the lower left, we identify the current mirror Q_5-Q_6 as forming an active load for the CB *pnp* pair Q_3-Q_4. Normally we would expect this pair to be of the EC type, but this is not the case here due to the notoriously low betas of lateral *pnp* BJTs. Op amps are intended to draw negligibly small currents at the input terminals, so in order to minimize these currents the *pnp* pair Q_3-Q_4 is operated in the CB mode, and the v_P and v_N inputs are buffered via the CC pair

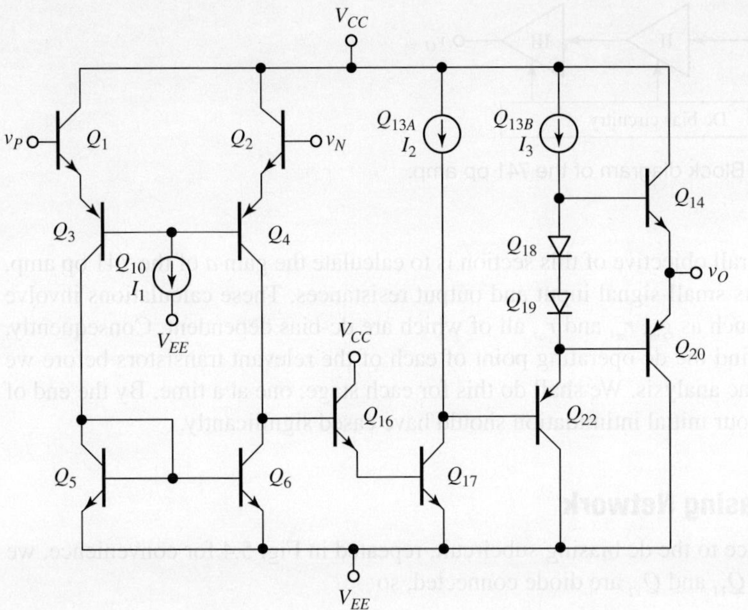

FIGURE 5.2 Simplified circuit schematic of the 741 op amp.

Q_1-Q_2. Since the bipolar process in use optimizes *npn* BJTs, Q_1 and Q_2 enjoy much higher betas than their *pnp* counterparts and therefore draw much lower base currents. We have thus identified our first subcircuit: a *differential-input, single-ended-output amplifier stage* made up of two CC-CB halves as well as a current-mirror load.

- Moving toward the right we identify a Darlington-type stage made up of the CC-CE pair Q_{16}-Q_{17}. The function of the CE amplifier Q_{17} is to provide *additional gain*, and that of the CC buffer Q_{16} is to provide *high input resistance* in order to avoid loading the first stage excessively. (Figure 5.1 shows also a capacitance C_c between this stage's input and output terminals. Its function is to stabilize the op amp against possible oscillations, a subject to be investigated at the end of Chapter 7. The present analysis is limited to low frequencies, where C_c acts as an open circuit and will thus be ignored.)

- Moving further to the right we find yet another CC *buffer*, namely, Q_{22}, whose function is to reduce loading of Q_{17}. The emitter network of Q_{22} includes the pair Q_{18}-Q_{19} to provide the two *pn*-junction voltage drops needed to bias the Class AB push-pull pair Q_{14}-Q_{20} forming the *output stage*. The *overload protection* circuitry has been omitted precisely because at this point we want to highlight only the essentials, leaving the details for later.

- It stands to reason to regard the 741 op amp as consisting of a cascade of *three stages*, as block-diagrammed in Fig. 5.3. (In fact a great many op amps conform to a similar three-stage schematization.) One further consideration is in order: to function, all active transistors must be operated in the forward-active region, so a *fourth* subcircuit is needed to properly bias the aforementioned stages. In Fig. 5.1 this subcircuit is made up of the current-mirror pairs Q_8-Q_9, Q_{10}-Q_{11}, and Q_{12}-Q_{13}.

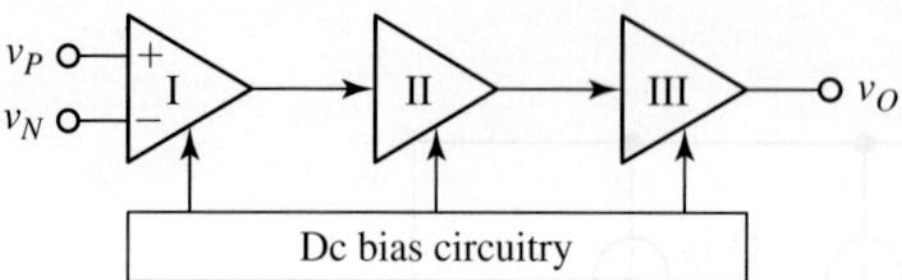

FIGURE 5.3 Block diagram of the 741 op amp.

The overall objective of this section is to calculate the gain a of the 741 op amp, along with its small-signal input and output resistances. These calculations involve parameters such as g_m, r_π, and r_o, all of which are dc-bias dependent. Consequently, we need to find the dc operating point of each of the relevant transistors before we attempt any ac analysis. We shall do this for each stage, one at a time. By the end of the process, our initial intimidation should have eased significantly.

The Dc Biasing Network

With reference to the dc biasing subcircuit, repeated in Fig. 5.4 for convenience, we observe that Q_{11} and Q_{12} are diode connected, so

$$I_{REF} = \frac{(V_{CC} - V_{EB12(on)}) - (V_{EE} + V_{BE11(on)})}{R_5} \tag{5.1}$$

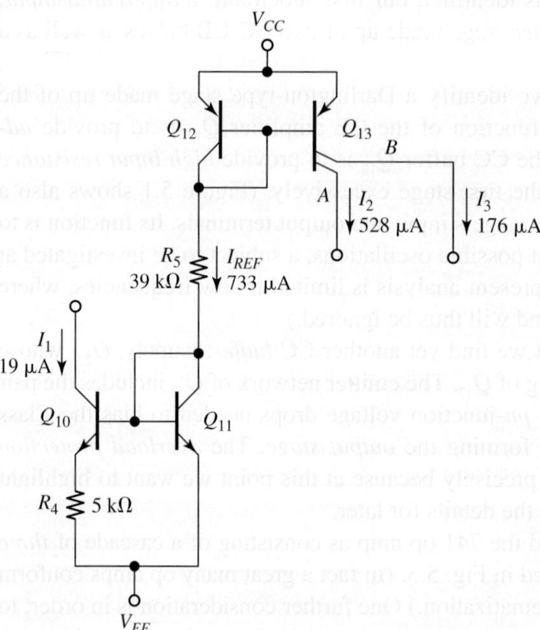

FIGURE 5.4 The dc biasing circuitry of the 741 op amp.

The 741 was designed to operate with nominal power supplies of ± 15 V, though it can function with lower values just as well, such as ± 5 V. In the following analysis we shall assume nominal supplies along with typical *pn* junction voltage drops of 0.7 V. With these values, Eq. (5.1) gives $I_{REF} = 733$ μA.

The pair Q_{10}-Q_{11} forms a Widlar current sink. Using the iterative technique presented in Chapter 4 in connection with this type of source, we get

$$I_1 = 19 \ \mu A \tag{5.2}$$

The Q_{12}-Q_{13} pair forms an ordinary current mirror, except that Q_{13} is equipped with *two* collectors instead of one. Two collectors are fabricated by making two separate *p*-type diffusions into the *n*-type base region. In the present case the areas are fabricated in a 3-to-1 ratio, so the larger of the two will collect ¾ and the smaller ¼ of the total collector current I_{C13}. (Alternatively, you can think of Q_{13} as consisting of *two* separate transistors Q_{13A} and Q_{13B} with their emitters and bases tied pairwise together and with their I_ss in a 3-to-1 ratio. Except that fabricating two separate BJTs would take up more chip area than a single BJT equipped with two collectors.) By current-mirror action, $I_{C13} = I_{REF}(1 - 2/\beta_{F13}) = 733(1 - 2/50) = 704$ μA, where $\beta_{F13} = 50$ is assumed. Letting $I_2 = 3I_3 = ¾ I_{C13}$ gives

$$I_2 = 528 \ \mu A \qquad I_3 = 176 \ \mu A \tag{5.3}$$

The 1st or Input Stage

The operation of this stage is best understood by examining it right after we apply power. With all BJTs still off, the first item to go on is the current I_{REF} of Fig 5.4, followed by I_1 through I_3. Initially all of I_1 in Fig. 5.5 will be pulled out of the bases of Q_3 and Q_4, causing their rapid turn on. But, as Q_3 and Q_4 conduct, so do Q_1 and Q_2, as the latter are in series with the former. Q_1 and Q_2 draw their collector currents from the diode-connected transistor Q_8, whose current is mirrored by Q_9 back to the very node where I_1 triggered the whole sequence of events. We have a negative-feedback situation whereby the circuit will automatically stabilize at the operating point where the current I_{C9} returned by Q_9, augmented by the current $2I_B$ coming from the bases of Q_3 and Q_4, will *equal* I_1.

We know from Chapter 4 that the current returned by Q_9, compared to that of Q_8, is afflicted by a fractional error of $-2/\beta_{Fp}$. This may not be negligible, as lateral *pnp* BJTs have mediocre betas. However, in the present arrangement this error is cancelled out by the current $2I_B$ returned by the bases of Q_3 and Q_4 themselves. (This error cancellation is similar to that occurring in Wilson current mirrors.) We thus conclude that the BJTs of each half of the input stage carry a current of $I_1/2$, or

$$I_{C1} = I_{C2} = I_{C3} = I_{C4} = I_{C5} = I_{C6} = \frac{I_1}{2} = 9.5 \ \mu A \tag{5.4}$$

The base currents drawn by Q_1 and Q_2 are

$$I_P = I_N = \frac{I_1/2}{\beta_{Fn}} = 47.5 \ nA \tag{5.5}$$

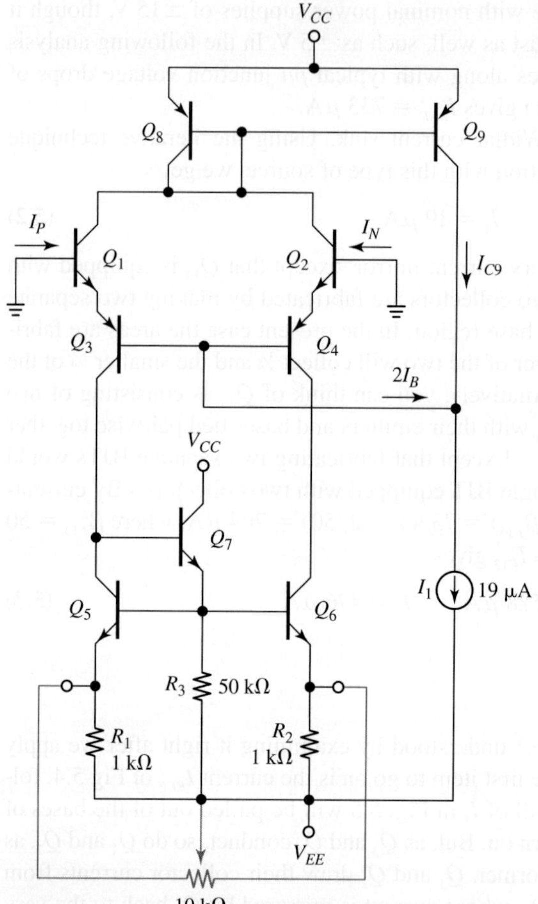

FIGURE 5.5 Circuit for the dc analysis of the input stage of the 741 op amp.

where $\beta_{Fn} = 200$ is assumed for the Q_1-Q_2 pair. Once powered, the op amp will automatically draw these currents from the surrounding circuitry. Looking back at Fig. 5.1 we note that both collectors of Q_5 and Q_6 are two *pn*-junction voltage drops above V_{EE}. Consequently, besides acting as a beta helper, Q_7 ensures *equal collector voltages* for Q_5 and Q_6 as well as for Q_3 and Q_4, thus rendering the two halves of the stage perfectly *balanced*.

We now wish to find the small-signal gain and the input/output resistances of the input stage. This task is simplified considerably if we exploit the half-circuit concept of Fig. 5.6*a*. By applying $v_{id}/2$ to Q_1 and $-v_{id}/2$ to Q_2 we force the common base node of Q_3 and Q_4 to remain at ac ground, so we can work with just one of the two halves, say, Q_1 and Q_3. Their small-signal emitter resistances r_{e1} and r_{e3} cause a voltage division by two to give, at their shared emitter terminals, the ac signal

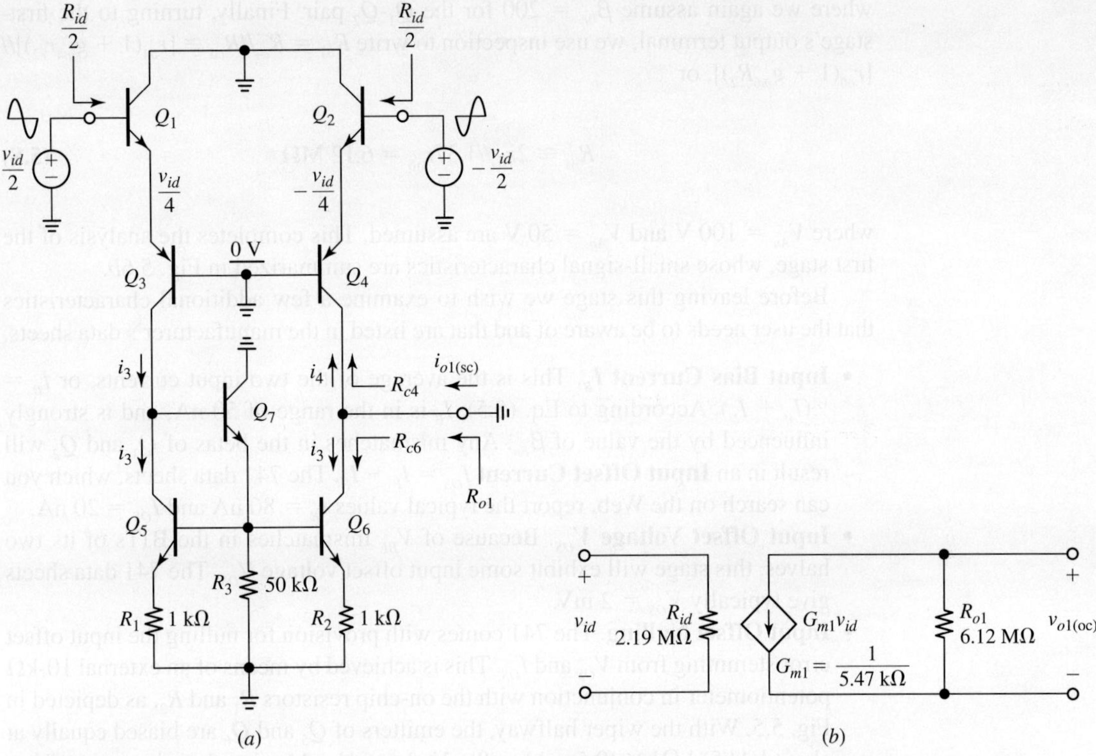

FIGURE 5.6 (a) Circuit for the ac analysis of the input stage, and (b) input-stage small-signal equivalent.

$(1/2)(v_{id}/2) = v_{id}/4$. So, the ac current *out* of Q_3's collector is $i_3 = g_{m3}v_{id}/4$, whereas the ac current *into* Q_4's collector is, by symmetry, $i_4 = g_{m4}v_{id}/4$. The current mirror made up of Q_5 and Q_6 reflects i_3 so that the net short-circuit ac current sunk from the output node is $i_{ol(sc)} = i_3 + i_4 = g_{m3}v_{id}/4 + g_{m4}v_{id}/4 = g_m v_{id}/2$, where we have dropped the subscripts as the two g_ms are identical. This allows us to write

$$G_{m1} = \frac{i_{ol(sc)}}{v_{id}} = \frac{g_m}{2} = \frac{9.5\ \mu A}{2 \times 26\ mV} = \frac{1}{5.47\ k\Omega} \tag{5.6}$$

where G_{m1} is the *first-stage transconductance*. Looking into Q_1's base terminal we find, by inspection, $R_{id}/2 = r_{\pi 1} + (\beta_{01} + 1)r_{e3} \cong r_{\pi 1} + (\beta_{01} + 1)r_{e1} = 2r_{\pi 1} = 2r_{\pi n}$, having used the fact that $r_{e3} \cong r_{e1}$. Consequently,

$$R_{id} \cong 4r_{\pi n} = 2.19\ M\Omega \tag{5.7}$$

where we again assume $\beta_{0n} = 200$ for the Q_1-Q_2 pair. Finally, turning to the first-stage's output terminal, we use inspection to write $R_{o1} = R_{c4}//R_{c6} \cong [r_{o4}(1 + g_{m4}r_{e2})]//[r_{o6}(1 + g_{m6}R_2)]$, or

$$R_{o1} \cong 2r_{o4}//1.37r_{o6} = 6.12 \text{ M}\Omega \tag{5.8}$$

where $V_{An} = 100$ V and $V_{Ap} = 50$ V are assumed. This completes the analysis of the first stage, whose small-signal characteristics are summarized in Fig. 5.6b.

Before leaving this stage we wish to examine a few additional characteristics that the user needs to be aware of and that are listed in the manufacturer's data sheets.

- **Input Bias Current I_B.** This is the average of the two input currents, or $I_B = \frac{1}{2}(I_P + I_N)$. According to Eq. (5.5), I_B is in the range of 50 nA, and is strongly influenced by the value of β_{Fn}. Any mismatches in the betas of Q_1 and Q_2 will result in an **Input Offset Current** $I_{OS} = I_P - I_N$. The 741 data sheets, which you can search on the Web, report the typical values $I_B = 80$ nA and $I_{OS} = 20$ nA.
- **Input Offset Voltage V_{OS}.** Because of V_{BE} mismatches in the BJTs of its two halves, this stage will exhibit some input offset voltage V_{OS}. The 741 data sheets give typically $V_{OS} = 2$ mV.
- **Input Offset Nulling.** The 741 comes with provision for nulling the input offset error stemming from V_{OS} and I_{OS}. This is achieved by means of an external 10-kΩ potentiometer in conjunction with the on-chip resistors R_1 and R_2, as depicted in Fig. 5.5. With the wiper halfway, the emitters of Q_5 and Q_6 are biased equally at about $[(1//5) \text{ k}\Omega] \times (9.5 \text{ }\mu\text{A}) = 8$ mV above V_{EE}. Moving the wiper away from its center position will make one transistor more conductive by bringing its emitter closer to V_{EE} while leaving the other still about 8 mV above V_{EE}. This external imbalance is adjusted empirically by the user until it *cancels out* any internal imbalance to give the appearance of offset-less behavior.
- **Input Voltage Range (IVR).** In negative-feedback operation the op amp forces v_N to track v_P, so the common-mode input voltage is $v_{IC} = \frac{1}{2}(v_P + v_N) \cong v_P$. We wish to find the *common-mode input voltage range*, that is, the range of values of v_{IC} over which the input stage will function properly, with all BJTs operating in the forward-active region or, at most, at the edge of saturation (EOS). Referring to Fig. 5.1 we observe that the *upper* limit is reached when Q_1 is driven to the EOS, so we use KVL to write $v_{P(\max)} \cong V_{CC} - V_{EB8(\text{on})} - V_{CE1(\text{EOS})} + V_{BE1(\text{on})}$. The *lower* limit is reached when Q_3 is driven to the EOS, so $v_{P(\min)} \cong V_{EE} + V_{BE5(\text{on})} + V_{BE7(\text{on})} + V_{EC3(\text{EOS})} + V_{BE1(\text{on})}$. Letting $v_P \rightarrow v_{IC}$ gives

$$v_{IC(\max)} \cong V_{CC} - V_{CE(\text{EOS})} \qquad v_{IC(\min)} \cong V_{EE} + 3V_{BE(\text{on})} + V_{EC(\text{EOS})} \tag{5.9}$$

(The above estimates ignore the small voltage drops across R_1, R_2, and R_8, and also make the simplifying assumption that all junction voltage drops be equal.) To get an idea, if we assume typical junction drops of 0.7 V and EOS drops of 0.2 V, the above estimates give $v_{IC(\max)} \cong V_{CC} - 0.2$ V and $v_{IC(\min)} \cong V_{EE} + 2.3$ V. With ±15-V supplies, the IVR is thus -12.7 V $\le v_{IC} \le +14.8$ V. Pushing v_{IC} outside the IVR will cause malfunction.

The 2nd or Intermediate Stage

With reference to Fig. 5.7 we have, by inspection, $I_{C17} = I_2$, or

$$I_{C17} = 528 \ \mu A \tag{5.10}$$

Moreover we have $I_{C6} = \alpha_{F6}(I_{R_9} + I_{B17})$, which can be approximated as

$$I_{C16} \cong \frac{V_{BE17} + R_8 I_2/\alpha_{F17}}{R_9} + \frac{I_2}{\beta_{F17}} \cong \frac{V_T \ln(I_2/I_{s17}) + R_8 I_2}{R_9} + \frac{I_2}{\beta_{F17}} \tag{5.11a}$$

Assuming $I_{s17} = 10$ fA and $\beta_{F17} = 250$ we get

$$I_{C16} = 16 \ \mu A \tag{5.11b}$$

This completes the dc analysis.

For the small-signal analysis refer to Fig. 5.8a, where C_c has been omitted because here we are interested only in low-frequency behavior. Starting out at the left and working our way toward the right we write, by inspection

$$R_{i2} = r_{\pi16} + (\beta_{016} + 1)\{R_9//[r_{\pi17} + (\beta_{017} + 1)R_8]\} \tag{5.12a}$$

Assuming $\beta_{016} = 200$ and $\beta_{017} = 250$ we get

$$R_{i2} = 4.63 \ M\Omega \tag{5.12b}$$

Using again inspection, we write

$$R_{o2} = r_{o13A}//R_{c17} \cong r_{o13A}//[r_{o17}(1 + g_{m17}R_8)] \tag{5.13a}$$

Letting $r_{o13A} = (50 \ V)/(528 \ \mu A) = 94.7 \ k\Omega$ and $r_{o17} = (100 \ V)/(528 \ \mu A) = 189.4 \ k\Omega$ we get

$$R_{o2} = 81.3 \ k\Omega \tag{5.13b}$$

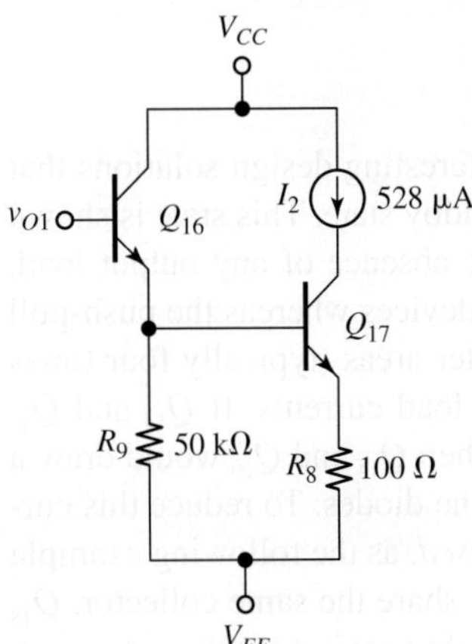

FIGURE 5.7 Circuit for the dc analysis of the intermediate stage of the 741 op amp.

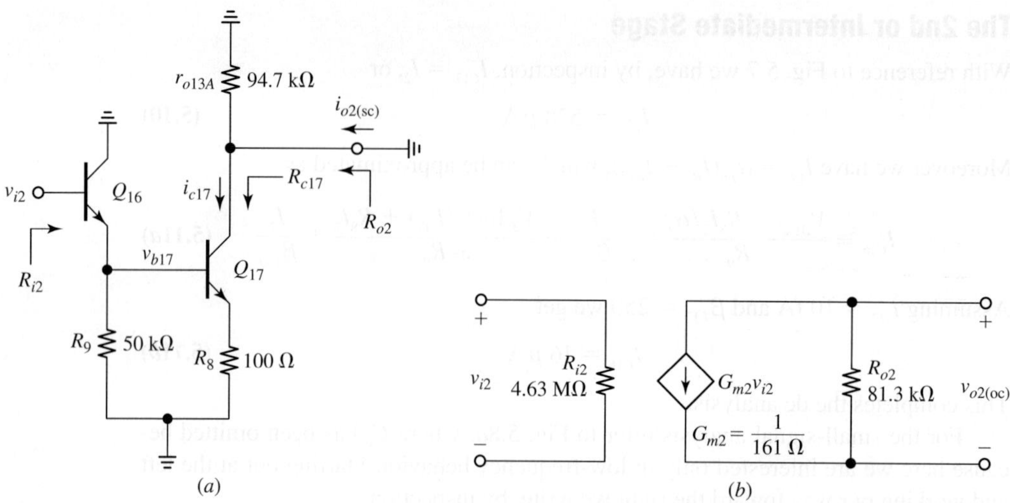

FIGURE 5.8 (a) Circuit for the ac analysis of the second stage, and (b) second-stage small-signal equivalent.

Finally, considering that Q_{16} is operating in the CC mode and Q_{17} in the CE-ED mode, we write

$$G_{m2} = \frac{i_{o2(sc)}}{v_{i2}} = \frac{v_{b17}}{v_{i2}} \times \frac{i_{c17}}{v_{b17}} = \frac{1}{1 + \dfrac{r_{e16}}{R_9 /\!/ [r_{\pi17} + (\beta_{017} + 1)R_8]}} \times \frac{g_{m17}}{1 + g_{m17}R_8}$$

$$= \frac{1}{161\ \Omega}$$

(5.14)

where G_{m2} is the *second-stage transconductance*. The small-signal characteristics of this stage are summarized in Fig. 5.8b, and this completes the second-stage analysis.

The 3rd or Output Stage

The output stage of the 741 op amp presents some interesting design solutions that are best appreciated by examining the circuit in its standby state. This state is shown in Fig. 5.9 for the case of the input at 0 V and in the absence of any output load. The biasing transistors Q_{18} and Q_{19} are minimum-size devices whereas the push-pull transistors Q_{14} and Q_{20} are fabricated with larger emitter areas (typically four times as large) to maintain good current gains also at high load currents. If Q_{14} and Q_{19} were diode-connected in the manner of Section 4.10, then Q_{14} and Q_{20} would draw a standby current roughly four times as large as that of the diodes. To reduce this current to a more acceptable level Q_{18} is suitably *underbiased*, as the following example will clarify. It is also worth mentioning that since they share the same collector, Q_{18} and Q_{19} are fabricated inside the same isolation region, thus saving precious die area.

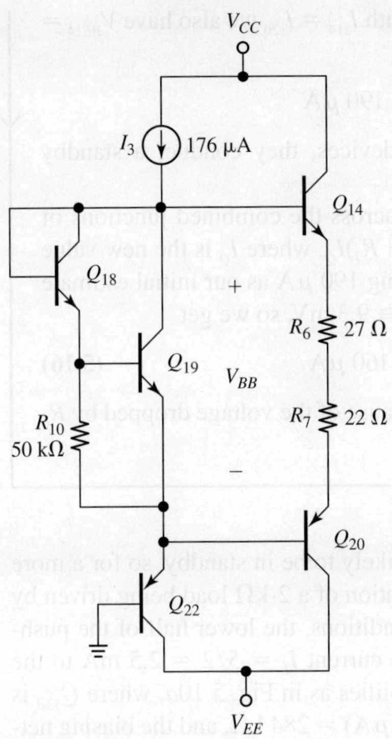

FIGURE 5.9 Dc analysis of the 741 output stage in standby.

In the circuit of Fig. 5.9 let $I_{s18} = I_{s19} = 2$ fA and $I_{s14} = I_{s20} = 8$ fA (four times as large). Moreover, whenever appropriate, assume negligible base currents for all BJTs.   **EXAMPLE 5.1**

(a) Use iterations to estimate the collector currents I_{C18} and I_{C19}. Hence, find V_{BB}, in mV.

(b) Estimate I_{C14} and I_{C20} under the simplifying assumption $R_6 = R_7 = 0$.

(c) Repeat part (b), but with R_6 and R_7 in place as shown. Comment on your findings.

Solution

(a) Start out with $V_{BE19} = 0.7$ V, so $I_{C18} \cong V_{BE19}/R_{10} \cong 0.7/50 = 14$ μA. By KCL, $I_{C19} = I_3 - I_{C18} \cong 176 - 14 = 162$ μA. By the BJT equation, $V_{BE19} = V_T \ln(I_{C19}/I_{s19}) = 0.026\ln[(162 \times 10^{-6})/(2 \times 10^{-15})] = 0.653$ V. Use this value for the improved estimate $I_{C18} \cong 0.653/50 \cong 13$ μA. Allowing for, say, 1 μA of base current for Q_{19}, we finally get

$$I_{C18} \cong 14 \text{ μA} \qquad I_{C19} \cong 162 \text{ μA} \qquad (5.15)$$

and no further iterations are needed. Applying again the BJT equation with $I_{s18} = I_{s19} = 2$ fA we get

$$V_{BB} = V_{BE18} + V_{BE19} = 589.4 + 653.1 \cong 1242 \text{ mV}$$

(b) In standby we have $I_{C14} = I_{C20}$. Moreover, with $I_{s14} = I_{s20}$ we also have $V_{BE14} = V_{EB20} = V_{BB}/2$ so

$$I_{C14} = I_{C20} = 8 \times 10^{-15}\exp(1242/52) \cong 190 \ \mu A$$

Even though Q_{14} and Q_{20} are large-area devices, they conduct a standby current that is comparable to I_3.

(c) With R_6 and R_7 in place, the voltage drop across the combined junctions of Q_{14} and Q_{20} will be reduced by $\Delta V = (R_6 + R_7)I_Q$, where I_Q is the new value of the quiescent current of Q_{14} and Q_{20}. Using 190 μA as our initial estimate for I_Q we find $\Delta V = (27 + 22)190 \times 10^{-6} \cong 9.3$ mV, so we get

$$I_Q = 8 \times 10^{-15}\exp[(1242 - 9.3)/52] \cong 160 \ \mu A \tag{5.16}$$

This is lower than the previous estimate because of the voltage dropped by R_6 and R_7.

In actual application the output stage is unlikely to be in standby, so for a more realistic ac analysis we consider the typical situation of a 2-kΩ load being driven by a voltage centered at $V_O = 5$ V. Under these conditions, the lower half of the push-pull stage will be off, leaving Q_{14} to source the current $I_L = 5/2 = 2.5$ mA to the load. The ac equivalent of the output stage simplifies as in Fig. 5.10a, where Q_{13B} is modeled with the resistance $r_{o13B} = (50\ \text{V})/(176\ \mu\text{A}) = 284$ kΩ, and the biasing network made up of Q_{18}, Q_{19}, and R_{10} with a single small-signal resistance $r_{bb} \cong 174 \ \Omega$ (see Exercise 5.1).

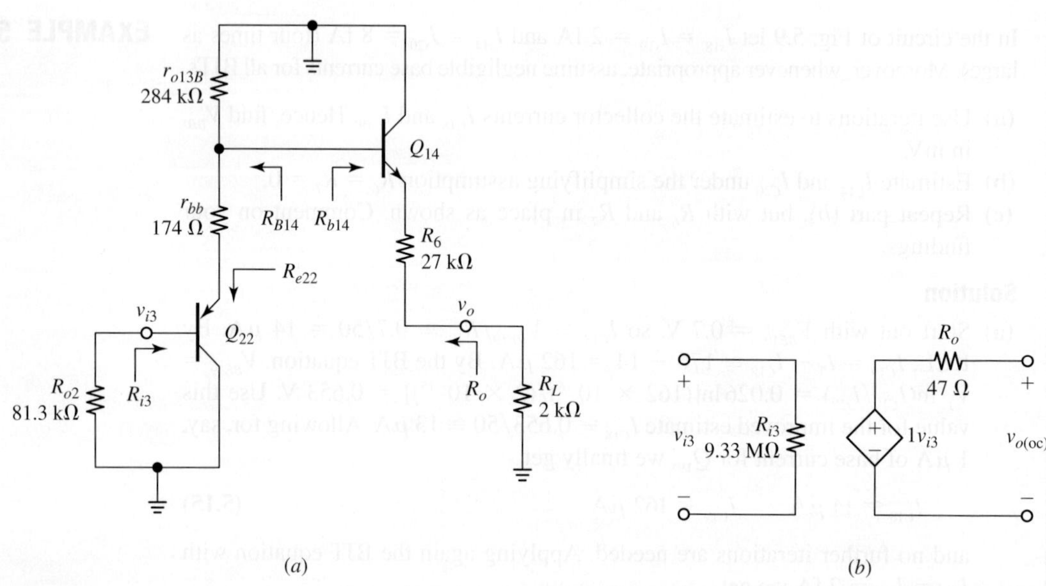

FIGURE 5.10 (a) Circuit for the ac analysis of the output stage, and (b) output-stage small-signal equivalent.

Exercise 5.1

Draw the small-signal equivalent of the biasing network of Fig. 5.9, consisting of Q_{18}, Q_{19}, and R_{10}, and show that the entire network acts as a single resistance

$$r_{bb} = \frac{r_{d18} + (R_{10}//r_{\pi19})}{1 + g_{m19}(R_{10}//r_{\pi19})} \cong 174\ \Omega \tag{5.17}$$

where $r_{d18} = V_T/I_{C18}$ is the ac resistance of diode-connected Q_{18}, and $g_{m19} = I_{C19}/V_T$ and $r_{\pi19} = \beta_{019}/g_{m19}$ are the small-signal parameters of Q_{19}. Assume the currents of Example 5.1 as well as $\beta_{019} = 200$.

To find the small-signal characteristics of the output stage refer again to Fig. 5.10a. Considering that we have two voltage-followers coupled via the small resistance r_{bb}, we can approximate

$$v_o \cong 1 \times v_{i3} \tag{5.18}$$

By inspection we also have

$$R_{i3} = r_{\pi22} + (\beta_{022} + 1)[r_{bb} + (r_{o13B}//R_{b14})] \tag{5.19a}$$

where R_{b14} is the ac resistance seen looking into the base of Q_{14},

$$R_{b14} = r_{\pi14} + (\beta_{014} + 1)(R_6 + R_L)$$

Assuming $I_{C14} = 2.5$ mA and $\beta_{014} = 250$ we get $R_{b14} \cong 511$ kΩ. Substituting into Eq. (5.19a) and assuming $\beta_{022} = 50$, we finally obtain

$$R_{i3} = 9.33\ M\Omega \tag{5.19b}$$

Using again inspection, we write

$$R_o = R_6 + \frac{R_{B14} + r_{\pi14}}{\beta_{014} + 1} \tag{5.20a}$$

where R_{B14} is the ac resistance presented to Q_{14}'s base by the circuitry upstream,

$$R_{B14} = r_{o13B}//(r_{bb} + R_{e22})$$

and R_{e22} is the ac resistance seen looking into Q_{22}'s emitter

$$R_{e22} = \frac{R_{o2} + r_{\pi22}}{\beta_{022} + 1} = \frac{81.3 + 50(26/0.176)}{50 + 1} = 1.74\ k\Omega$$

Substituting gives $R_{B14} = 1.9$ kΩ. Substituting in turn into Eq. (5.20a) we finally get

$$R_o = 47\ \Omega \tag{5.20b}$$

The small-signal characteristics of the output stage are summarized in Fig. 5.10b. Before leaving this stage we wish to investigate one additional characteristic that the user needs to know and that is reported in the data sheets.

- **Output Voltage Swing (OVS).** This is the range of values of v_O over which the output stage will function properly, with all active BJTs operating in the forward-active region or at most at the edge of saturation (EOS). Simple inspection of Fig. 5.1 reveals that the *upper* limit for v_O is reached when Q_{13B} is driven to the EOS, and the *lower* limit when Q_{17} is driven to the EOS. Using KVL we thus write $v_{O(\max)} \cong V_{CC} - V_{EC13(EOS)} - V_{BE14(on)}$ and $v_{O(\min)} \cong V_{EE} + V_{CE17(EOS)} + V_{EB22(on)} + V_{EB20(on)}$. These simplify as

$$v_{O(\max)} \cong V_{CC} - V_{EC(EOS)} - V_{BE(on)} \qquad v_{O(\min)} \cong V_{EE} + V_{CE(EOS)} + 2V_{EB(on)} \quad \textbf{(5.21)}$$

(The above estimates assume identical junction voltage drops for the BJTs, along with light output loading so that the voltage drops across R_6 and R_7 can be ignored.) To get an idea, if we assume junction drops of 0.7 V and EOS drops of 0.2 V, the above estimates give $v_{O(\max)} \cong V_{CC} - 0.9$ V and $v_{O(\min)} \cong V_{EE} + 1.6$ V. With ±15-V supplies, the OVS is thus -13.4 V $\leq v_O \leq +14.1$ V.

The Small-Signal Characteristics of the 741 Op Amp

We now use the three-stage cascade of Fig. 5.11 to find the overall small-signal gain a. By inspection

$$a = \frac{v_o}{v_{id}} \cong (-G_{m1})(R_{o1}//R_{i2})(-G_{m2})(R_{o2}//R_{i3}) = (-482) \times (-501)$$
$$= 241 \times 10^3 \text{ V/V} \qquad \textbf{(5.22)}$$

indicating that the first two stages contribute gains on the order of 500 V/V each. The 741 data sheets give typically $R_{id} = 2$ MΩ, $a = 200 \times 10^3$ V/V, and $R_o = 75\ \Omega$. Our calculations are influenced by the assumed values for the betas and the Early voltages, both of which depend on critical fabrication parameters such as the base width. Also, assuming infinite Early voltages to simplify the dc calculations may have underestimated dc currents by as much as 20-30%, especially in the case of *pnp* BJTs. Finally, R_o depends on the operating current of Q_{14}, which was arbitrarily assumed to be 2.5 mA,

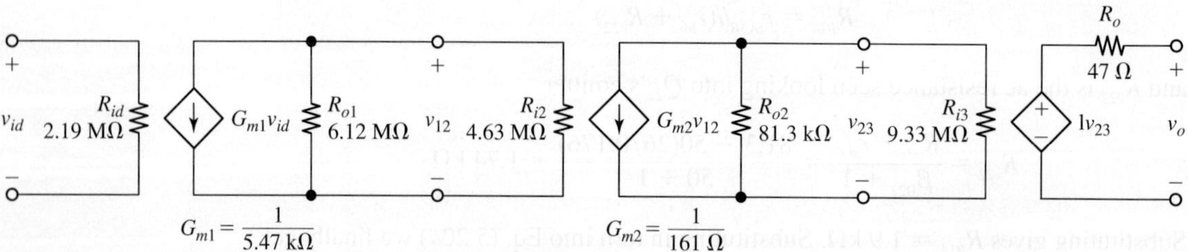

FIGURE 5.11 Small-signal model of the 741 op amp.

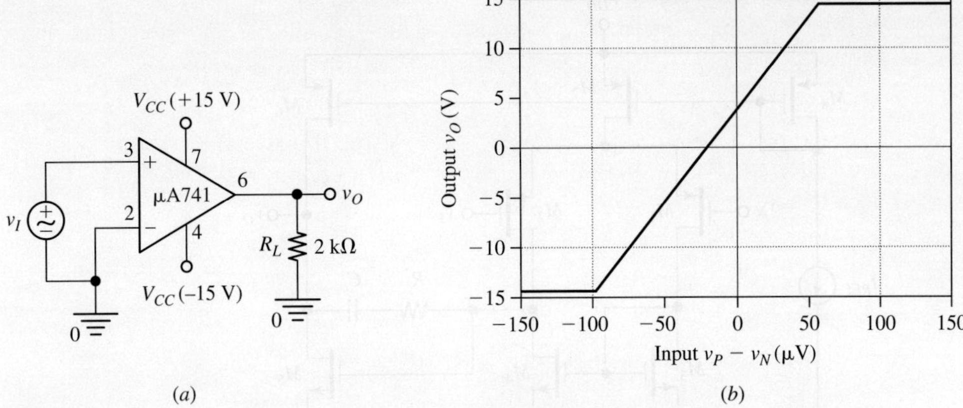

FIGURE 5.12 (a) PSpice circuit to display (b) the VTC of the 741 op amp.

so, the discrepancies between calculated and data-sheets values are not surprising. Yet, the mental process exercised in the estimation of these parameters is undoubtedly quite instructive—and presumably it has helped demystify our initial intimidation.

PSpice Simulation of the 741 Op Amp

To facilitate the simulation of op amps in an application setting, manufacturers usually provide SPICE macro-models of their device (see Appendix 5A). The PSpice circuit of Fig. 5.12a utilizes the uA741 macro-model available in PSpice's library to display the VTC as well as to calculate the ac gain a and the input and output resistances r_i and r_o. The results are as follows.

```
V(OUT)/VI = 1.992E+05
INPUT RESISTANCE AT VI = 9.963E+05
OUTPUT RESISTANCE AT V(OUT) = 1.517E+02
```

Figure 5.12b confirms that the output saturates in the vicinity of the supply voltages. It also shows a (systematic) input offset voltage of about 20 μV. This macro-model will be used in subsequent chapters to investigate other behavioral aspects, such as the frequency and transient responses as well as the stability of 741 circuits.

5.2 THE TWO-STAGE CMOS OPERATIONAL AMPLIFIER

This classic CMOS op amp topology (and variants thereof) is used especially in mixed-mode ICs. Shown in its basic form in Fig. 5.13, it comprises two gain stages and a dc biasing circuit as follows:

- The **1st** or **input stage** consists of the p-channel differential pair M_1-M_2 and the n-channel mirror load M_3-M_4. As we know, its voltage gain is

$$a_1 = -g_{m1}(r_{o2}//r_{o4}) \tag{5.23}$$

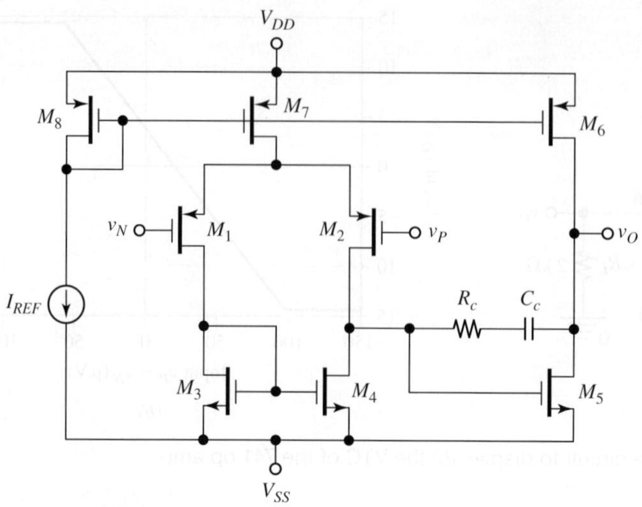

FIGURE 5.13 Two-stage CMOS op amp.

- The **2nd** or **output stage** consists of the CS amplifier M_5 and the active load M_6. (Also shown is an R_c-C_c network whose function is to stabilize the amplifier against possible oscillations in negative-feedback operation, a subject to be addressed in Chapter 7. The present analysis is restricted to low frequencies, where C_c acts as an open circuit, so the R_c-C_c network will be ignored.) This stage's low-frequency voltage gain is

$$a_2 = -g_{m5}(r_{o5}//r_{o6})$$ **(5.24)**

- The **dc biasing circuit** consists of the dual-output current mirror M_6-M_7-M_8, along with the I_{REF} current reference. The details of this reference have been omitted for simplicity, but this is typically a circuit shared among different op amps on the same chip. It generates a regulated current I_{REF} once, which is then replicated by a multiple-output current mirror to each of the other on-chip op amps. In Fig. 5.13, I_{REF} is replicated by M_7 to bias the M_1-M_2 pair, and by M_6 to actively load the CS stage M_5.

Thanks to the infinite resistance presented by M_5's gate, there is *no inter-stage loading*, so the overall gain is simply the *product* of the individual gains

$$a = \frac{v_o}{v_p - v_n} = a_1 \times a_2 = g_{m1}(r_{o2}//r_{o4})g_{m5}(r_{o5}//r_{o6})$$ **(5.25a)**

Adapting Eq. (4.152) to the present case, we express gain in the insightful alternative form

$$a = \frac{2}{V_{OV1}(\lambda_2 + \lambda_4)} \times \frac{2}{V_{OV5}(\lambda_5 + \lambda_6)}$$ **(5.25b)**

If the op amp is realized with FETs having the *same* channel length L and overdrive voltage V_{OV}, then gain takes on the concise form

$$a = \left[\frac{2L}{V_{OV}(\lambda'_n + \lambda'_p)}\right]^2 \tag{5.25c}$$

where, by Eq. (4.31), λ'_n and λ'_p are the process parameters characterizing channel-length modulation in the *n*FETs and *p*FETs, respectively. Clearly, the longer L and the lower V_{OV}, the higher the gain. The ac resistances between the two inputs and between the output and ground are, respectively,

$$R_i = \infty \qquad\qquad R_o = r_{o5}//r_{o6} \tag{5.26}$$

We note a close resemblance of the CMOS stages of Fig. 5.13 to the 1st and 2nd stages of the 741 op amp of Fig. 5.2. Actually, thanks to the infinite resistance presented by the gates, the CMOS stages are much simpler. We also note the absence of an output stage, even though R_o can be fairly high. Typically, an op amp of the type of Fig. 5.13 is likely to drive other on-chip CMOS circuits also presenting infinite input resistance (though not necessarily zero capacitance), so there is no need for a dedicated output stage. Only when meant to drive resistive loads, most likely off chip, will a two-stage op amp be equipped with a dedicated 3rd stage. This might be an output stage of the type discussed in Section 4.11.

Input Offset Voltage Considerations

It is important to realize that the W/L ratios of M_5 and M_6 cannot be specified arbitrarily, but must satisfy a specific constraint in order to avoid introducing gross input offset voltage errors. To find this constraint, note that since $V_{SG6} = V_{SG7}$, I_{D6} and I_{D7} must scale in proportion to their W/L ratios as

$$\frac{I_{D6}}{I_{D7}} = \frac{W_6/L_6}{W_7/L_7} \tag{5.27a}$$

Note also that at dc balance we have $V_{DS4} = V_{DS3}$. Consequently, $V_{GS5} = V_{DS4} = V_{DS3} = V_{GS3}$, indicating that I_{D5} and I_{D3} must scale in proportion to their W/L ratios as

$$\frac{I_{D5}}{I_{D3}} = \frac{W_5/L_5}{W_3/L_3} \tag{5.27b}$$

But, $I_{D5} = I_{D6}$ and $I_{D3} = I_{D7}/2$. Substituting into Eq. (5.27) and simplifying gives the important constraint

$$\frac{W_5/L_5}{W_3/L_3} = 2\frac{W_6/L_6}{W_7/L_7} \tag{5.28}$$

Once this constraint is met, any k and V_t mismatches in the M_1-M_2 and M_3-M_4 pairs will cause the 1st stage to exhibit an input offset voltage of the type of Eq. (4.161). Adapting to the present case, we have

$$V_{OS(1\text{st stage})} \cong \frac{V_{OVp}}{2}\sqrt{\left(\frac{\Delta k_p}{k_p}\right)^2 + \left(\frac{\Delta k_n}{k_n}\right)^2 + \left(\frac{\Delta V_{tp}}{0.5 V_{OVp}}\right)^2 + \left(\frac{\Delta V_{tn}}{0.5 V_{OVp}}\right)^2} \tag{5.29}$$

where subscripts p and n refer, respectively, to the M_1-M_2 and M_3-M_4 pairs. Even if the M_1-M_2 and M_3-M_4 pairs are perfectly matched, letting $v_P = v_N = 0$ is likely to result in $v_O \neq 0$ because of a possible imbalance between M_5 and M_6. Reflected to the input, the effect of this imbalance will result in an additional component for V_{OS} (this issue is explored further in the end-of-chapter problems).

Input Voltage Range (IVR)

In negative-feedback operation the op amp forces v_N to track v_P, so the common-mode input voltage is $v_{IC} = \frac{1}{2}(v_P + v_N) \cong v_P$. We wish to find the *common-mode input voltage range*, that is, the range of values of v_{IC} over which the input stage will function properly, with all FETs operating in saturation or at most at the edge of saturation (EOS). The *upper* limit is reached when M_7 is driven to the EOS, where $V_{SD7} = V_{OV7}$. Using inspection and KVL we thus write $v_{IC(\max)} = V_{DD} - V_{OV7} - V_{SG1}$, that is,

$$v_{IC(\max)} = V_{DD} - V_{OV7} - V_{OV1} - |V_{t1}| \tag{5.30a}$$

Similarly, the *lower* limit for v_{IC} is reached when M_1 is driven to the EOS, where $V_{SD1} = V_{OV1}$. Using inspection and KVL we thus write $v_{IC(\min)} = V_{SS} + V_{GS3} + V_{SD1} - V_{SG1}$. But, $V_{SD1} - V_{SG1} = V_{OV1} - (V_{OV1} + |V_{t1}|) = -|V_{t1}|$, so

$$v_{IC(\min)} = V_{SS} + V_{OV3} + V_{t3} - |V_{t1}| \tag{5.30b}$$

It is apparent that for this circuit $v_{IC(\max)}$ is more restrictive than $v_{IC(\min)}$.

Output Voltage Swing (OVS)

This is the range of values of v_O over which M_5 and M_6 operate in saturation or at most at the EOS. By inspection, we readily find

$$v_{O(\max)} = V_{DD} - V_{OV6} \qquad v_{O(\min)} = V_{SS} + V_{OV5} \tag{5.31}$$

In words, v_O can swing within a V_{OV} voltage drop of each supply rail.

EXAMPLE 5.2   (a) Suppose the two-stage CMOS op amp of Fig. 5.13 is fabricated in a process characterized by $k'_n = 2.5 k'_p = 100$ μA/V^2, $V_{tn} = -V_{tp} = 0.75$ V, $\lambda'_n = 0.1$ μm/V and $\lambda'_p = 0.05$ μm/V. Moreover, all FETs are fabricated with $L = 1$ μm and are designed to operate with $V_{OV} = 0.25$ V. If the circuit is powered from ± 2.5-V supplies and uses $I_{REF} = 100$ μA, specify suitable values for W_1 through W_8 to bias both stages at 100 μA (for simplicity assume $\lambda_n = \lambda_p = 0$ in the course of your dc calculations).

(b) Find the individual-stage gains, the overall gain, the output resistance, the IVR, and the OVS.

(c) Check with PSpice and account for any differences between calculated and simulated values.

Solution

(a) M_6, M_7, and M_8 must each satisfy the saturation-region condition

$$100\ \mu A = \frac{40\ \mu A}{2}\ \frac{W}{1\ \mu m} 0.25^2$$

which gives $W = 80\ \mu m$. So, $W_6 = W_7 = W_8 = 80\ \mu m$. M_1 and M_2 draw *half* as much current as M_7 but with the same overdrive voltage, so $W_1 = W_2 = W_7/2 = 40\ \mu m$. M_3 and M_4 must each satisfy the condition

$$50\ \mu A = \frac{100\ \mu A}{2}\ \frac{W}{1\ \mu m} 0.25^2$$

which gives $W_3 = W_4 = 16\ \mu m$. By Eq. (5.28), $(W_5/1)/(16/1) = 2(80/1)/(80/1)$, or $W_5 = 32\ \mu m$. In summary,

$$W_1 = W_2 = 40\ \mu m \qquad W_3 = W_4 = 16\ \mu m$$

$$W_5 = 32\ \mu m \qquad\qquad W_6 = W_7 = W_8 = 80\ \mu m$$

(b) By Eq. (5.25c) we have

$$a = \left[\frac{2 \times 1}{0.25(0.1 + 0.05)}\right]^2 = 53.3^2 = 2{,}844\ \text{V/V}$$

so $a_1 = a_2 = -53.3$ V/V. Also, $\lambda_n = \lambda'_n/1 = 0.1\ \text{V}^{-1}$ and $\lambda_p = \lambda'_p/1 = 0.05\ \text{V}^{-1}$, so $r_{o5} = 1/(0.1 \times 100 \times 10^{-6}) = 100\ \text{k}\Omega$, $r_{o6} = 1/(0.05 \times 100 \times 10^{-6}) = 200\ \text{k}\Omega$, and

$$R_o = 100//200 = 66.7\ \text{k}\Omega$$

Finally, Eqs. (5.30) and (5.31) give

$$v_{IC(\max)} = 2.5 - 2 \times 0.25 - 0.75 = 1.25\ \text{V}$$

$$v_{IC(\min)} = v_{O(\min)} = -2.5 + 0.25 = -2.25\ \text{V}$$

$$v_{O(\max)} = 2.25\ \text{V}$$

(c) A PSpice simulation using the circuit of Fig. 5.14a gives the VTC of Fig. 5.14b, which reveals a (systematic) input offset of 166 μV and confirms output saturation levels within a V_{OV} of the ± 2.5-V supplies. (The offset can be compensated for by specifying a dc component of $-166\ \mu$V for the ac input source v_p.) After directing PSpice to perform the small-signal analysis (.TF) we get $a = v_o/v_p = 4{,}271$ V/V and $R_o = 75\ \text{k}\Omega$. The discrepancy between calculated and simulated values stems primarily from the assumption $\lambda = 0$, especially while calculating g_{m1} and g_{m5}. For more accurate results we must multiply the calculated value of g_{m1} by $(1 + \lambda V_{SD1}) \cong (1 + 0.05 \times 2.5) = 1.125$

and that of g_{m5} by $(1 + \lambda V_{DS5}) \cong (1 + 0.1 \times 2.5) = 1.25$. With these corrections we get $a_1 \cong -53.3 \times 1.125 = -60$ V/V, $a_2 \cong -53.3 \times 1.25 = -66.7$ V/V, and $a \cong -4{,}000$ V/V, in better agreement with PSpice.

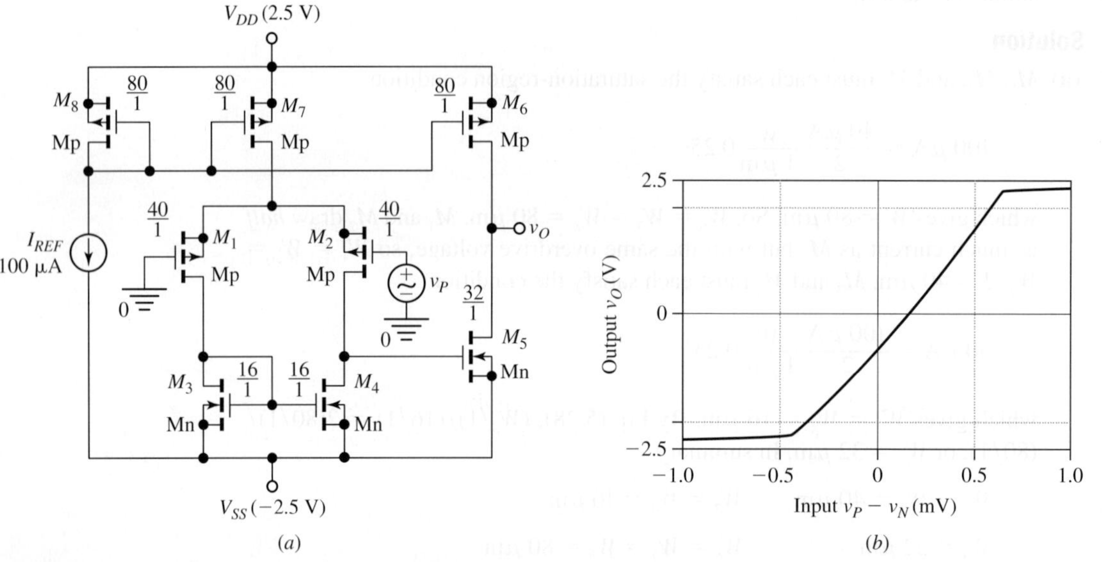

FIGURE 5.14 (a) PSpice circuit for Example 5.2, and (b) its VTC. The MOSFET parameters are as follows: $k'_n = 2.5k'_p = 100$ µA/V², $V_{tn} = -V_{tp} = 0.75$ V, and $\lambda_n = 2\lambda_p = 0.1$ V⁻¹.

Common-Mode Rejection Ratio (CMRR)

We estimate this parameter by adapting Eq. (4.158) to the present circuit,

$$\text{CMRR} = \frac{1 + g_{m3}r_{o3}}{1 + r_{o3}/r_{o1}}[1 + 2(g_{m1} + g_{mb1})r_{o7}] \tag{5.32}$$

If necessary, the CMRR can be improved by raising M_7's output resistance, for instance via the cascoding technique. The price is a reduction in the value of $v_{IC(\text{max})}$.

Power-Supply Rejection Ratio (PSRR)

The output of an amplifier should be unaffected by any variations in its power-supply voltages, such as ripple and supply noise induced by adjacent circuitry. However, a real-life amplifier will be somewhat sensitive also to these variations (besides the already familiar common-mode input voltage), so the small-signal output of an op amp with split supplies takes on the more general form

$$v_o = a_{dm}v_{id} + a_{cm}v_{ic} + a_{dd}v_{dd} + a_{ss}v_{ss} \tag{5.33}$$

where v_{dd} and v_{ss} are, respectively, the variations in the supply voltages V_{DD} and V_{SS}, and a_{dd} and a_{ss} are the gains with which the amplifier magnifies these variations, that is, $a_{dd} = v_o/v_{dd}$ and $a_{ss} = v_o/v_{ss}$. Ideally, both a_{dd} and a_{ss} should be zero (just like a_{cm} should be). To tell how insensitive a real-life amplifier is to power-supply variations we use a figure of merit called the *power-supply rejection ratio* (PSRR), which, in the case of split supplies, takes on the separate forms

$$\text{CMRR}_{dd} = \left| \frac{a_{dm}}{a_{dd}} \right| \qquad \text{CMRR}_{ss} = \left| \frac{a_{dm}}{a_{ss}} \right| \qquad (5.34)$$

where a_{dm} is the familiar differential-mode gain. To estimate the PSRRs of the two-stage CMOS op amp, refer to its ac equivalents of Fig. 5.15, with respect to which we make the following observations:

- Applying v_{dd} to the emitters of the M_1-M_2 pair via r_{o7} has the same effect as lifting their bases off ground and driving them with a common voltage of $-v_{dd}$ while holding the upper terminal of r_{o7} at ac ground. We can thus adapt the common-mode gain of Eq. (4.157c) to the present circuit. Ignoring the body effect for simplicity, we have

$$v_{ds4} = a_{cm(\text{MOS})}(-v_{dd}) \cong \frac{v_{dd}}{2g_{m4}r_{o7}}$$

Applying KCL at node v_o gives

$$\frac{v_{dd} - v_o}{r_{o6}} = g_{m5}\frac{v_{dd}}{2g_{m4}r_{o7}} + \frac{v_o}{r_{o5}}$$

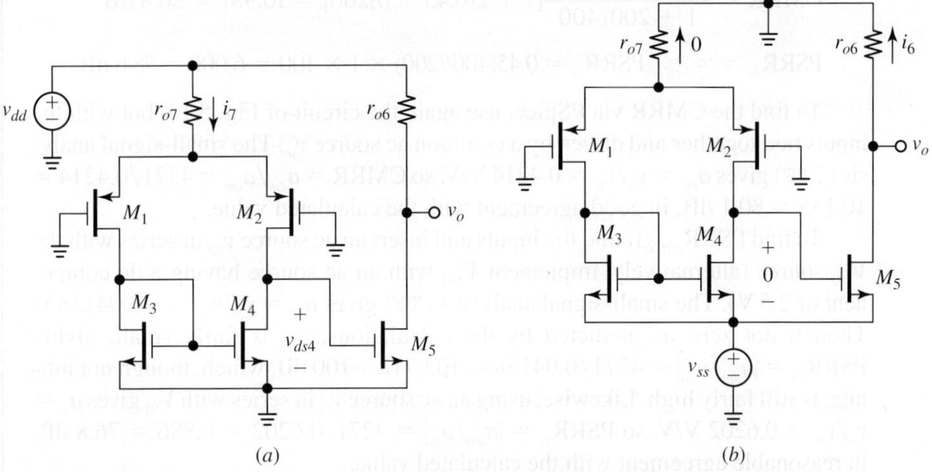

(a)　　　　　　　　(b)

FIGURE 5.15 Ac equivalents for the estimation of the supply rejection ratios to (a) V_{DD} and (b) V_{SS}.

Since $V_{OV5} = V_{OV4}$ and $V_{OV6} = V_{OV7}$, it follows that $r_{o6}/r_{o7} = 2g_{m4}/g_{m5}$, so the above expression gives $v_o(1/r_{o6} + 1/r_{o5}) = v_{dd}(1/r_{o6} - 1/r_{o6}) = 0$. This is possible only if $v_o = 0$, so

$$a_{dd} = \frac{v_o}{v_{dd}} = \frac{0}{v_{dd}} = 0 \text{ V/V} \tag{5.35a}$$

and

$$\text{PSRR}_{dd} = \infty \tag{5.35b}$$

Interestingly, the contributions by v_{dd} to v_o via r_{o6} and r_{o7} *cancel* each other out. However, because of various approximations made, PSRR_{dd} will in practice not be infinite, though we expect it to be reasonably high.

- Turning next to Fig. 5.15b, we observe that the balance condition of the input stage is unperturbed by v_{ss}, so we have $v_{ds4} = v_{ds3} = 0$. M_5's internal dependent source is now dormant, so we use the voltage divider rule to write

$$a_{ss} = \frac{v_o}{v_{ss}} = \frac{r_{o6}}{r_{o5} + r_{o6}} \tag{5.36a}$$

Substituting into Eq. (5.34), along with a_{dm} as given in Eq. (5.25a), we finally get

$$\text{PSRR}_{ss} = g_{m1}(r_{o2}//r_{o4})g_{m5}r_{o5} \tag{5.36b}$$

EXAMPLE 5.3   Estimate the CMRR and PSRRs of the CMOS op amp of Example 5.2 (ignore the body effect for M_1 and M_2). Compare with PSpice, and comment.

Solution

Using $g_{m1} = g_{m3} = (2 \times 50 \times 10^{-6}/0.25) \times 1.125 = 0.45$ mA/V and $g_{m5} = (2 \times 100 \times 10^{-6}/0.25) \times 1.25 = 1$ mA/V, we have, by Eqs. (5.32), (5.35b), and (5.36),

$$\text{CMRR} = \frac{1 + 0.45 \times 200}{1 + 200/400}[1 + 2(0.45 + 0)200] = 10{,}981 = 80.8 \text{ dB}$$

$$\text{PSRR}_{dd} = \infty \qquad \text{PSRR}_{ss} = 0.45(400//200) \times 1 \times 100 = 6{,}000 = 75.6 \text{ dB}$$

To find the CMRR via PSpice, use again the circuit of Fig. 5.14, but with the inputs tied together and driven by a common ac source v_{ic}. The small-signal analysis (.TF) gives $a_{cm} = v_o/v_{ic} = 0.4214$ V/V, so $\text{CMRR} = a_{dm}/a_{cm} = 4271/0.4214 = 10{,}135 = 80.1$ dB, in good agreement with the calculated value.

To find PSRR_{dd}, ground the inputs and insert an ac source v_{dd} in series with the V_{DD} source (alternatively, implement V_{DD} with an ac source having a dc component of 2.5 V). The small-signal analysis (.TF) gives $a_{dd} = v_o/v_{dd} = -0.04158$ V. Though not zero as predicted by the calculations, a_{dd} is fairly small, giving $\text{PSRR}_{dd} = |a_{dm}/a_{dd}| = 4271/0.04158 = 102{,}718 = 100$ dB, which, though not infinite, is still fairly high. Likewise, using an ac source v_{ss} in series with V_{SS} gives $a_{ss} = v_o/v_{ss} = 0.6202$ V/V, so $\text{PSRR}_{ss} = |a_{dm}/a_{ss}| = 4271/0.6202 = 6{,}886 = 76.8$ dB, in reasonable agreement with the calculated value.

Rewriting Eq. (5.33) in the form

$$v_o = a_{dm}\left(v_{id} + \frac{a_{cm}}{a_{dm}}v_{ic} + \frac{a_{dd}}{a_{dm}}v_{dd} + \frac{a_{ss}}{a_{dm}}v_{ss}\right)$$

$$= a_{dm}\left(v_{id} + \frac{v_{ic}}{\text{CMRR}} + \frac{v_{dd}}{\text{PSRR}_{dd}} + \frac{v_{ss}}{\text{PSRR}_{ss}}\right) \tag{5.37}$$

offers an instructive interpretation for the rejection ratios: (*a*) the voltages v_{ic}, v_{dd}, and v_{ss}, reflected to the input, get divided by the corresponding rejection ratios; (*b*) since the reflected voltages are in *series* with v_{id}, they act as separate *input offset-voltage components*. Clearly, the higher a rejection ratio, the smaller the corresponding input offset term.

5.3 THE FOLDED-CASCODE CMOS OPERATIONAL AMPLIFIER

The two-stage op amp just studied realizes voltage gain in the form $a = (G_{m1}R_{o1}) \times (G_{m2}R_{o2})$, that is, as the product of its individual stage gains. Regrouping as $a = G_{m1}(R_{o1}G_{m2}R_{o2})$ suggests an alternative implementation, namely, a *single stage* (G_{m1}) but having a *much higher* output resistance ($R_o = R_{o1}G_{m2}R_{o2}$). We start out with an active-loaded differential pair to implement G_{m1}, and then we *cascode* both the pair and the load to raise the output resistance. To avoid the notorious voltage-swing limitations of straight cascodes we use the *folded-cascode* scheme introduced at the end of Section 4.9. This results in the popular CMOS op amp alternative of Fig. 5.16, in

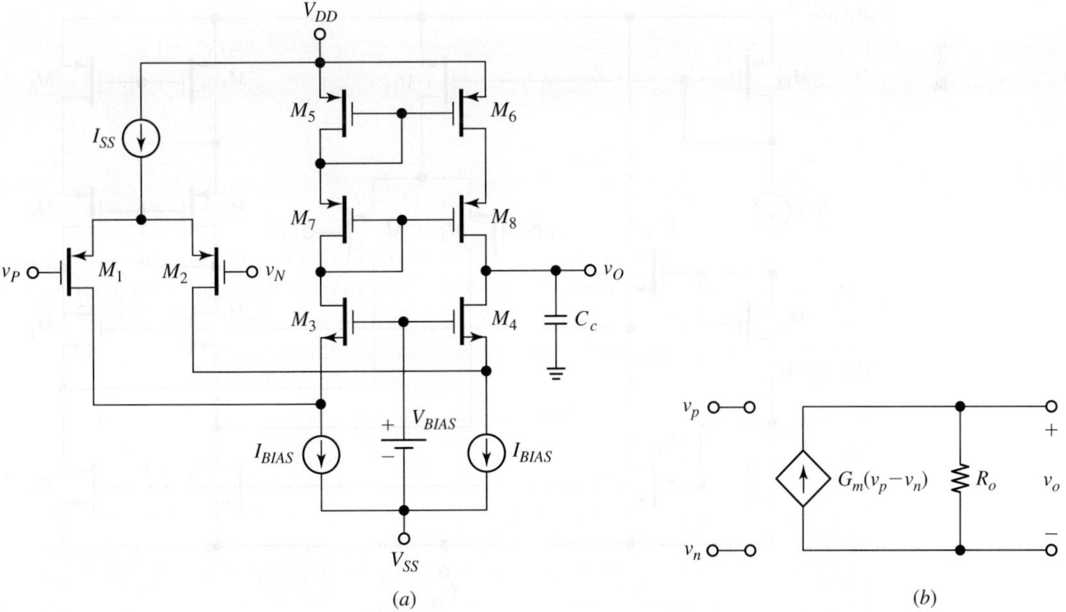

FIGURE 5.16 (*a*) Simplified diagram of the folded-cascode CMOS op amp and (*b*) its Norton equivalent.

connection with which we note the following:

- The heart of the circuit is the *p*-channel *differential pair* M_1-M_2, in turn *cascoded* by the *n*-channel CG pair M_3-M_4. As we know, in the folded arrangement the CG pair requires separate biasing, a function here provided by the two I_{BIAS} current sinks.
- The *active load* is the *cascode current mirror* made up of the M_5-M_6 and M_7-M_8 pairs. (The function of capacitor C_c is to stabilize the amplifier against possible oscillations in negative-feedback operation, a subject to be addressed in detail in Chapter 7.)
- The *dc biasing circuitry* comprises the sources I_{SS}, I_{BIAS}, and V_{BIAS}. Should the M_1-M_2 pair be over-driven, one of its FETs would go off. To prevent the corresponding half of the load from also turning off and then require a delay to go back on when the overdrive condition is removed, it is customary to specify $I_{BIAS} > I_{SS}$, such as $I_{BIAS} \cong 1.25 I_{SS}$. Shown in Fig. 5.17 is a possible realization of the dc biasing circuitry: M_9 and M_{10} sink the I_{BIAS} currents, M_{11} sources the I_{SS} current, and M_{12} through M_{16} provide the proper voltages to bias M_{11} as well as the M_3-M_4 and M_9-M_{10} pairs.

We now wish to find the element values G_m and R_o of the Norton equivalent depicted in Fig. 5.16*b*. To find G_m we need to find $i_{o(sc)}$, a task that we shall carry out using the half-circuit concept of Fig. 5.18. The differential pair responds to an input imbalance $v_{id} = v_p - v_n$ with the drain currents

$$i_1 = g_{m1}\frac{v_{id}}{2} \qquad i_2 = g_{m2}\frac{v_{id}}{2}$$

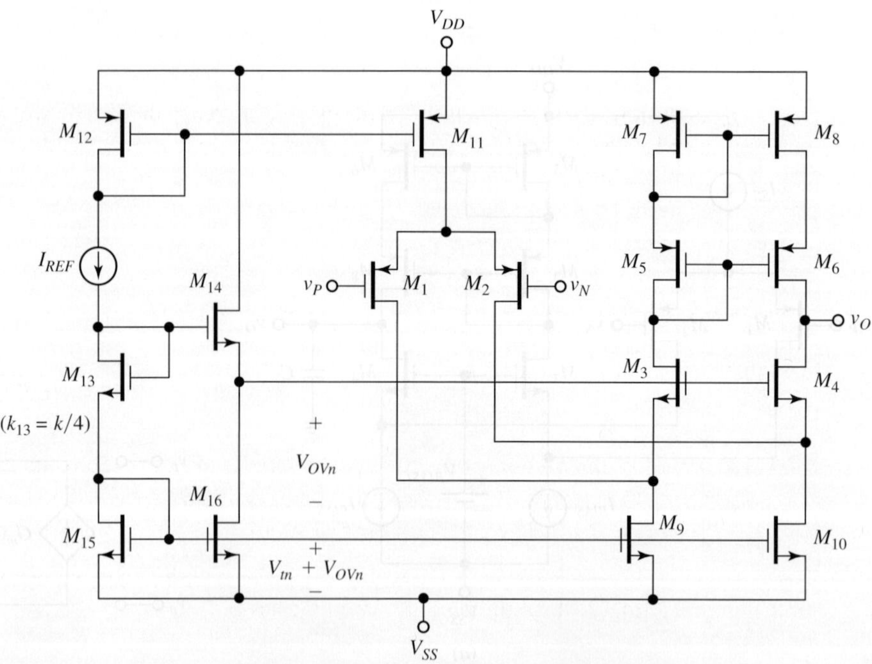

FIGURE 5.17 Detailed circuit schematic of the folded-cascode CMOS op amp.

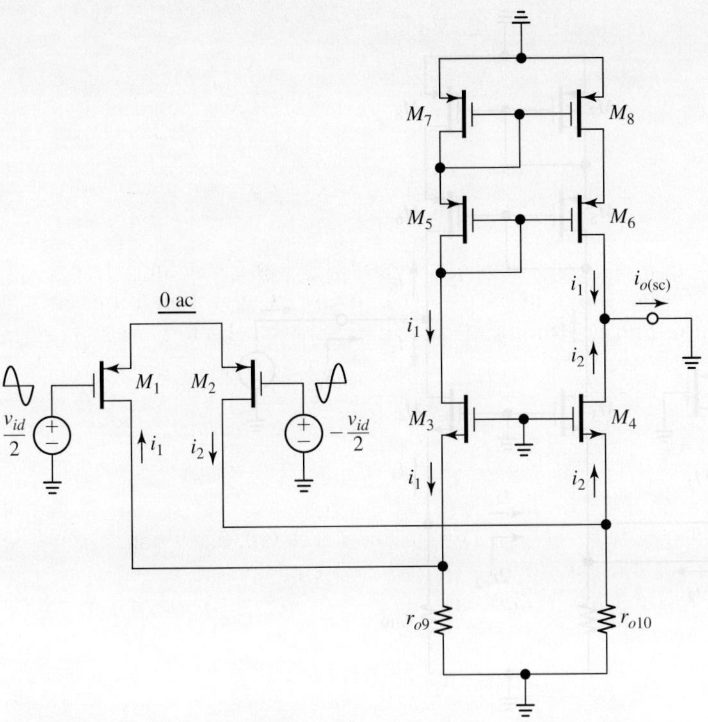

FIGURE 5.18 Ac model to find the short-circuit output current $i_{o(sc)}$.

Since $g_{m2} = g_{m1}$, it follows that $i_2 = i_1$. Once i_2 reaches M_4's source, it divides between the resistance seen looking into M_4's source and that seen looking into M_{10}'s drain. The former is $R_{s4} = [1/(g_{m4} + g_{mb4})]//r_{o4}$ and the latter is $R_{d10} = r_{o10}$. Since $R_{s4} \ll R_{d10}$, virtually all of i_2 will flow into M_4 and thence to the output ac short, as shown. Similar considerations hold for the current division experienced by i_1 at M_3's source. Virtually all of i_1 will come out of M_3 and will thus be mirrored into the output ac short, as shown. By KCL, $i_{o(sc)} = i_1 + i_2 = 2(g_{m1}v_{id}/2) = g_{m1}v_{id}$. Consequently,

$$G_m = \frac{i_{o(sc)}}{v_p - v_n} = g_{m1} \qquad (5.38)$$

To find the *small-signal output resistance* R_o, set the input sources to zero, apply a test voltage v, find the current i out of the test source, and let $R_o = v/i$. The test method is depicted in Fig. 5.19, where we observe that i consists of three components:

- The component $i_6 = v/R_{d6}$, where R_{d6} is the resistance seen looking into M_6's drain. Adapting Eq. (4.41) to the present circuit we have

$$i_6 = \frac{v}{r_{o6} + r_{o8} + (g_{m6} + g_{mb6})r_{o6}r_{o8}} \cong \frac{v}{(g_{m6} + g_{mb6})r_{o6}r_{o8}}$$

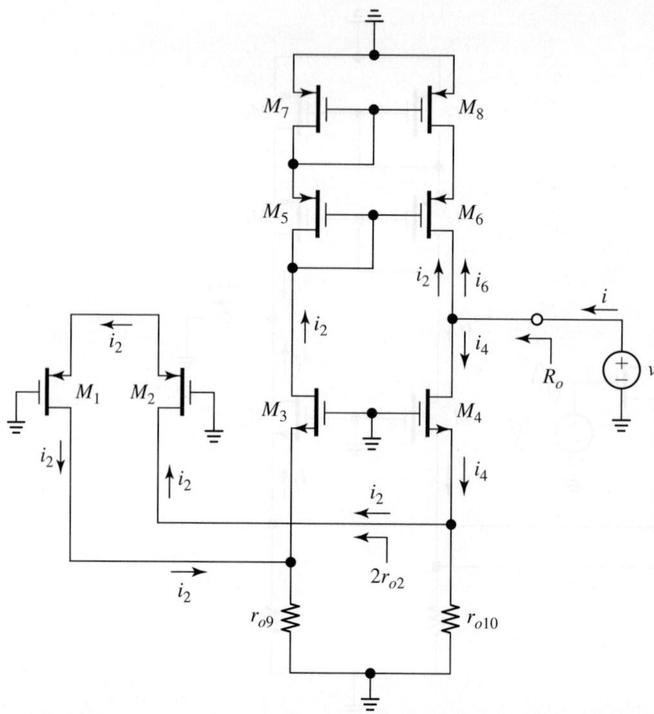

FIGURE 5.19 Ac model to find the output resistance R_o.

- The component $i_4 = v/R_{d4}$, where R_{d4} is the resistance seen looking into M_4's drain. Again adapting we get

$$i_4 = \frac{v}{r_{o4} + (2r_{o2}//r_{o10}) + (g_{m4} + g_{mb4})r_{o4}(2r_{o2}//r_{o10})} \cong \frac{v}{(g_{m4} + g_{mb4})r_{o4}(2r_{o2}//r_{o10})}$$

where $2r_{o2}$ is the resistance seen looking into M_2's drain.

- Upon exiting M_4, i_4 divides between $2r_{o2}$ and r_{o10} to give, by the current divider rule,

$$i_2 = \frac{r_{o10}}{2r_{o2} + r_{o10}}i_4 = \frac{2r_{o2}//r_{o10}}{2r_{o2}}i_4 \cong \frac{v}{(g_{m4} + g_{mb4})2r_{o2}r_{o4}}$$

This current continues through M_1 to M_3's source, where it experiences negligible current division to proceed through M_3 to the mirror, which then replicates it at the test source, as shown.

We now apply KCL to write

$$i \cong \frac{v}{(g_{m6} + g_{mb6})r_{o6}r_{o8}} + \frac{v}{(g_{m4} + g_{mb4})r_{o4}(2r_{o2}//r_{o10})} + \frac{v}{(g_{m4} + g_{mb4})2r_{o2}r_{o4}}$$

Combining the last two terms and simplifying, we write

$$i \cong \frac{v}{(g_{m6} + g_{mb6})r_{o6}r_{o8}} + \frac{v}{(g_{m4} + g_{mb4})r_{o4}(r_{o2}//r_{o10})} = \frac{v}{R_o}$$

where

$$R_o \cong \left[(g_{m6} + g_{mb6}) r_{o6} r_{o8} \right] // \left[(g_{m4} + g_{mb4}) r_{o4} (r_{o2} // r_{o10}) \right] \qquad (5.39)$$

Finally, the unloaded voltage gain is

$$a = \frac{v_o}{v_p - v_n} = g_{m1} R_o \qquad (5.40)$$

Input Voltage Range and Output Voltage Swing

In order for the circuit to function properly, all FETs must operate in saturation or at most at its edge (EOS). The permissible range of values for the common-mode input voltage v_{IC} defines the input voltage range (IVR). Using inspection and KVL in the circuit of Fig. 5.17 we readily find

$$v_{IC(max)} = V_{DD} - V_{OV11} - V_{OV1} - |V_{t1}| \qquad v_{IC(min)} = V_{SS} + V_{OV9} - |V_{t1}| \qquad (5.41)$$

Likewise, the limits of the output voltage swing (OVS) are

$$v_{O(max)} = V_{DD} - |V_{t8}| - V_{OV8} - V_{OV6} \qquad v_{O(min)} = V_{SS} + V_{OV10} + V_{OV4} \qquad (5.42)$$

We can eliminate the $|V_{t8}|$ term from $v_{O(max)}$ by using a wide-swing cascode mirror such as the p-channel version of the Sooch mirror discussed in Section 4.8.

EXAMPLE 5.4

(a) Suppose the folded-cascode op amp of Fig. 5.17 is fabricated in a process with $k'_n = 2.5 k'_p = 100 \ \mu A/V^2$, $V_{tn} = -V_{tp} = 0.75$ V, $\lambda'_n = 0.1 \ \mu m/V$ and $\lambda'_p = 0.05 \ \mu m/V$. Moreover, all FETs are fabricated with $L = 1 \ \mu m$ and are designed to operate with $V_{OV} = 0.25$ V. If the circuit is powered from ± 2.5-V supplies and uses $I_{REF} = 100 \ \mu A$, specify suitable values for W_1 through W_{16} for $I_{SS} = 100 \ \mu A$ and $I_{BIAS} = 125 \ \mu A$. (For simplicity assume $\lambda = 0$ and ignore the body effect in the course of your dc calculations.)

(b) Assuming $\chi = 0.1$ throughout, find R_o, a, the IVR, and the OVS.

Solution

(a) Each FET (except for M_{13}) must satisfy the saturation-region condition

$$I_D = \frac{k'}{2} \frac{W}{1 \ \mu m} 0.25^2$$

which allows us to perform the following calculations:

- Letting $I_D = 100 \ \mu A$ and $k' = 40 \ \mu A/V^2$ in the above expression gives $W_{11} = W_{12} = 80 \ \mu m$.
- Since M_1 and M_2 draw half as much dc current as M_{11} we have $W_1 = W_2 = W_{11}/2 = 80/2 = 40 \ \mu m$.
- Letting $I_D = 100 \ \mu A$ and $k' = 100 \ \mu A/V^2$ gives $W_{14} = W_{15} = W_{16} = 32 \ \mu m$. Also, $W_{13} = 32/4 = 8 \ \mu m$.

- Since M_9 and M_{10} draw 125 μA ($=1.25 \times 100$ μA), they must be 1.25 times as wide as M_{15}, so $W_9 = W_{10} = 1.25W_{15} = 1.25 \times 32 = 40$ μm.
- At dc balance M_3 and M_4 draw $125 - 50 = 75$ μA each, or $\tfrac{3}{4}I_{D15}$. Consequently, $W_3 = W_4 = \tfrac{3}{4}W_{15} = \tfrac{3}{4}32 = 24$ μm.
- M_5 through M_8 draw 75 μA, or $\tfrac{3}{4}I_{D11}$, so $W_5 = W_6 = W_7 = W_8 = \tfrac{3}{4}W_{11} = \tfrac{3}{4}80 = 60$ μm. The widths of the signal-processing FETs are

$$W_1 = W_2 = W_9 = W_{10} = 40 \ \mu\text{m} \qquad W_3 = W_4 = 24 \ \mu\text{m}$$

$$W_5 = W_6 = W_7 = W_8 = 60 \ \mu\text{m} \qquad W_{11} = 80 \ \mu\text{m}$$

(b) Proceeding as usual, we find

$$g_{m1} = 2\frac{I_{D1}}{V_{OV1}} = 2\frac{50 \times 10^{-6}}{0.25} = 0.4 \ \text{mA/V}$$

$$g_{m4} = g_{m6} = 2\frac{75 \times 10^{-6}}{0.25} = 0.6 \ \text{mA/V}$$

$$r_{o2} = \frac{1}{\lambda_2 I_{D2}} = \frac{1}{0.05 \times 50 \times 10^{-6}} = 400 \ \text{k}\Omega$$

$$r_{o4} = \frac{1}{0.1 \times 75 \times 10^{-6}} = 133 \ \text{k}\Omega$$

$$r_{o6} = r_{o8} = \frac{1}{0.05 \times 75 \times 10^{-6}} = 267 \ \text{k}\Omega$$

$$r_{o10} = \frac{1}{0.1 \times 125 \times 10^{-6}} = 80 \ \text{k}\Omega$$

Substituting into Eqs. (5.32) and (5.33) we get

$$R_o \cong [0.6(1 + 0.1)267 \times 267]//[0.6(1 + 0.1)133(400//80)]$$

$$\cong (49{,}050//5867) \ \text{k}\Omega = 5.22 \ \text{M}\Omega$$

$$a \cong 0.4 \times 10^{-3} \times 5.22 \times 10^6 = 2{,}088 \ \text{V/V}$$

Finally, use Eq. (5.41) and (5.42) to find

$$v_{IC(\max)} = 1.25 \ \text{V} \qquad v_{IC(\min)} = -3.0 \ \text{V}$$

$$v_{O(\max)} = 1.25 \ \text{V} \qquad v_{O(\min)} = -2 \ \text{V}$$

The two-stage and folded-cascode topologies, along with variants thereof, are in widespread use today. As mentioned, both topologies achieve an overall gain on the order of $(g_m r_o)^2$, though by different means. Yet the folded cascode requires more transistors and its OVS is more limited, so one may wonder what its advantages are compared to the two-stage version. To answer this question we need to study stability and frequency compensation, in Chapter 7. There we shall see that in negative-feedback operation, which is the preferred mode of operation of op amps, the folded cascode is *much easier to stabilize* against unwanted oscillations. This advantage alone warrants the additional transistor count!

5.4 VOLTAGE COMPARATORS

After the op amp, the voltage comparator is probably the most popular high-gain amplifier. Its function is to compare two analog inputs v_P and v_N and yield a binary-valued output such as

$$v_O = V_{OL} \quad \text{for } v_P < v_N \qquad (5.43a)$$

$$v_O = V_{OH} \quad \text{for } v_P > v_N \qquad (5.43b)$$

where V_{OL} and V_{OH} are prescribed logic levels such as the familiar TTL/CMOS-compatible voltages $V_{OL} = 0$ V and $V_{OH} = 5$ V. Aptly called a *decision circuit*, the comparator can also be viewed as a 1-bit analog-to-digital converter. Figure 5.20a shows the comparator's circuit symbol, whereas Fig. 5.20b shows the voltage transfer curve (VTC) implied by Eq. (5.43). As we know, the *slope* of a VTC represents *gain*, so the VTC of Fig. 5.20b implies an amplifier having *infinite gain* and *saturating* at V_{OL} and V_{OH}.

Infinite gain is physically impossible, so the VTC of a real-life comparator is more likely as in Fig. 5.20c, where we estimate voltage gain as

$$a \cong \frac{V_{OH} - V_{OL}}{V_{IH} - V_{IL}} \qquad (5.44)$$

with V_{IL} and V_{IH} representing the values of v_I at which $a = 1$ V/V. Accordingly, Eq. (5.43) changes to

$$v_O = V_{OL} \quad \text{for } v_P < (v_N + V_{IL}) \qquad (5.45a)$$

$$v_O = V_{OH} \quad \text{for } v_P > (v_N + V_{IH}) \qquad (5.45b)$$

Moreover, we have

$$v_O = a(v_P - v_N) \quad \text{for } (v_N + V_{IL}) < v_P < (v_N + V_{IH}) \qquad (5.45c)$$

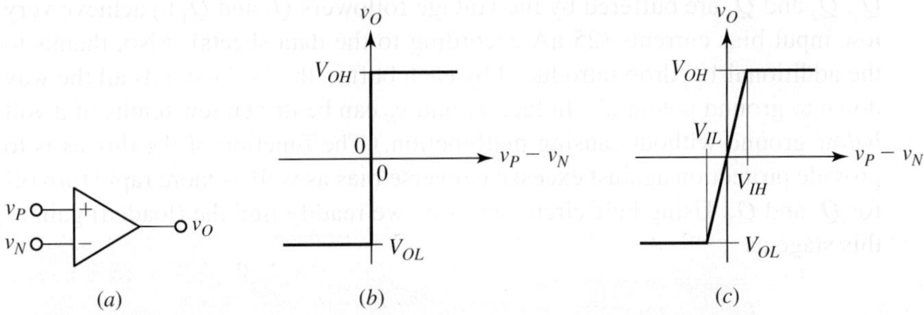

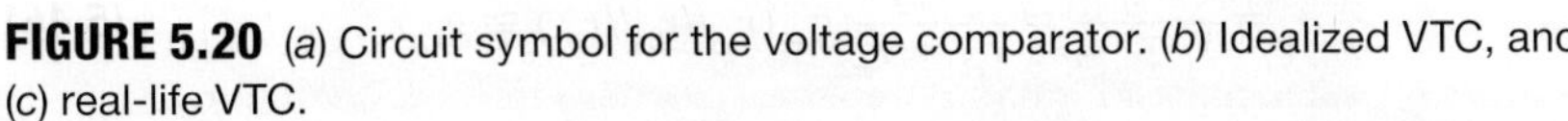

FIGURE 5.20 (a) Circuit symbol for the voltage comparator. (b) Idealized VTC, and (c) real-life VTC.

It is apparent that the real-life VTC of Fig. 5.20c is only an approximation to the ideal VTC of Fig. 5.20b. However, the closer V_{IL} and V_{IH} are to each other, the higher the gain and the closer the VTC to ideal.

As a high-gain amplifier, the voltage comparator bears strong similarities to the op amp (in fact, we use the same circuit symbol for both). However, they differ in two important respects:

- Op amps are intended to operate with *negative feedback*, whereas comparators are not. In Chapter 7 we shall see that in order to stave off possible oscillation, an op amp is equipped with a frequency-compensation network which, in its simplest form, consists of a mere capacitance, such as C_c in Figs. 5.1, 5.13 and 5.16. In Chapter 6 we shall see that C_c slows down the dynamics of the op amp considerably. Comparators, on the other hand, do not need to be frequency compensated because voltage comparison does not involve negative feedback (in fact, a compensation capacitance would only slow down the comparator unnecessarily). Freed of compensation requirements, comparators operate at full speed (voltage-comparator dynamics are investigated in Chapter 6).
- The output saturation voltages of op amps are not digitally compatible (for instance, a 741 op amp powered from ±15-V supplies saturates at about ±13 V, a far cry from TTL/CMOS logic levels). Conversely, voltage-comparator output stages are designed with this type of compatibility in mind.

If speed and logic compatibility are not of concern, then an op amp can indeed be used as a voltage comparator. However, most comparator applications require specialized circuits that have been optimized for this specific operation. To get a feel, let us examine some typical bipolar and CMOS voltage-comparator realizations.

The LM339 Voltage Comparator

This popular bipolar comparator, available in a quad package from most analog IC manufacturers, is depicted in simplified form in Fig. 5.21. We identify the following blocks:

- The **1st** or **input stage** consists of the EC pair Q_2-Q_3 and the active load Q_5-Q_6. Q_2 and Q_3 are buffered by the voltage followers Q_1 and Q_4 to achieve very low input bias currents (25 nA according to the data sheets). Also, thanks to the additional V_{EB} drop introduced by each buffer, the IVR extends all the way down to ground potential. (In fact, v_P and v_N can be driven few tenths of a volt *below* ground without causing malfunction.) The function of the diodes is to provide protection against excessive reverse bias as well as more rapid turn off for Q_2 and Q_3. Using half-circuit analysis we readily find the (loaded) gain of this stage as

$$a_1 = \frac{v_{b7}}{v_P - v_N} = \frac{r_{\pi2}}{r_{e1} + r_{\pi2}} g_{m3}(r_{o3}//r_{o6}//r_{\pi7}) \cong g_{m3}r_{\pi7} \qquad (5.46)$$

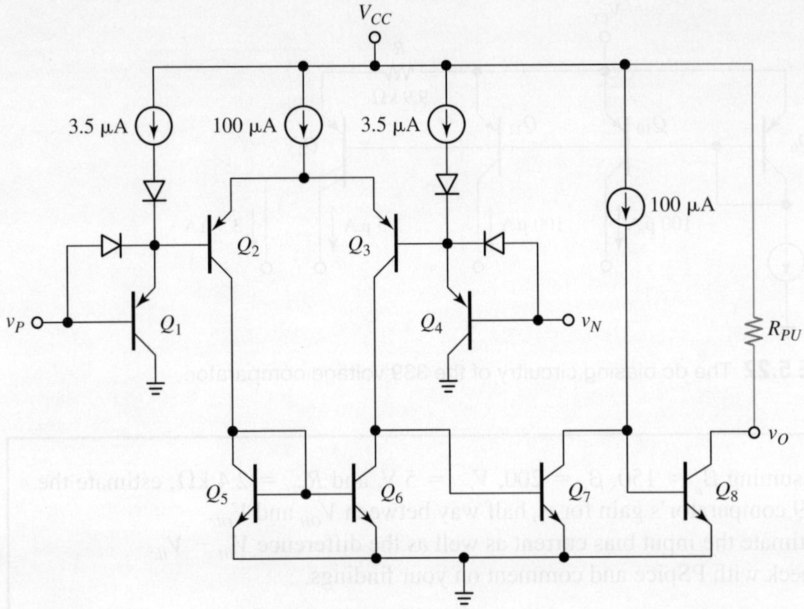

FIGURE 5.21 Simplified circuit schematic of the LM339 voltage comparator.

- The **2nd** or **intermediate stage** consists of the CE amplifier Q_7. Its function is to provide additional voltage gain, which we estimate as

$$a_2 = -\frac{v_{b8}}{v_{b7}} \cong -g_{m7}r_{\pi8} \qquad (5.47)$$

- The **3rd** or **output** stage consists of the so-called *open collector* CE transistor Q_8. The reason for leaving the collector uncommitted is so that the user can configure it externally for the desired logic levels at the output. The simplest form is to use a *pull-up resistor* R_{PU}, in which case the circuit yields $V_{OL} = V_{CE8(\text{sat})} \cong 0.2$ V when Q_8 is driven in saturation, and $V_{OH} = V_{CC}$ when Q_8 is driven in cutoff. With $V_{CC} = 5$ V the circuit yields TTL/CMOS logic levels of about 0 V and 5 V. The gain of this stage is estimated as

$$a_3 = -\frac{v_o}{v_{b8}} \cong -g_{m8}R_{PU} \qquad (5.48)$$

- The **dc biasing circuit**, shown in Fig. 5.22, consists of the multiple-output current mirror Q_9 through Q_{12}, along with the stabilized current reference I_{REF}, which in turn is shared among all four comparators on the chip. The current I_{REF} is replicated by Q_{10} to bias the Q_2-Q_3 pair, and by Q_{11} to actively load Q_7. Moreover, Q_{12} forms a Widlar current source synthesizing a 7-μA current that subsequently divides between the two collectors to provide the 3.5-μA bias currents for the Q_1 and Q_4 buffers.

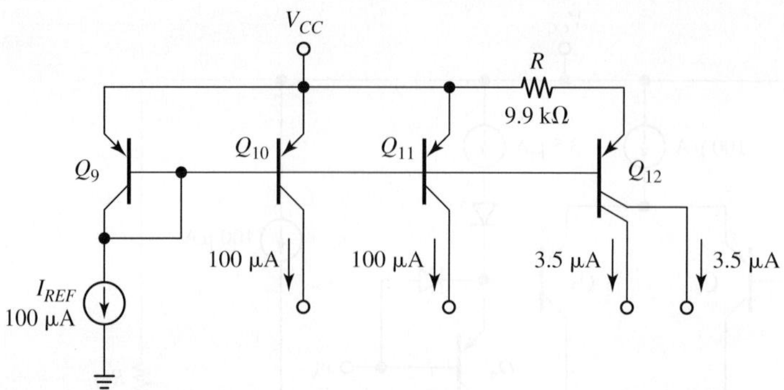

FIGURE 5.22 The dc biasing circuitry of the 339 voltage comparator.

EXAMPLE 5.5

(a) Assuming $\beta_p = 150$, $\beta_n = 200$, $V_{CC} = 5$ V, and $R_{PU} = 2.4$ kΩ, estimate the 339 comparator's gain for v_o half way between V_{OH} and V_{OL}.

(b) Estimate the input bias current as well as the difference $V_{IH} - V_{IL}$.

(c) Check with PSpice and comment on your findings.

Solution

(a) For $v_o = \frac{1}{2}(V_{OH} + V_{OL}) = \frac{1}{2}(5 + 0.2) = 2.6$ V we have $I_{C8} = (5 - 2.6)/2.4 = 1$ mA. Using Eqs. (5.46) through (5.48),

$$a_1 \cong \frac{0.05}{26}\beta_7\frac{26}{0.1} = \frac{\beta_7}{2} = 100 \text{ V/V}$$

$$a_2 \cong -\frac{0.1}{26}\beta_8\frac{26}{1} = -\frac{\beta_8}{10} = -20 \text{ V/V}$$

$$a_3 \cong -\frac{R_{PU}}{26\ \Omega} = -92.3 \text{ V/V}$$

The overall gain is $a = a_1 \times a_2 \times a_3 = 100 \times 20 \times 92.3 \cong 185$ V/mV.

(b) $I_B = I_{B1} = I_{E1}/(\beta_1 + 1) = (3.5\ \mu\text{A} + I_{B2})/(\beta_1 + 1) = [3.5\ \mu\text{A} + (50\ \mu\text{A})/(\beta_2 + 1)]/(\beta_1 + 1) = 25.4$ nA. By Eq. (5.44), $V_{IH} - V_{IL} = (V_{OH} - V_{OL})/a = (5 - 0.2)/(185 \times 10^3) \cong 26\ \mu$V.

(c) The circuit of Fig. 5.23a utilizes a 339 macro-model to display the VTC (see Appendix 5A).

From Fig. 5.23b we find $V_{IL} \cong 23\ \mu$V, $V_{IH} \cong 52\ \mu$V, $V_{OL} \cong 0.25$ V, and $V_{OH} = 5$ V. By Eq. (5.44), the gain is $a \cong (5 - 0.25)/[(52 - 23)10^{-6}] \cong 183$ V/mV. We also observe that the simulated VTC is shifted toward the right by about 37 μV. This represents the input offset voltage V_{OS} of the 339 macro-model for the given value of R_{PU}. Due primarily to mismatches between the two halves of its input stage, an actual 339 is likely to exhibit a much higher V_{OS}. You can easily search the Web for the LM339 data sheets, which list the following typical values at room temperature: $a = 200$ V/mV, $V_{OS} = 2$ mV, $I_B = 25$ nA, and $I_{OS} = 5$ nA. The above calculations indicate that I_B strongly depends on β_p, a_1 on β_n, and a_2 on R_{PU}.

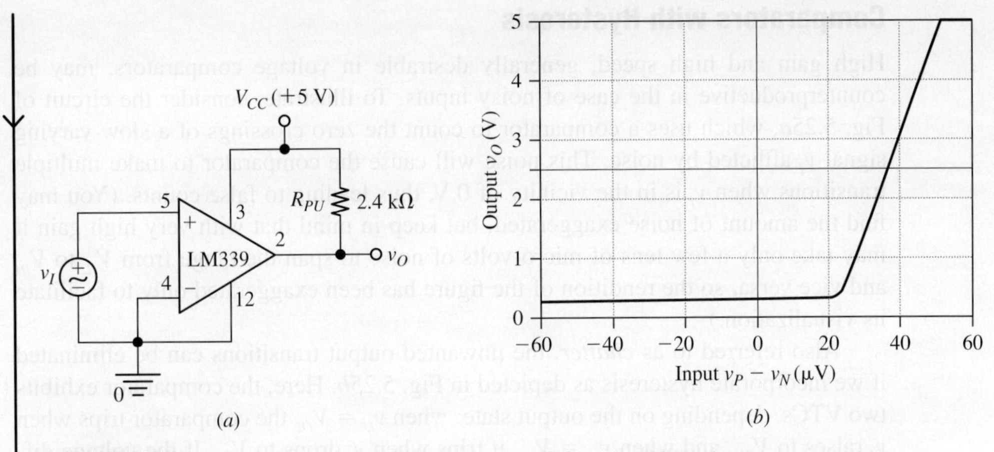

FIGURE 5.23 (a) PSpice circuit to display (b) the VTC of the 339 voltage comparator.

A CMOS Voltage Comparator

The circuit of Fig. 5.24 is the complementary version of the CMOS amplifier of Fig. 5.13, but *without* the compensation network R_c-C_c, which is unnecessary in voltage comparison. The circuit is also equipped with an output inverter to boost the gain as well as to provide a rail-to-rail output swing, or $V_{OL} = V_{SS}$ and $V_{OH} = V_{DD}$. If necessary, the channel widths W_9 and W_{10} can be made suitably large to boost the output current-drive capabilities.

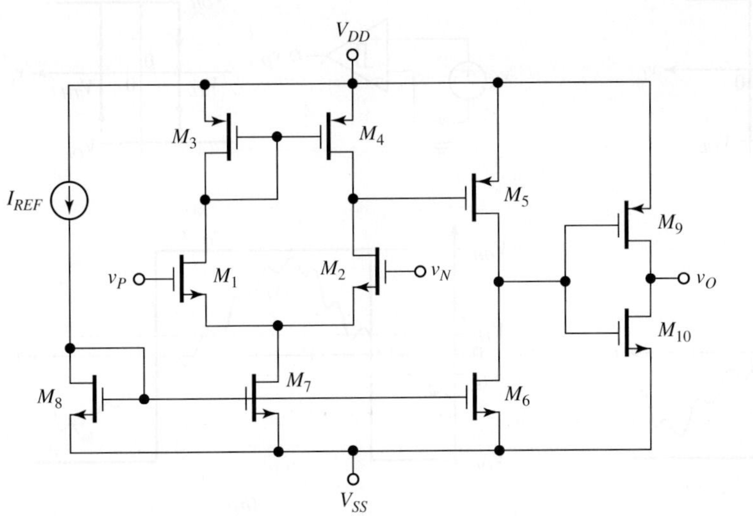

FIGURE 5.24 CMOS voltage comparator.

Comparators with Hysteresis

High gain and high speed, generally desirable in voltage comparators, may be counterproductive in the case of noisy inputs. To illustrate, consider the circuit of Fig. 5.25a, which uses a comparator to count the zero crossings of a slow varying signal v_I afflicted by noise. This noise will cause the comparator to make multiple transitions when v_I is in the vicinity of 0 V, thus leading to false counts. (You may find the amount of noise exaggerated, but keep in mind that with very high gain it may take only a few tens of micro-volts of noise to span the range from V_{IL} to V_{IH} and vice versa, so the rendition of the figure has been exaggerated only to facilitate its visualization.)

Also referred to as *chatter*, the unwanted output transitions can be eliminated if we incorporate hysteresis as depicted in Fig. 5.25b. Here, the comparator exhibits two VTCs, depending on the output state: when $v_O = V_{OL}$ the comparator trips when v_I raises to V_{TH}, and when $v_O = V_{OH}$ it trips when v_I drops to V_{TL}. If the voltage difference $V_{TH} - V_{TL}$, aptly called the *hysteresis width*, exceeds the maximum peak-to-peak amplitude of the input noise, then chatter will be eliminated, as exemplified in Fig. 5.25b. Hysteresis is introduced via *positive feedback*, either externally by the user[7] or internally by the IC designer. Here we are interested in the latter.

Figure 5.26 shows a popular CMOS realization of the comparator-with-hysteresis concept. Ignoring M_5 and M_6 for a moment, we observe that the M_4-M_7 mirror steers i_{D2} *toward* the output node v_O, whereas the M_3-M_8 and M_9-M_{10} mirrors steer i_{D1} *away from* node v_O. Consequently, the circuit acts as a differential amplifier with small-signal gain $a = v_o/(v_p - v_n) = g_{m1}(r_{o7}//r_{o10})$ and a VTC of the type of Fig. 5.20c (see also Fig. 4.66).

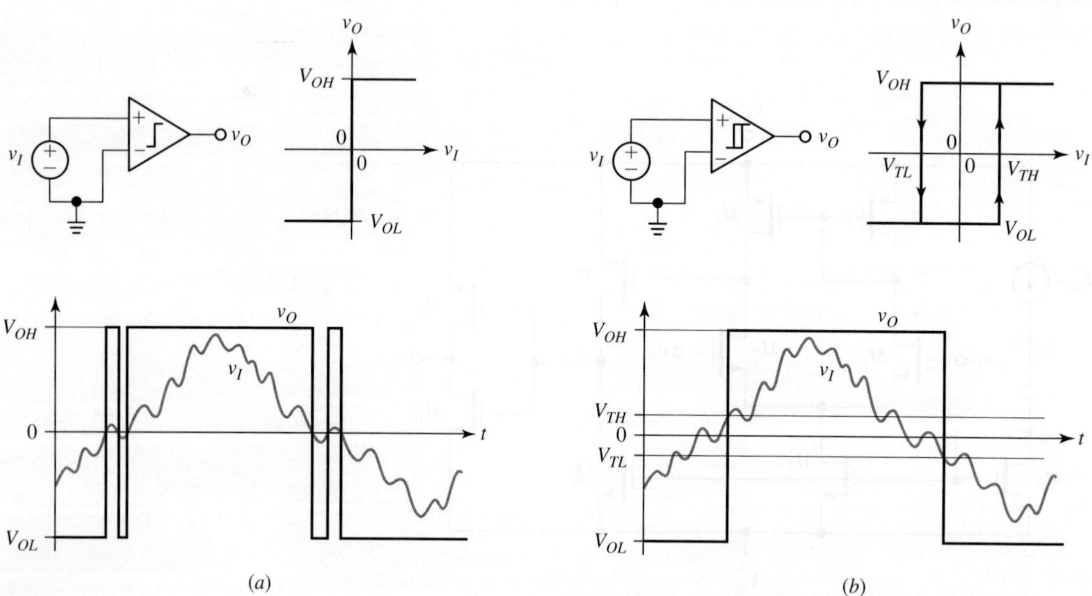

(a)   (b)

FIGURE 5.25 Illustrating (a) *comparator chatter*, and (b) its elimination via *hysteresis*.

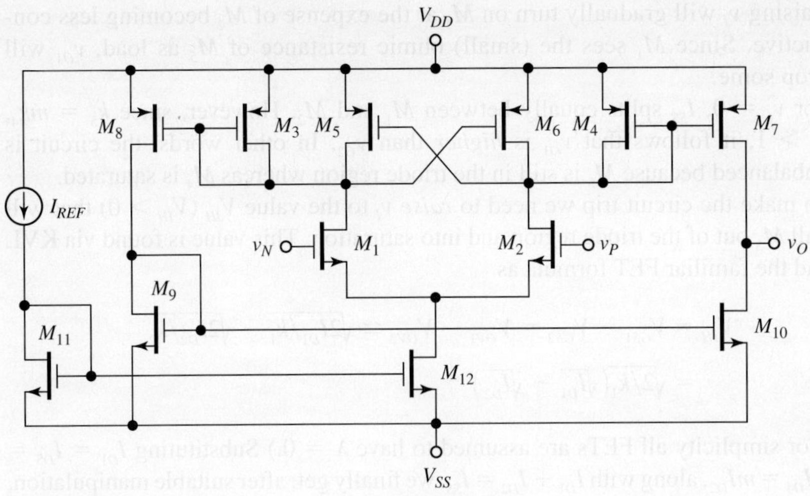

FIGURE 5.26 CMOS comparator with hysteresis.

Consider now the effect of having M_5 and M_6 present. Viewed as two cross-coupled inverters, these transistors introduce a flip-flop action that keeps the loads seen by M_1 and M_2 unbalanced. It is precisely this imbalance that causes hysteresis. To see how, refer to the reduced renditions of Fig. 5.27, where the cross-coupled FETs M_5 and M_6 are assumed to have W/L ratios that are m $(m > 1)$ times as *large* as those of the diode-connected FETs M_3 and M_4. We make the following observations:

- For v_I sufficiently negative as in Fig. 5.27a, M_1 is off, so M_3 and M_5 are also off. All of I_{SS} flows though M_2, causing v_{O2} to be low. This forces M_5 in the ohmic region, thus pulling v_{O1} to V_{DD}. With $v_{SG6} = 0$, M_6 is also off, resulting in $i_{D4} = i_{D2} = I_{SS}$.

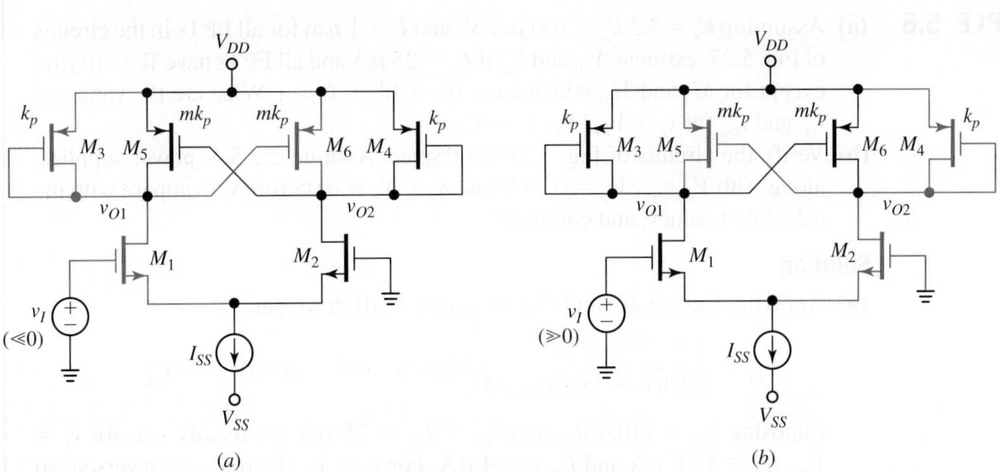

FIGURE 5.27 The central part of the circuit of Fig. 5.26 for the cases (a) $v_I \ll 0$ and (b) $v_I \gg 0$.

- Raising v_I will gradually turn on M_1 at the expense of M_2 becoming less conductive. Since M_1 sees the (small) ohmic resistance of M_5 as load, v_{O1} will drop some.
- For $v_I = 0$, I_{SS} splits equally between M_1 and M_2. However, since $k_5 = mk_4$, $m > 1$, it follows that v_{O1} is *higher* than v_{O2}. In other words, the circuit is unbalanced because M_5 is still in the triode region whereas M_4 is saturated.
- To make the circuit trip we need to *raise* v_I to the value V_{IH} ($V_{IH} > 0$) that will pull M_5 out of the triode region and into saturation. This value is found via KVL and the familiar FET formula as

$$V_{IH} = V_{GS1} - V_{GS2} = V_{OV1} - V_{OV2} \cong \sqrt{2I_{D1}/k_1} - \sqrt{2I_{D2}/k_2}$$

$$= \sqrt{2/k_1}\left(\sqrt{I_{D1}} - \sqrt{I_{D2}}\right)$$

(For simplicity all FETs are assumed to have $\lambda = 0$.) Substituting $I_{D1} = I_{D5} = mI_{D4} = mI_{D2}$, along with $I_{D1} + I_{D2} = I_{SS}$, we finally get, after suitable manipulation,

$$V_{IH} \cong \sqrt{\frac{2I_{SS}}{(m+1)k_1}}(\sqrt{m} - 1) \tag{5.49}$$

- For v_I sufficiently positive as in Fig. 5.27b the roles of the various FET pairs are interchanged, so we exploit the symmetry of the circuit to state that in order to make it trip in the opposite direction we now need to *lower* v_I to the value V_{IL} ($V_{IL} < 0$) such that

$$V_{IL} = -V_{IH} \tag{5.50}$$

EXAMPLE 5.6   (a) Assuming $k'_n = 2.5\, k'_p = 100\ \mu\text{A/V}^2$ and $L = 1\ \mu$m for all FETs in the circuits of Fig. 5.27, estimate V_{IH} and V_{IL} if $I_{SS} = 25\ \mu$A and all FETs have $W = 10\ \mu$m, except for M_3 and M_4, which have $W_3 = W_4 = 6\ \mu$m. What are the values of I_{D1} and I_{D2} for $v_I = V_{IH}$? For $v_I = V_{IL}$?

(b) Verify the circuits of Fig. 5.27 via PSpice. Assume ± 2.5-V power supplies, along with $V_{tn} = -V_{tp} = 0.75$ V and $\lambda'_n = \lambda'_p = 0.05\ \mu$m/V. Compare with the calculated values, and comment.

Solution

(a) Applying Eqs. (5.49) and (5.50) with $m = 10/6$ we get

$$V_{IH} \cong \sqrt{\frac{2 \times 25}{(10/6 + 1)(10/1)100}}\left(\sqrt{10/6} - 1\right) \cong 40\ \text{mV} = -V_{IL}$$

Imposing $I_{D1} = (10/6)I_{D2}$ and $I_{D1} + I_{D2} = 25\ \mu$A we readily get, for $v_I = V_{IH}$, $I_{D1} \cong 15.6\ \mu$A and $I_{D2} \cong 9.4\ \mu$A. For $v_I = V_{IL}$ the roles are reversed, so $I_{D1} \cong 9.4\ \mu$A and $I_{D2} \cong 15.6\ \mu$A.

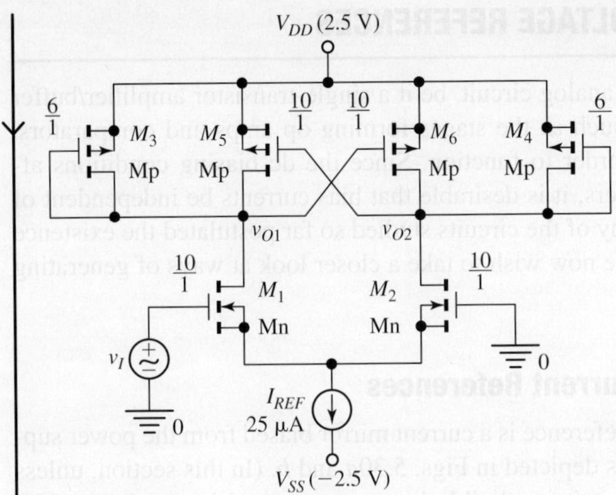

FIGURE 5.28 PSpice circuit to investigate the hysteresis of the comparator of Example 5.6.

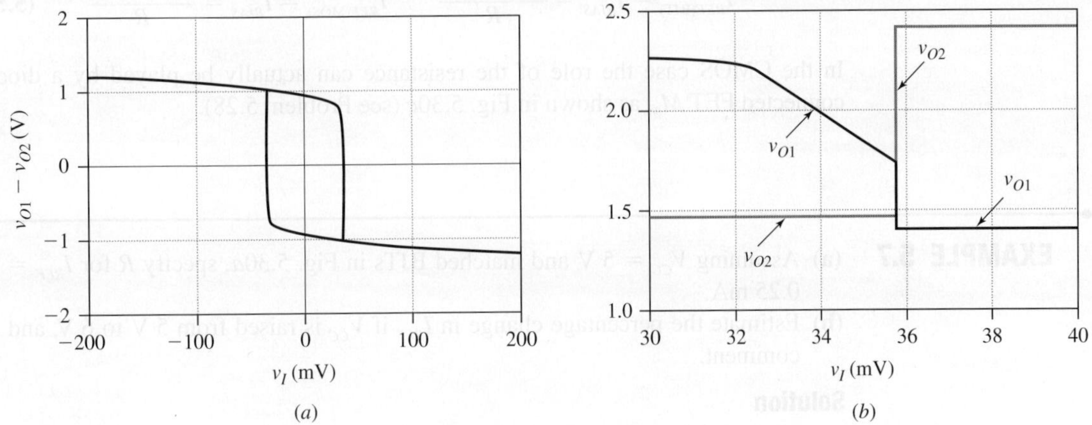

(a) (b)

FIGURE 5.29 Transfer characteristics of the comparator of Fig. 5.28: (a) hysteresis and (b) detail.

(b) Using the PSpice circuit of Fig. 5.28 we readily obtain the plots of Fig. 5.29, from which we find $V_{IH} = -V_{IL} = 35.7$ mV. This value differs from 40 mV because our hand calculations are based on $\lambda = 0$. Figure 5.29b provides an expanded view of the snapping action taking place as v_I is raised to V_{IH}. Once M_1 pulls M_5 out of the triode region, v_{O1} drops more rapidly until M_6 turns on. At this point M_6 pulls v_{O2} toward V_{DD} as fast as it can. This, in turn, shuts off M_5, causing a final and more rapid drop in v_{O1}. It is apparent that because of the flip-flop action provided by M_5 and M_6, v_{O1} and v_{O2} coexist only in complementary form (one is high whereas the other is low, and vice versa).

5.5 CURRENT AND VOLTAGE REFERENCES

It is apparent by now that an analog circuit, be it a single-transistor amplifier/buffer or a multi-transistor circuit such as the stages forming op amps and comparators, must be properly biased in order to function. Since the dc biasing conditions affect the small-signal parameters, it is desirable that bias currents be independent of power-supply variations. Many of the circuits studied so far postulated the existence of a stabilized current I_{REF}. We now wish to take a closer look at ways of generating such a current.

Power-Supply-Based Current References

Perhaps the simplest current reference is a current mirror biased from the power supply via a plain resistance R, as depicted in Figs. 5.30a and b. (In this section, unless stated to the contrary, we assume negligible base currents and ignore base-width and channel-length modulation effects as well as the body effect in MOSFETs.) By current-mirror action and Ohm's law we have, for the two circuits,

$$I_{REF(BJT)} = I_{BIAS} = \frac{V_{CC} - V_{BE}}{R} \qquad I_{REF(MOS)} = I_{BIAS} = \frac{V_{DD} - V_{GS}}{R} \qquad (5.51)$$

In the CMOS case the role of the resistance can actually be played by a diode-connected FET M_3, as shown in Fig. 5.30c (see Problem 5.28).

EXAMPLE 5.7 (a) Assuming $V_{CC} = 5$ V and matched BJTs in Fig. 5.30a, specify R for $I_{REF} = 0.25$ mA.

(b) Estimate the percentage change in I_{REF} if V_{CC} is raised from 5 V to 6 V, and comment.

Solution

(a) Assuming $V_{BE} = 0.7$ V, use Ohm's law to calculate $R = (5 - 0.7)/0.25 = 17.2$ kΩ.

(b) With $V_{CC} = 6$ V we get $I_{REF} = I_{BIAS} \cong (6 - 0.7)/17.2 = 0.308$ mA, indicating a percentage increase of $100(0.308 - 0.25)/0.25 \cong 23\%$, quite a change!

EXAMPLE 5.8 (a) Let the FETs of Fig. 5.30b have $V_t = 0.75$ V and $k' = 125$ μA/V². Assuming $V_{DD} = 5$ V, specify R and W/L so that both FETS draw 100 μA with $V_{OV} = 0.25$ V.

(b) Estimate the percentage change in I_{REF} if V_{DD} is raised from 5 V to 6 V.

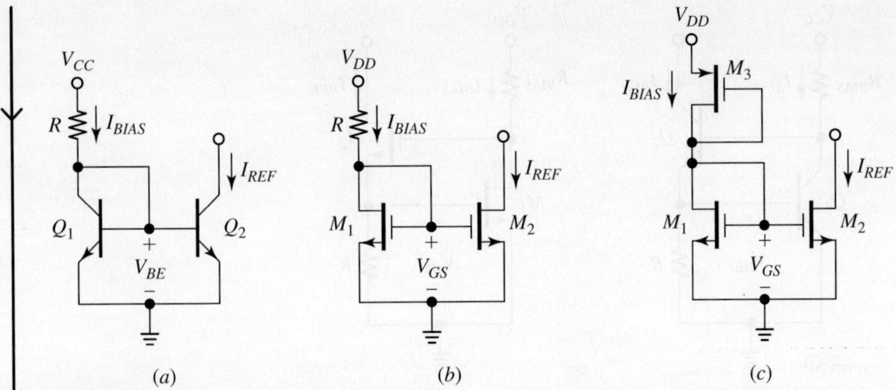

FIGURE 5.30 Power-supply-based current references: (a) bipolar, and (b) MOS. (c) All MOS version.

Solution

(a) Since $V_{GS} = 0.75 + 0.25 = 1$ V, Ohm's law gives $R = (5 - 1)/0.1 = 40$ kΩ. Moreover, imposing $0.1 = (0.125/2) \times (W/L) \times 0.25^2$ gives $W/L = 25.6$.

(b) From part (a) we have $k = k'(W/L) = 0.125 \times 25.6 = 3.2$ mA/V^2. Imposing

$$I_{BIAS} = \frac{6 - 0.75 - \sqrt{2I_{BIAS}/(3.2 \times 10^{-3})}}{40 \times 10^3}$$

and solving by iteration gives $I_{REF} = I_{BIAS} \cong 124$ μA, indicating a 24% increase, quite a change!

The above examples indicate a fairly strong dependence of I_{REF} upon the supply voltage. As an example, the biasing scheme of the 741 op amp of Fig. 5.4 gives, for ± 15-V supplies, $I_{REF} = (30 - 1.4)/39 = 733$ μA. However, should the user opt for a pair of ± 9-V supply batteries, I_{REF} would drop to $(18 - 1.4)/39 = 426$ μA, causing appreciable changes in most small-signal parameters. In the following we shall explore ways to reduce the dependence of I_{REF} on the power supply.

V_{BE}-Based and V_{GS}-Based Current References

The circuits of Fig. 5.31 can be viewed as modified Wilson current mirrors with the diode-connected transistor replaced by the current-setting resistance R. Ignoring base currents we have, for the BJT case

$$I_{REF(BJT)} = \frac{V_{BE}}{R} = \frac{V_T \ln(I_{BIAS}/I_s)}{R} \tag{5.52a}$$

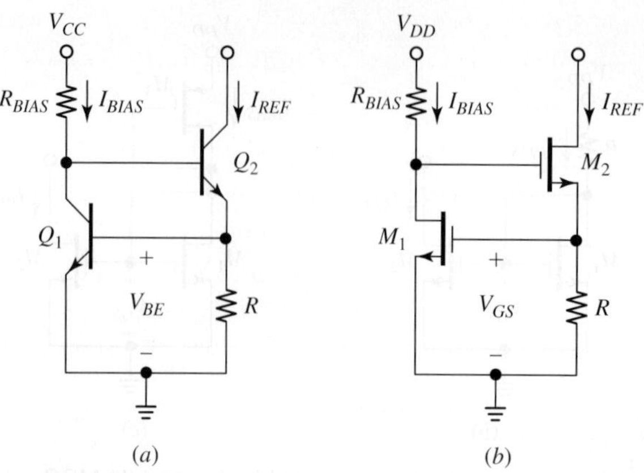

FIGURE 5.31 (a) V_{BE}-based and (b) V_{GS}-based current references.

where $I_{BIAS} = (V_{CC} - V_{BE2} - V_{BE})/R_{BIAS}$. Likewise we have, for the MOS case,

$$I_{REF(MOS)} = \frac{V_{GS}}{R} = \frac{V_t + \sqrt{2I_{BIAS}/k}}{R} \tag{5.52b}$$

where $I_{BIAS} = (V_{DD} - V_{GS2} - V_{GS})/R_{BIAS}$. In both cases I_{BIAS} is still strongly dependent on the supply voltage. However, its effect on I_{REF} is greatly mitigated by the fact that I_{BIAS} now appears in the argument of a *logarithm* in the BJT case, and of a *square root* in the MOS case. As we know, V_{BE} will remain fairly close to about 0.7 V even for appreciable variations in V_{CC}. Moreover, if the FETs are fabricated with sufficiently large W/L ratios to render the variations of the square-root term negligible compared to V_t, then V_{GS} too will remain fairly constant in the vicinity of V_t.

EXAMPLE 5.9   (a) Assuming $V_{CC} = 5$ V, $I_s = 1$ fA, and negligible base currents, specify R_{BIAS} and R in the circuit of Fig. 5.31a for $I_{REF} = 2I_{BIAS} = 250$ μA. Hence, estimate the percentage change in I_{REF} if V_{CC} is raised from 5 V to 6 V. Compare this change with Example 5.7, and comment.

(b) Assuming $V_{DD} = 5$ V, $V_t = 0.75$ V, and $k = 3.2$ mA/V^2, specify R_{BIAS} and R in the circuit of Fig. 5.31b for $I_{REF} = I_{BIAS} = 100$ μA. Hence, estimate the percentage change in I_{REF} if V_{DD} is raised from 5 V to 6 V, compare with Example 5.8, and comment.

Solution

(a) We have $V_{BE} = 0.026 \ln(125 \times 10^{-6}/10^{-15}) = 0.664$ V so $R = 0.664/0.25 = 2.66$ kΩ. By the 18-mV rule of thumb, $V_{BE2} = V_{BE} + 18$ mV $= 0.682$ V. So, $R_{BIAS} = (5 - 0.682 - 0.664]/(0.25/2) = 29.2$ kΩ. With $V_{CC} = 6$ V we get $I_{BIAS} \cong [6 - 2 \times 0.67)]/29.2 = 159$ μA, $V_{BE} = 0.026 \ln(159 \times 10^{-6}/10^{-15}) = 0.670$ V, and $I_{REF} = 0.670/2.66 = 252.5$ μA. This represents an increase in I_{REF} of $100(252.5 - 250)/250 \cong 1\%$, quite an improvement over the 23% increase of Example 5.7!

(b) From Example 5.8 we have $V_{GS} = 0.75 + 0.25 = 1$ V, so $R = 1/0.1 = 10\,\text{k}\Omega$ and $R_{BIAS} = [5 - 1 - 1)]/(0.1) = 30\,\text{k}\Omega$. As we raise V_{DD} from 5 V to 6 V we expect the change in V_{GS2} to be negligible compared to the change in V_{DD}, so we write $I_{BIAS} \cong [6 - 0.75 - (2I_{BIAS}/3.2)^{1/2} - 1]/30$. Solving by iterations gives $I_{BIAS} = 0.132$ mA. Consequently, $V_{GS} = 0.75 + (2 \times 0.132/3.2)^{1/2} = 1.037$ V and $I_{REF} = 1.037/10 = 103.7\,\mu\text{A}$, indicating a 3.7% increase in I_{REF}, an appreciable improvement over 24% of Example 5.8!

Imbalance-Based Current References

An elegant alternative for achieving power-supply independence is offered by the introduction of deliberate imbalances in device geometries, or in the dc biasing conditions, or both. In Fig. 5.32 the Q_1-Q_2 and M_1-M_2 pairs can be viewed as modified Widlar current mirrors with deliberate device imbalances. Specifically, Q_2 is fabricated with an emitter area m_n (≥ 1) times as large as the emitter area A_n of Q_1, so their saturation currents are related as $I_{s2} = m_n I_{s1}$. Likewise, M_2 is fabricated with a W/L ratio m_n times as large as that of M_1, so their device transconductance parameters are related as $k_2 = m_n k_1$. Also unbalanced are the current mirrors Q_3-Q_4 and M_3-M_4, for which $I_{s3} = m_p I_{s4}$ and $k_3 = m_p k_4$. These imbalances force the current in the left branch to be m_p times as large as that of the right branch, as shown.

- Turning first to the **bipolar current reference** of Fig. 5.32a, we observe that since Q_2 has a larger emitter area than Q_1, it will require a smaller V_{BE} drive to sustain a current that is also m_p times smaller than that of Q_1. We thus write

$$\Delta V_{BE} = V_{BE1} - V_{BE2} = V_T \ln\frac{m_p I_{REF}}{I_{s1}} - V_T \ln\frac{I_{REF}}{m_n I_{s1}} = V_T \ln(m_p m_n) \qquad (5.53)$$

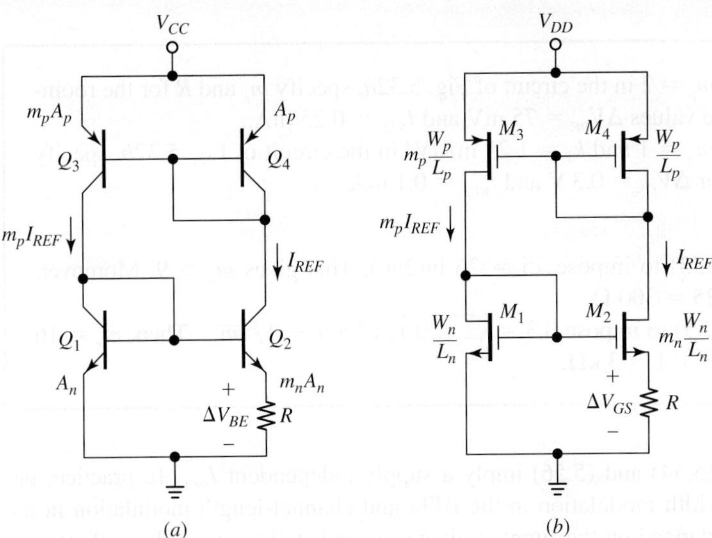

FIGURE 5.32 Imbalance-based current references: (a) bipolar and (b) CMOS.

where $V_T = kT/q$ is the familiar *thermal voltage*. Note that ΔV_{BE} is proportional to absolute temperature, or PTAT for short. We use precisely this voltage difference to establish the reference current as $I_{REF} = \Delta V_{BE}/R$, or

$$I_{REF(BJT)} = \frac{V_T \ln(m_p m_n)}{R} \qquad (5.54)$$

- Turning next to the **MOS current reference** of Fig. 5.32b, we likewise observe that since M_2 has a larger W/L ratio than M_1, it will require a smaller V_{GS} drive to sustain a current that is also m_p times smaller than that of M_1. Ignoring M_2's body effect we thus write

$$\Delta V_{GS} = V_{GS1} - V_{GS2} = \left(V_t + \sqrt{\frac{2m_p I_{REF}}{k_1}}\right) - \left(V_t + \sqrt{\frac{2I_{REF}}{m_n k_1}}\right)$$

$$= \sqrt{\frac{2I_{REF}}{k_1}}\left(\sqrt{m_p} - \frac{1}{\sqrt{m_n}}\right) \qquad (5.55)$$

We use precisely this voltage difference to establish the reference current as $I_{REF} = \Delta V_{GS}/R$. Eliminating ΔV_{GS} and simplifying we get

$$I_{REF(MOS)} = \frac{2}{k_1 R^2}\left(\sqrt{m_p} - \frac{1}{\sqrt{m_n}}\right)^2 \qquad (5.56)$$

We can specify a variety of values for m_p and m_n in the circuits of Fig. 5.32.
- In the special case $m_p = 1$ and $m_n > 1$ the voltages ΔV_{BE} and ΔV_{GS} arise from deliberate *fabrication mismatches* in the transistors of the Widlar mirrors.
- In the special case $m_n = 1$ and $m_p > 1$ the Widlar transistors are matched and ΔV_{BE} and ΔV_{GS} arise from deliberate *unbalanced current drive*.

EXAMPLE 5.10

(a) Assuming $m_p = 2$ in the circuit of Fig. 5.32a, specify m_n and R for the room-temperature values $\Delta V_{BE} = 75$ mV and $I_{REF} = 0.25$ mA.

(b) Assuming $m_p = 1$ and $k_1 = 1.25$ mA/V^2 in the circuit of Fig. 5.32b, specify m_n and R for $\Delta V_{GS} = 0.3$ V and $I_{REF} = 0.1$ mA.

Solution

(a) Use Eq. (5.53) to impose $75 = 26 \ln(2m_n)$. This gives $m_n \cong 9$. Moreover, $R = 75/0.25 = 300 \ \Omega$.

(b) Use Eq. (5.55) to impose $0.3 = \sqrt{2 \times 0.1/1.25}(1 - 1/\sqrt{m_n})$. Then, $m_n = 16$ and $R = 0.3/0.1 = 3$ kΩ.

Equations (5.54) and (5.56) imply a supply-independent I_{REF}. In practice, because of base-width modulation in the BJTs and channel-length modulation in the FETs, I_{REF} will depend on the supply voltage somewhat. To get an idea, refer to the small-signal equivalent of Fig. 5.33, and use the test method to find the change i

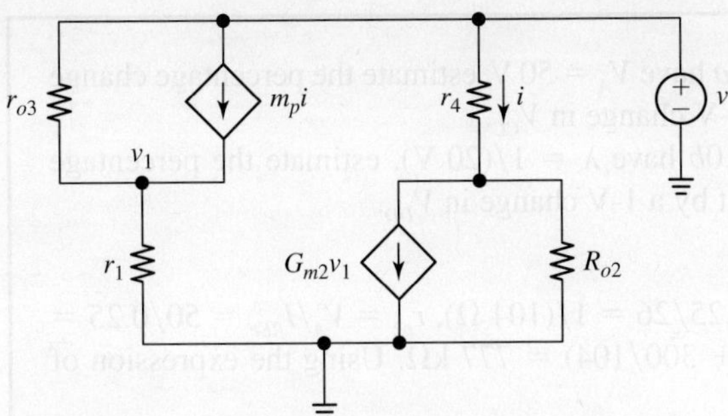

FIGURE 5.33 Small-signal model to find the power-supply dependence of a mismatch-based reference current.

experienced by I_{REF} due to a change v in the power supply. Here, the diode-connected transistors Q_1/M_1 and Q_4/M_4 are modeled by their (small) ac resistances $r_1 \cong 1/g_{m1}$ and $r_4 \cong 1/g_{m4}$, the mirror transistor Q_3/M_3 is modeled by a dependent source $m_p i$ with parallel resistance r_{o3}, and the degeneration resistance R has been absorbed into the model for transistor Q_2/M_2 by letting

$$G_{m2} \cong g_{m2}/(1 + g_{m2}R) \qquad R_{o2} \cong r_{o2}(1 + g_{m2}R)$$

(for simplicity we are ignoring M_2's body effect). Using nodal analysis along with KVL we get

$$\frac{v - v_1}{r_{o3}} + m_p i = \frac{v_1}{r_1} \qquad i = G_{m2}v_1 + \frac{v - r_4 i}{R_{o2}}$$

Typically $r_{o3} \gg r_1$ and $r_4/R_{o2} \ll 1$, so the above expressions simplify as

$$\frac{v}{r_{o3}} + m_p i \cong \frac{v_1}{r_1} \cong g_{m1}v_1 \qquad i \cong G_{m2}v_1 + \frac{v}{R_{o2}}$$

Eliminating v_1, collecting, and simplifying, we get

$$\frac{i}{v} = \frac{1 + (g_{m2}/g_{m1}) \times (r_{o2}/r_{o3})}{R_{o2}(1 - m_p G_{m2}/g_{m1})} \tag{5.57}$$

It is apparent that the larger R_{o2} the less dependent the current from the supply voltage. If desired, we can reduce this dependence further by cascoding Q_2 and M_2.

EXAMPLE 5.11   (a) If all BJTs of Example 5.10*a* have $V_A = 50$ V, estimate the percentage change in I_{REF} brought about by a 1-V change in V_{CC}.

(b) If all FETs of Example 5.10*b* have $\lambda = 1/(20$ V$)$, estimate the percentage change in I_{REF} brought about by a 1-V change in V_{DD}.

Solution

(a) We have $g_{m2} = I_{REF}/V_T = 0.25/26 = 1/(104\ \Omega)$, $r_{o2} = V_A/I_{REF} = 50/0.25 = 200$ kΩ, and $R_{o2} = 200(1 + 300/104) = 777$ kΩ. Using the expression of Exercise 5.2,

$$\frac{i}{v} = \frac{(1 + 50/50) \times [1 + 1/(300/104)]}{777} = \frac{1}{288\ \text{k}\Omega}$$

Letting $\Delta I_{REF} = (1$ V$)/(288$ k$\Omega) = 3.47\ \mu$A indicates a per-volt change of $100(3.47/250) \cong 1.4\%$.

(b) We have $g_{m1} = (2 \times 1.25 \times 0.1)^{1/2} = 0.5$ mA/V, $g_{m2} = (2 \times 16 \times 1.25 \times 0.1)^{1/2} = 2$ mA/V $= 4g_{m1}$, $r_{o2} = r_{o3} = 20/0.1 = 200$ kΩ, $G_{m2} = 2/(1 + 2 \times 3) = 1/(3.5$ k$\Omega)$, and $R_{o2} = 200(1 + 2 \times 3) = 1.4$ MΩ. By Eq. (5.57),

$$\frac{i}{v} = \frac{1 + (4) \times (1)}{1400[1 - 1 \times (1/3.5)/0.5)]} = \frac{1}{120\ \text{k}\Omega}$$

Letting $\Delta I_{REF} = (1$ V$)/(120$ k$\Omega) = 8.3\ \mu$A indicates a per-volt change of 8.3%.

Startup Circuits

The circuits of Fig. 5.32 are said to be *self-biased* or *bootstrapped* because each mirror biases and is in turn biased by the other. In particular, if one of the mirrors fails to turn on, so will the other and the circuit will remain in this unwanted state indefinitely. We therefore need a mechanism that, once the circuit receives power, will turn on at least one of the mirrors, forcing the circuit to evolve toward the desired state and remain there. Aptly referred to as *startup circuit*, in its simplest form it consists of a large resistance $R_{startup}$ to inject a small startup current into one of the mirrors. This is shown in Fig. 5.34*a* for the bipolar case. (It goes without saying that such a current must be much smaller than I_{REF} in order to avoid introducing an intolerable error.)

A more elegant approach is a startup circuit that intervenes only when the reference is in the unwanted state, but goes dormant once the reference has reached the desired state. In the CMOS example of Fig. 5.34*b* the diode-connected transistors M_7 and M_8 form a voltage divider to provide a suitable gate bias for M_9. Should M_1 be off, V_{GS9} is designed to be high enough to turn on M_9 and thus force both mirrors out of the cutoff state. Once M_1 turns fully on, V_{GS9} is designed to drop below V_{t9} so as to turn M_9 off, leaving the rest of the circuit undisturbed.

Since the current I_{REF} is used internally, provision must be made for its replication to the outside. In Fig. 5.34 this function is provided by Q_5/M_5 when the stabilized current is to be *sourced* to an external load, and by Q_6/M_6 when it is to be *sunk* from an external load.

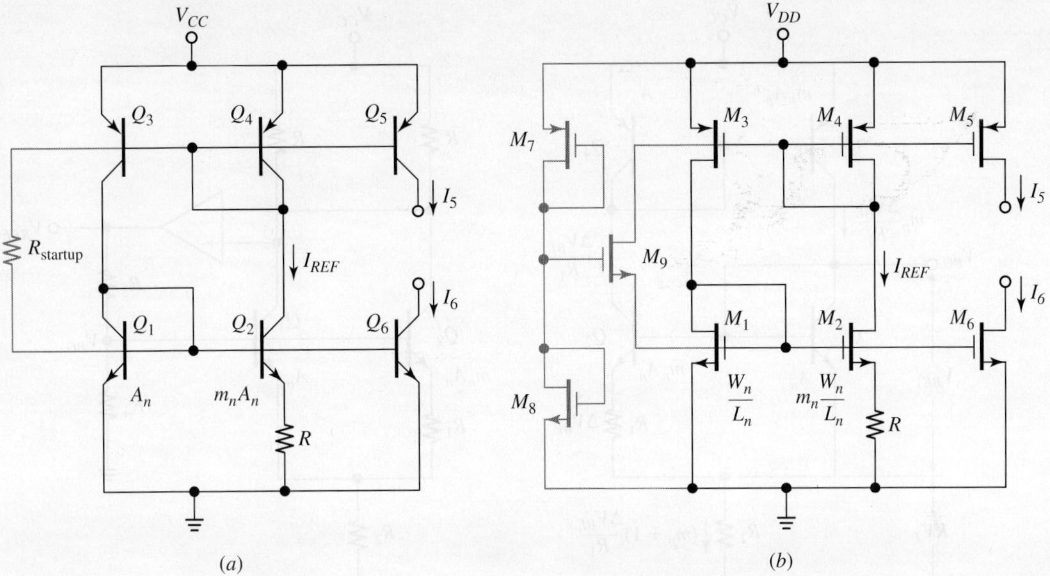

FIGURE 5.34 Mismatch-based current references with start-up circuit examples: (a) bipolar and (b) CMOS.

A final caveat is in order, namely, the circuits of Fig. 5.34 incorporate *positive feedback*, a situa-tion that may lead to instability if the loop gain exceeds unity (more on this in Chapter 7). With reference to Fig. 5.33 we observe that the current i is first amplified to $m_p i$ by Q_3, then it is converted to $v_1 \cong r_1 m_p i \cong m_p i / g_{m1}$ by Q_1, and finally it is returned by Q_2 as $G_{m2} v_1 \cong (m_p G_{m2} / g_{m1}) i$, so the overall amplification experienced by i around the loop, aptly called the loop gain T, is $T = m_p G_{m2} / g_{m1}$. To avert instability we must ensure that $T < 1$ (indeed, Example 5.11a has $T = 0.257$ A/A, and Example 5.11b has $T = 0.571$ A/A.)

Bandgap Voltage References

In electronic instrumentation and measurement as well as in data conversion the need arises for references that are not only supply independent, but also thermally stable. We make the following observations:

- The reference of Fig. 5.31a is based on V_{BE}, which *decreases* with temperature (recall the familiar rule of thumb). Adapting Eq. (1.66), we express the thermal coefficient of V_{BE} as

$$\text{TC}(V_{BE}) = \frac{\partial V_{BE}}{\partial T} = \frac{V_{BE} - (4 + m)V_T - V_{G0}}{T} \cong -2 \text{ mV/°C} \qquad (5.58)$$

- The reference of Fig. 5.31a is based on the difference $\Delta V_{BE} = V_T \ln(m_p m_n)$, which instead *increases* with temperature because $V_T = kT/q$ (recall that ΔV_{BE} is PTAT). The thermal coefficient of V_T is

$$\text{TC}(V_T) = \frac{\partial V_T}{\partial T} = \frac{k}{q} = \frac{1.381 \times 10^{-23}}{1.602 \times 10^{-19}} \cong +85 \text{ } \mu\text{V/°C} \qquad (5.59)$$

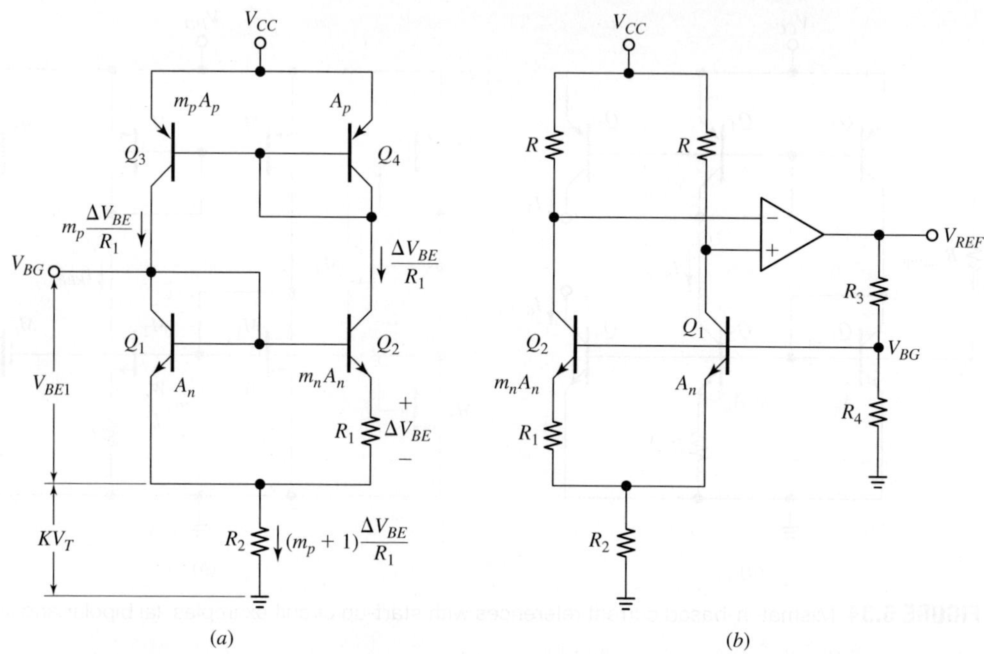

FIGURE 5.35 (a) Illustration of the bandgap voltage reference concept. (b) The Brokaw Cell realization.

- Now, if we could generate a voltage of value KV_T with

$$K = \left| \frac{\mathrm{TC}(V_{BE})}{\mathrm{TC}(V_T)} \right|$$

or $K \cong (2 \text{ mV/°C})/(85 \ \mu\text{V/°C}) = 23.5$, and we sum it to V_{BE} to obtain the composite voltage $V_{BG} = KV_T + V_{BE}$, then V_{BG} would be *temperature independent* because it consists of two components of *equal* but *opposing* thermal coefficients. This principle is depicted in Fig. 5.35a, which is obtained from Fig. 5.32a by the mere artifice of lifting the circuit off ground and inserting an additional series resistance R_2, as shown.

To investigate circuit behavior, note that the current through R_2 is the sum of the currents through Q_1 and Q_2, or $(m_p + 1)(\Delta V_{BE}/R_1)$. By Ohm's law, R_2 drops the voltage $R_2(m_p + 1)(\Delta V_{BE}/R_1)$ which, by Eq. (5.53), can be expressed as $(m_p + 1) \times (R_2/R_1) \times \ln(m_p m_n)V_T = KV_T$ (see Fig. 5.35a). It is precisely this voltage that is added to V_{BE1} to create the thermally stable voltage V_{BG}. To achieve this we must impose

$$K = (m_p + 1)\frac{R_2}{R_1}\ln(m_p m_n) = \frac{-\mathrm{TC}(V_{BE1})}{\mathrm{TC}(V_T)} = \frac{V_{G0} + (4 + m)V_T - V_{BE1}}{T(k/q)}$$

that is,

$$K = \frac{V_{G0} - V_{BE1}}{V_T} + 4 + m \tag{5.60}$$

When this condition is met, the desired voltage is

$$V_{BG} = KV_T + V_{BE1} = \left(\frac{V_{G0} - V_{BE1}}{V_T} + 4 + m\right)V_T + V_{BE1}$$

$$= V_{G0} + (4 + m)V_T \tag{5.61}$$

which is just a bit higher than the bandgap voltage V_{G0} ($=1.205$ V). Hence the reason for the designation *bandgap voltage reference*. Note the disappearance from Eq. (5.61) of both resistances, themselves functions of temperature. The proportionality constant K depends only on their *ratio*, which can be kept quite stable thanks to the thermal-tracking advantages of monolithic devices.

The bandgap concept has been realized in a variety of circuit forms. A popular version is the *Brokaw Cell* of Fig. 5.35b, so-called for its inventor Paul Brokaw. This realization uses an op amp to bring about a number of performance enhancements as follows. Fist, the op amp uses negative feedback to keep the collectors at the *same potential* over a range of V_{CC} values, thus ensuring a high degree of power-supply rejection. (Since the collector resistances are identical, we presently have $m_p = 1$, so $I_{C1} = I_{C2}$.) Second, the op amp can be used to amplify V_{BG} to a more manageable value, such as $V_{REF} = 2.50$ V, 5.0 V, and so forth, via the gain-setting resistances R_3 and R_4. In fact, the op amp acts as a noinverting amplifier with respect to V_{BG} to give, by Eq. (1.11),

$$V_{REF} = (1 + R_3/R_4)V_{BG} \tag{5.62}$$

Finally, the op amp provides low output resistance, preventing external loading of the cell. An additional advantage of the Brokaw Cell is that the voltage across R_2 can be used as a PTAT temperature sensor.

Suppose the Brokaw Cell of Fig. 5.35b is fabricated with $m_n = 8$. Assuming $m = -1.5$, $I_{s2} = 2$ fA, and $V_T = 26$ mV, specify suitable resistance values to achieve $V_{REF} = 5.0$ V with collector currents of 0.1 mA and 1-V voltage drops across the collector resistances R.

EXAMPLE 5.12

Solution

We have $\Delta V_{BE} = V_T \ln(m_p \times m_n) = 26\ln(1 \times 8) = 54$ mV, $R_1 = \Delta V_{BE}/I_{E2} \cong 54/0.1 = 0.54$ kΩ, and $R = 1/0.1 = 10$ kΩ. We also have $V_{BE1} = V_T\ln[(0.1 \times 10^{-3})/(2 \times 10^{-15})] = 640.5$ mV. Use Eq. (5.60) to impose

$$(1 + 1)\frac{R_2}{0.54}\ln(1 \times 8) = \frac{1{,}205 - 640.5}{26} + 4 - 1.5$$

and get $R_2 = 3.14$ kΩ. By Eq. (5.61) we have $V_{BG} = 1.205 + (4 - 1.5)0.026 = 1.27$ V. Finally, use Eq. (5.62) to impose $5.0 = (1 + R_3/R_4)1.27$ and get $R_3/R_4 = 2.94$. Use $R_4 = 10$ kΩ and $R_3 = 29.4$ kΩ.

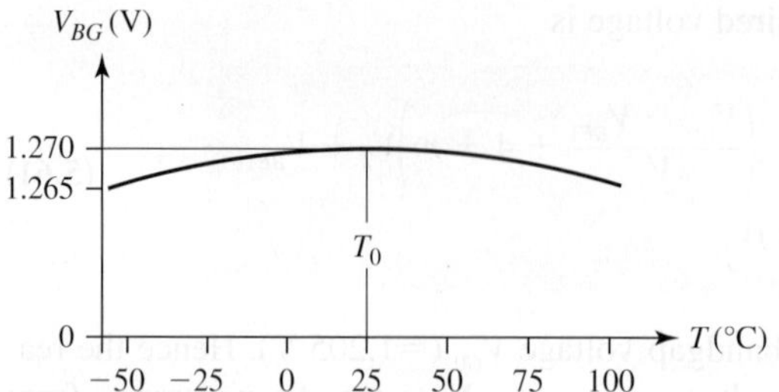

FIGURE 5.36 Thermal variation of a bandgap reference.

One last observation is in order. Equation (5.60) provides the value of K necessary to achieve $TC(V_{BG}) = 0$ at a *specific temperature* T_0, typically ambient temperature. Since K depends on V_{BE1} and V_T, themselves functions of temperature, $TC(V_{BG})$ will depart from zero at temperatures other than T_0. As depicted in Fig. 5.36, the plot of V_{BG} as a function of T exhibits a curvature[1] typical of this class of references. (The student is encouraged to look up the literature[1-3] for clever *curvature-correction techniques* designed to compensate for this higher-order effect.)

CMOS Bandgap References

The bandgap cell of Fig. 5.35a cannot be replicated in MOS form because the voltage ΔV_{GS} of Eq. (5.55) is not PTAT. We need BJTs to generate V_{BE} and ΔV_{BE}. Mercifully, the MOS structure of Fig. 3.2 does lend itself to the fabrication of so-called *well BJTs*. Figure 5.37a shows how the n^- well, normally used to host pMOSFETs, can be turned into a *pnp* BJT by the mere artifice of using a p^+ source/drain implant as the *emitter*, the n^- well itself as the *base* region, and the p^- body as the *collector*. As we know, the p^- substrate must be tied to the most negative voltage (MNV) to

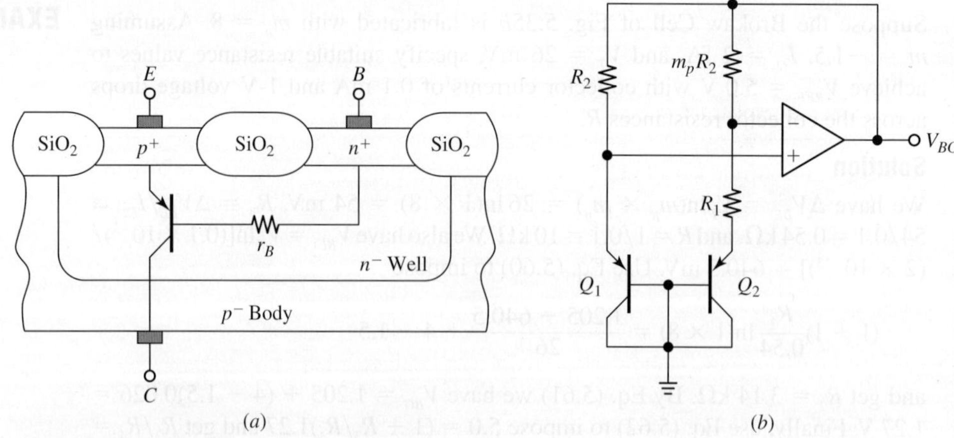

FIGURE 5.37 (a) *pnp* BJT fabricated in an n^- CMOS process, and (b) CMOS bandgap voltage reference.

prevent forward-biasing the body-source junctions of the nMOSFETs sharing the same p^- body. Consequently, the bandgap structure of Fig. 5.35a must be modified to reflect the MNV constraint. A popular solution is shown in Fig. 5.37b.

Typically the two BJTs are fabricated with equal emitter areas, so the imbalance necessary to create ΔV_{BE} is achieved by biasing the BJTs differently via unequal resistances. Through negative feedback, the op amp keeps its own input terminals at the same potential, so Q_2 is forced to operate at a current m_p *times smaller* than Q_1's. Adapting Eq. (5.53) we have $\Delta V_{EB} = V_T \ln(m_p)$, so the current through Q_2 is $\Delta V_{EB}/R_1$ and that through Q_1 is $m_p \Delta V_{EB}/R_1$. Finally, KVL gives $V_{BG} = V_{EB1} + R_2 m_p \Delta V_{EB}/R_1$, or

$$V_{BG} = V_{EB1} + KV_T \qquad\qquad K = m_p \frac{R_2}{R_1} \ln(m_p) \qquad\qquad (5.62)$$

A well BJT tends to exhibit high bulk resistance r_B across the long and lightly doped base region, so in order to minimize the voltage drop across r_B it is customary to bias well BJTs at suitably low currents.

EXAMPLE 5.13

Suppose the resistors of Fig. 5.36b are fabricated with $m_p = 10$. Assuming $m = -1.5$, $I_{s1} = 1$ fA, and $V_T = 26$ mV, specify suitable resistance values for a 100-μA current through Q_1.

Solution

We have $\Delta V_{EB} = 26\ln(10 \times 1) = 59.9$ mV, $R_1 = \Delta V_{EB}/I_{E2} = \Delta V_{EB}/(I_{E1}/10) = (59.9 \times 10^{-3})/(10 \times 10^{-6}) = 5.99$ kΩ, and $V_{EB1} = V_T \times \ln[(100 \times 10^{-6})/10^{-15}] = 658.5$ mV. The calculation of K yields

$$(10 + 1)\frac{R_2}{5.99}\ln(10 \times 1) = \frac{1{,}205 - 658.5}{26} + 4 - 1.5$$

This gives $R_2 = 5.56$ kΩ and $m_p R_2 = 55.6$ kΩ.

5.6 CURRENT-MODE INTEGRATED CIRCUITS

The voltages and currents in a linear circuit are mathematically equivalent because the laws governing voltages admit counterpart laws governing currents, and vice versa (familiar examples of this equivalence, known as *duality*, are Kirchhoff's voltage/current laws, the node/loop methods, and the Thévenin/Norton theorems). However, transistors, the basic ingredients of today's electronics, are nonlinear devices that process voltage and current differently. Specifically, the exponential characteristics of BJTs and the quadratic characteristics of FETs indicate inherently *wider dynamic ranges* for currents than for voltages: for instance, a 10-to-1 (20 dB) change in the overdrive voltage of a FET results a 100-to-1 (40 dB) change in the channel current. It also turns out that the manipulation of currents in a physical circuit is *inherently faster* than the manipulation of voltages: this is so because stray inductances, which oppose rapid changes in mesh currents, are of far lesser consequence than stray capacitances, which oppose rapid changes in node voltages (these aspects will be investigated in great detail in Chapter 6). These advantages in both range and speed provide a strong motivation for *current-mode circuits*, that is, circuits emphasizing the manipulation of currents over voltages. (Interestingly, the engineering community has traditionally favored the voltage viewpoint, as signified by the fact that voltage-mode circuits have reached commercial maturity and popularity well ahead of current-mode circuits, a classic example being the op amp, a voltage-in/voltage-out block. The development of current-mode analog ICs has been delayed also by technological reasons, such as the ability to fabricate monolithic *pnp* BJTs of comparable quality to their *npn* counterparts.) In this section we investigate current-mode circuits such as *transconductors*, *current conveyors* (CCs), *operational transconductance amplifiers* (OTAs), *current feedback amplifiers* (CFAs), and *Gilbert Cells*.

Transconductors

A transconductor is a voltage-in/current-out circuit. To prevent loading, a transconductor should exhibit sufficiently high (ideally infinite) resistances both at the input and at the output terminals. The simplest transconductor is the transistor itself. However, transistors operate over only one quadrant of their i_C-v_{BE} or i_D-v_{GS} characteristics. To cope with this drawback, we bias the transistor at a specified operating point in the active region, and we then achieve four-quadrant operation by effecting *variations*, both positive and negative, about this point. But we need to keep these variations suitably small in order to ensure approximately linear circuit operation, this being the basis of small-signal models.

We can achieve much more versatility and flexibility if we manage to (*a*) establish the operating point right at the origin of the *i-v* characteristic, and (*b*) lift the small-signal constraint altogether. A circuit meeting meets both demands is shown in Fig. 5.38 (though the version shown is bipolar, a CMOS version is readily obtained by replacing each BJT with a MOSFET of the corresponding type). We make the following observations:

- In order to handle emitter currents of either polarity, the class AB pair Q_1-Q_2 is used, with Q_1 *sourcing* current to and Q_2 *sinking* current from the circuitry external to node E (this is similar to the push-pull output stages of Chapter 4).
- The emitter followers Q_3-Q_4 generate the pair of junction voltage drops needed to bias the Q_1-Q_2 pair for class AB operation. They also provide a Darlington function to raise the input resistance as well as lower the input bias current of node B. It is apparent that Q_1 through Q_4 form a voltage buffer with approximately unity gain, as depicted in the simplified equivalent circuit at the right. Assuming matched voltage drops $V_{EB3} = V_{BE1}$ and $V_{BE4} = V_{EB2}$, we thus have

$$v_E = v_B \tag{5.63}$$

- The Wilson current mirrors Q_5-Q_6-Q_7 and Q_{10}-Q_9-Q_8 replicate the collector currents of Q_1 and Q_2, respectively, and convey them to the output node C, where they are subtracted from each other to give the output current $i_C = i_7 - i_8 = i_1 - i_2$.

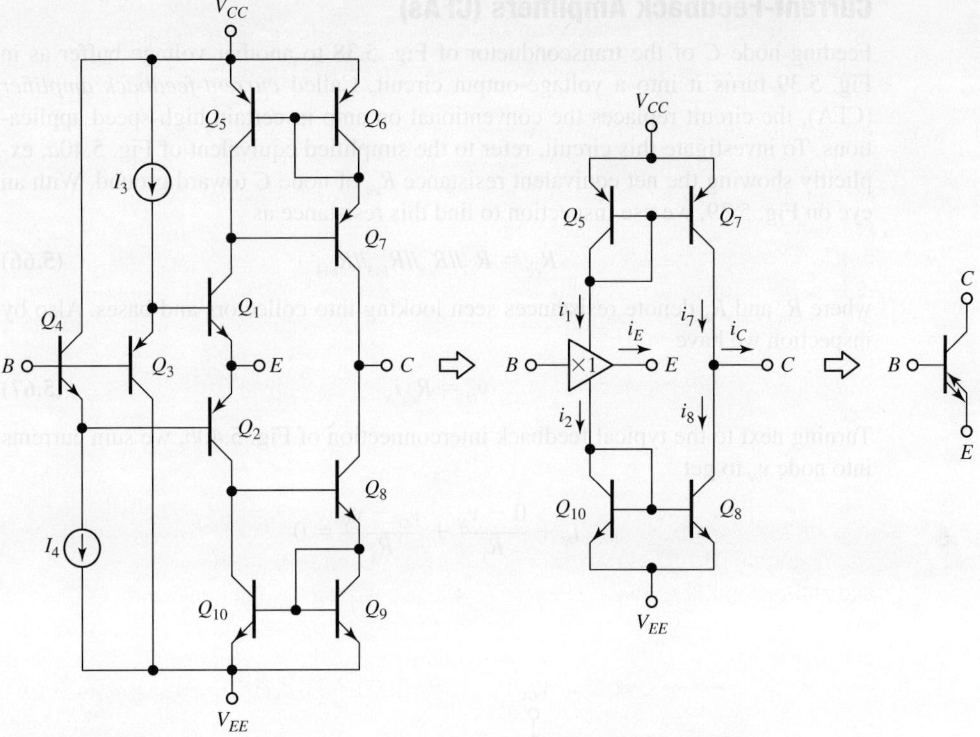

FIGURE 5.38 Bipolar transcoductor, simplified circuit representation, and circuit symbol used sometimes.

But, applying KCL at the voltage-buffer supernode we have, assuming negligible base currents, $i_1 = i_2 + i_E$, or $i_1 - i_2 = i_E$. Eliminating the difference $i_1 - i_2$ gives

$$i_C = i_E \qquad (5.64)$$

Finally, the output resistance at node C is the parallel combination of the Wilson-mirror resistances $R_{c7} \cong (\beta_{o7}/2)r_{o7}$ and $R_{c8} \cong (\beta_{o8}/2)r_{o8}$. Using subscripts n and p as usual, we write

$$R_c \cong \left(\frac{\beta_{on}}{2}r_{on}\right) /\!/ \left(\frac{\beta_{op}}{2}r_{op}\right) \qquad (5.65)$$

It is apparent that the transconductor can be regarded as a form of *idealized transistor* having (*a*) zero B-E voltage drop, (*b*) much higher input and output resistances than an ordinary transistor (this, thanks to the Darlington circuitry at the input and the Wilson circuitry at the output), and (*c*) full four-quadrant operation (for instance, if we terminate node E on a resistive load to ground, i_E and i_C will flow *out* of the transconductor for $v_B > 0$, but *into* the transconductor for $v_B < 0$). Also called *macro transistor*, *diamond transistor*, and *Current Conveyor II+*, the transconductor shown can be used in its own right in the basic configurations of CC, CB, and CE, or as a building block for other current-mode ICs.

Current-Feedback Amplifiers (CFAs)

Feeding node C of the transconductor of Fig. 5.38 to another voltage buffer as in Fig. 5.39 turns it into a voltage-output circuit. Called *current-feedback amplifier* (CFA), the circuit replaces the conventional op amp in certain high-speed applications. To investigate this circuit, refer to the simplified equivalent of Fig. 5.40a, explicitly showing the net equivalent resistance R_{eq} of node C toward ground. With an eye on Fig. 5.39, we use inspection to find this resistance as

$$R_{eq} = R_{c7}/\!/R_{c8}/\!/R_{b13}/\!/R_{b14} \tag{5.66}$$

where R_c and R_b denote resistances seen looking into collectors and bases. Also by inspection we have

$$v_O = R_{eq}i_N \tag{5.67}$$

Turning next to the typical feedback interconnection of Fig. 5.40b, we sum currents into node v_N to get

$$i_N + \frac{0 - v_N}{R_1} + \frac{v_O - v_N}{R_2} = 0$$

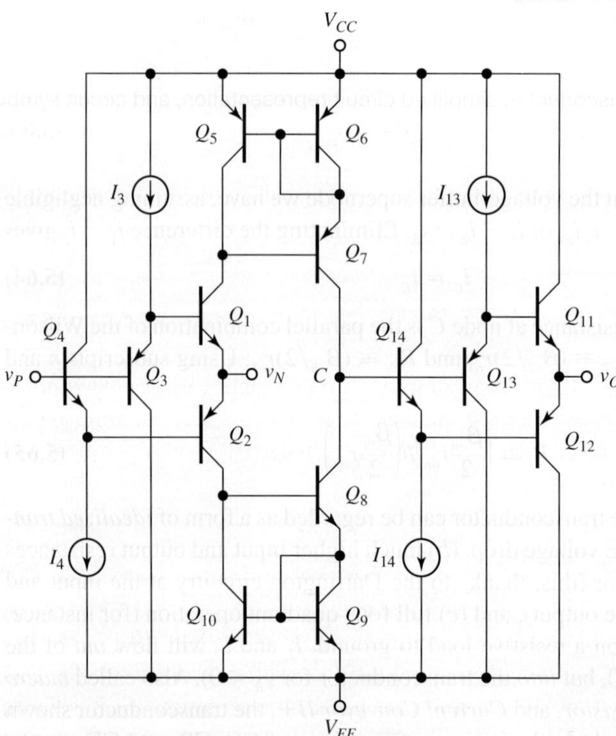

FIGURE 5.39 Current-feedback amplifier.

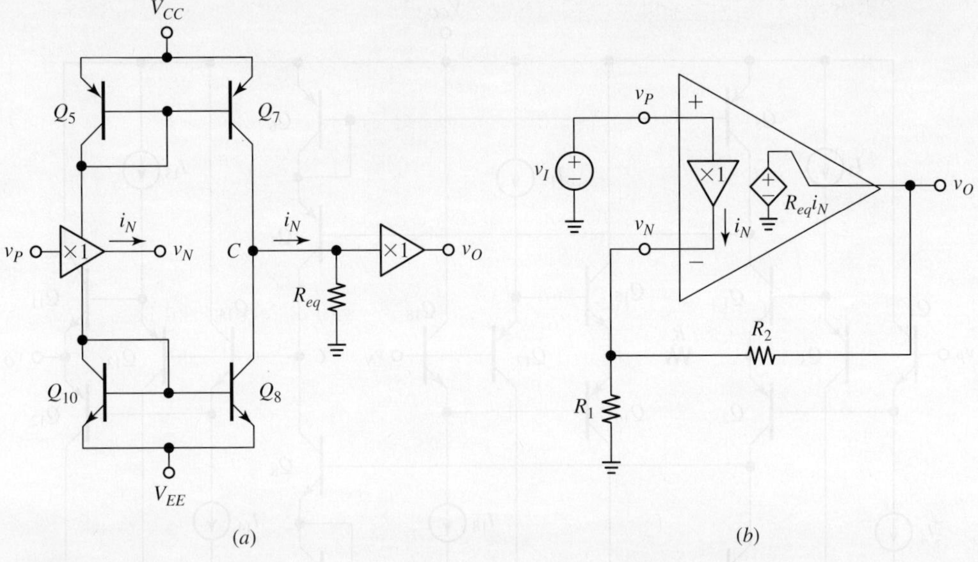

FIGURE 5.40 (*a*) Simplified circuit equivalent of the CFA. (*b*) CFA symbol and interconnection for operation as a noninverting amplifier.

Letting $v_N = v_P = v_I$, solving for i_N, and inserting into Eq. (5.67) gives the *closed-loop voltage gain*

$$A = \frac{v_O}{v_I} = \left(1 + \frac{R_2}{R_1}\right)\frac{1}{1 + R_2/R_{eq}} \tag{5.68}$$

In a well-designed circuit R_2 is on the order of $10^3\ \Omega$ whereas R_{eq} is on the order of $10^5 \sim 10^6\ \Omega$, so $R_2 \ll R_{eq}$. Under this condition, A tends to the expression already familiar from op amps,

$$A \cong 1 + \frac{R_2}{R_1} \tag{5.69}$$

The unique advantages of CFAs compared to ordinary op amps (also called voltage-feedback amplifiers or VFAs) are fast dynamics. These advantages, not immediately obvious from the present discussion but stemming from current-mode operation, will be addressed in Chapter 6.

CFA-Derived Voltage-Feedback Amplifiers

Another important application of the transconductor of Fig. 5.38 is as a building block for high-speed voltage-feedback amplifiers (VFAs). The VFA of Fig. 5.41 is obtained from the CFA of Fig. 5.39 by using a third voltage buffer (Q_{15}-Q_{16}-Q_{17}-Q_{18}) to turn node v_N into a high resistance input, and by inserting a resistance R between the outputs of the first buffer and this new buffer to generate the control current previously denoted as i_N. This current (assumed to flow from left to right) is now

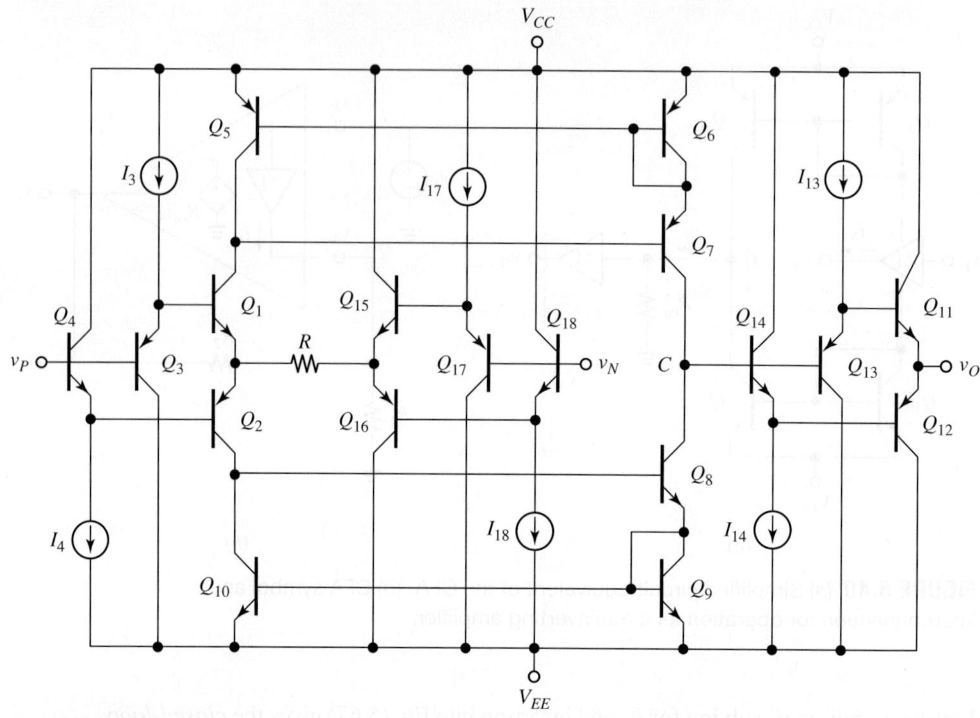

FIGURE 5.41 CFA-derived VFA.

$(v_P - v_N)/R$, and is conveyed to node C, where it produces the voltage $R_{eq}(v_P - v_N)/R$. This voltage is then buffered to the output node to give v_O, so the *open-loop voltage gain* of the resulting op amp is

$$a = \frac{v_O}{v_P - v_N} = \frac{R_{eq}}{R} \tag{5.70}$$

Again, a well-designed circuit has $a \gg 1$. Owing to its inherently fast current-mode operation, this op amp type is especially suited to high-speed applications.

Differential-Input Transconductors

The transconductor's versatility can be enhanced appreciably if we configure it to respond to *differential-type inputs*. In the popular rendition of Fig. 5.42, called *operational transconductance amplifier* (OTA), this feature is achieved by means of the differential pair Q_1-Q_2. To ensure high CMRR, the pair is biased via the high output-resistance Wilson mirror Q_3-Q_4-Q_5. Q_2's current is replicated and conveyed to the output node by the Wilson mirror Q_9-Q_{10}-Q_{11}; Q_1's current is first replicated by the mirror Q_6-Q_7-Q_8, and then conveyed to the output node by the mirror Q_{12}-Q_{13}-Q_{14}, where it gets *subtracted* from that conveyed by Q_9-Q_{10}-Q_{11}. Ignoring the diodes D_1 and D_2 for the time being, we adapt Eq. (4.73) to write

$$i_o = i_{C11} - i_{C12} = i_{C2} - i_{C1} = I \times \tanh\left(\frac{v_P - v_N}{2V_T}\right) \tag{5.71}$$

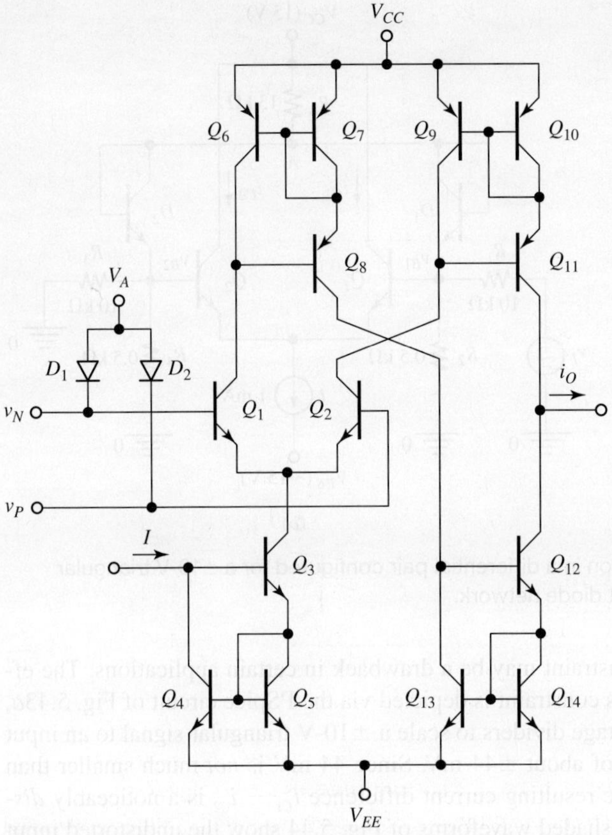

FIGURE 5.42 Operational transconductance amplifer (OTA).

where I is the current supplied by the user to bias the Q_1-Q_2 pair. The plot of i_O versus the difference $v_P - v_N$ is the familiar s-shaped curve of Fig. 4.44. This curve is *nonlinear*, but if we restrict the input voltage within the range $|v_P - v_N| \ll 4V_T$ ($\cong 100$ mV), then the small-signal approximation allows us to retain only the first term in the series expansion $\tanh x = x - x^3/3 + \cdots$ and write

$$i_o = \frac{I}{2V_T}(v_p - v_n) \tag{5.72}$$

We make some important observations:

- In small-signal operation, the *transconductance gain*

$$G_m = \frac{i_o}{v_p - v_n} = \frac{I}{2V_T} \tag{5.73}$$

  is *linearly proportional* to the bias current I, so it can be programmed externally by the user. In fact, BJTs easily allow for current ranges in excess of five decades (>100 dB), making OTAs particularly suited to a variety of programmable circuits[7,8] such as programmable amplifiers, programmable oscillators, and programmable filters, especially in wide-dynamic range applications like audio and electronic music.

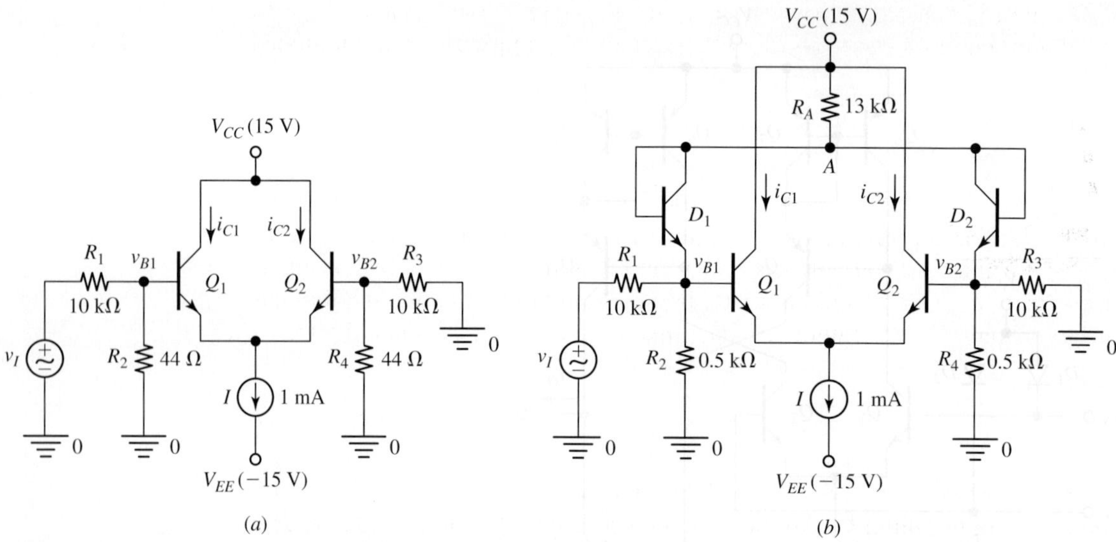

FIGURE 5.43 PSpice circuits to investigate distortion in a differential pair configured for a ±10-V triangular input: (*a*) without and (*b*) with the predistorting input diode network.

- The small-signal constraint may be a drawback in certain applications. The effect of exceeding this constraint is depicted via the PSpice circuit of Fig. 5.43*a*, which uses input voltage dividers to scale a ±10-V triangular signal to an input difference $v_{B1} - v_{B2}$ of about ±44-mV. Since 44 mV is *not* much smaller than $4V_T$ ($\cong 100$ mV), the resulting current difference $i_{C1} - i_{C2}$ is a noticeably *distorted* triangle. (The shaded waveforms of Fig. 5.44 show the undistorted input voltage as well as the distorted output current.)

- If we wish an undistorted current waveform, the input voltage must be *predistorted* according to the *inverse* of the hyperbolic tangent function. An approximation to this function is achieved by means of the diode-connected BJT pair D_1-D_2 of Fig. 5.43*b*. Its effect, depicted in Fig. 5.44 via the solid traces, is

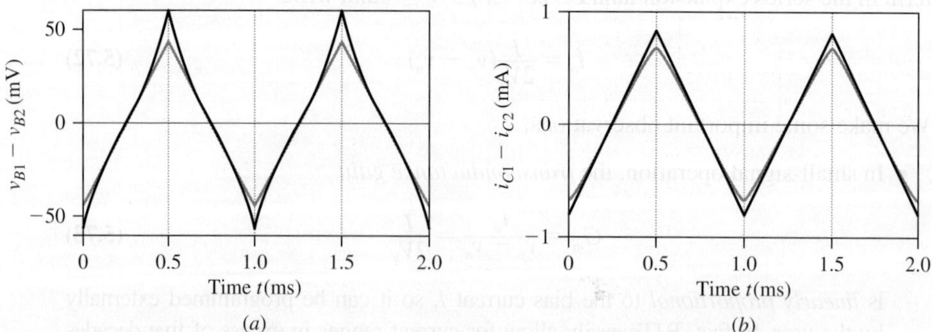

FIGURE 5.44 Waveforms for the PSpice circuits of Fig. 5.43 (the *shaded* traces pertain to the circuit of Fig. 5.43*a*, the *solid* traces to that of Fig. 5.43*b*.)

to stretch in approximately inverse hyperbolic-tangent fashion the input peaks from about ±44 mV to about ±60 mV, the values required for an undistorted current waveform.

Variable Transconductance Multipliers

The OTA of Fig. 5.42 is said to be an *analog multiplier* because, according to Eq. (5.72), i_o is proportional to the product $I \times (v_p - v_n)$. While the difference $v_p - v_n$ can assume either polarity, the current I must always flow out of the differential pair ($I \geq 0$), so the OTA is said to be a *two-quadrant multiplier*. Applications such as communications require full *four-quadrant analog multiplication*. This function is implemented using the current-mode circuit of Fig. 5.45. Popularly known as the *Gilbert Cell* for its inventor Barrie Gilbert,[8] this most elegant current-mode circuit consists of (*a*) two emitter-coupled (EC) differential pairs Q_1-Q_2 and Q_3-Q_4 driven by the same voltage v_X but in *phase opposition* to each other, so their output currents *subtract* pair-wise, and (*b*) a third EC pair (Q_5-Q_6) designed to *steer* the bias current I to the two differential pairs in ratios controlled by the voltage v_Y. Depending on the polarity of v_Y, the output of one pair will prevail over that of the other, allowing the output current difference $i_{O1} - i_{O2}$ to attain *either polarity* and thus provide *four-quadrant operation*.

To obtain a relationship between the output currents and the input voltages we use KCL to write

$$i_{O1} - i_{O2} = (i_{C1} + i_{C3}) - (i_{C2} + i_{C4}) = (i_{C1} - i_{C2}) - (i_{C4} - i_{C3})$$

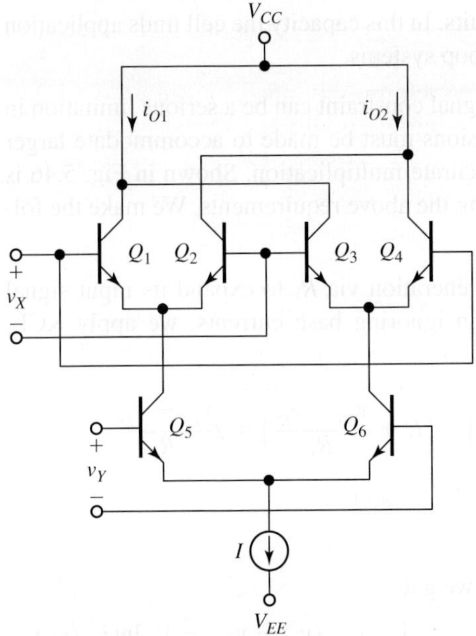

FIGURE 5.45 The Gilbert Cell.

Assuming negligible base currents, we adapt Eq. (5.71) and write

$$i_{O1} - i_{O2} = i_{C5}\tanh\left(\frac{v_X}{2V_T}\right) - i_{C6}\tanh\left(\frac{v_X}{2V_T}\right) = (i_{C5} - i_{C6})\tanh\left(\frac{v_X}{2V_T}\right)$$

that is,

$$i_{O1} - i_{O2} = I\tanh\left(\frac{v_Y}{2V_T}\right)\tanh\left(\frac{v_X}{2V_T}\right) \tag{5.74}$$

We identify three classes of applications for the cell:

- Both inputs are of the small-signal type, that is, $|v_x| \ll 4V_T$ and $|v_y| \ll 4V_T$, so we approximate $\tanh x \cong x$ and write

$$i_{o1} - i_{o2} \cong I \times \frac{v_x}{2V_T} \times \frac{v_y}{2V_T} \tag{5.75}$$

Clearly, in this mode the cell operates as a true *four-quadrant multiplier*.
- One of the inputs (v_x) is a small-signal continuous wave such as a sine wave, while the other input (v_y) is a square wave of sufficient magnitude to overdrive the Q_5-Q_6 pair so as to turn one of its BJTs off. Then, for $v_y > 0$, Q_6 is off and Q_5 steers all of I to the Q_1-Q_2 pair to give $i_{o1} - i_{o2} = +Iv_x/(2V_T) = +g_m v_x$. Conversely, for $v_y < 0$, Q_5 is off and Q_6 steers all of I to the Q_3-Q_4 pair to give $i_{o1} - i_{o2} = -Iv_x/(2V_T) = -g_m v_x$, the negative sign stemming from the anti-phase connection at the input. In this capacity the cell finds application in communications as *modulator/detector*.
- Both inputs are of the large-signal type. Then, when v_X and v_Y have the *same polarity* the cell gives $i_{O1} - i_{O2} = I - 0 = I$, whereas when v_X and v_Y have *opposite polarities* it gives $i_{O1} - i_{O2} = 0 - I = -I$. This function is similar to the exclusive-or function of digital circuits. In this capacity the cell finds application as *phase detector* in phase-locked loop systems.

As in the case of OTAs, the small-signal constraint can be a serious limitation in certain multiplier applications, so provisions must be made to accommodate larger input magnitudes while still assuring accurate multiplication. Shown in Fig. 5.46 is a popular Gilbert Cell application meeting the above requirements. We make the following observations:

- The Q_5-Q_6 pair exploits emitter degeneration via R_Y to expand its input signal range by the desired amount. Again ignoring base currents, we apply KCL, Ohm's law, and KVL to write

$$i_{C5} - i_{C6} = \left(I_Y + \frac{v_{E5} - v_{E6}}{R_Y}\right) - \left(I_Y + \frac{v_{E6} - v_{E5}}{R_Y}\right) = 2\frac{v_{E5} - v_{E6}}{R_Y}$$

$$= \frac{(v_{Y1} - v_{BE5}) - (v_{Y2} - v_{BE6})}{0.5R_Y}$$

Regrouping and using $v_{BE} = V_T\ln(i_C/I_s)$ we get

$$i_{C5} - i_{C6} = \frac{(v_{Y1} - v_{Y2}) - (v_{BE5} - v_{BE6})}{0.5R_Y} = \frac{(v_{Y1} - v_{Y2}) - V_T\ln(i_{C5}/i_{C6})}{0.5R_Y}$$

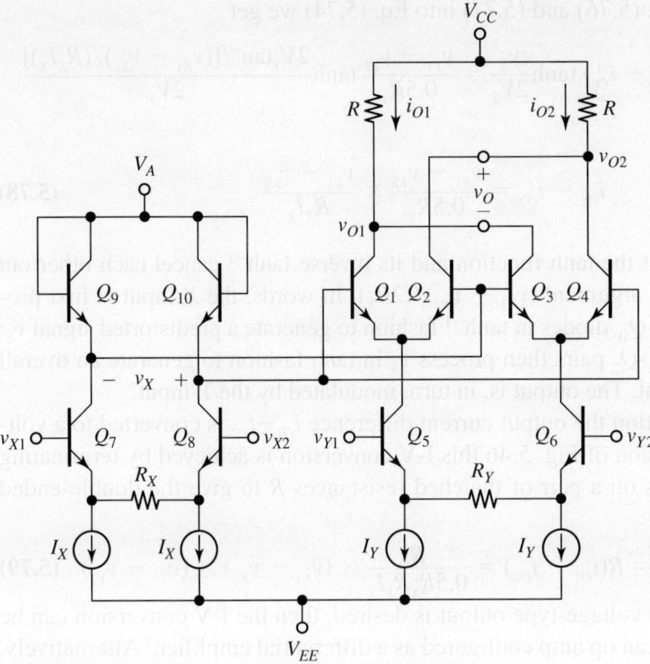

FIGURE 5.46 Four-quadrant analog multiplier.

In a well-designed circuit the last numerator term is negligible (see Problem 5.39), so we approximate

$$i_{C5} - i_{C6} = \frac{v_{Y1} - v_{Y2}}{0.5R_Y} \tag{5.76}$$

indicating that the emitter-degenerated Q_5-Q_6 pair acts as a linear *voltage-to-current* (V-I) *converter*.

- Likewise, the Q_7-Q_8 pair is a V-I converter giving $i_{C7} - i_{C8} = (v_{X1} - v_{X2})/(0.5R_X)$ and driving the diode-connected pair Q_9-Q_{10}. Using KVL and $v_{BE} = V_T \ln(i_C/I_s)$ we have

$$v_X = v_{E10} - v_{E9} = (V_A - v_{BE10}) - (V_A - v_{BE9}) = v_{BE9} - v_{BE10}$$

$$= V_T \ln\frac{i_{C9}}{i_{C10}} = V_T \ln\frac{i_{C7}}{i_{C8}}$$

But, KCL gives $i_{C7} = I_X + (v_{X1} - v_{X2})/R_X$ and $i_{C8} = I_X - (v_{X1} - v_{X2})/R_X$, so

$$v_X = V_T \ln\frac{I_X + (v_{X1} - v_{X2})/R_X}{I_X - (v_{X1} - v_{X2})/R_X} = V_T \ln\frac{1 + (v_{X1} - v_{X2})/(R_X I_X)}{1 - (v_{X1} - v_{X2})/(R_X I_X)}$$

Using the identity $\ln[(1 + x)/(1 - x)] = 2\tan^{-1}x$, which holds for $|x| < 1$, we have

$$v_X = 2V_T\tan^{-1}\frac{v_{X1} - v_{X2}}{R_X I_X} \tag{5.77}$$

- Substituting Eqs. (5.76) and (5.77) into Eq. (5.74) we get

$$i_{O1} - i_{O2} = (i_{C5} - i_{C6})\tanh\frac{v_X}{2V_T} = \frac{v_{Y1} - v_{Y2}}{0.5R_Y}\tanh\frac{2V_T\tan^{-1}[(v_{X1} - v_{X2})/(R_XI_X)]}{2V_T}$$

that is,

$$i_{O1} - i_{O2} = \frac{v_{Y1} - v_{Y2}}{0.5R_Y} \times \frac{v_{X1} - v_{X2}}{R_XI_X} \qquad (5.78)$$

It is apparent that the tanh function and its inverse $\tanh^{-1}$ cancel each other out to leave only the argument $(v_{X1} - v_{X2})/(2R_X)$. In words, the X-input is first processed by the Q_9-Q_{10} diodes in $\tanh^{-1}$ fashion to generate a predistorted signal v_X; the Q_1-Q_2 and Q_3-Q_2 pairs then process v_X in tanh fashion to generate an overall undistorted output. The output is, in turn, modulated by the Y-input.

- In actual application the output current difference $i_{O1}-i_{O2}$ is converted to a voltage. In the rendition of Fig. 5.46 this I-V conversion is achieved by terminating the cell's outputs on a pair of matched resistances R to give the double-ended output voltage

$$v_O = v_{O2} - v_{O1} = R(i_{O1} - i_{O2}) = \frac{R}{0.5R_XR_YI_X} \times (v_{X1} - v_{X2}) \times (v_{Y1} - v_{Y2}) \quad (5.79)$$

If a single-ended voltage-type output is desired, then the I-V conversion can be implemented via an op amp configured as a differential amplifier.[7] Alternatively, if a single-ended current output is desired, then we can use three Wilson current mirrors in the manner of the OTA of Fig. 5.42.

5.7 FULLY DIFFERENTIAL OPERATIONAL AMPLIFIERS

A common way of encoding analog information electronically is via single-ended voltages such as ground-referenced potentials. A classic example is the inverting amplifier of Fig. 5.47a, whose input v_I and output v_O are both referenced to ground potential. For large a, they are related as $v_O = (-R_2/R_1)v_I$. In actual application the circuit is likely to be part of a larger system, where it may be subject to various

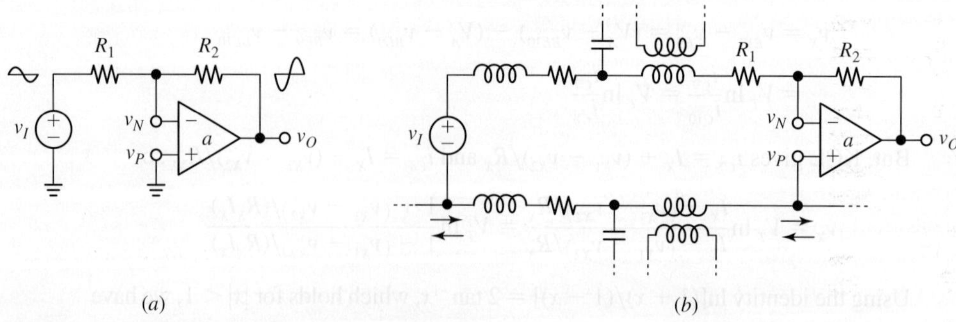

(a)   (b)

FIGURE 5.47 Basic op amp circuit: (a) idealized situation, and (b) in actual application, showing the stray elements of its interconnections.

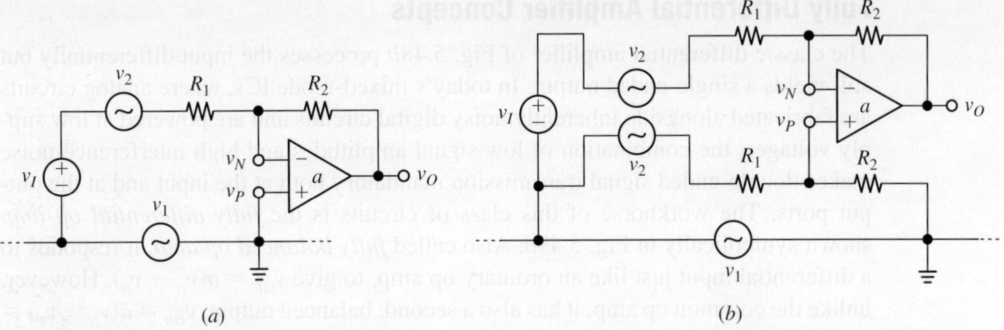

FIGURE 5.48 (a) Interference noise model for the inverting amplifier of Fig. 5.47. (b) Interconnecting the op amp as a *difference amplifier* to reject v_1, and using *balanced input lines* to cancel out the v_2 noise sources.

forms of interference noise that might obfuscate its useful signals. To begin with, the ground interconnection, far from being a perfect conductor, exhibits *distributed resistance, inductance,* and *capacitance,* and is shared with other subcircuits in the system. As the currents of the other circuits flow back and forth on the ground line, as illustrated via the lumped model of Fig. 5.47b, they produce undesirable voltage drops along the distributed line impedance. Also shown are *stray mutual inductances* and *stray internode capacitances*, which provide additional paths through which the surrounding circuits can inject interference noise into the line. We model the cumulative effect of all noise sources with a single *ground noise source* v_1 as in Fig. 5.48a. Interference noise pickup from neighboring circuits affects also the signal line connecting the source to the amplifier, so we model this via the *line noise source* v_2.

Given the noise drops along the ground line, the very concept of reference node becomes blurred. For the sake of discussion let us arbitrarily pick node v_P as our ground reference, as depicted in Fig. 5.48a. Then, using KVL we find $v_O = (-R_2/R_1) \times (v_1 + v_I + v_2)$, indicating that both noise sources get amplified by the same amount as the useful signal v_I. This is unacceptable especially in high-gain applications, where the input signal strength may be comparable to that of noise. The best way to dispose of ground noise is to treat v_1 as a common-mode signal and use a *differential amplifier* with a high common-mode rejection ratio (CMRR) to reject it. As depicted in Fig. 5.48b, this requires lifting node v_P off ground and driving it by means of an *additional dedicated line* as well as an additional R_1-R_2 pair, as shown. This second line will of course pick up noise just like the preexisting line, but if the two lines are identical and are laid out in close proximity to each other so as to form what is aptly referred to as *balanced lines*, then noise pickup will be *identical* on the two lines and the circuit will give[7] $v_O = (R_2/R_1)[(v_1 + v_2) - (v_1 + v_I + v_2)] = (-R_2/R_1)v_I$. Summarizing, v_1 is rejected thanks to the amplifier's CMRR, and the v_2 sources simply cancel each other out thanks to line balance.

Fully Differential Amplifier Concepts

The classic differential amplifier of Fig. 5.48*b* processes the input differentially but still yields a single-ended output. In today's mixed-mode ICs, where analog circuits are fabricated alongside inherently noisy digital circuits and are powered at low supply voltages, the combination of low signal amplitudes and high interference noise makes double-ended signal transmission mandatory both at the input and at the output ports. The workhorse of this class of circuits is the *fully differential op amp* shown symbolically in Fig. 5.49*a*. Also called *fully balanced op amp*, it responds to a differential input just like an ordinary op amp, to give $v_{OP} = a(v_P - v_N)$. However, unlike the common op amp, it has also a second, balanced output, $v_{ON} = a(v_N - v_P) = -v_{OP}$. Additionally, it comes with a control input for the user to specify the common-mode output voltage $V_{OC(\text{set})}$, as we shall see shortly.

Beside the noise-immunity advantages of balanced transmission, the fully differential op amp offers also (*a*) *wider output signal swing* and (*b*) *reduced second-harmonic distortion*. To see how, rewrite the outputs in more general forms accounting for inherent circuit nonlinearities, namely, $v_{OP} = a(v_P - v_N) + b(v_P - v_N)^2 + c(v_P - v_N)^3 \ldots$ and $v_{ON} = a(v_N - v_P) + b(v_N - v_P)^2 + c(v_N - v_P)^3 \ldots$ Then, the differential output reduces, after simplification, to $v_{OP} - v_{ON} = 2a(v_P - v_N) + 2c(v_P - v_N)^3 \ldots$, indicating a differential output signal swing *twice* as large as that of the individual single-ended outputs, as well as a cancellation of the second-harmonic components.

The fully differential op amp is configured for negative-feedback operation with two identical networks as depicted in Fig. 5.49*b* for the resistive case. (Both networks are of the negative-feedback type as each connects from one of the outputs to the input of the opposite polarity.) For the purpose of illustration, the external inputs v_{IP} and v_{IN} have been decomposed in terms of their differential-mode and common-mode components

$$v_{ID} = v_{IP} - v_{IN} \qquad v_{IC} = \frac{v_{IP} + v_{IN}}{2} \tag{5.80a}$$

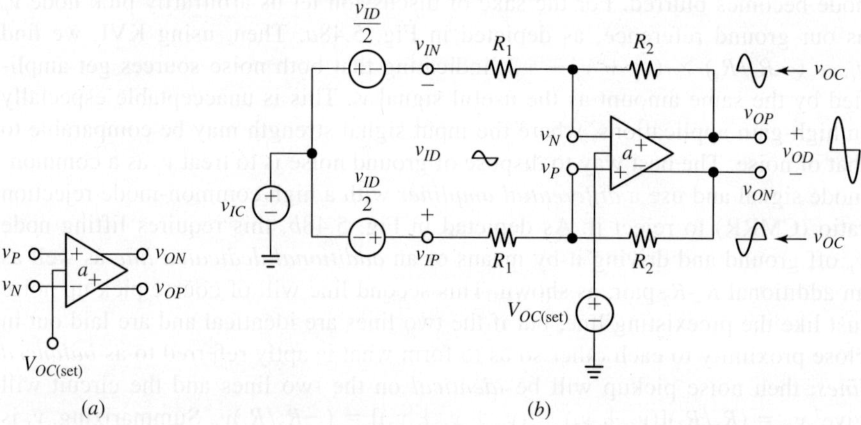

FIGURE 5.49 Fully differential op amp: (a) symbol and labels, and (b) interconnection for operation as a fully balanced amplifier.

Likewise, the differential-mode and common-mode components at the output are

$$v_{OD} = v_{OP} - v_{ON} \qquad v_{OC} = \frac{v_{OP} + v_{ON}}{2} \qquad (5.80b)$$

If the open-loop gain a is sufficiently large, the differential-mode components are related as

$$v_{OD} = \frac{R_2}{R_1}v_{ID} \qquad (5.81)$$

(Note that since the single-ended output signals have opposite phases, as shown, the differential output swing v_{OD} is *twice* as wide.) We also observe that whereas v_{IC} is established by the input sources, v_{OC} is unspecified and as such it needs to be *set* externally by the user.

Fully Differential Bipolar Op Amp

Figure 5.50 shows how the folded-cascode bipolar structure of Fig. 4.71a can be turned into a fully differential op amp. The first important change is in Q_5, formerly diode connected but now operated as a current-sink active load, just like Q_6. This

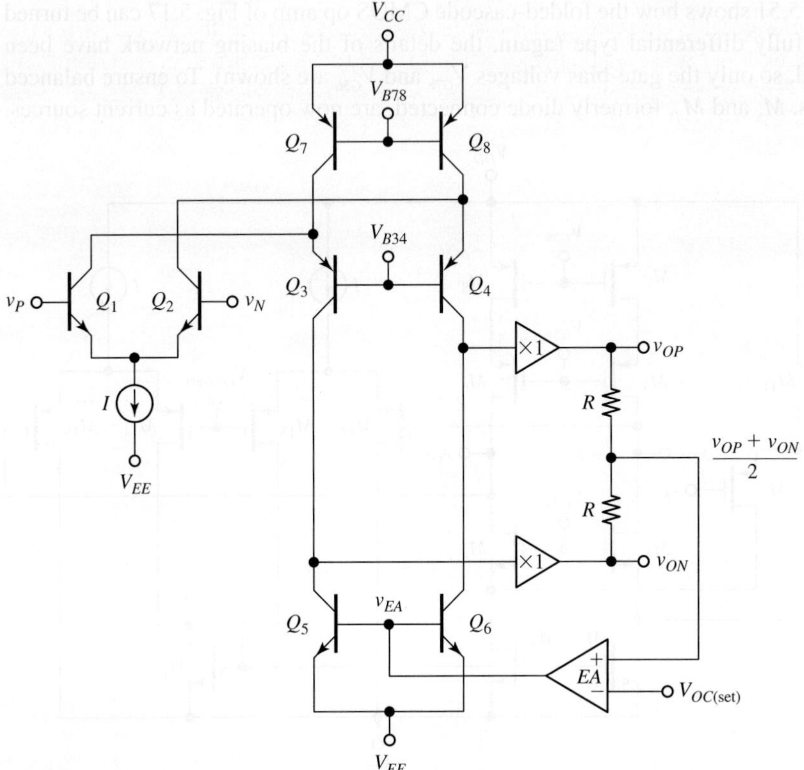

FIGURE 5.50 Simplified fully differential bipolar op amp.

results in a perfectly symmetric structure. (To avoid cluttering, the details of the biasing network have been omitted, so only the base-bias voltages V_{B78} and V_{B34} are shown (also, in actual implementation, Q_5 and Q_6 are likely to be cascoded to raise the gain, but here we are focusing only on the basics.) The high-resistance collector nodes of the Q_3-Q_5 and Q_4-Q_6 pairs are buffered via unity-gain push-pull stages to give v_{OP} and v_{ON}. The other important feature is the matched R-R pair and the *error amplifier EA*, forming what is referred to as the *common-mode feedback network* (CMFN). Its function is to properly center v_{OP} and v_{ON} up or down the permissible output voltage range (OVR), usually halfway in order to maximize the output voltage swing. Specifically, the resistance pair synthesizes the voltage $(v_{OP} + v_{ON})/2$, which the EA then compares against the externally supplied voltage $V_{OC(\text{set})}$, and adjusts its output v_{EA} to force $(v_{OP} + v_{ON})/2$ to approach $V_{OC(\text{set})}$. To see how, recall that v_{OP} and v_{ON} are the voltages at which the pull-up actions by Q_3 and Q_4 balance, respectively, the pull-down actions by Q_5 and Q_6. So, raising/lowering v_{EA} makes pull-down stronger/weaker than pull-up, shifting v_{OP} and v_{ON} down/up the OVR. In control-theory parlance, the negative-feedback loop comprising the R-R pair, the EA, the Q_5-Q_6 pair, and the buffers is said to form a *servo loop*.

Fully Differential CMOS Op Amps

Figure 5.51 shows how the folded-cascode CMOS op amp of Fig. 5.17 can be turned into a fully differential type (again, the details of the biasing network have been omitted, so only the gate-bias voltages V_{G78} and V_{G56} are shown). To ensure balanced outputs, M_5 and M_7, formerly diode connected, are now operated as current sources.

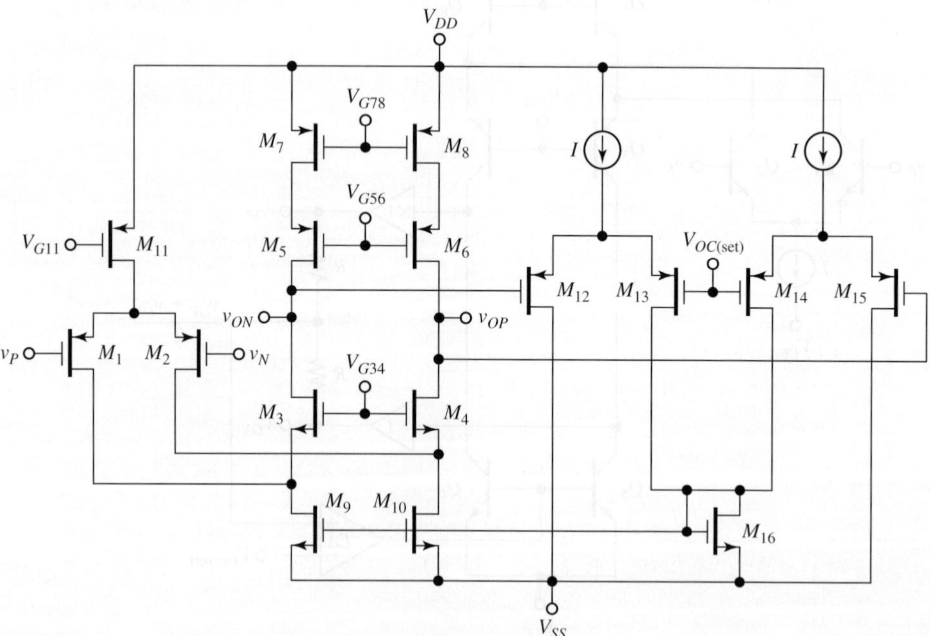

FIGURE 5.51 Fully-differential CMOS op amp of the folded-cascode type.

Moreover, the CMFN is the all-transistor network made up of M_{12} through M_{16}. To see how it functions, consider the following:

- Start out assuming $v_P = v_N$, so $v_{OP} = v_{ON}$. Also, assume $v_{OP} = v_{ON} = V_{OC(set)}$, so the two I sources split equally between the SC pairs to give $I_{D12} = I_{D13} = 0.5I$ and $I_{D14} = I_{D15} = 0.5I$. The diode-connected FET M_{16} draws $I_{D16} = I_{D13} + I_{D14} = I$, which M_9 and M_{10} then mirror to provide, respectively, the bias currents for M_3 and M_4 (refer also to Fig. 5.16). As we know, v_{OP} and v_{ON} represent the voltages at which the pull-up actions by M_5 and M_6 balance, respectively, the pull-down actions by M_3 and M_4.
- Suppose now that for some reason v_{OP} and v_{ON} try to *rise* above $V_{OC(set)}$. Then, M_{12} and M_{15} will become less conductive, while M_{13} and M_{14} will conduct more, thus raising I_{D16}. By mirror action this will raise I_{D9} and I_{D10} and thus boost the pull-down action by M_3 and M_4, with the result of forcing v_{OP} and v_{ON} to drop back to $V_{OC(set)}$. By symmetric reasoning, if v_{OP} and v_{ON} try to *drop* below $V_{OC(set)}$, M_{13} and M_{14} will conduct less, decreasing I_{D16} and thus reducing the pull-down action by M_3 and M_4. The pull-up action by M_5 and M_6 will now prevail, forcing v_{OP} and v_{ON} to rise back to $V_{OC(set)}$.
- Let us now apply an ac signal across the inputs so as to generate the symmetric swings $v_{OP} = V_{OC(set)} + v_{OD}/2$ and $v_{ON} = V_{OC(set)} - v_{OD}/2$. These swings will imbalance the differential pairs to give $i_{D15} = 0.5I - i$ and $i_{D14} = 0.5I + i$, and $i_{D12} = 0.5I + i$ and $i_{D13} = 0.5I - i$, for some current change i. However, M_{16} still draws

$$i_{D16} = i_{D13} + i_{D14} = 0.5I - i + 0.5I + i = I$$

indicating that the servo loop provided by the CMFN keeps the value of $(v_{OP} + v_{ON})/2$ near $V_{OC(set)}$ regardless of the ac signals present. Of course this holds true so long as none of the transistors is forced out of its active region.

EXAMPLE 5.14

Find the small-signal gain v_{od}/v_{id} of the op amp of Fig. 5.51, assuming the parameters of Example 5.4. Compare with the example and comment.

Solution

The circuit of Fig. 5.51 is perfectly symmetric about the vertical axis running down from node V_{G78}, so we work with the half-circuit ac equivalent of Fig. 5.52. By inspection, $R_o = R_{d5}//R_{d3}$, where R_{d5} and R_{d3} are the ac resistances seen looking into the drains of M_5 and M_3. Using again inspection we write

$$R_o = [r_{o5} + r_{o7} + (g_{m5} + g_{mb5})r_{o5}r_{o7}]//[r_{o3} + (r_{o9}//r_{o1}) + (g_{m3} + g_{mb3})r_{o3}(r_{o9}//r_{o1})]$$

This is formally identical to Eq. (5.39), so we recycle Example 5.4 to write $R_o \cong$ 5.22 MΩ. We also have $i_1 = g_{m1}(v_{id}/2) = (0.4$ mA/V$) \times (v_{id}/2)$. Following the line of reasoning of Example 5.4 we can state that most of i_1 flows out of M_3, so

$$-\frac{v_{od}}{2} \cong -R_o i_1 = -g_{m1}R_o\frac{v_{id}}{2}$$

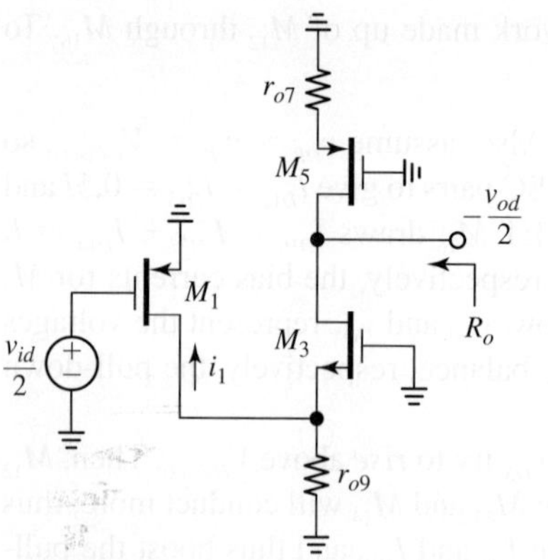

FIGURE 5.52 Half-circuit equivalent of the op amp of Fig. 5.51.

Consequently, $v_{od}/v_{id} = g_{m1}R_o \cong 0.4 \times 5,220 = 2,088$ V/V. The single-ended signal v_o of Example 5.4 now splits between the two opposite halves v_{op} and v_{on}, so the gain is unchanged.

Figure 5.53 shows how the two-stage CMOS op amp of Fig. 5.13 can be turned into a fully differential type (again, the biasing details have been omitted for simplicity). In this rendition the CMFN is now the complementary of that of Fig. 5.51 because it is designed to adjust the current of a *p*MOSFET, namely, the tail current supplied by M_9 (evidently there are various different ways of designing a CMFN).

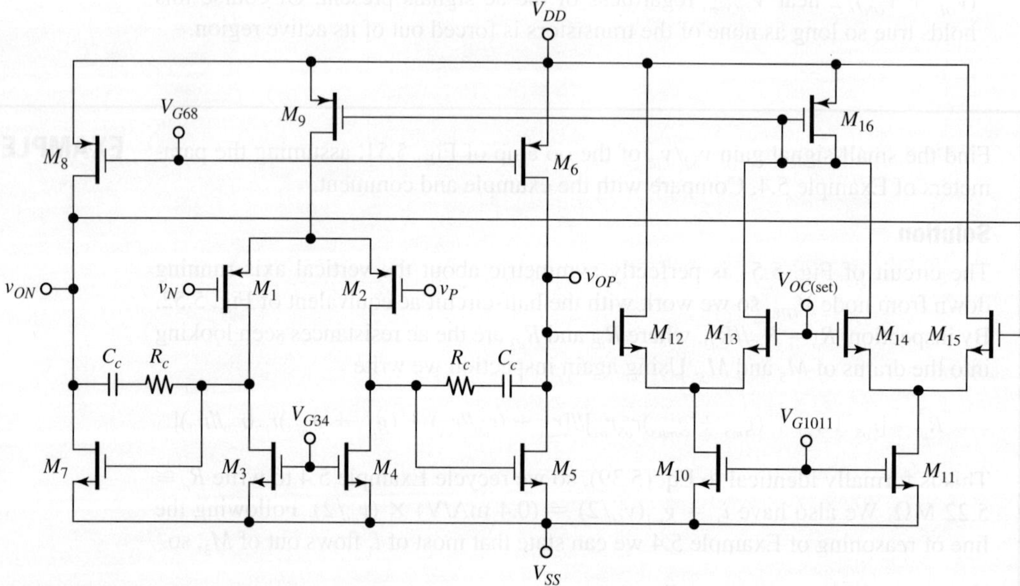

FIGURE 5.53 Fully-differential CMOS op amp of the two-stage type.

Expanding the Input Voltage Range

If we specify V_{G34} in the folded-cascode op amp of Fig. 5.51 so as to bias the M_9-M_{10} pair right at the edge of saturation ($V_{DS9} = V_{OV9}$ and $V_{DS10} = V_{OV10}$), then, the input voltage range (IVR) is specified by Eq. (5.41). For instance, suppose we have a single-supply system with $V_{SS} = 0$ and $V_{DD} = 3$ V, and let $|V_{t1}| = 0.75$ V and $V_{OV} = 0.25$ V throughout. Then, the common-mode input voltage $v_{IC} = \frac{1}{2}(v_P + v_N)$ is such that

$$v_{IC(min)} = V_{SS} + V_{OV9} - |V_{t1}| = 0 + 0.25 - 0.75 = -0.5 \text{ V}$$

$$v_{IC(max)} = V_{DD} - V_{OV11} - V_{OV1} - |V_{t1}| = 3 - 0.25 - 0.25 - 0.75 = 1.75 \text{ V}$$

While the lower limit is quite desirable because it allows us to lower the inputs all the way down to the ground rail, and even 0.5-V below, without causing any malfunction, the upper limit is generally too restrictive, especially in a low-supply situation such as the 3-V case considered here. Ideally we want the IVR to extend from *rail to rail* and possibly beyond by, say, a 0.5-V margin (from -0.5 V to 3.5 V in our example). Were we to use the dual op amp realization by interchanging *p*MOSFETs and *n*MOSFETs as well as interchanging the power supplies (see Fig. P5.21), then the *n*-type input pair would exhibit the dual constraints $v_{IC(min)} = 1.25$ V and $v_{IC(max)} = 3.5$ V. In other words, while a *p*-type input stage works well near V_{SS} but goes off near V_{DD}, an *n*-type works well near V_{DD} but goes off near V_{SS}, just the opposite.

   The above considerations provide the basis for the ingenious solution of Fig. 5.54, which uses both input-stage types to get the *best of both*. (The CMFN, not shown to

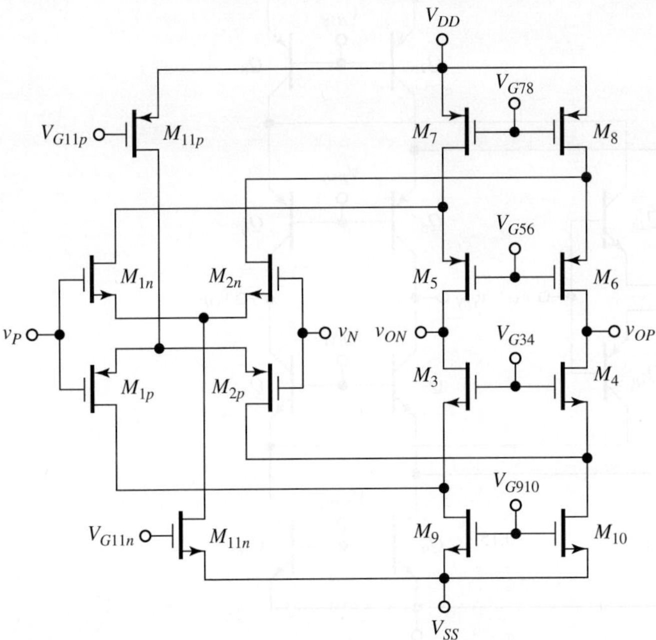

FIGURE 5.54 Using two complementary input stages (M_{1n}-M_{2n} and M_{1p}-M_{2p}) to achieve a rail-to-rail IVR (CMFN not shown).

avoid cluttering the schematic, can be made to adjust either V_{G78} or V_{G910}, or both.) Assuming again $V_{tn} = -V_{tp} = 0.75$ V and $V_{OV} = 0.25$ V throughout, we note the following:

- For $v_{IC} < V_{SS} + V_{OV11n} + V_{tn}$ ($=1$ V for $V_{SS} = 0$) the M_{1n}-M_{2n} pair is off whereas the M_{1p}-M_{2p} pair is fully operational, so we rewrite Eq. (5.38) as $G_m = g_{m1p}$.
- Raising v_{IC} above 1 V will gradually turn on the M_{1n}-M_{2n} pair until both pairs are fully operational once we pass the 1.25-V mark. For 1.25 V $\leq v_{IC} \leq 1.75$ V the two pairs work in tandem, giving $G_m = g_{m1p} + g_{m1n}$.
- Raising v_{IC} above 1.75 V will gradually turn off the M_{1p}-M_{2p} pair while the M_{1n}-M_{2n} pair continues to be fully operational.
- For $v_{IC} > V_{DD} - V_{OV11p} - |V_{tp}|$ ($=2$ V for $V_{DD} = 3$ V) the M_{1p}-M_{2p} pair is totally off, so the circuit gives $G_m = g_{m1n}$.
- So long as the M_7-M_8 and the M_9-M_{19} pairs are biased at the edge of saturation, the output voltage swing (OVS) will extend to within two V_{OV}s of either supply rail. If a true rail-to-rail OVS is desired, then an ad-hoc output stage is required, such as the CS push-pull output stage of Section 4.11.

Shown in Fig. 5.55 is the bipolar version of Fig. 5.54. To maximize the IVR, V_{B34} is set to bias the Q_9-Q_{10} pair slightly above the edge of saturation, say at $V_{CE9} = V_{CE10} \cong 0.25$ V. Likewise, V_{B56} is set so that $V_{EC7} = V_{EC8} \cong 0.25$ V. Then, assuming B-E voltage drops of 0.65 V, we have $v_{IC(min)} = V_{EE} + 2 \times 0.25 - 0.65 = V_{EE} - 0.15$ V

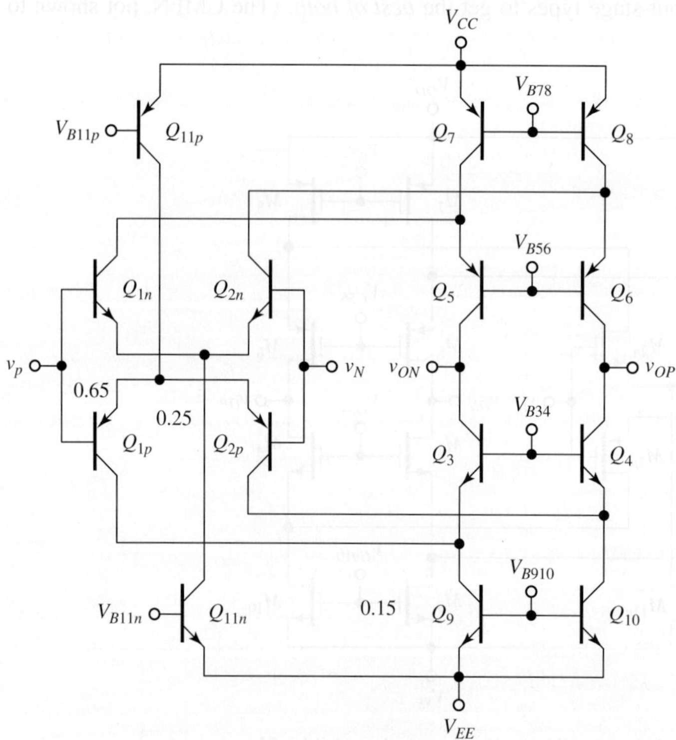

FIGURE 5.55 Bipolar op amp of with a rail-to-rail IVR (CMFN not shown).

and $v_{IC(\max)} = V_{DD} + 0.15$ V. In some embodiments the Q_7-Q_8 and Q_9-Q_{10} pairs are replaced by *resistor pairs*[1] to develop the voltage drops of about 0.25 V that are required to prevent the input pairs from saturating in the vicinity of the supply rails.

A notorious issue with the op amps of Figs. 5.54 and 5.55 is that near the lower end of the IVR the input offset voltage of the *p*-type pair dominates, whereas near the upper end that of the *n*-type pair dominates. So, as v_{IC} changes from one end to the other of the IVR, the net input offset voltage will generally change in magnitude and possibly in polarity, depending on the magnitudes and directions of the individual mismatches. In the bipolar case also the input bias current may pose problems: at the lower end it coincides with the base current of the *p*-type pair, flowing *out of* the op amp, but at the higher end it coincides with that of the *n*-type pair, flowing *into* the op amp. At intermediate IVR values both pairs are on and the base currents of the *p*-type pair tend to cancel those of the *n*-type pair. Consequently, input bias-current magnitude and direction are strong functions of v_{IC}.

5.8 SWITCHED-CAPACITOR CIRCUITS

Traditionally, electronic systems have been designed using off-the-shelf analog and digital ICs, along with passive components such as resistors and capacitors, all mounted on a common medium such as a printed-circuit board (PCB). Nowadays, for reasons of cost, miniaturization, power consumption, and reliability, the trend is to integrate both analog and digital functions on the *same silicon chip* to create circuits aptly called *mixed-signal* ICs. Given that digital electronics is dominated by CMOS technology, analog functions must be implemented using exclusively the components most easily available in this technology, namely, MOSFETs and capacitors (to keep die size within reason, capacitances must be limited to a few tens of picofarads or less). *Resistors and larger capacitors must therefore be avoided altogether*. Figure 5.56 shows the conceptual structure of these two basic components (the shaded capacitances represent parasitics that will be discussed later). The type of capacitor shown is referred to as *double-polysilicon capacitor* because it is made up of two polysilicon layers, the same material type as the gate (being of the n^+ type, these layers can for all purposes be regarded as metal plates).

The most common applications of the MOSFET are *amplification* and *switching*. We have already investigated switching in the design of the logic gates of Chapter 3, and amplification in the design of CMOS op amps such as the cascode op amp of Fig. 5.17. A unique advantage of CMOS op amps is that they exhibit the virtually infinite input resistances presented by the gate terminals of the input FETs. This makes it possible to drive the op amp with voltages stored in on-chip capacitors with very small leakage. Also, the reason for cascoding is to raise the resistance levels in order to achieve high dc gain. If we were to apply resistive feedback around such an op amp, output loading by the external resistances would be so severe as to invalidate the high-gain advantages accrued via cascoding. Mercifully, capacitive feedback (as opposed to resistive feedback) loads down the op amp only during transients. Once steady state is achieved, capacitors act as open circuits, thus preserving high dc gains. Another important advantage of cascode op amps, not immediately obvious at this

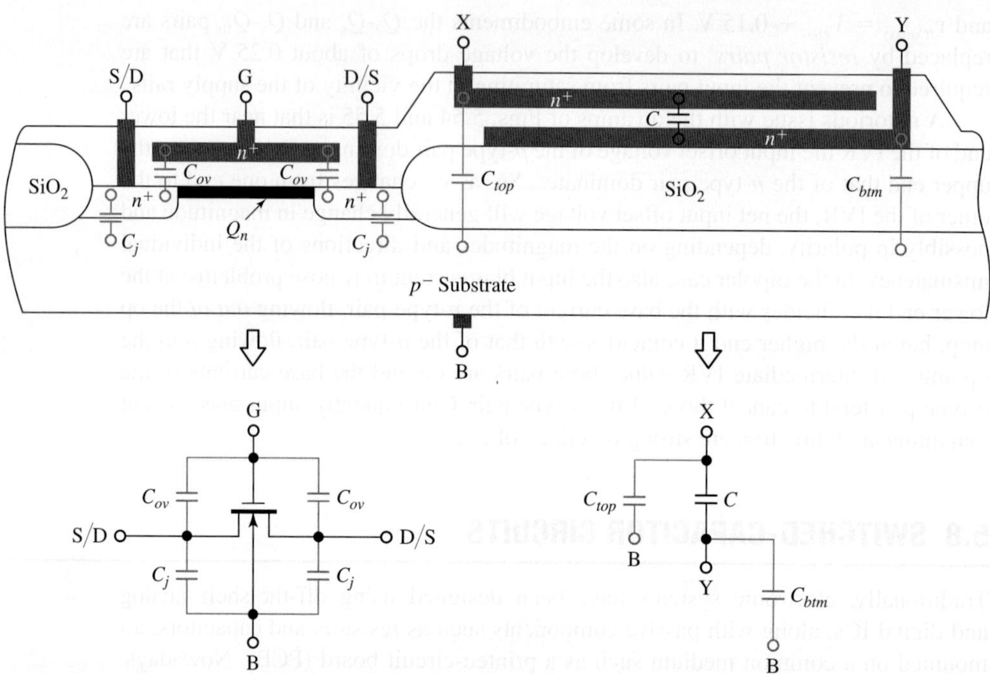

FIGURE 5.56 Conceptual structure of an *n*MOSFET and a double-poly capacitor, showing the parasitic overlap capacitances C_{ov} and parasitic top-plate/bottom-plate capacitances C_{top} and C_{btm}.

juncture but to be investigated in great detail in Chapter 7, is that capacitive loading reinforces their ability to stave off the possibility of oscillation, a feature not available in most other op amp types. In summary, the on-chip components available to the mixed-signal IC designer are (*a*) CMOS op amps, especially cascode types, (*b*) small capacitors (on the order of picofarads or less), and (*c*) MOSFET switches. Let us now address the switched-capacitor concept.

The Switched Capacitor

Figure 5.57 shows a basic switched-capacitor arrangement, along with the gate drive for the two FETs. We make the following observations:

- When ϕ is high, M_1 is on while M_2 is off because $\bar{\phi}$ is low. M_1 charges C toward v_1, and when v_C comes within v_{OV1} of v_1, M_1 enters the triode region, where it approximates a resistance r_{DS1} as in Fig. 5.58*a*. If ϕ is kept high for a duration sufficiently longer than the time constant $\tau_1 = r_{DS1}C$, we can say that by the time ϕ returns low we practically have $v_C = v_1$.
- When $\bar{\phi}$ is high the situation reverses as in Fig. 5.58*b*, so now M_2 charges C toward v_2. Assuming $\bar{\phi}$ is kept high long enough compared to $\tau_2 = r_{DS2}C$ to fully

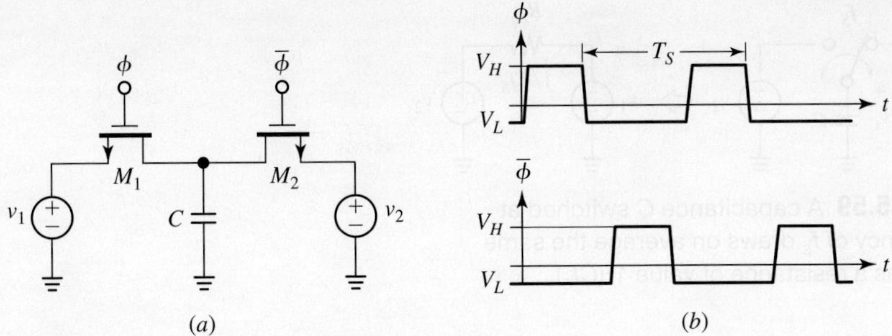

FIGURE 5.57 (a) Basic switched capacitor and (b) non-overlapping clock drive.

charge C to v_2, we can say that by the time $\bar{\phi}$ returns low there has been a charge transfer ΔQ from v_1 to v_2 such that

$$\Delta Q = C(v_1 - v_2) \tag{5.82}$$

(This, assuming $v_1 > v_2$. If $v_1 < v_2$, then charge transfer is from v_2 to v_1.)

- If the process is repeated at a rate of f_s $(=1/T_s)$ cycles/sec, then the charge transferred in one second is $f_s\Delta Q = f_s C(v_1 - v_2)$. But, charge/second represents *average current*, $i_{\text{avg}} = f_s C(v_1 - v_2)$, suggesting that the switched-capacitor block acts like an *equivalent resistance* $R_{eq} = (v_1 - v_2)/i_{\text{avg}}$, or

$$R_{eq} = \frac{1}{Cf_s} \tag{5.83}$$

This equivalence is depicted in Fig. 5.59, where the MOSFET pair of Fig. 5.57a is shown as a *break-before-make switch* of the single-pole double-throw (SPDT) type. We must prevent the MOSFETs from being on simultaneously as this would short the v_1 and v_2 sources together and invalidate Eq. (5.82). Hence, the need for *non-overlapping* gate drives in the manner of Fig. 5.57b.

A popular application of the simulated resistor of Eq. (5.82) is in the synthesis of filters, of which the integrator to be investigated next is the most basic building block.

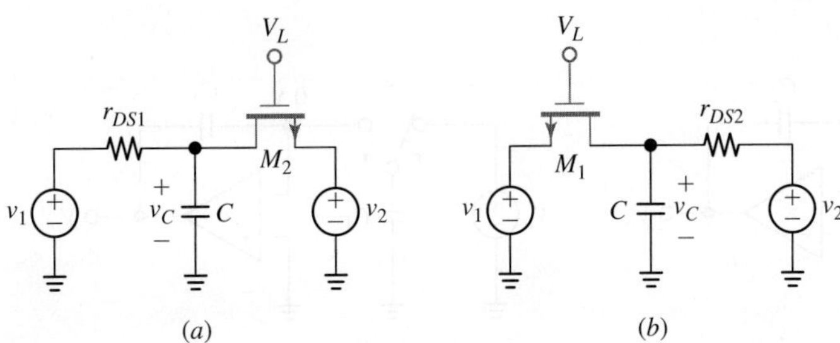

FIGURE 5.58 Circuit equivalent of the switched capacitor of Fig. 5.57a when (a) ϕ is high and (b) $\bar{\phi}$ is high.

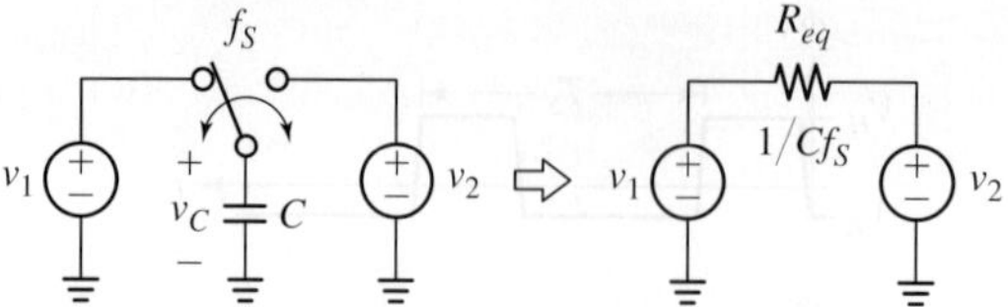

FIGURE 5.59 A capacitance C switched at a frequency of f_S draws on average the same current as a resistance of value $1/(Cf_S)$.

Switched-Capacitor Integrator

The classic integrator of Fig. 5.60a is also called *RC integrator* because we use the external *R-C* network to configure the op amp for integration. Applying KCL we get $(V_i - 0)/R_1 = (0 - V_o)/[1/(sC_2)]$. Letting $s \to j2\pi f$ and simplifying gives the *inverting-type integrator transfer function*

$$H_{RC}(jf) = \frac{V_o}{V_i} = -\frac{1}{jf/f_0} \tag{5.84}$$

where

$$f_0 = \frac{1}{2\pi R_1 C_2} \tag{5.85}$$

is called the *unity-gain frequency* because $|H_{RC}(jf_0)| = 1$. We now wish to rephrase the circuit of Fig. 5.60a in switched-capacitor (SC) form so we can fabricate it entirely on chip. To achieve this we replace R_1 with a *simulated* resistor as in Fig. 5.60b, and we scale C_2 to a small enough value so we can realistically fabricate it on chip. Using Eq. (5.83) to write $R_1 = 1/(C_1 f_S)$ and substituting into Eq. (5.85) we get

$$f_0 = \frac{1}{2\pi} \frac{C_1}{C_2} f_S \tag{5.86}$$

We make the following important observations:

- The *SC* integrator uses *no resistors*, a definite advantage because accurate and stable resistors are difficult to fabricate on chip. The *SC* integrator uses MOSFETs and capacitors instead.

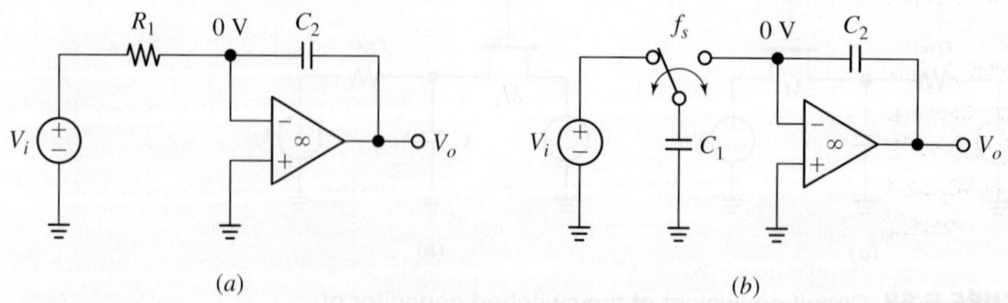

(a)   (b)

FIGURE 5.60 (a) The *RC* integrator, and (b) its switched-capacitor (SC) implementation.

- The unity-gain frequency f_0 is established by the product R_1C_2 in the RC integrator, but by the *ratio* C_2/C_1 in the SC integrator. Compared to products, ratios are much easier to control during fabrication and to maintain with temperature and time. Capacitance matching within 0.5% is easily achievable in MOS technology.[2] This is another important advantage of SC integrators.
- The unity-gain frequency f_0 of the SC integrator is directly proportional to the switching frequency f_S, so, if desired, we can program f_0 *automatically* over a wide range of values by suitably varying f_S.

EXAMPLE 5.15

(a) Specify component values for $f_0 = 10$ kHz in the integrator of Fig. 5.60*a*, and comment.

(b) Repeat, but for the circuit of Fig. 5.60*b* if $f_S = 1$ MHz. Do not exceed a total capacitance of 10 pF.

Solution

(a) Arbitrarily choose $C_2 = 1$ nF. Then Eq. (5.85) gives,

$$R_1 = \frac{1}{2\pi C_2 f_0} = \frac{1}{2\pi 10^{-9} \times 10^4} = 15.9 \text{ k}\Omega$$

Both values are easily available off the shelf, but C_2 could hardly be fabricated on chip. Moreover, because of component tolerances, f_0 may require tuning (via R_1), and will drift with temperature because so do R_1 and C_2.

(b) By Eq. (5.86) we have

$$\frac{C_2}{C_1} = \frac{f_S}{2\pi f_0} = \frac{10^6}{2\pi 10^4} = 15.9$$

Arbitrarily pick $C_1 = 0.5$ pF, so $C_2 = 7.958$ pF, for a total capacitance of less than 10 pF, which we can readily fabricate on chip.

Discrete-Time Considerations

The SC integrator is a *discrete-time* circuit because the current between the input source and the op amp flows in the form of charge packets. (Conversely, the RC integrator is a *continuous-time* circuit because so is the current flowing via R_1.) We now wish to investigate how charge discretization affects the transfer function of the SC integrator compared to its RC counterpart. To this end we use the PSpice circuit of Fig. 5.61 and turn our attention to the waveforms of Fig. 5.62. As shown, C_1 switches back and forth between v_I and the op amp's virtual ground (0 V) at a frequency of $f_S = 10$ MHz, so we divide the time axis into intervals of $T_S = 1/f_S = 1/10^7 = 100$ ns each. For $v_I > 0$, C_1's charge is dumped *into* C_2, causing a *negative-going* step in v_O. Conversely, for $v_I < 0$, charge is pulled *out* of C_2, yielding *a positive-going* step in v_O. Of particular significance are the *falling edges* of the v_{G1} pulse train because these are the instants at which C_1, after having been fully charged to v_I, is ready to be discharged into C_2. Labeling these instants sequentially as $\ldots n-1, n, n+1 \ldots$

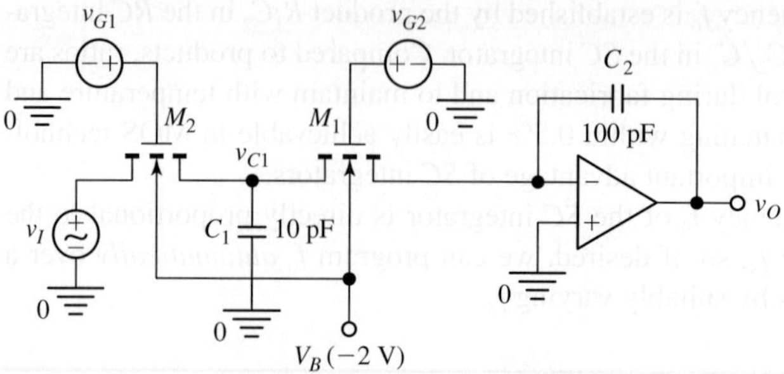

FIGURE 5.61 PSpice circuit to investigate the *SC* integrator in the time domain.

we can say that the change experienced by v_O from the instant $n - 1$ to the instant n is set by the charge $C_1 v_I(n - 1)$ accumulated in C_1 at the instant $n - 1$ according to

$$v_O(n) - v_O(n - 1) = \frac{-C_1 v_I(n - 1)}{C_2} \qquad (5.87)$$

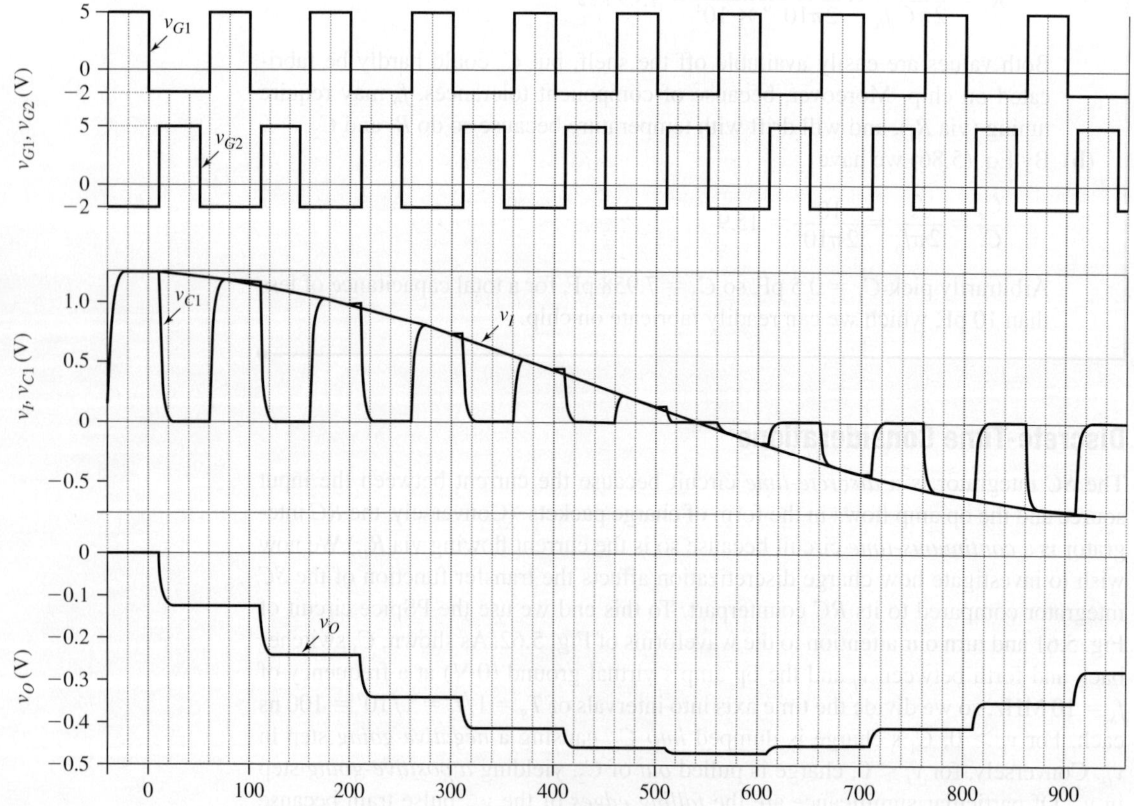

FIGURE 5.62 Waveforms of the PSpice circuit of Fig. 5.61.

This *discrete-time sequence* is best investigated via the Fourier transforms $V_i(j\omega)$ and $V_o(j\omega)$. A well-known property of these transforms states that delaying a signal by one clock period T_S is equivalent to multiplying its Fourier transform by $\exp(-j\omega T_S)$. This means that Eq. (5.87) transforms into

$$V_o(j\omega) - V_o(j\omega)e^{-j\omega T_S} = -\frac{C_1}{C_2}V_i(j\omega)e^{-j\omega T_S}$$

Collecting and simplifying we get

$$V_o(j\omega) = -\frac{C_1}{C_2}V_i(j\omega)\frac{e^{-j\omega T_S}}{1 - e^{-j\omega T_S}} = -\frac{C_1}{C_2}V_i(j\omega)\frac{e^{-j\omega T_S/2}}{e^{j\omega T_S/2} - e^{-j\omega T_S/2}}$$

Letting $\omega = 2\pi f$, $T_S = 1/f_S$, and using the identity $\sin x = (e^x - e^{-x})/2j$ we get, after some algebra,

$$H_{SC}(jf) = \frac{V_o(jf)}{V_i(jf)} = -\frac{1}{jf/f_0} \times \frac{\pi f/f_S}{\sin(\pi f/f_S)} \times e^{-j\pi f/f_S} \qquad (5.88)$$

where again

$$f_0 = \frac{1}{2\pi}\frac{C_1}{C_2}f_S \qquad (5.89)$$

We observe that in the limit $f \ll f_S$ we have $H_{SC}(jf) \to H_{RC}(jf)$, that is, the *SC* integrator approximates the *RC* integrator. Physically this makes sense because for $f_S \gg f$ charge transfer can be regarded as a continuous process. Otherwise, $H_{SC}(jf)$ will deviate from $H_{RC}(jf)$ by the *magnitude* and *phase* errors

$$\varepsilon_m(f) = \frac{\pi f/f_S}{\sin(\pi f/f_S)} - 1 \qquad \varepsilon_\phi(f) = -180°(f/f_S) \qquad (5.90)$$

Figure 5.63 compares the transfer function magnitudes for the case $f_S = 10f_0$.

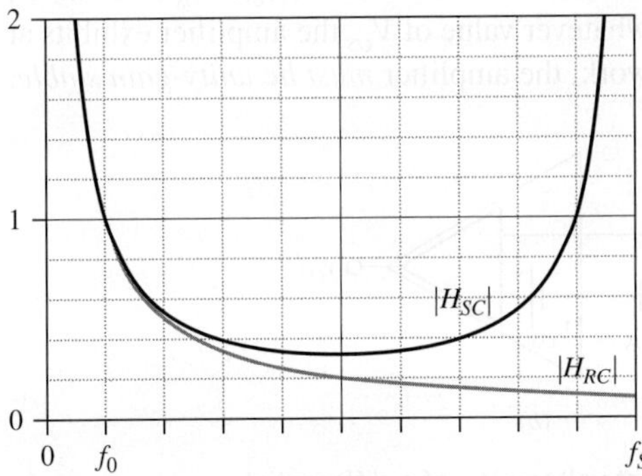

FIGURE 5.63 Comparing the transfer functions of the *RC* and *SC* integrators.

EXAMPLE 5.16   (a) Calculate the deviation of $H_{SC}(jf)$ from $H_{RC}(jf)$ in Fig. 5.63 at $f = f_0$ and at $f = 2f_0$. Compare and comment.

(b) Repeat, but for the circuit of Fig. 5.61. Compare with (a) and comment.

Solution

(a) In Fig. 5.63 we have $f_0/f_S = 1/10$, so Eq. (5.90) gives $\varepsilon_m(f_0) = (\pi/10)/\sin(\pi/10) - 1 = 1.0166 - 1 = 1.66\%$ and $\varepsilon_\phi(f_0) = -180/10 = -18°$. Similarly, $\varepsilon_m(2f_0) = 6.9\%$ and $\varepsilon_\phi(2f_0) = -36°$, indicating a doubling in ε_ϕ but a much greater increase in ε_m.

(b) By Eq. (5.89) the circuit of Fig. 5.61 has $f_0/f_S = (C_1/C_2)/(2\pi) = (10/100)/(2\pi) = 1/62.83$, so we now get $\varepsilon_m(f_0) = 0.0417\%$ and $\varepsilon_\phi(f_0) = -2.86°$. Also, $\varepsilon_m(2f_0) = 0.167\%$ and $\varepsilon_\phi(2f_0) = -5.73°$. The deviations are now appreciably smaller because the ratio f_0/f_S is 2π times as small.

Autozeroing Techniques

Because of component mismatches stemming from fabrication process variations, a differential amplifier such as an op amp or a voltage comparator exhibit some input offset voltage V_{OS}. As shown in Fig. 5.64a, we can visualize a real-life amplifier as consisting of an ideal, offsetless amplifier but equipped with an internal voltage source V_{OS} in series with one of its inputs to account for the internal imbalances (it doesn't matter which input, as V_{OS} is unpredictable both in magnitude and polarity). The presence of V_{OS} is likely to be a serious drawback in precision applications, so it is desirable to automatically compensate for its presence, a task referred to as *autozeroing*. CMOS circuits utilize a capacitor C as in Fig. 5.64b to store a voltage of equal magnitude but opposite polarity as V_{OS}. Being in series, the two voltages cancel each other out, giving the appearance of offset-free behavior. The amplifier alternates between two modes of operation known as the *autozeroing* and the *sampling* mode:

- During the autozero phase depicted in Fig. 5.65a, switches S_1 and S_2 disconnect the amplifier from nodes v_P and v_N while S_4 configures it as a unity-gain voltage follower. The circuit now buffers its own internal offset voltage V_{OS} to C via S_3, so with S_5 closed C will charge to whatever value of V_{OS} the amplifier exhibits at that moment. (For this scheme to work, the amplifier *must be unity-gain stable.*

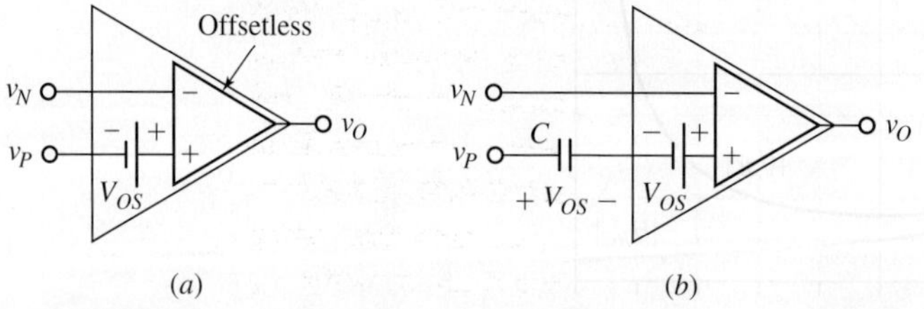

FIGURE 5.64 (a) Modeling the input offset voltage V_{OS} of a differential amplifier, and (b) nulling the effect of V_{OS} by using a capacitor to store a corrective voltage of equal magnitude but opposite polarity as V_{OS}.

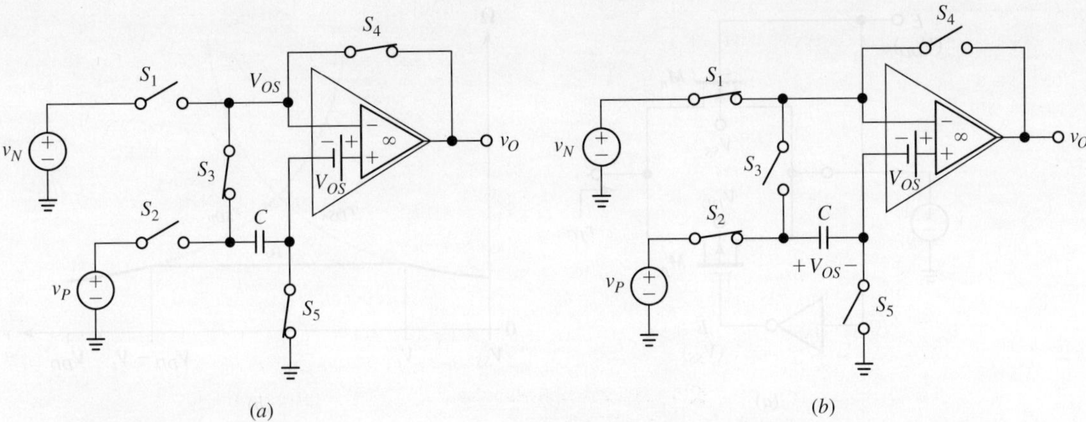

FIGURE 5.65 Switch positions during the (a) *autozeroing mode* and (b) during the *sampling mode*.

If necessary, a suitable frequency-compensation network must be switched into the amplifier during this phase.)

- Once we have stored V_{OS} in the capacitor, we open switches S_3 through S_5 as in Fig. 5.65b to place the capacitor voltage in series with the internal offset voltage so the two cancel each other out. At the same time, we close S_1 and S_2 to reconnect the amplifier to nodes v_P and v_N for normal operation, such as negative-feedback op amp operation or voltage comparison. The scheme shown is popular especially with voltage comparators, where the circuit is autozeroed before each comparison instance. Should V_{OS} drift with temperature or should C experience leakage between consecutive comparisons, autozeroing is renewed each time.

Transmission Gates

The *n*MOSFET switches of Fig. 5.57 will close convincingly only as long as the voltages at the source and drain terminals are *sufficiently lower* than V_H. As v_1 and v_2 increase, the overdrive voltages are reduced, thus increasing the channel resistances. If v_1 or v_2 rise above $V_H - V_{tn}$, the FET won't even turn on and the switch will therefore fail to close. If we were to use *p*MOSFET switches instead, they would work well for v_1 and v_2 sufficiently high, but fail to turn on if v_1 or v_2 drop below $V_L + |V_{tp}|$, just the opposite of *n*MOSFETs. The circuit of Fig. 5.66, popularly known as *transmission gate*, combines the *best of both* by using a pair of complementary FETs connected in parallel and driven in anti-phase.

When the switch-enable control E is low ($E = V_{SS}$) and its complement $\overline{E}$ is therefore high ($\overline{E} = V_{DD}$), both FETs are off. Driving E high and thus $\overline{E}$ low will tend to turn on both FETs. In fact, so long as $V_{SS} + |V_{tp}| < v < V_{DD} - V_{tn}$, both FETs are indeed on, ensuring a net transmission-gate resistance r_{TG} that, for the case of matched devices with $k_n = k_p = k$ and $V_{tn} = -V_{tp} = V_t$, is such that

$$r_{TG} = r_{DSn} /\!/ r_{SDp} = \frac{1}{kv_{OVn}} /\!/ \frac{1}{kv_{OVp}} = \frac{1}{k(V_{DD} - v - V_t)} /\!/ \frac{1}{k(v - V_{SS} - V_t)}$$

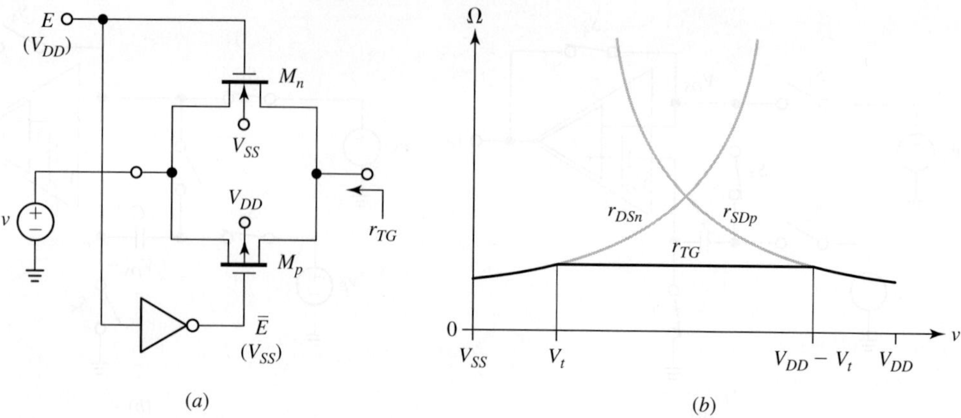

(a)   (b)

FIGURE 5.66 (a) Transmission gate and (b) its zero-current resistance r_{TG}, along with the individual channel resistances r_{DSn} and r_{SDp}.

Expanding and simplifying gives

$$r_{TG} = \frac{1}{k(V_{DD} - V_{SS} - 2V_t)} \tag{5.91}$$

Under the assumption of matched FETs, r_{TG} is *independent* of v over the given interval (see Fig. 5.66*b*). If v is driven outside the above interval, one of the FETs will go off but the other will become even more conductive, thus ensuring a *low* overall resistance over the entire range $V_{SS} < v < V_{DD}$.

Stray Capacitances and Charge Injection

Returning to Fig. 5.56, we observe the presence of a number of stray or parasitic capacitances:

- The top and bottom plates of C form themselves stray capacitances with the substrate, here denoted as C_{top} and C_{btm}. While C_{btm} is typically 10%-20% of C, C_{top} is considerably smaller.
- The sides of the gate of the MOSFET switch form the *overlap capacitances* C_{OV} with the source and drain regions, and these regions form in turn the *diffusion capacitances* C_j with the substrate.
- Additionally, we observe that in order to turn on/off the *n*MOSFET we need to build up or remove the electron charge Q_n in the channel (see Example 3.3). Q_n enters/exits the channel through the source and drain regions.

Depending on circuit topology, the stray capacitances may interfere with the intended charge transfer and cause errors. In the switch/capacitor example of Fig. 5.67 the bottom-plate capacitance C_{btm} plays no role because shorted to ground. Moreover, the overlap and junction capacitances associated with the source region, relabeled as C_{gs} and C_{sb}, affect the input source v_1 but have no impact on the useful capacitance C. However, the remaining parasitics do play a role. Specifically, C_{top} and C_{db} are in parallel with C, thus yielding a total capacitance $C_{tot} = C_{db} + C_{top} + C$. Moreover,

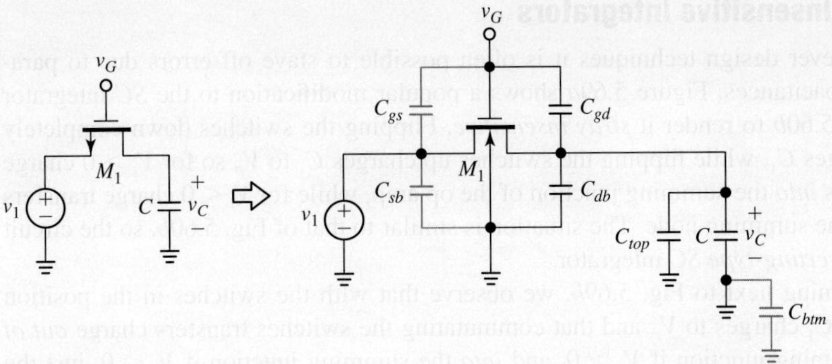

FIGURE 5.67 Showing the capacitive parasitics of a simple switch-capacitor pair (the capacitances shown in shaded form have no effect in this particular circuit).

C_{gd} couples the clock signal v_G to C_{tot}, a phenomenon referred to as *clock feedthrough*. Additionally, turning the switch off/on causes a fraction α $(0 < \alpha < 1)$ of the total electron charge Q_n in the channel to flow to/from C_{tot}, a phenomenon referred to as *charge injection*.

We are particularly interested in the trailing edge of v_G because at this instant, denoted as t_{OFF}, we witness (*a*) the removal of the charge Q_{gd} from C_{tot} and (*b*) the injection of the charge αQ_n into C_{tot} (see Fig. 5.68*a*). Since the electron charge Q_n is negative, the effect is equivalent to removing the positive charge $\alpha|Q_n|$ from C_{tot}, so the two undesired charge transfers reinforce each other. As a result, v_O will be affected by an error of $\Delta v_O \cong -(Q_{gd} + \alpha|Q_n|)/C_{tot}$, where $C_{gd} \ll C_{tot}$ has been assumed.

One way to reduce the above error is to deliberately introduce an opposing clock feedthrough and charge injection via a dummy transistor M_2 driven in anti-phase with respect to M_1 (see Fig. 5.68*b*). With proper selection of M_2's channel width W_2, M_2's feedthrough and injection will approximately cancel out those of M_1, resulting in a much smaller error in v_O (see Problem 5.50). Alternatively, we can implement the switch with a transmission gate and adjust the W/L ratios of the nFET and pFET so as to make their undesirable charge transfers approximately cancel each other out.

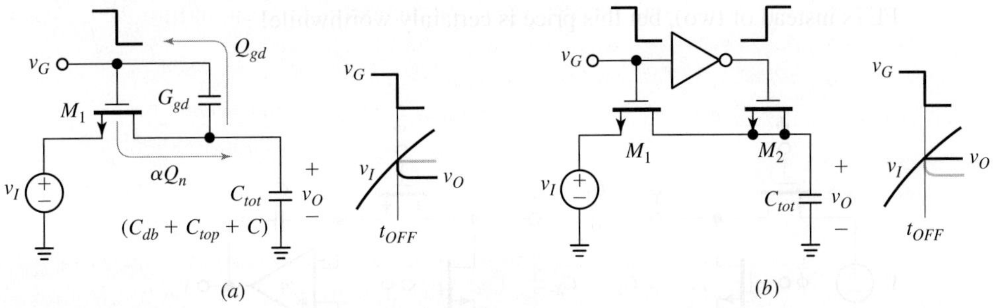

(*a*)   (*b*)

FIGURE 5.68 (*a*) Charge transfer (Q_{gd}) and charge injection (Q_n) at the instant t_{OFF} in which the M_1 switch is turned off. (*b*) Compensation via a second dummy transistor M_2 driven in anti-phase with respect to M_1.

Stray-Insensitive Integrators

With clever design techniques it is often possible to stave off errors due to parasitic capacitances. Figure 5.69a shows a popular modification to the *SC* integrator of Fig. 5.60b to render it *stray insensitive*. Flipping the switches down completely discharges C_1, while flipping the switches up charges C_1 to V_i, so for $V_i > 0$ charge transfers *into* the summing junction of the op amp, while for $V_i < 0$ charge transfers *out of* the summing node. The situation is similar to that of Fig. 5.60b, so the circuit is an *inverting-type SC* integrator.

Turning next to Fig. 5.69b, we observe that with the switches in the position shown, C_1 charges to V_i, and that commutating the switches transfers charge *out of* the summing junction if $V_i > 0$, and *into* the summing junction if $V_i < 0$, just the opposite of Fig. 5.69a. So, the circuit is a *noninverting-type SC* integrator, obtained from that of Fig. 5.69a by a mere clock-phase change for the second SPDT switch.

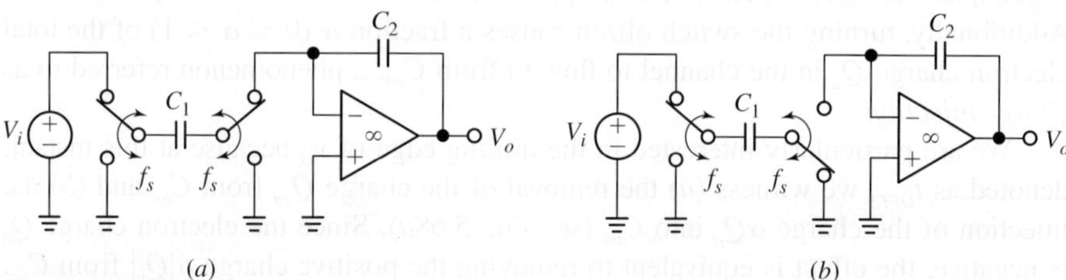

FIGURE 5.69 Stray-insensitive *SC* integrators: (a) inverting type and (b) noninverting type.

To appreciate the stray insensitivity of the above integrators, consider Fig. 5.70, where the total stray capacitances (top/bottom parasitics plus FET parasitics) associated of the two plates of C_1 have been denoted as C_X and C_Y. Since C_X is switched between ground and the op amp's virtual-ground input, it remains permanently discharged, so the circuit is insensitive to C_X. C_Y does intervene in the transfer of charge between V_i and ground, but without interfering with C_1, so the circuit is insensitive also to C_Y. The price for stray insensitivity is a more complex switch system (four FETs instead of two), but this price is certainly worthwhile!

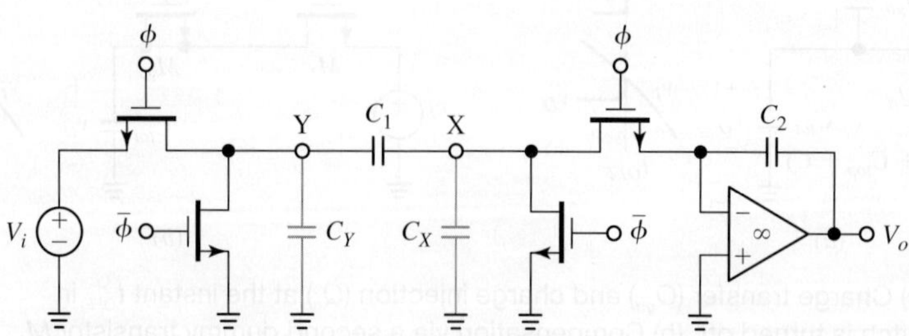

FIGURE 5.70 Investigating stray insensitivity.

SPICE Macro-Models

Used judiciously, computer simulators such as SPICE can be very powerful aids both to the designer and the user of analog ICs. The IC designer will simulate the product under development at the *transistor level*, using device parameters derived from direct measurements of the fabrication process, usually proprietary. The IC user, on the other hand, is more interested in a *behavioral* simulation of the IC, without having to worry about too many internal details. Besides, the transistor-level simulations of complex system comprising a multitude of individual ICs may be excessively time consuming, not to mention the fact that the numerical calculations may fail to converge.

To alleviate the user's task, IC manufacturers provide SPICE *macro-models* that can be downloaded directly from their websites for immediate use. PSpice's library contains the macro-models of a number of popular ICs such as the 741 op amp. Its netlist is as follows:

```
* connections:    non-inverting input
*                 | inverting input
*                 | | positive power supply
*                 | | | negative power supply
*                 | | | | output
*                 | | | | |
.subckt uA741     1 2 3 4 5
*
  c1     11    12    8.661E-12
  c2      6     7    30.00E-12
  dc      5    53    dx
  de     54     5    dx
  dlp    90    91    dx
  dln    92    90    dx
  dp      4     3    dx
  egnd   99     0    poly(2) (3,0) (4,0) 0 .5 .5
  fb      7    99    poly(5) vb vc ve vlp vln 0 10.61E6 -10E6 10E6
  +                  10E6 -10E6
  ga      6     0    11 12 188.5E-6
  gcm     0     6    10 99 5.961E-9
  iee    10     4    dc 15.16E-6
  hlim   90     0    vlim 1K
  q1     11     2    13 qx
  q2     12     1    14 qx
  r2      6     9    100.0E3
  rc1     3    11    5.305E3
  rc2     3    12    5.305E3
  re1    13    10    1.836E3
  re2    14    10    1.836E3
```

```
ree     10    99    13.19E6
ro1      8     5    50
ro2      7    99    100
rp       3     4    18.16E3
vb       9     0    dc 0
vc       3    53    dc 1
ve      54     4    dc 1
vlim     7     8    dc 0
vlp     91     0    dc 40
vln      0    92    dc 40
.model dx D(Is=800.0E-18 Rs=1)
.model qx NPN(Is=800.0E-18 Bf=93.75)
.ends
```

Far from including all the 24 BJTs of Fig. 5.1, the macro-model uses only *two* BJTs (q1 and q2) to emulate the input-stage front end, and a number of other simpler components (resistors, capacitors, diodes, and dependent/independent sources) to mimic op amp behavior at a macroscopic level (gain, output saturation, and dynamic characteristics such as input/output impedances, frequency response, and slew-rate limiting, to be covered in Chapter 6). Compared to the transistor-level simulation by the IC designer, the macro-model-level simulation by the user is generally much faster. However, both the designer and the user need to be aware that all models come with inherent limitations, so the results of any simulation need to be properly evaluated and eventually verified in the lab. (For instance, a notorious limitation of many op amp macro-models is their failure to mimic the input noise characteristics correctly.)

REFERENCES

1. P. R. Gray, P. J. Hurst, S. H. Lewis, and R. G. Meyer, *Analysis and Design of Analog Integrated Circuits*, 5/E, Wiley and Sons, 2009.
2. P. E. Allen and D. R. Holberg, *CMOS Analog Circuit Design*, 2/E, Oxford University Press, 2002.
3. D. A. Johns and K. Martin, *Analog Integrated Circuit Design*, Wiley and Sons, 1997.
4. B. Razavi, *Design of Analog CMOS Integrated Circuits*, McGraw-Hill, 2001.
5. A. S. Sedra and K. C. Smith, *Microelectronic Circuits*, 6/E, Oxford University Press, 2010.
6. R. C. Jaeger and T. N. Blalock, *Microelectronic Circuit Design*, 2/E, McGraw-Hill, 2004.
7. S. Franco, *Design with Operational Amplifiers and Analog Integrated Circuits*, 4/E, McGraw-Hill, 2014.
8. C. Toumazou, F. J. Lidgey, and D. G. Haigh, Eds., *Analogue IC Design: The Current-Mode Approach*, Peter Peregrinus Ltd., 1990.
9. H. Camenzind, *Designing Analog Chips*, www.designinganalogchips.com, 2005.

PROBLEMS

5.1 The μA741 Operational Amplifier

5.1 (*a*) Suppose the input stage of Fig. 5.6 is redesigned by dropping Q_1 and Q_2 and by operating Q_3 and Q_4 as an EC pair biased at $I_{EE} = 19$ μA. Show the modified circuit and use the device parameters of the text, along with $\beta_{F3} = \beta_{F4} = 50$, to recalculate I_P, I_N, R_{id}, G_{m1}, R_{o1}, and the unloaded gain $v_{o1(oc)}/(v_p - v_n)$.

(*b*) Assuming the B-E junctions of the *pnp* BJTs break down at 7 V and those of the *npn* BJTs break down at 20 V, what is the maximum voltage difference $(v_P - v_N)_{max}$ that can safely be applied to the modified circuit of (*a*) before breakdown occurs?

  (*c*) Compare all the above parameters with those of the original 741 design, and give reasons why the original design is generally preferable.

5.2 (*a*) Assuming $\beta_{F16} = 200$, verify that I_{B16} in the 741 circuit of Fig. 5.1 is negligibly small compared to I_{C4} and I_{C6}. This allows us to define the first-stage input offset voltage V_{OS1} as the voltage difference $V_P - V_N$ needed to make $I_{C4} = I_{C6}$ (not necessarily $= 9.5\ \mu\text{A}$) at $V_{C4} = V_{C6} = V_{BE16} + V_{BE17} + V_{EE}$.

  (*b*) Based on the above definition, investigate the effect of a 10% mismatch between I_{s1} and I_{s2}, assuming no other mismatches in the circuit.

  (*c*) Repeat, but for a 10% mismatch between I_{s3} and I_{s4}.

  (*d*) Repeat, but for a 10% mismatch between I_{s5} and I_{s6}.

  (*e*) Repeat, but for a 10% mismatch between R_1 and R_2.

5.3 (*a*) In the text Q_1 and Q_2 are assumed to be matched with $\beta_{F1} = \beta_{F2} = 200$. Consider a particular op amp sample having $\beta_{F1} = 175$ and $\beta_{F2} = 225$. What are I_P, I_N, I_B, and I_{OS}?

  (*b*) What is the range of allowed values for β_{F1} and β_{F2} if we wish to ensure $I_B \le 50$ nA and $I_{OS} \le 5$ nA?

  (*c*) Nominally, Q_3 and Q_4 have $\beta_{F3} = \beta_{F4} = 50$. Suppose that because of a fabrication glitch a certain 741 sample has $\beta_{F3} = \beta_{F4}/2 = 25$. Use the definition of V_{OS1} of Problem 5.2 (*a*) to show that this mismatch produces an input offset voltage, and calculate it.

5.4 (*a*) Based on the definition of V_{OS1} of Problem 5.2 (*a*), calculate V_{OS1} if R_2 is shorted out in Fig. 5.5. What is the new value of $I_{C6}\ (=I_{C4})$?

  (*b*) Repeat if R_2 is lowered from 1 kΩ to 0.5 kΩ.

  (*c*) Repeat (*a*) and (*b*), but for R_1 instead of R_2.

  (*d*) Having demonstrated that altering the ratio R_1/R_2 unbalances the circuit by creating an offset, we can exploit this feature for *nulling* an existing offset, that is, for ensuring $I_{C6} = I_{C4}$ with $V_P = V_N$. If $V_{OS1} = +5$ mV, which of R_1 or R_2 must be lowered to *null* V_{OS1}? What is the lowered resistance value, and what is the new value of $I_{C6}\ (=I_{C4})$?

5.5 In an attempt to simplify the 741 op amp, consideration is given to dropping the emitter follower

Q_{22} altogether and replacing its B-E junction by a plain wire.

  (*a*) What are the new values of I_{C17} and R_{i3}?

  (*b*) Recalculate the 2nd-stage parameters R_{i2}, G_{m2}, and R_{o2}, and compare them with those of the original circuit.

  (*c*) Find the 2nd-stage *loaded* voltage gain (that is, the gain in the presence of R_{i3}), compare with that of the original design, and comment.

5.6 (*a*) Using the BJT parameters of the text, find R_{10} so that $I_{C19} = 2I_{C18}$ in the 3rd stage. What is the resulting value of V_{BB}?

  (*b*) Using the BJT data of Example 5.1, find the standby currents I_{C14} and I_{C20} under the assumption that R_6 and R_7 can be ignored.

  (*c*) Refine your calculations with R_6 and R_7 in place.

5.7 (*a*) Find the power P_{OA} absorbed by an *unloaded* 741 op amp powered from ±15-V supplies and with the output near 0 V. Hence, verify that the sum of the power released by the supplies equals that absorbed by the op amp.

  (*b*) Find P_{OA} if the op amp is *sourcing* to the surrounding circuit a current of $I_O = 2$ mA at $V_O = 10$ V. Again, verify the conservation of power.

  (*c*) Repeat (*b*) if the op amp is *sinking* from the surrounding circuit a current of $I_O = 1.5$ mA at $V_O = -8$ V.

5.8 (*a*) What value must R_4 be changed to if we want to double I_1 in the Widlar current sink of Fig. 1.4?

  (*b*) How does doubling I_1 affect the parameters of the input stage?

  (*c*) How does it affect the overall characteristics of the op amp?

5.9 (*a*) Which parameters of the 741 op amp are affected if we reduce Q_{17}'s emitter resistance R_8 from 100 Ω to 0 Ω? How is the overall gain affected?

  (*b*) Provide an intuitive justification for your results.

5.10 (*a*) Investigate the change in its dc bias conditions if the 741 op amp is powered by a pair of ±9-V batteries instead of the usual ±15-V power supplies. Hence, assuming the betas of the BJTs are unaffected, give rough estimates of the changes intervening in:

  (*b*) R_{id}, G_{m1}, and R_{o1};

  (*c*) R_{i2}, G_{m2}, R_{o2};

  (*d*) R_{i3} and R_o.

  (*e*) How is the overall gain *a* affected? Justify your results.

5.11 Advanced considerations[1] indicate that reducing the 741 op amp's input-stage transconductance offers certain benefits, such as a slew rate increase and a reduction in the required value of the on-chip compensation capacitance (these issues will be addressed in Chapters 6 and 7). As we know, a common way to reduce transconductance is by adding suitable degeneration resistors in series with the emitters of Q_3 and Q_4. However, resistors are undesirable in IC technology, so a clever technique to lower transconductance is to rob Q_3 and Q_4 of collector current by fabricating each BJT with a double collector, and tying the extra collectors in the manner depicted in Fig. P5.11, where the original collectors are labeled as A and the extra ones as B. (This also results in simpler input-stage circuitry, as the collectors of Q_1 and Q are now tied directly to V_{CC}.)

(a) Assuming $\beta_{Fp} = 50$ and β_{Fn} is very large, and the A and B collectors are fabricated with equal areas, find I_1 so that we still have $I_{C1} = I_{C2} = 9.5\ \mu A$. What is the required value of R_4 in the Widlar sink of Fig. 5.4? What is of the new value of the input-stage transconductance G_{m1}?

(b) Repeat if the areas of the B collectors are fabricated three times as large as the areas of the A collectors.

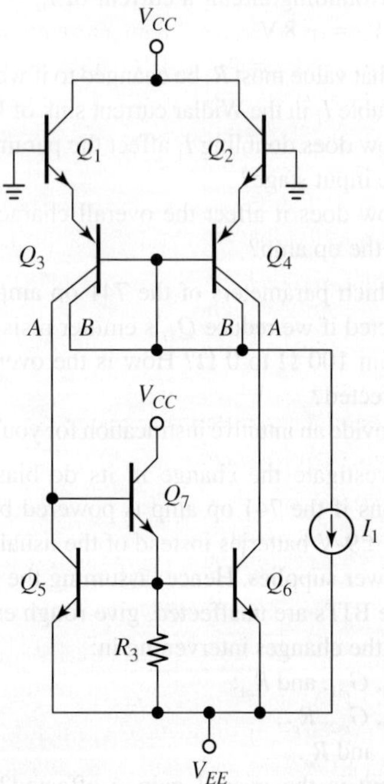

FIGURE P5.11

5.12 Some second-source manufacturers of the 741 op amp use the alternative circuit of P5.12 to establish the voltage drop V_{BB} needed to bias the push-pull pair. The circuit is called V_{BE} *multiplier* because V_{BB} can be adjusted to any (not necessarily integer) multiple of V_{BE} by suitable choice of the R_1/R_2 ratio.

(a) Assuming $I_{s1} = I_{s2} = 4I_{s3} = 8\ \text{fA}$ and Q_3 draws negligible base current, specify suitable values for R_1 and R_2 to achieve $V_{BB} = 1242\ \text{mV}$ as in Example 5.1, under the constraint that the $176\text{-}\mu A$ current split equally between R_1 and Q_3's collector. What is the quiescent current I_Q of the Q_1-Q_2 pair?

(b) Assuming $I_{s4} = I_{s3}$, find v_I (in millivolts) for $v_O = 0$.

(c) Find v_O (in millivolts) if $v_I = 0$ and $R_L = \infty$.

(d) Repeat (c), but with $R_L = 1\ \text{k}\Omega$.

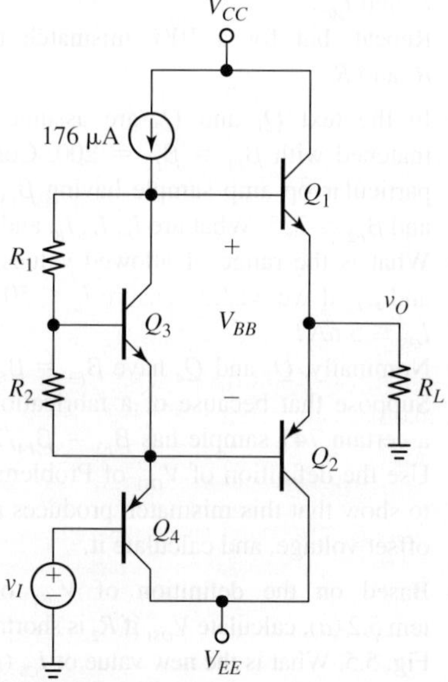

FIGURE P5.12

5.13 In Fig. 5.1 the overload protection circuitry for Q_{20} consists of R_7, Q_{21}, R_{11}, Q_{24}, and Q_{23}. Suppose the op amp is trying to swing v_O negative, but because of an overload condition, Q_4 steers *all* of I_{C8} ($=19\ \mu A$) toward the second stage.

(a) Assuming $I_{s21} = I_{s24} = I_{s23} = 2\ \text{fA}$, find the current drawn by Q_{20} (this is I_{SC} for the current *sinking* case).

Hint: find I_{C23}, I_{s24}, I_{C21}, and finally V_{EB21}.

(b) What value should R_7 be changed to if we want $I_{SC} = 15\ \text{mA}$?

5.14 (**a**) Assuming the inputs and output terminals are at 0-V dc in Fig. P5.14, estimate the small-signal gain $a = v_o/(v_p - v_n)$. Assume $\beta_0 = 100$ and $V_A = \infty$ for all BJTs, and ignore base currents during your dc calculations.

(**b**) Using matched *pnp* BJTs, design a circuit to provide the 0.1 mA and 1 mA bias currents.

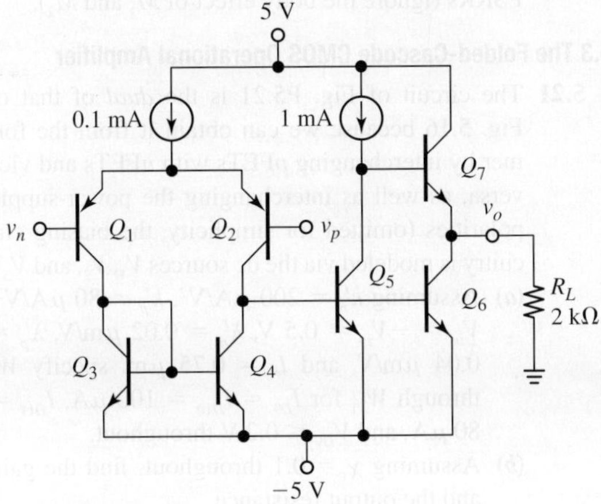

FIGURE P5.14

5.15 The circuit of Fig. P5.15 is known as a *current-differencing amplifier* because it responds to the difference of its input currents to give $v_o = z(i_p - i_n)$, where z is the transimpedance gain, in V/A. The Q_1-Q_2 mirror reverses the direction of i_p so that the current into Q_3's base is $i_{b3} = i_n - i_p$, and Q_3 is a CE amplifier whose output is buffered to the load R_L by the Darlington-like pair Q_4-Q_5. (Omitted for simplicity is the circuitry providing the sources V_1 and V_2 to bias the 200-μA current source Q_6 and the 1.3-mA current sink Q_7). Assuming $\beta_3 = 125$, $\beta_5 = 200$, $\beta_4 = 50$, and $V_{An} = V_{Ap} = 100$ V,

(**a**) find I_{B3} to sustain $V_O = 5$ V;

(**b**) find the voltage gain $a_v = v_o/v_{b3}$;

(**c**) find the transimpedance gain $z = v_o/(i_p - i_n)$.

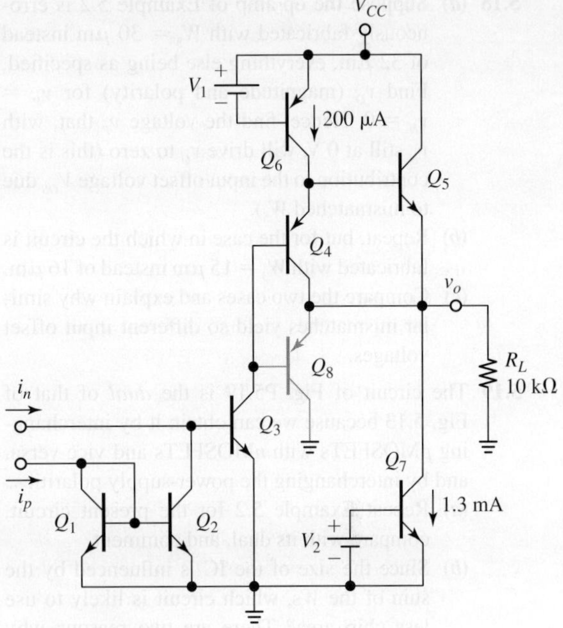

FIGURE P5.15

5.2 The Two-Stage CMOS Operational Amplifier

5.16 Suppose the two-stage op amp of Fig. 5.13 is fabricated in a process characterized by $k'_n = 200$ μA/V^2, $k'_p = 80$ μA/V^2, $V_{tn} = 0.65$ V, $V_{tp} = -0.75$ V, $\lambda'_n = 0.02$ μm/V, $\lambda'_p = 0.04$ μm/V, and $L = 0.75$ μm throughout. Moreover, the circuit is powered from ± 1.65-V supplies and uses $I_{REF} = 150$ μA. If $W_1 = W_2 = W_6 = W_7 = W_8 = 100$ μm, $W_3 = W_4 = W_5/2 = 30$ μm, find the individual-stage gains, the overall gain, the output resistance, the IVR, the OVS, the CMRR, and the PSRRs (ignore the body effect of M_1 and M_2).

5.17 The dc calculations of Example 5.2 assume $\lambda_n = \lambda_p = 0$ and as such they are inadequate to predict the input offset voltage.

(**a**) Following Example 4.7, calculate V_{OV6} and V_{OV5} for $v_P = v_N = 0$ but using the given non-zero values of λ_n and λ_p (you will find that V_{OV6} and V_{OV5} are *slightly less* than 250 mV. Why?). Hence, find v_o (magnitude and polarity).

(**b**) Find the voltage v_P that, with v_N still at 0 V, will drive v_o to zero (this is the contribution to the input offset voltage V_{OS} due to nonzero λ_5 and λ_6). Hence, compare with the PSpice value of Fig. 5.14*b*.

5.18 (*a*) Suppose the op amp of Example 5.2 is erroneously fabricated with $W_5 = 30\ \mu$m instead of 32 μm, everything else being as specified. Find v_O (magnitude and polarity) for $v_P = v_N = 0$. Hence, find the voltage v_P that, with v_N still at 0 V, will drive v_O to zero (this is the contribution to the input offset voltage V_{OS} due to mismatched W_5).

(*b*) Repeat, but for the case in which the circuit is fabricated with $W_3 = 15\ \mu$m instead of 16 μm.

(*c*) Compare the two cases and explain why similar mismatches yield so different input offset voltages.

5.19 The circuit of Fig. P5.19 is the *dual* of that of Fig. 5.13 because we can obtain it by interchanging *p*MOSFETs with *n*MOSFETs and vice versa, and by interchanging the power-supply polarities.

(*a*) Repeat Example 5.2 for the present circuit, compare with its dual, and comment.

(*b*) Since the size of the IC is influenced by the sum of the *W*s, which circuit is likely to use less chip area? There are two reasons why the version of Fig. 5.13 is more popular: (*a*) a *p*MOSFET SC pair exhibits lower flicker noise than an *n*MOSFET pair (see Chapter 7), and (*b*) the CS stage exhibits a higher g_m if M_5 is an *n*MOSFET instead of a *p*MOSFET, a feature that facilitates frequency compensation (again, see Chapter 7).

5.20 Suppose the op amp of Fig. P5.19 is fabricated in the process of Problem 5.16 with $L = 0.75\ \mu$m throughout. Assuming ± 1.5-V supplies, $I_{REF} = 200\ \mu$A, $W_1 = W_2 = W_7/2 = W_6/4 = W_8/4 = 20\ \mu$m, and $W_3 = W_4 = W_5/4 = 50\ \mu$m, find the individual-stage gains, the overall gain, the output resistance, the IVR, the OVS, the CMRR, and the PSRRs (ignore the body effect of M_1 and M_2).

5.3 The Folded-Cascode CMOS Operational Amplifier

5.21 The circuit of Fig. P5.21 is the *dual* of that of Fig. 5.16 because we can obtain it from the former by interchanging *p*FETs with *n*FETs and vice versa, as well as interchanging the power-supply polarities (omitted for simplicity, the biasing circuitry is modeled via the dc sources V_1, V_2, and V_3).

(*a*) Assuming $k'_n = 200\ \mu$A/V^2, $k'_p = 80\ \mu$A/V^2, $V_{tn} = -V_{tp} = 0.5$ V, $\lambda'_n = 0.02\ \mu$m/V, $\lambda'_p = 0.04\ \mu$m/V, and $L = 0.75\ \mu$m, specify W_1 through W_{11} for $I_{D9} = I_{D10} = 100\ \mu$A, $I_{D11} = 80\ \mu$A, and $V_{OV} \leq 0.2$ V throughout.

(*b*) Assuming $\chi = 0.1$ throughout, find the gain and the output resistance.

(*c*) If V_2 and V_3 are biasing the *p*FETs right at the EOS and the circuit is powered from ± 1.65-V supplies, find the IVR and the OVS.

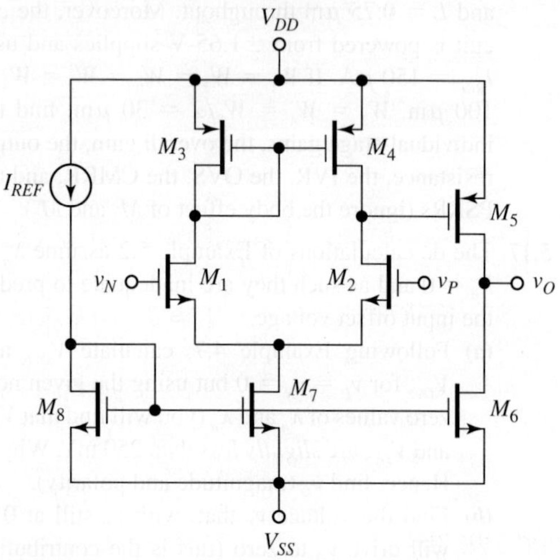

FIGURE P5.19

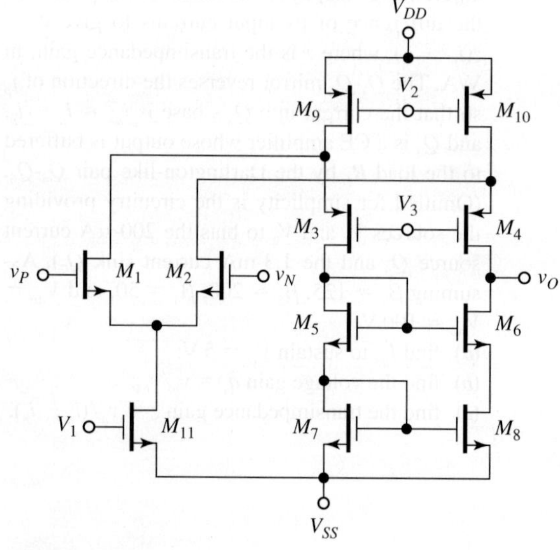

FIGURE P5.21

5.22 **(a)** If W_1 is 1% higher than W_2 in the folded-cascode op amp of Example 5.4, find the polarity and projected value of v_O for $v_P = v_N = 0$. Hence, find the voltage v_P that, with v_N still at 0 V, will drive v_O to zero (as we know, this is the input offset voltage V_{OS}).

(b) Repeat (a) if W_7 is 1% higher than W_8.

(c) Repeat (a) if W_5 is 1% higher than W_6.

(d) Repeat (a) if W_3 is 1% higher than W_4.

(e) Repeat (a) if W_9 is 1% higher than W_{10}.

(f) Compare all cases and comment.

5.23 Figure P5.23 shows how we can exploit the folded-cascode technique to realize a two-stage op amp in which *all* signal-processing transistors are *n*MOSFETs, which are faster than *p*MOSFETs because μ_n is two to three times higher than μ_p. Signal coupling from the differential pair M_1-M_2 to the current-mirror load M_3-M_4 takes place via the level-shifting voltage sources denoted as V_{LS} (omitted for simplicity, the biasing circuitry is modeled via the dc sources V_1 and V_2).

(a) Assuming $k'_n = 150\ \mu\text{A/V}^2$, $k'_p = 60\ \mu\text{A/V}^2$, $V_{tn} = -V_{tp} = 0.6$ V, $\lambda'_n = 0.02\ \mu\text{m/V}$, $\lambda'_p = 0.04\ \mu\text{m/V}$, and $L = 0.75\ \mu\text{m}$ throughout, specify suitable values for W_1 through W_9 for $I_{D9} = I_{D5} = 100\ \mu\text{A}$ and $V_{ov} = 0.25$ V throughout.

(b) Find the voltage gain and the output resistance.

(c) Assuming ± 1.65-V supplies, find the input voltage range and the output voltage swing, as well as the value of V_{LS} that will make $v_{IC(\text{max})} = V_{DD}$.

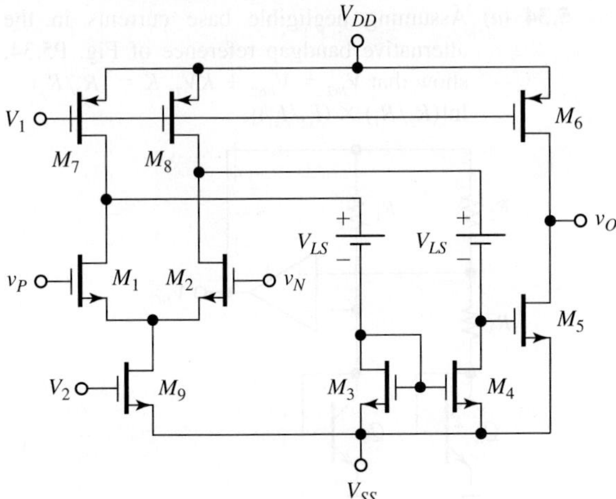

FIGURE P5.23

5.24 Shown in Fig. P5.24 is a possible circuit realization of the level-shifting sources V_{LS} of Fig. P5.23, along with the Thévenin equivalent.

(a) Find expressions for V_{Th} and R_{Th} (ignore the body effect, and use the test method to find R_{Th}).

(b) Assuming the process parameters of Problem 5.23, specify W_1 and W_2 for $R_{Th} = 3\ \text{k}\Omega$ and $V_{Th} = 1.85$ V for $I = 75\ \mu\text{A}$ and $I_2 = 50\ \mu\text{A}$.

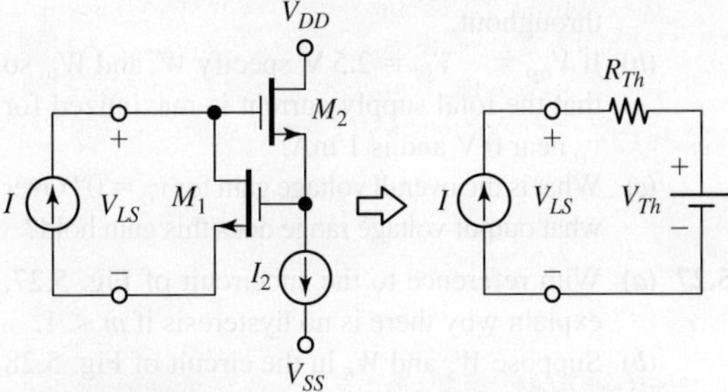

FIGURE P5.24

5.4 Voltage Comparators

5.25 Shown in Fig. P5.25 is a voltage comparator that allows for V_{OL} and V_{OH} to be set independently of the supplies V_{CC} and V_{EE}. The circuit consists of the differential pair Q_1-Q_2, the current mirrors Q_3-Q_4, Q_5-Q_6, and Q_7-Q_8, the open-collector output stage Q_9, and the external pull-up resistor R_{PU}.

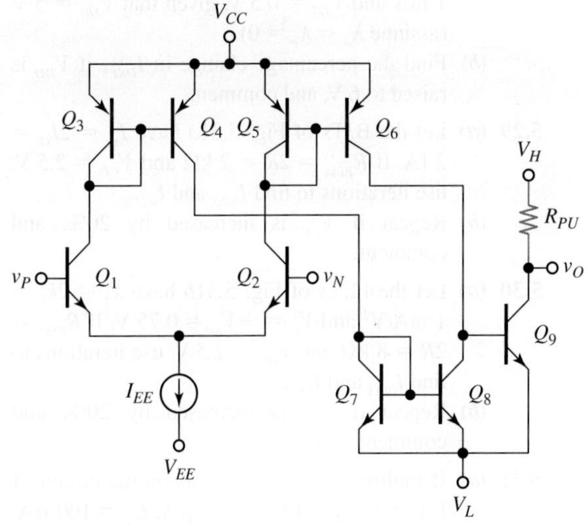

FIGURE P5.25

(a) Let $V_{CC} = -V_{EE} = 10$ V, $I_{EE} = 0.2$ mA, $V_H = 5$ V, $V_L = 0$ V, $\beta_{F9} = 200$, $V_{CE9(\text{sat})} = 0.2$ V, and $R_{PU} = 10\ \text{k}\Omega$. Assuming $r_o \gg r_\pi$ for simplicity, find V_{OH}, V_{OL}, and the gain a for $v_O = \frac{1}{2}(V_{OL} + V_{OH})$. Hence, estimate the difference $V_{IH} - V_{IL}$.

(b) If $v_N = 0$, find v_P for $v_O = \frac{1}{2}(V_{OL} + V_{OH})$.

5.26 In the CMOS comparator of Fig. 5.24 let $k'_n = 2.5k'_p = 120\ \mu\text{A/V}^2$, $V_{tn} = -V_{tp} = 0.7$ V, $\lambda_n = \lambda_p = 1/(30\ \text{V})$, and $L = 0.75\ \mu\text{m}$.

(a) If $I_{REF} = 100\ \mu\text{A}$, specify W_1 through W_7 for $I_{D6} = I_{D7} = I_{D8} = 100\ \mu\text{A}$ and $V_{OV} = 0.2$ V throughout.

(b) If $V_{DD} = -V_{SS} = 2.5$ V, specify W_9 and W_{10} so that the total supply current is maximized for v_O near 0 V and is 1 mA.

(c) What is the overall voltage gain for $v_O = 0$? Over what output voltage range does this gain hold?

5.27 **(a)** With reference to the subcircuit of Fig. 5.27, explain why there is no hysteresis if $m < 1$.

(b) Suppose W_5 and W_6 in the circuit of Fig. 5.28 are lowered from 10 μm to 4 μm, so that $m = 4/6$. Assuming $\lambda_n = \lambda_p = 0$ for simplicity, sketch and label the plots of v_{O1}, v_{O2}, and $v_{O1} - v_{O2}$ versus v_I. What is the value of the gain in the vicinity of $v_I = 0$?

5.5 Current and Voltage References

5.28 **(a)** Assuming $k'_n = 2.7k'_p = 160\ \mu\text{A/V}^2$ and $V_{tn} = -V_{tp} = 0.75$ V in the circuit of Fig. 5.30c, specify W/L ratios for $I_{REF} = 5I_{BIAS} = 1$ mA and $V_{GS} = 0.5$ V, given that $V_{DD} = 5$ V (assume $\lambda_n = \lambda_p = 0$).

(b) Find the percentage change in I_{DREF} if V_{DD} is raised to 6 V, and comment.

5.29 **(a)** Let the BJTs of Fig. 5.31a have $I_{s1} = 2I_{s2} = 2$ fA. If $R_{BIAS} = 2R = 2$ kΩ and $V_{CC} = 2.5$ V, use iterations to find I_{BIAS} and I_{REF}.

(b) Repeat if V_{CC} is increased by 20%, and comment.

5.30 **(a)** Let the FETs of Fig. 5.31b have $k_2 = 2k_1 = 1$ mA/V^2 and $V_{tn} = -V_{tp} = 0.75$ V. If $R_{BIAS} = 2R = 8$ kΩ and $V_{DD} = 2.5$ V, use iterations to find I_{BIAS} and I_{REF}.

(b) Repeat if V_{DD} is increased by 20%, and comment.

5.31 **(a)** If multimeter measurements on the circuit of Fig. 5.32a yield $I_{C1} = 250\ \mu\text{A}$, $I_{C2} = 100\ \mu\text{A}$, and $\Delta V_{BE} = 60$ mV, find R. If $I_{s1} = I_{s4} = 2$ fA, find I_{s2} and I_{s3}.

(b) If R is halved, which voltages and current change, and by how much? Which remain unchanged?

5.32 **(a)** If multimeter measurements on the circuit of Fig. 5.32b yield $I_{D1} = 100\ \mu\text{A}$, $I_{D2} = 50\ \mu\text{A}$, $\Delta V_{GS} = 250$ mV, and $R = 5$ kΩ, find k_1. If $k_4 = k_1$, find k_2 and k_3.

(b) If R is halved, which voltages and current change, and by how much?

5.33 **(a)** If the BJT of Fig. P5.33 exhibits $\beta_F = 200$ and $V_{BE} = 0.64$ V at $I_C = 1$ mA, specify suitable resistance values for $V_{REF} = 1.0$ V at $V_{CC} = 5$ V, under the constraints $I_C = 0.25$ mA and $I_{R_1} = 10I_B$.

(b) Use small-signal analysis to estimate the percentage change in V_{REF} in response to a 20% increase in V_{CC}.

(c) Assuming V_{BE} drifts by -2 mV/°C, estimate the drift of V_{REF}.

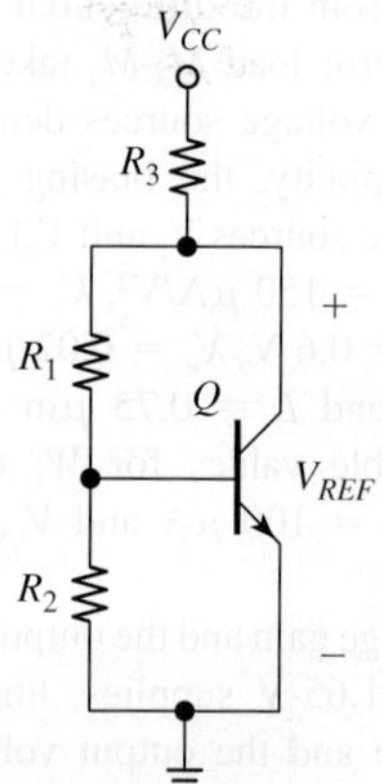

FIGURE P5.33

5.34 **(a)** Assuming negligible base currents in the alternative bandgap reference of Fig. P5.34, show that $V_{BG} = V_{BE2} + KV_T$, $K = (R_2/R_3) \times \ln[(R_2/R_1) \times (I_{s2}/I_{s1})]$.

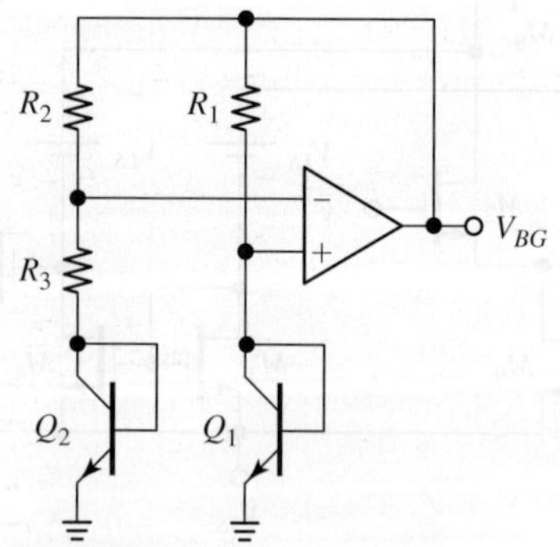

FIGURE P5.34

(b) If $I_{s2}(25\,°C) = 2I_{s1}(25\,°C) = 5$ fA, specify R_1 through R_3 for TC(V_{REF}) = 0 at $T = 25\,°C$ under the constraint $I_{C1} = 5I_{C2} = 0.2$ mA. What is the value of V_{BG}?

5.35 The circuit of Fig. P5.35 is called the Widlar band-gap reference for its inventor.

(a) Assuming matched BJTs with negligible base currents, show that $V_{BG} = V_{BE3} + KV_T$, $K = (R_2/R_3)\ln(I_{C1}/I_{C2})$.

(b) If $I_s(25\,°C) = 2$ fA for all BJTs, specify R_1 through R_3 for TC(V_{BG}) = 0 at $T = 25\,°C$ under the constraint $I_{C1} = I_{C3} = 5I_{C2} = 0.2$ mA.

(c) Specify R_4 through R_6 for $V_{REF} = 5.0$ V.

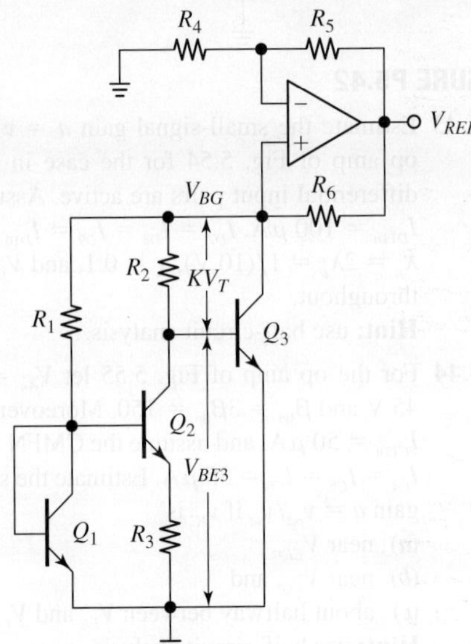

FIGURE P5.35

5.6 Current-Mode Integrated Circuits

5.36 Suppose the BJTs in the transconductor of Fig. 5.38 have $\beta = 200$ and $V_A = 50$ V. Moreover, let $I_3 = I_4 = 0.1$ mA, and suppose $I_{s1} = 10I_{s3}$ and $I_{s2} = 10I_{s4}$. The transconductor is now connected in the form shown concisely in Fig. P5.36, with the power supplies not shown to avoid cluttering. Calculate R_b, R_c, and R_e if $R_E = 250\ \Omega$ and $R_C = 10$ kΩ.

Hint: exploit the symmetry of the circuit about the horizontal line joining B, E, and C in Fig. 5.38.

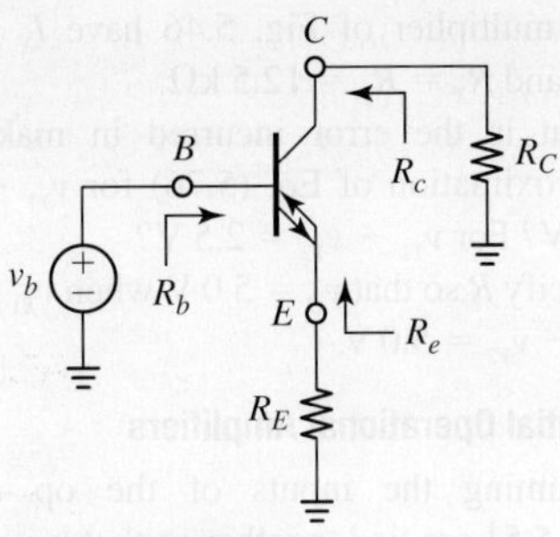

FIGURE P5.36

5.37 Figure P5.36 shows how easy it is to operate the transconductor of Fig. 5.38 as a CE ampli-fier. Since all biasing is done internally (to avoid cluttering, the power supplies are not explicitly shown), and the device is capable of true four-quadrant operation, the external resistors can be terminated directly to ground, as shown.

(a) Assuming the parameters of Problem 5.36, esti-mate the voltage gain v_c/v_b. What is its polarity?

(b) Repeat if $R_E = 250\ \Omega$ and $R_C = \infty$.

(c) Repeat if $R_E = 0$ and $R_C = \infty$, and comment.

5.38 Equation (5.68) was derived under the assumption of an ideal input voltage buffer. A real-life buffer will exhibit a small output resistance R_n, as shown in Fig. P5.38.

(a) Show that Eq. (5.68) still holds, provided the term $1 + R_2/R_{eq}$ is changed to $1 + [R_2 + R_n(1 + R_2/R_1)]/R_{eq}$.

(b) If a CFA with $R_{eq} = 750$ kΩ and $R_n = 40\ \Omega$ is configured as a noninverting amplifier with $R_2 = 9R_1 = 1.2$ kΩ, find its gain A. What is its percentage deviation from the ideal? How does it compare with the case $R_n = 0$?

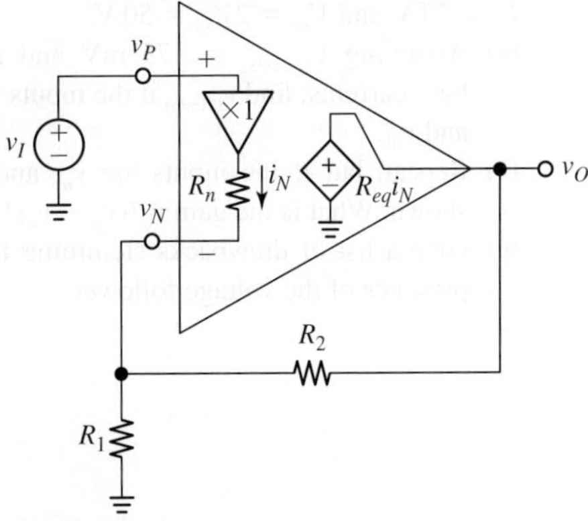

FIGURE P5.38

5.39 Let the multiplier of Fig. 5.46 have $I_X = I_Y = 0.5$ mA and $R_X = R_Y = 12.5$ kΩ.

(a) What is the error incurred in making the approximation of Eq. (5.76) for $v_{Y1} - v_{Y2} = 5.0$ V? For $v_{Y1} - v_{Y2} = 2.5$ V?

(b) Specify R so that $v_O = 5.0$ V when $v_{X1} - v_{X2} = v_{Y1} - v_{Y2} = 5.0$ V.

5.7 Fully Differential Operational Amplifiers

5.40 (a) Assuming the inputs of the op amp of Fig. 5.51 are tied together so that $v_{OP} = v_{ON} = v_{OC}$, show that $v_{OC} = V_{OC(set)}/(1 + 1/a_s)$, where a_s is the gain of the servo loop. (To find a_s, set $v_p = v_n = 0$, break M_{12}'s gate connection and ground M_{12}'s gate, break M_{15}'s gate connection and apply a test voltage v_{test} to M_{15}'s gate, and finally obtain $a_s = v_{op}/v_{test}$.

(b) Suppose the op amp is used in a single-supply system with $V_{DD} = 5$ V, $V_{SS} = 0$, and $V_{OC(set)} = 2.5$ V. Assuming the parameters of Example 5.4, along with $k_{12} = k_{13} = k_{14} = k_{15} = k_1$, $I = 100$ μA, and $k_{15} = k_9/1.25$, calculate a_s and, hence, v_{OC}.

5.41 (a) Assuming the parameters of Example 5.2, find the small-signal gain $a = v_{od}/v_{id}$ of the op amp of Fig. 5.53.

(b) If the op amp is used in a single-supply system with $V_{DD} = 5$ V and $V_{SS} = 0$, what is the value of $V_{OC(set)}$ that will maximize the OVS?

(c) If v_{id} is a sinusoidal wave, what is its maximum peak amplitude before the output clips?

5.42 Figure P5.42 shows how the IVR of the single-supply active-loaded Q_2-Q_3 pair can be extended *below* ground potential by using the matched voltage followers/shifters Q_1-Q_4. Let $I_{sn} = 5$ fA, $I_{sp} = 2$ fA, and $V_{An} = 2V_{Ap} = 50$ V.

(a) Assuming $V_{EC(EOS)} = 175$ mV and ignoring base currents, find $v_{IC(min)}$ if the inputs were v_{B2} and v_{B3}.

(b) Repeat, but if the inputs are v_{B1} and v_{B4}, as shown. What is the gain $v_o/(v_p - v_n)$?

(c) Give a list of drawbacks stemming from the presence of the voltage followers.

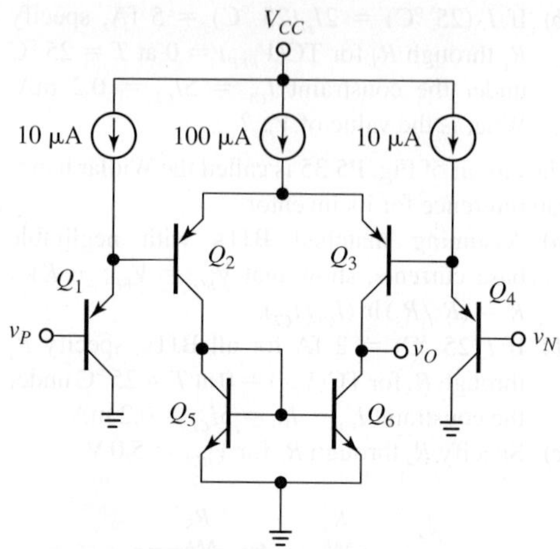

FIGURE P5.42

5.43 Estimate the small-signal gain $a = v_{od}/v_{id}$ of the op amp of Fig. 5.54 for the case in which both differential input pairs are active. Assume $I_{D11p} = I_{D11n} = 100$ μA, $I_{D7} = I_{D8} = I_{D9} = I_{D10} = 125$ μA, $\lambda_n = 2\lambda_p = 1/(10$ V$)$, $\chi = 0.1$, and $V_{OV} = 0.25$ V throughout.

Hint: use half-circuit analysis.

5.44 For the op amp of Fig. 5.55 let $V_{An} = 1.5V_{Ap} = 45$ V and $\beta_{0n} = 3\beta_{0p} = 150$. Moreover, let $I_{C11p} = I_{C11n} = 50$ μA, and assume the CMFN keeps $I_{C3} = I_{C4} = I_{C5} = I_{C6} = 35$ μA. Estimate the small-signal gain $a = v_{od}/v_{id}$ if v_{IC} is

(a) near V_{EE},

(b) near V_{CC}, and

(c) about halfway between V_{EE} and V_{CC}.

Hint: use half-circuit analysis.

5.8 Switched-Capacitor Circuits

5.45 Show that the switch-capacitor arrangement of Fig. P5.45 simulates a resistor with $R_{eq} = 1/(4Cf_S)$. List a possible advantage of this scheme, as well as possible disadvantages.

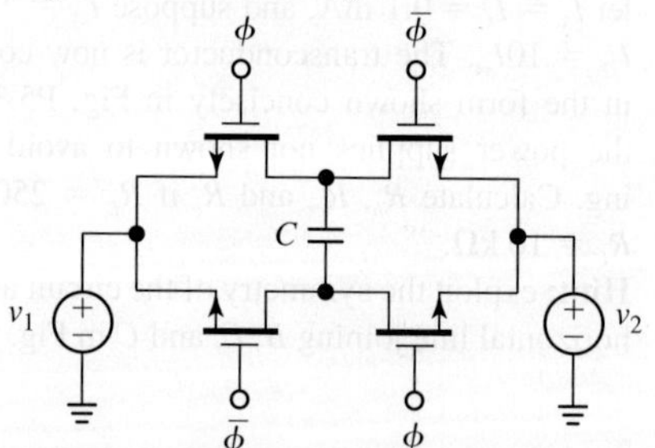

FIGURE P5.45

5.46 Figure P5.46 shows the autozeroing scheme for a high-gain amplifier designed to be operated as an inverting voltage comparator. Discuss its operation and show the circuit and all voltages during the autozeroing mode as well as in normal operation, in the manner of Fig. 5.65, given that $V_{OS} = 5$ mV.

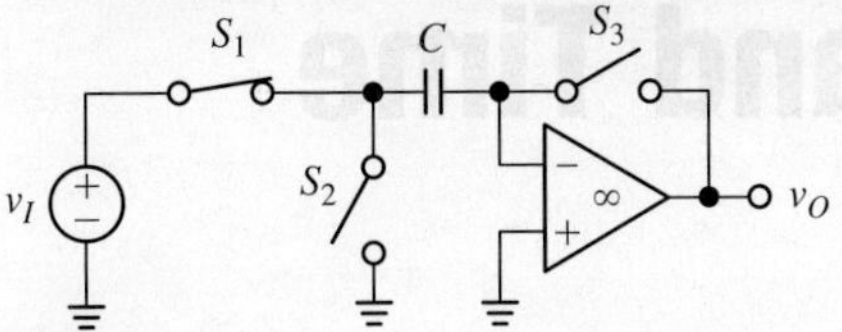

FIGURE P5.46

5.47 Derive an expression for the transmission-gate resistance r_{TG} of Fig. 5.66 for arbitrary values of k_n, k_p, V_{tn}, and V_{tp}.

(a) Assuming $V_{SS} = 0$ and $V_{DD} = 5.0$ V, sketch and label r_{TG} as a function of v over the range $0 \leq v \leq 5.0$ V if $k_n = 1.25k_p = 1$ mA/V^2, $V_{tn} = 0.75$ V, and $V_{tp} = -1$ V. What are the maximum and minimum values attained by r_{TG}?

5.48 (a) Assuming f_S in Fig. P5.48 is sufficiently high to make the switching process appear as virtually continuous at the signal frequency, find V_o as a function of V_1 and V_2. How would you name this circuit?

(b) If $f_S = 1$ MHz, specify suitable values for C_1 and C_2 for a unity-gain frequency of 5 kHz under the constraint $C_1 + C_2 \leq 20$ pF.

(c) Find the magnitude and phase errors at $f = 20$ kHz.

(d) What happens if we change the clock phase of the upper switch so that it is flipped to the right while the bottom switch is flipped to the left, and vice versa?

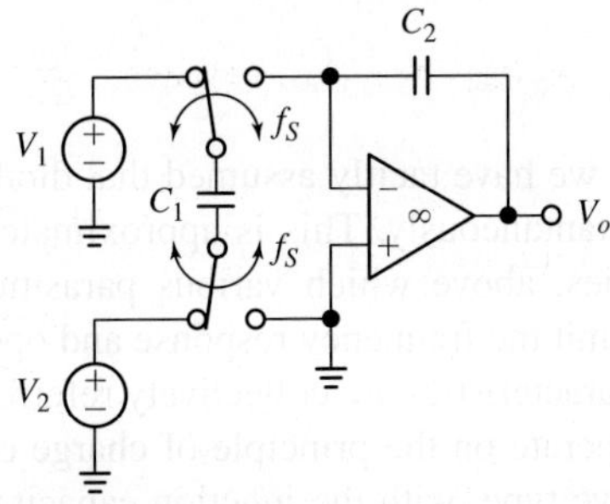

FIGURE P5.48

5.49 (a) Assuming f_S in Fig. P5.49 is sufficiently high to make the switching process appear as virtually continuous at the signal frequency, derive the transfer function $H = V_o/V_i$ and show that the circuit is a low-pass filter with gain.[7] What are the expressions for its dc gain and its -3 dB frequency?

(b) If $f_S = 1$ MHz, specify suitable capacitance values for a low-frequency gain of 2 V/V and a -3 dB frequency of 10 kHz under the constraint $C_1 + C_2 + C_3 = 30$ pF.

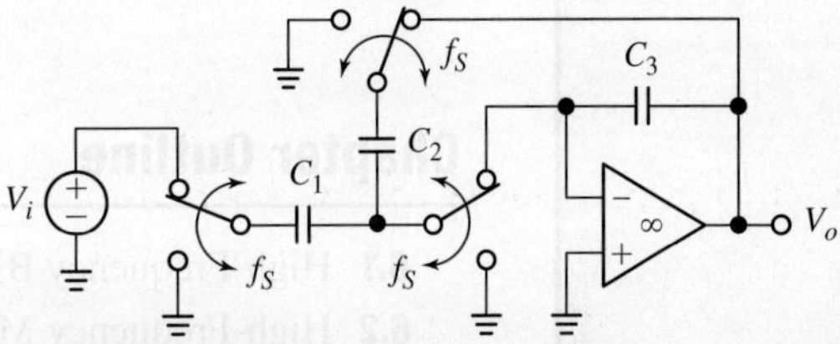

FIGURE P5.49

5.50 Suppose the FETs of Fig. 5.68b are fabricated in a process for which we can approximate $C_{gd} \cong 0.25W$ fF, $Q_n \cong 1.75WV_{OV}$ fC, and $k = 125W$ μA/V^2, where W is the channel width, in μm, and V_{OV} is the overdrive voltage, in V. Moreover, let $V_t = 0.5$ V, and assume the gate voltages alternate between ± 2.5 V.

(a) Specify W_1 so that $r_{DS1} = 500$ Ω for $v_I = 0$. What is r_{DS1} at $v_I = 0.5$ V? At $v_I = -0.5$ V?

(b) If $C_{tot} = 750$ fF, find the error in v_O due to clock feedthrough for $v_I = 0, 0.5$ V, and -0.5 V.

(c) Repeat part (b) but for charge injection with $\alpha = 0.5$.

(d) Specify W_2 so as to cancel out the clock feedthrough due to M_1.

(e) What can you say about charge injection cancellation if $\alpha = 0.5$? If $\alpha = 1$?

Frequency and Time Responses

Chapter Outline

I n our study of electronic circuits so far we have tacitly assumed that diodes and transistors react to external signals instantaneously. This is approximately true only up to certain operating frequencies, above which various parasitic reactances come into play, whose effect is to limit the frequency response and operating speed of a circuit (speed and frequency characteristics are collectively referred to as *dynamics*). Since diodes and transistors operate on the principle of charge control, their internal parasitics are of the capacitive type, with the junction capacitance C_j encountered in Chapter 1 being a familiar example. But, even without any electronic devices present, a circuit exhibits inherent reactive parasitics stemming from its dimensions and layout. In fact, viewing a circuit as a collection of nodes and branches, we can safely say that

- each node exhibits stray capacitance toward its nearby nodes, including the reference or ground node
- each branch exhibits stray self-inductance as well as inductive coupling to its nearby branches

A capacitance, whether intentional or parasitic, tends to *oppose* changes in the *voltage* across it, whereas an inductance tends to *oppose* changes in the *current* through it. These parasitics are generally small, indicating that below certain operating frequencies, stray capacitances act as open circuits and stray inductances act as short circuits and can thus be ignored. However, as a circuit's operating frequency or operating speed is increased, the roles of capacitances and inductances are reversed, so capacitance now approach short-circuit behavior and inductances open-circuit behavior. In the circuits of interest to us here it so happens that the most severe limitations stem from stray capacitances. To be able to investigate their effect upon the dynamics of a circuit we therefore need to augment the small-signal transistor models utilized so far with suitable capacitances.

Both BJTs and MOSFETs include *pn* junctions, so their models must incorporate the corresponding junction capacitances. The MOSFET includes also the capacitances formed by the gate with the channel as well as with the source and drain regions. There is no counterpart of these capacitances in the BJT. On the other hand, the BJT exhibits a capacitance associated with the minority charge buildup inside the base region that does not have a counterpart in the MOSFET. We thus have similarities but also differences in the physical origin of the parasitics of the two devices. Yet, the similarities are strong enough that once the student has mastered the high-frequency analysis of one of the two device types, a good deal of it can be adapted to the other.

CHAPTER HIGHLIGHTS

The chapter begins with the physical basis of the parasitic capacitances of BJTs and MOSFETs, after which suitable high-frequency models are developed for both devices.

These models are then applied to the study of the high-frequency behavior of the basic single transistor configurations (CE/CS, CC/CD, and CB/CG configurations), as well as differential pairs, both passive loaded and active loaded. Since the distinguishing feature of the CC/CD and CB/CG configurations is their ability to effect impedance transformation, particular attention is given to terminal impedances and modeling thereof, especially when it comes to inductive impedances due to their notorious tendency to cause ringing or even oscillations.

As circuit complexity increases, frequency analysis tends to become prohibitively difficult. The *open-circuit time-constant* (OCTC) technique discussed next alleviates the task by breaking it down into multiple but simpler sub-tasks, which we then use to estimate the circuit's bandwidth. The OCTC technique proves particularly useful in the analysis of multiple-transistor circuits such as the cascode configuration.

The frequency response illuminates only certain aspects of a circuit's dynamics. To complete the picture we need to know also the transient response, the dual counterpart. Op amps offer a classic example in which it proves useful to know both viewpoints side by side.

There are situations in which only the transient response is of interest, especially when it comes to highly nonlinear applications such as switches and logic gates. The chapter concludes with the switching transients of *pn* junctions and BJTs, the propagation delays of CMOS logic gates, and the response times of voltage comparators.

Frequency analysis requires a certain amount of dexterity with the manipulations of transfer functions and the construction of Bode plots. For convenience, these subjects are briefly reviewed in Appendix 6A, at the end of the chapter. The chapter makes abundant use of PSpice both as a software oscilloscope to display Bode plots and transient responses, and as a verification tool for hand calculations.

6.1 HIGH-FREQUENCY BJT MODEL

Recall that a BJT comprises two *pn* junctions. In junction-isolated integrated circuits (ICs) there is a third junction providing electric isolation between each BJT and its surrounding components on the chip (this junction is always reverse biased). As we know, each junction exhibits a voltage-dependent capacitance called the *junction capacitance*, whose characteristic is depicted in Fig. 1.41*b*. In the case of the *npn* BJT of Fig. 6.1, the three capacitances of interest are:

- The *base-emitter junction capacitance* C_{je},

$$C_{je}(v_{BE}) = \frac{C_{je0}}{\left(1 - v_{BE}/\phi_e\right)^{m_e}} \tag{6.1a}$$

where C_{je0} is the *zero-bias* value of C_{je}, v_{BE} is the base-emitter voltage, ϕ_e is the *built-in potential* of the base-emitter junction, and m_e is the junction's *grading coefficient*, typically between 1/2 for *abrupt* junctions, and 1/3 for *graded* junctions. In forward-active operation v_{BE} doesn't vary much from its typical value $V_{BE(on)}$ (=0.7 V), so it is common practice to approximate

$$C_{je}(V_{BE(on)}) \cong 2C_{je0} \tag{6.1b}$$

- The *base-collector junction capacitance* C_{jc}, more commonly denoted as C_μ in analog electronics,

$$C_\mu(v_{BC}) = \frac{C_{\mu0}}{\left(1 - v_{BC}/\phi_c\right)^{m_c}} \tag{6.2}$$

with similar meaning for the various parameters. In forward-active operation we usually have $v_{BC} < 0$, so it follows that $C_\mu < C_{\mu0}$.

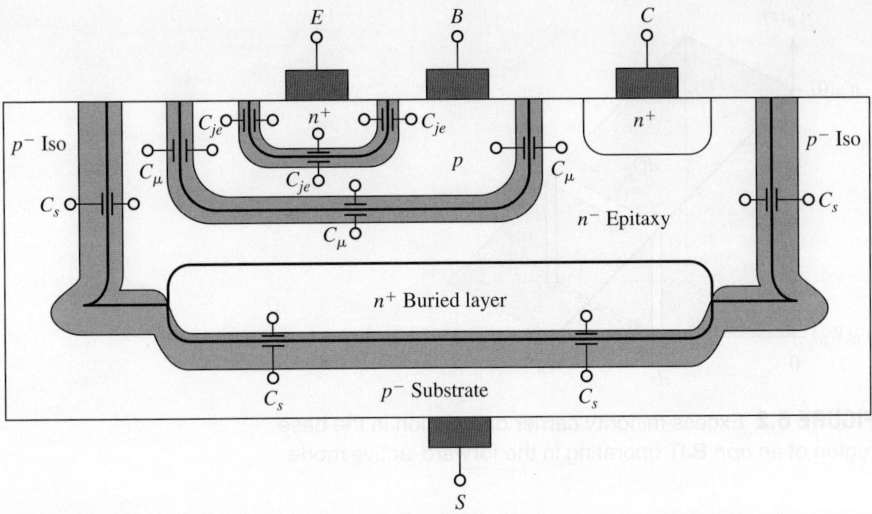

FIGURE 6.1 The junction capacitances of a monolithic *npn* BJT.

- The *collector-to-substrate junction capacitance* C_s, with a similar expression as the one above. To ensure isolation, this junction must always be in reverse bias, so to prevent it from ever turning on, the substrate S is internally tied to the most negative voltage (MNV) in the circuit. With $v_{SC} < 0$ this junction is nonconductive, yet it presents a stray capacitance C_s whose effect is to shunt high-frequency collector signals to the ac grounded substrate.

Figure 6.1 indicates that all three capacitances are of the distributed type and that their areas increase as we go from C_{je} to C_μ to C_s. However, the depletion-layer widths also increase in the same order due to progressively lighter doping levels from emitter to base to collector and to substrate. So, if we visualize each capacitance as $C_j = \varepsilon_{si} A / X_d$ as per Fig. 1.42*b*, it is not surprising that the three capacitance values do not differ that much from each other. Depending on transistor size, C_{je0}, $C_{\mu0}$, and C_{s0} may range from a few picofarads (1 pF = 10^{-12} F) down to ten femtofarads (1 fF = 10^{-15} F) or so.

The Base-Charging Capacitance C_b

We now wish to demonstrate that the BJT exhibits an *additional* capacitance C_b due to the injection of minority charges into its base region. Recall that in order to operate a BJT in the forward-active (FA) region we need to establish an *excess* of minority charge-carriers in its base (this charge, already discussed in connection with Fig. 2.8, is repeated here as Fig. 6.2 for convenience). To find this charge, henceforth denoted as Q_F and consisting of electrons in the *npn* BJT and holes in the *pnp* BJT, we consider the charge dQ_F contained in an infinitesimal slice of thickness dx located at some point x along the horizontal axis, and we then integrate dQ_F from 0 to the base width W_B. For an *npn* BJT with emitter area A_E, we first multiply the volume $A_E dx$

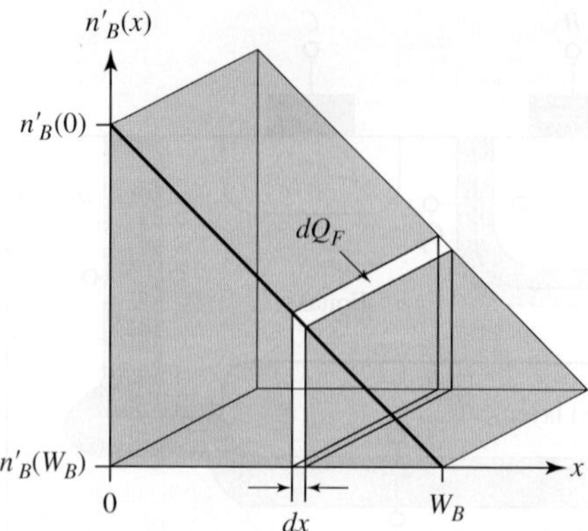

FIGURE 6.2 Excess minority carrier distribution in the base region of an *npn* BJT operating in the forward-active mode.

by the local excess electron density $n'_B(x)$ to find the number of electrons within the slice, then we multiply by the electron charge $-q$ to find dQ_F, finally we integrate

$$Q_F = -qA_E \int_0^{W_B} n'_B(x)dx = -qA_E \frac{W_B \times [n'_B(0) - n'_B(W_B)]}{2}$$

where we have used the fact that the integral is just the *area* of the triangle with base W_B and height $n'_B(0) - n'_B(W_B)$. We also know that the collector current I_C is proportional to the *slope* of the triangle,

$$I_C = A_E J_n = qA_E D_n \frac{dn'_B(x)}{dx} = qA_E D_n \left(-\frac{n'_B(0) - n'_B(W_B)}{W_B} \right)$$

where D_n is the electron diffusivity. Eliminating the difference $n'_B(0) - n'_B(W_B)$ and simplifying, we get

$$Q_F = \tau_F I_C \qquad (6.3)$$

where

$$\tau_F = \frac{W_B^2}{2D_n} \qquad (6.4)$$

is the *mean transit time* for electrons in the forward direction, so-called because it represents the average time an electron spends in crossing the base region. (To see why, rewrite as $I_C = Q_F/\tau_F$, and use the definition of current, $I = \Delta Q/\Delta t$, to view τ_F as the time taken by the charge Q_F to diffuse through the base.) For monolithic *npn* BJTs, τ_F is typically in the range of 10 to 100 ps (1 ps $= 10^{-12}$ s). Equation (6.4) applies to the *pnp* BJT as well, provided we replace D_n with D_p. By Einstein's equations, $D_p = (\mu_p/\mu_n)D_n$. Since $\mu_p < \mu_n$, it follows that $D_p < D_n$, indicating that *pnp* BJTs exhibit *longer* transit times than *npn* BJTs, a feature that makes the *npn* types inherently faster and thus better suited to high-speed applications.

Recall that a change v_{be} results in the change $i_c = g_m v_{be}$, and hence, by Eq. (6.3), in the change $q_f = \tau_F i_c$, that is,

$$q_f = \tau_F g_m v_{be}$$

Whenever a *voltage change* causes *charge redistribution* we have the phenomenon of *capacitance*, so we write $C_b = q_f / v_{be}$, that is,

$$C_b = \tau_F g_m = \tau_F \frac{I_C}{V_T} \tag{6.5}$$

where C_b is the *base-charging capacitance*, also called the *diffusion capacitance*. Note that C_b depends on the bias current I_C, just like g_m, r_π, and r_o do.

The High-Frequency BJT Model

We are now ready to incorporate all above information into a small-signal BJT model that will allow us to investigate the high-frequency behavior of bipolar ICs. The result is shown in Fig. 6.3, where small-signal voltages and currents, now frequency-dependent, are represented in terms of their Laplace's transforms (upper-case letters with lower-case subscripts). Proceeding from right to left, we make the following observations:

- First, we have the substrate capacitance C_s. Since the substrate of an *npn* BJT is connected to the MNV, which is a dc potential, the substrate appears as a signal ground in our model. (Oftentimes C_s is ignored in order to simplify the calculations.)
- Next, we note the base-collector junction capacitance C_μ. This capacitance is most intriguing, for there are situations in which its effect is negligible and it can thus be ignored, whereas there are others in which its role gets magnified because of a phenomenon known as the Miller effect, and thus becomes the dominant capacitance in the circuit.
- In parallel with r_π we have a capacitance C_π made up of two components,

$$C_\pi = C_{je} + C_b \tag{6.6}$$

namely, the approximately constant junction component C_{je} ($\cong 2C_{je0}$), and the bias-dependent component C_b ($= \tau_F I_C / V_T$).
- With reference to Fig. 6.1, we observe that as current enters the base terminal B and progresses to the thin base region separating emitter and collector, where the

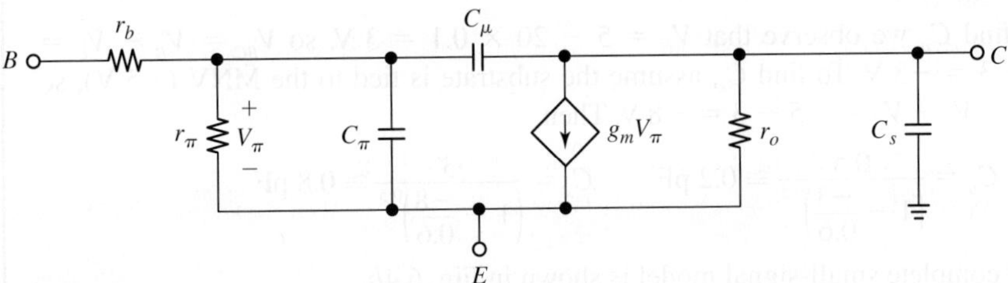

FIGURE 6.3 High-frequency small-signal model for the BJT.

central transistor action takes place, it encounters some distributed resistance. This is simply the bulk resistance r_b of the moderately-doped p-type base region. Being typically on the order of 10^2 Ω in monolithic BJTs, r_b has been ignored so far, since the voltage drop it produces in response to i_b is usually negligible compared to that produced by r_π. However, we will find that in general r_b cannot be ignored in high-frequency operation, as it limits the dynamics of certain BJT configurations, particularly the CE configuration.

EXAMPLE 6.1   Find the element values of the small-signal model of the BJT of Fig. 6.4a using the data of a typical high-voltage process: $\beta_0 = 150$, $V_A = 80$ V, $r_b = 250$ Ω; $\tau_F = 200$ ps; $C_{je0} = 1.0$ pF, $\phi_e = 0.8$ V, $m_e = 0.33$; $C_{\mu 0} = 0.5$ pF, $\phi_c = 0.6$ V, $m_c = 0.5$; $C_{s0} = 3.0$ pF, $\phi_s = 0.6$ V, and $m_s = 0.5$. Show your final circuit.

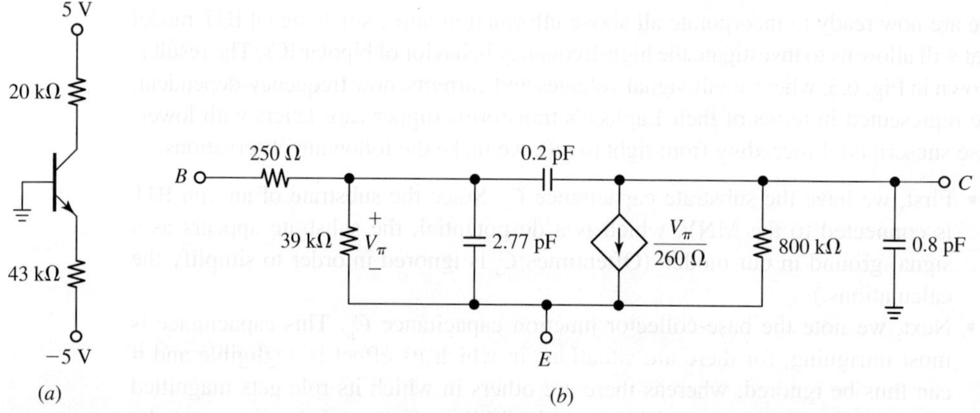

FIGURE 6.4   (a) Circuit of Example 6.1, and (b) the BJT's small-signal model values.

Solution

By inspection, $I_C \cong I_E = (5 - 0.7)/43 = 0.1$ mA. Proceeding as usual, we get $g_m = 1/(260\ \Omega)$, $r_\pi = 39$ kΩ, and $r_o = 800$ kΩ. We also have

$$C_b = \tau_F g_m = \frac{200 \times 10^{-12}}{260} \cong 0.77 \text{ pF}$$

$$C_{je} \cong 2C_{je0} = 2.0 \text{ pF}$$

$$C_\pi = C_b + C_{je} \cong 2.77 \text{ pF}$$

To find C_μ we observe that $V_C = 5 - 20 \times 0.1 = 3$ V, so $V_{BC} = V_B - V_C = 0 - 3 = -3$ V. To find C_s, assume the substrate is tied to the MNV (-5 V), so $V_{SC} = V_S - V_C = -5 - 3 = -8$ V. Then,

$$C_\mu = \frac{0.5}{\left(1 - \dfrac{-3}{0.6}\right)^{0.5}} \cong 0.2 \text{ pF} \qquad C_s = \frac{3}{\left(1 - \dfrac{-8}{0.6}\right)^{0.5}} \cong 0.8 \text{ pF}$$

The complete small-signal model is shown in Fig. 6.4b.

Specification of the BJT Frequency Response

It is common practice to specify the frequency capability of a BJT in terms of the *transition frequency* f_T, representing the frequency at which its small-signal current gain $|\beta(jf)|$ drops to unity. This frequency is used as a *figure of merit* for high-speed operation, and it can either be calculated or measured, using the ac concept of Fig. 6.5. Specifically, we apply a small-signal ac current i_b to the base, we find the ac current i_c with the collector at ac ground, and we take the ratio $\beta = I_c/I_b$, where I_b and I_c are the Laplace's transforms of i_b and i_c. Finally, we obtain f_T as the frequency such that $|\beta(jf_T)| = 1$, or 0 dB.

Turning to the equivalent circuit of Fig. 6.5b, we observe that shorting the collector to ground renders r_o and C_s irrelevant and places C_μ in parallel with C_π. We can thus apply Ohm's law and write

$$V_\pi = \left[r_\pi // \frac{1}{s(C_\pi + C_\mu)} \right] I_b = \frac{r_\pi I_b}{1 + sr_\pi(C_\pi + C_\mu)}$$

where s is the *complex frequency*. It turns out (see Exercise 6.1 below) that over the frequency range of interest the current fed forward via C_μ is negligible compared to $g_m V_\pi$, so we approximate

$$I_c \cong g_m V_\pi = \frac{g_m r_\pi}{1 + sr_\pi(C_\pi + C_\mu)} I_b$$

Letting $g_m r_\pi \to \beta_0$ and solving for the ratio I_c/I_b, we get

$$\beta(s) = \frac{I_c}{I_b} = \frac{\beta_0}{1 + sr_\pi(C_\pi + C_\mu)}$$

We are primarily interested in the transistor's *ac steady-state response*, also called the *frequency response*, so we let $s \to j\omega$ (or $s \to j2\pi f$) and get

$$\beta(j\omega) = \frac{\beta_0}{1 + j\omega/\omega_\beta} \tag{6.7}$$

where

$$\omega_\beta = \frac{1}{r_\pi(C_\pi + C_\mu)} \tag{6.8}$$

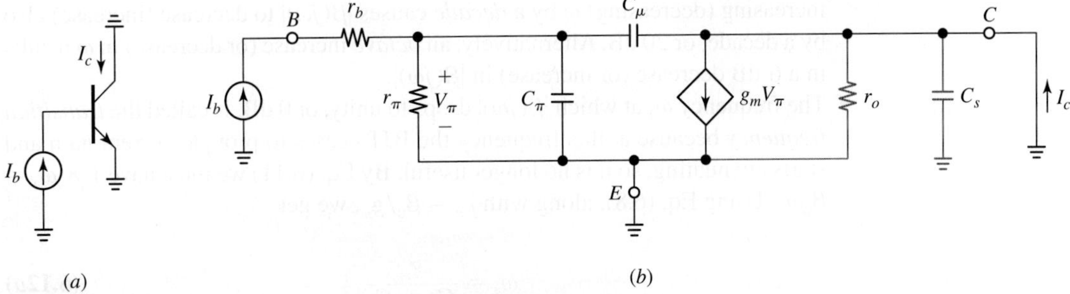

FIGURE 6.5 (a) Ac circuit to find f_T, and (b) its small-signal equivalent.

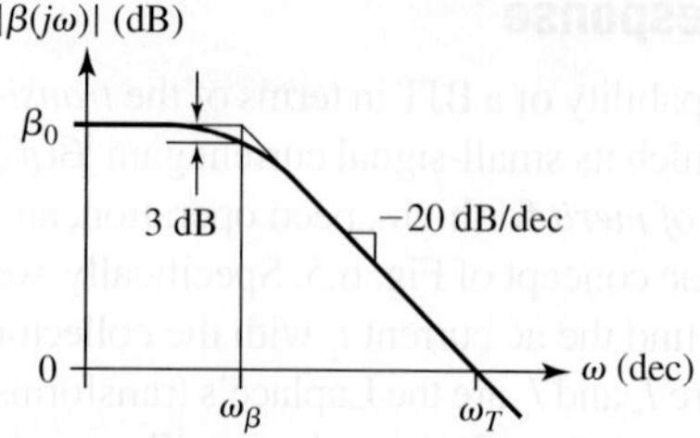

FIGURE 6.6 Bode plot of $|\beta(j\omega)|$.

Since the denominator of the $\beta(s)$ vanishes for $s = -\omega_\beta$, causing $\beta(s)$ to blow up to infinity, ω_β is aptly called *pole frequency* (see Bode plots in Appendix 6A). The magnitude of the current gain is

$$|\beta(j\omega)| = \frac{\beta_0}{\sqrt{1 + (\omega/\omega_\beta)^2}} \tag{6.9}$$

and is plotted on logarithmic scales, with ω in decades (or octaves), and magnitude in decibels. The resulting plot, known as *magnitude Bode plot* and shown in Fig. 6.6, is of such a common form that it warrants some useful observations:

- For $\omega \ll \omega\beta$, Eq. (6.9) predicts the *low-frequency asymptote*

$$|\beta(j\omega)| \to \beta_0 \tag{6.10}$$

  This is the frequency range over which we have been implicitly operating up to this chapter.

- For $\omega \gg \omega_\beta$, Eq. (6.9) predicts the *high-frequency asymptote*

$$|\beta(j\omega)| \to \frac{\beta_0}{\omega/\omega_\beta} = \frac{\beta_0 \omega_\beta}{\omega}$$

  Defining the *gain-bandwidth product* GBP $= |\beta| \times \omega$, we observe that for $\omega \gg \omega_\beta$ we have

$$\text{GBP} = |\beta(j\omega)| \times \omega = \beta_0 \omega_\beta \tag{6.11}$$

  that is, the GBP is *constant* with frequency. In other words, if we pick any point on the high-frequency asymptote and take the product of its ordinate $|\beta(j\omega)|$ by its abscissa ω, we always get the same value, namely, the GBP. In particular, increasing (decreasing) ω by a *decade* causes $|\beta(j\omega)|$ to decrease (increase) also by a decade, or 20 dB. Alternatively, an *octave* increase (or decrease) in ω results in a 6-dB decrease (or increase) in $|\beta(j\omega)|$.

- The frequency ω_T at which $|\beta(j\omega)|$ drops to unity, or 0 dB, is called the *transition frequency* because at this frequency the BJT ceases to provide current gain and starts attenuating, so it is no longer useful. By Eq. (6.11) we must have $1 \times \omega_T = \beta_0 \omega_\beta$. Using Eq. (6.8), along with $r_\pi = \beta_0/g_m$, we get

$$\omega_T = \frac{g_m}{C_\pi + C_\mu} \tag{6.12a}$$

or, alternatively,

$$f_T = \frac{g_m}{2\pi(C_\pi + C_\mu)} \qquad (6.12b)$$

In monolithic BJTs f_T ranges from a few hundred MHz to tens of GHz.

- For $\omega = \omega_\beta$ Eq. (6.9) predicts $|\beta(j\omega_\beta)| = \beta_0/\sqrt{2} = 0.707\beta_0$, that is, at $\omega = \omega_\beta$ the magnitude $|\beta|$ is down to 70.7% of its low-frequency value β_0. Since $1/\sqrt{2} = -3$ db, the pole frequency ω_β is also called the -3-dB frequency.

Exercise 6.1

The current fed forward via C_μ in Fig. 6.5b is $I_\mu = V_\pi/[1/(j\omega C_\mu)]$. Using the fact that $C_\mu \ll C_\pi$, show that for frequencies up to at least ω_T we have $|I_\mu| \ll |g_m V_\pi|$, thus justifying the approximation $I_c \cong g_m V_\pi$.

EXAMPLE 6.2

If a certain BJT exhibits $|\beta| = 200$ at $f = 1$ kHz and $|\beta| = 10$ at $f = 500$ MHz, find β_0, f_β, and f_T.

Solution

Since 1 kHz is such a low frequency, the first datum must lie on the low-frequency asymptote, so $\beta_0 = 200$. Since the second datum is much smaller than β_0, it must lie on the high-frequency asymptote, where the GBP is constant. So, $f_T = \text{GBP} = 10 \times 500 = 5$ GHz. Finally, $f_\beta = f_T/\beta_0 = 5000/200 = 25$ MHz.

It is instructive to take a closer look at the transition frequency f_T. Combining Eqs. (6.5), (6.6), and (6.12), we express this frequency in the insightful form

$$f_T = \frac{1/2\pi}{\dfrac{C_{je} + C_\mu}{I_C}V_T + \tau_F} \qquad (6.13)$$

which shows explicitly the dependence on the bias current I_C. At sufficiently low collector currents f_T is dominated by $C_{je} + C_\mu$, and it increases with I_C. For I_C sufficiently high, f_T saturates at

$$f_{T(\text{max})} = \frac{1}{2\pi\tau_F} \qquad (6.14)$$

indicating that τ_F poses the ultimate limit on f_T. Figure 6.7 shows a decline in f_T at high collector currents. This is due to the fact that τ_F increases with high-level injection and other high collector-current effects.[2]

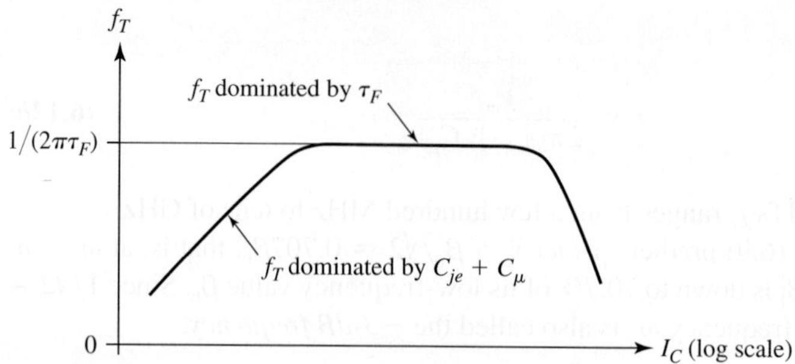

FIGURE 6.7 Dependence of f_T on the bias current I_C.

Using Eq. (6.4), along with Einstein's relation $D_n = \mu_n V_T$, we can also write, for an *npn* BJT,

$$f_{T(\max)} = \frac{1}{\pi}\frac{\mu_n}{W_B^2}V_T \tag{6.15}$$

(For a *pnp* BJT replace μ_n with μ_p.) It is apparent that for fast operation a BJT should be fabricated with a very *narrow* base, and it should be of the *npn* type as electrons are 2 to 3 times *more mobile* than holes.

EXAMPLE 6.3   Find f_T for the BJT of Example 6.1. How does this compare with $f_{T(\max)}$? Which parasitic dominates f_T in this example? Which dominates the least?

Solution

Equation (6.13) gives

$$f_T = \frac{1/2\pi}{\dfrac{(2+0.2)10^{-12}}{0.1}26 + 200 \times 10^{-12}} = \frac{1}{2\pi(520 + 52 + 200)10^{-12}} = 206 \text{ MHz}$$

Equation (6.14) gives

$$f_{T(\max)} = \frac{1}{2\pi200 \times 10^{-12}} = 796 \text{ MHz}$$

It is apparent that the main culprit in this example is C_{je}. This is followed by C_b, whereas C_μ has the least impact.

6.2 HIGH-FREQUENCY MOSFET MODEL

As depicted in Fig. 6.8, an integrated-circuit (IC) MOSFET presents a variety of internal capacitances:[1]

- The *gate-channel oxide capacitance* C_{gc}, also called the *intrinsic capacitance*,

$$C_{gc} = WLC_{ox} \tag{6.16}$$

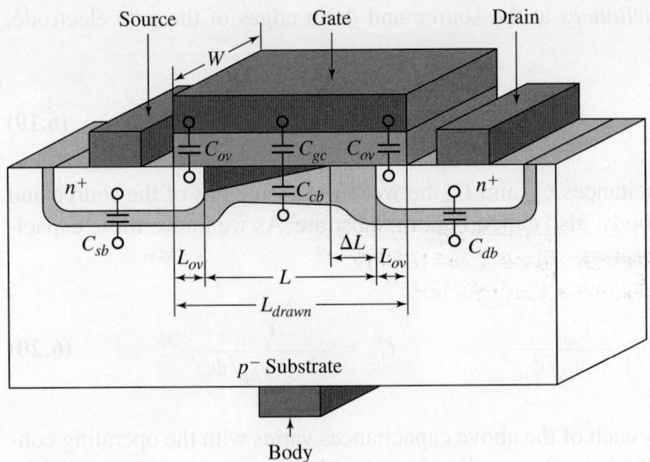

FIGURE 6.8 The capacitances in a saturated monolithic *n*MOSFET.

where W is the *channel width*, L is the distance between the inner edges of the source and drain diffusion regions, and C_{ox} is the *oxide capacitance per unit area*. Recall from Chapter 3 that

$$C_{ox} = \frac{\varepsilon_{ox}}{t_{ox}} = \frac{34.5}{t_{ox}} \text{ fF/}\mu\text{m}^2 \tag{6.17}$$

where t_{ox} is the *oxide thickness*, in nm. For instance, a process with $t_{ox} = 10$ nm gives $C_{ox} = 34.5/10 = 3.45$ fF/μm^2. The gate length as drawn on the mask before fabrication is denoted as L_{drawn}. During fabrication of the n^+ source and drain regions via ion implantation, ions diffuse *laterally*, resulting in some *overlap* between the inner edges of these regions and the outer edges of the gate electrode. Denoting the amount of overlap as L_{ov} (in PSpice this is denoted as Ld) we thus have

$$L = L_{drawn} - 2L_{ov} \tag{6.18}$$

Typically, L_{ov} is on the order of 10-20% of L_{drawn}. (Note that when calculating the device transconductance parameter $k = k'(W/L)$ we must use L as given above, and also use the multiplicative factor $(1 + \lambda v_{DS})$ to account for the channel length modulation ΔL. When referring to a particular fabrication process, engineers use L to denote what is actually L_{drawn}. This is also the convention used by PSpice, where statements of the type L=1.0u Ld=0.15u imply a fabrication process with $L_{drawn} = 1.0$ μm and $L_{ov} = 0.15$ μm, and thus $L = 1 - 2 \times 0.15 = 0.7$ μm. For consistency with previous chapters we shall continue using L to denote the distance between the inner edges of the source and drain regions.)

- The *channel-body depletion capacitance* C_{cb}. In saturation operation, which is the region of greatest interest in analog applications, this capacitance is shielded from the gate by the inversion layer and therefore plays a negligible role.

- The *overlap capacitances* at the source and drain edges of the gate electrode, each of which is

$$C_{ov} \cong WL_{ov}C_{ox} \tag{6.19}$$

- The junction capacitances C_{sb} and C_{db} between the n^+ regions of the source and drain, and the p^- body, also called bulk or substrate. As we know, these capacitances take on the forms

$$C_{sb} = \frac{C_{sb0}}{\left(1 + v_{SB}/\phi_0\right)^m} \qquad\qquad C_{db} = \frac{C_{db0}}{\left(1 + v_{DB}/\phi_0\right)^m} \tag{6.20}$$

The role played by each of the above capacitances varies with the operating conditions of the MOSFET.[2] In analog applications the FETs are operated in saturation, where the channel takes on the familiar tapered form (see Fig. 6.8), where ΔL is the SCL portion extending into the channel side. This asymmetry causes $(2/3)$ of C_{gc} to go to the source side, and none to the drain side.[2] Based on these considerations, the complete high-frequency model of the MOSFET is as in Fig. 6.9. As usual, small-signal voltages and currents, now frequency-dependent, are represented in terms of their Laplace's transforms (upper-case letters with lower-case subscripts). We shall see that the capacitances playing the biggest roles in the frequency response of a FET are C_{gs} and C_{gd}, which take on the forms

$$C_{gs} \cong \frac{2}{3}WLC_{ox} + WL_{ov}C_{ox} \qquad\qquad C_{gd} \cong WL_{ov}C_{ox} \tag{6.21}$$

The model includes also the stray capacitance C_{gb}, not immediately obvious from the structure of Fig. 6.8, to account for the capacitive coupling between the gate interconnections and the underlying substrate *outside* the active device area. In today's technology, the various capacitances appearing in a MOSFET's small-signal model are in the femtofarad range (1 fF $= 10^{-15}$ F).

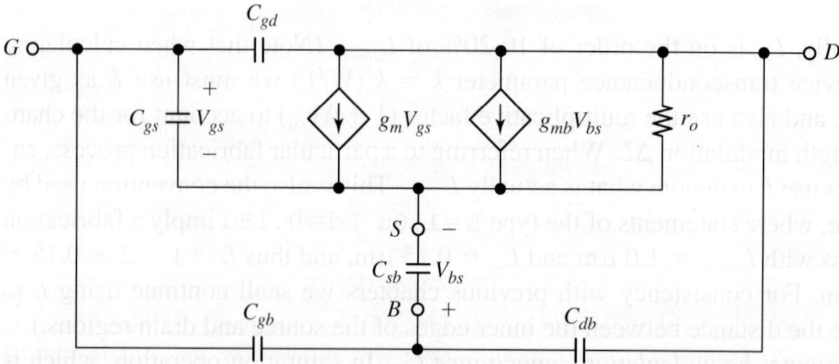

FIGURE 6.9 Complete high-frequency small-signal model for the MOSFET.

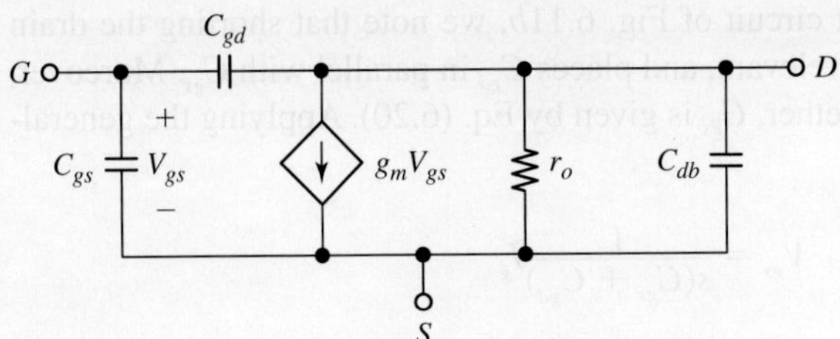

FIGURE 6.10 High-frequency small-signal model for a MOSFET with body and source tied together.

The model of Fig. 6.9 is certainly intimidating for hand calculations—though not necessarily so for PSpice simulations. In the cases in which body and source are tied together, the model simplifies as in Fig. 6.10, where the expression for C_{gs} in Eq. (6.21) is now changed as

$$C_{gs} \cong \frac{2}{3}WLC_{ox} + WL_{ov}C_{ox} + C_{gb} \tag{6.22}$$

Specification of the MOSFET Frequency Response

As in the BJT case, the frequency capability of a MOSFET is expressed in terms of the *transition frequency* f_T at which the magnitude of its small-signal current gain drops to unity. As we know, no current flows into the gate terminal at dc. However, as the operating frequency is increased, the stray capacitances associated with the gate terminal draw increasing current, thus decreasing the current gain of the FET. The transition frequency represents a *figure of merit* for high-frequency operation, and it can either be calculated or measured, using the ac concept of Fig. 6.11a. Specifically, we apply a small-signal ac current I_g to the gate terminal, we find the current I_d drawn by the FET with the drain at ac ground, we take the ratio I_d/I_g, and we finally determine the frequency ω_T at which $|I_d/I_g| = 1$, or 0 dB.

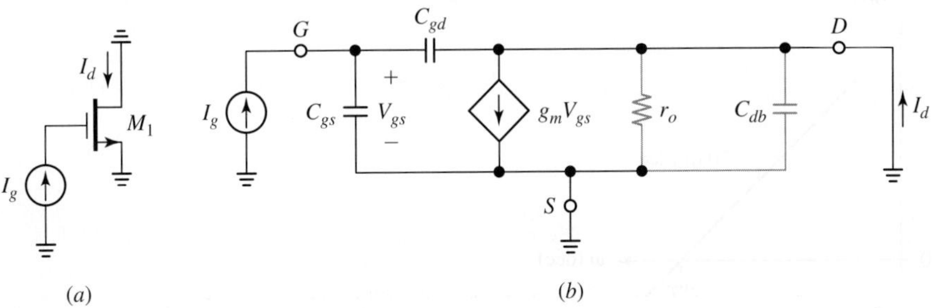

(a)   (b)

FIGURE 6.11 (a) Ac circuit to find f_T, and (b) its small-signal equivalent.

Turning to the equivalent circuit of Fig. 6.11b, we note that shorting the drain to ground renders r_o and C_{db} irrelevant, and places C_{gd} in parallel with C_{gs}. Moreover, with body and source tied together, C_{gs} is given by Eq. (6.20). Applying the generalized Ohm's law,

$$V_{gs} = \frac{1}{s(C_{gs} + C_{gd})} I_g$$

As in BJT case, one can verify that over our frequency range of interest, the current fed forward via C_{gd} is negligible compared to that of the dependent source, so we approximate

$$I_d \cong g_m V_{gs} = \frac{g_m}{s(C_{gs} + C_{gd})} I_g$$

Taking the ratio I_d/I_g and letting $s \to j\omega$, we get

$$\frac{I_d}{I_g} = \frac{1}{j\omega/\omega_T} \tag{6.23}$$

where

$$\omega_T = \frac{g_m}{C_{gs} + C_{gd}} \tag{6.24a}$$

or, alternatively,

$$f_T = \frac{g_m}{2\pi(C_{gs} + C_{gd})} \tag{6.24b}$$

(Note the formal similarity with Eq. (6.12) of the BJT.) Since g_m depends on the bias current I_D, so does f_T. Figure 6.12 shows the Bode plot of the MOSFET's current gain. At low frequencies this gain tends to infinity because the gate draws no dc current. But, at $f = f_T$, the current entering the gate equals that drawn by the drain. In current monolithic MOSFETs, f_T ranges from hundreds of MHz to tens of GHz.

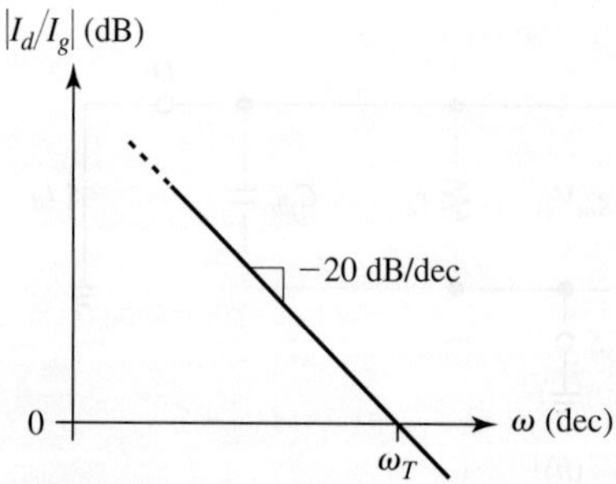

FIGURE 6.12 Bode plot of a MOSFET's current gain.

Exercise 6.2

The current fed forward via C_{gd} in Fig. 6.11b is $I_{gd} = V_{gs}/[1/(j\omega C_{gd})]$. Exploiting the fact that $C_{gd} \ll C_{gs}$, show that for frequencies up to at least ω_T we have $|I_{gd}| \ll |g_m V_{gs}|$, thus justifying the approximation $I_d \cong g_m V_{gs}$.

EXAMPLE 6.4

(a) Assuming V_{GS} has been adjusted for $I_D = 100\ \mu\text{A}$ in the circuit of Fig. 6.13a, find the element values in the small-signal model of the MOSFET, and show the final circuit. The *process parameters* are: $k' = 250\ \mu\text{A/V}^2$, $C_{ox} = 4\ \text{fF}/\mu\text{m}^2$, $\lambda' = 0.02\ \mu\text{m/V}$, $\gamma = 0.5\ \text{V}^{1/2}$, $\phi_p = -0.3\ \text{V}$, $\phi_0 = 0.6\ \text{V}$, and $m = 0.5$. The *device parameters* are $W = 10\ \mu\text{m}$, $L = 1.0\ \mu\text{m}$, $L_{ov} = 25\ \text{nm}$, $C_{sb0} = C_{db0} = 10\ \text{fF}$, and $C_{gb} = 5\ \text{fF}$.

(b) Estimate f_T.

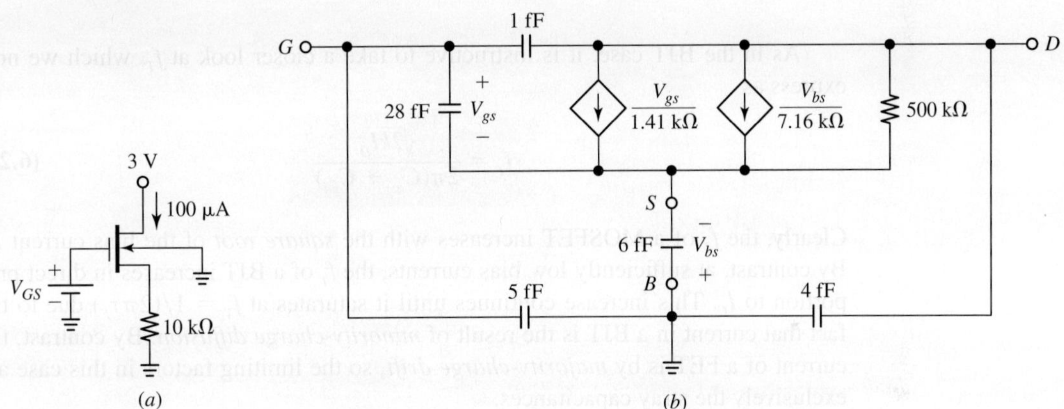

FIGURE 6.13 (a) Circuit of Example 6.2, and (b) the MOSFET's small-signal model values.

Solution

(a) The device transconductance parameter is $k = k'(W/L) = 0.25(10/1) = 2.5\ \text{mA/V}$, and the overdrive voltage is $V_{OV} = \sqrt{2I_D/k} = \sqrt{2 \times 0.1/2.5} = 0.283\ \text{V}$. Since $V_S = 10 \times 0.1 = 1\ \text{V}$ and $V_{DS} = 3 - 1 = 2\ \text{V}$, it follows that $V_{DS} > V_{OV}$, indicating a saturated FET. We have

$$\lambda = \lambda'/L = 0.02/1 = 0.02\ \text{V}^{-1}$$

$$r_o = \frac{1}{\lambda I_D} = \frac{1}{0.02 \times 0.1} = 500\ \text{k}\Omega$$

$$g_m = \sqrt{2kI_D} = \sqrt{2 \times 2.5 \times 0.1} = \frac{1}{1.41\ \text{k}\Omega}$$

$$g_{mb} = \frac{\gamma g_m}{2\sqrt{V_{SB} + 2|\phi_p|}} = \frac{0.5/1.41}{2\sqrt{1 + 2 \times 0.3}} = \frac{1}{7.16\ \text{k}\Omega}$$

$$C_{sb} = \frac{C_{sb0}}{(1 + v_{SB}/\phi_0)^m} = \frac{10}{(1 + 1/0.6)^{0.5}} \cong 6\,\text{fF}$$

$$C_{db} = \frac{C_{db0}}{(1 + v_{DB}/\phi_0)^m} = \frac{10}{(1 + 3/0.6)^{0.5}} \cong 4\,\text{fF}$$

$$C_{gs} \cong \frac{2}{3}WLC_{ox} + WL_{ov}C_{ox} = \frac{2}{3}10 \times 1 \times 4 + 10 \times 0.025 \times 4 \cong 27 + 1 = 28\,\text{fF}$$

$$C_{gd} = 1\,\text{fF}$$

The complete small-signal model is shown in Fig. 6.13b.
(b) Find f_T using Eq. (6.24b), but with C_{gs} as in Eq. (6.22), namely, $C_{gs} = 27 + 1 + 5 = 33$ fF. Thus

$$f_T = \frac{g_m}{2\pi(C_{gs} + C_{gd})} = \frac{1/(1.41 \times 10^3)}{2\pi(33 + 1)10^{-15}} = 3.31\,\text{GHz}$$

As in the BJT case, it is instructive to take a closer look at f_T, which we now express as

$$f_T = \frac{\sqrt{2kI_D}}{2\pi(C_{gs} + C_{gd})} \tag{6.25}$$

Clearly, the f_T of a MOSFET increases with the *square root* of the bias current I_D. By contrast, at sufficiently low bias currents, the f_T of a BJT increases in direct proportion to I_C. This increase continues until it saturates at $f_T = 1/(2\pi\tau_F)$ due to the fact that current in a BJT is the result of *minority-charge diffusion*. By contrast, the current of a FET is by *majority-charge drift*, so the limiting factors in this case are exclusively the stray capacitances.

Of all capacitances in a MOSFET, the dominant one is usually the first component in the right-hand side of Eq. (6.21). If we approximate Eq. (6.24b) as $f_T \cong g_m/(2\pi C_{gs})$ with $C_{gs} \cong (2/3)WLC_{ox}$, then

$$f_T \cong \frac{g_m}{2\pi(2/3)WLC_{ox}} = \frac{0.75}{\pi}\frac{g_m}{WLC_{ox}}$$

Letting $g_m = kV_{OV} = [(W/L)\mu_n C_{ox}]V_{OV}$ and simplifying, we can finally place an upper limit on the f_T of a nMOSFET for a given V_{OV} by writing

$$f_{T(\text{max})} = \frac{0.75}{\pi}\frac{\mu_n}{L^2}V_{OV} \tag{6.26}$$

(For a pMOSFET replace μ_n with μ_p.) It is apparent that for fast operation a MOSFET should have a very *short* channel, and it should be of the n type as electrons are 2 to 3 times as mobile as the holes of a p-type. Note the striking similarity to the BJT limit of Eq. (6.15), except for the replacement of the (fixed) thermal voltage V_T by the (user-imposed) overdrive voltage V_{OV}: the higher V_{OV} the faster the MOSFET is likely to operate.

6.3 FREQUENCY RESPONSE OF CE/CS AMPLIFIERS

With the high-frequency models in hand, we are now eager to investigate the frequency response of the most popular transistor configurations. The first to come to mind are the common-emitter (CE) and common-source (CS) configurations, the workhorses of voltage amplification. Their ac equivalents, shown in Figs. 6.14a and 6.15a, could refer to any of the *discrete* CE realizations of Chapters 2 and 3, so long as the operating frequencies are such that the ac coupling and bypassing capacitors act as ac short circuits. But, they could also represent the differential-mode half-circuits of the dc-coupled EC and SC pairs of Chapter 4. Consequently, the analysis we are about to undertake is quite general.

Replacing the transistors with their high-frequency small-signal models we obtain the ac circuits of 6.14b and 6.15b. (For the time being we are deliberately ignoring the output-node parasitics, namely, C_s for the BJT and C_{db} for the FET, so we can focus on the two remaining capacitances and develop valuable intuition in the process. These parasitics will be taken up later in this section.) The two circuits exhibit inevitable differences, but also formal similarities. In fact, using simple circuit transformations, we can reduce them to a *common* form, and then perform a *single* analysis on this common circuit to avoid duplication (mercifully, opportunities of this type will arise often as we proceed).

- Turning first to the **BJT equivalent** of Fig. 6.14b, we simplify its left side by applying Thévenin's theorem, and its right side by combining the two parallel resistances into a single one,

$$V_i = \frac{r_\pi}{R_{sig} + r_b + r_\pi}V_{sig} \qquad R_1 = (R_{sig} + r_b)//r_\pi \qquad R_2 = R_C//r_o \qquad \textbf{(6.27)}$$

  After this, the circuit of Fig. 6.14b reduces to that of Fig. 6.16, where the input-node capacitance C_1 plays the role of C_π, the feedback capacitance C_f plays that of C_μ, and V_1 that of V_π.
- Likewise, turning to the **MOS equivalent** of Fig. 6.15b and letting

$$V_i = V_{sig} \qquad R_1 = R_{sig} \qquad R_2 = R_D//r_o \qquad \textbf{(6.28)}$$

  we reduce it to the same equivalent of Fig. 6.16, where now the input-node capacitance C_1 plays the role of C_{gs}, the feedback capacitance C_f plays that of C_{gd}, and V_1 that of V_{gs}.

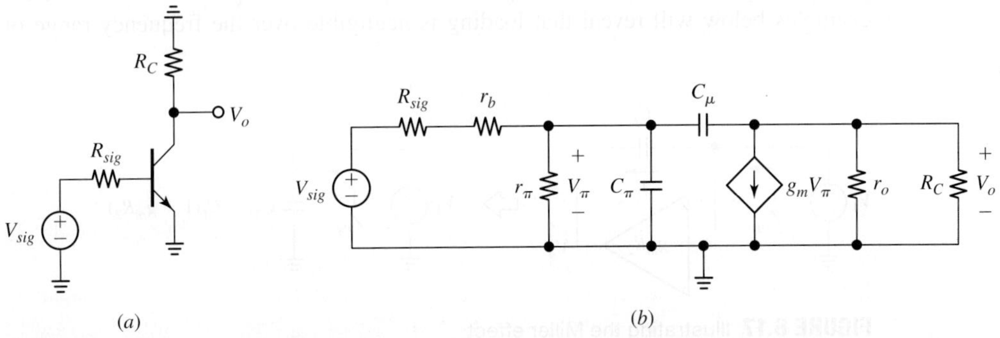

(a)   (b)

FIGURE 6.14 (a) Ac equivalent of the CE amplifier, and (b) its high-frequency small-signal model.

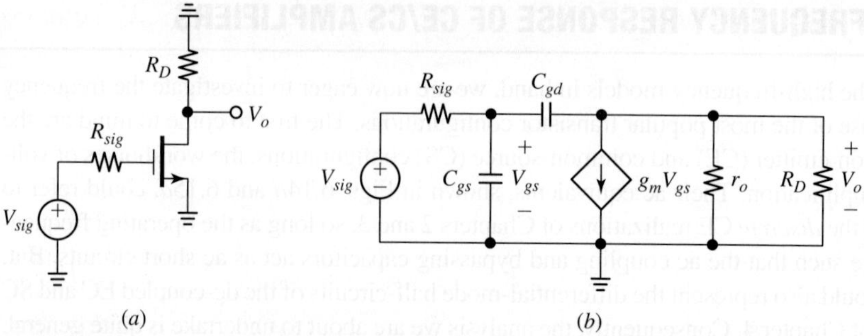

FIGURE 6.15 (a) Ac equivalent of the CS amplifier, and (b) its high-frequency small-signal model.

Let us investigate the common circuit of Fig. 6.16, and then adapt our results to the BJT and the FET circuits of Figs. 6.14 and 6.15 with the aid of Eqs. (6.27) and (6.28), respectively. The analysis of this circuit is facilitated further if we take advantage of the *Miller effect*, to be discussed next. The results, though not exact, will prove quite insightful, as we shall see.

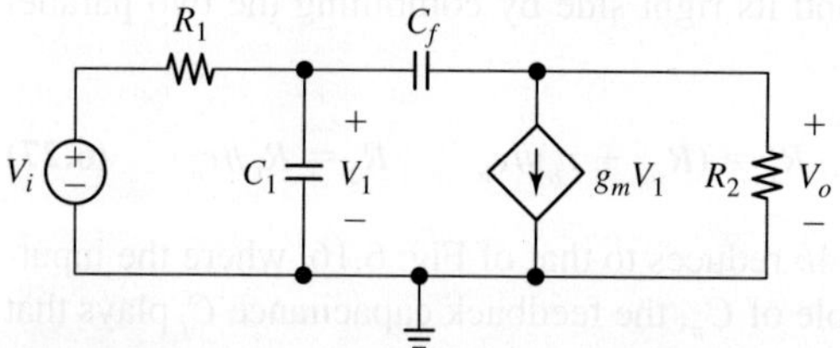

FIGURE 6.16 General model for the CE and CS amplifiers.

The Miller Effect

With C_f absent, the circuit of Fig. 6.16 gives $V_o = -g_m R_2 V_1$, indicating that we can model its portion from V_1 to V_o with an inverting amplifier as in Fig. 6.17. With C_f *present* there will be some loading at the output of the amplifier; however, actual examples below will reveal that loading is negligible over the frequency range of

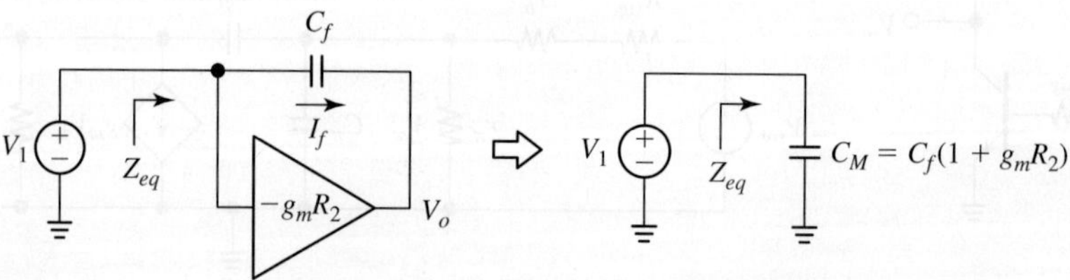

FIGURE 6.17 Illustrating the Miller effect.

interest. We now wish to find the equivalent impedance Z_{eq} seen looking toward the right by the V_1 source of Fig. 6.17. By Ohm's law, the current supplied by V_1 is

$$I_f = \frac{V_1 - V_o}{1/(sC_f)} \cong sC_f[V_1 - (-g_mR_2V_1)] = sC_f(1 + g_mR_2)V_1 = sC_MV_1 \qquad \textbf{(6.29)}$$

Where

$$C_M = C_f(1 + g_mR_2) \qquad \textbf{(6.30)}$$

Letting $Z_{eq} = V_1/I_f = 1/(sC_M)$, we conclude that the block consisting of the amplifier and its feedback capacitance appears to the V_1 source as a mere *equivalent capacitance* C_M toward ground. This capacitance is $(1 + g_mR_2)$ times as large as C_f. This intriguing phenomenon is called the *Miller effect* for John M. Miller, who first described it in 1920. The term $(1 + g_mR_2)$ is called the *Miller multiplier* and C_M the *Miller capacitance*. In general, $C_M \gg C_f$.

EXAMPLE 6.5

To better understand the Miller effect, let us investigate the process of *changing* the voltage across a 1-pF capacitor from 0 V to 1 mV, first for the case in which the capacitor is *grounded*, then for the case in which it is placed in the *feedback path* of an amplifier with a gain of -99 V/V. Let us compare the two cases and comment.

Solution

With reference to Fig. 6.18a we note that applying $\Delta V = 1$ mV causes a charge transfer of $\Delta Q = C\Delta V = 10^{-12} \times 10^{-3} = 10^{-15}$ C $= 1$ fC. Consider next the case in which the (initially discharged) capacitor is in the feedback path of the amplifier as in Fig. 6.18b. As we raise the left plate from 0 V to 1 mV, the amplifier will lower the right plate from 0 V to -99 mV, causing a 100-mV total change *across* the capacitor. The charge transfer is now $\Delta Q = C\Delta V = 10^{-2} \times (100 \times 10^{-3}) = 100$ fC. Even though the physical capacitance of Fig. 6.18b is still 1 pF, the charge transfer is 100 times that of Fig. 6.18a. Yet, the applied voltage is still 1 mV. So if we regroup the terms as $\Delta Q = (10^{-12} \times 100) \times 10^{-3} = (100$ pF$)(1$ mV$) = 100$ fC, we can state that things go *as if* the input source were driving a fictitious capacitance 100 times as large, or $C_M = 100$ pF!

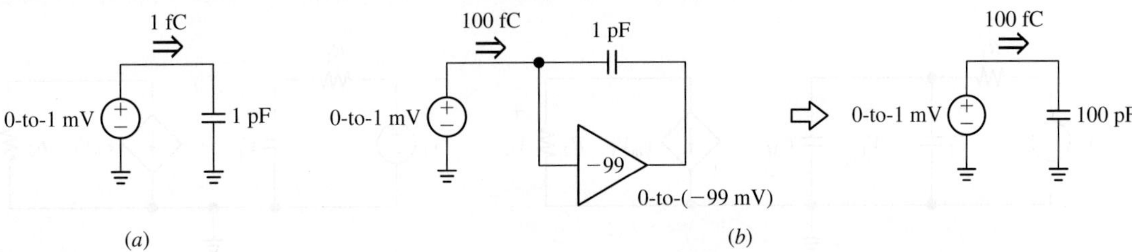

FIGURE 6.18 Quantitative illustration of the Miller effect.

Analysis Using the Miller Approximation

Thanks to the Miller effect, the circuit of Fig. 6.16 simplifies as in Fig 6.19*a*. In fact, we can combine the two parallel capacitances into a single *total capacitance* C_t,

$$C_t = C_1 + C_M \tag{6.31}$$

and work with the even simpler circuit of Fig. 6.19*b*. Using the ac voltage divider formula,

$$V_o = -g_m V_1 R_2 = -g_m R_2 \frac{1/sC_t}{R_1 + 1/sC_t} V_i = \frac{-g_m R_2}{1 + sR_1 C_t} V_i$$

so the voltage gain of the circuit is

$$a(s) = \frac{V_o}{V_i} = \frac{-g_m R_2}{1 + sR_1 C_t} \tag{6.32}$$

The value of s that makes the denominator vanish and therefore causes $a(s)$ to blow up to infinity is referred to as a *pole*. This value is

$$s = -\frac{1}{R_1 C_t} \tag{6.33}$$

indicating a *real* and *negative* pole. Letting $s \to j\omega$ in Eq. (6.32) gives the *frequency response*, which we express in the standard for of Eq. (6A.1) of Appendix 6A as

$$a(j\omega) = \frac{a_0}{1 + j\omega/\omega_p} \tag{6.34}$$

where

$$a_0 = -g_m R_2 \tag{6.35a}$$

is the value of a in the limit $\omega \to 0$, aptly called the *low-frequency gain*, and

$$\omega_p = \frac{1}{R_1 C_t} \tag{6.35b}$$

is called the *pole frequency.*

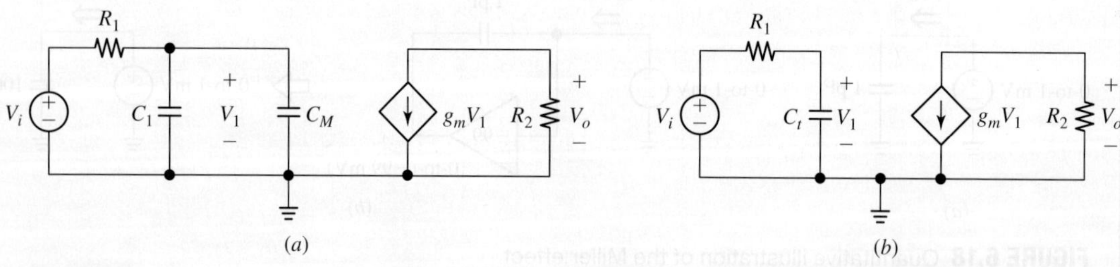

FIGURE 6.19 Equivalent-circuit simplifications using the Miller approximation.

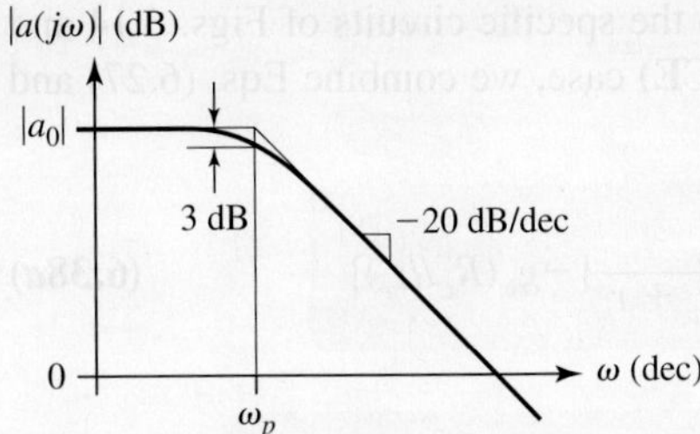

FIGURE 6.20 Magnitude gain plot for the circuit of Fig. 6.19b.

The magnitude Bode plot is shown in Fig. 6.20 (see also Bode plots in Appendix 6A). It is instructive to justify this plot using *physical insight*. With reference to Fig. 6.19b, we observe that the frequency response is governed by the ac voltage divider formed by C_t with R_1, the equivalent resistance seen by C_t itself. The impedance presented by C_t is $Z_t(j\omega) = 1/(j\omega C_t)$, and depending on how its magnitude $|Z_t(j\omega)| = 1/(\omega C_t)$ compares with R_1, we have three significant cases:

- At *low frequencies* we have $|Z_t(\omega)| \gg R_1$, indicating that C_t approximates an *open circuit* compared to R_1. Consequently, $V_1 \to V_i$, and thus $|a| \to a_0$. This is the situation we have been dealing with before embarking upon the present chapter.
- At *high frequencies* we have $|Z_t(\omega)| \ll R_1$, indicating that C_t now approximates a *short circuit* compared to R_1. Consequently, $V_1 \to 0$, and $|a|$ rolls off with frequency, as shown.
- The *borderline* between the two limiting cases occurs when $\omega = \omega_p$. Rewriting Eq. (6.35b) as $1/(\omega_p C_t) = R_1$, we see that at this frequency we have

$$|Z_t(\omega_p)| = R_1 \tag{6.36}$$

We now have a physical interpretation for ω_p: this is the frequency at which the capacitance's impedance *equals*, in magnitude, the equivalent resistance seen by the capacitance itself. In the MOS case of Fig. 6.15 this resistance is simply R_{sig}, but in the bipolar case of Fig. 6.14 it is $r_\pi//(R_{sig} + r_b)$. A circuit designer will always use physical insight to check the results of mathematical derivations as well as to develop a feel for the workings of the circuit at hand.

- Because of the gain roll-off with frequency, an amplifier is in effect a *low-pass filter*, this being the reason why ω_p is also variously known as *corner frequency*, *cutoff frequency*, or also *break frequency*. At $\omega = \omega_p$, $|V_1|$ is down to $1/\sqrt{2}$ (=0.707, or -3-dB) of its low-frequency value, so ω_p is also called the -3-*dB frequency*. Since the power of an ac signal is proportional to the square of its magnitude, another name of ω_p is *half-power frequency*. The gain-bandwidth product is

$$GBP = |a_0| \times f_p \tag{6.37}$$

We are now ready to apply our results to the specific circuits of Figs. 6.14 and 6.15. Turning first to the **common-emitter (CE)** case, we combine Eqs. (6.27) and (6.35) to write

$$a_{0(BJT)} = \frac{V_o}{V_{sig}}\bigg|_{\omega \to 0} = \frac{r_\pi}{R_{sig} + r_b + r_\pi}[-g_m(R_C//r_o)] \qquad (6.38a)$$

$$\omega_{p(BJT)} = \frac{1}{\{r_\pi//(R_{sig} + r_b)\} \times \{C_\pi + C_\mu[1 + g_m(R_C//r_o)]\}} \qquad (6.38b)$$

Turning next to the **common-source (CS)** case, we combine Eqs. (6.28) and (6.35) to write

$$a_{0(FET)} = \frac{V_o}{V_{sig}}\bigg|_{\omega \to 0} = -g_m(R_D//r_o) \qquad (6.39a)$$

$$\omega_{p(FET)} = \frac{1}{R_{sig}\{C_{gs} + C_{gd}[1 + g_m(R_D//r_o)]\}} \qquad (6.39b)$$

To develop a better feel, let us look at some actual examples.

EXAMPLE 6.6   Let the CE amplifier of Fig. 6.14 use a BJT with $\beta_0 = 200$, $V_A = 50$ V, $r_b = 200\ \Omega$, and $C_\mu = 0.5$ pF. The BJT is biased at $I_C = 1$ mA, where it exhibits $f_T = 500$ MHz. Moreover, let $R_{sig} = 1$ kΩ and $R_C = 5$ kΩ.

(a) Estimate the amplifier's low-frequency gain as well as its -3-dB frequency. What is the gain-bandwidth product of this amplifier?

(b) Verify that loading of the output node by the feedback capacitance is negligible over the frequency range of interest ($f \leq f_p$), thus validating the Miller approximation.

Solution

(a) Proceeding as usual, we find $g_m = 1/(26\ \Omega)$, $r_\pi = 5.2$ kΩ, and $r_o = 50$ kΩ. Moreover,

$$R_1 = r_\pi//(R_{sig} + r_b) = 5.2//(1 + 0.2) = 0.975\ \text{k}\Omega$$

$$R_2 = R_C//r_o = 5//50 = 4.55\ \text{k}\Omega$$

The low-frequency gain is

$$a_0 = \frac{r_\pi}{R_{sig} + r_b + r_\pi}(-g_m R_2) = \frac{5.2}{1 + 0.2 + 5.2}(-4.55/0.026)$$

$$\cong 0.81 \times (-175) = -142\ \text{V/V}$$

By Eq. (6.12b) we have

$$C_\pi = \frac{g_m}{2\pi f_T} - C_\mu = \frac{1/26}{2\pi \times 500 \times 10^6} - 0.5 \times 10^{-12} \cong 12 \text{ pF}$$

The Miller capacitance is

$$C_M = C_\mu[1 + g_m(R_C//r_o)] = 0.5 \times 10^{-12}[1 + 175] = 88 \text{ pF}$$

indicating a Miller multiplier of 176. The total capacitance is thus

$$C_t = C_\pi + C_M = 12 + 88 = 100 \text{ pF}$$

Clearly, the Miller capacitance plays a dominant role in this amplifier. Together, R_1 and C_t create a pole frequency at

$$f_p = \frac{1}{2\pi R_1 C_t} = \frac{1}{2\pi \times 975 \times 100 \times 10^{-12}} = 1.63 \text{ MHz}$$

The gain-bandwidth product is

$$\text{GBP} = |a_0| \times f_p = 142 \times 1.63 \cong 230 \text{ MHz}$$

(b) By Eq. (6.29), the current fed forward via C_μ is maximized at the upper edge of the frequency band of interest, where

$$I_f(jf_p) \cong j2\pi f_p C_M V_\pi = j2\pi \times 1.63 \times 10^6 \times 88 \times 10^{-12} V_\pi = jV_\pi/(1110\ \Omega)$$

On the other hand, the current drawn by the dependent source is

$$g_m V_\pi = V_\pi/(26\ \Omega)$$

The ratio of the two currents is thus

$$\frac{|I_f(jf_p)|}{|g_m V_\pi|} = \frac{26}{1110} \ll 1$$

This confirms the validity of the approximation $V_o \cong -g_m R_2 V_\pi$ for the BJT.

EXAMPLE 6.7

Repeat Example 6.6, but for the CS amplifier of Fig. 6.15. Assume the MOSFET has $k = 8 \text{ mA/V}^2$, $\lambda = 1/(50\text{ V})$, and $C_{gd} = 0.1 \text{ pF}$, and is biased at $I_D = 1 \text{ mA}$, where it exhibits $f_T = 500 \text{ MHz}$. Moreover, let $R_{sig} = 10 \text{ k}\Omega$ and $R_D = 5 \text{ k}\Omega$.

Solution

(a) Proceeding as usual we find $g_m = 4 \text{ mA/V}$ and $r_o = 50 \text{ k}\Omega$, so the low-frequency gain is

$$a_0 = -g_m(R_D//r_o) = -4(5//50) = -4 \times 4.55 = -18.2 \text{ V/V}$$

By Eq. (6.24b) we have

$$C_{gs} = \frac{g_m}{2\pi f_T} - C_{gd} = \frac{4 \times 10^{-3}}{2\pi 500 \times 10^6} - 0.1 \times 10^{-12} \cong 1.17 \text{ pF}$$

The Miller capacitance is

$$C_M = C_{gd}[1 + g_m(R_D /\!/ r_o)] = 0.1 \times 10^{-12}[1 + 18.2] \cong 1.92 \text{ pF}$$

indicating a Miller multiplier of 19.2. (In general, this multiplier is lower in FETs than in BJTs because a FET has notoriously a lower g_m.) The total capacitance is thus

$$C_t = C_{gs} + C_M = 1.17 + 1.92 = 3.09 \text{ pF}$$

so the Miller capacitance plays a dominant role also in this amplifier. The resistance seen by C_t is now R_{sig}. Together, they create a pole frequency at

$$f_p = \frac{1}{2\pi R_{sig} C_t} = \frac{1}{2\pi \times 10^4 \times 3.09 \times 10^{-12}} = 5.15 \text{ MHz}$$

The gain-bandwidth product is

$$\text{GBP} = |a_0| \times f_p = 18.2 \times 5.2 \cong 94 \text{ MHz}$$

(b) By Eq. (6.29), the current fed forward via C_{gd} at the upper edge of the frequency band of interest is

$$I_f(jf_p) \cong j2\pi f_p C_M V_{gs} = j2\pi \times 5.2 \times 10^6 \times 1.92 \times 10^{-12} V_{gs} = jV_{gs}/(16 \text{ k}\Omega)$$

whereas the current drawn by the dependent source is

$$g_m V_{gs} = V_{gs}/(0.25 \text{ k}\Omega)$$

Consequently, the ratio of the two currents is

$$\frac{|I_f(jf_p)|}{|g_m V_{gs}|} = \frac{0.25}{16} \ll 1$$

thus confirming the validity of the approximation $V_o \cong -g_m R_2 V_{gs}$ for the MOSFET.

A More Accurate Analysis

To assess the accuracy of the Miller approximation and also to gain additional insight into circuit behavior, let us perform the exact analysis of the small-signal circuit of Fig. 6.16. Since we are at it, we may as well generalize by including also the output-node capacitance C_2, as in Fig. 6.21. As we know, the collector of a monolithic BJT exhibits the collector-to-substrate capacitance C_s, and the drain of a FET exhibits the drain-to-body capacitance C_{db}. Moreover, in actual application, the output node is likely to be loaded by an external capacitance C_L, so, in general, $C_2 = C_s + C_L$ for the BJT, and $C_2 = C_{db} + C_L$ for the FET.

Applying KCL at the node to the left of C_f gives

$$\frac{V_i - V_1}{R_1} = \frac{V_1}{1/(sC_1)} + \frac{V_1 - V_o}{1/(sC_f)}$$

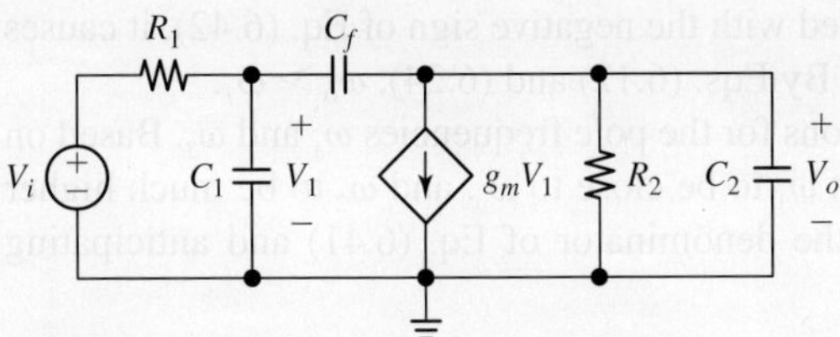

FIGURE 6.21 Ac circuit for a more accurate analysis of the CE/CS amplifiers.

Likewise, KCL at the node to the right of C_f gives

$$\frac{V_1 - V_o}{1/sC_f} = g_m V_1 + \frac{V_o}{R_2} + \frac{V_o}{1/sC_2}$$

Eliminating V_1 and solving for the ratio V_o/V_i we get, after a bit of algebraic labor,

$$a(s) = \frac{V_o}{V_i}$$

$$= \frac{(-g_m R_2) \times (1 - sC_f/g_m)}{1 + s\{R_1[C_1 + C_f(1 + g_m R_2)] + R_2(C_f + C_2)\} + s^2 R_1 R_2(C_1 C_f + C_1 C_2 + C_f C_2)} \tag{6.40}$$

The denominator is a quandratic polynomial in s, so $a(s)$ admits *two* poles. Denoting the corresponding pole frequencies as ω_1 and ω_2, we express gain more concisely in the standard form of Eq. (6A.1) in the Appendix,

$$a(s) = a_0 \frac{1 - s/\omega_0}{(1 + s/\omega_1)(1 + s/\omega_2)} \tag{6.41}$$

where

$$a_0 = -g_m R_2 \tag{6.42}$$

is the familiar *low-frequency gain*, and

$$\omega_0 = \frac{g_m}{C_f} \tag{6.43}$$

is the *zero frequency* of $a(s)$. Physically, the current fed forward via C_f at this frequency *equals* the current drawn by the dependent source, resulting in a net current of zero through the parallel combination of R_2 and C_2. Consequently, V_o drops to zero, implying a *gain of zero* at this frequency. As a physical check, when $V_o = 0$ we have $I_f = (V_\pi - 0)/(1/sC_f) = sC_f V_\pi$, so imposing $sC_f V_\pi = g_m V_\pi$ yields $s = g_m/C_f$. In the s plane this zero is located on the *positive real axis*. Note that for $\omega > \omega_0$ the current through C_f *exceeds* that of the dependent source, indicating gain-polarity *reversal*. This provides a physical justification for the presence of the negative sign in the numerator of Eq. (6.41); combined with the negative sign of Eq. (6.42), it causes gain to become positive for $\omega > \omega_0$. By Eqs. (6.12) and (6.24), $\omega_0 \gg \omega_T$.

We now wish to derive expressions for the pole frequencies ω_1 and ω_2. Based on the Miller approximation, we expect ω_1 to be close to ω_p, and ω_2 to be much higher than ω_p. Consequently, expanding the denominator of Eq. (6.41) and anticipating $\omega_2 \gg \omega_1$, we write

$$a(s) = a_0 \frac{1 - s/\omega_0}{1 + s(1/\omega_1 + 1/\omega_2) + s^2/(\omega_1\omega_2)} \cong a_0 \frac{1 - s/\omega_0}{1 + s/\omega_1 + s^2/(\omega_1\omega_2)} \tag{6.44}$$

Equating the coefficients of s in the denominators of Eqs. (6.40) and (6.44), we readily find

$$\omega_1 = \frac{1}{R_1[C_1 + C_f(1 + g_m R_2 + R_2/R_1)] + R_2 C_2} \tag{6.45}$$

We observe that in the limit $C_2 \to 0$ this expression differs from that of ω_p derived earlier only in the denominator term R_2/R_1. But, $R_2/R_1 \ll g_m R_2$, so $\omega_1 \cong \omega_p$, confirming that the Miller approximation is an excellent one, considering also how quicker is its derivation. Likewise, equating the coefficients of s^2 in the denominators of Eqs. (6.40) and (6.44),

$$\omega_2 = \frac{1}{R_1 R_2 (C_1 C_f + C_1 C_2 + C_f C_2)\omega_1} \tag{6.46}$$

The next examples will confirm that $\omega_2 \gg \omega_1$, indicating that the frequency response of Fig. 6.20, though approximate, provides a good indication of the actual response over the frequency range of interest.

EXAMPLE 6.8   (a) Find f_0, f_1, and f_2 for the CE amplifier of Example 6.6. Compare with the example and comment.

(b) Repeat, but taking into account a substrate capacitance $C_s = 1$ pF.

(c) Verify with PSpice.

Solution

(a) For the BJT, Eq. (6.43) predicts a zero frequency at

$$f_0 = \frac{g_m}{2\pi C_\mu} = \frac{1/2\pi}{26 \times 0.5 \times 10^{-12}} \cong 12 \text{ GHz}$$

Moreover, with $R_1 = 975\ \Omega$, $R_2 = 4.55$ kΩ, and $C_2 = 0$, Eqs. (6.45) and (6.46) predict pole frequencies at

$$f_1 = \frac{1/2\pi}{R_1[C_\pi + C_\mu(1 + g_m R_2 + R_2/R_1)]}$$

$$= \frac{1/2\pi}{975[12 + 0.5(1 + 175 + 4.55/0.975)]10^{-12}} = 1.56 \text{ MHz}$$

and

$$f_2 = \frac{1/(2\pi)^2}{R_1 R_2 C_\pi C_\mu f_1}$$

$$= \frac{1/(2\pi)^2}{975 \times 4.55 \times 10^3 \times 12 \times 10^{-12} \times 0.5 \times 10^{-12} \times 1.56 \times 10^6}$$

$$\cong 600 \text{ MHz}$$

Both f_0 and f_2 are well above f_1, so they are of irrelevant consequence in this example and can be ignored. The first pole frequency ($f_1 = 1.56$ MHz) is slightly lower than that predicted via the Miller approximation ($f_p = 1.63$ MHz), indicating that the Miller estimate suffices for all practical pusposes.

(b) Recalculating with $C_2 = C_s = 1$ pF we get $f_0 = 12$ GHz, $f_1 = 1.55$ MHz, and $f_2 \cong 1.3$ GHz. The effect of C_s is negligible in this example.

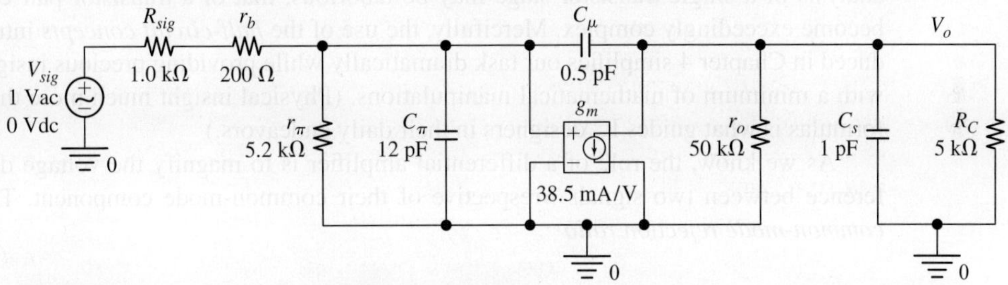

FIGURE 6.22 PSpice circuit to display the gain of the CE amplifier of Example 6.8.

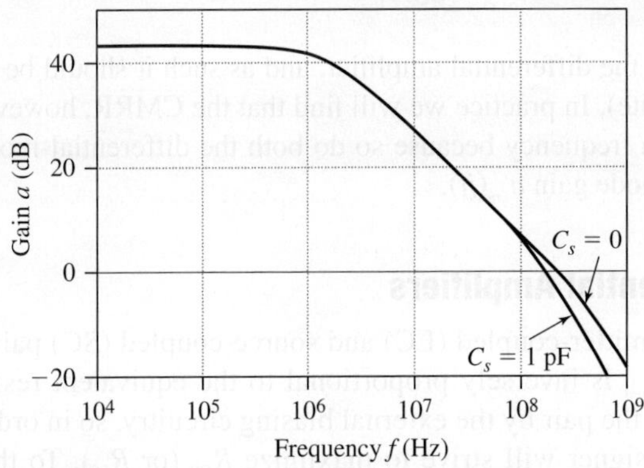

FIGURE 6.23 Gain plot of the circuit of Fig. 6.22.

(c) Using the PSpice circuit of Fig. 6.22 we get the gain plot of Fig. 6.23. Using the cursor facilty, we find $|a_0| = 43.057$ dB, or 142 V/V, and $f_{-3dB} = 1.522$ MHz, in agreement with the calculations. The plot confirms the minimal role played by C_s in this example.

Remark: an IC designer will simulate an amplifier using a *PSpice model* for the transistor. But here, for pedagogical purposes, it is more convenient to work with the homebrew model depicted in Fig. 6.22.

Exercise 6.3

Find f_0, f_1, and f_2 for the CS amplifier of Example 6.7, if $C_{db} = 0.1$ pF. Compare with the results obtained there and comment.

Ans. $f_0 = 6.4$ GHz, $f_1 = 5.0$ MHz, and $f_2 \cong 1$ GHz; $f_p = 5.15$ MHz ($\cong f_1$).

6.4 FREQUENCY RESPONSE OF DIFFERENTIAL AMPLIFIERS

Given the importance of the differential amplifier as an analog building block, it is only appropriate that we investigate its frequency response in detail. If the frequency analysis of a single-transistor stage may be laborious, that of a transistor pair can become exceedingly complex. Mercifully, the use of the *half-circuit concepts* introduced in Chapter 4 simplifies our task dramatically while providing precious insight with a minimum of mathematical manipulations. (Physical insight much more than formulas is what guides IC designers in their daily endeavors.)

As we know, the role of a differential amplifier is to magnify the voltage difference between two signals irrespective of their common-mode component. The *common-mode rejection ratio*

$$\text{CMRR} = \left| \frac{a_{dm}(jf)}{a_{cm}(jf)} \right| \tag{6.47}$$

constitutes a figure of merit of the differential amplifier, and as such it should be as large as possible (ideally, infinite). In practice we will find that the CMRR, however high initially, deteriorates with frequency because so do both the differential-mode gain $a_{dm}(jf)$ and the common-mode gain $a_{cm}(jf)$.

Resistive-Loaded Differential Amplifiers

Figure 6.24 shows the basic emitter-coupled (EC) and source-coupled (SC) pairs. Recall from Chapter 4 that a_{cm} is inversely proportional to the equivalent resistance R_{EE} (or R_{SS}) presented to the pair by the external biasing circuitry, so in order to maximize the CMRR a designer will strive to maximize R_{EE} (or R_{SS}). To this end, the current sinking transistor Q_3 (or M_3), whose biasing details have been omitted for simplicity, is likely to be part of a very high output-resistance topology such as the Wilson or the cascode types. As a rule, the impedance Z_{EE} (or Z_{SS}) presented to the EC (or SC) pair consists of a *resistive* component R_{EE} (or R_{SS}) in parallel with a *capacitive* component C_{EE} (or C_{SS}). As we are about to see, it is precisely the capacitive component that causes $a_{cm}(jf)$, and thus CMRR, to deteriorate with frequency.

To investigate the CMMR, we first need to find the differential-mode and common-mode gains $a_{dm}(jf)$ and $a_{cm}(jf)$. We shall do so using the differential-mode and common-mode half circuits. Though the analysis will be carried out for the

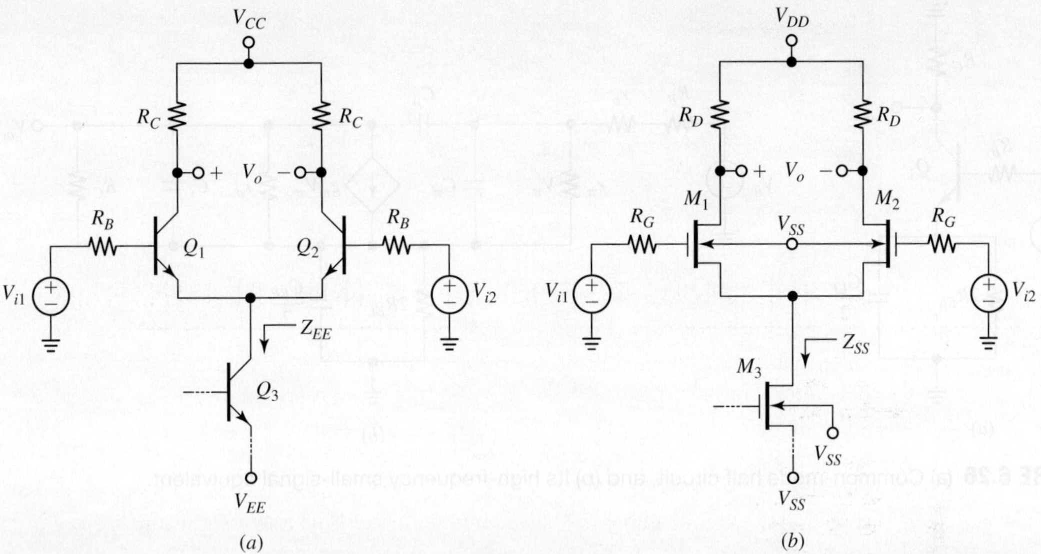

FIGURE 6.24 (a) The EC and (b) SC pairs with resistive loads.

EC pair, the results are readily adapted to the SC pair as well. To find $a_{dm}(jf)$ we use the half-circuit equivalent of Fig. 6.25. This is the familiar CE configuration, whose gain contains a *dominant pole* due primarily to the Miller capacitance. Adapting Eq. (6.45),

$$f_{p(dm)} \cong \frac{1/2\pi}{[(R_B + r_b)//r_\pi] \times \{C_\pi + C_\mu[1 + g_m(R_C//r_o)]\} + (R_C//r_o) \times C_s} \qquad (6.48)$$

To find $a_{cm}(jf)$ we use the half-circuit equivalent of Fig. 6.26. Note that as we split the impedance Z_{EE} into two identical parts, R_{EE} must be *doubled* to give $(2R_{EE})//(2R_{EE}) = R_{EE}$, whereas C_{EE} must be *halved* to give $(C_{EE}/2)//(C_{EE}/2) = C_{EE}/2 + C_{EE}/2 = C_{EE}$.

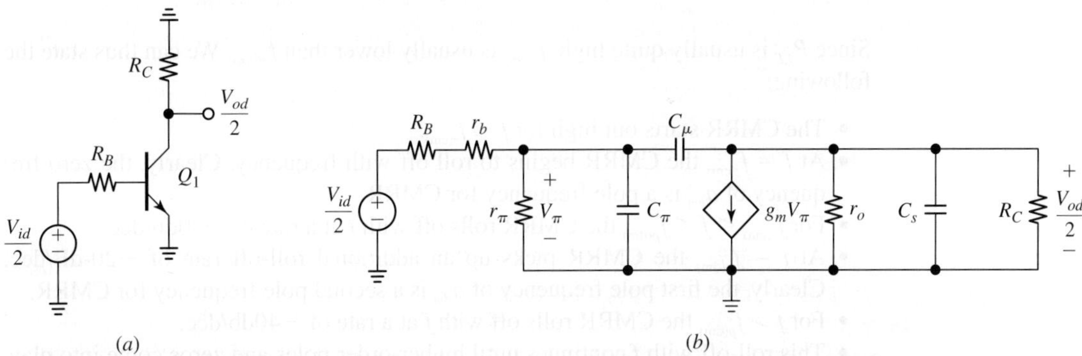

FIGURE 6.25 (a) Differential-mode half circuit, and (b) its high-frequency small-signal equivalent.

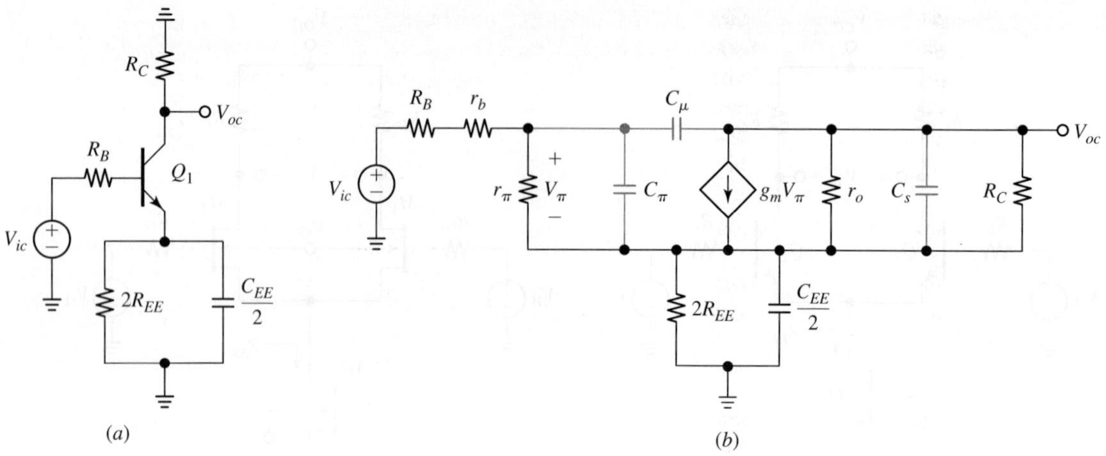

FIGURE 6.26 (a) Common-mode half circuit, and (b) its high-frequency small-signal equivalent.

The circuit of Fig. 6.26 is the familiar CE-ED configuration, but with degeneration now provided by the impedance $2Z_{EE} = (2R_{EE})//[1/s(C_{EE}/2)]$. At sufficiently low frequencies $C_{EE}/2$ acts as an open circuit compared to $2R_{EE}$, so $a_{cm}(jf)$ starts out low and CMRR starts out high. As the operating frequency is increased, the impedance provided by $C_{EE}/2$ decreases, causing Z_{EE} to decrease as well. This, in turn, causes $a_{cm}(jf)$ to increase and CMRR to decrease with frequency. Clearly, $a_{cm}(jf)$ exhibits a *zero frequency* $f_{z(cm)}$. This is the frequency at which the impedance provided by $C_{EE}/2$ *equals*, in magnitude, $2R_{EE}$. This condition yields the familiar result $f_{z(cm)} = 1/[2\pi(2R_{EE})(C_{EE}/2)]$, or

$$f_{z(cm)} = \frac{1}{2\pi R_{EE} C_{EE}} \tag{6.49}$$

Since R_{EE} is usually quite high, $f_{z(cm)}$ is usually lower than $f_{p(dm)}$. We can thus state the following:

- The CMRR starts out high for $f \ll f_{z(cm)}$.
- At $f = f_{z(cm)}$ the CMRR begins to roll off with frequency. Clearly, the zero frequency of a_{cm} is a pole frequency for CMRR.
- For $f_{z(cm)} < f < f_{p(dm)}$ the CMRR rolls off with f at a rate of -20db/dec.
- At $f = f_{p(dm)}$ the CMRR picks up an additional roll-off rate of -20-dB/dec. Clearly, the first pole frequency of a_{dm} is a second pole frequency for CMRR.
- For $f > f_{p(dm)}$ the CMRR rolls off with f at a rate of -40db/dec.
- This roll-off with f continues until higher-order poles and zeros come into play, by which point the CMRR has already deteriorated to fairly low values.

EXAMPLE 6.9

(a) Let the EC pair of Fig. 6.24*a* use BJTs having $\beta_0 = 200$, $V_A = 50$ V, $r_b = 200\ \Omega$, $C_\pi = 25$ pF, $C_\mu = 0.3$ pF, and $C_s = 1$ pF. Moreover, let $R_B = 2$ kΩ and $R_C = 10$ kΩ, and let the emitter-biasing current sink have $I_{EE} = 1$ mA, $R_{EE} = 1$ MΩ, and $C_{EE} = 1.5$ pF. Estimate the low-frequency value of the CMRR as well as its two principal poles.

(b) Use PSpice to display the Bode plots of $|a_{dm}|$, $|a_{cm}|$, and $|a_{dm}/a_{cm}|$, and comment.

Solution

(a) We have $g_m = (200/201) \times (0.5)/26 = 19.1$ mA/V, $r_\pi = 10.5$ kΩ, $r_o = 100$ kΩ, $2R_{EE} = 2$ MΩ, and $C_{EE}/2 = 0.75$ pF. Also, $(R_B + r_b)//r_\pi = 1.82$ kΩ and $R_C//r_o = 9.09$ kΩ. Then, Eqs. (6.48) and (6.49) give

$$f_{p(dm)} \cong \frac{1}{2\pi 1.82 \times 10^3 [25 + 0.3 \times (1 + 19.1 \times 9.09) + (9.09/1.82) \times 1] 10^{-12}}$$

$$= 1.06 \text{ MHz}$$

and

$$f_{z(cm)} \cong \frac{1}{2\pi \times 1 \times 10^6 \times 1.5 \times 10^{-12}} = 106 \text{ kHz}$$

At low frequencies, CMRR $= |a_{dm0}/a_{cm0}| \cong |-g_m(R_C//r_o)/(-R_C/2R_{EE})| = 174/0.005 = 34{,}800 = 90.8$ dB.

(b) To generate the Bode plot of $|a_{dm}|$ we reuse the PSpice circuit of Fig. 6.22, but with the present parameter values. To generate that of $|a_{dm}|$ we use again the same circuit, but after lifting the emitter terminal off ground and inserting the parallel combination of a 2-MΩ resistance and a 0.75-pF capacitance between emitter and ground, in accordance with Fig. 6.26. The plots of $|a_{dm}|$, $|a_{cm}|$, and $|a_{dm}/a_{cm}|$ are shown in Fig. 6.27. The PSpice values $|a_{dm0}/a_{cm0}| = 90.1$ dB, $f_{p(dm)} = 1.05$ MHz, and $f_{z(cm)} = 161$ kHz are in fair agreement with the calculated values.

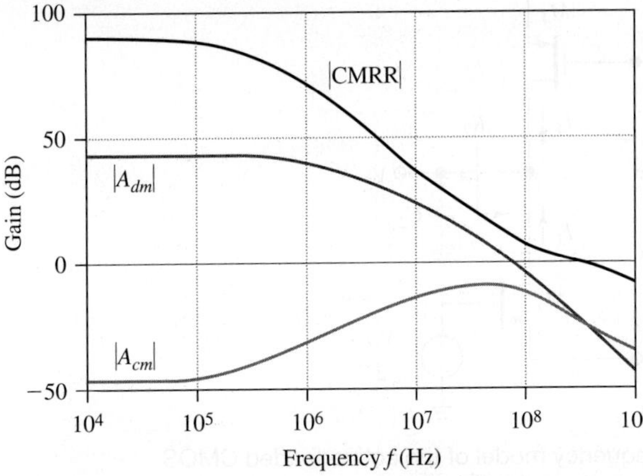

FIGURE 6.27 Bode plots for the EC pair of Example 6.9.

The results obtained above for the EC pair are readily adapted to the **SC pair**. The dominant pole of $a_{dm}(jf)$, being also the second pole of the CMRR, is now

$$f_{p(dm)} \cong \frac{1/2\pi}{R_{sig}\{C_{gs} + C_{gd}[1 + g_m(R_D//r_o)]\} + (R_D//r_o)C_{db}} \tag{6.50}$$

whereas the dominant zero of $a_{cm}(jf)$, being also the first pole of CMRR, is

$$f_{z(cm)} = \frac{1}{2\pi R_{ss}C_{ss}} \tag{6.51}$$

The Bode plots are qualitatively similar to those of Fig. 6.27.

Active-Loaded Differential Amplifiers

With an active load, the circuit loses its symmetry and the additional parasitics introduced by the load transistors tend to complicate the analysis. It is nevertheless possible to gain quick insight into the predominant characteristics of the circuit if we are willing to make suitable approximations. As shown in the CMOS version of Fig. 6.28, the circuit possesses two significant nodes, denoted as V_1 and V_o. With a balanced input drive, the third node, denoted as V_{ss}, would be at ac ground if the drains of M_1 and M_1 were terminated equally. In practice M_1's drain is terminated on the resistance $(1/g_{m3})//r_{o3} \cong 1/g_{m3}$ while M_2's drain is terminated on the resistance r_{o4}, such that $r_{o4} \gg 1/g_{m3}$. This imbalance results in $V_{ss} \neq 0$. Let us nevertheless continue to assume an ac ground at V_{ss} so that we can apply the half-circuit concept to simplify our analysis.

The net capacitances associated with the nodes V_1 and V_o are, respectively,

$$C_1 \cong C_{gs3} + C_{gs4} + C_{db3} + C_{db1} + C_{gd1} \tag{6.52a}$$

$$C_2 \cong C_{gd4} + C_{db4} + C_{gd2} + C_{db2} + C_L \tag{6.52b}$$

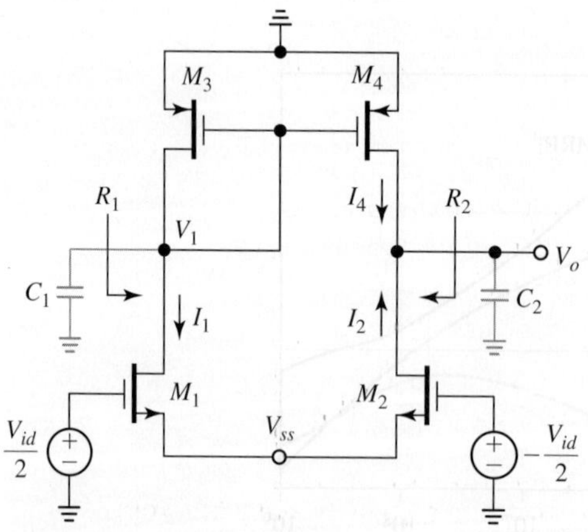

FIGURE 6.28 High-frequency model of the active-loaded CMOS differential amplifier (to facilitate analysis, all stray capacitances have been moved outside of the FETs and lumped together into two equivalent capacitances C_1 and C_2 as shown).

where C_L is the capacitance of the external load, if any. It can be proved that these capacitances see, respectively, the equivalent resistances

$$R_1 = \frac{r_{op}//(2r_{on} + r_{op})}{1 + g_{mp}r_{op}} \tag{6.53a}$$

$$R_2 = r_{op}//r_{on} \tag{6.53b}$$

and as such yield two pole frequencies at $\omega_1 = 1/(R_1C_1)$ and $\omega_2 = 1/(R_2C_2)$.

Exercise 6.4

Prove Eq. (6.53).

Hint: replace each transistor with its small-signal model and use the test method, exploiting also the fact that M_1 and M_2 are matched, and so are M_3 and M_4.

By the superposition principle, $V_o = V_{o4} + V_{o2}$, where V_{o4} is due to M_4's response to $+V_{id}/2$, and V_{o2} to M_2's response to $-V_{id}/2$. By inspection, $V_{o2} = [R_2//(1/sC_2)]I_2$, where $I_2 = g_{m2}V_{id}/2$. Expanding, we get

$$V_{o2} = \frac{g_{m2}R_2}{1 + sR_2C_2}\frac{V_{id}}{2} \tag{6.54a}$$

Similarly, $V_{o4} = [R_2//(1/sC_2)]I_4$, where $I_4 = g_{m4}V_1$. But,

$$V_1 \cong \frac{(1/g_{m3})I_1}{1 + sR_1C_1} = \frac{(1/g_{m3})}{1 + sR_1C_1}g_{m1}\frac{V_{id}}{2}$$

Substituting, and exploiting the fact that $g_{m4}/g_{m3} = 1$ because of matching, we get

$$V_{o4} \cong \frac{1}{1 + sR_1C_1}\frac{g_{m2}R_2}{1 + sR_2C_2}\frac{V_{id}}{2} \tag{6.54b}$$

It is interesting to note that V_{id} contributes to V_o via the *shorter* signal path made up of M_2 as well as via the *longer* signal path made up of M_1-M_3-M_4. Both paths converge at the common output-node pole formed by R_2 and C_2. However, the longer path is *slower* because it includes also the additional pole due to R_1 and C_1. Letting $V_o = V_{o4} + V_{o2}$, collecting and simplifying, we finally get

$$a_{dm}(s) = \frac{V_o}{V_{id}} = g_{m2}R_2\frac{1 + s0.5R_1C_1}{1 + sR_1C_1}\frac{1}{1 + sR_2C_2}$$

Letting $s \to 2\pi f$ and $g_{m1} = g_{m2} = g_{mn}$, we express gain in the insightful form

$$a_{dm}(jf) = a_{dm0}\frac{1 + jf/f_0}{(1 + jf/f_1)(1 + jf/f_2)} \tag{6.55}$$

where

$$a_{dm0} = g_{mn}(r_{op}//r_{op}) \tag{6.56}$$

$$f_1 = \frac{1}{2\pi R_1 C_1} \qquad f_2 = \frac{1}{2\pi R_2 C_2} \qquad f_0 = 2f_1 \tag{6.57}$$

It is apparent that beside the aforementioned pole frequencies f_1 and f_2, $a_{dm}(jf)$ exhibits also a zero frequency f_0 stemming from the direct signal path via M_2. At sufficiently high frequencies the *slower* signal path via M_1-M_3-M_4 is shunted by C_1, leaving only the *faster* path via M_2. Regardless, the overall frequency response is dominated by f_2 due to the fact that $R_2 \gg R_1$.

EXAMPLE 6.10   In the circuit of Fig. 6.28 let all FETs have $g_m = 1$ mA/V, $r_o = 50$ kΩ, $C_{gs} = 50$ fF, and $C_{gd} = C_{db} = 5$ fF. Moreover, assume the circuit is terminated on a load capacitance $C_L = 0.25$ pF. Estimate all the parameters intervening in the calculation of $a_{dm}(jf)$. Hence, verify with PSpice, and comment.

Solution

Plugging the given data into Eqs. (6.52) and (6.53) gives $C_1 = 115$ fF, $R_1 = 735\ \Omega$, $C_2 = 270$ fF, and $R_2 = 25$ kΩ. Plugging in turn into Eqs. (6.56) and (6.57) we get

$$a_{dm0} = 25 \text{ V/V} \qquad f_2 = 23.6 \text{ MHz} \qquad f_1 = 1.88 \text{ GHz} \qquad f_0 = 3.27 \text{ GHz}$$

A PSpice simulation yields the gain plot of Fig. 6.29a. Using the cursor facility we find $a_{dm0} = 24.7$ V/V and $f_{-3\text{ dB}} = 23.3$ MHz, in good agreement with the calculated values. Figure 6.29b shows the magnitude plot of the amplifier's short-circuit transconductance $G_m = I_{o(sc)}/V_{id}$. Short-circuiting the output node eliminates the pole frequency f_2, so the plot evidences only the pole-zero frequency pair f_1 and f_0, both of which are in the GHz range.

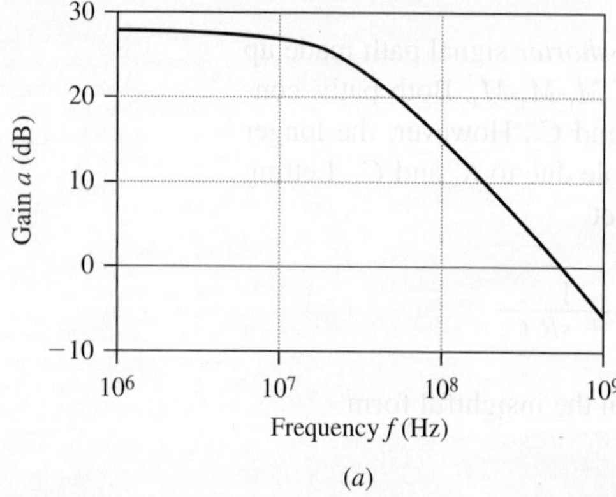

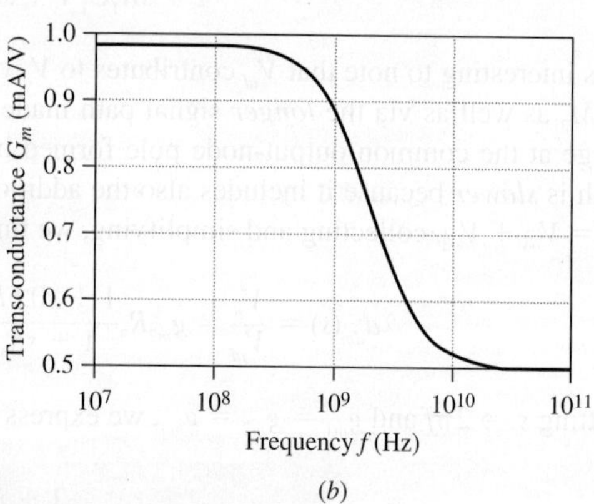

(a)   (b)

FIGURE 6.29 Frequency response of the active-loaded CMOS amplifier of Example 6.10.

6.5 BIPOLAR VOLTAGE AND CURRENT BUFFERS

Recall that the role of a *voltage* buffer is to provide *unity voltage gain* with *high input impedance* and *low output impedance*, whereas that of a *current* buffer is to provide *unity* current gain with *low input impedance* and *high output impedance*. We are about to see that the CC and CB configurations approach the above characteristics over a fairly wide frequency band because they are exempt from the Miller effect, the main frequency bottleneck of CE amplifiers. However, the BJT's stray capacitances do come into play at high frequencies, where they tend to degrade both the gain and the terminal impedances. In particular, impedances that start out *high* tend to *decrease* with frequency, thus exhibiting *capacitive* behavior, and impedances that start out *low* tend to *increase* with frequency (at least up to a point), thus exhibiting *inductive* behavior. Moreover, gain may depart appreciably from unity at high frequencies.

If the study of voltage amplifiers focuses on the frequency behavior of gain because gain is the most important amplifier parameter, the study of buffers must emphasize the frequency behavior of the terminal impedances because impedance transformation is the primary function of a buffer.

Frequency Characteristics of the Emitter Follower

Figure 6.30 shows the ac equivalent of the emitter follower, along with its high-frequency model. Since its right plate is grounded, C_μ is exempt from the Miller multiplication, so we expect the CC to be an *inherently fast* configuration. In fact, to develop a quick (if only approximate) feel for the circuit, let us ignore C_μ altogether (the general case with C_μ present will be addressed in Section 6.7). We wish to investigate the frequency dependence of the impedance $Z_i(j\omega)$ seen by the signal source, the source-to-load voltage gain $a(j\omega) = V_o/V_{sig}$, and the impedance $Z_o(j\omega)$ seen by the load.

The starting point is offered by Eqs. (2.83), (2.84), and (2.86), provided we make the substitutions

$$\beta_0 \to \beta(j\omega) \qquad r_\pi \to z_\pi(j\omega) \tag{6.58}$$

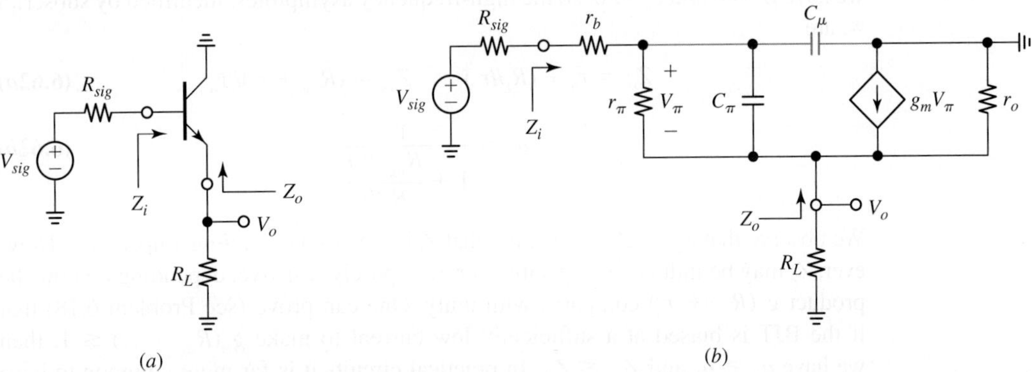

(a)

(b)

FIGURE 6.30 (a) The emitter follower and (b) its high-frequency small-signal model.

where

$$\beta(j\omega) = \frac{\beta_0}{1 + j\omega/\omega_\beta} \tag{6.59a}$$

and

$$z_\pi(j\omega) = r_\pi//\frac{1}{j\omega C_\pi} = \frac{r_\pi}{1 + j\omega r_\pi C_\pi}$$

By Eq. (6.8), $\omega_\beta = 1/[r_\pi(C_\pi + C_\mu)]$. As long as $C_\mu \ll C_\pi$, we can approximate $\omega_\beta \cong 1/(r_\pi C_\pi)$ and write

$$z_\pi(j\omega) \cong \frac{r_\pi}{1 + j\omega/\omega_\beta} = \frac{\beta(j\omega)}{g_m} \tag{6.59b}$$

With the above substitutions, the emitter-follower parameters become

$$Z_i(j\omega) = r_b + z_\pi(j\omega) + [\beta(j\omega) + 1](R_L//r_o)$$

$$Z_o(j\omega) = \frac{R_{sig} + r_b + z_\pi(j\omega)}{\beta(j\omega) + 1}//r_o \tag{6.60a}$$

$$a(j\omega) = \frac{V_o}{V_{sig}} = \frac{1}{1 + \dfrac{R_{sig} + r_b + z_\pi(j\omega)}{[\beta(j\omega) + 1](R_L//r_o)}} \tag{6.60b}$$

At sufficiently *low frequencies*, where C_π acts as an *open circuit*, we have $\beta \to \beta_0$ and $z_\pi \to r_\pi$, so the above expressions tend to their familiar low-frequency forms, which we identify with subscript 0,

$$Z_{i0} = r_b + r_\pi + (\beta_0 + 1)(R_L//r_o) \qquad Z_{o0} = \frac{R_{sig} + r_b + r_\pi}{\beta_0 + 1}//r_o \tag{6.61a}$$

$$a_0 = \frac{1}{1 + \dfrac{R_{sig} + r_b + r_\pi}{(\beta_0 + 1)(R_L//r_o)}} \tag{6.61b}$$

On the other hand, at sufficiently *high frequencies*, where C_π acts as a *short circuit*, we have $\beta \to 0$ and $z_\pi \to 0$, so the high-frequency asymptotes, identified by subscript ∞, are

$$Z_{i\infty} = r_b + (R_L//r_o) \qquad Z_{o\infty} = (R_{sig} + r_b)//r_o \tag{6.62a}$$

$$a_\infty = \frac{1}{1 + \dfrac{R_{sig} + r_b}{R_L//r_o}} \tag{6.62b}$$

We observe that $Z_{i\infty} \ll Z_{i0}$, indicating that Z_i is always a *capacitive* impedance. However, Z_o may be inductive, capacitive, or even purely resistive, depending on how the product $g_m(R_{sig} + r_b)$ compares with unity. One can prove (see Problem 6.18) that if the BJT is biased at a sufficiently low current to make $g_m(R_{sig} + r_b) \leq 1$, then we have $a_\infty \geq a_0$ and $Z_{o\infty} \leq Z_{o0}$. In practical circuits it is far more common to have $g_m(R_{sig} + r_b) \gg 1$, in which case $a_\infty < a_0$ and $Z_{o\infty} \gg Z_{o0}$, indicating an *inductive* Z_o.

Figure 6.31 shows typical magnitude plots of Z_i, a, and Z_o for $C_\mu = 0$ and $g_m(R_{sig} + r_b) \gg 1$. In accordance with Appendix 6A, these parameters must take on the forms

$$Z_i(j\omega) = Z_{i0}\frac{1 + j\omega/\omega_{zi}}{1 + j\omega/\omega_{pi}} \qquad\qquad a(j\omega) = a_0\frac{1 + j\omega/\omega_{za}}{1 + j\omega/\omega_{pa}}$$

$$Z_o(j\omega) = Z_{o0}\frac{1 + j\omega/\omega_{zo}}{1 + j\omega/\omega_{po}} \qquad\qquad (6.63)$$

We now seek to estimate, for each of the above expressions, the zero and pole frequencies ω_z and ω_p, also known as the *break frequencies*. This task is facilitated by the fact that letting $\omega \to \infty$ in the above expressions we obtain the following constraints

$$\frac{Z_{i0}}{Z_{i\infty}} = \frac{\omega_{zi}}{\omega_{pi}} \qquad\qquad \frac{a_0}{a_\infty} = \frac{\omega_{za}}{\omega_{pa}} \qquad\qquad \frac{Z_{o\infty}}{Z_{o0}} = \frac{\omega_{po}}{\omega_{zo}} \qquad\qquad (6.64)$$

Consequently, for each expression we need only to estimate *one* of its two break frequencies. The *other* can be found via the appropriate constraint of Eq. (6.64). Though the exact derivations are left as an exercise in Problem 6.19, here we wish to pursue quick estimations to gain basic insight. Thus, according to Eq. (6.60), each of $Z_i(j\omega)$, $Z_o(j\omega)$, and $a(j\omega)$ contains the terms $z_\pi(j\omega)$ and $\beta(j\omega) + 1$. These terms affect the frequency response up to ω_T, beyond which $z_\pi(j\omega)$ becomes negligible compared to the other resistances in the circuit, and $\beta(j\omega)$ becomes negligible compared to 1. So, we expect each curve in Fig. 6.31 to make the transition to its high-frequency asymptote in the vicinity of ω_T, thus giving

$$\omega_{zi} \cong \omega_{za} \cong \omega_{po} \cong \omega_T \qquad\qquad (6.65)$$

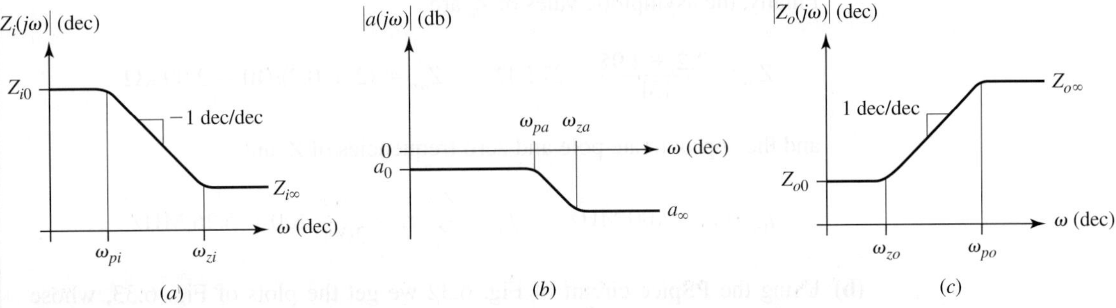

FIGURE 6.31 Typical emitter-follower characteristics for $C_\mu = 0$: (a) input impedance Z_i, (b) voltage gain a, and (c) output impedance Z_o. The plots of a and Z_o are for the case $g_m(R_{sig} + r_b) \gg 1$.

EXAMPLE 6.11   (a) Let the BJT of Fig. 6.30 have $\beta_0 = 150$, $V_A = 80$ V, $r_b = 200\ \Omega$, and $C_\mu = 1$ pF, and suppose it is biased at $I_C = 2$ mA, where it has $f_T = 400$ MHz. Moreover, let $R_{sig} = 2$ kΩ and $R_L = 5$ kΩ. Ignoring C_μ, provide quick estimates for the asymptotic values as well as the pole and zero frequencies of a, Z_i, and Z_o.

(b) Verify with PSpice and compare with the calculated values.

(c) Re-run PSpice with $C_\mu = 1$ pF and use physical insight to justify the ensuing changes in the plots.

Solution

(a) Proceeding as usual we find $g_m = 1/(13\ \Omega)$, $r_\pi = 1.95$ kΩ, and $r_o = 40$ kΩ, so that $R_{sig} + r_b = 2 + 0.2 = 2.2$ kΩ and $R_L//r_o = 5//40 = 4.44$ kΩ. The asymptotic values of gain are

$$a_0 = \frac{1}{1 + \dfrac{2.2 + 1.95}{151 \times 4.44}} = 0.994 \text{ V/V} = -0.05 \text{ dB}$$

$$a_\infty = \frac{1}{1 + \dfrac{2.2}{4.44}} = 0.669 \text{ V/V} = -3.5 \text{ dB}$$

and its zero and pole frequencies are

$$f_{za} \cong f_T = 400 \text{ MHz} \qquad f_{pa} \cong \frac{a_\infty}{a_0} f_{za} = \frac{0.669}{0.994} 400 = 269 \text{ MHz}$$

which are fairly high and close to each other. The asymptotic vales of Z_i are

$$Z_{i0} = 0.2 + 1.95 + 151 \times 4.44 = 673 \text{ k}\Omega \qquad Z_{i\infty} = 0.2 + 4.44 = 4.64 \text{ k}\Omega$$

and the approximate zero and pole frequencies of Z_i are

$$f_{zi} \cong f_T = 400 \text{ MHz} \qquad f_{pi} \cong \frac{Z_{i\infty}}{Z_{i0}} f_T = \frac{4.64}{673} 400 = 2.76 \text{ MHz}$$

Finally, the asymptotic vales of Z_o are

$$Z_{o0} = \frac{2.2 + 1.95}{151} = 27.5\ \Omega \qquad Z_{o\infty} = (2 + 0.2)//40 = 2.09 \text{ k}\Omega$$

and the approximate pole and zero frequencies of Z_o are

$$f_{po} \cong f_T = 400 \text{ MHz} \qquad f_{zo} \cong \frac{Z_{o0}}{Z_{o\infty}} f_T = \frac{27.5}{2090} 400 = 5.26 \text{ MHz}$$

(b) Using the PSpice circuit of Fig. 6.32 we get the plots of Fig. 6.33, whose asymptotic values and break frequencies for $C_\mu = 0$ are in good agreement with the values estimated above.

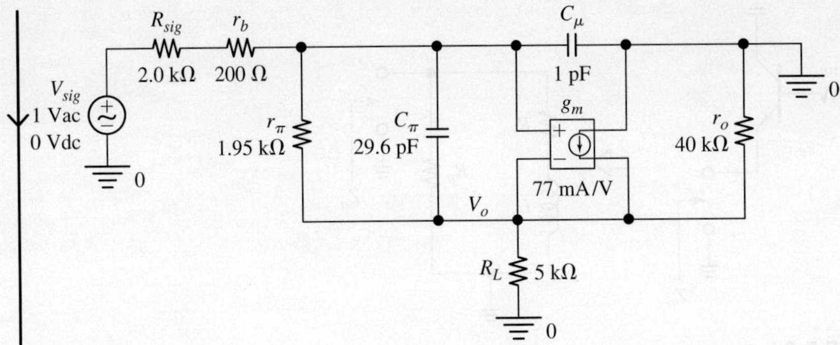

FIGURE 6.32 PSpice circuit to display the Bode plots of the emitter follower of Example 6.11.

(c) Re-running PSpice with $C_\mu = 1$ pF indicates the presence of an additional pole frequency for each plot. Using physical insight we estimate the additional pole frequency of Z_i in the vicinity of $1/(2\pi Z_{i0}C_\mu) = 236$ MHz, that of a in the vicinity of $1/[2\pi(R_{sig} + r_b)C_\mu] = 72$ MHz, and that of Z_o in the vicinity of $1/\{2\pi[(R_{sig} + r_b)//R_L]C_\mu\} = 104$ MHz. Above this frequency Z_o changes from inductive to capacitive.

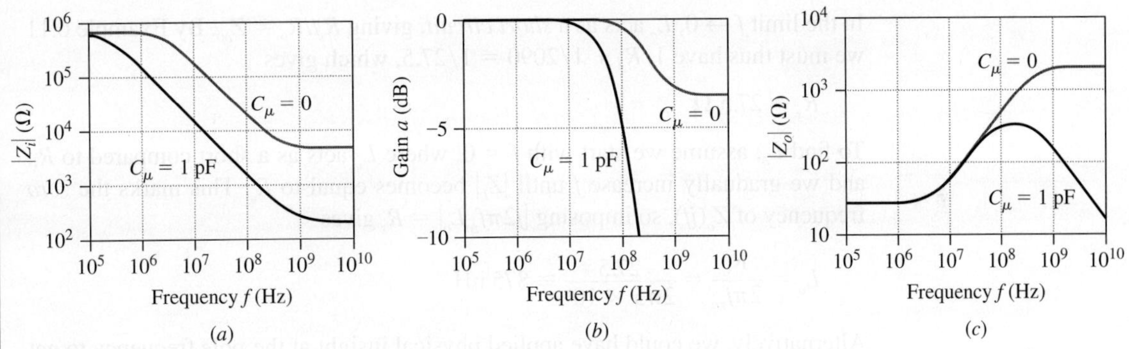

FIGURE 6.33 Gain and impedance plots for the emitter follower of Example 6.11.

Inductive behavior may pose problems when an emitter follower drives a *capacitive load* because of the tendency by capacitive and inductive impedances to *resonate* with each other. Depending on the damping conditions, the emitter follower may exhibit undesirable ringing or even oscillations. To better assess the situation, it is often convenient to model Z_o in terms of a suitable network, such as the one of Fig. 6.34, consisting of an equivalent inductance L_o with a series resistance R_s and a parallel resistance R_p. Their values are found by matching the asymptotic values and break frequencies of the equivalent network to those of Z_o.

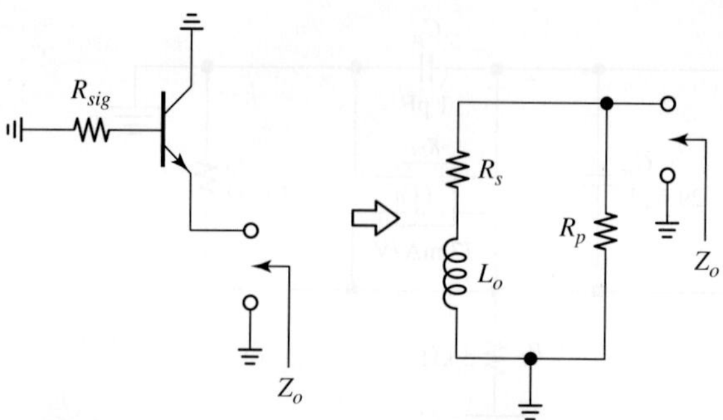

FIGURE 6.34 Equivalent network for the output impedance Z_e of an emitter follower for the case $C_\mu = 0$.

EXAMPLE 6.12 Find R_p, R_s, and L_o for the emitter follower of Example 6.11. Again, ignore C_μ.

Solution

In the limit $f \rightarrow \infty$, L_o acts as an *open circuit*, resulting in $R_p = Z_{o\infty}$. By Example 6.11 we must thus have

$$R_p = 2.09 \text{ k}\Omega$$

In the limit $f \rightarrow 0$, L_o acts as a *short circuit*, giving $R_s//R_p = Z_{o0}$. By Example 6.11 we must thus have $1/R_s + 1/2090 = 1/27.5$, which gives

$$R_s \cong 27.5 \ \Omega$$

To find L_o, assume we start with $f = 0$, where L_o acts as a short compared to R_s, and we gradually increase f until $|Z_L|$ becomes equal to R_s. This marks the *zero* frequency of $Z_o(jf)$, so imposing $|j2\pi f_{zo}L_o| = R_s$ gives

$$L_o = \frac{R_s}{2\pi f_{zo}} = \frac{27.5}{2\pi 5 \times 10^6} \cong 875 \text{ nH}$$

Alternatively, we could have applied physical insight at the *pole* frequency to get $L_o = R_p/(2\pi f_{po})$.

Frequency Characteristics of Bipolar Current Buffers

Figure 6.35 shows the ac equivalent of a bipolar current buffer, along with its high-frequency model. Like the CC configuration, the CB configuration is *inherently fast* because C_μ is exempt from the Miller effect. We wish to find the frequency dependence of the impedance $Z_i(j\omega)$ seen looking into the emitter, the current gain $a(j\omega) = I_o/I_i$, and the impedance $Z_o(j\omega)$ seen looking into the collector.

To gain quick—if approximate—insight into Z_i and a it is convenient to ignore r_o and C_μ for then the input port is *isolated* from the output port and can

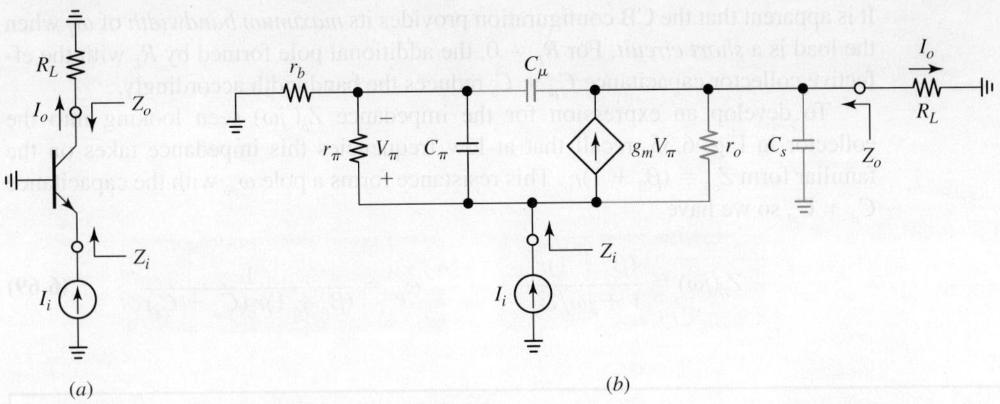

(a)   (b)

FIGURE 6.35 (a) The bipolar current buffer, and (b) its high-frequency small-signal model.

thus be analyzed separately. To find $Z_i(j\omega)$ we simply recycle the expression for $Z_o(j\omega)$ derived earlier for the CC configuration, but with $R_{sig} = 0$. The result is, for $r_o \to \infty$,

$$Z_i(j\omega) = Z_{i0}\frac{1 + j\omega/\omega_{zi}}{1 + j\omega/\omega_{pi}} \qquad (6.66a)$$

where

$$Z_{i0} \cong \frac{r_b + r_\pi}{\beta_0 + 1} \qquad \omega_{pi} \cong \omega_T \qquad \omega_{zi} = \frac{Z_{i0}}{r_b}\omega_{pi} \qquad (6.66b)$$

Again, one can readily prove that this impedance is inductive for $g_m r_b > 1$, capacitive for $g_m r_b < 1$, and purely resistive for $g_m r_b = 1$.

Using Eq. (6.59a) and expanding, we readily obtain the *short-circuit current gain*

$$\alpha(j\omega) = \frac{I_{o(sc)}}{I_i} = \frac{\beta(j\omega)}{\beta(j\omega) + 1} \cong \frac{\alpha_0}{1 + j\omega/\omega_T} \qquad (6.67)$$

where $I_{o(sc)}$ is the collector current in the limit $R_L \to 0$, and $\alpha_0 = \beta_0/(\beta_0 + 1)$. Once it reaches the collector node, $I_{o(sc)}$ divides among R_L, C_μ, and C_s. Since r_b is small, we can treat C_μ as if its left plate was directly grounded, so we can lump C_μ with C_s. Using the current-divider formula we get, for $r_o \to \infty$,

$$I_o \cong \frac{1/[j\omega(C_\mu + C_s)]}{R_L + 1/[j\omega(C_\mu + C_s)]}\alpha(j\omega)I_i = \frac{\alpha(j\omega)}{1 + j\omega R_L(C_\mu + C_s)}I_i$$

Substituting Eq. (6.67), we finally obtain the *overall current gain*

$$a(j\omega) = \frac{I_i}{I_i} \cong \frac{\alpha_0}{(1 + j\omega/\omega_L)(1 + j\omega/\omega_T)} \qquad \omega_L \cong \frac{1}{R_L(C_\mu + C_s)} \qquad (6.68)$$

It is apparent that the CB configuration provides its *maximum bandwidth* of ω_T when the load is a *short circuit*. For $R_L \neq 0$, the additional pole formed by R_L with the effective collector capacitance $C_\mu + C_s$ reduces the bandwidth accordingly.

To develop an expression for the impedance $Z_o(j\omega)$ seen looking into the collector in Fig. 6.35, recall that at low frequencies this impedance takes on the familiar form $Z_{o0} = (\beta_0 + 1)r_o$. This resistance forms a pole ω_{po} with the capacitance $C_\mu + C_s$, so we have

$$Z_o(j\omega) = \frac{(\beta_0 + 1)r_o}{1 + j\omega/\omega_{po}} \qquad \omega_{po} = \frac{1}{(\beta_0 + 1)r_o(C_\mu + C_s)} \qquad (6.69)$$

EXAMPLE 6.13   The CB amplifier of Fig. 6.35a uses a BJT with $\beta_0 = 200$, $V_A = 50$ V, $r_b = 250\ \Omega$, $C_\mu = 0.5$ pF, and $C_s = 1$ pF. Also, the BJT is biased at $I_C = 1$ mA, where it exhibits $f_T = 500$ MHz. If $R_L = 5\ \text{k}\Omega$, estimate expressions for $a(jf)$, $Z_i(jf)$, and $Z_o(jf)$.

Solution

We have $r_\pi = 5.2\ \text{k}\Omega$, $r_o = 50\ \text{k}\Omega$, and $C_\mu + C_s = 0.5 + 1 = 1.5$ pF. Plugging into the above formulas we get $\alpha_0 = 0.995$, $f_L \cong 21$ MHz, $Z_{i0} \cong 27\ \Omega$, $f_{zi} \cong 54$ MHz, $Z_{o0} \cong 10$ MΩ, and $f_{po} \cong 10.6$ kHz. Consequently,

$$a(jf) \cong \frac{0.995}{[1 + jf/(21\ \text{MHz})][1 + jf/(500\ \text{MHz})]}$$

indicating that the band-limiting bottleneck for $a(jf)$ is the pole f_L. We also have

$$Z_i(jf) \cong (27\ \Omega)\frac{1 + jf/(54\ \text{MHz})}{1 + jf/(500\ \text{MHz})} \qquad Z_o(jf) = \frac{10\ \text{M}\Omega}{1 + jf/(10.6\ \text{kHz})}$$

indicating an inductive Z_i and a capacitive Z_o (as we know, because of C_μ and C_s, Z_i eventually will turn capacitive.

Remark: Even an apparently simple circuit such as a buffer may prove too complex for a thorough paper-and-pencil analysis. A reasonable approach is (a) to start out with a simplied but more manageable circuit version (such as Fig. 6.30b, where we ignore C_μ to focus on the more important C_π), (b) develop a basic feel for this simplified circuit, and (c) then use PSpice to investigate higher-order effects (such as the effect of $C_\mu = 1$ pF in Fig. 6.33). Regardless, we still need hand analysis in order to anticipate the results of computer simulation, and thus provide a form of check. This is how engineers proceed on their daily jobs.

6.6 MOS VOLTAGE AND CURRENT BUFFERS

The voltage/current buffer considerations at the beginning of Section 6.5, which you are encouraged to review, hold also for MOSFETs, so our analysis will proceed along the lines of the bipolar case, with special emphasis on the frequency behavior of the terminal impedances.

Frequency Characteristics of the Source Follower

Figure 6.36 shows the ac equivalent of the source follower, along with its high-frequency model. Since its right plate is grounded, C_{gd} is exempt from the Miller multiplication, so we expect the CD to be an *inherently fast* configuration. In fact, to develop a quick (if only approximate) feel for the circuit, let us ignore all capacitances except for C_{gs}, which is the dominant capacitance in the circuit. We wish to investigate the frequency dependence of the impedance $Z_i(j\omega)$ seen by the signal source, the source-to-load voltage gain $a(j\omega) = V_o/V_{sig}$, and the impedance $Z_o(j\omega)$ seen by the load.

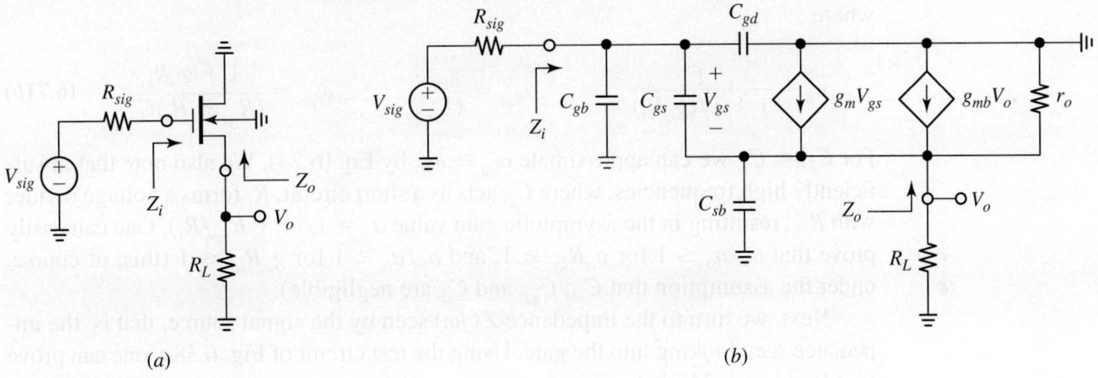

(a)   (b)

FIGURE 6.36 (a) The source follower and (b) its high-frequency small-signal model.

To find the voltage gain, refer to the simplified equivalent of Fig. 6.37, obtained from that of Fig. 6.36b by lumping R_L, r_o, and $1/g_{mb}$ into a single equivalent resistance,

$$R_1 = R_L//r_o//\frac{1}{g_{mb}} \tag{6.70}$$

(Though C_{gd} and C_{gb} are being ignored in the present analysis, they too have been lumped together as they are in parallel in the original circuit). Applying Ohm's law, KCL, and the voltage divider rule, we write

$$V_o = R_1\left(\frac{V_{gs}}{1/(j\omega C_{gs})} + g_m V_{gs}\right) = R_1(j\omega C_{gs} + g_m)V_{gs}$$

$$V_{gs} = \frac{1/j\omega C_{gs}}{R_{sig} + 1/(j\omega C_{gs})}(V_{sig} - V_o) = \frac{1}{1 + j\omega R_{sig}C_{gs}}(V_{sig} - V_o)$$

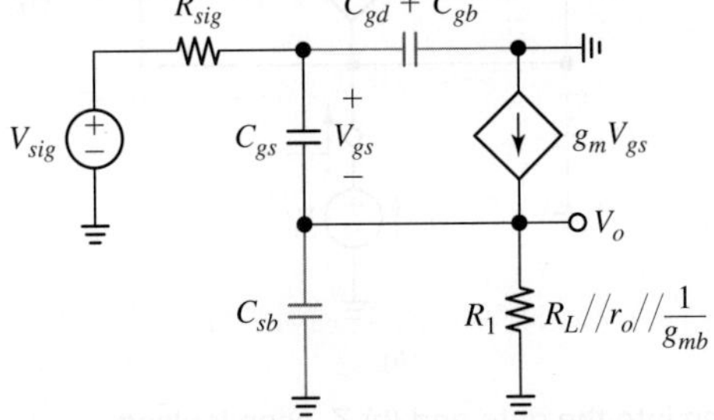

FIGURE 6.37 Simplified equivalent of the circuit of Fig. 6.36b.

Eliminating V_{gs} and collecting gives

$$a(j\omega) = \frac{V_o}{V_{sig}} = \frac{g_m R_1 + j\omega R_1 C_{gs}}{1 + g_m R_1 + j\omega(R_{sig} + R_1)C_{gs}}$$

With a bit of algebraic manipulation we put gain in the more insightful form advocated in Appendix 6A,

$$a(j\omega) = a_0 \frac{1 + j\omega/\omega_{za}}{1 + j\omega/\omega_{pa}} \tag{6.71a}$$

where

$$a_0 = \frac{1}{1 + 1/(g_m R_1)} \qquad \omega_{za} = \frac{g_m}{C_{gs}} \qquad \omega_{pa} = \frac{1 + g_m R_1}{(R_{sig} + R_1)C_{gs}} \tag{6.71b}$$

For $C_{gd} \ll C_{gs}$ we can approximate $\omega_{za} \cong \omega_T$, by Eq. (6.24). We also note that at sufficiently high frequencies, where C_{gs} acts as a short circuit, R_1 forms a voltage divider with R_{sig}, resulting in the asymptotic gain value $a_\infty = 1/(1 + R_{sig}/R_1)$. One can easily prove that $a_\infty/a_0 \leq 1$ for $g_m R_{sig} \geq 1$, and $a_\infty/a_0 \geq 1$ for $g_m R_{sig} \leq 1$ (this, of course, under the assumption that C_{gb}, C_{gd}, and C_{sb} are negligible).

Next, we turn to the impedance $Z_i(j\omega)$ seen by the signal source, that is, the impedance seen looking into the gate. Using the test circuit of Fig. 6.38a, one can prove (see Problem 6.24) that

$$Z_i(j\omega) = \frac{V_i}{I_i} = \frac{1}{j\omega C_1} + R_1 \tag{6.72a}$$

where

$$C_1 = \frac{C_{gs}}{1 + g_m R_1} \tag{6.72b}$$

indicating that Z_i appears as an equivalent capacitance C_1 in series with the resistance R_1. As depicted in Fig. 6.39a, C_1 dominates at low frequencies, leading to the familiar dc limit $Z_{i0} = \infty$. At high frequencies, where C_{gs} acts as a short circuit, R_1 dominates,

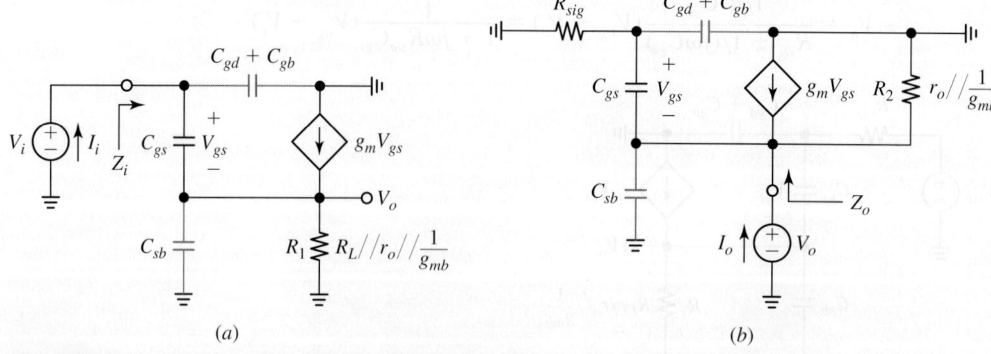

(a)   (b)

FIGURE 6.38 Finding (a) the impedance Z_i seen looking into the gate and (b) Z_o seen looking into the source.

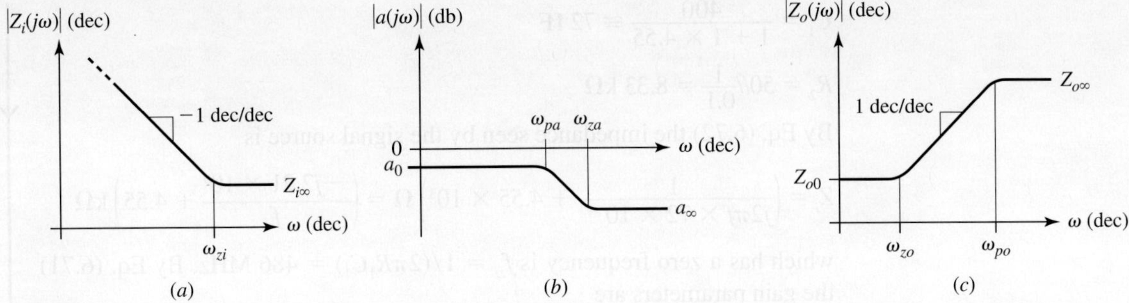

FIGURE 6.39 Typical source-follower characteristics for $C_{gb} = C_{gd} = C_{sb} = 0$: (a) input impedance Z_i, (b) voltage gain a, and (c) output impedance Z_o. The plots of a and Z_o are for the case $g_m R_{sig} \gg 1$.

giving $Z_{i\infty} = R_1$. The zero frequency is $\omega_{zi} = 1/(R_1 C_1)$. For $g_m R_1 \gg 1$ we can approximate $\omega_{zi} \cong \omega_T$.

Finally, to find the impedance $Z_o(j\omega)$ seen by the load, we use the test circuit of Fig. 6.38b, where

$$R_2 = r_o // \frac{1}{g_{mb}} \tag{6.73}$$

The result (see Problem 6.24) is

$$Z_o(j\omega) = Z_{o0} \frac{1 + j\omega/\omega_{zo}}{1 + j\omega/\omega_{po}} \tag{6.74a}$$

where Z_{o0} is the familiar *low-frequency resistance* seen looking into the source terminal, and ω_{zo} and ω_{po} are the zero and pole frequencies,

$$Z_{o0} = \frac{1}{g_m} // R_2 \qquad \omega_{zo} = \frac{1}{R_{sig} C_{gs}} \qquad \omega_{po} = \frac{1 + g_m R_2}{(R_{sig} + R_2) C_{gs}} \tag{6.74b}$$

By inspection, the high-frequency asymptotic value is $Z_{o\infty} = R_{sig} // R_1$. One can easily verify that as long as C_{gb}, C_{gd}, and C_{sb} can be ignored, Z_o is inductive for $g_m R_{sig} > 1$ and capacitive for $g_m R_{sig} < 1$.

EXAMPLE 6.14

(a) The source follower of Fig. 6.36 uses a FET with $g_m = 1$ mA/V, $g_{mb} = 0.1$ mA/V, $r_o = 50$ kΩ, and $C_{gs} = 400$ fF. Moreover, $R_{sig} = R_L = 10$ kΩ. Assuming $C_{gb} = C_{gd} = C_{sb} = 0$, find the frequency characteristics of Z_i, a, and Z_o.

(b) Verify with PSpice both for $C_{gb} = C_{gd} = C_{sb} = 0$ and for $C_{gb} = C_{gd} = C_{sb} = 25$ fF, and comment.

Solution

We have

$$f_T \cong \frac{g_m}{2\pi C_{gs}} = \frac{10^{-3}}{2\pi 400 \times 10^{-12}} \cong 400 \text{ MHz}$$

$$R_1 = 10//50//\frac{1}{0.1} = 4.55 \text{ k}\Omega$$

$$C_1 = \frac{400}{1 + 1 \times 4.55} \cong 72 \text{ fF}$$

$$R_2 = 50 // \frac{1}{0.1} = 8.33 \text{ k}\Omega$$

By Eq. (6.72) the impedance seen by the signal source is

$$Z_i = \left(\frac{1}{j2\pi f \times 72 \times 10^{-15}} + 4.55 \times 10^3 \right) \Omega = \left(\frac{-j2.21 \times 10^9}{f} + 4.55 \right) \text{k}\Omega$$

which has a zero frequency is $f_{zi} = 1/(2\pi R_1 C_1) = 486$ MHz. By Eq. (6.71) the gain parameters are

$$a_0 = \frac{1}{1 + 1/(1 \times 4.55)} = 0.820 \text{ V/V} = -1.72 \text{ dB}$$

$$a_\infty = \frac{1}{1 + 10/4.55} = 0.312 \text{ V/V} = -10.1 \text{ dB}$$

$$f_{za} = 400 \text{ MHz}$$

$$f_{pa} = \frac{1 + 1 \times 4.55}{2\pi(10 + 4.55)10^3 \times 400 \times 10^{-15}} = 152 \text{ MHz}$$

By. Eq. (6.74) the parameters of the impedance seen by the load are

$$Z_{o0} = \frac{1}{1} // 8.33 = 0.893 \text{ k}\Omega \qquad Z_{o\infty} = 10 // 8.33 = 4.55 \text{ k}\Omega$$

$$f_{zo} = \frac{1/(2\pi)}{10^4 \times 400 \times 10^{-15}} \cong 40 \text{ MHz}$$

$$f_{po} = \frac{1 + 1 \times 8.33}{2\pi(10 + 8.33)10^3 \times 400 \times 10^{-15}} = 202 \text{ MHz}$$

It is apparent that Z_o is inductive, at least up to a point.

(b) Adapting the PSpice circuit of Fig. 6.32 to the present case we obtain the frequency plots of Fig. 6.40. The asymptotic values and break frequencies are in good agreement with the estimated ones under the assumption $C_{gb} = C_{gd} = C_{sb} = 0$. With $C_{gb} = C_{gd} = C_{sb} = 25$ fF, the high-frequency asymptotic values tend to zero, turning Z_o from inductive to capacitive at high frequencies.

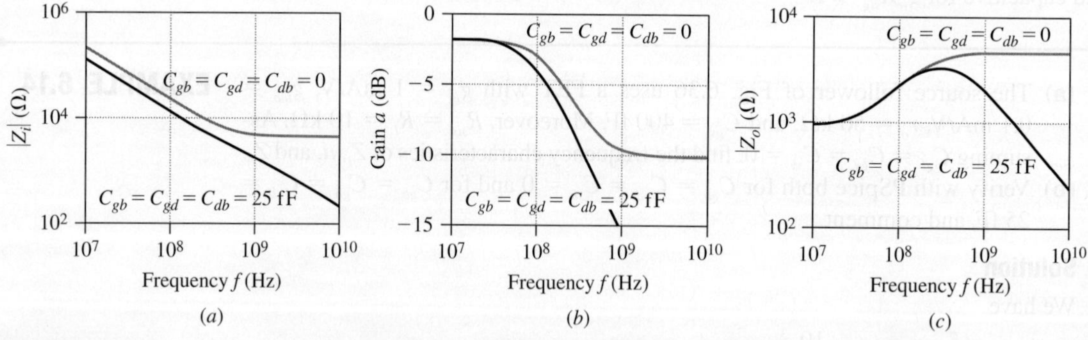

FIGURE 6.40 Gain and impedance plots for the source follower of Example 6.14.

Frequency Characteristics of MOS Current Buffers

Figure 6.41 shows the ac equivalent of the MOS current buffer, along with its high-frequency model. Like the CD configuration, the CG configuration is *inherently fast* because C_{gd} is exempt from the Miller effect. We wish to find the impedance $Z_i(j\omega)$ seen looking into the source terminal, the current gain $a(j\omega) = I_o/I_i$, and the impedance $Z_o(j\omega)$ seen looking into the drain. To this end, refer to the more compact equivalent of Fig. 6.42, obtained by lumping together the capacitances as shown.

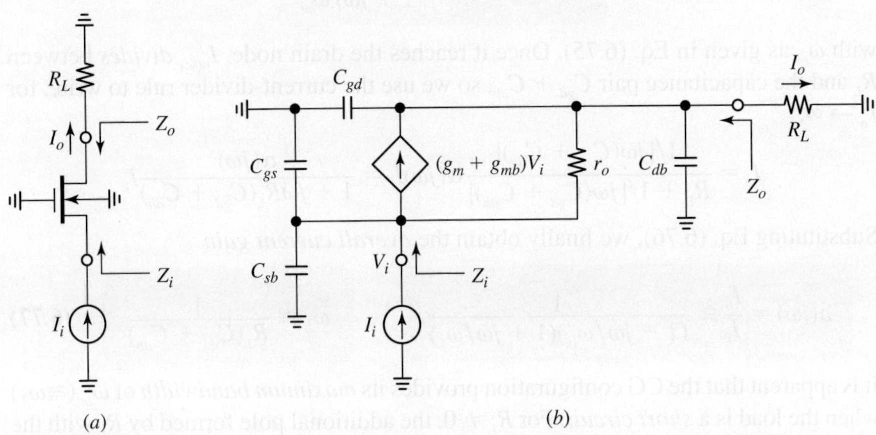

FIGURE 6.41 (a) The MOS current buffer and (b) its high-frequency small-signal model.

To gain quick insight into the frequency dependence of Z_i and a it is convenient to ignore r_o for then the input port is isolated from the output port and can thus be analyzed separately. KCL at the input node gives, for $r_o \to \infty$,

$$I_i = (g_m + g_{mb})V_i + \frac{V_i}{1/[j\omega(C_{gs} + C_{sb})]} = (g_m + g_{mb})\left[1 + \frac{j\omega(C_{gs} + C_{sb})}{g_m + g_{mb}}\right]V_i$$

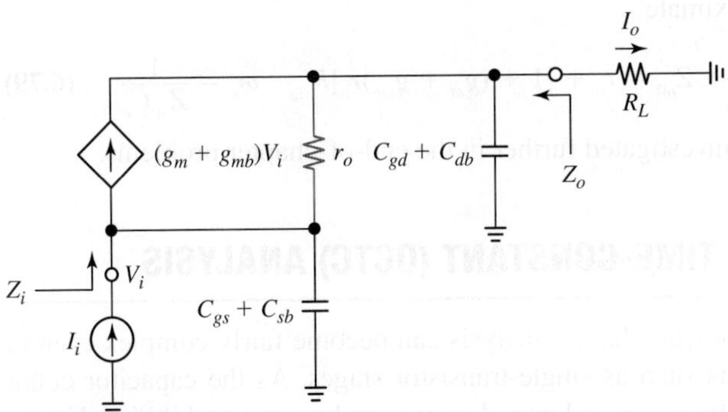

FIGURE 6.42 Compact rendition of the MOS current buffer.

Consequently, the impedance seen by the signal source is

$$Z_i = \frac{V_i}{I_i} = \frac{1}{g_m + g_{mb}} \times \frac{1}{1 + j\omega/\omega_{pi}} \qquad \omega_{pi} = \frac{g_m + g_{mb}}{C_{gs} + C_{sb}} \qquad (6.75)$$

The current into a short-circuit load ($R_L = 0$) is, for $r_o \to \infty$, $I_{o(sc)} = (g_m + g_{mb})V_i = (g_m + g_{mb})Z_i I_i$, so the *short-circuit current gain* is

$$\alpha(j\omega) = \frac{I_{o(sc)}}{I_i} = \frac{1}{1 + j\omega/\omega_{pi}} \qquad (6.76)$$

with ω_{pi} as given in Eq. (6.75). Once it reaches the drain node, $I_{o(sc)}$ *divides* between R_L and the capacitance pair $C_{gd} + C_{db}$, so we use the current-divider rule to write, for $r_o \to \infty$,

$$I_o = \frac{1/[j\omega(C_{gd} + C_{db})]}{R_L + 1/[j\omega(C_{gd} + C_{db})]}\alpha(j\omega)I_i = \frac{\alpha(j\omega)}{1 + j\omega R_L(C_{gd} + C_{db})}I_i$$

Substituting Eq. (6.76), we finally obtain the *overall current gain*

$$a(j\omega) = \frac{I_o}{I_i} \cong \frac{1}{(1 + j\omega/\omega_{pi})(1 + j\omega/\omega_L)} \qquad \omega_L \cong \frac{1}{R_L(C_{gd} + C_{db})} \qquad (6.77)$$

It is apparent that the CG configuration provides its *maximum bandwidth* of ω_{pi} ($\cong \omega_T$) when the load is a *short circuit*. For $R_L \neq 0$, the additional pole formed by R_L with the effective drain capacitance $C_{gd} + C_{db}$ reduces the bandwidth accordingly.

It is left as an exercise (see Problem 6.28) to prove that the impedance seen by the load is

$$Z_o(j\omega) = \frac{1}{j\omega C_o} \times \frac{1 + j\omega/\omega_{zo}}{1 + j\omega/\omega_{po}} \qquad C_o = C_{gd} + C_{db} + \frac{C_{gs} + C_{sb}}{1 + (g_m + g_{mb})r_o} \qquad (6.78a)$$

$$\omega_{zo} = \frac{1 + (g_m + g_{mb})r_o}{r_o(C_{gs} + C_{sb})} \qquad \omega_{po} = \omega_{zo} + \frac{1}{r_o(C_{gd} + C_{db})} \qquad (6.78b)$$

The frequency dependence of Z_o is dominated by C_o, and $Z_{o0} \to \infty$ at dc because the signal source has been assumed ideal. A practical signal source will have $R_{sig} < \infty$, in which case we can approximate

$$Z_o(j\omega) \cong \frac{Z_{o0}}{1 + j\omega/\omega_o} \qquad Z_{o0} = r_o + [1 + (g_m + g_{mb})r_o]R_{sig} \qquad \omega_o = \frac{1}{Z_{o0}C_o} \qquad (6.79)$$

The CG configuration is investigated further in the end-of-chapter problems.

6.7 OPEN-CIRCUIT TIME-CONSTANT (OCTC) ANALYSIS

The preceding sections indicate that ac analysis can become fairly complex even in the case of simple circuits such as single-transistor stages. As the capacitor count increases, exact analysis by paper and pencil may soon become prohibitive. Yet, in everyday practice a designer must be able to come up with quick, if approximate, estimates of a circuit's most salient ac characteristics, such as its -3-dB frequency, and identify what changes need to be made in case the circuit fails to meet the specifications. Then, we turn to computer simulation to verify the modified design.

If the circuit contains a single pole, its -3-dB frequency is the pole frequency itself. Even if it contains additional poles and/or zeros but at sufficiently higher frequencies, $\omega_{-3\,\mathrm{dB}}$ will still be close to the frequency of the lowest pole, aptly referred to as the *dominant pole*. Eloquent examples were offered by the common-emitter/common-source amplifiers of Section 6.3, where we used the Miller approximation to speed up the estimation of $\omega_{-3\,\mathrm{dB}}$. We wonder whether there is a quick way to estimate $\omega_{-3\,\mathrm{dB}}$ also in a multi-pole circuit. A widely used such technique is the *Open-Circuit Time-Constant (OCTC) Analysis Technique* pioneered by P. E. Gray and C. L. Searle in 1969.[2]

To develop an intuitive feel for this technique, start out with a circuit containing a single capacitor C_k. As we know, a pole arises at the frequency ω_k at which the impedance of C_k *equals*, in magnitude, the equivalent resistance R_k presented to C_k by the surrounding circuit, a condition that we express as $1/(\omega_k C_k) = R_k$. This gives $\omega_k = 1/\tau_k$, where $\tau_k = R_k C_k$ is the *time constant* formed by C_k and R_k. As a function of frequency, C_k starts out as an *open circuit* for $\omega \ll \omega_k$, it presents an impedance *equal* in magnitude to R_k at $\omega = \omega_k$, and it becomes a *short circuit* for $\omega \gg \omega_p$.

What if the circuit contains more than just one capacitance? If there is a dominant pole, we can say that at that pole frequency all capacitances still act as *open circuits*, except for the capacitance responsible for that one pole, which will exhibit an impedance *equal* in magnitude to the resistance presented by the surrounding circuit. To find which capacitance is responsible for the dominant pole we need to test one capacitance at a time, assuming all remaining capacitances are acting as open circuits. We find the equivalent resistance seen by the capacitance under scrutiny, and we calculate the corresponding time constant. We repeat this procedure for each capacitance present, and finally we let $\omega_{-3\,\mathrm{dB}} \cong 1/\tau_D$, where τ_D is by far the longest and thus *dominant time constant*.

What if there isn't a definite dominant time constant in the circuit? The information gathered is still useful, for the OCTC technique states that we can estimate the -3-dB frequency as[2]

$$\omega_{-3\,\mathrm{dB}} \cong \frac{1}{R_1 C_1 + R_2 C_2 + \cdots + R_n C_n} \qquad (6.80)$$

where R_i is the *equivalent resistance* seen by the capacitance C_i ($i = 1, 2, \ldots n$) with *all other capacitances open-circuited*. Note that this technique provides no information about higher-order poles and possibly zeros. It gives only an estimate of the -3-dB frequency, also called the *half-power bandwidth*, but via n simple time-constant calculations. Moreover, by showing explicitly which time constant contributes most heavily to the -3-dB frequency, it pinpoints the parameters that need to be altered if the design fails to meet specific bandwidth requirements. Some examples will better illustrate the OCTC technique.

OCTC Analysis of the CE/CS Amplifiers

As our first application of the open-circuit time-constant (OCTC) technique, let us return to the circuit of Fig. 6.21 (repeated in Fig. 6.43), representing the common-emitter/common-source voltage amplifier. With three capacitances present

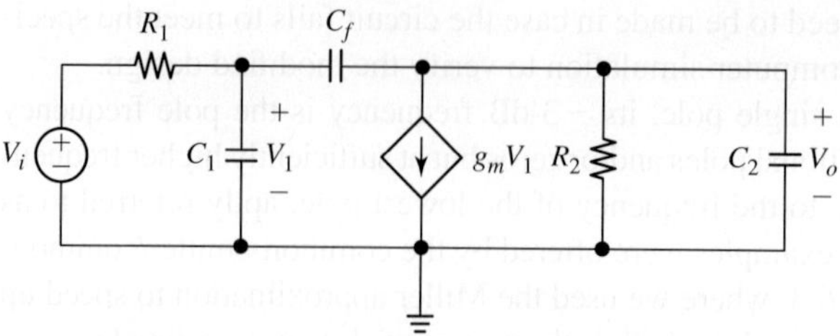

FIGURE 6.43 Revisiting the CE/CS amplifier.

$(C_1, C_2,$ and $C_f)$, we need to find three open-circuit equivalent resistances. With reference to Fig. 6.44a, it is apparent that the resistances seen by C_1 and C_2 are just R_1 and R_2. However, to find R_f, we can no longer rely on mere inspection, as the presence of the dependent source mandates using the *test method* instead. With reference to Fig. 6.44b we have, by Ohm's law, $v_1 = R_1 i$. By KCL, R_2 must supply the current $i + g_m v_1 = i + g_m R_1 i$, so $v_2 = -R_2(i + g_m R_1 i) = -R_2(1 + g_m R_1)i$. By KVL,

$$v = v_1 - v_2 = R_1 i - [-R_2(1 + g_m R_1)i]$$

so, taking the ratio $R_f = v/i$ we get

$$R_f = R_1 + R_2 + g_m R_1 R_2 \tag{6.81}$$

Finally, we estimate the 3-dB frequency using Eq. (6.80),

$$\omega_{-3\,\text{dB}} \cong \frac{1}{R_1 C_1 + R_f C_f + R_2 C_2} = \frac{1}{R_1 C_1 + (R_1 + R_2 + g_m R_1 R_2)C_f + R_2 C_2} \tag{6.82}$$

Aside from a rearrangement of its denominator terms, Eq. (6.82) is *identical* to Eq. (6.45), which was obtained via the much more laborious exact analysis. Of course, exact analysis provides information also about the higher order pole and zero frequencies, while the OCTC method estimates only $\omega_{-3\,\text{dB}}$. But this is often all the designer wants to know, so if we consider the much simpler calculations required

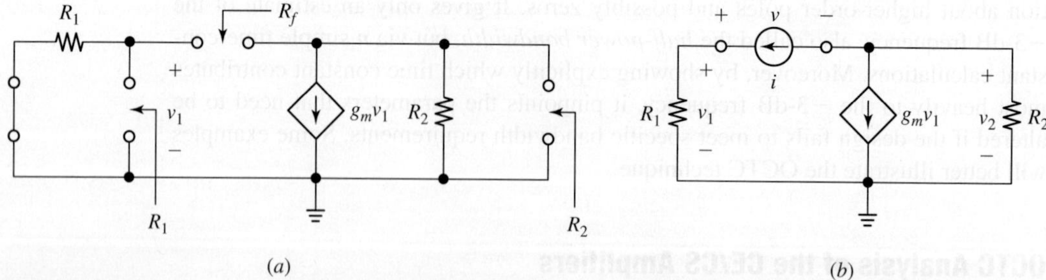

(a)   (b)

FIGURE 6.44 (a) The open-circuit resistances seen by the capacitances in the generalized voltage amplifier of Fig. 6.43. (b) Using the test method to find R_f.

by the OCTC technique, the latter is a powerful tool indeed. As a final remark, we observe that if we define the multiplier term

$$M = 1 + g_m R_2 + \frac{R_2}{R_1} \qquad (6.83)$$

then the time constant τ_f associated with C_f is expressed as $\tau_f = (R_1 M)C_f$ in the OCTC approximation, but as $\tau_f = R_1(MC_f)$ in the Miller approximation. We have two different viewpoints for the same result!

Reconsider the CE amplifier of Example 6.8, for which $g_m = 1/(26\ \Omega)$, $R_1 = 0.975\ \text{k}\Omega$, $R_2 = 4.55\ \text{k}\Omega$, $C_\pi = 12\ \text{pF}$, $C_\mu = 0.5\ \text{pF}$ and $C_s = 1\ \text{pF}$.

EXAMPLE 6.15

(a) Estimate $f_{-3\ \text{dB}}$ via the OCTC method, and comment on which capacitor contributes the most and which the least to the dominant-pole frequency.

(b) Propose a way to increase the circuit's bandwidth without reducing its low-frequency gain a_0.

Solution

(a) By inspection, the resistances seen by C_π and C_s are, respectively, $R_\pi = 0.975\ \text{k}\Omega$ and $R_s = 4.55\ \text{k}\Omega$. Adapting Eq. (6.81) to the present case we get $R_\mu = 0.975 + 4.55 + 0.975 \times 4.55/0.026 = 176\ \text{k}\Omega$. With resistances in $\text{k}\Omega$ (10^3) and capacitances in pF (10^{-12}), the time constants in Eq. (6.82) will be in ns (10^{-9}),

$$f_{-3\ \text{dB}} \cong \frac{1/2\pi}{(0.975 \times 12 + 176 \times 0.5 + 4.55 \times 1)\ \text{ns}}$$

$$= \frac{1/2\pi}{(11.7 + 88 + 4.6)\ \text{ns}} = 1.53\ \text{MHz}$$

This is in excellent agreement with Example 6.8. As expected, the main band-limiting culprit is the time constant associated with C_μ, whereas that associated with C_s counts the least.

(b) An obvious way to increase the bandwidth is to reduce R_μ as it intervenes in the longest time constant. By Eq. (6.81), this resistance depends both on R_1 and R_2. To avoid significantly perturbing the low-frequency gain a_0, don't change R_2. This leaves R_1, which can be reduced by driving the amplifier with a *low output-impedance signal source*. In the limit $R_{sig} \to 0$ we have $R_1 = r_\pi // r_b$, so

$$R_\pi = R_1 = 5.2 // 0.2 = 0.193\ \text{k}\Omega$$

$$R_\mu = 0.193 + 4.55 + 0.193 \times 4.55/0.026 = 38.4\ \text{k}\Omega.$$

Consequently, in the limit $R_{sig} \to 0$ we get

$$f_{-3\ \text{dB}} \cong \frac{1/2\pi}{(0.193 \times 12 + 38.4 \times 0.5 + 4.55 \times 1)\ \text{ns}}$$

$$= \frac{1/2\pi}{(2.3 + 19.2 + 4.6)\ \text{ns}} \cong 6\ \text{MHz}$$

indicating a bandwidth expansion of almost two octaves. (It can be seen that with $R_{sig} = 0$, a_0 is raised from -142 V/V to -169 V/V. Usually this is not a problem, but if it is, we can always tweak with the value of R_2 to re-establish the original gain.)

Remark: If the available signal source doesn't have a sufficiently low R_{sig}, we can interpose a *voltage buffer* between source and amplifier. A buffer provides high input impedance, low output impedance, and close-to-unity gain over a much wider bandwidth than the voltage amplifier under consideration, so it will serve the required function of impedance translation quite well.

EXAMPLE 6.16   Reconsider the CS amplifier of Exercise 6.3, for which $g_m = 4$ mA/V, $R_1 = 10$ kΩ, $R_2 = 4.55$ kΩ, $C_{gs} = 1.17$ pF, and $C_{gd} = 0.1$ pF. Assuming $C_{db} = 0.2$ pF, estimate $f_{-3\,\text{dB}}$ via the OCTC method.

Solution

By inspection, $R_{gs} = 10$ kΩ and $R_{db} = 4.55$ kΩ. Adapting Eq. (6.81) to the present case, $R_{gd} = 10 + 4.55 + 4 \times 10 \times 4.55 = 196$ kΩ. With resistances in kΩ and capacitances in pF, the time constants will be in ns,

$$f_{-3\,\text{dB}} \cong \frac{1/2\pi}{(10 \times 1.17 + 196 \times 0.1 + 4.55 \times 0.2)\ \text{ns}}$$

$$= \frac{1/2\pi}{(11.7 + 19.6 + 0.91)\ \text{ns}} = 4.9\ \text{MHz}$$

in close agreement with Exercise 6.3.

OCTC Analysis of Voltage Buffers

The **emitter follower** of Fig. 6.30b, repeated for convenience in Fig. 6.45, was analyzed under the assumption $C_\mu = 0$ to keep the derivations manageable. We now wish to estimate its -3-dB frequency via the OCTC technique, but with C_μ present. To find the open-circuit equivalent resistances, use the circuit of Fig. 6.46a with $R_1 = R_{sig} + r_b$ and $R_2 = R_L//r_o$, as shown. With reference to the node common to R_1 and r_π we observe that looking to the left we see R_1, and looking down toward the right we see $r_\pi + (\beta_0 + 1)R_2$, so the resistance seen by C_μ is

$$R_\mu = R_1//[r_\pi + (\beta_0 + 1)R_2] \tag{6.84}$$

To find R_π we need to apply the test method of Fig. 6.46b. By KCL, the current into R_1 is $i_1 = i - v/r_\pi$ and that into R_2 is $i_2 = v/r_\pi + g_m v - i$, so we can write

$$v = R_1 i_1 - R_2 i_2 = R_1\left(i - \frac{v}{r_\pi}\right) - R_2\left(\frac{v}{r_\pi} + g_m v - i\right)$$

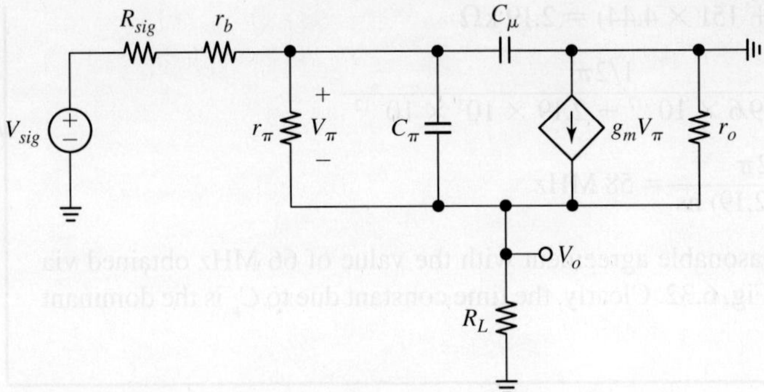

FIGURE 6.45 Revisiting the emitter follower.

Collecting and taking the ratio $R_\pi = v/i$ we get, after some algebra,

$$R_\pi = r_\pi // \frac{R_1 + R_2}{1 + g_m R_2} \tag{6.85}$$

Finally,

$$\omega_{-3\,dB} \cong \frac{1}{R_\pi C_\pi + R_\mu C_\mu}$$

$$= \frac{1}{\{r_\pi // [(R_1 + R_2)/(1 + g_m R_2)]\} C_\pi + \{R_1 // [r_\pi + (\beta_0 + 1)R_2]\} C_\mu} \tag{6.86}$$

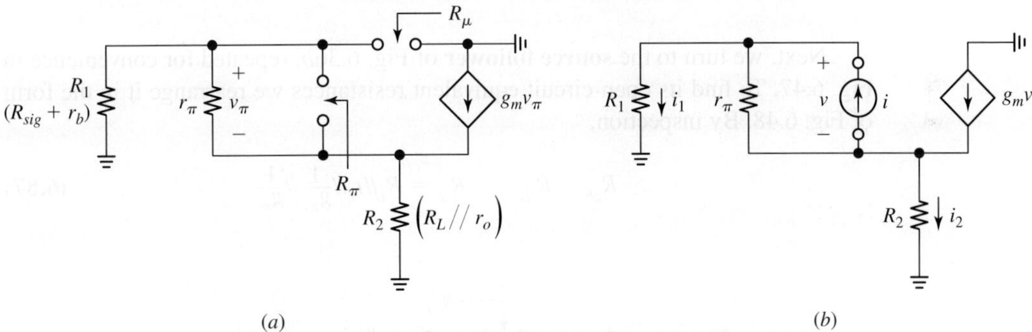

(a) (b)

FIGURE 6.46 (a) The open-circuit resistances seen by the capacitances of the *emitter follower* of Fig. 6.45. (b) Using the test method to find R_π.

Reconsider the emitter follower of Example 6.11, for which $\beta_0 = 150$, $g_m = 1/(13\ \Omega)$, $r_\pi = 1.95\ \text{k}\Omega$, $R_1 = 2.2\ \text{k}\Omega$, $R_2 = 4.44\ \text{k}\Omega$, $C_\pi = 29.6\ \text{pF}$, and $C_\mu = 1\ \text{pF}$. Estimate $f_{-3\,dB}$ via the OCTC method, comment.   **EXAMPLE 6.17**

Solution

By Eqs. (6.84) through (6.86), we have

$$R_\pi = 1.95 // \frac{2.2 + 4.44}{1 + 4.44/0.013}\ \text{k}\Omega = 19.2\ \Omega$$

$$R_\mu = 2.2//(1.95 + 151 \times 4.44) = 2.19 \text{ k}\Omega$$

$$f_{-3\,\text{dB}} \cong \frac{1/2\pi}{19.2 \times 29.6 \times 10^{-12} + 2.19 \times 10^3 \times 10^{-12}}$$

$$= \frac{1/2\pi}{(0.57 + 2.19)\ \text{ns}} = 58 \text{ MHz}$$

This estimate is in reasonable agreement with the value of 66 MHz obtained via the PSpice circuit of Fig. 6.32. Clearly, the time constant due to C_μ is the dominant one in this case.

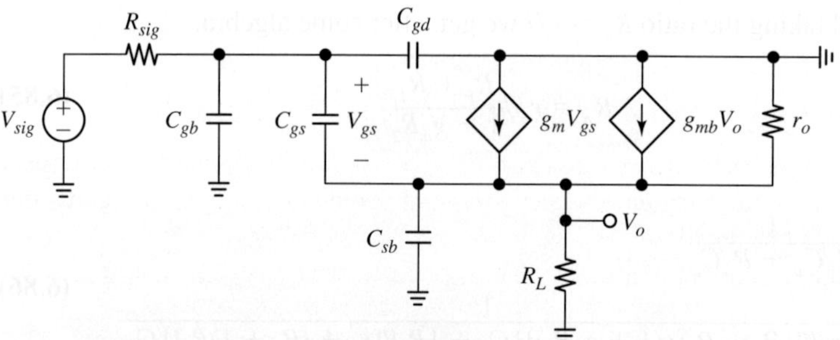

FIGURE 6.47 Revisiting the source follower.

Next, we turn to the **source follower** of Fig. 6.36*b*, repeated for convenience in Fig. 6.47. To find its open-circuit equivalent resistances we rearrange it in the form of Fig. 6.48. By inspection,

$$R_{gb} = R_{sig} \qquad R_{sb} = R_L//r_o//\frac{1}{g_{mb}}//\frac{1}{g_m} \tag{6.87}$$

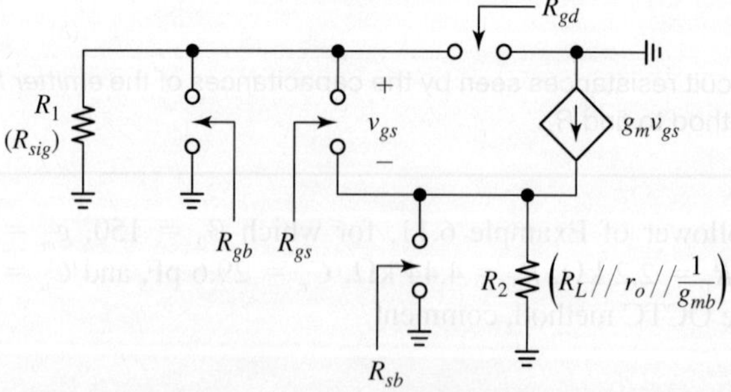

FIGURE 6.48 The open-circuit resistances seen by the capacitances in the *source follower* of Fig. 6.47.

To find R_{gs} and R_{gd} we note this circuit's similarity to its bipolar counterpart of Fig. 6.46, except that now $r_\pi \to \infty$. To save labor we thus recycle Eqs. (6.84) and (6.85) by letting $r_\pi \to \infty$ and $\beta_0 \to \infty$, and get

$$R_{gd} = R_1 = R_{sig} \qquad R_{gs} = \frac{R_1 + R_2}{1 + g_m R_2} = \frac{R_{sig} + [R_L /\!/ r_o /\!/ (1/g_{mb})]}{1 + g_m[R_L /\!/ r_o /\!/ (1/g_{mb})]} \qquad (6.88)$$

We now have all the parameters needed to estimate $f_{-3\,\text{dB}}$.

Exercise 6.5

Reconsider the source follower of Example 6.14, having $g_m = 1$ mA/V, $g_{mb} = 0.1$ mA/V, $r_o = 50$ kΩ, $C_{gs} = 400$ fF, $C_{gb} = C_{gd} = C_{sb} = 25$ fF, and $R_{sig} = R_L = 10$ kΩ. Use the OCTC technique to estimate $f_{-3\,\text{dB}}$.

Ans. 122 MHz.

Voltage Amplifiers with Emitter/Source Degeneration

Recall that degeneration *reduces gain* in a CE/CS amplifier. Consequently, we expect a reduced Miller multiplier and, hence, a *higher* value for $f_{-3\,\text{dB}}$. Gain-bandwidth tradeoffs arise all the time in electronics, as we shall frequently see. We wish to use the OCTC technique to estimate the -3-dB frequency of a voltage amplifier with degeneration.

Consider first the **CE-ED amplifier** shown in ac form in Fig. 6.49a. As long as r_o is fairly large compared with the resistances external to the BJT, we can ignore r_o altogether to simplify our analysis. Turning to the open-circuit equivalent of Fig. 6.49b, we adapt Eq. (6.85) to obtain an expression for R_π,

$$R_\pi \cong r_\pi /\!/ \frac{R_B + R_E}{1 + g_m R_E} \qquad (6.89)$$

Also, by inspection,

$$R_s \cong R_C \qquad (6.90)$$

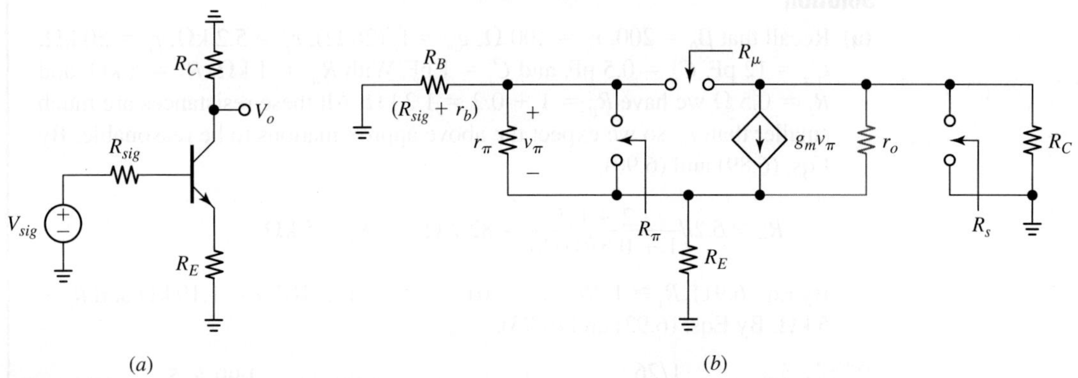

FIGURE 6.49 (a) The CE-ED amplifier, and (b) the open-circuit resistances seen by its capacitances.

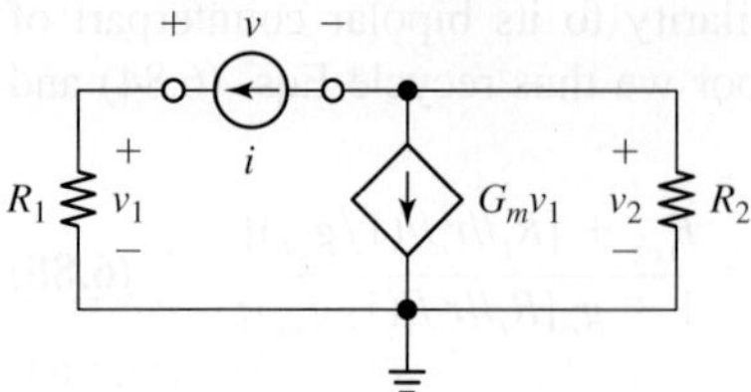

FIGURE 6.50 Equivalent circuit to find R_μ in Fig. 6.49b.

To find R_μ we use the test method, but after having reduced the circuit of Fig. 6.49b to the more compact form of Fig. 6.50. Here, R_1 and R_2 represent the equivalent resistances seen looking from the positive and from the negative terminals of the test source, and G_m represents the degenerated transconductance. Adapting the expressions tabulated in Fig. 4.9, we have, for $R_C \ll r_o$,

$$R_1 \cong (R_{sig} + r_b)//[r_\pi + (\beta_0 + 1)R_E] \qquad R_2 \cong R_C \qquad (6.91)$$

$$G_m = \frac{g_m}{1 + g_m R_E} \qquad (6.92)$$

But the circuit of Fig. 6.50 is formally identical to that of Fig. 6.44b, so we adapt Eq. (6.81) to the present case and write

$$R_\mu \cong R_1 + R_2 + G_m R_1 R_2 \qquad (6.93)$$

We now have all the parameters needed to estimate $f_{-3\,\text{dB}}$.

EXAMPLE 6.18   (a) Investigate the effect of adding an emitter-degeneration resistance $R_E = 500\ \Omega$ to the CE amplifier of Example 6.8. Compare with the example, and comment.
(b) Verify via PSpice.

Solution

(a) Recall that $\beta_0 = 200$, $r_b = 200\ \Omega$, $g_m = 1/(26\ \Omega)$, $r_\pi = 5.2\ \text{k}\Omega$, $r_o = 50\ \text{k}\Omega$, $C_\pi = 12\ \text{pF}$, $C_\mu = 0.5\ \text{pF}$, and $C_s = 1\ \text{pF}$. With $R_{sig} = 1\ \text{k}\Omega$, $R_C = 5\ \text{k}\Omega$, and $R_E = 0.5\ \Omega$ we have $R_B = 1 + 0.2 = 1.2\ \text{k}\Omega$. All these resistances are much smaller than r_o, so we expect the above approximations to be reasonable. By Eqs. (6.89) and (6.90),

$$R_\pi \cong 5.2//\frac{1.2 + 0.5}{1 + 0.5/0.026} = 82.7\ \Omega \qquad R_s \cong 5\ \text{k}\Omega$$

By Eq. (6.91), $R_1 \cong 1.2//(5.2 + 201 \times 0.5) = 1.2//105.7 = 1.19\ \text{k}\Omega$ and $R_2 \cong 5\ \text{k}\Omega$. By Eqs. (6.92) and (6.93),

$$G_m = \frac{1/26}{1 + 500/26} = \frac{1}{526\ \Omega} \qquad R_\mu = 1.19 + 5 + \frac{1.99 \times 5}{0.526} = 17.5\ \text{k}\Omega$$

Finally, expressing resistances in kΩ and capacitances in pF, we get

$$f_{-3\,dB} = \frac{1/2\pi}{R_\pi C_\pi + R_\mu C_\mu + R_s C_s} \cong \frac{1/2\pi}{(1 + 8.75 + 5)\ \text{ns}} = 10.8\ \text{MHz}$$

Compared to the case $R_E = 0$ of Example 6.15, there has been an order-of-magnitude reduction both in τ_π and in τ_μ, causing $f_{-3\,dB}$ to increase from about 1.53 MH to 10.8 MHz. On the other hand, the low-frequency gain has dropped from $a_0 = -142$ V/V (43 dB) of Example 6.8 to the present value

$$a_0 = \frac{105.7}{1.2 + 105.7} \times \frac{-5}{0.526} \cong -9.4\ \text{V/V (19.5 dB)}$$

(b) To verify via PSpice, reuse the circuit of Fig. 6.22, but after lifting the emitter terminal off ground to insert $R_E = 500\ \Omega$ between emitter and ground. The simulation yields the Bode plots of Fig. 6.51, which are in good agreement with the results found via the OCRC technique.

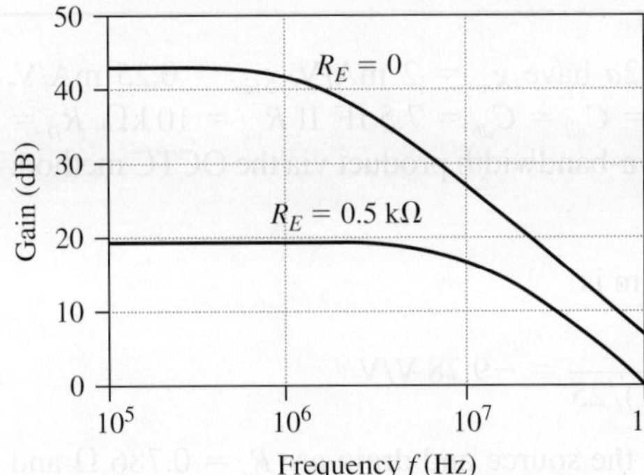

FIGURE 6.51 Gain plots for the CE amplifier of Example 6.8 ($R_E = 0$), and the CE-ED amplifier of Example 6.18 ($R_E = 0.5$ kΩ).

Next, we turn to the **CS-SD amplifier** of Fig. 6.52a. With reference to its open-circuit equivalent of Fig. 6.52b we find, by inspection,

$$R_{gb} = R_{sig} \qquad R_{sb} = R_S/\!/R_s \qquad R_{db} = R_D/\!/R_d \tag{6.94}$$

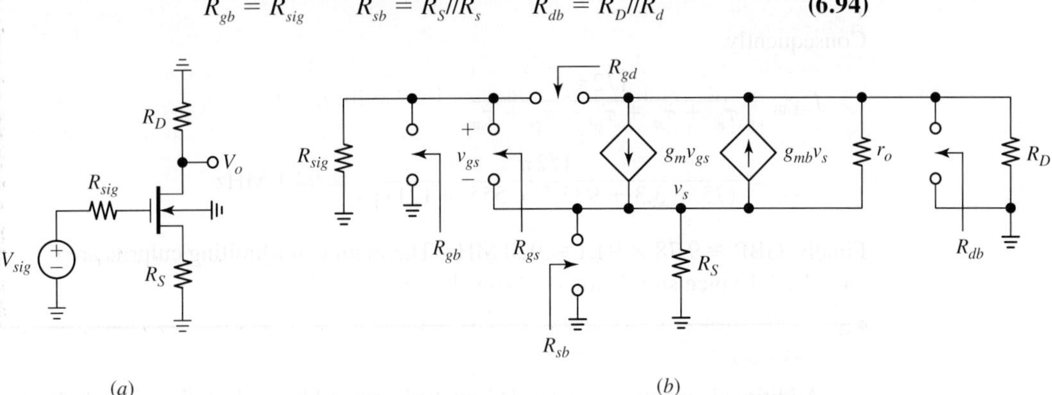

FIGURE 6.52 (a) The CS-SD amplifier and (b) the open-circuit resistances seen by its capacitances.

where R_s and R_d are the resistances seen looking into the source and the drain, both of which are tabulated in Fig. 4.23. To find R_{gd} we simply adapt Eqs. (6.92) and (6.93) to the MOS case,

$$R_{gd} \cong R_{sig} + R_{db} + \frac{g_m}{1 + (g_m + g_{mb})R_S} R_{sig} R_{db} \tag{6.95}$$

Finally, to find R_{gs} we apply the test method as usual. The result, whose derivation is left as an exercise for the reader, is

$$R_{gs} = \frac{R_{sig} + R_S + g_{mb}R_{sig}R_S + (R_{sig}R_S + R_{sig}R_D + R_D R_S)/r_o}{1 + (g_m + g_{mb})R_S + (R_D + R_S)/r_o} \tag{6.96}$$

We now have all the parameters needed to estimate $f_{-3\,\mathrm{dB}}$.

EXAMPLE 6.19   Let the CS-SD amplifier of Fig. 6.52a have $g_m = 2$ mA/V, $g_{mb} = 0.25$ mA/V, $r_o = 25$ kΩ, $C_{gs} = 100$ fF, $C_{gb} = C_{sb} = C_{gd} = C_{db} = 7.5$ fF. If $R_{sig} = 10$ kΩ, $R_D = 20$ kΩ, and $R_S = 1$ kΩ, estimate its gain-bandwidth product via the OCTC method.

Solution

By Eq. (4.39a), the low-frequency gain is

$$a_0 = \frac{-2 \times 20}{1 + (2 + 0.25)1 + (20 + 1)/25} = -9.78 \text{ V/V}$$

and the resistances seen looking into the source and drain are $R_s = 0.786 \, \Omega$ and $R_d = 82.3$ kΩ, so Eq. (6.94) gives

$$R_{gb} = 10 \text{ k}\Omega \qquad R_{sb} = 1//0.786 = 0.44 \text{ k}\Omega \qquad R_{db} = 20//82.3 = 16.1 \text{ k}\Omega$$

Likewise, Eqs. (6.95) and (6.96) give

$$R_{gd} \cong 125 \text{ k}\Omega \qquad R_{gs} = 5.55 \text{ k}\Omega$$

Consequently,

$$f_{-3\,\mathrm{dB}} = \frac{1/2\pi}{\tau_{gb} + \tau_{sb} + \tau_{gd} + \tau_{gs} + \tau_{db}}$$

$$\cong \frac{1/2\pi}{(75 + 3.3 + 937.5 + 555 + 121) \text{ ps}} \cong 94.1 \text{ MHz}$$

Finally, GBP $\cong 9.78 \times 94.1 \cong 920$ MHz. The main band-limiting culprits are C_{gd} and C_{gs}. A PSpice simulation confirms the above results.

Additional examples of OCTC analysis are addressed in the end-of-chapter problems.

6.8 FREQUENCY RESPONSE OF CASCODE AMPLIFIERS

In Chapter 4 we investigated the use of cascoding as a means to raise the unloaded voltage gain dramatically. We are now ready to appreciate another inherent advantage of cascoding, namely, an increase in the gain-bandwidth product (GBP).

The Bipolar Cascode

Shown in Fig. 6.53a is the ac equivalent of the bipolar cascode configuration, highlighting the capacitances intervening in the frequency response. We wish to estimate the -3-dB frequency via OCTC analysis. The node labeled as V_x plays an important role in this circuit, so we start out by finding the open-circuit equivalent resistance R_x between this node and ground. With reference to Fig. 6.53b we note that $R_x = r_{o1}//R_{e2}$, where R_{e2} is the resistance seen looking into Q_2's emitter. This is obtained by adapting Eq. (4.10) to the present case. The result is

$$R_x = r_{o1}//\left[\left(\frac{r_{b2} + r_{\pi2}}{\beta_{02} + 1}//r_{o2}\right) \times \frac{1 + R_C/r_{o2}}{1 + R_C/[(\beta_{02} + 1)r_{o2} + r_{b2} + r_{\pi2}]}\right] \quad (6.97)$$

(Note that for large r_{o1} and β_{02} we have $R_x \cong r_{e2} + r_{b2}/(\beta_{02} + 1)$ for $R_C \to 0$, and $R_x \cong r_{\pi2} + r_{b2}$ for $R_C \to \infty$).

We observe that Q_1 is similar to the circuit of Fig. 6.43, but with $R_2 = R_x$. We can thus recycle the results developed there and write,

$$R_{\pi1} = (R_{sig} + r_{b1})//r_{\pi1} \qquad R_{\mu1} = R_{\pi1} + R_x + g_{m1}R_{\pi1}R_x \qquad R_{s1} = R_x \quad (6.98)$$

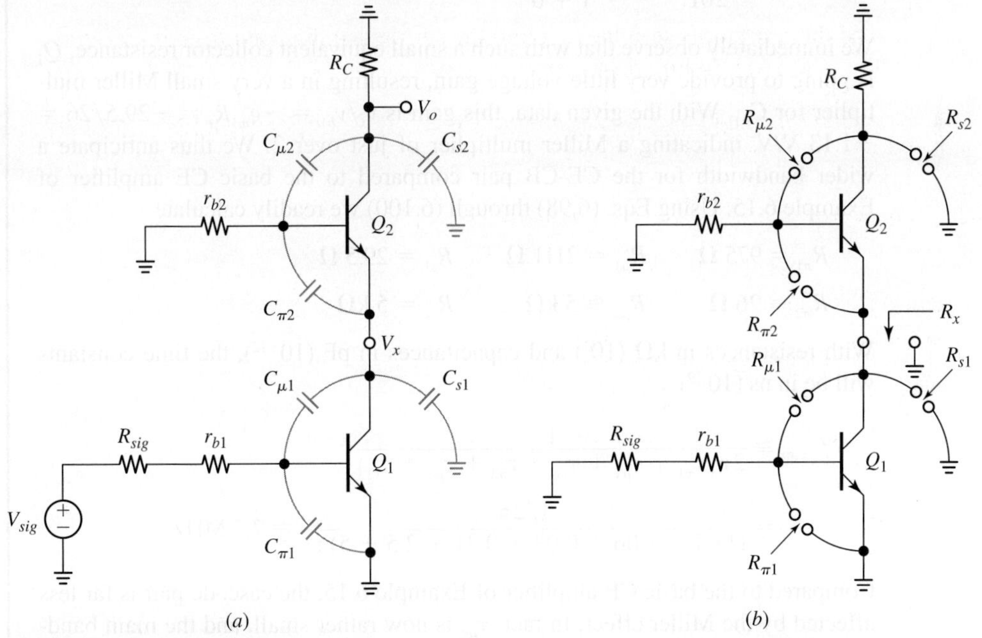

FIGURE 6.53 (a) The bipolar cascode configuration, with all relevant stray capacitances explicitly shown outside of the BJTs. (b) Circuit for the calculation of the open-circuit equivalent resistances.

Likewise, Q_2 is similar to the circuit of Fig. 6.49b, but with $R_B = r_{b2}$ and $R_E = r_{o1}$. We can thus recycle the results developed there and write,

$$R_{\pi2} \cong r_{\pi2} /\!/ \frac{r_{b2} + r_{o1}}{1 + g_{m2}r_{o1}} \cong r_{e2} \qquad R_{\mu2} \cong R_1 + R_2 + G_{m2}R_1R_2 \qquad R_{s2} \cong R_C \quad \textbf{(6.99)}$$

where

$$R_1 \cong r_{b2}/\!/[r_{\pi2} + (\beta_{02} + 1)r_{o1}] \cong r_{b2} \quad R_2 \cong R_C \quad G_{m2} = \frac{g_{m2}}{1 + g_{m2}r_{o1}} \cong \frac{1}{r_{o1}} \quad \textbf{(6.100)}$$

It is now a straightforward matter to apply Eq. (6.80) to estimate the $f_{-3\,dB}$ for the bipolar cascode.

EXAMPLE 6.20   The CE amplifier of Example 6.15 was found to have $f_{-3\,dB} = 1.53$ MHz with $R_{sig} = 1$ kΩ and $R_C = 5$ kΩ. What happens if we buffer it to R_C via a CB stage, thus implementing a cascoded pair? Compare and comment. (For simplicity assume identical BJT parameters, namely, $\beta_0 = 200$, $r_b = 200\ \Omega$, $g_m = 1/(26\ \Omega)$, $r_\pi = 5.2$ kΩ, $r_o = 50$ kΩ, $C_\pi = 12$ pF, $C_\mu = 0.5$ pF and $C_s = 1$ pF. Also, assume $r_\mu = \infty$.)

Solution

Applying Eq. (6.97) with suitable approximations we get

$$R_x \cong \frac{200 + 5200}{201} \times \frac{1 + 5/50}{1 + 0} = 29.5\ \Omega$$

We immediately observe that with such a small equivalent collector resistance, Q_1 is going to provide very little voltage gain, resulting in a very small Miller multiplier for $C_{\mu1}$. With the given data, this gain is $v_x/v_{b1} = -g_{m1}R_x = -29.5/26 = -1.13$ V/V, indicating a Miller multiplier of just over 2. We thus anticipate a wider bandwidth for the CE-CB pair compared to the basic CE amplifier of Example 6.15. Using Eqs. (6.98) through (6.100) we readily calculate

$$R_{\pi1} = 975\ \Omega \qquad R_{\mu1} = 2111\ \Omega \qquad R_{s1} = 29.5\ \Omega$$

$$R_{\pi2} \cong 26\ \Omega \qquad R_{\mu2} \cong 5\ \text{k}\Omega \qquad R_{s2} = 5\ \text{k}\Omega$$

With resistances in kΩ (10^3) and capacitances in pF (10^{-12}), the time constants will be in ns (10^{-9}),

$$f_{-3\,dB} \cong \frac{1}{2\pi(\tau_{\pi1} + \tau_{\mu1} + \tau_{s1} + \tau_{\pi2} + \tau_{\mu2} + \tau_{s2})}$$

$$= \frac{1/2\pi}{(11.7 + 1.06 + 0.03 + 0.31 + 2.5 + 5)\ \text{ns}} \cong 7.7\ \text{MHz}$$

Compared to the basic CE amplifier of Example 6.15, the cascode pair is far less affected by the Miller effect. In fact, $\tau_{\mu1}$ is now rather small, and the main band-limiting culprits are $C_{\pi1}$ and C_{s2}.

EXAMPLE 6.21

Investigate the limiting case in which the cascode circuit of Example 6.20 is terminated on an ideal current-source load so that $R_C \to \infty$. Compare the two situations, verify with PSpice, and comment.

Solution

Letting $R_C \to \infty$ in Eq. (6.97) raises R_x to

$$R_x = r_{o1}/\!/(r_{b2} + r_{\pi 2}) = 50/\!/(0.2 + 5.2) = 4.87 \text{ k}\Omega$$

This increase has no effect on $R_{\pi 1}$, but raises $R_{\mu 1}$ and R_{s1}. Using again Eqs. (6.98) through (6.100) we get

$$R_{\pi 1} = 975 \ \Omega \qquad R_{\mu 1} \cong 188 \text{ k}\Omega \qquad R_{s1} = 4.87 \text{ k}\Omega$$

Also, adapting Eq. (4.11) to the present case we find, for negligible r_b,

$$R_{s2} = R_{c2} \cong r_{o2}\left(1 + \beta_{02}\frac{r_{o1}}{r_{o1} + r_{\pi 2}}\right) = 9.1 \text{ M}\Omega \qquad R_{\mu 2} \cong R_{s2} = 9.1 \text{ M}\Omega$$

With Q_2's collector terminated on an ac open circuit, we now have

$$R_{\pi 2} = r_{\pi 2}/\!/(r_{b2} + r_{o1}) = 4.95 \text{ k}\Omega$$

Finally,

$$f_{-3 \text{ dB}} \cong \frac{1/2\pi}{\tau_{\pi 1} + \tau_{\mu 1} + \tau_{s1} + \tau_{\pi 2} + \tau_{\mu 2} + \tau_{s2}}$$

$$\cong \frac{1/2\pi}{(12 + 94 + 5 + 59 + 4550 + 9100) \text{ ns}} \cong 11.5 \text{ kHz}$$

We observe that letting $R_C \to \infty$ increases R_x significantly, causing Q_1 to provide a much higher gain, namely, $v_x/v_{b1} = -g_{m1}R_x = -4870/26 = -187$ V/V. Consequently, we now have a Miller multiplier of 188, which raises $\tau_{\mu 1}$ from about 1 ns to 94 ns. The other major effect of letting $R_C \to \infty$ is the dramatic increase in the output resistance R_o, namely, from 5 kΩ to 9.1 MΩ. While raising the unloaded gain a_0, this also reduces the pole frequency $f_{-3 \text{ dB}}$ in proportion.

Using the PSpice circuit of Fig. 6.54 we obtain the frequency plots of Fig. 6.55. It is apparent that in the limit $R_C \to \infty$ the main band-limiting culprits are $C_{\mu 2}$ and C_{s2}, which establish a dominant pole at

$$f_p = \frac{1}{2\pi R_o(C_{s2} + C_{\mu 2})} = 11.7 \text{ kHz}.$$

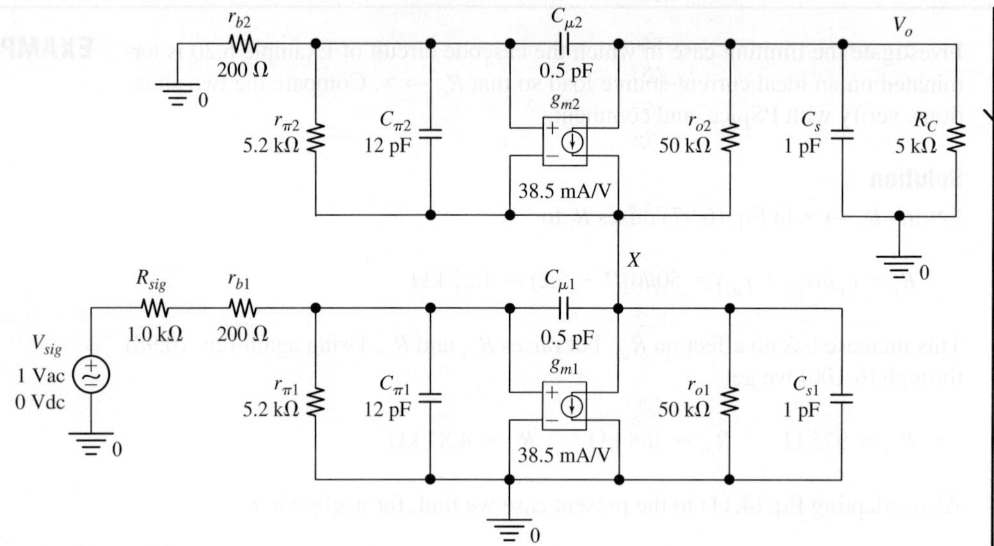

FIGURE 6.54 PSpice circuit to display gain vs. frequency for the CE-CB amplifier.

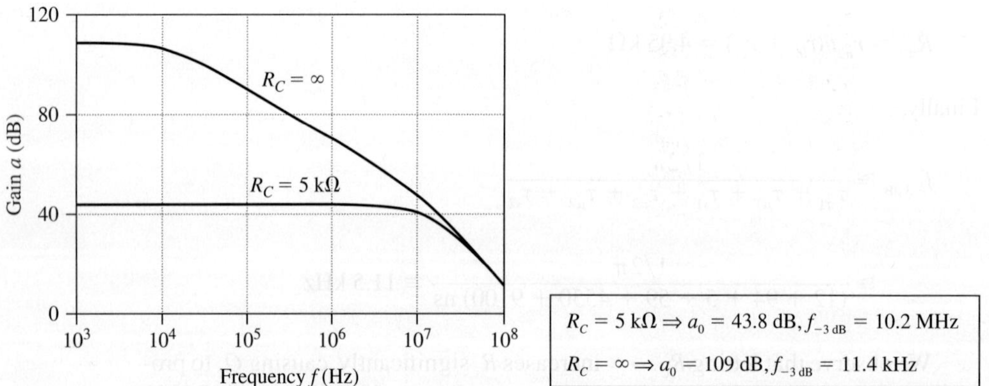

FIGURE 6.55 Gain plots for the CE-CB amplifier of Example 6.20 ($R_C = 5$ kΩ) and of Example 6.21 ($R_C = \infty$).

The MOS Cascode

Shown in Fig. 6.56a is the ac equivalent of the MOS cascode configuration, along with all capacitances affecting its frequency response. We wish to estimate the -3-dB frequency via the OCTC technique. As in the bipolar case, the node labeled as V_x plays an important role in this circuit, so we start out by finding the open-circuit equivalent resistance R_x between this node and ground. With reference to

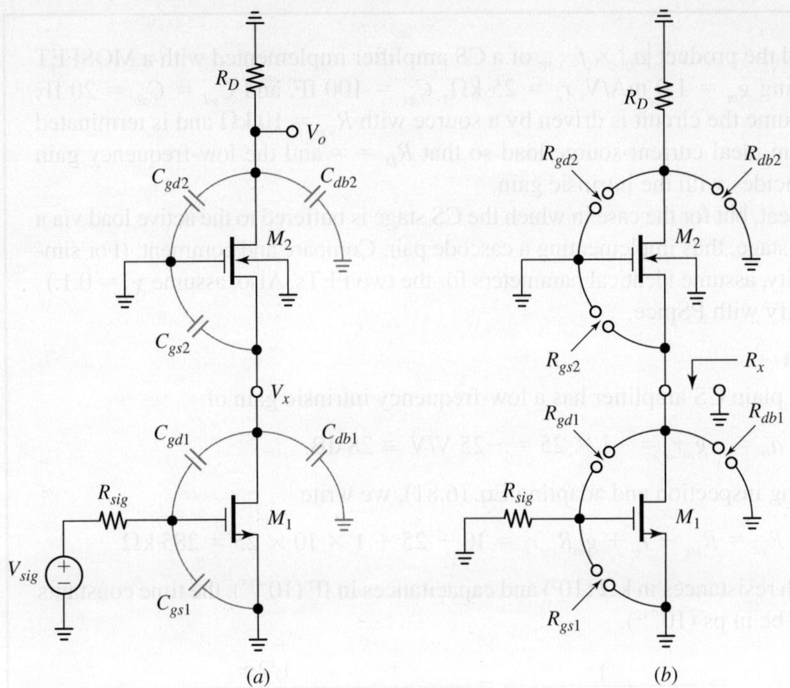

FIGURE 6.56 (a) The MOS cascode configuration, with all relevant stray capacitances explicitly shown outside of the FETs. (b) Circuit for the calculation of the open-circuit equivalent resistances.

Fig. 6.56b we note that $R_x = R_{d1}//R_{s2}$, where R_{d1} and R_{s2} are the resistances seen looking into M_1's drain and M_2's source, respectively. The former is simply r_{o1}, and the latter is obtained by adapting Eq. (4.40a) to the present case. The result can be written as

$$R_x = r_{o1}//\frac{r_{o2} + R_D}{1 + (g_{m2} + g_{mb2})r_{o2}} \qquad (6.101)$$

(Note that for $R_D \to 0$ we have $R_x \cong 1/(g_{m2} + g_{m2})$, whereas for $R_D \to \infty$ we have $R_x \cong r_{o1}$.) Next, using inspection and also adapting Eq. (6.81) to the present case we write

$$R_{gs1} = R_{sig} \qquad R_{gd1} = R_{sig} + R_x + g_{m1}R_{sig}R_x \qquad R_{db1} = R_x \qquad (6.102)$$

Finally, using again inspection and adapting Eq. (4.41) to the present case we write

$$R_{gs2} = R_x \qquad R_{gd2} = R_D//[r_{o1} + r_{o2} + (g_{m2} + g_{mb2})r_{o1}r_{o2}] \qquad R_{db2} = R_{gd2} \qquad (6.103)$$

It is now a straightforward matter to apply Eq. (6.80) to estimate the -3-dB frequency of the MOS cascode.

EXAMPLE 6.22   (a) Find the product $|a_0| \times f_{-3\,dB}$ of a CS amplifier implemented with a MOSFET having $g_m = 1.0$ mA/V, $r_o = 25$ kΩ, $C_{gs} = 100$ fF, and $C_{gd} = C_{db} = 20$ fF. Assume the circuit is driven by a source with $R_{sig} = 10$ kΩ and is terminated on an ideal current-source load so that $R_D = \infty$ and the low-frequency gain coincides with the intrinsic gain.

   (b) Repeat, but for the case in which the CS stage is buffered to the active load via a CG stage, thus implementing a cascode pair. Compare and comment. (For simplicity, assume identical parameters for the two FETs. Also, assume $\chi_2 = 0.1$.)

   (c) Verify with PSpice.

Solution

(a) The plain CS amplifier has a low-frequency intrinsic gain of

$$a_0 = -g_m r_o = -1 \times 25 = -25 \text{ V/V} \cong 28 \text{ dB.}$$

Using inspection and adapting Eq. (6.81), we write

$$R_{gd} = R_{sig} + r_o + g_m R_{sig} r_o = 10 + 25 + 1 \times 10 \times 25 = 285 \text{ k}\Omega$$

With resistances in kΩ (10^3) and capacitances in fF (10^{-15}), the time constants will be in ps (10^{-12}),

$$f_{-3\,dB} \cong \frac{1}{2\pi(\tau_{gs} + \tau_{gd} + \tau_{db})} = \frac{1/2\pi}{(10 \times 100 + 285 \times 20 + 25 \times 20) \text{ ps}}$$

$$= \frac{1/2\pi}{(1 + 5.7 + 0.5) \text{ ns}} \cong 22 \text{ MHz}$$

Finally, $|a_0| \times f_{-3\,dB} = 25 \times 22 = 550$ MHz.

(b) Cascoding will raise the intrinsic gain to

$$a_0 = -g_{m1} r_{o1}[1 + (g_{m2} + g_{mb2})r_{o2}] = -25(1 + 1.1 \times 25)$$

$$= -712.5 \text{ V/V} = 57 \text{ dB}$$

To apply the OCTC method we note that with an ideal current-source load we have $R_D = \infty$, so Eq. (6.102) gives $R_x = r_{o1} = 25$ kΩ. Using inspection, along with Eqs. (6.102) and (6.103), we now have

$$R_{gs1} = 10 \text{ k}\Omega \quad R_{db1} = R_{gs2} = 25 \text{ k}\Omega \quad R_{gd1} = 285 \text{ k}\Omega$$

$$R_{gd2} = R_{db2} = 737.5 \text{ k}\Omega$$

With resistances in kΩ (10^3) and capacitances in pF (10^{-12}), the time constants will be in ns (10^{-9}),

$$f_{-3\,dB} \cong \frac{1/2\pi}{\tau_{gs1} + \tau_{gd1} + \tau_{db1} + \tau_{gs2} + \tau_{gd2} + \tau_{db2}}$$

$$= \frac{1/2\pi}{(1 + 5.7 + 0.5 + 2.5 + 14.7 + 14.7) \text{ ns}} = 4.1 \text{ MHz}$$

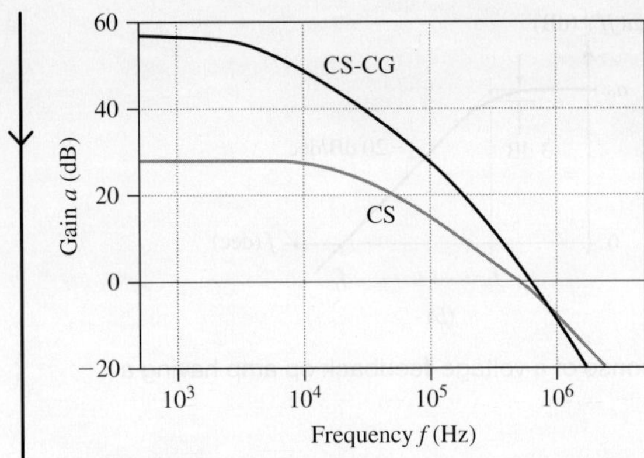

FIGURE 6.57 Gain plots for the Cs and the CS-CB amplifiers of Example 6.22.

We now have $|a_0| \times f_{-3\,\text{dB}} = 712.5 \times 4.1 = 2.9$ GHz. Cascoding has raised a_0 while lowering $f_{-3\,\text{dB}}$. Yet their product is still higher than in part (*a*). The main band-limiting culprits are now C_{gd2} and C_{db2}, which form a pole with the high output resistance R_o of the cascode structure.

(c) Using a PSpice circuit of the type of Fig. 6.54 we obtain the plots of Fig. 6.57, which are in excellent agreement with the hand calculations.

We summarize the advantages of cascoding by stating that the CE/CS stages Q_1/M_1 overcome their inherent Miller-effect drawbacks by delegating the task of voltage amplification to the CB/CG stages Q_2/M_2, which are inherently much faster because they are immune from the Miller effect. The result is a dramatic increase in the gain-bandwidth product.

6.9 FREQUENCY AND TRANSIENT RESPONSES OF OP AMPS

Most op amps are designed for a gain that is dominated by a single low-frequency pole (the reason, to be investigated in detail in Chapter 7, is to prevent possible oscillations in negative-feedback operation). For the traditional voltage-feedback op amp of Fig. 6.58*a* the gain $a(jf)$ exhibits the frequency profile of Fig. 6.58*b* and takes on the mathematical form

$$a(jf) = \frac{V_o}{V_p - V_n} \cong \frac{a_0}{1 + jf/f_b} \tag{6.104}$$

where V_o, V_p, and V_n are the Laplace transforms of the small signals v_o, v_p, and v_n, a_0 is the *dc gain*, and f_b is the *dominant-pole frequency*. Gain is approximately constant up to f_b. At f_b it is 3-dB below its dc value. Above f_b it rolls off with frequency at a uniform rate of -20 dB/dec until it drops to unity, or 0 dB, at the *transition frequency f_t*.

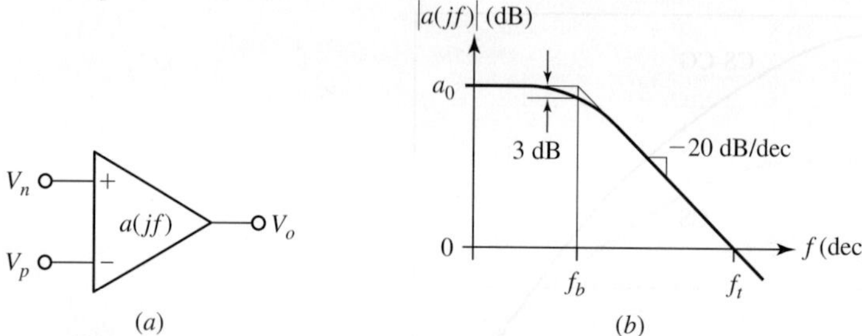

FIGURE 6.58 The frequency response of a voltage-feedback op amp having a dominant pole.

Exploiting the constancy of the gain-bandwidth product (GBP) we write $a_0 \times f_b = 1 \times f_t$. Consequently, a dominant-pole op amp has a *constant* GBP,

$$\text{GBP} = f_t = a_0 \times f_b \tag{6.105}$$

The frequency response of an op amp is easily visualized via PSpice, as demonstrated in Fig. 6.59 for the case of the 741 op amp. Using PSpice's cursor facility on the magnitude plot we find $a_0 = 105.3$ dB, $f_b = 5.2$ Hz, and $f_t = 871$ MHz. Above f_t the rolloff rate increases, indicating the presence of additional root frequencies there. (Higher-order roots will be addressed in greater detail in Chapter 7.)

Let us now investigate how the dominant pole is established in the three representative op amps discussed in Chapter 5, namely, the 741 *bipolar* and the *two-stage* and *folded cascode* CMOS op amps.

- Figure 6.60 shows the portion of the **741 op amp** involved in frequency compensation. The goal is to establish a dominant pole at a *sufficiently low* frequency f_b to force gain to drop to unity *before* the phase shift by higher-order poles

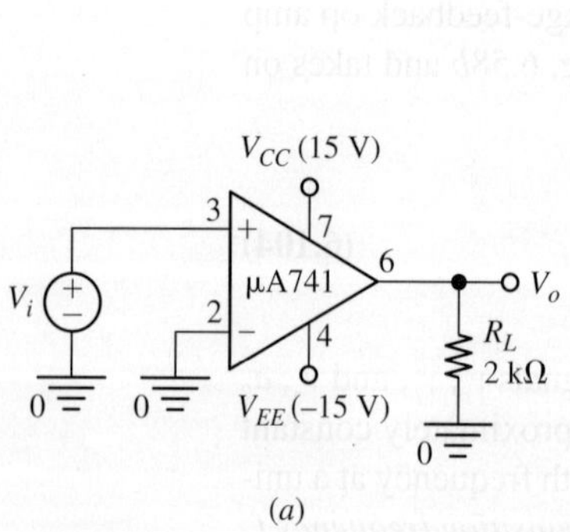

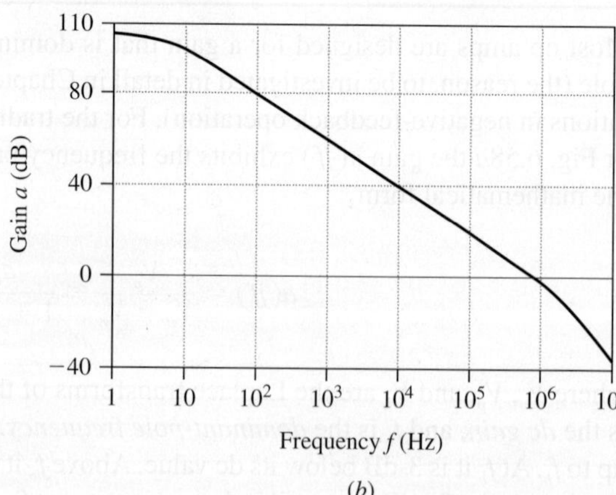

FIGURE 6.59 Using the 741 macromodel available in PSpice's library to plot the frequency response.

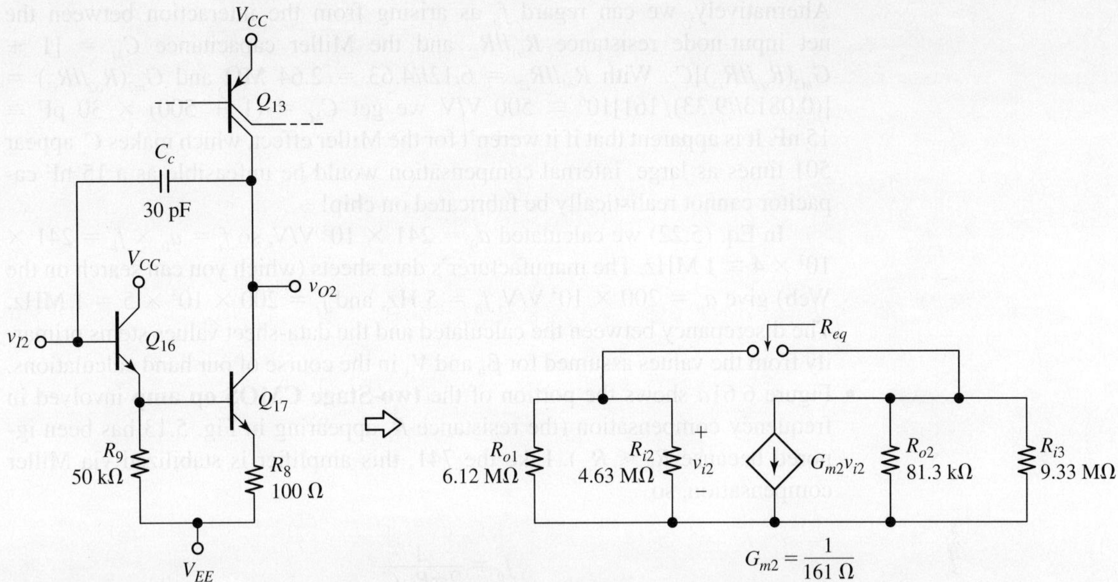

FIGURE 6.60 The 2nd stage of the 741 op amp and its ac equivalent for the calculation of the resistance R_{eq} as seen by the frequency-compensation capacitor C_c.

destabilizes the op amp in negative-feedback operation. This pole is obtained by placing a small capacitance C_c ($=30$ pF) across the 2nd stage and taking advantage of the Miller effect to achieve the much larger effective value needed to establish this low-frequency pole. Since C_c is small enough that it can be fabricated on chip, the 741 is said to be *internally compensated*. (In fact, the 741 was the *first* op amp in this category, whereas previous op amps had to be compensated externally by the user. Once internal compensation became a technological reality, the op amp took off to become the most widely used analog IC.)

Using OCTC analysis we get

$$f_b = \frac{1}{2\pi R_{eq} C_c}$$

where R_{eq} is the equivalent resistance seen by C_c. This resistance is depicted in the ac equivalent shown at the right. Taking advantage of the similarity to Fig. 6.44*a*, we adapt Eq. (6.81) and write

$$R_{eq} = (R_{o1}//R_{i2}) + (R_{o2}//R_{i3}) + G_{m2}(R_{o1}//R_{i2})(R_{o2}//R_{i3})$$

Substituting the values shown we get

$$R_{eq} = \left[(6.12//4.63) + (0.0813//9.33) + \frac{(6.12//4.63) + (0.0813//9.33)}{161}\right]10^6 = 1.32 \text{ G}\Omega$$

$$f_b = \frac{1}{2\pi \times 1.32 \times 10^9 \times 30 \times 10^{-12}} = 4 \text{ Hz}$$

Alternatively, we can regard f_b as arising from the interaction between the net input-node resistance $R_{o1}//R_{i2}$ and the Miller capacitance $C_M = [1 + G_{m2}(R_{o2}//R_{i3})]C_c$. With $R_{o1}//R_{i2} = 6.12//4.63 = 2.64$ MΩ and $G_{m2}(R_{o2}//R_{i3}) = [(0.0813//9.33)/161]10^6 \cong 500$ V/V we get $C_M = (1 + 500) \times 30$ pF $\cong$ 15 nF. It is apparent that if it weren't for the Miller effect, which makes C_c appear 501 times as large, internal compensation would be unfeasible as a 15-nF capacitor cannot realistically be fabricated on chip!

In Eq. (5.22) we calculated $a_0 = 241 \times 10^3$ V/V, so $f_t = a_0 \times f_b = 241 \times 10^3 \times 4 \cong 1$ MHz. The manufacturer's data sheets (which you can search on the Web) give $a_0 = 200 \times 10^3$ V/V, $f_b = 5$ Hz, and $f_t = 200 \times 10^3 \times 5 = 1$ MHz. The discrepancy between the calculated and the data-sheet values stems primarily from the values assumed for β_0 and V_A in the course of our hand calculations.

- Figure 6.61a shows the portion of the **two-Stage CMOS op amp** involved in frequency compensation (the resistance R_c appearing in Fig. 5.13 has been ignored because $R_c \ll R_{eq}$). Like the 741, this amplifier is stabilized via Miller compensation, so

$$f_b = \frac{1}{2\pi R_{eq} C_c}$$

where R_{eq} is the equivalent resistance seen by C_c,

$$R_{eq} \cong R_{o1} + R_{o2} + g_{m5}R_{o1}R_{o2}$$

- Figure 6.61b shows the portion of the **folded-cascode CMOS op amp** involved in frequency compensation. This configuration differs from the two op amps just

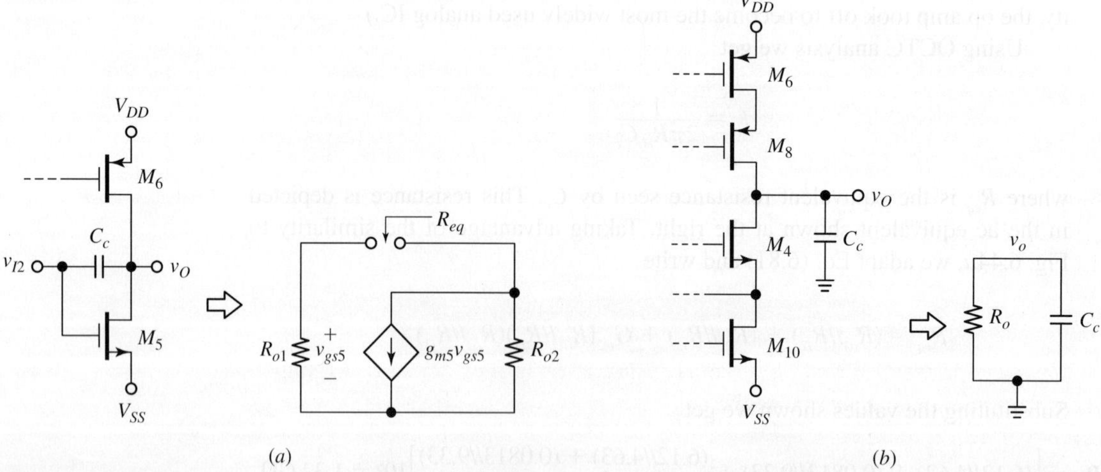

FIGURE 6.61 Portions of the (a) *two-stage* and (b) *folded-cascode* CMOS op amps involved in frequency compensation, along with the ac equivalents showing the ac resistance seen by the compensation capacitance.

investigated because the dominant pole is established by the output resistance R_o and the net output-node capacitance C_c as

$$f_b = \frac{1}{2\pi R_o C_c}$$

Here, C_c is the sum of the capacitances associated with the drains of M_4 and M_8 and any external load capacitance C_L. Moreover, by Eq. (5.39),

$$R_o \cong [(g_{m6} + g_{mb6})r_{o6}r_{o8}]//[(g_{m4} + g_{mb4})r_{o4}(r_{o2}//r_{o10})]$$

Looking at the folded-cascode schematic of Fig. 5.16, we expect additional poles due to the stray capacitances of the other nodes along the signal path. However, each of these nodes exhibits an equivalent resistance on the order of $1/(g_m + g_{mb})$, which is much lower than R_o, so the output-node pole dominates the entire response. This is why the folded cascode is often regarded as a *single stage* op amp!

EXAMPLE 6.23

(a) For the two-stage CMOS op amp of Example 5.2, find f_b and f_t if $C_c = 3$ pF.
(b) Repeat, but for the folded-cascode CMOS op amp of Example 5.4 if $C_c = 2.5$ pF.

Solution

(a) From Example 5.2 we have $a_0 = 2{,}844$ V/V, $R_{o1} = r_{o2}//r_{o4} = 400//200 = 133$ kΩ, $R_{o2} = r_{o5}//r_{o6} = 100//200 = 66.7$ kΩ, and $g_{m5} = 2 \times 0.1/0.25 = 1/(1.25 \text{ k}\Omega)$. We thus write $R_{eq} = 133 + 66.7 + 133 \times 66.7/1.25 = 7.31$ MΩ, so

$$f_b = \frac{1}{2\pi \times 7.31 \times 10^6 \times 3 \times 10^{-12}} = 7.34 \text{ kHz}$$

$$f_t = a_0 \times f_b = 2{,}844 \times 7.34 \times 10^3 = 20.9 \text{ MHz}$$

(b) From Example 5.4 we have $a_0 = 2{,}088$ V/V and $R_o = 5.22$ MΩ, so

$$f_b = \frac{1}{2\pi \times 5.22 \times 10^6 \times 2.5 \times 10^{-12}} \cong 12.2 \text{ kHz}$$

$$f_t = 2{,}088 \times 12.2 \times 10^3 = 25.5 \text{ MHz}$$

The Transient Response

Op amps are characterized both in the *frequency* and the *time* domains. An important time-domain characteristic is the *transient response*, which shows how an op amp circuit reacts to an input *voltage step*. The circuit commonly used is the *unity-gain voltage follower* of Fig. 6.62a because this is most difficult configuration to stabilize, as we shall see in Chapter 7. To facilitate the transient analysis, refer to the simplified renditions of Fig. 6.63, where the second stage and the compensation

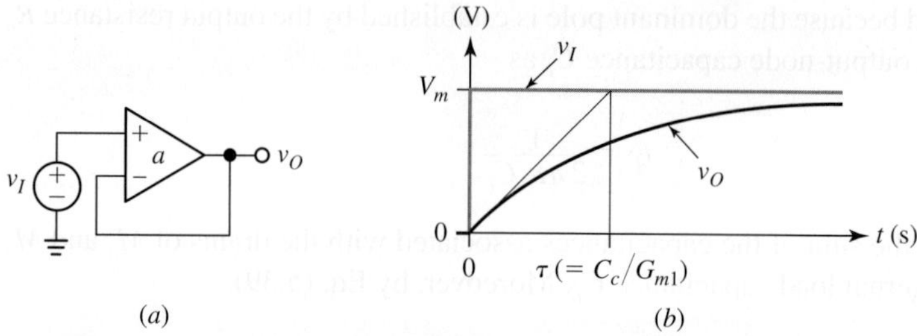

FIGURE 6.62 (a) Voltage follower and (b) *small-signal transient response* for a dominant-pole op amp.

capacitance have been combined together to form an integrator. As long as the gain a_2 of this stage is sufficiently high, the integrator's input terminal will approximate a virtual ground, so we have $C_c d(v_O - 0)/dt = i_C$, or

$$C_c \frac{dv_O}{dt} = i_C \tag{6.106}$$

We make the following considerations:

- In dc steady state both circuits yield $i_C = 0$ and $v_O \cong v_I$. Moreover, the input-stage bias current I_1 splits *equally* between the two halves of the differential pair.
- Applying a positive voltage step at the input will make Q_2/M_2 less conductive, so a greater portion of I_1 will be diverted to Q_1/M_1 to be subsequently replicated at the integrator's input by the current mirror. This results in $i_C > 0$ and causes C_c to charge up as per Eq. (6.106). As long as the step amplitude V_m is *sufficiently*

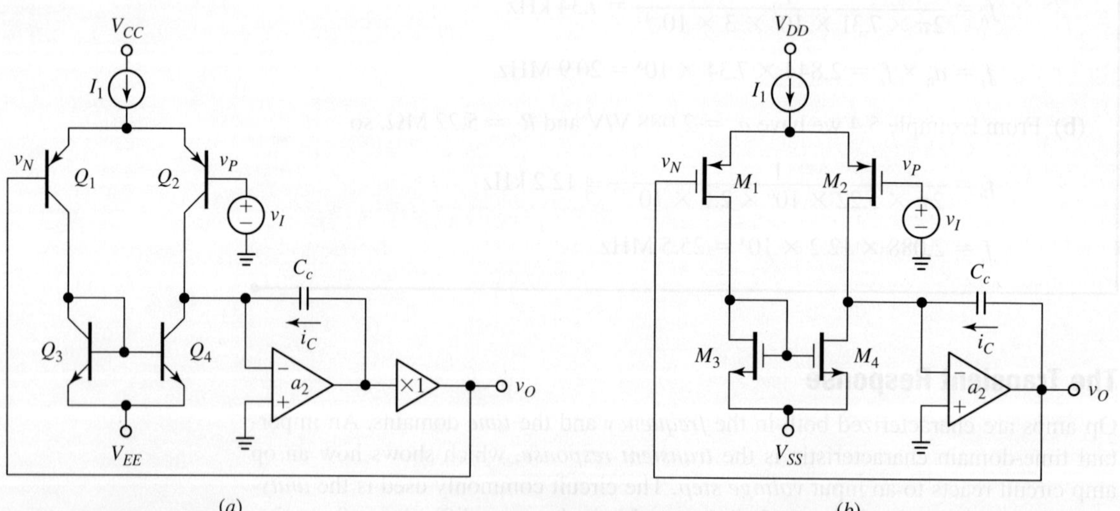

FIGURE 6.63 Simplified equivalents of (a) the 741 op amp and (b) the two-stage CMOS op amp (the folded-cascode CMOS op amp is similar, except that there is no second stage and C_c is connected to ground.)

small, we can use the small-signal approximation to write $i_C = G_{m1}(v_P - v_N) = G_{m1}(V_m - v_O)$, where G_{m1} is the *input-stage transconductance*. Substituting into Eq. (6.106) and rearranging gives

$$\tau \frac{dv_O}{dt} + v_O = V_m \qquad (6.107)$$

where

$$\tau = \frac{C_c}{G_{m1}} \qquad (6.108)$$

Recall from basic circuit courses that the solution to the above differential equation is an *exponential transient* governed by the *time constant* τ. For $v_O = 0$, Eq. (6.107) reduces to $\tau dv_O/dt = V_m$, indicating an initial slope of $dv_O/dt = V_m/\tau$ (see Fig. 6.62b). As the transient dies out, v_O eventually settles at V_m.

- We can obtain an insightful alternative expression for τ by noting that for $f = f_t$ we have

$$V_o(jf_t) = \frac{1}{j2\pi f_t C_c} I_c = \frac{1}{j2\pi f_t C_c} G_{m1}(V_p - V_n)$$

But, we also have $V_o(jf_t) = a(jf_t) \times (V_p - V_n) = (1/j) \times (V_p - V_n)$, so $G_{m1}/(2\pi f_t C_c) = 1$. Combining with Eq. (6.108) gives

$$\tau = \frac{1}{2\pi f_t} \qquad (6.109)$$

This provides a link between the *frequency-domain* parameter f_t and the *time-domain* parameter τ. The 741 op amp has $\tau = 1/(2\pi \times 10^6) = 159$ ns.

Slew-Rate (SR) Limiting

If we raise V_m further, the small-signal approximation ceases to hold. This is so because the i_C characteristic as a function of the difference $v_P - v_N$ takes on the familiar *s-shaped* form of Section 4.5, so the transient response becomes a *sluggish exponential*. Still, v_O will settle at V_m once the transient has died out. For V_m sufficiently large, *all* of I_1 will be diverted to Q_2/M_2, so i_C will *saturate* at $i_{C(max)} = I_1$, causing v_O to ramp up at the *maximum possible rate*. This rate is called the *slew rate* (SR) and is expressed in V/μs. By Eq. (6.106),

$$SR = \left. \frac{dv_O}{dt} \right|_{max} = \frac{I_1}{C_c} \qquad (6.110)$$

The 741 op amp has $I_1 = 19$ μA, so SR $= (19\ \mu A)/(30\ pF) = 0.633 \times 10^6$ V/s $= 0.633$ V/μs, which is fairly close to the more conservative data-sheet value of 0.5 V/μs.

It is of interest to know the step amplitude $V_{m(onset)}$ that marks the onset of SR limiting. This occurs for $V_{m(onset)}/\tau = SR$, or $V_{m(onset)} = SR/(2\pi f_t)$. The 741 has $V_{m(onset)} = 0.5 \times 10^6/(2\pi 10^6) \cong 80$ mV.

Exercise 6.6

Find τ, SR, and $V_{m(\text{onset})}$ for (a) the two-stage and (b) the folded-cascode CMOS op amps of Example 6.23. Assume $I_1 = 100\ \mu\text{A}$ for both circuits.

Ans. (a) 7.73 ns, 33.3 V/μs, 257 mV; (b) 6.24 ns, 40 V/μs, 250 mV.

We can readily visualize transient responses via PSpice. The circuit of 6.64a uses the 741 macro-model available in PSpice's Library to display the response to an input step of magnitude $V_m = 10\ \text{mV}$. Recall from Fig. 5.2 that the differential input stage of the 741 involves four base-emitter junctions, so each junction is subjected to a step of $10/4 = 2.5\ \text{mV}$, which is quite adequate for the small-signal approximation. The actual response, shown in Fig. 6.64b, differs somewhat from the exponential form predicted by single-pole analysis, and also exhibits overshoot. This is due to the presence of additional pole frequencies above f_t, as per Fig. 6.59 (more on this in Chapter 7).

Using the 60-mV rule of thumb we can state that raising V_m to $2 \times 60 = 120\ \text{mV}$ will make the current of one side of the differential stage *ten* times as large as the other side, thus bringing the 741 on the verge of slew-rate limiting. The pulse response of Fig. 6.65a is based on $V_m = 1.0\ \text{V}$, a convincingly large overdrive. Consequently, v_O ramps up at a constant rate of about 0.5 V/μs, taking about 2 μs to approach the 1.0-V plateau. As v_O approaches the plateau, the op amp ceases to slew-rate limit and v_O completes the last part of the transient in small-signal fashion, according to Fig. 6.64b.

Slew-rate limiting is a form of *nonlinear distortion* that limits the useful frequency range for large-signal operation. This distortion is illustrated in Fig. 6.65b for the case of a sinusoidal input

$$v_I = V_m \sin(2\pi ft)$$

with $V_m = 10\ \text{V}$ and $f = 15\ \text{kHz}$. The slope of a sinusoid is steepest at the 0-V crossings, where we have

$$\left|\frac{dv_I}{dt}\right|_{\max} = \left|-2\pi f V_m \cos(2\pi ft)\right|_{\max} = 2\pi f V_m$$

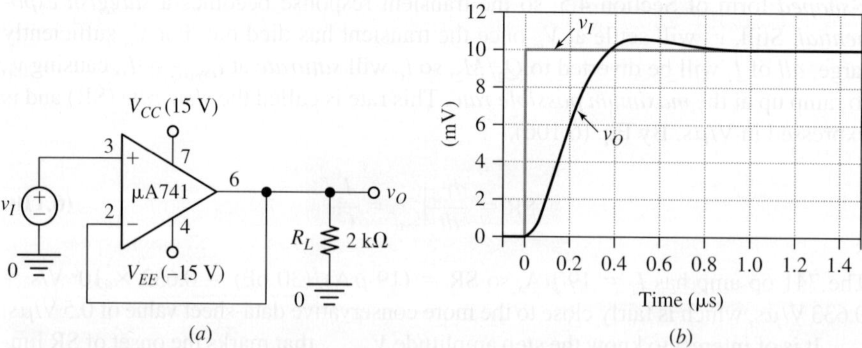

FIGURE 6.64 (a) PSpice circuit to display the 741 transient responses. (b) Small-signal step response.

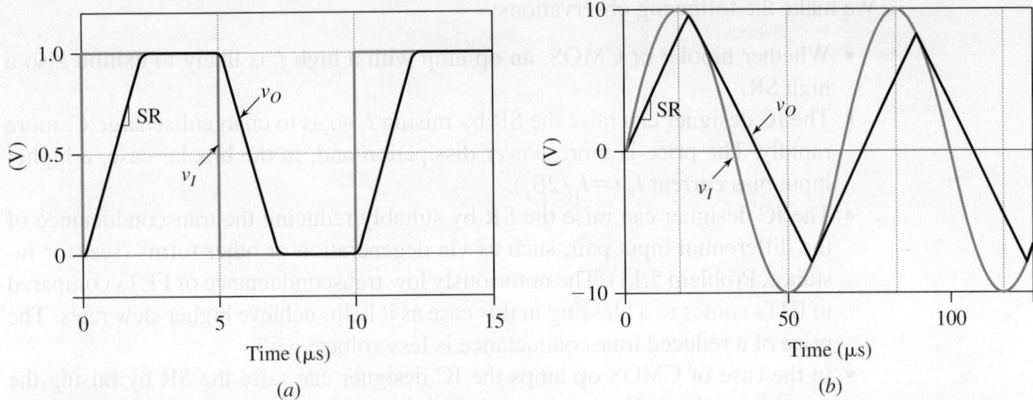

FIGURE 6.65 Slew-rate limited responses of the 741 follower to (a) a pulse and (b) a sinusoid.

If we want to avoid slew-rate limiting, this slope must be less than the slew rate,

$$2\pi f V_m \leq SR \qquad (6.111)$$

For instance, a 741 op amp with $V_m = 10$ V requires that its frequency f meet the condition

$$f \leq \frac{SR}{2\pi V_m} = \frac{0.5/10^{-6}}{2\pi 10} \cong 8 \text{ kHz}$$

Figure 6.65b shows the effect of exceeding the above limit with $f = 15$ kHz (> 8 kHz). It is apparent that as soon as the slope of v_I exceeds the SR, v_O ceases to track v_I and ramps up or down at a fixed rate of ± 0.5 V/μs. We can avoid slew-rate limiting either by lowering f to 8 kHz or less while keeping $V_m = 10$ V, or by suitably lowering V_m while keeping $f = 15$ kHz. In fact, we can turn around Eq. (6.111) and find $V_m \leq$ SR/$(2\pi f) = (0.5/10^{-6})/(2\pi \times 15 \times 10^3) \cong 5.3$ V. Likewise, if we wish the 741 follower to give an undistorted sine wave over the entire audio range, whose upper limit is $f = 20$ kHz, then we must ensure that $V_m \leq (0.5/10^{-6})/(2\pi \times 20 \times 10^3) \cong 4$ V.

Insightful Expressions for the Slew Rate

Combining Eqs. (6.108) through (6.110) we can express the slew rate in the insightful alternative form

$$SR = 2\pi \frac{I_1}{G_{m1}} f_t \qquad (6.112)$$

where f_t is the op amp's transition frequency, G_{m1} is the transconductance of the differential input pair, and I_1 is the pair's bias current. In the case of CMOS op amps we use $G_{m1} = 2(I_1/2)/V_{OV1} = I_1/V_{OV1}$ to obtain yet another insightful form,

$$SR_{CMOS} = 2\pi V_{OV1} f_t \qquad (6.113)$$

We make the following observations:

- Whether bipolar or CMOS, an op amp with a high f_t is likely to exhibit also a high SR.
- The IC designer can raise the SR by raising I_1 so as to charge/discharge C_c more rapidly. The price is more power dissipation and, in the bipolar case, a higher input bias current I_B ($=I_1/2\beta_F$).
- The IC designer can raise the SR by suitably reducing the transconductance of the differential input pair, such as via degeneration or other forms (see, for instance, Problem 5.11). The notoriously low transconductance of FETs compared to BJTs comes as a blessing in this case as it helps achieve higher slew rates. The price of a reduced transconductance is less voltage gain.
- In the case of CMOS op amps the IC designer can raise the SR by raising the overdrive voltage V_{OV1} of the differential input pair (another way of signifying low G_{m1}). This is another reason why preference is given to pMOSTETs in the differential pair: for similar dimensions and biasing conditions, the lower mobility of holes compared to electrons makes V_{OVp} two to three times higher than V_{OVn}.

Current-Feedback Amplifiers

The current-feedback amplifier (CFA) circuit of Fig. 5.39 reveals the presence of a number of internal low-resistance nodes along with a single high-resistance node denoted as node C. We anticipate the frequency response to be dominated by the pole associated with this one node. Denoting this node's total stray capacitance to ground as C_{eq}, as depicted in Fig. 6.66a, we have $V_o = z(s)I_n$, where $z(s) = R_{eq}//[1/(sC_{eq})]$. Expanding and letting $s \to j2\pi f$, we readily find the gain of the CFA as

$$z(jf) = \frac{V_o}{I_n} = \frac{R_{eq}}{1 + jf/f_b} \qquad (6.114)$$

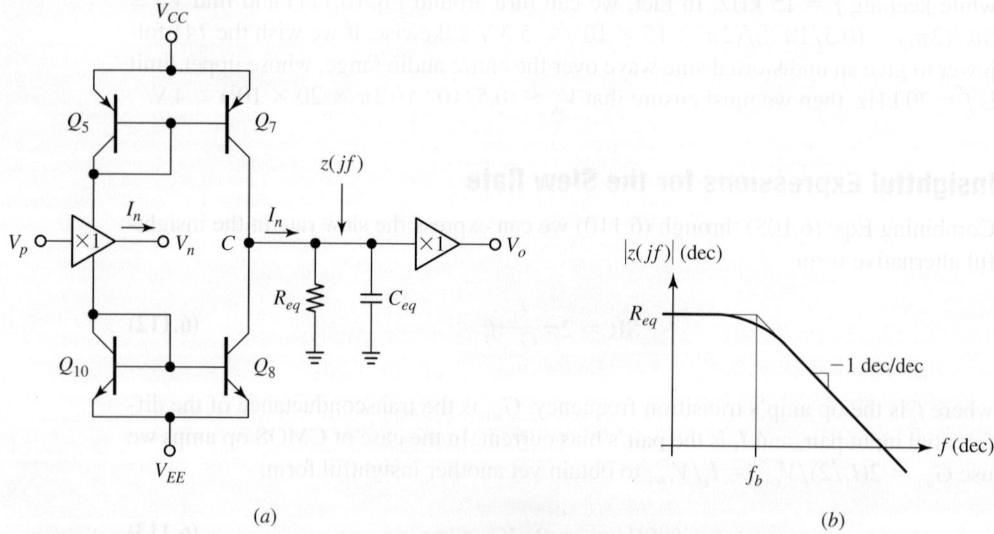

FIGURE 6.66 The frequency response of a current-feedback op amp (CFA).

where R_{eq} represents the dc value of gain, and

$$f_b = \frac{1}{2\pi R_{eq}C_{eq}} \tag{6.115}$$

represents the frequency at which gain drops to $1/\sqrt{2}$ ($=70.7\%$) of its low-frequency value. Above f_b gain rolls off by one Ω-decade for every Hz-decade, or by -1 dec/dec, as shown. Since it has the dimensions of impedance, z is also referred to as *transimpedance gain*, and the CFA as *transimpedance amplifier*.

We observe that the current dumped into C_{eq} at the onset of a step in V_p depends on the external resistance on which node V_n is terminated *as well as the step magnitude*. Consequently, there are no current-saturation effects in CFAs, and, hence, *no slew-rate-limiting*. For instance, with $C_{eq} \sim 1$ pF and $I_n \sim 1$ mA the initial slope is $dV_o/dt \sim 10^{-3}/10^{-12} = 10^9 = 1{,}000$ V/μs. To fully appreciate the dynamic advantages of the CFA we need to investigate its frequency response in negative-feedback operation, in Chapter 7.

6.10 DIODE SWITCHING TRANSIENTS

Up to now diodes have been assumed to turn on or off instantaneously. An actual *pn* junction diode takes time to switch from one state to the other because charge must be transferred in or out of the device to effect the switching, and charge transfers cannot occur instantaneously. The transient behavior of a *pn* junction is governed by the *charge-control equation*[1, 8]

$$i_D = \frac{dQ_j}{dt} + \frac{dQ_F}{dt} + \frac{Q_F}{\tau_F} \tag{6.116}$$

where i_D is the diode current, Q_j is the charge of the junction capacitance C_j, and Q_F is the excess minority charge in forward bias. In words, the current i_D supplied to a diode in the forward mode goes toward (a) charging up the capacitance C_j, (b) building up the excess charge Q_F, and (c) sustaining the charge Q_F built up to that point. Once all transients have died out, the diode reaches its dc steady state, where $i_D = Q_F/\tau_F$. Consequently, the time-constant τ_F represents the ratio Q_F/i_D in the forward steady state.

Recall that most practical junctions are fabricated with one side much more heavily doped than the other. For a junction with $N_A \gg N_D$ we can approximate $Q_F \cong Q_p$; moreover, $\tau_F \cong \tau_p = L_p^2/D_p$ in the case of a long-base diode, $\tau_F = W_n^2/(2D_p)$ in the short-base case (refer to Figs. 1.43 and 1.45). Likewise, for a junction with $N_D \gg N_A$ we have $Q_F \cong Q_n$, and $\tau_F \cong \tau_n = L_n^2/D_n$ for a long-base diode, $\tau_F = W_p^2/(2D_n)$ for a short-base diode. In the long-base case τ_F is called the minority-carrier *mean recombination time*, or also the minority-carrier *mean lifetime*, whereas in the short-base case it is called the minority-carrier *mean transit time* and it is denoted as τ_T (this is also the symbol used by PSpice).

Figure 6.67 shows a PSpice circuit to display the switching characteristics of a diode having the parameters shown in the table. Figure 6.68 shows all relevant

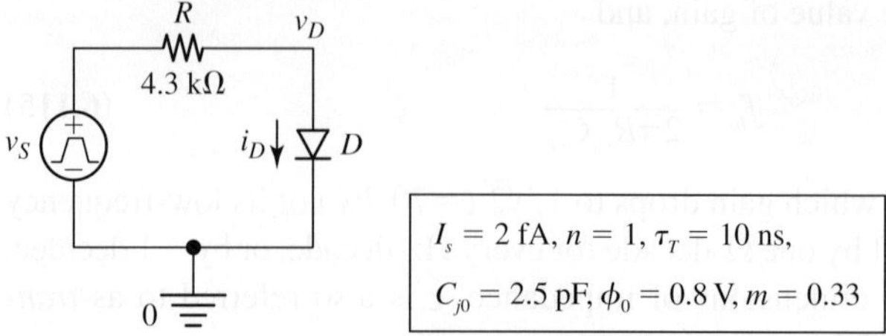

FIGURE 6.67 PSpice circuit to visualize the switching transients of a *pn* junction diode.

waveforms for the case in which the driving source v_S is switched from V_R ($= -2$ V) to V_F ($=5$ V) at $t = t_0 = 0$ ns, and back to -2 V at $t = t_2 = 50$ ns, with the diode in steady state prior to $t = t_0$. We make the following observations.

- Right after the leading edge of v_S there is no excess charge Q_F yet, so Eq. (6.116) simplifies as

$$i_D \cong \frac{dQ_j}{dt} \tag{6.117}$$

indicating that initially all of i_D goes toward charging C_j. As C_j charges up, v_D increases until the diode reaches the *edge of conduction* (EOC) at the instant t_1 when $v_D \cong 0.6$ V. Recall from Chapter 1 that the junction capacitance is

$$C_j(v_D) = \frac{C_{j0}}{(1 - v_D/\phi_0)^m} \tag{6.118}$$

where C_{j0} is the zero-bias value of C_j, ϕ_0 is the built-in potential, and m the grading coefficient. Since C_j is nonlinear, its charging process is rather complex, but we can estimate its charging time $t_1 - t_0$ ($=t_1$) by approximating Eq. (6.117) as

$$i_{D(\text{avg})} \cong \frac{\Delta Q_j}{\Delta t} = \frac{C_{j(\text{eq})}\Delta v_D}{t_1 - t_0}$$

where $i_{D(\text{avg})}$ is the average of $i_D(t_0^+) = [5 - (-2)]/4.3 = 1.63$ mA and $i_D(t_1) = (5 - 0.6)/4.3 = 1.02$ mA, that is, $i_{D(\text{avg})} = (1.63 + 1.02)/2 = 1.3$ mA, and $\Delta v_D = v_D(t_1) - v_D(t_0) = 0.6 - (-2) = 2.6$ V. Making the crude approximation $C_{j(\text{eq})} \cong C_{j0} = 2.5$ pF (see Problem 6.49 for a better estimate) we get $t_1 \cong 2.5 \times 10^{-12} \times 2.6/(1.3 \times 10^{-3}) = 5$ ns. This well agrees with $t_1 = 4.77$ ns measured via the cursor facility of PSpice.

- Following t_1, the diode is brought from the EOC ($v_D \cong 0.6$ V) to full conduction ($v_D \cong 0.7$ V), and this is when Q_F comes into play. The change in v_D from 0.6 V to 0.7 V is small enough that we can ignore the term dQ_j/dt and simplify Eq. (6.116) as

$$I_F \cong \frac{dQ_F}{dt} + \frac{Q_F}{\tau_F} \tag{6.119}$$

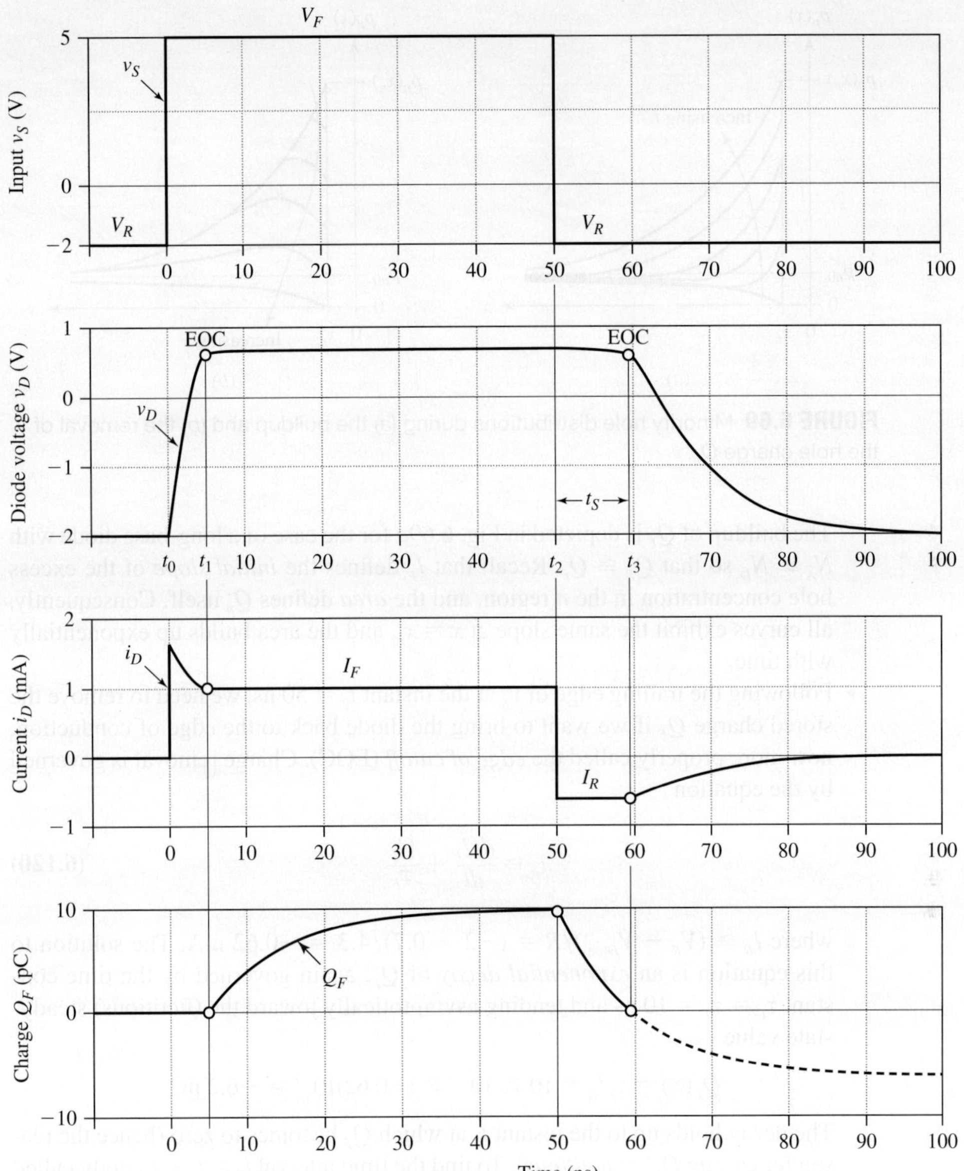

FIGURE 6.68 Relevant waveforms for the circuit of Fig. 6.67.

where $I_F \cong (V_F - V_{D(\mathrm{on})})/R = (5 - 0.7)/4.3 = 1$ mA. The solution to this equation is an *exponential buildup* of Q_F, governed by the time constant $\tau_F = \tau_T = 10$ ns and tending asymptotically toward the steady-state value

$$Q_{F(\mathrm{ss})} = \tau_F I_F = 10 \times 10^{-9} \times 10^{-3} = 10 \text{ pC}$$

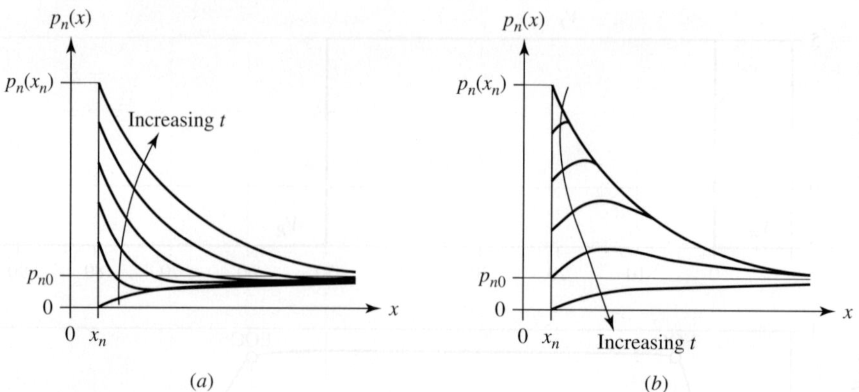

FIGURE 6.69 Minority hole distributions during (a) the buildup and (b) the removal of the hole charge Q_p.

The buildup of Q_F is depicted in Fig. 6.69a for the case of a long-base diode with $N_A \gg N_D$ so that $Q_F \cong Q_p$. Recall that I_F defines the *initial slope* of the excess hole concentration in the n region, and the *area* defines Q_p itself. Consequently, all curves exhibit the same slope at $x = x_n$, and the area builds up exponentially with time.

- Following the trailing edge of v_S at the instant $t_2 = 50$ ns, we need to remove the stored charge Q_F if we want to bring the diode back to the edge of conduction, now more properly called the *edge of cutoff* (EOC). Charge removal is governed by the equation

$$I_R \cong \frac{dQ_F}{dt} + \frac{Q_F}{\tau_F} \tag{6.120}$$

where $I_R \cong (V_R - V_{D(\text{on})})/R = (-2 - 0.7)/4.3 = -0.62$ mA. The solution to this equation is an *exponential decay* of Q_F, again governed by the time constant $\tau_F = \tau_T = 10$ ns and tending asymptotically toward the (fictitious) steady-state value

$$Q_F(\infty) = \tau_F I_R = 10 \times 10^{-9} \times (-0.62)10^{-3} = -6.2 \text{ pC}$$

The decay holds up to the instant t_3 at which Q_F becomes to zero (hence the reason for calling $Q_F(\infty)$ *fictitious*). To find the time interval $t_S = t_3 - t_2$, aptly called the *storage time*, we use[7]

$$t_S = \tau_F \ln \frac{Q_F(t_2) - Q_F(\infty)}{Q_F(t_3) - Q_F(\infty)} = \tau_F \ln \frac{\tau_F I_F - \tau_F I_R}{0 - \tau_F I_R}$$

that is,

$$t_S = \tau_F \ln \frac{I_F - I_R}{-I_R} \tag{6.121}$$

Presently, $t_S = (10 \text{ ns}) \ln[(1 + 0.62)/0.62] = 9.6$ ns. This well agrees with the measured PSpice value of 9.5 ns. The removal of Q_F is depicted in Fig. 6.69b, where we note that the initial slope is now defined by I_R, which is negative.

- Once all the stored charge has been removed, the diode will retrace the initial voltage transient, but in *reverse*. Were C_j linear, the reverse transient would be an exponential transient from $+0.6$ V to -2 V governed by the time constant $\tau = RC_j$. In practice this transient is pseudo exponential, a bit slower at the beginning where $C_j \cong 2C_{j0}$, but getting faster as C_j decreases in reverse bias, where $C_j < C_{j0}$.

Looking at the waveforms of v_D and i_D we observe that the diode starts conducting right away when we turn it on. However, when we try to turn it off, it *continues to remain on* for t_S nanoseconds. During this time it acts as a $\sim$0.7-V battery and the reverse current, far from dropping instantaneously to zero as it would in the case of an ideal diode, remains at I_R ($|I_R| \gg 0$), as confirmed by the waveform of i_D. It is apparent that the storage time t_S can be a drawback, especially in high-speed applications.

According to Eq. (6.120) t_S depends on the intrinsic *pn*-junction parameter τ_F as well as on the user-provided drives I_F and I_R. Short-base diodes are preferable in high-speed applications because they have $\tau_T \ll \tau_F$. We can also reduce the logarithmic term of Eq. (6.121) by driving the diode with a large reverse current I_R to wipe out the stored charge more rapidly. However, other design constraints may limit the reverse drive, indicating that the circuit designer must learn to live with storage-time limitations.

If a diode has $t_S = 25$ ns with $I_F = 10$ mA and $I_R = -2$ mA, find t_S if $I_F = 4$ mA and $I_R = -5$ mA. Comment.   **EXAMPLE 6.24**

Solution

Imposing 25 ns $= \tau_F \ln[(10 + 2)/2]$ gives $\tau_F \cong 14$ ns. (This provides a means of finding τ_F experimentally via storage-time and current-drive measurements.) Now $t_S = 14 \ln[(4 + 5)/5] = 8.23$ ns. Lowering I_F from 10 mA to 4 mA reduces the amount of stored charge, and raising $-I_R$ from 2 mA to 5 mA wipes out the stored charge more rapidly. However, because of the logarithmic dependence, the reduction in t_S is not that dramatic.

Schottky-Barrier Diodes

Schottky-barrier diodes (SBDs) overcome the charge-storage limitation of *pn* junctions by avoiding minority charges altogether. The two diode types are compared in Fig. 6.70. A conventional monolithic diode, shown in Fig. 6.70a, consists of a p-n^- junction and corresponding metal electrodes, for instance made of aluminum (Al). Consider now the Al-n^- junction of Fig. 6.70b. Because of the lightly doped n^- region (typically $N_D \leq 10^{-16}/\text{cm}^3$), a space-charge layer (SCL) with rectifying properties forms just below the Al electrode. Applying a positive bias to the Al electrode

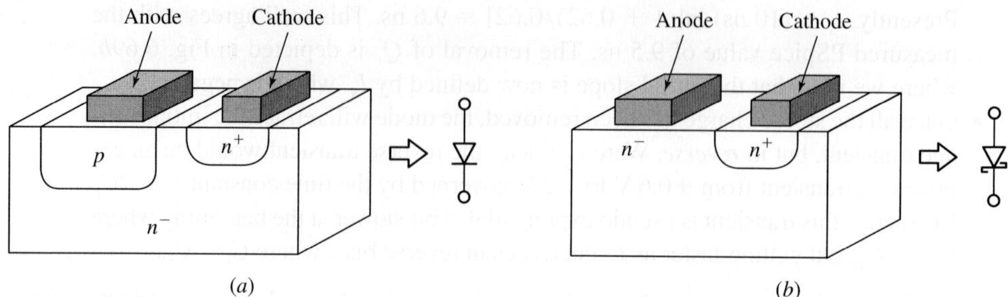

FIGURE 6.70 (a) Ordinary *pn* diode structure, and (b) Schottky barrier diode (SBD) structure.

relative to the n^- region will overcome the ensuing electrostatic barrier and cause electrons to flow from the n^- region, through the SCL, to the Al electrode. Since conduction is exclusively via electrons, which are the (only) charge carriers in the Al metal and are the majority carriers in the n^- material, there is *no minority-charge storage* in SBDs (aptly enough, SBDs are also called *hot-carrier diodes* because of this.) However, the SBD exhibits a junction capacitance C_j just like the conventional *pn* diode, indicating similar behavior during turn-on from t_0 to t_1 as well as during turnoff past t_3. But, in the SBD case t_3 coincides with t_2 because $t_S = 0$.

SBDs exhibit an exponential *i-v* characteristic just like ordinary *pn* diodes, except that the saturation current I_s of a SBD is some 5 orders of magnitude higher than that of a *pn* diode of comparable dimensions. Using the 60-mV/decade rule, we conclude that SBDs have typically $V_{D(\text{on})} \cong 0.7 - 5 \times 0.06 = 0.4$ V. The advantages of (a) a *lower voltage drop* and (b) the *absence of minority-charge storage effects* makes SBDs preferable to *pn* diodes in applications such as switching power supplies and high-speed diode circuits. Figure 6.70*b* shows also the circuit symbol in common use for the SBD.

Before concluding, we observe that both structures of Fig. 6.70 include an Al-n^+ junction at the cathode side. This junction too results in the formation of an SCL below the cathode electrode; however, because of heavy doping, this SCL is so narrow that electrons can easily tunnel across it in either direction to form what is known as an *ohmic contact*. Were the n^+ region absent, the "cathode" electrode would form another SBD with the n^- material underneath, resulting in a back-to-back diode pair that would serve no useful purpose. As we know, ohmic contacts play an important role in connection with the collector of bipolar transistors as well as the tubs of CMOS transistors.

6.11 BJT SWITCHING TRANSIENTS

The analysis of BJT transients draws heavily from the diode treatment of the previous section, except that the BJT comprises *two* junctions and the analysis is therefore more complex. When used as a switch, a BJT usually alternates between the *cutoff* (CO) and *saturation* (Sat) modes, with brief transitions through the *forward-active* mode. In saturation both junctions are forward biased, so this mode is a combination of forward-active (FA) and *reverse-active* (RA) operation. The transient behavior of the *npn* BJT is governed by the *charge-control equations*[1,8]

$$i_B = \frac{Q_F}{\tau_{BF}} + \frac{Q_R}{\tau_{BR}} + \frac{d}{dt}(Q_F + Q_R + Q_{je} + Q_{jc}) \tag{6.122}$$

$$i_C = \frac{Q_F}{\tau_F} - Q_R\left(\frac{1}{\tau_R} + \frac{1}{\tau_{BR}}\right) - \frac{d}{dt}(Q_R + Q_{jc}) \tag{6.123}$$

where i_B and i_C are the currents into the base and into the collector terminals; Q_F and Q_R are the FA and RA excess minority-carrier charges in the base; Q_{je} and Q_{jc} are the charges of the base-emitter (BE) and base-collector (BC) capacitances C_{je} and C_{jc} ($=C_\mu$ in ac analysis). The time constants τ_F and τ_{BF} are called, respectively, the *mean transit time* and the *mean lifetime* of the base minority carriers in FA operation. According to Eq. (6.4) the *npn* BJT has

$$\tau_F = \frac{W_B^2}{2D_n} \tag{6.124}$$

where W_B is the base width and D_n is the electron diffusivity. The two time constants are related as[1]

$$\tau_{BF} = \beta_F \tau_F \tag{6.125}$$

where β_F is the familiar FA current gain. Similar terminology holds for the time constants τ_R and τ_{BR} ($=\beta_R\tau_R$), except that they pertain to RA operation, where the current gain is β_R ($\ll \beta_F$). Equation (6.122) describes the effect of i_B on the various charges, whereas Eq. (6.123) describes the effect of these charges on i_C, indicating a relationship of cause-and-effect between the two currents. Also, once we know i_B and i_C, we can find the emitter current via KVL as $i_E = i_B + i_E$.

However intimidating the above equations might appear, we shall use them in simple and intuitive ways to investigate the response of a basic BJT inverter/switch to an input pulse. To this end we use the PSpice circuit of Fig. 6.71 to display all

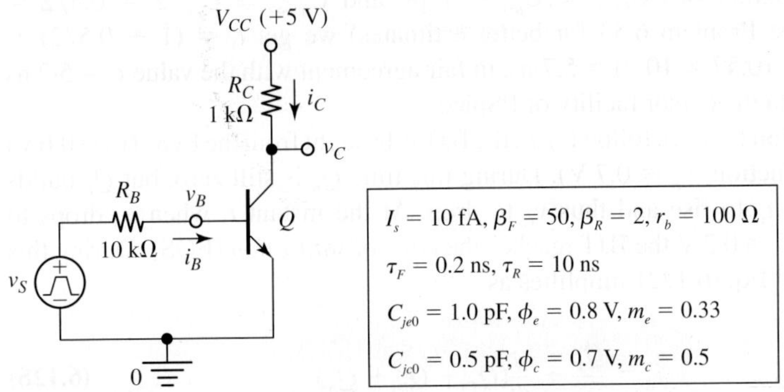

FIGURE 6.71 PSpice circuit to visualize the switching transients of a BJT inverter/switch.

relevant waveforms for the case of a BJT inverter with the parameters shown in the table, and then we apply the above equations to calculate[8] the various transient components. As shown in Fig. 6.72, the driving source v_S is switched from V_R ($= -2$ V) to V_F ($=5$ V) at $t = t_0 = 0$ ns and back to -2 V at $t = t_3 = 100$ ns. Assuming the BJT is in steady state prior to $t = t_0$, we identify the following operating regions and corresponding time intervals.

- **Cutoff Region** (t_0 to t_1): right after the leading edge of v_S there are no base excess charges Q_F and Q_R yet, so Eqs. (6.122) and (6.123) simplify as

$$i_B = \frac{d}{dt}(Q_{je} + Q_{jc}) \qquad i_C = -\frac{dQ_{jc}}{dt} \tag{6.126}$$

The first equation states that i_B merely goes toward charging up the capacitances C_{je} and C_{jc}, and the second that the portion of i_B flowing through C_{jc} *exits* (hence the "$-$" sign in the second tem above) the collector terminal. This current then flows into R_C to produce the initial voltage bump visible in the plot of v_C. As C_{je} and C_{jc} charge up, v_B rises until the BJT reaches the *edge of conduction* (EOC) at the instant t_1 when $v_B \cong 0.6$ V. As we know, the junction capacitances are

$$C_{je} = \frac{C_{je0}}{(1 - v_{BE}/\phi_e)^{m_e}} \qquad C_{jc} = \frac{C_{jc0}}{(1 - v_{BC}/\phi_c)^{m_c}} \tag{6.127}$$

where C_{je0} and C_{jc0} are their zero-bias values, ϕ_e and ϕ_c are the built-in potentials of the two junctions, and m_e and m_c are their grading coefficients. Following the diode treatment of the previous section we estimate the charging time $t_1 - t_0 (= t_1)$ by approximating the first of Eq. (6.126) as

$$i_{B(avg)} \cong \frac{\Delta Q_{je} + \Delta Q_{jc}}{\Delta t} = \frac{C_{je(eq)}\Delta v_{BE} + C_{jc(eq)}\Delta v_{BC}}{t_1 - t_0}$$

where $i_{B(avg)}$ is the average of $i_B(t_0^+) = [5 - (-2)]/10 = 0.7$ mA and $i_B(t_1) = (5 - 0.6)/10 = 0.44$ mA, that is, $i_{B(avg)} = (0.7 + 0.44)/2 = 0.57$ mA; moreover, $\Delta v_{BE} = v_{BE}(t_1) - v_{BE}(t_0) = 0.6 - (-2) = 2.6$ V $= \Delta v_{BC}$. Making the crude approximations $C_{je(eq)} \cong C_{je0} = 1$ pF and $C_{jc(eq)} \cong C_{je0}/2 = 0.5/2 = 0.25$ pF (see Problem 6.53 for better estimates) we get $t_1 = (1 + 0.5/2) \times 10^{-12} \times 2.6/(0.57 \times 10^{-3}) = 5.7$ ns, in fair agreement with the value $t_1 = 5.2$ ns measured via the cursor facility of PSpice.

- **Active Region** (t_1 to t_2): following t_1, the BJT is brought from the EOC ($v_B \cong 0.6$ V) to full conduction ($v_B \cong 0.7$ V). During this time Q_R is still zero, but Q_F builds up, causing i_C to rise and thus v_C to drop. At the instant t_2 when v_C drops to $v_C = V_{CE(EOS)} \cong 0.2$ V the BJT reaches the *edge of saturation* (EOS). During this time interval Eq. (6.122) simplifies as

$$i_B = \frac{Q_F}{\tau_{BF}} + \frac{d}{dt}(Q_F + Q_{je} + Q_{jc}) \tag{6.128}$$

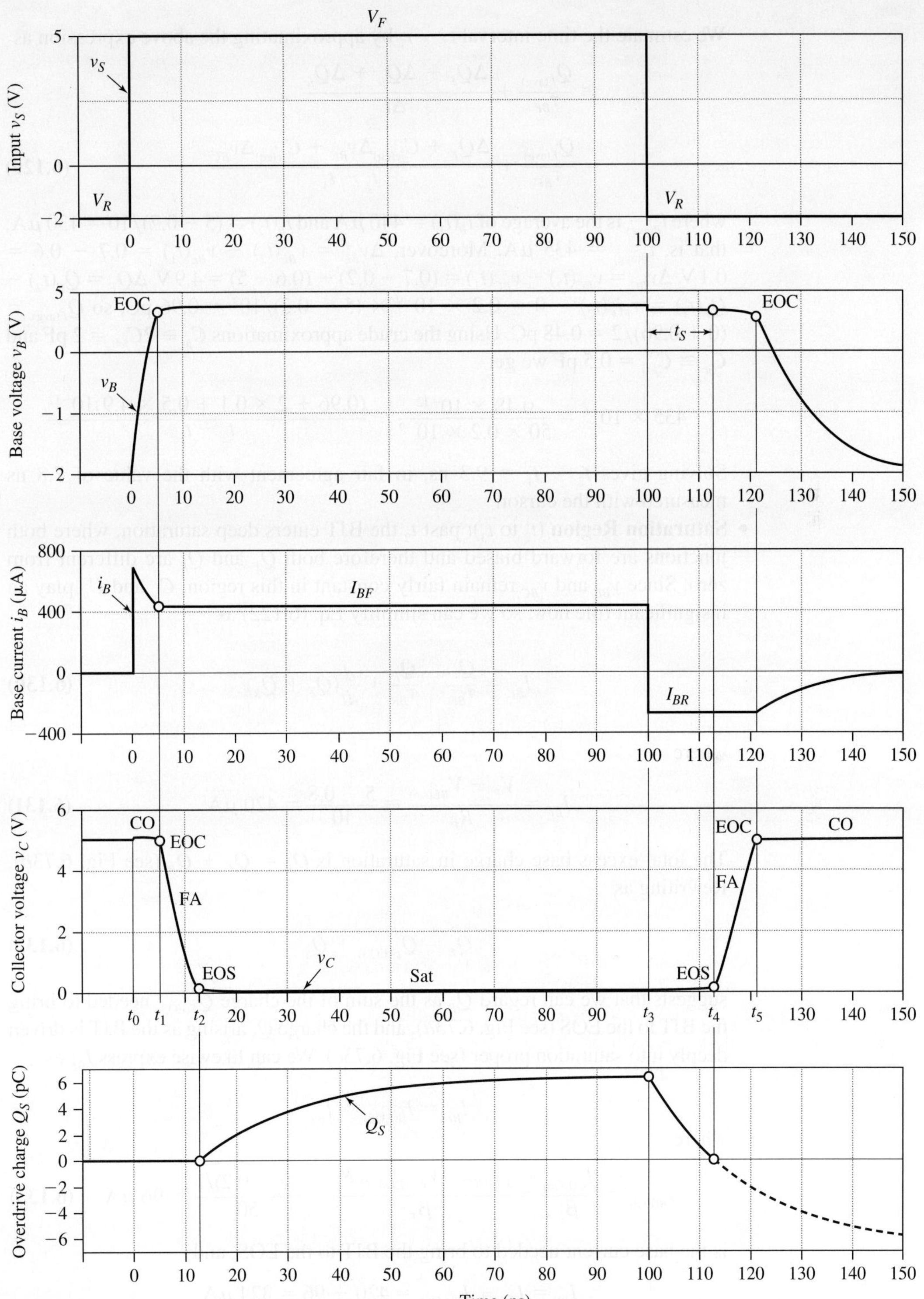

FIGURE 6.72 Plot of all relevant waveforms for the BJT inverter/switch of Fig. 6.71.

We estimate the time interval $t_2 - t_1$ by approximating the above expression as

$$i_{B(\text{avg})} \cong \frac{Q_{F(\text{avg})}}{\tau_{BF}} + \frac{\Delta Q_F + \Delta Q_{je} + \Delta Q_{jc}}{\Delta t}$$

$$= \frac{Q_{F(\text{avg})}}{\tau_{BF}} + \frac{\Delta Q_F + C_{je(\text{eq})}\Delta v_{BE} + C_{jc(\text{eq})}\Delta v_{BC}}{t_2 - t_1} \tag{6.129}$$

where $i_{B(\text{avg})}$ is the average of $i_B(t_1) = 440\ \mu\text{A}$ and $i_B(t_2) = (5 - 0.7)/10 = 430\ \mu\text{A}$, that is, $i_{B(\text{avg})} = 435\ \mu\text{A}$. Moreover, $\Delta v_{BE} = v_{BE}(t_2) - v_{BE}(t_1) = 0.7 - 0.6 = 0.1\ \text{V}$, $\Delta v_{BC} = v_{BC}(t_2) - v_{BC}(t_1) = (0.7 - 0.2) - (0.6 - 5) = 4.9\ \text{V}$, $\Delta Q_F = Q_F(t_2) - Q_F(t_1) = \tau_F i_C(t_2) - 0 = 0.2 \times 10^{-9} \times (5 - 0.2)/10^3 = 0.96\ \text{pC}$, so $Q_{F(\text{avg})} = (0 + 0.96)/2 = 0.48\ \text{pC}$. Using the crude approximations $C_{je} \cong 2C_{je0} = 2\ \text{pF}$ and $C_{jc} \cong C_{jc0} = 0.5\ \text{pF}$ we get

$$435 \times 10^{-6} = \frac{0.48 \times 10^{-12}}{50 \times 0.2 \times 10^{-9}} + \frac{(0.96 + 2 \times 0.1 + 0.5 \times 4.9)10^{-12}}{t_2 - t_1}$$

Solving gives $t_2 - t_1 = 9.3$ ns, in fair agreement with the value of 7.8 ns measured with the cursor.

- **Saturation Region** (t_2 to t_3): past t_2 the BJT enters deep saturation, where both junctions are forward biased and therefore both Q_F and Q_R are different from zero. Since v_{BE} and v_{BC} remain fairly constant in this region, C_{je} and C_{jc} play an insignificant role now, so we can simplify Eq. (6.122) as

$$I_{BF} = \frac{Q_F}{\tau_{BF}} + \frac{Q_R}{\tau_{BR}} + \frac{d}{dt}(Q_F + Q_R) \tag{6.130}$$

where

$$I_{BF} = \frac{V_F - V_{BE(\text{sat})}}{R_B} \cong \frac{5 - 0.8}{10} = 420\ \mu\text{A} \tag{6.131}$$

The total excess base charge in saturation is $Q_B = Q_F + Q_R$ (see Fig. 6.73b). Rewriting as

$$Q_B = Q_{F(\text{EOS})} + Q_S \tag{6.132}$$

suggests that we can regard Q_B as the sum of the charge $Q_{F(\text{EOS})}$ needed to bring the BJT to the EOS (see Fig. 6.73a), and the charge Q_S arising as the BJT is driven deeply into saturation proper (see Fig. 6.73c). We can likewise express I_{BF} as

$$I_{BF} = I_{B(\text{EOS})} + I_{BS}$$

where

$$I_{B(\text{EOS})} = \frac{I_{C(\text{EOS})}}{\beta_F} = \frac{(V_{CC} - V_{CE(\text{EOS})})/R_C}{\beta_F} \cong \frac{(5 - 0.2)/1}{50} = 96\ \mu\text{A} \tag{6.133}$$

is the base current needed to bring the BJT to the EOS, and

$$I_{BS} = I_{BF} - I_{B(\text{EOS})} = 420 - 96 = 324\ \mu\text{A}$$

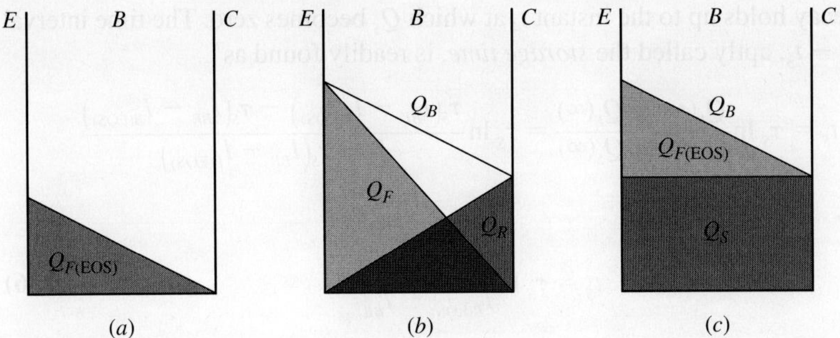

FIGURE 6.73 Excess minority charges in the base: (*a*) at the EOS, and (*b*), (*c*) in deep saturation.

is the amount of base current *above* $I_{B(\text{EOS})}$ that drives the BJT in deep saturation (note that the BJT is indeed saturated because $\beta_{sat} = 4.9/0.42 = 11.7 \ll \beta_F$). Aptly called the *overdrive base current*, I_{BS} causes the *overdrive base charge* Q_S to build up according to the charge-control equation[8]

$$I_{BS} = \frac{Q_S}{\tau_S} + \frac{dQ_S}{dt} \tag{6.134}$$

where the time constant τ_S is a linear combination[8] of τ_{BF} and τ_{BR},

$$\tau_S = \frac{(\beta_R + 1)\tau_{BF} + \beta_F \tau_{BR}}{\beta_F + \beta_R + 1} \tag{6.135}$$

Using the data tabulated in Fig. 6.71 we get $\tau_S = [(2 + 1)50 \times 0.2 + 50 \times 2 \times 10]/(50 + 2 + 1) = 19.4$ ns. The solution to Eq. (6.134) is an *exponential buildup* of Q_S, governed by the time constant τ_S and tending asymptotically toward the steady-state value

$$Q_{S(\text{ss})} = \tau_S I_{BS} = 19.4 \times 10^{-9} \times 324 \times 10^{-6} = 6.3 \text{ pC}$$

- **Storage Time** (t_3 to t_4): once v_S is switched back to -2 V at $t = t_3 = 100$ ns, we need to remove the overdrive charge Q_S if we want to bring the BJT back to the EOS. The removal of Q_S is still governed by Eq. (6.134) provided we now use

$$I_{BS} = I_{BR} - I_{B(\text{EOS})} = \frac{V_R - V_{BE(\text{sat})}}{R_B} - I_{B(\text{EOS})} = \frac{-2 - 0.7}{10 \times 10^3} - (96 \ \mu\text{A})$$

$$= -270 - 96 = -366 \ \mu\text{A}$$

(Cognizant of the small step decrease in v_B, which is $\Delta v_B = r_b \Delta I_B$, we are now assuming $V_{BE(\text{sat})} \cong 0.7$ V instead of the usual 0.8 V.) Since I_{BS} is now negative, the solution to Eq. (6.134) is an *exponential decay* in Q_S, still governed by the time constant τ_S but tending asymptotically toward the (fictitious) steady-state value

$$Q_S(\infty) = \tau_S I_{BS} = 19.4 \times 10^{-9} \times (-366 \times 10^{-6}) = -7.1 \text{ pC}$$

The decay holds up to the instant t_4 at which Q_S becomes zero. The time interval $t_S = t_4 - t_3$, aptly called the *storage time*, is readily found as[7]

$$t_S = \tau_S \ln \frac{Q_S(t_3) - Q_S(\infty)}{Q_S(t_4) - Q_S(\infty)} = \tau_S \ln \frac{\tau_S(I_{BF} - I_{B(\text{EOS})}) - \tau_S(I_{BR} - I_{B(\text{EOS})})}{0 - \tau_S(I_{BR} - I_{B(\text{EOS})})}$$

that is,

$$t_S = \tau_S \ln \frac{I_{BF} - I_{BR}}{I_{B(\text{EOS})} - I_{BR}} \tag{6.136}$$

Presently we get $t_S = (19.4 \text{ ns}) \ln[(420 + 270)/(96 + 270)] = 12.3$ ns, which matches PSpice exactly.

- **Active Region Again** (t_4 to t_5): after removing Q_S to bring the BJT back to the EOS, we need to remove Q_F to bring it back to the edge of conduction, now more properly called the *edge of cutoff* (EOC). The removal of Q_F is still governed by Eq. (6.128), but with $i_B = (V_R - V_{BE})/R_B$. Using $i_{B(\text{avg})} = (-2 - 0.65)/10 = -265$ μA, we suitably recycle Eq. (6.129) to write

$$-265 \times 10^{-6} = \frac{0.48 \times 10^{-12}}{50 \times 0.2 \times 10^{-9}} - \frac{(0.96 + 2 \times 0.1 + 0.5 \times 4.9)10^{-12}}{t_5 - t_4}$$

Solving gives $t_5 - t_4 = 11.5$ ns, in fair agreement with the PSpice value of 9.2 ns.

- **Recovery Region** ($t > t_5$): once all the excess base charge has been wiped out, the BJT goes through a recovery phase to return C_{je} and C_{jc} to the steady-state conditions preceding t_0. During this phase v_B makes a pseudo exponential transistion from $v_B = V_{BE(\text{EOC})} \cong 0.6$ V to $v_B = V_R = -2$ V, as shown.

It is apparent that a BJT inverter/switch takes time to turn on and off. In particular, during turn-off, the BE junction *continues to remain on* from t_3 to t_5, during which time it acts as a 0.7-V battery. Of special significance is the storage time t_S from t_3 to t_4, during which the BJT remains *saturated*. This is usually a serious drawback, especially in high-speed logic or in power-switch applications.

Schottky-Clamped BJTs

We can eliminate the storage time t_S altogether by placing a Schottky barrier diode (SBD) in parallel with the base-collector (BC) junction, as shown in Fig. 6.74a for the case of the *npn* BJT. Recall that the saturation current of a SBD is typically five orders of magnitude higher than that of an ordinary *pn* junction, indicating that the forward voltage drop across a SBD is *lower* than that across a *pn* junction by about $5 \times (60 \text{ mV}) = 0.3$ V, giving $V_{SBD(\text{on})} \cong 0.7 - 0.3 \cong 0.4$ V typical. Because of the SBD clamp, the BC junction is subject to the constraint $v_{BC} \leq 0.4$ V, which is insufficient to allow the BC junction to turn convincingly on. Hence, the BJT will always have $Q_R = 0$ and, as such, it will never saturate.

Shown in Fig.11.4b is the monolithic realization of a Schottky-clamped *npn* BJT. Comparing with the structure of Fig. 2.1, we observe that all it takes to Schottky-clamp the BJT is to extend the metal electrode of the base over the n^- collector

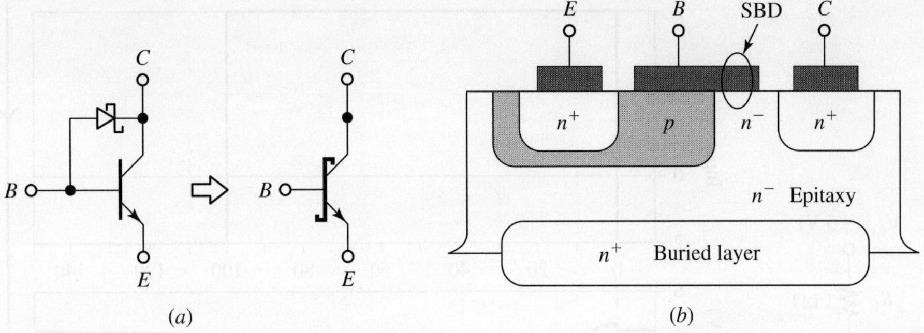

FIGURE 6.74 (a) Schottky-clamped BJT and its circuit symbol. (a) Planar process fabrication.

region, where it forms the SBD. (Also shown at the right is the metal-n^+ structure, which provides an ohmic contact between the metal electrode of the collector and the n^- epitaxial layer underneath.)

(a) Assuming the BJT of Fig. 6.71 is equipped with a SBD clamp having $V_{SBD(on)} \cong 0.4\,\text{V}$, find the steady-state currents I_B, I_C, and I_{SBD} when $v_S = V_F = 5\,\text{V}$, and comment on your findings.

EXAMPLE 6.25

(b) Rerun the PSpice circuit of Fig. 6.71, but using a SBD. Comment on your findings.

Solution

(a) The current through R_B is $I(R_B) = (V_F - V_B)/R_B \cong (5 - 0.7)/10 = 0.43$ mA. By KVL, $V_C = V_B - V_{SBD(on)} \cong 0.7 - 0.4 = 0.3$ V, so the current through R_C is $I(R_C) = (V_{CC} - V_C)/R_C \cong (5 - 0.3)/1 = 4.7$ mA. Moreover, the BJT is operating in the FA region because $V_{CE} \cong 0.3$ V, which is 0.1-V higher than $V_{CE(EOS)}$ ($\cong 0.2$ V). Consequently, we can write $I_C = \beta_F I_B = 50 I_B$. We need two more equations to solve for the three unknowns. These are provided by KCL at the base and at the collector nodes, where $I_B = I(R_B) - I_{SBD} = 0.43 - I_{SBD}$, and $I_C = I(R_C) + I_{SBD} = 4.73 + I_{SBD}$. Solving, we get

$$I_B = 100.6\ \mu\text{A} \qquad I_C = 5.03\ \text{mA} \qquad I_{SBD} = 0.33\ \text{mA}$$

Because of the SBD, only enough current is allowed into the base to keep the BJT about 0.1-V away from the EOS. The overdrive current is diverted by the SBD to the collector, thus preventing the BJT from ever saturating.

(b) The clamped BJT of Fig. 6.75a uses a SBD with the following PSpice model:

```
.model DSBD D(IS=1nA CJO=0.25pF VJ=1.6V M=0.4 EG=0.7)
```

The waveforms of Fig. 6.75b confirm the elimination of the storage time. The rise and fall times of v_C are a bit longer than those of the unclamped circuit of Fig. 6.71 because of the SBD's junction capacitance C_j, which is in parallel with C_{jc}. Moreover, v_C is clamped at about 0.3 V, a bit higher than the 0.1 V of the circuit of Fig. 6.71.

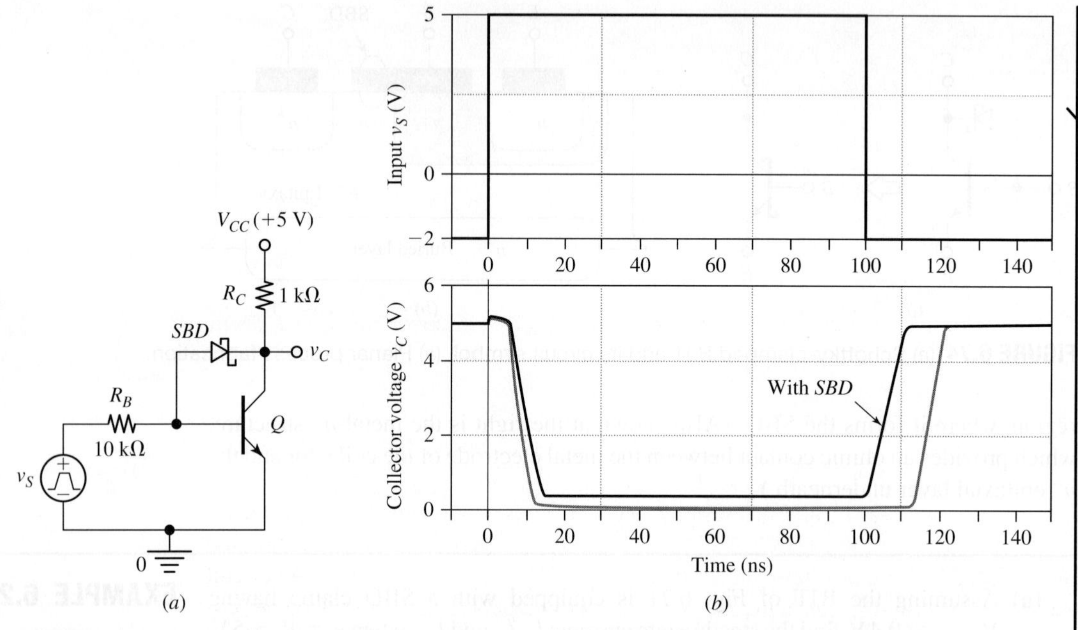

FIGURE 6.75 (a) Schottky-clamped BJT inverter/switch, and (b) input and output waveforms.

6.12 TRANSIENT RESPONSE OF CMOS GATES AND VOLTAGE COMPARATORS

In binary-output circuits such as logic gates and voltage comparators it is of interest to know how rapidly the output changes state in response to a sudden change at the input. It takes time for the stray capacitances of the transistors and their interconnections to charge/discharge in response to an input step. As a rule, the smaller the capacitances and the higher the currents available to charge/discharge them, the faster the response.

Propagation Delays in Logic Gates

The dynamics of a logic gate are characterized via the *propagation delays*, usually specified for the case of an inverter (the most basic representative of a logic family) driving *n* similar inverters, or for a *fanout* of *n* in logic-design jargon (see Fig. 6.76). A propagation delay is the amount of time it takes for the output to accomplish 50% of its transition from one output level to the other following an input-step edge. Denoting the output levels as V_{OL} and V_{OH}, we define the 50%-point as

$$V_{50\%} = \frac{V_{OL} + V_{OH}}{2} \tag{6.137}$$

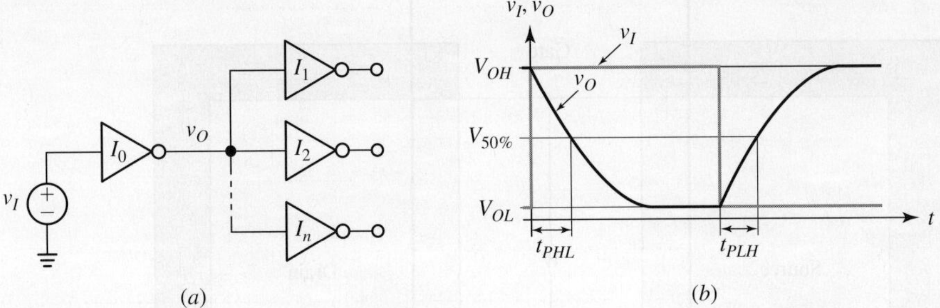

FIGURE 6.76 (a) Logic inverter I_0 with a fanout of n. (b) The propagation delays t_{PHL} and t_{PLH}.

The time it takes for v_O to rise from V_{OL} to $V_{50\%}$ is denoted as t_{PLH}, and the time it takes to drop from V_{OH} to $V_{50\%}$ is denoted as t_{PHL}. (Because of inherent internal circuit dissymmetries, t_{PLH} and t_{PHL} are not necessarily identical.) In the following we investigate the propagation delays of CMOS gates, presently the predominant digital technology. These gates have $V_{OL} = 0$ and $V_{OH} = V_{DD}$, so $V_{50\%} = V_{DD}/2$. As circuit complexity increases, transient analysis by paper and pencil can easily become prohibitive, so we shall find the propagation delays via PSpice, and then use simplified hand analysis to obtain approximate estimates, both as a check on PSpice and as a means to gain insight into the inner workings of the gate.

Transient Analysis of CMOS Gates via PSpice

In order to display the transient response, PSpice calculates all parasitic capacitances in the gate, so we need to provide PSpice with suitable *process* and *device* parameters. To this end, refer to the conceptual rendition of Fig. 6.77, which is similar to the basic structure of Fig. 6.8, except for an additional detail hitherto omitted for simplicity; these are the p^+ *channel-stop implants* surrounding the n^+ source and drain regions on three sides other than that facing the channel. The function of these implants is to *electrically isolate* adjacent FETs sharing the same bulk. (The portion of the bulk between the source/drain regions of two adjacent FETs forms a spurious channel that might become accidentally conductive. The p^+ implants are designed to greatly increase the V_t of these spurious channels and thus prevent them from accidentally turning on. Typically, the doping of the p^+ implants is an *order of magnitude* higher than that of the p^- bulk, or $N_{\text{implant}} \cong 10N_{\text{bulk}}$.) Because of the channel stops, each of the junction capacitances C_{sb} and C_{db} consists of a *bottom* component $C_{j(\text{btm})}$ associated with the p^--n^+ junction at the bottom of the source and drain region, and a *sidewall* component $C_{j(\text{sw})}$ associated with the p^+-n^+ junction around the region's perimeter. The net junction capacitance of the *drain region* is expressed as

$$C_{db} = \frac{A_d \times C_{j0(\text{btm})}}{\left(1 - v_{BD}/\phi_{0(\text{btm})}\right)^{m_{\text{btm}}}} + \frac{P_d \times C_{j0(\text{sw})}}{\left(1 - v_{BD}/\phi_{0(\text{sw})}\right)^{m_{\text{sw}}}} \qquad (6.138)$$

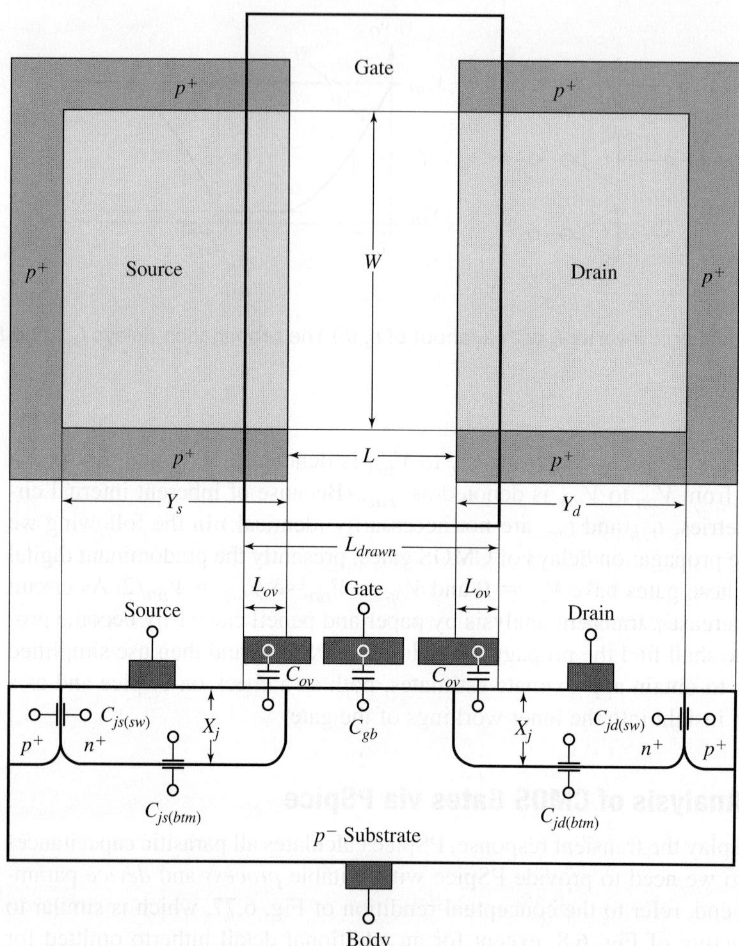

FIGURE 6.77 Conceptual views (top as well as cross sectional) of an *n*MOSFET. Note the p^+ channel-stop implants surrounding the source and drain regions on the three sides other than the side facing the channel.

where

- A_d is the *area* of the drain's bottom junction and $C_{j0(\text{btm})}$ is the zero-bias *junction capacitance per unit area*. Moreover, $\phi_{0(\text{btm})}$ is the bottom-junction built-in potential and m_{btm} is the grading coefficient.
- P_d is the *perimeter* of the drain's sidewall junction, $C_{j0(\text{sw})}$ is the zero-bias *junction capacitance per unit perimeter*, $\phi_{0(\text{sw})}$ is the sidewall-junction built-in potential, and m_{sw} is the grading coefficient.

For the geometry depicted at the top of Fig. 6.77 we have $A_d = Y_d \times W$ and $P_d = 2Y_d + W$. Moreover, adapting Eq. (1.47*b*) to the present case, and simplifying owing to the fact that $N_{A(\text{bulk})} \ll N_{D(\text{drain})}$ and $N_{A(\text{implant})} \ll N_{D(\text{drain})}$, we have

$$C_{j0(\text{btm})} \cong \sqrt{\frac{\varepsilon_{si} q N_{A(\text{bulk})}}{2\phi_{0(\text{btm})}}} \qquad\qquad C_{j0(\text{sw})} \cong X_j \sqrt{\frac{\varepsilon_{si} q N_{A(\text{implant})}}{2\phi_{0(\text{sw})}}} \qquad (6.139)$$

where X_j is the *depth* of the drain region, also shown in the figure. The process parameters of Eq. (6.139) apply also to the *source region*, the only possible differences being in its area A_s and perimeter P_s, depending on device layout geometry. The above expressions are readily adapted to the case of the *p*MOSFET, as the following example illustrates.

EXAMPLE 6.26

(a) Assuming an *n*MOSFET with the process parameters $t_{ox} = 20$ nm, $\mu_n = 600$ cm²/Vs, $V_t = 0.7$ V, $\lambda' = 0.1$ μm/V, $L_{ov} = 0.15$ μm, $X_j = 0.25$ μm, $N_{D(\text{poly})} = 10^{20}$ cm^{-3}, $N_{A(\text{bulk})} = 3 \times 10^{15}$ cm^{-3}, $N_{A(\text{implant})} = 10 N_{A(\text{bulk})}$, $m_{\text{btm}} = 0.5$, and $m_{sw} = 0.33$, find its process-related capacitances. Hence, find A_d, P_d, A_s, and P_s if $L_{drawn} = 1.0$ μm, $W = 2.0$ μm, and $Y_d = Y_s = 2.5$ μm.

(b) Assuming a *p*MOSFET with the process parameters $t_{ox} = 20$ nm, $\mu_p = 250$ cm²/Vs, $V_t = -0.7$ V, $\lambda' = 0.05$ μm/V, $L_{ov} = 0.2$ μm, $X_j = 0.3$ μm, $N_{A(\text{poly})} = 10^{20}$ cm^{-3}, $N_{D(\text{bulk})} = 1.8 \times 10^{16}$ cm^{-3}, $N_{D(\text{implant})} = 10 N_{D(\text{bulk})}$, $m_{\text{btm}} = 0.5$, and $m_{sw} = 0.33$, find its process-related capacitances. Hence, find A_d, P_d, A_s, and P_s if $L_{drawn} = 1.0$ μm, $W = 4.0$ μm, and $Y_d = Y_s = 3.0$ μm.

Solution

(a) For the *n*MOSFET we have

$$C_{ox} = \frac{\varepsilon_{ox}}{t_{ox}} = \frac{34.5}{20} = 1.725 \ \frac{\text{fF}}{\mu\text{m}^2}$$

$$C_{ov} = C_{ox} L_{ov} = 1.725 \times 0.15 = 0.259 \ \frac{\text{fF}}{\mu\text{m}} = 0.259 \ \frac{\text{nF}}{\text{m}}$$

$$\phi_{0(\text{btm})} = V_T \ln \frac{N_{A(\text{bulk})} N_{D(\text{poly})}}{n_i^2} = 0.026 \ln \frac{3 \times 10^{15} \times 10^{20}}{2 \times 10^{20}} = 0.909 \ \text{V}$$

$$\phi_{0(\text{sw})} = V_T \ln \frac{N_{A(\text{implant})} N_{D(\text{poly})}}{n_i^2} = 0.026 \ln \frac{30 \times 10^{15} \times 10^{20}}{2 \times 10^{20}} = 0.968 \ \text{V}$$

$$C_{j0(\text{btm})} \cong \sqrt{\frac{1.04 \times 10^{-12} \times 1.602 \times 10^{-19} \times 3 \times 10^{15}}{2 \times 0.909}} = 16.6 \ \frac{\text{nF}}{\text{cm}^2}$$

$$= 0.166 \ \frac{\text{fF}}{\mu\text{m}^2} = 166 \ \frac{\mu\text{F}}{\text{m}^2}$$

$$C_{j0(\text{sw})} \cong (0.25 \times 10^{-4}) \sqrt{\frac{1.04 \times 10^{-12} \times 1.602 \times 10^{-19} \times 30 \times 10^{15}}{2 \times 0.968}}$$

$$= 12.7 \ \frac{\text{nF}}{\text{cm}} = 0.127 \ \frac{\text{fF}}{\mu\text{m}} = 0.127 \ \frac{\text{nF}}{\text{m}}$$

Finally, $A_s = A_d = Y_d \times W = 2.5 \times 2 = 5~\mu\text{m}^2 = 5 \times 10^{-12}~\text{m}^2$, and $P_s = P_d = 2Y_d + W = 2 \times 2.5 + 2 = 7~\mu\text{m}$.

(b) For the pMOSFET we likewise find

$$C_{ox} = 1.725~\frac{\text{fF}}{\mu\text{m}^2} \qquad C_{ov} = 1.725 \times 0.2 = 0.345~\frac{\text{fF}}{\mu\text{m}} = 0.345~\frac{\text{nF}}{\text{m}}$$

$$\phi_{0(\text{btm})} = V_T \ln \frac{N_{D(\text{bulk})} N_{A(\text{poly})}}{n_i^2} = 0.026 \ln \frac{1.8 \times 10^{16} \times 10^{20}}{2 \times 10^{20}} = 0.955~\text{V}$$

$$\phi_{0(\text{sw})} = V_T \ln \frac{N_{D(\text{implant})} N_{A(\text{poly})}}{n_i^2} = 0.026 \ln \frac{18 \times 10^{16} \times 10^{20}}{2 \times 10^{20}} = 1.01~\text{V}$$

$$C_{j0(\text{btm})} \cong \sqrt{\frac{1.04 \times 10^{-12} \times 1.602 \times 10^{-19} \times 1.8 \times 10^{16}}{2 \times 0.955}}$$

$$= 39.6~\frac{\text{nF}}{\text{cm}^2} = 0.396~\frac{\text{fF}}{\mu\text{m}^2} = 396~\frac{\mu\text{F}}{\text{m}^2}$$

$$C_{j0(\text{sw})} \cong (0.3 \times 10^{-4})\sqrt{\frac{1.04 \times 10^{-12} \times 1.602 \times 10^{-19} \times 18 \times 10^{16}}{2 \times 1.01}}$$

$$= 36.6~\frac{\text{nF}}{\text{cm}} = 0.366~\frac{\text{fF}}{\mu\text{m}} = 0.366~\frac{\text{nF}}{\text{m}}$$

Finally, $A_s = A_d = Y_d \times W = 3 \times 4 = 12~\mu\text{m}^2 = 12 \times 10^{-12}~\text{m}^2$, and $P_s = P_d = 2Y_d + W = 2 \times 3 + 4 = 10~\mu\text{m}$.

We are now ready to input the above data to PSpice for the transient analysis of a CMOS inverter based on the above MOSFETs and using $V_{DD} = 3.3$ V. The circuit, shown in Fig. 6.78, has been created by recycling that of Fig. 3.65, and then by suitably editing the FET models and the circuit's netlist. The example shown is for a fanout of 1, but it can easily be adapted to a higher fanout. The circuit includes also the wire capacitance C_w to model the parasitic capacitance of the interconnections.

Following the instructions of Appendix 3A, we create the PSpice models as follows:

```
.model  Mn NMOS(Level=1 Tox=20n Uo=600 Vto=0.7 Lambda=0.1
+       Ld=0.15u Gamma=0.18 phi=0.64 Cj=166u Mj=0.5 Cjsw=0.127n
+       Mjsw=0.33 Pb=0.909 Cgso=0.259n Cgdo=0.259n)


.model  Mp PMOS(Level=1 Tox=20n Uo=250 Vto=-0.7 Lambda=0.05
+       Ld=0.2u Gamma=0.42 phi=0.73 Cj=396u Mj=0.5 Cjsw=0.366n
+       Mjsw=0.33 Pb=0.955 Cgso=0.345n Cgdo=0.345n)
```

A few remarks are in order. First, note that the dimensions are V, A, m, and s, with the exception of mobilities, which must be expressed in cm^2/Vs (also, doping densities, when specified, must be expressed in atoms/cm^3). In PSpice the zero-bias bottom and sidewall capacitances are denoted as `Cj` and `Cjsw`, and both use the same built-in potential `Pb`. Moreover, the unit-length overlap capacitances associated with the source

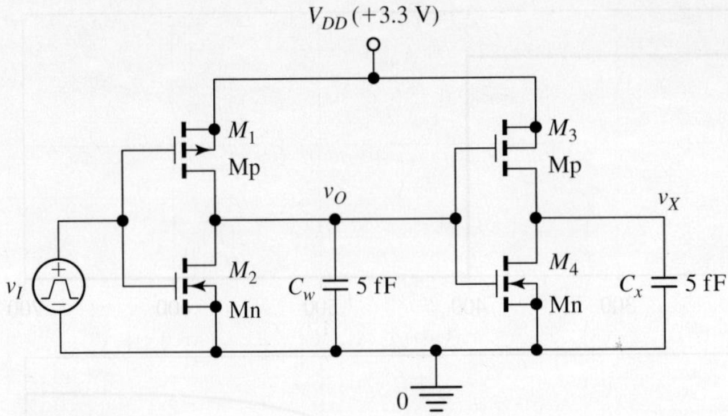

FIGURE 6.78 PSpice circuit to display the transient response of CMOS inverter I_0 with a fanout of 1.

and drain are denoted as Cgso and Cgdo. PSpice calculates automatically the parasitic capacitances of each FET in accordance with its instantaneous region of operation.

Once the *process parameters* have been entered in the model statements, we need to enter the *device parameters* in the netlist. To this end, use **PSpice → Create Netlist** to direct PSpice to generate the netlist, and then use **PSpice → View Netlist** to visualize it. The result is

```
* source CKT_of_Fig_6.78
V_VDD    VDD 0 3.3Vdc
C_Cw     0 VO 5fF
C_Cx     0 VX 5fF
M_M1     VO IN VDD VDD Mp
M_M2     VO IN 0 0 Mn
M_M3     VX VO VDD VDD Mp
M_M4     VX VO 0 0 Mn
V_VI     IN 0 PULSE 0 3.3V 100ps 0.1ps 0.1ps 400ps 1ns
```

Next, type in the individual transistor dimensions L, W, A_s, P_s, A_d, and P_d as follows:

```
* source CKT_of_Fig_6.78
V_VDD    VDD 0 3.3Vdc
C_Cw     0 VO 5fF
C_Cx     0 VX 5fF
M_M1     VO IN VDD VDD Mp      L=1u W=4u As=12p Ps=10u Ad=12p
+                             Pd=10u
M_M2     VO IN 0 0 Mn          L=1u W=2u As=5p Ps=7u Ad=5p
+                             Pd=7u
M_M3     VX VO VDD VDD Mp      L=1u W=4u As=12p Ps=10u Ad=12p
+                             Pd=10u
M_M4     VX VO 0 0 Mn          L=1u W=2u As=5p Ps=7u Ad=5p
+                             Pd=7u
V_VI     IN 0 PULSE 0 3.3V 100ps 0.1ps 0.1ps 400ps 1ns
```

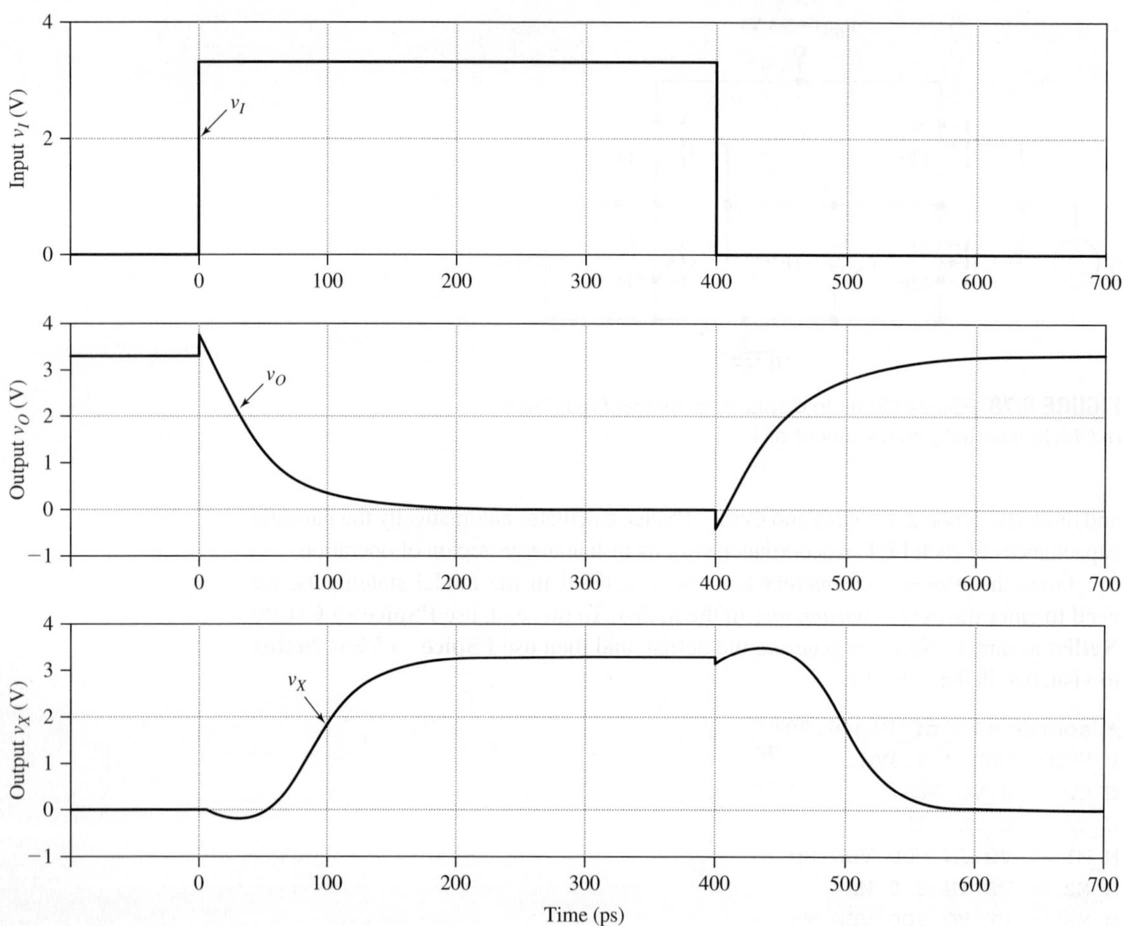

FIGURE 6.79 Waveforms for the PSpice circuit of Fig. 6.78.

Finally, use **File → Save** to save it, and **PSpice → Run** to launch PSpice. This gives the waveforms of Fig. 6.79. Using the cursor facility we measure the delays and find $t_{PHL} \cong 39.6$ ps and $t_{PLH} \cong 43.3$ ps.

Hand Calculation of CMOS Gate Delays

No matter how powerful the computational tools at hand, a scrupulous engineer will always try to anticipate/check the results of simulation via hand calculations, even if the latter may provide only gross approximations. Shown in Fig. 6.80 are all parasitic capacitances for the case of a CMOS inverter with a fanout of 1. To facilitate hand analysis we lump all stray capacitances into a single equivalent capacitance C_{eq} at the output node of the driving inverter I_0. Using inspection, we express the net capacitance between node v_O and ground as

$$C_{eq} = C_0 + C_w + C_1 \tag{6.140}$$

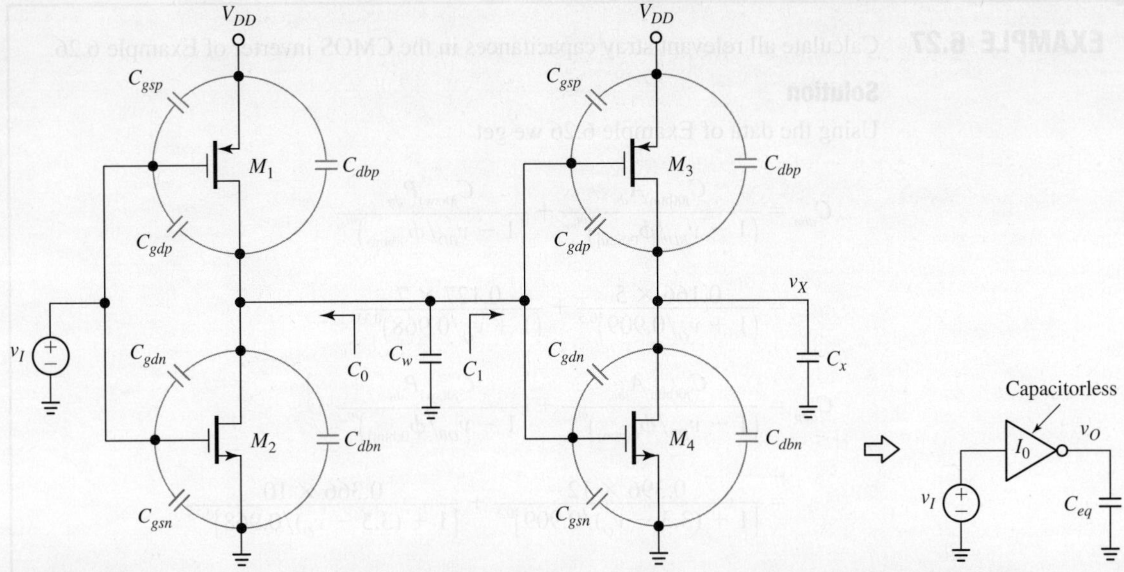

FIGURE 6.80 Showing all stray capacitances for a CMOS inverter with a fanout of 1. To simplify hand calculations, I_0 can be regarded as a parasitics-free inverter driving a suitable equivalent capacitance C_{eq}.

where:

- C_0 is the equivalent capacitance seen looking into the output terminal of the inverter I_0 made up of M_1 and M_2. We have

$$C_0 = C_{dbn} + C_{dbp} + 2(C_{gdn} + C_{gdp}) \qquad \textbf{(6.141)}$$

with the factor of 2 stemming from the Miller effect (as v_I changes from 0 to V_{DD}, v_O changes from V_{DD} to 0, subjecting each of C_{gdn} and C_{gdp} to a voltage change of $2V_{DD}$, in effect doubling both capacitances). Note that C_{gsn} and C_{gsp} are not connected to node v_O, so they do not contribute to C_0 (they only load down the input source v_I).

- C_w is the capacitance of the wire interconnecting the two inverters (C_w increases with the fanout.)

- C_1 is the equivalent capacitance seen looking into the input terminal of the load inverter made up of M_3 and M_4. According to Fig. 6.79, the output v_X of this inverter doesn't change significantly during I_0's t_{PLH} or t_{PHL}, so we can ignore C_{dbn} and C_{dbp} and approximate $C_1 \cong C_{gsn} + C_{gdn} + C_{gsp} + C_{gdn}$. Regardless of how the gate-channel capacitance splits between source and drain, we simply have

$$C_1 \cong C_{ox}\big(W_n \times L_{n(\text{drawn})} + W_p \times L_{p(\text{drawn})}\big) \qquad \textbf{(6.142)}$$

where $L_{n(\text{drawn})}$ and $L_{p(\text{drawn})}$ are the drawn channel lengths, as depicted in Fig. 6.77.

EXAMPLE 6.27   Calculate all relevant stray capacitances in the CMOS inverter of Example 6.26.

Solution

Using the data of Example 6.26 we get

$$C_{dbn} = \frac{C_{j0(\text{btm})}A_{dn}}{\left(1 - v_{BD}/\phi_{0(\text{btm})}\right)^{m_{\text{btm}}}} + \frac{C_{j0(\text{sw})}P_{dn}}{\left(1 - v_{BD}/\phi_{0(\text{btm})}\right)^{m_{\text{sw}}}}$$

$$= \frac{0.166 \times 5}{\left(1 + v_O/0.909\right)^{0.5}} + \frac{0.127 \times 7}{\left(1 + v_O/0.968\right)^{0.33}}$$

$$C_{dbp} = \frac{C_{j0(\text{btm})}A_{dn}}{\left(1 - v_{DB}/\phi_{0(\text{btm})}\right)^{m_{\text{btm}}}} + \frac{C_{j0(\text{sw})}P_{dn}}{\left(1 - v_{DB}/\phi_{0(\text{btm})}\right)^{m_{\text{sw}}}}$$

$$= \frac{0.396 \times 12}{\left[1 + (3.3 - v_O)/0.909\right]^{0.5}} + \frac{0.366 \times 10}{\left[1 + (3.3 - v_O)/0.968\right]^{0.33}}$$

that is,

$$C_{dbn} = \frac{0.83 \text{ fF}}{\left(1 + v_O/0.909\right)^{0.5}} + \frac{0.89 \text{ fF}}{\left(1 + v_O/0.968\right)^{0.33}}$$

$$C_{dbp} = \frac{4.75 \text{ fF}}{\left(4.63 - v_O/0.909\right)^{0.5}} + \frac{3.66 \text{ fF}}{\left(4.41 - v_O/0.968\right)^{0.33}}$$

Moreover,

$$2(C_{gdn} + C_{gdp}) = 2(C_{ovn} \times W_n + C_{ovp} \times W_p) = 2(0.259 \times 2 + 0.345 \times 4)$$

$$= 3.80 \text{ fF}$$

and

$$C_1 = C_{ox}\left(W_n \times L_{n(\text{drawn})} + W_p \times L_{p(\text{drawn})}\right) = 1.725(2 \times 1 + 4 \times 1)$$

$$= 10.35 \text{ fF}$$

Taking also $C_w = 5$ fF into account and combining according to Eq. (6.140) we finally obtain

$$C_{eq} = \frac{0.83 \text{ fF}}{\left(1 + \dfrac{v_O}{0.909}\right)^{0.5}} + \frac{0.89 \text{ fF}}{\left(1 + \dfrac{v_O}{0.968}\right)^{0.33}} + \frac{4.75 \text{ fF}}{\left(4.63 - \dfrac{v_O}{0.909}\right)^{0.5}}$$

$$+ \frac{3.66 \text{ fF}}{\left(4.41 - \dfrac{v_O}{0.968}\right)^{0.33}} + 19.15 \text{ fF}$$

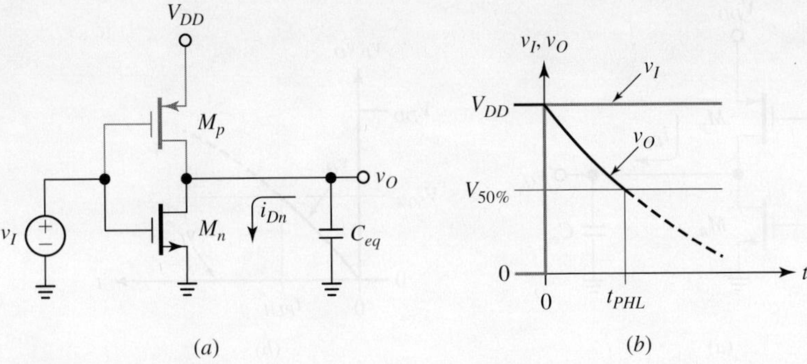

FIGURE 6.81 (a) Equivalent circuit for the estimation of t_{PHL} and (b) waveforms.

We now wish to find quick estimates for the propagation delays. To estimate t_{PHL} refer to Fig. 6.81, showing the situation *after* the transition of v_I from 0 to V_{DD}. With M_p off, M_n pulls the current i_{Dn} out of C_{eq}, thus discharging it. Given the various approximations already made, we make one more and estimate the discharge from V_{DD} to $V_{50\%}$ $(=0.5V_{DD})$ via the rule $C\Delta V = I\Delta t$ (*cee delta vee equals aye delta tee*), where $C = C_{eq}$, $\Delta V = 0.5V_{DD}$, $\Delta t = t_{PHL}$, and $I = i_{Dn(avg)}$ is the average of i_{Dn} over the t_{PHL} interval. We thus have

$$t_{PHL} \cong \frac{0.5V_{DD}C_{eq}}{i_{Dn(avg)}} \tag{6.143a}$$

Considering that at the beginning of the t_{PHL} interval M_n is in saturation and at the end it is in the triode region, we write

$$i_{Dn(avg)} = \frac{1}{2}k'_n\frac{W_n}{L_n}\left\{\frac{1}{2}(V_{DD} - V_{tn})^2(1 + \lambda_n V_{DD}) + \left[(V_{DD} - V_{tn})\frac{V_{DD}}{2} - \frac{1}{2}\left(\frac{V_{DD}}{2}\right)^2\right]\left(1 + \lambda_n\frac{V_{DD}}{2}\right)\right\} \tag{6.143b}$$

Similar considerations hold for the estimation of t_{PLH}. Shown in Fig. 6.82 is the situation *after* the transition of v_I from V_{DD} to 0. Now M_n is off while M_p sources the current i_{Dp} to C_{eq}, thus charging it. Adapting the above equations we write

$$t_{PLH} \cong \frac{0.5V_{DD}C_{eq}}{i_{Dp(avg)}} \tag{6.144a}$$

where

$$i_{Dp(avg)} = \frac{1}{2}k'_p\frac{W_p}{L_p}\left\{\frac{1}{2}(V_{DD} + V_{tp})^2(1 + \lambda_p V_{DD}) + \left[(V_{DD} + V_{tp})\frac{V_{DD}}{2} - \frac{1}{2}\left(\frac{V_{DD}}{2}\right)^2\right]\left(1 + \lambda_p\frac{V_{DD}}{2}\right)\right\} \tag{6.144b}$$

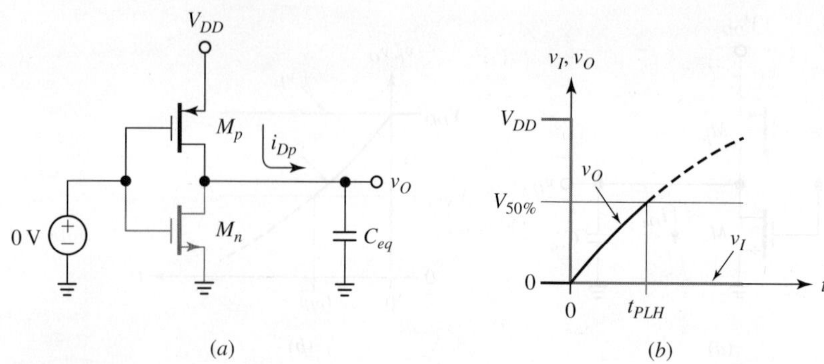

FIGURE 6.82 (a) Equivalent circuit for the estimation of t_{PLH} and (b) waveforms.

EXAMPLE 6.28   (a) Using the data of Example 6.27, estimate the propagation delays of the CMOS inverter of Fig. 6.78 and compare with PSpice.

(b) What would be the delays with a fanout of 0 and $C_w = 0$?

Solution

(a) We have $k'_n = 600 \times (10^{-2})^2 \times 1.725 \times 10^{-15}/(10^{-6})^2 = 103.5 \ \mu\text{A/V}^2$, $L_n = 1.0 - 2 \times 0.15 = 0.7 \ \mu\text{m}$, $k'_p = 250 \times 1.725 \times (0.1) = 43.1 \ \mu\text{A/V}^2$, and $L_p = 1.0 - 2 \times 0.2 = 0.6 \ \mu\text{m}$. By Eqs. (6.143$b$) and (6.144$b$),

$$
i_{Dn(\text{avg})} = \frac{1}{2} 103.5 \frac{2}{0.7} \left\{ \frac{1}{2}(3.3 - 0.7)^2(1 + 0.1 \times 3.3) \right.
$$

$$
\left. + \left[(3.3 - 0.7)\frac{3.3}{2} - \frac{1}{2}\left(\frac{3.3}{2}\right)^2 \right]\left(1 + 0.1\frac{3.3}{2} \right) \right\} = 1.17 \ \text{mA}
$$

$$
i_{Dp(\text{avg})} = \frac{1}{2} 43.1 \frac{4}{0.6} \left\{ \frac{1}{2}(3.3 - 0.7)^2(1 + 0.05 \times 3.3) \right.
$$

$$
\left. + \left[(3.3 - 0.7)\frac{3.3}{2} - \frac{1}{2}\left(\frac{3.3}{2}\right)^2 \right]\left(1 + 0.05\frac{3.3}{2} \right) \right\} = 1.02 \ \text{mA}
$$

A problem with C_{eq} is that its components C_{dbn} and C_{dbp} depend on v_O. We can simplify our analysis by computing them at the beginning and at the end of the propagation interval under consideration, and then using their average value. Thus, at the beginning of t_{PHL} we have $v_O = 3.3$ V, where we calculate $C_{dbn}(3.3) = 0.93$ fF and $C_{dbp}(3.3) = 8.41$ fF. At the end of t_{PHL} we have $v_O = 1.65$ V, where we calculate $C_{dbn}(1.65) = 1.15$ fF and $C_{dbp}(1.65) = 5.47$ fF. The average of their sum is thus $0.5(0.93 + 8.41 + 1.15 + 5.47) = 7.98$ fF. Consequently, $C_{eq} = 7.98 + 19.15 = 27.13$ fF, and

$$
t_{PHL} \cong \frac{1.65 \times 27.13 \times 10^{-15}}{1.17 \times 10^{-3}} = 38.3 \ \text{ps} \ (39.6 \ \text{ps with PSpice})
$$

Likewise, at the beginning of t_{PLH} we have $v_O = 0$, where $C_{dbn}(0) = 1.72$ fF and $C_{dbp}(0) = 4.45$ fF. At the end of t_{PLH} we have $v_O = 1.65$ V, so we recycle $C_{dbn}(1.65) = 1.15$ fF and $C_{dbp}(3.3) = 5.47$ fF. The average of their sum is thus $0.5(1.72 + 4.45 + 1.65 + 5.47) = 6.65$ fF. Consequently, $C_{eq} = 6.65 + 19.15 = 25.80$ fF, and

$$t_{PLH} \cong \frac{1.65 \times 25.80 \times 10^{-15}}{1.02 \times 10^{-3}} = 41.7 \text{ ps (43.3 ps with PSpice)}$$

(b) With a fanout of 0, only C_0 will appear in the calculations, so we need to recalculate but with C_{eq} reduced by the sum $C_1 + C_w (= 10.35 + 5 = 15.35$ fF). So, for t_{PHL} we have $C_{eq} = 27.13 - 15.35 = 11.78$ fF, and we use proportionality to find $t_{PHL} = 38.3(11.78/27.13) \cong 16.6$ ps (20 ps with PSpice). Likewise, using $C_{eq} = 28.80 - 15.35 = 13.45$ fF we get $t_{PLH} = 41.7(13.45/25.80) \cong 21.7$ ps (18.7 ps with PSpice).

Power Dissipation of CMOS Logic Gates

As a CMOS inverter is sitting in either state ($v_O = 0$ or $v_O = V_{DD}$), one of its two transistors is off, drawing only leakage current. We say that the *static* power dissipation of a CMOS gate is virtually zero. However, when the output is switched from one state to the other, energy is expended to charge/discharge the stray capacitances, thus resulting in nonzero *dynamic* power dissipation.

Specifically, to switch v_O from 0 to V_{DD}, M_p must expend some energy E_p in order to charge C_{eq} to V_{DD} (see Fig. 6.82a). Once charged, C_{eq} holds the energy $E_C = (1/2)C_{eq}V_{DD}^2$. Likewise, to return v_O from V_{DD} to 0, M_n must expend some energy E_n in order to discharge C_{eq} to 0 (see Fig. 6.81a). By the energy conservation principle we must have $E_n = E_C$. Also, by symmetry, $E_p = E_n$. The amount of energy dissipated by the gate during a complete cycle is thus $E_{cycle} = E_p + E_n = 2E_C = C_{eq}V_{DD}^2$. If the gate is operated at an average frequency of f_{avg} cycles/sec, the average power P dissipated by the gate equals, by definition, the amount of energy dissipated in 1 sec, or $P = E_{cycle} \times f_{avg}$. Consequently, we have

$$P = C_{eq}V_{DD}^2 f_{avg} \qquad (6.145)$$

It is apparent that the higher the operating frequency, the higher the dissipation. Moreover, power scales *quadratically* with the power supply.

Using the data of Example 6.28, estimate P for $f_{avg} = 1$ kHz. What happens if f_{avg} is raised to 100 MHz?   **EXAMPLE 6.29**

Solution

By Eq. (6.145), $P = 25.80 \times 10^{-15} \times 3.3^2 \times 10^3 = 0.281$ nW. Now $P = (0.281$ nW$) \times (10^8/10^3) = 28.1$ μW.

Transient Response of Voltage Comparators

Being *digital-ouput* devices, comparators are characterized via *propagation delays* just like logic gates. However, being *analog-input* devices, the input test conditions are specified differently. As shown in Fig. 6.83, the input is usually a pulse with a 100-mV baseline and an overdrive V_{OV} designed to *barely exceed* a level that will cause the comparator to trip (typically, V_{OV} is in the mV range). Given the circuit complexity of a comparator, hand analysis is generally prohibitive, so computer simulation becomes a necessity. The IC designer will simulate a comparator at the transistor level, whereas the user will most likely simulate it using the macro-model supplied by the manufacturer.

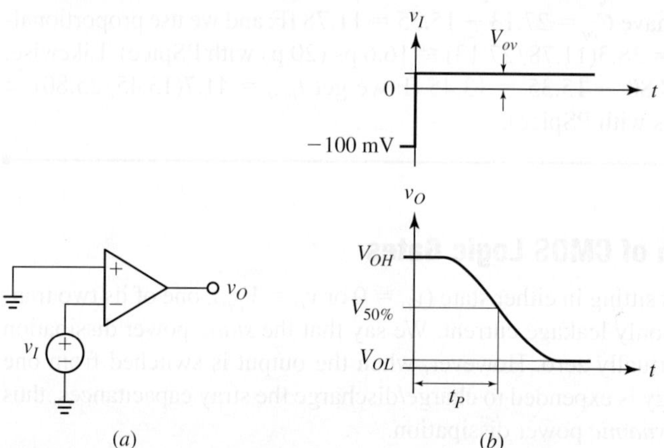

FIGURE 6.83 (a) Test circuit to investigate (b) the transient response of a voltage comparator.

Shown in Fig. 6.84 is a PSpice circuit to display the transient response of the popular LM339 comparator, using its SPICE macro-model. Figure 6.85 shows the responses to the *leading* as well as the *trailing* edge of the input pulse (note the dissymmetry reflecting internal circuit dissymmetries, particularly in the output stage). The overdrives used are $V_{OV} = 5\text{mV}$, 20 mV, and 100 mV. As a rule, the higher the overdrive, the shorter the propagation delays.

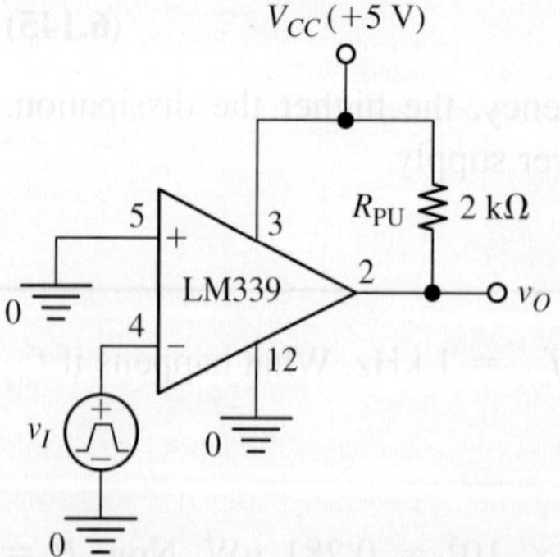

FIGURE 6.84 PSpice circuit to display the transient responses of the LM339 voltage comparator for different input overdrives.

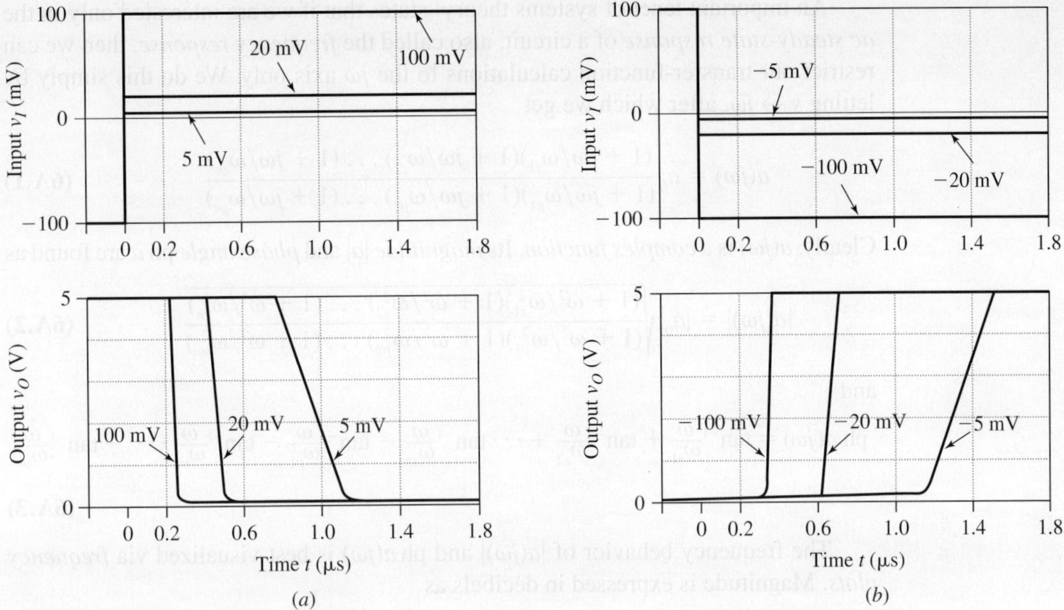

FIGURE 6.85 The transient responses of the PSpice circuit of Fig. 6.84 for different input overdrives.

APPENDIX 6A

Transfer Functions and Bode Plots

The frequency characteristics of circuits are represented mathematically via *transfer functions* and are visualized graphically via *Bode plots*. A transfer function is a function of the *complex frequency s*, the most common examples being *gain* and *input/output impedances*. Gain is the ratio of the *Laplace's transforms* of the output and input signals, or $a(s) = S_o(s)/S_i(s)$. The transfer functions of interest to us can always be put, through suitable algebraic manipulations, in the following insightful *standard form*

$$a(s) = a_0 \frac{(1 + s/\omega_{z1})(1 + s/\omega_{z2}) \ldots (1 + s/\omega_{zn})}{(1 + s/\omega_{p1})(1 + s/\omega_{p2}) \ldots (1 + s/\omega_{pn})}$$

where a_0 is the value of $a(s)$ in the limit $s \to 0$ and is thus referred to as the *low-frequency gain*. (This is the gain that we have been dealing with in the previous chapters). Since the numerator vanishes for $s = -\omega_{z1}$, $s = -\omega_{z2}$, ... $s = -\omega_{zn}$, the ω_zs are referred to as the *zero frequencies* of $a(s)$. These frequencies are *real*, and may be *positive*, *negative*, or even *infinite*. The denominator vanishes for $s = -\omega_{p1}$, $s = -\omega_{p2}$, ... $s = -\omega_{pn}$, causing $a(s)$ to blow up to infinity. The ω_ps are referred to as the *pole frequencies* of $a(s)$, and in this chapter they are *real* and *positive*. Poles and zeros are jointly referred to as *roots*.

An important tenet of systems theory[7] states that if we are interested only in the *ac steady-state response* of a circuit, also called the *frequency response*, then we can restrict our transfer-function calculations to the $j\omega$ axis only. We do this simply by letting $s \rightarrow j\omega$, after which we get

$$a(j\omega) = a_0 \frac{(1 + j\omega/\omega_{z1})(1 + j\omega/\omega_{z2}) \ldots (1 + j\omega/\omega_{zn})}{(1 + j\omega/\omega_{p1})(1 + j\omega/\omega_{p2}) \ldots (1 + j\omega/\omega_{pn})} \tag{6A.1}$$

Clearly, $a(j\omega)$ is a *complex function*. Its *magnitude* $|a|$ and *phase angle* ph a are found as

$$|a(j\omega)| = |a_0| \sqrt{\frac{(1 + \omega^2/\omega_{z1}^2)(1 + \omega^2/\omega_{z2}^2) \ldots (1 + \omega^2/\omega_{zn}^2)}{(1 + \omega^2/\omega_{p1}^2)(1 + \omega^2/\omega_{p2}^2) \ldots (1 + \omega^2/\omega_{pn}^2)}} \tag{6A.2}$$

and

$$\text{ph } a(j\omega) = \tan^{-1}\frac{\omega}{\omega_{z1}} + \tan^{-1}\frac{\omega}{\omega_{z2}} + \cdots \tan^{-1}\frac{\omega}{\omega_{zn}} - \tan^{-1}\frac{\omega}{\omega_{p1}} - \tan^{-1}\frac{\omega}{\omega_{p2}} \cdots - \tan^{-1}\frac{\omega}{\omega_{pn}}$$

$$\tag{6A.3}$$

The frequency behavior of $|a(j\omega)|$ and ph $a(j\omega)$ is best visualized via *frequency plots*. Magnitude is expressed in decibels as

$$|a(j\omega)|_{\text{dB}} = 20\log_{10}|a(j\omega)| \tag{6A.4}$$

and ph $a(j\omega)$ is expressed in degrees, and both functions are plotted versus ω on a logarithmic scale. The most common frequency intervals are decades ($\omega = \ldots 10^{-2}$, 10^{-1}, 10^0, 10^1, $10^2\ldots$ rad/s), though octave intervals are also used ($\omega = \ldots 2^{-2}$, 2^{-1}, 2^0, 2^1, $2^2\ldots$ rad/s), especially in connection with audio circuits. Following is a worth-remembering list of frequently occurring gains as well as their decibel values:

$$|1|_{\text{dB}} = 0 \text{ dB} \quad |2^{\pm 1/2}|_{\text{dB}} = \pm 3 \text{ dB} \quad |2^{\pm n}|_{\text{dB}} = \pm 6n \text{ dB} \quad |10^{\pm n}|_{\text{dB}} = \pm 20n \text{ dB}$$

$$\tag{6A.5}$$

Note that *positive* decibels imply *amplification* and *negative* decibels imply *attenuation*, with the borderline between the two being 0 dB, or unity gain. Moreover, given two transfer functions $a_1(j\omega)$ and $a_2(j\omega)$, we have, by familiar properties of logarithms,

$$|a_1 \times a_2|_{\text{dB}} = |a_1|_{\text{dB}} + |a_2|_{\text{dB}} \tag{6A.6a}$$

$$|a_1(j\omega)/a_2(j\omega)|_{\text{dB}} = |a_1|_{\text{dB}} - |a_2|_{\text{dB}} \tag{6A.6b}$$

that is, the magnitude plot of a product (ratio) is simply the sum (difference) of the individual magnitude plots. In particular, $|1/a|_{\text{dB}} = -|a|_{\text{dB}}$, that is, the magnitude plot of the reciprocal is simply the magnitude plot of the original, but reflected about the horizontal axis.

Given a root ω_0, we will make use of the following approximations:

$$(\omega \ll \omega_0) \Rightarrow (1 + j\omega/\omega_0) \rightarrow 1 \tag{6A.7a}$$

$$(\omega \gg \omega_0) \Rightarrow (1 + j\omega/\omega_0) \rightarrow j\omega/\omega_0 \tag{6A.7b}$$

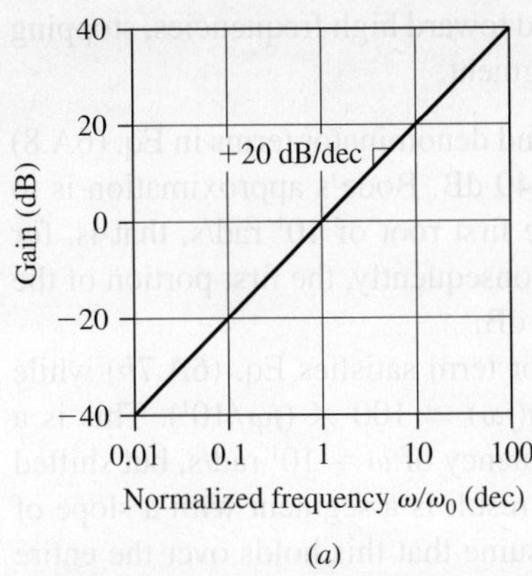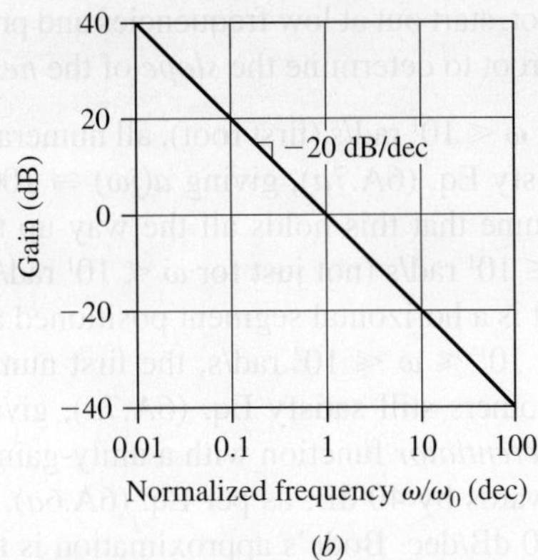

FIGURE 6A.1 Frequency plots of the (a) *differentiator* $j\omega/\omega_0$ and (b) *integrator* $1/(j\omega/\omega_0)$ functions.

The function $j\omega/\omega_0$ is called the *differentiator* function, and its reciprocal $1/(j\omega/\omega_0)$ is called the *integrator* function. These functions have, respectively, a zero and a pole at the origin, and their magnitudes are, $|j\omega/\omega_0|_{dB} = 20\log(\omega/\omega_0)$ and $|1/(j\omega/\omega_0)|_{dB} = -20\log(\omega/\omega_0)$. With a logarithmic frequency scale, these equations are of the type $y = \pm 20x$, that is, *straight lines* with a slope of $+20$ dB/decade ($+6$ dB/octave) in the differentiator case, and -20 dB/decade (-6 dB/octave) in the integrator case. These curves are shown in Fig. 6A.1. Since the two functions are the reciprocal of each other, the plot of one is obtained from the other by a simple reflection about the 0-dB axis. Both curves intersect the 0-dB axis at $\omega = \omega_0$, so ω_0 is aptly called the *unity-gain frequency.*

Bode Plots

To speed up drawing frequency plots by hand, Hendrik W. Bode (1905–1982) proposed a *piecewise linear approximation* consisting of suitably sloped straight segments joined together at the various root frequencies. This technique assumes the approximations of Eqs. (6A.7) to be valid not only far away from a given root, but also in the vicinity of it. It turns out that this technique is quite adequate if the roots are widely spaced, say by a decade or more. Even if they aren't, it still provides valuable insight into the frequency behavior of a circuit.

To illustrate the technique, consider the function

$$a(j\omega) = 100\frac{(1 + j\omega/10^1)(1 + j\omega/10^5)}{(1 + j\omega/10^2)(1 + j\omega/10^3)(1 + j\omega/10^4)} \tag{6A.8}$$

having a dc gain of 100, two zero frequencies at 10^1 and 10^5 rad/s, and three pole frequencies at 10^2, 10^3, and 10^4 rad/s. (For simplicity the roots have been assumed to have nicely rounded values, spaced a decade apart from each other.) To construct the

Bode plot, start out at low frequencies and proceed toward high frequencies, stopping at each root to determine the *slope* of the *next* segment.

- For $\omega \ll 10^1$ rad/s (first root), all numerator and denominator terms in Eq. (6A.8) satisfy Eq. (6A.7*a*), giving $a(j\omega) \cong 100 = 40$ dB. Bode's approximation is to assume that this holds all the way up to the first root of 10^1 rad/s, that is, for $\omega \le 10^1$ rad/s (not just for $\omega \ll 10^1$ rad/s). Consequently, the first portion of the plot is a horizontal segment positioned at 40 dB.

- For $10^1 \ll \omega \ll 10^2$ rad/s, the first numerator term satisfies Eq. (6A.7*b*) while all others still satisfy Eq. (6A.7*a*), giving $a(j\omega) \cong 100 \times (j\omega/10^1)$. This is a *differentiator* function with a unity-gain frequency of $\omega = 10^1$ rad/s, but shifted upwards by 40 dB, as per Eq. (6A.6*a*). The result is a segment with a slope of $+20$ dB/dec. Bode's approximation is to assume that this holds over the entire interval $10^1 \le \omega \le 10^2$ rad/s, not just well away from its extremes.

- Proceeding in similar manner, we can say that for $10^2 \le \omega \le 10^3$ rad/s, the first numerator and the first denominator terms satisfy Eq. (6A.7*b*) while all others still satisfy Eq. (6A.7*a*). Consequently, $a(j\omega) \cong 100 \times (j\omega/10^1)/(j\omega/10^2) = 1000 = 60$ dB. This is again a horizontal segment but positioned at 60 dB.

- For $10^3 \le \omega \le 10^4$ rad/s, the first numerator term, and the second and third denominator terms satisfy Eq. (6A.7*b*) while all others still satisfy Eq. (6A.7*a*), so $a(j\omega) \cong 100 \times (j\omega/10^1)/[(j\omega/10^2)(j\omega/10^3)] = 1000/(j\omega/10^3)$. This is an *integrator* function with a unity-gain frequency of $\omega = 10^3$ rad/s, but shifted upwards by 60 dB, as per Eq. (6A.6*a*). The result is a segment with a slope of -20 dB/dec.

- Likewise, for $10^4 \le \omega \le 10^5$ rad/s we write $a(j\omega) \cong 1000/[(j\omega/10^3)(j\omega/10^4)]$, indicating that another *integrator* term has kicked in at $\omega = 10^4$ rad/s, causing an additional -20-dB change in slope, for a net slope of -40 dB/dec over this frequency interval. We can say that over this frequency interval our transfer function exhibits *double-integrator* behavior.

- For $\omega \ge 10^5$ rad/s, all numerator and denominator terms satisfy Eq. (6A.7*b*), thus giving, after simplifications, $a(j\omega) \cong 1/(j\omega/10^5)$. This is again an *integrator* function, now with a unity-gain frequency of 10^5 rad/s. Above this breakpoint gain rolls off with frequency with a slope of -20 dB/dec.

To get an idea of the errors incurred in using linearized magnitude plots, consider the gain at $\omega = 10^1$ rad/s (first root). By Eq. (6A.2) we have

$$|a(j10)| = 100\sqrt{\frac{[1 + (10/10^1)^2][1 + (10/10^5)^2]}{[1 + (10/10^2)^2][1 + (10/10^3)^2][1 + (10/10^4)^2]}}$$

$$\cong 100\sqrt{2} = (40 + 3)\ \text{dB} = 43\ \text{dB}$$

indicating that our linearized plot *underestimates* magnitude by 3 dB at the first root frequency. Likewise, you can verify that $|a(j10^2)| \cong 1000/\sqrt{2} = 60 - 3 = 57$ dB, indicating a 3-dB *overestimate* at the second root . A quick look at Fig. 6A.2 confirms that the piecewise linear approximation is fairly close to the exact plot, shown in shaded form.

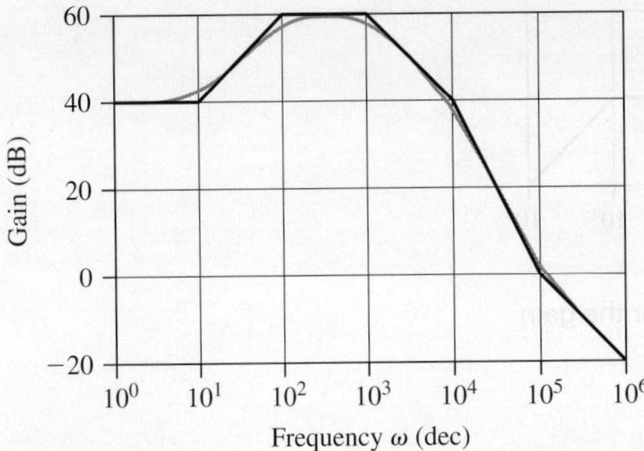

FIGURE 6A.2 Linearized Bode plot for the gain of Eq. (6A.8). The shaded curves show the exact plot.

The procedure for drawing linearized Bode plots can be speeded up considerably as follows:

- Starting out at low frequencies, draw the *low-frequency* asymptote up to the *first nonzero root*. This asymptote will be horizontal if the transfer function has no roots at the origin, or will have a slope of ± 20-dB/dec for each zero/pole at the origin.
- As you move toward the right and hit a root frequency, *change* the present slope by either $+20$ dB/dec or by -20 dB/dec, depending on whether this root is, respectively, a zero ($+$) or a pole ($-$).
- Proceed toward the right till all roots have been exhausted.

As another example, consider the function

$$a(j\omega) = 10\frac{j\omega(1 + j\omega/10^3)}{(1 + j\omega/10^1)(1 + j\omega/10^2)(1 + j\omega/10^4)} \qquad (6A.9)$$

To draw the linearized magnitude plot, proceed as follows:

- At low frequencies all terms within parentheses reduce to unity, so the low-frequency asymptote is $a(j\omega) \cong 10j\omega = j\omega/10^{-1}$. This is a differentiator function with a unity-gain frequency of $\omega = 10^{-1}$ rad/s, so the asymptote is a straight line with a slope of $+20$ dB/dec and a 0-dB axis intercept at 10^{-1} rad/s.
- Coming from the left, draw this asymptote till you hit the first nonzero root at 10^1 rad/s.
- Since 10^1 rad/s is a pole, change the present slope by -20 db/dec, that is, change it from $+20$ dB/dec to $+20 - 20 = 0$ dB/dec. This yields a horizontal segment till the next root at 10^2 rad/s.
- Since 10^2 rad/s is a pole, change the slope from 0 to -20 dB/dec, and continue till the next root at 10^3 rad/s.
- Since 10^3 rad/s is a zero, change the slope from -20 dB/dec back to $-20 + 20 = 0$ dB/dec, and continue till the next root at 10^4 rad/s.
- Since 10^4 rad/s is a pole, change the slope to -20 dB/dec, and draw the final asymptote accordingly. The final plot is shown in Fig. 6A.3.

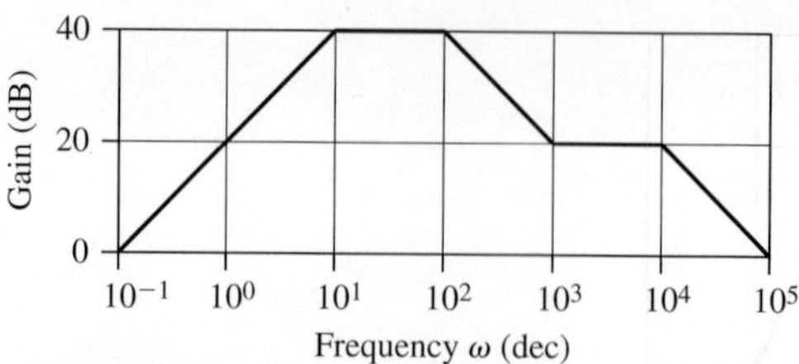

FIGURE 6A.3 Linearized Bode plot for the gain of Eq. (6A.9)

Impedance Plots

Just like gain, impedances are plotted using logarithmic scales. However, impedances are expressed in Ohms (not in dBs!), so while the horizontal axis is still marked in frequency decades (or octaves), the vertical axis is now marked in resistance decades (or octaves). To give an idea, consider Fig. 6A.4a, showing the magnitude plots of the impedances

$$Z_R = R = 10^3 \ \Omega \qquad Z_C = \frac{1}{j\omega C} = \frac{1}{j\omega 10^{-6}} = -j\frac{10^6}{\omega}$$

The plot of $|Z_R|$ (=R) is just a *horizontal* line positioned at $10^3 \ \Omega$, whereas that of $|Z_C|$ (= $10^6/\omega$) is a slanted line with a slope of $-(1$ resistance-decade)/(frequency-decade), or simply -1 dec/dec. Moreover, the two curves intersect each other at $\omega_0 = 1$ krad/s.

Knowing the individual plots of $|Z_R|$ and $|Z_C|$, we find it instructive to construct the magnitude plots of their *series* and *parallel* combinations Z_s and Z_p using mere

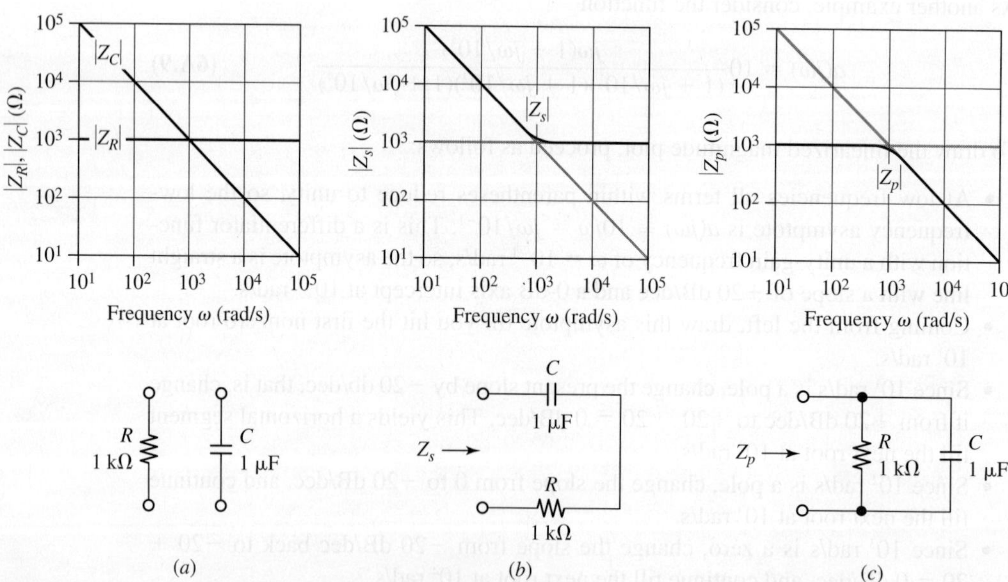

FIGURE 6A.4 Magnitude plots of (a) the individual impedances Z_R and Z_C, (b) their series combination $Z_s = Z_R + Z_C$, and (c) their parallel combination $Z_p = Z_R//Z_C$.

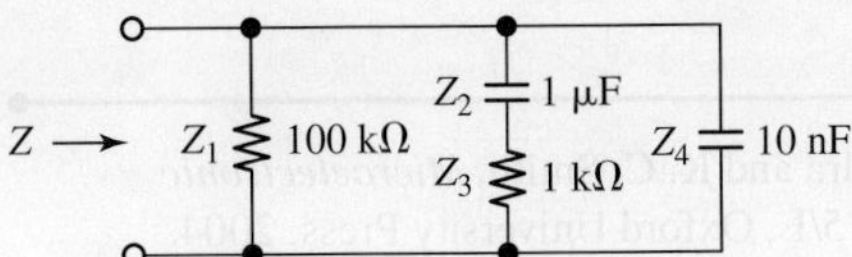

FIGURE 6A.5 An impedance network.

inspection. To this end, recall that in a *series* combination the *larger* of the two impedances dominates, whereas in a *parallel* combination the *smaller* of the two dominates. We make the following observations:

- At *low frequencies*, where $|Z_C| \gg |Z_R|$, we have $Z_s \cong Z_C$ and $Z_p \cong Z_R$.
- At *high frequencies*, where $|Z_C| \ll |Z_R|$, the opposite holds, namely, $Z_s \cong Z_R$ and $Z_p \cong Z_C$.
- The individual impedances exhibit *equal* magnitudes ($=1\ \text{k}\Omega$ in the example) at a special frequency that we shall call ω_0 ($=1$ krad/s in the example), so this frequency marks the *breakpoint* between the low-frequency and the high-frequency asymptotes. Imposing $|Z_C(j\omega_0)| = |Z_R|$, or $1/(\omega_0 C) = R$, we readily find $\omega_0 = 1/(RC) = 1/(10^3 \times 10^{-6}) = 1$ krad/s. At this frequency we have $|Z_s(j\omega_0)| = R\sqrt{2}$ ($=1.414\ \text{k}\Omega$ in our example), and $|Z_p(j\omega_0)| = R/\sqrt{2}$ ($= 0.707\ \text{k}\Omega$ in the example).

As an additional example, let us apply the above intuitive reasoning to plot the magnitude of the equivalent impedance Z presented by the network of Fig. 6A.5. First, plot the individual impedances as in Fig. 6A.6*a*. Then, starting at the low-frequency end and gradually moving toward higher frequencies, construct the plot of $|Z|$ as in Fig 6A.6*b*, based on the following observations.

- At sufficiently low frequencies, where Z_2 and Z_4 act as open circuits, Z_1 dominates.
- Next, Z_2 kicks in $\omega = 10^1$ rad/s because $|Z_2| = |Z_1|$ there.
- Z_2 dominates till $\omega = 10^3$ rad/s, where Z_3 kicks in and dominates till $\omega = 10^5$ rad/s.
- At this final point Z_4 kicks in to dominate for the rest of the frequency spectrum.

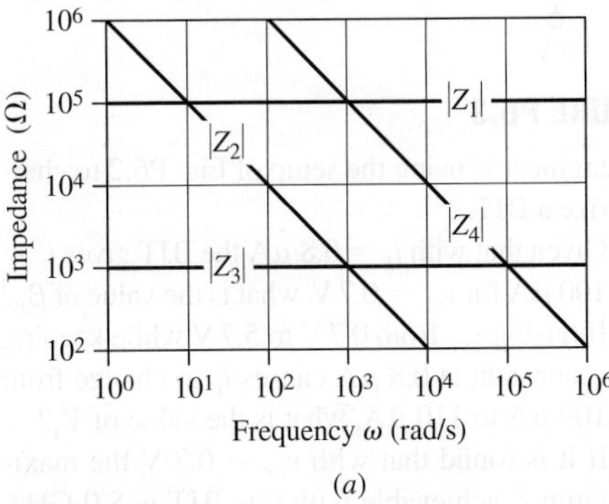

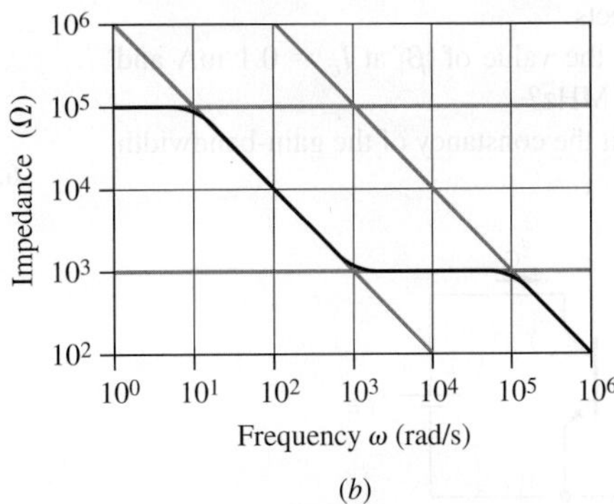

FIGURE 6A.6 Magnitude plots of (a) the individual impedances of Fig. 6A.5, and (b) the overall equivalent impedance Z.

REFERENCES

1. R. S. Muller and T. I. Kamins, *Device Electronics for Integrated Circuits*, 2/E, J. Wiley and Sons, 1986.
2. P. R. Gray, P. J. Hurst, S. H. Lewis, and R. G. Meyer, *Analysis and Design of Analog Integrated Circuits*, 5/E, Wiley and Sons, 2009.
3. P. E. Allen and D. R. Holberg, *CMOS Analog Circuit Design*, 2/E, Oxford University Press, 2002.
4. R. T. Howe and C. G. Sodini, *Microelectronics: An Integrated Approach*, Prentice Hall, 1997

5. A. S. Sedra and K. C. Smith, *Microelectronic Circuits*, 5/E, Oxford University Press, 2004.
6. R. C. Jaeger and T. N. Blalock, *Microelectronic Circuit Design*, 2/E, McGraw-Hill, 2004.
7. S. Franco, *Electric Circuits Fundamentals*, Oxford University Press, 1995.
8. D. A Hodges and H. G. Jackson, *Analysis and Design of Digital Integrated Circuits*, 2/E, McGraw-Hill, 1988.

PROBLEMS

6.1 High-Frequency BJT Model

6.1 A bipolar IC designer is using a planar process with *npn* BJTs having $C_{je0} = 1.0$ pF, $\tau_F = 0.3$ ns, and $C_{\mu0} = 0.4$ pF, and lateral *pnp* BJTs having $C_{je0} = 0.5$ pF, $\tau_F = 25$ ns, and $C_{\mu0} = 1.5$ pF. Both devices have $|\phi_c| = 0.55$ V and $m_c = 1/2$.

(a) Compare their f_Ts at $I_C = 1$ mA and base-collector reverse voltages of 5 V, and comment.

(b) Repeat if I_C is lowered to 0.01 mA, compare with (a), and comment.

6.2 Suppose the BJT of Fig. P6.2 has $C_\mu = 40$ fF and its current gain $\beta = I_c/I_b$ is measured at $f = 500$ MHz for two different dc-bias conditions.

(a) If it is found that $|\beta| = 13.5$ at $I_C = 1.0$ mA, and $|\beta| = 9.2$ at $I_C = 0.25$ mA, estimate C_{je} and τ_F. Assume these parameters are dc-bias independent and there are no high-level injection effects.

(b) What is the value of $|\beta|$ at $I_C = 0.1$ mA and $f = 200$ MHz?

Hint: exploit the constancy of the gain-bandwidth product.

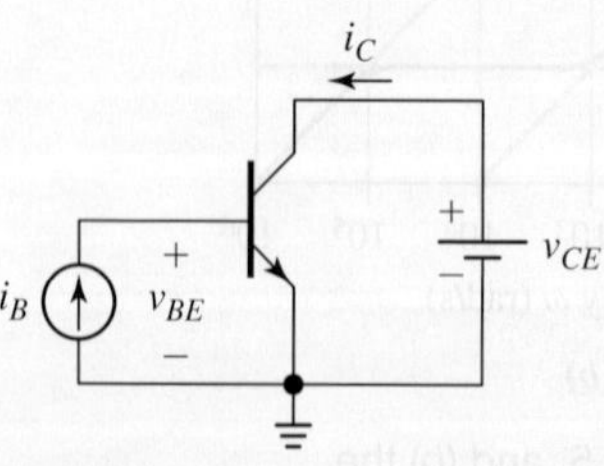

FIGURE P6.2

6.3 (a) Sketch the high-frequency model of the *pnp* BJT of Fig. P6.3, and find its element values if $\beta_0 = 75$, $V_A = 50$ V, $\tau_F = 25$ ps, $r_b = 300\ \Omega$, $C_{je0} = 0.5$ pF, $\phi_e = 0.8$ V, $m_e = 1/3$, $C_{\mu0} = 0.3$ pF, $\phi_c = 0.6$ V, and $m_c = 1/2$. What is the value of f_T?

(b) How do the various element values change if the 4.3-kΩ resistance is increased to 10 kΩ? What is the new value of f_T?

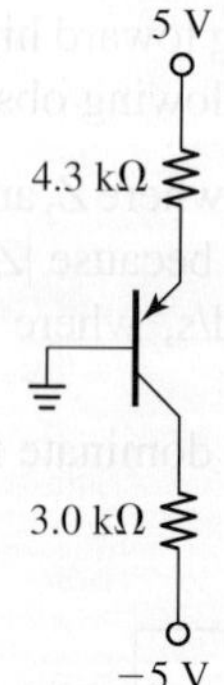

FIGURE P6.3

6.4 An engineer is using the setup of Fig. P6.2 to characterize a BJT.

(a) Given that with $i_B = 0.8\ \mu$A the BJT gives $i_C = 100\ \mu$A for $v_{CE} = 0.7$ V, what is the value of β_F?

(b) If raising v_{CE} from 0.7 V to 5.7 V while keeping i_B constant at 0.8 μA causes i_C to change from 100 μA to 110 μA, what is the value of V_A?

(c) If it is found that with $v_{CE} = 0.7$ V the maximum f_T achievable with this BJT is 8.0 GHz, what is the value of τ_F? What is the effective base width W_B if $D_n = 10$ cm^2/s?

(d) If it is found that with $v_{CE} = 0.7$ V $f_T = 4.0$ GHz at $i_C = 0.1$ mA, what is the value of the sum $(C_{je} + C_\mu)$?

(e) If raising v_{CE} from 0.7 V to 5.7 V while adjusting i_B so as to keep i_C constant at 100 μA causes f_T to increase from 4.0 GHz to 4.30 GHz, what are the values of C_{je0} and $C_{\mu0}$? Assume $C_{je} = 2C_{je0}$, $\phi_c = 0.6$ V, and $m_c = 1/2$.

(f) Suppose the collector terminal is disconnected so as to leave only the B-E junction active. If it is found that with $i_B = 100$ μA the BJT gives $v_{BE} = 650$ mV and with $i_B = 200$ μA it gives $v_{BE} = 700$ mV, what is the value of r_b?
Hint: by a well-known rule of thumb, doubling i_B should require only an 18-mV increase in v_{BE}. The additional voltage drop is due to i_B flowing through the bulk resistance r_b appearing in series with the *pn* junction formed by the base and emitter regions.

(g) Sketch and label the high-frequency equivalent of the BJT at the operating point $Q(I_C, V_{CE}) = Q(0.5$ mA, 3 V$)$.

6.5 Let the BJT of Fig. P6.5 have $r_b = 300$ Ω, $\beta_0 = 100$, and $f_T = 600$ MHz.

(a) Assuming $C_\mu = 0$ and $V_A = \infty$, sketch and label the high-frequency equivalent circuit. Hence, use physical insight to predict the asymptotic values of Z in the limits $f \to 0$ and $f \to \infty$.

(b) Use the test method to obtain an expression for $Z(f)$, put it in the standard form of Eq. 6A.1, and sketch and label the frequency plot of $|Z(jf)|$ (use logarithmic scales). What are the values of its zero and pole frequencies?

(c) Use physical insight to discuss how the above frequency plot will change if $C_\mu = 0.5$ pF.
Hint: the effect of $C_\mu \ne 0$ is to introduce an additional high-frequency pole, thus causing a -1-dec/dec bend in the high-frequency asymptote (refer also to Appendix 6A).

6.2 High-Frequency MOSFET Model

6.6 (a) Show that if the capacitance $(2/3)WLC_{ox}$ is much greater than all other parasitic capacitances in a FET, then

$$f_{T(\text{max})} \cong \frac{1.5}{\pi L}\sqrt{\frac{\mu_n I_D}{2C_{ox}WL}}$$

indicating that faster operation is achieved with *small* FETs operated at *high* currents.

(b) If $\mu_n = 500$ cm²/Vs and $C_{ox} = 3.5$ fF/μm², estimate $f_{T(\text{max})}$ for a FET with $W/L = (10\ \mu\text{m})/(1\ \mu\text{m})$ that is operating at $I_D = 100$ μA.

(c) Repeat part (b) if I_D is doubled.

(d) Repeat part (b) if both W and L are halved, assuming everything else remains the same.

6.7 A CMOS IC designer is working with nMOSFETs characterized by $k' = 125$ μA/V² and $\lambda = (0.04\ \mu\text{m})/L$ V^{-1}.

(a) Assuming the designer wants to bias a particular device at $I_D = 200$ μA with $V_{OV} = 0.25$ V, what is the required value of W if $L = 1$ μm? What is the value of the intrinsic gain $a_{\text{intrinsic}} = g_m r_o$?

(b) Given that $C_{ox} = 2.5$ fF/μm², $L_{ov} = 0.1L$, and $C_{gb} = 5$ fF, what are the values of C_{gs}, C_{gd}, and f_T?

(c) Dissatisfied with the estimated value of f_T, the designer decides to reduce W to ¼ of the value of part (a) so as to reduce C_{gs}, C_{gd} and thus raise f_T. What is the required V_{OV} to retain the same operating point at $I_D = 200$ μA? What are the new values of $a_{\text{intrinsic}}$, C_{gs}, C_{gd}, and f_T? Comment on your findings.

6.8 (a) Find the operating point of the pMOSFET of Fig. P6.8 if $k' = 50$ μA/V², $W = 10$ μm, $L = 1$ μm, $V_{t0} = -0.5$ V, $\gamma = 0.445$ V$^{1/2}$, $2\phi_n = 0.6$ V, and $\lambda = 0.05$ V^{-1}.
Hint: assume $\lambda = 0$ for dc analysis, and use iterations.

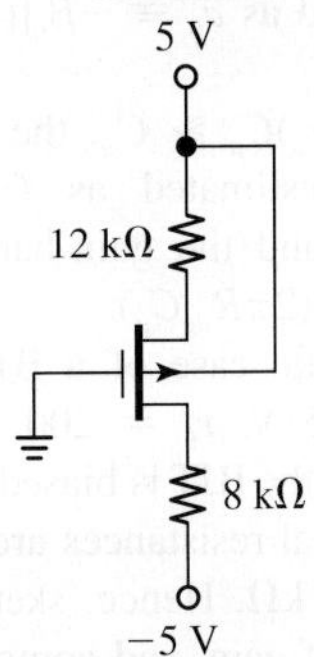

FIGURE P6.8

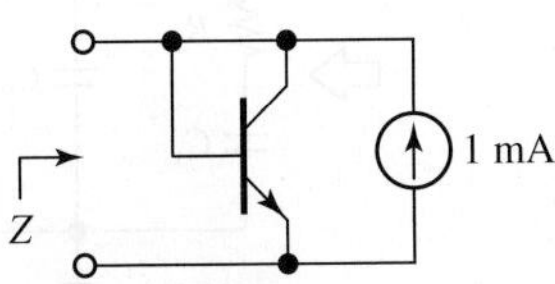

FIGURE P6.5

(*b*) Find the element values in the small-signal model of the MOSFET and show the final circuit, given that $C_{ox} = 3.6$ fF/μm^2, $L_{ov} = 55$ nm, $C_{sb0} = C_{db0} = 25$ fF, $C_{gb} = 5$ fF, $\phi_0 = 0.6$ V, and $m = 0.5$. (*c*) Estimate f_T.

6.9 Let the diode-connected FET of Fig. P6.9 have $V_t = 0.5$ V, $\phi_0 = 0.6$ V, and $m = 0.5$.

(*a*) Given that with $I_D = 0.28$ mA it gives $V_{DS} = 1.5$ V, and with $I_D = 1.20$ mA it gives $V_{DS} = 2.5$ V, what are the values of k and λ?

(*b*) Given that at $(I_D, V_{DS}) = (0.28$ mA, 1.5 V$)$ the FET has $C_{gs} = (2/3)WLC_{ox} + WL_{ov}C_{ox} + C_{gb} = (30 + 2 + 3)$ fF and $C_{db} = 5$ fF, sketch and label its small-signal model. Hence, find the small-signal impedance $Z(jf)$, and sketch and label the frequency plot of $|Z(jf)|$ using logarithmic scales.

(*c*) Repeat part (*b*), but for the case in which the FET is operated at $(I_D, V_{DS}) = (1.20$ mA, 2.5 V$)$. **Hint:** increasing V_{DS} reduces both L and C_{db}, so you can utilize the data of part (*a*) to calculate the percentage reductions in L and in C_{db} as V_{DS} is raised from 1.5 V to 2.5 V.

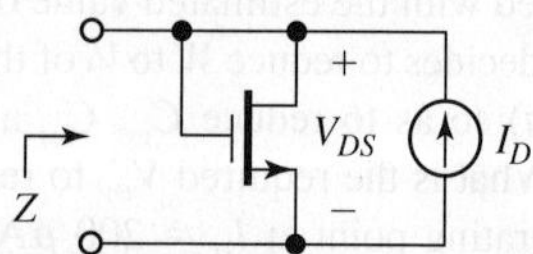

FIGURE P6.9

6.3 Frequency Response of CE/CS Amplifiers

6.10 This problem investigates the frequency response of the CE amplifier of Fig. 6.14*a* to a source with a *high* resistance R_{sig}.

(*a*) Show that if $R_{sig} \gg r_b + r_\pi$, the low-frequency gain can be estimated as $a_0 \cong -\beta_0[(R_C//r_o)/R_{sig}]$.

(*b*) Show that if $g_m(R_C//r_o)C_\mu \gg C_\pi$, the -3-dB frequency can be estimated as $f_{-3\,\text{dB}} \cong 1/[2\pi\beta_0(R_C//r_o)C_\mu]$, and the gain-bandwidth product as GBP $\cong 1/(2\pi R_{sig}C_\mu)$.

(*c*) Investigate the specific case of a BJT with $\beta_0 = 125$, $V_A = 75$ V, $r_b = 200$ Ω, and $C_\mu = 1.0$ pF. Assume the BJT is biased at $I_C = 1$ mA, and the external resistances are $R_{sig} = 30$ kΩ and $R_C = 10$ kΩ. Hence, sketch and label the Bode plot of gain, and comment on your results.

6.11 The CE amplifier of Fig. P6.11 utilizes the feedback-bias scheme. Assume the BJT is biased at $I_C = 1$ mA, and has $\beta_0 = 150$, $V_A = 75$ V, $r_b = 300$ Ω, $f_T = 500$ MHz, and $C_\mu = 0.3$ pF. Estimate a_0 and $f_{-3\,\text{dB}}$ if $R_{sig} = 1.0$ kΩ, $R_C = 10$ kΩ, and $R_B = 100$ kΩ. **Hint:** beware that both C_μ and R_B are subject to the Miller effect.

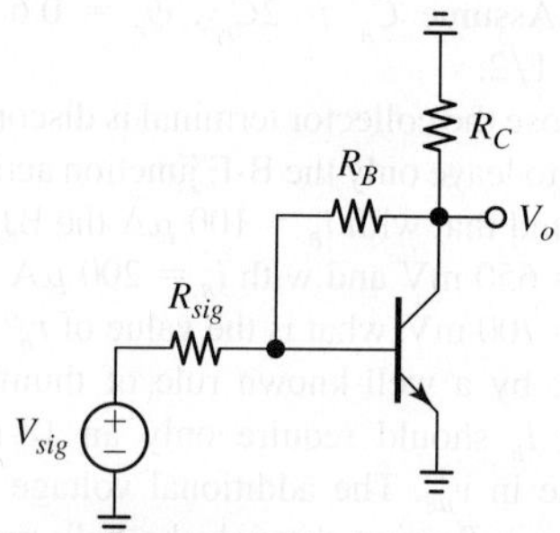

FIGURE P6.11

6.12 (*a*) Show that as long as $C_\mu \ll C_\pi$, the small-signal impedance Z_c seen looking into the collector of the BJT of Fig. P6.12 can be modeled with the all-passive network shown at the right. What are the expressions for R_x, C_x, and C_y? **Hint:** apply the test method to the high-frequency equivalent of the circuit at the left, and exploit the condition $C_\mu \ll C_\pi$ to simplify your calculations.

(*b*) Calculate R_x, C_x, and C_y if the BJT has $g_m = 1/(40\,\Omega)$, $\beta_0 = 150$, $r_o = 100$ kΩ, $r_b = 300$ Ω, $f_T = 400$ MHz, $C_\mu = 0.45$ pF, $C_s = 0.55$ pF, and the input source has $R_{sig} = 2.0$ kΩ. (*c*) Sketch and label the frequency plot of $|Z_c(jf)|$ from 1 kHz to 10 GHz (use logarithmic scales). **Hint:** first sketch and label the plot of the series combination of $|Z_x(jf)|$, where $Z_x = R_x + 1/sC_x$. Next, use physical insight to assess the effect of adding r_o in parallel. Finally, use again physical insight to assess the effect of adding C_y.

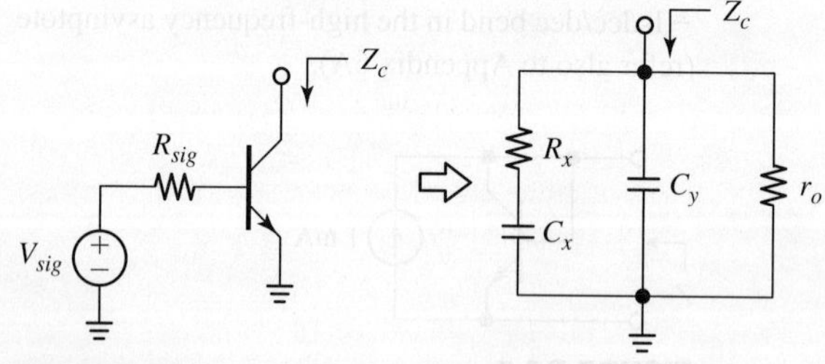

FIGURE P6.12

6.13 This problem investigates the frequency response of the CS amplifier of Fig. 6.15a to a source with a *low* resistance R_{sig}, ideally $R_{sig} \to 0$.

(a) Redraw the high-frequency equivalent of Fig. 6.15b, but with $R_{sig} = 0$, and also add a load capacitance C_L in parallel with R_D for a more general analysis. Writing a node equation at the output node, show that

$$\frac{V_o}{V_{sig}} = a_0 \frac{1 - jf/f_z}{1 + jf/f_p},$$

where $a_0 = -g_m(R_D /\!/ r_o)$, $f_z = \dfrac{g_m}{2\pi C_{gd}}$,

and $f_p = \dfrac{1}{2\pi(R_D /\!/ r_o)(C_{gd} + C_L)}$

indicating that the pole frequency is now established by the net resistance and capacitance of the *output* (rather than the input) node.

(b) Investigate the specific case of a FET with $g_m = 1.8$ mA/V, $r_o = 50$ kΩ, and $C_{gd} = 0.5$ pF, driving $R_D = 10$ kΩ and $C_L = 2$ pF.

(c) Sketch and label the Bode plot of gain, and estimate the GBP of this amplifier.

6.14 The CS amplifier of Fig. P6.14 utilizes the feedback-bias scheme. Assume the FET has $g_m = 1.5$ mA/V, $r_o = 50$ kΩ, $C_{gs} = 2.0$ pF, and $C_{gd} = 0.2$ pF. Estimate a_0 and $f_{-3\,\text{dB}}$, and the GBP if $R_{sig} = 100$ kΩ, $R_D = 10$ kΩ, and $R_G = 3.0$ MΩ.

Hint: beware that both C_{gd} and R_B are subject to the Miller effect.

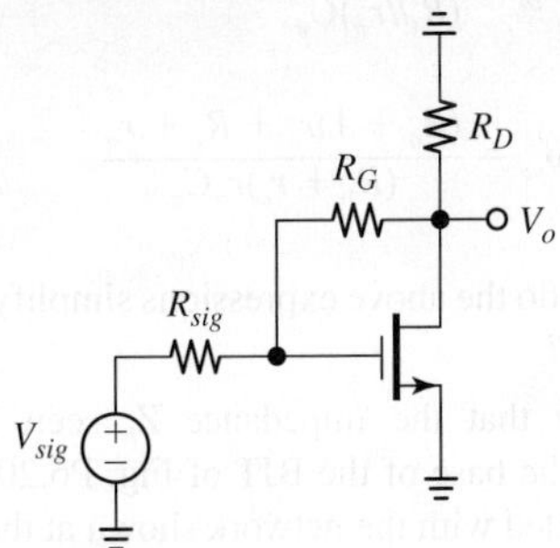

FIGURE P6.14

6.15 (a) Show that as long as $C_{gd} \ll C_{gs}$, the small-signal impedance Z_d seen looking into the drain of the FET of Fig. P6.15 can be modeled with the network shown at the right. What are the expressions for R_x, C_x, and C_y?

Hint: apply the test method to the high-frequency equivalent of the circuit at the left, and exploit the condition $C_{gd} \ll C_{gs}$ to simplify your calculations.

(b) Calculate R_x, C_x, and C_y if the FET has $g_m = 2.0$ mA/V, $r_o = 50$ kΩ, $C_{gs} = 100$ fF, $C_{gd} = 10$ fF, $C_{db} = 20$ fF, and the input source has $R_{sig} = 10$ kΩ. (c) Sketch and label the frequency plot of $|Z_d(jf)|$ from 1 MHz to 10 GHz (use logarithmic scales).

Hint: first plot the impedance $|Z_x| = |R_x + 1/sC_x|$. Next, use physical insight to assess the effect of adding r_o in parallel. Finally, use again physical insight to assess the effect of adding C_y.

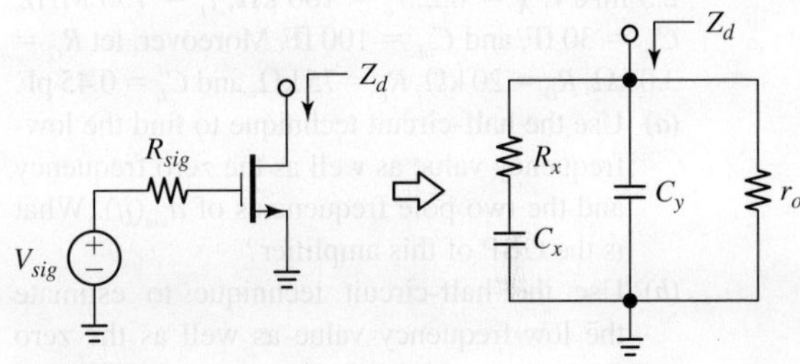

FIGURE P6.15

6.4 Frequency Response of Differential Amplifiers

6.16 Shown in Fig. P6.16 is the ac equivalent of a bipolar differential amplifier driving a load consisting of R_L and C_L. Let the BJTs have $g_m = 1/(50\ \Omega)$, $\beta_0 = 250$, $r_o = 120$ kΩ, $r_b = 250\ \Omega$, $f_T = 400$ MHz, $C_\mu = 0.45$ pF, and $C_s = 1$ pF. Moreover, let $R_B = 2.0$ kΩ, $R_C = 10$ kΩ, $R_L = 50$ kΩ, and $C_L = 4.5$ pF.

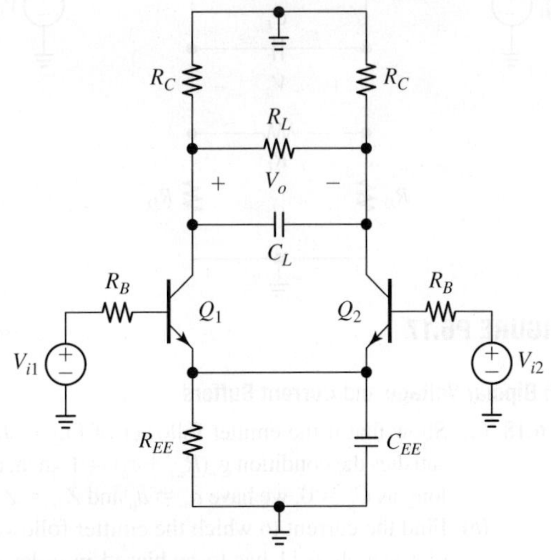

FIGURE P6.16

(a) Use the half-circuit technique to find the low-frequency value as well as the zero frequency and the two pole frequencies of $a_{dm}(jf)$. What is the GBP of this amplifier?

(b) Use the half-circuit technique to estimate the low-frequency value as well as the zero frequency of $a_{cm}(jf)$ if $R_{EE} = 3$ MΩ and $C_{EE} = 0.35$ pF.

6.17 Shown in Fig. P6.17 is the ac equivalent of a CMOS differential amplifier driving a load consisting of R_L and C_L. Let all FETs have $g_m = 2.5$ mA/V, $\chi = 0.2$, $r_o = 100$ kΩ, $f_T = 750$ MHz, $C_{gd} = 30$ fF, and $C_{db} = 100$ fF. Moreover, let $R_G = 3.0$ kΩ, $R_D = 20$ kΩ, $R_L = 75$ kΩ, and $C_L = 0.45$ pF.

(a) Use the half-circuit technique to find the low-frequency value as well as the zero frequency and the two pole frequencies of $a_{dm}(jf)$. What is the GBP of this amplifier?

(b) Use the half-circuit technique to estimate the low-frequency value as well as the zero frequency of $a_{cm}(jf)$ if $R_{SS} = 1$ MΩ and $C_{SS} = 50$ fF.

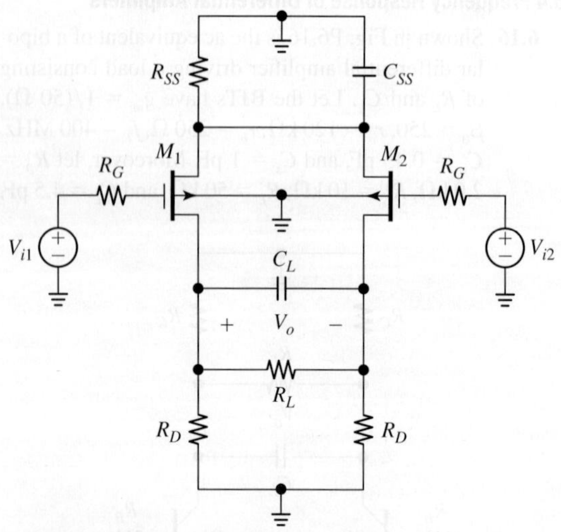

FIGURE P6.17

6.5 Bipolar Voltage and Current Buffers

6.18 **(a)** Show that if the emitter follower of Fig. 6.30a satisfies the condition $g_m(R_{sig} + r_b) = 1$, then, as long as $C_\mu = 0$, we have $a_\infty = a_0$ and $Z_{o\infty} = Z_{o0}$.

(b) Find the current to which the emitter follower of Example 6.11 has to be biased in order to achieve the condition of part (a). **(c)** Sketch

and label the magnitude plot of $a(jf)$ and $Z_o(jf)$ for the follower of part (b) if $C_\mu = 1$ pF. Assume z_π is negligible at the break frequency of each plot. Verify with PSpice.

6.19 **(a)** Assuming $C_\mu = 0$, show that the voltage gain of the emitter follower of Fig. 6.30a has the pole-zero frequency pair

$$\omega_{pa} = \frac{(\beta_0 + 1)R_2 + R_1 + r_\pi}{(R_1 + R_2)r_\pi C_\pi}$$

$$\omega_{za} = \frac{1}{r_e C_\pi}$$

where $R_1 = R_{sig} + r_b$ and $R_2 = R_L//r_o$.
Hint: expand the gain expression of Eq. (6.60) and put it in the standard form of Eq. (6.63).

(b) Show that the impedance Z_i has the pole-zero frequency pair

$$\omega_{pi} = \frac{1}{r_\pi C_\pi}$$

$$\omega_{zi} = \frac{(\beta_0 + 1)R_2 + r_b + r_\pi}{(r_b + R_2)r_\pi C_\pi}$$

(c) Show that the impedance Z_o has the zero-pole frequency pair

$$\omega_{zo} = \frac{1}{(R_1//r_\pi)C_\pi}$$

$$\omega_{po} = \frac{(\beta_0 + 1)r_o + R_1 + r_\pi}{(R_1 + r_o)r_\pi C_\pi}$$

How do the above expressions simplify if β_0 is large?

6.20 **(a)** Show that the impedance Z_b seen looking into the base of the BJT of Fig. P6.20 can be modeled with the network shown at the right.

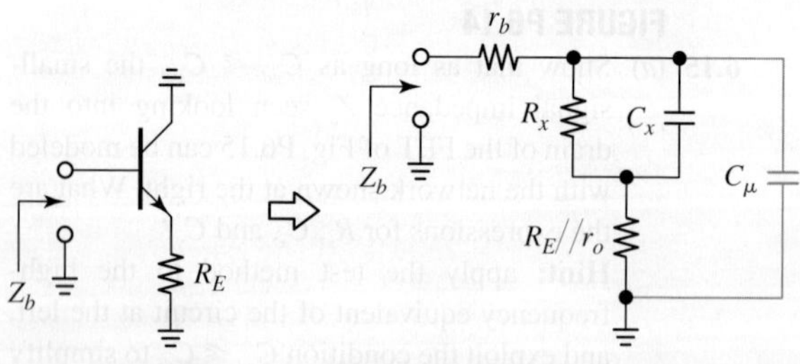

FIGURE P6.20

Hint: apply the test method to the high-frequency equivalent of the circuit shown at the left, and express Z_b as $Z_b = r_b + [R_x /\!/ (1/sC_x)] + (R_E /\!/ r_o)$. Do this first with $C_\mu = 0$, and then add C_μ to your circuit at the end. What are the expressions for R_x and C_x?

(b) Calculate all element values using the data of the emitter follower of Example 6.11.

6.21 The BJT of Fig. P6.21 is a lateral *pnp* with $V_{EB(on)} = 0.7\,\text{V}$, $\beta_F = 50$, $r_b = 250\,\Omega$, $V_A = 50\,\text{V}$, $C_{je} = 0.6\,\text{pF}$, $C_\mu = 0.3\,\text{pF}$, and $\tau_F = 15\,\text{ns}$.

(a) Assuming $C_\mu = 0$, find the gain $a(jf)$.

(b) Find the element values of the inductive network modeling $Z_o(jf)$.

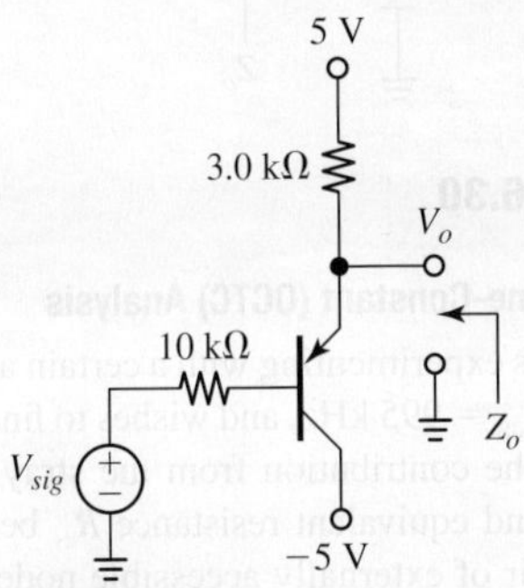

FIGURE P6.21

6.22 The BJT in the current buffer of Fig. 6.35a has $\beta_0 = 125$ and $C_\pi = 20\,\text{pF}$, and is biased at $I_C = 2\,\text{mA}$.

(a) Assuming $r_b = 0$, sketch and label the frequency plots of $|a_{sc}(jf)|$ and $|Z_i(jf)|$ from 1 MHz to 10 GHz (use logarithmic scales).

(b) Repeat, but with $r_b = 265\,\Omega$. Compare and comment.

6.23 This problem investigates the CB configuration as a voltage amplifier. Assuming the collector in Fig. P6.23 is terminated on an ideal active load, sketch and label the magnitude plots of the voltage gain $a(j\omega) = V_o/V_i$ and the impedances $Z_i(j\omega)$ and $Z_o(j\omega)$ if the BJT has $g_m = 25\,\text{mA/V}$, $r_\pi = 6\,\text{k}\Omega$, $r_o = 100\,\text{k}\Omega$, $C_\pi = 10\,\text{pF}$, $C_\mu = 0.25\,\text{pF}$, and $C_s = 0.5\,\text{pF}$. Assume $r_b = 0$ for simplicity.

Hint: write $Z_i = Z_{i1} /\!/ Z_{i2}$, where $Z_{i1} = z_\pi$ and $Z_{i2} = R_{i2} + 1/(sC_{i2})$ is the impedance presented to the driving source by the rest of the circuit. Obtain expressions for R_{i2} and C_{i2}, and use the impedance plotting techniques described in Appendix 6A.

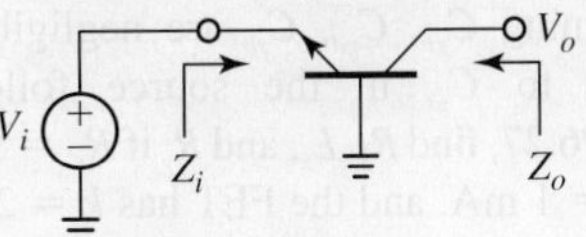

FIGURE P6.23

6.6 MOS Voltage and Current Buffers

6.24 (a) Derive the MOS buffer's input-impedance characteristics of Eq. (6.72).

(b) Repeat, but for the output-impedance characteristics of Eq. (6.74).

6.25 (a) Sketch and label the frequency plot of $|Z_o(jf)|$ for the source follower of Example 6.14a (use logarithmic scales).

(b) Repeat part (a), but for $R_{sig} = 100\,\Omega$. (c) Repeat part (a), but for $R_{sig} = 1\,\text{k}\Omega$. Compare the three cases, and comment.

6.26 (a) Assuming C_{gb}, C_{gd}, C_{sb} are negligible compared to C_{gs}, show that the input impedance Z_i of the source follower of Fig. P6.26 can be modeled with R_G in parallel with the series combination of a suitable capacitance C_x and resistance R_x, as shown. What are the expressions for C_x and R_x?

(b) Find C_x and R_x if $R_G = 100\,\text{k}\Omega$ and $R_s = 2\,\text{k}\Omega$, and the FET has $k = 4\,\text{mA/V}^2$, $\lambda = 0.05\,\text{V}^{-1}$, $\chi = 0.125$, and $C_{gs} = 3\,\text{pF}$, and is biased at $I_D = 2\,\text{mA}$.

(c) Sketch and label the frequency plot of $|Z_i(jf)|$ from 100 kHz to 10 GHz (use logarithmic scales).

(d) Now consider the case $C_{gb} = 50\,\text{fF}$, $C_{gd} = 100\,\text{fF}$, and $C_{sb} = 500\,\text{fF}$. To account for the presence of these capacitances, the model needs to be augmented with C_y and C_z, as shown. What does C_y model, and what is its value? What does C_z model, and what is its value?

(e) Discuss how C_y and C_z affect the plot of part (c). **Hint:** find $Z_{i\infty}$.

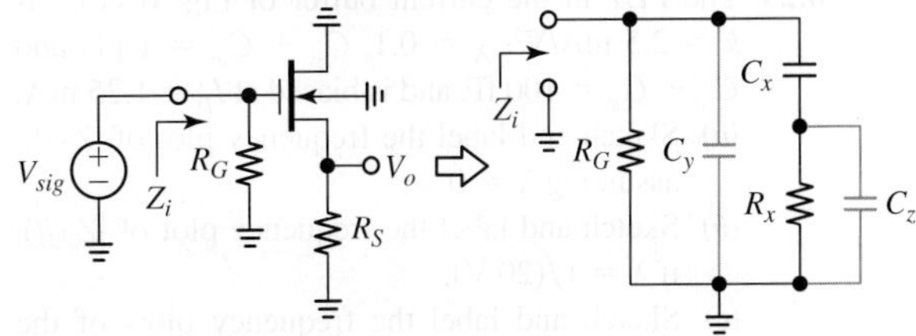

FIGURE P6.26

6.27 **(a)** Assuming C_{gb}, C_{gd}, C_{sb} are negligible compared to C_{gs} in the source follower of Fig. P6.27, find R_x, L_x, and R_y if $R_G = 5$ kΩ and $I_{BIAS} = 1$ mA, and the FET has $k = 2$ mA/V^2, $\lambda = 0.05$ V^{-1}, $\chi = 0.125$, and $C_{gs} = 1$ pF.

(b) Sketch and label the frequency plot of $|Z_o(jf)|$ from 1 MHz to 100 GHz (use logarithmic scales).

(c) Use physical insight to discuss how the plot is affected if $C_{gb} = C_{gd} = 50$ fF and $C_{sb} = 100$ fF. **Hint:** find $Z_{o\infty}$.

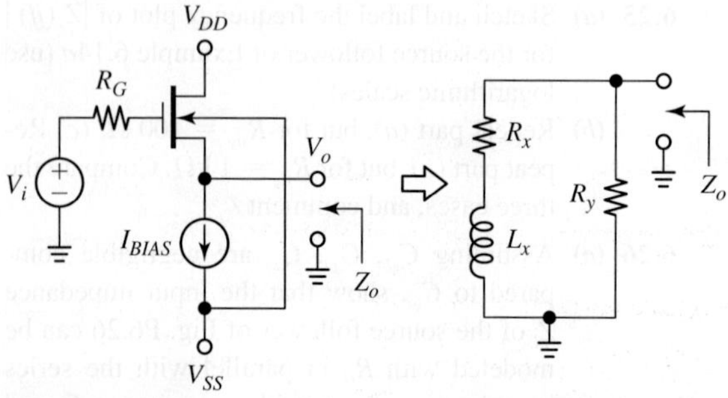

FIGURE P6.27

6.28 **(a)** For the MOS current buffer equivalent of Fig. 6.41b we can write $Z_o = Z_{o1}//Z_{o2}$, where Z_{o2} is the impedance presented by $C_{gd} + C_{db}$, and $Z_{o1} = R_{o1} + 1/(sC_{o1})$ is the impedance presented by the rest of the circuit. Obtain expressions for R_{o1} and C_{o1}, and then prove Eq. (6.78).

(b) Assuming $g_m + g_{mb} = 0.5$ mA/V, $r_o = 20$ kΩ, $C_{gs} + C_{sb} = 450$ fF, and $C_{gd} + C_{db} = 100$ fF, sketch and label the magnitude plots of $Z_{o1}(jf)$, $Z_{o2}(jf)$, and $Z_o(jf)$ using the impedance-plotting techniques of Appendix 6A.

(c) How does the plot of $Z_o(jf)$ change if the signal source, instead of being ideal, has a parallel resistance $R_{sig} = 30$ kΩ?

6.29 The FET in the current buffer of Fig. 6.41a has $k = 2.5$ mA/V^2, $\chi = 0.1$, $C_{gs} + C_{sb} = 1$ pF, and $C_{gd} + C_{db} = 100$ fF, and is biased at $I_D = 1.25$ mA.

(a) Sketch and label the frequency plot of $|Z_i(jf)|$ assuming $\lambda = 0$.

(b) Sketch and label the frequency plot of $|Z_o(jf)|$ if $\lambda = 1/(20$ V$)$.

(c) Sketch and label the frequency plots of the current gain $|a(jf)|$ for the cases $R_L = 0$ and $R_L = 1.0$ kΩ, compare and comment.

6.30 This problem investigates the CG configuration as a voltage amplifier. Referring to the ac equivalent of Fig. P6.30, and assuming $g_m + g_{mb} = 0.35$ mA/V, $r_o = 40$ kΩ, $C_{gs} + C_{sb} = 750$ fF, and $C_{gd} + C_{db} = 150$ fF, sketch and label the magnitude plots of the unloaded voltage gain V_o/V_i and the terminal impedances Z_i and Z_o.

Hint: write $Z_i = Z_{i1}//Z_{i2}$, where Z_{i1} is the impedance of $C_{gs} + C_{sb}$, and Z_{i2} is the impedance of $C_{gd} + C_{db}$ reflected to the input. Hence, use the impedance plotting techniques described in Appendix 6A.

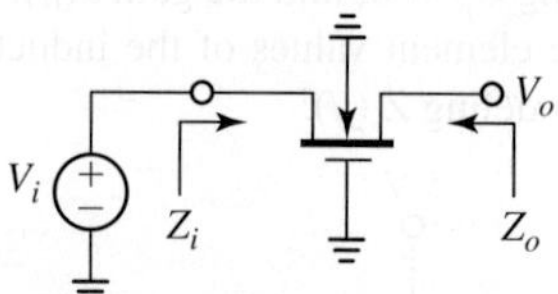

FIGURE P6.30

6.7 Open-Circuit Time-Constant (OCTC) Analysis

6.31 A student is experimenting with a certain amplifier that has $f_{-3\text{ dB}} = 995$ kHz, and wishes to find experimentally the contribution from the stray capacitance C_{xy} and equivalent resistance R_{xy} between a specific pair of externally accessible nodes X and Y. Writing $f_{-3\text{ dB}} = 1/[2\pi(R_{xy}C_{xy} + \tau_{rest})]$, where τ_{rest} is the sum of the time constants due to the rest of the stray capacitances, the student decides to measure $f_{-3\text{ dB}}$ under different loading conditions to indirectly find R_{xy} and C_{xy}.

(a) If connecting an external capacitance $C_{ext} = 10$ pF between nodes X and Y lowers the bandwidth to $f_{-3\text{ dB}} = 612$ kHz, what is the value of R_x?

(b) If the student connects in parallel with C_{ext} also a resistance $R_{ext} = R_x$ (to avoid perturbing the dc operating conditions of the circuit, the student uses a 0.1-μF capacitance in series with R_x) and finds that $f_{-3\text{ dB}} = 884$ kHz, what is the value of C_x? What is the value of τ_{rest}?

6.32 Because of the distributed nature of r_b, we can better approximate high-frequency BJT behavior by splitting C_μ into two parts, as shown in Fig. P6.32. Assuming $R_{sig} = 1$ kΩ, $r_b = 300$ Ω, $\beta_0 = 100$, $g_m = 1/(15$ $\Omega)$, $r_o//R_L = 10$ kΩ, $C_\pi = 10$ pF, and $C_{\mu1} = C_{\mu2} = 0.5$ pF, estimate $f_{-3\text{ dB}}$ via OCTC analysis. Comment on your findings, and specify under what conditions it is acceptable to lump $C_{\mu1}$ with $C_{\mu2}$.

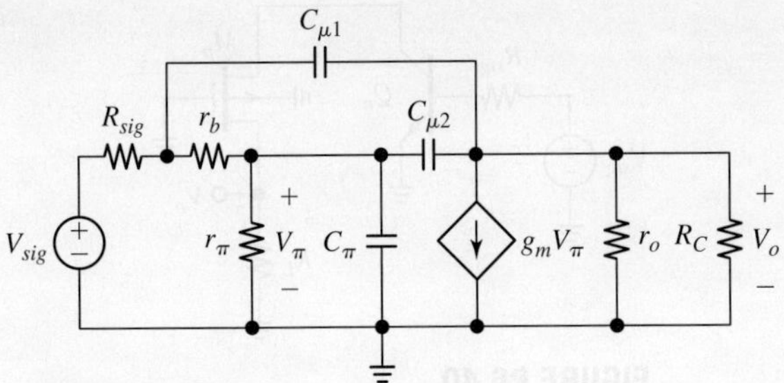

FIGURE P6.32

6.33 Find the low-frequency gain $a_0 = v_o/v_{sig}$ and use OCTC analysis to estimate $f_{-3\,dB}$ for the feedback-bias amplifier shown in ac form in Fig. P6.33, given that $R_{sig} = 1\ k\Omega$, $R_F = 100\ k\Omega$, and $r_o//R_L = 5\ k\Omega$, and that the BJT has $g_m = 1/(25\ \Omega)$, $r_\pi = 5\ k\Omega$, $C_\pi = 15\ pF$, $C_\mu = 0.5\ pF$, and $C_s + C_L = 2\ pF$. Assume $r_b \ll r_\pi$.

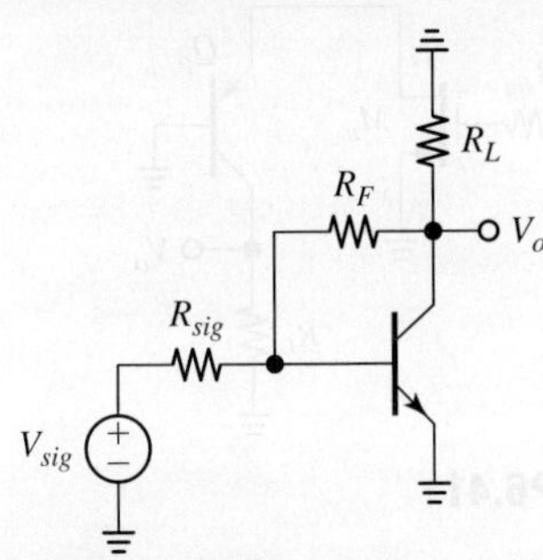

FIGURE P6.33

6.34 Find the low-frequency gain $a_0 = v_o/i_i$ and use OCTC analysis to estimate $f_{-3\,dB}$ for the I-V converter of Fig. P6.34 assuming $k = 2\ mA/V^2$, $\lambda = 0.04\ V^{-1}$, $C_{gs} = 1\ pF$, $C_{gd} = 0.1\ pF$, and $C_{db} + C_L = 1\ pF$.

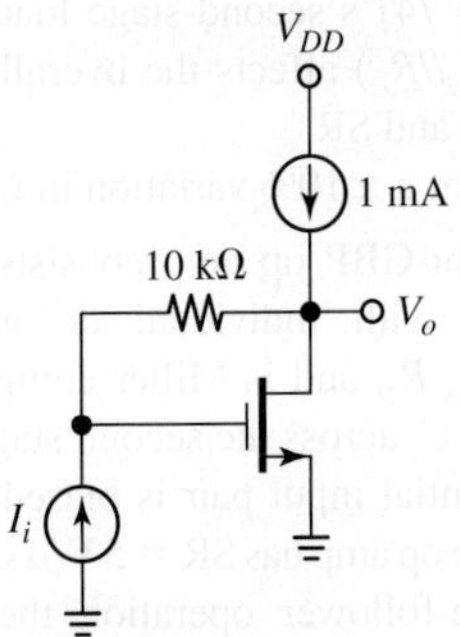

FIGURE P6.34

6.35 This problem investigates the CB configuration as an I-V converter. With reference to the ac equivalent of Fig. P6.35, find the low-frequency gain $a_0 = v_o/i_{sig}$ and use OCTC analysis to estimate $f_{-3\,dB}$ if $R_{sig} = R_L = 10\ k\Omega$. Assume the BJT is biased at $I_C = 0.5\ mA$ and has $\beta_0 = 150$, $r_b = 250\ \Omega$, $V_A = 50\ V$, $C_\pi = 10\ pF$, $C_\mu = 0.25\ pF$, and $C_s + C_L = 1\ pF$.

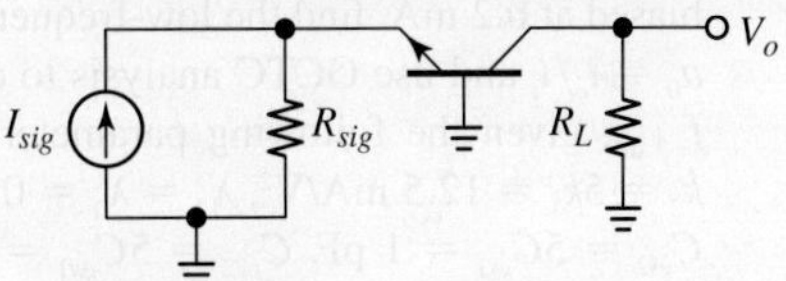

FIGURE P6.35

6.36 This problem investigates the CG configuration of Fig. P6.36 as a voltage amplifier and assumes $r_o \neq \infty$. Find the low-frequency gain $a_0 = v_o/v_{sig}$ and use OCTC analysis to estimate $f_{-3\,dB}$ if $R_{sig} = 3\ k\Omega$ and $R_L = 30\ k\Omega$. Assume the FET has $g_m = 1.25\ mA/V$, $\chi = 0.2$, $r_o = 15\ k\Omega$, $C_{gs} = 250\ fF$, and $C_{gd} + C_{db} = 100\ fF$.

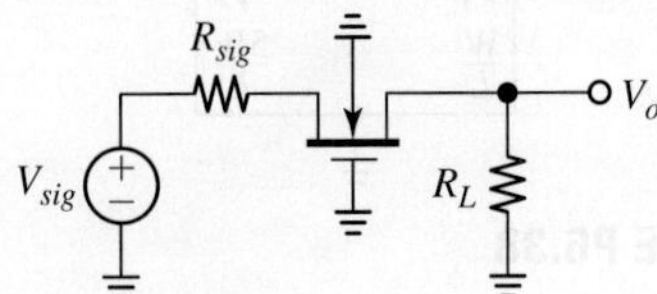

FIGURE P6.36

6.37 Shown in Fig. P6.37 is the ac equivalent of a bipolar current mirror operating as a high-frequency current amplifier. Since the emitter area of Q_2 is four times that of Q_1, the mirror provides a nominal gain of 4 A/A. Draw the high-frequency small-signal equivalent; then, assuming the diode-connected transistor Q_1 is biased at 0.25 mA, find the low-frequency gain $a_0 = i_o/i_i$ and use OCTC analysis to estimate $f_{-3\,dB}$, given the following parameter values: $\beta_{01} = \beta_{02} = 250$, $\tau_{F1} = \tau_{F2} = 0.25\ ns$, $C_{je2} = 4C_{je1} = 4\ pF$, $C_{\mu 2} = 4C_{\mu 1} = 1\ pF$, and $C_{s2} = 4C_{s1} = 6\ pF$. Explain what makes this circuit a high-frequency type.

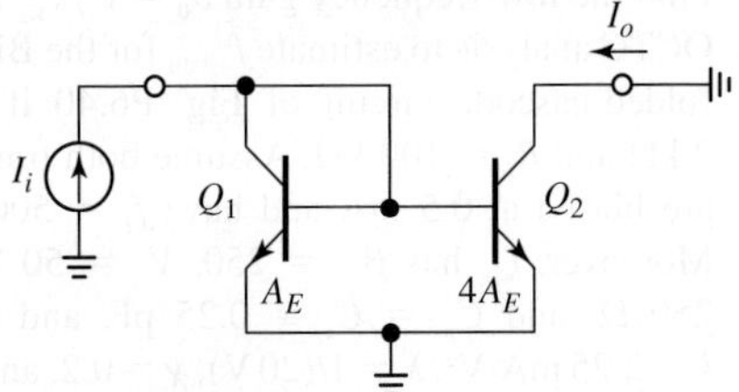

FIGURE P6.37

6.38 (*a*) Shown in Fig. P6.38 is the ac equivalent of a MOS current mirror operating as a high-frequency current amplifier. Since the channel width of M_2 is five times that of M_1, the mirror provides a nominal gain of 5 A/A. Draw the high-frequency small-signal equivalent; then, assuming the diode-connected transistor M_1 is biased at 0.2 mA, find the low-frequency gain $a_0 = i_o/i_i$ and use OCTC analysis to estimate $f_{-3\,\text{dB}}$, given the following parameter values: $k_2 = 5k_1 = 12.5\ \text{mA/V}^2$, $\lambda_2 = \lambda_1 = 0.05\ \text{V}^{-1}$, $C_{gs2} = 5C_{gs1} = 1\ \text{pF}$, $C_{gd2} = 5C_{gd1} = 200\ \text{fF}$, $C_{db2} = 5C_{db1} = 50\ \text{fF}$. Explain what makes this circuit a high-frequency type.

(*b*) Repeat, if M_2's drain is terminated on a load $R_L = 2\ \text{k}\Omega$. Compare with part (*a*), and comment.

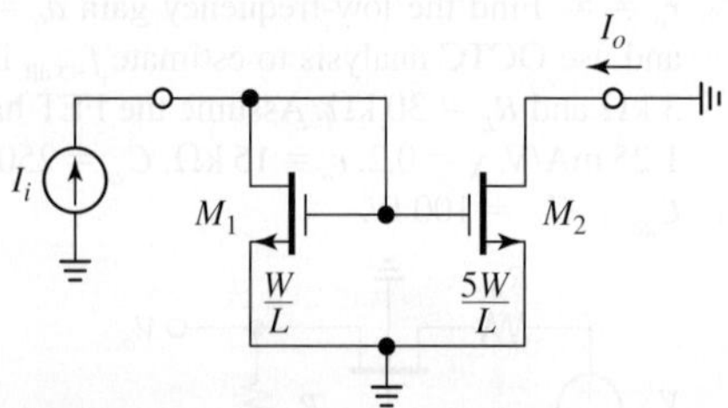

FIGURE P6.38

6.39 Reconsider the EC pair of Example 6.9, but with the inclusion of a pair of degeneration resistances $R_{E1} = R_{E2} = 200\ \Omega$ in series with the emitters, in the manner of Fig. P4.44. Assuming the dc bias conditions and the values of the internal capacitances are unaffected by the addition of R_{E1} and R_{E2}, use OCTC analysis, along with half-circuit techniques, to investigate how the inclusion of degeneration affects $a_{dm}(jf)$ $a_{cm}(jf)$, and $|a_{dm}(jf)/a_{cm}(jf)|$. Compare with the example, and comment.

6.8 Frequency Response of Cascode Amplifiers

6.40 Find the low-frequency gain $a_0 = v_o/v_{sig}$ and use OCTC analysis to estimate $f_{-3\,\text{dB}}$ for the BiCMOS folded-cascode circuit of Fig. P6.40 if $R_{sig} = 2\ \text{k}\Omega$ and $R_L = 100\ \text{k}\Omega$. Assume both transistors are biased at 0.5 mA and have $f_T = 500\ \text{MHz}$. Moreover, Q_n has $\beta_{01} = 250$, $V_A = 50\ \text{V}$, $r_b = 250\ \Omega$, and $C_\mu = C_s = 0.25\ \text{pF}$, and M_p has $k = 2.25\ \text{mA/V}^2$, $\lambda = 1/(20\ \text{V})$, $\chi = 0.2$, and $C_{gd} = C_{db} = 50\ \text{fF}$.

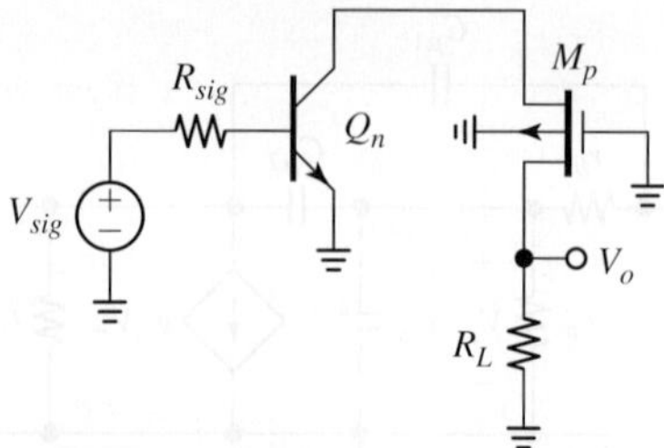

FIGURE P6.40

6.41 Find the low-frequency gain $a_0 = v_o/v_{sig}$ and use OCTC analysis to estimate $f_{-3\,\text{dB}}$ for the BiCMOS folded-cascode circuit of Fig. P6.41 if $R_{sig} = 10\ \text{k}\Omega$ and $R_L = 100\ \text{k}\Omega$. Assume both transistors are biased at 0.25 mA and have $f_T = 400\ \text{MHz}$. Moreover, M_n has $k = 1.28\ \text{mA/V}^2$, $\lambda = 1/(15\ \text{V})$, and $C_{gd} = C_{db} = 25\ \text{fF}$, and Q_p has $\beta_{01} = 200$, $V_A = 50\ \text{V}$, $C_\mu = C_s = 0.25\ \text{pF}$.

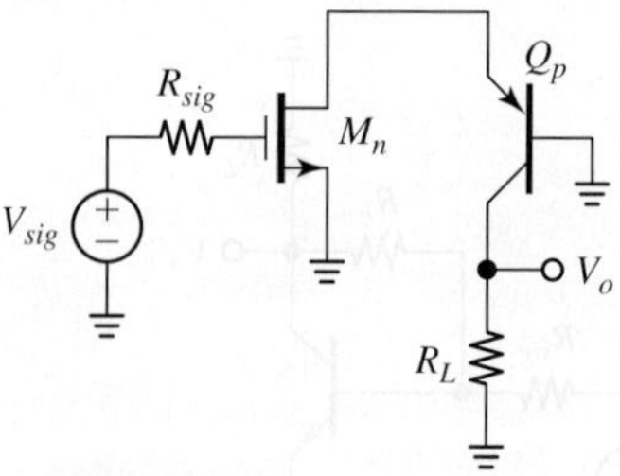

FIGURE P6.41

6.9 Frequency and Transient Responses of Op Amps

6.42 If the gain of a constant-GBP op amp has a magnitude of 80 dB at $f = 10\ \text{Hz}$ and a phase angle of $-58°$ at $f = 320\ \text{Hz}$, estimate a_0, f_b, f_t, and the GBP. What are the gain magnitude and phase at $f = 440\ \text{Hz}$?

6.43 (*a*) Discuss how a fabrication process variation of $\pm10\%$ in the 741's second-stage loaded gain $a_{20} = G_{m2}(R_{o2}/\!/R_{i3})$ affects the overall parameters a_0, f_b, f_t, and SR.

(*b*) Repeat, but for a $\pm10\%$ variation in C_c.

6.44 A certain constant-GBP op amp consists of two inverting stages with individual dc gains of $-G_{m1}R_1$ and $-G_{m2}R_2$, and is Miller compensated via a capacitance C_c across the second stage.

(*a*) If the differential input pair is biased at $I_1 = 50\ \mu\text{A}$ and the op amp has SR $= 5\ \text{V}/\mu\text{s}$, find C_c.

(*b*) If in voltage-follower operation the small-signal transient response is governed by the time constant $\tau = 40$ ns, find the step magnitude $V_{m(onset)}$ corresponding to the onset of SR limiting.

(c) If the overall dc gain is $a_0 = 100$ dB, find f_b as well as the equivalent resistance R_{eq} seen by C_c.

(d) If the dc gain of the second stage is 1.6 times as large as that of the first stage, find the effective capacitance resulting from the Miller effect, as well as the first-stage parameters G_{m1} and R_1.

6.45 A two-stage CMOS op amp of the type of Fig. 5.13 has $C_c = 2$ pF and is fabricated in a process characterized by $k'_p = 65$ μA/V^2, $\lambda_n = 0.02$ V^{-1}, and $\lambda_p = 0.05$ V^{-1}. If SR $= 40$ V/μs and $f_t = 25$ MHz, find the bias current I_{SS} of the SC input pair, the overdrive voltage V_{OV} and the W/L ratio of the individual SC transistors, and the 1st-stage dc gain a_{10}.

6.46 (a) A student is characterizing the folded-cascode op amp of Fig. 5.16 using an oscilloscope. If loading the output node with a capacitance $C_L = 3$ pF causes SR to drop from 60 V/μs to 24 V/μs, find the differential input pair's bias current I_{SS} as well as the stray output-node capacitance C_o.

(b) If loading the output with a resistance $R_L = 10$ MΩ causes the dc gain a_0 to drop from 5000 V/V to 2500 V/V, find the output resistance R_o.

(c) Find the overdrive voltage V_{OV} of the differential-pair transistors, the corner frequency f_b, and the GBP for the unloaded case.

(d) Find a_0, GBP, and SR if the amplifier is loaded with $C_L = 2$ pF.

(e) Repeat part (d), but for the case of a load $R_L = 15$ MΩ.

(f) Repeat part (d), but for the case in which the 2-pF and 15-MΩ loads are present *simultaneously*.

6.47 Suppose that because of some design error the folded-cascode op amp of Example 5.4 is fabricated with $W_9 = W_{10} = 24$ μm instead of 40 μm.

(a) Assuming $C_c = 2.5$ pF, discuss how this error affects a_0, f_b, and f_t, compare with Example 6.23b, and comment.

(b) Find all drain currents during positive SR limiting ($v_P \gg v_N$), compare with Exercise 6.6b.

(c) Repeat (b), but for negative SR limiting ($v_P \ll v_N$), and compare.

(d) Discuss the main functional difference between the correct and the erroneous circuit.

6.48 An IC designer is considering the folded-cascode op amp of the type of Fig. 5.16 under the following constraints: $I_{BIAS} = 1.2I_{SS}$, and all FETS must operate with the same overdrive voltage V_{OV}. The op amp is to have a dc gain of 7,500 V/V and a slew rate of 20 V/μs for a 6-pF output load.

(a) If $\lambda_n = \lambda_p = 1/(22$ V$)$, find V_{OV}, I_{SS}, and I_{BIAS} (for simplicity ignore the body effect and assume $\lambda = 0$ in the course of dc calculations).

(b) What is the GBP for the given load?

(c) What happens if C_L is doubled?

6.10 Diode Switching Transients

6.49 Because of its dependence on v_D, the junction capacitance C_j is nonlinear. To simplify the calculations it is often convenient to work with the *equivalent capacitance* $C_{j(eq)}$ that, in response to a voltage change $\Delta V_D = V_{D2} - V_{D1}$, displaces the *same amount* of charge $\Delta Q_j = Q_j(V_{D2}) - Q_j(V_{D1})$ as C_j, or

$$\Delta Q_j = C_{j(eq)} \Delta V_D = \int_{V_{D1}}^{V_{D2}} C_j(v_D)\,dv_D$$

(a) Calculate the above integral using Eq. (6.118), and show that

$$C_{j(eq)} = \frac{-C_{j0}\phi_0}{(V_{D2} - V_{D1}) \times (1 - m))}$$

$$\times \left[\left(1 - \frac{V_{D2}}{\phi_0}\right)^{1-m} - \left(1 - \frac{V_{D1}}{\phi_0}\right)^{1-m} \right]$$

(b) With reference to Fig. 6.68, calculate $C_{j(eq)}$ for v_D changing from $V_{D1} = -2$ V to $V_{D2} = +0.6$ V, and compare with the approximation $C_j \cong C_{j0}$ made in the text.

6.50 (a) To what value must we lower V_R in Fig. 6.68 if we want to halve the storage time t_S?

(b) Suppose we reduce the duration of the pulse of Fig. 6.68 so that v_S switches from V_F back to V_R at $t_2 = 15$ ns (instead of $t_2 = 50$ ns, as shown). Recalculate the storage time t_S and comment.

6.51 In Fig. P6.51 let $R = 3$ kΩ and let the diode have $V_{D(on)} = 0.7$ V. Suppose v_S in Fig. P6.51a, after having been at $+4$ V for a sufficiently long time, is switched to -4 V, and the storage time is measured to be 25 ns. If D and R are swapped as shown in Fig. P6.51b and v_S, after having been at $+5$ V for a long time, is switched to -5 V at $t = 0$, sketch and label v_S and v_O vs. time for $t \geq 0$ and comment (for simplicity assume $C_j = $ constant $= 5$ pF).

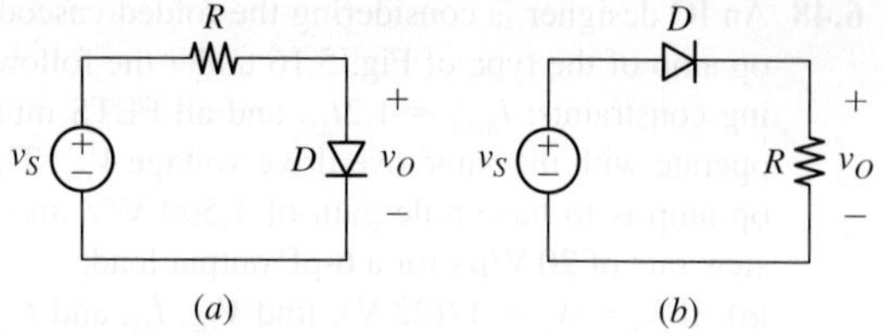

FIGURE P6.51

6.52 In Fig. P6.52 let $R = 2$ kΩ and $v_2 = 2.5$ V, and let the diode have $V_{D(on)} = 0.7$ V and $\tau_F = 20$ ns.
   (**a**) If v_1, after having been at 0 V for a long time, is switched to $+5$ V at $t = 0$, sketch and label v_1 and v_o vs. time for $t \geq 0$ and comment (for simplicity assume $C_j = $ constant $= 5$ pF).
   (**b**) Repeat part (**a**), but for the case in which v_1 is switched to $+2.5$ V (instead of $+5$ V).

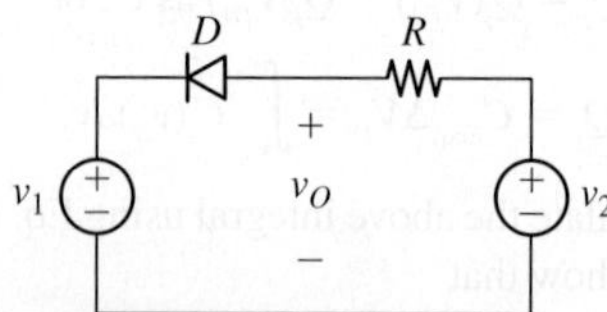

FIGURE P6.52

6.11 BJT Switching Transients

6.53 Suppose the BJT of Fig. P6.53 has $\tau_F = 2$ ns, $C_{je0} = 5$ pF, $\phi_e = 0.8$ V, $m_e = 0.33$, $C_{jc0} = 1$ pF, $\phi_c = 0.75$ V, $m_c = 0.33$, $V_{BE(EOC)} = 0.6$ V, and $V_{CE(EOS)} = 0.2$ V. Moreover, let $V_{CC} = -V_{BB} = 5$V, $R_1 = 3$ kΩ, $R_2 = 10$ kΩ, and $R_C = 1$ kΩ.
   (**a**) If v_S in P6.53a, after having been at 0 V for a long time, is switched to $+5$ V, find the time t_{EOC} it takes to bring the BJT to the EOC (use the formula of Problem 6.49 to calculate $C_{je(eq)}$ and $C_{jc(eq)}$).
   (**b**) We can make $t_{EOC} \to 0$ by adding a capacitance C as in Fig. P6.53b and adjusting it to the value C_{EOC} that, as v_S switches from 0 V to 5 V, will inject into the base the charge needed to bring the BJT *right* to the EOC. What is the required C_{EOC}?
   (**c**) In fact, we can go one step further, and increase the capacitance to the value C_{EOS} that will bring the BJT from CO, through the FA region, *right* to the EOS. What is the required C_{EOS}?

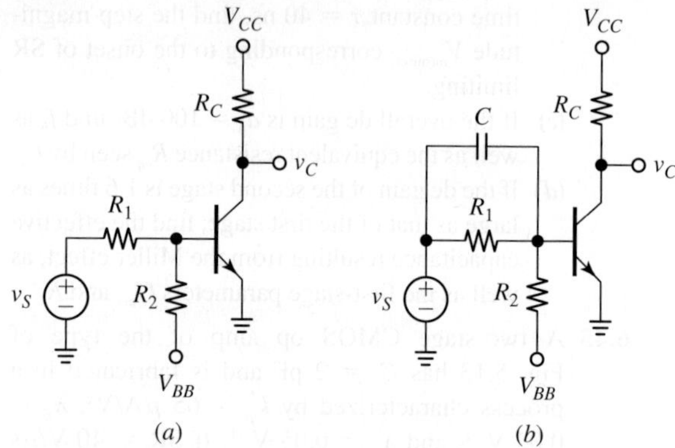

FIGURE P6.53

6.54 Let the BJT of Fig. P6.53a have $V_{BE(sat)} = 0.7$ V, $V_{CE(EOS)} = 0.2$ V, $V_{CE(sat)} = 0.1$ V, and $\beta_F = 75$. Moreover, let $V_{CC} = 5$V, $V_{BB} = -2$ V, $R_1 = 10$ kΩ, $R_2 = 20$ kΩ, and $R_C = 1.2$ kΩ.
   (**a**) If v_S, after having been at 5 V for a long time, is switched to 0 V and the storage time is $t_S = 30$ ns, find the time constant τ_S.
   (**b**) We can make $t_S \to 0$ by adding a capacitance C as in Fig. P6.53b and adjusting it to the value C_{EOS} that, as v_S switches from 5 V to 0 V, will pull out of the base the charge Q_S needed to bring the BJT *right* to the EOS. What is the required C_{EOS}?
   (**c**) Find t_S if C is made equal to $C_{EOS}/2$.

6.55 The circuit of Fig. P6.55, known as *Baker clamp* and much in use before SBDs became more popular, uses two regular diodes to prevent the BJT from saturating. Assuming $V_{BE(on)} = V_{D(on)} = 0.7$ V, find all voltages and current in the circuit for $v_S = 0$ V and $v_S = 5$ V.

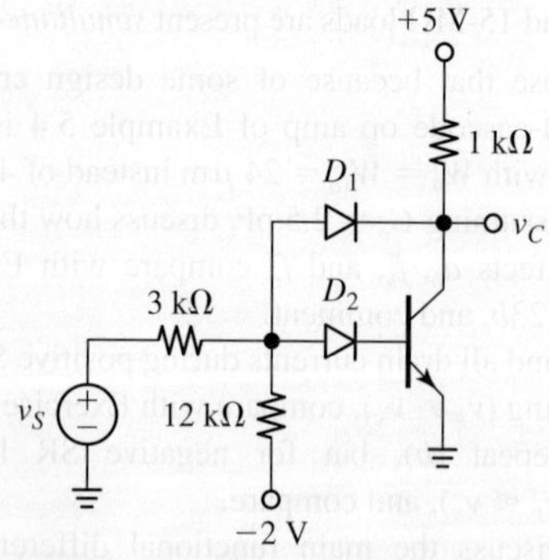

FIGURE P6.55

6.56 The BJT in Fig. P6.56 is designed to operate within the FA region, where Δv_{BE} is small enough that we can ignore ΔQ_{je}. Moreover, R_3 is made deliberately small so that Δv_{BC} is also small enough that we can ignore ΔQ_{jc}. Under these conditions, Eq. (6.128) simplifies as $i_B = Q_F/\tau_{BF} + dQ_F/dt$.

  (*a*) Given that in response to a 1-V input step the circuit yields an exponential transient with a 0.1-V magnitude, find β_F, V_{OH}, and V_{OL}.

  (*b*) Suppose we now add a capacitor and adjust its value until the output transition becomes the perfect step shown in shaded form. If this occurs for $C = 2$ pF, find τ_{BF} and τ_F.

  (*c*) Sketch and label v_O vs. time if C is lowered to 1 pF. (*d*) Repeat part (*c*) if C is raised to 3 pF.

  (*d*) Sketch and label v_O vs. time if $C = 2$ pF and R_1 is removed from the circuit.

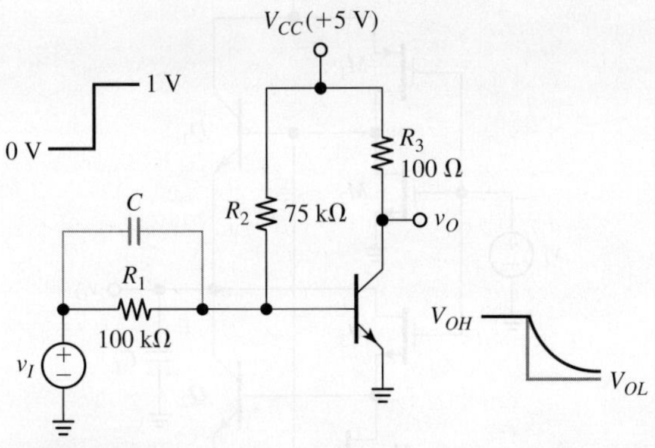

FIGURE P6.56

6.12 Transient Response of CMOS Gates and Voltage Comparators

6.57 If the CMOS inverter of Fig. P6.57 has the current transfer curve shown at the right, estimate t_{PHL} and t_{PLH} under the assumption that all stray capacitances can be modeled with $C_{eq} = 1$ pF.

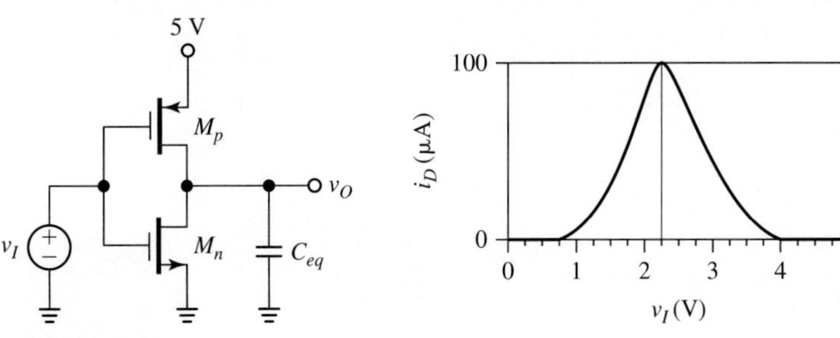

FIGURE P6.57

6.58 (*a*) Shown in Fig. P6.58 are three CMOS inverters connected to form a ring. Sketch v_1, v_2, and v_3 vs. time (for simplicity assume approximately triangular waveforms), and verify that the circuit oscillates (hence the name *ring oscillator*). Find a relationship between the frequency of oscillation f_{osc} and the average propagation delay t_p of each gate.

  (*b*) What happens if a fourth inverter is inserted in the loop? What if the number of inverters is five? What conclusions do you draw?

  (*c*) If it is found that loading the gates with three external capacitances $C_1 = C_2 = C_3 = 2$ pF as shown in shaded form causes f_{osc} to drop from 300 MHz to 100 MHz, estimate the net stray capacitance C_{stray} of each node as well as the average current supplied by each gate to charge/discharge C_{stray} during consecutive half periods.

  (*d*) Estimate the frequency of oscillation if the circuit drives an external 5-pF capacitive load connected to v_3. Compare with PSpice and comment.

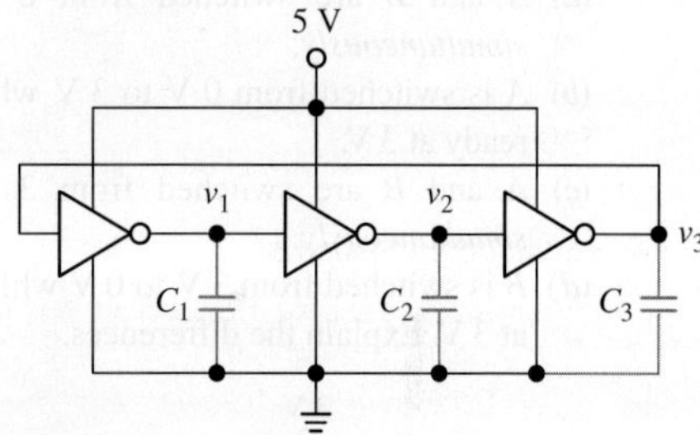

FIGURE P6.58

6.59 For the inverter of Fig. P6.59, known as a *pseudo CMOS inverter*, find V_{OH}, V_{OL}, t_{PLH} and t_{PHL} for $C_{eq} = 0.75$ pF. Assume $V_{tn} = -V_{tp} = 0.5$ V, $k_n = 6.25k_p = 500$ μA/V^2, and $\lambda_n = \lambda_p = 0$. Comment.

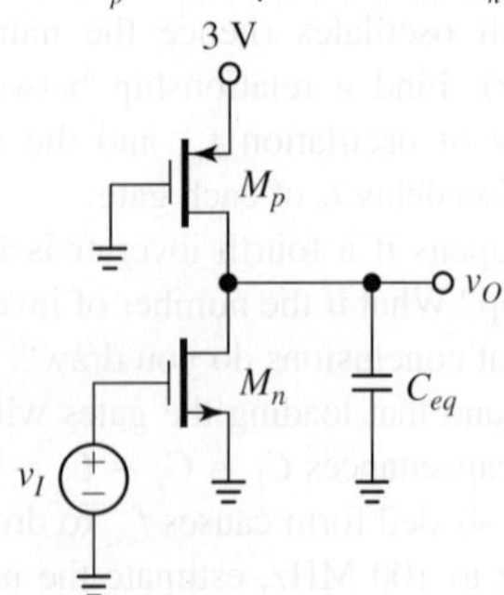

FIGURE P6.59

6.60 Let the FETs of the NAND gate of Fig. P6.60 have $V_{tn} = -V_{tp} = 0.6$ V, $k_n = 2.5k_p = 100$ μA/V^2, and $\lambda_n = \lambda_p = 0$. Assuming that all stray capacitances can be modeled with $C_{eq} = 1$ pF, estimate the propagation delays for the following cases:

(*a*) *A* and *B* are switched from 0 V to 3 V simultaneously.

(*b*) *A* is switched from 0 V to 3 V while *B* is already at 3 V.

(*c*) *A* and *B* are switched from 3 V to 0 V simultaneously.

(*d*) *B* is switched from 3 V to 0 V while *A* is kept at 3 V. Explain the differences.

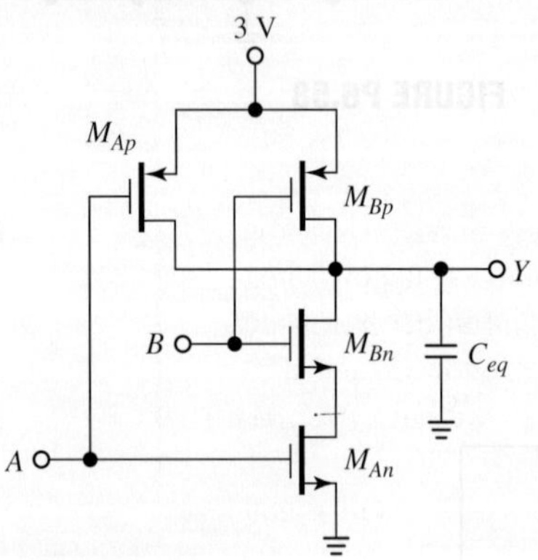

FIGURE P6.60

6.61 The biCMOS inverter of Fig. P6.61 takes advantage of the best of both technologies (high input impedance of MOSFETs and high current-drive capabilities of BJTs) to handle large capacitive loads. As v_I is switched to 0V, M_1 pulls Q_1's base to 5 V, giving $V_{OH} \cong 5 - 0.7 = 4.3$ V. At the same time, M_4 pulls Q_2's base to 0 V to rapidly turn it off. As v_I is switched to 5 V, M_2 pulls Q_1's base to 0 V to rapidly turn it off. At the same time, M_3 turns Q_2 on in Darlington fashion while also clamping its collector at $V_{OL} = V_{BE2} + V_{DS3} \cong 0.7 + 0 = 0.7$ V, thus preventing it from saturating. Assuming the FETs have $V_t = 1$ V and $k = 100$ μA/V^2, and the BJTs have $V_{BE(on)} = 0.7$ V and $\beta_F = 75$, estimate t_{PLH} and t_{PHL} for $C_L = 25$ pF.

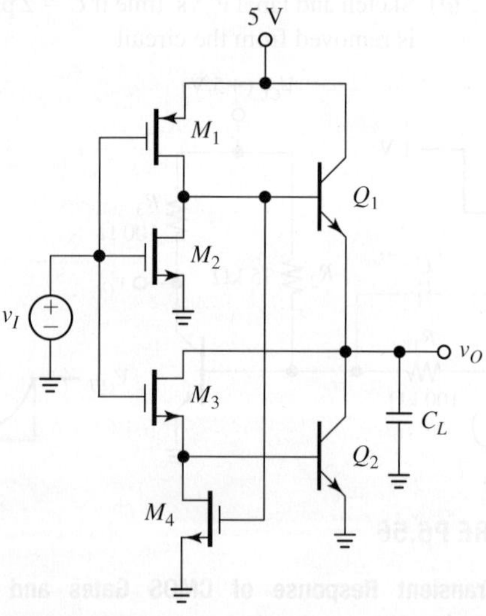

FIGURE P6.61

Feedback, Stability, and Noise

Chapter Outline

Feedback in electronics refers to the situation whereby a signal derived from the output port of an amplifier is returned to the input port, where it is combined with the externally applied input signal to create a new signal to be processed by the amplifier itself. The most common forms of combination are addition and subtraction. When the feedback signal is *added* to the external signal, we have *positive feedback*, and when it is *subtracted* we have *negative feedback*.

In positive feedback the returned signal is designed to *reinforce* the input signal in such a way as to deliberately drive the amplifier in saturation. Also referred to as *regenerative feedback*, it is used in the synthesis of digital circuits such as flip flops and Schmitt triggers.

In negative feedback the returned signal is designed to *oppose* (rather than reinforce) the input signal, this being the reason why it is also referred to as *degenerative feedback*. This type of feedback is far more interesting than positive feedback because of the many potential advantages it offers. First, it tends to *stabilize the gain* against parameter variations and drift of the components making up the amplifier itself. Second, it tends to *reduce distortion* as well as certain types of *noise*. Third, it can be used to *control* the *input* and *output resistances* in such a way as to reduce the unwanted effects of loading. Finally, it can be exploited to *control* the amplifier's *dynamics*, like extending the bandwidth (broadbanding) and speeding up its transient response.

Like many inventions, negative feedback comes with a price and a risk:

- As we move along we'll see that in order to fully realize the benefits available from this type of feedback we need to start out with a *much higher gain* than that finally demanded by the application at hand. (The student who has been exposed to the operational amplifier, the most popular amplifier type intended for negative-feedback operation, already knows this.) However, in today's integrated-circuit technology, high-gain amplifiers such as op amps are manufactured readily and inexpensively, so price is generally not much of an issue.

- Far more serious is the fact that negative feedback poses the *risk of oscillation*. As the signal propagates through the amplifier, it experiences unavoidable delays, collectively referred to as *phase lag*. If, by the time it returns to the input, the signal has acquired a shift of $-180°$, feedback turns from negative to positive. Moreover, if the signal is at least as strong as when it started out, feedback becomes *regenerative*, resulting in high-frequency oscillation. Though this effect is exploited on purpose in the design of oscillators, it is otherwise undesirable as it may render a circuit totally useless. Mercifully, a variety of techniques have been developed to tame unwanted oscillations. Generally known as *frequency compensation techniques*, they constitute one of the most fascinating aspects of systems theory as applied to electronics and control.

Negative feedback was conceived in 1928 by Harold Black in his quest to reduce distortion in telephone repeaters. For instance, when using a gain-of-ten voltage amplifier, we expect the circuit to respond to a given input v_I with the output $v_O = 10v_I$. In practice, due to the nonlinearities of the devices making up the amplifier (vacuum tubes then, transistors today), such a relationship holds only in small-signal operation. In large-signal operation an amplifier generally yields a much-distorted output, as we have already seen when transistor amplifiers are being overdriven as in Sections 2.5 and 3.6. We can model this situation by regarding the actual output as consisting of the *desired* component $10v_I$ plus an *unwanted* component (or *noise*) v_U,

$$v_O = 10v_I + v_U$$

In his attempt to reduce v_U, Harold Black reasoned that if (*a*) we take a *fraction* of the actual output equal to the *reciprocal* of the desired gain, or $(1/10)v_O$ in the present example, (*b*) we *subtract* this fraction from the input to create a *new signal* v_E (subsequently named *error signal*)

$$v_E = v_I - \frac{1}{10}v_O = v_I - \frac{10v_I + v_U}{10} = \frac{-v_U}{10}$$

(*c*) we feed this new signal v_E to the amplifier (now renamed *error amplifier*), and (*d*) we substantially *increase* the gain (now called *open-loop gain*) so that the amplifier can sustain v_O with a *vanishingly small* v_E, (or $v_E \to 0$), then also v_U $(= -10v_E)$ will be small, resulting in the virtual *elimination* of the distortion v_U from the output to give

$$v_O \to 10v_I$$

It is fascinating that such a terse and disarmingly simple line of reasoning would result in one of the most important inventions of electronics!

Like other revolutionary inventions, negative feedback wasn't immediately accepted by the engineering community because of the risk of oscillation that it posed. However, once this risk became better understood and suitable cures were developed to tame unwanted oscillations, negative feedback became a cornerstone not only in electronic circuit design, but also in such disparate disciplines as automatic control and modeling of biological systems. The student has already been exposed informally to negative feedback in a variety of different situations: op amp circuits use negative feedback; the feedback bias technique stabilizes dc biasing for transistors; emitter/source degeneration is a negative-feedback example designed to stabilize gain against transistor parameter variations. After our informal exposure to negative feedback, we are now ready to tackle it in a thorough and systematic fashion.

CHAPTER HIGHLIGHTS

The chapter begins with basic negative-feedback concepts and terminology, emphasizing the *loop gain* as a central parameter of a negative feedback system. The curative properties of feedback are illustrated in a variety of situations such as distortion reduction, noise reduction, and broadbanding.

Next, the chapter introduces the four basic negative-feedback topologies and discusses the effect of feedback on gain as well as the input and output resistances. Prerequisite courses have already exposed the student to negative feedback via operational amplifiers, if only informally. This is the right juncture to capitalize on basic op-amp background and expand it to illustrate the different feedback topologies in a more systematic fashion.

In real-life feedback circuits the basic amplifier and the feedback network tend to load each other, so we need suitable methods to investigate the various topologies in the presence of loading. The first such method, known as *two-port* analysis, is illustrated via a variety of circuit examples, ranging from full-blown op amps and current-feedback amplifiers, to multi-transistor configurations, to single-transistor stages (the student will finally be able to appreciate the stabilizing effect of already familiar single-transistor feedback schemes such as emitter/source degeneration and feedback bias).

A powerful alternative to two-port analysis, known as *return-ratio* analysis, is also illustrated in a variety of circuit examples. Directly related to this type of analysis are Blackman's impedance formula and injection methods, which are particularly useful in laboratory measurements and in the course of computer simulations.

The chapter progresses to the study of stability and frequency compensation techniques. After introducing graphical as well as experimental and computer tools to assess the stability of a negative-feedback circuit, the chapter investigates the internal frequency compensation of the most common op amps discussed in Chapter 5: the bipolar 741-type op amp, and the two-stage and cascode CMOS op amps.

The chapter concludes with noise in integrated circuits. After an introduction to basic noise properties, analytical tools, and noise types, the noise models of diodes and transistors are discussed. The chapter ends with the noise analysis of important circuit configurations such as op amp circuits and differential pairs, both bipolar and CMOS.

The chapter makes abundant use of PSpice both as a verification tool for paper-and-pencil calculations and as a software oscilloscope to display critical waveforms, especially when investigating the curative properties on distortion, the complex issues of stability and frequency compensation, or understanding the noise performance of a circuit.

7.1 NEGATIVE-FEEDBACK BASICS

Figure 7.1 shows the structure of a negative feedback system. Its main ingredients are an *error amplifier* and a *feedback network*. The system receives an external *input signal* s_i (which in an electronic circuit is typically a voltage or a current) and produces in turn an *output signal* s_o (again, a voltage or a current). The feedback network senses s_o to produce a *scaled* version of it, called the *feedback signal* s_f, such that

$$s_f = bs_o \tag{7.1}$$

where b is called the *feedback factor*. The feedback signal is then fed to an input *summer*, where it is *subtracted* from the input signal to produce a signal called the *error signal* s_ε,

$$s_\varepsilon = s_i - s_f \tag{7.2}$$

This, in turn, is fed to the error amplifier, thus closing a signal-propagation *loop* around the amplifier.

As implied by Eq. (7.2), the aim of *negative feedback* is to *reduce* the input signal s_i to a *smaller* signal s_ε. Were we to *add* (rather than subtract) s_f to s_i, then s_ε would be *greater* than s_i. After undergoing further magnification by the amplifier, the signal would return to the summer even greater, continuously feeding upon itself until the

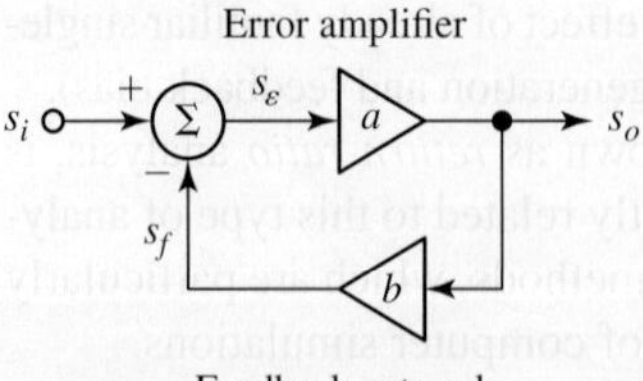

FIGURE 7.1 Block diagram of a negative-feedback circuit.

amplifier is ultimately driven in saturation. Aptly called *positive feedback*, this form of feedback is used in the synthesis of highly nonlinear circuits such as flip flops and Schmitt triggers. This chapter will deal with negative feedback only.

We now wish to obtain a relationship between the system's output s_o and input s_i. By definition, the error amplifier yields

$$s_o = as_\varepsilon \qquad (7.3)$$

where a is the amplifier's gain. Were we to break the feedback loop to make $s_f = 0$, then the amplifier would yield $s_o = as_i$, indicating that a is the gain by which s_i would be magnified in the absence of any feedback loop. Consequently, a is called the *open-loop gain*. Combining the above equations,

$$s_o = a(s_i - s_f) = a(s_i - bs_o)$$

Collecting and solving for s_o gives

$$s_o = As_i \qquad (7.4)$$

where

$$A = \frac{a}{1 + ab} \qquad (7.5)$$

is the gain by which the overall negative-feedback system amplifies the input s_i. Aptly called the *closed-loop gain*, A $(=s_o/s_i)$ should not be confused with the open loop gain a $(=s_o/s_\varepsilon)$. In fact, to underscore the distinction, we shall use *upper-case* letters to indicate *closed-loop* parameters, such as gain and (later) input and output resistances (A, R_i, R_o), and *lower-case* letters to indicate the parameters of the basic error amplifier, aptly called *open-loop* parameters (a, r_i, r_o).

As a signal propagates around the loop, starting, say, at the amplifier's input, it undergoes first magnification by a as it goes through the amplifier, then attenuation by b as it returns through the feed-back network, and finally inversion $(-)$ as it goes through the summer Σ. The overall gain around the loop is thus $-ab$. The *negative* of this overall gain is (somewhat improperly) called the *loop gain L*,

$$L = ab \qquad (7.6)$$

As we shall see, L plays a central role in a feedback system. Manipulating Eq. (7.5) as

$$A = \frac{1}{b}\frac{ab}{1 + ab} = \frac{1}{b}\frac{1}{1 + 1/(ab)}$$

allows us to express the closed-loop gain in the insightful alternative form

$$A = \frac{1}{b} \times \frac{1}{1 + 1/L} \qquad (7.7)$$

Of particular interest is the condition $L \gg 1$, for then Eq. (7.7) can be approximated as

$$A \cong \frac{1}{b}\left(1 - \frac{1}{L}\right) \cong \frac{1}{b} \tag{7.8}$$

This result alone underscores two of the most important benefits accruing from the use of negative feedback under the condition $L \gg 1$, namely:

- The closed-loop gain A is virtually independent of the open loop gain a. This is highly desirable as the open-loop gain a is usually an ill-defined parameter that depends on the parameters of the transistors making up the amplifier. As we know, these parameters vary with the dc biasing conditions, drift with temperature and time, and vary from device to device due to fabrication process variations.
- We can tailor A to a wide variety of applications by a suitable choice of the feedback network. This network is usually implemented with passive components such as resistors and capacitors. By using components of adequate quality, we can make A as predictable, accurate, and stable as needed.

If we regard $1/L$ as an error term in Eq. (7.8), it is apparent that L gives a measure of how close the *actual* gain A is to the *ideal* gain $1/b$. Specifically, the larger L the better. Since $L = ab$, it follows that to ensure a suitably large L for a given b we need an amplifier with an *adequately high gain a*. In other words, we need to start out with a high open-loop gain a to achieve a much lower but much more stable and predictable closed-loop gain A. Since gain drops from a to $a/(1 + L)$, we are in affect *throwing away gain* by the *amount of feedback* $(1 + L)$. Considering the benefits as well as the fact that modern integrated circuit (IC) technology allows for high gains to be achieved readily and inexpensively, this price is well worth paying.

As we move along, we shall refer to the limit $L \rightarrow \infty$ as representing the *ideal situation*. The corresponding closed-loop gain is then

$$A_{\text{ideal}} = \lim_{T \to \infty} A = \frac{1}{b} \tag{7.9}$$

Though the ideal condition is physically unattainable, a circuit designer will strive to approach it within a specified degree of accuracy by ensuring an adequately high loop gain L, and hence, by using an amplifier with a correspondingly high open-loop gain a $(=L/b)$.

EXAMPLE 7.1   **(a)** An engineer is asked to design a voltage amplifier having a closed-loop gain of 10 V/V with an error of 1% or less. What values of a and b are needed? What is the resulting value of A?

    (b) To be on the safe side, the engineer decides to use an amplifier with a gain a ten times as large as that calculated in part (a). What is the resulting value of A?

Solution

(a) Impose 10 V/V = $1/b$, or $b = 0.1$ V/V. For a 1% error we need $1/L = 1/100$, so $a = L/b = 100/0.1 = 1000$ V/V. Moreover, $A \cong (1/b) \times (1 - 1/L) = 10(1 - 1/100) = 9.9$ V/V.

(b) Now $L = 1000$, so $A \cong 10(1 - 1/1000) = 9.99$ V/V, even closer to the ideal value of 10 V/V.

The Error Signal s_ε and the Feedback Signal s_f

Additional properties of negative feedback are readily found by writing

$$s_\varepsilon = \frac{s_o}{a} = \frac{As_i}{a} = \frac{a}{1+L}\frac{s_i}{a}$$

or

$$s_\varepsilon = \frac{s_i}{1+L} \tag{7.10}$$

Moreover,

$$s_f = bs_o = bAs_i = b\frac{1}{b}\frac{1}{1+1/L}s_i$$

which gives

$$s_f = \frac{s_i}{1+1/L} \tag{7.11}$$

These (equivalent) results indicate that for a *sufficiently large loop gain* (ideally, for $L \to \infty$), the error signal becomes vanishingly *small* (ideally, $s_\varepsilon \to 0$), causing the feedback signal to closely *follow* the input signal ($s_f \to s_i$). These properties are worth keeping in mind as we attempt to develop a quick (if approximate) feel for the inner workings of a negative-feedback circuit.

Gain Desensitivity

Given that the open-loop gain a is an ill-defined parameter because of production variations as well as environmental changes, we wish to investigate the impact of these uncertainties upon the closed-loop gain A. To this end, let us differentiate A with respect to a in Eq. (7.5),

$$\frac{dA}{da} = \frac{1 \times (1+ab) - a \times b}{(1+ab)^2} = \frac{1}{(1+ab)^2} = \frac{1}{1+ab} \times \frac{a}{1+ab}\frac{1}{a} = \frac{1}{1+ab} \times \frac{A}{a}$$

Multiplying both sides by $100\,da/A$ and replacing differentials (d) with small differences (Δ), we obtain

$$100\frac{\Delta A}{A} \cong \frac{1}{1+L}\left(100\frac{\Delta a}{a}\right) \qquad (7.12)$$

This result indicates that the *percentage variation* in the closed-loop gain ($100 \times \Delta A/A$) due to a given percentage variation in the open-loop gain ($100 \times \Delta A/A$) is approximately ($1 + L$) times as small. With L sufficiently large, even an outlandish variation in a will have a minimal effect upon A! To reflect this stabilizing effect, the amount of feedback ($1 + L$) is also called the *gain desensitivity*. Once again we observe that the size of L offers a measure of how close a negative-feedback system is to ideal.

EXAMPLE 7.2   Suppose the open-loop gain a of the amplifier of Example 7.1a has a tolerance of $\pm 20\%$. Estimate the tolerance of the closed-loop gain A. Repeat, but for Example 7.1b, and comment on your findings.

Solution

In Example 7.1a we have $L = 100$, so the approximate tolerance of A is $(\pm 20\%)/(1 + 100) \cong \pm 0.2\%$. In Example 7.1b, L is ten times as large, so the tolerance of A will be about ten times as small, or $\pm 0.02\%$. In either case negative feedback has a dramatic stabilizing effect upon the closed-loop gain A.

A Classic Example: The Non-Inverting Op Amp Configuration

A circuit example conforming exactly to the diagram of Fig. 7.1, and thus embodying all the features discussed so far, is the familiar non-inverting op amp configuration of Fig. 7.2. The op amp combines the roles of error amplifier as well as summer, the latter thanks to the fact that the op amp responds to the *difference* between its input voltages. The feedback network is a plain *voltage divider*, giving

$$b = \frac{v_f}{v_o} = \frac{R_1}{R_1 + R_2} = \frac{1}{1 + R_2/R_1} \qquad (7.13)$$

Op amps are deliberately designed to have very high open-loop gains so as to ensure high loop gains, and hence, nearly ideal behavior in negative-feedback operation. In the *ideal* limit $a \to \infty$, the circuit would give $L \to \infty$ and thus $v_\varepsilon \to 0$ and $v_f \to v_i$. The closed-loop gain would then take on the *ideal* value

$$A_{\text{ideal}} = \frac{v_o}{v_i} = \frac{1}{b} = 1 + \frac{R_2}{R_1} \qquad (7.14)$$

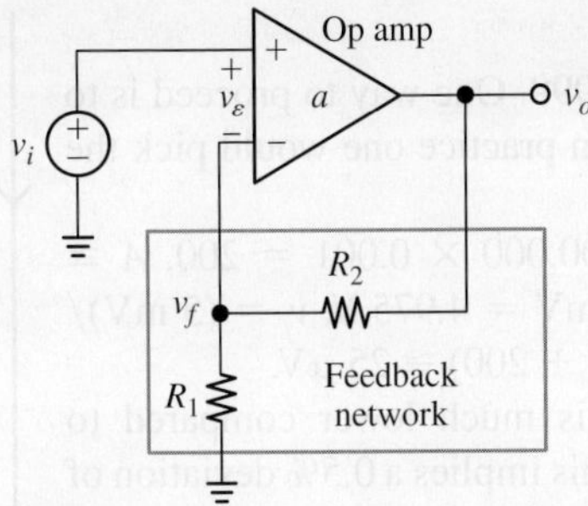

FIGURE 7.2 The non-inverting op amp circuit as a classic example of a negative-feedback system.

EXAMPLE 7.3

(a) Let the op amp of Fig. 7.2 be the popular 741-type, whose data sheets report the typical gain $a = 200{,}000$ V/V. Find the closed-loop gain if $R_1 = 1.0$ kΩ and $R_2 = 3.0$ kΩ.

(b) Find v_o, v_f, and v_ε if $v_i = 2.0$ V. Comment on your results.

(c) The data sheets also report that because of fabrication process variations, the gain a can be as low as 50,000 V/V. How does this impact the results found in part (a)? Comment.

Solution

(a) We have $b = 1/(1 + 3) = 1/4$, $L = ab = 200{,}000/4 = 50{,}000$, and $A \cong 4(1 - 1/50{,}000) = 3.99992$ V/V. Thanks to the high loop gain, A is very close to $A_{\text{ideal}}(= 4.0$ V/V$)$.

(b) We have $v_o = Av_i$, $= 3.99992 \times 2.0 = 7.99984$ V, $v_f = v_i/(1 + 1/L) = 2.0/(1 + 1/50{,}000) = 1.99996$ V, and $v_\varepsilon = v_i/(1 + L) \cong 2.0/50{,}000 = 40 \times 10^{-6}$ V $= 40$ μV. For practical purposes, we can state that $v_o \cong 8$ V, $v_f \cong 2$ V, and $v_\varepsilon \cong 0$. We observe that just as the voltage divider *divides down* v_o by 4 to give v_f, the op amp performs the inverse operation, namely, it *multiplies up* v_i by 4 to give v_o.

(c) We now have $L = ab = 50{,}000/4 = 12{,}500$, so $A \cong 4(1 - 1/12{,}500) = 3.99968$ V/V. The change in A is insignificant (-0.006%), and so are the changes in v_o and v_f, both of which continue to be extremely close to their ideal values of 8.0 V and 2.0 V, respectively. However, due to the drop in a, we now have $v_\varepsilon = v_o/a \cong 8/50{,}000 = 160$ μV, higher than in part (b) but still truly negligible compared to the other voltages in the circuit.

EXAMPLE 7.4

(a) Suitably modify the circuit of Fig. 7.2 so that it amplifies a transducer signal $v_i = 5$ mV with a closed loop gain of 1000 V/V.

(b) Assuming a 741-type op amp, estimate A, v_o, v_f, and v_ε.

(c) Compare with Example 7.3a and comment.

Solution

(a) Imposing $1000 = 1 + R_2/R_1$ we get $R_2/R_1 = 999$. One way to proceed is to leave $R_1 = 1.0\ \text{k}\Omega$ and make $R_2 = 999\ \text{k}\Omega$. (In practice one would pick the closest standard value of $1.0\ \text{M}\Omega$.)

(b) We now have $b \cong 0.001\ \text{V/V}$, $L = ab \cong 200{,}000 \times 0.001 = 200$, $A \cong 1000(1 - 1/200) = 995\ \text{V/V}$, $v_o = 995 \times 5\ \text{mV} = 4.975\ \text{V}$, $v_f = (5\ \text{mV})/(1 + 1/200) = 4.975\ \text{mV}$, and $v_\varepsilon \cong (5\ \text{mV})/(1 + 200) \cong 25\ \mu\text{V}$.

(c) Due to the much higher gain A sought, b is much lower compared to Example 7.3a, so the loop gain drops to 200. This implies a 0.5% deviation of A and v_f from their ideal values—still a fairly small deviation. The op amp's input is always $v_\varepsilon = v_o/a$, that is, the departure of v_ε from its ideal value of 0 V depends only on v_o and a, regardless of the loop gain L.

A Single-Transistor Example of a Negative-Feedback System

If Fig. 7.2 illustrates feedback around a complex circuit such as an op amp, consisting of *many* transistors, Fig. 7.3 depicts the opposite extreme of feedback around just *one* transistor. The latter is a CG amplifier utilizing the voltage divider R_1-R_2 as its feedback network. As long as $(R_1 + R_2) \gg R_D$, we can write

$$v_o \cong -g_m(R_D//r_o)v_{gs} = -g_m(R_D//r_o) \times (v_g - v_s) = g_m(R_D//r_o) \times (v_s - v_g)$$

where the body effect has been ignored. This expression is of the type

$$v_o = a(v_i - v_f) = a(v_i - bv_o)$$

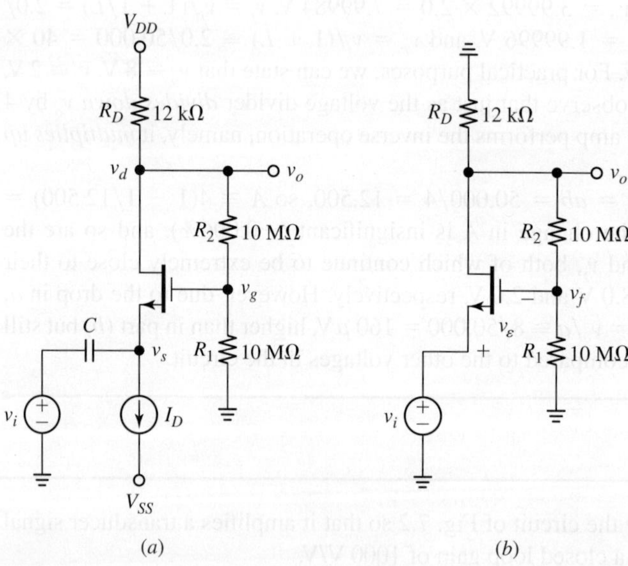

FIGURE 7.3 (a) A single-transistor circuit as an example of a negative-feedback system, and (b) its ac equivalent.

provided we let $v_s \to v_i$, $v_g \to v_f$, and $v_{sg} \to v_\varepsilon$, as depicted in Fig. 7.3b. Also, the open-loop gain is $a = g_m(R_D//r_o)$, and the feedback factor is

$$b = \frac{v_g}{v_d} = \frac{R_2}{R_1 + R_2} = \frac{1}{1 + R_2/R_1}$$

Just like the op amp example above, the present circuit conforms exactly to the diagram of Fig. 7.1.

Let the FET of Fig. 7.3 have $g_m = 2$ mA/V and $r_o = 60$ kΩ. Estimate L and A, and comment.   **EXAMPLE 7.5**

Solution

We have $a = g_m(R_D//r_o) = 2(12//60) = 20$ V/V, $b = 1/(1 + 10/10) = 1/2$, $L = ab = 20/2 = 10$, and

$$A = \frac{1}{b}\frac{1}{1 + 1/L} = 2\frac{1}{1 + 1/10} = 1.82 \text{ V/V}$$

Given the notoriously low voltage gains achievable with FETs, it is not surprising that the loop gain is so low compared to that of an op amp, resulting in a noticeable departure of A from its ideal value of 2 V/V. Yet, it is instructive to investigate the given single-transistor circuit from a negative-feedback standpoint!

Exercise 7.1

As we move along we shall find that negative feedback affects not only the gain, but also the input and output resistances. Using the test method, show that for $R_1 + R_2 \gg R_D$ in Fig. 7.3, the resistance R_i seen looking into the input terminal and the resistance R_o seen looking into the output terminal are

$$R_i = \frac{1}{g_m}(1 + L) \qquad R_o = \frac{R_D//r_o}{1 + L}$$

7.2 EFFECT OF FEEDBACK ON DISTORTION, NOISE, AND BANDWIDTH

Equation (7.3) implies a relationship of *linear proportionality* between output and input, with the proportionality constant representing the open-loop gain a. A practical amplifier, such as an IC op amp, is made up of transistors, which are inherently nonlinear devices. Moreover, the amplifier cannot swing its output beyond its own power-supply voltages. Consequently, the voltage transfer curve (VTC) of a practical amplifier is not a straight line, but a *nonlinear curve* of the type exemplified in Fig. 7.4a (top). As long as operation is restricted in the vicinity of the origin, the curve can be regarded as approximately linear, and its slope, representing the gain a, is the steepest there. However, as we move away from the origin, slope progressively decreases until the VTC

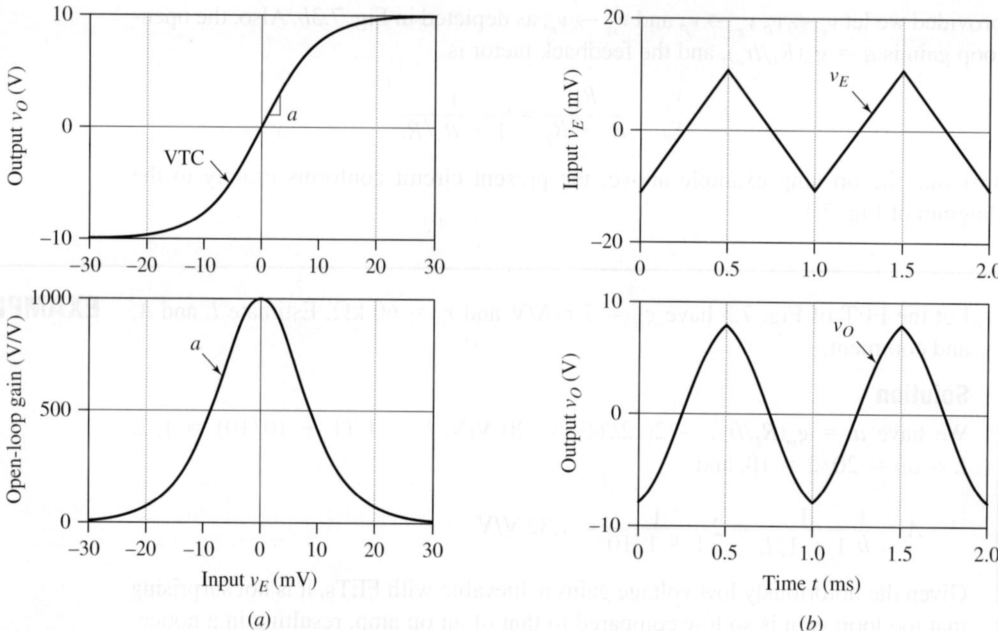

FIGURE 7.4 (a) The voltage transfer curve (VTC) of a practical error amplifier (*top*), and its slope (*bottom*), representing the open-loop gain a. (b) Input (*top*) and output (*bottom*) waveforms.

eventually flattens out (or saturates), making the gain a drop to zero. (In the example shown v_O saturates at ± 10 V.) The open loop gain is now more aptly defined as

$$a = \frac{dv_O}{dv_E} \qquad (7.15)$$

where v_O is the instantaneous voltage at the output and v_E the instantaneous error voltage at the input. As shown in Fig. 7.4a (bottom), the gain a reaches its maximum of 1000 V/V at the origin, progressively decreasing away from the origin, and finally dropping to zero as the amplifier is driven in saturation.

Because of its nonlinear VTC, a practical amplifier will generally yield a *distorted* output. This is depicted in Fig. 7.4b for the case of a triangular wave at the input (top). The output (bottom) can be regarded as a magnified version of the input, but with significantly compressed peaks because of the decreased gain there. What are we to do with a nonlinear device of this sort in demanding applications such as high-fidelity (hi-fi) audio or precision instrumentation, where distortion is intolerable? As we shall see next, this is another instance where negative feedback comes to our rescue.

Once a nonlinear amplifier is placed inside a negative-feedback loop, a closed-loop VTC results, whose slope represents the *closed loop gain A*,

$$A = \frac{dv_O}{dv_I} \qquad (7.16)$$

where v_O and v_I are the instantaneous output and input voltages. Rewriting Eq. (7.15) as

$$\frac{1}{a} = \left[\frac{dv_O}{d(v_I - v_F)} \right]^{-1} = \frac{dv_I - d(bv_O)}{dv_O} = \frac{1}{A} - b$$

and rearranging, we get the familiar result

$$A = \frac{a}{1 + ab} = A_{\text{ideal}} \times \frac{1}{1 + 1/L} \tag{7.17}$$

where $A_{\text{ideal}} = 1/b$, and $L = ab$ is the familiar *loop gain*. This indicates that as long as the open-loop gain a is sufficiently large to ensure an adequately *high loop gain L*, the closed-loop gain A will be fairly close to A_{ideal}, even though the gain a decreases as we move away from the origin. Consequently, negative feedback can *linearize* the VTC of an amplifier dramatically!

To illustrate, consider the PSpice circuit of Fig. 7.5, where an error amplifier with the nonlinear VTC of Fig. 7.4a is placed inside a negative-feedback loop with $b = R_1/(R_1 + R_2) = \frac{1}{4}$. The linearizing properties of negative feedback are demonstrated in Fig. 7.6. Compared to the open-loop VTC of Fig. 7.4a, the closed-loop VTC of Fig. 7.6a is far *more linear*, and the gain A is close to its ideal value of $1/b$ ($= 4$ V/V in this example) over a much *wider range* of output voltages. So long as we restrict circuit operation within this range, the output will be a faithful (magnified-by-four) replica of the input. This is shown in Fig. 7.6b (top) for the case in which v_I is a triangular waveform with peak values of ± 2 V, so v_O is a fairly undistorted triangular wave with peak values of ± 8 V. Even more revealing is the error signal v_E displayed in Fig. 7.6b (bottom), because it shows the effort required of the error amplifier to make $v_O = 4v_I$. It is apparent that in order to compensate for the decrease in its open-loop gain away from the origin, the amplifier suitably pre-distorts its own error signal! Historically, it was precisely to reduce output distortion that Harold Black conceived negative feedback in the first place.

EVALUE 10*((exp(2E2*V(%IN+, %IN−))−1)/(exp(2E2*V(%IN+, %IN−))+1))

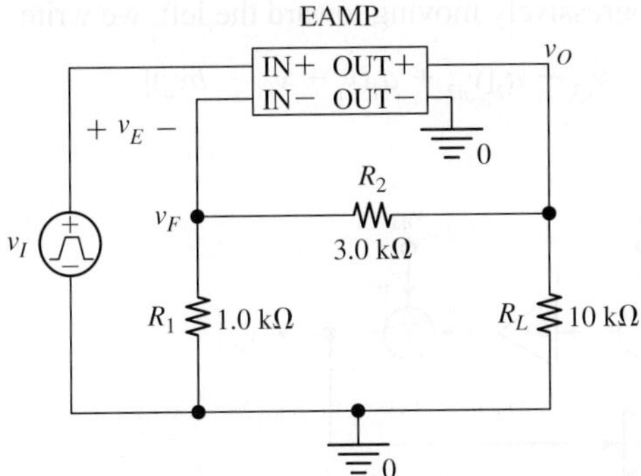

FIGURE 7.5 PSpice circuit to display the waveforms of a negative-feedback system utilizing an error amplifier with the nonlinear VTC of Fig. 7.4a.

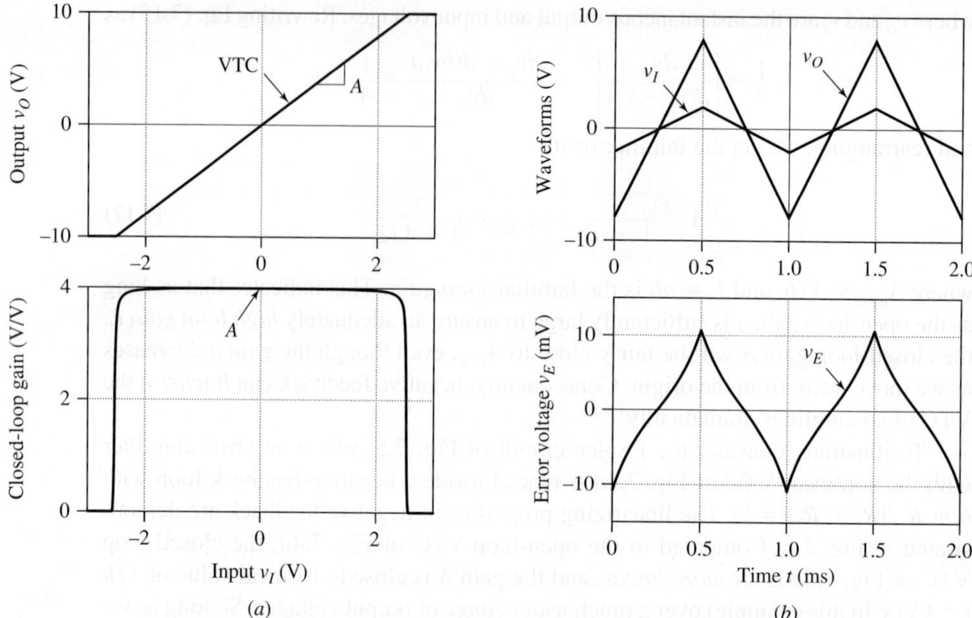

FIGURE 7.6 Illustrating the linearizing properties of negative feedback: (*a*) the closed-loop VTC (*top*) and its slope (*bottom*), representing the closed-loop gain *A*; (*b*) the input and output waveforms (*top*) and the error waveform (*bottom*).

Effect of Negative Feedback on Noise

We now wish to investigate the effect of negative feedback upon disturbances. Henceforth referred to as *noise*, disturbances may enter the amplifier at the input node (v_{n1}), at some intermediate node (v_{n2}), or at the output node (v_{n3}). As depicted in Fig. 7.7, we use summers to model the points of entry of the various noise components. Moreover, to model the entry of intermediate noise, we split the amplifier into two stages with individual gains a_1 and a_2, respectively (clearly, the overall gain is $a = a_1 \times a_2$). Starting out at the right and progressively moving toward the left, we write

$$v_o = v_{n3} + a_2[v_{n2} + a_1(v_i + v_{n1} - bv_o)]$$

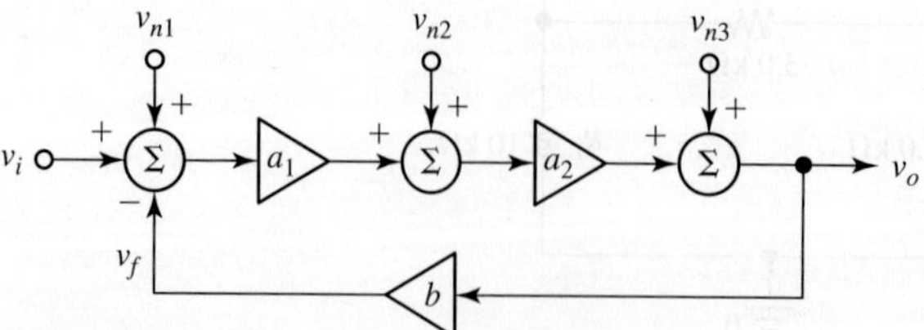

FIGURE 7.7 Model to investigate the effect of negative feedback upon voltage noise.

where we have used $v_f = bv_o$. Collecting, letting $a = a_1 \times a_2$, and solving for v_o, we get, after minor algebra,

$$v_o = A\left(v_i + v_{n1} + \frac{v_{n2}}{a_1} + \frac{v_{n3}}{a_1 \times a_2}\right) \qquad (7.18)$$

where A is the closed-loop gain of Eq. (7.17). We observe that a negative-feedback circuit amplifies all three noise terms with the same gain A as the useful signal v_i. However, while v_{n1} is unchanged, v_{n2} is *divided* by a_1, and v_{n3} is *divided* by $a_1 \times a_2$. We summarize this by saying that in a negative-feedback circuit a noise component, *reflected* to the input, gets *divided* by the gain(s) of the stage(s) *preceding* it. This property is often exploited to reduce the effect of a given noise source, such as hum creeping into the power stage of an audio system. If we precede this stage by an additional amplifier with suitably high gain, and we close a negative-feedback loop around the composite circuit, we can make the hum component, reflected to the input, as small as desired compared to the audio signal v_i, thus raising the *signal-to-noise ratio* to an acceptable level.

As a dramatic demonstration of the curative properties of negative feedback upon noise, consider the situation of Fig. 7.8a, where a source v_I is buffered to a load R_L via the Class AB push-pull stage Q_1-Q_2. Ideally, the buffer should give

$$v_{O(ideal)} = v_I$$

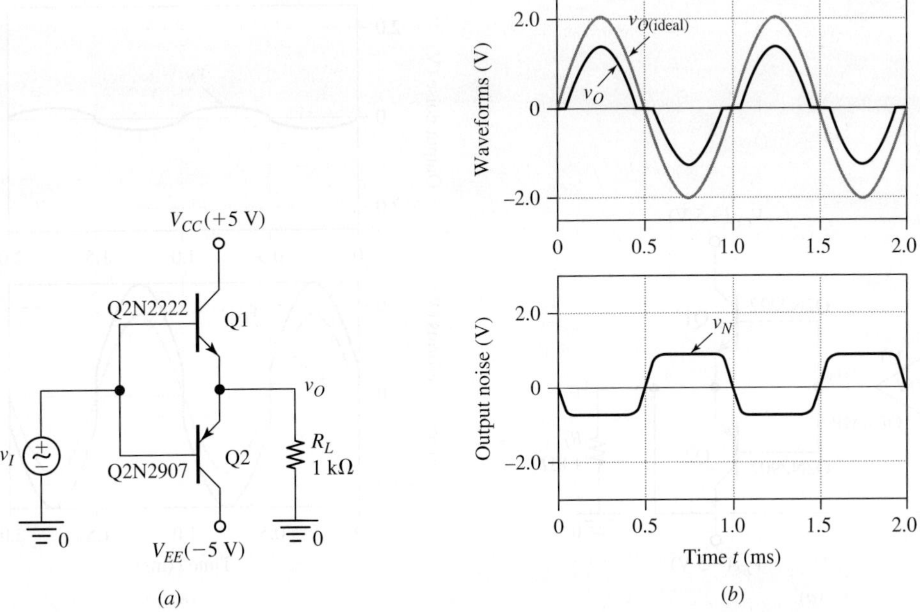

FIGURE 7.8 (a) PSpice circuit simulating a source v_I driving a load R_L via a push-pull stage Q_1-Q_2. (b) Input and output waveforms (*top*), and output noise $v_N = v_O - v_{O(ideal)}$ (*bottom*).

However, since it takes about 0.7 V for each BJT to turn on, the circuit gives $v_O = 0$ for $-0.7\ \text{V} < v_I < 0.7\ \text{V}$, $v_O \cong v_I - 0.7\ \text{V}$ for $v_I > 0.7\ \text{V}$, and $v_O \cong v_I + 0.7\ \text{V}$ for $v_I < 0.7\ \text{V}$. The waveforms, shown in Fig. 7.8b (top), reveal a much distorted output. In fact, things go as if the non-linearity of the push-pull stage resulted in the injection of the voltage noise

$$v_N = v_O - v_{O(\text{ideal})}$$

depicted in Fig. 7.8b (bottom). We can reduce v_N by preceding the buffer by a suit-able error amplifier and then closing a negative-feedback loop around the composite circuit. The example depicted in Fig. 7.9a uses an amplifier with gain of only $a = 10\ \text{V/V}$, and a plain wire as the feedback network to yield $b = 1\ \text{V/V}$. The ensuing reduction in distortion can be appreciated by comparing the waveforms of Fig. 7.9b (top and middle) with their counterparts of Fig. 7.8b (top and bottom). One can read-ily see that the insertion of the amplifier makes the 0.7-voltage drops appear *as if* they were reduced to about $(0.7\ \text{V})/a$, or 0.07 V in our example.

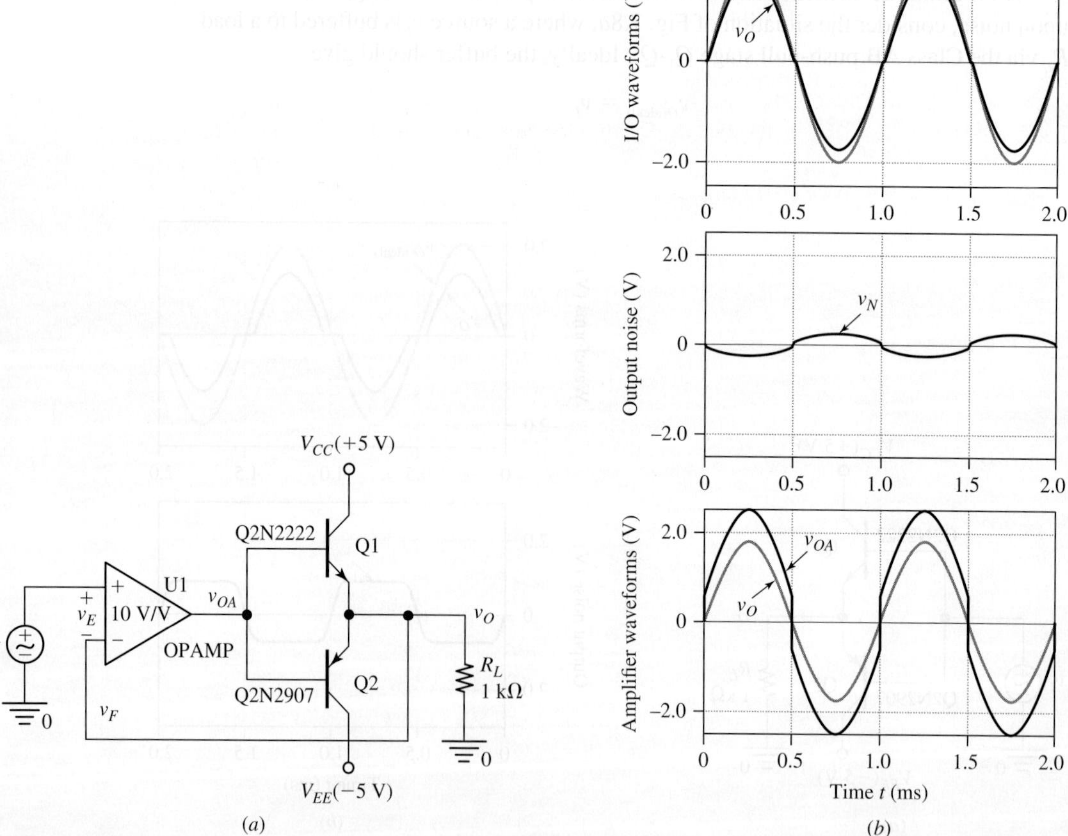

(a)

(b)

FIGURE 7.9 (a) Preceding the push-pull stage of Fig. 7.8a by an amplifier, and closing a negative-feedback loop around the whole circuit in order to reduce output distortion. (b) Input and output waveforms (*top*), output noise $v_N = v_O - v_{O(\text{ideal})}$ (*middle*), and the op amp output waveform v_{OA} (*bottom*).

To fully appreciate the role of the amplifier, it is instructive to display also its output v_{OA}, which is shown in Fig. 7.9b (bottom). In its attempt to make v_O follow v_I, the amplifier will have to swing its output v_{OA} about 0.7 V *above* v_O during the *positive* alternations, and 0.7 V *below* v_O during the *negative* alternations. In this example we have used an amplifier with a gain of only 10 V/V, but a unit with much higher gain will reduce distortion in proportion, making v_O a much more faithful replica of v_I. Recall that we have already observed this behavior in connection with the *superdiode* circuit of Section 1.10. Now we are simply revisiting an already-familiar concept, but from a negative-feedback perspective.

Remark: The goal of the negative-feedback circuit example of Fig. 7.9a is to implement an amplifier with gain $A = 1$ V/V and reduced distortion. To this end, we insert an error amplifier with gain $a = 10$ V/V. In so doing we are in effect throwing away a gain-of-ten to achieve only a gain-of-one, but this price is well worth the ensuing reduction in distortion. Needless to say, it is important to distinguish between the *basic error amplifier* and the *overall amplifier* resulting from operating the former in conjunction with the negative feedback network (a mere wire in the present example).

Effect of Negative Feedback on the Frequency Bandwidth

Negative feedback has a profound effect also on the frequency response. In fact, if used carelessly, it may lead to unwanted oscillation, in which case suitable measures need to be taken to stabilize the system. Though the subject of stability will be treated in detail later in this chapter, here we examine the effect of feedback upon two op amp representatives, the voltage-feedback amplifier (VFA) and the current-feedback amplifier (CFA).

Let us start out with the noninverting VFA configuration of Fig. 7.2, repeated in Fig. 7.10a for convenience. As seen in Chapter 6, the *open-loop gain* of the VFA is of the type

$$a(jf) \cong \frac{a_0}{1 + jf/f_b} \tag{7.19}$$

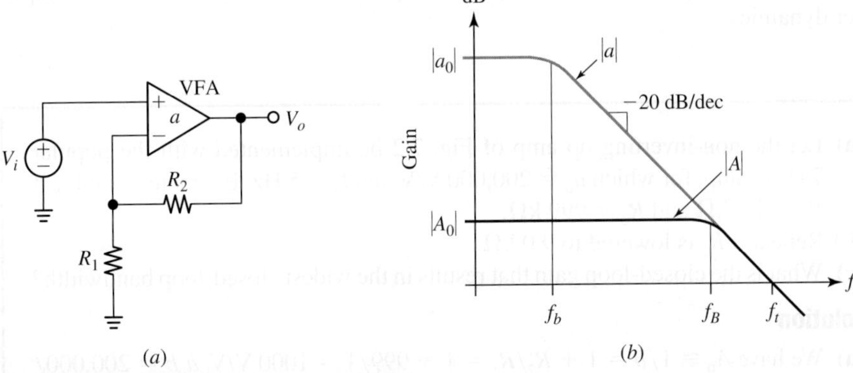

FIGURE 7.10 (a) Noninverting VFA configuration. (b) Visualizing the open-loop response $|a|$ and the closed-loop response $|A|$, both in dB.

where f is the input-signal frequency, a_0 is the *dc gain*, and f_b is the *open-loop bandwidth*. As depicted in Fig. 7.10b, the gain a is high ($\cong a_0$) from dc up to f_b, after which it rolls off with f at the rate of -20 dB/decade. As we know, an important figure of merit of such an amplifier is its *gain-bandwidth product*

$$GBP = a_0 \times f_b = f_t \tag{7.20}$$

For instance, the popular 741 op amp has $a_0 = 200{,}000$ V/V and $f_b = 5$ Hz, so $GBP = 200{,}000 \times 5 = 1$ MHz. The *closed-loop gain* is, by Eqs. (7.6) and (7.7),

$$A(jf) = \frac{V_o}{V_i} = \frac{1}{b}\frac{1}{1 + \dfrac{1 + jf/f_b}{ba_0}} = \frac{1}{b} \times \frac{1}{1 + 1/(ba_0)} \times \frac{1}{1 + \dfrac{jf}{(1 + ba_0)f_b}}$$

where $b = 1/(1 + R_2/R_1)$, by Eq. (7.13). This is put in the more insightful form

$$A(jf) = \frac{A_0}{1 + jf/f_B} \tag{7.21}$$

where

$$A_0 = \left(1 + \frac{R_2}{R_1}\right) \times \frac{1}{1 + 1/(ba_0)} \cong 1 + \frac{R_2}{R_1} \tag{7.22}$$

is the already familiar *closed-loop dc gain*, and

$$f_B = (1 + ba_0)f_b \cong \frac{a_0}{A_0}f_b = \frac{f_t}{A_0} \tag{7.23}$$

is the *closed-loop bandwidth*. (Again, note the use of *lower-case* letters to designate *open-loop* parameters, and *upper-case* letters to designate their *closed-loop* counterparts.) With reference to Fig. 7.10b we observe that negative feedback, while *reducing* the dc gain from a_0 to $A_0 \cong 1 + R_2/R_1$, also *expands* the bandwidth from f_b to $f_B \cong (a_0/A_0)f_b$, so the closed-loop GBP ($= A_0 \times f_B = f_t$) remains constant. *Gain-bandwidth tradeoff* is exploited on purpose by the circuit designer to control amplifier dynamics.

EXAMPLE 7.6   (a) Let the non-inverting op amp of Fig. 7.2 be implemented with the popular 741 op amp, for which $a_0 = 200{,}000$ V/V and $f_b = 5$ Hz. Estimate A_0 and f_B if $R_1 = 1.0$ kΩ and $R_2 = 999$ kΩ.

(b) Repeat if R_2 is lowered to 9.0 kΩ.

(c) What is the closed-loop gain that results in the widest closed-loop bandwidth?

Solution

(a) We have $A_0 \cong 1/b = 1 + R_2/R_1 = 1 + 999/1 = 1000$ V/V, $a_0b = 200{,}000/1000 = 200$, and $f_B = (1 + a_0b)f_b = (1 + 200)5 \cong 1.0$ kHz.

(b) We now have $A_0 \cong 10$ V/V, $a_0 b = 20{,}000$, and $f_B \cong 100$ kHz. Compared to part (a), A_0 has been reduced by two decades while f_B has been increased by two decades.

(c) The widest bandwidth is achieved with $b = 1$, that is, when we configure the op amp as a *unity-gain voltage follower* by replacing R_2 with a wire and removing R_1 altogether. Then, $A_0 \cong 1$ V/V, $a_0 b = a_0 \times 1 = a_0$, and $f_B = (1 + a_0)f_b \cong a_0 f_b = f_t = 1$ MHz.

Next, let us turn to the CFA amplifier of Fig. 7.11*a*. Recall from Eq. (6.114) that the CFA's *open loop transimpedance gain* is of the type

$$z(jf) \cong \frac{R_{eq}}{1 + jf/f_b} \tag{7.24}$$

where f is the input-signal frequency, R_{eq} is the dc gain, and f_b is the open-loop bandwidth (the magnitude plot of z is repeated in Fig. 7.11*b* for convenience). To find the closed-loop gain $A(jf)$, we adapt the expression for A derived in Section 5.6, but with R_{eq} replaced by $z(jf)$. The result is

$$A(jf) = \frac{V_o}{V_i} = \left(1 + \frac{R_2}{R_1}\right) \times \frac{1}{1 + \dfrac{R_2(1 + jf/f_b)}{R_{eq}}}$$

$$= \left(1 + \frac{R_2}{R_1}\right) \times \frac{1}{1 + R_2/R_{eq}} \times \frac{1}{1 + \dfrac{jf}{(1 + R_{eq}/R_2)f_b}}$$

We put this in the more insightful form

$$A(jf) = \frac{A_0}{1 + jf/f_B} \tag{7.25}$$

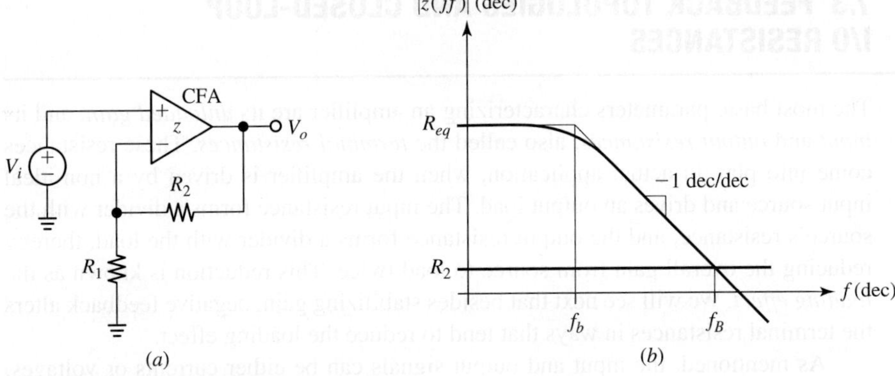

FIGURE 7.11 (a) Noninverting CFA configuration. (b) Visualizing the open-loop response $|z|$ and the closed-loop bandwidth f_B.

where

$$A_0 = \left(1 + \frac{R_2}{R_1}\right) \times \frac{1}{1 + R_2/R_{eq}} \cong 1 + \frac{R_2}{R_1} \qquad (7.26)$$

is the already familiar *closed-loop dc gain*, and

$$f_B = \left(1 + \frac{R_{eq}}{R_2}\right) f_b \cong \frac{R_{eq}}{R_2} \times \frac{1}{2\pi R_{eq} C_{eq}} = \frac{1}{2\pi R_2 C_{eq}} \qquad (7.27)$$

is the *closed-loop bandwidth* (both approximations exploit the fact that in a well-designed circuit we have $R_2 \ll R_{eq}$). Using simple geometry we visualize f_B as the frequency at which $|z|$ drops to R_2, as illustrated in Fig. 7.11b. This frequency is established by the user via R_2 regardless of the closed-loop gain A_0, which is set separately via R_1. Consequently, CFA circuits are *not* subject to the gain-bandwidth tradeoff of their VFA counterparts. Along with the absence of slew-rate limiting, this is a fundamental advantage of CFAs compared to VFAs. (See also Problem 7.10 for higher-order effects.)

EXAMPLE 7.7   Suppose the CFA of Fig. 7.11 has $R_{eq} = 750$ kΩ and $C_{eq} = 2.21$ pF. If the data sheets recommend using $R_2 = 1.2$ kΩ, specify R_1 for a closed-loop dc gain of 10 V/V. What is the closed-loop bandwidth?

Solution

Imposing $1 + 1,200/R_1 = 10$ gives $R_1 = 133.3$ Ω. Moreover,

$$f_B \cong \frac{1}{2\pi R_2 C_{eq}} = \frac{1}{2\pi \times 1200 \times 2.21 \times 10^{-12}} = 60 \text{ MHz}$$

7.3 FEEDBACK TOPOLOGIES AND CLOSED-LOOP I/O RESISTANCES

The most basic parameters characterizing an amplifier are its *unloaded gain*, and its *input* and *output resistances*, also called the *terminal resistances*. These resistances come into play in actual application, when the amplifier is driven by a non-ideal input source and drives an output load. The input resistance forms a divider with the source's resistance, and the output resistance forms a divider with the load, thereby reducing the overall gain from source to load twice. This reduction is known as the *loading effect*. We will see next that besides stabilizing gain, negative feedback alters the terminal resistances in ways that tend to reduce the loading effect.

As mentioned, the input and output signals can be either currents or voltages, so we have four possible amplifier types: (*a*) the *voltage* amplifier, giving $v_o = Av_i$, A in V/V; (*b*) the *current* amplifier, giving $i_o = Ai_i$, A in A/A; (*c*) the *transresistance* amplifier, giving $v_o = Ai_i$, A in V/A; and (*d*) the *transconductance* amplifier, giving $i_o = Av_i$, A in A/V. When negative feedback is applied around an amplifier, four different topologies arise. We wish to investigate how negative feedback affects gain as well as the terminal resistances of each configuration.

The Series-Shunt Configuration

We start out by investigating the application of negative feedback around a *voltage amplifier*, the first of the aforementioned amplifier types. As shown in Fig. 7.12*a*, the amplifier's output port is modeled with a *Thévenin equivalent* consisting of the dependent source av_ε and the *series* output resistance r_o. The input port, assumed to play a purely passive role, is modeled with the input resistance r_i. The quantities a, r_i, and r_o are referred to as *open-loop parameters*, and are thus denoted with *lower-case* letters. The role of the feedback network is to sense the output voltage v_o and produce a scaled version of it, or $v_f = bv_o$, such that $-v_f$ is then summed to the input v_i to generate the error signal $v_\varepsilon = v_i - v_f$. Note that the operation of *voltage sensing* at the output is performed in parallel, or *shunt*, just as in ordinary voltmeter measurements (in the lab we always measure voltage in parallel, never in series!). However, the operation of *voltage summing* at the input is performed in *series*, which is how we connect together different voltage sources when we want to add or subtract their voltages—never connect different voltage sources in parallel!

To be able to focus on the effect of negative feedback upon a, r_i, and r_o alone, irrespective of the surrounding circuitry's details, we deliberately assume the *absence of any loading* from both ports of the amplifier (the presence of loading encountered in practical circuits will be taken up in the next section). Thus, to prevent loading at the amplifier's input port, assume the v_i and bv_o sources have *zero* series resistances so we can write $v_\varepsilon = v_i - v_f$ (had these resistances been different from zero, we would have had voltage division there). Likewise, to prevent loading at the amplifier's output port, assume this port is left *open circuited*, and the feedback-network's presents *infinite* resistance to said port so we can write $v_o = av_\varepsilon$ (had this not been the case,

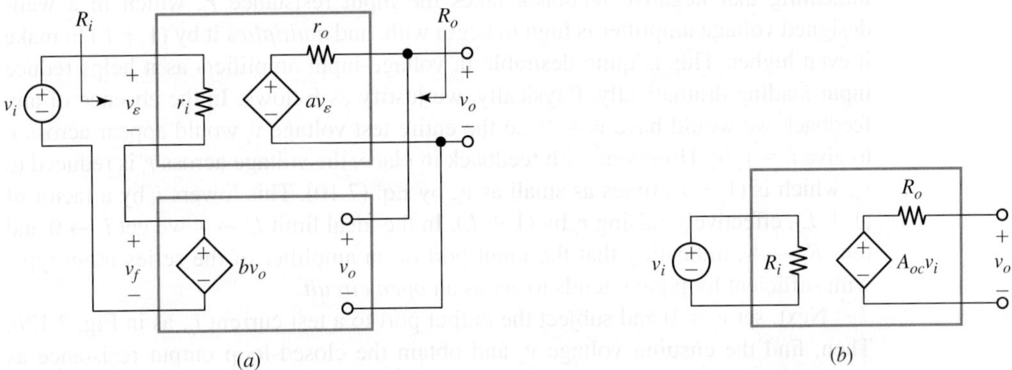

FIGURE 7.12 The series-shunt configuration or voltage amplifier: (*a*) error amplifier (*top*) and idealized feedback network (*bottom*); (*b*) equivalent circuit.

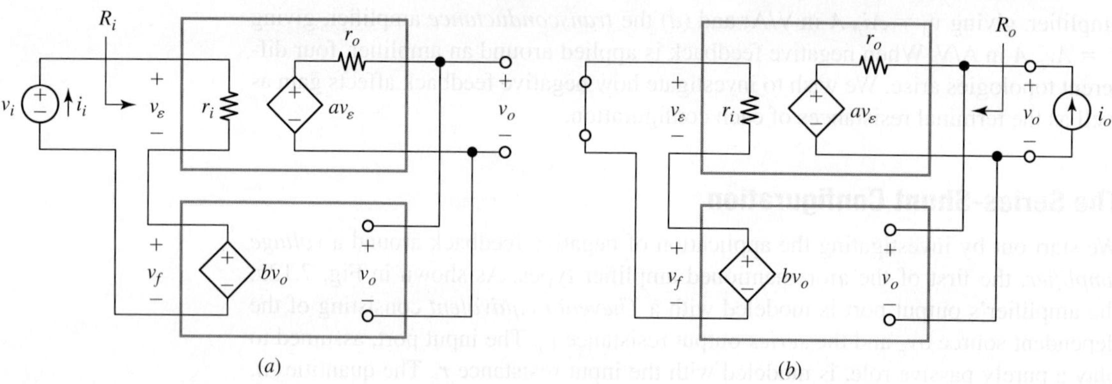

FIGURE 7.13 (a) Test circuit to find the closed-loop parameters $A = v_o/v_i$ and R_i for the series-shunt configuration. (b) Test circuit to find R_o.

we would have had voltage division at the output port as well). With reference to Fig. 7.13a, we have, by inspection, $v_o = av_\varepsilon = a(v_i - v_f) = a(v_i - bv_o)$. Collecting yields the familiar result

$$A_{oc} = \frac{v_o}{v_i} = \frac{1}{b}\frac{1}{1 + 1/L} \tag{7.28}$$

where $L = ab$. As we know, negative feedback stabilizes gain by making $A_{oc} \to 1/b$ in the limit $L \to \infty$.

To find the effect upon r_i, apply a test voltage v_i as in Fig. 7.13a, find the ensuing current i_i, and obtain the closed-loop input resistance as the ratio $R_i = v_i/i_i$. Thus, Ohm's law gives $i_i = v_\varepsilon/r_i$. But Eq. (7.10) predicts $v_\varepsilon = v_i/(1 + L)$, so

$$R_i = \frac{v_i}{i_i} = r_i(1 + L) \tag{7.29}$$

indicating that negative feedback takes the input resistance r_i, which in a well-designed voltage amplifier is high to begin with, and *multiplies* it by $(1 + L)$ to make it even higher. This is quite desirable in voltage-input amplifiers as it helps reduce input loading dramatically. Physically, we justify as follows. In the absence of any feedback we would have $v_f = 0$, so the entire test voltage v_i would appear across r_i to give $i_i = v_i/r_i$. However, with feedback in place, the voltage across r_i is reduced to v_ε, which is $(1 + L)$ times as small as v_i, by Eq. (7.10). This lowers i_i by a factor of $(1 + L)$, effectively raising r_i by $(1 + L)$. In the ideal limit $L \to \infty$ we get $i_i \to 0$ and thus $R_i \to \infty$, indicating that the input port of an amplifier of the series-input type, with sufficient loop gain, tends to act as an *open circuit*.

Next, set $v_i = 0$ and subject the output port to a test current i_o, as in Fig. 7.13b. Then, find the ensuing voltage v_o and obtain the closed-loop output resistance as $R_o = v_o/i_o$. By KVL and Ohm's law,

$$v_o = av_\varepsilon + r_o i_o = a(-v_f) + r_o i_o = -abv_o + r_o i_o$$

Collecting and simplifying gives

$$R_o = \frac{v_o}{i_o} = \frac{r_o}{1 + L} \qquad (7.30)$$

indicating that negative feedback takes the output resistance r_o, which in a well-designed voltage amplifier is low to begin with, and *divides* it by $(1 + L)$ to make it even lower. This is highly desirable in voltage-output amplifiers as it helps reduce output loading dramatically. Physically, we justify as follows. In the absence of feedback ($b = 0$) the circuit of Fig. 7.13b would give $av_\varepsilon = 0$ and thus $v_o = r_o i_o$. However, with feedback in place, the amplifier, in its attempt to drive v_ε close to zero, adjusts its dependent source so that v_o ($= v_f/b = -v_\varepsilon/b$) becomes $(1 + L)$ times as small, effectively lowering r_o by $(1 + L)$. In the ideal limit $L \to \infty$ we get $v_o \to 0$ and thus $R_o \to 0$, indicating that in the absence of any input signal, the output port of an amplifier of the output-shunt type, with sufficient loop gain, tends to act as a *short circuit*.

In light of the above results, it is apparent that the series-shunt configuration of Fig. 7.12a admits the voltage-amplifier equivalent of Fig. 7.12b, with the closed-loop parameters A, R_i, and R_o as given in Eqs. (7.28) through (7.30). Note the use of upper-case letters to distinguish these parameters from their open-loop counterparts a, r_i, and r_o.

An op amp with $a = 10^5$ V/V, $r_i = 1$ MΩ, and $r_o = 100$ Ω is operated in the series-shunt mode with $b = 0.01$ V/V. Estimate A_{oc}, R_i, and R_o, and comment.   **EXAMPLE 7.8**

Solution

We have $L = 10^5 \times 0.01 = 10^3$. Thus, $A_{oc} \cong 100(1 - 1/10^3) = 99.9$ V/V, $R_i = 10^6(1 + 10^3) \cong 10^9$ $\Omega = 1$ GΩ, and $R_o = 100/(1 + 10^3) \cong 0.1$ Ω. Compared with the resistances surrounding an op amp, which are typically in the kΩ-range, R_i effectively appears as an open circuit and R_o as a short circuit.

The Shunt-Series Configuration

We now turn our attention to the application of negative feedback around a *current amplifier*, the dual of the voltage amplifier. As shown in Fig. 7.14a, the amplifier's output port is modeled with a *Norton equivalent* consisting of the dependent source ai_ε and the *parallel* output resistance r_o. The input port, assumed to play a purely passive role, is modeled with the input resistance r_i. The role of the feedback network is to sense the output current i_o and produce a scaled version of it, or $i_f = bi_o$, such that i_f is then summed to the input i_i to generate the error signal $i_\varepsilon = i_i - i_f$. The operation of *current sensing* at the output port is performed in *series*, just as in laboratory current measurements, where we break the circuit to insert the ammeter in series (recall that currents are always measured in series, voltages always in parallel!). However, the operation of *current summing* at the input is performed in *parallel*, or *shunt*, as this is how we connect together different current sources when we want to add or subtract their currents—never connect different current sources in series!

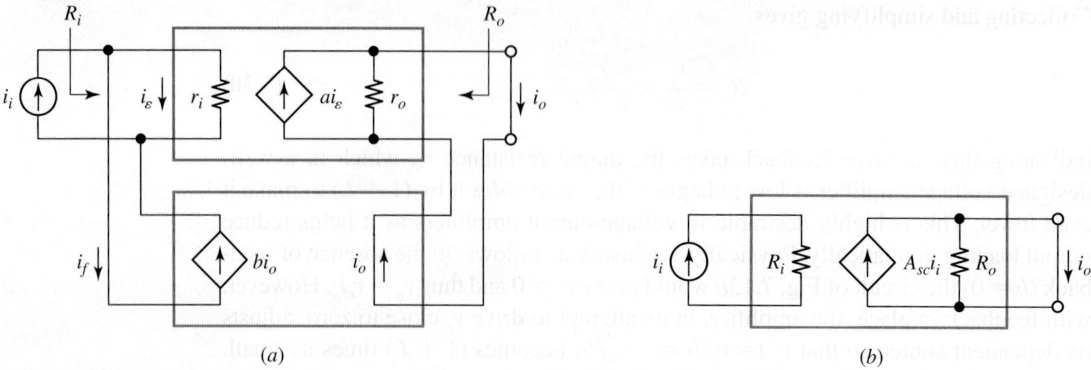

FIGURE 7.14 The shunt-series configuration or current amplifier: (*a*) error amplifier (*top*) and idealized feedback network (*bottom*). (*b*) Equivalent circuit.

To be able to focus on the effect of negative feedback upon the open-loop parameters a, r_i, and r_o alone, irrespective of the surrounding circuitry's details, we again deliberately assume the *absence of any loading* from both ports of the amplifier (the presence of loading encountered in practical circuits will be taken up in the next section). Thus, to prevent loading at the amplifier's input port, assume the i_i and bi_o sources have *infinite* parallel resistance so we can write $i_\varepsilon = i_i - i_f$ (had these sources had non-infinite resistances, we would have had current division at the input port). Likewise, to prevent loading at the amplifier's output port, assume this port is *short circuited*, and the feedback-network's presents *zero* resistance to said port so we can simply write $i_o = ai_\varepsilon$ (had this not been the case, we would have had current division at the amplifier's output port as well). With reference to Fig. 7.14*a*, we have, by inspection, $i_o = ai_\varepsilon = a(i_i - i_f) = a(i_i - bi_o)$. Collecting yields the familiar result

$$A_{sc} = \frac{i_o}{i_i} = \frac{1}{b}\frac{1}{1 + 1/L} \tag{7.31}$$

where $L = ab$. As we know, negative feedback stabilizes gain by making $A_{sc} \to 1/b$ in the limit $L \to \infty$.

To find the effect upon r_i, apply a test current i_i as in Fig. 7.15*a*, find the ensuing voltage v_i, and obtain the closed-loop input resistance as $R_i = v_i/i_i$. Thus, Ohm's law gives $v_i = r_i i_\varepsilon$. But Eq. (7.10) predicts $i_\varepsilon = i_i/(1 + L)$, so

$$R_i = \frac{v_i}{i_i} = \frac{r_i}{1 + L} \tag{7.32}$$

indicating that negative feedback takes the input resistance r_i, which in a well-designed current amplifier is usually low to begin with, and *divides* it by $(1 + L)$ to make it even lower. This is highly desirable in current-input amplifiers, as it helps reduce input loading dramatically. Physically, we justify as follows. In the absence of any feedback, the entire test current i_i would flow through r_i, giving $v_i = r_i i_i$. However, with feedback in place, the current through r_i reduces to the error current

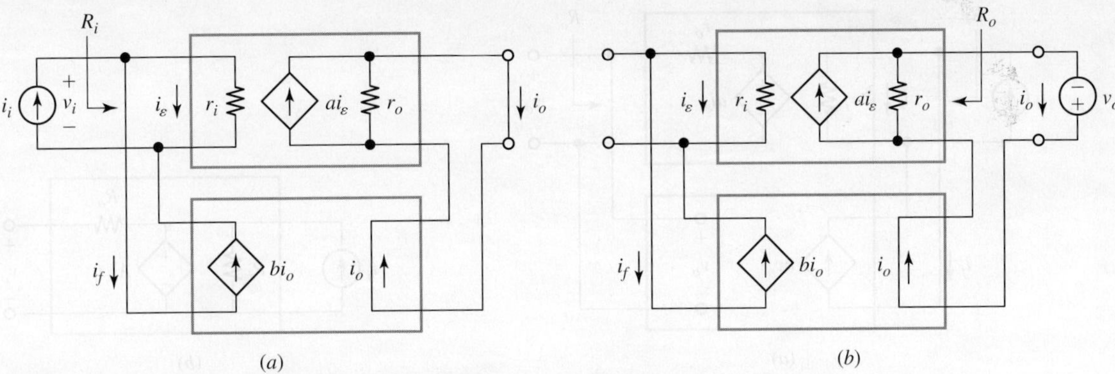

(a)   (b)

FIGURE 7.15 (a) Test circuit to find the closed-loop parameters $A = v_o/v_i$ and R_i for the shunt-series configuration. (b) Test circuit to find R_o.

i_ε, which is $(1 + L)$ times as small as i_i, by Eq. (7.10). This reduces v_i by a factor of $(1 + L)$, effectively dividing r_i by $(1 + L)$. In the ideal limit $L \to \infty$ we get $v_i \to 0$ and $R_i \to 0$, indicating that the input port of an amplifier of the input-shunt type, with sufficient loop gain, tends to act as a *short circuit*.

Next, set $i_i = 0$ and subject the output port to a test voltage v_o as shown in Fig. 7.15b. Then, find the ensuing current i_o and obtain the closed-loop output resistance as $R_o = v_o/i_o$. By KCL and Ohm's law,

$$i_o = ai_\varepsilon + v_o/r_o = a(-i_f) + v_o/r_o = -abi_o + v_o/r_o$$

Collecting and simplifying gives

$$R_o = \frac{v_o}{i_o} = r_o(1 + L) \tag{7.33}$$

indicating that negative feedback takes the output resistance r_o, which in a well-designed current amplifier is usually high to begin with, and *multiplies* it by $(1 + L)$ to make it even higher. This is quite desirable in current-output amplifiers as it helps reduce output loading dramatically. Physically, we justify as follows. In the absence of any feedback we would have $i_o = v_o/r_o$. However, with feedback in place, the amplifier, in its attempt to drive i_ε close to zero, adjusts its dependent source so that $i_o \, (= i_f/b = -i_\varepsilon/b)$ becomes $(1 + L)$ times as small, effectively raising r_o by $(1 + L)$. In the ideal limit $L \to \infty$ we get $i_o \to 0$ and thus $R_o \to \infty$, indicating that in the absence of any input, the output port of an amplifier of the output-series type, with sufficient loop gain, tends to act as an *open circuit*.

In light of the above results, it is apparent that the shunt-series configuration of Fig. 7.14a admits the current-amplifier equivalent of Fig. 7.14b, with the closed-loop parameters A_{sc}, R_i, and R_o as given in Eqs. (7.31) through (7.33). Note that what is good for a current amplifier ($R_i \to 0$ and $R_o \to \infty$) is just the opposite of what is good for a voltage amplifier ($R_i \to \infty$ and $R_o \to 0$). This is yet another manifestation of the duality principle.

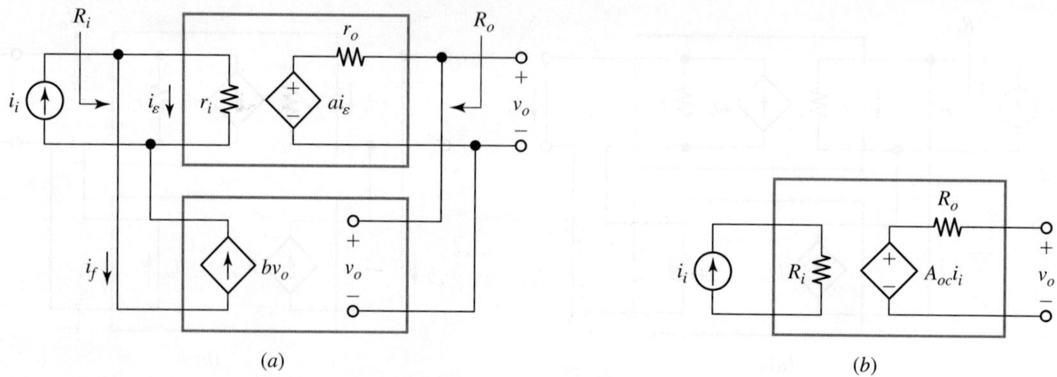

FIGURE 7.16 The shunt-shunt configuration or transresistance amplifier: (a) error amplifier (*top*) and idealized feedback network (*bottom*). (b) Equivalent circuit.

The Shunt-Shunt Configuration

Figure 7.16a shows the application of negative feedback around a *transresistance amplifier*. The input signal is a current, so input current summing is done in *shunt* fashion, as in the case of the current amplifier of Fig. 7.14a. On the other hand, the output signal is a voltage, so the amplifier's output port is modeled with the *Thévenin* source ai_ε, and output voltage sensing is done in *shunt* fashion, as in the case of the voltage amplifier of Fig. 7.12a. The transresistance amplifier is thus of the *shunt-shunt* type.

To find the closed-loop gain A_{oc} we write $v_o = ai_\varepsilon = a(i_i - i_f) = a(i_i - bv_o)$. Collecting yields the familiar result

$$A_{oc} = \frac{v_o}{i_i} = \frac{1}{b} \frac{1}{1 + 1/L} \tag{7.34}$$

where $L = ab$. To find the closed-loop resistances R_i and R_o, we use the test-signal techniques already utilized above. Actually, since the input port is similar to that of the shunt-series configuration, and the output port is similar to that of the series-shunt configuration, we recycle the results developed in connection with Figs. 7.15a and 7.13b, and write

$$R_i = \frac{r_i}{1 + L} \qquad\qquad R_o = \frac{r_o}{1 + L} \tag{7.35}$$

In the limit $L \to \infty$ the shunt-shunt configuration gives $A_{oc} \to 1/b$, $R_i \to 0$, and $R_o \to 0$.

The Series-Series Configuration

Figure 7.17a shows the application of negative feedback around a *transconductance amplifier*. The input signal is a voltage, so input voltage summing is done in *series* fashion, as in the case of the voltage amplifier of Fig. 7.12a. Conversely, the output signal is a current, so the amplifier's output port is modeled with the *Norton* source

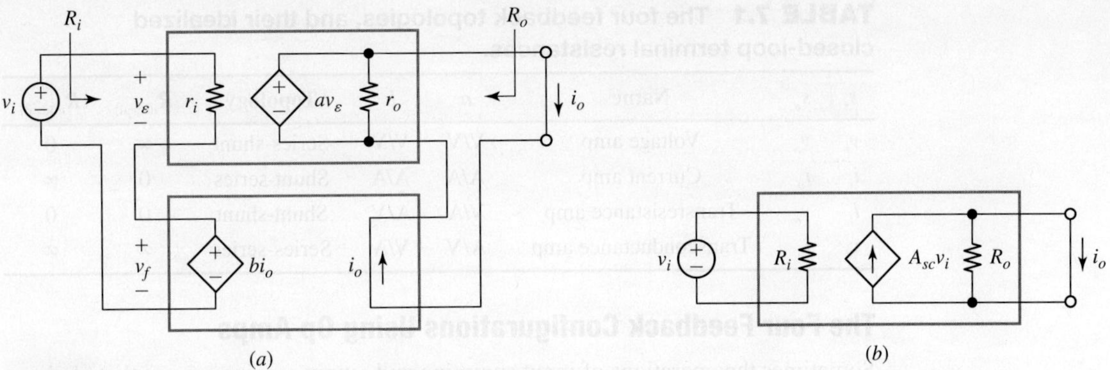

FIGURE 7.17 The series-series configuration or transconductance amplifier: (a) error amplifier (*top*) and idealized feedback network (*bottom*). (b) Equivalent circuit.

av_ε, and output current sensing is done in *series* fashion, as in the case of the current amplifier of Fig. 7.14a. The transconductance amplifier is thus of the *series-series* type, the dual type of the shunt-shunt amplifier.

To find the closed-loop parameters we proceed in the usual fashion, obtaining

$$A_{sc} = \frac{i_o}{v_i} = \frac{1}{b}\frac{1}{1 + 1/L} \tag{7.36}$$

$$R_i = r_i(1 + L) \qquad R_o = r_o(1 + L) \tag{7.37}$$

where $L = ab$. In the limit $L \to \infty$, the series-series configuration gives $A_{sc} \to 1/b$, $R_i \to \infty$, and $R_o \to \infty$.

Summary

In light of the above results, we can state that the closed-loop gain A of each of the four negative-feedback configurations can be expressed in the form

$$A = A_{ideal}\frac{1}{1 + 1/L} \tag{7.38}$$

where $A_{ideal} = 1/b$ is the closed-loop gain in the limit $L \to \infty$, $L = ab$ is the loop gain, a is the open-loop gain, and b is the feedback factor. Moreover, the closed-loop terminal resistances $R_{i/o}$ can be expressed in terms of the open-loop resistances $r_{i/o}$ as

$$R_{i/o} = r_{i/o}(1 + L)^{\pm 1} \tag{7.39}$$

with $+1$ for the *series* cases, and -1 for *shunt* cases. Table 7.1 summarizes the four negative-feedback topologies, along with the terminal resistances in the idealized limit $L \to \infty$.

TABLE 7.1 The four feedback topologies, and their idealized closed-loop terminal resistances.

s_i	s_o	Name	a	b	Topology	$R_{i(\text{ideal})}$	$R_{o(\text{ideal})}$
v_i	v_o	Voltage amp	V/V	V/V	Series-shunt	∞	0
i_i	i_o	Current amp	A/A	A/A	Shunt-series	0	∞
i_i	v_o	Transresistance amp	V/A	A/V	Shunt-shunt	0	0
v_i	i_o	Transconductance amp	A/V	V/A	Series-series	∞	∞

The Four Feedback Configurations Using Op Amps

Sometimes the operations of input summing and output sensing are not that obvious. So, to develop a quick feel, let us examine actual circuit examples using the op amp, the most popular building block designed for negative-feedback operation. Even though the op amp is, strictly speaking, a voltage-type amplifier (to stress this, we use the symbol a_v to denote its voltage gain), it can be used in any of the four feedback topologies discussed above, giving further credence to the designation *operational*.

- **Series-Shunt Configuration** (Fig. 7.18a): we have already examined this configuration in connec-tion with Fig. 7.2. With sufficiently high loop gain, as is generally the case with op amp circuits, the input error voltage v_ε will be

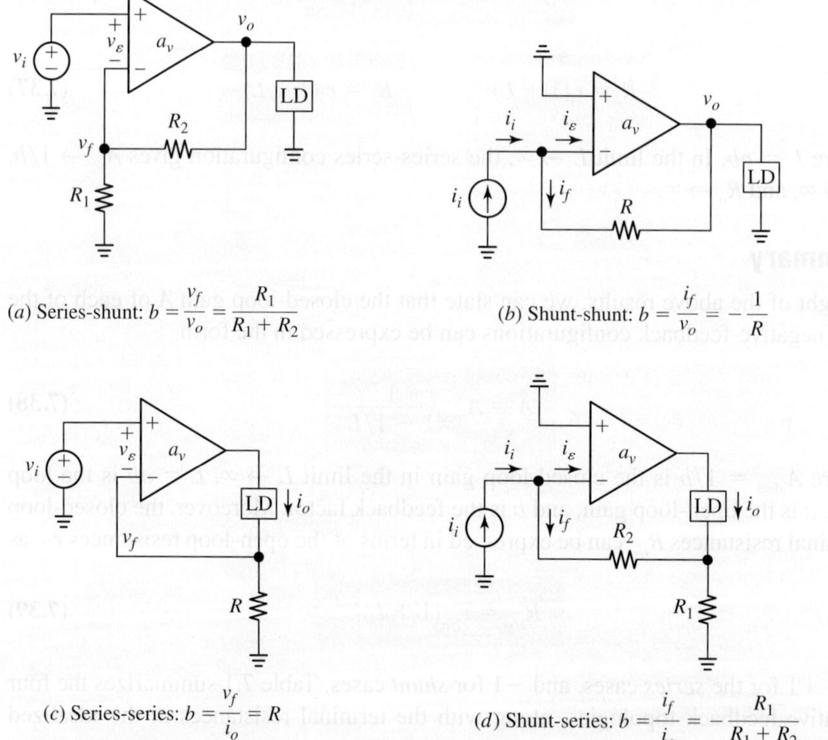

(a) Series-shunt: $b = \dfrac{v_f}{v_o} = \dfrac{R_1}{R_1 + R_2}$

(b) Shunt-shunt: $b = \dfrac{i_f}{v_o} = -\dfrac{1}{R}$

(c) Series-series: $b = \dfrac{v_f}{i_o} = R$

(d) Shunt-series: $b = \dfrac{i_f}{i_o} = -\dfrac{R_1}{R_1 + R_2}$

FIGURE 7.18 Illustrating the four basic negative-feedback topologies using op amps.

vanishingly small, indicating a vanishingly small current through the op amp's internal resistance r_i. Consequently, for all practical purposes the op amp's input port appears as an *open circuit* to the feedback network. This allows us to apply the voltage divider rule and write

$$b = \frac{v_f}{v_i} = \frac{R_1}{R_1 + R_2} = \frac{1}{1 + R_2/R_1} \tag{7.40}$$

In the ideal op-amp limit $a_v \to \infty$ this circuit gives $A_{oc} = v_o/v_i \to 1/b = 1 + R_2/R_1$, $R_i \to \infty$, and $R_o \to 0$.

- **Shunt-Shunt Configuration** (Fig. 7.18*b*): the student will recall from basic op amp theory that in the ideal limit $a_v \to \infty$ the inverting-input node, also called a *current-summing junction*, acts as a *virtual ground*. The natural input signal is in this case a current, so the error signal is $i_\varepsilon = i_i - i_p$, as shown. Moreover, with the inverting-input node at 0 V, we have, by Ohm's law, $i_f = (0 - v_o)/R$, so

$$b = \frac{i_f}{v_o} = -\frac{1}{R} \tag{7.41}$$

In the ideal op-amp limit $a_v \to \infty$, this circuit gives $A_{oc} = v_o/i_i \to 1/b = -R$. Moreover, the resistance seen by the input source is $R_i \to 0$, and that seen by the output load is $R_o \to 0$.

The shunt-shunt configuration forms the basis of the popular *inverting voltage amplifier* of Fig. 7.19*a*. Even though this is a voltage-in, voltage-out amplifier, from a feedback viewpoint it is a *shunt-shunt* configuration. This becomes more apparent if we perform a source transformation to convert from the voltage v_i to the current $i_i = v_i/R_1$. Then, the closed-loop voltage gain is

$$A_v = \frac{v_o}{v_i} = \frac{v_o}{i_i} \times \frac{i_i}{v_i} = (-R_2) \times \frac{1}{R_1} = -\frac{R_2}{R_1} \tag{7.42}$$

- **Series-Series Configuration** (Fig. 7.18*c*): the natural signals for this circuit are a voltage at the input and a current at the output. The task of the feedback network is to sense the output current i_o in *series*, and convert it to the feedback voltage v_f to be summed to the input v_i in *series*. Amazingly, both operations are performed by means of a single resistance, R. As discussed earlier, the op amp's input port appears for all practical purposes as an *open circuit* to the feedback network. We can thus use Ohm's law and write $v_f = Ri_o$, so

$$b = \frac{v_f}{i_o} = R \tag{7.43}$$

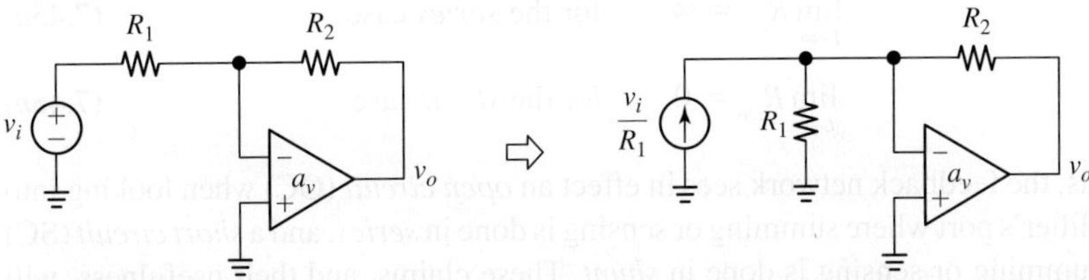

FIGURE 7.19 The popular inverting voltage amplifier is actually a shunt-shunt amplifier, as demonstrated by the source transformation.

In the ideal op-amp limit $a_v \to \infty$, this circuit gives $A_{sc} = i_o/v_i \to 1/b = 1/R$. Moreover, the resistance seen by the input source is $R_i \to \infty$, and that seen by the output load is $R_o \to \infty$.

- **Shunt-Series Configuration** (Fig. 7.18d): the natural signals for this circuit are a current at the input and a current at the output. Given that the inverting-input node is at 0 V, we can apply the current-divider rule and write

$$-i_f = \frac{R_1}{R_1 + R_2}i_o$$

so

$$b = \frac{i_f}{i_o} = \frac{-R_1}{R_1 + R_2} = \frac{-1}{1 + R_2/R_1} \tag{7.44}$$

In the ideal op-amp limit $a_v \to \infty$, this circuit gives $A_{sc} = i_o/i_i \to -(1 + R_2/R_1)$. Moreover, the resistance seen by the input source is $R_i \to 0$, and that seen by the output load is $R_o \to \infty$.

7.4 PRACTICAL CONFIGURATIONS AND THE EFFECT OF LOADING

The feedback networks of Section 7.3 were deliberately assumed idealized so we could focus on the effects of negative feedback upon the amplifier's parameters a, r_i, and r_o alone. In a practical circuit the feedback network introduces *loading* both at the amplifier's input and output. Moreover, the amplifier and the feedback network are sometimes interwoven with each other, making the separation between the two not always obvious. A negative-feedback circuit can always be analyzed in its entirety via standard techniques such as nodal or loop analysis. However, as circuit complexity increases, this approach soon becomes prohibitive.

Mercifully, suitable approximations can be made that allow us to *separate* the circuit into a basic amplifier incorporating the effects of loading, and a distinct feedback network. With this separation in hand, we can then apply Eqs. (7.38) and (7.39) to find the closed-loop parameters A, R_i, and R_o. These approximations exploit a priori the fact that with a sufficiently high loop gain T, a terminal resistance $R_{i/o}$ satisfies

$$\lim_{L \to \infty} R_{i/o} = \infty \qquad \text{for the } series \text{ case} \tag{7.45a}$$

$$\lim_{L \to \infty} R_{i/o} = 0 \qquad \text{for the } shunt \text{ case} \tag{7.45b}$$

In words, the feedback network sees in effect an *open circuit* (OC) when looking into an amplifier's port where summing or sensing is done in *series*, and a *short circuit* (SC) when summing or sensing is done in *shunt*. These claims, and their usefulness, will become clearer as we proceed.

Series-Shunt Circuits

In Fig. 7.18a we have investigated the non-inverting op amp configuration as a classic example of a series-shunt circuit. We now re-examine it with an eye on the interaction between the error amplifier's internal resistances r_i and r_o and the external resistances R_1 and R_2 making up the feedback network. This interaction results in *loading* both at the amplifier's input and output. To be able to focus on loading by the feedback network alone, we continue to assume a driving source v_i with zero series resistance, and an unloaded (open-circuited) output port, as shown in Fig. 7.20.

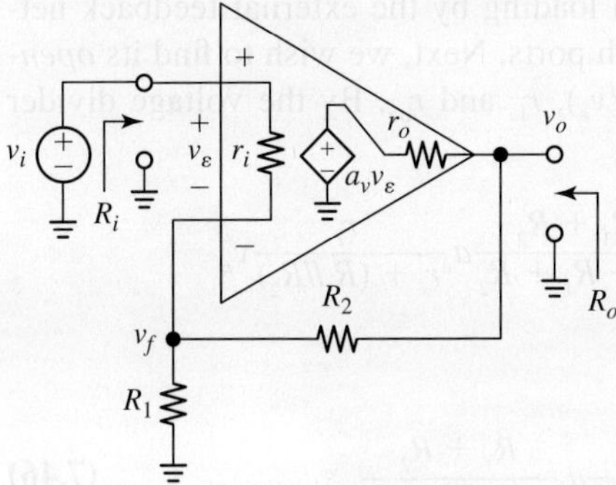

FIGURE 7.20 The non-inverting op amp with its internal resistances r_i and r_o explicitly shown.

To investigate loading at the *input port*, note that the external resistance seen by this port would appear to be $R_1//(R_2 + r_o)$. But we know a priori that in feedback operation the system will present a very low resistance ($R_o \rightarrow 0$) at the output node due to shunt-type sensing there. This swamps out the effect of r_o, causing the external resistance seen by the amplifier's input port to be in effect $R_1//R_2$. This is depicted in the amplifier's equivalent of Fig. 7.21a, where R_2 is shown *short-circuited* (SC) to ground.

To investigate loading at the *output port*, note that the external resistance seen by this port would appear to be $R_2 + (R_1//r_i)$. But we know a priori that in feedback

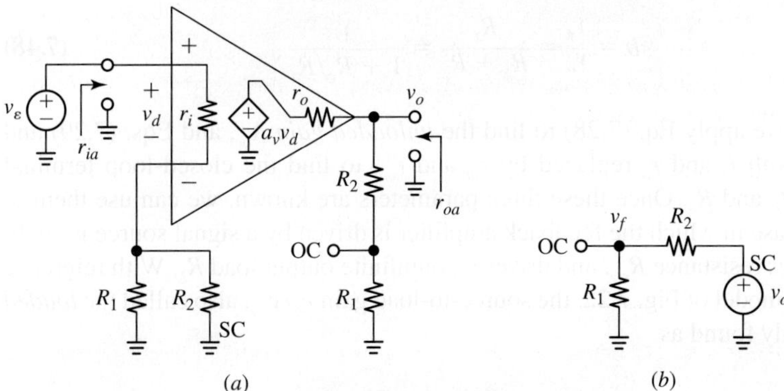

FIGURE 7.21 Decomposing the circuit of Fig. 7.20 into (a) the error amplifier and (b) the feedback network.

operation the system will present a very high resistance ($R_i \to \infty$) at the input node due to series-type summing there. This stems from the fact that $v_\varepsilon \to 0$, making the current through r_i truly negligible and thus swamping out the effect of r_i. For practical purposes, the amplifier's input port appears as an open circuit to the feedback network. This too is depicted in Fig. 7.21a, where the node common to R_1 and R_2 is shown as *open-circuited* (OC), indicating that the external resistance seen by the amplifier's output port is in effect $R_2 + R_1$.

With the feedback path interrupted both at the output (via short circuiting) and at the input (via open circuiting), the circuit of Fig. 7.21a represents the error amplifier in *open-loop* operation, but with loading by the external feedback network specifically taken into account at both ports. Next, we wish to find its *open-loop parameters*, now denoted as $a\ (=v_o/v_\varepsilon)$, r_{ia} and r_{oa}. By the voltage divider rule we have

$$v_o = \frac{R_1 + R_2}{r_o + R_1 + R_2}a_v v_d = \frac{R_1 + R_2}{r_o + R_1 + R_2}a_v\frac{r_i}{r_i + (R_1//R_2)}v_\varepsilon$$

so that

$$a = \frac{v_o}{v_\varepsilon} = \frac{r_i}{r_i + (R_1//R_2)}a_v\frac{R_1 + R_2}{r_o + R_1 + R_2} \qquad (7.46)$$

Note that $a < a_v$ because of input and output loading. We also have, by inspection

$$r_{ia} = r_i + (R_1//R_2) \qquad r_{oa} = r_o//(R_2 + R_1) \qquad (7.47)$$

Next, we seek a representation for the feedback network that is *separate* from the basic amplifier. Referring back to Fig. 7.20 we observe that at the sensing side this network is driven by a voltage source v_o expected to exhibit extremely low series resistance R_o, and that at the summing side it drives the amplifier's input port which, for practical purposes, appears as an open circuit (OC). The circuit for the calculation of b is thus as in Fig. 7.21b, from which we readily find

$$b = \frac{v_f}{v_o} = \frac{R_1}{R_1 + R_2} = \frac{1}{1 + R_2/R_1} \qquad (7.48)$$

Finally, we apply Eq. (7.28) to find the *unloaded gain* A_{oc}, and Eqs. (7.29) and (7.30), but with r_i and r_o replaced by r_{ia} and r_{oa}, to find the closed-loop terminal resistances R_i, and R_o. Once these three parameters are known, we can use them in the general case in which the feedback amplifier is driven by a signal source v_{sig} with nonzero series resistance R_{sig}, and drives a noninfinite output load R_L. With reference to the circuit model of Fig. 7.22, the source-to-load gain v_o/v_{sig}, also called the *loaded gain*, is readily found as

$$\frac{v_o}{v_{sig}} = \frac{R_i}{R_{sig} + R_i}A_{oc}\frac{R_L}{R_o + R_L} \qquad (7.49)$$

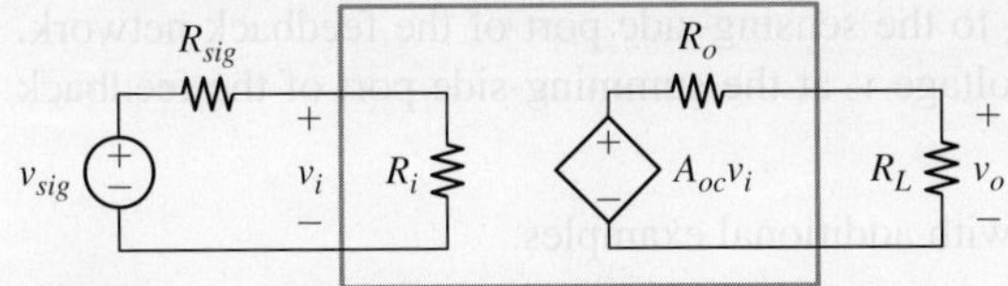

FIGURE 7.22 Modeling a closed-loop voltage amplifier driven by a source v_{sig} and driving an external load R_L.

EXAMPLE 7.9

(a) In the circuit of Fig. 7.20 let the op amp be a mediocre type with $a_v = 1000$ V/V, $r_i = 10$ kΩ, and $r_o = 1.0$ kΩ. If $R_1 = 1.0$ kΩ and $R_2 = 9.0$ kΩ, find R_i, R_o, and the unloaded gain A_{oc}.

(b) Find the *source-to-load gain* $A = v_o/v_{sig}$ if the feedback amplifier is driven by a signal source v_{sig} having a series resistance $R_{sig} = 20$ kΩ, and drives a load $R_L = 2.0$ kΩ.

Solution

(a) Using Eqs. (7.46) through (7.48) we have

$$a = \frac{10}{10 + (1.0//9.0)} 1000 \frac{1.0 + 9.0}{1.0 + 1.0 + 9.0} = 0.917 \times 1000 \times 0.909$$

$$= 833.3 \text{ V/V} \ (<1000 \text{ V/V})$$

$$r_{ia} = 10 + (1.0//9.0) = 10.9 \text{ k}\Omega \qquad r_{oa} = 1.0//(9.0 + 1.0) = 0.909 \text{ k}\Omega$$

$$b = \frac{1.0}{1.0 + 9.0} = \frac{1}{10} \text{ V/V}$$

Thus, $L = ab = 833.3/10 = 83.33$, so

$$A_{oc} = \frac{10}{1 + 1/83.33} = 9.881 \text{ V/V}$$

$$R_i = 10.9(1 + 83.33) \cong 920 \text{ k}\Omega$$

$$R_o = \frac{909}{1 + 83.33} \cong 10.8 \ \Omega$$

(b) By Eq. (7.49), the source-to-load gain is

$$\frac{v_o}{v_{sig}} = \frac{920}{20 + 920} 9.88 \frac{2000}{10.8 + 2000} = 0.979 \times 9.881 \times 0.995$$

$$= 9.618 \text{ V/V} \ (<A_{oc})$$

The line of reasoning developed for the above example can be generalized to the following **Series-Shunt Procedure**:

- To calculate loading at the *input* of the basic amplifier, *short-circuit* (SC) the sensing-side port of the feedback network.
- To calculate loading at the *output* of the basic amplifier, *open-circuit* (OC) the summing-side port of the feedback network.

- To find b, apply a voltage v_o to the sensing-side port of the feedback network, find the *open-circuit* (OC) voltage v_f at the summing-side port of the feedback network, and let $b = v_f/v_o$.

We shall illustrate the procedure with additional examples.

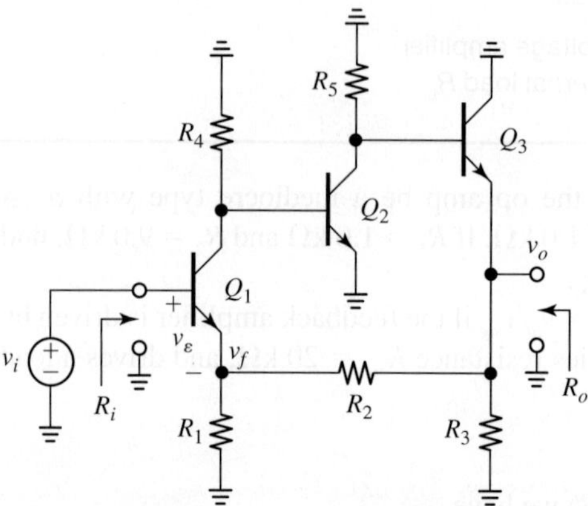

FIGURE 7.23 Series-shunt feedback triple.

In the feedback circuit of Fig. 7.23, Q_1 responds to the difference $v_\varepsilon = v_i - v_f$, so we have series-type summing. Q_2 is a CE stage designed to provide additional voltage gain and thus boost the system's loop gain L. Q_3 is an emitter follower designed to provide current gain at low output resistance. Finally, we note that v_o is shunted to the feedback network consisting of R_1 and R_2, indicating shunt-type sensing. Functionally, this circuit is similar to the non-inverting op amp of Fig. 7.20. Applying the aforementioned Series-Shunt Procedure, we end up with the error amplifier and feedback network of Fig. 7.24.

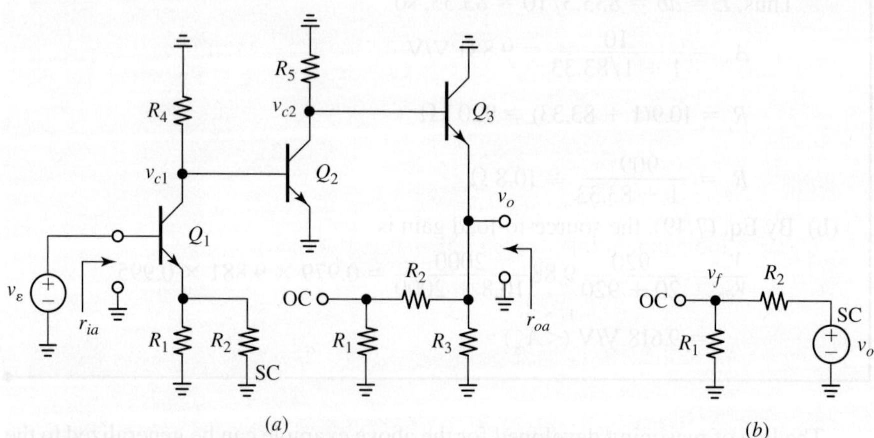

(a)   (b)

FIGURE 7.24 Decomposing the circuit of Fig. 7.23 into (a) the error amplifier and (b) the feedback network.

EXAMPLE 7.10   (a) In the circuit of Fig. 7.23 let $R_1 = 1.0\ \text{k}\Omega$, $R_2 = 3.0\ \text{k}\Omega$, and $R_3 = R_4 = R_5 = 10\ \text{k}\Omega$. Moreover, let all BJTs have $g_m = 1/(25\ \Omega)$, $r_\pi = 5\ \text{k}\Omega$, and $r_o = \infty$. Find R_i, R_o, and the unloaded gain A_{oc}.

(b) Verify your results via PSpice.

Solution

(a) For the BJTs we have $\beta_0 = g_m r_\pi = 5000/25 = 200$. Using Fig. 7.24a, we find, by inspection,

$$r_{ia} = r_{\pi 1} + (\beta_{01} + 1)(R_1//R_2) = 5 + 201(1//3) \cong 156 \text{ k}\Omega$$

$$r_{oa} = \frac{R_5 + r_{\pi 3}}{\beta_{03} + 1}//R_3//(R_1 + R_2) = \frac{10 + 5}{201}//10//(1.0 + 3.0) \cong 73 \ \Omega$$

To find the overall voltage gain a, examine one stage at a time. We note that Q_1 is a CE-ED stage, so we use the popular rule of thumb stating that the gain is the negative of the ratio of the total collector-node resistance to the total emitter-node resistance, or

$$\frac{v_{c1}}{v_\varepsilon} = -\frac{R_4//r_{\pi 2}}{r_{e1} + (R_1//R_2)} = -\frac{10//5}{0.025 + (1.0//3.0)} = -4.3 \text{ V/V}$$

One can readily verify that the resistance R_{b3} seen looking into Q_3's base is such that $R_{b3} \gg R_5$, so the gain provided by Q_2 is

$$\frac{v_{c2}}{v_{c1}} = -g_{m2}(R_5//R_{b3}) \cong -g_{m2}R_5 = -10/0.025 = -400 \text{ V/V}$$

We expect the gain of the voltage follower Q_3 to be very close to unity,

$$\frac{v_o}{v_{c2}} \cong 1 \text{ V/V}$$

Combining the above results, we finally obtain

$$a = \frac{v_o}{v_\varepsilon} = \frac{v_{c1}}{v_\varepsilon} \times \frac{v_{c2}}{v_{c1}} \times \frac{v_o}{v_{c2}} \cong (-4.3)(-400)(1) = 1720 \text{ V/V}$$

We also have, from Fig. 7.24b,

$$b = \frac{v_f}{v_o} = \frac{R_1}{R_1 + R_2} = \frac{1.0}{1.0 + 3.0} = \frac{1}{4} \text{ V/V}$$

so the loop gain is $L = ab = 1720/4 = 430$. Finally,

$$A_{oc} = \frac{v_o}{v_i} = \frac{4}{1 + 1/430} = 3.991 \text{ V/V}$$

$$R_i = 156 \times 10^3(1 + 430) = 67 \text{ M}\Omega \qquad R_o = \frac{73}{1 + 430} = 0.17 \ \Omega$$

Remark: Note that if Q_1 weren't part of the negative-feedback loop, the resistance seen looking into its emitter would be r_{e1}, or $25 \ \Omega$ in our example, a rather low value. However, in feedback operation the situation changes dramatically. To get a feel, consider the case $v_i = 1.0$ V. Then, the base current is $i_{b1} = v_i/R_i = 1/(67 \text{ M}\Omega) \cong 15$ nA, and the emitter current is $i_{e1} = (\beta_{01} + 1)i_{b1} = 201(15 \text{ nA}) \cong 3 \ \mu\text{A}$. On the other hand, the current through R_1 is $i_1 = v_f/R_1 \cong v_i/R_1 = (1 \text{ V})/(1 \text{ k}\Omega) = 1$ mA. Given that $3 \ \mu\text{A} \ll 1$ mA, there is no question that for all practical purposes Q_1's emitter appears as an open circuit (OC) to R_1 and R_2!

(b) The PSpice circuit of Fig. 7.25 uses three VCCSs to model the BJTs in small-signal operation, and it includes the 0-V dummy source V_0 to sense the current out of Q_1's emitter. After directing PSpice to perform the dc analysis as well as the transfer function (.TF) analysis, we get $A_{oc} = 3.99$ V/V, $R_i = 65.13$ MΩ, and $R_o = 0.1739$ Ω. Moreover, the current through the test source V_0 for $v_i = 1.0$ V is 3.086 μA. All data are amazingly close to those calculated by hand!

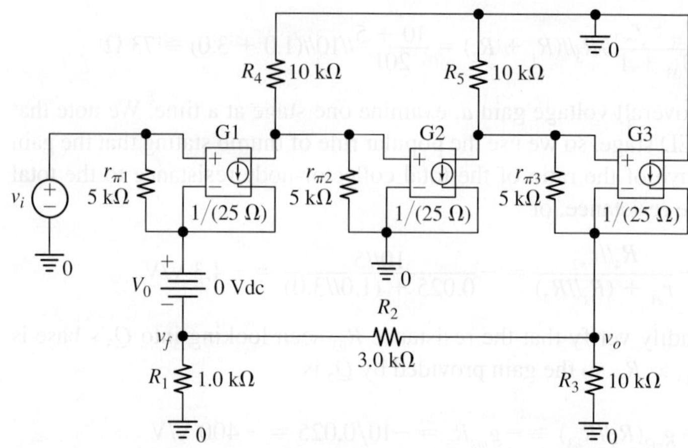

FIGURE 7.25 PSpice circuit to verify the series-shunt triple of Example 7.10.

EXAMPLE 7.11   Assuming $g_m = 1$ mA/V and $r_o = 50$ kΩ for all FETs in the circuit of Fig. 7.26, along with $\chi_5 = 0.25$, find the unloaded gain A_{oc} and the output resistance R_o ($R_i = \infty$ because the input source is looking right into M_1's gate).

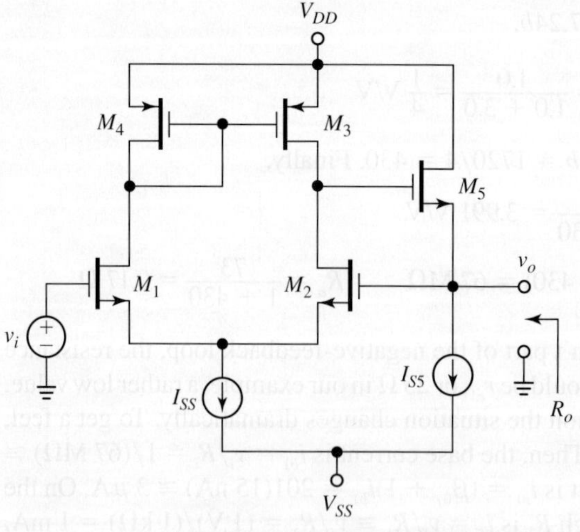

FIGURE 7.26 Series-shunt circuit of Example 7.11.

Solution

This is a unity-gain non-inverting amplifier whose decomposition appears as in Fig. 7.27. The voltage gain of the first stage is $g_m(r_{op}//r_{on}) = g_m r_o/2$ and that of the M_5 source follower is $1/(1 + \chi_5)$, so we write

$$a = \frac{v_o}{v_\varepsilon} = \frac{g_m r_o/2}{1 + \chi_5} = \frac{1 \times 50/2}{1 + 0.25} = 20 \text{ V/V} \qquad b = 1.0 \text{ V/V} \qquad L = 20$$

We also have $r_{oa} = 1/[g_{m5}(1 + \chi_5)] = 1/1.25 = 800 \ \Omega$. Consequently,

$$A_{oc} = \frac{v_o}{v_i} = \frac{1}{1 + 1/20} = 0.9524 \text{ V/V} \qquad R_o = \frac{800}{1 + 20} = 38.1 \ \Omega$$

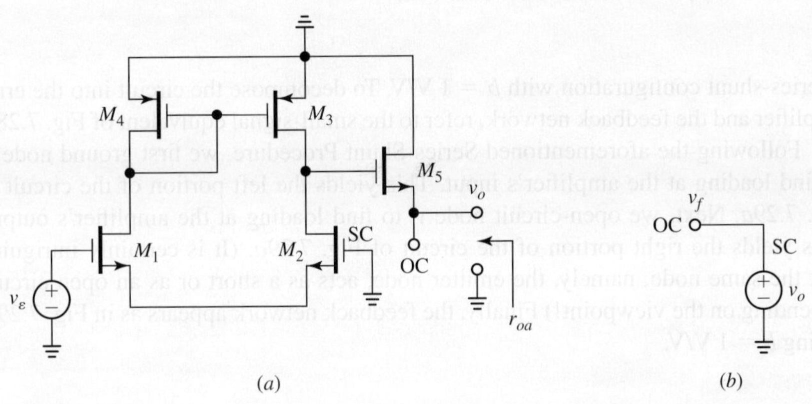

FIGURE 7.27 Decomposing the circuit of Fig. 7.26 into (a) the error amplifier and (b) the feedback network.

As our last example, let us investigate the *emitter follower* of Fig. 7.28a. On a variety of occasions it has been said that a resistance in series with the emitter of a BJT or with the source of a FET introduces *degeneration*, and that, among other things, it stabilizes the dc bias point. We now wish to reexamine this issue from the viewpoint of negative feedback. The BJT responds to the signal $v_\varepsilon = v_i - v_o$, indicating

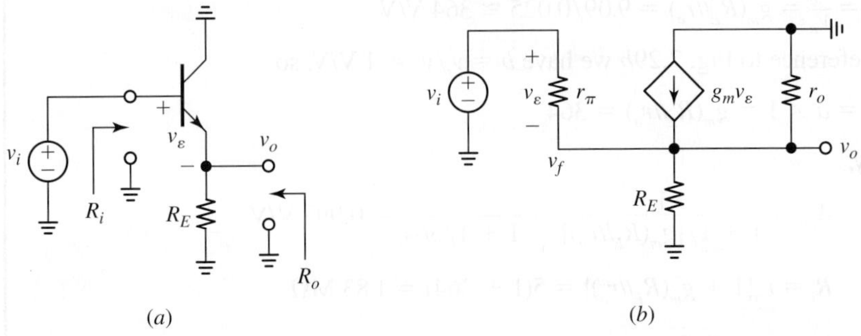

FIGURE 7.28 The emitter follower as a negative-feedback system: (a) ac equivalent and (b) small-signal representation.

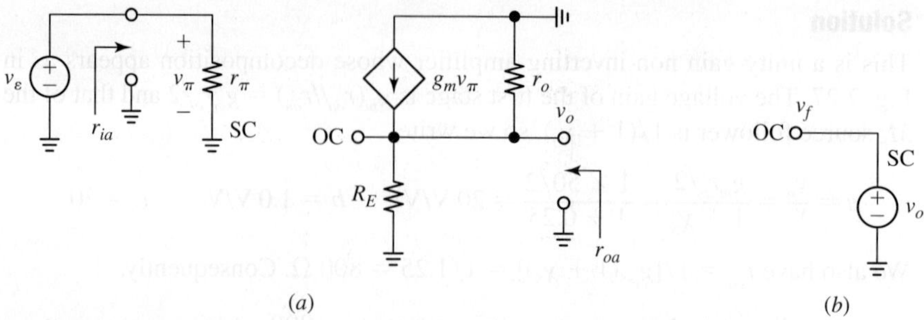

(a)   (b)

FIGURE 7.29 Decomposing the feedback circuit of Fig. 7.28 into (a) the error amplifier and (b) the feedback network.

a series-shunt configuration with $b = 1$ V/V. To decompose the circuit into the error amplifier and the feedback network, refer to the small-signal equivalent of Fig. 7.28b.

Following the aforementioned Series-Shunt Procedure, we first ground node v_o to find loading at the amplifier's input. This yields the left portion of the circuit of Fig. 7.29a. Next, we open-circuit node v_f to find loading at the amplifier's output. This yields the right portion of the circuit of Fig. 7.29a. (It is certainly intriguing that the same node, namely, the emitter node, acts as a short or as an open circuit, depending on the viewpoint!) Finally, the feedback network appears as in Fig. 7.29b, giving $b = 1$ V/V.

EXAMPLE 7.12   In the emitter follower of Fig. 7.28a let $R_E = 10$ kΩ. Moreover, let the BJT have $g_m = 1/(25\ \Omega)$, $r_\pi = 5$ kΩ, and $r_o = 100$ kΩ. Find R_i, R_o, and the unloaded gain A_{oc}. Comment on your findings.

Solution

With reference to Fig. 7.29a we have, by inspection,

$$r_{ia} = r_\pi = 5\ \text{k}\Omega \qquad r_{oa} = R_E /\!/ r_o = 10 /\!/ 100 = 9.09\ \text{k}\Omega$$

$$a = \frac{v_o}{v_\varepsilon} = g_m(R_E /\!/ r_o) = 9.09/0.025 \cong 364\ \text{V/V}$$

With reference to Fig. 7.29b we have $b = v_f/v_o = 1$ V/V, so

$$L = a \times 1 = g_m(R_E /\!/ r_o) = 364$$

Finally,

$$A_{oc} = \frac{1}{1 + 1/[g_m(R_E /\!/ r_o)]} = \frac{1}{1 + 1/364} = 0.997\ \text{V/V}$$

$$R_i = r_\pi[1 + g_m(R_E /\!/ r_o)] = 5(1 + 364) = 1.83\ \text{M}\Omega$$

$$R_o = \frac{R_E /\!/ r_o}{1 + g_m(R_E /\!/ r_o)} = \frac{1}{g_m} /\!/ R_E /\!/ r_o \cong \frac{1}{g_m} \cong r_e = 25\ \Omega$$

The interested student can, after suitable manipulations, compare the above expressions with their exact counterparts of Section 2.9. The minor differences (such as β_0 instead of $\beta_0 + 1$) stem from the fact that the method used here is based on Eqs. 7.45a and b, and as such it is an approximation—though a fairly good one in this case, as confirmed by the numerical results.

Remark: It is interesting to note that a feedback circuit can in turn be part of another more complex feedback circuit: an example is the emitter follower subcircuit (Q_3) of the series-shunt triple of Fig. 7.23.

Shunt-Shunt Circuits

In Fig. 7.30 we re-examine the shunt-shunt op amp configuration of Fig. 7.18b, but with an eye on the interaction between the amplifier's internal resistances r_i and r_o and the external feedback network, in this case consisting of R. To be able to focus on loading by the feedback network alone, we continue to assume a driving source i_i with infinite parallel resistance, and an unloaded (open-circuited) output port.

As we know, because of the high gain a_v, the differential input voltage v_d is going to be vanishingly small, indicating that the inverting-input node will be kept at virtual-ground potential. Thus, to calculate loading at the amplifier's output port, we regard R as if its left terminal were short-circuited (SC) to ground. To find loading at the input, we note that sensing at the output is of the shunt type, just as in the case of the series-shunt circuit of Fig. 7.20. Consequently, to calculate loading at the amplifier's input port, we regard R as if its right terminal were short-circuited (SC) to ground. Both situations are depicted in Fig. 7.31a, which shows the error amplifier in open-loop operation, but with loading by the external feedback network specifically taken into account at both ports. Next, we wish to find its open-loop parameters, now denoted as a ($= v_o / i_\varepsilon$), r_{ia} and r_{oa}. By the voltage divider rule and Ohm's law,

$$v_o = \frac{R}{r_o + R} a_v v_d = \frac{R}{r_o + R} a_v (r_i /\!/ R)(-i_\varepsilon)$$

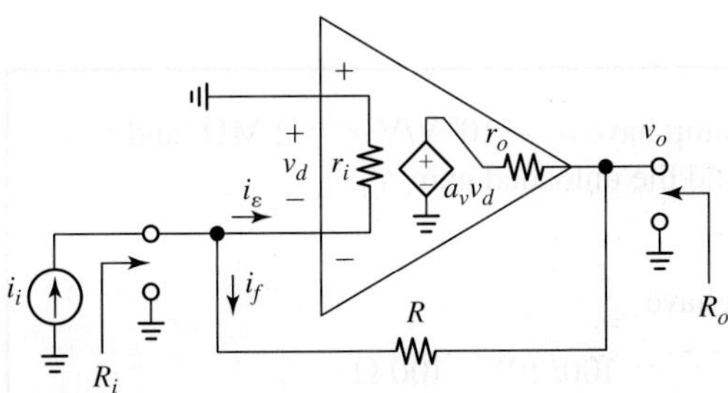

FIGURE 7.30 The transresistance op amp with its internal resistances r_i and r_o explicitly shown.

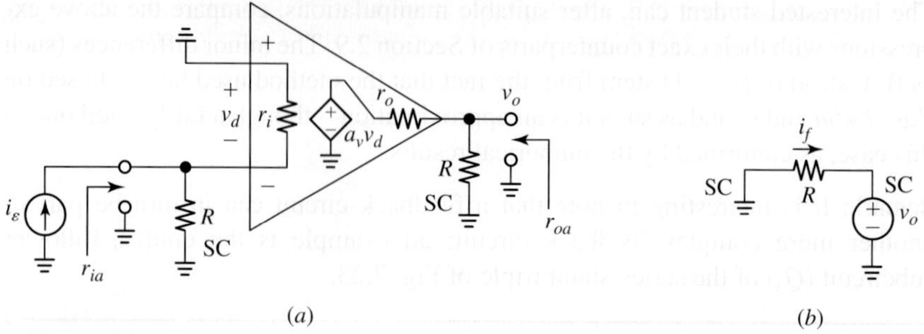

FIGURE 7.31 Decomposing the circuit of Fig. 7.30 into (a) basic amplifier and (b) feedback network.

so that

$$a = \frac{v_o}{i_\varepsilon} = -(r_i//R)a_v\frac{R}{r_o + R} \tag{7.50}$$

Note that a now has the dimensions of V/A, and is negative. Moreover, by inspection, we have

$$r_{ia} = r_i//R \qquad r_{oa} = r_o//R \tag{7.51}$$

Next, we seek a representation for the feedback network that is *separate* from the basic amplifier. Referring back to Fig. 7.30, we observe that at the sensing side this network is driven by a voltage source v_o expected to exhibit extremely small series resistance R_o, and that at the summing side it drives the amplifier's input port which, for practical purposes, appears as a short circuit (SC). The circuit for the calculation of b is thus as in Fig. 7.31b, from which we readily find $i_f = (0 - v_o)/R$, so

$$b = \frac{i_f}{v_o} = -\frac{1}{R} \tag{7.52}$$

The dimensions of b are A/V, the reciprocal of those of a. Moreover, b is negative, just like a. Both properties are consistent with the fact that $L = ab$ must be *dimensionless* and *positive*. We can now apply Eq. (7.34) to find the *unloaded gain* A_{oc}, and Eq. (7.35), but with r_i and r_o replaced by r_{ia} and r_{oa}, to find the closed-loop terminal resistances R_i, and R_o.

EXAMPLE 7.13   In the circuit of Fig. 7.30 let the op amp have $a_v = 10^5$ V/V, $r_i = 2$ MΩ, and $r_o = 100\ \Omega$. If $R = 1.0$ MΩ, find R_i, R_o, and the unloaded gain A_{oc}.

Solution

Using Eqs. (7.50) through (7.52) we have

$$r_{ia} = (2//1) \times 10^6 = 667\ \text{k}\Omega \qquad r_{oa} = 100//10^6 = 100\ \Omega$$

$$a = -(2//1)10^6 \times 10^5\frac{10^6}{100 + 10^6} = -6.66 \times 10^{10}\ \text{V/A} \qquad b = -10^{-6}\ \text{A/V}$$

$$L = (-6.667 \times 10^{10})(-10^{-6}) = 66,667$$

$$A_{oc} = \frac{-10^6}{1 + 1/66,667} \cong -1.0 \text{ V/}\mu\text{A}$$

$$R_i = \frac{667 \text{ k}\Omega}{66,667} = 10 \ \Omega$$

$$R_o = \frac{100 \ \Omega}{66,667} = 1.5 \text{ m}\Omega$$

The line of reasoning developed for the above example can be generalized to the following **Shunt-Shunt Procedure**:

- To calculate loading at the *input* of the basic amplifier, *short-circuit* (SC) the sensing-side port of the feedback network.
- To calculate loading at the *output* of the basic amplifier, *short-circuit* (SC) the summing-side port of the feedback network.
- To find b, apply a voltage v_o to the sensing-side port of the feedback network, find the *short-circuit* current i_f flowing into the summing-side port of the feedback network, and let $b = i_f/v_o$.

We shall illustrate the procedure with additional examples.

The feedback circuit of Fig. 7.32 is similar to that of Fig. 7.23, except that the input source is now a current that has been moved to Q_1's emitter, while the base has been grounded. Thus, Q_1 is now operating in the CB mode. As we know, the resistance seen looking into its emitter is low to begin with, and we anticipate that feedback will make it even lower. Applying the aforementioned Shunt-Shunt Procedure, we end up with the basic amplifier and feedback network of Fig. 7.33.

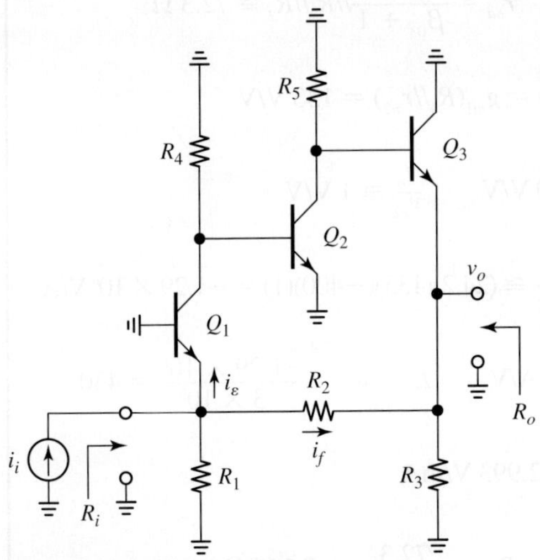

FIGURE 7.32 Shunt-shunt feedback triple.

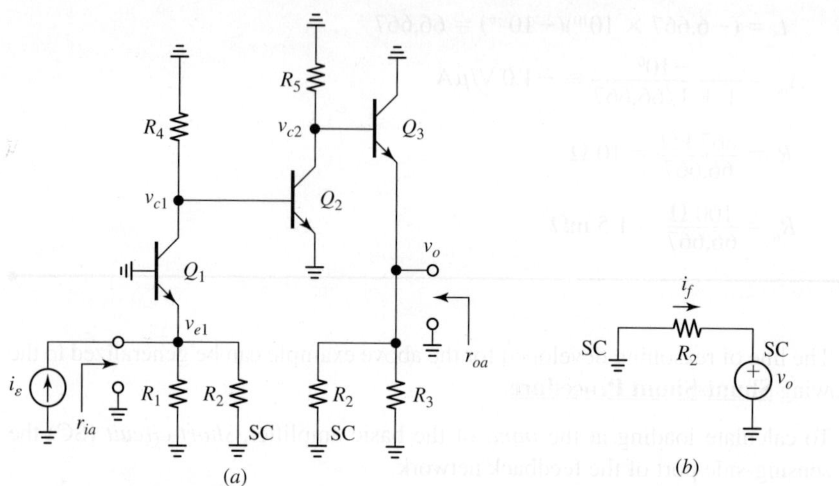

FIGURE 7.33 Decomposing the circuit of Fig. 7.32 into (a) the error amplifier and (b) the feedback network.

EXAMPLE 7.14   (a) Let the circuit of Fig. 7.32 have the same parameters of Example 7.10, namely, $R_1 = 1.0$ kΩ, $R_2 = 3.0$ kΩ, and $R_3 = R_4 = R_5 = 10$ kΩ. Moreover, let all BJTs have $g_m = 1/(25\ \Omega)$, $r_\pi = 5$ kΩ, and $r_o = \infty$. Find R_i, R_o, and the unloaded gain A_{oc}.

(b) Compare with Example 7.10, and justify similarities as well as differences.

Solution

(a) Retracing similar steps as in Example 7.10 using the circuit of Fig. 7.33 we find

$$r_{ia} = r_{e1}//R_1//R_2 = 24.2\ \Omega \qquad r_{oa} = \frac{R_5 + r_{\pi 3}}{\beta_{03} + 1}//R_2//R_3 \cong 72.3\ \Omega$$

$$\frac{v_{e1}}{i_\varepsilon} = r_{ia} = 24.2\ \text{V/A} \qquad \frac{v_{c1}}{v_{e1}} = g_{m1}(R_4//r_{\pi 2}) = 133\ \text{V/V}$$

$$\frac{v_{c2}}{v_{c1}} = -g_{m2}(R_5//R_{b3}) = -400\ \text{V/V} \qquad \frac{v_o}{v_{c2}} \cong 1\ \text{V/V}$$

$$a = \frac{v_o}{i_\varepsilon} = \frac{v_{e1}}{i_\varepsilon} \times \frac{v_{c1}}{v_{e1}} \times \frac{v_{c2}}{v_{c1}} \times \frac{v_o}{v_{c2}} \cong (24.2)(133)(-400)(1) = -1.29 \times 10^6\ \text{V/A}$$

$$b = \frac{i_f}{v_o} = -\frac{1}{R_2} = -\frac{1}{3 \times 10^3}\ \text{A/V} \qquad L = ab = \frac{-1.29 \times 10^6}{-3 \times 10^3} = 430$$

$$A_{oc} = \frac{v_o}{i_i} = \frac{-3 \times 10^3}{1 + 1/430} = -2.993\ \text{V/mA}$$

$$R_i = \frac{24.2}{1 + 430} = 0.056\ \Omega \qquad R_o = \frac{72.3}{1 + 430} \cong 0.168\ \Omega$$

(b) We immediately observe that even though a and b are quite different in the two examples, both in values and dimensions, their product $L = ab$ is the *same* (430). However, some of the closed-loop parameters have changed drastically because of the different feedback topology. In both cases, output sensing is of the shunt type, so R_o is virtually unchanged. By contrast, R_i changes from 67 MΩ in the input-series circuit of Example 7.10, to 0.056 Ω in the input-shunt circuit of the present example—quite a change! Also, the closed-loop gain has changed in value, dimensions, and even polarity, as it should be.

As our last example, we turn to the ac circuit of Fig. 7.34a, representing the familiar *feedback-bias* scheme. As we know, this scheme offers a good dc biasing alternative for a transistor. By inspection we find that this circuit is a shunt-shunt type. We thus apply the aforementioned Shunt-Shunt Procedure to end up with the decompositions of Fig. 7.34b and c.

(a) In the shunt-shunt circuit of Fig. 7.34a let $R_B = 100$ kΩ and $R_C = 10$ kΩ. Moreover, let the BJT have $g_m = 1/(25\ \Omega)$, $r_\pi = 5$ kΩ, and $r_o = 100$ kΩ. Find R_i, R_o, and the unloaded gain A_{oc}.
(b) Repeat if R_B is lowered to 5.0 kΩ. Comment.

EXAMPLE 7.15

Solution
(a) With reference to Figs. 7.34b and c we have, by inspection,

$$r_{ia} = R_B//r_\pi = 100//5 = 4.76\ \text{k}\Omega \qquad r_{oa} = R_C//R_B//r_o = 10//100//100 = 8.33\ \text{k}\Omega$$

$$\frac{v_b}{i_\varepsilon} = r_{ia} = 4.76\ \text{V/mA} \qquad \frac{v_o}{v_b} = -g_m r_{oa} = -\frac{8.333}{0.025} = -333\ \text{V/V}$$

$$a = \frac{v_o}{i_\varepsilon} = \frac{v_b}{i_\varepsilon} \times \frac{v_o}{v_b} = -g_m r_{ia} r_{oa} = (4.76 \times 10^3)(-333) = -1.59 \times 10^6\ \text{V/A}$$

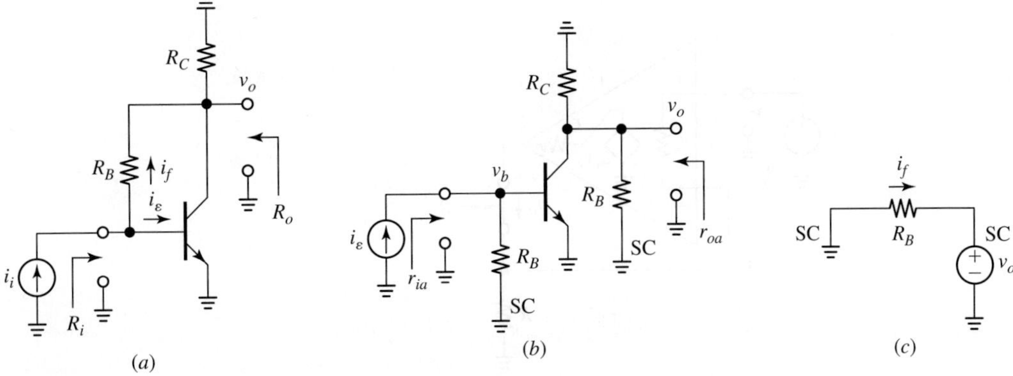

FIGURE 7.34 (a) The feedback-bias BJT configuration as a shunt-shunt circuit. (b) Error amplifier and (c) feedback network.

$$b = \frac{i_f}{v_o} = -\frac{1}{R_B} = -\frac{1}{10^5} \text{ A/V} \qquad L = ab = \frac{-1.59 \times 10^6}{-10^5} = 15.9$$

$$A_{oc} = \frac{v_o}{i_i} = -\frac{10^5}{1 + 1/15.9} = -94 \text{ V/mA}$$

$$R_i = \frac{4.76}{1 + 15.9} \cong 0.28 \text{ k}\Omega \qquad\qquad R_o = \frac{8.33}{1 + 15.9} \cong 0.5 \text{ k}\Omega$$

(b) With $R_B = 5.0$ kΩ we get $r_{ia} = 2.5$ kΩ, $r_{oa} = 3.23$ kΩ, $a = -3.23 \times 10^5$ V/A, and $b = -1/(5 \times 10^3)$ A/V. Consequently,

$$L = 65.5 \qquad A_{oc} = -4.925 \text{ V/mA} \qquad R_i = 38.2 \ \Omega \qquad R_o = 49.2 \ \Omega$$

Note the significant increase in the value of L, from 15.9 to 65.5. Can you justify this intuitively?

Series-Series Circuits

In Fig. 7.35 we re-examine the series-series op amp configuration of Fig. 7.18c, but with the usual eye on the interaction between the amplifier's internal resistances r_i and r_o and the external feedback network, in this case consisting of R. To be able to focus on loading by the feedback network alone, we continue to assume the absence of any external loading, meaning a driving source v_i with zero series resistance, and a short-circuit output load (note that an unloaded output port is an open circuit for a voltage-output device, but a short circuit for a current-output type).

As already seen in connection with the circuit of Fig. 7.20, the op amp's input port appears as an open circuit to the feedback network, so to calculate loading at the amplifier's output we regard the wire to the left of R as an open circuit (OC). To find loading at the input, we note that sensing at the output is of the series type, where we anticipate a high closed-loop resistance. Consequently, to calculate loading at the amplifier's input port, we regard R as if its terminal facing the load was open-circuited (OC).

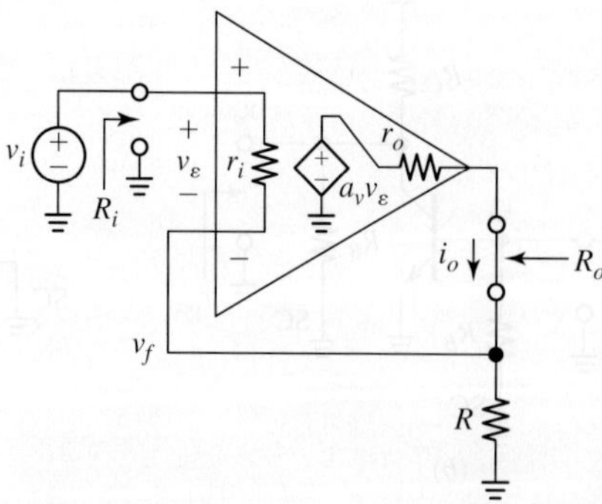

FIGURE 7.35 The transconductance op amp with its internal resistances r_i and r_o explicitly shown.

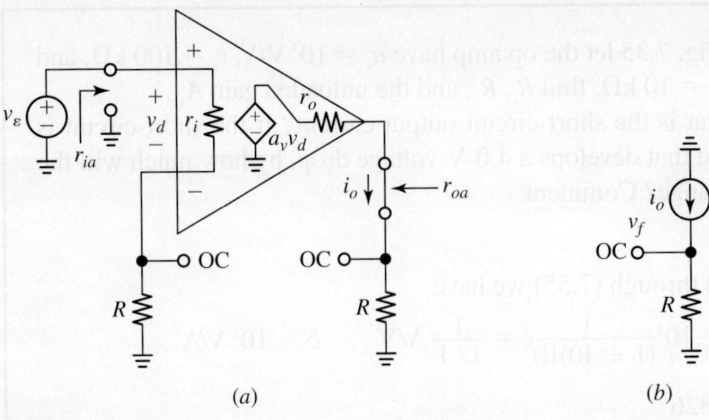

FIGURE 7.36 Decomposing the circuit of Fig. 7.35 into (a) the error amplifier, and (b) the feedback network.

The two situations are depicted in Fig. 7.36a, which represents the error amplifier in open-loop operation, but with loading by the external feedback network specifically taken into account at both ports. Next, we wish to find its open-loop parameters, now denoted as a, r_{ia} and r_{oa}. By Ohm's law and the voltage divider rule,

$$i_o = \frac{a_v v_d}{r_o + R} = \frac{1}{r_o + R} a_v \frac{r_i}{r_i + R} v_\varepsilon$$

so

$$a = \frac{i_o}{v_\varepsilon} = \frac{r_i}{r_i + R} a_v \frac{1}{r_o + R} \tag{7.53}$$

Note that a now has the dimensions of A/V. Moreover, the resistances seen by the input source v_ε and by the short-circuit load are, respectively,

$$r_{ia} = r_i + R \qquad r_{oa} = r_o + R \tag{7.54}$$

Next, we seek a representation for the feedback network that is *separate* from the basic amplifier. Referring back to Fig. 7.35, we observe that at the sensing side R is driven by a current source i_o expected to exhibit extremely high parallel resistance R_o, and that at the summing side it drives the amplifier's input port which, for practical purposes, appears as an open circuit (OC). The circuit to calculate b is thus as in Fig. 7.36b, from which we readily find $v_f = Ri_o$, so

$$b = \frac{v_f}{i_o} = R \tag{7.55}$$

The dimensions of b are V/A, the reciprocal of those of a, as it must be in order to make L dimensionless. As usual, we now apply Eq. (7.36) to find the *unloaded gain* A_{sc}, and Eq. (7.37), but with r_i and r_o replaced by r_{ia} and r_{oa}, to find the closed-loop terminal resistances R_i and R_o.

EXAMPLE 7.16   (a) In the circuit of Fig. 7.35 let the op amp have $a_v = 10^3$ V/V, $r_i = 100$ kΩ, and $r_o = 1.0$ kΩ. If $R = 10$ kΩ, find R_i, R_o, and the unloaded gain A_{sc}.

(b) If $v_i = 5.0$ V, what is the short-circuit output current? If the short-circuit is replaced by a load that develops a 4.0-V voltage drop, by how much will the output current change? Comment.

Solution

(a) Using Eqs. (7.53) through (7.55) we have

$$a = \frac{100}{100 + 10}10^3\frac{1}{(1 + 10)10^3} = \frac{1}{12.1} \text{ A/V} \qquad b = 10^4 \text{ V/A}$$

$$L = \frac{10^4}{12.1} = 826$$

$$r_{ia} = 100 + 10 = 110 \text{ kΩ} \qquad r_{oa} = 1 + 10 = 11 \text{ kΩ}$$

$$A_{sc} = \frac{1/10^4}{1 + 1/826} = 99.88 \text{ } \mu\text{A/V} \qquad R_i = 110(1 + 826) = 91 \text{ MΩ}$$

$$R_o = 11(1 + 826) = 9.1 \text{ MΩ}$$

(b) The short-circuit output current is $i_o = A_{sc}v_i = 99.88 \times 5.0 = 499.4$ μA (ideally, 500 μA). Visualizing the output port via its Norton equivalent, we note that with a 4.0-V load-voltage increase, the current will *decrease* by (4.0 V)/(9.1 MΩ) = 0.44 μA, an insignificant amount. This, thanks to the high output resistance brought about by negative feedback!

The line of reasoning developed for the above example can be generalized to the following **Series-Series Procedure**:

- To calculate loading at the *input* of the basic amplifier, *open-circuit* (OC) the sensing-side port of the feedback network.
- To calculate loading at the *output* of the basic amplifier, *open-circuit* (OC) the summing-side port of the feedback network.
- To find b, apply a current i_o to the sensing-side port of the feedback network, find the *open-circuit* voltage v_f at the summing-side port of the feedback network, and let $b = v_f/i_o$.

Let us illustrate the procedure using the single BJT circuit of Fig. 7.37, here operated in the series-series mode to act as a voltage-to-current (V-I) converter. The circuit is similar to the CC of Fig. 7.28, except that the output is now the short-circuit collector current instead of the open-circuit emitter voltage. To decompose the feedback circuit into the basic error amplifier and the feedback network, refer to the small-signal equivalent of Fig. 7.37b, where we note that the current being sensed is not i_o itself, but i_o/α_0.

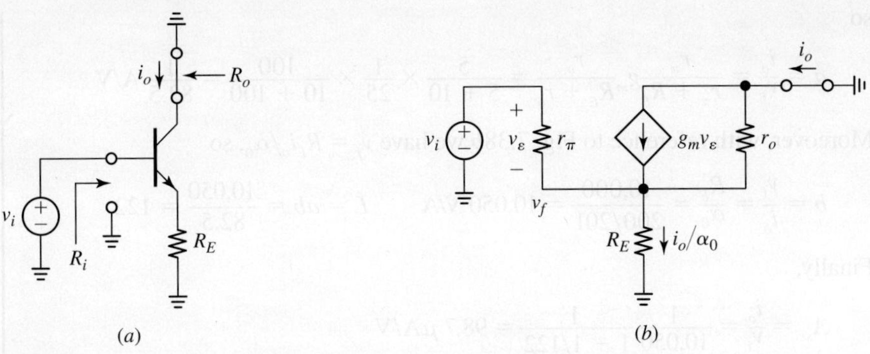

FIGURE 7.37 The single-BJT voltage-to-current (V-I) converter as a series-series feedback system. (*a*) Ac equivalent and (*b*) small-signal representation.

Following the aforementioned Series-Series Procedure, we open-circuit at the right-hand side of R_E to find loading at the input, thus obtaining the left portion of the circuit of Fig. 7.38*a*. Likewise, we open-circuit at the left-hand side of R_E to find loading at the output, thus obtaining the right portion of the circuit of Fig. 7.38*a*. Finally, we drive the feedback network R_E with the current i_o/a_0 to find v_f.

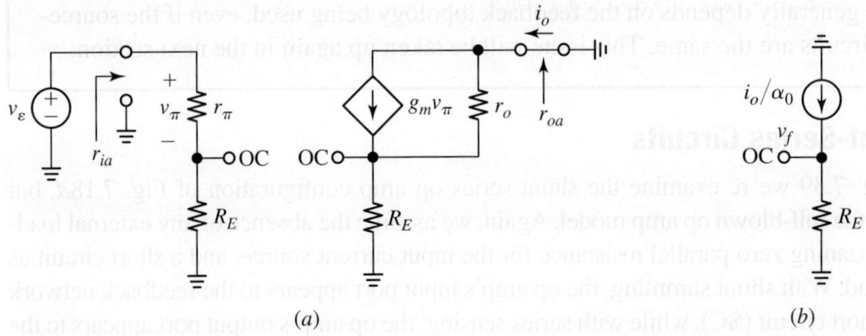

FIGURE 7.38 Decomposing the circuit of Fig. 7.37 into (*a*) the error amplifier and (*b*) the feedback network.

In the V-I converter of Fig. 7.37*a* let $R_E = 10$ kΩ. Moreover, let the BJT have $g_m = 1/(25\ \Omega)$, $r_\pi = 5$ kΩ, and $r_o = 100$ kΩ. Find R_i, R_o, and the unloaded gain A_{sc}. Comment on your findings.

Solution

With reference to Fig. 7.38*a* we have, by inspection,

$$r_{ia} = r_\pi + R_E = 5 + 10 = 15\ \text{k}\Omega \qquad r_{oa} = r_o + R_E = 100 + 10 = 110\ \text{k}\Omega$$

$$i_o = i_{R_E} = \frac{r_o}{R_E + r_o} g_m v_\pi = \frac{r_o}{R_E + r_o} g_m \frac{r_\pi}{r_\pi + R_E} v_\varepsilon$$

so

$$a = \frac{i_o}{v_\varepsilon} = \frac{r_\pi}{r_\pi + R_E} g_m \frac{r_o}{R_E + r_o} = \frac{5}{5 + 10} \times \frac{1}{25} \times \frac{100}{10 + 100} = \frac{1}{82.5} \text{ A/V}$$

Moreover, with reference to Fig. 7.38b we have $v_f = R_E i_o / \alpha_0$, so

$$b = \frac{v_f}{i_o} = \frac{R_E}{\alpha_0} = \frac{10{,}000}{200/201} = 10{,}050 \text{ V/A} \qquad L = ab = \frac{10{,}050}{82.5} = 122$$

Finally,

$$A_{sc} = \frac{i_o}{v_i} = \frac{1}{10{,}050} \frac{1}{1 + 1/122} = 98.7 \ \mu\text{A/V}$$

$$R_i = 15(1 + 122) = 1.85 \text{ M}\Omega \qquad R_o = 110(1 + 122) = 13.5 \text{ M}\Omega$$

Remark: The student should compare the above (approximate) results with those provided by the exact expressions of Section 2.6, and verify that the differences are reasonably small. Even though the V-I converter of Fig. 7.37a is similar to the emitter follower of Fig. 7.28 and uses the same component values, the loop gains are quite different ($L = 364$ for the emitter follower, $L = 122$ for the V-I converter). On the other hand, the series-shunt circuit of Example 7.10 and its shunt-shunt counterpart of Example 7.14 have the same value of L (430). It must be said that L generally depends on the feedback topology being used, even if the source-less circuits are the same. This issue will be taken up again in the next section.

Shunt-Series Circuits

In Fig. 7.39 we re-examine the shunt-series op amp configuration of Fig. 7.18d, but using the full-blown op amp model. Again, we assume the absence of any external loading, meaning zero parallel resistance for the input current source, and a short circuit as the load. With shunt summing, the op amp's input port appears to the feedback network as a short circuit (SC), while with series sensing, the op amp's output port appears to the feedback network as an open circuit (OC). Consequently, we end up with the situation

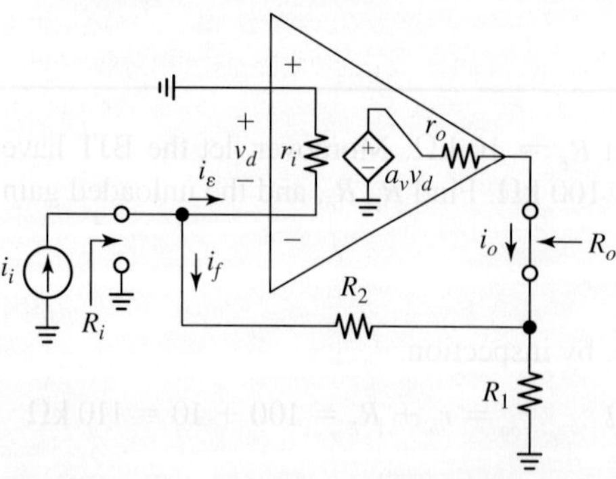

FIGURE 7.39 The op amp as a current amplifier with its internal resistances r_i and r_o explicitly shown.

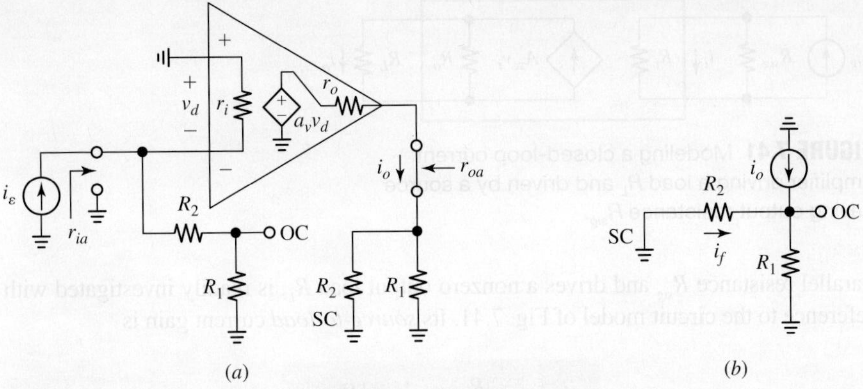

FIGURE 7.40 Decomposing the circuit of Fig. 7.39 into (a) the error amplifier and (b) the feedback network.

of Fig. 7.40a, showing the error amplifier in open-loop operation, but with loading by the feedback network specifically taken into account at both ports. Next, we wish to find its open-loop parameters, now denoted as a, r_{ia} and r_{oa}. By inspection

$$r_{ia} = r_i//(R_2 + R_1) \qquad r_{oa} = r_o + (R_2//R_1) \qquad (7.56)$$

Moreover,

$$i_o = \frac{a_v v_d}{r_{oa}} = \frac{1}{r_{oa}} a_v r_{ia}(-i_\varepsilon)$$

so that

$$a = \frac{i_o}{i_\varepsilon} = -\frac{r_{ia}}{r_{oa}} a_v \qquad (7.57)$$

Note that a now has the dimensions of A/A, and it is negative.

Next, we seek a representation for the feedback network that is *separate* from the basic amplifier. Referring back to Fig. 7.39 we observe that at the sensing side the feedback network is driven by a current source i_o expected to exhibit extremely high parallel resistance R_o, whereas at the summing side it drives the amplifier's input port which, for practical purposes, appears as a short circuit (SC). The circuit for the calculation of b is thus as in Fig. 7.40b. Using the current divider rule, we have

$$b = \frac{i_f}{i_o} = -\frac{R_1}{R_1 + R_2} = -\frac{1}{1 + R_2/R_1} \qquad (7.58)$$

Note that b is negative and is in A/A, just like a. As usual, we now apply Eq. (7.31) to find the *unloaded gain* A_{sc}, and Eqs. (7.32) and (7.33), but with r_i and r_o replaced by r_{ia} and r_{oa}, to find the closed-loop terminal resistances R_i and R_o. Moreover, the general case in which a feedback current amplifier is driven by a signal source i_{sig} with noninfinite

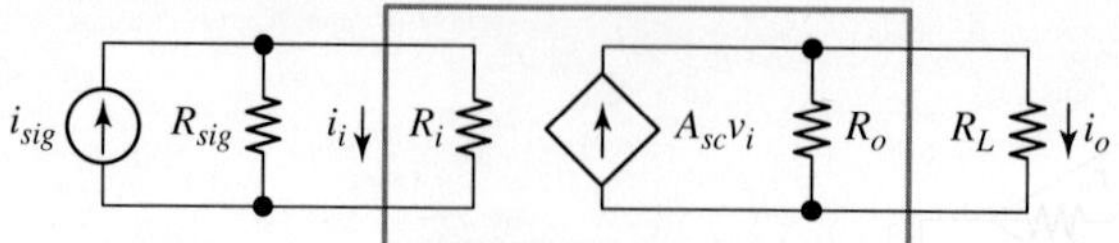

FIGURE 7.41 Modeling a closed-loop current amplifier driving a load R_L and driven by a source having output resistance R_{sig}.

parallel resistance R_{sig} and drives a nonzero output load R_L, is readily investigated with reference to the circuit model of Fig. 7.41. Its *source-to-load* current gain is

$$\frac{i_o}{i_{sig}} = \frac{R_{sig}}{R_{sig} + R_i} A_{sc} \frac{R_o}{R_o + R_L} \tag{7.59}$$

EXAMPLE 7.18

(a) In the circuit of Fig. 7.39 let the op amp have $a_v = 10^3$ V/V, $r_i = 100$ kΩ, and $r_o = 100$ Ω. If $R_1 = 1.0$ kΩ and $R_2 = 9.0$ kΩ, find R_i, R_o, and the unloaded gain A_{sc}.

(b) Find the *signal-to-load gain* i_o/i_{sig} if the feedback amplifier is driven by a signal source i_{sig} having a parallel resistance $R_{sig} = 1.0$ kΩ, and drives a load $R_L = 1.0$ kΩ.

Solution

(a) Using Eqs. (7.56) through (7.58) we have

$$r_{ia} = 100//(9.0 + 1.0) = 9.09 \text{ k}\Omega \qquad r_{oa} = 0.1 + (9.0//1.0) = 1.0 \text{ k}\Omega$$

$$a = -\frac{9.09}{1.0}10^3 = -9{,}090 \text{ A/A}$$

$$b = -\frac{1}{1 + 9/1} = -0.1 \text{ A/A}$$

$$L = ab = 909$$

$$A_{sc} \cong -10\left(1 - \frac{1}{909}\right) = -9.989 \text{ A/A}$$

$$R_i = \frac{9100}{1 + 909} = 10 \ \Omega$$

$$R_o = 1(1 + 909) = 910 \text{ k}\Omega$$

(b) Refer to the equivalent representation of Fig. 7.41, where we note that we have current division both at the input and at the output. The signal-to-load gain is readily found as

$$\frac{i_o}{i_{sig}} = \frac{1.0}{1.0 + 0.010}(-9.989)\frac{910}{910 + 2.0} = -9.868 \text{ A/A}$$

The line of reasoning developed for the above example can be generalized to the following **<u>Shunt-Series Procedure</u>**:

- To calculate loading at the *input* of the basic amplifier, *open-circuit* (OC) the sensing-side port of the feedback network.
- To calculate loading at the *output* of the basic amplifier, *short-circuit* (SC) the summing-side port of the feedback network.
- To find b, apply a current i_o to the sensing-side port of the feedback network, find the *short-circuit* current i_f through the summing-side port of the feedback network, and let $b = i_f/i_o$.

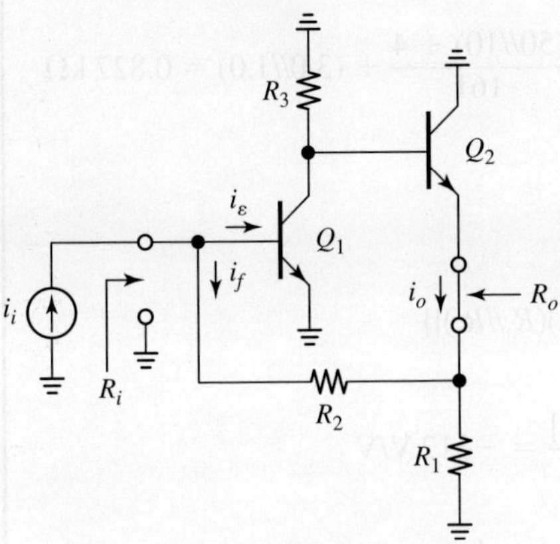

FIGURE 7.42 Shunt-series feedback pair.

Let us illustrate the procedure using the shunt-series pair of Fig. 7.42, also referred to as a current amplifier pair. Applying the aforementioned Shunt-Series Procedure, we end up with the basic amplifier and feedback network of Fig. 7.43.

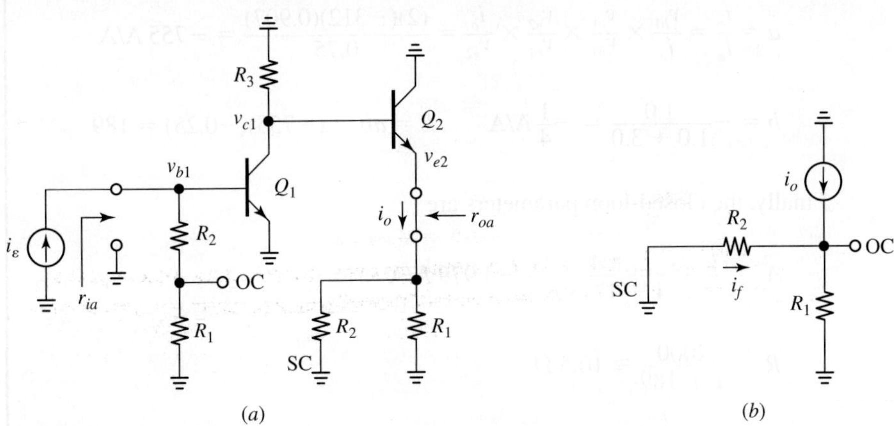

FIGURE 7.43 Decomposing the circuit of Fig. 7.42 into (a) the error amplifier and (b) the feedback network.

EXAMPLE 7.19   (a) In the shunt-series pair of Fig. 7.42 let $R_1 = 1.0$ kΩ, $R_2 = 3.0$ kΩ, and $R_3 = 10$ kΩ. Moreover, let the BJTs have $g_m = 1/(25\ \Omega)$, $r_\pi = 4.0$ kΩ, and $r_o = 50$ kΩ. Find R_i, R_o, and the unloaded gain A_{sc}.

(b) Verify your results via PSpice.

Solution

(a) We have $\beta_0 = g_m r_\pi = 4/0.025 = 160$. With reference to Fig. 7.43 we have, by inspection,

$$r_{ia} = (R_2 + R_1)//r_{\pi 1} = (3.0 + 1.0)//4 = 2.0\ \text{k}\Omega$$

$$r_{oa} = \frac{(r_{o1}//R_3) + r_{\pi 2}}{\beta_{02} + 1} + (R_2//R_1) \cong \frac{(50//10) + 4}{161} + (3.0//1.0) = 0.827\ \text{k}\Omega$$

$$\frac{v_{b1}}{i_\varepsilon} = r_{ia} = 2.0\ \text{k}\Omega$$

$$\frac{v_{c1}}{v_{b1}} = -g_{m1}\{R_3//r_{o1}//[r_{\pi 2} + (\beta_{02} + 1)(R_1//R_2)]\}$$

$$= -\frac{10//50//[4 + 161(1.0//3.0)]}{0.025} = -312\ \text{V/V}$$

$$\frac{v_{e2}}{v_{c1}} = \frac{1}{1 + \dfrac{(R_3//r_{o2}) + r_{\pi 2}}{(\beta_{02} + 1)(R_1//R_2)}} = \frac{1}{1 + \dfrac{10//50 + 4}{161(1.0//3.0)}} = 0.907\ \text{V/V}$$

$$\frac{i_o}{v_{e2}} = \frac{1}{R_1//R_2} = \frac{1}{1.0//3.0} = \frac{1}{0.75\ \text{k}\Omega}$$

$$a = \frac{i_o}{i_\varepsilon} = \frac{v_{b1}}{i_\varepsilon} \times \frac{v_{c1}}{v_{b1}} \times \frac{v_{e2}}{v_{c1}} \times \frac{i_o}{v_{e2}} = \frac{(2)(-312)(0.907)}{0.75} = -755\ \text{A/A}$$

$$b = -\frac{1.0}{1.0 + 3.0} = -\frac{1}{4}\ \text{A/A} \qquad L = ab = (-755)(-0.25) = 189$$

Finally, the closed-loop parameters are

$$A_{sc} = \frac{i_o}{i_i} \cong \frac{-4}{1 + 1/189} = -3.979\ \text{A/A}$$

$$R_i = \frac{2000}{1 + 189} \cong 10.5\ \Omega$$

$$R_o = 0.827(1 + 189) = 157\ \text{k}\Omega$$

(b) The PSpice circuit, shown in Fig. 7.44, includes the 0-V dummy source V_L to measure the short-circuit load current. After directing PSpice to perform the transfer function (.TF) analysis we get $A_{sc} = -3.980$ A/A, $R_i = 9.877\ \Omega$ and $R_o = 167.4$ kΩ, all in fair agreement with our calculations above.

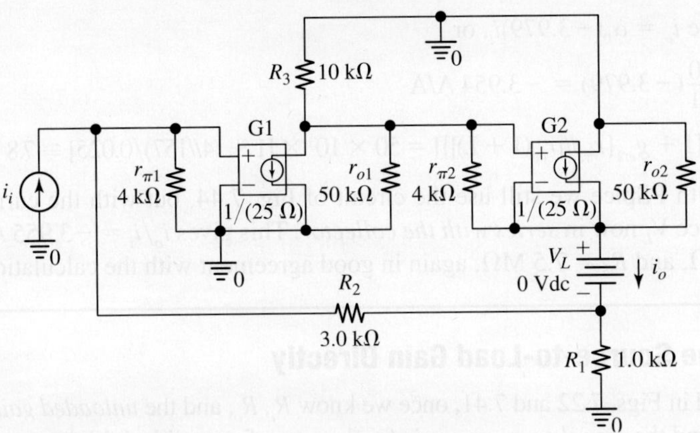

FIGURE 7.44 PSpice circuit to verify Example 7.19.

If an even higher output resistance is desired, the load can be moved to the collector side, which usually presents a much higher ac resistance than the emitter. This results in the alternative of Fig. 7.45a. Note that the load is now *outside* the feedback loop; however, the B-E port of Q_2 is still *inside* the loop, so the current being regulated is still Q_2's emitter current i_{e2}. This current is related to the load current as $i_{e2} = i_o/\alpha_0$. Moreover, as depicted in Fig. 7.45b, negative feedback in effect places in series with Q_2's emitter a resistance equal to that found in connection with Fig. 7.42, namely, $r_{oa}(1 + L)$.

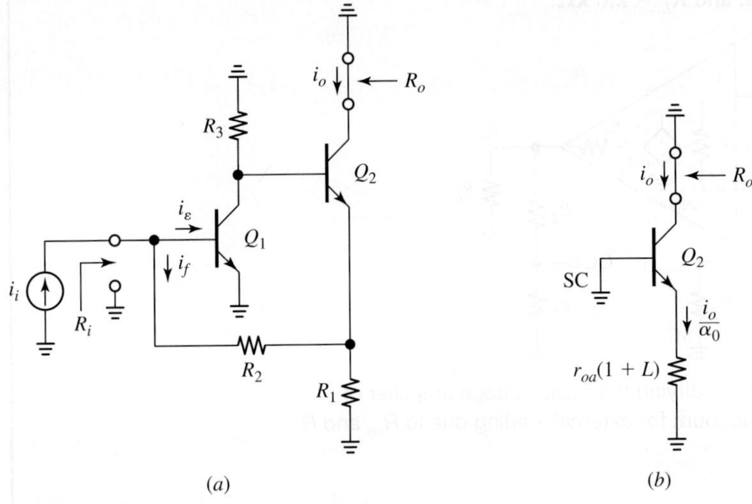

(a)　　　　　　　　　　　　　　　　(b)

FIGURE 7.45 (a) Shunt-series feedback pair with the load in series with Q_2's collector. (b) Equivalent circuit for finding i_o and R_o.

EXAMPLE 7.20   Assuming the same parameters of Example 7.19, along with $r_\mu = \infty$, find the unloaded gain i_o/i_i as well as R_o for the circuit of Fig. 7.45a. Hence, verify via PSpice.

Solution

We now have $i_o = \alpha_0(-3.979)i_i$, or

$$\frac{i_o}{i_i} = \frac{160}{161}(-3.979) = -3.954 \text{ A/A}$$

$$R_o \cong r_{o2}[1 + g_{m2}\{r_{\pi2}//[r_{oa}(1 + L)]\}] = 50 \times 10^3 \times [1 + (4//157)/0.025] \cong 7.8 \text{ M}\Omega$$

To verify with PSpice we still use the circuit of Fig. 7.44, but with the current-sensing source V_L now in *series with the collector*. This gives $i_o/i_i = -3.955$ A/A, $R_i = 9.877\ \Omega$, and $R_o = 7.5$ MΩ, again in good agreement with the calculations.

Finding the Source-to-Load Gain Directly

As illustrated in Figs. 7.22 and 7.41, once we know R_i, R_o, and the *unloaded gain* A_{oc} or A_{sc}, we can find the *signal-to-source gain* for the case of a non-ideal driving source and an arbitrary output load. However, the need often arises to find the signal-to-load gain *directly*, without having to go through the intermediate calculations of R_i and R_o. This is achieved by *including* also the *signal source resistance* R_{sig} and the *load resistance* R_L in the circuit of the basic amplifier when calculating its open-loop gain a. Then, the source-to-load gain is simply $A = a/(1 + ab)$. We shall demonstrate for the cases of voltage and current amplifiers, but the technique is applicable also to the other two amplifier types.

EXAMPLE 7.21   Use direct calculation to find the *signal-to-load gain* $A = v_o/v_{sig}$ of Example 7.9b, for which $a_v = 1000$ V/V, $r_i = 10$ kΩ, $r_o = 1.0$ kΩ, $R_1 = 1.0$ kΩ, $R_2 = 9.0$ kΩ, $R_{sig} = 20$ kΩ, and $R_L = 2.0$ kΩ.

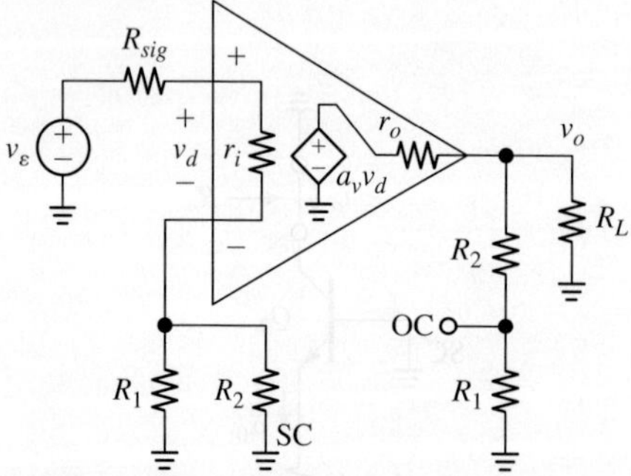

FIGURE 7.46 Modifying the basic voltage amplifier of Fig. 7.21a to account for external loading due to R_{sig} and R_L.

Solution

Redraw the circuit of Fig. 7.21a, but with R_{sig} and R_L also included. This results in the circuit of Fig. 7.46, where we note that R_{sig} appears in series and R_L in shunt, in accordance with the present topology type. Thus, Eq. (7.46) changes to

$$a = \frac{v_o}{v_\varepsilon} = \frac{r_i}{R_{sig} + r_i + (R_1 /\!/ R_2)} a_v \frac{(R_1 + R_2)/\!/R_L}{r_o + [(R_1 + R_2)/\!/R_L]}$$

or

$$a = \frac{10}{20 + 10 + (1.0/\!/9.0)} 1000 \frac{(1.0 + 9.0)/\!/2.0}{1 + [(1.0 + 9.0)/\!/2.0]}$$

$$= 0.324 \times 1000 \times 0.625 = 202 \text{ V/V}$$

By Eq. (7.48) we still have $b = 0.1$ V/V, so $L = 202 \times 0.1 = 20.2$ and

$$\frac{v_o}{v_{sig}} = \frac{1/b}{1 + 1/L} = \frac{10}{1 + 1/20.2} = 9.528 \text{ V/V}$$

in reasonable agreement with the result of Example 7.9*b*.

Use direct calculation to find the *signal-to-load gain* $A = i_o/i_{sig}$ of the circuit of   **EXAMPLE 7.22**
Example 7.18*b*.

Solution

Redraw the circuit of Fig. 7.40*a*, but with R_{sig} and R_L also included. The result is shown in Fig. 7.47, where we note that R_{sig} appears in shunt and R_L in series, in accordance with the present feedback topology. Thus, Eq. (7.57) changes to

$$a = \frac{i_o}{i_\varepsilon} = -\frac{R_{sig}/\!/r_i/\!/(R_1 + R_2)}{r_o + R_L + (R_1/\!/R_2)} a_v = -\frac{1.0/\!/100/\!/(1.0 + 9.0)}{1.0 + 2.0 + (1.0/\!/9.0)} 10^3 = -231 \text{ A/A}$$

With b still given as in Eq. (7.58), or $b = -0.1$ A/A, we get

$$\frac{i_o}{i_{sig}} = \frac{a}{1 + ab} = \frac{-231}{1 + (-231)(-0.1)} = -9.585 \text{ A/A}$$

in reasonable agreement with the result of Example 7.18*b*.

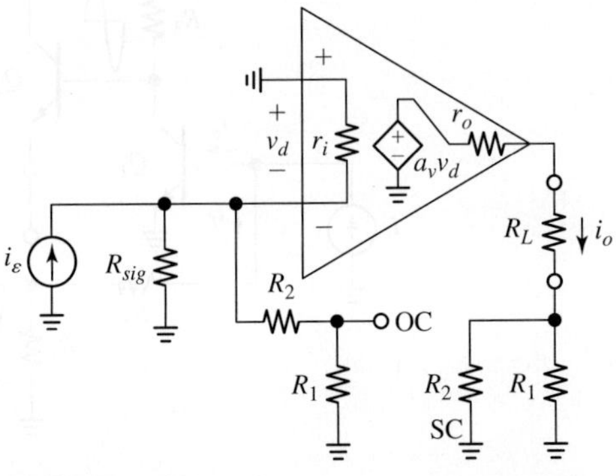

FIGURE 7.47 Modifying the basic current amplifier of Fig. 7.40*a* to account for external loading due to R_{sig} and R_L.

Identifying Feedback Type and Topology

The analytical tools developed so far can be applied to any feedback circuit provided feedback is itself *negative*. To assess feedback polarity, apply a stimulus s_i at the input, follow its effect around the loop, and determine whether the returned signal s_f opposes or reinforces s_i. Only if it *opposes* is feedback *negative*. The procedure is readily visualized via signal "bumps" as exemplified in Fig. 7.48.

In Fig. 7.48a we have frozen the ac input v_i during a *positive* alternation, signified by a positive bump. This bump is *amplified* and *inverted* by Q_1 and then by Q_2, and is finally buffered by Q_4 back to the input port via the voltage divider R_2-R_1. With two inversions, the returned v_f bump has the same polarity as the original v_i bump. Since Q_1 responds to the *difference* $v_i - v_f$, it is apparent that v_f tends to neutralize the effect of v_i, so feedback is negative. Had R_2 been returned to Q_1's base instead of the emitter, v_f would have reinforced v_i, so feedback would have been positive.

In Fig. 7.48b a positive current bump in i_i is converted by Q_1 to a negative voltage bump at the collector, which is then buffered by Q_2 back to Q_1's base via R_2. With a negative bump at its right side, R_2 will sink current away from to Q_1's base. Since Q_1 responds to the *difference* $i_i - i_f$, it is apparent that i_f tends to neutralize the effect of i_i, so feedback is negative. Had we inserted an additional *inverting* stage between Q_1 and Q_2 in the manner of Fig. 7.48a, the direction of i_f would have been reversed, causing i_f to reinforce i_i, so feedback would have been positive.

To identify the feedback topology we must determine the type of signal *sensing* at the output as well as the type of signal *summing* at the input. To identify the type of output sensing, proceed like this:

- To test for *voltage sensing*, set $v_o \to 0$ by short-circuiting the output port. Then, if the signal returned to the input port also drops to zero, sensing is of the *shunt* type.
- To test for *current sensing*, set $i_o \to 0$ by open-circuiting the output port. Then, if the signal returned to the input port also drops to zero, sensing is of the *series* type.

As an example, short-circuiting the output port in Fig. 7.48a yields $v_f \to 0$, confirming voltage feedback. Conversely, open-circuiting the output port in Fig. 7.48b yields

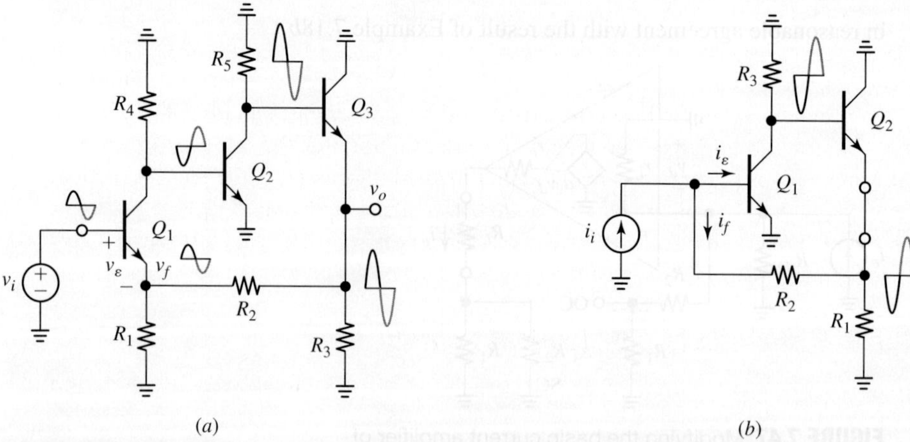

(a)   (b)

FIGURE 7.48 Using voltage "bumps" to investigate signal propagation around the feedback loop of (a) the series-shunt triple and (b) the shunt-series pair studied above.

$i_f \rightarrow 0$, confirming current feedback. In general one would apply both tests, one to identify and the other to corroborate.

To identify the type of input summing, proceed as follows:

- To test for *voltage summing*, look for an input *mesh* comprising the *series* combination of the input voltage, the voltage controlling the amplifier, and the voltage fed back from that across the output load. If such a mesh can be found, then summing is of the *series* type. If this test fails, then test for current summing.
- To test for *current summing*, look for an input *node* comprising the *shunt* combination of the input current, the current controlling the amplifier, and the current fed back from that through the output load. If such a node can be found, then summing is of the *shunt* type. If this test fails, then test for voltage summing.

In Fig. 7.48*a* we identify a mesh comprising the series combination of v_i, v_ε, and v_f, indicating series summing. In Fig. 7.48*b* we identify a node comprising i_i, i_ε, and i_f, indicating shunt summing. No such node is present in Fig. 7.48*a*, just as no mesh counterpart is present in Fig. 7.48*b*.

It also helps exercising intuition to anticipate the expected resistance level: a high resistance implies series type, and a low resistance implies shunt type. The student is urged to apply the sensing and summing tests to all circuits investigated so far. With enough practice, you will soon be able to identify a topology by mere inspection!

7.5 RETURN-RATIO ANALYSIS

There is no doubt that the loop gain plays a central role in a negative-feedback system. Since it gives a measure of how close circuit behavior comes to ideal, the designer often needs to come up with a quick estimate for this gain to see whether the circuit under development meets the design specifications or needs improvement. *Two-port techniques* require that we find *a* and *b* *separately* via suitable circuit manipulations and approximations, and then estimate the loop gain as $L = ab$ (as seen in the previous section, both *a* and *b* vary with the feedback topology in use). Unlike two-port methods, *direct methods* are based on the idea of injecting a stimulus into the feedback loop and then observing the system's response to find the overall loop characteristics (as such, they are topology independent). As we inject a stimulus we must be careful (*a*) not to upset the existing *dc bias conditions* and (*b*) not to alter the local *loading conditions*. Two techniques are available for the direct estimation of the feedback loop characteristics. The first method, discussed in the present section, is suited to hand calculations on circuit schematics involving dependent sources. The second method, discussed in the next section, is suited to test and measurement environments such as lab measurements or transistor-level computer simulations.

The Return Ratio of a Dependent Source

Given a *dependent source* inside a single-loop feedback circuit, we find its *return ratio* as follows:

- Set the external driving source s_i to zero, that is, replace s_i with a *short circuit* if it is a *voltage* source, or with an *open circuit* if it is a *current* source.

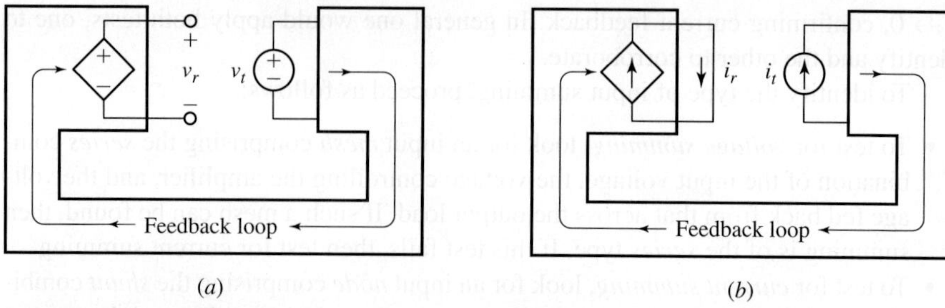

FIGURE 7.49 Finding the return ratio T of a dependent source of the (a) voltage type and (b) current type.

- Break the feedback loop *right downstream* of the dependent source, and leave this source *open-circuited* if it is of the *voltage* type, or *short-circuited* it if it is of the *current* type.
- If the dependent source is of the *voltage* type, drive the circuit downstream with a *test voltage* v_t having the *same polarity* as the dependent source (see Fig. 7.49a). Hence, find the *open-circuit voltage* v_r returned by the dependent source, and obtain the source's return ratio, denoted as T, as

$$T = -\frac{v_r}{v_t} \tag{7.60a}$$

- If the dependent source is of the *current* type, drive the circuit downstream with a *test current* i_t having the *same direction* as the dependent source (see Fig. 7.49b). Hence, find the *short-circuit current* i_r returned by the dependent source, and obtain the source's return ratio as

$$T = -\frac{i_r}{i_t} \tag{7.60b}$$

Comparing T and L

Intuitively, it would seem that T is the same as L (in fact, it is common practice to refer also to T as the loop gain, further adding to the confusion.) Though they may coincide in some some cases, T and L are generally *different* (when necessary to distinguish between the two, we shall refer to T as the *return-ratio loop gain*, and to L as the *two-port loop gain*). To understand the difference, recall that L as defined in Eq. (7.6) is predicated on two premises, namely,

- *Forward* signal transmission occurs exclusively through the *amplifier* (in Fig. 7.1 we signify this by representing the amplifier with an arrowhead in the forward direction)
- *Reverse* signal transmission occurs exclusively through the *feedback network* (in Fig. 7.1 we signify this by representing the feedback network with an arrowhead in the reverse direction)

When this is the case, the amplifier and feedback network are said to be *unilateral*. However, most real-life feedback networks are *bilateral*. Add to this the fact that in the two-port approach we use short-circuit (SC) and open-circuit (OC) approximations to expedite the calculation of a and b, and we have additional reasons for possible differences between T and L. At this juncture it must be stressed that the calculation of T does not predicate any premises and thus yields *exact* results.

We can better appreciate the similarities and differences between T and L by examining the manner in which each intervenes in the expression for the closed-loop gain A. Let the gain of the dependent source in question be denoted as k (for instance, $k = a_v$ for an op amp, $k = \beta_0$ for a BJT, $k = g_m$ for a FET or a BJT, $k = z$ for a CFA). As we know, the closed-loop gain A is related to the loop gain L as

$$A = \frac{s_o}{s_i} = \frac{A_{\text{ideal}}}{1 + 1/L} \tag{7.61}$$

where

$$A_{\text{ideal}} = \lim_{k \to \infty} \frac{s_o}{s_i} \tag{7.62}$$

is the value of A in the idealized limit of the dependent source having *infinite gain*. We visualize Eq. (7.61) via the familiar block diagram of Fig. 7.50a, which has been derived from Fig. 7.1 by letting $b \to 1/A_{\text{ideal}}$ and $a \to ab/b = L/b = LA_{\text{ideal}}$. As we know, a signal starting at the error amplifier's input and going around the loop clockwise undergoes an overall amplification of $(LA_{\text{ideal}}) \times (1/A_{\text{ideal}}) \times (-1)$, or $-L$.

By contrast, the dependence of the closed-loop gain A upon the return ratio T takes on the form[1, 4]

$$A = \frac{s_o}{s_i} = \frac{A_{\text{ideal}}}{1 + 1/T} + \frac{a_{\text{ft}}}{1 + T} \tag{7.63}$$

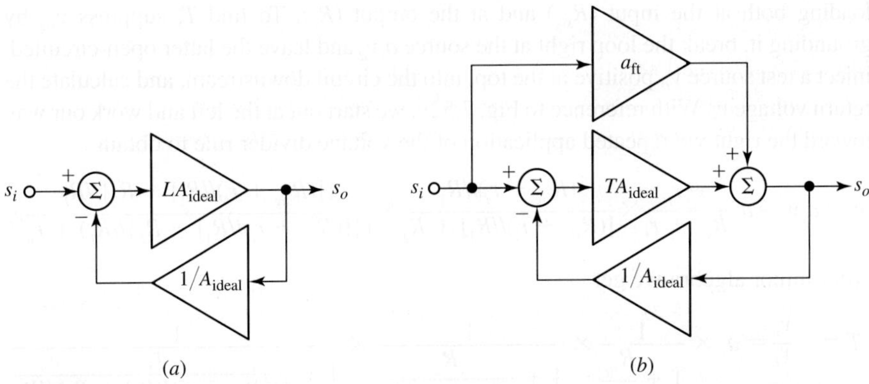

FIGURE 7.50 Block diagrams illustrating the roles of (a) the loop gain L, and (b) the return ratio T.

where

$$a_{ft} = \lim_{k \to 0} \frac{s_o}{s_i} \tag{7.64}$$

is the *feedthrough gain*, that is, the gain with the dependent source *set to zero*. This gain stems from forward transmission *around* the dependent source. We visualize Eq. (7.63) via the block diagram of Fig. 7.50*b*, which is a generalization of that of Fig. 7.50*a* because it includes also the feedthrough path shown at the top.

Compared to two-port analysis, return-ratio analysis is *more insightful* as it splits A into two separate components, both stemming from forward transmission, but one through the error amplifier and the other through the feedback network. Two-port analysis, on the other hand, attempts at making the entire feedback system conform to the simpler diagram of Fig. 7.50*a*, this being the reason why it gives only approximate results compared to the exact results of the return-ratio method.[1,4] It is apparent from Eq. (7.63) that if the condition

$$|a_{ft}| \ll |TA_{ideal}| \tag{7.65}$$

is met, the feedthrough component of A in Eq. (7.63) can be ignored, implying that Eq. (7.63) becomes *formally identical* to Eq. (7.61). Even so, T and L may still differ from each other due to the approximations at the basis of L. As such, T and L tend to provide different values for A, though this difference may be very small for large T and L.

As an alternative viewpoint, we can regard the signal component $a_{ft}s_i$ that manages to creep to the output as a form of *output noise*. Reflected to the input, this noise gets divided by the gain TA_{ideal}, thus resulting in the equivalent input noise component $s_{ni} = a_{ft}s_i/(TA_{ideal})$. It is clear that as long as Eq. (7.65) is met, then $|s_{ni}| \ll |s_i|$, so the fact that the feedback network is not unilateral is of little consequence in this case.

Return-Ratio Calculation Examples

The above concepts are best illustrated via practical examples. Let us start with the familiar op amp circuit of Fig. 7.51, now shown in a more general setting that includes loading both at the input (R_{sig}) and at the output (R_L). To find T, suppress v_{sig} by grounding it, break the loop right at the source $a_v v_d$ and leave the latter open-circuited, inject a test source v_t, positive at the top, into the circuit downstream, and calculate the return voltage v_r. With reference to Fig. 7.52*a*, we start out at the left and work our way toward the right via repeated application of the voltage divider rule to obtain

$$v_r = a_v v_d = a_v \frac{-r_i}{R_{sig} + r_i} \times \frac{(R_{sig} + r_i)//R_1}{[(R_{sig} + r_i)//R_1] + R_2} \times \frac{\{[(R_{sig} + r_i)//R_1] + R_2\}//R_L}{(\{[(R_{sig} + r_i)//R_1] + R_2\}//R_L) + r_o} v_t$$

After minor algebra we get

$$T = -\frac{v_r}{v_t} = a_v \times \frac{1}{1 + \dfrac{R_{sig}}{r_i}} \times \frac{1}{1 + \dfrac{R_2}{(R_{sig} + r_i)//R_1}} \times \frac{1}{1 + \dfrac{r_o}{\{[(R_{sig} + r_i)//R_1] + R_2\}//R_L}}$$

$$\tag{7.66}$$

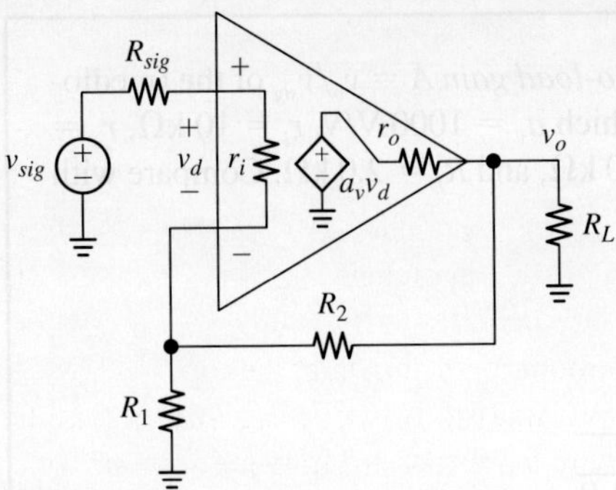

FIGURE 7.51 Non-inverting op amp with input and output loading.

To find a_{ft}, let $a_v \to 0$, so that the dependent-source side of r_o goes to 0 V, or ground. This gives us the situation of Fig. 7.52b. Starting out at the right, and applying the voltage divider rule twice, we get

$$v_o = \frac{r_o/\!/R_L}{R_2 + (r_o/\!/R_L)} \times \frac{R_1/\![R_2 + (r_o/\!/R_L)]}{R_{sig} + r_i + \{R_1/\![R_2 + (r_o/\!/R_L)]\}} v_{sig}$$

or

$$a_{\mathrm{ft}} = \lim_{a_v \to 0} \frac{v_o}{v_{sig}} = \frac{1}{1 + \dfrac{R_2}{r_o/\!/R_L}} \times \frac{1}{1 + \dfrac{R_{sig} + r_i}{R_1/\![R_2 + (r_o/\!/R_L)]}} \qquad (7.67)$$

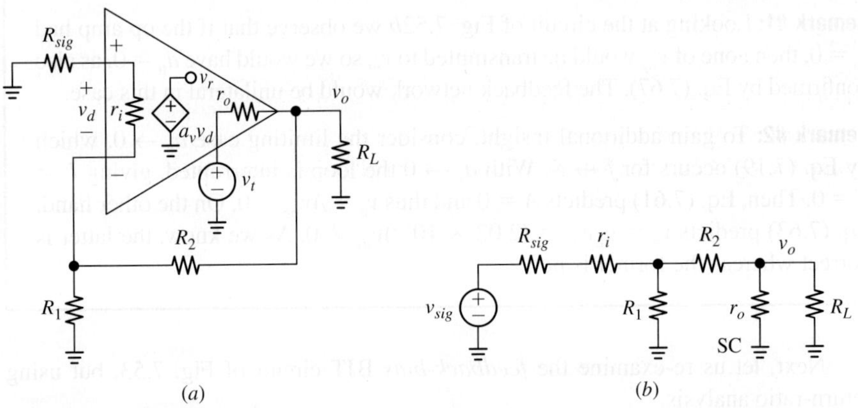

(a)

(b)

FIGURE 7.52 Modifying the circuit of Fig. 7.51 to find (a) its return ratio T, and (b) the forward transmission a_{ft} around the dependent source.

EXAMPLE 7.23   Use return-ratio analysis to find the *signal-to-load gain* $A = v_o/v_{sig}$ of the (mediocre) op amp circuit of Example 7.21, for which $a_v = 1000$ V/V, $r_i = 10$ kΩ, $r_o = 1.0$ kΩ, $R_1 = 1.0$ kΩ, $R_2 = 9.0$ kΩ, $R_{sig} = 20$ kΩ, and $R_L = 2.0$ kΩ. Compare with Example 7.21 and comment.

Solution

Using Eqs. (7.66) and (7.67), we get

$$T = 10^3 \times \frac{1}{1 + \dfrac{20}{10}} \times \frac{1}{1 + \dfrac{9.0}{(20 + 10)//1.0}}$$

$$\times \frac{1}{1 + \dfrac{1.0}{\{[(20 + 10)//1.0] + 9.0\}//2.0}} = 10^3 \times \frac{1}{3} \times \frac{1}{10.3} \times \frac{1}{1.6} = 20.2$$

and

$$a_{\text{ft}} = \frac{1}{1 + \dfrac{9.0}{1.0//2.0}} \times \frac{1}{1 + \dfrac{20 + 10}{1.0//[9.0 + (1.0//2.0)]}}$$

$$= \frac{1}{14.5} \times \frac{1}{34.1} = 2.02 \times 10^{-3} \text{ V/V}$$

Also, we know from basic op amp theory that in the limit $a_v \to \infty$ we have $A_{\text{ideal}} = 1 + R_2/R_2 = 1 + 9.0/1.0 = 10$ V/V, so Eq. (7.63) gives

$$A = \frac{10}{1 + 1/20.2} + \frac{2.02 \times 10^{-3}}{1 + 20.2} = 9.528 + 0.000095 \cong 9.528 \text{ V/V}$$

The value of T found here matches the value of L of Example 7.21, so the effect of feedthrough is truly negligible in this example. Indeed, the numerical decomposition of A reveals that the term due to a_{ft} accounts for only 0.001% of the overall gain A. Alternatively, the fact that $TA_{\text{ideal}} = 20.2 \times 10 = 202$ V/V and $a_{\text{ft}} = 2.02 \times 100^{-3}$ indicates that Eq. (7.65) is met abundantly.

Remark #1: Looking at the circuit of Fig. 7.52*b* we observe that if the op amp had $r_o = 0$, then none of v_{sig} would be transmitted to v_o, so we would have $a_{\text{ft}} = 0$, as also confirmed by Eq. (7.67). The feedback network would be unilateral in this case.

Remark #2: To gain additional insight, consider the limiting case $a_v \to 0$, which by Eq. (7.19) occurs for $f \to \infty$. With $a_v \to 0$ the loop is interrupted, giving $L = T = 0$. Then, Eq. (7.61) predicts $A = 0$ and thus $v_o = Av_{sig} = 0$. On the other hand, Eq. (7.63) predicts $v_o = a_{\text{ft}}v_{sig} = (2.02 \times 10^{-3})v_{sig} \neq 0$. As we know, the latter is correct whereas the former is not.

Next, let us re-examine the *feedback-bias* BJT circuit of Fig. 7.53, but using return-ratio analysis.

To find T, refer to the ac equivalent of Fig. 7.54*a*, where the loop has been broken right at the upper terminal of the dependent source, which must be short-circuited in

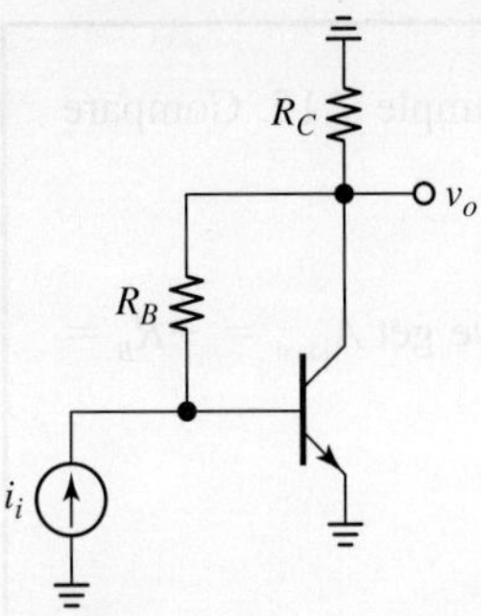

FIGURE 7.53 The feedback-bias BJT configuration.

order to provide a path for its current to flow. Next, subject the rest of the circuit to a test source i_t flowing downward like the dependent source. Using the voltage divider rule and Ohm's law we get

$$i_r = g_m v_\pi = g_m \frac{r_\pi}{r_\pi + R_B} v_o = g_m \frac{r_\pi}{r_\pi + R_B}[(R_B + r_\pi)//R_C//r_o](-i_t)$$

Letting $g_m v_\pi \to \beta_0$ we obtain, after minor manipulation,

$$T = -\frac{i_r}{i_t} = \frac{\beta_0}{1 + \dfrac{R_B + r_\pi}{R_C//r_o}} \tag{7.68}$$

(Note that $T < \beta_0$.) To find a_{ft}, suppress the dependent source to obtain the situation of Fig. 7.54b. Using the voltage divider rule and Ohm's law, we get

$$a_{ft} = \lim_{g_m \to 0} \frac{v_o}{i_i} = \frac{r_o//R_C}{R_B + (r_o//R_C)} \times \{r_\pi//[R_B + (r_o//R_C)]\} \tag{7.69}$$

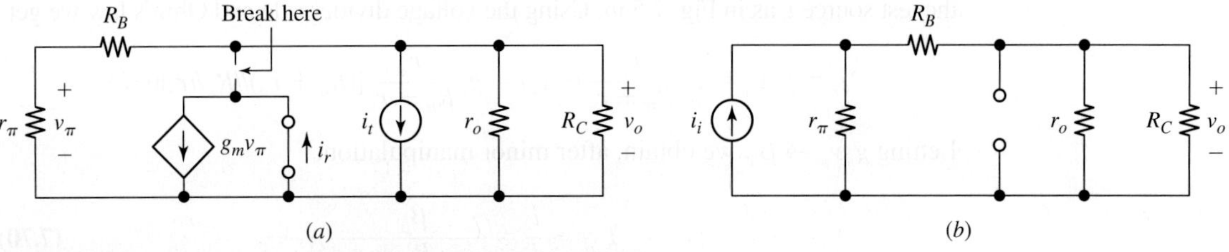

(a)　　　　　　　　　　　　　　　(b)

FIGURE 7.54 Ac models of the circuit of Fig. 7.53 to find (a) T and (b) a_{ft}.

EXAMPLE 7.24   Find T, a_{ft}, and A for the feedback-bias BJT circuit of Example 7.15. Compare with the results found there, and comment.

Solution

(a) Using the component values given in Example 7.15a we get $A_{ideal} = -R_B = -100$ V/mA,

$$T = \frac{200}{1 + \dfrac{100 + 5}{10//100}} = 15.9$$

$$a_{ft} = \frac{100//10}{100 + (100//10)} \times \{5//[100 + (100//10)]\} = 0.4 \text{ V/mA}$$

$$A = \frac{-100}{1 + 1/15.9} + \frac{0.40}{1 + 15.9} = -94.08 + 0.024 = -94.1 \text{ V/mA}$$

In Example 7.15a we found $L = 15.9$ and $A = -94.0$ V/mA. In this case T and L coincide, and feedthrough accounts for only 0.025% of A, a truly negligible amount.

(b) Lowering R_B from 100 kΩ to 5.0 kΩ gives

$$A_{ideal} = -5.0 \text{ V/mA} \qquad T = 95.2$$

$$a_{ft} = 2.4 \text{ V/mA} \qquad A = -4.948 + 0.025 = -4.923 \text{ V/mA}$$

In Example 7.15b we found $L = 65.5$ and $A = -4.925$ V/mA. Even though the values of A are still very close, T and L differ appreciably, and feedthrough has increased from 0.025% to about 0.5% of A. Note that the value of A obtained via T is exact, whereas that obtained via L is only approximate.

Next, consider the generalized BJT circuit of Fig. 7.55a, shown with the external driving source already suppressed. Depending on where we apply the input and where we obtain the output, different topologies can be realized with just this one circuit! Thus, applying a voltage to the base or a current to the emitter we have, respectively, series or shunt summing. Using as output the emitter voltage or the collector current we have, respectively, shunt or series sensing. Yet, T is an *intrinsic* characteristic of the circuit that must be *invariant* irrespective of the topology in use. On the contrary, both L and a_{ft} will generally vary with the topology. To find T, break the loop right downstream of the controlled source, short-circuit the latter, and then subject the circuit downstream to the test source i_t as in Fig. 7.55b. Using the voltage divider rule and Ohm's law we get

$$i_r = g_m v_\pi = g_m \frac{r_\pi}{R_B + r_\pi}(-v_e) = g_m \frac{r_\pi}{R_B + r_\pi}[(R_B + r_\pi)//R_E//r_o](-i_t)$$

Letting $g_m v_\pi \to \beta_0$, we obtain, after minor manipulation,

$$T = -\frac{i_r}{i_t} = \frac{\beta_0}{1 + \dfrac{R_B + r_\pi}{R_E//r_o}} \tag{7.70}$$

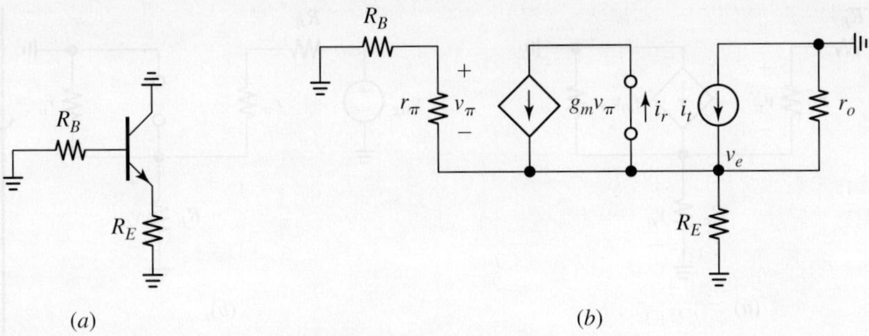

FIGURE 7.55 (a) A generalized BJT ac circuit, and (b) its small-signal model to find T.

(Note that $T < \beta_0$.) Interestingly enough, there is a formal similarity with Eq. (7.68), provided we let $R_C \to R_E$. Can you explain why?

EXAMPLE 7.25

(a) In the generalized ac circuit of Fig. 7.55a, let $R_B = 0$ and $R_E = 10\ \text{k}\Omega$. More-over, let the BJT have $g_m = 1/(25\ \Omega)$, $r_\pi = 5\ \text{k}\Omega$, and $r_o = 100\ \text{k}\Omega$, as in previous examples. Find T.

(b) Repeat, but with $R_B = 30\ \text{k}\Omega$.

(c) Under what conditions is T maximized, and what is its value in this example?

Solution

(a) Applying Eq. (7.70) we get

$$T = \frac{200}{1 + \dfrac{5}{10//100}} = 129$$

(b) Changing R_B from zero to 30 kΩ lowers the return ratio to $T = 41$.

(c) Equation (7.70) indicates that T is maximized when $R_B = 0$ and $R_E = \infty$, implying the use of an ideal current sink to provide emitter bias for the CB transistor. Then, $T_{max} = \beta_0/(1 + r_\pi/r_o) = 190.5$.

EXAMPLE 7.26

(a) Using the value of T calculated in Example 7.25a, find the gain A for the case in which the generalized BJT circuit of Fig. 7.55a is operated in the *series-shunt* mode as in Example 7.12, with the output being the emitter voltage. Compare and comment.

(b) Find the gain A if the generalized BJT circuit of Fig. 7.55a is operated in the *series-series* mode as in Example 7.17, with the output being now the collec-tor current. Compare and comment.

Solution

From Example 7.25a we know that $T = 129$ regardless of the feedback topology.

(a) For *series-shunt* operation, refer to Fig. 7.56. To find A_{ideal}, let $g_m \to \infty$ in Fig. 7.56a. In this limit the dependent source will require a vanishingly small control voltage v_π to sustain a finite output v_o. So, both the voltage across and

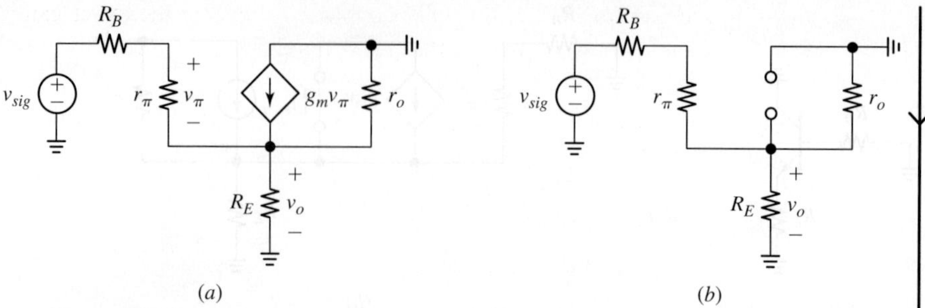

FIGURE 7.56 (a) Ac model for the BJT of Fig. 7.55a in series-shunt operation. (b) Circuit to find a_{ft}.

the current through r_π will approach zero, giving, for $R_B = 0$, $v_o = v_b = v_{sig}$, or $A_{ideal} = v_o/v_{sig} = 1$ V/V. To find a_{ft}, suppress the BJT's dependent source by replacing it with an open circuit. This results in the situation of Fig. 7.56b, where we have, by the voltage divider rule,

$$a_{ft} = \lim_{g_m \to 0} \frac{v_o}{v_{sig}} = \frac{R_E /\!/ r_o}{R_B + r_\pi + (R_E /\!/ r_o)} = \frac{1}{1 + \dfrac{R_B + r_\pi}{R_E /\!/ r_o}} = \frac{1}{1 + \dfrac{0 + 5}{10 /\!/ 100}} = \frac{1}{1.55}$$

Using Eq. (7.63) we finally get

$$A = \frac{1}{1 + 1/129} + \frac{1/1.55}{1 + 129} = 0.992 + 0.005 = 0.997 \text{ V/V}$$

This result matches exactly that of Example 7.12. To test for the accuracy of both methods, recall from Section 2.9 that the exact gain for $R_B = 0$ is

$$A_{exact} = \frac{1}{1 + \dfrac{r_\pi}{(\beta_0 + 1)(R_E /\!/ r_o)}} = \frac{1}{1 + \dfrac{5}{201(10 /\!/ 100)}} = 0.997 \text{ V/V}$$

Viewing the above expression as $A_{exact} = 1/(1 + 1/L_{exact})$, we readily find

$$L_{exact} = \frac{(\beta_0 + 1)(R_E /\!/ r_o)}{r_\pi} = \frac{201(10 /\!/ 100)}{5} = 365.5$$

which is in excellent agreement with $L = 364$ of Example 7.12. Thus, whether we use Eq. (7.61) with $L = 364$ (as we did in Example 7.12), or we use Eq. (7.63) with $T = 129$ and $a_{ft} = 1/1.55$ (as we are doing presently), we obtain consistent results for A.

(b) For *series-series* operation refer to Fig. 7.57. As we know, negative feedback regulates the emitter current i_e. The output current i_o is outside the feedback loop, but it is related to the former as $i_o = \alpha_0 i_e$. Given that $i_e = v_e/R_E$, we can recycle the value of v_e/v_{sig} found in part (a) and write

$$\frac{i_o}{v_{sig}} = \frac{i_o}{i_e} \times \frac{i_e}{v_e} \times \frac{v_e}{v_{sig}} = \alpha_0 \times \frac{1}{R_E} \times A = \frac{200}{201} \times \frac{1}{10^4} \times 0.997 = 99.2 \ \mu\text{A/V}$$

Comparing the gain of 98.7 μA/V of Example 7.17 against the exact gain of 99.2 μA found here we conclude that the discrepancy stems from the approximations at the basis of the circuits of Fig. 7.38.

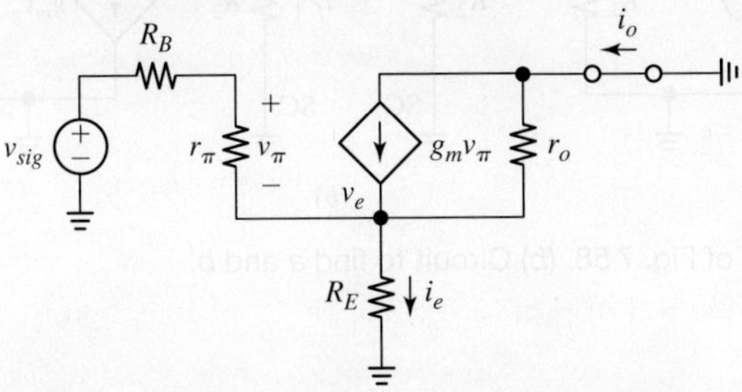

FIGURE 7.57 Ac model for the BJT of Fig. 7.55a in series-series operation.

Remark: As mentioned, L generally varies with the feedback topology, even though the sourceless circuit is the same. However, T remains invariant.

We conclude our list of examples with the circuit of Fig. 7.58a, which uses a CMOS inverter to implement an inverting amplifier with ideal voltage gain $A_v = v_o/v_i = -R_2/R_1$. As we know, this is a shunt-shunt configuration, so to prepare the circuit for two-port analysis we perform the input source transformation of Fig. 7.58b. To find A_v, we first find the transresistance gain $A = v_o/i_i$, then we let

$$A_v = \frac{v_o}{v_i} = \frac{v_o}{i_i} \times \frac{i_i}{v_i} = \frac{A}{R_1} \tag{7.71}$$

As we know, the two FETs of a CMOS inverter reinforce each other in providing transconductance, acting like a single equivalent FET with equivalent parameters

$$g_m = g_{mn} + g_{mp} \qquad r_o = r_{on}//r_{op}$$

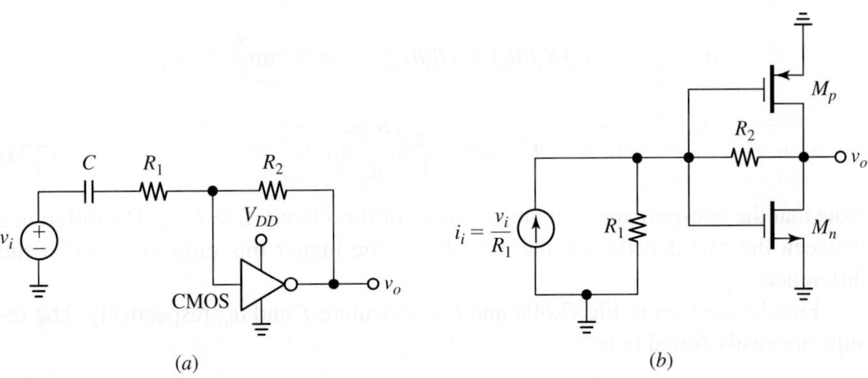

FIGURE 7.58 (a) Using a CMOS inverter as an inverting amplifier. (b) Its shunt-shunt ac equivalent.

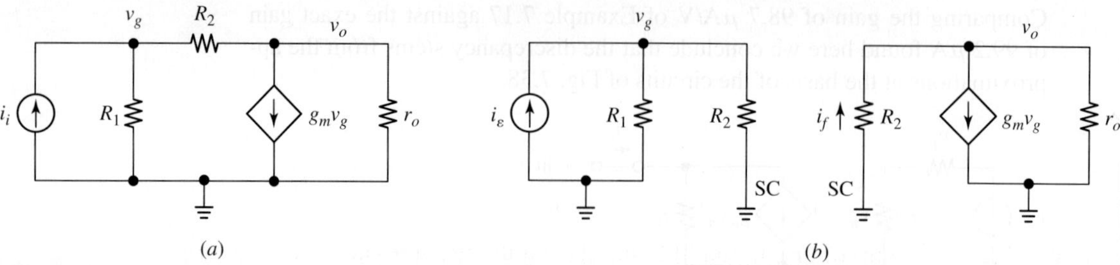

FIGURE 7.59 (a) Small-signal equivalent of the circuit of Fig. 7.58. (b) Circuit to find a and b.

This is shown in the small-signal equivalent of Fig. 7.59a. This circuit is simple enough that we can find its exact gain A *directly*. We will also find A via *loop-gain analysis* and via *return-ratio analysis*, so we can compare the three methods and get a better feel for similarities and differences.

> ### Exercise 7.2
> Use node analysis to prove that the circuit of Fig. 7.59a gives
>
> $$A = \frac{v_o}{i_i} = -R_2 \cfrac{1}{1 + \cfrac{(1 + R_2/R_1)(1 + R_2/r_o)}{g_m R_2 - 1}} \tag{7.72}$$
>
> so that, expressing as $A = (-R_2)/(1 + 1/L_{\text{exact}})$ we get, after minor manipulation,
>
> $$L_{\text{exact}} = \frac{g_m(R_2/\!/r_o) - 1/(1 + R_2/r_o)}{1 + R_2/R_1} \tag{7.73}$$

Next, estimate L via two-port analysis. With reference to Fig. 7.59b we readily find

$$a = \frac{v_o}{i_\varepsilon} = -g_m(R_1/\!/R_2) \times (R_2/\!/r_o) \qquad b = \lim_{i_\varepsilon \to 0} \frac{i_f}{v_o} = -\frac{1}{R_2}$$

$$L = ab = \frac{g_m(R_2/\!/r_o)}{1 + R_2/R_1} \tag{7.74}$$

Note that the two-port approximation gives, for this circuit, $L > L_{\text{exact}}$. The difference between the two depends on the ratio R_2/r_o: the higher this ratio, the smaller the difference.

Finally, we turn to Fig. 7.60a and b to calculate T and a_{ft}, respectively. The results are easily found to be

$$T = -\frac{i_r}{i_t} = \frac{g_m r_o}{1 + (R_2 + r_o)/R_1} \qquad a_{\text{ft}} = \lim_{g_m \to 0} \frac{v_o}{i_i} = \frac{r_o}{1 + (R_2 + r_o)/R_1} \tag{7.75}$$

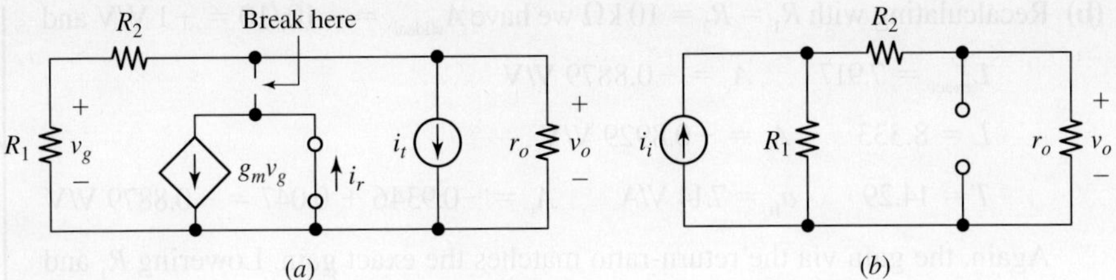

FIGURE 7.60 Small-signal circuits to find (a) T and (b) a_{ft} for the circuit of Fig. 7.58.

EXAMPLE 7.27

(a) In the circuit of Fig. 7.58a let $g_m = 2$ mA/V and $r_o = 50$ kΩ. Assuming the circuit has been configured for unity voltage-gain operation using $R_1 = R_2 = 100$ kΩ, find $A_v = v_o/v_i$ via exact analysis, via two-port analysis, and via return-ratio analysis. Compare the three cases, and comment.

(b) Repeat, but for the case $R_1 = R_2 = 10$ kΩ. Compare with part (a), and comment.

Solution

(a) We have $A_{v(ideal)} = -100/100 = -1$ V/V. Applying Eqs. (7.71) through (7.73) we get

$$L_{exact} = \frac{2(100//50) - 1/(1 + 100/50)}{1 + 100/100} = 33.33 - 0.17 = 33.17$$

$$A_v = \frac{-100/100}{1 + 1/33.17} = -0.9707 \text{ V/V}$$

Applying Eqs. (7.74) and (7.61) gives

$$L = 2\frac{100//50}{1 + 100/100} = 33.33 \qquad A_v = \frac{-100/100}{1 + 1/33.3} = -0.9709 \text{ V/V}$$

Applying Eq. (7.75) we find

$$T = \frac{2 \times 50}{1 + (100 + 50)/100} = 40.0 \qquad a_{ft} = \frac{50}{1 + (100 + 50)/100} = 20 \text{ V/A}$$

so Eq. (7.63) gives

$$A_v = \frac{1}{100}\left(\frac{-100}{1 + 1/40} + \frac{20}{1 + 40}\right) = \frac{1}{100}(-97.56 + 0.49) = -0.9707 \text{ V/V}$$

As mentioned, two-port analysis for this circuit slightly overestimates the value of L compared to the exact value, but in this case the effect upon A_v is negligible. Note also that $T > L$ as a_{ft} and $A_{v(ideal)}$ have opposite polarities. Forward transmission through the feedback network accounts for about 0.5% of A_v.

(b) Recalculating with $R_1 = R_2 = 10\ \text{k}\Omega$ we have $A_{v(\text{ideal})} = -10/10 = -1$ V/V and

$$L_{\text{exact}} = 7.917 \qquad A_v = -0.8879\ \text{V/V}$$

$$L = 8.333 \qquad A_v = -0.8929\ \text{V/V}$$

$$T = 14.29 \qquad a_{\text{ft}} = 7.14\ \text{V/A} \qquad A_v = -0.9346 + 0.047 = -0.8879\ \text{V/V}$$

Again, the gain via the return-ratio matches the exact gain. Lowering R_1 and R_2 simultaneously by an order of magnitude leaves $A_{v(\text{ideal})}$ unchanged but alters the loading as well as the forward transmission conditions appreciably. Indeed, T is now almost twice as large as L and forward transmission through the feedback network now accounts for 5% of A_v.

The Feedback Factor β

By analogy with the two-port loop gain L, which takes on the form of a product $L = ab$, it is convenient to express also the return-ratio loop gain T as a product,

$$T = a\beta \qquad\qquad (7.76)$$

where β, the counterpart of b, shall be called the *return-ratio feedback factor* (this, to distinguish it from b, which shall be called the *two-port feedback factor*). The derivation of β is similar to that of T of Eq. (7.60), except that the return signal is now the signal controlling the dependent source (alternatively, we can find T and then let $\beta = T/a$). Just as T and L are generally different, so are β and b, though they may coincide in particular cases. Let us illustrate via the noninverting op amp of Fig. 7.20, repeated for convenience in Fig. 7.61a. With reference to Fig. 7.61b, we readily find β using the voltage divider twice. So, we write

$$\beta = -\frac{v_d}{v_t} = \frac{1}{1 + \dfrac{R_2}{r_i /\!/ R_1}} \times \frac{1}{1 + \dfrac{r_o}{r_i /\!/ R_1 + R_2}} \qquad a = a_v$$

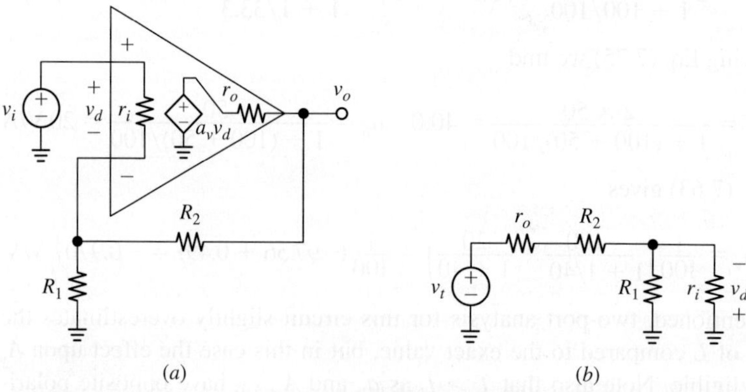

(a) (b)

FIGURE 7.61 (a) Noninverting amplifier and (b) circuit to find β.
($a_v = 1000$ V/V, $r_i = 10$ kΩ, $r_o = 1.0$ kΩ, $R_1 = 1.0$ kΩ, and $R_2 = 9.0$ kΩ.)

On the other hand, according to Eqs. (7.48) and (7.46), we have, respectively,

$$b = \frac{1}{1 + R_2/R_1} \qquad a = \frac{r_i}{r_i + (R_1//R_2)} a_v \frac{R_1 + R_2}{r_o + R_1 + R_2}$$

Plugging in the component values of Fig. 7.61 we get

$$L = ab = 833.3 \times \frac{1}{10} = 83.33 \qquad T = a\beta = 1000 \times \frac{1}{12} = 83.33$$

We now make the following generalizations:

- Two-port analysis makes the feedback factor b depend exclusively on the feedback network such that $b = 1/A_{\text{ideal}}$ ($=1/10$ in the above example). Loading by the feedback network is absorbed by the amplifier, causing gain to drop from $a_v = 1000$ to $a = 833.3$ in the example.
- Return-ratio analysis leaves the amplifier gain unchanged at a_v ($= 1000$ in the example). Loading by amplifier is absorbed by the feedback network, causing the feedback factor to drop from $b = 1/10$ to $\beta = 1/12$ in the example. Consequently, $A_{\text{ideal}} \neq 1/\beta$ (to avoid confusion, always calculate A_{ideal} in the limit $k \to \infty$, or $a_v \to \infty$ in the case of an op amp circuit).
- In this particular case we have $T = L$, though in general T and L differ. As mentioned, even when T and L are equal, they contribute differently to A as per Eqs (7.61) and (7.63).

As we embark upon the study of stability, in Section 7.7, we shall use return-ratio analysis because it shifts loading to β and it makes easier to investigate stability graphically, starting with the Bode plot of the gain a.

7.6 BLACKMAN'S IMPEDANCE FORMULA AND INJECTION METHODS

The return ratio is a powerful tool also for the calculation of the *closed-loop resistance* R between any pair of nodes of a negative-feedback circuit — not just the nodes of the input and output ports. Such a resistance is given by *Blackman's impedance formula*[5]

$$R = r_0 \frac{1 + T_{sc}}{1 + T_{oc}} \tag{7.77}$$

where

$$r_0 = \lim_{k \to 0} R \tag{7.78}$$

is the resistance between the given node pair with the *dependent source set to zero*, and T_{sc} and T_{oc} are the return ratios of the same source with the two nodes *short-circuited* and *open-circuited*, respectively. Blackman's formula holds regardless of the feedback topology in use. Oftentimes either T_{sc} or T_{oc} is zero, indicating that negative feedback will either *raise* r_0 by $(1 + T_{sc})$ (this is the case of a series topology), or *lower* r_0 by $(1 + T_{oc})$ (shunt topology).

EXAMPLE 7.28   Consider the feedback-bias shunt-shunt circuit of Fig. 7.62, which has already been the focus of our attention in Examples 7.15 and 7.24. Assuming the parameter values of Example 7.15b ($R_B = 5.0$ kΩ, $R_C = 10$ kΩ, $g_m = 1/(25\ \Omega)$, $r_\pi = 5$ kΩ, and $r_o = 100$ kΩ), use Blackman's impedance formula to find (a) R_i and (b) R_o. Compare with the example and comment.

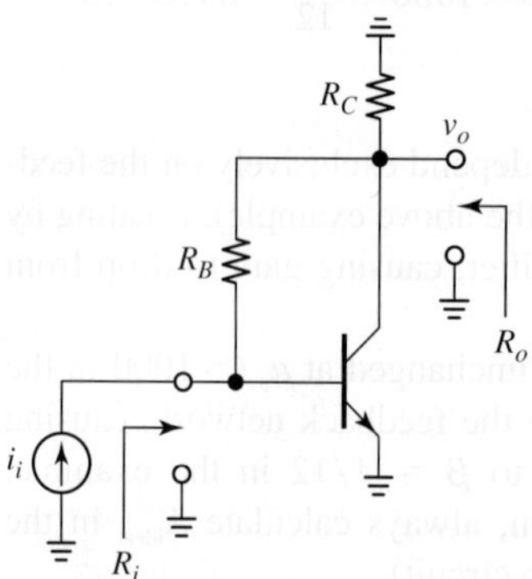

FIGURE 7.62 Feedback-bias BJT circuit of Example 7.28.

Solution

To find r_{i0} and r_{o0}, we let $g_m \to 0$ while suppressing the input source i_i and leaving the output port open-circuited. This yields the situation of Fig. 7.63a. To find the return ratios we use Fig. 7.63b.

(a) By inspection,

$$r_{i0} = r_\pi \,//\,[R_B + (r_o//R_C)] = 5//[5 + (100//10)] = 3.69\ \text{k}\Omega$$

Short-circuiting the input port in Fig. 7.63b yields $v_\pi = 0$, and thus $i_r = 0$, so $T_{sc} = 0$. Leaving the input port open-circuited results in the same situation of Example 7.24b, so we recycle the result obtained there and write $T_{oc} = 95.2$. Consequently,

$$R_i = r_{i0}\frac{1 + T_{sc}}{1 + T_{oc}} = 3.69\frac{1 + 0}{1 + 95.2} = 38.4\ \Omega$$

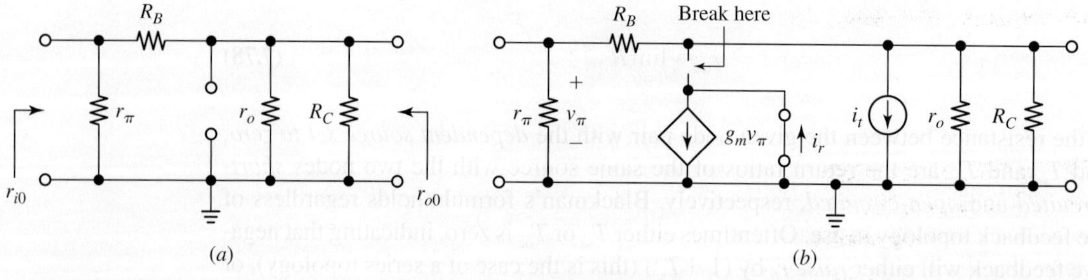

FIGURE 7.63 Ac models of the circuit of Fig. 7.62 for finding (a) r_{i0} and r_{o0}, and (b) the return ratios.

(b) Referring again to Fig. 7.63a, we have, by inspection,

$$r_{o0} = R_C /\!/ r_o /\!/ (R_B + r_\pi) = 4.76 \text{ k}\Omega$$

Short-circuiting the output port in Fig. 7.63b gives again $v_\pi = 0$, and thus $T_{sc} = 0$. Likewise, leaving it open-circuited gives again $T_{oc} = 95.2$. Consequently,

$$R_o = r_{o0}\frac{1 + T_{sc}}{1 + T_{oc}} = 4.76\frac{1 + 0}{1 + 95.2} = 49.5 \ \Omega$$

The values of 38.2 Ω and 49.2 Ω found in Example 7.15b are fairly close to the present (exact) ones.

EXAMPLE 7.29

The Wilson current mirror, shown in ac form in Fig. 7.64a, is a shunt-series feedback circuit based on Q_3. This is better understood via the small-signal equivalent of Fig. 7.64b, where we note that diode-connected Q_2 senses Q_3's emitter current, and current mirror Q_1 replicates this current with a feedback factor of unity and then sinks it from the input summing node. We wish to use Blackman's impedance formula in connection with the dependent source modeling Q_1 to find expressions for (a) R_i and (b) R_o.

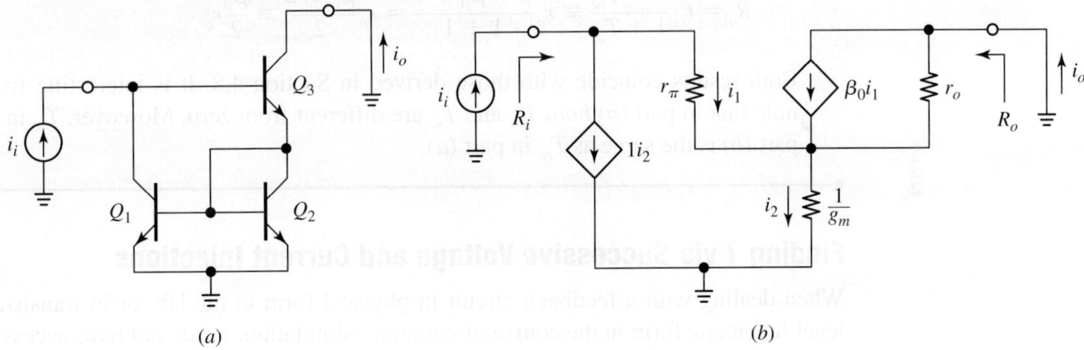

FIGURE 7.64 (a) The Wilson current mirror, and (b) its small-signal equivalent.

Solution

To find r_{i0} and r_{o0}, we suppress the input source i_i and set the dependent source modeling Q_1 to zero. This yields the situation of Fig. 7.65a. By inspection,

$$r_{i0} = r_\pi + (\beta_0 + 1)\left(\frac{1}{g_m}/\!/r_o\right) \cong 2r_\pi \qquad r_{o0} = r_o + \frac{1}{g_m} \cong r_o$$

To find the return ratios, refer to Fig. 7.65b.

(a) With the input terminal open circuited as in Fig. 7.65b we have $i_r = 1i_2 \cong (\beta_0 + 1)i_1 = -(\beta_0 + 1)i_t$, where we have ignored r_o because it is in parallel with $1/g_m$, and $r_o \gg 1/g_m$. Consequently, $T_{oc} = -i_r/i_t = (\beta_0 + 1)$. Shorting the

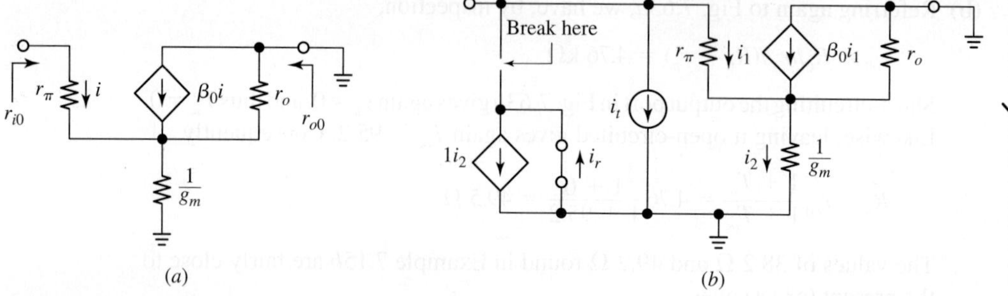

FIGURE 7.65 Modifying the circuit of Fig. 7.64b to find (a) r_{i0} and r_{o0}, and (b) to find the return ratios.

input terminal to ground will cause i_t to flow through this short, giving $i_1 = i_2 = 0$. So, $i_r = 0$ and $T_{sc} = 0$. Then,

$$R_i = r_{i0}\frac{1 + T_{sc}}{1 + T_{oc}} = 2r_\pi\frac{1 + 0}{1 + \beta_0 + 1} \cong \frac{2}{g_m}$$

(b) With the output terminal shorted to ground as in Fig. 7.65b we can recycle the result of (a) and write $T_{sc} = (\beta_0 + 1)$. Open-circuiting the output terminal will cause the current $\beta_0 i_1$ to flow through r_o, yielding $i_r = 1i_2 = i_1 = -i_t$. So, $T_{oc} = -i_r/i_t = 1$, and

$$R_o = r_{o0}\frac{1 + T_{sc}}{1 + T_{oc}} \cong r_o\frac{1 + \beta_0 + 1}{1 + 1} = r_o\frac{\beta_0 + 2}{2} \cong \frac{\beta_0}{2}r_o$$

Both results coincide with those derived in Section 4.8. It is interesting to note that in part (b) *both* T_{oc} and T_{sc} are different from zero. Moreover, T_{sc} in part (b) is the same as T_{oc} in part (a).

Finding *T* via Successive Voltage and Current Injections

When dealing with a feedback circuit in physical form in the lab, or in transistor-level schematic form in the course of computer simulation, we do not have access to any internal dependent sources, so we need alternative techniques for finding *T*. An elegant technique, devised by R. D. Middlebrook, is the *successive voltage* and *current injection technique*[6] depicted in Fig. 7.66. Its steps are as follows.

- First, set all external signal sources to zero so as to put the circuit in its *dormant* state.
- *Break* the feedback loop and insert a *series* test source v_t as shown in Fig. 7.66a. The perturbation introduced by v_t will cause a signal v_f to propagate in the *forward* direction, in turn causing the feedback system to respond with a *return* signal v_r. Measure the *voltage ratio*

$$T_v = -\frac{v_r}{v_f} \tag{7.79a}$$

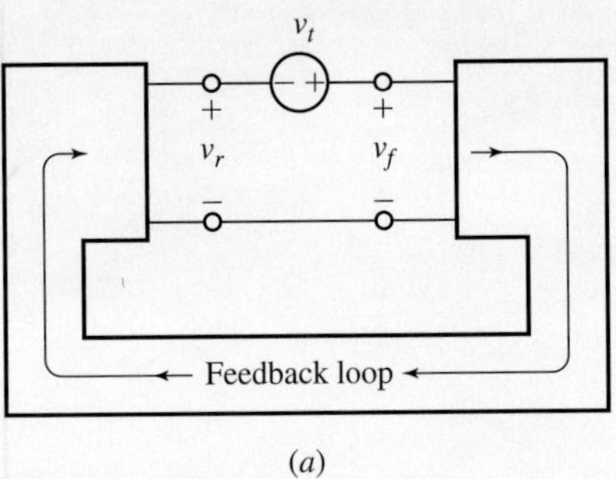

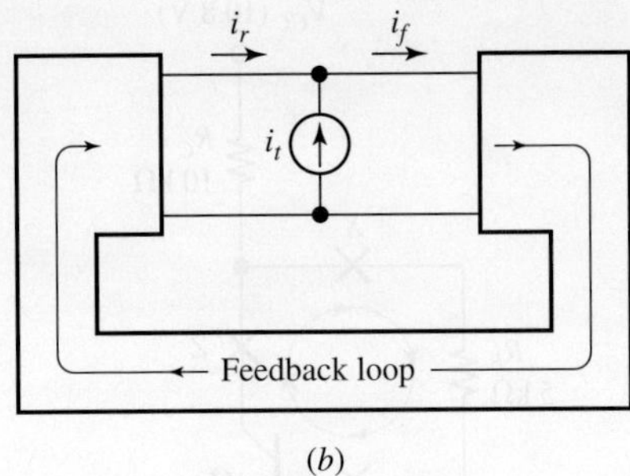

FIGURE 7.66 Illustrating the (a) voltage injection and (b) current injection method.

- Remove the test source v_t and apply between the *same* set of wires a *shunt* test source i_t as shown in Fig. 7.66b. The perturbation introduced by i_t will cause a signal i_f to propagate in the *forward* direction, in turn causing the feedback system to respond with a *return* signal i_r. Measure the *current ratio*

$$T_i = -\frac{i_r}{i_f} \tag{7.79b}$$

- It has been proved[6] that the loop's return ratio T is such that

$$\frac{1}{1+T} = \frac{1}{1+T_v} + \frac{1}{1+T_i} \tag{7.80}$$

Solving for T we get

$$T = \frac{T_v T_i - 1}{T_v + T_i + 2} \tag{7.81}$$

It should be noted that T is independent of the particular point of the loop at which the two injections are made.

Depending on the point of signal injection, the ac signals v_t and i_t may need to be kept suitably small to prevent the circuit from behaving nonlinearly. Note that both test setups preserve the existing dc biasing conditions as well as the existing loading conditions. An actual example will better illustrate.

EXAMPLE 7.30

Shown in Fig. 7.67 is the feedback-bias BJT that we have already investigated in Examples 7.15 and 7.24. Lacking any input signal source, the circuit is already in its ac dormant state. Assuming the BJT has $I_s = 2$ fA, $\beta_F = 200$, and $V_A = 100$ V, use PSpice to find the return ratio of this circuit, compare with Example 7.24, and comment.

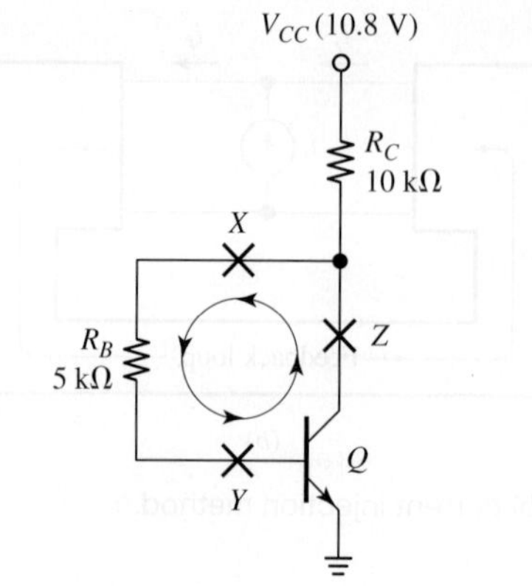

FIGURE 7.67 The feedback-bias BJT configuration.

Solution

In this circuit rendition feedback signal flow is *counterclockwise*, and there are three points where we can apply our double injection, namely, X, Y, and Z. In the laboratory one would choose a point where signal levels are the highest and therefore easier to measure, especially in the presence of noise. In the PSpice circuits of Fig. 7.68 we have arbitrarily chosen point Y, but the procedure can readily be repeated for the other two. (Note that the current injection circuit requires the use of two 0-V dummy sources V_1 and V_2 to sense the currents i_r and i_f.) The results of the simulation are as follows:

At point X:	$T_v = 179.8$	$T_i = 201.1$	$T = 94.4$
At point Y:	$T_v = 353.4$	$T_i = 129.4$	$T = 94.3$
At point Z:	$T_v = 1983$	$T_i = 99.26$	$T = 94.4$

It is interesting that T_v and T_i change with the point of injection whereas T is invariant (within roundoff errors). Also, T agrees well with the value found in Example 7.24*b* via hand calculations.

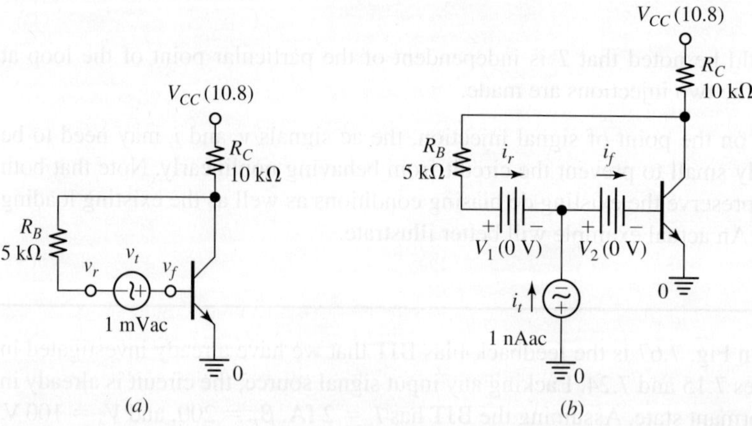

FIGURE 7.68 Applying (*a*) *voltage* and (*b*) *current* injection to the feedback-bias circuit of Fig. 7.67.

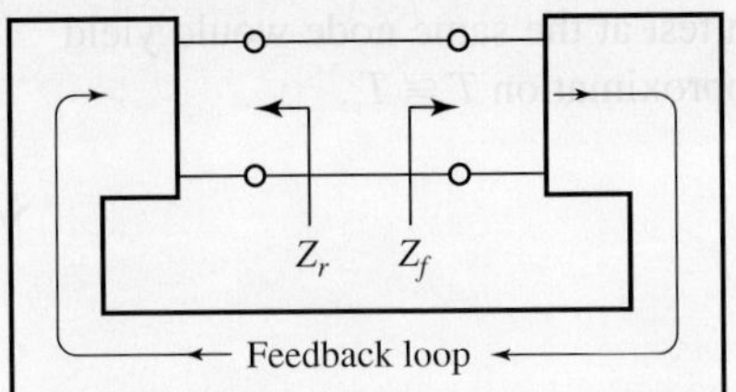

FIGURE 7.69 The impedances Z_f and Z_r seen looking in the forward and the return directions.

Single-Injection Approximations

According to Eq. (7.80), the terms $(1 + T_v)$ and $(1 + T_i)$ combine like *resistances in parallel*. If one happens to be much larger than the other, the smaller will prevail and we can estimate T more quickly by limiting ourselves to just *one* signal injection, namely, the one that results in the much smaller return ratio. For instance, had we anticipated the condition $T_v \gg T_i$ at point Z in the circuit of Fig. 7.67, we could have skipped the voltage injection to approximate $T \cong T_i \cong 99$, for an error of less than 5%. The choice of where to break the feedback loop is facilitated by the fact that T_v and T_i are constrained as[6]

$$\frac{1 + T_v}{1 + T_i} = \frac{Z_r}{Z_f} \tag{7.82}$$

where Z_f and Z_r are the impedances seen looking in the *forward* and in the *return* directions from the point of signal injection, as depicted in Fig. 7.69. For instance, if we select point Z in Fig. 7.67, then Z_r is the impedance seen looking into the collector, which is much higher than the impedance Z_f seen looking toward the node common to R_C and R_B. Consequently, Eq. (7.82) predicts $(1 + T_v) \gg (1 + T_i)$, as we already know from Example 7.30.

EXAMPLE 7.31

Using the 741 macro-model available in PSpice's library, find T for the case in which the 741 is configured to function as an *inverting amplifier* with a gain of -100 V/V by means of two resistances $R_1 = 1.0$ kΩ and $R_2 = 100$ kΩ. Pick a point where only one injection is sufficient.

Solution

After setting the input source to zero to put the circuit in its dormant state, we end up with the situation of Fig. 7.70. The point of injection has been chosen right at the op amp's inverting input, where $Z_f \, (=r_i)$ is in the MΩ range, while $Z_r \, [=R_1 // (R_2 + r_o)]$ is less than 1 kΩ. With a disparity on the order of three orders of magnitude it is hardly worth performing the current-injection test, so we limit ourselves to a mere voltage test, as shown. The simulation yields

$$T_v = 1970$$

The student can verify that a current-injection test at the same node would yield $T_i = 1.983 \times 10^6$ ($\gg T_v$), thus validating the approximation $T \cong T_v$.

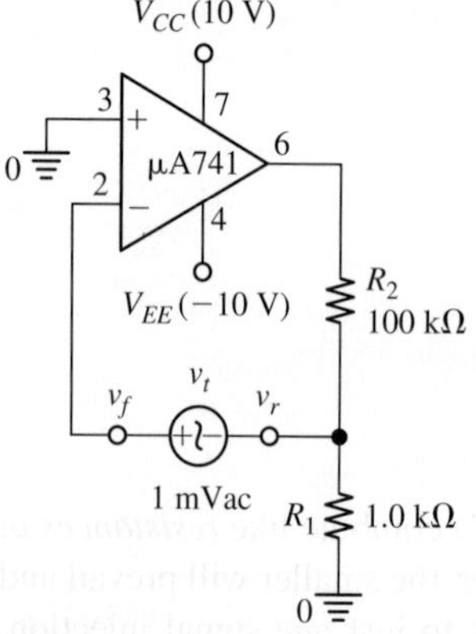

FIGURE 7.70 Circuit of Example 7.31, where only voltage injection suffices.

7.7 STABILITY IN NEGATIVE-FEEDBACK CIRCUITS

For all its benefits, negative feedback poses the potential risk of *oscillation*. Because of this, negative feedback was met with skepticism when first proposed by Harold Black, in 1928. However, once this risk became better understood and suitable cures were developed to tame unwanted oscillations, negative feedback established itself as the cornerstone of electronics and control we know today. To develop a basic feel for stability (or lack thereof), let us assume *negligible feedthrough*, so both the amplifier and the feedback network can be regarded as unilateral as in Fig. 7.71. Moreover, we take the viewpoint of the IC designer, whose concern is to ensure that the amplifier is stable when operated with *purely resistive* feedback networks. With this type of feedback, the feedback factor β is *frequency independent* and its phase angle is thus zero. (If the feedback network contains reactive elements such as capacitors, then β too will be frequency dependent and the responsibility for ensuring stability is now shifted to the user.[7])

Let us start out by expressing the closed-loop gain in the more insightful form

$$A(jf) = \frac{S_o}{S_i} = A_{\text{ideal}} \times D(jf)$$ (7.83)

where $A_{\text{ideal}} = \lim_{a \to \infty} A$, and the quantity

$$D(jf) = \frac{1}{1 + 1/T(jf)}$$ (7.84)

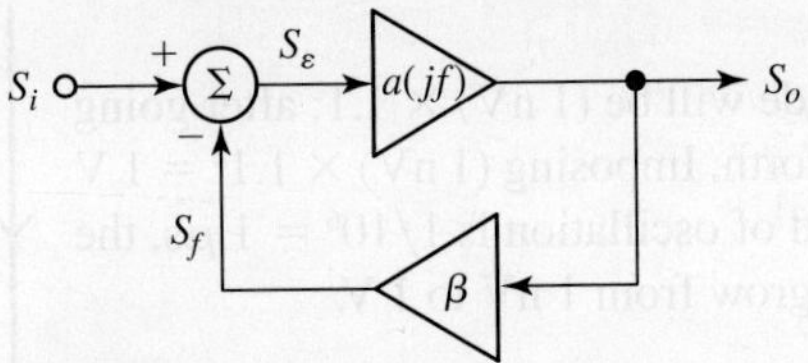

FIGURE 7.71 Negative-feedback system with unilateral amplifier and feedback network, and a frequency independent feedback factor β.

is called the *discrepancy function* because it gives a measure of the deviation of $A(jf)$ from A_{ideal}. Moreover, $T(jf) = \beta a(jf)$ is the familiar loop gain. As a signal S_ε propagates around the loop and returns to the summer Σ as S_f, it experiences a frequency-dependent *phase shift*, which we shall denote as ph $T(jf)$. If this shift reaches $-180°$, then feedback turns from negative to *positive*. Denoting the frequency at which this happens as $f_{-180°}$, it follows that $T(jf_{-180°})$ is a *real*, *negative* number such as -0.5, -1, -2, etc. We have three possibilities:

- $|T(jf_{-180°})| < 1$. Even though feedback is *positive* at $f_{-180°}$, S_ε gets *attenuated* each time it goes around the loop, so if we set $S_i \to 0$, any signal present in the loop will eventually decay to zero, and the system is said to be *stable*.
- $T(jf_{-180°}) \to -1$. By Eq. (7.84) we have $D(jf_{-180°}) \to \infty$ and so $A(jf_{-180°}) \to \infty$, indicating that the system can sustain an output signal $S_o \neq 0$ with a *vanishingly small* input $S_i \to 0$. Once a signal component $S_\varepsilon(jf_{-180°})$ is initiated in the circuit (for instance by electronic noise, which is always present in one form or another), this component will emerge from the feedback network as $S_f(jf_{-180°}) = -S_\varepsilon(jf_{-180°})$, and then undergo one more inversion at the summer Σ to reappear at the amplifier's input as the original signal $S_\varepsilon(jf_{-180°})$ itself! A circuit's ability to sustain a signal at $f_{-180°}$ qualifies it as an *oscillator*, and such the circuit is regarded as *unstable*.
- $|T(jf_{-180°})| > 1$. Suppose we change $T(jf_{-180°})$ from -1 to a *more negative* value, say, -1.1. This implies a 10% increase in magnitude each time a signal goes around the loop, thus resulting in oscillations of growing magnitude. The oscillations will grow until some inherent circuit nonlinearity, such as op amp saturation, will limit T to -1 *exactly* and thus sustain steady oscillation. Though this mechanism can be exploited on purpose to design oscillators, when it comes to amplifiers, buffers, V-I and I-V converters, voltage and current sources, and active filters, oscillations must be avoided by all means. As we shall see, unwanted oscillations are tamed via suitable techniques generally referred to as *frequency compensation*.

Suppose the circuit of Fig. 7.71 is unstable and such that its own internal noise has initiated a growing oscillation at 1 MHz. If amplitude increases by 10% from one oscillation cycle to the next, find the number of cycles as well as the amount of time it takes for the oscillation to grow from 1 nV to 1 V.

EXAMPLE 7.32

Solution

After going around the loop once, the amplitude will be $(1 \text{ nV}) \times 1.1$; after going around twice, it will be $(1 \text{ nV}) \times 1.1^2$, and so forth. Imposing $(1 \text{ nV}) \times 1.1^n = 1 \text{ V}$ and solving, we get $n \cong 217$. Since one period of oscillation is $1/10^6 = 1\ \mu s$, the signal takes about $217 \times (1\ \mu s) = 217\ \mu s$ to grow from 1 nV to 1 V.

Graphical Visualization of the Loop Gain *T*

Given the impact of the loop gain T upon the stability of a circuit, we seek a quick way to visualize the frequency plots of the magnitude $|T(jf)|$ and the phase angle $\mathrm{ph}\,T(jf)$. By definition,

$$|T|_{\mathrm{dB}} = 20\log|T| = 20\log|a\beta| = 20\log\left|\frac{a}{1/\beta}\right| = 20\log|a| - 20\log\left|\frac{1}{\beta}\right|$$

that is,

$$|T(jf)|_{\mathrm{dB}} = |a(jf)|_{\mathrm{dB}} - \left|\frac{1}{\beta}\right|_{\mathrm{dB}} \tag{7.85a}$$

Similarly, $\mathrm{ph}\,T(jf) = \mathrm{ph}\,a(jf) - \mathrm{ph}\,\beta$. So long as β is *frequency independent* we have $\mathrm{ph}\,\beta = 0$, so

$$\mathrm{ph}\,T(jf) = \mathrm{ph}\,a(jf) \tag{7.85b}$$

Based on these relationships, we proceed as follows:

- On the decibel plot of $|a(jf)|$, draw the decibel plot of the *noise gain* $1/\beta$ as depicted in Fig. 7.72a, top. Since β is assumed to be frequency independent and

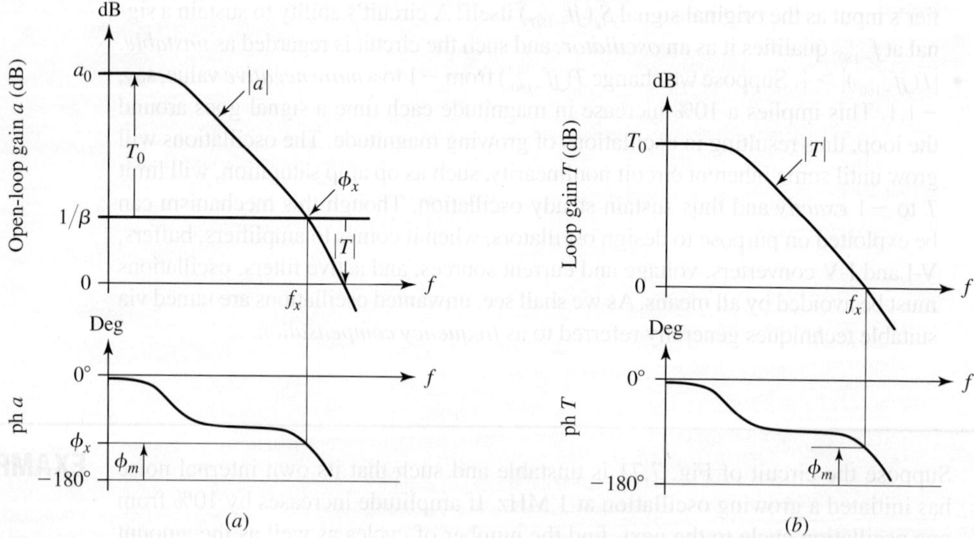

FIGURE 7.72 Visualizing the loop gain $|T(jf)|$, the crossover frequency f_x, and the phase margin ϕ_m.

$0 < \beta \leq 1$, the $1/\beta$ curve will be a *horizontal line* positioned somewhere *above* the 0-dB axis (or right on the 0-dB axis if $\beta = 1$).

- Visualize the plot of $|T(jf)|$ as the *difference* between the plot of $|a(jf)|$ and the $1/\beta$ curve. Equivalently, visualize the plot of $|T(jf)|$ as the plot of $|a(jf)|$, but with the $1/\beta$ curve as its *new* 0-dB axis, as depicted in Fig. 7.72b, top.
- The plot of $|T(jf)|$ starts out at the typically high dc value $T_0 = \beta a_0$, and then it rolls off with frequency with the same profile as $|a(jf)|$.
- The frequency at which the $1/\beta$ curve intersects the $|a(jf)|$ curve is aptly called the *crossover frequency f_x*. At this frequency we have $|T(jf_x)| = 0$ dB, or $|T(jf_x)| = 1$ V/V, so we can write $T(jf_x) = 1\exp(j\phi_x)$, where $\phi_x = \text{ph}\,a(jf_x)$, as depicted in Fig. 7.72a (bottom).
- For $f > f_x$ the decibel value of $|T(jf)|$ becomes negative, indicating loop signal *attenuation*. For $|T(jf)| \ll 1$ we can approximate Eq. (7.84) as $|D(jf)| \cong |1/(1/T)| = |T(jf)|$.

The Phase Margin ϕ_m

In light of the discussion at the beginning of this section, we want the crossover to occur *well before* the dreaded condition $T = 1e^{j(-180°)} = -1$ occurs, which is the recipe for oscillation. The degree of stability of a circuit is quantified via the *phase margin*, defined as $\phi_m = \phi_x - (-180°)$, that is,

$$\phi_m = 180 + \phi_x \tag{7.86}$$

This margin is depicted in Fig. 7.72b (bottom). As we move along, we shall be interested in the value of the discrepancy function at f_x. Letting $T(jf_x) = 1\exp(j\phi_x) = 1\exp[j(\phi_m - 180°)] = -1\exp(j\phi_m)$ in Eq. (7.84), we get

$$|D(jf_x)| = \left|\frac{1}{1 + 1/(-1e^{j\phi_m})}\right| = \left|\frac{1}{1 - e^{-j\phi_m}}\right| = \left|\frac{1}{1 - (\cos\phi_m - j\sin\phi_m)}\right|$$

$$= \frac{1}{\sqrt{(1 - \cos\phi_m)^2 + (\sin\phi_m)^2}}$$

where Euler's formula has been used. Expanding and using the identity $\cos^2\phi_m + \sin^2\phi_m = 1$ gives

$$|D(jf_x)| = \frac{1}{\sqrt{2(1 - \cos\phi_m)}} \tag{7.87}$$

An Illustrative Example

Let us illustrate the above concepts using the feedback circuit of Fig. 7.73 as a vehicle. The circuit utilizes a three-stage amplifier consisting of a pair of transconductance stages with dc gains $a_{10} = G_{m1}R_1 = 10^3$ V/V and $a_{20} = G_{m2}R_2 = 10^3$ V/V, followed by a unity-gain buffer, so $a_0 = a_{10}a_{20} = 10^6$ V/V. The pole frequencies of

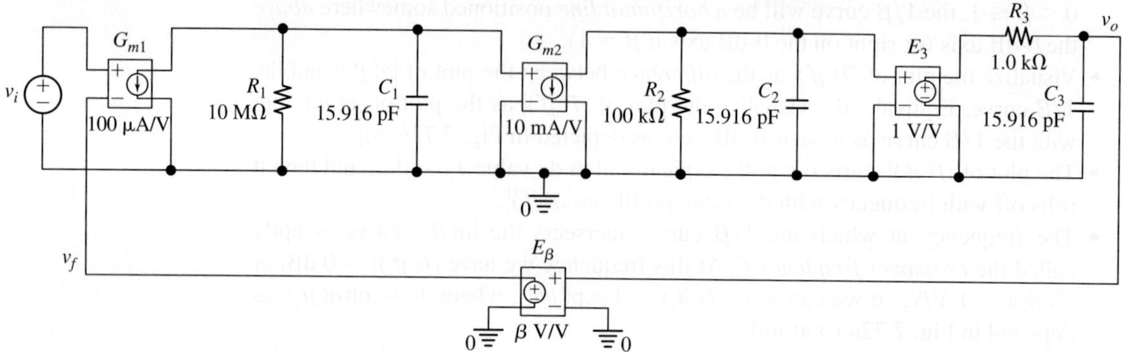

FIGURE 7.73 PSpice circuit to investigate a three-pole op amp under different amounts of feedback.

the three stages are $f_1 = 1/(2\pi R_1 C_1) = 1$ kHz, $f_2 = 1/(2\pi R_2 C_2) = 100$ kHz, and $f_3 = 1/(2\pi R_3 C_3) = 10$ MHz, so the open-loop gain is

$$a(jf) = \frac{10^6}{(1 + jf/10^3)(1 + jf/10^5)(1 + jf/10^7)} \tag{7.88}$$

Magnitude and phase are calculated as

$$|a(jf)| = \frac{10^6}{\sqrt{[1 + (f/10^3)^2] \times [1 + (f/10^5)^2] \times [1 + (f/10^7)^2]}} \tag{7.89a}$$

$$\text{ph } a(jf) = -[\tan^{-1}(f/10^3) + \tan^{-1}(f/10^5) + \tan^{-1}(f/10^7)] \tag{7.89b}$$

and are plotted via PSpice, as depicted in Fig. 7.74a (magnitude is in dB, as marked at the left, and phase in degrees, as marked at the right).

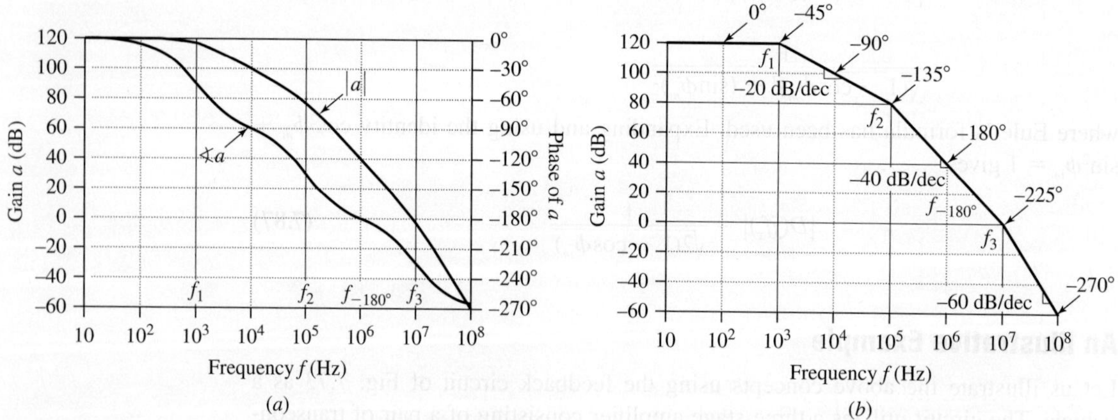

FIGURE 7.74 (a) Frequency plots of magnitude $|a|$ and phase $\angle a$ for the error amplifier of Fig. 7.73. (b) Linearized magnitude plot associating phase with slope.

If the roots are widely spaced apart as in the present example, we can combine magnitude and phase in the more concise and visually intuitive form of Fig. 7.74b. Specifically, we draw a linearized magnitude plot using segments with progressively steeper slopes, and we mark significant phase values using the correspondence

$$\text{Phase(in °)} \leftrightarrow 4.5 \times \text{Slope(in dB/dec)} \qquad (7.90)$$

Thus, from dc to f_1 we draw a segment with a slope of 0 dB/dec, for which Eq. (7.90) implies a phase of 0°. From f_1 to f_2 we draw a segment with a slope of -20 dB/dec, implying a phase of $4.5 \times (-20)$, or $-90°$. Right at f_1 the slope is -10 dB/dec, so the corresponding phase is $4.5 \times (-10)$, or $-45°$. Likewise, the segment from f_2 to f_3 has a slope of -40 dB/dec, which implies a phase of $-180°$. The phase at f_2 is $-135°$. Past f_3 slope approaches -60 dB/dec and phase approaches $-270°$.

We now wish to investigate the closed-loop response under increasing amounts of feedback, which we implement by varying the feedback factor β. We have the following significant cases:

- Starting out with $\beta = 10^{-5}$, we observe that if we draw the $1/\beta$ line ($1/\beta = 10^5 = 100$ dB) in Fig. 7.74b, it will intersect the gain curve at $f_x \cong 10$ kHz, where $\phi_x \cong -90°$. Consequently, $\phi_m \cong 180° - 90° = 90°$. Figure 7.75 shows that the closed-loop gain starts out as $A_0 = (1/\beta)/(1 + 1/T_0) = 10^5/(1 + 1/10) = 90{,}909$ V/V (about 10% lower than $A_{\text{ideal}} = 100{,}000$ V/V because of low T_0), and it exhibits a dominant pole frequency of f_x. Shown in Fig. 7.76a is the response to an input step of β V ($=10 \ \mu$V). This is an approximately *exponential transient*, so once the input is stepped back to zero, $v_o(t)$ will decay in approximately exponential fashion, indicating a stable circuit.

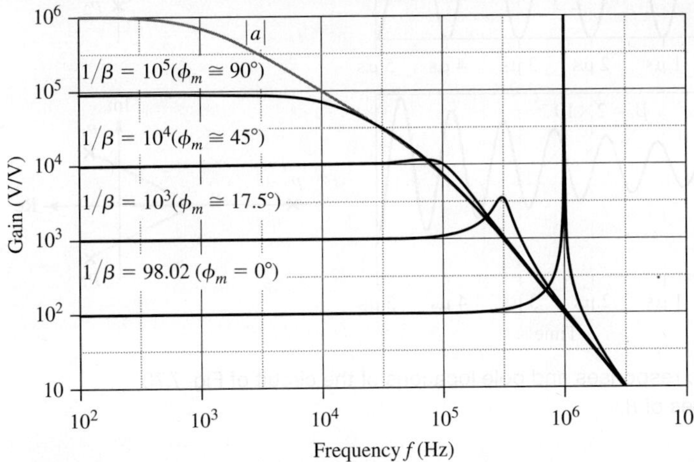

FIGURE 7.75 Closed-loop responses of the circuit of Fig. 7.73 for different amounts of feedback. Raising β lowers the $1/\beta$ curve, shifting the crossover frequency towards regions of greater phase shift and thus lower phase margin. This, in turn, increases the amount of peaking as well as ringing (see Fig. 7.76).

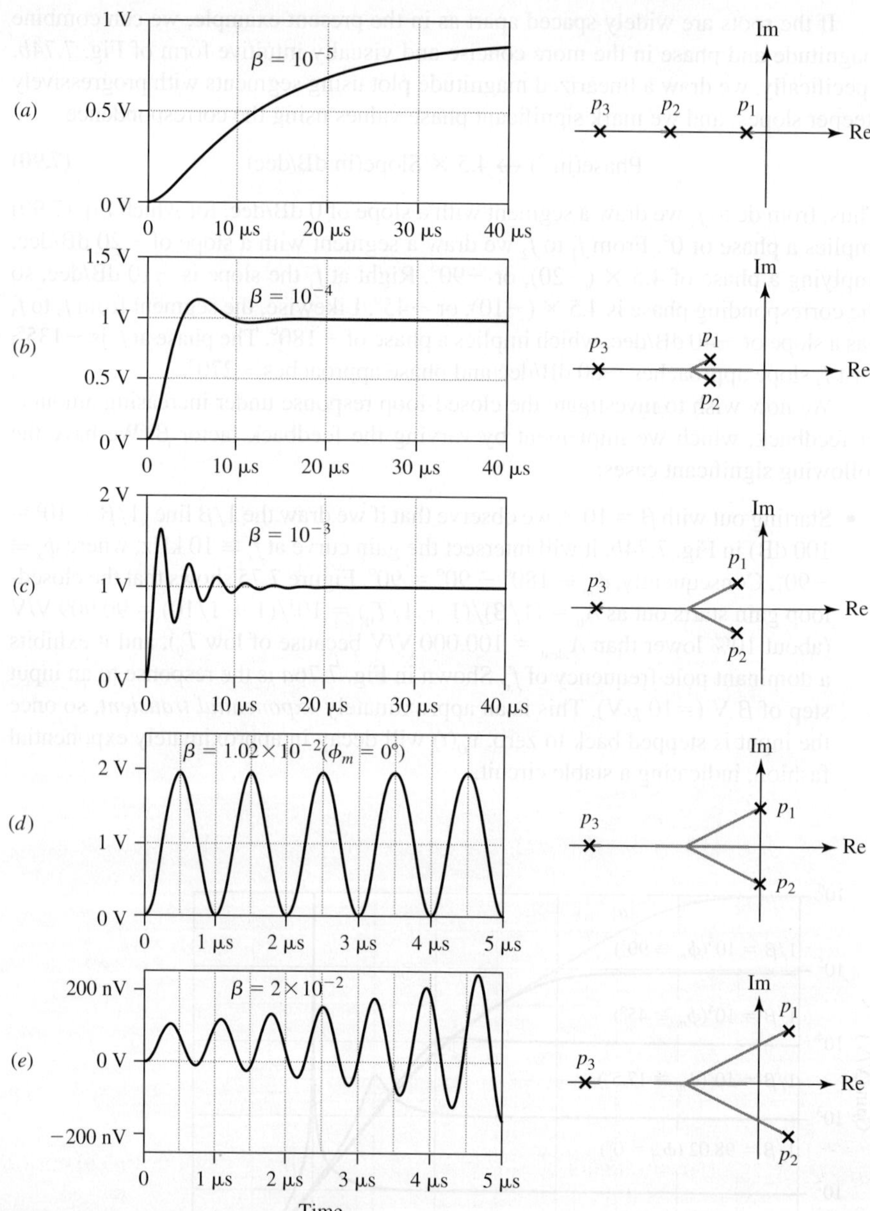

FIGURE 7.76 Step responses and pole locations of the circuit of Fig. 7.73 for increasing values of β.

- Raising β to 10^{-4} lowers the $1/\beta$ line to 80 dB in Fig. 7.74b, giving $f_x \cong 100\,\text{kHz}$ and $\phi_x \cong -135°$, so $\phi_m \cong 180° - 135° = 45°$. The closed-loop gain now exhibits a bit of *peaking* just before f_x, after which it rolls off with frequency as $|a(jf)|$. By Eq. (7.87) we have $|D(jf_x)| = 1/\sqrt{2(1 - \cos45°)} = 1.307$, so

Eq. (7.83) predicts $|A(jf_x)| = 10^4 \times 1.307 = 13{,}070$ V/V, which is 30.7% *higher* than A_{ideal}. Recall from systems theory that *peaking* in the frequency domain is accompanied by *ringing* in the time domain. This is confirmed by Fig. 7.76*b*, showing the transient response to an input step of β V ($= 0.1$ mV). Once we step the input back to zero, $v_O(t)$ will decay to zero, albeit with a bit of ringing. We conclude that a circuit with $\phi_m = 45°$ is still a stable circuit, though its (moderate) amount of peaking and ringing may be undesirable in certain applications.

- For $\beta = 10^{-3}$ the $1/\beta$ line is lowered further to 60 dB in Fig. 7.74*b*, so f_x is now the *geometric mean* of 100 kHz and 1 MHz, or $f_x \cong \sqrt{10^5 \times 10^6} = 316$ kHz, and ϕ_x is the *arithmetic mean* of $-145°$ and $-180°$, or $\phi_x \cong -162.5°$, so $\phi_m \cong 180° - 162.5° = 17.5°$. With a reduced phase margin, both peaking and ringing are more pronounced. In fact we now have $|D(jf_x)| = 1/\sqrt{2(1 - \cos 17.5°)} = 3.29$, so Eq. (7.83) predicts $|A(jf_x)| \cong 1{,}000 \times 3.29 = 3{,}290$ V/V, or almost $3.3 \times A_{ideal}$! Figure 7.76*c* shows the response to an input step of β V ($=1$ mV). Once we step the input back to zero, $v_O(t)$ will decay with quite a bit of ringing. We conclude that a circuit with $\phi_m = 17.5°$, though still stable, is likely to be unacceptable in most applications because of excessive peaking and/or ringing.

- Cursor measurements on the PSpice plots of Fig. 7.74*a* give $f_{-180°} = 1.006$ MHz, where $|a(jf_{-180°})| = 98.02$ V/V, so if we let $\beta = 1/98.02 = 1.02 \times 10^{-2}$, we get $f_x = 1.006$ MHz and $\phi_x = -180°$, or $\phi_m = 0°$. Consequently, $|D(jf_x)| \to \infty$, indicating oscillatory bevavior. This is confirmed by the response to a 10-mV input step of Fig. 7.76*d*.

- Raising β to 2×10^{-2} lowers the $1/\beta$ curve further, pushing f_x into a region of additional phase shift and thus $\phi_m < 0°$. All it takes now is internal noise to trigger a growing oscillation. Using an input step of just 1 nV to simulate noise, we obtain the response of Fig. 7.76*e*.

It is instructive to visualize circuit behavior also in terms of its poles in the complex plane. Letting $jf \to s/(2\pi)$ in Eq. (7.88), substituting into Eq. (7.84) and then into Eq. (7.83), gives

$$A(s) = \frac{10^6}{\beta 10^6 + \left(1 + \dfrac{s}{2\pi 10^3}\right)\left(1 + \dfrac{s}{2\pi 10^5}\right)\left(1 + \dfrac{s}{2\pi 10^7}\right)}$$

The roots of the denominator are the poles of $A(s)$. Using a scientific calculator or similar, we find the poles tabulated in Table 7.2. These data are best visualized in the *s*-plane as shown (not to scale) on the right side of Fig. 7.76. Starting out with no feedback ($\beta = 0$) and gradually increasing β brings the two lowest poles closer together, until they become coincident and then split apart to become complex conjugate and move toward the imaginary axis (see the shaded root locus). Once on the imaginary axis, they result in sustained oscillation, and once they spill into the right half of the complex plane, they result in a growing oscillation.

Looking at Fig. 7.75, we observe that if we can tolerate the amounts of peaking and ringing that come with, say, $\phi_m = 45°$, then we must restrict operation to

TABLE 7.2 Poles of the closed-loop gain of the circuit of Fig. 7.73.

β	$p_1(\text{s}^{-1})$	$p_2(\text{s}^{-1})$	$p_3(\text{s}^{-1})$
0	$2\pi(-1.0\text{ k})$	$2\pi(-100\text{ k})$	$2\pi(-10\text{ M})$
10^{-5}	$2\pi(-12.4\text{ k})$	$2\pi(-88.5\text{ k})$	$2\pi(-10\text{ M})$
10^{-4}	$2\pi(-50\text{ k} + j87.2\text{ k})$	$2\pi(-50\text{ k} - j87.2\text{ k})$	$2\pi(-10\text{ M})$
10^{-3}	$2\pi(-45.4\text{ k} + j313\text{ k})$	$2\pi(-45.4\text{ k} - j313\text{ k})$	$2\pi(-10.01\text{ M})$
1.02×10^{-2}	$2\pi(0 + 1.0\text{ M})$	$2\pi(0 - 1.0\text{ M})$	$2\pi(-10.1\text{ M})$
2×10^{-2}	$2\pi(46.7\text{ k} + 1.34\text{ M})$	$2\pi(46.7\text{ k} - 1.34\text{ M})$	$2\pi(-10.19\text{ M})$

$1/\beta \geq 10^4$ V/V. What if we want to operate the amplifier with lower closed-loop gains, such as $1/\beta = 50$ V/V, or $1/\beta = 2$ V/V? As is, the circuit will simply oscillate for these values of β! Mercifully, some clever frequency-compensation techniques have been developed that allow us to stabilize an amplifier for virtually any closed-loop gain we wish, including what by now we recognize as the most difficult configuration to stabilize, namely, the voltage follower, for which $\beta = 1$.

EXAMPLE 7.33 (a) What is the minimum allowable noise gain $1/\beta$ if we want to operate the amplifier of Fig. 7.73 with a phase margin of 60°?
(b) Find $|D(jf_x)|$ and comment.
(c) Verify with PSpice.

Solution

(a) By Eq. (7.86) we have $\phi_x = \phi_m - 180° = 60° - 180° = -120°$. Figure 7.74$a$ indicates that $f_{-120°}$ occurs a bit below 100 kHz. Start out with the initial estimate $f_{-120°} = 50$ kHz and iterate using Eq. (7.89b) until you settle at $f_{-120°} = 59.2$ kHz. Next, use Eq. (7.89a) to find $|a(jf_{-120°})| \cong 14{,}534$ V/V. This is the minimum value of $1/\beta$ for $\phi_m \geq 60°$ with this particular amplifier.
(b) Proceeding in the usual manner we get

$$|D(jf_x)| \cong \frac{1}{\sqrt{2(1 - \cos 60°)}} = 1$$

Since $T_0 = 10^6/14{,}534 = 68.8$, we also have $D_0 = 1/(1 + 1/68.8) = 0.986$, indicating a very small amount of peaking.
(c) Using the circuit of Fig. 7.73 with $\beta = 1/14{,}534 = 68.8 \times 10^{-6}$ V/V we get the plots of Fig. 7.77. It is fair to say that, aside for a slight amount of peaking and ringing, which is acceptable in many applications, the condition $\phi_m = 60°$ is *almost as good* as $\phi_m = 90°$ in terms of stability, yet it expands the range of possible closed-loop gains by decreasing the lower limit of acceptable values for $1/\beta$ from 100,000 V/V to 14,534 V/V, or by almost 17 dB.

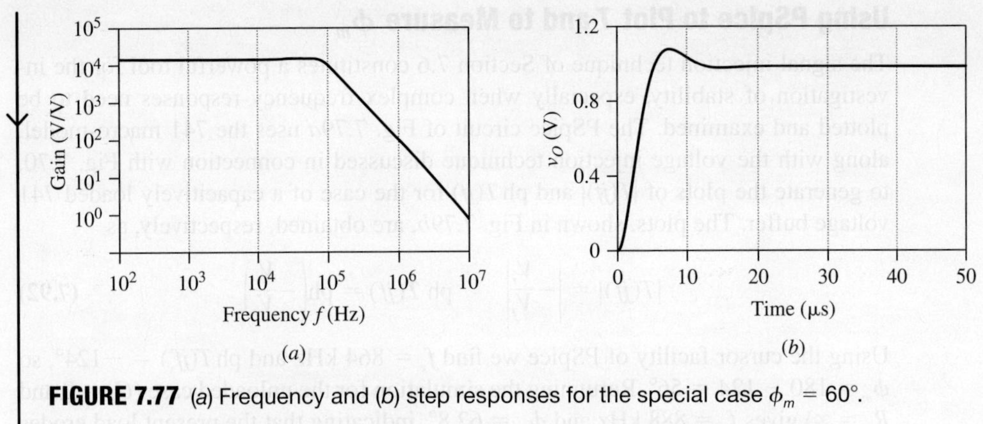

FIGURE 7.77 (a) Frequency and (b) step responses for the special case $\phi_m = 60°$.

Peaking and Ringing as Functions of the Phase Margin ϕ_m

Figure 7.78 depicts peaking and ringing by the circuit of Fig. 7.73 as functions of the phase margin ϕ_m. These characteristics are quantified in terms of the gain peaking GP and the overshoot OS, defined as

$$GP = 20 \log_{10}|A_p - A_0| \qquad OS(\%) = 100\frac{V_p - V_\infty}{V_\infty} \qquad (7.91)$$

Peaking occurs for ϕ_m less than about 65°, and ringing for ϕ_m less than about 75°.

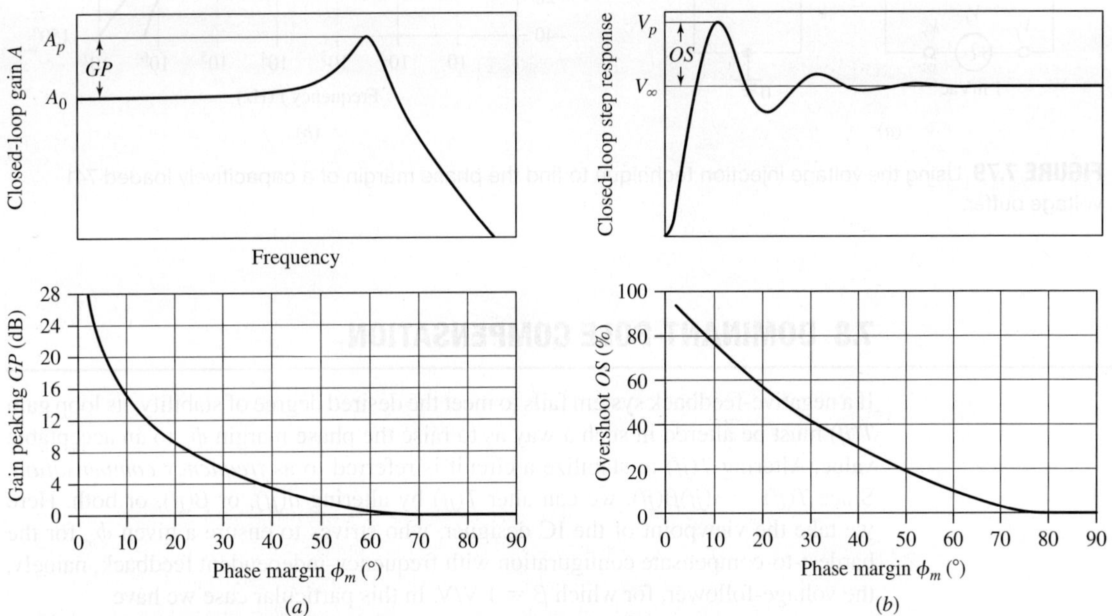

FIGURE 7.78 (a) Peaking and (b) ringing as functions of the phase margin ϕ_m.

Using PSpice to Plot *T* and to Measure ϕ_m

The signal injection technique of Section 7.6 constitutes a powerful tool for the investigation of stability, especially when complex frequency responses need to be plotted and examined. The PSpice circuit of Fig. 7.79*a* uses the 741 macro-model, along with the voltage injection technique discussed in connection with Fig. 7.70, to generate the plots of $|T(jf)|$ and ph $T(jf)$ for the case of a capacitively loaded 741 voltage buffer. The plots, shown in Fig. 7.79*b*, are obtained, respectively, as

$$|T(jf)| = \left| -\frac{V_r}{V_f} \right| \qquad \text{ph } T(jf) = \text{ph}\left(-\frac{V_r}{V_f} \right) \tag{7.92}$$

Using the cursor facility of PSpice we find $f_x = 864$ kHz and ph $T(jf_x) = -124°$, so $\phi_m = 180 - 124 = 56°$. Rerunning the simulation for the unloaded case ($C_L = 0$ and $R_L = \infty$) gives $f_x = 888$ kHz and $\phi_m = 62.8°$, indicating that the present load erodes ϕ_m by 6.8° (this is due to the high-frequency pole formed by C_L with the output resistance r_o of the op amp[7]).

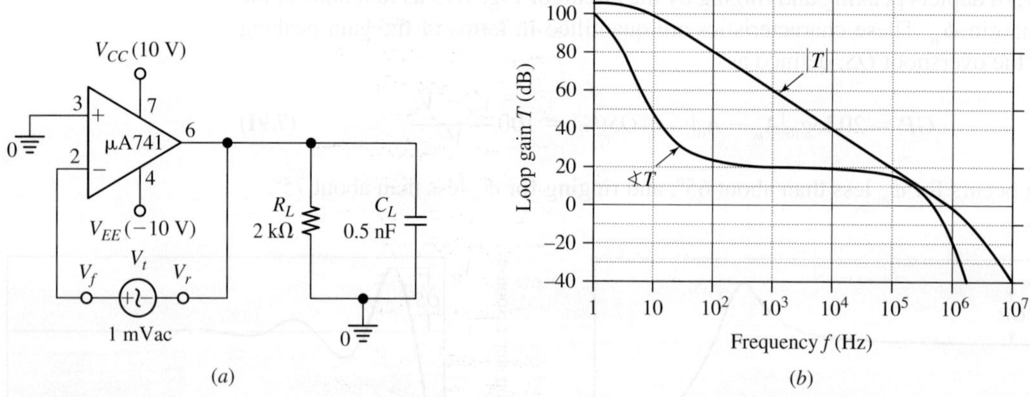

FIGURE 7.79 Using the voltage injection technique to find the phase margin of a capacitively loaded 741 voltage buffer.

7.8 DOMINANT-POLE COMPENSATION

If a negative-feedback system fails to meet the desired degree of stability, its loop gain $T(jf)$ must be altered in such a way as to raise the phase margin ϕ_m to an acceptable value. Altering $T(jf)$ to stabilize a circuit is referred to as *frequency compensation*. Since $T(jf) = a(jf)\beta(jf)$, we can alter $T(jf)$ by altering $a(jf)$, or $\beta(jf)$, or both. Here we take the viewpoint of the IC designer, who strives to ensure a given ϕ_m for the hardest-to-compensate configuration with frequency-independent feedback, namely, the voltage-follower, for which $\beta = 1$ V/V. In this particular case we have

$$T(jf) = a(jf)\beta = a(jf) \times 1 = a(jf) \tag{7.93}$$

that is, $T(jf)$ *coincides* with $a(jf)$. If the op amp utilizes frequency-dependent feedback, then it may be necessary for the user to take additional measures[7] to stabilize it (see also the end-of-chapter problems).

A popular compensation technique involves *lowering* the first pole frequency f_1 to a new value f_D such that the compensated response is dominated by this lone pole *all the way* up to the crossover frequency f_x, which, for $\beta = 1$ V/V, coincides with the already-familiar *transition frequency* f_t. The phase shift at f_x is then $\phi_x = -90° + \phi_{x(HOR)}$, where $-90°$ is the phase shift due to f_D, and $\phi_{x(HOR)}$ is the *combined phase shift* due to the *higher-order roots* (poles and possibly zeros) at f_x. (For instance, the loaded 741 op amp of Fig. 7.79 has $\phi_{x(HOR)} = 90° + \phi_x = 90° - 124° = -34°$.) The phase margin after compensation is $\phi_m = 180° + \phi_x = 180° - 90° + \phi_{x(HOR)}$, that is,

$$\phi_m = 90° + \phi_{x(HOR)} \tag{7.94}$$

By making f_D sufficiently low, we can keep $\phi_{x(HOR)}$ as small as needed. For instance, for $\phi_m \geq 60°$ we need to ensure $\phi_{x(HOR)} \geq -30°$. An obvious drawback of dominant-pole compensation is a drastic gain reduction above f_D, but this is the price we must pay for the sake of stability. Though there are other, more sophisticated techniques[7] that preserve gain over a wider portion of the frequency spectrum, here we limit ourselves to the dominant-pole types. Two popular techniques for shifting f_1 to f_D ($\ll f_1$) are the *shunt-capacitance* technique and the *Miller-compensation* technique.

Shunt-Capacitance Compensation

A simple technique for lowering a pole frequency f_1 is to deliberately *increase* the capacitance of the node responsible for f_1 itself. In the case of the 3-pole amplifier example of the previous section we simply add a *shunt capacitance* C_{shunt} in parallel with C_1 to *lower* its pole frequency from $f_1 = 1/(2\pi R_1 C_1)$ to

$$f_D = \frac{1}{2\pi R_1(C_1 + C_{shunt})} \tag{7.95}$$

To visualize the required location of f_D, refer to the linearized plot of Fig. 7.80 and proceed as follows:

- Identify the frequency f_x at which the phase of $a(jf)$ is $\phi_x = \phi_m - 180°$, where ϕ_m is the desired phase margin (e.g. 45°, 60°, etc.) This is going to be the *transition frequency f_t* after compensation. (We can read out ϕ_x from the phase plot of $a(jf)$, if available, or from slope readings on the magnitude plot of $a(jf)$, in the manner illustrated in Fig. 7.74b.)
- Starting at f_x on the 0-dB axis of the decibel plot of $|a(jf)|$, draw a line with a slope of -20 dB/dec till it intersects the a_0 curve. The corresponding frequency is going to be f_D. Exploiting the constancy of the gain-bandwidth product, we write $a_0 \times f_D = 1 \times f_x$, which allows us to calculate the new pole location as

$$f_D = \frac{f_x}{a_0} \tag{7.96}$$

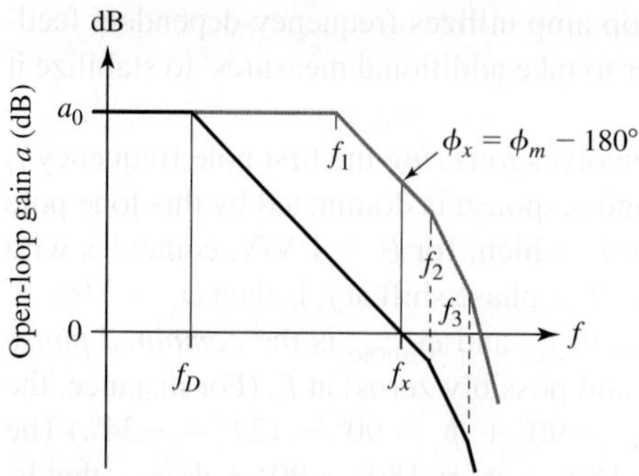

FIGURE 7.80 Graphical method for locating the dominant pole f_D for the case of shunt-capacitance compensation (the shaded curve shows gain before compensation).

If the first two poles are widely spaced, and any higher-order root frequencies (such as f_3 in our example) are sufficiently high, a graphically and computationally convenient starting point is $\phi_x = -135°$, for then $f_x = f_2$, giving $f_D = f_2/a_0$. The corresponding phase margin is only $45°$, but we can always lower f_D further to increase ϕ_m, for instance to $60°$, which Example 7.33 has demonstrated to be often adequate.

EXAMPLE 7.34   **(a)** Find the capacitance C_{shunt} that will stabilize the amplifier of Fig. 7.72 for $\phi_m = 45°$ when used with $\beta = 1$. Verify with PSpice.

(b) To what value must we increase C_{shunt} if we want to achieve $\phi_m = 60°$?

Solution

(a) Letting $f_x = f_2 = 100\ \text{kHz}$ and $a_0 = 10^6\ \text{V/V}$ in Eq. (7.96) gives $f_D = 10^5/10^6 = 0.1\ \text{Hz}$. By Eq. (7.95),

$$C_{shunt} = \frac{1}{2\pi R_1 f_D} - C_1 = \frac{1}{2\pi 10^7 \times 0.1} - 15.9 \times 10^{-12} = 159.1\ \text{nF}$$

Using the PSpice circuit of Fig. 7.81 we get the curves of Fig. 7.82, which show unequivocally the large amount of gain that needs to be sacrificed for the sake of shunt-capacitance stabilization. But it is better to have a stable circuit with sacrificed gain than an unstable one!

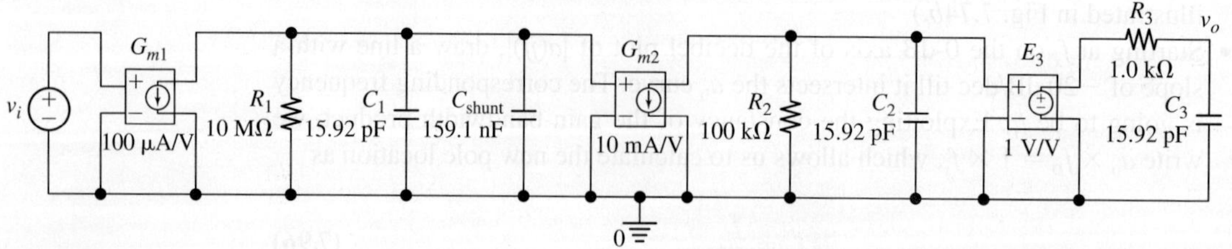

FIGURE 7.81 Three-pole amplifier with shunt-capacitance compensation.

(b) For $\phi_m = 60°$ the crossover must be placed at the frequency f_x where $\phi_x = \phi_m - 180° = 60° - 180° = -120°$. Example 7.33 indicated that $f_x = 59.2$ kHz, so Eq. (7.96) gives $f_D = 0.0592$ Hz, and Eq. (7.95) gives $C_{\text{shunt}} \cong 269$ nF. Re-running PSpice with $C_{\text{shunt}} = 269$ nF and using the cursor facility, we measure $f_x = 59.4$ kHz and $\phi_x = -118°$, in fairly good agreement with the predicted values.

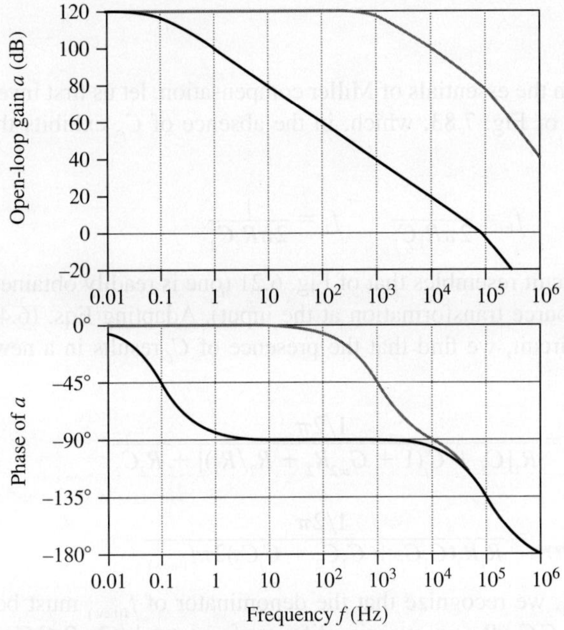

FIGURE 7.82 Magnitude and phase plots with shunt-capacitance compensation for $\phi_m = 45°$. The shaded curves show the response before compensation ($C_{\text{shunt}} = 0$).

Miller Compensation

The low value required of f_D for dominant-pole compensation results in a relatively large C_{shunt}. This may not necessarily be a problem if C_{shunt} is a discrete capacitor external to the op amp. However, in integrated circuits it is desirable to fabricate the compensation circuitry directly on chip, and capacitors higher than a few tens of picofarads would take up too much chip area. A clever way around this limitation is to start out with a small enough capacitance C_f that can be fabricated on chip, and then use the Miller effect to make it appear as large as needed for frequency compensation. We shall see that two additional benefits accrue from this scheme, namely, pole splitting as well as higher slew rates.

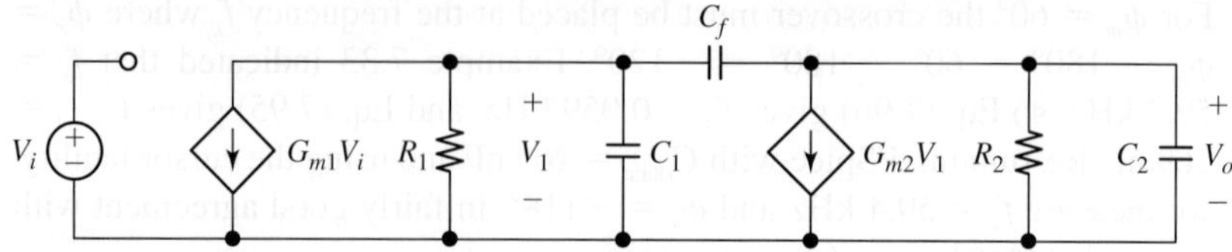

FIGURE 7.83 Two-pole amplifier with Miller compensation.

In order to focus on the essentials of Miller compensation, let us first investigate the two-pole amplifier of Fig. 7.83, which, in the absence of C_f, exhibits the pole frequencies

$$f_1 = \frac{1}{2\pi R_1 C_1} \qquad f_2 = \frac{1}{2\pi R_2 C_2} \tag{7.97}$$

With C_f present, the circuit resembles that of Fig. 6.21 (one is readily obtained from the other via a mere source transformation at the input). Adapting Eqs. (6.45) and (6.46) to the present circuit, we find that the presence of C_f results in a new pole-frequency pair

$$f_{1(\text{new})} = \frac{1/2\pi}{R_1[C_1 + C_f(1 + G_{m2}R_2 + R_2/R_1)] + R_2 C_2}$$

$$f_{2(\text{new})} = \frac{1/2\pi}{R_1 R_2(C_1 C_f + C_1 C_2 + C_f C_2)2\pi f_{1(\text{new})}}$$

Anticipating $f_{1(\text{new})} \ll f_1$, we recognize that the denominator of $f_{1(\text{new})}$ must be dominated by the term $R_1 C_f G_{m2} R_2$, so we simplify as $f_{1(\text{new})} \cong 1/(2\pi R_1 C_f G_{m2} R_2) = 1/(2\pi R_1 C_1 G_{m2} R_2 C_f/C_1)$. Using Eq. (7.97), along with straightforward algebra, we express the new pole pair in the more insightful form

$$f_{1(\text{new})} \cong \frac{f_1}{(G_{m2} R_2 C_f)/C_1} \qquad f_{2(\text{new})} \cong \frac{(G_{m2} R_2 C_f)f_2}{C_1 + C_f(1 + C_1/C_2)} \tag{7.98}$$

Since f_1 is *divided* by $G_{m2}R_2 C_f$ whereas f_2 is multiplied by the same term, we conclude that increasing C_f moves the first pole *down* in frequency and the second pole *up* in frequency, a phenomenon aptly called *pole splitting*. Depicted in Fig. 7.84, pole splitting is highly beneficial because it pushes the second pole and its phase lag *up in frequency*, making the placement of the dominant pole less stringent compared to shunt-capacitance compensation.

Recall from Eq. (6.43) that the presence of C_f results also in the creation of a right-half-plane (RHP) *zero* at $s = G_{m2}/C_f$. Noting that the *dc gain* in Fig. 7.83 is $a_0 = (-G_{m1}R_1) \times (-G_{m2}R_2)$, we summarize by stating that with Miller compensation gain takes on the form

$$a(jf) = a_0 \frac{1 - jf/f_0}{(1 + jf/f_{1(\text{new})})(1 + jf/f_{2(\text{new})})} \tag{7.99}$$

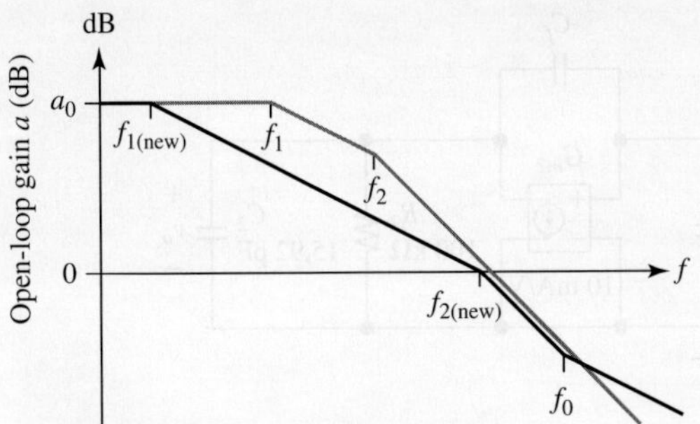

FIGURE 7.84 Miller compensation and pole splitting for the two-pole amplifier of Fig. 7.83 (the shaded curve shows the gain before compensation).

where

$$a_0 = G_{m1}R_1G_{m2}R_2 \qquad\qquad f_0 = \frac{G_{m2}}{2\pi C_f} \tag{7.100}$$

$$f_{1(new)} \cong \frac{1/2\pi}{R_1[C_1 + C_f(1 + G_{m2}R_2 + R_2/R_1)] + R_2C_2} \cong \frac{1}{2\pi R_1 G_{m2}R_2 C_f} \tag{7.101}$$

$$f_{2(new)} \cong \frac{G_{m2}/2\pi}{C_1 + C_2(1 + C_1/C_f)} \tag{7.102}$$

Also, the *gain-bandwidth product* after compensation is GBP $= a_0 \times f_{1(new)}$. Combining Eqs. (7.100) and (7.101), we readily find

$$\text{GBP} \cong \frac{G_{m1}}{2\pi C_f} \tag{7.103}$$

Magnitude and phase are readily found as

$$|a(jf)| = a_0\sqrt{\frac{1 + (f/f_0)^2}{\left[1 + \left(f/f_{1(new)}\right)^2\right]\left[1 + \left(f/f_{2(new)}\right)^2\right]}} \tag{7.104a}$$

$$\text{ph } a(jf) = -\tan^{-1}(f/f_0) - \tan^{-1}\left(f/f_{1(new)}\right) - \tan^{-1}\left(f/f_{2(new)}\right) \tag{7.104b}$$

Remark: Note that a RHP zero contributes *phase lag*, just like a pole in the left half plane! (As we shall see, this can be an issue in two-stage CMOS op amps.)

Pole splitting is readily visualized via PSpice. The circuit of Fig 7.85 utilizes only *the first two stages* of our three-pole working example to show how

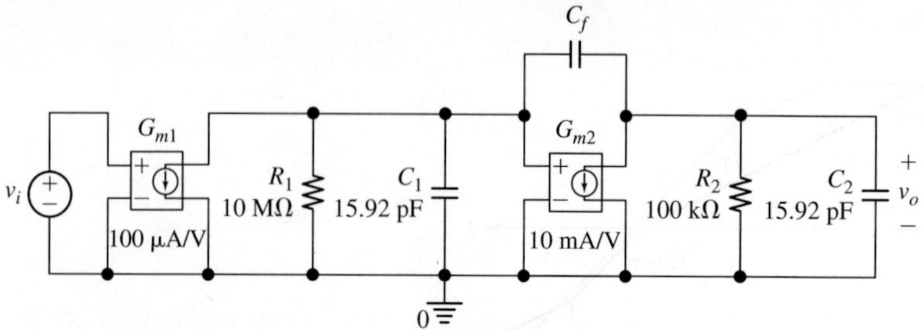

FIGURE 7.85 Using the first two stages of the amplifier of Fig. 7.81 to investigate *pole splitting*.

increasing C_f causes the poles to split apart from their initial values $f_1 = 1$ kHz and $f_2 = 100$ kHz (see Fig. 7.86). The phase plot indicates that the maximum phase shift before compensation is $-180°$ (with each pole contributing a maximum of $-90°$), but after compensation it becomes $-270°$ because of the additional $-90°$ phase lag due to the RHP zero.

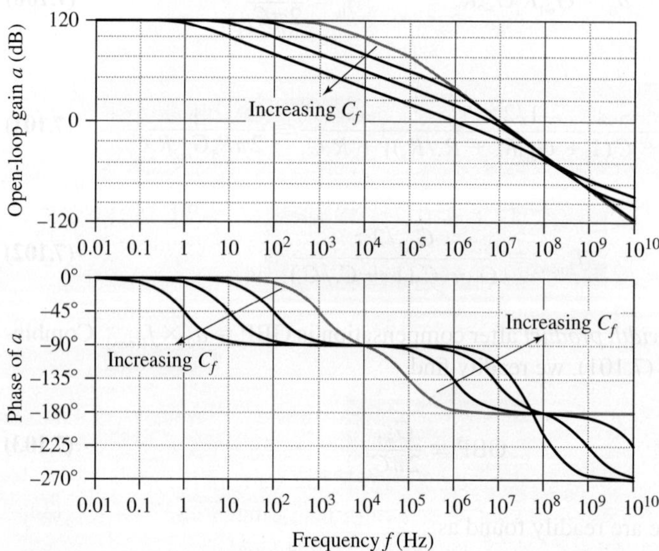

FIGURE 7.86 Pole splitting as a function of C_f for the two-stage amplifier of Fig. 7.85. The values used are $C_f = 0.143$ pF, 1.59 pF, and 15.9 pF, which lower f_1 from 1 kHz to $f_{1(\text{new})} = 100$ Hz, 10 Hz, and 1 Hz, respectively. The gray curves show the response before compensation ($C_f = 0$).

EXAMPLE 7.35

(a) Estimate the pole and zero frequencies of the two-stage amplifier of Fig. 7.85 if $C_f = 5$ pF.

(b) Estimate C_f for a phase margin of 60°.

(c) Verify part (b) with PSpice and comment.

Solution

(a) By Eq. (7.100) we have $f_0 = 10^{-2}/(2\pi \times 5 \times 10^{-12}) = 318$ MHz. Moreover, using $G_{m2}R_2 = 10^3$, we calculate Eq. (7.98) as

$$f_{1(\text{new})} \cong \frac{10^3}{10^3 \times 5/15.92} = 3.18 \text{ Hz}$$

$$f_{2(\text{new})} \cong \frac{10^3 \times 5 \times 10^5}{16.92 + 5(1 + 15.92/15.92)} = 18.6 \text{ MHz}$$

It is apparent that the first pole has moved *down* from 1 kHz to 3.18 Hz, whereas the second pole has moved *up* from 100 kHz to 18.6 MHz. Moreover, $f_0 \gg f_{2(\text{new})}$.

(b) By Eq. (7.94) the combined phase shift at f_x due to $f_{2(\text{new})}$ and f_0 must be $\phi_{x(\text{HOR})} = \phi_m - 90° = 60° - 90° = -30°$. Since in this particular example f_0 is so high, we can ignore its phase contribution at f_x and impose $-30° = -\tan^{-1}[f_x/f_{2(\text{new})}]$, which gives $f_x = (\tan 30°) \times f_{2(\text{new})} = 0.577 \times f_{2(\text{new})}$. Exploiting the constancy of the gain-bandwidth product we write

$$a_0 \times f_{1(\text{new})} = 1 \times f_x = 0.577 \times f_{2(\text{new})}$$

Using Eqs. (7.102) and (7.103) we express this as

$$\frac{G_{m1}}{2\pi C_f} = 0.577 \times \frac{G_{m2}/2\pi}{C_1 + C_2(1 + C_1/C_f)}$$

Substituting the given values of G_{m1}, G_{m2}, C_1, and C_2, and solving for C_f finally gives $C_f = 2.388$ pF. Plugging into Eqs. (7.101) and (7.102) we get $f_{1(\text{new})} = 6.66$ Hz and $f_{2(\text{new})} = 11.54$ MHz, so $f_x = 0.577 \times 11.54 = 6.66$ MHz.

(c) Running the PSpice circuit of Fig. 7.85 with $C_f = 2.388$ pF gives $f_x \cong 5.9$ MHz and $\phi_m \cong 62°$. Considering all the approximations made, the calculated values are in reasonable agreement with PSpice.

Turning to the full-blown *three-pole* amplifier of Fig. 7.81 we observe that the phase lag due to f_3 will erode the phase margin calculated for the *two-pole* version of Fig. 7.85, so C_f will have to be increased slightly if the same phase margin is to be retained. In particular, we use PSpice to find empirically that with $C_f = 4.7$ pF the three-pole amplifier of Fig. 7.81 exhibits $f_x \cong 3.2$ MHz and $\phi_m \cong 62°$.

Shunt-Capacitance and Miller Compensation Comparison

We wish to compare the shunt-capacitance and the Miller compensation schemes for the case of the three-pole amplifier operated with $\beta = 1$ and $\phi_m \cong 60°$. Running the PSpice circuit of Fig. 7.87, first with $C_{\text{shunt}} = 269$ nF (and $C_f = 0$), and then with $C_f = 4.7$ pF (and $C_{\text{shunt}} = 0$), we obtain the closed-loop frequency and step responses of Fig. 7.88. It is apparent that the Miller scheme results in much faster dynamics, thanks to the pole splitting effect, which pushes the second pole toward higher frequencies, thus relaxing the constraints on the dominant pole location. Also, the

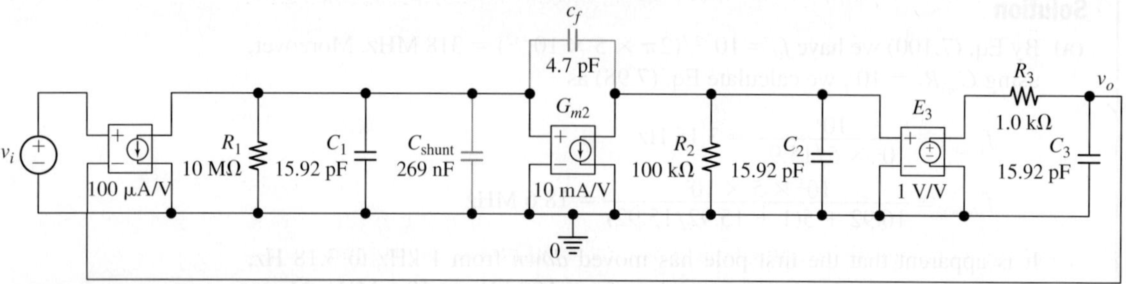

FIGURE 7.87 PSpice circuit to compare the shunt-capacitance and the Miller compensation schemes for the case $\beta = 1$ and $\phi_m \cong 60°$.

Miller-multiplication effect allows for C_f to be much smaller than C_{shunt} so it can be fabricated on chip. Finally, not immediately apparent from the amplifier model used above, is the fact that the slew rate SR is likely to be much higher with the Miller compensation scheme as the much smaller C_f can get charged/discharged much more rapidly.

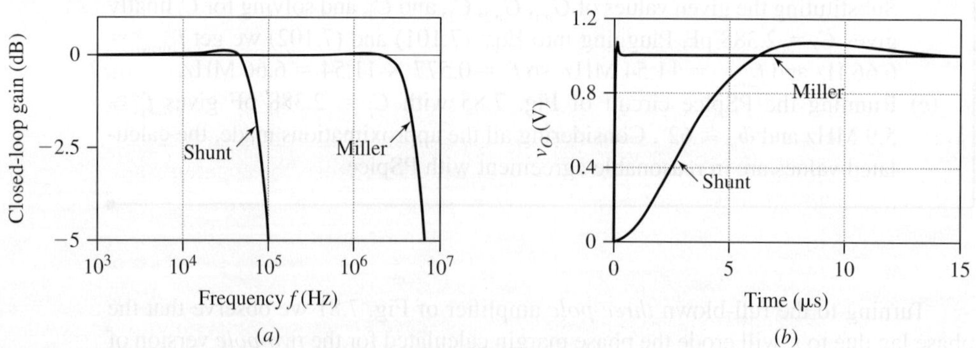

FIGURE 7.88 (a) Frequency and (b) step responses for the unity-gain amplifier of Fig. 7.87.

7.9 FREQUENCY COMPENSATION OF MONOLITHIC OP AMPS

We are now ready to apply the techniques of the previous section to the compensation of the monolithic op amp representatives of Chapter 5, namely, the 741 *bipolar op amp* and the CMOS op amps of the *two-stage* and the *folded-cascode* types. Historically, the 741 was the first op amp to incorporate frequency compensation on chip, in the mid 1960s, a feature that contributed to making the 741 one of the most popular ICs ever. MOS analog technology reached commercial maturity later, and the first significant articles on CMOS op amp frequency compensation started to appear only in the early 1980s.

Frequency Compensation of the 741 Op Amp

With two dozen transistors, an uncompensated 741 op amp exhibits a multitude of pole and zero frequencies, including complex-conjugate pole pairs,[1] whose cumulative phase lag far exceeds $-180°$. As depicted in the 741 schematic of Fig. 5.1, the 741 is Miller compensated via the capacitance $C_c = 30$ pF across the intermediate stage consisting of the CC-CE pair Q_{16}-Q_{17}. After compensation, the roots undergo a drastic rearrangement in the complex plane,[1] such that their combined phase lag is pushed sufficiently above the transition frequency f_t ($\cong 1$ MHz). To estimate the dominant pole frequency, refer to the ac equivalent of Fig. 5.11, repeated for convenience in Fig. 7.89 along with the Miller capacitance C_c. (In anticipation of the dominant role played by C_c, the much smaller stray capacitances of the nodes at either side of C_c have been omitted for simplicity.) This circuit is similar to that of Fig. 7.83, provided we let $C_f = C_c = 30$ pF, $R_1 = R_{o1}//R_{i2} = 6.12//4.63 = 2.64$ MΩ, and $R_2 = R_{o2}//R_{i3} = 81.3//9330 = 80.6$ kΩ. Adapting Eq. (7.101) and retaining only the dominant denominator term, we estimate the -3-dB frequency f_b as

$$f_b \cong \frac{1/2\pi}{R_1 C_c(1 + G_{m2}R_2)} = \frac{1/2\pi}{2.64 \times 10^6 \times 30 \times 10^{-12}(1 + 80.6/0.161)} = 4 \text{ Hz}$$

$$(7.105)$$

Considering all the approximations made, this is close enough to data-sheet value of 5 Hz.

A simulation with the 741 macro-model available in PSpice's library gives, for the case of ± 10-V power supplies, $a_0 = 199{,}220$ V/V, $f_b = 5.0$ Hz, $f_t = 888$ kHz, and ph$a(jf_t) = -117.2°$, so $\phi_m = 62.8°$. As we know from Chapter 6, C_c also sets the slew rate at a nominal value of 0.5 V/μs. The Miller effect boosts the apparent value of C_c by $1 + 80.6/0.161 = 501.6$, making it appear as an equivalent capacitance of 501.6×30 pF $= 150$ nF! Were we to use shunt-capacitance compensation, such a large value could not be fabricated on chip and it would also result in a much lower slew rate.

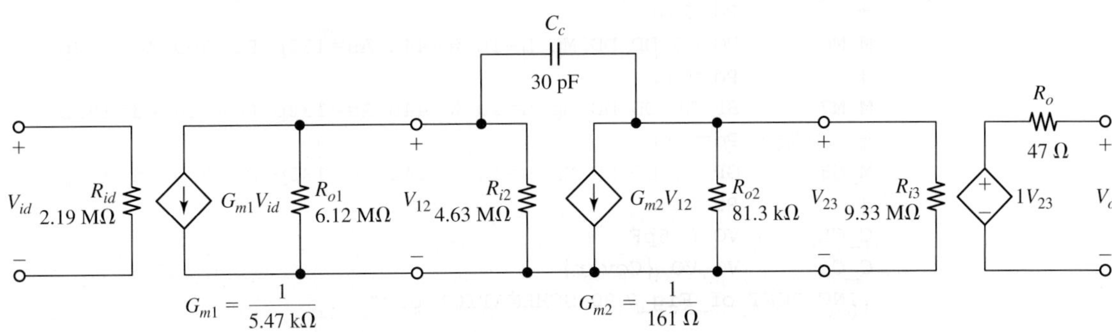

FIGURE 7.89 Ac equivalent of the 741 op amp for the estimation of the dominant pole f_D.

Frequency Compensation of the Two-Stage CMOS Op Amp

We wish to investigate the frequency compensation of the two-stage CMOS op amp of Fig. 7.90, assuming a 1-pF load. We assume the process parameters of Section 6.12, which are based on $L_{\text{drawn}} = 1$ μm and are listed in the following PSpice models:

```
.model    Mn NMOS(Level=1 Tox=20n Uo=600 Vto=0.7 Lambda=0.1
+         Ld=0.15u Gamma=0.6 phi=0.75 Cj=166u Mj=0.5 Cjsw=0.127n
+         Mjsw=0.33 Pb=0.909 Cgso=0.259n Cgdo=0.259n)


.model    Mp PMOS(Level=1 Tox=20n Uo=250 Vto=-0.7 Lambda=0.05
+         Ld=0.2u Gamma=0.5 phi=0.7 Cj=396u Mj=0.5 Cjsw=0.366n
+         Mjsw=0.33 Pb=0.955 Cgso=0.345n Cgdo=0.345n)
```

Following the procedure of Section 6.12, we specify the individual transistor dimensions A_s, P_s, A_d, and P_d so as to reflect the device geometries of Fig. 7.90. Here, the Ws have been calculated on the basis of $V_{OV} = 0.25$ V for all FETs as well as the effective channel lengths $L_n = L_{\text{drawn}} - 2L_{ovn} = 1 - 2 \times 0.15 = 0.7$ μm and $L_p = L_{\text{drawn}} - 2L_{ovp} = 1 - 2 \times 0.2 = 0.6$ μm. The outcome is the following PSpice netlist

```
* source CKT_of_Fig_7.90
V_VDD      DD 0 2.5V
V_VSS      0 SS 2.5V
I_IREF     GP SS DC 100uA
V_Vi       IN 0 DC 0Vdc AC 1Vac
M_M1       GN 0 SP SP Mp L=1u W=22u As=66p Ps=28u Ad=66p
+          Pd=28u
M_M2       V1 IN SP SP Mp L=1u W=22u As=66p Ps=28u Ad=66p
+          Pd=28u
M_M3       GN GN SS SS Mn L=1u W=12u As=30p Ps=17u Ad=30p
+          Pd=17u
M_M4       V1 GN SS SS Mn L=1u W=12u As=30p Ps=17u Ad=30p
+          Pd=17u
M_M5       VO V1 SS SS Mn L=1u W=22U As=55p Ps=27u Ad=55p
+          Pd=27u
M_M6       VO GP DD DD Mp L=1u W=44u As=132p Ps=50u Ad=132p
+          Pd=50u
M_M7       SP GP DD DD Mp L=1u W=44u As=132p Ps=50u Ad=132p
+          Pd=50u
M_M8       GP GP DD DD Mp L=1u W=44u As=132p Ps=50u Ad=132p
+          Pd=50u
C_CL       VO 0 5pF
C_Cc       V1 VO {Ccvar}
.INC "CKT_of_Fig_7.90-SCHEMATIC1.par"
```

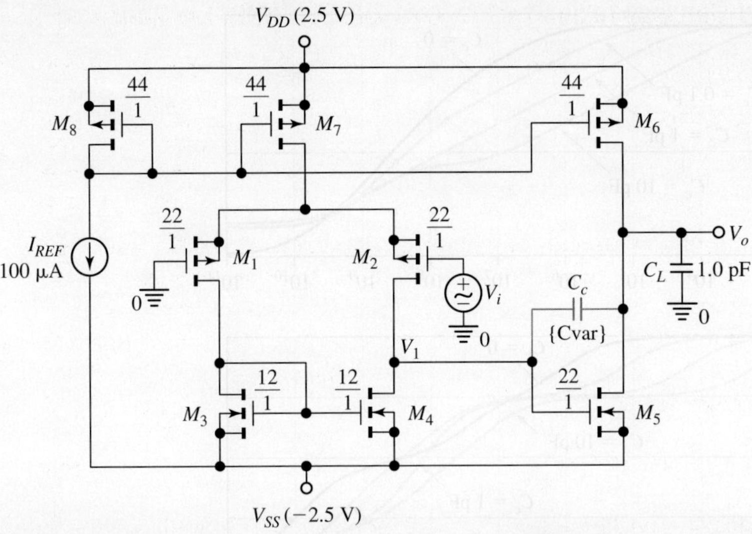

FIGURE 7.90 PSpice circuit to plot the frequency response a two-stage CMOS op amp using conventional Miller compensation.

The small-signal (.TF) analysis yields the following low-frequency parameter values:

$$a_0 = 4245 \text{ V/V} \qquad g_{m1} = 0.425 \text{ mA/V} \qquad r_{o2}/\!/r_{o4} = 143.3 \text{ k}\Omega$$

$$g_{m5} = 0.9344 \text{ mA/V} \qquad r_{o5}/\!/r_{o6} = 74.59 \text{ k}\Omega$$

The ac analysis gives the open-loop responses of Fig. 7.91. Using the cursor facility of PSpice on the uncompensated curves ($C_c = 0$), we find that gain crosses the 0-dB axis at $f_x \cong 366$ MHz, where $\phi_x \cong -184°$, indicating an amplifier in dire need of frequency compensation.

The logical candidate for dominant-pole compensation is the CS stage M_5, where we can take advantage of the Miller effect to multiply a small compensation capacitance C_c placed between its gate and drain terminals for the purpose of establishing a dominant pole at a suitably low frequency. To investigate further, refer to the ac equivalent of Fig. 7.92, where

$$R_1 = r_{o2}/\!/r_{o4} \qquad R_2 = r_{o5}/\!/r_{o6} \qquad (7.106)$$

and C_1 and C_2 are the net capacitances associated with the gate and drain terminals of M_5,

$$C_1 = C_{gd2} + C_{db2} + C_{gd4} + C_{db4} + C_{gs5} \qquad C_2 = C_{db5} + C_{db6} + C_{gd6} + C_L \qquad (7.107)$$

(Usually the load capacitance C_L dominates all the output-node parasitic capacitances, so we approximate $C_2 \cong C_L$ and ignore C_1 compared to C_2.) Moreover, the feedback capacitance is

$$C_f = C_c + C_{gd5} \qquad (7.108)$$

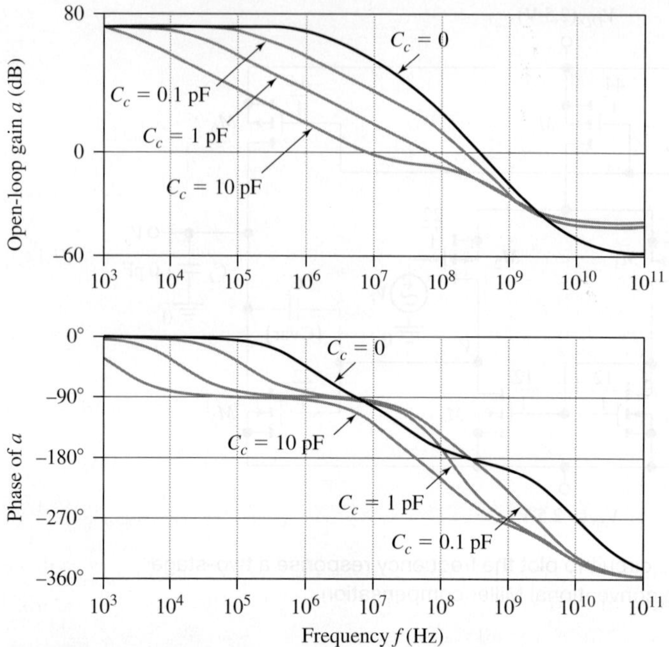

FIGURE 7.91 Magnitude and phase plots for the op amp of Fig. 7.90. The dark curves show the uncompensated ($C_c = 0$) response. The shaded curves indicate that conventional Miller compensation *fails* to convincingly bend the phase curves toward $-90°$ in the vicinity of crossover.

(Usually $C_c \gg C_{gd5}$, so we approximate $C_f \cong C_c$.) Ignoring the stray capacitance C_i, we observe that this circuit is similar to that of Fig. 7.83, so we adapt the formulas developed there (after dropping the subscripts "new" to simplify the notation), and state that in the presence of C_c, gain takes on the form

$$a(jf) = a_0 \frac{1 - jf/f_0}{(1 + jf/f_1)(1 + jf/f_2)}$$

(7.109)

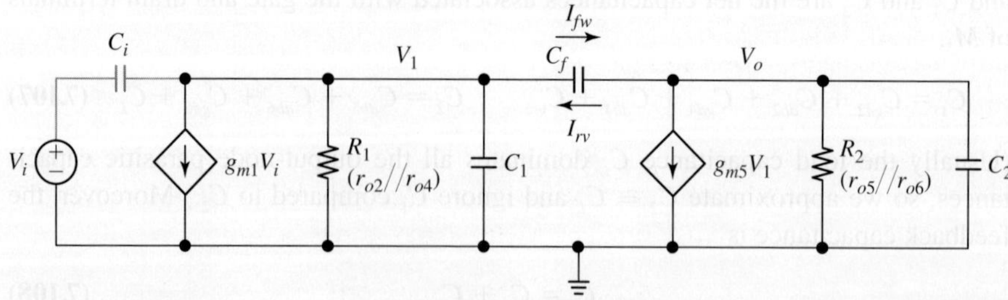

FIGURE 7.92 Approximate ac equivalent of the two-stage CMOS amplifier of Fig. 7.90.

where

$$a_0 = g_{m1}R_1g_{m5}R_2 \tag{7.110}$$

$$f_1 \cong \frac{1/2\pi}{R_1[C_1 + C_f(1 + g_{m5}R_2 + R_2/R_1)] + R_2C_2} \cong \frac{1}{2\pi R_1 g_{m5}R_2C_f} \cong \frac{1}{2\pi R_1 g_{m5}R_2C_c} \tag{7.111}$$

$$f_2 \cong \frac{g_{m5}/2\pi}{C_1 + C_2(1 + C_1/C_f)} \cong \frac{g_{m5}}{2\pi C_2} \cong \frac{g_{m5}}{2\pi C_L} \tag{7.112}$$

$$f_0 = \frac{g_{m5}}{2\pi C_f} \cong \frac{g_{m5}}{2\pi C_c} \tag{7.113}$$

Moreover, the gain-bandwidth product over the frequency region dominated by f_1 is GBP $= a_0 \times f_1$, or

$$\text{GBP} = \frac{g_{m1}}{2\pi C_f} \cong \frac{g_{m1}}{2\pi C_c} \tag{7.114}$$

(Beside confirming the pole pair and RHP zero, the plots of Fig. 7.91 reveal the existence of an additional RHP zero in the 10-GHz range, stemming from forward transmission from V_i to V_1 via the parasitics of the input stage. However, given its high frequency, its impact on the phase margin will be negligible and it shall be ignored. Moreover, the first stage introduces an additional pole-zero pair spaced an octave apart, as per Fig. 6.29b. This too will be ignored for the sake of simplicity.)

Equations (7.111) through (7.113) indicate that increasing C_c will downshift both f_1 and f_0 by about the same amount, while leaving f_2 approximately unchanged. In fact, for $C_c = C_L (=1$ pF), f_0 overlaps f_2, and for $C_c = 10C_L (=10$ pF), f_0 is a decade below f_2 (this is confirmed also by the magnitude plots of Fig. 7.91). We make the important observation that regardless of the value of C_c, f_1 and f_0 are generally not separated widely enough to prevent the phase lag of the RHP zero from eroding the phase margin. This *inherent* limitation of MOSFETs stems from their notoriously low g_ms (g_{m5} in the present case, as evidenced by the fact that the separation f_0/f_1 is proportional to g_{m5}^2). Were it possible to raise g_{m5} by, say, a decade without altering the capacitances, the GBP would remain unchanged, while both f_0 and f_2 would move up and away from f_x by a decade, raising the phase margin significantly (see Problem 7.64).

EXAMPLE 7.36

(a) Estimate the phase margin of the amplifier of Fig. 7.90 for the special case $C_c = C_L (= 1$ pF$)$.

(b) Comment on your results.

Solution

(a) Equations (7.111) through (7.113) give $f_1 = 1/(2\pi \times 143.3 \times 10^3 \times 0.9344 \times 10^{-3} \times 74.59 \times 10^3 \times 1 \times 10^{-12}) = 15.9$ kHz and $f_0 = f_2 = 0.9344 \times 10^{-3}/(2\pi \times 1 \times 10^{-12}) = 149$ MHz, so the gain is

$$a(jf) = \frac{4245[1 - jf/(149 \times 10^6)]}{[1 + jf/(15.9 \times 10^3)] \times [1 + jf/(149 \times 10^6)]}$$

We easily find that the magnitude

$$|a(jf)| = 4245\sqrt{\frac{1 + f^2/(149 \times 10^6)^2}{[1 + f^2/(15.9 \times 10^3)^2][1 + f^2/(149 \times 10^6)^2]}}$$

$$= \frac{4245}{\sqrt{1 + f^2/(15.9 \times 10^3)^2}}$$

drops to 1 V/V at $f_x = 67.5$ MHz, where

$$\phi_x = -\tan^{-1}\frac{67.5 \times 10^6}{15.9 \times 10^3} - \tan^{-1}\frac{67.5 \times 10^6}{149 \times 10^6} - \tan^{-1}\frac{67.5 \times 10^6}{149 \times 10^6}$$

$$\cong -90° - 24.4° - 24.4° \cong -139°$$

Consequently, $\phi_m = 180 - 139 = 41°$, which is generally not sufficiently high. (PSpice gives $f_x = 63.5$ MHz, $\phi_x = -141.6°$, and $\phi_m = 38.4°$, in reasonable agreement with the calculations.)

(b) For $C_c = C_L$ the RHP zero and the second pole *cancel* each other from the magnitude expression, leaving only the dominant pole (this is confirmed by the magnitude curve corresponding to $C_c = 1$ pF in Fig. 7.91). However, their individual phase lags, far from canceling out, *reinforce* each other!

To figure ways around the effect of low g_{m5}, we need to take a closer look at the physical basis of the zero frequency f_0. With reference to Fig. 7.92, we decompose the current through C_f into a *forward* component $I_{fw}(jf) = V_1/(1/j2\pi f C_c)$ and a *reverse* component $I_{rv}(jf) = V_o/(1/j2\pi f C_c)$, or

$$I_{fw}(jf) = j2\pi f C_c V_1 \qquad I_{rv}(jf) = j2\pi f C_c V_o \qquad (7.115)$$

We make the following observations:

- $I_{rv}(jf)$ is responsible for making C_c appear magnified by the Miller multiplier when reflected to the input node V_1. As such, $I_{rv}(jf)$ is *desirable* because it establishes the dominant pole.

- $I_{fw}(jf)$ is responsible for creating the RHP transmission zero at the output node V_o. As such, it is *undesirable* because the presence of this zero erodes the phase margin and halts the gain roll-off that is needed to stabilize the circuit. Recall from Section 6.3 that f_0 is the frequency at which the condition $|I_{fw}(jf_0)| = g_{m5}V_1$ is met, which causes V_o to drop to zero (hence the name). Rewriting as $|j2\pi f_0 C_f V_1| = g_{m5}V_1$ gives the familiar result $f_0 = g_{m5}/(2\pi C_f)$. For $f > f_0$ we have $|I_{fw}(jf)| > g_{m5}V_1$, indicating a *polarity reversal* of V_o that turns feedback from negative to positive. This is not necessarily a problem if $f_0 \gg f_x$, which is usually the case with bipolar amplifiers. However, in MOS amplifiers f_0 tends to be much lower due to the aforementioned low g_ms of FETs.

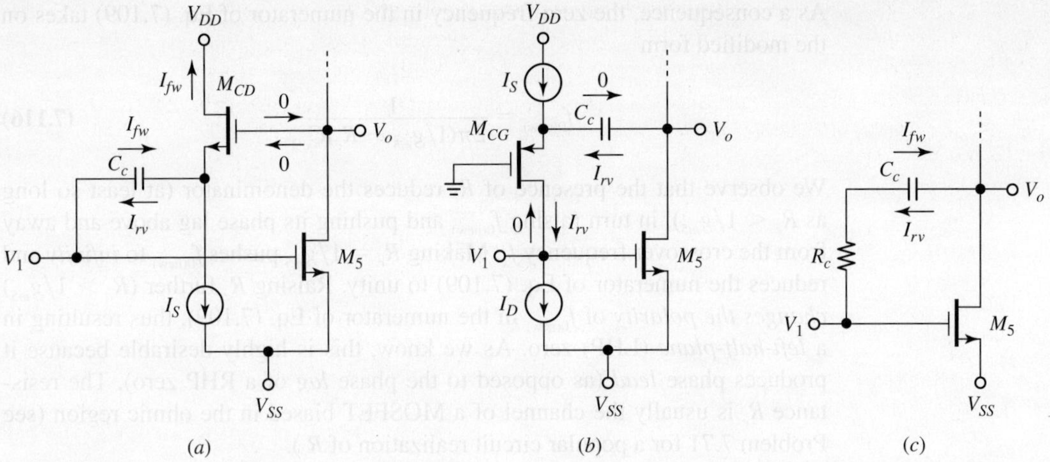

FIGURE 7.93 The RHP zero can be eliminated via (*a*) a voltage buffer or (*b*) a current buffer, or it can be relocated via (*c*) a suitable series resistance R_c.

The best way to cope with the undesirable RHP zero is to eliminate it altogether or at least to shift it to a less harmful location. Figure 7.93 shows *three* different techniques for achieving this goal.

- The scheme[8] of Fig. 7.93*a* exploits the unilateral nature of a CD *voltage buffer* to transmit I_{rv} to node V_1 while diverting I_{fw} to V_{DD} and thus preventing it from reaching node V_o, where it would create the RHP zero (see Problem 7.69). A drawback of this scheme is a reduction in the OVS, as $v_{O(\min)}$ is raised from $V_{SS} + V_{OV5}$ to $V_{SS} + V(I_S)_{\min} + V_{GS(CD)}$, where $V(I_S)_{\min}$ is the minimum allowable voltage drop across the I_S current sink, and $V_{GS(CD)}$ is the gate-source voltage drop of the CD buffer.
- The scheme[9] of Fig. 7.93*b* uses a CG *current buffer* to transmit I_{rv} to node V_1 while inhibiting the formation of I_{fw} because of the high resistance seen looking into the buffer's drain (see Problem 7.70). A drawback of this scheme is that I_S and I_D must be closely matched to avoid creating an intolerable offset error (see Ref. [10] for an ingenious way around this).
- The scheme[11] of Fig. 7.93*c* utilizes a series resistance R_c to curb I_{fw} at high frequencies. At low frequencies, where $R_c \ll |1/(j2\pi f C_c)|$, C_c still dominates, giving $I_{rv}(jf) \cong j2\pi f C_c V_o$ to sustain the dominant pole f_1. But at high frequencies, the presence of R_c changes the location of the transmission zero because we now have $I_{fw}(s) = V_1/(R_c + 1/sC_c)$. V_o will drop to zero at the complex frequency s_0 such that $I_{fw}(s_0) = g_{m5}V_1$, or

$$\frac{V_1}{R_c + 1/(s_0 C_c)} = g_{m5}V_1$$

Solving for s_0, we get the *s-plane zero*

$$s_0 = \frac{1}{(1/g_{m5} - R_c)C_c}$$

As a consequence, the zero frequency in the numerator of Eq. (7.109) takes on the modified form

$$f_{0(new)} = \frac{1}{2\pi(1/g_{m5} - R_c)C_c}$$

(7.116)

We observe that the presence of R_c reduces the denominator (at least so long as $R_c < 1/g_{m5}$), in turn raising $f_{0(new)}$ and pushing its phase lag above and away from the crossover frequency f_x. Making $R_c = 1/g_{m5}$ pushes $f_{0(new)}$ to *infinity* and reduces the numerator of Eq. (7.109) to unity. Raising R_c further ($R_c > 1/g_{m5}$) *changes the polarity* of $f_{0(new)}$ in the numerator of Eq. (7.109), thus resulting in a *left-half-plane* (LHP) zero. As we know, this is highly desirable because it produces phase *lead* (as opposed to the phase *lag* of a RHP zero). The resistance R_c is usually the channel of a MOSFET biased in the ohmic region (see Problem 7.71 for a popular circuit realization of R_c).

EXAMPLE 7.37

(a) Find R_c and C_c to compensate the two-stage op amp of Fig. 7.90 for $\phi_m = 75°$ with $f_0 = \infty$. Hence, estimate the open-loop gain $a(jf)$, the gain-bandwidth product GBP, and the slew rate SR.

(b) Compare with PSpice, and comment.

(c) Assuming $\beta = 1$, plot the closed-loop frequency response as well as the transient response to a pulse alternating between -100 mV and $+100$ mV. Is the pulse response slew-rate limited?

Solution

(a) The pole frequency due to C_L is still $f_2 = g_{m5}/(2\pi C_L) \cong 149$ MHz. For $\phi_m = 75°$ we need $\phi_x = 75 - 180 = -105°$. With $-90°$ coming from the dominant pole frequency f_1, the cumulative phase contribution by f_2 and higher-order root frequencies must be $90 - 105 = -15°$. Ignoring the higher-order roots, we require that $-15° \cong -\tan^{-1}(f_x/f_2)$, or

$$f_x \cong f_2 \times \tan 15° = 149 \times 0.268 = 39.8 \text{ MHz}$$

The required dominant pole frequency is

$$f_1 = \frac{f_x}{a_0} = \frac{39.8 \times 10^6}{4245} = 9.39 \text{ kHz}$$

so Eq. (7.111) gives

$$C_c \cong \frac{1}{2\pi R_1 g_{m5} R_2 f_1}$$

$$= \frac{1}{2\pi \times 143.3 \times 10^3 \times 0.9344 \times 10^{-3} \times 74.59 \times 10^3 \times 9.39 \times 10^3}$$

$$\cong 1.70 \text{ pF}$$

Finally, to move the zero frequency to infinity we need

$$R_c = 1/g_{m5} = 1/(0.9344 \times 10^{-3}) = 1.07 \text{ k}\Omega$$

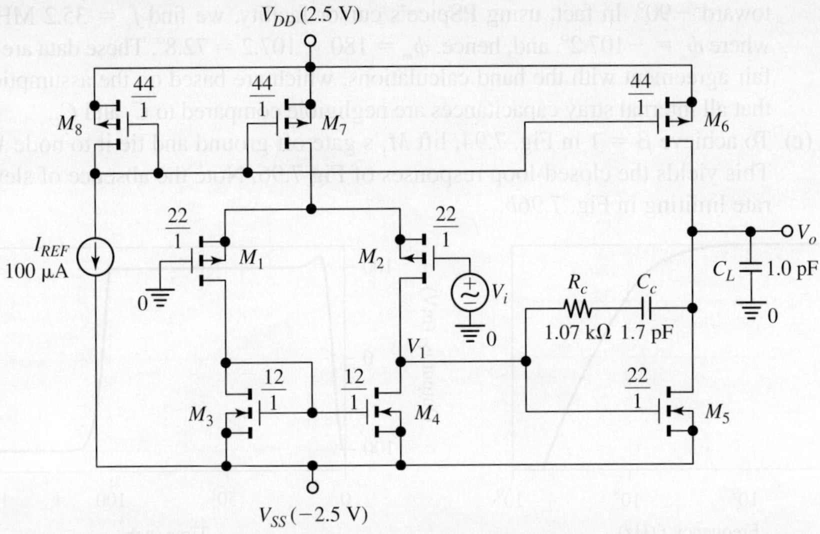

FIGURE 7.94 Frequency compensation for op amp of Example 7.37.

Ignoring higher-order roots, the gain after compensation is approximately

$$a(jf) \cong \frac{4245}{[1 + jf/(9.39 \times 10^3)] \times [1 + jf/(149 \times 10^6)]}$$

We also have GBP $\cong f_x = 39.8$ MHz and SR $= I_{D7}/C_c = (100\,\mu\text{A})/(1.70\,\text{pF}) \cong$ 59 V/μs.

(b) Using the PSpice circuit of Fig. 7.94, we obtain the plots of Fig. 7.95. Compared to Fig. 7.91, the phase curve at crossover is now far more convincingly bent

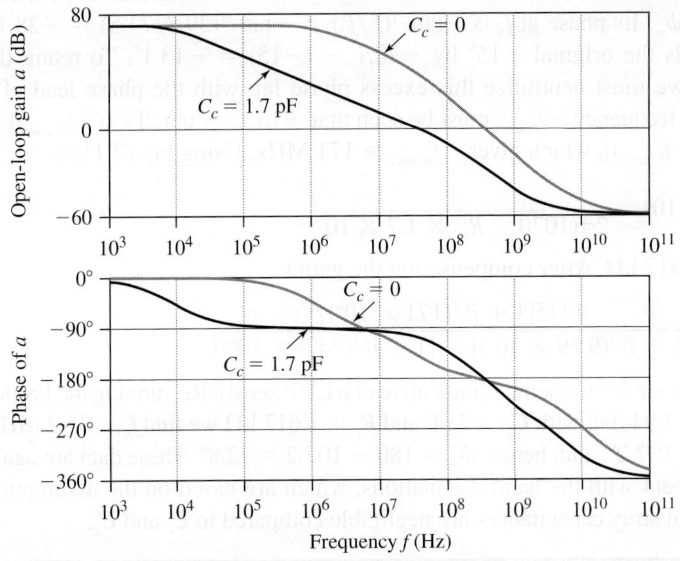

FIGURE 7.95 Magnitude and phase plots for the op amp of Fig. 7.94.

toward $-90°$. In fact, using PSpice's cursor facility, we find $f_x = 35.2$ MHz, where $\phi_x = -107.2°$, and, hence, $\phi_m = 180 - 107.2 = 72.8°$. These data are in fair agreement with the hand calculations, which are based on the assumption that all internal stray capacitances are negligible compared to C_c and C_L.

(c) To achieve $\beta = 1$ in Fig. 7.94, lift M_1's gate off ground and tie it to node V_o. This yields the closed-loop responses of Fig 7.96. Note the absence of slew-rate limiting in Fig. 7.96*b*.

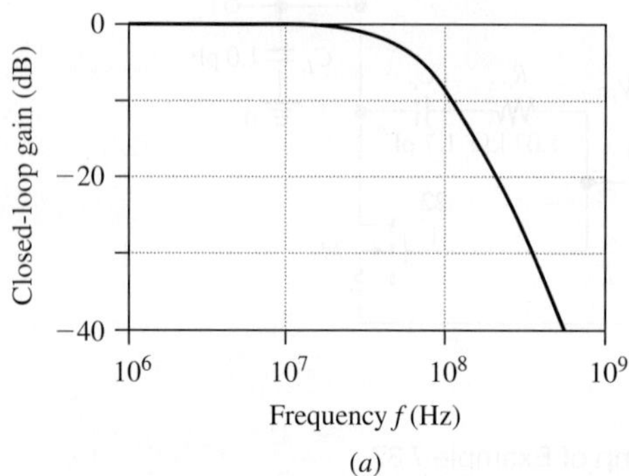
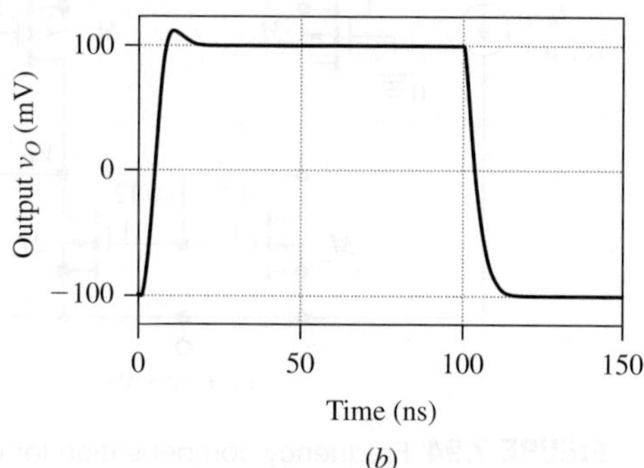

FIGURE 7.96 (*a*) Frequency (*b*) pulse responses of the CMOS op amp of Fig. 7.94 configured for negative-feedback operation with $\beta = 1$.

EXAMPLE 7.38   If C_L is doubled in the circuit of Example 7.37, find R_c so as to retain the *same phase margin* without increasing C_c. What is the expression for $a(jf)$ after compensation? Verify with PSpice and comment.

Solution

Doubling C_L will halve f_2 to 74.5 MHz, bringing it an octave closer to f_x, where it will erode ϕ_m. Its phase at f_x is $-\tan^{-1}(f_x/f_2) = -\tan^{-1}(39.8/74.5) = -28.1°$, which exceeds the original $-15°$ by $-28.1 - (-15) = -13.1°$. To retain the original ϕ_m, we must neutralize this excess phase lag with the phase lead of a LHP zero. Its frequency $-f_{0(new)}$ must be such that $+13.1° = \tan^{-1}[f_x/(-f_{0(new)})] = \tan^{-1}[39.8/(-f_{0(new)})]$, which gives $-f_{0(new)} \cong 171$ MHz. Using Eq. (7.116)

$$-171 \times 10^6 = \frac{1}{2\pi(1070 - R_c) \times 1.7 \times 10^{-12}}$$

gives $R_c = 1.617$ kΩ. After compensation the gain is

$$a(jf) \cong \frac{4245[1 + jf/(171 \times 10^6)]}{[1 + jf/(9.39 \times 10^3)] \times [1 + jf/(74.5 \times 10^6)]}$$

(note the $+$ sign in the numerator, indicative of a LHP zero!). Re-running the PSpice circuit of Fig. 7.94, but with $C_L = 2$ pF and $R_c = 1.617$ kΩ we find $f_x = 35.2$ MHz, where $\phi_x = -107.2°$, and, hence, $\phi_m = 180 - 107.2 = 72.8°$. These data are again in fair agreement with the hand calculations, which are based on the assumption that all internal stray capacitances are negligible compared to C_c and C_L.

Frequency Compensation of the Folded-Cascode CMOS Op Amp

Figure 7.97 shows the ac model of the folded-cascode op amp, along with the most relevant capacitances intervening in its frequency response. Each capacitance is the result of lumping together all parasitic capacitances associated with the corresponding node (the output node includes also the external load capacitance). According to the OCTC procedure of Section 6.7, the -3-dB frequency is

$$f_b \cong \frac{1/2\pi}{R_1C_1 + R_2C_2 + \cdots + R_5C_5 + R_o(C_6 + C_c)} \qquad (7.117)$$

where R_1 through R_5 are the open-circuit equivalent resistances seen by C_1 through C_5, and R_o is the op amp's output resistance. By Eq. (5.69), this resistance is

$$R_o \cong [(g_{m6} + g_{mb6})r_{o6}r_{o8}]//[(g_{m4} + g_{mb4})r_{o4}(r_{o2}//r_{o10})] \qquad (7.118)$$

The resistances R_1 through R_5 are all *source resistances*, on the order of $1/(g_m + g_{mb})$. As such they are much smaller than R_o, and even though the time constants R_1C_1 through R_5C_5 are likely to differ from one another, they all tend to be negligible compared to that associated with R_o, so we approximate

$$f_b \cong \frac{1}{2\pi R_o(C_6 + C_c)} \qquad (7.119)$$

According to Eq. (7.94), the phase margin can be expressed as $\phi_m = 90° + \phi_{x(\text{HOR})}$, where $\phi_{x(\text{HOR})}$ is the *combined phase shift* due to all *higher-order roots*. Consequently,

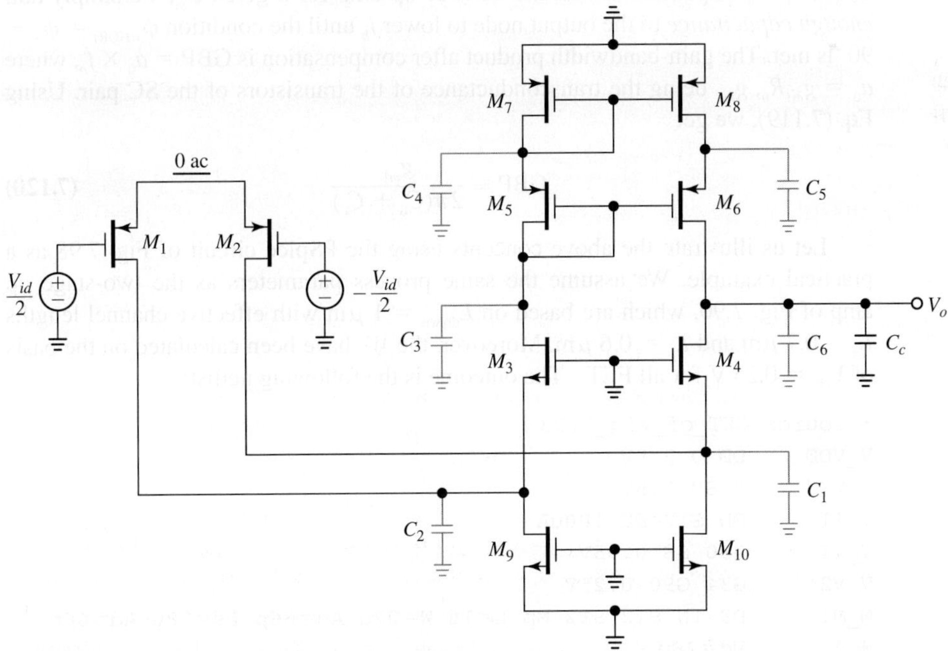

FIGURE 7.97 Ac model of the folded-cascode op amp.

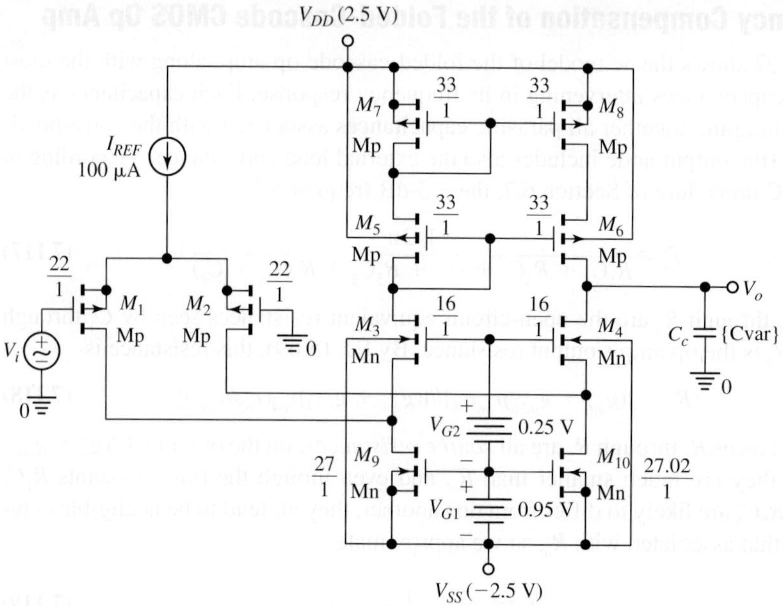

FIGURE 7.98 PSpice circuit to plot the gain of a folded-cascode CMOS op amp with a variable load C_c.

in order to compensate the folded-cascode op amp for a given ϕ_m, we simply *add enough capacitance* to the output node to lower f_b until the condition $\phi_{x(\text{HOR})} = \phi_m - 90°$ is met. The gain-bandwidth product after compensation is GBP $= a_0 \times f_b$, where $a_0 = g_{m1}R_o$, g_{m1} being the transconductance of the transistors of the SC pair. Using Eq. (7.119), we get

$$\text{GBP} \cong \frac{g_{m1}}{2\pi(C_6 + C_c)} \tag{7.120}$$

Let us illustrate the above concepts using the PSpice circuit of Fig. 7.98 as a practical example. We assume the same process parameters as the two-stage op amp of Fig. 7.90, which are based on $L_{\text{drawn}} = 1$ μm with effective channel lengths $L_n = 0.7$ μm and $L_p = 0.6$ μm. Moreover, the Ws have been calculated on the basis of $V_{OV} = 0.25$ V for all FETs. The outcome is the following netlist:

```
* source CKT_of_Fig_7.98
V_VDD      DD 0 2.5V
V_VSS      0 SS 2.5V
I_I1       DD S12 DC 100uA
V_V1       G90 SS 0.95V
V_V2       G34 G90 0.25V
M_M1       D9 IN S12 S12 Mp L=1u W=22u As=66p Ps=28u Ad=66p
+          Pd=28u
M_M2       D10 0 S12 S12 Mp L=1u W=22u As=66p Ps=28u Ad=66p
+          Pd=28u
```

```
M_M3       G56 G34 D9 SS Mn L=1u W=16U As=40p Ps=21u As=40p
+          Ps=21u
M_M4       OUT G34 D10 SS Mn L=1u W=16U As=40p Ps=21u As=40p
+          Ps=21u
M_M5       G56 G56 G78 DD Mp L=1u W=33u As=99p Ps=39u Ad=99p
+          Pd=39u
M_M6       OUT G56 S6 DD Mp L=1u W=33u As=99p Ps=39u Ad=99p
+          Pd=39u
M_M7       G78 G78 DD DD Mp L=1u W=33u As=99p Ps=39u Ad=99p
+          Pd=39u
M_M8       S6 G78 DD DD Mp L=1u W=33u As=99p Ps=39u Ad=99p
+          Pd=39u
M_M9       D9 G90 SS SS Mn L=1u W=27U As=68p Ps=33u Ad=68p
+          Pd=33u
M_M10      D10 G90 SS SS Mn L=1u W=27.02U As=68p Ps=33u Ad=68p
+          Pd=33u
C_Cc       V1 VO {Ccvar}
.INC "CKT_of_Fig_9.11-SCHEMATIC1.par"
```

The small-signal (.TF) analysis gives the following low-frequency parameter values

$$a_0 = 2679 \text{ V/V} \qquad R_o = 6.391 \text{ M}\Omega \qquad g_{m1} = 419 \text{ }\mu\text{A/V} \qquad \textbf{(7.121}a\textbf{)}$$

The ac analysis gives the open-loop responses of Fig. 7.99. Using PSpice's cursor facility, we find that the uncompensated response, corresponding to $C_c = 0$, has

$$f_b = 425 \text{ kHz} \qquad f_x = 749.5 \text{ MHz} \qquad \phi_x = -131.5° \qquad \phi_m = 48.5° \qquad \textbf{(7.121}b\textbf{)}$$

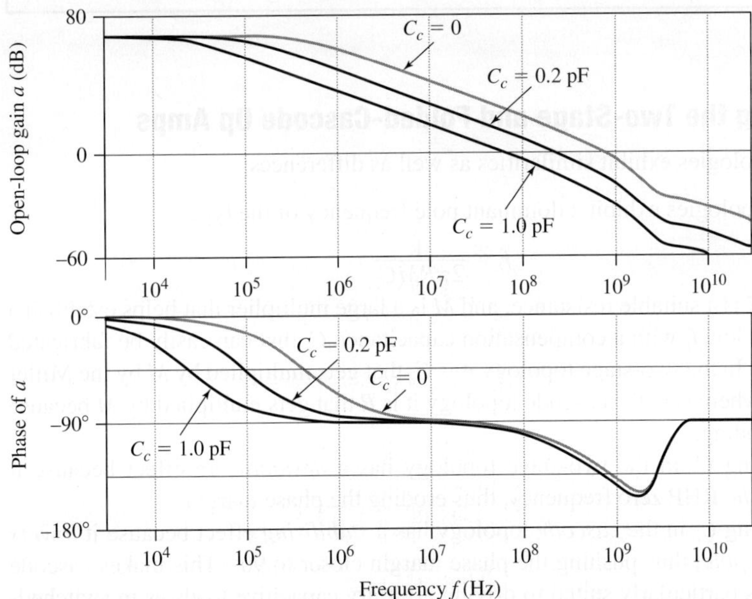

FIGURE 7.99 Magnitude and phase plots for the op amp of Fig. 7.98 for different values of C_c.

To increase ϕ_m we deliberately add capacitance to the output node and obtain the curves shown for the cases $C_c = 0.2$ pF and 1 pF. As expected, the price for more phase margin is bandwidth reduction.

EXAMPLE 7.39   (a) Using the above cursor data, estimate the output-node parasitic capacitance C_6.
(b) Find C_c for $\phi_m = 75°$ using the fact that the cursor gives, for the uncompensated response, $f_{-105°} = 186$ MHz. What are the ensuing bandwidth f_b and GBP? Check with PSpice and comment.

Solution

(a) For $C_c = 0$ we use Eq. (7.119) to impose

$$425 \times 10^3 = \frac{1}{2\pi \times 6.391 \times 10^6 \times (C_6 + 0)}$$

This gives $C_6 = 58.6$ fF.

(b) Adapting Eq. (7.96) to the present case we write $f_b = f_{-105°}/a_0 = 186 \times 10^6/2679 \cong 69.4$ kHz. Using again Eq. (7.119),

$$69.4 \times 10^3 = \frac{1}{2\pi \times 6.391 \times 10^6 \times (58.6 \times 10^{-15} + C_c)}$$

which gives $C_c = 300$ fF and GBP $= 186$ MHz. Rerunning PSpice with $C_c = 0.3$ pF we get $f_b = 69.3$ kHz, GBP $= f_x = 173$ MHz, $\phi_x = -108°$, and $\phi_m = 72°$, all in reasonable agreement with the calculated values.

Comparing the Two-Stage and Folded-Cascode Op Amps

The two topologies exhibit similarities as well as differences:

- Both topologies exhibit a dominant pole frequency of the type

$$f_b \propto \frac{1}{2\pi RMC_c}$$

  where R is a suitable resistance, and M is a large multiplier that helps establish a suitably low f_b with a compensation capacitance C_c that can easily be fabricated on chip. In the two-stage topology it is C_c that gets multiplied by M by the Miller effect, whereas in the cascode topology it is R that gets multiplied by M because of cascoding.

- Increasing C_c in the two-stage topology has a *destabilizing* effect because it *lowers the* RHP zero frequency, thus eroding the phase margin.

- Increasing C_c in the *cascode* topology has a *stabilizing* effect because it *lowers the first pole*, thus pushing the phase margin closer to 90°. This makes cascode op amps particularly suited to driving arbitrary capacitive loads as in switched-capacitor filters.

7.10 NOISE

Any disturbance that tends to obscure a signal of interest is generally referred to as noise. A familiar example is *hum* ("*hmmm*") in a poorly designed audio system. This noise is injected into the circuit from the outside (in this case from the utility power). Another example is the *hiss* ("*shhh*") produced by a low quality audio amplifier and best evidenced when we turn the volume all the way up with no audio input. This is an example of *intrinsic noise*, so called because it is generated internally by the components (resistors, diodes, and transistors) making up the circuit (conversely, noise injected from the outside is called *extrinsic* noise). Though we can virtually eliminate extrinsic noise via proper design, layout, and shielding, intrinsic noise is always present in a circuit. It can be reduced via proper component and circuit topology selection, and filtering, but it can never be eliminated entirely.

The most common example of intrinsic noise is *resistor noise*, which is the result of thermal agitation of the charge carriers (electrons in ordinary resistors and n-type materials, holes in p-type materials), and is present even if the resistor is sitting in the drawer. Though the voltage across an unconnected resistor averages to zero, its instantaneous value is constantly fluctuating about zero as depicted in Fig. 7.100.

Basic Noise Properties

Denoting a noise voltage as $e_n(t)$ and a noise current as $i_n(t)$, we are interested in their *root-mean-square* (rms) *values* E_n and I_n over a time interval t_1 to t_2, respectively defined as

$$E_n = \sqrt{\frac{1}{t_2 - t_1} \int_{t_1}^{t_2} e_n^2(t)\, dt} \qquad I_n = \sqrt{\frac{1}{t_2 - t_1} \int_{t_1}^{t_2} i_n^2(t)\, dt} \qquad (7.122)$$

Their physical meanings are that if we play $e_n(t)$ *across* or $i_n(t)$ *through* a resistor R, the power dissipated by R is E_n^2/R or RI_n^2. Alternatively, we say that E_n^2 and I_n^2 represent the power dissipated by $e_n(t)$ and $i_n(t)$ in a 1-Ω resistance. Though $e_n(t)$ and $i_n(t)$ cannot be integrated analytically because they are random variables, E_n and I_n are easily measured with a true rms multimeter.

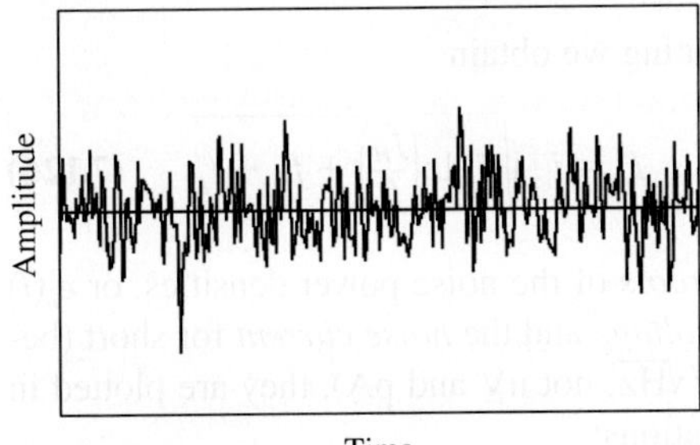

FIGURE 7.100 Oscilloscope display of (suitably magnified) resistor noise.

In the course of noise analysis we often deal with noise voltages in *series* or noise currents in *parallel*, so we are interested in the rms value of the *combined* noise. For the case of a pair of noise voltages $e_{n1}(t)$ and $e_{n2}(t)$ in series, the overall rms value is

$$E_n^2 = \frac{1}{t_2 - t_1} \int_{t_1}^{t_2} [e_{n1}(t) + e_{n2}(t)]^2 \, dt = E_{n1}^2 + \frac{2}{t_2 - t_1} \int_{t_1}^{t_2} e_{n1}(t) \times e_{n2}(t) \, dt + E_{n2}^2$$

having expanded the quadratic term and then used Eq. (7.122) twice. In most cases of interest $e_{n1}(t)$ and $e_{n2}(t)$ are *uncorrelated*, so their product averages to zero, giving $E_n^2 = E_{n1}^2 + E_{n2}^2$. We readily generalize this result to the case of N uncorrelated noise voltages in series or N uncorrelated noise currents in parallel by stating that they combine in *rms fashion* as

$$E_n = \sqrt{E_{n1}^2 + E_{n2}^2 \cdots + E_{nN}^2} \qquad I_n = \sqrt{I_{n1}^2 + I_{n2}^2 \cdots + I_{nN}^2} \qquad \textbf{(7.123)}$$

Noise Spectra

The physical meaning of the rms value of noise is similar to the rms value of an ac signal, except that the power of a sinusoid is concentrated at just one frequency whereas noise power is usually spread all over the frequency spectrum. The frequency distribution of noise power is specified via the *noise power densities* $e_n^2(f)$ and $i_n^2(f)$, each representing the average noise power over a 1-Hz bandwidth as a function of frequency f. We use the power densities to calculate analytically the rms values over an arbitrary frequency interval f_L to f_H as

$$E_n = \sqrt{\int_{f_L}^{f_H} e_n^2(f) \, df} \qquad I_n = \sqrt{\int_{f_L}^{f_H} i_n^2(f) \, df} \qquad \textbf{(7.124)}$$

Since E_n is in V and f in Hz, it follows that $e_n^2(f)$ is in V^2/Hz; similarly, $i_n^2(f)$ is in A^2/Hz.

As an example, the power densities of *integrated-circuit* (IC) *noise* take on the analytical forms

$$e_n^2(f) = e_{nw}^2 \left(\frac{f_{ce}}{f} + 1 \right) \qquad i_n^2(f) = i_{nw}^2 \left(\frac{f_{ci}}{f} + 1 \right) \qquad \textbf{(7.125)}$$

so substituting into Eq. (7.124) and integrating we obtain

$$E_n = e_{nw} \sqrt{f_{ce} \ln\left(\frac{f_H}{f_L}\right) + f_H - f_L} \qquad I_n = i_{nw} \sqrt{f_{ci} \ln\left(\frac{f_H}{f_L}\right) + f_H - f_L} \qquad \textbf{(7.126)}$$

Data sheets usually show the *square roots* of the noise power densities, or $e_n(f)$ and $i_n(f)$. Called, if improperly, the *noise voltage* and the *noise current* for short (beware that their units are nV/$\sqrt{\text{Hz}}$ and pA/$\sqrt{\text{Hz}}$, not nV and pA), they are plotted in Fig. 7.101. We make the following observations:

- For $f \gg f_{ce}$ we have $e_n \to e_{nw}$, and for $f \gg f_{ci}$ we have $i_n \to i_{nw}$, indicating high-frequency asymptotes that are *constant with frequency*. This type of noise is

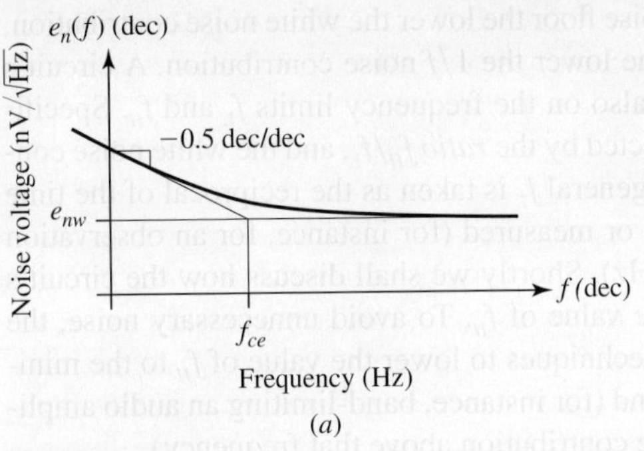
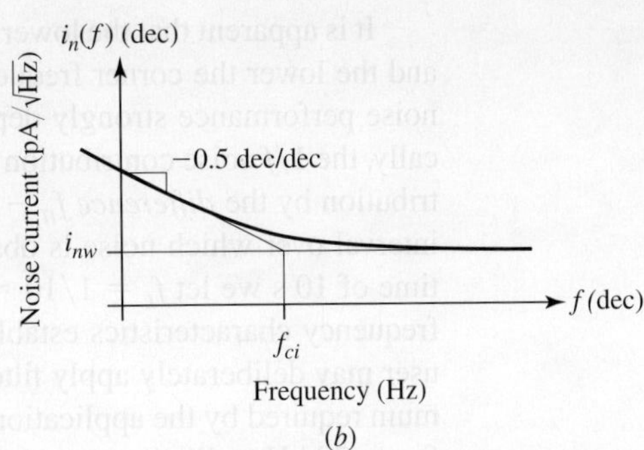

FIGURE 7.101 Typical spectral densities of integrated circuit noise: (*a*) noise voltage and (*b*) noise current.

called *white noise* by analogy with white light, which contains all frequency components, and e_{nw} and i_{nw} are aptly called *white-noise floors*.

- For $f \ll f_{ce}$ we have, by Eq. (7.125), $e_n^2(f) \propto 1/f$, and for $f \ll f_{ci}$ we have $i_n^2(f) \propto 1/f$, indicating low-frequency asymptotes with slopes of -1 dec/dec. (By contrast, the plots of the square roots of Fig. 7.101 exhibit slopes of -0.5 dec/dec). Noise with a power density inversely proportional to frequency f is aptly called $1/f$ *noise*.
- The frequencies f_{ce} and f_{ci}, representing the borderlines between $1/f$ noise and white noise are called the *corner frequencies* by analogy with filters. As a practical example, the data sheets of the popular 741 op amp give

$$e_{nw} = 20 \text{ nV}/\sqrt{\text{Hz}} \quad f_{ce} = 200 \text{ Hz} \quad i_{nw} = 0.5 \text{ pA}/\sqrt{\text{Hz}} \quad f_{ci} = 2 \text{ kHz} \quad \textbf{(7.127)}$$

EXAMPLE 7.40

(a) Find E_n for the 741 op amp over the *audio range* (20 Hz to 20 kHz).
(b) Repeat, but for the *wide-band range* of 0.1 Hz to 1 MHz, compare with (*a*), and comment.

Solution

(a) Using Eq. (7.126),

$$E_n = 20 \times 10^{-9}\sqrt{200 \ln\left(\frac{20 \times 10^3}{20}\right) + 20 \times 10^3 - 20}$$

$$\cong 20 \times 10^{-9}\sqrt{1{,}382 + 20 \times 10^3} = 2.92 \ \mu\text{V rms.}$$

(b) Likewise,

$$E_n = 20 \times 10^{-9}\sqrt{200 \ln\left(\frac{10^6}{0.1}\right) + 10^6 - 0.1}$$

$$\cong 20 \times 10^{-9}\sqrt{3{,}224 + 10^6} = 20.0 \ \mu\text{V rms.}$$

Increasing the bandwidth increases both the $1/f$ noise and the white noise contributions. However, due to its logarithmic dependence, the $1/f$ noise contribution increases less noticeably than the white noise contribution, which increases with the square root of the bandwidth.

It is apparent that the lower the noise floor the lower the white noise contribution, and the lower the corner frequency the lower the $1/f$ noise contribution. A circuit's noise performance strongly depends also on the frequency limits f_L and f_H. Specifically, the $1/f$ noise contribution is affected by the *ratio* f_H/f_L, and the white noise contribution by the *difference* $f_H - f_L$. In general f_L is taken as the reciprocal of the time interval over which noise is observed or measured (for instance, for an observation time of 10 s we let $f_L = 1/10 = 0.1$ Hz). Shortly we shall discuss how the circuit's frequency characteristics establish the value of f_H. To avoid unnecessary noise, the user may deliberately apply filtering techniques to lower the value of f_H to the minimum required by the application at hand (for instance, band-limiting an audio amplifier to 20 kHz will eliminate any noise contribution above that frequency).

Noise Types

Noise stems from a number of different physical mechanisms. Following are the types of noise most commonly found in electronic devices:

- **Thermal noise.** As mentioned, this form of noise stems from *thermal agitation* of the charge carriers in conductors. It is white, and its power density is proportional to absolute temperature T. Also called *Johnson noise* for John B. Johnson who first investigated it in 1928, it is present in all resistors, whether intentional resistors or parasitic resistors like the bulk resistance of a *pn* junction, the base-region resistance of a BJT, or the channel resistance of a FET. We model an actual resistor with a noiseless resistance R in *series* with a noise *voltage* e_{nr}, as depicted in Fig. 7.102a. Or, we can perform a source transformation and use a noiseless resistance R in *parallel* with a noise *current* $i_{nr} = e_{nr}/R$, as in Fig. 7.102b. The power densities are given by[1]

$$e_{nr}^2 = 4kTR \qquad\qquad i_{nr}^2 = \frac{4kT}{R} \tag{7.128}$$

where $k = 1.38 \times 10^{-23}$ J/K is Boltzmann's constant. For instance, at room temperature a 1-kΩ resistor has $e_{nr} = \sqrt{4 \times 1.38 \times 10^{-23} \times 300 \times 10^3} \cong 4$ nV/$\sqrt{\text{Hz}}$ and $i_{nr} = 4 \times 10^{-9}/10^3 = 4$ pA/$\sqrt{\text{Hz}}$. The rms voltage over a 1-MHz bandwidth is $E_n = 4 \times 10^{-9} \times \sqrt{10^6} = 4$ μV rms.

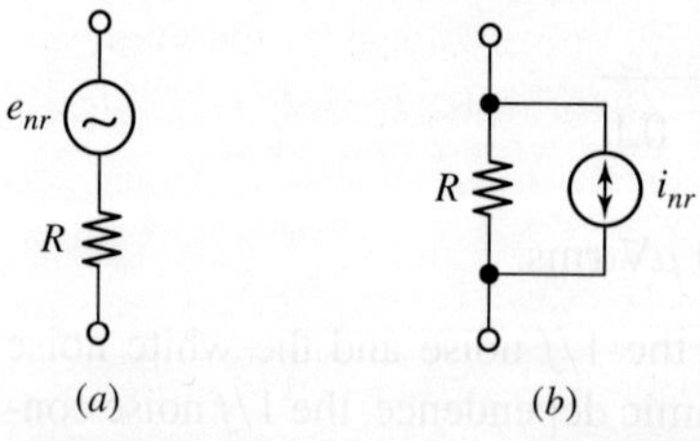

FIGURE 7.102 Resistor noise models, consisting of a noiseless resistance R and (a) a series noise voltage e_{nr} or (b) a parallel noise current i_{nr}.

- **Shot Noise.** As charge carriers flow across a potential barrier such as that of a *pn* junction, they produce a current $i(t)$ that is constantly fluctuating around its average value I because of the discrete nature of charge. The power density of these fluctuations is proportional to I according to[1]

$$i_n^2 = 2qI \tag{7.129}$$

where $q = 1.602 \times 10^{-19}$ C is the electron charge.

- **Flicker Noise.** Also known as *contact noise*, this form of noise has various origins, depending on the device type. In transistors it is generally due to *traps* that capture and release charge carriers randomly as they flow to produce current. The ensuing fluctuations exhibit a power density of the type[1]

$$i_n^2 = \frac{KI^a}{f^b} \tag{7.130}$$

where K is a device constant called the *flicker coefficient*, I is the average current, and a and b are additional device constants with $0.5 < a < 2$ and $b \cong 1$. Since its power density is (approximately) inversely proportional to frequency f, flicker noise is also called $1/f$ *noise*. Another name is *pink noise*, by analogy with pink light, whose power is denser at low frequencies.

- **Other Forms of Noise.** Another form of low-frequency noise is *burst noise*, so called because of its appearance when observed on the oscilloscope. Also called *popcorn noise* because of the sound it produces when played through a loud-speaker, it manifests itself in the presence of heavy metal-ion contaminants, such as gold doping[1] (we will ignore this form of noise here). Another form of noise is *breakdown noise*, so called because produced by *pn* junctions when operated in breakdown.

Noise Models of Semiconductor Devices

We now wish to develop noise models for semiconductor devices using noiseless devices but equipped with suitable external noise sources, in the manner already seen for the resistor.

- **Diode Noise Model.** To account for its shot and flicker noise, we model an actual diode with a noiseless *pn* junction but having a parallel *noise current* i_{nd} as depicted in Table 7.3. Also shown is a series noise voltage e_{nr} modeling the thermal noise of the diode's bulk resistance r_S (see Chapter 1). The power densities of these sources are, respectively,[1]

$$i_{nd}^2 = 2qI_D + \frac{KI_D^a}{f} \tag{7.131a}$$

$$e_{nr}^2 = 4kTr_S \tag{7.131b}$$

TABLE 7.3 Noise models and noise power densities for semiconductor devices (the devices in the models are assumed *noiseless* and noise is accounted for by suitable noise sources, as shown).

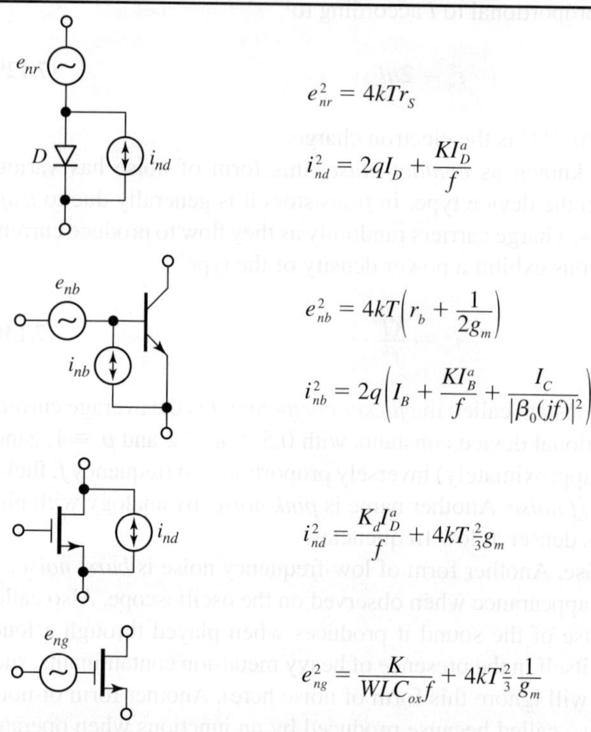

- **BJT Noise Model.** A forward-biased BJT with base current I_B and collector current I_C exhibits shot-noise power densities of $2qI_B$ at the base and $2qI_C$ at the collector. Dividing $2qI_C$ by g_m^2, $g_m = I_C/V_T = qI_C/kT$, reflects it to the base, where it is added to $4kTr_b$, the thermal noise due to the base's bulk resistance r_b (see Chapter 2). The result is the overall base power density[1]

$$e_{nb}^2 = 4kT\left(r_b + \frac{1}{2g_m}\right) \tag{7.132a}$$

Likewise, dividing $2qI_C$ by $|\beta_0(jf)|^2$ reflects it to the base, where it is added to the base's shot noise as well as to the base's flicker noise, to give the overall base power density[1]

$$i_{nb}^2 = 2q\left(I_B + \frac{KI_B^a}{f} + \frac{I_C}{|\beta_0(jf)|^2}\right) \tag{7.132b}$$

where K and a are process-related parameters. The BJT's noise model and relative power densities are summarized in Table 7.3. By Eq. (7.132a), a BJT produces the same amount of voltage noise as a resistance $R_{eq} = r_b + 1/(2g_m)$. For instance, a BJT with $r_b = 250\ \Omega$ operating at $I_C = 0.1$ mA has $R_{eq} = 250 + (26/0.1)/2 = 380\ \Omega$, giving $e_{nb} \cong 2.5$ nV/$\sqrt{\text{Hz}}$.

- **MOSFET Noise Models.** The noise characteristics of a saturated MOSFET are dominated by two noise types: (*a*) *flicker noise*, due to the dangling bonds in the channel near the interface with the oxide, which act as traps, randomly capturing and releasing charge carriers as they flow from source to drain; (*b*) *thermal noise*, due to the channel's resistance. The noise power density of the drain current is given by[1]

$$i_{nd}^2 = \frac{K_d I_D^a}{f} + 4kT\tfrac{2}{3}g_m \qquad (7.133a)$$

where K_d and a are suitable parameters. Alternatively, we reflect the drain noise to the gate as $e_{ng}^2 = i_{nd}^2/g_m^2$. The result is expressed in the insightful form[1]

$$e_{ng}^2 = \frac{K}{WLC_{ox}f} + 4kT\tfrac{2}{3}\frac{1}{g_m} \qquad (7.133b)$$

where K is a process-related constant on the order of 10^{-24} V²F. The two MOSFET noise models and relative power densities are summarized in Table 7.3. Flicker noise is inversely proportional to the area $W \times L$, so the IC designer has the option of specifying large-area MOSFETs to keep flicker noise below a prescribed level. Also, K tends to be lower in *p*MOSFETs than in *n*MOSFETs as holes are less likely to get trapped than electrons. This is another reason why *p*-channel input stages are preferable. By Eq. (7.133), a MOSFET produces the same amount of voltage noise as a resistance $R_{eq} = 2/(3g_m)$. For instance, a FET with $g_m = 1$ mA/V has $R_{eq} = 667\ \Omega$, giving $e_{ng} \cong 3.3$ nV/$\sqrt{\text{Hz}}$.

Noise Dynamics

An integrated circuit such as an amplifier consists of a variety of devices, both active and passive, each contributing noise to the output, so finding the individual contributions may be an arduous task. Mercifully, the noise characteristics of the entire circuit can be modeled with just a pair of input noise sources as depicted in Fig. 7.103a. The source e_n models the *short-circuit* input noise because short-circuiting the input port forces i_n to flow through e_n, so i_n has no effect on the output noise, which is thus due exclusively to e_n. Conversely, the source i_n models the *open-circuit* input noise because open-circuiting the input port eliminates the effect of e_n and noise is now due exclusively to i_n flowing through the circuit's input impedance.

Once we embed an amplifier in a circuit, we are interested in its *total rms output noise* E_{no} for the case of a voltage-output type, or in I_{no} for a current-output type (by *total* we mean over the frequency interval $f_L < f < \infty$). To this end, we first combine the effects of e_n and i_n into a single equivalent input source $e_{ni}(f)$ for a voltage-input type, or $i_{ni}(f)$ for a current-input type. Next, we multiply this input source by the

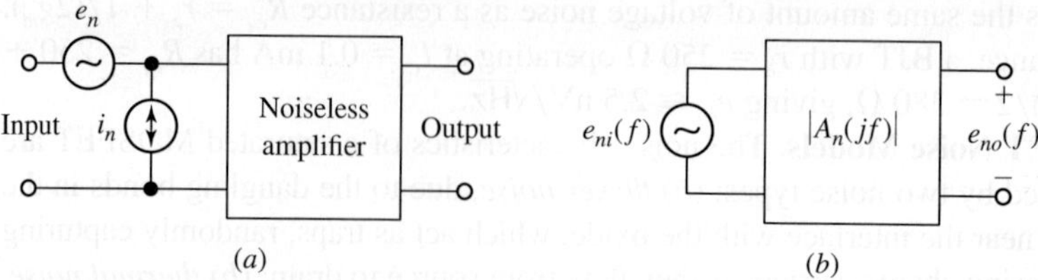

FIGURE 7.103 (a) Noise model of an amplifier. (b) Circuit for the calculation of the output noise voltage e_{no} of a *voltage amplifier* with total input noise voltage e_{ni} and noise gain A_n.

circuit's noise gain to obtain the output noise. For example, in the case of the voltage-type amplifier exemplified in Fig. 7.103b we write $e_{no}(f) = |A_n(jf)|e_{ni}(f)$, where $A_n(jf)$ is the noise gain, in V/V. Finally, we adapt Eq. (7.124) to find the *total rms output noise* above some prescribed frequency f_L as

$$E_{no} = \sqrt{\int_{f_L}^{\infty} |A_n(jf)|^2 e_{ni}^2(f)\, df} \tag{7.134}$$

Most noise gains of interest are dominated by a single pole, or

$$A_n(jf) = \frac{A_{n0}}{1 + jf/f_B} \tag{7.135}$$

where A_{n0} is the dc gain and f_B is the -3-dB frequency. We are interested in two special cases, namely, the case of *white* input noise and the case of $1/f$ input noise.

- **White-Noise Equivalent Bandwidth (NEB).** Assuming $e_{ni}(f) = e_{niw}$, we combine Eqs. (7.134) and (7.135) and obtain (see Problem 7.74)

$$E_{no} = A_{n0}e_{niw}\sqrt{\int_{f_L}^{\infty} \frac{df}{1 + (f/f_B)^2}} = A_{n0}e_{niw}\sqrt{f_B\left(\tan^{-1}\infty - \tan^{-1}\frac{f_L}{f_B}\right)} = A_{n0}e_{niw}\sqrt{\text{NEB}}$$

where

$$\text{NEB} = f_B\left(\frac{\pi}{2} - \tan^{-1}\frac{f_L}{f_B}\right) \tag{7.136}$$

is the *white-noise equivalent bandwidth*. Situations of practical interest are such that $f_L \ll f_B$, indicating that we can approximate $\tan^{-1}(f_L/f_B) \cong f_L/f_B$ in Eq. (7.136). Consequently,

$$\text{NEB} \cong 1.57f_B - f_L \tag{7.137}$$

We observe that if $|A_n(jf)|$ rolled off *sharply* (brick-wall type) at f_B, then f_H would coincide with f_B itself. However, because the rolloff is gradual (1st order), the noise above f_B provides an additional contribution of 57%.

- **Brick-Wall Equivalent for Flicker Noise.** Assuming $e_{ni}^2(f) = KI/f$, we again combine Eqs. (7.134) and (7.135) and obtain (see Problem 7.74)

$$E_{no} = A_{n0}\sqrt{\int_{f_L}^{\infty} \frac{KI\,df}{f[1 + (f/f_B)^2]}} = A_{n0}\sqrt{KI \ln \frac{f_B\sqrt{1 + (f_L/f_B)^2}}{f_L}} = A_{n0}\sqrt{KI \ln \frac{f_H}{f_L}}$$

where

$$f_H = f_B\sqrt{1 + (f_L/f_B)^2} \tag{7.138}$$

In words, passing above-f_L-flicker-noise through a 1st-order low-pass filter with -3-dB of f_B is equivalent to passing it through a brick-wall filter with a cut-off frequency of f_H as given above. Situations of practical interest are such that $f_L \ll f_B$, so we approximate the above expression as

$$f_H \cong f_B \tag{7.139}$$

(a) A pn junction with $r_s \cong 0$, $K = 5 \times 10^{-17}$ A, and $a = 1$ is forward-biased at $I_D = 100$ μA by means of a noiseless current source. Find the rms noise voltage E_n across its terminals from 0.1 Hz to 1 MHz.

(b) Find a capacitance C that, when placed across its terminals, will limit E_n to 1.0 μV rms.

EXAMPLE 7.41

Solution

(a) By Eq. (7.131a) we have

$$i_{nd}^2 = 2 \times 1.602 \times 10^{-19} \times 100 \times 10^{-6} + \frac{5 \times 10^{-17} \times 100 \times 10^{-6}}{f}$$

$$= (5.66 \text{ pA})^2\left(\frac{156}{f} + 1\right)$$

The noise voltage across the junction is $e_{nd} = r_d i_{nd}$, where $= r_d = 26/0.1 = 260$ Ω, so

$$e_{nd}^2 = (260 \times 5.66 \times 10^{-12})^2\left(\frac{156}{f} + 1\right) = (1.472 \text{ nV})^2\left(\frac{156}{f} + 1\right)$$

Using Eq. (7.126) we get

$$E_n \cong (1.472 \text{ nV})\sqrt{156 \ln\left(\frac{10^6}{0.1}\right) + 10^6} \cong (1.472 \text{ nV})\sqrt{10^6} = 1.47 \text{ } \mu\text{V rms}$$

indicating that flicker noise is negligible in this case.

(b) Placing a capacitor C in parallel with the junction establishes a pole frequency at $f_B = 1/(2\pi r_d C)$. Since flicker noise is negligible, we impose

$$(1.472 \text{ nV})\sqrt{1.57 f_B} = 1.0 \text{ } \mu\text{V}$$

to obtain $f_B = 294$ kHz. Finally, $C = 1/(2\pi r_d f_B) = 1/(2\pi \times 260 \times 294 \times 10^3) \cong 2.1$ nF.

An Op Amp Circuit Example

The op amp circuit of Fig. 7.104a is a classic benchmark for noise calculations. No input signals are shown, so whatever emerges from the output is just noise. (The circuit is fairly general in that, depending on where we apply the input or inputs, it could be an inverting amplifier, a noninverting amplifier, a summing or a difference amplifier, a buffer, an I-V converter, and so forth.) We wish to find the total output noise E_{no} assuming an op amp with a constant GBP product of f_t and the spectral densities of Eq. 7.126. To this end we redraw the circuit as in Fig. 7.104b, showing all noise sources explicitly. Since the op amp is a differential-input device, both noise currents i_{nn} and i_{np} must be shown (however, the individual noise voltages are lumped together into a single source e_n in series with just one of the inputs). Additionally, we must show the noise of each resistor (we use either the series or the parallel model, depending on which one makes the calculations easier). Next, we combine the effects of all sources into a single input source e_{ni} as shown in Fig. 7.104c. Finally, we apply Eq. (7.134) to find E_{no}.

The feedback factor is $\beta = R_1/(R_1 + R_2)$, so $A_{n0} = 1/\beta$ and $f_B = \beta f_t$, or

$$A_{n0} = 1 + \frac{R_2}{R_1} \qquad f_B = \frac{f_t}{1 + R_2/R_1} \tag{7.140}$$

To find the contributions to e_{ni} by e_3, e_n, and i_{np}, we observe that R_3 and i_{np} produce the noise voltage $R_3 i_{np}$, so their power densities combine as

$$e_3^2 + e_n^2 + R_3^2 i_{np}^2 = 4kTR_3 + e_n^2 + R_3^2 i_{np}^2$$

where Eq. (7.128) has been used. To find the contributions to e_{ni} by i_1, i_2, and i_{nn}, set e_3, e_n, and i_{np} to zero. This forces the inverting node to virtual ground, causing

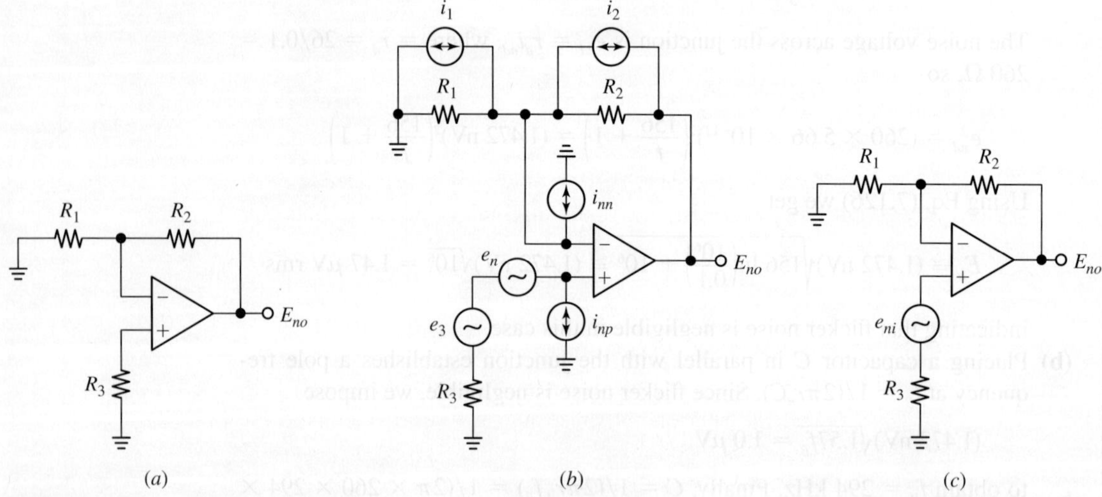

FIGURE 7.104 (a) Generalized resistive-type op amp circuit. (b) Redrawing the circuit with all noise sources explicitly shown. (c) Noise model after all sources have been combined into a single noise voltage e_{ni}.

i_1, i_2, and i_{nn} to flow through R_2 and thus yield the power density $R_2^2(i_1^2 + i_2^2 + i_{nn}^2)$ at the output. To find its contribution to e_{ni}, we must reflect the output power density to the noninverting input by dividing it by A_{n0}^2. So, the contributions by i_1, i_2, and i_{nn} are

$$\frac{R_2^2(i_1^2 + i_2^2 + i_{nn}^2)}{A_{n0}^2} = \frac{R_2^2(i_1^2 + i_2^2 + i_{nn}^2)}{(1 + R_2/R_1)^2} = (R_1 /\!/ R_2)^2(i_1^2 + i_2^2 + i_{nn}^2)$$

$$= 4kT(R_1 /\!/ R_2) + (R_1 /\!/ R_2)^2 i_{nn}^2$$

where Eq. (7.128) has been used again. Combining all contributions gives the total input power density

$$e_{ni}^2 = 4kT[R_3 + (R_1 /\!/ R_2)] + e_n^2 + R_3^2 i_{np}^2 + (R_1 /\!/ R_2)^2 i_{nn}^2 \tag{7.141}$$

Finally, we substitute into Eq. (7.134) and write

$$E_{no} = \sqrt{E_r^2 + E_n^2 + E_{np}^2 + E_{nn}^2} \tag{7.142}$$

where E_r, E_n, E_{np} and E_{nn} represent, respectively, the noise contributions by the resistances, e_n, i_{np}, and i_{nn}. Applying Eq. (7.126) with $\text{NEB}_{white} = 1.57 f_B$ and $\text{NEB}_{flicker} \cong f_B$, and ignoring f_L compared to $1.57 f_B$, gives

$$E_r \cong A_{n0}\sqrt{4kT[R_3 + (R_1 /\!/ R_2)] \times 1.57 f_B} \tag{7.143a}$$

$$E_n \cong A_{n0}e_{nw}\sqrt{f_{ce}\ln\frac{f_B}{f_L} + 1.57 f_B} \tag{7.143b}$$

$$E_{np} \cong A_{n0}R_3 i_{npw}\sqrt{f_{cip}\ln\frac{f_B}{f_L} + 1.57 f_B} \tag{7.143c}$$

$$E_{nn} \cong A_{n0}(R_1 /\!/ R_2)i_{nnw}\sqrt{f_{cin}\ln\frac{f_B}{f_L} + 1.57 f_B} \tag{7.143d}$$

The above derivations for voltage-type amplifiers are easily generalized to other amplifier types such as current, transresistance, and transconductance amplifiers (see the end-of-chapter problems).

(a) If the circuit of Fig. 7.104 uses a 741 op amp with $R_1 = 20\ \text{k}\Omega$, $R_2 = 180\ \text{k}\Omega$, and $R_3 = 18\ \text{k}\Omega$, find the rms output noise above 0.1 Hz. Compare the different noise components and comment.

(b) What happens if all resistances are simultaneously scaled to one-tenth of their original value? How would you scale the resistances in order to minimize output noise? What would the minimum be?

EXAMPLE 7.42

Solution

(a) We have $A_{n0} = 1 + 180/20 = 10$ V/V and $f_B = 10^6/10 = 10^5$ Hz. Using Eq. (7.143) with the data of Eq. (7.127), and then substituting into Eq. (7.142), we find

$$E_r = 10\sqrt{4 \times 1.38 \times 10^{-23} \times 300 \times (18 + 18)10^3 \times 1.57 \times 10^5}$$

$$= 96.7 \; \mu V \text{ rms}$$

$$E_n = 10 \times 20 \times 10^{-9}\sqrt{200 \times \ln\frac{10^5}{0.1} + 1.57 \times 10^5} = 79.9 \; \mu V \text{ rms}$$

$$E_{np} = E_{nn} = 10 \times 18 \times 10^3 \times 0.5$$

$$\times 10^{-12}\sqrt{2 \times 10^3 \times \ln\frac{10^5}{0.1} + 1.57 \times 10^5} = 38.7 \; \mu V \text{ rms}$$

$$E_{no} = \sqrt{96.7^2 + 79.9^2 + 2 \times 38.7^2} \cong 137 \; \mu V \text{ rms}$$

It is apparent that resistor noise dominates in this circuit, followed by op amp voltage noise, in turn followed by op amp current noise.

(b) Reducing all resistances to $1/10$ of their original values leaves A_{n0}, f_B, and E_n unchanged while reducing both E_{np} and E_{nn} to $1/10$, or to $3.87 \; \mu V$ rms, which is negligible compared to E_n. However, because of the square-root dependence, E_r is reduced only to $1/\sqrt{10}$, or to $96.7/\sqrt{10} = 30.6 \; \mu V$ rms, giving $E_{no} \cong \sqrt{30.6^2 + 79.9^2} \cong 85 \; \mu V$ rms. We can minimize noise by scaling all resistances further, until the condition $E_r \ll E_n$ is met, at which point $E_{no} \cong E_n \cong 80 \; \mu V$ rms. Of course, the price for smaller resistances is an increase in power dissipation.

High-gain amplifiers such as op amps and voltage comparators utilize a differential pair as the input stage. Since its noise adds directly to the useful input signals, its noise performance is usually critical (by contrast, a subsequent stage tends to be less critical because its noise, reflected to the input, gets divided by the gains of the preceding stages, as discussed in Section 7.2). It is therefore appropriate that we investigate the noise performance of differential pairs in detail.

Noise in CMOS Differential Stages

We wish to model the noise characteristics of a CMOS differential stage with a single input source e_n as in Fig. 7.105a, given the individual noise sources of Fig. 7.105b. To find e_n, we first calculate the short-circuit noise current i_{no} at the output, then we divide it by the circuit's transconductance g_m to reflect it to the input and thus get $e_n = i_{no}/g_m$. We use the superposition principle to find the power density contribution by each FET acting alone and then we add up all contributions to obtain i_{no}^2.

By inspection, the contributions to i_{no}^2 by M_1 and M_2 are, respectively, $g_{m1}^2 e_{n1}^2$ and $g_{m2}^2 e_{n2}^2$ (note that polarity inversion by one of the inputs is inconsequential because of the randomness of noise). Next, we find the contributions by M_3 and M_4 by setting

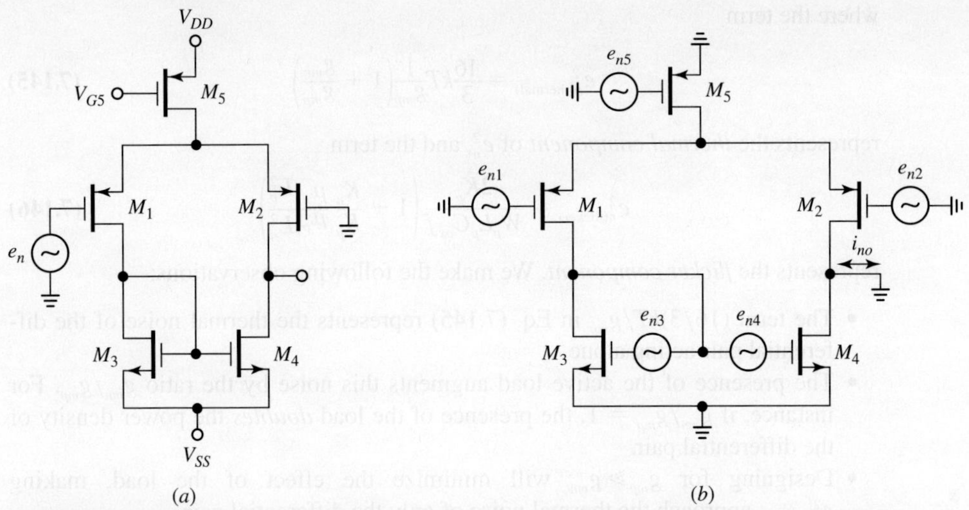

FIGURE 7.105 (a) Noise model of an active-loaded CMOS differential stage (all FETs in the model are assumed noiseless). (b) Noise circuit to find the individual FET contribution to the output noise current i_{no}.

$e_{n1} = e_{n2} = e_{n5} = 0$. These conditions place M_3's gate at ac ground, so M_3 does not contribute noise directly. Rather, it contributes indirectly via M_4 because e_{n3} appears in series with e_{n4} and e_{n3} is referenced to ac ground. Consequently, M_4 contributes to i_{no}^2 the power density $g_{m4}^2(e_{n3}^2 + e_{n4}^2)$. Finally, we observe that e_{n5} appears as a common-mode signal, so in a well-designed differential pair its contribution tends to be negligible compared to the others because the common-mode gain is usually very small. Summarizing, we have

$$i_{no}^2 = g_{m1}^2 e_{n1}^2 + g_{m2}^2 e_{n2}^2 + g_{m4}^2(e_{n3}^2 + e_{n4}^2) = g_{mp}^2(e_{np}^2 + e_{np}^2) + g_{mn}^2(e_{nn}^2 + e_{nn}^2)$$

$$= 2(g_{mp}^2 e_{np}^2 + g_{mn}^2 e_{nn}^2)$$

where the subscripts have been changed to p and n to identify p-channel and n-channel parameters. Dividing i_{no}^2 by g_{mp}^2 reflects it to the input, giving

$$e_n^2 = 2\left(e_{np}^2 + \frac{g_{mn}^2}{g_{mp}^2}e_{nn}^2\right)$$

Combining with Eq. (7.133b) we obtain the more explicit expression

$$e_n^2 = 2\left[\frac{K_p}{W_p L_p C_{ox} f} + 4kT\frac{2}{3}\frac{1}{g_{mp}} + \frac{g_{mn}^2}{g_{mp}^2}\left(\frac{K_n}{W_n L_n C_{ox} f} + 4kT\frac{2}{3}\frac{1}{g_{mn}}\right)\right]$$

where K_p and K_n are the flicker-noise coefficients of the p-type and n-type FETs. Regrouping terms and using $g_{mn}^2/g_{mp}^2 = (2k_n I_{Dn})/(2k_p I_{Dp}) = k_n/k_p = (\mu_n W_n/L_n)/(\mu_p W_p/L_p)$, we put the above expression in the more insightful form

$$e_n^2 = e_{n(\text{thermal})}^2 + e_{n(\text{flicker})}^2 \tag{7.144}$$

where the term

$$e_{n(\text{thermal})}^2 = \frac{16}{3}kT\frac{1}{g_{mp}}\left(1 + \frac{g_{mn}}{g_{mp}}\right) \tag{7.145}$$

represents the *thermal component* of e_n^2, and the term

$$e_{n(\text{flicker})}^2 = \frac{2K_p}{W_pL_pC_{ox}f}\left(1 + \frac{K_n}{K_p}\frac{\mu_n}{\mu_p}\frac{L_p^2}{L_n^2}\right) \tag{7.146}$$

represents the *flicker component*. We make the following observations:

- The term $(16/3)kT/g_{mp}$ in Eq. (7.145) represents the thermal noise of the differential pair acting alone.
- The presence of the active load augments this noise by the ratio g_{mn}/g_{mp}. For instance, if $g_{mn}/g_{mp} = 1$, the presence of the load *doubles* the power density of the differential pair.
- Designing for $g_{mp} \gg g_{mn}$ will minimize the effect of the load, making $e_{n(\text{thermal})}^2$ approach the thermal noise of only the differential pair.
- If desirable, we can reduce the differential-pair noise by designing for a suitably large g_{mp}.
- The term $2K_p/(W_pL_pC_{ox}f)$ in Eq. (7.146) represents the flicker noise of the differential pair alone.
- The presence of the active load augments this noise by the ratio $(K_n\mu_nL_p^2)/(K_p\mu_pL_n^2)$. (Interestingly, this augmentation depends only on the channel lengths, regardless of the channel widths.)
- Designing for $K_p\mu_pL_n^2 \gg K_n\mu_nL_p^2$ will minimize the effect of the load, in turn making $e_{n(\text{flicker})}^2$ approach the flicker noise of only the differential pair.
- If desirable, we can reduce the differential-pair noise by designing for a suitably large area W_pL_p.

Noise in Bipolar Differential Pairs

We wish to find an expression for the sources modeling the noise characteristics of the bipolar differential pair of Fig. 7.106a. To find the short-circuit noise e_n, refer to the circuit of Fig. 7.106b, whose inputs have been grounded to blank out the noise currents and leave active only the noise voltages. Assuming $r_o \gg R_C$, the voltage noise gain is g_mR_C, so we write, by inspection,

$$e_{no}^2 = g_{m1}^2R_C^2e_{n1}^2 + g_{m2}^2R_C^2e_{n2}^2 + e_{r1}^2 + e_{r2}^2 = 2[g_m^2R_C^2e_{n1}^2 + 4kTR_C]$$

having ignored the source e_{n3} because of the high CMRR of this circuit. Reflecting e_{no}^2 to the input gives

$$e_n^2 = \frac{e_{no}^2}{g_m^2R_C^2} = 2\left[e_{n1}^2 + \frac{4kT}{g_m^2R_C}\right]$$

Substituting the expression of Eq. (7.132a) for e_{n1}^2 we get, after minor algebra,

$$e_n^2 = 2\left[4kT\left(r_b + \frac{1}{2g_m} + \frac{1}{(g_mR_C)g_m}\right)\right]$$

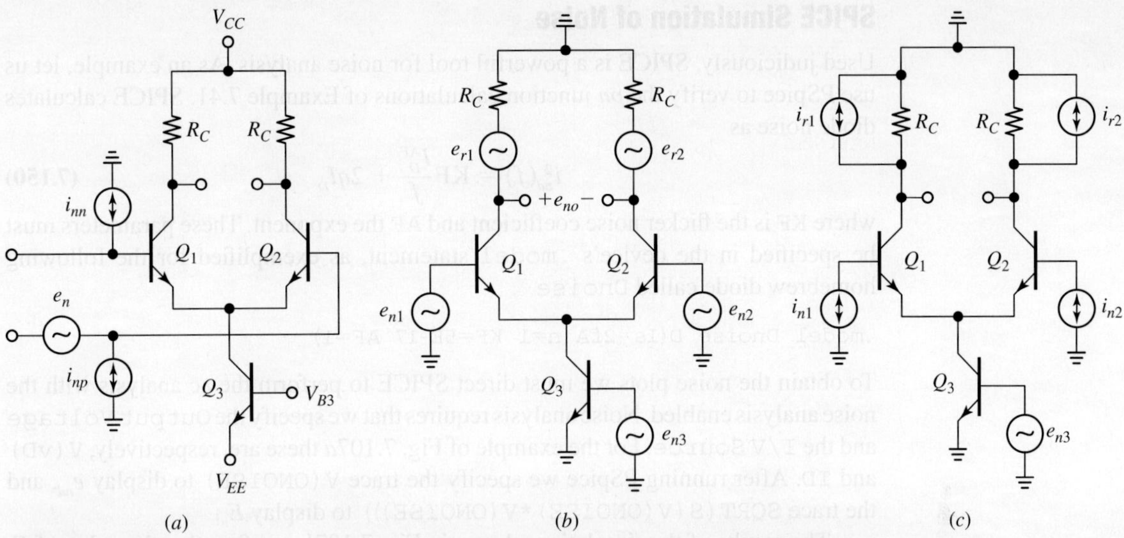

FIGURE 7.106 (a) Noise model of a resistive-loaded bipolar differential pair (the transistors and resistors in the model are assumed noiseless). Noise circuits to find (a) e_n and (b) i_{nn} and i_{np}.

But a well-designed differential amplifier has $g_m R_C \gg 2$, so the above expression reduces to

$$e_n^2 \cong 2 \times 4kT\left(r_b + \frac{1}{2g_m}\right) = 2e_{nb}^2 \tag{7.147}$$

It is apparent that the noise of the resistive load tends to play an insignificant role, though this is not necessarily true in the case of an active load (see Problem 7.85).

To find the open-circuit noise currents i_{nn} and i_{np}, refer to the circuit of Fig. 7.106c, whose inputs have been left floating to blank out the effect of the noise voltages and enable only that of the noise currents. We find i_{nn} by reflecting i_{r1} to Q_1's base and then combining it with i_{n1}. Since the current noise gain is $|\beta_0(jf)|$, we get

$$i_{nn}^2 = i_{n1}^2 + \frac{i_{r1}^2}{|\beta_0(jf)|^2} = i_{n1}^2 + \frac{4kT/R_C}{|\beta_0(jf)|^2}$$

Substituting the expression of Eq. (7.132b) for i_{n1}^2 we get

$$i_{nn}^2 = 2q\left(I_B + \frac{KI_B^a}{f} + \frac{I_C}{|\beta_0(jf)|^2}\right) + \frac{4kT/R_C}{|\beta_0(jf)|^2} = 2q\left(I_B + \frac{KI_B^a}{f} + \frac{I_C + 2V_T/R_C}{|\beta_0(jf)|^2}\right) \cong i_{nb}^2 \tag{7.148}$$

having exploited the fact that in a well-designed differential circuit the condition $R_C I_C \gg 2V_T$ holds. It is apparent that the noise of a resistive load plays an insignificant role also in the case of current noise. By symmetry, similar considerations hold for i_{np}, so we summarize by writing

$$i_{np}^2 = i_{nn}^2 \cong 2q\left(I_B + \frac{KI_B^a}{f} + \frac{I_C}{|\beta_0(jf)|^2}\right) \tag{7.149}$$

This completes the noise analysis of the bipolar pair.

SPICE Simulation of Noise

Used judiciously, SPICE is a powerful tool for noise analysis. As an example, let us use PSpice to verify the *pn* junction calculations of Example 7.41. SPICE calculates diode noise as

$$i_{nd}^2(f) = \text{KF}\frac{I_D^{\text{AF}}}{f} + 2qI_D \tag{7.150}$$

where KF is the flicker noise coefficient and AF the exponent. These parameters must be specified in the device's `.model` statement, as exemplified for the following homebrew diode called `Dnoise`

```
.model Dnoise D(Is=2fA n=1 KF=5E-17 AF=1)
```

To obtain the noise plots we must direct SPICE to perform the ac analysis with the noise analysis enabled. Noise analysis requires that we specify the Output Voltage and the I/V Source. For the example of Fig. 7.107*a* these are, respectively, V(vD) and ID. After running PSpice we specify the trace V(ONOISE) to display e_{nd}, and the trace SQRT(S(V(ONOISE)*V(ONOISE))) to display E_n.

The results of the simulation, shown in Fig. 7.107*b*, confirm that the value of E_n strongly depends on our choice of f_H. For $f_H = 1$ MHz PSpice gives $E_n = 1.466$ μV rms, in excellent agreement with hand calculations. Placing a 2.1-nF capacitance in parallel with the diode filters out high-frequency noise, causing E_n to settle asymptotically to 1.0 μV rms, as desired.

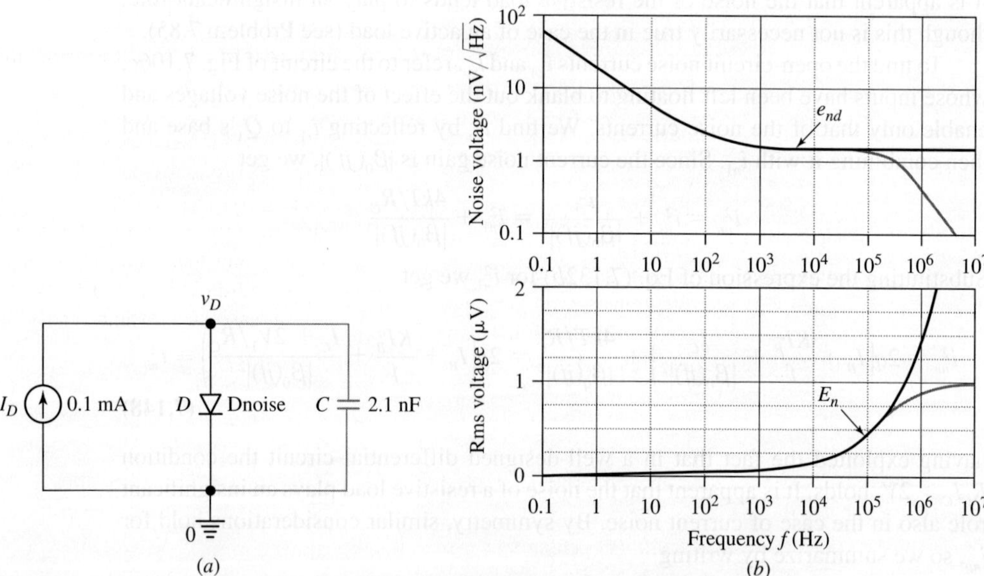

FIGURE 7.107 (a) PSpice circuit to plot the noise characteristics of the diode of Example 7.41. (b) Frequency plots of the noise voltage e_{nd} (*top*) and the rms voltage E_n (*bottom*).

REFERENCES

1. P. R. Gray, P. J. Hurst, S. H. Lewis, and R. G. Meyer, *Analysis and Design of Analog Integrated Circuits*, 5/E, John Wiley and Sons, 2009.
2. A. S. Sedra and K. C. Smith, *Microelectronic Circuits*, 6/E, Oxford University Press, 2010.
3. R. C. Jaeger and T. N. Blalock, *Microelectronic Circuit Design*, 2/E, McGraw-Hill, 2004.
4. P. J. Hurst, "A Comparison of Two Approaches to Feedback Circuit Analysis," IEEE Trans. on Education, Vol. 35, No. 3, August 1992, pp. 253–261.
5. R. B. Blackman, "Effect of Feedback on Impedance," *Bell Sys. Tech. J.*, Vol. 23, pp. 269–277, October 1943.
6. R. D. Middlebrook, "Measurement of Loop Gain in Feedback Systems," *Int. J. Electronics*, Vol. 38, no. 4, pp. 485–512, 1975.
7. S. Franco, *Design with Operational Amplifiers and Analog Integrated Circuits*, 4/E, McGraw-Hill, 2014.
8. Y. P. Tsividis and P. R. Gray, "An Integrated NMOS Operational Amplifier with Internal Frequency Compensation," *IEEE J. Solid-State Circuits*, Vol. SC-11, pp. 748–753, December 1976.
9. B. K. Ahuja, "An Improved Frequency Compensation Technique for CMOS Operational Amplifiers," *IEEE J. Solid-State Circuits*, Vol. SC-18, pp. 629–633, December 1983.
10. D. B. Ribner and M.A. Copeland, "Design Techniques for Cascoded CMOS Op Amps with Improved PSRR and Common-Mode Input Range," *IEEE J. Solid-State Circuits*, pp. 919–925, December 1984.
11. D. Senderowicz, D. A. Hodges and P. R. Gray, "A High-Performance NMOS Operational Amplifier," *IEEE J. Solid-State Circuits*, Vol. SC-13, pp. 760–768, December 1978.
12. H. Camenzind, *Designing Analog Chips*, www.designinganalogchips.com, 2005.

PROBLEMS

7.1 Negative-Feedback Basics

7.1 (a) If it is found that with $v_i = 1.0$ V the circuit of Fig. 7.1 gives $v_o = 10.0$ V while requiring $v_\varepsilon = 25$ mV, what are the values of a, b, and L?

    (b) If a drops to 50% of the value found in part (a), what are the new values of v_o and v_ε if v_i is still 1.0 V?

    (c) How would you adjust the value of b to make up for the drop in a and still ensure $v_o = 10.0$ V with $v_i = 1.0$ V? Would v_ε change?

7.2 (a) Suppose a certain class of error amplifiers offers open-loop gains that, due to fabrication and environmental variations as well as aging, can drop to as little as 10% of their nominal values. Specify a and b such that $A = 100.0$ V/V for a nominal open-loop gain of a, and A drops only by 1% when a drops to 10% of its nominal value.

    (b) What is A if a has twice its nominal value? What is the % change of A?

7.3 By varying the position of the potentiometer's wiper W in the circuit of Fig. P7.3 we can select any value of b from 0 (W all the way down) to 1 (W all the way up). For instance, with W set 1/4 of the way up so that the portion of the potentiometer's resistance below W is $R_{\text{below}} = 2.5$ kΩ and that above W is $R_{\text{above}} = 7.5$ kΩ, we get $b = 2.5/(2.5 + 7.5) = 0.25$.

    (a) Let $v_i = 1.0$ V. If it is found that $v_o = 1.0$ V with W all the way down, what is v_o with W all the way up? With W halfway?

    (b) If v_i is held constant and it is found that the circuit gives $v_o = -6$ V with W all the way up, and $v_o = -11$ V with W halfway, find a and v_i.

    (c) Derive an expression for b in terms of $1/A$ and $1/a$, and then find the wiper's position that will result in $v_o/v_i = 10$ V/V in the following two cases: $a = \infty$ and $a = 100$ V/V.

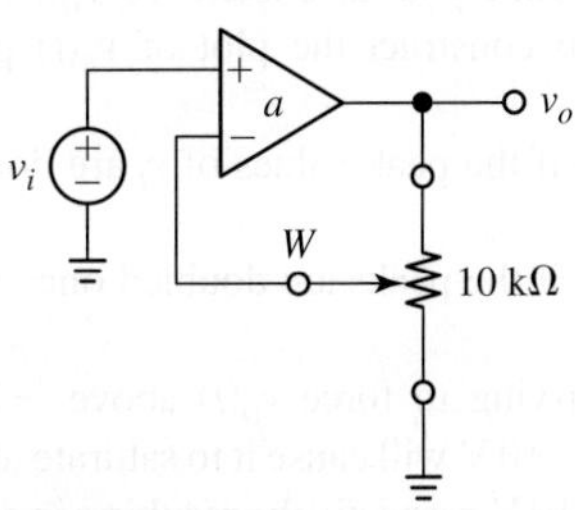

FIGURE P7.3

7.4 **(a)** Let the circuit of Fig. P7.4 have $a = 150$ V/V, $R_1 = 10$ kΩ, $R_2 = 20$ kΩ, $R_3 = 30$ kΩ, and $R_4 = 35$ kΩ. Find v_o and v_ε if $v_i = 0.2$ V.

(b) Change the value of R_4 so as to make the actual closed-loop gain equal in value to A_{ideal}. If $v_i = 0.2$ V, what is the new value of v_ε?

(c) Estimate v_o and v_ε if a drops to 60% of its nominal value.

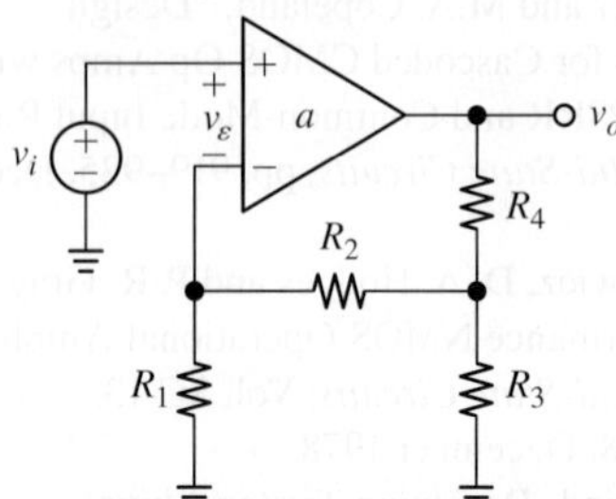

FIGURE P7.4

7.5 **(a)** An engineer is to design an amplifier with a gain of 10^3 V/V and a $\pm 0.25\%$ accuracy, or $A = 10^3$ V/V $\pm$ 0.25%, using amplifier stages with $a = 10^4$ V/V $\pm$ 25% each. Verify that a single stage won't do it. Hence, consider cascading two or more identical stages, each using local feedback to desensitize its own gain and thus give $A = A_1 \times A_2 \times \cdots$. What is the minimum number of stages required, and what is the net tolerance of the overall gain?

(b) Repeat, but for the tighter specification $A = 10^3$ V/V $\pm$ 0.1%.

7.2 Effect of Feedback on Distortion, Noise, and Bandwidth

7.6 **(a)** Given that the op amp of Fig. P7.6a has the open-loop VTC of Fig. P7.6b, sketch and label $v_I(t)$, $v_O(t)$, and $v_E(t)$ if $R_1 = R_2 = 10$ kΩ and $v_I(t)$ is a 500-Hz triangular wave with peak values of ± 2 V.

Hint: Once you have sketched $v_O(t)$, use the VTC to construct the plot of $v_E(t)$ point by point.

(b) Repeat if the peak values of v_I are doubled to ± 4 V.

(c) Repeat if the peaks are doubled once again to ± 8 V.

Hint: trying to force $v_O(t)$ above $+10$ V or below -10 V will cause it to saturate at $+10$ V or at -10 V, respectively, resulting in a *clipped* output waveform.

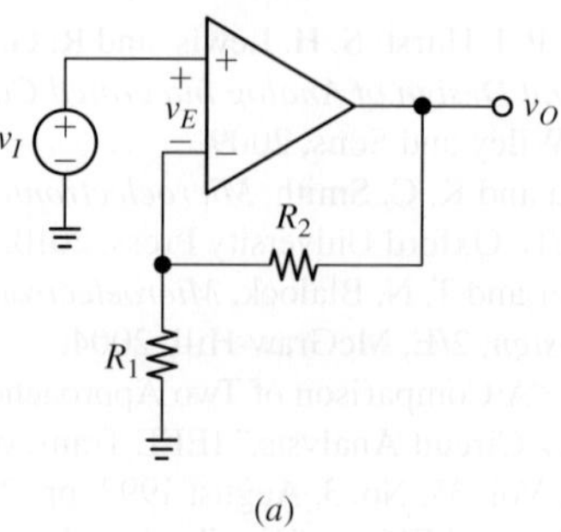

(a)

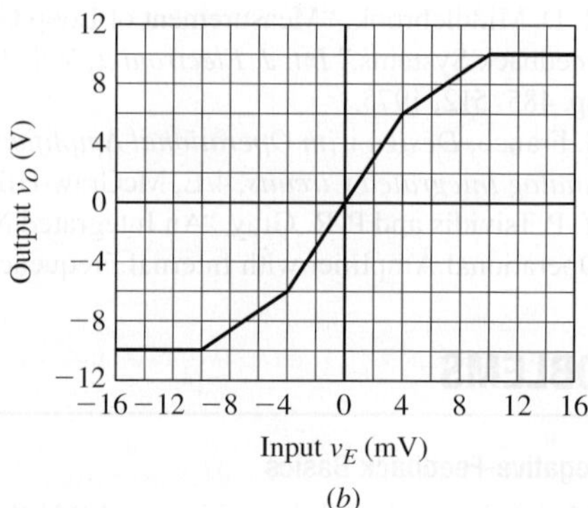

(b)

FIGURE P7.6

7.7 **(a)** Suppose a novel voltage buffer has been invented, which exhibits $R_i \cong \infty$ and $R_o \cong 0$, and accepts an input voltage v_I to give the output voltage

$$v_O = v_I - \left(\frac{v_I}{10 \text{ V}}\right)^3$$

Sketch and label $v_I(t)$ and $v_O(t)$ if $v_I(t)$ is a voltage source producing a triangular wave with peak values of ± 10 V.

(b) To eliminate the effects of the nonlinearity stemming from the cubic term, let us interpose an op amp between the source v_I and this buffer, in a manner similar to Fig. 7.9. Assuming the op amp has $a = \infty$, sketch and label $v_I(t)$ and $v_O(t)$ as well as the op amp's output $v_{OA}(t)$.

(c) Find the gain a sufficient to keep the output error down to 1 mV when $v_I(t)$ peaks out. Hence, sketch and label $v_E(t)$.

7.8 Suppose a certain audio power amplifier gives

$$v_O = 0.9v_I + (250\text{ mV})\cos(2\pi120t)$$

where the second term at the right-hand side represents 120-Hz hum that the amplifier picks up from its own crudely-designed power supply. Rather than switching to a cleaner but more expensive supply, an engineer decides to reduce the output hum by exploiting the curative properties of negative feedback.

(*a*) Using a suitable op amp, design a feedback circuit that accepts an input $v_I(t)$ and gives $v_O(t) = 1.0v_I(t) + (50\ \mu\text{V})\cos(2\pi120t)$, and show your circuit. What is the gain a required of the op amp at 120 Hz? What is its required GBP?

(*b*) Repeat, but for $v_O(t) = 10v_I(t) + (50\ \mu\text{V}) \times \cos(2\pi120t)$.

7.9 An engineer is trying to design a 60-dB audio amplifier (that is, an amplifier with $A_0 = 1000$ V/V and $f_B \geq 20$ kHz) using an op amp with $f_t = 1$ MHz. Realizing that just one op amp won't do it because it would give $f_B = f_t/A_0 = 10^6/10^3 = 1$ kHz $\ll 20$ kHz, the engineer tries *cascading two* stages with lower individual gains A_{10} and A_{20} but wider individual bandwidths f_{B1} and f_{B2}.

(*a*) If the engineer decides to impose $A_{10} = A_{20} = \sqrt{1000}$ V/V, what are the bandwidths f_{B1} and f_{B2}?

(*b*) What is the overall -3-dB frequency of the cascade combination of the two stages? **Hint:** by definition, $f_{-3\text{dB}}$ is such that $|A(jf_{-3\text{dB}})| = A_0/\sqrt{2}$, where $A(jf) = A_1(jf) \times A_2(jf)$.

(*c*) What if the engineer had cascaded two stages of unequal gains but still such that $A_{01} \times A_{02} = 1000$ V/V, say, $A_{10} = 20$ V/V and $A_{20} = 50$ V/V? Would the overall -3-dB frequency still meet the specifications? Explain!

7.10 The CFA closed-loop bandwidth of Eq. (7.27) was derived under the assumption of an ideal input voltage buffer across the inputs. However, a real-life buffer will exhibit a small output resistance R_n, as shown in Fig. P7.10.

(*a*) Using the results of Problem 5.38, show that Eq. (7.27) still holds, provided we let $R_2 \rightarrow R_2 + R_n(1 + R_2/R_1)$.

(*b*) What is the actual closed-loop bandwidth of the circuit of Example 7.7?

(*c*) How would you lower R_2 to retain the same bandwidth (60 MHz), and R_1 to retain the same gain (10 V/V)?

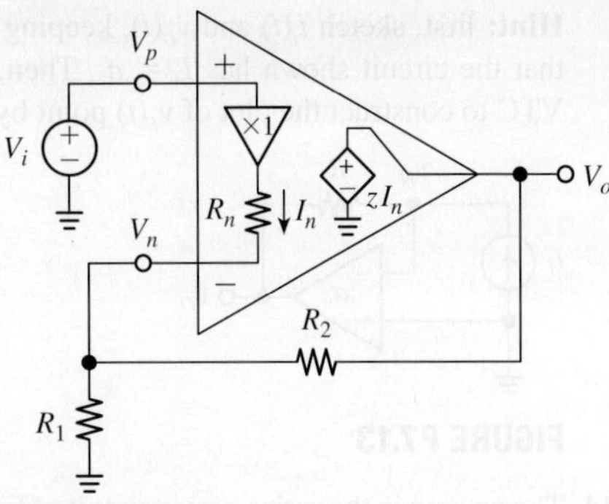

FIGURE P7.10

7.3 Feedback Topologies and Closed-Loop I/O Resistances

7.11 (*a*) Given that with $v_i = 10.0$ mV the series-shunt configuration of Fig. 7.12a yields $v_f = 9.5$ mV and $v_o = 5.0$ V, find the values (and units) of a, b, L, and A (use two methods to calculate A).

(*b*) If gain a drops to 50% of the value found above, how are v_f and v_o affected? Cross check!

(*c*) Repeat part (*a*), but for the series-series configuration of Fig. 7.17a, assuming the same situation at the summing side, but $i_o = 5.0$ mA at the sensing side.

(*d*) If a doubles, how are v_f and i_o affected? Cross check!

7.12 (*a*) Given that with $i_i = 10.0\ \mu$A the shunt-series configuration of Fig. 7.14a yields $i_f = 9.5\ \mu$A and $i_o = 5.0$ mA, find the values (and units) of a, b, L, and A (use two methods to calculate A).

(*b*) If a doubles, how are i_f and i_o affected? Cross check!

(*c*) Repeat part (*a*), but for the shunt-shunt configuration of Fig. 7.16a, assuming the same situation at the summing side, but $v_o = 5.0$ V at the sensing side.

(*d*) Suppose that because of an internal amplifier nonlinearity the actual gain is $1.5a$ for $i_i > 0$ and $0.75a$ for $i_i < 0$, where a is the nominal gain found above. If i_i is a sinewave with peak values of $\pm10.0\ \mu$A, what are the peak values of i_f and v_o?

7.13 Suppose the op amp utilized in the shunt-shunt circuit of Fig. P7.13 has the open-loop VTC of Fig. P7.6b. Sketch and label $i_I(t)$, $v_O(t)$, and $v_N(t)$ if $R = 10$ kΩ and $i_I(t)$ is a triangular wave with peak values of ±0.8 mA.

Hint: first, sketch $i_I(t)$ and $v_O(t)$, keeping in mind that the circuit shown has $L = a_v$. Then, use the VTC to construct the plot of $v_E(t)$ point by point.

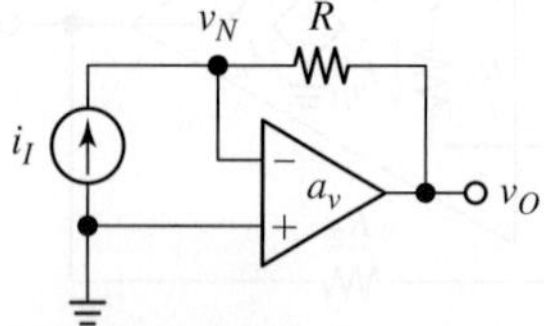

FIGURE P7.13

7.14 The op amp in the series-series circuit of Fig. P7.14 drives a nonlinear load that has been modeled with two diodes and a resistor R_1. Let $R_1 = 1\ \text{k}\Omega$ and $V_{D1(on)} = V_{D2(on)} = 0.75\ \text{V}$.

(a) Assuming $a_v = \infty$, sketch and label $v_I(t)$, $i_O(t)$, and $v_{OA}(t)$ if $R_2 = 2\ \text{k}\Omega$ and $v_I(t)$ is a triangular wave with peak values of $\pm 3\ \text{V}$. Comment on the waveform of $v_{OA}(t)$.

Hint: consider first the case in which D_1 and D_2 are both off, then the case of D_1 on, and finally the case of D_2 on.

(b) How are the waveforms of $i_O(t)$ and $v_{OA}(t)$ affected if the op amp is no longer ideal, but has $a_v = 10^3\ \text{V/V}$? How is the waveform of $v_E(t)$?

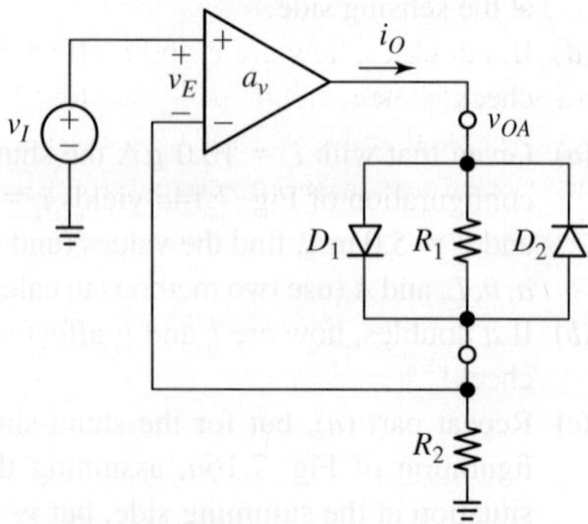

FIGURE P7.14

7.15 The circuit of Fig. P7.15 is called a *current mirror* because with $R_1 = R_2$ it gives, in the limit $a_v \to \infty$, $i_O = i_I$.

(a) Using KVL, KCL, and the op amp relationship $v_{OA} = a_v v_E$, show that the circuit gives $i_O = Ai_I - v_O/R_o$, where A and R_o are suitable functions of a_v, R_1, and R_2. Clearly, A is the closed-loop gain, and R_o is the circuit's output resistance as seen by the load.

(b) Calculate A and R_o if $R_1 = 2R_2 = 20\ \text{k}\Omega$ and $a_v = \infty$. Repeat, but for $a_v = 10^3\ \text{V/V}$, and comment.

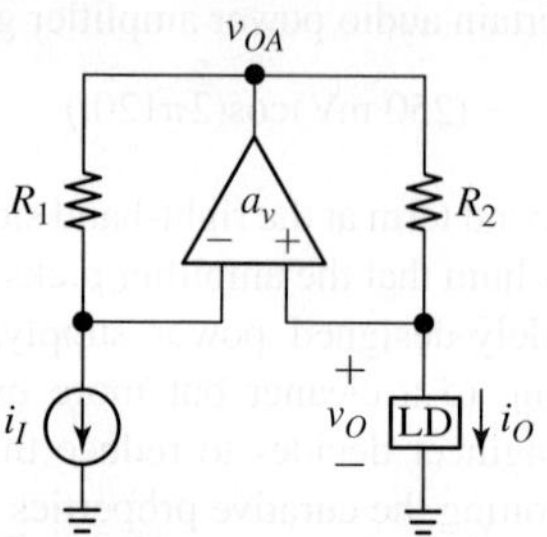

FIGURE P7.15

7.4 Practical Configurations and the Effect of Loading

7.16 Assuming the op amp in the feedback circuit of Fig. P7.4 has the open-loop parameters $r_i = 100\ \text{k}\Omega$, $r_o = 100\ \Omega$, and $a_v = 10^4\ \text{V/V}$, use the series-shunt procedure to estimate R_i, R_o, and the unloaded gain $A_{oc} = v_o/v_i$, given that $R_1 = 10\ \text{k}\Omega$, $R_2 = R_3 = 30\ \text{k}\Omega$, and $R_4 = 120\ \text{k}\Omega$.

7.17 Assuming the output node of the feedback amplifier of Fig. P7.17 is near 0 V dc, estimate R_i, R_o, and $A_{oc} = v_o/v_i$. Assume $V_{BEn(on)} = V_{EBp(on)} = 0.7\ \text{V}$, $\beta_{0n} = 200$, $\beta_{0p} = 100$, and $V_{An} = V_{Ap} = \infty$.

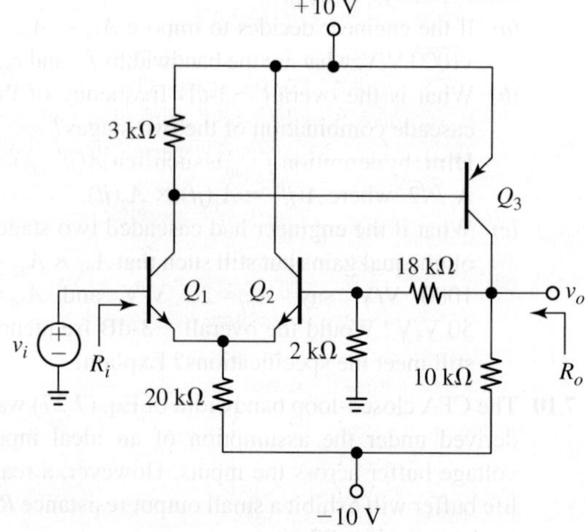

FIGURE P7.17

7.18 Assuming the output node of the feedback amplifier of Fig. P7.18 is near 0 V dc, find R_i, R_o, and the unloaded gain $A_{oc} = v_o/v_i$ if $g_{m1} = g_{m2} = g_{m5}/2 = 0.5\ \text{mA/V}$, and $r_{o1} = r_{o3} = 4r_{o5} = 120\ \text{k}\Omega$.

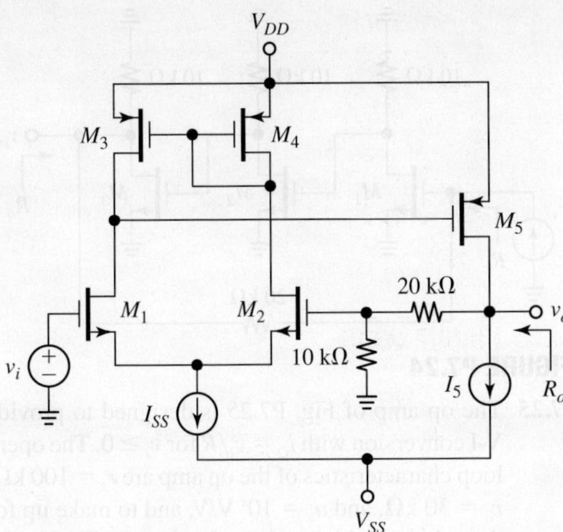

FIGURE P7.18

7.19 The emitter follower uses local series-shunt feedback to achieve high input resistance and low output resistance. These characteristics can be improved further if we apply additional negative feedback, in the manner depicted in Fig. P7.19. In this circuit, aptly known as *super emitter follower*, Q_1 is the emitter follower proper, and Q_2 provides additional feedback around Q_1.

 (*a*) Assuming $\beta_{01} = \beta_{02} = \beta_0$, $r_{o1} = r_{o1} = r_o$, and $r_{\mu 1} = r_{\mu 2} = \infty$, use the series-shunt procedure to estimate R_i and R_o, and verify that the presence of Q_2 raises R_i and lowers R_o by a factor of $\beta_0 \times r_o/(r_o + r_\pi)$.

 (*b*) Assuming $\beta_0 = 100$ and $V_A = 50$ V, calculate R_i and R_o, compare with the values they would have if Q_1 were operating alone at 1 mA, and comment on your results.

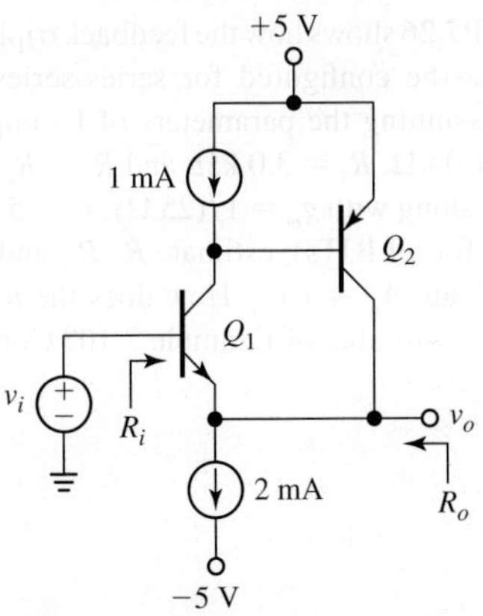

FIGURE P7.19

7.20 Shown in Fig. P7.20 is the *source follower*, the MOS counterpart of the emitter follower of Fig. 7.28a. Using the series-shunt procedure, develop expressions for R_i, R_o, and the unloaded gain A, and compare them with already-known results. Hence, calculate their values if $R_S = 20$ kΩ, $g_m = 1.25$ mA/V, $\chi = 0.1$, and $r_o = 50$ kΩ.

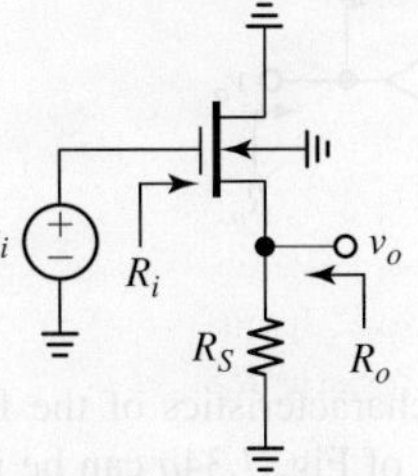

FIGURE P7.20

7.21 The source follower uses local series-shunt feedback to achieve a low output resistance R_o. This resistance can be lowered further if we apply additional negative feedback, in the manner depicted in Fig. P7.21. In this circuit, aptly known as *super source follower*, M_1 is the source follower proper, and M_2 provides additional feedback around M_1. Using the series-shunt procedure, estimate R_o, and verify that the presence of M_2 lowers the output resistance $1/g_{m1}$ approximately by the factor $g_{m2}r_{o1}$. Hence, calculate R_o if both FETs have $g_m = 1$ mA/V and $r_o = 25$ kΩ. How does the present procedure handle the contribution from g_{mb1}? Explain!

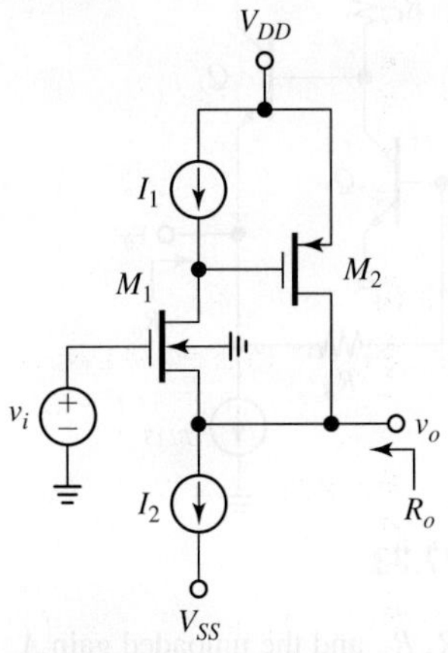

FIGURE P7.21

7.22 Assuming the op amp in the circuit of Fig. P7.22 has the open-loop parameters $r_i = 2$ MΩ, $r_o = 100$ Ω, and $a_v = 10^5$ V/V, use the shunt-shunt procedure to estimate R_i, R_o, and the unloaded gain $A_{oc} = v_o/i_i$ if $R_1 = 1\,\text{M}\Omega$, $R_2 = 2\,\text{k}\Omega$, and $R_3 = 18\,\text{k}\Omega$.

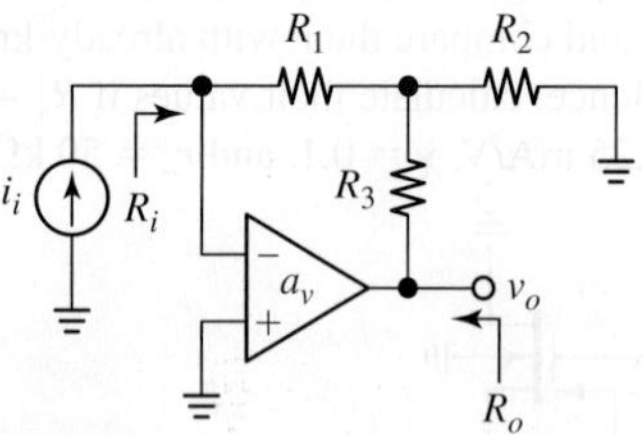

FIGURE P7.22

7.23 The closed-loop characteristics of the feedback-bias configuration of Fig. 7.34a can be improved appreciably if we buffer the amplifier's output with an emitter follower to reduce loading by the feedback network. In Fig. P7.23, Q_1 is the amplifier proper and Q_2 is the buffer. To appreciate the improvement quantitatively, let us use the same component values of Example 7.15b ($R_B = 5.0\,\text{k}\Omega$ and $R_C = 10\,\text{k}\Omega$). Moreover, assume both BJTs have $g_m = 1/(25\,\Omega)$, $r_\pi = 5\,\text{k}\Omega$, and $r_o = 100\,\text{k}\Omega$. Estimate R_i, R_o, and the unloaded gain $A_{oc} = v_o/i_i$. Compare with Example 7.15b, and comment.

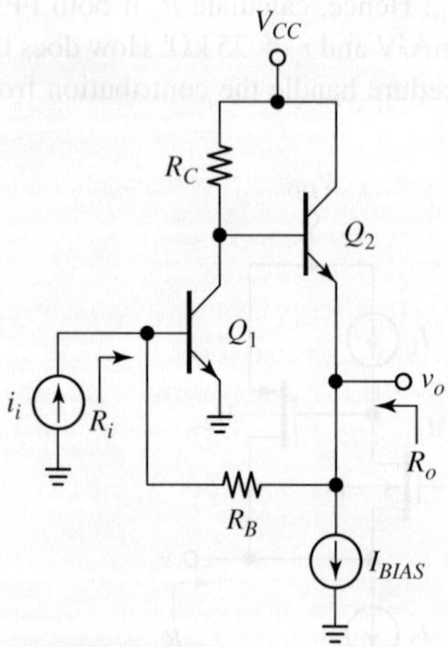

FIGURE P7.23

7.24 Estimate R_i, R_o, and the unloaded gain $A_{oc} = v_o/i_i$ for the shunt-shunt circuit of Fig. P7.24 if all FETs have $g_m = 1.5\,\text{mA/V}$ and $r_o = 30\,\text{k}\Omega$.

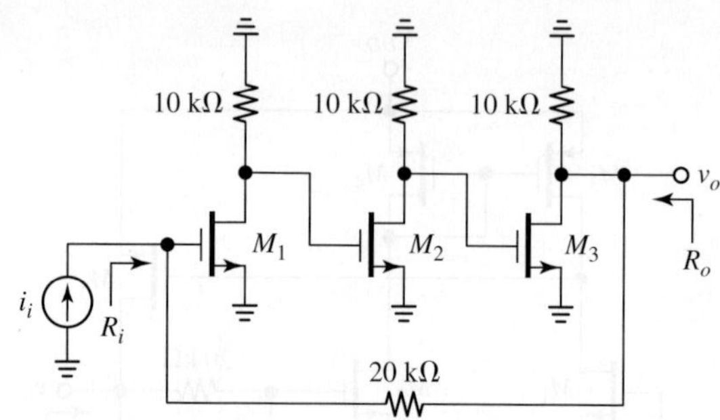

FIGURE P7.24

7.25 The op amp of Fig. P7.25 is designed to provide V-I conversion with $i_O = v_I/R$ for $v_I \geq 0$. The open-loop characteristics of the op amp are $r_i = 100\,\text{k}\Omega$, $r_o = 30\,\text{k}\Omega$, and $a_v = 10^4$ V/V, and to make up for the relatively high value of r_o, we buffer it with a BJT booster, as shown. Using the series-series procedure, estimate R_i, R_o, and the unloaded gain $A_{sc} = i_o/v_I$ at $v_I = 1$ V if $R = 1.0\,\text{k}\Omega$ and the BJT has $\beta_0 = 200$ and $V_A = 50$ V.

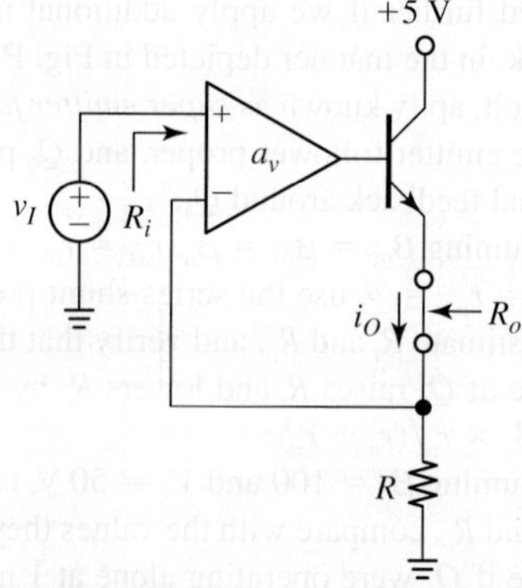

FIGURE P7.25

7.26 Figure P7.26 shows how the feedback triple of Fig. 7.23 can be configured for series-series operation. Assuming the parameters of Example 7.10 ($R_1 = 1.0\,\text{k}\Omega$, $R_2 = 3.0\,\text{k}\Omega$, and $R_3 = R_4 = R_5 = 10\,\text{k}\Omega$, along with $g_m = 1/(25\,\Omega)$, $r_\pi = 5\,\text{k}\Omega$, and $r_o = \infty$ for all BJTs), estimate R_i, R_o, and the unloaded gain $A_{sc} = i_o/v_i$. How does the loop gain compare with that of Example 7.10? Comment.

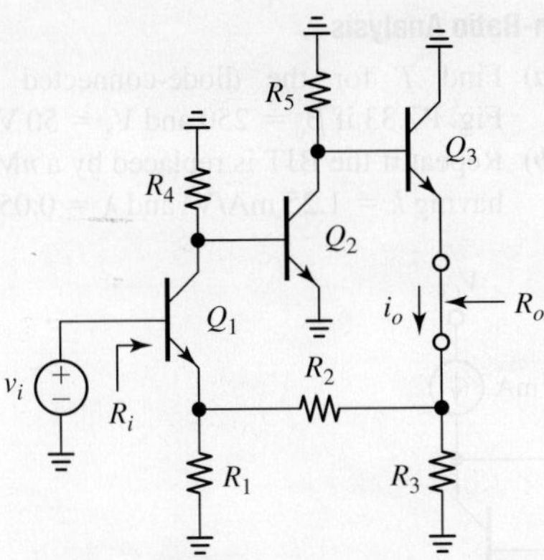

FIGURE P7.26

7.27 Using the series-series procedure, estimate R_o as well as the unloaded gain $A_{sc} = i_o/v_i$ of the circuit of Fig. P7.27. Assume 0-V dc across the 10-kΩ sensing resistance, as well as the following transistor parameters: $k_1 = k_2 = k_5/4 = 0.5$ mA/V^2, $\lambda_2 = \lambda_3 = \lambda_5 = 0.05$ V^{-1}, and $\chi_5 = 0.1$.

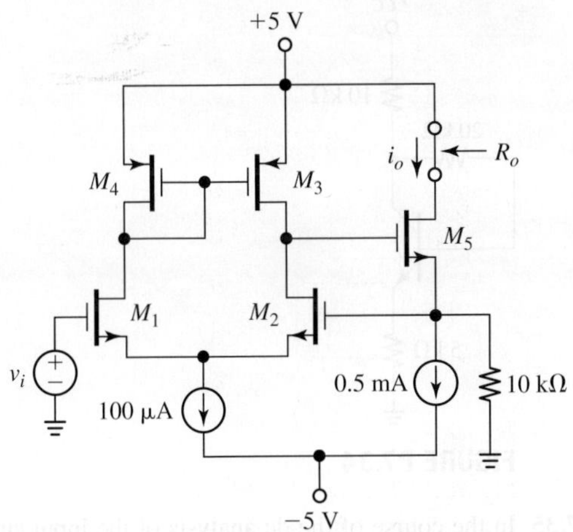

FIGURE P7.27

7.28 The feedback circuit of Fig. P7.28 uses a CMOS op amp to implement a current amplifier with $i_O = 100i_I$ for $i_I \geq 0$. The open-loop characteristics of the op amp are $r_i = \infty$, $r_o = 100$ kΩ, and $a_v = 10^4$ V/V, and to make up for the relatively high value of r_o, we buffer the op amp with a pMOSFET booster, as shown. Assuming the FET has $k = 0.5$ mA/V^2, $\lambda = 0.05$ V^{-1}, and $\chi = 0.1$, use the shunt-series procedure to estimate R_i, R_o, and the unloaded gain $A_{sc} = i_O/i_I$ at $i_I = 10$ μA.

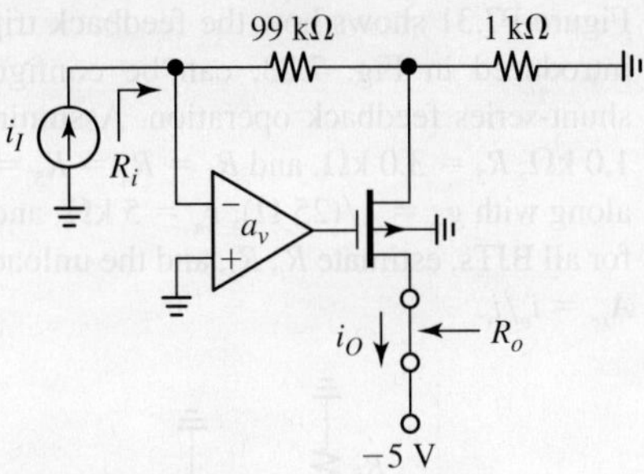

FIGURE P7.28

7.29 Assuming $V_{BE(\text{on})} = 0.7$ V, $\beta_0 = 100$, $V_A = 75$ V, and $r_\mu = \infty$ for the current buffer of Fig. P7.29, use the shunt-series procedure to estimate R_i, R_o, and the unloaded gain $A_{sc} = i_o/i_i$.

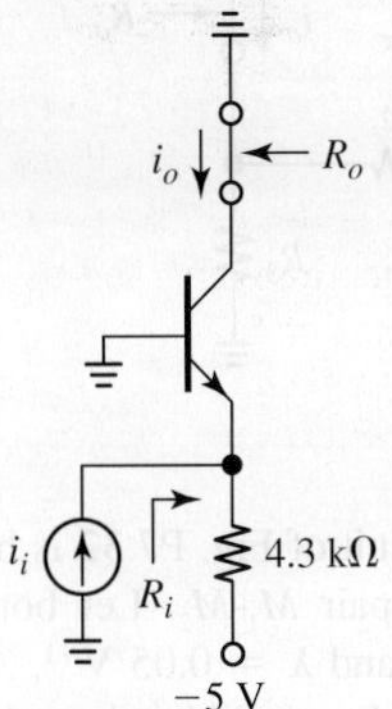

FIGURE P7.29

7.30 A diode-connected FET can be viewed as a shunt-series feedback circuit with unity feedback factor ($b = 1$). A load placed inside the feedback loop as in Fig. P7.30 will see a resistance $R_o \gg R_s$ (this, so long as the voltage developed by the load is small enough to prevent v_{DS} from dropping below V_{OV}). Use the shunt-series procedure to estimate R_o and the gain $A_{sc} = i_o/i_i$, if $R_s = 10$ kΩ and I_{BIAS} has been adjusted for $I_D = 1$ mA. Assume $k = 1.5$ mA/V^2 and $\lambda = 0.02$ V^{-1}.

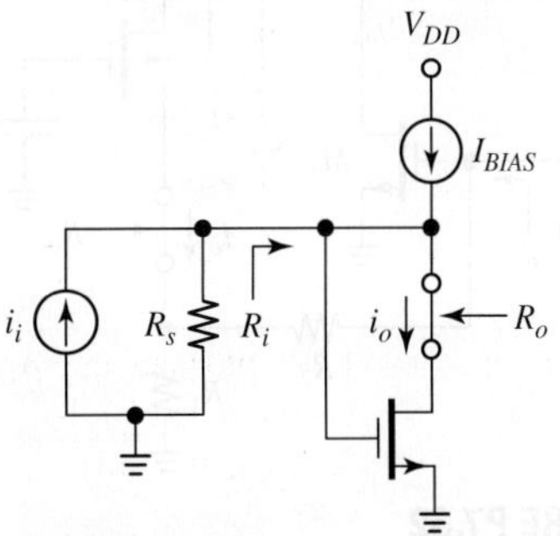

FIGURE P7.30

7.31 Figure P7.31 shows how the feedback triple, first introduced in Fig. 7.23, can be configured for shunt-series feedback operation. Assuming $R_3 = 1.0\ \text{k}\Omega$, $R_2 = 3.0\ \text{k}\Omega$, and $R_1 = R_4 = R_5 = 10\ \text{k}\Omega$, along with $g_m = 1/(25\ \Omega)$, $r_\pi = 5\ \text{k}\Omega$, and $r_o = \infty$ for all BJTs, estimate R_i, R_o, and the unloaded gain $A_{sc} = i_o/i_i$.

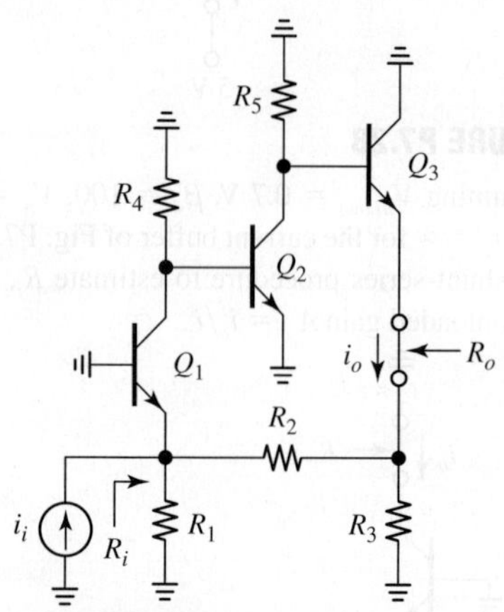

FIGURE P7.31

7.32 The shunt-series circuit of Fig. P7.32 is based on the folded cascode pair M_1-M_2. Let both FETs have $k = 2\ \text{mA/V}^2$ and $\lambda = 0.05\ \text{V}^{-1}$. Also, let $\chi_2 = 0.2$. Assuming I_{BIAS} and V_{BIAS} have been adjusted so that $I_{D1} = I_{D2} = 1\ \text{mA}$, estimate R_i, R_o, and the unloaded gain $A_{sc} = i_o/v_i$ if $R_1 = 3\ \text{k}\Omega$ and $R_2 = 12\ \text{k}\Omega$.

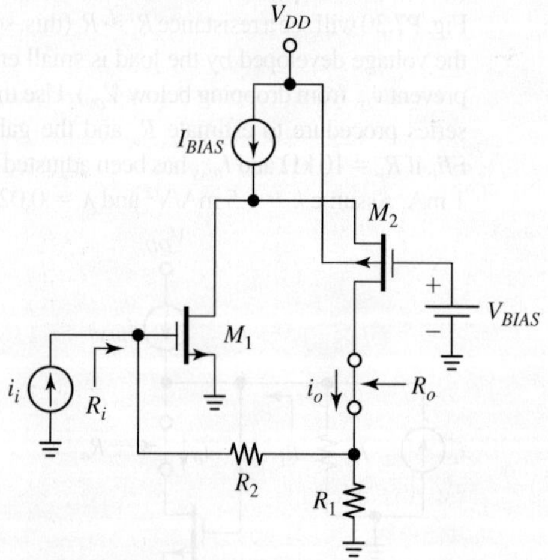

FIGURE P7.32

7.5 Return-Ratio Analysis

7.33 **(a)** Find T for the diode-connected BJT of Fig. P7.33 if $\beta_0 = 250$ and $V_A = 50$ V.

(b) Repeat if the BJT is replaced by a nMOSFET having $k = 1.25\ \text{mA/V}^2$ and $\lambda = 0.05\ \text{V}^{-1}$.

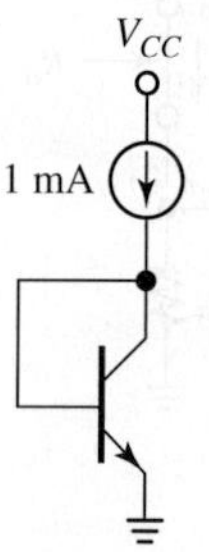

FIGURE P7.33

7.34 Find T for the circuit of Fig. P7.34 if V_{CC} has been adjusted for $I_C = 0.5$ mA. Assume $\beta_0 = 200$ and $V_A = \infty$.

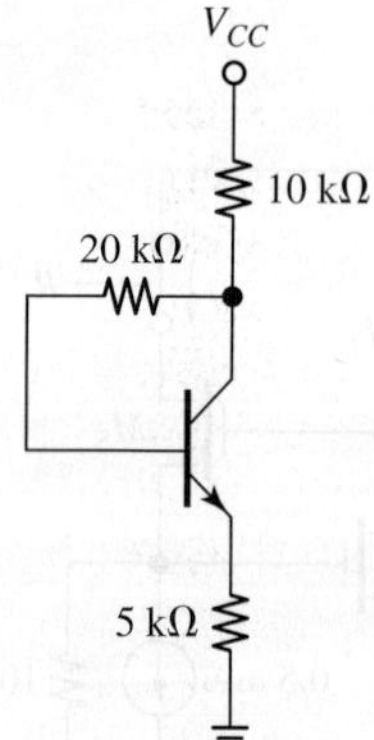

FIGURE P7.34

7.35 In the course of the dc analysis of the input stage of the 741 op amp, repeated in Fig. P7.35 for convenience, it was stated that the biasing scheme adopted utilizes negative feedback. Assuming $\beta_{Fp} = 50$ and very large β_{Fn}, find T.

Hint: break the loop at pint X, inject a test current i_t into Q_8, and find the amount of current i_r returned by the Q_1-Q_2 pair.

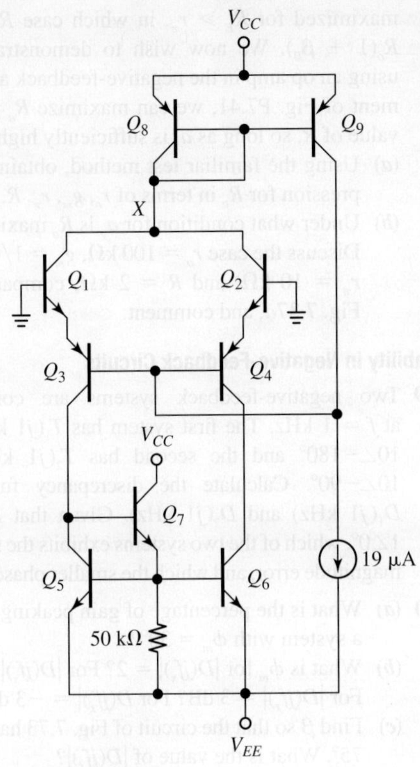

FIGURE P7.35

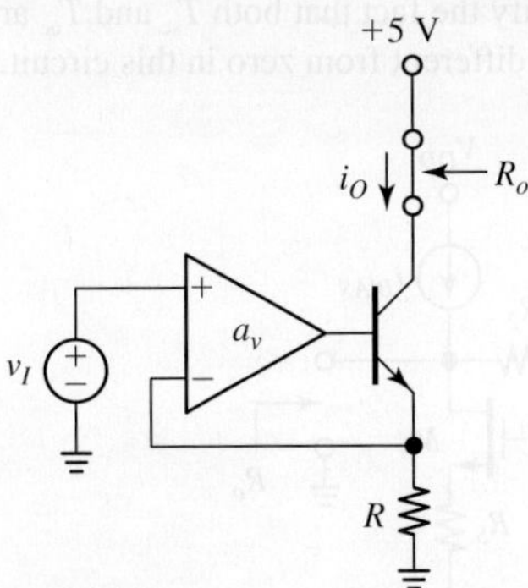

FIGURE P7.41

7.42 As we know, one way of increasing the output resistance of MOS cascodes is via the telescopic technique. Telescoping, however, reduces the output voltage swing. A clever way to overcome this limitation is to place the top FET inside a negative-feedback loop of the series-output type as depicted in Fig. P7.42. Assuming a CMOS op amp with $a_v = 10^3$ V/V and FETs with $g_m = 0.5$ mA/V, $r_o = 20$ kΩ, and $\chi = 0.1$, find R_o via Blackman's impedance formula. What is the unloaded gain $A_{oc} = v_o/v_i$? **Hint:** select the dependent source modeling the op amp for your return-ratio calculations.

7.36 Using return-ratio analysis, find the gain v_o/v_i of the ordinary source follower of Problem 7.20.

7.37 Using return-ratio analysis, find the gain v_o/v_i of the op amp circuit of Problem 7.22.

7.38 With reference to the shunt-shunt BJT pair of Problem 7.23, find the gain v_o/i_i via the return ratio of the dependent source modeling Q_1.

7.6 Blackman's Impedance Formula and Injection Methods

7.39 Use Blackman's formula to find R_i and R_o for the bipolar V-I converter of Example 7.17.

7.40 Use Blackman's formula to find R_o for the super source follower of Problem 7.21. Assume both FETs have $g_m = 1$ mA/V and $r_o = 25$ kΩ. Also, for M_1 assume $\chi = 0.2$. **Hint:** select the dependent source modeling Q_2 for your return-ratio calculations.

7.41 Use Blackman's formula to find R_o for the V-I converter of Fig. P7.41 for $v_I = 1$ V and $R = 1$ kΩ. For the op amp assume $a_v = 10^4$ V/V, $r_i = \infty$, and $r_o = 0$, and for the BJT assume $\beta_0 = 100$, $V_A = 50$ V, and $r_\mu = \infty$. **Hint:** select the dependent source modeling the op amp for your return-ratio calculations.

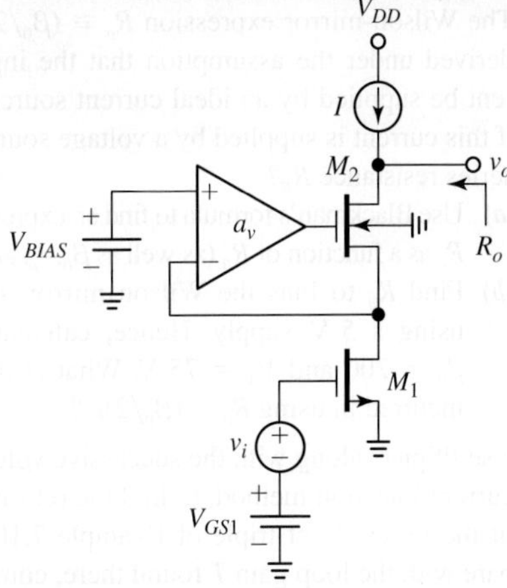

FIGURE P7.42

7.43 (*a*) Assuming R_1 and R_2 are so large that their current can be ignored compared to I_{BIAS} in the circuit of Fig. P7.43, use Blackman's formula to obtain an expression for R_o as a function of g_m, r_o, R_S, and β, where $\beta = R_1/(R_1 + R_2)$.

(**b**) Discuss the limiting cases $R_1 \to \infty$, and $R_2 \to \infty$; compare with known results.

(**c**) Justify the fact that both T_{sc} and T_{oc} are generally different from zero in this circuit.

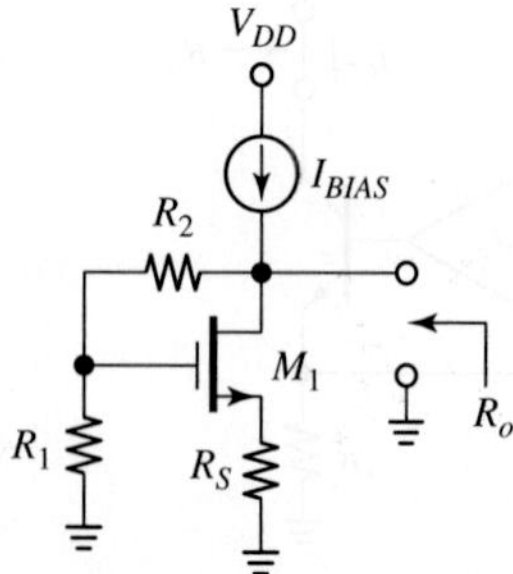

FIGURE P7.43

7.44 Use Blackman's formula to find R_i and R_o for the shunt-series BJT pair of Fig. 7.45a. Use the data of Example 7.20.

Hint: select the dependent source modeling Q_1 for your return-ratio calculations.

7.45 Use Blackman's formula to find R_i and R_o for the shunt-series MOS pair of Fig. P7.32. Use the data of Problem P7.32.

Hint: select the dependent source modeling M_1 for your return-ratio calculations, and verify via PSpice.

7.46 The Wilson-mirror expression $R_o \cong (\beta_0/2)r_o$ was derived under the assumption that the input current be supplied by an ideal current source. What if this current is supplied by a voltage source via a series resistance R_B?

(**a**) Use Blackman's formula to find an expression for R_o as a function of R_B (as well as β_0, r_o, and r_π).

(**b**) Find R_B to bias the Wilson mirror at 1 mA using a 5-V supply. Hence, calculate R_o if $\beta_0 = 200$ and $V_A = 75$ V. What is the error incurred in using $R_o \cong (\beta_0/2)r_o$?

7.47 Use PSpice, along with the successive voltage and current injection method, to find the return ratio T of the series-shunt triple of Example 7.10. Compare with the loop gain T found there, comment.

7.48 The single-BJT V-I converter of Fig. 7.37a can be analyzed *directly* via the test method to give the familiar expression $R_o = r_o[1 + g_m(r_\pi/\!/R_E)]$, under the assumption $r_\mu \to \infty$. As we know, R_o is

maximized for $R_E \gg r_\pi$, in which case $R_{o(max)} = R_o(1 + \beta_0)$. We now wish to demonstrate that using an op amp in the negative-feedback arrangement of Fig. P7.41, we can maximize R_o for *any* value of R, so long as a_v is sufficiently high.

(**a**) Using the familiar test method, obtain an expression for R_o in terms of r_o, g_m, r_π, R, and a_v.

(**b**) Under what condition for a_v is R_o maximized? Discuss the case $r_o = 100$ kΩ, $g_m = 1/(50\ \Omega)$, $r_\pi = 10$ kΩ, and $R = 2$ kΩ, compare with Fig. 7.37a, and comment.

7.7 Stability in Negative-Feedback Circuits

7.49 Two negative-feedback systems are compared at $f = 1$ kHz. The first system has $T_1(j1\ \text{kHz}) = 10\angle-180°$ and the second has $T_2(j1\ \text{kHz}) = 10\angle-90°$. Calculate the discrepancy functions $D_1(j1\ \text{kHz})$ and $D_2(j1\ \text{kHz})$. Given that $D_{ideal} = 1\angle0°$, which of the two systems exhibits the smaller magnitude error, and which the smaller phase error?

7.50 (**a**) What is the percentage of gain peaking at f_x of a system with $\phi_m = 30°$?

(**b**) What is ϕ_m for $|D(jf_x)| = 2$? For $|D(jf_x)| = 10$? For $|D(jf_x)| = 3$ dB? For $D(jf_x) = -3$ dB?

(**c**) Find β so that the circuit of Fig. 7.73 has $\phi_m = 75°$. What is the value of $|D(jf_x)|$?

7.51 A voltage amplifier with open-loop gain $a(jf) = a_0/[(1 + jf/f_1) \times (1 + jf/f_2)]$ is operated in negative feedback with $\beta = 0.1$ V/V.

(**a**) If the closed-loop dc gain is $A_0 = 9$ V/V, find a_0; hence, develop a standard-form expression for $A(jf)$ in terms of f_1 and f_2.

(**b**) If it is found that the phase and magnitude of $A(jf)$ at $f = 10$ kHz are $-90°$ and 90/11 V/V, what are the values of f_1 and f_2?

(**c**) Find the crossover frequency f_x and, hence, the phase margin ϕ_m.

(**d**) Find ϕ_m if β is raised to 1 V/V.

7.52 An amplifier with open-loop gain $a(s) = 100/[(1 + s/10^3) \times (1 + s/10^5)]$ is placed in a negative-feedback loop.

(**a**) Derive an expression for the closed-loop gain $A(s)$ as a function of the feedback factor β, and find the value of β that causes the poles of $A(s)$ to be *coincident*. What is their common value?

(**b**) Find the crossover frequency f_x and, hence, the phase margin ϕ_m.

7.53 An alternative way of quantifying the stability of a negative-feedback system is via *the gain margin* g_m, defined as $g_m = -20 \log |T(jf_{-180°})|$, where $f_{-180°}$ represents the frequency at which ph $T = -180°$. (As we know, for our system to be convincingly stable, we want $|T(jf)|$ to have *dropped well below* 1 at $f_{-180°}$)

(a) If an amplifier with dc gain $a_0 = 10^5$ and three pole frequencies $f_1 = 1$ kHz, $f_2 = 1$ MHz, and $f_3 = 10$ MHz is to be operated in negative feedback, find β for $\phi_m = 60°$. What is the corresponding value of g_m?

(b) Find β for $g_m = 20$ dB. What is the corresponding value of ϕ_m?

7.54 Even though we have focused on frequency-independent feedback, we can easily generalize to frequency-dependent cases because stability is determined by $T(jf) = a(jf)\beta(jf)$ regardless of whether frequency dependence is due to $a(jf)$, or $\beta(jf)$, or both. A classic example is the op amp *differentiator* of Fig. P7.54. Suppressing V_i, we find the feedback factor as $\beta_v(jf) = V_n/V_o = 1/(1 + jf/f_0)$, $f_0 = 1/(2\pi RC)$. Ideally, a differentiator gives $A_{\text{ideal}}(jf) = V_o/V_i = -jf/f_0$, but because of $T(jf) \neq \infty$, the actual gain $A(jf)$ will depart from the ideal.

(a) Assuming a single-pole op amp with $a_v(jf) = a_{v0}/(1 + jf/f_b)$, sketch and label the linearized Bode plots of $|a_v(jf)|$ and $|1/\beta_v(jf)|$ if $a_{v0} = 10^5$ V/V, $f_b = 10$ Hz, $R = 10$ kΩ, and $C = 15.9$ nF. Hence, use visual inspection for an initial estimation of the frequency f_x at which the two curves intersect.

(b) Obtain an expression for $T(jf) = a_v(jf)\beta_v(jf)$, use trial and error for a more accurate calculation of f_x, ϕ_x, and ϕ_m, and verify that the circuit is on the verge of oscillation.

(c) Calculate $D(jf_x)$ and $A(jf_x)$, and compare with $A_{\text{ideal}}(jf_x)$. What are the magnitude and phase errors at this frequency?

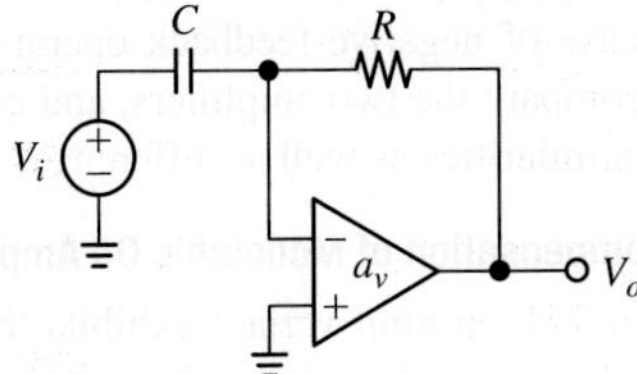

FIGURE P7.54

7.8 Dominant-Pole Compensation

7.55 The negative-feedback loop of Fig. P7.55 is made up of a ring of three CMOS *inverters*. Assume the FETs have $g_{mn} = g_{mp} = 1$ mA/V and $r_{on} = r_{op} = 30$ kΩ,

and the stray capacitances of each inverter can be modeled by a single equivalent capacitance $C_o = 1$ pF from the output node to ground.

(a) Obtain an expression for $T(jf)$, find f_x, ϕ_x, and ϕ_m, and verify that the loop is unstable, this being the reason why the circuit is called a *ring oscillator*.

(b) Find an external shunt capacitance C_{shunt} that, when connected between the output terminal and ground of any one of the inverters will stabilize the loop for $\phi_m = 45°$.

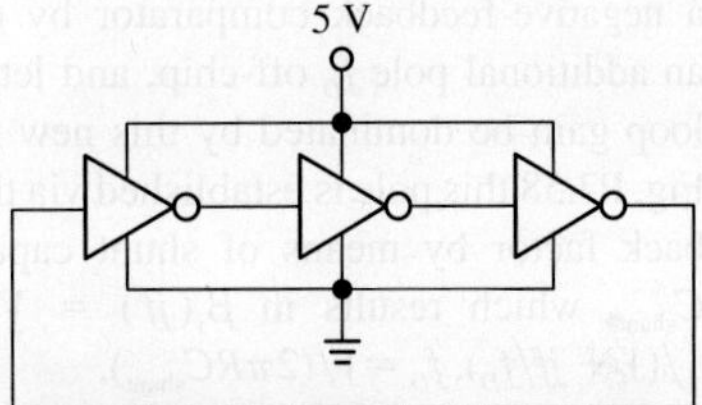

FIGURE P7.55

7.56 (a) Consider a negative-feedback loop with dc gain $T_0 = 10^3$, a dominant pole at 1 kHz, and a pair of coincident poles at 250 kHz. Using trial and error, find the crossover frequency f_x and, hence, the phase margin ϕ_m, and verify that this loop is unstable.

(b) One way of stabilizing it is to reduce T_0 so that the ensuing downward shift of the magnitude plot will lower f_x to a frequency region of less phase lag. Find the value to which T_0 must be reduced for $\phi_m = 45°$.

Hint: since the dominant-pole contribution to ϕ_x is $-90°$, the coincident poles must contribute about $-45/2 = -22.5°$ each.

(c) As we know, another way is to reduce f_1. To what value must f_1 be reduced for $\phi_m = 45°$?

(d) Repeat (b) and (c), but for $\phi_m = 60°$.

7.57 An amplifier has dc gain $a_0 = 10^5$ V/V and three pole frequencies $f_1 = 100$ kHz, $f_2 = 1$ MHz, and $f_3 = 10$ MHz, arising at three nodes with equivalent resistances R_1, R_2, and R_3.

(a) Sketch and label the linearized Bode plot of gain magnitude for a visual estimation of f_x and ϕ_x for operation with $\beta = 1$ V/V, and verify that the circuit is unstable.

(b) One way of stabilizing it is to place a compensating resistance R_c in parallel with R_2 so as raise f_2 and lower a_0. Sketch the new Bode plot if $R_c = R_2/99$, and verify that this causes a two-decade drop in a_0 as well as a two-decade increase in f_2. How do f_x and ϕ_m change?

(c) Find the ratio R_c/R_2 that results in $\phi_m = 60°$.

7.58 (a) A student is trying to use a voltage comparator as a unity-gain buffer ($\beta_v = 1$). If the comparator has a dc gain of 10^4 V/V and three pole frequencies at 1 MHz, 20 MHz, and 50 MHz, find f_x, ϕ_x, and ϕ_m, and verify that the circuit is unstable.

(b) Comparators do not have on-chip compensation because they are meant for open-loop operation. However, the user can still stabilize a negative-feedback comparator by creating an additional pole f_D off-chip, and letting the loop gain be dominated by this new pole. In Fig. P7.58 this pole is established via the feedback factor by means of shunt capacitance C_{shunt}, which results in $\beta_v(jf) = V_n/V_o = 1/(1 + jf/f_D), f_D = 1/(2\pi RC_{\text{shunt}})$.

(a) Find f_D for $\phi_m = 60°$. Hence, assuming $R = 30$ kΩ, find the required value of C_{shunt}. What is the expression for $T(jf)$ after compensation?

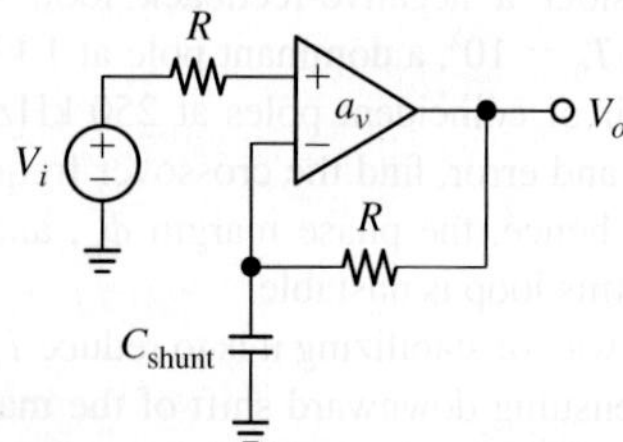

FIGURE P7.58

7.59 The op amp differentiator of Problem 7.54 can be stabilized by inserting a series resistance R_c, as in Fig. P7.59, to create a LHP transmission zero at $f_z = 1/(2\pi R_c C)$. As we know, a LHP zero contributes phase *lead*, and as such it can be positioned so as to *raise* ϕ_m.

(a) Assuming the component values of Problem 7.54, sketch and label the linearized Bode plots of $|a_v(jf)|$ and $|1/\beta_v(jf)|$ before compensation (i.e. with $R_c = 0$). Locate the frequency f_x at which the two curves intersect and find the value of R_c that makes $f_z = f_x$ (this will result in $\phi_m \cong 45°$). How does the $|1/\beta_v(jf)|$ curve look after compensation?

Hint: the presence of R_c results in the high-frequency asymptote $1/\beta_{v\infty} = 1 + R/R_c$.

(b) Obtain an expression for $\beta_v(jf) = V_n/V_o$ with $V_i = 0$, and then use trial and error to find the actual values of f_x, ϕ_x, and ϕ_m.

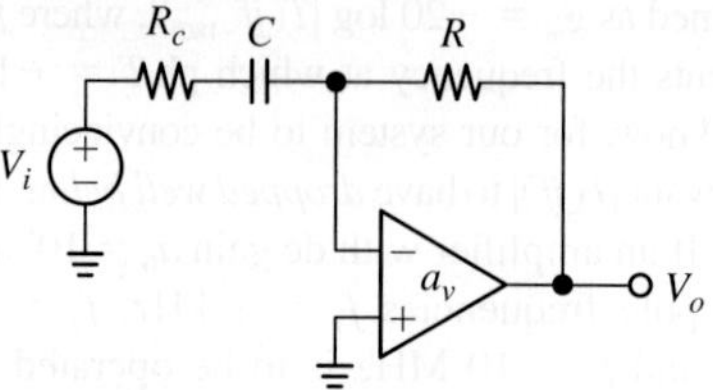

FIGURE P7.59

7.60 An amplifier with a dc gain of 20,000 V/V and three pole frequencies at 100 kHz, 3 MHz, and 5 MHz is operated in negative feedback with $\beta = 1$ V/V.

(a) Sketch and label the linearized Bode plot of gain magnitude for a visual estimation of f_x and ϕ_x, and verify that the circuit is unstable.

(b) Given that the first two poles are produced at the input and output nodes of an internal transistor amplifier with a dc gain of -200 V/V and input-node and output-node resistances $R_1 = 100$ kΩ and $R_2 = 10$ kΩ, find the capacitance C_c that, connected across the terminals of this internal stage, stabilizes the circuit for $\phi_m = 45°$.

(c) Calculate g_m, C_1, and C_2. What are the values of the new pole frequencies as well as the RHP zero frequency? Use these values to obtain an expression for $T(jf)$.

7.61 Consider two amplifiers having, respectively, open-loop gains

$$a_1(jf) = \frac{10^4(1 + jf/10^5)}{(1 + jf/10^3)(1 + jf/10^7)}$$

$$a_2(jf) = \frac{10^4(1 - jf/10^5)}{(1 + jf/10^3)(1 + jf/10^7)}$$

Sketch and label their linearized Bode plots (magnitude as well as phase). Find the phase margins for the case of negative-feedback operation with $\beta = 1$, compare the two amplifiers, and comment on their similarities as well as differences.

7.9 Frequency Compensation of Monolithic Op Amps

7.62 A certain 741 op amp variant exhibits the resistance and transconductance values of Fig. 5.11, but with different frequency characteristics because of a different fabrication process. Before compensation the 2nd stage contributes a pole pair with its stray input and output capacitances $C_1 = 2$ pF and $C_2 = 4$ pF, and the rest of the circuitry contributes a 3rd pole frequency at 7.5 MHz.

(a) Find the Miller capacitance C_c that will ensure $\phi_m = 75°$ for negative-feedback operation with $\beta = 1$. What is the slew rate SR?

(b) Obtain an expression for the loop gain $T(jf)$.

7.63 (a) A PSpice simulation of the 741 op amp gives $a_0 = 200$ V/mV, $f_b = 5$ Hz, $f_t = 888$ kHz, and ph $a(jf_t) = -117.2°$, so the phase margin for operation with $\beta = 1$ V/V is $\phi_m = 62.8°$. Assuming all higher-order roots can be modeled with a single pole frequency f_p of fixed value, find f_p.

(b) To what value should the 30-pF capacitance be lowered if we wish to ensure the same phase margin for operation with $1/\beta \geq 5$ V/V? What is the corresponding slew rate? (Sometimes manufacturers offer under-compensated op amps for improved dynamics so long as their operation is restricted to closed-loop gains above a prescribed value, such as 5 V/V in the present example.)

7.64 A two-stage op amp of the type of Fig. 7.92 has the following loop gain

$$T(jf) = \frac{3160(1 - jf/10^6)}{(1 + jf/10^2)(1 + jf/10^8)}$$

(a) Sketch and label the linearized Bode plots of its magnitude and phase, and estimate f_x, ϕ_x, and ϕ_m. Have four copies of the plots at hand, and show separately on each copy the modifications brought about by increasing one of the following parameters by a factor of ten while leaving all others unchanged:

(b) g_{m1},

(c) g_{m5},

(d) C_L,

(e) C_c.

Estimate ϕ_m in each of the cases and comment.

7.65 In deriving Eqs. (7.111) through (7.113) for the two-stage CMOS op amp, we assumed the internal parasitic capacitances to be *negligible* compared to the external capacitances C_c and C_L. We wish to verify by performing a series of tests on the basic PSpice op amp of Fig. 7.90 that will allow for the indirect estimation of its internal parasitics (all tests are shall be performed with $C_c = 0$, so $C_f = C_{gd5}$).

(a) For the first test we set $V_i = 0$, short V_o to ground to inhibit the Miller affect by M_5, inject a test current I_t into node V_1, and plot the impedance $Z_1 = V_1/I_t$. If Z_1 exhibits a pole frequency of 16.2 MHz, estimate the sum $C_i + C_1 + C_f$ in the ac equivalent of Fig. 7.92.

(b) Now repeat part (a), but with node V_o open circuited to enable the Miller effect. If this causes the dominant pole frequency of Z_1 to drop from 16.2 MHz to 2.25 MHz, estimate C_f.

(c) Next, terminate node V_1 on a large (1 μF) capacitance to establish an ac ground over the frequencies of interest, inject a test current I_t into node V_o, and plot the impedance $Z_o = V_o/I_t$. If Z_o exhibits a pole frequency of 25.3 MHz, estimate $C_2 + C_f$ in the ac equivalent of Fig. 7.92.

(d) Finally, activate the input source V_i and short V_o to ground. If the gain $a_1 = V_1/V_i$ exhibits a high-frequency asymptote of -15 dB, estimate C_i and C_1 in Fig. 7.92.

(e) Using the above stray capacitances, recalculate f_1 and f_2 for Example 7.37, compare, and comment.

7.66 (a) With reference to the two-stage CMOS op amp of Example 7.37, what is the maximum C_L for which $\phi_m = 60°$?

(b) Repeat, but for $\phi_m = 45°$.

(c) Find R_c that will restore $\phi_m = 75°$ for the value of C_L of part (b).

(d) If R_c is doubled in the circuit of Example 7.37, what is the maximum C_L for which $\phi_m = 60°$? Compare with part (b) and comment.

7.67 A two-stage CMOS op amp of the type of Fig. 7.94 has $g_{m5} = 2.5g_{m1}$ and its transmission zero is at infinity.

(a) If GBP $=37.1$ MHz and loading the op amp with $C_L = 3$ pF results in $\phi_m = 65°$, find R_c and C_c.

(b) If SR $= 40$ V/μs, find I_{D7}, V_{OV1}, and k_1. If $V_{OV5} = V_{OV1}$, what are k_5 and I_{D5}?

(c) Assuming $\lambda_n = \lambda_p = \lambda$, find f_1 and λ if $a_0 = 10$ V/mV.

7.68 (a) A two-stage CMOS op amp of the type of Fig. 7.94 has $g_{m1} = 0.5$ mA/V, $g_{m2} = 1.25$ mA/V, and $I_{D7} = 80$ μA. If $C_c = 2$ pF, find the GBP and the SR. What are k_1 and V_{OV1}?

(b) If $R_c = 1.5$ kΩ, find the maximum C_L for which $\phi_m = 70°$.

7.69 Figure P7.69 shows the ac equivalent of the compensation scheme of Fig. 7.93a (to ease analysis, the CD buffer is shown as ideal.) Find the gain $a(s) = V_o/V_i$, verify that the transmission zero has been moved to infinity, and obtain approximate expressions for its poles f_1, and f_2 under the assumption $f_1 \ll f_2$. Compare with Eqs. (7.111) and (7.112) and comment.

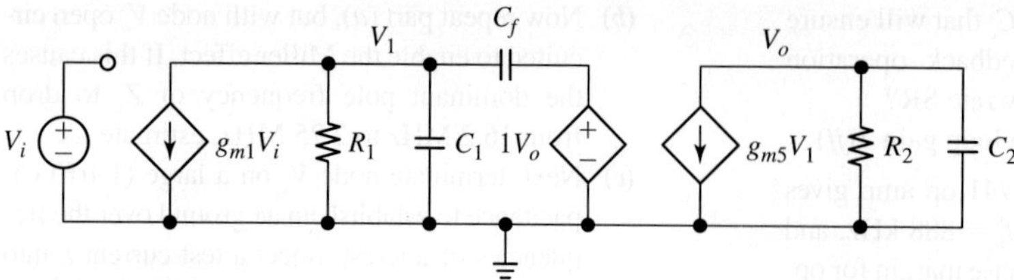

FIGURE P7.69

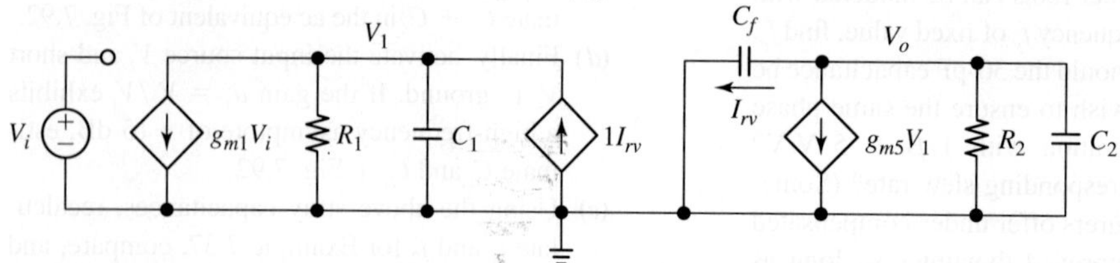

FIGURE P7.70

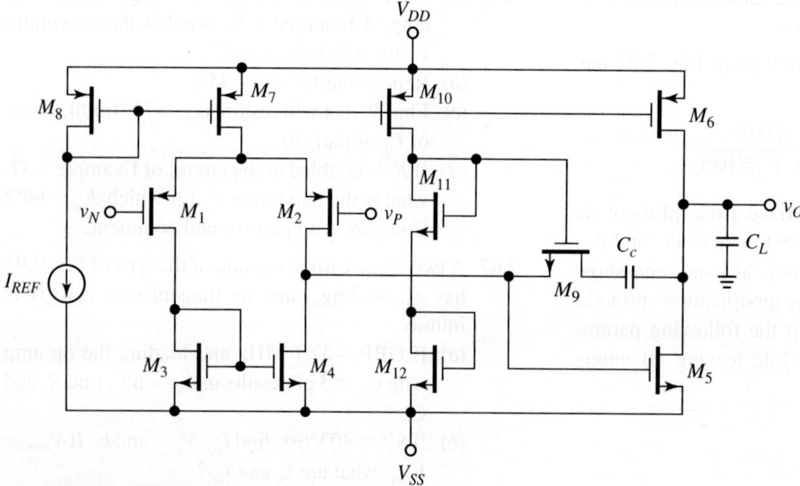

FIGURE P7.71

7.70 Figure P7.70 shows the ac equivalent of the compensation scheme of Fig. 7.93b (to ease analysis, the CG buffer is shown as ideal.) Find the gain $a(s) = V_o/V_i$, verify that the transmission zero has been moved to infinity, and obtain approximate expressions for its poles f_1 and f_2 under the assumption $f_1 \ll f_2$. Compare with Eqs. (7.112) and (7.113) and comment.

7.71 Shown in Fig. P7.71 is a popular scheme for synthesizing the compensation resistance R_c. This is simply the channel resistance of M_9, in turn biased by the current source M_{10} and the diode-connected pair M_{11} and M_{12}.

(a) If this scheme is used in the PSpice circuit of Fig. 7.74 of Example 7.37, specify suitable W/L ratios for M_9 through and M_{12} to implement the resistance $R_c = 1.07$ kΩ under the constraint $I_{D10} = 50$ μA. Hence, run a PSpice simulation, compare and comment.

(b) Repeat, but for Example 7.38.

7.72 We wish to design a folded-cascode op amp of the type of Fig. 5.16 under the constraints that $I_{BIAS} = 1.25I_{SS}$ and that all FETs have the same overdrive voltage V_{OV}. The op amp is to have a dc gain of 8,000 V/V, and when terminated on an

8-pF external load, its gain-bandwidth product must be 16 MHz.

(a) If $\lambda_n = \lambda_p = 0.05$ V^{-1}, find V_{OV}, I_{SS}, I_{BIAS}, and SR (for simplicity ignore the body effect and assume $\lambda = 0$ in the course of dc calculations).

(b) If the biasing circuit has $I_{REF} = I_{BIAS}$ and the entire op amp is powered from ± 2.5-V supplies, what is its power dissipation?

7.73 As mentioned in connection with the folded-cascode op amp of Fig. 7.97, the phase shift $\phi_{x(HOR)}$ is due to the capacitances C_1 through C_5, each of which sees a source resistance on the order of $1/(g_m + g_{mb})$.

(a) If we make the simplifying assumption that the effect of all these roots can be modeled, at least in the vicinity of f_x, with a *single* pole frequency f_2, use the values of a_0 and f_b of Eq. (7.121) to find the value f_2 that will ensure the value of f_x of Eq. (7.121b).

(b) If we make the additional assumptions that the effect of C_1 through C_5 can be modeled with a *single* equivalent capacitance C_{eq}, and that the various source resistances can be modeled with the *common* value $R_{eq} = 1/(g_m + g_{mb}) \cong 1.5$ kΩ, estimate C_{eq}.

(c) Develop a relationship between C_c and C_{eq} in terms of g_{cm1}, R_{eq}, C_6, and ϕ_m, where ϕ_m is the desired phase margin for operation with $\beta = 1$.

(d) Estimate the required value C_c for $\phi_m = 60°$. What is the corresponding GBP?

(e) Repeat, but for $\phi_m = 75°$, compare with Example 7.39, and comment.

7.10 Noise

7.74 **(a)** Use the identity $\int dx/(x^2 + 1) = \tan^{-1}x$ to prove Eq. (7.136).

(b) Look up the integral tables, for instance on the Web, and prove Eq. (7.138).

(c) Find NEB if $A_n(jf)$ contains two (instead of one) pole frequencies, both at $f = f_B$. Compare with Eq. (7.136) and justify the difference in terms of Bode plot areas.

(d) What approximation would you use in this case for f_H for flicker noise?

7.75 **(a)** Two IC noise sources e_{n1} and e_{n2} of the type of Eq. (7.125) are in *series* with each other. Show that the equivalent noise e_n of the combination of the two sources is still of the type of Eq. (7.125) by deriving expressions for its white-noise floor e_{nw} and corner frequency f_{ce}.

(b) If e_{n1} has $e_{nw1} = 30$ nV/$\sqrt{\text{Hz}}$ and $f_{ce1} = 400$ Hz, and e_{n2} has $e_{nw2} = 40$ nV/$\sqrt{\text{Hz}}$ and $f_{ce2} = 100$ Hz, find e_{nw} and f_{ce}.

7.76 **(a)** Show that the total rms voltage across the parallel combination of a resistance R and a capacitance C is $E_n = \sqrt{kT/C}$ regardless of R. Can you justify the independence of R?

(b) What capacitance C is required for $E_n = 1$ μV rms? What resistance R is required so that NEB$_{white} = 100$ kHz?

7.77 **(a)** The noise of an IC current source is measured at two different frequencies and is found to be $i_n(250$ Hz$) = 6.71$ pA/$\sqrt{\text{Hz}}$ and $i_n(2500$ Hz$) = 3.55$ pA/$\sqrt{\text{Hz}}$. What are the values of i_{nw} and f_{ci}?

(b) If the source is fed to a 1-kΩ resistance, find the values of e_{nw} and f_{ce} for the noise voltage across its terminals.

(c) Find the total rms noise E_n above 0.01 Hz if a 10-nF capacitor is placed across the resistor.

7.78 A diode with $I_s = 2$ fA, $r_S \cong 0$, $K = 10^{-16}$ A, and $a = 1$ is forward-biased at $I_D = 100$ μA by a 3.3-V supply via a series resistance R. The supply itself exhibits noise that can be assumed white with a density of $e_{ns} = 100$ nV/$\sqrt{\text{Hz}}$.

(a) Combine e_{ns}, e_{nr}, and the effect of i_{nd} to find the overall noise voltage $e_n(f)$ across the diode, and express it in the form of Eq. (7.125). What are the values of e_{nw} and f_{ce}?

(b) Repeat if R is changed so as to bias the diode at $I_D = 1$ mA.

7.79 **(a)** Find the output noise voltage e_{no} of a CMOS inverter that is powered from ± 3.3-V supplies and is driven by an input voltage source with a dc component of 0 V and a series resistance $R_s = 1$ kΩ. Assume matched devices with $k = 1$ mA/V^2, $V_t = 0.7$ V, $\lambda = 1/(33.8$ V), and $f_{ce} = 10$ kHz for the input noise voltage of both devices.

(b) Find the total rms output noise E_{no} above 1 Hz if the inverter output is loaded with a capacitance $C_L = 10$ pF (ignore all internal parasitics).

7.80 **(a)** For the I-V converter of Fig. P7.80a find the equivalent input noise current i_{ni} depicted in Fig. P7.80b. Assume a 741 op amp with $R = 100$ kΩ.

(b) What is the total rms output voltage E_{no}?

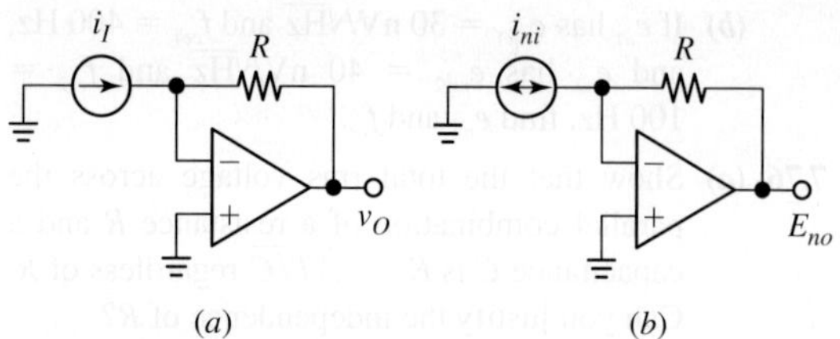

FIGURE P7.80

7.81 A certain bandgap voltage reference is buffered to the outside via an op amp voltage follower. The reference's output noise e_{n1} has $e_{nw1} = 100 \text{ nV}/\sqrt{\text{Hz}}$ and $f_{ce1} = 20$ Hz, and the op amp's input noise e_{n2} has $e_{nw2} = 25 \text{ nV}/\sqrt{\text{Hz}}$ and $f_{ce2} = 200$ Hz. Moreover, the op amp is dominant-pole compensated for GBP = 1 MHz via an external capacitance C_c.

(a) Calculate the total noise above 0.1 Hz at the output of the follower.

(b) Repeat if C_c is increased by a factor of 100.

7.82 (a) Derive an expression for the short-circuit noise current i_{no} at the output of the bipolar current mirror depicted in Fig. 4.56a. Assume the input source is noiseless.

(b) Repeat, but for the MOS mirror of Fig. 4.58a. What are the expressions for the noise floor and the corner frequency of i_{no}^2? How are they affected if the input current is doubled?

7.83 (a) Assuming $r_o \gg R_D$, derive an expression for the short-circuit noise e_{ni} of the CS amplifier of Fig. P7.83a.

(b) Find the open-circuit noise i_{ni} of Fig. P7.83b.

(c) One might think that with two input sources noise is counted twice. To prove that this is not the case, refer to Fig. P7.83c, where R is assumed noiseless for simplicity. Prove that the combined power density e_{ng}^2 at the gate is independent of R, indicating that for $R \to 0$ only e_{ni} suffices, for $R \to \infty$ only i_{ni} suffices, but for intermediate values of R both sources are necessary in order to make the output noise e_{no} come out right. (In a practical circuit one would try to keep R suitably small to make its thermal noise contribution negligible).

7.84 Recalculate the input noise voltage e_n of the CMOS differential pair of Fig. 7.105 for the case of a *passive* load implemented with a pair of identical resistors R_D. Compare with the active-load case and comment.

7.85 Recalculate the input noise voltage e_n of the bipolar differential pair of Fig. 7.106 for the case of an *active* load implemented with a *pnp* current mirror. Compare with the passive-load case and comment.

7.86 We conclude with an open problem based on the circuit of Fig. P7.86, which uses the three Main ingredients of this book, namely, BJTs, MOSFETs, and ICs. Make up as many problems as you can think of, from basic to noise, and then use typical device parameter values for your calculations and PSpice simulations.

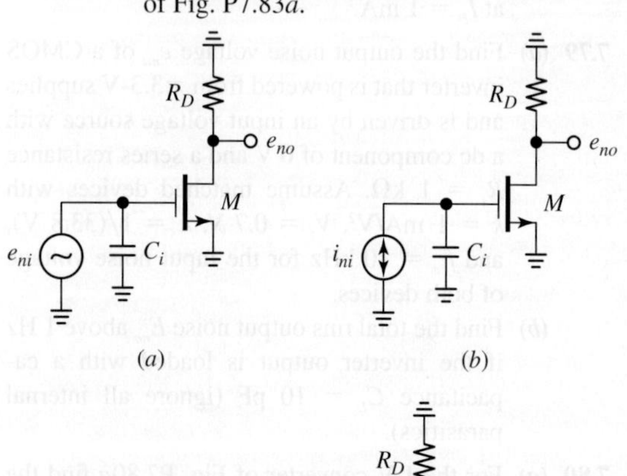

FIGURE. P7.83

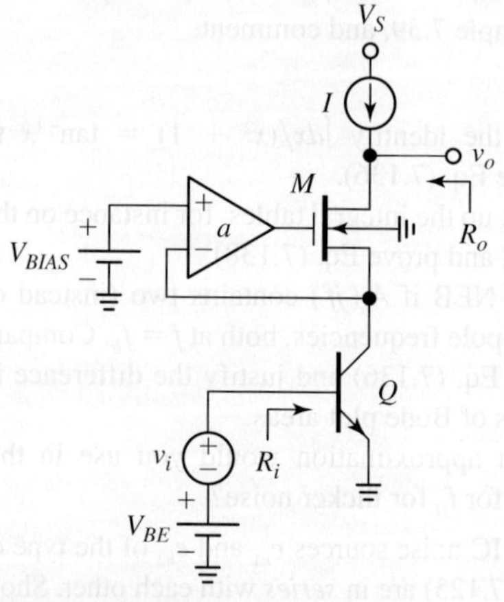

FIGURE. P7.86

Index